P9-CQU-655

Sociobiology

By the Same Author

Life on Earth
with Thomas Eisner, Winslow R. Briggs, Richard E. Dickerson, Robert
L. Metzenberg, Richard D. O'Brien, Millard Susman, William E. Boggs

The Insect Societies

A Primer of Population Biology
with William H. Bossert

The Theory of Island Biogeography
with Robert H. MacArthur

Sociobiology THE NEW SYNTHESIS

Edward O. Wilson

The Belknap Press of Harvard University Press
Cambridge, Massachusetts, and London, England

Copyright © 1975 by the President and Fellows of Harvard College
All rights reserved
Third printing 1976
Library of Congress Catalog Card Number 74-83910
ISBN 0-674-81621-8
Printed in the United States of America

Acknowledgments

Modern sociobiology is being created by gifted investigators who work primarily in population biology, invertebrate zoology, including entomology especially, and vertebrate zoology. Because my training and research experience were fortuitously in the first two subjects and there was some momentum left from writing *The Insect Societies*, I decided to learn enough about vertebrates to attempt a general synthesis. The generosity which experts in this third field showed me, patiently guiding me through films and publications, correcting my errors, and offering the kind of enthusiastic encouragement usually reserved for promising undergraduate students, is a testament to the communality of science.

My new colleagues also critically read most of the chapters in early draft. The remaining portions were reviewed by population biologists and anthropologists. I am especially grateful to Robert L. Trivers for reading most of the book and discussing it with me from the time of its conception. Others who reviewed portions of the manuscript, with the chapter numbers listed after their names, are Ivan Chase (13), Irven DeVore (27), John F. Eisenberg (23, 24, 25, 26), Richard D. Estes (24), Robert Fagen (1–5, 7), Madhav Gadgil (1–5), Robert A. Hinde (7), Bert Hölldobler (8–13), F. Clark Howell (27), Sarah Blaffer Hrdy (1–13, 15–16, 27), Alison Jolly (26), A. Ross Kiester (7, 11–13), Bruce R. Levin (4, 5), Peter R. Marler (7), Ernst Mayr (11–13), Donald W. Pfaff (11), Katherine Ralls (15), Jon Seger (1–6, 8–13, 27), W. John Smith (8–10), Robert M. Woollacott (19), James Weinrich (1–5, 8–13), and Amotz Zahavi (5).

Illustrations, unpublished manuscripts, and technical advice were supplied by R. D. Alexander, Herbert Bloch, S. A. Boorman, Jack Bradbury, F. H. Bronson, W. L. Brown, Francine and P. A. Buckley, Noam Chomsky, Malcolm Coe, P. A. Corning, Iain Douglas-Hamilton, Mary Jane West Eberhard, John F. Eisenberg, R. D. Estes, O. R. Floody, Charles Galt, Valerius Geist, Peter Haas, W. J. Hamilton III, Bert Hölldobler, Sarah Hrdy, Alison Jolly, J. H. Kaufmann, M. H. A. Keenleyside, A. R. Kiester, Hans Kummer, J. A. Kurland, M. R. Lein, B. R. Levin, P. R. Levitt, P. R. Marler, Ernst Mayr, G. M. McKay, D. B. Means, A. J. Meyerriecks, Martin Moynihan, R. A. Paynter, Jr., D. W. Pfaff, W. P. Porter, Katherine Ralls, Lynn Riddiford, P. S. Rodman, L. L. Rogers, Thelma E. Rowell, W. E. Schevill, N. G. Smith, Judy A. Stamps, R. L. Trivers, J. W. Truman, F. R. Walther, Peter Weygoldt, W. Wickler, R. H. Wiley, E. N. Wilmsen, E. E. Williams, and D. S. Wilson.

Kathleen M. Horton assisted closely in bibliographic research, checked many technical details, and typed the manuscript through two intricate drafts. Nancy Clemente edited the manuscript, providing many helpful suggestions concerning organization and exposition.

Sarah Landry executed the drawings of animal societies presented in Chapters 20–27. In the case of the vertebrate species, her compositions are among the first to represent entire societies, in the correct demographic proportions, with as many social interactions displayed as can plausibly be included in one scene. In order to make the drawings as accurate as possible, we sought and were generously given the help of the following biologists who had conducted research on the sociobiology of the individual species: Robert T. Bakker (reconstruction of the appearance and possible social behavior of dinosaurs), Brian Bertram (lions), Iain Douglas-Hamilton (African elephants), Richard D. Estes (wild dogs, wildebeest), F. Clark Howell (reconstructions of primitive man and the Pleistocene mammal fauna), Alison Jolly (ring-tailed lemurs), James Malcolm (wild dogs), John H. Kaufmann (whip-tailed wallabies), Hans Kummer (hamadryas baboons), George B. Schaller (gorillas), and Glen E. Woolfenden (Florida scrub jays). Elso S. Barghoorn, Leslie A. Garay, and Rolla M. Tryon added advice on the depiction of the surrounding vegetation. Other drawings in this book were executed by Joshua B. Clark, and most of the graphs and diagrams by William G. Minty.

Certain passages have been taken with little or no change from *The Insect Societies*, by E. O. Wilson (Belknap Press of Harvard University Press, 1971); these include short portions of Chapters 1, 3, 6, 8, 9, 13, 14, 16, and 17 in the present book as well as a substantial portion of Chapter 20, which presents a brief review of the social insects. Other excerpts have been taken from *A Primer of Population Biology*, by E. O. Wilson and W. H. Bossert (Sinauer Associates, 1971), and *Life on Earth*, by E. O. Wilson et al. (Sinauer Associates, 1973). Pages 106–117 come from my article "Group Selection and Its Significance for Ecology" (*BioScience*, vol. 23, pp. 631–638, 1973), copyright © 1973 by the President and Fellows of Harvard College. Other passages have been adapted from various of my articles in *Bulletin of the Entomological Society of America* (vol. 19, pp. 20–22, 1973); *Science* (vol. 163, p. 1184, 1969; vol. 179, p. 466, 1973; copyright © 1969, 1973, by the American Association for the Advancement of Science); *Scientific American* (vol. 227, pp. 53–54, 1972); *Chemical Ecology* (E. Sondheimer and J. B. Simeone, eds., Academic Press, 1970); *Man and Beast: Comparative Social Behavior* (J. F. Eisenberg and W. S. Dillon, eds., Smithsonian Institution Press, 1970). The quotations from the Bhagavad-Gita are taken from the Peter Pauper Press translation. The editors and publishers are thanked for their permission to reproduce these excerpts.

I wish further to thank the following agencies and individuals for permission to reproduce materials for which they hold the copyright: Academic Press, Inc.; Aldine Publishing Company; American Association for the Advancement of Science, representing *Science*; *American Midland Naturalist*; *American Zoologist*; Annual Reviews, Inc.; Associated University Presses, Inc., representing Bucknell University Press; Ballière Tindall, Ltd.; Professor George W. Barlow; Blackwell Scientific Publications, Ltd.; E. J. Brill Co.; Cambridge University Press; Dr. M. J. Coe; Cooper Ornithological Society, representing *The Condor*; American Society of Ichthyologists and Herpetologists, representing *Copeia*; Deutsche Ornithologen-Gesellschaft, representing *Journal für Ornithologie*; Dr. Iain Douglas-Hamilton (Ph.D. thesis, Oxford University); Dowden, Hutchinson and Ross, Inc.; Duke University Press and the Ecological Society of America, representing *Ecology*; Dr. Mary Jane West Eberhard; Professor Thomas Eisner; *Evolution*; Dr. W. Faber; W. H. Freeman and Company, representing *Scientific American*; Gustav Fischer Verlag; Harper and Row, Publishers, Inc., including representation for *Psychosomatic Medicine*; Dr. Charles S. Henry; the Herpetologists' League, representing *Herpetologica*; Holt, Rinehart and Winston, Inc.; Dr. J. A. R. A. M. van Hooff; Houghton Mifflin Company; Indiana University Press; *Journal of Mammalogy*; Dr. Heinrich Kutter; Professor James E. Lloyd; Macmillan Publishing Company, Inc.; Professor Peter Marler; McGraw-Hill Book Company; Masson et Cie, representing *Insectes Sociaux*; Dr. L. David Mech; Methuen and Co., Ltd.; Museum of Zoology, University of Michigan; Dr. Eugene L. Nakamura; *Nature*, for Macmillan (Journals), Ltd.; Professor Charles Noirot; Pergamon Press, Inc.; Professor Donald W. Pfaff; Professor Daniel Otte; Plenum Publishing Corporation; *The Quarterly Review of Biology*; Dr. Katherine Ralls; Random House, Inc.; Professor Carl W. Rettenmeyer; the Royal Society, London; *Science Journal*; Dr. Neal G. Smith; Springer-Verlag New York, Inc.; Dr. Robert Stumper; University of California Press; The University of Chicago Press, including representation of *The American Naturalist*; Walter de Gruyter and Co.; Dr. Peter Weygoldt; Professor W. Wickler; John Wiley and Sons, Inc.; Worth Publishers, Inc.; The Zoological Society of London, representing *Journal of Zoology*; Zoologischer Garten Köln (Aktiengesellschaft).

Finally, much of my personal research reported in the book has been supported continuously by the National Science Foundation during the past sixteen years. It is fair to say that I would not have reached the point from which a synthesis could be attempted if it had not been for this generous public support.

E. O. W.

Cambridge, Massachusetts
October 1974

Contents

Arjuna to Lord Krishna: *Although these are my enemies, whose wits are overthrown by greed, see not the guilt of destroying a family, see not the treason to friends, yet how, O Troubler of the Folk, shall we with clear sight not see the sin of destroying a family?*

Lord Krishna to Arjuna: *He who thinks this Self to be a slayer, and he who thinks this Self to be slain, are both without discernment; the Soul slays not, neither is it slain.*

Part I Social Evolution

Chapter 1 **The Morality of the Gene**

Camus said that the only serious philosophical question is suicide. That is wrong even in the strict sense intended. The biologist, who is concerned with questions of physiology and evolutionary history, realizes that self-knowledge is constrained and shaped by the emotional control centers in the hypothalamus and limbic system of the brain. These centers flood our consciousness with all the emotions—hate, love, guilt, fear, and others—that are consulted by ethical philosophers who wish to intuit the standards of good and evil. What, we are then compelled to ask, made the hypothalamus and limbic system? They evolved by natural selection. That simple biological statement must be pursued to explain ethics and ethical philosophers, if not epistemology and epistemologists, at all depths. Self-existence, or the suicide that terminates it, is not the central question of philosophy. The hypothalamic-limbic complex automatically denies such logical reduction by countering it with feelings of guilt and altruism. In this one way the philosopher's own emotional control centers are wiser than his solipsist consciousness, "knowing" that in evolutionary time the individual organism counts for almost nothing. In a Darwinist sense the organism does not live for itself. Its primary function is not even to reproduce other organisms; it reproduces genes, and it serves as their temporary carrier. Each organism generated by sexual reproduction is a unique, accidental subset of all the genes constituting the species. Natural selection is the process whereby certain genes gain representation in the following generations superior to that of other genes located at the same chromosome positions. When new sex cells are manufactured in each generation, the winning genes are pulled apart and reassembled to manufacture new organisms that, on the average, contain a higher proportion of the same genes. But the individual organism is only their vehicle, part of an elaborate device to preserve and spread them with the least possible biochemical perturbation. Samuel Butler's famous aphorism, that the chicken is only an egg's way of making another egg, has been modernized: the organism is only DNA's way of making more DNA. More to the point, the hypothalamus and limbic system are engineered to perpetuate DNA.

In the process of natural selection, then, any device that can insert a higher proportion of certain genes into subsequent generations will come to characterize the species. One class of such devices promotes prolonged individual survival. Another promotes superior mating performance and care of the resulting offspring. As more complex social behavior by the organism is added to the genes' techniques for replicating themselves, altruism becomes increasingly prevalent and eventually appears in exaggerated forms. This brings us to the central theoretical problem of sociobiology: how can altruism, which by definition reduces personal fitness, possibly evolve by natural selection? The answer is kinship: if the genes causing the altruism are shared by two organisms because of common descent, and if the altruistic act by one organism increases the joint contribution of these genes to the next generation, the propensity to altruism will spread through the gene pool. This occurs even though the altruist makes less of a solitary contribution to the gene pool as the price of its altruistic act.

To his own question, "Does the Absurd dictate death?" Camus replied that the struggle toward the heights is itself enough to fill a man's heart. This arid judgment is probably correct, but it makes little sense except when closely examined in the light of evolutionary theory. The hypothalamic-limbic complex of a highly social species, such as man, "knows," or more precisely it has been programmed to perform as if it knows, that its underlying genes will be proliferated maximally only if it orchestrates behavioral responses that bring into play an efficient mixture of personal survival, reproduction, and altruism. Consequently, the centers of the complex tax the conscious mind with ambivalences whenever the organisms encounter stressful situations. Love joins hate; aggression, fear; expansiveness, withdrawal; and so on; in blends designed not to promote the happiness and survival of the individual, but to favor the maximum transmission of the controlling genes.

The ambivalences stem from counteracting pressures on the units of natural selection. Their genetic consequences will be explored formally later in this book. For the moment suffice it to note that what is good for the individual can be destructive to the family; what preserves the family can be harsh on both the individual and the tribe to which its family belongs; what promotes the tribe can weaken the family and destroy the individual; and so on upward through the permutations of levels of organization. Counteracting selection on these different units will result in certain genes being multiplied and fixed, others lost, and combinations of still others held in static proportions. According to the present theory, some of the genes will produce emotional states that reflect the balance of counteracting selection forces at the different levels.

I have raised a problem in ethical philosophy in order to characterize the essence of sociobiology. Sociobiology is defined as the systematic study of the biological basis of all social behavior. For the present it focuses on animal societies, their population structure, castes, and communication, together with all of the physiology underlying the social adaptations. But the discipline is also concerned with the social behavior of early man and the adaptive features of organization in the more primitive contemporary human societies. Sociology *sensu stricto*, the study of human societies at all levels of complexity, still stands apart from sociobiology because of its largely structuralist and nongenetic approach. It attempts to explain human behavior primarily by empirical description of the outermost phenotypes and by unaided intuition, without reference to evolutionary explanations in the true genetic sense. It is most successful, in the

way descriptive taxonomy and ecology have been most successful, when it provides a detailed description of particular phenomena and demonstrates first-order correlations with features of the environment. Taxonomy and ecology, however, have been reshaped entirely during the past forty years by integration into neo-Darwinist evolutionary theory—the "Modern Synthesis," as it is often called—in which each phenomenon is weighed for its adaptive significance and then related to the basic principles of population genetics. It may not be too much to say that sociology and the other social sciences, as well as the humanities, are the last branches of biology waiting to be included in the Modern Synthesis. One of the functions of sociobiology, then, is to reformulate the foundations of the social sciences in a way that draws these subjects into the Modern Synthesis. Whether the social sciences can be truly biologicized in this fashion remains to be seen.

This book makes an attempt to codify sociobiology into a branch of evolutionary biology and particularly of modern population biology. I believe that the subject has an adequate richness of detail and aggregate of self-sufficient concepts to be ranked as coordinate with such disciplines as molecular biology and developmental biology. In the past its development has been slowed by too close an identification with ethology and behavioral physiology. In the view presented here, the new sociobiology should be compounded of roughly equal parts of invertebrate zoology, vertebrate zoology, and population biology. Figure 1-1 shows the schema with which I closed *The Insect Societies*, suggesting how the amalgam can be achieved. Biologists have always been intrigued by comparisons between societies of invertebrates, especially insect societies, and those of vertebrates. They have dreamed of identifying the common properties of such disparate units in a way that would provide insight into all aspects of social evolution, including that of man. The goal can be expressed in modern terms as follows: when the same parameters and quantitative theory are used to analyze both termite colonies and troops of rhesus macaques, we will have a unified science of sociobiology. This may seem an impossibly difficult task. But as my own studies have advanced, I have been increasingly impressed with the functional similarities between invertebrate and vertebrate societies and less so with the structural differences that seem, at first glance, to constitute such an immense gulf between them. Consider for a moment termites and monkeys. Both are formed into cooperative groups that occupy territories. The group members communicate hunger, alarm, hostility, caste status or rank, and reproductive status among themselves by means of something on the order of 10 to 100 nonsyntactical signals. Individuals are intensely aware of the distinction between groupmates and nonmembers. Kinship plays an important role in group structure and probably served as a chief generative force of sociality in the first place. In both kinds of society there is a well-marked division of labor,

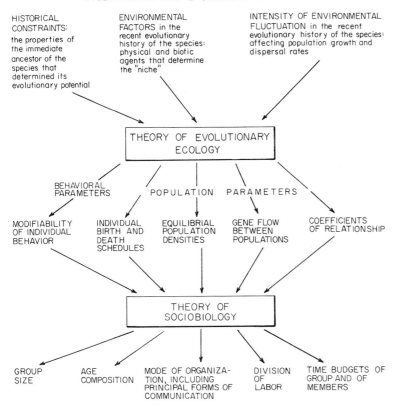

Figure 1-1 The connections that can be made between phylogenetic studies, ecology, and sociobiology.

although in the insect society there is a much stronger reproductive component. The details of organization have been evolved by an evolutionary optimization process of unknown precision, during which some measure of added fitness was given to individuals with cooperative tendencies—at least toward relatives. The fruits of co-operativeness depend upon the particular conditions of the environment and are available to only a minority of animal species during the course of their evolution.

This comparison may seem facile, but it is out of such deliberate oversimplification that the beginnings of a general theory are made. The formulation of a theory of sociobiology constitutes, in my opinion, one of the great manageable problems of biology for the next twenty or thirty years. The prolegomenon of Figure 1-1 guesses part of its future outline and some of the directions in which it is most likely to lead animal behavior research. Its central precept is that the evolution of social behavior can be fully comprehended only through an understanding, first, of demography, which yields the vital infor-

mation concerning population growth and age structure, and, second, of the genetic structure of the populations, which tells us what we need to know about effective population size in the genetic sense, the coefficients of relationship within the societies, and the amounts of gene flow between them. The principal goal of a general theory of sociobiology should be an ability to predict features of social organization from a knowledge of these population parameters combined with information on the behavioral constraints imposed by the genetic constitution of the species. It will be a chief task of evolutionary ecology, in turn, to derive the population parameters from a knowledge of the evolutionary history of the species and of the environment in which the most recent segment of that history unfolded. The most

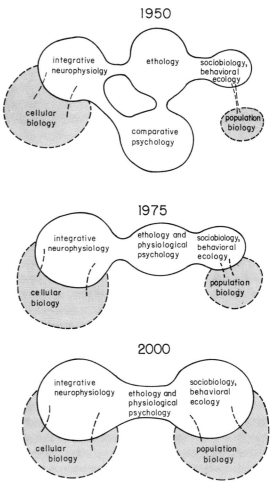

Figure 1-2 A subjective conception of the relative number of ideas in various disciplines in and adjacent to behavioral biology to the present time and as it might be in the future.

important feature of the prolegomenon, then, is the sequential relation between evolutionary studies, ecology, population biology, and sociobiology.

In stressing the tightness of this sequence, however, I do not wish to underrate the filial relationship that sociobiology has had in the past with the remainder of behavioral biology. Although behavioral biology is traditionally spoken of as if it were a unified subject, it is now emerging as two distinct disciplines centered on neurophysiology and on sociobiology, respectively. The conventional wisdom also speaks of ethology, which is the naturalistic study of whole patterns of animal behavior, and its companion enterprise, comparative psychology, as the central, unifying fields of behavioral biology. They are not; both are destined to be cannibalized by neurophysiology and sensory physiology from one end and sociobiology and behavioral ecology from the other (see Figure 1-2).

I hope not too many scholars in ethology and psychology will be offended by this vision of the future of behavioral biology. It seems to be indicated both by the extrapolation of current events and by consideration of the logical relationship behavioral biology holds with the remainder of science. The future, it seems clear, cannot be with the ad hoc terminology, crude models, and curve fitting that characterize most of contemporary ethology and comparative psychology. Whole patterns of animal behavior will inevitably be explained within the framework, first, of integrative neurophysiology, which classifies neurons and reconstructs their circuitry, and, second, of sensory physiology, which seeks to characterize the cellular transducers at the molecular level. Endocrinology will continue to play a peripheral role, since it is concerned with the cruder tuning devices of nervous activity. To pass from this level and reach the next really distinct discipline, we must travel all the way up to the society and the population. Not only are the phenomena best described by families of models different from those of cellular and molecular biology, but the explanations become largely evolutionary. There should be nothing surprising in this distinction. It is only a reflection of the larger division that separates the two greater domains of evolutionary biology and functional biology. As Lewontin (1972a) has truly said: "Natural selection of the character states themselves is the essence of Darwinism. All else is molecular biology."

Chapter 2   **Elementary Concepts of Sociobiology**

Genes, like Leibnitz's monads, have no windows; the higher properties of life are emergent. To specify an entire cell, we are compelled to provide not only the nucleotide sequences but also the identity and configuration of other kinds of molecules placed in and around the cell. To specify an organism requires still more information about both the properties of the cells and their spatial positions. And once assembled, organisms have no windows. A society can be described only as a set of particular organisms, and even then it is difficult to extrapolate the joint activity of this ensemble from the instant of specification, that is, to predict social behavior. To cite one concrete example, Maslow (1936) found that the dominance relations of a group of rhesus monkeys cannot be predicted from the interactions of its members matched in pairs. Rhesus monkeys, like other higher primates, are intensely affected by their social environment—an isolated individual will repeatedly pull a lever with no reward other than the glimpse of another monkey (Butler, 1954). Moreover, this behavior is subject to higher-order interactions. The monkeys form coalitions in the struggle for dominance, so that an individual falls in rank if deprived of its allies. A second-ranking male, for example, may owe its position to protection given it by the alpha male or support from one or more close peers (Hall and DeVore, 1965; Varley and Symmes, 1966). Such coalitions cannot be predicted from the outcome of pairwise encounters, let alone from the behavior of an isolated monkey.

The recognition and study of emergent properties is holism, once a burning subject for philosophical discussion by such scientists as Lloyd Morgan (1922) and W. M. Wheeler (1927), but later, in the 1940's and 1950's, temporarily eclipsed by the triumphant reductionism of molecular biology. The new holism is much more quantitative in nature, supplanting the unaided intuition of the old theories with mathematical models. Unlike the old, it does not stop at philosophical retrospection but states assumptions explicitly and extends them in mathematical models that can be used to test their validity. In the sections to follow we will examine several of the properties of societies that are emergent and hence deserving of a special language and treatment. We begin with a straightforward, didactic review of a set of the most basic definitions, some general for biology, others peculiar to sociobiology.

Society: a group of individuals belonging to the same species and organized in a cooperative manner. The terms *society* and *social* need to be defined broadly in order to prevent the exclusion of many interesting phenomena. Such exclusion would cause confusion in all further comparative discussions of sociobiology. Reciprocal communication of a cooperative nature, transcending mere sexual activity, is the essential intuitive criterion of a society. Thus it is difficult to think of a bird egg, or even a honeybee larva sealed in its brood cell, as a member of the society that produced it, even though it may function as a true member at other stages of its development. It is also not satisfying to view the simplest aggregations of organisms, such as swarms of courting males, as true societies. They are often drawn together by mutually attractive stimuli, but if they interact in no other way it seems excessive to refer to them by a term stronger than aggregation. By the same token a pair of animals engaged in simple courtship or a group of males in territorial contention can be called a society in the broadest sense, but only at the price of diluting the expression to the point of uselessness. Yet aggregation, sexual behavior, and territoriality are important *properties* of true societies, and they are correctly referred to as social behavior. Bird flocks, wolf packs, and locust swarms are good examples of true elementary societies. So are parents and offspring if they communicate reciprocally. Although this last, extreme example may seem at first trivial, parent-offspring interactions are in fact often complex and serve multiple functions. Furthermore, in many groups of organisms, from the social insects to the primates, the most advanced societies appear to have evolved directly from family units. Another way of defining societies is by delimiting particular groups. Since the bond of the society is simply and solely communication, its boundaries can be defined in terms of the curtailment of communication. Altmann (1965) has expressed this aspect: "A society . . . is an aggregation of socially intercommunicating, conspecific individuals that is bounded by frontiers of far less frequent communication."

The definition of a society as a cooperating group of conspecific organisms is about the same as that used, more or less explicitly, by writers as early as Alverdes (1927), Allee (1931), and Darling (1938). There has, nevertheless, always been some ambiguity about the cut-off point or, to be more precise, the level of organization at which we cease to refer to a group as a society and start labeling it as an aggregation or nonsocial population.

Aggregation: a group of individuals of the same species, comprised of more than just a mated pair or a family, gathered in the same place but not internally organized or engaged in cooperative behavior. Winter congregations of rattlesnakes and ladybird beetles, for example, may provide superior protection for their members, but unless they are organized by some behavior other than mutual attraction they are better classified as aggregations rather than true societies. Students of fish behavior attending the 11th International Ethological Congress at Rennes, France, recommended a formal adoption of essentially this distinction between an association and a school of fish (Shaw, 1970). However, they further specified that an aggregation is a group whose members are brought together by extrinsic conditions rather than by social attraction to one another. This addendum seems to me to be gratuitous and an impracticably fine distinction.

Colony: in strict biological usage, a society of organisms which are highly integrated, either by physical union of the bodies or by divi-

sion into specialized zooids or castes, or by both. In the vernacular and even in some technical descriptions, a colony can mean almost any group of organisms, especially if they are fixed in one locality. In sociobiology, however, the word is best restricted to the societies of social insects, together with the tightly integrated masses of sponges, siphonophores, bryozoans, and other "colonial" invertebrates.

Individual: any physically distinct organism. Although pondering the definition of an individual might strike one as a waste of time, it is actually a substantial philosophical problem. G. C. Williams (1966a), for example, has suggested that from the standpoint of evolutionary theory, "the concept of an 'individual' implies genetic uniqueness." This recommendation overlooks identical twins, who must be treated as separate entities even by the most detached theoretician. In his definition Williams, like many others before him, was concerned with elucidating the status of the clonal zooids of siphonophores and other invertebrate colonies, some of which have been reduced in evolution to the status of accessory organs attached to other, more complete organisms. The distinction between the individual and the colony can be especially baffling in the sponges (Hartman and Reiswig, 1971). In "solitary" forms such as *Sycon,* each organism possesses a single terminal oscule. Water is passed through the exhalant vent of the oscule after being depleted of oxygen and food. Thus in colonial sponges the oscules seem to be the best markers of the individual organisms. However, in the encrusting colonial species, the outer channels of adjacent water systems run together, so that water flowing from the boundary chambers can be captured by either water system. As a result it is difficult or impossible to map water systems precisely onto particular oscules, hence impracticable to make any clean distinction between individuals. Furthermore, some colonies pump water in a rhythmic fashion, so that in this sense the entire sponge behaves as though it were one individual.

Group: a set of organisms belonging to the same species that remain together for any period of time while interacting with one another to a much greater degree than with other conspecific organisms. The word *group* is thus used with the greatest flexibility to designate any aggregation or kind of society or subset of a society. The expression is especially useful in accommodating descriptions of certain primate societies in which there exists a hierarchy of levels of organization constructed of nested subsets of individuals belonging to a single large congregation. Here, for example, is the hierarchy of groups recognized by Kummer (1968) in his study of the hamadryas baboon:

Troop: a large group that gathers in the protective shelter of a sleeping rock, consisting of one or more bands that support each other in alerting and defending against predators

Band: a group headed by one or more males that maintains itself apart during foraging trips and occasionally fights with other bands (the band can be broken down into one or more two-male teams, the unit defined below)

Two-male team: an older and younger male, the latter initially in the role of a tolerated "apprentice"; the two operate closely together but maintain their own harems and off-spring

One-male unit: the older or younger male of the two-male team, together with his family

Clearly, no single set of hamadryas baboons is "the" society. The problem of designating a social unit by fixed criteria becomes still more acute when analyzing the rapid formation, breakup, and reformation of *casual groups* or *subgroups* (Cohen, 1971), examples of which include clusters of grooming monkeys, regurgitating ants, and conversationalists at a cocktail party. In many such cases, not even a hierarchy of groups can be clearly defined.

Yet the ambiguity of the expression *group* becomes felicitous when the nature of the organization is still unknown or there is no desire to specify it. In this context we are permitted the use of terms of venery (Lipton, 1968), which are solely for purposes of taxonomy and convey no information on social organization. Of largely medieval origin, many of these words still enjoy everyday use, while others are little more than amusing relicts: a school of fish, a pride of lions, a swarm of bees, a gang of elk, a pace of asses, a troop of kangaroos, a route of wolves, a skulk of foxes, a sleuth of bears, a crash of rhinoceroses, a trip (or herd) of seals, a pod of sea otters, a siege of herons, a herd of cranes, a tok of capercaillies, a murmuration of starlings, an exaltation of larks, a bouquet of pheasants, a murder of crows, a building of rooks, a knot of toads, a smack of jellyfish, and so forth. There is no reason why any of these terms cannot be employed when it is expedient to do so, even in technical descriptions of behavior.

Population: a set of organisms belonging to the same species and occupying a clearly delimited area at the same time. This unit—the most basic but also one of the most loosely employed in evolutionary biology—is defined in terms of genetic continuity. In the case of sexually reproducing organisms, the population is a geographically delimited set of organisms capable of freely interbreeding with one another under natural conditions. The special population used by model builders is the *deme*, the smallest local set of organisms within which interbreeding occurs freely. The idealized deme is panmictic, that is, its members breed completely at random. Put another way, panmixia means that each reproductively mature male is equally likely to mate with each reproductively mature female, regardless of their location within the range of the deme. Although not likely to be attained in absolute form in nature, especially in social organisms,

panmixia is an important simplifying assumption made in much of elementary quantitative theory.

In sexually reproducing forms, including the vast majority of social organisms, a *species* is a population or set of populations within which the individuals are capable of freely interbreeding under natural conditions. By definition the members of the species do not interbreed freely with those of other species, however closely related they may be genetically. The existence of natural conditions is a basic part of the definition of the species. In establishing the limits of a species it is not enough merely to prove that genes of two or more populations can be exchanged under experimental conditions. The populations must be demonstrated to interbreed fully in the free state. To illustrate the point, let us consider a familiar case with some surprising implications. Lions (*Panthera leo*) and tigers (*Panthera tigris*) are genetically closely related, despite their marked differences in outward appearance. They are sometimes crossed in zoos to produce hybrids, called "tiglons" (tiger as father) and "ligers" (lion as father). But this breeder's accomplishment does not prove them to belong to the same species. The ability to hybridize under a suitable experimental environment can be said to be a necessary condition under the biological species concept, but not a sufficient one. The important question is whether the two forms cross freely where they occur together in the wild. Lions and tigers did coexist over most of India until the 1800's, when lions began to be reduced even more quickly than tigers by intensive hunting and deterioration of the environment. Now lions are nearly extinct, limited to a few hundred individuals in the Gir Forest in the state of Gujarat. There can be no doubt that lions and tigers were fully isolated reproductively during their coexistence, for no tiglons or ligers have ever been found in India. Suppose that lions and tigers had been shown to be wholly intersterile under experimental conditions. This could reasonably have been interpreted to mean that they are distinct species, because the condition could be assumed to hold in nature also. But the opposite evidence means nothing, since many other genetic devices in addition to mere intersterility might (and obviously do) operate to isolate them in nature. In fact lions and tigers differ strongly in their behavior and in the habitats they prefer. The lion is more social, living in small groups called prides, and it prefers open country. The tiger is solitary and is found more frequently in forested regions. These differences between the two species, which almost certainly have a genetic basis, could be great enough to account for their failure to hybridize.

A population that differs significantly from other populations belonging to the same species is referred to as a *geographic race* or *subspecies*. Subspecies are separated from other subspecies by distance and geographic barriers that prevent the exchange of individuals, as opposed to the genetically based "intrinsic isolating mechanisms" that hold species apart. Subspecies, insofar as they can be distinguished with any objectivity at all, show every conceivable degree of differentiation from other subspecies. At one extreme are the populations that fall along a cline—a simple gradient in the geographic variation of a given character. In other words, a character that varies in a clinal pattern is one that changes gradually over a substantial portion of the entire range of the species. At the other extreme are subspecies consisting of easily distinguished populations that are differentiated from one another by numerous genetic traits and exchange genes across a narrow zone of intergradation.

The main obstacle in dealing with the population as a unit, one that extends into theoretical sociobiology, is the practical difficulty of deciding the limits of particular populations. There are some extreme cases which for special reasons present no problem. All 200 to 800 desert pupfish constituting the species *Cyprinodon diabolis* live in a single thermal spring at Devil's Hole, Nevada. Each year all 50 or so of the living whooping cranes (*Grus americana*) fly from their nesting ground in Canada to their winter home at the Aransas National Wildlife Refuge, Texas, where they are watched and counted to the last fledgling by anxious wildlife managers. But very few populations, let alone species, are so restricted. The eastern highland gorilla (*Gorilla gorilla beringei*), for example, generally regarded as a subspecific equivalent of the lowland gorilla, occupies a relatively narrow range. The 10,000 or so individuals that constitute it have been grouped by Emlen and Schaller (1960) into about 60 populations, each occupying 25 to 250 square kilometers of mountainous country in Central Africa. In the center of the distribution there is a large area in which the species appears to be sparse but continuously distributed. In fact, the true limits of these "populations" are unknown, since the rate at which gorillas move from one area to another to breed is not known. To express this in the language of population genetics, we do not know the rate of gene flow. Lacking that crucial parameter, we can conclude very little more about the population structure of mountain gorillas. *G. gorilla beringei* is not at all unusual in this regard. On the contrary, it is much better known at the present time than the vast majority of the more than 10 million living plant and animal species and subspecies.

What is the relation between the population and the society? Here we arrive unexpectedly at the crux of theoretical sociobiology. The distinction between the two categories is essentially as follows: the population is bounded by a zone of sharply reduced gene flow, while the society is bounded by a zone of sharply reduced communication. Often the two zones are the same, since social bonds tend to promote gene flow among the members of the society to the exclusion of outsiders. For example, detailed field studies by Stuart and Jeanne Altmann (1970) on the yellow baboons (*Papio cynocephalus*) of Amboseli show that in this species the society and the deme are essentially the same thing. The baboons are internally organized by

dominance hierarchies and are usually hostile toward outsiders. Gene exchange occurs between troops by the emigration from one to another of subordinate males, who typically leave their home troop after the loss of a fight or during competition for estrous females. Using the Altmanns' data, Cohen (1969b) estimated the immigration rate into one large troop to be 8.043×10^{-3} individuals per group per day, a degree of flow that is many orders of magnitude below that which occurred between subgroups belonging to the same troop.

In open-group species the relation between the population and the society can be vastly more complex. The chimpanzee (*Pan troglodytes*) provides an extreme example of this type of organization, a fact that has intrigued and puzzled every investigator who has conducted extensive field studies to date (Reynolds and Reynolds, 1965; Reynolds, 1966; Goodall, 1965; Itani, 1966; Sugiyama, 1968, 1972; Izawa, 1970). A local population of chimpanzees is a weakly strung nexus of troops, the members of which know one another to some extent. Troop membership changes frequently, and the residents are friendly even to strangers who enter the area from outside the nexus. Apparently the limits of personal acquaintanceship, and hence of the society by broadest definition, are set either by the existence of physical barriers that prevent migration of chimpanzees or by great distance, over which personal contacts become too tenuous to be socially significant. Sugiyama (1968) has labeled such societies "regional populations," but the expression is redundant (populations are generally defined as being regional) and ambiguous with reference to other usages of the population unit in biology. A better expression would be *group complex* or simply *group*. Open groups are known in a few ant species, including the Argentine ant *Iridomyrmex humilis* and certain members of *Pseudomyrmex*, *Crematogaster*, *Myrmica*, and *Formica* (Wilson, 1971a). The "colonies" occupy discrete nest sites, but, unlike those of the great majority of other ant species, they exchange members freely and accept back queens from any part of the local population following the nuptial flights. I have labeled such populations "unicolonial," to distinguish them from the multicolonial populations that represent the more general and primitive state in ants and other social insects.

Communication: action on the part of one organism (or cell) that alters the probability pattern of behavior in another organism (or cell) in an adaptive fashion. This definition conforms well both to our intuitive understanding of communication and to the procedure by which the process is mathematically analyzed (see Chapter 8).

Coordination: interaction among units of a group such that the overall effort of the group is divided among the units without leadership being assumed by any one of them. Coordination may be influenced by a unit in a higher level of the social hierarchy, but such outside control is not essential. The formation of a fish school, the exchange of liquid food back and forth by worker ants, and the encirclement of prey by a pride of lions are all examples of coordination among organisms at the same organizational level.

Hierarchy: in ordinary sociobiological usage, the dominance of one member of a group over another, as measured by superiority in aggressive encounters and order of access to food, mates, resting sites, and other objects promoting survivorship and reproductive fitness. Technically, there need be only two individuals to make such a hierarchy, but chains of many individuals in descending order of dominance are also frequent. More generally, a hierarchy can be defined without reference to dominance as a system of two or more levels of units, the higher levels controlling at least to some extent the activities of the lower levels in order to achieve the goal of the group as a whole (Mesarović et al., 1970). Hierarchies without dominance are common in social insect colonies and occur in certain facets of the behavior of such highly coordinated mammals as higher primates and social canids. The more advanced animal societies are in general organized at one or at most two hierarchical levels and consist of individuals tightly connected by relatively few kinds of social bonds and communicative signals. Human societies, in contrast, are typically organized through many hierarchical levels and are comprised of numerous individuals loosely joined by very many kinds of social bonds and an extremely rich language. Human societies also differ from animal societies in their tendency to differentiate into large numbers of highly organized subgroups (families, clubs, committees, corporations, and so on) with overlapping memberships.

Regulation: in biology, the coordination of units to achieve the maintenance of one or more physical or biological variables at a constant level. The result of regulation is termed *homeostasis.* The most familiar form of homeostasis is physiological: a properly tuned organism maintains constant values in pH, in concentrations of dissolved nutrients and salts, in proportions of active enzymes and organelles, and so forth, which fall close to the optimal values for survival and reproduction. Like a man-designed machine system, physiological homeostasis is self-regulated by internal feedback loops that increase the values of important variables when they fall below certain levels and decrease them when they exceed other, higher values. At a higher level, social insects display marked homeostasis in the regulation of their own colony populations, caste proportions, and nest environment. This form of steady-state maintenance has aptly been termed social homeostasis by Emerson (1956a). A still higher level of regulation is genetic homeostasis, defined as the automatic resistance of evolving populations to selection which proceeds at a rate fast enough to make deep inroads into genetic variability (Lerner, 1954; Mayr, 1963).

The Multiplier Effect

Social organization is the class of phenotypes furthest removed from the genes. It is derived jointly from the behavior of individuals and the demographic properties of the population, both of which are themselves highly synthetic properties. A small evolutionary change in the behavior pattern of individuals can be amplified into a major social effect by the expanding upward distribution of the effect into multiple facets of social life. Consider, for example, the differing social organizations of the related olive baboon (*Papio anubis*) and hamadryas baboon (*P. hamadryas*). These two species are so close genetically that they interbreed extensively where their ranges overlap and could reasonably be classified as no more than subspecies. The hamadryas male is distinguished by its proprietary attitude toward females, which is total and permanent, whereas the olive male attempts to appropriate females only around the time of their estrus. This difference is only one of degree, and would scarcely be noticeable if one's interest were restricted in each species to the activities of a single dominant male and one consort female. Yet this trait alone is enough to account for profound differences in social structure, affecting the size of the troops, the relationship of troops to one another, and the relationship of males within each troop (Kummer, 1971).

Even stronger multiplier effects occur in the social insects. Termites are notable for the fact that their behavioral diversity generally exceeds morphological diversity at the species level (Noirot, 1958–1959). The structure of nests alone can be used to distinguish species within the higher termites. Certain species of the African genus *Apicotermes*, for example, can be most easily distinguished from their closest relatives on this basis, and in one instance (*A. arquieri* versus *A. occultus*) the taxonomic diagnosis is based exclusively on the nest (Emerson, 1956b). Comparable examples have recently been discovered in the halictine bee genus *Dialictus* (Knerer and Atwood, 1966) and the wasp genus *Stenogaster* (Sakagami and Yoshikawa, 1968). Emerson (1938) was the first to point out that such variation in the fine details of nest structure provides an opportunity to study the evolution of instinct, since each nest is a frozen product of social behavior that can be literally weighed, measured, and geometrically analyzed. The nests are often very complex even by vertebrate standards, the extreme example being the immense structures erected by *Macrotermes* and other fungus-growing termites in Africa (Figure 2-1). The labyrinthine internal structure of these termitaries has been designed in the course of evolution to guide a regular flow of air from the central fungus gardens, where it is heated and rises by convection, upward and outward to a flat, peripheral system of capillarylike chambers, where it is cooled and freshened by proximity to the outside air. In *M. natalensis* the architecture is so efficient that the tempera-

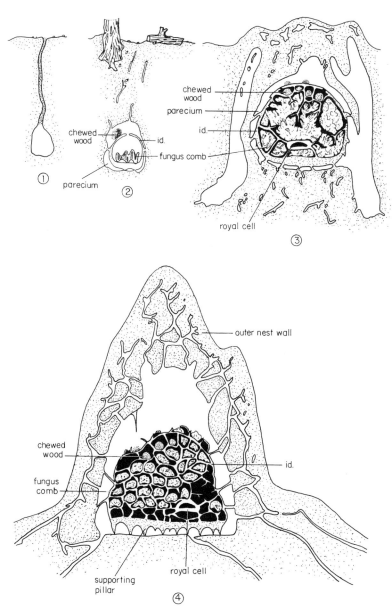

Figure 2-1 Development of the nest of the African fungus-growing termite *Macrotermes bellicosus,* from the initial chamber dug by the newly mated queen and king (*1*), through intermediate periods of growth as worker and soldier castes are added (*2, 3*), to the fully mature form (*4*). The wall (*id.*) of the fungus garden (*idiothèque* of Grassé and Noirot) surrounds numerous chambers that contain masses of finely chewed wood used as substrate for the symbiotic fungus; the parecium is the air space surrounding the fungus garden. At maturity the nest may rise 5 meters or more from the ground and contain over 2 million inhabitants. (From Wilson, 1971a; based on Grassé and Noirot, 1958.)

ture within the fungus garden remains within one degree of 30°C and the carbon dioxide concentration varies only slightly, around 2.6 percent (Lüscher, 1961). The construction of termitaries, and formed nests of other social insects, is coordinated by the perception of work previously accomplished, rather than by direct communication. Even if the work force is constantly renewed, the nest structure already completed determines, by its location, its height, its shape, and probably also its odor, what further work will be done. This principle is nicely exemplified in the construction of a single foundation arch by *M. bellicosus* as the first step in the erection of a fungus garden. When workers of this species are separated from the remainder of the colony and placed in a container with some building material consisting of pellets of soil and excrement, each first explores the container individually. Next, pellets are picked up, carried about, and put down in a seemingly haphazard fashion. Although crude passageways may begin to take shape, the termites, for the most part, still act independently of each other. Finally, seemingly by chance, two or three pellets get stuck on top of each other. This little structure proves much more attractive to the termites than do single pellets. They quickly begin to add more pellets on top, and a column starts to grow. If the column is the only one in the vicinity, construction on it will cease after a while. If another column is located nearby, however, the termites continue adding pellets, and, after a certain height is reached, they bend the column at an angle in the direction of the neighboring column. When the tilted growing ends of the two columns meet, the arch is finished, and the workers move away.

The *Macrotermes* workers give every appearance of accomplishing their astonishing feat by means of what computer scientists call dynamic programming. As each step of the operation is completed, its result is assessed, and the precise program for the next step (out of several or many available) is chosen and activated. Thus no termite need serve as overseer with blueprint in hand. The opportunities for the multiplier effect to operate in the evolution of such a system are obviously very great. A slight alteration of the termites' response to a particular structure will tend to be amplified to a much greater alteration in the final product. The magnitude of the diversity in termite nests, then, is probably a reflection of a much lower degree of diversity in individual behavior patterns. The latter patterns, upon further analysis, may prove to be no more differentiated than the morphological characteristics by which termite species can be distinguished.

Multiplier effects can speed social evolution still more when an individual's behavior is strongly influenced by the particularities of its social experience. This process, called socialization, becomes increasingly prominent as one moves upward phylogenetically into more intelligent species, and it reaches its maximum influence in the higher primates. Although the evidence is still largely inferential, socialization appears to amplify phenotypic differences among primate species. As an example, take the diverging pathways of development of social behavior observed in young olive baboons (*Papio anubis*) as opposed to Nilgiri langurs (*Presbytis johnii*) (Eimerl and DeVore, 1965; Poirier, 1972). The baboon infant stays close to its mother during the first month of its life, and the mother discourages the approach of other females. But afterward the growing infant associates freely with adults. It is even approached by the males, who often draw close to the mother, smacking their lips in the typical conciliatory signal in order to be near the youngster. From the age of nine months, male baboons progressively lose the protection of their mothers, who reject them with increasing severity. As a result they come to mingle with other members of the troop even more quickly and freely. The social structure of the olive baboon is consistent with this program of socialization. Adult males and females mingle freely, and peripheral groups of males and solitary individuals are rare or absent. Langur social development is far more sex oriented than that of baboons. The infant is relinquished readily to other adult females, who pass it around. But it has little contact with adult males, who are chased away whenever they disturb the youngster. Juvenile males begin to associate with adult males only after eight months, while young females do not permit contact until the onset of sexual activity at the age of three years. The young males spend most of their free time playing. As the play-fighting becomes rougher and requires more room, they tend to drift to the periphery of the group, well apart from the infants and adults. The langur society reflects this form of segregated rearing. Adult males and females tend to remain apart. Groups of peripheral males are common, and they often interact aggressively with the dominant males within the troops in an attempt to penetrate and gain ascendancy.

Socialization can also amplify genetically based variation of individual behavior within troops. The temperament and rank of a higher primate is strongly influenced by its early experiences with its peers and its mother. In his early studies of the Japanese macaque (*Macaca fuscata*), Kawai (1958) was the first to show that the mother's dominance rank has an influence on the ultimate status of her offspring, and the result has since been abundantly confirmed by other investigators. Japanese macaque troops tend to array themselves concentrically around provisioning sites maintained by the human observers, with the dominant males, adult females, and infants and juveniles at the center and subadults and low-ranking males around the periphery. A young male whose mother is highly dominant may never have to leave the center for a period of temporary exile, but instead will probably graduate smoothly to the status of dominant male. A similar form of maternal influence has been described in olive baboons by Ransom and Rowell (1972). Insofar as such primate capabilities have a genetic basis—and there will almost certainly be some degree of

heritability—the initial differences in developmental tendencies will be amplified into the striking divergences in status and roles that provide much of the social structure.

The Evolutionary Pacemaker and Social Drift

The multiplier effect, whether purely genetic in basis or reinforced by socialization and other forms of learning, makes behavior the part of the phenotype most likely to change in response to long-term changes in the environment. It follows that when evolution involves both structure and behavior, behavior should change first and then structure. In other words, behavior should be the evolutionary pacemaker. This is an old idea, with roots extending back at least as far as the sixth edition of Darwin's *Origin of Species* (1872) and the principle of *Funktionswechsel* expressed by Anton Dohrn (1875). Dohrn postulated that the function of an organ, which in retrospect we can view to be most clearly expressed in its behavior, is continually changing and dichotomizing over many generations according to the experience of the organism. Changes in the structure of the organ represent accommodations to these functional shifts. Among recent zoologists, Wickler (1967a,b) has most explicitly argued the same point of view with reference to behavior, citing many examples from birds and fishes. Among the tetraodontiform fishes, to take one of the simpler and clearer cases, a number of species are able to inflate themselves tremendously with water or air as a protective device against predators. In young porcupine fishes of the genus *Diodon*, the median fins disappear into pouches of the skin that fold inward during inflation. The inflated stage has become irreversible in the diodontid genus *Hyosphaera*, while the tetraodontid globe fish, *Kanduka michiei*, not only is permanently inflated but also has lost the dorsal fin and reduced the anal fin to vestigial form. Social behavior also frequently serves as an evolutionary pacemaker. The entire process of ritualization, during which a behavior is transformed by evolution into a more efficient signaling device, typically involves a behavioral change followed by morphological alterations that enhance the visibility and distinctiveness of the behavior.

The relative lability of behavior leads inevitably to *social drift*, the random divergence in the behavior and mode of organization of societies or groups of societies. The term *random* means that the behavioral differences are not the result of adaptation to the particular conditions by which the habitats of one society differ from those of other societies. If the divergence has a genetic basis, the hereditary component of social drift is simply the same as genetic drift, an evolutionary phenomenon whose potential has been thoroughly investigated by the conventional models of mathematical population genetics (see Chapter 4). The component of divergence based purely on differences in experience can be referred to as tradition drift

(Burton, 1972). The amount of variance within a population of societies is the sum of the variances due to genetic drift, tradition drift, and their interaction. In any particular case the genetic and tradition components will be difficult to tease apart and to measure. Even if the alteration in social structure of a group is due to a behavioral change in a single key individual, we cannot be sure that this member was not predisposed to the act by a distinctive capability or temperament conferred by a particular set of genes. And then, how can the relative contributions of the genetic component be estimated? Burton has described an example of social drift in the Gibraltar population of the Barbary ape (*Macaca sylvanus*) which she suggests may be due to tradition drift. In the late 1940's infants were handled by both adult females, particularly siblings of the mother, and adult males. At the present time the lending of infants is confined mostly to the adult males, who use them as conciliation devices in interactions with other males. In the 1940's the Gibraltar population consisted of two strains, namely, the monkeys derived from the original population that occupied the island prior to World War II, and those derived from African imports made to secure the population. The mixed population probably had greater genetic variability and was in a position to evolve to a limited degree in a few generations, but it is impossible to judge to what extent evolution occurred and influenced the behavioral trait in question. Equal uncertainty extends even to the famous cultural innovations of the Japanese macaques (*M. fuscata*) of Koshima Island. At the age of 18 months, the female monkey "genius" Imo invented potato washing in the sea, a skill which then spread through the Koshima troop. At the age of four years she invented the flotation method of separating wheat grains from sand (Kawai, 1965a). Did Imo's achievements result from a rare genetic endowment, likely to occur in only some of the macaque troops picked at random? Or was she well within the range of variation of most of the local populations, so that any troop first encountering the sea and certain foods under the same conditions as those on Koshima might have responded with the inventions? If the former, the drift could be said to be primarily genetic drift; if the latter, it was primarily tradition drift.

To find an example of unalloyed tradition drift, we might have to travel phylogenetically all the way up to human cultural evolution. Cavalli-Sforza (1971) and Cavalli-Sforza and Feldman (1973) have suggested that in human social evolution the equivalent of an important mutation is a new idea. If it is acceptable and advantageous, the idea will spread quickly. If not, it will decline in frequency and be forgotten. Tradition drift in such instances, like purely genetic drift, has stochastic properties amenable to mathematical analysis. Probabilities can presumably be written first for the interaction between the two or more people who play the active and passive roles in the transmission, and then for the acceptance by each passive individual.

It is possible that a formal theory of tradition drift can be created that roughly parallels the sophisticated one already in existence for genetic drift.

The Concept of Adaptive Demography

All true societies are differentiated populations. When cooperative behavior evolves it is put to service by one kind of individual on behalf of another, either unilaterally or mutually. A male and a female cooperate to hold a territory, a parent feeds its young, two nurse workers groom a honeybee queen, and so forth. This being the case, the behavior of the society as a whole can be said to be defined by its demography. The breeding females of a bird flock, the helpless infants of a baboon troop, and the middle-aged soldiers of a termite colony are examples of demographic classes whose relative proportions help determine the mass behavior of the group to which they belong.

The proportions of the demographic classes also affect the fitness of the group and, ultimately, of each individual member. A group comprised wholly of infants or aging males will perish—obviously. Another, less deviant, group has a higher fitness that can be defined as a higher probability of survival, which can be translated as a longer waiting time to extinction. Either measure has meaning only over periods of time on the order of a generation in length, because a deviant population allowed to reproduce for one to several generations will go far to restore the age distribution of populations normal for the species. Unless the species is highly opportunistic, that is, unless it follows a strategy of colonizing empty habitats and holding on to them only for a relatively short time, the age distribution will tend to approach a steady state. In species with seasonal natality and mortality, which is to say nearly all animal species, the age distribution will undergo annual fluctuation. But even then the age distribution can be said to approach stability, in the sense that the fluctuation is periodic and predictable when corrected for season.

A population with a stable age distribution is not ipso facto well adjusted to the environment. It can be in a state of gradual decline, destined ultimately for extinction; or it can be increasing, in which case it may still be on its way to a population crash that leads to a decimation of numbers, strong deviation in the age distribution, and possibly even extinction. Only if its growth is zero when averaged out over many generations can the population have a chance of long life. There is one remaining way to be a success. A population headed for extinction can still possess a high degree of fitness if it succeeds in sending out propagules and creates new populations elsewhere. This is the basis of the opportunistic strategy, to be described in greater detail in Chapter 4.

We can therefore speak of a "normal" demographic distribution as the age distribution of the sexes and castes that occurs in populations with a high degree of fitness. But to what extent is the demographic distribution itself really adaptive? This is a semantic distinction that depends on the level at which natural selection acts to sustain the distribution. If selection operates to favor individuals but not groups, the demographic distribution will be an incidental effect of the selection. Suppose, for example, that a species is opportunistic, and females are strongly selected for their capacity to produce the largest number of offspring in the shortest possible time. Theory teaches that evolution will probably proceed to reduce the maturation time, increase the reproductive effort and progeny size, and shorten the natural life span. The demographic consequence will be a flattening of the age pyramid. A squashed age distribution is a statistical property of the population. It is a secondary effect of the selection that occurred at the individual level, contributes nothing of itself to the fitness of either the individual or the population, and therefore cannot be said to be adaptive in the usual sense of the word.

Now consider a colony of social insects. The demographic distribution, expressed in part by the age pyramid, is vital to the fitness of the colony as a whole and particularly of the progenitrix queen, with reference to whom the nonreproductive members can be regarded as a somatic extension. If too few soldiers are present at the right moment, the colony may be demolished by a predator; or if too few nurse workers of the appropriate age are not always available, the larvae may starve to death. Thus the demographic distribution is adaptive, in the sense that it is tested directly by natural selection. It can be shaped by altering growth thresholds, so that a lower or higher proportion of nymphs or larvae reaching a certain weight, or detecting a sufficient amount of a certain odorous secretion, is able to metamorphose into a given caste. It can also be shaped by changing the periods of time an individual spends at a certain task. For example, if each worker has a shortened tenure as a nurse, the percentage of colony members who are active nurses at any moment will be less. Finally, the demographic distribution can be changed by altering longevity: if soldiers die sooner, their caste will be less well represented numerically in each moment of time.

With reference to social behavior, the two most important components of a demographic distribution are age and size. In Figure 2-2 I have represented age-size frequency distributions as they might appear in two societies (*A* and *B*) subjected to little selection at the level of the society, as opposed to the distribution in a society (*C*) in which such group selection has been a major force. All can agree that demography is more interesting when it is adaptive. The patterns are likely to be not only more complex but more meaningful. Nonadaptive demography follows from a study of the behavior and life cycles of individuals; but adaptive demography must be analyzed holistically before the behavior and life cycles of the individuals take on meaning.

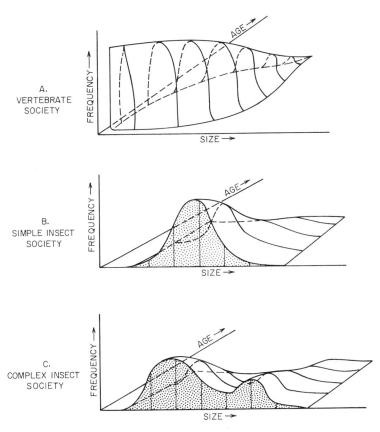

Figure 2-2 The age-size frequency distributions of three kinds of animal societies. These examples are based on the known general properties of real species but their details are imaginary. *A*: The distribution of the "vertebrate society" is nonadaptive at the group level and therefore is essentially the same as that found in local populations of otherwise similar, nonsocial species. In this particular case the individuals are shown to be growing continuously throughout their lives, and mortality rates change only slightly with age. *B*: The "simple insect society" may be subject to selection at the group level, but its age-size distribution does not yet show the effect and is therefore still close to the distribution of an otherwise similar but nonsocial population. The age shown is that of the imago, or adult instar, during which most or all of the labor is performed for the colony; and no further increase in size occurs. *C*: The "complex insect society" has a strongly adaptive demography reflected in its complex age-size curve: there are two distinct size classes, and the larger is longer lived.

In an adaptive setting, the moments of demographic frequency distributions take on new significance. The means still reflect the rough adjustment of the size and age of individual organisms to the exigencies of the environment. The variance and higher moments acquire a directly adaptive significance because of their reflection of caste structure. These and other aspects of demography will be discussed in Chapters 4 and 14.

The Kinds and Degrees of Sociality

All previous attempts to classify animal societies have failed. The reason is very simple: classification depends on the qualities chosen to specify the sets, and no two authors have agreed on which qualities of sociality are essential. The more kinds of social traits employed, the more complex the classification and the more likely its author to come into serious conflict with other classifiers. The pioneering system of Espinas (1878), for example, and the one that W. M. Wheeler (1930) derived from it have at least the virtue of simplicity. They were based upon whether the associations are active or passive, primarily reproductive, nutritive, or defensive, and colonial or free-ranging. From this elementary mixture Wheeler produced 5 basic kinds of societies. In contrast, Deegener (1918), who paid close attention to fine details of food habits, life cycles, orientation cues, and so on, proposed no less than 40 categories—and still more, if certain imprecise subdivisions are also recognized. Unfortunately, Deegener felt compelled to provide a full terminology for his classification. One form of concunnubium, he noted, is the amphoterosynhesmia, a swarm of both sexes gathered for reproductive purposes. If this does not discourage all but the most dedicated lexicographic scholars, consider the syncheimadium, a hibernating aggregation, or the polygynopaedium, an association of mothers and daughters each of which is reproducing parthenogenetically.

Deegener's *reductio ad absurdum* served as fair warning that classification based upon all relevant traits is a bottomless pit. It is to be avoided only by turning to the social qualities themselves and cataloging *them*, according to our intuitive idea of the way they can clarify social process as opposed to static consociations of individuals. The usefulness of such a list is twofold. First, by explicitly distinguishing and labeling discrete traits, we identify certain phenomena that have been hitherto understudied. Second, the list can be consulted for help in the preparation of sociograms (complete descriptions of the social behavior) of particular species. Recently a growing number of authors have reflected on the abstract qualities of social organization, among them Thompson (1958), Crook (1970a), Mesarović et al. (1970), Brereton (1971), Cohen (1971), and Wilson (1971a). From the suggestions expressed in these articles, and from my own further study of the literature of social systems, I have compiled the following set of ten qualities of sociality, which I believe can be both measured and ultimately incorporated into models of particular social systems (see also Figure 2-3):

1. *Group size.* Joel Cohen (1969, 1971) has shown the existence

of orderly patterns in the frequency distribution of group size among primate troops. In the case of closed, relatively stable groups, much (but not all) of the information can be accounted for with stochastic models that assume constant gain rates through birth and immigration and constant loss rates through death and emigration. Orderliness also occurs in the frequency distributions of casual subgroups in monkeys and man, and can be predicted in good part by reference to the variation of the attractiveness of groups of different sizes and of the attractiveness and joining tendency of individual group members.

2. *Demographic distributions.* The significance of these frequency distributions and the degree of their stability were discussed in the previous section on adaptive demography.

3. *Cohesiveness.* Intuitively we expect that the closeness of group members to one another is an index of the sociality of the species. This is true, first, because the effectiveness of group defense and group feeding is enhanced by tight formations and, second, because the widest range of communication channels can be brought into play at close range. There is indeed a correlation between physical cohesiveness and the magnitude of the other nine social parameters listed here but it is only loose. Honeybee colonies, for example, are more cohesive than nesting aggregations of solitary halictid bees. But chimpanzee troops and human societies are much less cohesive than fish schools and herds of cattle.

4. *Amount and pattern of connectedness.* The network of communication within a group can be patterned or not. That is, different kinds of signals can be directed preferentially at particular individuals or classes of individuals; or else, in the unpatterned case, all signals can be directed randomly for periods of time at any individuals close enough to receive them. In unpatterned networks, such as fish schools and temporary roosting flocks of birds, the number of arcs per node in the network, meaning the number of individuals contacted by the average member per unit of time, provides a straightforward measure of the sociality. This is a number that increases with the cohesiveness of the group or, in the case of animals that communicate over distances exceeding the diameter of the aggregation, the size of the group. In the case of patterned networks, the situation is radically different. Hierarchies with multiple levels can be constructed with relatively few arcs (see Figure 2-3). Provided the members are also performing separate functions, the degree of coordination and efficiency of the group as a whole can be vastly increased over an unpatterned network containing a comparable number of members, even if the degree of connectedness (the number of arcs per member) is much lower. All higher forms of societies, those recognized to possess a strong development of the other nine social qualities, are characterized by an advanced degree of patterning in connectedness. They are not always characterized by a large amount of connectedness, however.

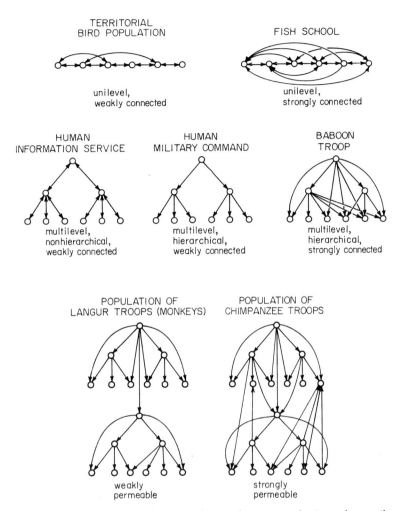

TERRITORIAL
BIRD POPULATION

unilevel,
weakly connected

FISH SCHOOL

unilevel,
strongly connected

HUMAN
INFORMATION SERVICE

multilevel,
nonhierarchical,
weakly connected

HUMAN
MILITARY COMMAND

multilevel,
hierarchical,
weakly connected

BABOON
TROOP

multilevel,
hierarchical,
strongly connected

POPULATION OF
LANGUR TROOPS (MONKEYS)

weakly
permeable

POPULATION OF
CHIMPANZEE TROOPS

strongly
permeable

Figure 2-3 Seven social groups depicted as networks in order to illustrate variation in several of the qualities of sociality. The traits of the social groups are abstracted, and the details are imaginary.

5. *Permeability.*   To say that a society is closed means that it communicates relatively little with nearby societies of the same species and seldom if ever accepts immigrants. A troop of langurs (*Presbytis entellus*) is an example of a society with low permeability. Exchanges between troops consist mostly of aggressive encounters over territory, and, at least in the dense populations of southern India, immigration is mostly limited to the intrusion of males who usurp the position of the dominant male (Ripley, 1967; Sugiyama, 1967). At the opposite extreme are the very permeable troops of chimpanzees, in which groups temporarily fuse and exchange members freely. All other things being equal, an increase in permeability should result in an increase in gene flow through entire populations and a reduced

degree of genetic relationship between any two members chosen at random from within a single society. The consequences of these relationships for social evolution will be explored in Chapters 4 and 5. Increased permeability is also associated with a reduction in the stability of such interpersonal relationships within the society as dominance hierarchies, coalitions, and kin groups. Whether the permeability is the cause or the effect of the correlation can be determined only by analysis of particular cases.

6. *Compartmentalization.*   The extent to which the subgroups of a society operate as discrete units is another measure of the complexity of the society. When confronted with danger, a herd of wildebeest flees as a disorganized mob, with the mothers turning individually to defend themselves and their calves only if overtaken. Zebra herds, in contrast, sort out into family groups, with each dominant stallion maneuvering to place himself between the predator and his harem. When the danger passes, the families merge again into a single formation. The colonies of certain ant species, including *Oecophylla smaragdina* and members of the *Formica exsecta* group, greatly increase their size and complexity by building new nests that are modular replications of the original mother nests. The subunits remain in contact through the continuous exchange of individuals, but they are also capable of independent existence and can become mother nests themselves to initiate new episodes of colonization.

7. *Differentiation of roles.*   Specialization of members of a group is a hallmark of advance in social evolution. One of the theorems of ergonomic theory is that for each species (or genotype) in a particular environment there exists an optimum mix of coordinated specialists that performs more efficiently than groups of equal size consisting wholly of generalists (Wilson, 1968a; see also Chapter 14). It is also true that under many circumstances mixes of specialists can perform qualitatively different tasks not easily managed by otherwise equivalent groups of generalists, whereas the reverse is not true. Packs of African wild dogs, to cite one case, break into two "castes" during hunts: the adult pack that pursues, and the adults that remain behind at the den with the young. Without this division of labor, the pack could not subdue a sufficient number of the large ungulates that constitute its chief prey (Estes and Goddard, 1967). The development of elaborate caste systems is correlated in ants with an increase in colony size and an enlargement of the communication repertory (Wilson, 1953, 1961). In a wholly different environment, the species of marine invertebrates with the largest colonies are also generally those with the greatest differentiation of the zooids.

8. *Integration of behavior.*   The obverse of differentiation is integration: a set of specialists cannot be expected to function as well as a group of generalists unless they are in the correct proportions and their behaviors coordinated. The following example is among the most striking known in the social insects. Minor workers of the

tropical ant *Pheidole fallax* forage singly for food outside the nest. When they discover a food particle too large to carry home, they lay an odor trail back to the nest. The trail pheromone is produced by a hypertrophied Dufour's gland and released through the sting when the tip of the abdomen is dragged over the ground. The trail attracts and guides both the other minor workers and members of the soldier caste, all of whom then assist in the cutting up and transport of the food. But the soldiers are specialized for yet another function: they defend the food from intruders, especially members of other ant colonies. Their behavior includes the release of skatole, a fetid liquid manufactured in the enlarged poison gland. The soldiers do not possess a visible Dufour's gland and cannot lay odor trails of their own; the minor workers have ordinary poison glands which do not secrete skatole. Together the two castes perform the same task, perhaps with greater efficiency, as do the workers of other myrmicine ant species that constitute a single caste. But either caste would be less effective if their efforts were not coordinated and if each were required to perform alone. In fact, the soldier caste would be quite incompetent at foraging (Law et al., 1965).

9. Information flow. Norbert Wiener said that sociology, including animal sociobiology, is fundamentally the study of the means of communication. Indeed, many of the social qualities I am listing here could, with varying degrees of effort, be subsumed under communication. The magnitude of a communication system can be measured in three ways: the total number of signals, the amount of information in bits per signal, and the rate of information flow in bits per second per individual and in bits per second for the entire society. These measures will be exemplified and evaluated in Chapter 8.

10. Fraction of time devoted to social behavior. The allocation of individual effort to the affairs of the society is one fair measure of the degree of sociality. This is the case whether effort is measured by the percentage of the entire day devoted to it, by the fraction of time devoted to it out of all the time spent engaged in any activity, or by the fraction of energy expended. Social effort reflects, but is not an elementary function of, cohesiveness, differentiation, specialization, and rate of information flow. R. T. Davis and his coworkers (1968) detected a rough correlation with these several traits within the primates. Lemurs (*Lemur catta*), generally regarded to have a somewhat simple social organization, devote approximately 20 percent of their time to social behavior, while pig-tailed macaques (*Macaca nemestrina*) and stump-tailed macaques (*M. speciosa*), which by other criteria are relatively sophisticated social animals, invest about 80 and 90 percent of their time, respectively, in social acts. Intermediate degrees of commitment are shown by New World monkeys and, more surprisingly, by the rhesus macaque (*M. mulatta*). Strong differences were also recorded in the times devoted to different kinds of social behavior (Figure 2-4).

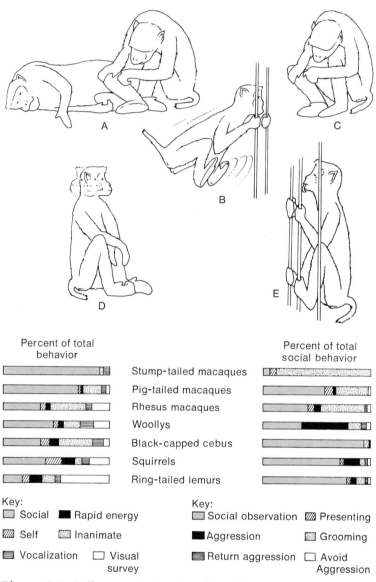

Key:

☐ Social	■ Rapid energy
▨ Self	▦ Inanimate
☰ Vocalization	☐ Visual survey

☐ Social observation	▨ Presenting
■ Aggression	▦ Grooming
☰ Return aggression	☐ Avoid Aggression

Figure 2-4 Differences in the time devoted to social behavior (left) and the different categories of social behavior (right) in seven species of primates. Major categories of behavior include (*A*) social behavior; (*B*) rapid energy expenditure; (*C*) self-directed behavior; (*D*) visual survey; and (*E*) manipulation of inanimate objects or cage. (From Jolly, 1972a; after Davis et al., 1968.)

The worker castes of higher social insects are nearly as fully committed to social existence as it is possible to conceive. Except for self-grooming and feeding, virtually all of their behavior is oriented toward the welfare of the colony. In most cases even feeding is to some degree social. The workers repeatedly regurgitate to one another, evening out the quantity and quality of food stored in their crops. The queen honeybee uses even self-grooming to a social end. By rubbing her legs over her own head and body she spreads queen substance (9-ketodecenoic acid) and mixes it with other attractant pheromones. As workers lick the surface of her body they pick up the queen substance, which proceeds to affect their behavior and physiology in several ways beneficial to both the queen and the colony as a whole (Wilson, 1971a).

There is still one more way of measuring the degree of sociality of a species, which might conveniently be called *minimum specification*. In a word, this criterion defines the complexity of a system as the number of its constituent units that need to be characterized in order to specify the system. This number will usually fall very short of the actual number of units present. Herbert A. Simon (1962), when characterizing the limits of complexity in general systems, observed, "Most things are only very weakly connected with most other things; for a tolerable description of reality only a tiny fraction of all probable interactions needs to be taken into account." Paul A. Weiss (1970) independently extended the same insight, as follows: "I have tried to translate the formula, 'the whole is more than the sum of its parts' into a mandate for action: a call for spelling out the irreducible minimum of supplementary information that is required beyond the information derivable from the knowledge of the ideally separated parts in order to yield a complete and meaningful description of the ordered behavior of the collective."

The criterion of minimum specification might be usefully extended to sociobiology as the number of individuals which, on the average, must be put together in order to observe the full behavioral repertory of the species. The criterion is not a simple quality of the society but rather a number derived as a compound function of most of the ten qualities of social structure previously cited. Consider the two species represented in Figure 2-5. The isolated individual of a solitary species typically has a larger behavioral repertory than the isolated member of a highly social form. Only a few more individuals need be added to evoke the remainder of the full potential of a solitary species: sexual behavior, territoriality, and even density-dependent responses such as emigration. With the addition of individuals to the group of the social species, the expressed behavioral repertory climbs more slowly. To reach its limit, every caste and adult age group must be added. The final result is a repertory larger than that of the solitary species.

Classifications of societies, as distinguished from classifications of social qualities, are feasible if confined to particular groups of organisms and based upon the sets of qualities which experience has proved to be most relevant to social evolution in the groups. A case in point is the classification of insect societies developed by Wheeler (1928)

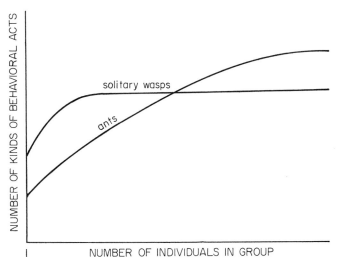

Figure 2-5 An application of the criterion of minimum specification to the characterization of social complexity in two species of insects. One solitary wasp has a larger behavioral repertory than one ant, and a smaller group of wasps is required for the display of the entire repertory of the species. As the ant group is enlarged, the full repertory is approached more slowly, but it is ultimately larger. These qualitative statements are correct, but the details of the curves shown are imaginary.

and Michener (1969) and refined by Wilson (1971a); this rather intricate system will be presented in full in Chapter 20.

The Concept of Behavioral Scaling

In the early years of vertebrate sociobiology it was customary for observers to assume that social structures, no less than ethological fixed action patterns, are among the invariant traits by which species can be diagnosed. If a short-term field study revealed no evidence of territoriality, dominance hierarchies, or some other looked-for social behavior, then the species as a whole was characterized as lacking the behavior. Even so skillful a field zoologist as George Schaller could state confidently on the basis of relatively few data that the gorilla "shares its range and its abundant food resources with others of its kind, disdaining all claims to a plot of land of its own."

Experience has begun to stiffen our caution about generalizing beyond single populations of particular species and at times other than the period of observation. Largely because of the amplifier effect, social organization is among the most labile of traits. The following case involving Old World monkeys is typical. Troops of vervets (*Cercopithecus aethiops*) observed by Struhsaker (1967) at the Amboseli Masai Game Reserve, Kenya, are strictly territorial and maintain rigid dominance hierarchies by frequent bouts of fighting. In contrast, those studied by J. S. Gartlan (cited by Thelma Rowell, 1969a) in Uganda had no visible dominance structure at the time of observation, exchange of males occurred frequently between troops, and fighting was rare.

In some cases differences of this sort are probably due to geographic variation of a genetic nature, originating in the adaptation of local populations to peculiarities in their immediate environment. Some fraction can undoubtedly also be credited to tradition drift. But a substantial percentage of cases do not represent permanent differences between populations at all: the societies are just temporarily at different points on the same *behavorial scale*. Behavioral scaling is variation in the magnitude or in the qualitative state of a behavior which is correlated with stages of the life cycle, population density, or certain parameters of the environment. It is a useful working hypothesis to suppose that in each case the scaling is adaptive, meaning that it is genetically programmed to provide the individual with the particular response more or less precisely appropriate to its situation at any moment in time. In other words, the entire scale, not isolated points on it, is the genetically based trait that has been fixed by natural selection (Wilson, 1971b). To make this notion clearer, and before we take up concrete examples, consider the following imaginary case of aggressive behavior programmed to cope with varying degrees of population density and crowding. At low population densities, all aggressive behavior is suspended. At moderate densities, it takes a mild form such as intermittent territorial defense. At high densities, territorial defense is sharp, while some joint occupancy of land is also permitted under the regime of dominance hierarchies. Finally, at extremely high densities, the system may break down almost completely, transforming the pattern of aggressive encounters into homosexuality, cannibalism, and other symptoms of "social pathology." Whatever the specific program that slides individual responses up and down the aggression scale, each of the various degrees of aggressiveness is adaptive at an appropriate population density— short of the rarely recurring pathological level. In sum, it is the total *pattern* of aggressive responses that is adaptive and has been fixed in the course of evolution.

In the published cases of scaled social response, the most frequently reported governing parameter is indeed population density. True threshold effects suggested in the imaginary example do exist in nature. Aggressive encounters among adult hippopotami, for example, are rare where populations are low to moderate. However, when populations in the Upper Semliki near Lake Edward became so dense that there was an average of one animal to every 5 meters of riverbank, males began to fight viciously, sometimes even to the death (Verheyen, 1954). When snowy owls (*Nyctea scandiaca*) live at normal population densities, each bird maintains a territory about 5,000 hectares in extent and it does not engage in territorial defense. But

when the owls are crowded together, particularly during the times of lemming highs in the Arctic, they are forced to occupy areas covering as few as 120 hectares. Under these conditions they defend the territories overtly, with characteristic sounds and postures (Frank A. Pitelka, in Schoener, 1968a). A similar density threshold for the expression of territorial defense has been reported in European weasels (*Mustela nivalis*) by Lockie (1966). A second class of aggression scaling effects, to be discussed in some detail in Chapter 13, is the transitions that occur in many vertebrate species from territorial to dominance behavior as the density passes a critical value.

Not all density-dependent social responses consist of aggressive behavior. When populations of European voles (*Microtus*) reach certain high densities, the females join in little nest communities, defend a common territory, and raise their young together (Frank, 1957). In a basically similar way, flocks of wild turkeys (*Meleagris gallopavo*) gradually increase in size as population density increases (Leopold, 1944).

Group size itself can affect the intensity of aggressive behavior in ways that can be reliably dissociated from the parallel influence of population density. Blue monkeys (*Cercopithecus mitis*) of the Budongo Forest, Uganda, are organized into troops of highly variable size. When one large group encounters another at a rich food source such as a fruiting fig tree, the adults threaten and chase one another until one group retreats from the area. Small parties, however, coalesce peacefully when they meet at feeding places (Aldrich-Blake, 1970). It is tempting to speculate that territorial behavior develops in the troops only when their size becomes so large that they have to compete with other large troops for sufficient food. In other words, they resort to aggression only if it is profitable. Aggressiveness also increases as a function of group size in colonies of many kinds of social insects. For example, newly established colonies of harvesting ants (*Pogonomyrmex badius*), which consist of only a few tens of workers, flee when their nest is broken open. But mature colonies, with populations of about 5,000 workers, pour out of the nest and attack any intruder in reach.

Jenkins (1961) has reported a case of variable social responses that apparently depends on the nature of the habitat. Partridges (*Perdix perdix*) living in thick vegetation seldom interact with one another, even when their populations are highly concentrated. In poor cover, however, their home ranges expand and the birds interact almost continuously.

The availability and quality of food can also move groups along behavioral scales. Well-fed honeybee colonies are very tolerant of intruding workers from nearby hives, letting them penetrate the nest and even take supplies without opposition. But when the same colonies are allowed to go without food for several days, they attack every intruder at the nest entrance. In general, primates also become increasingly intolerant of strangers and aggressive toward other group members during times of food shortages. Arthur N. Bragg (1955–1956) has reported a remarkable case of social scaling in the amphibians. Tadpoles of the spadefoot toads (*Scaphiopus*) are opportunistic, developing rapidly in short-lived rain pools. When nutritional conditions are good, the tadpoles live singly. When food is short, they become social, forming fishlike schools. By working in unison, the tadpoles also stir up the bottom more efficiently, with the result that each is rewarded with a larger quantity of food.

Even the way food is distributed in the environment can evoke strong variation in social behavior. The workers of higher ants, particularly those belonging to the dominant subfamilies Myrmicinae, Dolichoderinae, and Formicinae, forage singly outside the nest. If the food they encounter is widely dispersed as particles small enough to be carried home by a solitary ant, no recruitment occurs. When the food occurs in larger masses the workers of many species return home laying an odor trail behind them. By this means enough nestmates are eventually recruited either to transport the food masses or to protect them from the workers of other colonies. N. Chalmers (cited by Thelma Rowell, 1969b) found that mangabeys (*Cercocebus albigena*) have more aggressive interactions when they are feeding on large fruit growing singly than when the food is evenly dispersed in the trees. According to Rowell, forest-dwelling baboons (*Papio anubis*) in Uganda display a parallel scale of aggression. In most cases their food is spread out and abundant, and aggressive behavior is rare. But when they encounter piles of elephant dung old enough to contain sprouting seedlings of the kind prized as food, they exchange threats in attempts to gain possession.

Many forms of social behavior are episodic, in extreme cases limited to narrow periods in the day, season, or life cycle. Courtship behavior and parental care, as well as the maintenance of territories and dominance hierarchies specifically linked to these behaviors, are usually seasonal. Beyond the many vertebrate examples made familiar in the ethological literature (Marler and Hamilton, 1966; Hinde, 1970), there are other cases, especially among the invertebrates, that follow unusual temporal patterns. The smallest spiny lobsters (*Jasus lalandei*), measuring less than 4 centimeters in length, take solitary shelter during the day in separate holes in the back of caves and ledges. Those somewhat larger (4-9 centimeters) form aggregations in the caves and beneath the ledges. The biggest lobsters, 9 centimeters or more in length, usually take single possession of similar larger shelters, which they then defend as territories (Fielder, 1965). Bombardier beetles (*Brachinus*, family Carabidae) strongly aggregate during most of the year, orienting toward one another by odor cues. In the spring, when the businesses of individual courtship and egg-laying intervene, the aggregations break up and the beetles hide separately (Wautier, 1971).

Perhaps the most dramatic and instructive examples of behavioral scaling are the shifts in social behavior that occur on a daily basis in certain bird species. During the breeding season in East Africa, males of two species of paradise widow birds (genus *Steganura*) and of the straw-tailed widow bird (*Tetraenura fischeri*) are strongly territorial throughout the day, using elaborate and beautiful displays to exclude their conspecific rivals. Just before sunset, however, they quit and join the females and other males to form foraging groups away from the territories. The oilbirds (*Steatornis caripensis*) of South America nest on ledges in caves. Couples form permanent pair bonds and defend the scarce small spaces suitable for building nests. However, in the evening all flock together in feeding groups that search for the scattered oil palms and other fruit-bearing trees on which the species depends (Snow, 1961). Such patterns are in fact commonplace among colonially nesting birds, including many seabirds. They show that evolution can easily program social behavior to switch from one major state to another according to even a diel rhythm.

The Dualities of Evolutionary Biology

The theories of behavioral biology are riddled with semantic ambiguity. Like buildings constructed hastily on unknown ground, they sink, crack, and fall to pieces at a distressing rate for reasons seldom understood by their architects. In the special case of sociobiology, the unknown substratum is usually evolutionary theory. We should therefore set out to map the soft areas in the relevant parts of evolutionary biology. With remarkable consistency the most troublesome evolutionary concepts can be segregated into a series of dualities. Some are simple two-part classifications, but others reflect more profound differences in levels of selection and between genetic and physiological processes.

Adaptive versus *nonadaptive traits.*   A trait can be said to be adaptive if it is maintained in a population by selection. We can put the matter more precisely by saying that another trait is nonadaptive, or "abnormal," if it reduces the fitness of individuals that consistently manifest it under environmental circumstances that are usual for the species. In other words, deviant responses in abnormal environments may not be nonadaptive— they may simply reflect flexibility in a response that is quite adaptive in the environments ordinarily encountered by the species. A trait can be switched from an adaptive to a nonadaptive status by a simple change in the environment. For example, the sickle-cell trait of human beings, determined by the heterozygous state of a single gene, is adaptive under living conditions in Africa, where it confers some degree of resistance to falciparum malaria. In Americans of African descent, it is nonadaptive, for the simple reason that its bearers are no longer confronted by malaria.

The pervasive role of natural selection in shaping all classes of traits in organisms can be fairly called the central dogma of evolutionary biology. When relentlessly pressed, this proposition may not produce an absolute truth, but it is, as G. C. Williams disarmingly put the matter, the light and the way. A large part of the contribution of Konrad Lorenz and his fellow ethologists can be framed in the same metaphor. They convinced us that behavior and social structure, like all other biological phenomena, can be studied as "organs," extensions of the genes that exist because of their superior adaptive value.

How can we test the adaptation dogma in particular instances? There exist situations in which social behavior temporarily manifested by animals seems clearly to be abnormal, because it is possible to diagnose the causes of the deviation and to identify the response as destructive or at least ineffectual. When groups of hamadryas baboons were first introduced into a large enclosure in the London Zoo, social relationships were highly unstable and males fought viciously over possession of the females, sometimes to the death (Zuckerman, 1932). But these animals had been thrown together as strangers, and the ratio of males to females was higher than in the wild. Kummer's later studies in Africa showed that under natural conditions hamadryas societies are stable, with the basic unit composed of several adult females and their offspring dominated by one or two males. When C. R. Carpenter introduced rhesus macaques into the seminatural environment of Cayo Santiago, a small island off the south coast of Puerto Rico, the social structure was at first chaotic. Several ordinarily aberrant behaviors, including masturbation, female homosexuality, and copulation of members belonging to different groups, were commonplace (Carpenter, 1942a). In subsequent years the social structure stabilized and the deviant behaviors became rare. The Cayo Santiago colony converged in its social behavior toward the native populations of India.

For each such case of temporary maladaptation, many others appear to us to occupy a gray zone of uncertainty. Sometimes seemingly abnormal behavior proves on closer inspection to be adaptive after all. Consider specific cases of homosexuality, which we are conditioned to think of as necessarily abnormal. In the macaques pseudo-copulation is a common ritual used to express rank among males, with the dominant individual mounting the subordinate. In the South American leaf fish *Polycentrus schomburgkii* homosexuality is an imitation of female color change and behavior by subordinate males as they approach territorial males. True females ready to spawn enter the territories, turn upside down, and deposit their eggs on the lower surfaces of objects in the water. During the spawnings the pseudofemales often enter at the same time. In this way they evidently attempt to fool the resident males and to "steal" a fertilization by depositing their own sperm around the newly laid eggs (Barlow, 1967). If the interpretation is correct, we have here a case of transvestism evolved to serve heterosexuality!

Furthermore, what is adaptive social behavior for one member of the family may be nonadaptive for another. The Indian langur males who invade troops, overthrow the leaders, and destroy their offspring are clearly improving their own fitness, but at a severe cost to the females they take over as mates. When male elephant seals fight for possession of harems, they are being very adaptive with respect to their own genes, but they reduce the fitness of the females whose pups they trample underfoot.

Monadaptive versus *polyadaptive traits.* Social evolution is marked by repeated strong convergence of widely separate phylogenetic groups. The confusion inherent in this circumstance is worsened by the still coarse and shifting nature of our nomenclatural systems. Ideally, we should try to have a term for each major functional category of social behavior. This semantic refinement would result in most kinds of social behavior being recognized as monadaptive, that is, possessing only one function. In our far from perfect language, however, most behaviors are artificially construed as polyadaptive. Consider the polyadaptive nature of "aggression," or "agonistic behavior," in monkeys. Males of langurs, patas, and many other species use aggression to maintain troop distance. Similar behavior is also employed by a diversity of species, including langurs, to establish and to sustain dominance hierarchies. Male hamadryas baboons use aggression to herd females and discourage them from leaving the harems. Aggression, in short, is a vague term used to designate an array of behaviors, with various functions, that we intuitively feel resemble human aggression.

Some social behavior patterns nevertheless remain truly polyadaptive even after they have been semantically purified. Allogrooming in rhesus monkeys, for examples, serves the typically higher primate function of conciliation and bond maintenance. Yet it retains a second, apparently more primitive, cleansing function, because monkeys kept in isolation often develop severe infestations of lice. In some bird species, flocking behavior undoubtedly serves the dual function of predator evasion and improvement in foraging efficiency.

Reinforcing versus *counteracting selection.* A single force in natural selection acts on one or more levels in an ascending hierarchy of units: the individual, the family, the troop, and possibly even the entire population or species. If affected genes are uniformly favored or disfavored at more than one level, the selection is said to be reinforcing. Evolution, meaning changes in gene frequency, will be accelerated by the additive effects inserted at multiple levels. This process should offer no great problem to mathematicians. By contrast, the selection might be counteracting in nature: genes favored by a selective process at the individual level could be opposed by the same process at the family level, only to be favored again at the population level, and so on in various combinations. The compromise gene frequency is of general importance to the theory of social evolution, but it is mathematically difficult to predict. It will be considered formally in Chapters 5 and 14.

Ultimate versus *proximate causation.* The division between functional and evolutionary biology is never more clearly defined than when the proponents of each try to make a pithy statement about causation. Consider the problem of aging and senescence. Contemporary functional biologists are preoccupied with four competing theories of aging, all strictly physiological: rate-of-living, collagen wear, autoimmunity, and somatic mutation (Curtis, 1971). If one or more of these factors can be firmly implicated in a way that accounts for the whole process in the life of an individual, the more narrowly trained biochemist will consider the problem of causation solved. However, only the proximate causation will have been demonstrated. Meanwhile, as though dwelling in another land, the theoretical population geneticist works on senescence as a process that is molded in time so as to maximize the reproductive fitness in particular environments (Williams, 1957; Hamilton, 1966; J. M. Emlen, 1970). These specialists are aware of the existence of physiological processes but regard them abstractly as elements to be jiggered to obtain the optimum time of senescence according to the schedules of survivorship and fertility that prevail in their theoretical populations. This approach attempts to solve the problem of ultimate causation.

How is ultimate causation linked to proximate causation? Ultimate causation consists of the necessities created by the environment: the pressures imposed by weather, predators, and other stressors, and such opportunities as are presented by unfilled living space, new food sources, and accessible mates. The species responds to environmental exigencies by genetic evolution through natural selection, inadvertently shaping the anatomy, physiology, and behavior of the individual organisms. In the process of evolution, the species is constrained not only by the slowness of evolutionary time, which by definition covers generations, but also by the presence or absence of preadapted traits and certain deep-lying genetic qualities that affect the rate at which selection can proceed. These prime movers of evolution (see Chapter 3) are the ultimate biological causes, but they operate only over long spans of time. The anatomical, physiological, and behavioral machinery they create constitutes the proximate causation of the functional biologist. Operating within the lifetimes of organisms, and sometimes even within milliseconds, this machinery carries out the commands of the genes on a time scale so remote from that of ultimate causation that the two processes sometimes seem to be wholly decoupled.

Most psychologists and animal behaviorists trained in the conventional psychology departments of universities are nonevolutionary in their approach. Yet, like good scientists everywhere, they are always probing for deeper, more general explanations. What they should produce are specific assessments of ultimate causation rooted in pop-

ulation biology. What they typically produce instead are the nebulous independent variables of theoretical psychology—attraction-withdrawal thresholds, drive, deep-set aggregative or cooperative tendencies and so forth. And this approach creates confusion, because such notions are ad hoc and can seldom be linked either to neurophysiology or evolutionary biology and hence to the remainder of science.

The ambiguities created are embedded in the very meaning of cause and effect. Instances will be given through the remainder of the book, and concrete aspects of the underlying genetic theory will be discussed in Chapters 4 and 5. For the moment, let us view just one older example in order to illustrate the subtlety of the matter. Allee and Guhl (1942) conducted an experiment in which they daily replaced the oldest resident of a flock of seven white leghorn chickens. Similar flocks were left undisturbed to serve as controls. The experimental group, with a daily turnover of greater than 10 percent, naturally remained in a state of turmoil. The members pecked one another more, ate less food, and consistently lost weight, while the control chickens thrived. Allee and Guhl drew the plausible conclusion that organization in chicken flocks enhances group survival and therefore serves as the basis of natural selection. However, consider what might be the ultimate cause and effect in this case. Dominance hierarchies—the peck orders in these chickens—very likely evolve at the individual level, since it is more advantageous to live in a flock as a subordinate than to live alone. Once in a flock, the chicken would find it fruitful to employ aggression in an effort to ascend the pecking order—but judiciously, so as not to be the object of unnecessary and destructive amounts of retaliation. Hence order in the chicken society is viewed in the second hypothesis as the result of aggression, and not as its cause. In other words, aggression and dominance orders have not evolved as proximate devices to provide an orderly society; rather the order is a by-product of the tempering and compromise of aggressive behavior by individuals who join groups for other reasons.

Ideal versus *optimum permissible traits.* When organisms are thought of as machine analogs, their evolution can be viewed as a gradual perfecting of design. In this conception there exist ideal traits for survival in particular environments. There would be the ideal hammer bill and extrusible tongue for woodpeckers, the ideal caste system for army ants, and so forth. But we know that such traits vary greatly from species to species, even those belonging to the same phyletic group and occupying the same narrowly defined niche. In particular, it is disconcerting to find frequent cases of species with an advanced state or intermediate states of the same character.

Take the theoretical problem created by the primitively social insect species: Why have they progressed no further? Two extreme possibilities can be envisioned (Wilson, 1971a). First, there is what might be termed the "disequilibrium case." This means that the species is still actively evolving toward a higher social level. The situation can arise if social evolution is so slow that the species is embarked on a particular adaptive route but is still in transit (see Figure 2-6A). Bossert (1967) has shown that if the species perches on a knife ridge leading up an adaptive peak, it will move slowly if at all. The reason is that progress toward the optimum phenotype can occur only if the species moves precisely along the narrow path defined by the ridge. If it deviates, either by genetic drift or by

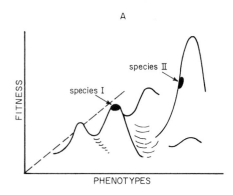

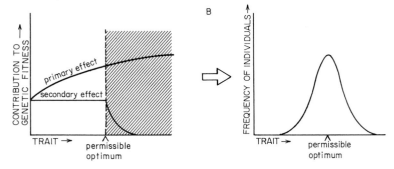

Figure 2-6 Concepts of optimization in evolutionary theory. *A*: An adaptive landscape: a surface of phenotypes (imaginary in this case) in which the similarity of the underlying genotypes is indicated by the nearness of the points on the surface and their relative fitness by their elevation. Species I is at equilibrium on a lower adaptive peak; it is characterized by an optimal permissible trait which is less perfect than the ideal conceivable trait. Species II is in disequilibrium because it is still evolving toward another permissible optimum. *B*: The species shown here is at equilibrium at the permissible optimum of a particular trait. Although the primary function of the trait would be ever more improved by an indefinite intensification of the trait, its secondary effect begins to reduce the fitness conferred on the organism when the trait exceeds a certain value. The threshold value is by definition the permissible optimum.

selection induced by a temporary perturbation in environmental conditions, it will descend rapidly down the steep slope on either side of the ridge. Recovery from this slip is unpredictable and may even lead to a position farther down the ridge. Or if the ridge dips slightly at some point of its ascent, the species could be stalled indefinitely. Disequilibrium can also be produced even if social evolution is rapid, provided that extinction rates of evolving species are so high that only a few species ever make the optimum phenotype and consequently most are in transit at any given time.

Implicit in a disequilibrium hypothesis is the assumption that the advanced social state, or some particular advanced social state, is the *summum bonum*, the solitary peak defined by the ideal trait toward which the species and its relatives are climbing. The opposite extreme is the "equilibrium case," in which species at different levels of social evolution are more or less equally well adapted. There can be multiple adaptive peaks, corresponding to "primitive," "intermediate," and "advanced" stages of sociality. In somewhat more concrete terms, the equilibrium hypothesis envisions lower levels of sociality as compromises struck by species under the influence of opposing selection pressures. The imaginary species represented in Figure 2-6B is favored by an indefinite intensification of the primary effect of the trait. But the evolution of the trait cannot continue forever, because a secondary effect begins to reduce the fitness of the organism when the trait exceeds a certain value. The species equilibrates at this, the permissible optimum. For example, among males of mountain sheep and other harem-forming ungulates dominance rank is strongly correlated with the size of the horns. The upper limit of horn size must therefore be set by other effects, presumably mechanical stress and loss of maneuverability caused by excessive horn size, together with the energetic cost of growing and maintaining the horns.

Potential versus *operational factors.* Experimental biologists break down causation in complex processes by artificially increasing the intensity of each suspected factor in turn while attempting to hold all other factors constant. By this means they draw up a list of factors to which the system is sensitive and estimate curves of the system's response to each factor in turn. Notice that the factors thus identified are only *potential*: they may or may not actually operate under natural conditions. For example, extended experiments have revealed that caste in ants can be influenced by at least six factors: larval nutrition, winter chilling, posthibernation temperature, queen influence, egg size, and queen age (Brian, 1965; Wilson, 1971a). The next question is, what is the relative importance of each in nature? Which factors, to put it another way, are truly operational? No answer is likely to be forthcoming without elaborate field studies that monitor all of the factors simultaneously.

The significance of this distinction between potential and operational factors is often missed by sociobiologists. To take an especially confusing case, the potential role of social behavior in population control has been repeatedly documented in controlled experiments in which captive populations were allowed to grow until physiological or social pathology brought the growth to a halt. Other potential factors were deliberately eliminated from contention during the experiments. Food and water were administered ad libitum, and parasites and predators were excluded. The conclusion has been frequently drawn from these results that social behavior is an important population control mechanism. That may turn out to be true in particular cases, but it cannot be proved solely by laboratory experiments. Ecologists are familiar with the process of intercompensation, which is the operation of only one or a small number of control factors at a time, with other mechanisms coming into play only if the primary ones are removed by an amelioration of environmental conditions. Whether social behavior is a primary control remains to be field tested in most particular cases (see Chapter 4).

Preferred versus *realized niche.*   Another special but equally important case of potential versus operational factors is implicit in the definition of the niche. Laboratory experiments are sometimes used to define the niche as a Hutchinsonian hyperspace, the space framed by the limits of each environmental parameter within which the species can exist and reproduce. The experiments can also be used to establish the preferred niche, which is the portion of the hyperspace in which the fitness is maximum and to which laboratory animals usually also move if given a choice along a series of environmental gradients. It should nevertheless be kept in mind that the preferred niche can differ from the actual portion of the hyperspace occupied by the species in nature. In marginal habitats the preferred niche can even be wholly lacking. Moreover, competing species tend to displace one another into portions of the habitat in which each is the best competitor; and these competitive strongholds are not necessarily the preferred portion of the niche. Hence the local ecological distribution of a given species, and along with it the population density and even the manifested form of its social behavior, often depends to some extent on what portion of the total geographical range the population occupies and on the presence or absence of particular competitors. Such facts alone account for some of the striking geographic variation recorded in field studies of social behavior.

Deep versus *shallow convergence.*   At this stage of our knowledge it is desirable to begin an analysis of evolutionary convergence per se, for the reason that an analogy recognized between two behaviors in one case may be a much more profound and significant phenomenon than an analogy recognized in another case. It will be useful to make a rough distinction between instances of evolutionary convergence that are deep and those that are shallow. The primary defining qualities of deep convergence are two: the complexity of the

adaptation and the extent to which the species has organized its way of life around it. The eye of the vertebrate and the eye of the cephalopod mollusk constitute a familiar example of a very deep convergence. Other characteristics associated with deep convergence, but not primarily defining the phenomenon, are degree of remoteness in phylogenetic origin, which helps determine the summed amount of evolution the two phyletic lines must travel to the point of convergence, and stability. Very shallow convergence is often marked by genetic lability. Related species, and sometimes populations within the same species, differ in the degree to which they show the trait, and some do not possess it at all.

Among the deepest and therefore most interesting cases of convergence in social behavior is the development of sterile worker castes in the social wasps, most of which belong to the family Vespidae, and in the social bees, which have evolved through nonsocial ancestors ultimately from the wasp family Sphecidae. The convergence of worker castes of ants and termites is even more profound, in that the adult forms have become flightless and reduced their vision as adaptations to a subterranean existence. Also, their phylogenetic bases are considerably farther apart: the ants originated from tiphiid wasps, and the termites from primitive social cockroaches. An example of an intermediate depth of convergence is the independent origin of communal arena displays in at least seven groups of birds. To pass to this unusual form of courtship, a species not only must establish breeding stations separate from the feeding and nesting areas, but also must reduce pair bonds to the brief period of actual mating. Also, the males must become polygamous and give up any role in the construction and defense of the nest (Gilliard, 1962). A second example of moderately deep convergence is the attainment by the most social of the marsupials, the whiptail wallaby *Macropus parryi*, of a social system similar in many details to that of the open-country ungulates and primates found elsewhere in the world. Each mob is territorial, or at least occupies a nearly exclusive home range, and contains 30 to 40 individuals of mixed sex and age. The males establish a linear dominance hierarchy by ritualized fighting, with their rank determining their access to estrous females (Kaufmann and Kaufmann, 1971). Finally, numerous examples of shallow convergence can be listed from the evolution of territoriality and dominance hierarchies, an aspect of the subject that will be explored in detail in Chapter 13.

Grades versus *clades.*   Evolution consists of two simultaneously occurring processes: while all species are evolving vertically through time, some of them split into two or more independently evolving lines. In the course of vertical evolution a species, or a group of species, ultimately passes through series of stages in certain morphological, physiological, or behavioral traits. If the stages are distinct enough they are referred to as evolutionary grades. Phylogenetically remote lines can reach and pass through the same grades, in which case we speak of the species making up these lines as being convergent with respect to the trait. Different species often reach the same grade at different times. A separate evolving line is referred to as a clade, and a branching diagram that shows how species split and form new species is called a cladogram (Simpson, 1961; Mayr, 1969). The full phylogenetic tree contains the information of the cladogram, plus some measure of the amount of divergence between the branches, plotted against a time scale. Sociobiologists are interested in both the evolutionary grades of social behavior and the phylogenetic relationships of the species within them. An excellent paradigm from the literature of social wasps is provided in Figure 2-7.

Instinct versus *learned behavior.* In the history of biology no distinction has produced a greater semantic morass than the one between instinct and learning. Some recent writers have attempted to skirt the issue altogether by declaring it a nonproblem and refusing to continue the instinct-learning dichotomy as part of modern language. Actually, the distinction remains a useful one, and the semantic difficulty can be cleared up rather easily.

The key to the problem is the recognition that instinct, or innate

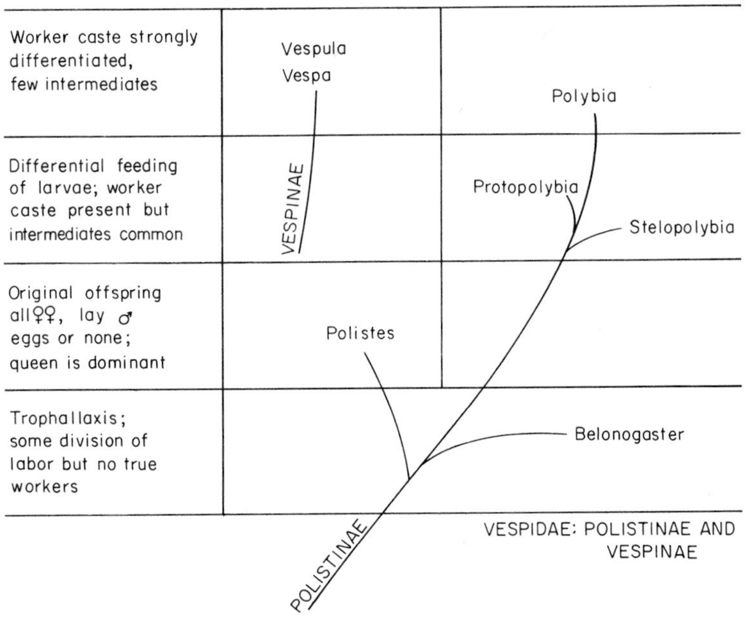

Figure 2-7 Cladograms of two groups of social wasps, the subfamilies Polistinae and Vespinae of the family Vespidae, are projected against the evolutionary grades of social behavior. The grades, which ascend from less advanced to more advanced states, are labeled on the left. The clades, or separate branches, are genera of the wasps. (Redrawn from Evans, 1958.)

behavior, as it is often called, has been intuitively defined in two very different ways:

1. An innate behavioral difference between two individuals or two species is one that is based at least in part on a genetic difference. We then speak of differences in the hereditary component of the behavior pattern, or of innate differences in behavior, or, most loosely, of differences in instinct.

2. An instinct, or innate behavior pattern, is a behavior pattern that either is subject to relatively little modification in the lifetime of the organism, or varies very little throughout the population, or (preferably) both.

The first definition can be made precise, since it is just a special case of the usual distinction made by geneticists between inherited and environmentally imposed variation. It requires, however, that we identify a difference between two or more individuals. Thus, by the first definition, blue eye color in human beings can be proved to be genetically different from brown eye color. But it is meaningless to ask whether blue eye color alone is determined by heredity or environment. Obviously both the genes for blue eye color and the environment contributed to the final product. The only useful question with reference to the first definition is whether human beings that develop blue eye color instead of brown eye color do so at least in part because they have genes different from those that control brown eye color. The same reasoning can be extended without change to different patterns of social behavior. Let us put it into practice by considering an actual problem in primate social organization. Titis (*Callicebus moloch*) and squirrel monkeys (*Saimiri sciureus*) occur together in South American forests but have very different social structures. The *Callicebus* are organized in small family groups consisting of one adult male, one adult female, and one or two young. Each group occupies a small area exclusively and frequently threatens neighboring groups. *Saimiri* groups, in contrast, consist of large and variable numbers of adult males, adult females, and young. They occupy an ill-defined home range which is not defended against neighboring groups. Mason (1971) combined single test monkeys with other individuals and variably composed groups belonging to the same species in order to isolate the classes of interactions that constitute the organization. The results of the experiments, which were conducted in cages and a large outdoor enclosure, are summarized in Table 2-1. The forms of the basic interactions in *Callicebus* and *Saimiri* are different, and provide a somewhat deeper explanation of the organizational differences. But are they themselves innate? Probably they are, but the hypothesis has not yet been put to a definitive test. Aggressive behavior in primates is strongly dependent on hormones, and endocrine schedules are known to differ among species, almost certainly on a genetic basis. The next step of our procedure

Table 2-1 The grouping tendencies of two South American monkeys; +, attraction; −, avoidance; ±, ambivalence. (From W. A. Mason, 1971.)

	Response to strangers		Response to familiar companions	
Species	Male strangers	Female strangers	Male companions	Female companions
Squirrel monkey (*Saimiri* sp.)				
Male	±	±	+++	++
Female	+	+++	++	++++
Titi (*Callicebus* sp.)				
Male	±	±	++	+++
Female	−	−	++++	±

would be to track the divergences between *Callicebus*, *Saimiri*, and other primates down to variation in the causative elements of endocrine physiology, learning schedules, microhabitat preferences, and other controlling and biassing factors, and finally to determine on what genetic foundation, if any, the variation rests.

The second intuitive definition of instinct can be most readily grasped by considering one of the extreme examples that fits it. The males of moth species are characteristically attracted only to the sex pheromones emitted by the females of their own species. In some cases they may be "fooled" by the pheromones of other, closely related species, but rarely to the extent of completing copulation. The sex pheromone of the silkworm moth (*Bombyx mori*) is 10, 12-hexadecadienol. The male responds only to this substance, and it is more sensitive by several orders of magnitude to one particular geometric isomer (*trans*-10-*cis*-12-hexadecadienol) than to the other isomers. Moreover, the discrimination takes place at the level of the sensilla trichodea, the hairlike olfactory receptors distributed over the antennae. Only when these organs encounter the correct pheromone do they send nervous impulses to the brain, triggering the efferent flow of commands that initiate the sexual response. In not the remotest sense is learning involved in such a machinelike response, which is typical of much of the behavior of arthropods and other invertebrates. Very few invertebrate zoologists feel self-conscious about alluding to this behavior as innate or instinctive, and they have in mind both the first and second definitions. At the opposite extreme, we have the plastic qualities of human speech and vertebrate social organization, and no one feels correct in labeling these traits instinctive by the second definition. A moment's reflection on the intermediate

cases reveals that they cannot be classified by a strict criterion comparable to the presence or absence of a genetic component used in the first definition. Therefore, the second definition can never be precise, and it really has informational content only when applied to the extreme cases.

Reasoning in Sociobiology

Much of what passes for theory in studies of animal behavior and sociobiology is semantic maneuvering to obtain a maximum congruence of classifications. This process is useful but better described as concept formation. Real theory is postulational-deductive. To formulate it, we first identify the parameters, then we define the relations between them as precisely as we can, and finally we construct models in order to relentlessly extend and to test the postulates. Good theory is either quantitative or at least cleanly qualitative in the sense that it produces easily recognized inequalities. Its results are often nonobvious or even counterintuitive. The important thing is that they exceed the capacity of unaided intuition. Good theory produces results that attract our attention as scientists and stimulate us to match them with phenomena not easily classified by previous schemes. Above all, good theory is testable. Its results can be translated into hypotheses subject to falsification by appropriate experiments and field studies.

Just as the experimental biologist assesses each potential factor controlling a process by varying it while holding all others constant, the theoretician predicts the importance of each parameter by varying it in the model while holding other parameters constant. By this means certain parameters are identified as being candidates for major roles while others are virtually eliminated from immediate consideration. Even then, the relative importance of the parameters cannot be guessed until their true values are measured in natural systems. Insofar as theory is consistent and correct, it provides a view of all possible worlds. Field biology identifies which of these worlds actually exist.

Theory can be pursued at either the phenomenological or the fundamental level. Of these the physicist Julian Schwinger said, "The true role of the fundamental theory is not to confront the raw data, but to explain the relatively few parameters of the phenomenological theory in terms of which the great mass of raw data has been organized." The aim of the fundamental theorist is to identify the minimal set of parameters by which the equations directly describing the data can be derived. The two levels are already emerging in some sociobiological research. Joel Cohen's models of casual group size in primates are a part of phenomenological theory; they can eventually be related to the fundamental theory of population genetics by explanations of the evolution of particular intensities of attractiveness of individuals and groups. Another effort in phenomenological theory attempts to explain population cycles as the interplay of population growth, emigration, and density-dependent social behavior. Fundamental theory in this and other topics is constructed at the next level down. It derives the demographic parameters that determine population growth and the individual behavioral scales that yield emigration and social responses as elements of strategies that maximize genetic fitness. In general, phenomenological theory aims at equations that predict the quantitative data of demography and of territory size, the ecological and physiological correlates of dominance hierarchies, role differentiation, and other features of social organization. Fundamental theory attempts to derive these equations from the first principles of population genetics and ecology.

Paradoxically, the greatest snare in sociobiological reasoning is the ease with which it is conducted. Whereas the physical sciences deal with precise results that are usually difficult to explain, sociobiology has imprecise results that can be too easily explained by many different schemes. Sociobiologists of the past have lost control by their failure to discriminate properly among the schemes. They have not yet employed the techniques of postulational-deductive model building. Nor, by and large, have they utilized the procedure of strong inference, which is standard in most of the physical sciences and biology. The steps of strong inference were summarized by John R. Platt (1964) as follows:

1. Devising alternative hypotheses (in population biology and sociobiology this step will often be taken with the aid of mathematical models).

2. Devising a crucial experiment or field study with alternate possible outcomes, each of which will, as nearly as possible, exclude one or more of the hypotheses.

3. Carrying out the experiment so as to get a clean result.

1'. Recycling the procedure, making subhypotheses or sequential hypotheses to refine the possibilities that remain; and so on.

In sociobiology, it is still considered respectable to use what might be called the advocacy method of developing science. Author X proposes a hypothesis to account for a certain phenomenon, selecting and arranging his evidence in the most persuasive manner possible. Author Y then rebuts X in part or in whole, raising a second hypothesis and arguing his case with equal conviction. Verbal skill now becomes a significant factor. Perhaps at this stage author Z appears as an *amicus curiae*, siding with one or the other or concluding that both have pieces of the truth that can be put together to form a third hypothesis—and so forth seriatim through many journals and over years of time. Often the advocacy method muddles through to the answer. But at its worse it leads to "schools" of thought that encapsulate logic for a full generation.

The advocacy method has been pursued remorselessly by many writers in the reconstruction of human social evolution. Here, for example, are Lionel Tiger and Robin Fox arguing (in *The Imperial Animal*) the social carnivore theory with brilliant clarity:

The main features of the hunting economy can be succinctly described.

The primate base provides for (a) a rudimentary sexual division of labor, (b) foraging by the males, (c) the cooperation of males in the framework of (d) competition between males.

It is small-scale, face-to-face, and personalized.

It is based on a sexual division of work requiring males to hunt and females to gather.

It is based on tool and weapon manufacture.

It is based on a division of skills and the integration of these skills through networks of exchange (of goods, services, and women).

These are networks of alliances and contracts—deals—among men.

It involves foresight, investment, judgment, risk taking—a strong element of gambling.

It involves social relationships based on a credit system of indebtedness and obligation.

It involves a redistributive system operating through the channels of exchange and generosity; exploitation is constrained in the interest of group survival.

It bases status on accumulative skill married to distributive control—again in the interest of the group as a whole.

It is important to see all these factors as integrated into the hunt. They are social, intellectual, and emotional devices that go to make up an efficient hunting economy, in the same way that muscles, joint articulation, eyesight, intelligence, etc., go to make up the efficient hunting body. They are the anatomy and physiology of the hunting body social. It is a system of the savannas and the hunting range, and it is the context of our social, emotional, and intellectual evolution.

What is wrong with this argument? It is of course ex post facto, but that alone does not make it wrong. Tiger and Fox might even be completely right. What really matters with respect to the scientific as opposed to the literary content is that the statement is not formulated in a way deliberately to make it falsifiable. No theory should be so loved that its authors try to move it out of harm's way. Quite the contrary: a theory that cannot be mortally threatened has little value in science. Most of the art of science consists of formulating falsifiable propositions in just this spirit. The good researcher does not grieve over the death of a particular hypothesis. Since he has attempted to set up multiple working hypotheses, he is committed to the survival of no one of them, but rather is interested to see how simply they can be formulated and how decisively they can be made to compete.

It was perhaps inevitable that the advocacy approach to human evolution should also produce a feminist theory. This has been duly supplied by Elaine Morgan in *The Descent of Woman* (1972). Her proposition is based on Sir Alister Hardy's idea that the human species was forced into becoming temporarily aquatic during the Pliocene drought. Man, according to this scenario, became erect to wade, lost his hair to swim better, and developed sensitive fingers to grope in the murky water. Pack hunting, male dominance, and other "antifeminist" phenomena have no place in Morgan's scheme. Her theory is advocated with the same intensity of conviction that characterized the earlier and radically different expositions of Robert Ardrey, Desmond Morris, and Tiger and Fox. *The Descent of Woman* was favorably reviewed in respectable popular magazines and newspapers, was adopted by the Book of the Month Club, and became a best seller. It does not matter much that it contains numerous errors and is far less critical in its handling of the evidence than the earlier popular books. The important point is that the argument could be accepted as serious scholarship by a large part of the educated public. For this frustrating circumstance, rival expositors have only themselves to blame. When the advocacy method is substituted for strong inference, "science" becomes a wide-open game in which any number can play.

Strong inference is not wholly unknown in sociobiology, however. It has been employed deliberately and with variable success in investigations of the adaptiveness of survivorship schedules in hemileucine moths (Blest, 1963), peculiarities in the social structure of rare ant species (Wilson, 1963), the adaptive significance of different degrees of reproductive effort in fishes (Williams, 1966a,b), the roles of species-specific plumage in birds (Hamilton and Barth, 1962), the function of territory in birds (Hailman, 1960; Fretwell, 1972), and the function of flocking behavior in desert finches (Cody, 1971). Sometimes a phenomenon allows only one reasonable explanation. The pseudopenis of the female hyena is a unique structure used conspicuously as part of the greeting ceremonies of these dangerously aggressive animals. Wolfgang Wickler has suggested that the organ evolved as a mimic of the true male penis to permit females to participate in the conciliatory communication within packs, which is based principally on penile displays. Kruuk (1972) has stated flatly that "it is impossible to think of any other purpose for this special female feature than for use in the meeting ceremony"; and he is probably right.

The single greatest difficulty encountered in the construction of multiple hypotheses is making them competitive instead of compatible. An example of a set of compatible hypotheses is the following group of explanations advanced by various authors for the role of cicada aggregations: they bring the sexes together for mating; they permit loud enough singing to confuse and repel predatory birds; they saturate the local predators with a superabundance of prey and thus permit the escape of much of the population. Not only are these propositions difficult to disentangle and to test in the form just given; they all may be true. If more than one is true, some method must eventually be devised to assess their relative importance. The subject

thereby gains one order of magnitude in difficulty. For a set of hypotheses that compete more cleanly, consider aunting in primates: it permits juvenile females to practice handling infants before their own primiparity; or it allies females with individuals of higher rank; or it results in the improved survival of infants genetically related to the aunts. Each one of these hypotheses is potentially subject to disproof in a straightforward way (see Chapter 16).

Compatibility of hypotheses leads easily to the Fallacy of Affirming the Consequent (Northrop, 1959). In scientific practice the fallacy takes the form of constructing a particular model from a set of postulates, obtaining a result, noting that approximately the predicted result does exist in nature, and concluding thereby that the postulates are true. The difficulty is that a second set of postulates, inspiring a different model, can often lead to the same result. It is even possible to start with the same conditions, construct wholly different models from them, and still arrive at the same result. I have presented just such a case from theoretical population biology in Figure 2-8. The way around the fallacy is to devise competing hypotheses such that all but one can be decisively defeated.

When carried to an extreme, the Fallacy of Affirming the Consequent generates what Garrett Hardin (1956) has called a panchreston—a word, or "concept," covering a wide range of different phenomena and loaded with a different meaning for each user, a word that attempts to "explain" everything but explains nothing. The history of the word *trophallaxis* illustrates vividly the process of creating a panchreston. The phenomenon on which it was based was the

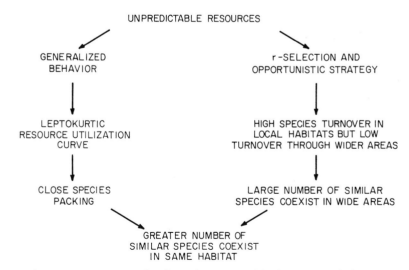

Figure 2-8 An example of two distinct models that start with the same condition and arrive at the same prediction. Either model "tested" by itself would have led to the Fallacy of Affirming the Consequent. (Based on Roughgarden, 1974, and personal communication.)

donation of salivary secretions by larvae of social wasps to their adult winged sisters. Emile Roubaud (1916) attributed a basic significance to this feeding bond. He saw it as the "raison d'être of the colonies of the social wasps," a case of association caused by trophic exploitation of the larvae by the adults. In later applying the name trophallaxis to the bond, Wheeler (1918) agreed with Roubaud's interpretation and extended it to ants. But then, stung by criticisms from Erich Wasmann and A. Reichensperger, who were promoting Wasmann's rival theory of symphilic instincts as the cause of social evolution, Wheeler (1928) proceeded to stretch and qualify the trophallaxis concept to the point of virtual uselessness: "There is no doubt that the glandular secretions of social insects are emitted in greater volume at times of excitement, but since even the persisting individual, caste, colony and nest odours are important means of recognition and communication, there is no reason why the odours should not be included with the gustatory stimuli as trophallactic." Caught up in the spirit of the idea, he went on to say, "If we compare the distribution of food in the colony regarded as a superorganism with the circulating blood current ('internal medium') in the individual insect or Vertebrate, trophallaxis, as the reciprocal exchange of food between the individuals of the colony, may be compared with the chemical exchanges between the tissue elements and the blood and between the various cells themselves." These two statements, of course, imply very different definitions, and the ambiguity persists through Wheeler's protean writings on the subject. If we select the broadest definition allowed by Wheeler, illustrated in the first of the two statements, trophallaxis must be the equivalent of all of chemical communication in the modern sense. In 1946, T. C. Schneirla, having misunderstood Wheeler (a forgivable mistake), extended trophallaxis to include tactile stimuli also. It remained for LeMasne (1953) to suggest the *reductio ad absurdum* by defining trophallaxis as synonymous with communication: "By this extension, all the life of the society is encompassed by the trophallaxis concept." In more recent years trophallaxis as a term has been rescued from oblivion by the tendency to utilize it in close to the original sense, to mean simply the exchange of alimentary liquid, either mutually or unilaterally. Many panchrestons still becloud the literature of behavioral biology, including drive, instinct, aggression, approach-withdrawal, altruism, and others. In most cases the term should not be thrown out of the biological literature—to try to do so causes even more confusion—but rather refined by narrower, more operational definitions, like that suggested for trophallaxis.

Another potentially misleading thought process in sociobiology can be conveniently designated the Fallacy of Simplifying the Cause. This fallacy is the a priori rule of choosing the simplest possible explanation of a biological phenomenon. One manifestation is "Morgan's Canon," proposed by the British comparative psychologist Lloyd Morgan in 1896. This law states that the behavior patterns of an animal should not be described in terms of anthropomorphic or higher psychic activity such as love, gentleness, deceit, and so forth, but instead interpreted exclusively by the simplest mechanisms known to work. Morgan's Canon helped inaugurate an era of reductionism in which even the most complex behavior patterns were broken down into a very few categories of response, such as reflexes, tropisms, and operant reinforcement. Although the trend had the salutary effect of curbing anthropomorphism, it went too far. Later animal behaviorists such as Bierens de Haan (1940) and Hediger (1955) correctly argued that behavior is based on complex mechanisms, and the goal of its study is to explain the mechanisms as correctly, not as simply, as possible. We are still permitted to share the pleasant view of Edward A. Armstrong, expressed in *Bird Display and Behaviour*, that "it is a thing to be thankful for, that Nature, having strictly practical ends in view, has achieved the creation of a wealth of beauty in carrying them out."

A more sophisticated variant of the same fallacy has been urged by G. C. Williams (1966a) for the construction of evolutionary hypotheses:

The ground rule—or perhaps *doctrine* would be a better term—is that adaptation is a special and onerous concept that should be used only where it is really necessary. When it must be recognized, it should be attributed to no higher a level of organization than is demanded by the evidence. In explaining adaptation, one should assume the adequacy of the simplest form of natural selection, that of alternative alleles in Mendelian populations, unless the evidence clearly shows that this theory does not suffice.

Williams' Canon was a healthy reaction to the excesses of explanation invoking group selection and higher social structure in populations that had been precipitated by earlier writings, particularly V. C. Wynne-Edwards' *Animal Dispersion in Relation to Social Behaviour* (1962). Nevertheless, Williams' distaste for group-selection hypotheses wrongly led him to urge the loading of the dice in favor of individual selection. As we shall see in Chapter 5, group selection and higher levels of organization, however intuitively improbable they may seem, are at least theoretically possible under a wide range of conditions. The goal of investigation should not be to advocate the simplest explanation, but rather to enumerate all of the possible explanations, improbable as well as likely, and then to devise tests to eliminate some of them.

Such testing is going to be time-consuming. Sociobiology, particularly evolutionary sociobiology, is not a science whose ideas can be checked by quick and elegant laboratory experiments. For example, one of the sufficiently thorough ethological studies that can be cited is the analysis of courtship displays in the goldeneye duck (*Bucephala clangula*) by Benjamin Dane and his associates (Dane et al., 1959;

Dane and Van der Kloot, 1964). These biologists examined 22,000 feet of film taken in the field to compile what is surely an exhaustive list of displays, then measured the duration of each display and the transition probabilities of all display pairs in each phase of the courtship. In a notable three-and-a-half year study of the Serengeti lions, George Schaller spent 2,900 hours and traveled 149,000 kilometers while locating and monitoring several prides on a nearly daily basis. Entomological problems can be just as demanding. In order to work out the orderly changes that occur in the labor programs of honeybee workers as they age, Sekiguchi and Sakagami (1966) spent 720 hours collecting data on 2,700 individually marked bees, while Lindauer (1952) watched a single worker for a total of 176 hours and 45 minutes.

The sociobiology of primates can be even more difficult and time-consuming. Our current knowledge has come principally from an exceptional effort in field studies which has been accelerating during the past 25 years. Prior to 1950 no more than 50 man-months had been devoted to such studies. By 1966 the cumulative field time had reached 1500 man-months, involved hundreds of investigators, and was increasing exponentially. The amount of research conducted in the 4 years from 1962 through 1965 alone exceeded that of all the research before it (Altmann, 1967b). In most cases, we can expect on the order of 100 man-hours of observation to bring a rough idea of the group organization and the communicative signals on which it is based. One thousand hours, approximately a year of daily field trips, bring a sound idea of the nature of individual relationships, seasonal change, and even behavioral ontogeny and socialization. Poirier (1970a), for example, attained this level for the Nilgiri langur (*Presbytis johnii*) with 1,250 hours, while T. W. Ransom reached even greater depth by devoting 2,555 hours during 15 months to one troop of olive baboons (Ransom and Rowell, 1972). The data yielded by such efforts become clinical in detail: each individual can be recognized, its idiosyncrasies recorded, and the development of its social status to some extent charted. Then the fine structure of the communication network begins to emerge. As we shall see in subsequent chapters, this new level of information is vital to the future development of sociobiology. In 1938 F. Fraser Darling expressed the matter with both accuracy and feeling as follows: "How surely it has been borne upon me that the glimpses of minutes, hours, days or even weeks, which a life of bird watching as a hobby have given, are inadequate for an interpretation or solution of the deeper problems of evolution, natural selection and survival in the bird world! We need time, time, time and a sense of timelessness. Our pictures of behaviour must be detailed in time equally with those of space."

Chapter 3   **The Prime Movers of Social Evolution**

In this chapter we will take an excursion into what can be termed the natural history of sociobiology, as opposed to its basic theory. Because natural history is sometimes so diverting, to the point of making one forget the main thrust of the theory, let me explain briefly the rationale for the next three chapters together. Then the reader can choose whether to skim the present chapter or to read it closely. In either case he must plan to make a careful study of Chapters 4 and 5 in order to gain a solid understanding of the foundations of sociobiology.

This chapter contains the following main argument. The major determinants of social organization are the demographic parameters (birth rates, death rates, and equilibrium population size), the rates of gene flow, and the coefficients of relationship. In both an evolutionary and functional sense these deeper factors, to be analyzed more formally in Chapter 4, orchestrate the joint behaviors of group members. But as population biologists come to understand them better, they see that the chain of causation has been traced only one link down. What, we then ask, determines the determinants? These prime movers of social evolution can be divided into two broad categories of very diverse phenomena: *phylogenetic inertia* and *ecological pressure*.

Phylogenetic inertia, similar to inertia in physics, consists of the deeper properties of the population that determine the extent to which its evolution can be deflected in one direction or another, as well as the amount by which its rate of evolution can be speeded or slowed. Environmental pressure is simply the set of all the environmental influences, both physical conditions such as temperature and humidity and the living part of the environment, including prey, predators, and competitors, that constitute the agents of natural selection and set the direction in which a species evolves.

Social evolution is the outcome of the genetic response of populations to ecological pressure within the constraints imposed by phylogenetic inertia. Typically the adaptation defined by the pressure is narrow in extent. It may be the exploitation of a new kind of food, or the fuller use of an old one, superior competitive ability against perhaps one formidable species, a stronger defense against a particularly effective predator, the ability to penetrate a new, difficult habitat, and so on. Such a unitary adaptation is manifest in the choice and interplay of the behaviors that make up the social life of the species. As a consequence, social behavior tends to be idiosyncratic. That is why any current discussion of the prime movers must take the form of natural history. The remainder of this chapter, then, consists of a survey of the many kinds of phylogenetic inertia and ecological pressure, together with a first attempt to assess their relative importance.

Phylogenetic Inertia

High inertia implies resistance to evolutionary change, and low inertia a relatively high degree of lability. Inertia includes a great deal of what evolutionists have always called preadaptation—the fortuitous predisposition of a trait to acquire functions other than the ones it originally served—but there are aspects of the process involved that fall outside the ordinary narrow usage of that term. Furthermore, as I hope to establish here, there is an advantage to continuing the physical analogy into at least the initial stages of evolutionary behavioral analysis.

Sociobiologists have found examples of phylogenetic diversity that are the outcome of inertial differences between evolving lines. One of the most striking is the restricted appearance of higher social behavior within the insects. Of the 12 or more times that true colonial life (eusociality) has originated in the insects, only once—in the termites—is this event known to have occurred outside the single order Hymenoptera, that is, in insects other than ants, bees, and wasps. W. D. Hamilton (1964) has argued with substantial logic and documentation that this peculiarity stems from the haplodiploid mode of sex determination used by the Hymenoptera and a few other groups of organisms, in which fertilized eggs produce females and unfertilized eggs produce males. One consequence of haplodiploidy is that females are more closely related to their sisters than they are to their own daughters. Therefore, all other things being equal, a female is more likely to contribute genes to the next generation by rearing a sister than by rearing a daughter. The likely result in evolution is the origin of sterile female castes and of a tight colonial organization centered on a single fertile female. This in fact is the typical condition of the hymenopterous societies. (For a full critique of the advantages and difficulties of this idea, see Wilson, 1971a, and Lin and Michener, 1972, as well as Chapter 20.)

Haplodiploid bias is an example of inertia that stems from a trait basic to the biology of a particular group of organisms. Another biasing trait is the tendency of some lower invertebrates, notably the sponges, the coelenterates, and the bryozoans (Ectoprocta), to form aggregations by asexual budding, a reproductive mode associated with their simple body organization. The aggregation habit is most pronounced in two dominant marine groups with sessile habits: the corals, which form the bulk of the tropical reefs, and the sponges and bryozoans, which constitute major elements of the encrusting communities of benthic organisms everywhere in the sea. This principal adaptation was established by no later than the early Paleozoic, and its consequence was the production of tight groups of genetically identical individuals. Altruism is easy for genetically identical individuals; in fact, in them such behavior is technically not even altruism. Furthermore, the primitive body forms of these animals enable them to unite physically with each other, to specialize individual function, and to divide labor at the cost of relatively few basic alterations in anatomy and behavior. The result, if this view of cause and effect is right, is the extraordinary "superorganisms" formed by colonies of the more advanced phyletic lines (Chapter 19).

An important component of inertia is the genetic variability of a population or, more precisely, the amount of phenotypic variability referable to genetic variation. The rate at which a population responds to selection depends exactly on the amount of this variability. Inertia in this case is measured by the rate of change of relative frequencies of genes that already exist in the population. If an environmental change renders old features of social organization inferior to new ones, the population can evolve relatively quickly to the new mode provided the appropriate genotypes can be assembled from within the existing gene pool. The population will proceed to the new mode at a rate that is a function of the product of the degree of superiority of the new mode, referred to as the intensity of selection, and the amount of phenotypic variability that has a genetic basis. Imagine some nonterritorial population faced with an environmental change that makes territoriality strongly advantageous. Suppose that a small fraction of the individuals occasionally display the rudiments of territorial behavior, and that this tendency has a genetic basis. We can expect the population to evolve relatively quickly, say over the order of 10 to at most 100 generations, to arrive at a primarily territorial mode of organization. Now consider a second population in identical circumstances, but with the occasional display of territorial behavior having no genetic basis—any genotype in the population is equally likely to develop it. In other words, genetic variability in the trait is zero. In this second case, the species will not evolve in the direction of territorial behavior.

There are some intriguing cases in which populations have failed to alter their social behavior to what seems to be a more adaptive form. The gray seal (*Halichoerus grypus*) has extended its range in recent years from the North Atlantic ice floes, where it breeds in pairs or in small groups, southward to localities where it breeds in large, crowded rookeries along rocky shores. Under the new circumstances the females might be expected to adopt the habit, characteristic of other colonial pinnipeds, of limiting their attention strictly to their own pups. But this has not occurred. Instead, mothers fail to discriminate between pups during mammary feeding, and many of the weaker young die of starvation (E. A. Smith, 1968). A second case of indiscriminate feeding that is possibly maladaptive has been recorded in the Mexican freetail bat *Tadarida mexicana* by Davis et al. (1962). Mothers give their milk not only to the young of other broods but also occasionally to other adults. The spotted hyenas of the Serengeti, unlike their relatives in the Ngorongoro Crater, subsist on game that is migratory during large parts of the year. Yet this population still behaves as though it were dealing with fixed ungulate populations, in ways that seem adapted to an environment like that of Ngorongoro Crater rather than to the unstable conditions of the Serengeti. The cubs are immobile and dependent on the mother over long periods of time, and they are not whelped at the most favorable season of the year. Several specific behavior patterns of the hyenas are clearly connected with an obsolete territorial system. They include stereotyped forms of scent marking, "border patrols," and direct aggression toward intruders (Kruuk, 1972).

We are led to ask whether the gray seal and hyena populations have failed to adapt because the required social alterations are not within their immediate genetic grasp. Or do they have the capacity and are evolving, but have not yet had sufficient time? A third possibility is that the requisite genetic variability is present but the populations cannot evolve further because of gene flow from nearby populations adapted to other circumstances. The last hypothesis, that of genetic "swamping," is basically the explanation offered by Kummer (1971) to account for maladaptive features in the social organization of baboon populations living just beyond the limits of the species' preferred habitats. Sugiyama (1967) has offered a similar hypothesis to account for the large amount of group fighting and social instability he observed in the langurs (*Presbytis entellus*) of South India. These monkeys are leaf-eating colobines, members of a group that are otherwise almost exclusively arboreal and organized under one-male dominance. The Indian langurs show evidence of having only recently adapted to life on the ground, but they have retained the one-male system, a form of organization that is less stable in ground-dwelling communities.

Success or failure in evolving a particular social mechanism often depends simply on the presence or absence of a particular *preadaptation*—a previously existing structure, physiological process, or behavior pattern which is already functional in another context and available as a stepping stone to the attainment of a new adaptation. Avicularia and vibracula, two of the more bizarre forms of specialized individuals found in bryozoan colonies, occur only within the ectoproct order Cheilostomata. The reason is simple: only the cheilostomes possess the operculum, a lidlike cover that protects the mouth of the organism. The essential structures of the specialized castes, the beak of the avicularium, which is used to fight off enemies, and the seta of the vibraculum, were both derived in evolution from the operculum (Ryland, 1970). Passerine birds accommodate the increased demands of territorial defense and reproduction in the breeding season by raising their total energy expenditure. But the same option is closed to the hummingbirds, whose hovering flight is already energetically very costly. Instead, hummingbirds maintain a nearly constant energy expenditure and simply devote less time during the breeding period to their nonsocial activities (Stiles, 1971). Social parasitism is rampant in the ants but virtually absent in the bees and termites. The reason appears to be simply that ant queens often return to nests of their own species after nuptial flights, predisposing them to enter nests of other species as well, while those of bees and termites do not (Wilson, 1971a).

When defined broadly, preadaptation can be viewed as a pervasive force in the histories of all species, creating multiplier effects that as a rule reach all the way to social behavior. Each organism, to be more specific, must find a place to live. It must occupy a space from which it can extract energy and avoid its predators, while moving within the humidity and temperature ranges it can tolerate. An evolving species squeezes and shapes its physiology to this end. Its behavior schedules are therefore determined by the particular opportunities presented to it by the environment. Consider the special case of a cold-blooded desert vertebrate, the desert iguana *Dipsosaurus dorsalis.* This creature's life is ruled to an unusual degree by fluctuations in temperature. It prefers a minimum 38.5°C for full activity but cannot tolerate temperatures greater than 43°C for long periods of time. Relying on this basic information, Porter et al. (1973) set out to measure the thermal regime of the lizard's environment in fine detail in order to delineate the yearly and daily schedules permitted to it. To a large degree they were successful (Figure 3-1), supporting the conjecture that the lizard makes the fullest use it can of the habitat within the constraints of thermoregulation. Sexual, territorial, and any other forms of social behavior are confined to the time-habitat envelopes defined by the temperature requirements. Limits are also automatically placed on the forms of communication, the seasonal and daily timing of reproductive events, and so forth. One ultimate result is a predisposition (that is, predadaptation) to certain modes of social organization. In general, we can hope to understand these constraints fully only when the most important governing factors are identified and analyzed. Additional techniques for microclimate analysis with special reference to animal behavior have been provided by Bartlett and Gates (1967), Porter and Gates (1969), and Gates (1970).

The kind of food on which the species feeds can also guide the evolution of social behavior. In Chapter 2 it was established that dispersed, predictable food sources tend to lead to territorial behavior, while patchily distributed sources unpredictable through time favor colonial existence. A second rule is that large, dangerous prey promote high degrees of cooperative and reciprocally altruistic behavior. Still another very general relation concerns the position on the trophic ladder: herbivores maintain the highest population densities and smallest home ranges, while top carnivores such as wolves and tigers are scarcest and utilize the largest home ranges. The reason is the substantial leakage of energy through respiration as the energy is passed up the food chains from plants to herbivores and thence to

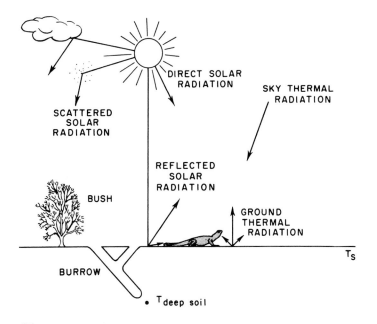

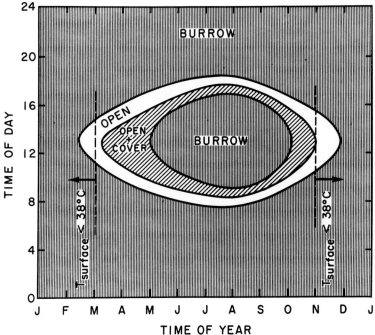

Figure 3-1 Predicting the activity schedule of a cold-blooded animal. The upper figure indicates the solar energy flows to the desert iguana *Dipsosaurus dorsalis* and to its immediate environment. The lower figure shows the predicted times and places at which the lizard can stay within its preferred temperature range (*T*) of 38.5°–43°C. A close approximation to the real schedules indicates that thermoregulation is a strongly governing requirement in the life of the species. (Modified from Porter et al., 1973.)

carnivores and top carnivores. In fact, only about 10 percent of the energy is transferred successfully from one trophic level to the next. The exact measurement used to make this generalization is ecological efficiency, defined as follows:

$$\text{Ecological efficiency} = \frac{\begin{array}{l}\text{the calories produced by}\\ \text{the population that are}\\ \text{consumed by its predator}\end{array}}{\begin{array}{l}\text{the calories that the}\\ \text{population consumes when}\\ \text{feeding on its own prey}\end{array}}$$

Suppose that we were studying a very simple ecosystem, consisting of a field of clover, the mice that eat the clover, and the cats that eat the mice. According to the "10 percent rule" of ecological efficiency, we would expect that for every 100 calories of clover eaten by the mice per unit of time, about 10 calories of mice would be eaten by the cats in the same unit of time. The ecological efficiency of the mice, with reference to the cats, is therefore 10 percent. Measurements in diverse ecosystems have shown that the ecological efficiencies actually vary from about 5 to 20 percent. Most are close enough to 10 percent to make this figure useful for rough first approximations. And even with this qualification, the rule is close enough to account for an important general feature of the organization of ecosystems: food chains seldom have more than four or five links. The explanation is that a 90 percent reduction (approximately) in productivity results in only $(1/10)^4 = 0.0001$ of the energy removed from the green plants being available to the fifth trophic level. In fact, the top carnivore that is utilizing only 0.0001 as many calories as produced by the plants on which it ultimately depends must be both sparsely distributed and far-ranging in its activities. Wolves must travel many miles each day to find enough energy. The ranges of tigers and other big cats often cover hundreds of square kilometers, while polar bears and killer whales travel back and forth over even greater distances. This demanding existence has exerted a strong evolutionary influence on the details of social behavior.

Finally, to complete this link between behavioral ecology and sociobiology, competitive interactions with other species are capable of constraining the social evolution of populations. The following example, provided by J. H. Brown (1971), illustrates one of the forms this relation takes. On the lower mountain sides of Nevada, clothed in sparse piñon-juniper woods, the cliff chipmunk *Eutamias dorsalis* is able to exclude the Uinta chipmunk *E. umbrinus* by territorial behavior. But at higher elevations, where the piñon and juniper become so dense that the branches interlock, *umbrinus* excludes *dorsalis*. The reason for this reversal is that in thick vegetation territorial behavior is less effective; *dorsalis* wastes a great deal of its time in fruitless pursuits of the less aggressive *umbrinus*, which is able to escape easily into the vegetation and go about its business. Under these circumstances *umbrinus* is able to outcompete *dorsalis* for food. Faced with opposing selection pressures generated by competitors and food, or competitors and reproductive opportunity, each species must "choose" the appropriate repertory of behavioral responses in order to persist.

The components of phylogenetic inertia include many *antisocial factors*, the selection pressures that tend to move the population to a less social state (Wilson, 1972a). Social insects and probably other highly colonial organisms have to contend with the "reproductivity effect": the larger the colony, the lower its rate of production of new individuals per colony member (Michener, 1964a; Wilson, 1971a). Large colonies, in other words, usually produce a higher total of new individuals in a given season, but the number of such individuals divided by the number already present in the colony is less. Ultimately, this means that social behavior can evolve only if large colonies survive at a significantly higher rate than small colonies and if individuals protected by colonies survive better than those left unprotected. Otherwise, the lower reproductivity of larger colonies will cause natural selection to reduce colony size and perhaps to eliminate social life altogether.

In mammals the principle antisocial factor appears to be chronic food shortage. Adult male coatis (*Nasua narica*) of Central American forests join the bands of females and juveniles only while large quantities of food are ripening on the trees, at which time mating takes place. In other seasons, when food is scarcer, the males are actively repulsed by the bands. The females and young begin to forage cooperatively for invertebrates on the forest floor, while the solitary males concentrate on somewhat larger prey (Smythe, 1970a). The moose (*Alces americana*), unlike many of the other great horned ungulates, is essentially solitary in its habits. Not only do the bulls stay apart outside the rutting season, but the cows drive away the yearling calves at just about the age when these young animals are able to fend off wolves, the principal predators of moose. Geist (1971a) has argued persuasively that this curtailment of social behavior, which would otherwise confer added protection against wolves, has been forced in evolution by the species' opportunistic feeding strategy. Moose depend to a great extent on second-growth forage, particularly that emerging after fires. This food source is patchy in distribution and subject to periodic shortages, especially when the winter snow is high. Parallel examples implicating food supply can be cited from the rodents and primates. In the latter group the general rule seems to be that adult males are added—and societies grow larger and more complex—only where particular auxiliary roles for males, such as defense or aid in parental care, become overridingly important to the fitness of the offspring.

A third potentially antisocial force is sexual selection. When cir-

cumstances favor the evolution of polygamy (see Chapter 15), sexual dimorphism increases. Typically the males become larger, more aggressive, and conspicuous by virtue of their exaggerated display behavior and secondary anatomical characteristics. The result is that the males are less likely to be closely integrated into the society formed by the females and juveniles. This is the apparent explanation for the female-centered societies that characterize deer, African plains antelopes, mountain sheep, and certain other ungulates whose males fight to establish harems during the rutting season. In the elephant seals, sea lions, and other strongly dimorphic pinnipeds, the large size and aggressive behavior of the males occasionally results in accidental injury or death to the young. Size dimorphism can also lead to different energetic requirements and sleeping sites, which have an even more disruptive effect. A larger individual requires a larger home range and is likely to need a different foraging regime to maintain its energy requirements. It is also likely to feed on a larger variety of food items (Schoener, 1971). The adult male of the orang-utan, for example, weighs almost twice as much as the adult female. In a study of a free-ranging population in Borneo, Peter S. Rodman (personal communication) noted important differences in the feeding behavior of the two sexes. The average length of feeding bouts for the male was 50 minutes, and for the female 35 minutes. The male averaged only 0.62 moves per hour to the female's 0.90 moves per hour. During an average 12-hour day a male fed in 8 episodes, moving 7 times, while the female fed in 8 episodes and moved 11 times. The female visits more fruit trees and feeds for a shorter time in each. By virtue of her smaller size she appears to be able to choose fruit in a more suitable stage of development. All of these differences contribute to the separation of the sexes in the orang-utan, which is essentially a solitary species. But which is the cause and which is the effect in this relation? It is equally conceivable that the prime inertial force is the advantage to a family of diversifying the diet and feeding rhythms of the different members. The divergence would cause sexual dimorphism, which in turn would lead to polygamy and social disruption. The alternatively possible cause-effect relations are visualized in Figure 3-2.

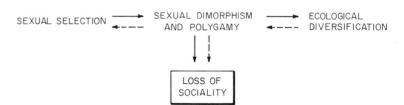

Figure 3-2   The two possible alternative pathways of cause and effect in the evolution of a solitary condition in the orang-utan and similar polygamous animals.

A fourth, possibly widespread antisocial factor is the loss of efficiency and individual fitness through inbreeding. Social organization, by closing groups off from one another, tightening the association of kin, and reducing individual movement, tends to restrict gene flow within the population as a whole. The result is increasing inbreeding and homozygosity (see Chapter 4). We have little information on the importance of this factor in real populations. If significant at all, it almost certainly varies greatly in effect from case to case, because of the idiosyncratic qualities of social organization and gene flow that characterize the biology of individual species.

The magnitude of phylogenetic inertia can be roughly gauged by comparing the evolutionary responses of closely related phyletic lines to divergent selective pressures. At the microevolutionary end of the scale, where low inertia is first detected, the analysis can be performed on laboratory populations. The results will be only partially applicable, since they can measure the heritability of the trait but not the adaptiveness of the newly evolved character states in nature. Comparative field studies of closely related species occupying different habitats can provide insight, under the right circumstances, into microevolutionary inertia, with none of the natural parameters altered. This approach will be stressed in reviews of many of the special topics to be taken up in later chapters. With increasing inertia, that is, with diminishing lability of the trait, evolutionary divergence between related phyletic lines can be detected only by comparisons of higher taxonomic categories. The genus may prove adequate, as in the analysis of socialization in *Papio* versus *Presbytis* (Chapter 2). Or the family may first reveal divergence, as in the case of social parasitism rampant in ants, belonging to the family Formicidae, but rare in bees, belonging to the family Apidae. At the level of the order, we have the marked tendency of the Hymenoptera to produce eusocial forms as opposed to the Diptera, which are exclusively solitary; and so forth.

Different categories of behavior vary enormously in the amount of phylogenetic inertia they display. Among those characterized by relatively low degrees of inertia are dominance, territoriality, courtship behavior, nest building, and taxes. Behaviors possessing high inertia include complex learning, feeding responses, oviposition, and parental care. In the case of low inertial systems, large components of the behavior can be added or discarded, or even the entire category evolved or discarded, in the course of evolution from one species to another. At least four aspects of a behavioral category, or any particular evolving morphological or physiological system underlying behavior, determine the inertia:

1. *Genetic variability.* This property of populations can be expected to cause differences between populations in low inertial social categories.

2. *Antisocial factors.* The processes are idiosyncratic in their occurrence and can be expected to generate inertia at various levels.

3. *The complexity of the social behavior.* The more numerous the components constituting the behavior, and the more elaborate the physiological machinery required to produce each component, the greater the inertia.

4. *The effect of the evolution on other traits.* To the extent that efficiency of other traits is impaired by alterations in the social system, the inertia is increased. Thus, if installment of territorial behavior cuts too far into feeding time or exposes individuals to too much predation, the evolution of the territorial behavior will be slowed or stopped.

Ecological Pressure

The natural history of sociobiology has begun to yield a very interesting series of ecological correlations. Some environmental factors tend to induce social evolution, others do not. Moreover, the form of social organization and the degree of complexity of the society is strongly influenced by only one or a very few of the principal adaptations of the species: the food on which it specializes, the degree to which seasonal change of its habitat forces it to migrate, its most dangerous predator, and so forth. To examine this generalization properly, let us next review the factors that have been identified as principal selective forces in field studies of particular social species.

Defense against Predators

An Ethiopian proverb says, "When spider webs unite, they can halt a lion." Defensive superiority is the adaptive advantage of cooperative behavior reported most frequently in field studies, and it is the one that occurs in the greatest diversity of organisms. It is easy to imagine the steps by which social integration of populations can be made increasingly complex by the force of sustained predation. The mere concentration of members of the same species in one place makes it more difficult for a predator to approach any one member without detection. Flying foxes (*Pteropus*), which are really large fruit bats, form dense sleeping aggregations in trees. Each male has his own resting position determined by dominance interactions with other males. The lower, more perilous branches of the trees serve as warning stations for the colony as a whole. Any predator attempting to climb the tree launches the entire colony into the air and out of reach (Neuweiler, 1969). In his study of arctic ground squirrels (*Spermophilus undulatus*), Ernest Carl (1971) was personally able to stalk isolated individuals to within 3 meters—close enough, in all probability, for a predator such as the red fox to make the rush and kill. But he found it impossible to close in on groups. From distances as great as 300 meters the *Spermophilus* set up waves of alarm calls, which increased in intensity and duration as the intruder came closer. By noting the quality and source of the alarm calls, Carl was even able to judge the shifting positions of predators as they passed through the *Spermophilus* colonies. Individual ground squirrels can probably do no less. Similar observations were made by King (1955) on the black-tail prairie dog (*Cynomys ludovicianus*). These rodents live in particularly dense, well-organized communities, the so-called towns, and it is probably one reward of their population structure that they suffer only to a minor degree from predation.

Birds increase resistance to predators under a variety of circumstances by forming flocks (Goss-Custard, 1970). Several kinds of wading birds respond to the alarm calls of their own species by bunching and flying off. A high-velocity bullet fired over a diffuse group of redshanks (*Tringa totanus*) causes them to congregate in agitation. The same response is shown by eider ducklings (*Somateria mollissima*) when attacked by predatory gulls. Several explanations have been advanced for the evolution of such behavior. First, it is as obvious with birds as with rodents that the efficiency of a group in detecting predators is superior to that of an individual. Provided an adequate alarm communication exists, group membership increases the probability that any given individual will survive the attack of a given predator. The flock members can furthermore "relax" and increase their efficiency in other activities. Murton (1968) showed that wood pigeons (*Columba palumbus*) collect food at a slower rate when alone than when in flocks because they spend more time looking around, evidently to guard against approaching predators. Second, birds flying or swimming in flocks may simply be more difficult to attack without injury to the predator. Flying groups of starlings (*Sturnus vulgaris*) respond to the sight of a peregrine falcon or sparrow hawk by drawing close together in a dense formation (Figure 3-3). Tinbergen (1951) pointed out that a dense formation is dangerous to the falcon, which normally takes prey by stooping at great speed (said to exceed 240 kilometers per hour); it runs a fatal risk if it collides with any birds other than the target because except for its talons, its body is fragile. The falcon accomplishes its purpose by carrying out a series of sham attacks until one or a few birds momentarily lose contact with the flock by inferior maneuvering. Then a real swoop is carried through. The response can be even more specific than that envisaged by Tinbergen. When flying above a sparrow hawk and hence out of danger, the starling flock remains dispersed. Only when the hawk flies above them do the birds assume a tight formation (Mohr, 1960).

Still another social way of avoiding predators is to utilize marginal individuals of the group as a shield. Since predators tend to seize the first individual they encounter, there is a great advantage for each individual to press toward the center of its group. The result in evolution would be a "herd instinct" that centripetally collapses populations into local aggregations. Francis Galton was the first to com-

prehend the effects of such an elementary natural selection for geometric pattern. In 1871 he described the behavior of cattle exposed to lions in the Damara country of South Africa:

> Yet although the ox had so little affection for, or individual interest in, his fellows, he cannot endure even a momentary severance from his herd. If he be separated from it by strategem or force, he exhibits every sign of mental agony; he strives with all his might to get back again and when he succeeds, he plunges into its middle, to bathe his whole body with the comfort of close companionship.

The result of centripetal movement is some of the most visually impressive but least organized of all forms of social behavior. Centripetal movement generates not only herds of cattle but also fish and squid schools, bird flocks, heronries, gulleries, terneries, locust swarms, and many other kinds of elementary motion groups and nesting associations (Figure 3-4). In more recent years the idea of the "selfish herd" has been developed persuasively, principally by means of circumstantial evidence and plausibility arguments, by G. C. Williams (1964, 1966a) and W. D. Hamilton (1971a).

Eibl-Eibesfeldt (1962) and Kühlmann and Karst (1967), among others, have postulated that special group movements have evolved to evade attacking predators. These maneuvers include streaming swiftly back and forth in parallel formation and splitting into subgroups that diverge and circle back to the rear. It is difficult, however, to judge to what degree these group patterns stem from coordination and to what degree they are the mere outcome of selfish evasive maneuvering by individual fish.

One potential variation on the selfish herd strategy is the utilization of a "protector" that consumes part of the population but more than compensates by excluding other predators. The widespread coral fish *Pempheris oualensis* forms schools of a few hundred or thousand individuals that find shelter during the day in well-shaded holes, coral passages, and caves facing the open sea. They share these hiding places with one or a few kinds of predatory fish, mostly the serranid *Cephalopholis argus*, which feed on them in limited amounts (Fishelson et al., 1971). Since the predators are territorial, the *Pempheris* gain to some extent by schooling and thus restricting their exposure during the daytime to only one or a few of their enemies. By jointly saturating the favored predators with more than they can consume, the individual members of the school are favored with an increased probability of survival. It is tempting to speculate that a convergent adaptation to that of the *Pempheris* is represented by the sleeping clusters of insects. Certain species of sand wasps, for example, congregate in large numbers each evening on the ends of flowerheads or branches (Evans, 1966). The sites are difficult for most predators to reach. The fact that many equally suitable sites occur in the vicinity suggests that the clustering enhances the protection of individual wasps, either through the concentration of repellent substances, or

Figure 3-3 Starlings fly in their usual loose formation when above a hawk but draw together into a tight flock when the hawk is above them. A stooping hawk must strike its prey with its talons first; if it passes through a dense flock it risks hitting a bird with a more fragile part of its body. (Original drawing by J. B. Clark; based on Mohr, 1960.)

Figure 3-4 A school of baitfish (*Stolephorus purpureus*) splits and streams away when attacked by a large kawakawa (*Euthynnus affinus*), a member of the tuna family. The adaptive value of moving from the edge of the school toward the center is obvious. (From E. L. Nakamura, 1972.)

through restriction by geography to a smaller number of predators, or both.

A close equivalent of herding and schooling is the "Fraser Darling effect," defined as the stimulation of reproductive activity at a social level beyond mere sexual pairing. In his study of colonial seabirds off the English coast, Darling (1938) noticed that "although the immediate mate of the opposite sex may be the most potent excitatory individual to reproductive condition, other birds of the same species, or even similar species, may play a decisive part if they are gregarious at the breeding season. Without the presence of others the individual pairs of birds may not complete the reproductive cycle to the limit of rearing young to the fledgling stage." Thus the essential effect deduced by Darling is the enhancement of reproduction by stimulation from animals other than the mate. Darling presented some data suggesting that small colonies of herring gulls (*Larus argentatus*) start laying eggs at a later date and have a longer breeding season than large colonies. As a consequence, their chicks are exposed to more cumulative predation by such enemies as herons and great black-backed gulls, whose densities and levels of activity tend to remain constant(see Figure 3-5, upper half). This distinction holds except for the very smallest colonies, where the sheer limitation of numbers of adults causes irregular egg-laying periods of brief duration. The hurrying and shortening of breeding activity in large colonies was attributed by Darling to social facilitation. Unfortunately, the time relation has proved not to be so simple. Coulson and White (1956) found Darling's data on herring gull colonies of various sizes not to be statistically significant. In their own detailed study of the kittiwake *Rissa tridactyla* (1960), they established that social facilitation of the Darling type does occur—the denser the local concentration, the earlier the onset of breeding. However, the effect extends only over about 2 meters. As a consequence, the larger the populations, the greater the spread of local densities, and hence the longer the breeding time of the population as a whole. The kittiwake is unusual in nesting along cliffs. It is therefore subject to less predation, and its nests tend to be arrayed in rows—both of which factors contribute to the peculiarities found by Coulson and White.

The Darling effect has also been documented in the red-winged blackbird *Agelaius phoeniceus* by H. M. Smith (1943), the tri-colored blackbird *A. tricolor* by Orians (1961a), the African village weaver-bird *Ploceus cucullatus* by Collias et al. (1971) and Hall (1970), and Viellot's blackweaver *Melanopteryx nigerrimus* by Hall (1970). In each case the result is the lengthening of the breeding period in larger colonies, but also synchronization and an increased peaking of reproductive output. The result in all these birds, then, including the kittiwake, is synchronization of breeding activity in *local* neighborhoods, coupled sometimes with longer, more productive breeding seasons (see lower half of Figure 3-5).

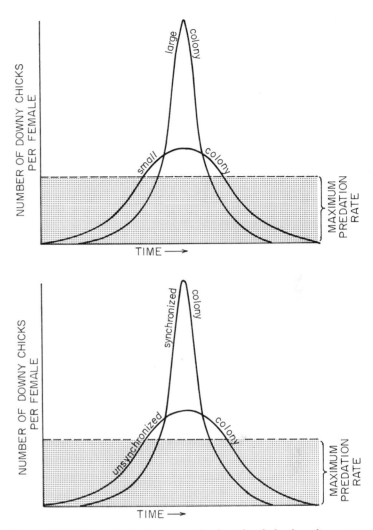

Figure 3-5 The relation between the length of the breeding season and the amount of mortality in chicks due to predation in colonial birds. The upper figure represents F. Fraser Darling's original hypothesis. Larger colonies were postulated to have a shorter breeding season and therefore to suffer less cumulative mortality. The lower figure represents modification of the hypothesis to accommodate the results of more recent field studies.

Let us suppose that the adaptive role attributed to the effect by Darling is correct, or at least the most plausible hypothesis among several conceivable. How might the effect have evolved? Notice that by "crowding" their reproductive effort into the time span when most of the other birds are producing chicks, the pair confront the waiting predators at the time the predators are most probably well fed and hence likely to ignore any particular chick. The pairs most insensitive to the Darling effect will tend to start too early or too late; their chicks will be the equivalent of the cattle living dangerously on the margin of a herd. The absence of the effect, meaning the absence of synchrony among the pairs, is the equivalent of what happens when the members of the herd scatter and expose themselves to increased predation. This inference is supported by the independent studies of Patterson (1965) on the population of the black-headed gull *Larus ridibundus* at Ravenglass in England. In 1962 most eggs were laid between the sixth and fifteenth days after the first eggs appeared, and of these, 11 percent gave rise to fledged young. But of the smaller number of eggs laid five days before and five days following this period, only 3.5 percent produced fledged young. Comparable results were obtained in 1963. The chief predators of the chicks, carrion crows and herring gulls, were simply saturated by the brief superabundance of their small prey.

Synchronized breeding, of unknown physiological origin, also occurs in social ungulates. The reproductive cycle of the wildebeest (*Connochaetes taurinus*) is characterized by sharp peaks of mating and birth. Mating occurs during a short interval in the middle of the long rainy season. Calving begins abruptly about eight months later and continues at a fairly constant rate for two to three weeks, during which 80 percent of the births occur. The remaining 20 percent occur at a slowly declining rate over the following four to five months. The synchronization of birth is even more precise than these data suggest: the majority of births occur in the forenoon, in large aggregations on calving grounds usually located on short grass (Estes, 1966). When a cow is thrown slightly out of phase, she is able to interrupt delivery at any stage prior to the emergence of the calf's head, thus giving her another chance to join the mass parturition. The synchronization almost certainly has among its results the saturation of local predators and the increased survival rate of the newborn calves. To this benefit is added an extraordinary precocity on the part of the calves: they are able to stand and to run within an average of seven minutes after their birth. And they *must* be able to do this, because the cows will defend them only if both are overtaken in flight. Synchronized calving has also been reported in the African buffalo *Syncerus caffer*, while in the barren-ground caribou *Rangifer arcticus* the calving ground is the single most fixed point in the annual migratory circuit of the species (Lent, 1966; Sinclair, 1970). The idea that synchronized birth in these and other mammals represents an adaptation specifically evolved to thwart predation is an attractive hypothesis, but it has not yet been subjected to adequate testing.

Crowding in time is also manifested in the en masse exits of cave crickets, cave bats, oilbirds, swallows, and other animals that take communal refuge in shelters. These animals emerge abruptly at certain times of the day or night in order to feed. Predators waiting near the exits find it difficult to cope with more than a small fraction of the prey. In the extreme case of the nursery populations of the Mexican freetail bat in the caves of the central United States, the emerging swarms often contain millions of individuals. At a distance a swarm resembles a continuous spiraling black rope rising from the mouth of the cave. There are hundreds of individuals per meter of cross section, each bat accelerating up to speeds of 90 kilometers per hour. Predators are further confounded by the fact that the bats are migratory, remaining at the nursery caves only in the late spring and summer (Davis et al., 1962). Nevertheless, it can only be speculated whether exit swarms are a device evolved primarily in response to predator pressure, or merely one of the secondary consequences of the cavernicolous habit of the freetail bats—which is itself the primary adaptation to escape predation.

Moving in a group can reduce the individual's risk of encountering a hungry predator for the simple reason that aggregation makes it difficult for a particular predator to find any prey at all. Suppose that a large fish has no way of tracking smaller fish and feeds only when it encounters the prey in the course of random searching. Brock and Riffenburgh (1960) have pursued a basic geometric and probability model to prove formally what intuition suggests, that as a prey population coalesces into larger and larger schools, the average distance between the schools increases, and there is a corresponding decrease in the frequency of the detection of schools by a randomly moving predator. Since one predator consumes no more than a fixed average number of prey at each encounter, the school size need only exceed this number in order for some of its members to escape. Thus above a certain level increase in school size confers a mounting degree of protection on its members. The same conclusion applies to herds, flocks, and other constantly moving groups. It loses force to the degree that the hunted group settles down, follows predictable migratory paths, can be tracked from place to place by the predators, or is easier to detect in the first place.

Perhaps the ultimate strategy of predator evasion has been achieved by the periodical cicadas (*Magicicada*) of eastern North America. The behavior and evolutionary relationships of these amazing insects has recently been reanalyzed by Alexander and Moore (1962) and their population ecology and adaptation by Lloyd and Dybas (1966a,b). Six species of *Magicicada* are now known; three emerge as adults every 13 years and three as adults every 17 years. The insects spend the long intervals between appearances as vegetarian nymphs burrowing underground. Although the nymphs go their own way over the years, their emergence as adults is tightly synchronized:

On some years practically all of the population in a given forest emerges on the same night, or on two or three different nights. There is almost always one night of maximum emergence. In 1957, Alexander witnessed such an emergence in Clinton County, Ohio. In a woods that during the afternoon had contained only scattered nymphal skins and no singing individuals, and in which no live adults had been found during a two-hour search, nymphs began to emerge in such tremendous numbers just past dusk that the noise of their progress through the oak leaf litter was the dominant sound across the forest. Thousands of individuals simultaneously ascended the trunk of each large tree in the area, and the next morning foliage everywhere was covered with newly molted adults. The numbers of subsequently emerging adults were negligible in comparison. In this case, it was literally true that the periodical cicadas had emerged as adults within a few hours from eggs laid across a period of several weeks seventeen years before. (Alexander and Moore, 1962: 39)

The geographical distribution of the swarms is highly patchy, and this fact alone must further reduce the total number of predators that can find them. The swarms are immense, often composed of millions of individuals. Since the separate insects are large to start with, predator satiation must occur quickly. It is also possible, as suggested by Simmons et al. (1971), that the extremely loud noise produced by the swarms repels some birds or at least interferes with their communication system in ways that reduce their effectiveness as predators. But far more impressive than the escape in space is, of course, the escape in time (Figure 3-6). No ordinary predator species can hope to adapt specifically to a prey that gluts it for a few days or weeks and then disappears for years. The only way to solve the problem would be to track the cicadas through time, entering dormancy for 13 or 17 years or molding the life cycle to pursue the cicada nymphs underground. No species is known to have turned the trick, although the possibility that one or more exists has not been wholly excluded.

For certain kinds of animals a potential bonus of living in groups is the enhancement of repellent powers. If a predator is more likely to be turned away by the defense systems presented by two individuals side by side than by that of a single individual, then (all other things being equal) aggregation will be favored in evolution. Many of the insects with the most formidable chemical defenses do in fact congregate in conspicuous aggregations. Included are a diversity of species from ladybird and bombardier beetles to "stink bugs" (that is, various hemipterans) and acraeine, danaiine, heliconiine, and nymphaline butterflies (Cott, 1957; Eisner, 1970; Wautier, 1971; Benson and Emmel, 1973). Such organisms are often marked by unusual anatomical projections, such as protrusible horns, together with striking color patterns that render them conspicuous. They may also wave their appendages, bob their bodies up and down, or engage in other distinctive behavior patterns. All such advertising traits used by dangerous animals are referred to by zoologists as *aposematism*. Experiments with insects and other arthropods have shown that vertebrate predators remember the aposematic characteristics after one or a few unpleasant experiences and emphatically avoid the animals afterward (Eisner, 1970; Eisner and Meinwald, 1966; Brower, 1969). It

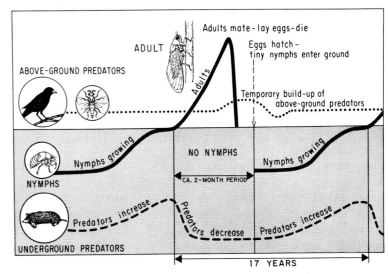

Figure 3-6 Predator escape by aggregation in time and space by the 17-year periodical cicadas, as hypothesized by Lloyd and Dybas (1966a). Significant numbers of adult cicadas appear above ground to lay eggs only once every 17 years. Birds and parasitoid wasps, their chief above-ground predators, increase in population that year, but the effects have vanished by the next bonanza 17 years later. Underground, moles can increase their populations somewhat over a few years by feeding on the long-lived cicada nymphs, but this benefit is taken away abruptly at the time of adult emergence and perhaps for a few years afterward, when the young nymphs of the new generation are too small to be useful as food.

is tempting to speculate that groups would be able to "teach" and "remind" the local predators more effectively than the same number of individuals diffusely scattered.

Substantial evidence exists of the greater effectiveness of group defense. In experiments on two European butterflies, the small tortoiseshell *Aglais urticae* and the peacock butterfly *Inachis io*, Erna Mosebach-Pukowski (1937) found that caterpillars in crowds were eaten less frequently than solitary ones. A study of ascalaphid neuropterans by Charles Henry (1972) has revealed what is virtually a controlled evolutionary experiment on the efficiency of group defense. The adults of these insects superficially resemble dragonflies and are sometimes popularly called owlflies. The female of *Ululodes mexicana* lay eggs in packets on the sides of twigs, then deposits a set of highly modified eggs called repagula ("barriers") farther down the stem. The repagula form a sticky barrier that prevents ants and other crawling predatory insects from reaching the nearby hatching larvae. Thus protected, the larvae quickly scatter from the oviposition site. A second ascalaphid species, *Ascaloptynx furciger*, employs a very different strategy. The modified eggs are used as food by the young owlfly larvae. They are not sticky and do not prevent predators from attacking the larvae. Unlike *Ululodes*, however, the *Ascaloptynx* larvae strongly aggregate and present potential enemies with a bristling mass of sharp, snapping jaws (Figure 3-7). The response is seen only when the *Ascaloptynx* are threatened by larger insects. Smaller insects such as fruit flies are treated as prey and captured by larvae who approach them singly. Henry's experiment demonstrated that larvae can be subdued by predators such as ants if they are caught alone, but that when defending en masse they are relatively safe. When properly searched for, similar phenomena will probably be found to be widespread among the arthropods. Among the likeliest possibilities are the dense aggregations formed by juveniles of spiny lobsters, spider crabs, and king crabs (Powell and Nickerson, 1965; Števčić, 1971).

3 mm

Figure 3-7 The mass defensive response of newly hatched owlfly larvae (*Ascaloptynx furciger*). When confronted by insect predators who crawl up the stem toward them, the larvae bunch together, turn to face the enemy, raise their heads, and rapidly and repeatedly snap their jaws. (From Henry, 1972.)

Cooperative behavior within the group, the essential ingredient that turns an aggregation into a society, can improve defensive capability still further. Among the bees, cooperative defense seems also to have been a principal element in the evolution to complex sociality. Bees are influenced by the reproductivity effect, which as we have already seen is a component of phylogenetic inertia that slows or reverses social evolution in primitively social insects. The effect has been overcome in halictid bees, according to Michener (1958), by the improved defense against parasitic and predatory arthropods that associations of little groups of nestmates provide. Several observers besides Michener have witnessed guard bees protecting their nest entrances against ants and mutillid wasps. Lin (1964) found that groups of *Dialictus zephyrus* females are more effective than solitary individuals in repelling mutillids. Michener and Kerfoot (1967) obtained indirect evidence that groups of *Pseudaugochloropsis* females survive longer than solitary ones, but whether improved nest defense is responsible remains moot. The structure of halictid bee nests makes them particularly convenient for communal defense. Even where multiple clusters of brood cells exist, each under the control of a reproductive female, the entire underground complex can ordinarily be reached only by a single entrance gallery not much wider than the body of a bee. By taking turns at guard duty, the bees can free each other for foraging trips without ever leaving the entrance untended.

Social ungulates that move in large amorphous herds, such as the wildebeest and Thomson's gazelle, do not cooperate in active defense against lions and other predators (Kruuk, 1972; Schaller, 1972). They depend chiefly on flight to escape. But ungulates that form small discrete units, comprised of one or more harems and other kinship groups, are more aggressive toward predators and mutually assist one another. Sometimes they move in complex patterns resembling military maneuvers. One of the most striking is the celebrated perimeter defense thrown up by musk oxen (*Ovibos moschatus*) against wolves. The following account by Tener (1954) is based on his observations on Ellesmere Island:

A herd of 14 musk-oxen that had been feeding undisturbed for several hours on the western slope of Black Top Ridge were seen to form a defensive group. Two wolves, one white and one grey were then noted lying down together 50 yards from the herd. Occasionally one of the wolves circled the herd and then returned to lie down. Eventually 10 of the musk-oxen lay down, while four remained standing facing the wolves. The calf in the herd kept close to the cows, grazing near the resting adults until the white wolf suddenly dashed around the four standing adults and toward the calf that was now outside the group of animals lying down. The calf immediately ran to the centre of the herd and all the musk-oxen rose to their feet. The one adult bull charged the wolf in an attempt to gore it but the wolf nimbly turned aside and trotted off to its mate. Both wolves left the vicinity about half an hour later, heading towards the eastern end of the fiord.

This singular behavior appears to be an adaptation specifically aimed at thwarting wolves, which are the principal natural predators of the musk oxen. When a man comes closer than about 100 meters to the massed group, the musk oxen break their line and run. Essentially the same formation is assumed by the eland (*Taurotragus oryx*), a giant African antelope (Figure 3-8), and the water buffalo (*Bubalus bubalis*) of Asia (Eisenberg and Lockhart, 1972). Their defensive array calls to mind one of Clausewitz's rules of war: "The side surrounded by the enemy is better off than the side that surrounds."

Elk (*Cervus canadensis*) frequently graze in a "windrow" formation, spread out in staggered rows that present a broad front to the wind. This formation allows the elk to catch the scent of predators from one direction while maintaining continuous visual surveillance in nearly all directions (Figure 3-9). Sometimes "calf pools" are formed in the meadows, with one or two cows staying with the calves while the others wander away for intervals to graze. When a human observer approaches, he is treated with yet another antipredator response: the leading cow turns and approaches him with a high-stepping gait, while the rest of the gang moves in the opposite direction in rapid single file (Margaret Altmann, 1956). When a solitary red deer (*C. elaphus*) rests, it faces upwind. When a group rests

Figure 3-9 The windrow formation of grazing elk. (Based on Margaret Altmann, 1956.)

together, they form a rough circle facing outward, so that all approaches are watched simultaneously (Darling, 1937).

Remarkably parallel accounts have been published on the social defense of the killer whale (*Orcinus orca*). For example, when a pack was surrounded by a net near Garden Bay, British Columbia, a large bull herded the cows together as they began to show excited behavior (Martinez and Klinghammer, 1970). Jacques-Yves Cousteau and Phillipe Diolé (1972), aboard the research vessel *Calypso*, described the role of another male in the following vivid terms:

The school is composed of an enormous male (at least three tons, 25 to 30 feet long, with a dorsal fin four-and-one-half feet high), a female almost as large as the male but with a smaller fin, seven or eight medium-sized females and six or eight calves. This is a nomadic school, comprising females and young, and with a single male taking the position of lord and master of the group.

At the beginning of the chase, the killer whales are very sure of themselves, diving every three or four minutes and reappearing about a half-mile away. Ordinarily, this would be enough to lose any marine attacker and to shake off a whaler. But the Zodiac is doing 20 knots on a sea of glass and is capable of turning on a dime. A few seconds after the grampuses surface to breathe, they hear the Zodiac's wasplike buzzing coming up from the rear.

After a while, the mammals try a new tactic. They surface every two

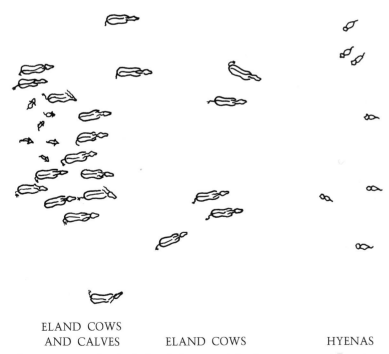

ELAND COWS AND CALVES   ELAND COWS   HYENAS

Figure 3-8 A herd of eland threatened by hyenas array themselves in a protective formation around the calves. (From Kruuk, 1972.)

or three minutes now and increase their speed. But the Zodiac keeps up with them.

The time has come for evasive tactics; the whales dart to the right at 90 degrees, then to the left and back again; then they make simulated turns at 180 degrees. Finally, they play their trump card: The male remains visible, swimming along at 15 or 20 knots and occasionally leaping out of the water. He is accompanied only by the largest female. His purpose, obviously, is to lure the Zodiac into following him—while the rest of the school escapes in the opposite direction.

One anecdote does not prove the existence of an adaptive behavior. Nevertheless, the degree of sophistication implied in this account is consistent with other observations of coordinated hunting behavior in the killer whale which will be reviewed later.

Primates also display defensive behavior parallel to that of the social ungulates. The gelada "baboon" (*Theropithecus gelada*), actually a large ground-dwelling cercopithecoid monkey, shows behavior notably similar to that of the wildebeest and Thomson's gazelle. In the rugged highlands of Ethiopia, it forms amorphous herds that travel as much as 8 kilometers a day in search of food. Single males defend their harems, mostly against other males, but there is no cooperative organization within the herd as a whole against outside dangers (Crook, 1966). The patas monkey (*Erythrocebus patas*) is an example of a species with small, nonherding troops dominated by single males. The defensive role in the patas troop is assumed almost exclusively by the male. He acts constantly as a watchful guardian, moving far away from his group when surveying a new feeding area or when approached too closely by a human observer. Diversionary tactics are occasionally used: the male crashes noisily through the bushes close to the observer and then far from the other members of the group, who remain hidden quietly in the vegetation (Hall, 1967). In higher primate species with multimale groups, organized defense is the rule. In fact, we can bring this generalization in line with current primatological theory by putting the proposition the other way around—the multimale unit may have evolved in order to provide coordinated, hence superior defense. The generalization was first illustrated by C. R. Carpenter's observation of a howler monkey infant (*Alouatta villosa*) threatened by an ocelot. The infant cried out and three adult males separated from the troop to come to its aid (Carpenter, 1934). Later, Chance (1955, 1961) explicitly suggested that groups of monkeys larger than the nuclear family have evolved as antipredator devices. DeVore (1963b) noted that the process has progressed furthest in species living in open habitats, specifically the grasslands and savannas of Africa, and suggested the following chain of causation: the more terrestrial the species, the larger its home range and the greater its exposure to predators, thus the larger the group and the more specialized the males for defensive fighting. DeVore further viewed this primary adaptation as enhancing the sexual dimorphism of the species as well as the aggressive behavior and dominance hierarchies

of the males. This ecological view, which has been modified and refined by Chance, Hall, Crook, Denham, and others, will be taken up in detail in Chapter 26.

The defensive maneuvers of a troop of large terrestrial primates is one of the natural world's most impressive sights. This is particularly true when the response is what the ornithologists call *mobbing*: the joint assault on a predator too formidable to be handled by a single individual in an attempt to disable it or at least drive it from the vicinity, even though the predator is not engaged in an attack on the group (Hartley, 1950). When presented with a stuffed leopard, for example, a troop of baboons goes into an aggressive frenzy. The dominant males dash forward, screaming and charging and retreating repeatedly in short rushes. When the "predator"does not react, the males grow more confident, slashing at the hind portions of the dummy with their long canines and dragging it for short distances. After a while other members of the troop join in the attack. Finally, the troop calms down and continues on its way (DeVore, 1972). Chimpanzees show a similar response to leopard models. When a stuffed leopard is dragged out from behind a blind into the presence of a troop, the chimpanzees first view it in silence, then burst into loud yelling and barking, while scrambling about in all directions. A majority begin to charge the leopard, waving sticks of broken-off saplings, throwing them in the direction of the leopard, and stamping the ground with their hands and feet. Some of the chimps charge upright on their hind legs. Near the leopard they seize saplings still rooted in the ground and lash them back and forth, sometimes striking the leopard in the process. These noisy attacks alternate with periods of quiet, during which the chimpanzees seek each other out for kissing, touching, and mock homosexual and heterosexual copulations. Diarrhea and intense body scratching also occur. Aggression gradually gives way to inquisitiveness, and the chimpanzees finally approach the model to investigate and pull at it (Kortlandt and Kooij, 1963).

Mobbing behavior occurs in a few other social mammals. Herds of axis deer (*Axis axis*) occasionally follow tigers and leopards for short distances while barking at them, although flight is the usual response (Schaller, 1967). Agoutis (*Dasyprocta punctata*) mob snakes and other potential predators that remain immobile (Smythe, 1970a). Janzen (1970) observed a band of coatis attacking a large boa that had just struck and coiled around one of their companions. The assault was accompanied by a loud, shrill chattering. The altruistic effort did no good; the victim was crushed to death within six minutes of the strike. Such interactions of coatis with predators are rarely observed, and it is not known whether these raccoonlike animals really mob boas and other predators, that is, attack them while they are quiescent.

Mobbing in birds is a well-defined behavioral pattern that occurs irregularly in a wide diversity of taxonomic groups, from certain hummingbirds, vireos, and sparrows to jays, thrushes, vireos, warblers, blackbirds, sparrows, finches, towhees, and still others (S. A. Altmann, 1956). It is apparently absent in other species of hummingbirds, vireos, and sparrows, and at least some doves. The attacks are normally directed at predatory birds, particularly hawks and owls, when they passively intrude into the territorial or roosting areas of the smaller birds. The mobbing calls are high-pitched, loud, and easy for human observers to localize. As Marler (1959) pointed out, the mobbing calls of different bird species are strongly convergent. In the majority of cases they are loud clicks, 0.1 second or less in duration and spread over at least 2 or 3 kiloherz of frequencies in the 0-8 kiloherz range. These two properties combine to provide a biaural receptor system, which birds possess as well as human beings, with an instant fixation on the sound source. Thus alerted birds are able to fly toward the predator being harassed, and sizable mobs are quickly assembled. Furthermore, different species respond to one another's calls, since all make nearly the same sound, and mobbing becomes a cooperative venture. Altmann's account of birds attracted to stuffed owls in California and Nevada may be taken as typical of the attacking behavior:

Wren-tits (*Chamaea fasciata*) stayed in the dense shrubbery when mobbing. They fluffed out their feathers and made a sound like a spinning wooden ratchet-wheel. Where the dense shrubbery was continuous around the owl, they approached to within a few inches of the specimen. But when the owl was on a perch surrounded by a small clear space without undergrowth, the Wren-tits approached only as close as they could without entering the clearing; then they called toward the owl from that position. The Wren-tits sometimes continued their agitation for two or three hours.

Flocks of Brewer Blackbirds (*Euphagus cyanocephalus*) circled around the tree that sheltered the owl or stood on the ground facing the owl, repeating a harsh, nasal, call note. Red-winged Blackbirds (*Agelaius phoeniceus*) behaved quite differently. On the one occasion I tested their reactions to Screech Owls, they sat in the same tree as the owl, the males calling *teeyee* and the females, *chack*. Some of the females and the males with yellow-orange epaulets (yearlings?) fluttered in the air in front of the owl. One of the adult males flew straight at the owl from a distance of 30 feet, swerving sharply a foot in front of the owl, then it flew back to the tree from which it came. Another of the adult males perched silently a foot behind the owl, then leaped out at it, clawing at the top of the owl's head.

One of the most spectacular methods of attack was that used by the Anna Hummingbird (*Calypte anna*). They flew around the owl, two or three inches from its head, facing it and making little jabbing motions in their flight . . . The bills of the hummingbirds seemed, in all cases, to be directed at the eyes of the owl. While circling around the owl in this manner, they called a short, repeated, high-pitched note. (Altmann, 1956)

As Altmann's description implies, mobbing of some species has a vicious intent, and it can result in injury or possibly even death to the predator. Gersdorf (1966) has described how starlings launch massive attacks against sparrow hawks in Germany. Sometimes the predator is chased out over open water or into the reeds along the waterside. On rare occasions the hawks are even killed. Many other aspects of mobbing behavior, especially the visual cues used in predator recognition and the development and properties of the mobbing call, have been subjected to careful experimental studies by Rand (1941), Hartley (1950), Hinde (1954), Andrew (1961a–d), Curio (1963), and others.

Organized defense by instinctive behavior attains its greatest heights in the social insects. The reason is altruism: because the workers are reproductive neuters devoted to the sustenance of the queen and maximum production of her offspring, their own brothers and sisters, they can afford to throw their lives away. And if the colony welfare is threatened they do just that, with impressive efficiency. The result has been the evolution of elaborate communication systems devoted primarily or exclusively to group defense, together with special soldier castes programmed for no function other than combat.

The alarm systems of insect colonies are chiefly chemical in nature. Beekeepers know, for example, that when one honeybee worker stings an intruder, her nestmates often move in swiftly to join the attack. The signal provoking such mass assaults is an odorous chemical secretion released from the vicinity of the stings. One of the active components has been identified as isoamyl acetate, the same substance as the essence of the odor of bananas, which the bee secretes from glandular cells lining the sting pouch. The barbed sting of the honeybee worker catches in the skin of its victim, and when the bee atempts to fly away it often leaves behind its sting along with the attached poison gland and parts of its viscera. The isoamyl acetate is exposed, probably along with other, unidentified alarm pheromones. It evaporates rapidly and attracts other workers to the source (Ghent and Gary, 1962; Shearer and Boch, 1965). When a worker of the subterranean formicine ant *Acanthomyops claviger* is strongly disturbed, for example placed under attack by a member of a rival colony or an insect predator, it reacts by simultaneously discharging the reservoirs of its mandibular and Dufour's glands. After a brief delay, other workers resting a short distance away display the following response: they raise and extend their antennae, then sweep them in an exploratory fashion through the air; they open their mandibles; and they begin to walk, then run, in the general direction of the disturbance. Workers sitting a few millimeters away begin to react within seconds, while those a few centimeters distant may take a minute or longer. In other words, the signal obeys the laws of gas diffusion. Experiments have implicated an array of hydrocarbons, ketones, and terpenes as the alarm pheromones. Undecane and the mandibular gland substances (the latter all terpenes) evoke the alarm response at concentrations of 10^9–10^{12} molecules per cubic centimeter. These same substances are individually present in amounts ranging from as low as 44 nanograms to as high as 4.3 micrograms per ant; altogether they total about 8 micrograms. Released in gaseous form during experiments, similar quantities of the synthetic pheromones produce the same responses. Apparently the *A. claviger* workers rely entirely on these pheromones for alarm communication. Their system seems designed to bring workers to the aid of a distressed nestmate over distances of up to 10 centimeters. Unless the signal is then reinforced by additional emissions, it dies out within a few minutes. The alerted workers approach their target in a truculent manner. This overall defensive strategy is in keeping with the structure of the *Acanthomyops* colonies, which are large in size and often densely concentrated in the constricted subterranean galleries. It seems that it would not pay for the colonies to try to disperse when their nests are invaded, and, consequently, the workers have evolved so as to meet danger head on (Regnier and Wilson, 1968).

A different strategy is employed in the chemical alarm-defense system of the related ant *Lasius alienus*. Colonies of this species are smaller and normally nest under rocks or in pieces of rotting wood on the ground; such nest sites give the ants ready egress when the colonies are seriously disturbed. *L. alienus* produce mostly the same volatile substances as *Acanthomyops claviger*, and from the same glands. When they smell the pheromones, the *Lasius* workers scatter and run frantically in a comparatively unoriented fashion. They are more sensitive to undecane, the principal conponent, than are the *Acanthomyops* workers, being activated by only 10^7–10^{10} molecules per cubic centimeter. It can be concluded that, in contrast to *A. claviger*, *L. alienus* utilizes an "early warning" system and subsequent evacuation in coping with serious intrusion (Regnier and Wilson, 1969).

Chemical alarm systems of one design or another are widespread in the higher social Hymenoptera. Maschwitz (1964, 1966a) found evidence of alarm pheromones in all 23 of the more highly social species he surveyed in Europe. Several well-formed exocrine glands were implicated: the mandibular gland in the honeybee and many species of ants, the poison gland in *Vespa* and a few ant species, and Dufour's gland and the anal gland in still other ant species. Thus a social alarm-defense system has evolved repeatedly in these insects, utilizing various combinations of glandular sources and volatile substances in different phyletic lines. In contrast, the more primitively social Hymenoptera, in particular the bumblebees and wasps of the genus *Polistes*, show no evidence of utilizing such pheromones.

Termites organize their colony defense by both chemical and sound communication. Some of the phylogenetically more advanced termite groups produce volatile substances that act as straightforward alarm signals reminiscent of the ant pheromones: for example, pinenes from

the cephalic glands of the nasute soldiers of *Nasutitermes* and limonene from the same glands in the soldiers of *Drepanotermes* (Moore, 1968). Some termites utilize chemical odor trails to assemble workers at points of stress and danger inside the nest. As Lüscher and Müller (1960) and Stuart (1960) independently discovered, nymphs of the primitive species *Zootermopsis nevadensis* guide other nymphs through the rotten wood galleries by means of substances streaked from the sternal gland. Subsequently, Stuart (1963, 1969) found that the trails are laid primarily or exclusively to breaches in the wall of the nest. Virtually all dangerous situations in the life of the colony, including attacks by ants and other predators, can be translated to this single proximate stimulus—a breach in the wall. Termite nymphs are extremely sensitive to the increased light intensities and to microcurrents of air associated with such an event, and when thus disturbed they run back into the interior of the nest laying an odor trail behind them (Figure 3-10). The pheromone is an attractant that "compels" the outward march of the nymphs encountering it, and it is adequate in itself to guide them to their destination. When recruited nymphs arrive at the damaged portion of the nest, they set about repairing it. If the breach is too extensive to be repaired at once, the newcomers remain in an alarmed state and lay trails of their own back into the interior of the nest. In this fashion a repair crew is built up in numbers sufficient for the work to be done. Once the repair is completed, alarm ceases, trails are no longer laid, and the activity dies out.

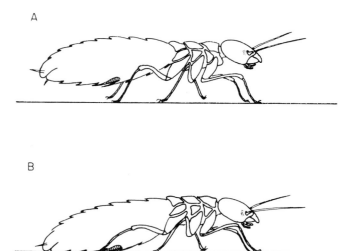

Figure 3-10 The nymph of the termite *Zootermopsis nevadensis*, on being alarmed, lays odor trails into the interior of the nest. The location of the gland that secretes the trail pheromone is indicated on the lower surface of the abdomen, in both the resting (*A*) and the trail-laying termite (*B*). (From Stuart, 1969.)

Sound communication in termites has been less securely documented. According to Howse (1964), the agitated soldier of *Zootermopsis angusticollis* alerts other colony members by sound transmitted through the wall of the nest. The sound is generated in a crude fashion: the soldier vibrates the forward part of its body by convulsively rocking its head upward and then down to a normal position again, over and over about 24 times a second. With each upward thrust the forelegs are lifted off the floor and the head is banged against the ceiling of the nest; the overall effect to the human ear is a faint rustling sound. The signals are transmitted through the substratum of the nest, not the air, and are picked up by the subgenual organs, specialized stretch receptors located in the legs. A systematic review of this and other forms of alarm-defense communication systems in social insects has been provided by Wilson (1971a).

A corollary development of increased efficiency in group defense is the narrowing of individual conformity. Predators counterrespond to social defensive mechanisms by watching for deviant individuals who, for reasons of health, inexperience, or whatever, fail to participate and, by failing, increase their vulnerability. In his field studies of muskrats (*Ondatra zibethica*), Paul Errington (1963) learned that minks concentrate on the muskrats that are excluded from the territorial aggregations and hence deprived of secure retreats. The same general effect has been independently documented in other rodent species and in several kinds of birds (Jenkins et al., 1963; Lack, 1966; Watson, 1967; Watson and Moss, 1971). Among mountain sheep, moose, and the antelopes and other ungulates of the African plains, the principal victims of predators are the young, aged, and infirm individuals who experience difficulty staying close to the herds (Murie, 1944; Mech, 1970; Kruuk, 1972). This phenomenon is probably of general occurrence whenever death by predation is more than negligible. Furthermore, there is abundant evidence that predators respond strongly to deviant individuals in the social groups they watch. Students of fish behavior and ecology have observed that it is difficult to tag fish or to introduce distinctive mutants in the presence of predators. Predatory fish are stimulated by any change in appearance and attack altered individuals preferentially. The preference for the simple property of oddity in prey has been demonstrated convincingly by Mueller (1971), who conducted experiments with sparrow hawks (*Falco sparverius*) and broad-winged hawks (*Buteo platypterus*). Eight tamed birds were simultaneously presented with sets of ten mice of which one (or none) had been dyed gray and the remainder left white. All of the hawks showed a preference for the oddly colored mice, but only if it was one particular color: four showed an oddity choice if the odd mouse was white, while the remaining four reacted only if the odd mouse was gray. Thus the oddity factor is combined with a preference for a particular color, a possible example of what L. Tinbergen (1960) has termed the "spe-

cific searching image" of predators. The two factors might interact in the following way. If the specific searching image results from previous successful experiences, which in turn are the outcome of pursuing odd individuals, the predators will tend to stay with a particular odd class. Thus they could adapt quickly to the class of helpless juveniles, the sick and the old, the dispossessed, and so forth. This strategy of choice could be a highly efficient one for the predator.

Increased Competitive Ability

The same social devices used to rebuff predators can be used to defeat competitors. Gangs of elk approaching salt licks are able to drive out other animals, including porcupines, mule deer, and even moose, simply by the intimidating appearance of the massed approach of the group (Margaret Altmann, 1956). Observers of the African wild dog (*Lycaon pictus*) have noted that coordinated pack behavior is required not only to capture game but also to protect the prey from hyenas immediately after the kill. The wild dogs and hyenas in turn each compete with lion prides.

Elsewhere (Wilson, 1971a) I have characterized as "bonanza strategists" a class of subsocial beetle species adapted to exploit food sources that are very rich but at the same time scattered and ephemeral: dung (*Platystethus* among the Staphylinidae; and Scarabaeidae), dead wood (Passalidae, Platypodidae, Scolytidae), and carrion (*Necrophorus* among the Silphidae). When individuals "strike it rich" by discovering such a food source, they are assured of a supply more than sufficient to rear their brood. They must, however, exclude others who are seeking to utilize the same bonanza. Territorial behavior is commonplace in all of these groups. Sometimes, as in *Necrophorus*, fighting leads to complete domination of the food site by a single pair. It is probably no coincidence that the males, and to a lesser extent the females, of so many of the species are equipped with horns and heavy mandibles—a generalization that extends to other bonanza strategists that are not subsocial, for example, the Lucanidae, the Ciidae, and many of the solitary Scarabaeidae. By the same token there is an obvious advantage to remaining in the vicinity of the food site to protect the young.

Within the higher social insects, group action is the decisive factor in aggressive encounters between colonies. It is a common observation that ant queens in the act of founding colonies as well as young colonies containing workers—the weaker units—are destroyed in large numbers by other, larger colonies belonging to the same species. Newly mated queens of *Formica fusca*, for example, are captured and killed as they run past the nest entrances (Donisthorpe, 1915); the same fate befalls a large percentage of the colony-founding queens of the Australian meat ant *Iridomyrmex detectus* and red imported fire ant *Solenopsis invicta*. Queens of *Myrmica* and *Lasius* are harried by ant colonies, including those belonging to their own species, and finally they are either driven from the area or killed (Brian, 1955, 1956a,b; Wilson, 1971a). As a corollary, colony-founding ant queens and juvenile colonies are more abundant where mature colonies are scarce or absent. Brian, who has studied this effect in the British fauna in some detail, discovered a striking inverse correlation in various habitats between the density of adult colonies and of foundress queens of *Myrmica* and *Formica*. Similar dispersing effects have been recorded in other social insects. In stable habitats of southwestern Australia, mature colonies of the termite *Coptotermes brunneus* are spaced about 90 meters apart. In the intervening areas, colony-founding queens are caught and destroyed. Also, the mature colonies compete intensely for the limited foraging space in the few available trees (Greaves, 1962). A similar pattern of strong territoriality has been described in the South African *Hodotermes mossambicus* by Nel (1968). It is true of termites generally that when more than one couple belonging to the same species succeeds in founding colonies together, they coexist peaceably or even combine forces for a while. But within a few months at most, fighting and cannibalism ensue, until finally only a single couple—and, hence, one effective colony—survives (Nutting, 1969). Colonies of the Japanese paper wasp *Polistes fadwigae* located 3.5 meters apart steal and eat one another's larvae. If they are brought by the experimenter to within 5 centimeters of each other, the dominant females fight until a new dominance order is achieved, and the colonies fuse (Yoshikawa, 1963). Honeybee workers from different colonies fight at the same food dishes when the sugar supply begins to be used up (Kalmus, 1941). Under more natural conditions, honeybee colonies placed together have been shown by use of radioactive tagging to restrict one another's foraging areas as a function of the degree of crowding (Levin and Glowska-Konopacka, 1963).

Territorial fighting among mature colonies of both the same and differing species is common but not universal in ants. It has been recorded in very diverse genera of which the following form only a partial list: *Pseudomyrmex, Myrmica, Pogonomyrmex, Leptothorax, Solenopsis, Pheidole, Tetramorium, Iridomyrmex, Azteca, Anoplolepis, Oecophylla, Formica, Lasius, Camponotus*. The most dramatic battles known within species are those conducted by the common pavement ant *Tetramorium caespitum*. First described by the Reverend Henry C. McCook (1879) from observations in Penn Square, Philadelphia, these "wars" can be witnessed in abundance on sidewalks and lawns in towns and cities of the eastern United States throughout the summer. Masses of hundreds or thousands of the small dark brown workers lock in combat for hours at a time, tumbling, biting, and pulling one another, while new recruits are guided to the melee along freshly laid odor trails. Although no careful study of this phenomenon has been undertaken, it appears superficially to be a contest between adjacent colonies in the vicinity of

their territorial boundaries. Curiously, only a minute fraction of the workers are injured or killed.

Territorial wars between colonies of different ant species occur only occasionally in the cold temperate zones. Colonies of *Myrmica* and *Formica*, for example, sometimes overrun and capture nest sites belonging to other species of the same genus (Brian, 1952a; Scherba, 1964). By contrast, intense aggression is very common in the tropics and warm temperate zones. Certain pest species, particularly *Pheidole megacephala, Solenopsis invicta*, and *Iridomyrmex humilis*, are famous for the belligerency and destructiveness of their attacks on native ant faunas wherever they have been introduced by human commerce (Haskins, 1939; Wilson and W. L. Brown, 1958; Haskins and Haskins, 1965; Wilson and Taylor, 1967). They even go so far as to eliminate some of the species, especially those closest to them taxonomically and ecologically. In the case of *I. humilis*, only the smallest, least aggressive ant species remain unaffected. Some of the battles between species are epic in their proportions. E. S. Brown (1959) has provided the following account of war between colonies of the introduced African ant *Anoplolepis longipes* and the defending colonies of two native species, *Oecophylla smaragdina* and *I. myrmecodiae*, in the Solomon Islands:

[The] invading *Anoplolepis* ants move on to the base of the trunk, which evokes the descent of large numbers of *Oecophylla* to ring the trunk in defensive formation just above them. It then becomes a ding-dong struggle, the dividing line between the two species sometimes moving up or down a few feet from day to day; any ant wandering alone into the other species' territory is usually surrounded and overcome. Eventually one species will get the better of the other, but this may not happen for several days or weeks . . .
Anoplolepis had advanced on to the base of the trunk of a palm occupied by *Iridomyrmex*, which had descended in force from the trunk and formed a complete phalanx of countless individuals, almost completely covering the trunk over about 2 ft. of its length. After a few days this defensive formation was still intact, but had retreated higher up the trunk; eventually it was driven from the trunk altogether, and later *Anoplolepis* took possession of the crown.

The outcome of such encounters must depend on a complex of factors: size and numbers of individuals, aggressiveness, secureness of the nest site, and so forth. Furthermore, the aggression may take the form of more subtle techniques. Brian (1952a,b) found that the takeover of nest sites by various species of Scottish ants is usually gradual and may involve any of several methods. *Myrmica scabrinodis*, for example, seizes nests of *M. ruginodis* either by direct siege, causing total evacuation of the *ruginodis*, by gradual encroachment of the nest, chamber by chamber, or by occupation following greater tenacity in the face of adverse physical conditions, particularly severe cold, that drive the other species away temporarily.

In the case of competition within the same species, we should expect to find that groups generally prevail over individuals, and larger groups over small ones. Consequently, competition, when it comes into play, should be a powerful selective force favoring not only social behavior but also large group size. Lindburg (1971) demonstrated a straightforward case of this relationship in a local population of free-ranging rhesus monkeys (*Macaca mulatta*) he studied in northern India. The population was divided into five troops, most of which had overlapping home ranges and therefore came into occasional contact. In the pairwise aggressive encounters that occurred, one group usually retreated, and this was almost invariably the smaller one. The same selective pressures should operate to favor coalitions or cliques with societies. The phenomenon does occur commonly in wolves and those primate species, such as baboons and rhesus monkeys, in which dominance hierarchies play an important role in social organization. In other words, coalitions are known in aggressive animals that have a sufficiently high degree of intelligence to remember and exploit cooperative relationships.

Increased Feeding Efficiency

We have finished considering the remarkably diverse ways in which aggregation and cooperative behavior can prevent individual organisms from being turned into energy by predators. Let us next examine the equally diverse ways by which social behavior can assist in converting other organisms into energy. There are two major categories of social feeding: *imitative foraging* and *cooperative foraging.* In imitative foraging the animal simply goes where the group goes, and eats what it eats. The pooled knowledge and efficiency of such a feeding assemblage exceeds that of an otherwise similar but independently acting group of individuals, but the outcome is a byproduct of essentially selfish actions on the part of each member of the assemblage. In cooperative foraging there is some measure of at least temporarily altruistic restraint, the behaviors of the group members are often diversified, and the modes of communication are typically complex. Some of the most advanced of all societies, possibly including those of primitive man, are based upon a strategy of cooperative hunting. One can reflect upon the fact that the qualities we intuitively associate with higher social behavior—altruism, differentiation of group members, and integration of group members by communication—are the same ones that evolve in a straightforward way to implement cooperative foraging.

Imitative foraging is based on an array of responses between animals that range from the simplest undirected stimulation of searching or feeding behavior to the most specific and elaborate imitation of one animal's movements by another. The classification of these various forms of coaction has evolved through the experiments and writings of Thorpe (1963a), Klopfer (1957, 1961), and Alcock (1969), whose synthesis is the one presented here.

True imitation: the copying of a novel or otherwise improbable act. Examples include the learning of particular song dialects by certain bird species and the cultural transmission of potato washing in Japanese macaques.

Social facilitation: an ordinary pattern of behavior that is initiated or increased in pace or frequency by the presence or actions of another animal. In order to provide the facilitating stimulus, the other animal need not be engaged in the act it causes. In some cases it does nothing at all except appear on the scene—the "audience effect."

Facilitation may produce only temporary results, or it may lead in an incidental manner to learning. For example, the observer animal might discover food in a particular spot as a result of having its attention drawn to that place and, thus rewarded, learn to look for food there even after the first animal is gone.

Observational learning (sometimes termed *empathic learning*): unrewarded learning that occurs when one animal watches the activities of another. In order to prove that observational learning has occurred, it is necessary to demonstrate that the observer was not rewarded while with its companion but altered its behavior later (in the absence of the companion) as a result of what it saw and remembered. Thus, a bird that saw a companion attacked by a snake and increased its avoidance of the same kind of snake in subsequent encounters could be said to have achieved pure observational learning. Technically, observational learning can be classified as either imitation or social facilitation, depending on the complexity and novelty of the behavior that is repeated. A great deal of human behavior, obviously, is based on observational learning that is imitative in nature.

The advantages of imitative foraging have been elucidated in a few instances. Turner (1964) described how chaffinches (*Fringilla coelebs*) commence feeding on familiar food if they see other chaffinches eating. Also, they occasionally enter new microhabitats and try new foods if they see others doing so; this is especially true of the young, who are less wary. As a result, chaffinch flocks can locate and switch to new feeding places more readily than birds acting separately. Primates appear to go out of their way to gain such information. Yellow baboons (*Papio cynocephalus*) sometimes touch muzzles in what appears to be an effort on the part of one animal to smell the contents of the mouth of the other (see Figure 3-11). The exchange is more frequent when the second baboon has food in its mouth. Altmann and Altmann (1970) have reasonably hypothesized that information about new food sources can be spread in this manner through the troop. Similar behavior has been recorded by Hall (1963a) in chacma baboons (*P. ursinus*) and by Struhsaker (1967a) in vervet monkeys (*Cercopithecus aethiops*).

Kummer (1971) has argued that the intensity of social facilitation in feeding, and from this the degree of coordination in group behav-

Figure 3-11 Muzzling in yellow baboons, an interaction hypothesized to spread information on new food sources through the troops. (From Altmann and Altmann, 1970.)

ior, increases with the severity of the environment to which the society is adapted. Troops of chimpanzees or tamarins live in forest habitats where food, water, and safe retreats are always a short distance away. Consequently, each member of the troop can eat, drink, and sleep when it pleases, and coordination with other members of the group is weak. But a troop of hamadryas baboons, which exists in a harsh environment where shelter is far removed from the sources of food and water, must operate with a high degree of synchrony. A baboon that stops to take a drink while the remainder of its troop continues the march is likely to lose contact and fall victim to a waiting predator. Conversely, a baboon that neglects to drink when its companions do so, because it is not yet thirsty, is likely to grow thirsty before the next drinking halt—unless it separates from the group and risks death from predation.

The conformist benefits from the pooled knowledge of its companions. Kummer's hamadryas and the Altmann's yellow baboons at Amboseli traveled directly back and forth to water sources that were not within sight of the sleeping places. They evidently operated on the basis of prior knowledge, one would assume within the memories of the adult leaders. In the Central Valley of California, enormous flocks of starlings leave their roosts and fly in straight lines to food sources as distant as 80 kilometers. The lengths of the flights are greatest in winter, when food is in shortest supply (W. J. Hamilton

III and Gilbert, 1969). By following a flock the individual starling has the greatest chance of locating adequate amounts of food on a given day, since it is utilizing the knowledge of the most experienced birds in the group. Also, it will expend the least amount of energy reaching the food. Theoretically, the prime factor for colonial roosting and nesting, as Horn (1968) has shown in an elegant geometric analysis, is that the food supply be considerably variable in space and time. That is, food must appear in unpredictable, irregular patches in the environment. If it occurs in patches but is available in certain spots permanently or at predictable intervals, individuals will simply roost or nest as closely as possible around those spots, and fly singly to them. But if the food is evenly distributed through the environment and concentrated enough to more than repay the energy expended in its defense, the individuals will stake out separate territories from which they exclude other birds (see Figure 3-12). Clumping in roosting sites does not preclude setting up "microterritories" that preserve the individual's exclusive access to a particular resting spot or nest site within the colony. The important feature of such colonial life is that the group be concentrated enough to forage more or less as a unit. Horn's principle is easily extensible to many kinds of colonial birds, from blackbirds and swallows to herons, ibises, spoonbills, and various seabirds. Terns, for example, are an extreme example of seabirds that nest in aggregations and forage in groups for highly unpredictable food patches. Their food consists of schools of small fish that move near the surface of the ocean. Notice that colonial flocking is favored *both* by the increase in feeding efficiency and by superiority in defense against predators. The most careful investigators of social behavior in birds, including Fisher (1954), J. M. Cullen (1960), Orians (1961a,b), Brown (1964), Kruuk (1964), Crook (1965), Patterson (1965), Ward (1965), Horn (1968), and Brereton (1971), have documented the operation of one or both of these prime factors. But the difficulty of putting both on the same scales of mortality and reproductive success has so far prevented any assessment of their relative contributions to social evolution.

Flocks are not just more expert at finding food than unorganized groups. They are also more likely to harvest it efficiently. The efficiency that counts to the individual member is not the depth to which a given patch of food is cropped by the group, but rather the food intake per animal in each unit of time. Insectivorous birds, such as cattle egrets, anis, parulid warblers, and tyrannid flycatchers, potentially benefit from foraging in flocks because the group as a whole can beat up a higher proportion of flying insects per bird than can scattered individuals. For the same reason, ant thrushes follow swarms of *Eciton* army ants in Central and South America, and cattle egrets, snowy egrets, and grackles attend cattle and other large, grazing mammals to catch the insects that they stir up (Short, 1961; Heatwole, 1965; Willis, 1967). A. L. Rand (1953) found that the feeding rates

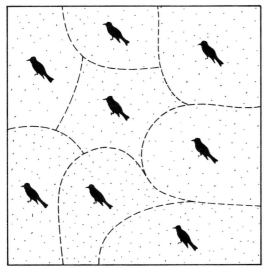

FEEDING TERRITORIES

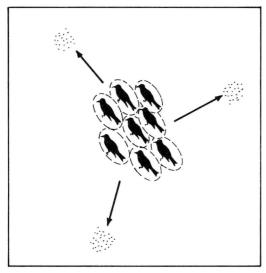

FEEDING FLOCKS

Figure 3-12 Horn's principle of group foraging. If food is more or less evenly distributed through the environment and can be defended economically, it is energetically most efficient to occupy exclusive territories (above). But if food occurs in unpredictable patches, the individuals should collapse their territories to roosting spots or nest sites, and forage as a group (below).

of groove-billed anis (*Crotophaga sulcirostris*) following cattle are higher than those feeding alone.

In the Mohave Desert of California large mixed flocks of birds forage slowly from November to May through the low, scrubby vegetation. The peak of flock diversity is reached in April, when a typical flock contains 50 to 200 individuals consisting chiefly of Brewer's, chipping, black-throated, and white-crowned sparrows, together with a motley assortment of other sparrows, juncos, grosbeaks, phoebes, woodpeckers, cactus wrens, vireos, warblers, kinglets, and *Empidonax* flycatchers. Most members of the assemblage are seed eaters. According to Cody (1971), the flocks move predictably along certain zones at relatively constant speeds. They also display momentum; that is, they pursue straight courses over longer distances than do solitary individuals. From the results of computer simulations, Cody concluded that under a wide range of conditions the flocks make more efficient use of both nonrenewable and renewable resources. Consider first the nonrenewable resources, such as the fruits of toyon (*Heteromeles arbutifolia*) and *Rhus laurina*. The flock reduces each patch of food more thoroughly than an individual would. Consequently, as it progresses it behaves like a giant mower, leaving a pattern of well-trimmed areas juxtaposed to relatively untouched areas. Wheeling and looping back over periods of days, the flock can easily distinguish and avoid previously exploited bushes and devote its full time to ones that hold full crops. In contrast, scattered individuals reduce such nonrenewable resources gradually and evenly. As the season passes, the time required to find each food item steadily increases, even though the total remaining crop may be equal to that in a similar area occupied by flocks. A different line of reasoning applies to renewable resources, such as grass seeds and flying insects. Because of its momentum, the flock will give each patch of vegetation a longer average rest between visits. Consequently it will obtain a higher average yield with each pass. Cody has gone so far as to suggest that the velocity and turning rate of the flocks have evolved to bring flocks back to previously visited patches at just about the time the plants bear a new full crop.

Another kind of feeding efficiency has been achieved by the larvae of the jack-pine sawfly (*Neodiprion pratti banksianae*). These caterpillarlike insects feed in tight groups on their coniferous hosts. Ghent (1960) discovered that the chief advantage of aggregation comes in the first stadium, when the larvae are very small and weak and experience great difficulty chewing holes in the tough pine needles on which they depend. In Ghent's experiments, 80 percent of the larvae isolated from their fellows died, while only 53 percent of those allowed to remain together died. The effect is a statistical one that improves with group size: even when a larva belongs to a group, it attempts individually to establish its own feeding site. When one does cut through into the succulent inner tissue, whether by luck, superior strength, or greater skill at finding a weak spot, the other larvae are quickly attracted to the spot by the odor of the volatile compounds among the salivary secretions and plant substances released into the air. Soon the breach is widened, and all of the larvae are able to feed.

It can be easily seen that if foraging in masses increases the yield of food, cooperative tactics by the same masses can improve it still more. Several groups of mammals have developed relatively sophisticated cooperative hunting maneuvers, in each case as an adaptation to help overcome unusually large or swift mammalian prey. In his pioneering study of the wolves of Mount McKinley National Park, Murie (1944) found that these carnivores could capture their principal large prey, Dall's sheep, only with difficulty. On a typical day a pack trots from one herd to another in search of a weak or sick individual, or a stray surprised on terrain in which it is at a disadvantage. A lone wolf can trap a healthy sheep only with great difficulty if it is on a slope; the sheep outdistances the wolf easily by racing it up the slope. Two or more wolves are able to hunt with greater success because they spread out and often are able to maneuver the sheep into a downhill race or force it onto flat land. Under both circumstances they hold the advantage. Where wolves hunt moose, as they do for example in the Isle Royale National Park of Michigan, cooperative hunting is required both to trap and to disable the prey (Mech, 1970).

The most social canids of all are the African wild dogs, *Lycaon pictus*. These relatively small animals are superbly specialized for hunting the large ungulates of the African plains, including gazelles, zebras, and wildebeest. The packs, often under the guidance of a lead dog, take aim on a single animal and chase it at a dead run. They pursue the target relentlessly, sometimes through crowds of other ungulates who either stand and watch or scatter away for short distances. The *Lycaon* do not ordinarily stalk their prey while in the open, although they sometimes use cover to approach animals more closely. Estes and Goddard (1967) watched a pack race blindly over a low crest in the apparent hope of surprising animals on the other side—in this one instance no quarry was there. Fleeing prey frequently circle back, a tactic that can help shake off a solitary pursuer. This maneuver, however, tends to be fatal when employed against a wild dog pack: the dogs lagging behind the leader simply swerve toward the turning animal and cut the loop. Once they have caught up to the prey, the dogs seize it on all sides and swiftly tear it to pieces. As soon as the prey is disabled, the dogs must be prepared to fight off hyenas, which habitually follow them and attempt to steal their food. The twin problems provided by the large size of the prey and the competition from hyenas make it unlikely that a wild dog could survive for long on its own. Estes and Goddard in fact estimate that the minimum pack size is four to six adults. The hyena has hunting habits similar to those of the African wild dogs and a com-

parably strong commitment to social life. While observing a total of 34 zebra hunts, Kruuk (1972: 185) gained the impression that the frequency of success was correlated with the number of hyenas taking part in the chase. His data were too few to be statistically significant, however.

Lions also hunt socially. As several members of a pride approach a prey together, they usually fan out along a broad front, sometimes extending laterally for as much as 200 meters. This coordination appears to be deliberate: the lions in the center halt or slow their advance while those on the flanks walk rapidly to their positions; then all move forward together. Schaller (1972) cites the following episode as typical:

At 1845 five lionesses and a male see a herd of some 60 wildebeest 2.7 km away—just black dots moving against the yellow-grey plains. The lions walk slowly toward them. At dusk the wildebeest bunch up. The last light has faded at 1930 when the lions stop, .3 km from the herd. The lionesses fan out there and advance at a walk in a front 160 m wide, moving downwind, the male 60 m behind them. They crouch when 200 m from the herd, and I can see only an occasional head as they stalk closer; the male remains standing. Five minutes later a female on the left flank rushes and catches a wildebeest, but I am unable to see the details. Two lionesses converge on her. The herd bolts to the right and two lionesses and the male run at an angle toward it, pursuing about 100 m without success. The wildebeest is on its back while one lioness clamps its muzzle shut with her teeth, a second bites it in the lower neck, and a third in the chest. Then the male bounds up and with one bite tears open the groin.

There can be no doubt that group action is also required to subdue certain especially difficult prey. Schaller witnessed an incident in which a lone lioness rushed a bull buffalo and seized him by the neck. The buffalo continued to walk or trot along until the lioness released her hold, whereupon he charged her and chased her into a tree!

Killer whales (*Orcinus orca*) are the wolves of the sea—large social predators that hunt in packs to catch even larger mammalian prey. Those prowling along the coasts of California and Mexico feed mostly on sea lions, porpoises, and whales (Brown and Norris, 1956; Martinez and Klinghammer, 1970). One pack of 15 to 20 was seen pursuing a school of about 100 porpoises, probably *Delphinus bairdi*. The killer whales encircled the porpoises, then gradually constricted the circle to crowd the porpoises inward. Suddenly one whale charged into the porpoises and ate several of the trapped animals while its companions held the line. Then it traded places with another whale, who fed for awhile. This procedure continued until all of the porpoises were consumed. The *Orcinus* use different tactics to subdue other kinds of whales larger than themselves. They attack en masse: some seize the pectoral fins, immobilizing the victim, while others bite at the lower jaw and tear flesh from it. The tongue is the most favored organ, however. If the whale does not stick out its tongue in its distress, the *Orcinus* force its mouth open by prying at it with their heads and pull the tongue out themselves. Some large predatory fish also hunt in schools. They have been observed to encircle schools of smaller fish and to drive them into tight spaces (Eibl-Eibesfeldt, 1962). The amount of cooperation in such maneuvering, if any exists at all, appears to be far below that displayed by killer whales.

The ultimate developments in cooperative foraging are found in the higher social insects. Members of the worker caste, bound together by their neuter status and altruistic commitment to the reproductive castes, are very sensitive to recruitment signals from their fellow colony members. Some large-eyed ants run toward sudden movements, including those of their nestmates, and thus become collectively involved in the trapping and killing of prey (Wilson, 1962a, 1971a). When individual workers of the harvesting ant *Pogonomyrmex badius* attack large, active insect prey in the vicinity of the nest, they discharge the alarm pheromone 4-methyl-3-heptenone from their mandibular glands. This substance both attracts and excites other workers within distances of 10 centimeters or so (just as it does in the presence of dangerous stimuli), with the result that the prey is more quickly subdued. Thus, in *P. badius*, and probably in other predaceous ant species that employ alarm pheromones, recruitment is a felicitous by-product of alarm communication (Wilson, 1958d). A parallel relation between two quite different behavioral functions exists in the social life of the honeybee, where the Nasanov gland pheromones are used in some instances to assemble workers that have become lost while foraging, or while participating in colony swarming, and in other instances to recruit nestmates to newly discovered pollen and nectar sources (Renner, 1960; Butler and Simpson, 1967).

There is some evidence to suggest that social insects leave chemical "signposts" around food discoveries, although few studies have been conducted to characterize the phenomenon. Glass feeding dishes that have been visited by worker honeybees are preferred by newcomer bees over unvisited dishes, even when each container holds identical food (Chauvin, 1960). The substance the workers deposit can be extracted and is said to come from Arnhardt's glands in the tarsi. The same pheromone may be responsible for the odor trails sometimes laid by walking honeybees, as described by Lecomte (1956) and Butler et al. (1969). The trails, whether laid deliberately or not, serve as rudimentary guides for worker bees that have landed on the correct hive but are still searching for the hive entrance.

The odor of food brought into the nests can also influence the behavior of nestmates and thereby serve as a primitive form of recruitment communication. Honeybee workers recognize the odor of food sources both from the smells adhering to the bodies of successful foragers and from the scent of nectar regurgitated to them. If they have had experience in the field with flowers or honeydew bearing

the same odor, they will then revisit the site searching for food. The response can be induced in the absence of waggle dancing or other forms of communication. Russian apiarists have used the principle to guide bees to crop plants they wish pollinated. To take a typical example, the colonies are trained to red clover by being fed with sugar water in which clover blossoms have been soaked for several hours. After this exposure, the foraging workers search preferentially for red clover in the vicinity of the hive. The same method has been used to increase pollination rates of vetch, alfalfa, sunflowers, and fruit trees (von Frisch, 1967). Free (1969) has recently been able to demonstrate that the odor of food stores has a similar effect on bumblebees.

The next step up the ladder of sophistication in chemical recruitment techniques is tandem running (Hingston, 1929; Wilson, 1959a; Hölldobler, 1971a). When a worker of the little African myrmicine ant *Cardiocondyla venustula* finds a food particle too large to carry, it returns to the nest and makes contact with another worker, which it leads from the nest. The interaction follows a stereotyped sequence. First the leader remains perfectly still until touched on the abdomen by the follower ant. Then it runs for a distance of approximately 3 to 10 millimeters, or about one to several times its body length, coming again to a complete halt. The follower ant, now in an excited state apparently due to a secretion released by the leader, runs swiftly behind, makes fresh contact, and "drives" the leader forward. After each contact and subsequent forward drive of the leader, the follower may press immediately behind and move it again. More commonly, it circles widely about in a hurried movement that lasts for several seconds and may take it as far as a centimeter from the path set by the leader. In a short time, however, the circling brings the follower once again into contact with the leader. Eventually the two arrive at the food particle. Tandem running also occurs in the large formicine genus *Camponotus*, where it has evolved independently in several phyletic lines.

The most elaborate of all the known forms of chemical recruitment is the odor trail system. Trail communication has evidently evolved, at least in some groups of ants, from tandem running. In several species of *Camponotus* and the slave-maker species *Harpagoxenus americanus*, an intermediate form of communication is employed. The leader ant does not wait to be touched, but instead runs outward from the nest to the target it previously discovered. As it proceeds it emits a pheromone that persists in the form of a short-lived odor trail. Depending on the species, from 1 to 20 or more workers follow single file behind the leader, and the entire group arrives at the target at more or less the same time.

It is only a short step in evolution from trail-guided processions such as these to typical trail communication, where in the absence of the trail layer followers are guided over long distances by odor alone. One well-analyzed case, that of the fire ants of the genus

Solenopsis, can serve as a paradigm (Wilson, 1959c, 1962a; Wilson and Bossert, 1963; Hangartner, 1969a). When workers of the red imported fire ant *S. invicta* (referred to as *S. saevissima* in earlier literature) leave their nest in search of food they may follow preexisting odor trails for a short while, but they eventually separate from one another and begin to explore singly. When alone they maintain knowledge of the location of the nest by sun-compass orientation; that is, they are aware of the angle subtended by lines drawn from the nest to their position and in the direction of the sun. When a foraging worker finds a particle of food too large to carry, it heads home at a slower, more deliberate pace. At frequent intervals the sting is extruded, and its tip drawn lightly over the ground surface, much as a pen is used to ink a thin line. As the sting touches the surface, a pheromone flows down from the Dufour's gland. Each worker possesses only a small fraction of a nanogram of the trail substance at any given moment. It follows that the pheromone must be a very potent attractant. In 1959 I showed that it is possible to induce the complete recruitment process in the fire ant with artificial trails made from extracts or smears of single Dufour's glands. Such trails induce following by dozens of individuals over a meter or more. When the concentrated pheromone is allowed to diffuse from a glass rod held in the air near the nest, workers mass beneath it, and they can be led along by the vapor alone if the rod is moved slowly enough (see Figure 3-13). When large quantities of the substance were allowed to evaporate near the entrances of artificial nests, they drew out most of the inhabitants, including workers carrying larvae and pupae and, on a single occasion, the mother queen.

If the trail-laying worker encounters another worker, she turns toward it. She may do nothing more than make an abrupt rush at it before moving on again, but sometimes the reaction is stronger: she climbs partly on top of the worker and, in some instances, shakes her own body lightly but vigorously in a vertical plane. The vibrating movement, which is unique to these encounters, has also been described in *Monomorium* and *Tapinoma* by Szlep and Jacobi (1967) and in *Camponotus* by Hölldobler (1971a). Hölldobler has been able to demonstrate experimentally that the movement stimulates other workers to follow the trail just laid. In *Solenopsis*, however, the movement does not appear to be essential, since contacted workers do not exhibit trail-following behavior different from the behavior of those not contacted. Moreover, the pheromone is by itself sufficient to induce immediate and full trail-following behavior when laid down in artificial trails.

Workers of some species of stingless bees (Meliponini) are able to communicate the location of food finds by chemical trail systems basically similar to those of the ants (Lindauer and Kerr, 1958, 1960). When a foraging worker of *Trigona postica* finds a feeding site, for example, she first makes three or more normal collecting flights

Figure 3-13 The response of fire ant workers to evaporated trail substance. Above: before the start of the experiment, air is being drawn into the nest (by suction tubing inserted to the left) from the direction of the still untreated glass rod. Below: within a short time after the glass rod has been dipped into Dufour's gland concentrate and replaced, a large fraction of the worker force leaves the nest and moves in the direction of the rod. (From Wilson, 1962a.)

straight back and forth between the hive and the site. Then she begins to stop in her homeward flight every two or three meters, settling onto a blade of grass, a pebble, or a clump of earth, opening her mandibles, and depositing a droplet of secretion from her mandibular glands. Other bees now leave the nest and begin to follow the odor trail outward.

Nedel (1960) subsequently found that the mandibular glands of the trail-laying Trigonas are greatly enlarged in comparison with those of other bee species. Furthermore, after being emptied, the gland reservoirs are refilled in as little time as 20 minutes. According to Kerr, Ferreira, and Simões de Mattos (1963) the overall *Trigona* trails are polarized; that is, larger quantities of scent are laid down nearer the food source. In three species studied by these investigators in Brazil, the odor spots retained their activity for periods ranging from 9 to 14 minutes. The alerting stimulus in *Trigona* communication, the action that arouses other workers before they move out along the odor trail, is believed by Kerr and his coworkers and by Esch (1965, 1967a,b) to be a buzzing sound made by successful foragers shortly after returning to the nest. According to Esch, the length of a particular pulse increases with the distance of the journey in a precise manner.

Most species of stingless bees nest and forage in tropical forests, and odor trail communication seems ideally suited for recruitment in this habitat. The individual forager bee can best thread its way through tree trunks and understory vegetation if guided point by point by frequently repeated cues. Odor trails also have the advantage of leading up and down tree trunks as well as over the ground, thus transmitting the three-dimensional information that is greatly needed in tall tropical forests. There can be no question concerning the superiority of trail communication as a recruitment device. The *Trigona* colonies that use it are able to assemble crowds of workers at new food sources far more quickly than colonies belonging to other species.

The waggle dance of the honeybee is in a sense the *ne plus ultra* of foraging communication, since it utilizes symbolic messages to direct workers to targets prior to leaving for the trip. It also operates over exceptionally long distances, exceeding the reach of any other known animal communication with the possible exception of the songs of whales. The waggle dance will be described in more detail in another context, in Chapter 8.

Penetration of New Adaptive Zones

Occasionally a social device permits a species to enter a novel habitat or even a whole new way of life. One case is provided by the staphylinid beetle *Bledius spectabilis*, which has evolved a complexity of maternal care rarely attained in the Coleoptera. The change has permitted the species to penetrate one of the harshest of all environments available to any insect: the intertidal mud of the European coast, where the beetle must subsist on algae and face extreme hazards from both the high salinity and periodic shortages of oxygen. The female constructs unusually wide tunnels in her brood nest, which are kept ventilated by tidal water movements and by renewed burrowing activity on the part of the female. If the mother is taken away, her brood soon perishes from lack of oxygen. The female also protects the eggs and larvae from intruders, and from time to time forages outside the nest for a supply of algae (Bro Larsen, 1952).

Termite colonies, which are among the most elaborate and successful of all societies, appear to have a peculiar, not to say bizarre raison d'être. Termites are unusual among the insects in their ability to digest cellulose, which they do with the aid of symbiotic intestinal microorganisms. Moreover, the exchange must be repeated each time a termite sheds its integument in order to grow, because the microorganisms are pulled out with the extension of the integument that lines the hind gut. It is very likely that termite social behavior received its initial impetus from this particular bond, which in turn evolved as part of a dietary specialization. The great ecological success of termites comes from a combination of their ability to feed on cellulose and the social organization that allows them to dominate logs, leaf litter, and other cellulose-rich parts of the environment.

Increased Reproductive Efficiency

Mating swarms, which rank with the most dramatic visual phenomena of the insect world, are formed by a diversity of species belonging to such groups as the mayflies, cicadas, coniopterygid neuropterans, mosquitoes and other nematoceran flies, empidid dance flies, braconid wasps, termites, and ants. They normally occur only during a short period of time at a certain hour of the day or night. Their primary function is to bring the sexes together for nuptial displays and mating (Kessel, 1955; Downes, 1958; Alexander and Moore, 1962; Chiang and Stenroos, 1963; Nielsen, 1964). Termites and some ants fill the air with diffuse clouds of individuals that mate either while traveling through the air or after falling to the ground. Nematoceran flies, dance flies, and some ant species typically gather in concentrated masses over prominent landmarks such as a bush, tree, or patch of bare earth. It is plausible (but unproved) that swarming is most advantageous to members of rare species and to those living in environments where the optimal time for mating is unpredictable. Newly mated ant queens and royal termite couples, for example, require soft, moist earth in which to excavate their first nest cells and to rear the first brood of workers. In drier climates their nuptial swarms usually occur immediately after heavy rains first break a prolonged dry spell. A second potential function of the swarms is to promote outcrossing. If mature individuals of scarce species began sexual activity immediately after emerging, or in response to very local microclimatic events, rather than traveling relatively long distances to join swarms, the amount of inbreeding would be much greater. A third reproductive function of the swarms, originally suggested by Downes, is to provide a premating isolating mechanism. The very specificity of the rendezvous in time and space reduces the chance that adults of different species will mingle and hybridize.

Immelmann (1966) hypothesized a special reproductive requirement as one of the prime movers guiding Australian wood swallows (*Artamus*) to an advanced social life. These desert-dwelling birds feed, bathe, roost, and nest in tight communal groups. They also groom and feed one another, and attack predators en masse. Perhaps the dominant feature of their environment is its great unpredictability. Rains come to the vast Central Desert at highly irregular intervals, bringing upsurges in the insect populations that are needed by the birds to rear healthy broods. By living in such tight associations, the swallows are in a position to stimulate one another and to synchronize gonadal development and sexual behavior with a minimum of delay.

The difficulty with Immelmann's plausibility argument is that other, equally plausible hypotheses can be erected. Thus wood swallows might also benefit from their improved defensive posture and greater efficiency in locating food. Perhaps multiple functions are served. This kind of possibility was impressed on me while studying the mating behavior of the small formicine ant *Brachymyrmex obscurior* in the Florida Keys. The winged males leave the nests in late afternoon to hover in swarms over open patches of ground. The females fly into the swarms and within seconds each is attached to one of the males. The process is fast and efficient. It undoubtedly enhances outcrossing in a population of insects that otherwise would, by virtue of the organization of ants into closed social units, find the free transmission of genes difficult. But the nuptial flight system is also effective in thwarting predators. After the *Brachymyrmex* swarms developed, numerous nighthawks (*Chordeiles minor*) invariably appeared on the scene and began feeding on the flying ants. These predators were hopelessly saturated. They were able to capture only a negligible fraction of the insects in the short intervals between the beginning of the swarm and the time the fecundated queens returned safely to earth.

To find an unambiguous example of reproductive efficiency as the ultimate cause of sociality, we must turn to a radically different kind of organism, the cellular slime molds (Bonner, 1967). In good times these organisms exist as single-celled amebas that creep through freshwater films, engulfing bacteria and reproducing by simple fission. Using laboratory cultures, E. G. Horn (1971) found that each species of two representative genera, *Dictyostelium* and *Polysphodylium*, is specialized to feed on certain kinds of bacteria and can exclude other species when it competes for its favored strains in isolation. Thus there is a premium on the rate at which the amebas can feed and reproduce. We can infer that the advantage favors the solitary condition for each ameba, because one-celled organisms can grow and reproduce faster on a diet of bacteria than can their multicellular equivalents. At certain times, presumably when the environment deteriorates, the amebas aggregate into a slug-shaped mass called a pseudoplasmodium. This newly formed society (or is it really an organism?) travels about for awhile. Then the cells differentiate, building up a stalk on the end of which is a swollen body containing thousands of tiny spores. The spores are released to disseminate through the air. If a spore falls on moist soil, it germinates as a single-celled ameba to initiate a new life cycle. The functions of the stalk and sporangium, the final productions of the colonial phase of the life cycle, are clearly reproduction and dissemination. In fact, the entire form of these structures, and hence their very sociality, seems designed to disperse spores. Remarkably convergent life cycles have evolved in the plasmodial slime molds, or myxomycetes, as well as in the procaryotic myxobacteria, which are phylogenetically extremely remote from each other.

Increased Survival at Birth

Evolving animal species are faced with two broad options in designing their birth process. First, they can invest time after the formation of zygotes by incubating the eggs, by bearing live young, or by otherwise assisting the embryos through the birth process. Failing one of these relatively involved procedures, they can simply deposit the eggs and gamble that the young will hatch and survive. In both alternatives the major risk comes from predators. We find that animals taking the second option, the simple ovipositors, also generally make an effort to conceal the eggs. The techniques include burying the eggs deep in the soil, inserting them into crevices, placing them on specially constructed stalks, and encrusting them with secretions that harden into an extra shell. The procedures improve the survival of the embryos but they make it more difficult for the newly hatched young to reach the outer world. In at least two recorded instances group behavior on the part of the newly born increases the survival of individuals.

The female green turtle (*Chelonia mydas*) journeys every second or third year to the beach of her birth to lay between 500 and 1000 eggs. The entire lot is parceled out at up to 15 intervals in clutches of about 100. Each clutch is deposited in a deep, flask-shaped hole excavated by the mother turtle, who then pulls sand in to bury it. In watching this process, Archie Carr and his coworkers gained the impression that mass effort on the part of the hatchlings is required to escape from the nests. They tested the idea by digging up clutches and reburying the eggs in lots of 1 to 10. Of 22 hatchlings reburied singly, only 6, or 27 percent, made it to the surface. Those that came out were too unmotivated or poorly oriented to crawl down to the sea. When allowed to hatch in groups of 2, the little turtles emerged at a strikingly higher rate—84 percent—and they journeyed to the water in a normal manner. Groups of 4 or more achieved virtually perfect emergence. Observations of the process through glass-sided nests revealed that emergence does depend on goup activity. The first young to hatch do not start digging at once but lie still until others have appeared. Each hatching adds to the working space, because the young turtles and crumpled egg shells take up less room between them than the unhatched, spherical eggs. The excavation then proceeds by a witless division of labor. By relatively uncoordinated digging and squirming, the hatchlings in the top layer scratch down the ceiling, while those around the side undercut the walls, and those on the bottom trample and compact the sand that falls down from above. Gradually the whole mass of individuals moves upward to the surface.

Once out on the sandy beach, the hatchling turtles mutually stimulate one another in the trip down to the water's edge. The groups tend to stop at frequent intervals, increasing their risk from desiccation and predation. But broodmates coming up from behind stimulate a stalled group to move off abruptly, "like toy turtles wound up and all let go together." Furthermore, stray individuals tend to change direction to join the group, and therefore reach the sea in a shorter average time (Carr and Ogren, 1960; Carr and Hirth, 1961). Hendrickson (1958) has also speculated that the metabolic heat of massed eggs speeds the development of the turtle embryos and improves their chances of hatching. Carr and Hirth did indeed find a gain of 2.3°C in their nests, but the improvement of embryo and hatchling fitness could not be tested with their data.

The invertebrate equivalent of turtle hatching is found in the Australian sawfly *Perga affinis* (Carne, 1966). The eggs of this species are laid in pods within the tissue of leaf blades. When the larvae hatch, they must rupture the overlying leaf tissue in order to escape and thus to survive. Usually only one or two larvae in a pod succeed in making it to the outside, and they are followed from the exit holes by their brother and sister larvae. It frequently happens that none of the progeny from a small pod succeeds in escaping, in which case all die. In one large sample of infested leaves studied by Carne, the mortality of pods containing fewer than 10 eggs was 66 percent; in those containing more than 30 eggs, only 43 percent. The *Perga* larvae also stay together when they leave the host tree to pupate. In order to cocoon, they must dig into the soil. Since their morphology is poorly adapted for burrowing, most are not able to penetrate the crust, and they face death by desiccation unless they can use the entrance burrow of a successful larva. In larger aggregations at least one larva usually succeeds in breaking through, with the result that other individuals are also able to cocoon. But in small groups, complete failure and total mortality are commonplace.

Improved Population Stability

Under a variety of special circumstances, social behavior increases the stability of populations. Specifically, it acts either as a buffer to absorb stress from the environment and to slow population decline, or as a control preventing excessive population increase, or both. The primary result is the damping of amplitude in the fluctuation of population numbers around a consistent, predictable level. One secondary result of such regulation is that in a fixed period of time the population has a lesser probability of extinction than another, otherwise comparable population lacking regulation. In other words, the

regulated population persists longer. Does a longer population survival time really benefit the individual belonging to it, whose own life span may be many orders of magnitude shorter than that of the population? Or has the regulation originated solely by selection at the level of the population, without reference to individual fitness? The third possibility is that the population stability is an epiphenomenon—an accidental by-product of individual selection with no direct adaptive value of its own.

These alternative explanations of the relation between social organization and population regulation will be explored in some detail in Chapters 4 and 5. Suffice it for the moment simply to note what the relation is. Territories are areas controlled by animals who exclude strangers. Members of a population who cannot obtain a territory wander singly or in groups through less desirable habitats, consequently suffering a relatively high rate of mortality. They constitute an excess that drains off quickly. Since the number of possible territories is relatively constant from year to year, the population remains correspondingly stable.

The reproductive caste structure of insect societies provides an additional means of population regulation. The effective population size in the true genetic sense is the number of fertile nest queens plus, in the case of termites, the consort males. The workers can be regarded as extensions of these individuals. Once a habitat is populated by mature colonies of social insects, the total number of workers can vary radically without altering the number of colonies, and hence without changing the effective population size. The reason is that it is possible for a cutback in numbers of individuals (workers), even a drastic one, to reduce the average size of colonies without changing the number of colonies. Thus the reduction does not endanger the existence of the population; it may not even alter its distribution in the area. When conditions ameliorate, the colonies serve as nuclei in the rapid restoration of the populations of workers. This inference is supported by the data of Pickles (1940), who for a period of four years kept careful records of both the nest populations and biomasses of ant species in a bracken heath in northern England. The number of nests of three species increased gradually by a factor of approximately two, while the number of workers fluctuated to a much stronger degree. The most interesting example was that of *Formica fusca*. In 1939, the number of workers of this species descended to low levels, but the number of nests actually rose, so that the chances of the species vanishing from the study area remained very remote.

Modification of the Environment

Manipulation of the physical environment is the ultimate adaptation. If it were somehow brought to perfection, environmental control would insure the indefinite survival of the species, because the genetic structure could at last be matched precisely to favorable conditions and freed from the capricious emergencies that endanger its survival. No species has approached full environmental control, not even man. Yet in a lesser sense all adaptations modify the environment in ways favorable to the individual. Social adaptations, by virtue of their great power and sophistication, have achieved the highest degree of modification.

At a primitive level, animal aggregations alter their own physical environment to an extent disproportionately greater than the extent achieved by isolated individuals, and sometimes even in qualitatively novel ways. This general effect was documented in detail by G. Bohn, A. Drzewina, W. C. Allee, and other biologists in the 1920's and 1930's (Allee, 1931, 1938). Consider the following two examples from the flatworms. *Planaria dorotocephala*, like most protistans and small invertebrates, is very vulnerable to colloidal suspensions of heavy metals. Kept at a certain marginal concentration of colloidal silver in 10 cubic centimeters of water, a single planarian shows the beginning of head degeneration within 10 hours. But lots of 10 worms or more maintained in the same concentration and volume survive for at least 36 hours with no externally obvious effects. The greater resistance of the group is due to the smaller amount of the toxic substance that each worm has to remove from its immediate vicinity in order to lower the concentration of the substance to a level beneath the lethal threshold. When single marine turbellarians of a certain species (*Procerodes wheatlandi*) are placed in small quantities of fresh water, they soon die and disintegrate. Groups survive longer and sometimes indefinitely. The effect is due to the higher rate at which calcium is emitted in groups, either by secretion from healthy individuals or by the disintegration of those unfortunate enough to succumb first. The group is therefore exposed to a dangerously hypotonic condition for a shorter period of time.

The existence of environmental imperatives in the evolution of aggregation behavior is nevertheless brought into question by the even more commonplace occurrence of adverse effects due to crowding in populations. More important, the value of many of the particular laboratory cases is compromised by the uncertainty of whether they ever occur in nature. Resistance of groups to colloidal silver suspensions may be an accidental result, an epiphenomenon with no direct relevance to the ecology of planarians. The restoration of calcium ions, however, may have meaning to *Procerodes*, which lives in tide pools, an environment occasionally subject to dilution from heavy rains. Somewhat more plausible is the hypothesized protective role of aggregations in woodlice of the genus *Oniscus*. These land-dwelling isopod crustaceans live in microenvironments where excessive drying is a constant peril. They are strongly attracted to one another and show a marked tendency to bunch together in tight piles. Experiments have demonstrated that groups of woodlice lose water much more slowly, and survive longer in dry air, than do isolated individuals in otherwise identical conditions (Allee, 1926; Friedlander, 1965).

Clearly, each group phenomenon must be judged on its own merits and as fully as possible with reference to the natural environment in which it evolved. In the case of more complex forms of social behavior, the adaptiveness of environmental modification becomes more easily identified. Colonies of black-tail prairie dogs drastically alter the vegetation of the prairie habitats in which they occur (King, 1955). Grasses are largely replaced by yellow sorrel, stickweed, nightshade, and a wide variety of other weeds that can tolerate the activities of the rodents. Many of the weeds are used by the prairie dogs for food, although several grass species are also eaten. In the immediate vicinity of the burrows, the ground is covered by heaps of subsoil, which especially favors the growth of several plants used by the rodents for food. Fetid marigold (*Boebera papposa*), scarlet mallow (*Sphaeralcea coccinea*), black nightshade (*Solanum nigrum*), and several others are limited almost exclusively to this habitat. Sage (*Artemisia frigida*) is a competitive species that tends to dominate weedy associations. The prairie dogs, which do not eat sage, prevent it from flourishing by clipping it close to the ground. Evidently these highly social rodents modify the environment in a way favorable to them. The precise cause-effect relation is obscure, because one could argue with equal force that the rodents have evolved so as to make use of the vegetation their social activities incidentally come to favor. But the outcome is the same, and the environmental modification is correctly viewed as adaptive.

Adaptive design in environmental control attains its clearest expression in the biology of the higher social insects. The complex architecture of the great nests of fungus-growing termites functions as an air-conditioning machine, the basic principles of which are illustrated in Figure 3-14.

Thermoregulation in honeybee colonies attains equal precision, but it is based more upon minute-to-minute behavioral responses by the worker bees (Ribbands, 1953; Lindauer, 1954, 1961). The honeybee colony makes an important first step toward thermoregulation by selecting a nest site, such as a hollow tree trunk or artificial hive, that tightly encloses the brood combs and the majority of the adult workers at all times. The workers use various plant gums, collectively referred to as propolis, to seal off all crevices and openings except for a single entrance hole. This procedure not only keeps enemies out, but, just as important, holds in heat and moisture. From late spring to fall, when the workers are foraging and the brood is present and growing, the interior temperature of the hive is almost always between 34.5° and 35.5°C—in other words, just below the normal body temperature of man. In winter the temperature of the clustered bees falls below this level, but it is still held very high (between 20°C and 30°C) most of the time and is almost never allowed to fall below 17°C. On one remarkable occasion, the temperature of the adult bee clusters was observed to be 31°C at the same time the air temperature outside the hive was −28°C, a difference of 59°C! The ability of

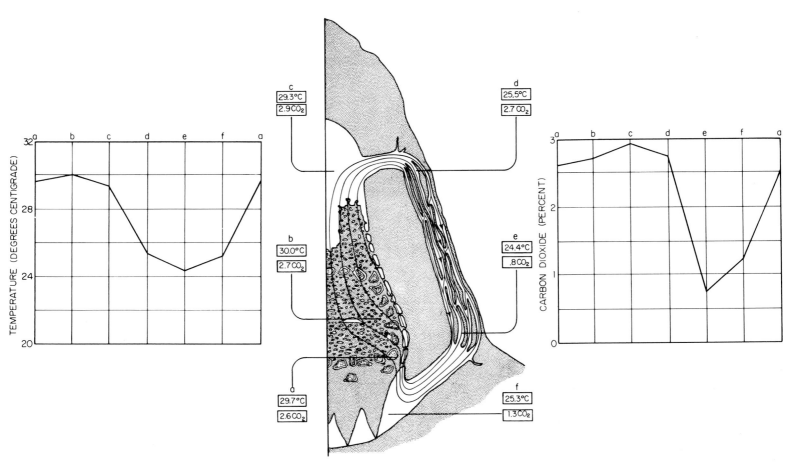

Figure 3-14 Air flow and microclimatic regulation in a nest of the African fungus-growing termite *Macrotermes bellicosus.* Half of a longitudinal section of the nest is shown here. At each of the positions indicated, the temperature (in degrees C) is shown in the upper rectangle and the percentage of carbon dioxide appears in the lower rectangle. As air warms in the central core of the nest (*a, b*) from the metabolic heat of the huge colony inside, it rises by convection to the large upper chamber (*c*) and then out to a flat, capillarylike network of chambers (*d*) next to the outer nest wall. In the outer chambers the air is cooled and refreshed. As this occurs, it sinks to the lower passages of the nest beneath the central core (*e, f*). The graphs at the side show how the temperature and carbon dioxide change during circulation. These changes are brought about by the diffusion of gases and the radiation of heat through the thin, dry walls of the ridge. (Modified from Lüscher, 1961. From "Air-conditioned Termite Nests," by M. Lüscher. © 1961 by Scientific American, Inc. All rights reserved.)

the bees to withstand high temperatures is equally impressive. Martin Lindauer placed a hive in full sunlight on a lava field near Salerno, Italy, where the surface temperature reached 70°C. As long as he permitted workers to take all the water they wanted from a nearby fountain, they were able to maintain the temperature inside the hive at the desired 35°C.

How do the worker bees do it? First, they are able to generate a respectable amount of heat as a by-product of metabolism. The amount produced varies greatly according to the age and activity of the individual bees, the humidity and temperature of the hive, and the time of the year. However, under most conditions each bee gener-

ates at least 0.1 calorie per minute at 10°C (M. Roth, in Chauvin, 1968). Presumably a colony of moderate size, containing 20,000 or more workers, is capable of producing thousands of calories per minute.

The honeybee colony makes use of this natural output of heat, which is about average at a rate per gram for insects generally, together with several ingenious behavioral devices, to hold the hive temperature at the preferred levels. The winter temperature of the hive, as we have just seen, is less closely regulated than the summer temperature. The mechanisms used in cold weather are first the formation of clusters and second the adjustment of cluster tightness, which is achieved as the outside temperature drops. The workers bunch closer together and the total cluster size correspondingly shrinks. Clusters begin to be formed when the hive temperature around the bees falls below 18°C. The clusters raise the temperature surrounding the bodies of the bees to some undetermined level. By the time the hive temperature has dropped to 13°C, and the temperature of the outside air has fallen much lower than that, most of the bees have formed into a very compact cluster that covers part of the brood combs like a warm, living blanket. The outer zone of the cluster is composed of several layers of bees who sit quietly with their heads pointed inward. Those composing the inner zone are more active. They move about restlessly, feed on the honey stores, and from time to time shake their abdomens and breathe more rapidly. Direct measurements have shown that the central bees generate most of the heat, while the outer bees serve as an insulating shell. Together they prevent the temperature of the inner zones of the cluster from falling below 20°C even when the air immediately surrounding the cluster inside the hive approaches the freezing mark.

Temperature control on summer days is even more sophisticated and precise. As summer heat drives the inner hive temperature upward past 30°C or thereabouts, the temperature of the air immediately surrounding the adult workers and the brood starts to rise above the preferred 35°C level. At first the workers cool the hive by fanning with their wings to circulate air over the brood combs and then out the nest entrance. When the hive temperature exceeds about 34°C, this simple device no longer suffices. Now water evaporation is added by an elaborate series of behavioral acts. Water is carried into the nest by the workers and distributed as hanging droplets over the brood cells. Other workers regurgitate droplets onto their tongues and then extend the tongues outward, spreading the water into films from which evaporation is rapid. Other workers fan their wings to drive the moist air away from the brood cells and out of the nest.

Temperature and humidity control is a general phenomenon in all major groups of social insects, including the termites, ants, bees, and wasps, and it is most advanced in those species with the largest colonies. The diverse mechanisms have been recently reviewed by Wilson (1971a).

The Reversibility of Social Evolution

Two broad generalizations have begun to emerge that will be reinforced in subsequent chapters: the ultimate dependence of particular cases of social evolution on one or a relatively few idiosyncratic environmental factors; and the existence of antisocial factors that also occur in a limited, unpredictable manner. If the antisocial pressures come to prevail at some time after social evolution has been initiated, it is theoretically possible for social species to be returned to a lower social state or even to the solitary condition. At least two such cases have been suggested. Michener (1964b, 1965) observed that allodapine bees of the genus *Exoneurella* are a little less than fully social, since the females disperse from the nest before being joined by their daughters. This condition appears to have been derived from the behavior still displayed by the closely related genus *Exoneura*, in which the mother and daughters remain in association. Michener (1969) also noted that reversals may have occurred in the primitively social species of the halictine sweat bees. The most likely selective force, inferred from field studies on the halictines, is the relaxation of pressure from such nest parasites as mutillid wasps. The second case is from the vertebrates. In the ploceine weaverbirds, as in most other passerine groups, the species that nest in forests and feed primarily on insects are solitary in habit, or at most territorial. According to Crook (1964), these species have evolved from other ploceines that live in savannas, eat seeds—and, like many other passerine groups similarly specialized, nest in colonial groups, in a few cases of very large size.

Chapter 4   **The Relevant Principles of Population Biology**

In 1886 August Weismann expressed metaphorically the central dogma of evolutionary biology:

It is true that this country is not entirely unknown, and if I am not mistaken, Charles Darwin, who in our time has been the first to revive the long-dormant theory of descent, has already given a sketch, which may well serve as a basis for the complete map of the domain; although perhaps many details will be added, and many others taken away. In the principle of natural selection, Darwin has indicated the route by which we must enter this unknown land.

Sociobiology will perhaps be regarded by history as the last of the disciplines to have remained in the "unknown land" beyond the route charted by Darwin's *Origin of Species.* In the first three chapters of this book we reviewed the elementary substance and mode of reasoning in sociobiology. Now let us proceed to a deeper level of analysis based at last on the principle of natural selection. The ultimate goal is a stoichiometry of social evolution. When perfected, the stoichiometry will consist of an interlocking set of models that permit the quantitative prediction of the qualities of social organization—group size, age composition, and mode of organization, including communication, division of labor, and time budgets—from a knowledge of the prime movers of social evolution discussed in Chapter 3.

To anticipate the form such an advance is likely to take, it will be useful to review briefly the recent history of the remainder of evolutionary biology. In the 1920's neo-Darwinism was born as a synthesis of Darwinian natural-selection theory and the new population genetics. Simultaneously Alfred Lotka, Vito Volterra, and others were creating the foundations of mathematical population ecology. When the publication of Ronald Fisher's *The Genetical Theory of Natural Selection* (1930), Sewall Wright's *Evolution in Mendelian Populations* (1931), and J. B. S. Haldane's *The Causes of Evolution* (1932) closed this pioneering decade, a respectable number of new ideas had been generated that constituted an extensive, albeit untested, framework on which a mature science might have been built. But evolutionary biology did not and could not proceed in this straightforward manner. It was necessary first for the science to pass through a period of about 30 years of consolidation of information, innovation in empirical research, and slow forward progress. These achievements are sometimes referred to as the Modern Synthesis or, rather loftily, as "the modern synthetic theory of evolution." Actually, very little theory in the strict sense was created between 1930 and 1960 beyond that already laid down in the 1920's. What really happened was that most of the several branches of evolutionary biology—systematics, comparative morphology and physiology, paleontology, cytogenetics, and ethology, to be exact—were reformulated in the *language* of early population genetics. The greatest accomplishment of this period was the elucidation, through excellent empirical research, of the nature of genetic variation within species and of the means by which species multiply. Other topics were clarified and extended, but some of the apparent new understanding of the Modern Synthesis was false illumination created by the too-facile use of a bastardized genetic lexicon: "fitness," "genetic drift," "gene migration," "mutation pressure," and the like. So many problems seemed to be solved by invoking these concepts, and so few really were. Stagnation inevitably followed. Reliance was placed increasingly on a few authoritative treatises in each of the respective fields that contained, in appropriately transmuted form, the magical genetic language. It thus happened that almost a whole generation of young evolutionists (roughly, those maturing in 1945–60) cut themselves off from the central theory. Having never grasped the true relation between theory and empiricism in the first place, they were willing to submit to authority rather than to advance the science by altering the central theory. In the new phase of evolutionary biology, dating from about 1960, evolutionists are attempting to produce a theory that can predict particular biological events in ecological and evolutionary time. This great task requires such profound changes in attitude and working methods that it can rightfully be called post-Darwinism. Its ultimate success cannot be predicted, but there is little question that the future of sociobiology will be heavily invested in it. If the reader will provisionally allow that much prophecy, we can proceed with a brief review of current theoretical population biology, arranged and exemplified in a way that stresses applications to sociobiology. This synopsis assumes a knowledge of elementary evolutionary theory and genetics at the level usually provided by beginning courses in biology. It also requires familiarity with mathematics through elementary probability theory and calculus.

Microevolution

The process of sexual reproduction creates new genotypes each generation but does not in itself cause evolution. More precisely, it creates new combinations of genes but does not change gene frequencies. If, in the simplest possible case, the frequencies of two alleles a_1 and a_2 on the same locus are p and q, respectively, and they occur in a Mendelian population within which sexual breeding occurs at random, $p + q = 1$ by definition; and the frequencies of the diploid genotypes can be written as the binomial expansion

$$(p + q)^2 = 1$$
$$p^2 + 2pq + q^2 = 1$$

where p^2 is the frequency of a_1a_1 individuals (a_1 homozygotes), $2pq$ is the frequency of a_1a_2 individuals (heterozygotes), and q^2 is the frequency of a_2a_2 individuals (a_2 homozygotes). The same result, usually called the *Hardy-Weinberg Law*, can be obtained in an intui-

tively clearer manner by noting that where breeding is random, the chance of getting an a_1a_1 individual is the product of the frequencies of the a_1 sperm and a_1 eggs, or $p \times p = p^2$. Likewise, a_2a_2 individuals must occur with frequency $q \times q = q^2$; and heterozygotes are generated by p sperm mating with q eggs (yielding a_1a_2 individuals) plus q sperm mating with p eggs (yielding a_2a_1 individuals), for a total of $2pq$. This result holds generation after generation. Thus sexual reproduction allows individuals to produce offspring with a diversity of genotypes, all similar to but different from its own. Yet the process does not alter the frequencies of the genes; it does not cause evolution.

Microevolution, which is evolution in its slightest, most elemental form, consists of changes in gene frequency. By experiments and field studies microevolution is known to be caused by one or a combination of the following five agents: mutation pressure, segregation distortion (meiotic drive), genetic drift, gene flow, and selection. Each is briefly described below.

1. *Mutation pressure:* the increase of allele a_1 at the expense of a_2 due to the fact that a_2 mutates to a_1 at a higher rate than a_1 mutates to a_2. Because mutation rates are mostly 10^{-4}/organism (or cell)/generation or less, mutation pressure is not likely to compete with the other evolutionary forces, which commonly alter gene frequencies at rates that are orders of magnitude higher.

2. *Segregation distortion:* the unequal representation of a_1 and a_2 in the initial production of gametes by heterozygous individuals. Segregation distortion, also known as meiotic drive, can be due to mechanical effects in the cell divisions of gametogenesis, in which one allele or the other is favored in the production of the fully formed gametes. This process, however, is difficult to distinguish from gamete selection, a true form of natural selection due to the differential mortality of cells during the period between the reductional division of meiosis and zygote formation. True segregation distortion appears to be sufficiently rare to be of minor general importance.

3. *Genetic drift:* the alteration of gene frequencies through sampling error. To gain an immediate intuitive understanding of what this means, consider the following simple experiment in probability theory. Suppose we were asked to take a random sample of 10 marbles from a very large bag containing exactly half black and half white marbles. Despite the 1:1 ratio in the bag, we could not expect to draw exactly 5 white and 5 black marbles each time. In fact, we expect from the binomial probability distribution that the probability of obtaining a perfect ratio is only

$$\frac{10!}{5!5!}\left(\frac{1}{2}\right)^{10} = 0.246$$

There is, however, a small probability—$2(\frac{1}{2})^{10} = 0.002$—of drawing a sample of either all white or all black marbles. This thought experiment is analogous to sampling in a small population of sexually reproducing organisms. In a 2-allele Mendelian system, a stable population of N parental individuals produces a large number of gametes whose allelic frequencies closely reflect those of the parents; this gamete pool is comparable to the bag of marbles. From the pool, approximately $2N$ gametes are drawn to form the next generation of N individuals. If $2N$ is small enough, and if the sampling is not overly biased by the operation of other forces such as selection, the proportions of a_1 and a_2 alleles (comparable to the black and white marbles) can change considerably from generation to generation by sampling error alone. In theory, three circumstances have been envisaged in which genetic drift can play an effective role in the evolution of small natural populations, including closed social groups. In *continuous drift*, the population remains small in size, and sampling error is effective each generation. In *intermittent drift*, the population is only occasionally reduced to a size small enough to allow drift to operate. Reduction can be effective in one of two ways: (a) if mortality is random at the time of the reduction, the sample of survivors can have a different genetic composition because of chance alone (the "bottleneck effect"); (b) if the population remains small over at least two generations, the process of continuous drift is initiated. The third process contributing to drift is the *founder effect:* new populations are often started by small numbers of individuals, which carry only a fraction of the genetic variability of the parental population and hence differ from it. If chance operates in deciding which genotypes are included among the founder individuals (and chance almost certainly does play a role), new populations will tend to differ from the parent population and from one another. The founder effect, or founder principle, as it has also been called, is of potential importance in the origin of species (Mayr, 1970).

We will now consider the way in which the effect of genetic drift can be roughly estimated. We are interested in the amount of change, Δq, in one generation in the frequency of some allele, a, due to chance alone. (The opposing allele will be designated A). Since a statistical, rather than a deterministic process is involved, it is necessary to calculate the *distribution* of Δq in a large series of populations of the same size. If the distribution is truly random, the *mean* of Δq among the populations will be zero, since the sum of all Δq in the positive direction (gains in gene frequency) is equal in absolute value to the sum of all Δq in the negative direction (losses in gene frequency). Each population has one Δq. When we sum up the Δq for all populations, the sum of the gains should equal the sum of the losses, yielding zero. What is interesting, then, is the dispersion of Δq among all the populations, measured by the variance. The distribution of q is binomial. The variance of a binomial sample about the mean q is pq/N, where N is the size of the sample. In the case of a Mendelian population, there are N organisms formed from $2N$ gametes. The latter figure is the size of the sample, since we are

dealing with $2N$ alleles with a probability p of A and a probability q of a. Therefore

$$\text{Variance of } \Delta q \text{ in one generation} = \frac{pq}{2N}$$

and

$$\text{Standard deviation of } \Delta q \text{ in one generation} = \sigma_{\Delta q} = \sqrt{\frac{pq}{2N}}$$

By the central limit theorem of probability, as N becomes large, Δq becomes normally distributed with mean 0 and standard deviation $\sigma_{\Delta q}$. Referring to tables of the normal distribution, we find that two thirds of the time Δq will be less than $\sigma_{\Delta q}$ in magnitude, and only about once in several hundred trials will it be greater than $3\sigma_{\Delta q}$. Notice that these values are the *maximum* that can be expected to be due to genetic drift, since they are calculated from a model in which no other evolutionary factors are operating. In real populations, these other factors are usually, if not invariably, important, and they diminish the effects of genetic drift in proportion to their intensity. The model, therefore, gives us an estimation of the upper limit of evolution by genetic drift.

It should now be clear why genetic drift is an appropriate term for the process of random change in gene frequencies. Evolution by this means in any given population has no predictable direction; if it is allowed to continue for several generations, the gene frequency would appear to drift about without approaching any particular value. The changes from one generation to the next follow what is called a *random walk* in probability theory. The ultimate fate of any given allele is that it is either lost ($q = 0$) or fixed ($q = 1$), as shown in Figure 4-1.

The most important result of genetic drift is the loss of heterozygosity in the populations. Sewall Wright has deduced the following theorem: in the absence of any other evolutionary force (selection, mutation, migration, meiotic drive), fixation and loss each proceed at a rate of about $1/(4N)$ per locus per generation. This function is useful in that it states the magnitude of rates of fixation and loss. The time to fixation or extinction of any given allele is therefore roughly $4N$ generations on the average.

What are "large" and "small" populations with reference to the potential of random fluctuation? Using the equations already given, we can develop a preliminary intuitive idea.

a. *Small.* If N is on the order of 10 or 100, alleles can be lost at a rate of about 0.1 or 0.01 per locus per generation. Also, $\sigma_{\Delta q}$ can be 0.1 or more of pq. Clearly, genetic drift is a factor of potential significance in populations of this size.

b. *Intermediate.* If N is on the order of 10,000, alleles can be lost at the most on the order of 10^{-4}/generation; $\sigma_{\Delta q}$ can be as high as

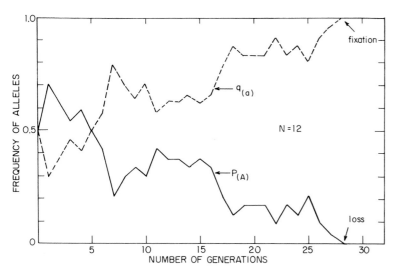

Figure 4-1 Continuous genetic drift, simulated with the aid of a computer, led to fixation of allele a_1 and loss of a_2 in a population consisting of only 12 individuals. In general, the smaller the population, the more rapid will be the drift to these end points. (From Wilson and Bossert, 1971.)

0.01 of pq. If allowed to operate freely, drift can force microevolution only to a moderate degree.

  c. *Large.* When N is 100,000 or greater, the maximum potential gene loss is negligible, while $\sigma_{\Delta q}$ is now only about 0.001 of pq. A very slight sampling bias due to other evolutionary agents will practically cancel these modest effects.

  In short, we would not expect genetic drift to be a factor of any importance in the present-day evolution of such dominant species as English sparrows and herring gulls, but it is conceivably critical for whooping cranes (1970 population: about 57) and North American ivory-billed woodpeckers (1970 population: under 20, if any at all). It has happened in the past that when the population of a vanishing species or subspecies, such as the European bison and North American heath hen, dropped to a few hundreds or tens of individuals, there was an apparent decrease in viability and fertility that hastened the decline. The effect has been attributed to the increase of deleterious genes through "inbreeding," that is, genetic drift. The extinction rates of animal and plant species endemic to small islands are higher than those of related species on larger land masses. This "evolutionary trap" effect has been attributed in part to genetic drift, but other features common to insular endemics, namely, small population size itself and a greater tendency to specialization, may be more important.

  Inbreeding due to consanguineous matings has the effect of reducing the effective population size and hence inducing genetic drift. Reciprocally, genetic drift caused by a small absolute population size increases the incidence of inbreeding, if we measure inbreeding as the probability that two alleles combined in a diploid organism are identical by common descent. Thus genetic drift, consanguinity, and inbreeding are united processes. Because of their importance in social evolution, they will be given special attention later in this chapter.

  4. *Gene flow.* Aside from selection, the quickest way by which gene frequencies can be altered is by gene flow: the immigration of groups of genetically different individuals into the population. Suppose that a population (which we will label α), containing a frequency q_α of a certain allele, receives some fraction m of its individuals in the next generation from a second population (called β) with a frequency q_β of the same allele. The frequency of the allele in population α can be expected to change to a new value that is the frequency of the allele in the nonimmigrant part of the population (q_α) times the proportion of individuals that are not immigrants $(1 - m)$, plus the frequency of the same allele among the immigrants (q_β) times the proportion of individuals in the original population that are immigrants (m). The altered frequency (q'_α) is thus

$$q'_\alpha = (1 - m)q_\alpha + mq_\beta$$

and the amount of change in one generation is

$$\Delta q = q'_\alpha - q_\alpha = -m(q_\alpha - q_\beta)$$

By inserting a few imaginary but plausible figures for $q_\alpha - q_\beta$ and m and noting the resulting Δq, one can see that only a small difference in gene frequencies (of the magnitude that often separates semi-isolated populations and social groups), together with a moderate immigration of individuals, is needed to effect a significant change. Two categories of gene flow can be usefully distinguished: intraspecific gene flow between geographically separate populations or societies of the same species; and interspecific hybridization. The former occurs almost universally within sexually reproducing plant and animal species and is a major determinant of the patterns of geographic variation. Interspecific hybridization occurs during breakdowns of normal species-isolating barriers. Ordinarily it is temporary, or at least rapidly shifting in nature. Although much less common than gene flow within species, it has a greater per generation effect because of the larger number of gene differences that normally separates species.

  5. *Selection.* Selection, whether artificial selection as deliberately practiced on populations by man or natural selection as it occurs

everywhere beyond the conscious intervention of man, is overwhelmingly the most important force in evolution and the only one that assembles and holds together particular ensembles of genes over long periods of time. Selection is defined simply as the change in relative frequency in genotypes due to differences in the ability of their phenotypes to obtain representation in the next generation. Under natural conditions the variation in competence can stem from many causes: different abilities in direct competition with other genotypes; differential survival under the onslaught of parasites, predators, and changes in the physical environment; variable reproductive competence; variable ability to penetrate new habitats; and so forth. The production of a superior variant in any or all of such categories represents *adaptation*. The devices of adaptation, together with genetic stability in constant environments and the ability to generate new genotypes to cope with fluctuating environments, constitute the *components of fitness* (Thoday, 1953). Natural selection means only that one genotype is increasing at a greater rate than another; stated in the conventional symbols of population growth, dn_i/dt (the instantaneous rate of growth of n_i) varies among the i genotypes. The absolute growth rate is meaningless in this regard. All of the tested genotypes may be increasing or decreasing in absolute terms while nonetheless differing in their relative increase or decrease. Acting upon genetic novelties created by mutation, natural selection is the agent that molds virtually all of the characteristics of species.

A selective force may act on the variation of a population in several radically different ways. The principal ensuing patterns are illustrated in Figure 4-2. In the diagrams, the phenotypic variation, measured along the horizontal axis, is given as normally distributed, with the frequencies of individuals measured along the vertical axis. Normal distributions are common but not universal among continuously varying characters (such as size, maturation time, and mental qualities). Stabilizing selection, sometimes also called optimizing selection, consists of a disproportionate elimination of extremes, with a consequent reduction of variance; the distribution "pulls in its skirts," as shown in the lefthand pair. This pattern of selection occurs in all populations. Variance is enlarged each generation by mutation pressure, recombination, and possibly also by immigrant gene flow; stabilizing selection constantly reduces the variance about the optimum "norm" best adapted to the local environment. Balanced genetic polymorphism (as opposed to social caste polymorphism) is sometimes effected by a special, very simple kind of stabilizing selection. In a simple two-allele system, the heterozygote a_1a_2 is favored over the homozygotes a_1a_1 and a_2a_2, and each generation sees a reduction in homozygotes. But the gene frequencies remain constant, and as a result the same diploid frequencies recur in each following generation, in a Hardy-Weinberg equilibrium, prior to the action of selection. True disruptive selection (often called diversifying selection) is

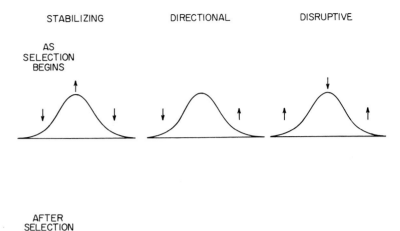

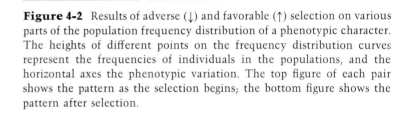

Figure 4-2 Results of adverse (↓) and favorable (↑) selection on various parts of the population frequency distribution of a phenotypic character. The heights of different points on the frequency distribution curves represent the frequencies of individuals in the populations, and the horizontal axes the phenotypic variation. The top figure of each pair shows the pattern as the selection begins; the bottom figure shows the pattern after selection.

a rarer phenomenon, or at least one less well known. It is caused by the existence of two or more accessible adaptive norms along the phenotypic scale, perhaps combined with preferential mating between individuals of the same genotype. Recent experimental evidence suggests that it might occasionally result in the creation of new species. Directional selection (or dynamic selection, as it is sometimes called) is the principal pattern through which progressive evolution is achieved.

Discussion of evolution by natural selection often seems initially circular: the fitter genotypes are those that leave more descendants, which, because of heredity, resemble the ancestors; and the genotypes that leave more descendants have greater Darwinian fitness. Expressed this way, natural selection is not a mechanism but simply a restatement of history. We cannot predict the future by making such a restatement, but must wait to see which genotypes will be more fit in the future. MacArthur (1971) has pointed out that some of the basic laws of population genetics turn rather trivially on the same tautological statement. The statement can be converted into the following equation in which n_x is the number of copies of a particular allele x in a population, N is the total number of genes of all alleles at this locus, and the frequency of gene x is $p_x = n_x/N$ by definition:

$$\frac{dp_x}{dt} = \frac{d}{dt}\left(\frac{n_x}{N}\right) = \frac{N(dn_x/dt) - n_x(dN/dt)}{N^2}$$

$$= \frac{n_x}{N}\left\{\frac{dn_x/dt}{n_x}\right\} - \frac{n_x}{N}\left\{\frac{dN/dt}{N}\right\}$$

$$= p_x[r_x - \overline{r}]$$

The r's are the fitnesses and are defined by the terms in braces. In particular, the entire population of alleles is by definition growing at the rate

$$\frac{dN}{dt} = \overline{r}N$$

and

$$\overline{r} = \frac{dN/dt}{N}$$

where $\overline{r}$ is the average rate of increase of all the alleles at the same locus. The x alleles are increasing as

$$\frac{dn_x}{dt} = r_x n_x$$

$$r_x = \frac{dn_x/dt}{n_x}$$

The theory of evolution by natural selection is embodied in this definition of the different rates of increase of alleles on the same locus. Wright's theorem about adaptive peaks can be derived from it in a straightforward way. Much of the remainder of theoretical population genetics is devoted to the complications introduced by sexual reproduction together with the alternation between haploid and diploid states. We will touch on some of these specialized developments as we go along. For the moment suffice it to note that the central formulations of the subject start with given constant r's. The forces that determine the r's fall within the province of ecology. When pursued that far the subject becomes experimental, realistic, and vastly richer and more interesting in detail. Ideally the analysis begins by breaking the r's down demographically, deriving each from the individual survivorship and fertility schedules of the separate genotypes. The procedure leads quickly to a consideration of particular biological phenomena, including those of social behavior. This is the bridge by which we will cross over from genetics to ecology later in the chapter.

Heritability

The phenotypic variation upon which natural selection acts has four sources (Milkman, 1970): purely genetic, based on allele differences between individuals; purely environmental, originating from variations in conditions exogenous to the organisms during their lifetimes; stochastic-genetic, based on developmental deviations caused by somatic mutations within the lifetimes of particular organisms; and historical, derived from deviant cytoplasmic traits transmitted over a period of two or more generations without benefit of special instruction from the hereditary nucleic acids. The last two contributors of population variation are probably insignificant in most instances. Stochastic-genetic activity, for example, is likely to be important only when the phenotype it affects lies close to a developmental threshold, so that one of the frequent changes in tautomers that normally occur in the base pairs of nucleic acid can shift it to another phenotype. Such activity is intrinsic to certain molecular components of the gene and it must occur no matter how constant the environment. However, this "developmental noise," as C. H. Waddington called it, is unlikely to be of much consequence in natural environments, where its effects are easily swamped by purely genetic and environmental variation. Historical sources of variation are likewise of less than general significance because they occur principally in microorganisms, as in biochemical capabilities of bacteria and the conformation of the cortex and siliceous shells of protozoans. Moreover, they persist for only a few generations at most.

We are thus left with a prevailing residue of phenotypic variation that is based jointly on clearly separable genetic effects, ultimately due to allele differences, and on purely exogenous, environmental effects. One should bear in mind that selection acts on phenotypes and not directly on the genes. But for evolution to occur it is necessary for phenotypic distributions of the kind schematized in Figure 4-2 to be determined at least in part by genetic variation. If phenotypic variation were not so determined, each new generation, being genetically uniform with respect to the phenotype, would spring back to the original distribution that existed before the selection occurred. The proportion of the total variance in the phenotype of a given character that is attributable to the average effects of genes in a particular environment is called the *heritability* of the character. It is symbolized as h^2 (which stands for the heritability and not its square) and can be estimated as follows. The total *phenotypic variance* (V_P) of a trait is the dispersion in the entire population of the trait and is the sum of the *genetic variance* (V_G) and *environmental variance* (V_E). Variance as used here is the standard measure of the dispersion of the individual data around the mean. To take an extremely simple case, suppose that we had two populations consisting of three individuals each. The three individuals in the first population measure 0, 2, and 4, respectively, in a given trait, and those in the second population measure 1, 2, and 3. Note that the first population has a greater dispersion than the second, even though both have a mean of 2. The variance is the average squared difference between the individual values and the mean value. The variance of the first population (0, 2, 4) is

$$\frac{(2-0)^2 + (2-2)^2 + (2-4)^2}{3} = \frac{8}{3}$$

while the variance of the second population (1, 2, 3) is

$$\frac{(2-1)^2 + (2-2)^2 + (2-3)^2}{3} = \frac{2}{3}$$

The advantage of the variance is that it can be partitioned into components that, provided they are independent of one another, are additive. When correlation between two contributing factors does exist, it can be estimated from the data in most cases as the covariance and simply subtracted from the summed variances. Thus, insofar as the genetic and environmental variances of a given trait are independent of each other, they can be summed in a straightforward way to yield the total phenotypic variance.

The genetic variance is the variance due to differences among genes affecting the same trait, and the environmental variance is the variance due to differing environments as they affect individual development. *Heritability in the broad sense* (h_B^2) is the proportion that genetic variance contributes to the total phenotypic variance:

$$h_B^2 = \frac{V_G}{V_P} = \frac{V_G}{V_G + V_E}$$

A heritability score of 1 means that all of the variation in the population is due to the differences between genotypes, and no variation is caused in the same genotype by the influence of the environment. A score of 0 means that all of the variation is caused by the environment; in other words, genetic differences among individuals have no influence on that particular trait. Heritability is a useful concept but one that must be used with great care. Notice that its magnitude depends on the character selected for measurement. Different characters in the same population vary drastically in their heritability scores. Notice also that heritability depends on the environment in which the population lives. The same population, with an unchanged genetic constitution, can yield a different heritability score for a given characteristic if placed in a new environment. Furthermore, it is possible to partition the genetic variance into three components. In the case of simple additive inheritance,

$$V_G = V_A + V_D + V_I$$

where

V_A is the variance due to the additive effect of genes contributing to the various individual genotypes. Some of the genes cause more of the characteristic (such as size, aggressiveness, or grouping tendency) to develop, some less; and the sum of the effects of the combination of such genes assembled in each individual helps to determine the degree to which the characteristic develops. Variation due to different combinations of these additive genes is V_A.

V_D is the variance due to dominance deviations, that is, differences in the degrees to which given genes are dominant over others at the same locus.

V_I is the variance due to epistatic interactions, that is, the various forms of suppression or enhancement among genes located at different loci. For example, the presence of b_1 at a given locus might suppress the contributions to the characteristic controlled by a_1 on a second locus, whereas the presence of b_2 might not.

From these three components of genetic variance, it is possible to separate out a narrower measure of heritability that permits a direct estimate of the rate at which evolution can occur. This *heritability in the narrow sense* (h_N^2) is defined as follows:

$$h_N^2 = \frac{V_A}{V_P}$$

The speed with which a trait is evolving in a population increases as the product of its heritability (in the narrow sense) and the intensity of the selection process. To be somewhat more precise, $R = h_N^2 S$, where R is the response of the population to selection, h_N^2 is heritability in the narrow sense, and S is a parameter determined in part by the proportion of the population included in the selection process. The system, as Mather and Harrison (1949) demonstrated long ago in their classic experimental study of selection for cheta number in *Drosophila*, responds in a linear fashion until either the genetic variance reaches zero or (much more likely) other genes on linked loci are altered to the point that the fitnesses of the participating organisms are lowered significantly. An example illustrating the relation between heritability and evolutionary plasticity is given in Figure 4-3. By means of similar selection experiments, moderate or high degrees of heritability have been documented in many familiar elements of social behavior: group size and dispersion, the openness of groups to strangers, dispersal tendency and capacity, the readiness to explore newly opened space, aggressiveness, fighting ability, the tendency to assume low or high rank in dominance hierarchies, song

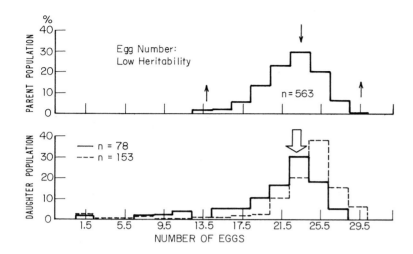

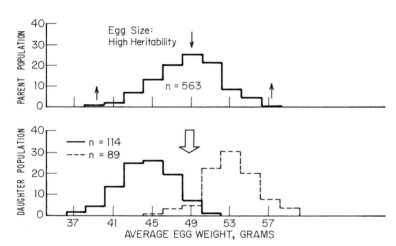

Figure 4-3 The relation between heritability and evolutionary plasticity in two reproductive traits in chickens. Disruptive selection was practiced in an attempt to separate low- and high-yield groups. (Modified from Lerner, 1958.)

Polygenes and Linkage Disequilibrium

The models of classical population genetics have a defect that has grown increasingly troublesome in recent years: they are for the most part based on single-locus systems and simulate competition between alleles. Real selection, however, is not directed at genes but at individual organisms, containing on the order of ten thousands of genes or more. Even when a trait can be precisely delimited for special study, it is ordinarily under the control of polygenes, that is, genes affecting the same character but located at two or more loci. So long as the loci are not linked and do not interact strongly enough to produce particularly favorable or unfavorable combinations, the classical theory is not seriously threatened. When these two conditions do coexist, however, a stable linkage disequilibrium can come into existence. Such a disequilibrium is just that in which the frequency of a gametic combination, such as $a_1b_3c_2$ representing three loci, is not the same as the product of the frequencies of the alleles a_1, b_3, and c_2. Recent work, summarized and extended greatly by Franklin and Lewontin (1970), indicates that linkage disequilibria are far more common than was indicated by earlier studies of the subject based on two-locus theory. When many loci are polymorphic, a condition which empirical research has now demonstrated to be very widespread, relatively small amounts of interaction between loci can generate sufficiently tight linkage disequilibria to make the entire chromosome respond to selection as a unit. Thus future population genetics seems destined to concentrate more on whole chromosomes, their recombination properties, the intensity of epistatic interactions among linked loci, and the effects of homozygosity on chromosomes of various length. This branch of theory will of necessity be developed concurrently with one-locus theory, the simplicity of which is still required for some of the conceptually difficult evolutionary processes. The simpler theory, for example, provides an entrée into the first analyses of group selection, to be reviewed in the following chapter. The adjustment of one-locus theory, on which this and most branches of theoretical population genetics still rest, to the new locus-interaction theory is a task for the future. It remains for sociobiologists to exploit both levels as opportunity provides.

The Maintenance of Genetic Variation

Early neo-Darwinist theory envisaged a simple process whereby raw genetic variation is created by mutation and then tested by natural selection. The reservoir of variation found in a population at any given moment of time was seen as being due to the presence of disfavored alleles in the process of being replaced by favorable mutations, or mutant alleles being sustained at low equilibrium levels at a point of balance between the selection opposing them and their

(in birds), the tendency to mate with similar or dissimilar individuals, and others. A principal conclusion of a 20-year study of dogs conducted at the Jackson Laboratory in Maine was that virtually every behavioral trait possesses sufficient heritability to respond rapidly to selection (Scott and Fuller, 1965). This malleability is the basis for man's success in creating such an imposing array of dog breeds, each specialized for a particular purpose within man's own social scheme.

renewal by fresh mutations. It came to be appreciated in time that although all new genetic variation originates ultimately by mutation in the conventional sense, its maintenance at levels higher than mutational equilibrium can be achieved by several distinct processes. Their effects, classified broadly as genetic polymorphism, are reviewed briefly below.

Transient polymorphism. Two alleles on the same locus can coexist at high frequencies during the long time it takes one to replace the other by natural selection. The two can coexist for even longer periods of time if they are selectively neutral and their relative frequencies shift randomly by genetic drift. The number of such genes originating with neither positive nor negative selective values may be small, and the chances of one coming into existence and increasing to fixation are certainly much smaller. But given enough time, all neutral genes can constitute a large pool.

The remaining cases are jointly referred to as balanced polymorphism.

Heterozygote superiority. If the heterozygote has greater fitness than either homozygote, it is easy to see that neither allele can be eliminated by selection alone. In fact, the frequency of one gene will be $s_2/(s_1 + s_2)$, where s_1 is the selection coefficient of its homozygote (the greater proportion by which the homozygote is eliminated in comparison to the heterozygote), and s_2 is the selection coefficient of the opposing homozygote.

Frequency-dependent selection. If the less frequent of the two alleles is favored in selection, the two will strike a balance at some intermediate frequency. Selection of this kind can arise if a parasite or predator repeatedly shifts its preference to attack a disproportionate number of individuals belonging to the more common type (Moment, 1962; Owen, 1963). It can also occur if there is sufficiently strong disassortative mating, in which individuals preferentially select others that possess a different allele. As a consequence, the rare genotypes are able to reproduce at a higher rate and to increase their abundance until they attain the frequency at which they are no longer scarce enough to gain the advantage.

Disruptive selection. Genetic polymorphism, or at least multimodality in continuously varying traits, can result if sufficiently strong selection is directed in a sustained fashion against intermediate types. One mechanism that could easily occur in nature is assortative mating, in which individuals show strong preference for those with a like phenotype (Karlin, 1969; Crow and Kimura, 1970).

Spatially heterogeneous environment with migration. Consider the case of two Mendelian populations far enough apart to have different environments and hence to favor different sets of genotypes, but close enough to exchange substantial numbers of individuals. As a result, each population can harbor significant numbers of the less favored allele. Provided the environments and migration rates are not too inconstant, the polymorphism will be balanced (Karlin and McGregor, 1972).

Cyclical selection. If selection is strong enough, a regular alternation in favor of first one allele and then the other can maintain balanced polymorphism. A probable example is the coexistence of alleles in some rodent populations that are associated with behavioral traits favored at certain parts of the population density cycle and disfavored at others. The basis for selection can include advantages accorded at different times to aggressive behavior and migration tendency (Krebs et al., 1973). A general theory allowing for a balance between slow-breeding and fast-breeding genotypes has been developed by Roughgarden (1971).

Counteracting selection at different levels. Under a variety of conditions, it is possible for altruist genes to be maintained in a state of balanced polymorphism with competing "selfish" genes. In simplest terms, the group selection favoring altruism and the individual selection opposing it are of sufficiently comparable intensities, and the populations appropriately structured, to lead to an equilibrium frequency of intermediate value (Boorman and Levitt, 1972, 1973a; Eshel, 1972). (See Chapter 5).

During the past ten years the use of high-resolution electrophoresis, by which proteins are allowed to separate in a strong electric field and then stained to pinpoint their location, has revealed a far larger amount of genetic polymorphism than geneticists had earlier believed possible. In their pioneering study of *Drosophila pseudoobscura*, Lewontin and Hubby (1966) discovered that about 30 percent of all loci in a single population have two or more alleles maintained in a polymorphic state; and each individual in the population is heterozygous for about 12 percent of its loci on the average. The revelation of such extensive variation has put a strain on the classical theory. If the polymorphism is balanced by means of heterozygote superiority, the stabilizing selection required to maintain so many genes would seem at first to create an intolerable load for the population. Thirty percent of the loci in *D. pseudoobscura* means, by conservative estimate, at least 2,000 loci. How can enough selection occur to keep 2,000 loci polymorphic? Consider the following model to see how these numbers create a dilemma. Assume for purposes of illustration that the alleles have equal frequencies, and suppose that this balance is maintained by removing 10 percent of the homozygotes at each locus in each generation. The reduced fitness per locus (the "genetic load" per locus) would therefore be

$$\frac{W_{\max} - \bar{W}}{W_{\max}} = \frac{1 - (0.5 \times 0.9 + 0.5 \times 1)}{1} = 0.05$$

where $W_{\max}$ is the fitness of the heterozygote (1 by definition) and $\bar{W}$ is the mean fitness of the three possible diploid genotypes in the population. If there are 2,000 such polymorphic loci, the relative population fitness would be reduced to

$$(1 - 0.05)^{2000} = 10^{-46}$$

Virtually all other reasonable numbers for homozygote fitness and allele frequencies put into this model give similarly impossible genetic loads. For example, if only 2 percent of the homozygotes are eliminated each generation, the fitness would still be cut to 10^{-9}. The population would have to become extinct many times over to achieve such a level of polymorphism. The way out of the difficulty may be through the selection for heterozygosity per se (King, 1967; Milkman, 1967, 1970; Sved et al., 1967). Instead of summing thousands of selective processes as though they were independent events, one views the individual as the object of selection. It is reasonable to assume that the alleles at different loci interact in favorable or unfavorable ways to produce the final product. Many in fact contribute as polygenes to the very same character; and different complexes will be held together in stable linkage disequilibria. If individuals that are heterozygous for a certain fraction of the loci or more are generally superior, as the experiments of Wallace (1968) and others have indicated, a relatively small set of selection episodes in each generation could sustain a large number of polymorphisms. This process, which has been referred to as truncation selection, is a notable departure from the conventional view of microevolution by competition between alleles.

A second, competing hypothesis to account for the maintenance of such high levels of variation suggests that the polymorphisms are transient, being based on selectively neutral genes that are spreading or receding through the population by genetic drift (Crow and Kimura, 1970). The probability that a particular mutant will ultimately be fixed is μ, the rate at which this kind of mutant appears in the population as a whole. This remarkably simple result is obtained as follows. Once the individual mutant gene comes into being, it constitutes exactly $1/2N$ of all the genes at its locus in the population. Since it is neutral, it has the same chance as every other gene present at the moment of its origin of having its descendants fixed at some future date in all $2N$ positions in the population. In other words, the chance that the descendants of a particular neutral mutant will be fixed to the exclusion of all other genes is $1/2N$. It follows that the probability that some neutral gene that arose in a given generation will be fixed is the total number that arose $(2N\mu)$ times the probability that one in particular will be fixed $(1/2N)$; this product is μ. Also, the average interval between the origination of successful mutants is $1/\mu$. These calculations are not inconsistent with the estimated rates of amino acid substitutions in such proteins as hemoglobin, cytochrome c, and fibrinopeptides. Extended to other enzyme

systems, and hence many loci, genetic drift of neutral alleles could account for much of the observed genetic variation in populations. Whether it does in fact, or whether the variation is based primarily or wholly on balanced polymorphism, is a problem whose solution will not come easily.

Phenodeviants and Genetic Assimilation

The student of social evolution is especially concerned with rare events that give small segments of populations unusual opportunities to innovate and thus, perhaps, to increase their fitness and affect the future of the species to a disproportionate degree. One such phenomenon found by geneticists is the appearance of *phenodeviants*, scarce aberrant individuals that appear regularly in populations because of the segregation of certain unusual combinations of individually common genes (Lerner, 1954; Milkman, 1970). Examples include pseudotumors and missing or defective crossveins in the wings of *Drosophila*, crooked toes in chickens, and diabetes in mammals. The traits often appear in larger numbers when stocks are being intensively selected for some other trait or are being inbred (the two processes usually amount to the same thing). They are often highly variable, and deliberate selection can further modify their penetrance and expressivity. The appearance of phenodeviants is generally part of the genetic load that slows evolution in other traits. Yet, clearly, they also represent potential points of departure for new pathways in evolution.

Closely related to phenodeviation is the special sequence of events referred to by Waddington as *genetic assimilation*. An extreme theoretical example is presented in Figure 4-4. Suppose that in each generation a few individuals possess unusual combinations of genes that gave them the potential to develop a trait in certain environments, but under ordinary circumstances the species does not encounter conditions that favor the development of the trait. When finally the environment changes long enough to permit the manifestation of the trait in some members of the species, the trait confers superior fitness. In the new circumstances, the genes that provide the potential also increase in proportion. In time they may become so common that most individuals contain a sufficient number to develop the trait *even in the old environment*. If the environment now returns to that original state, all or a substantial number of the individuals will still develop the trait spontaneously. On first inspection, genetic assimilation may seem to be just a sophisticated form of Lamarckism, but it is not. As far back as 1896 James Mark Baldwin recognized that the capacity to develop in one direction or another in various environments is subject to genetic control, and hence to evolution in the strict Darwinist sense.

Because behavior, and especially social behavior, has the greatest

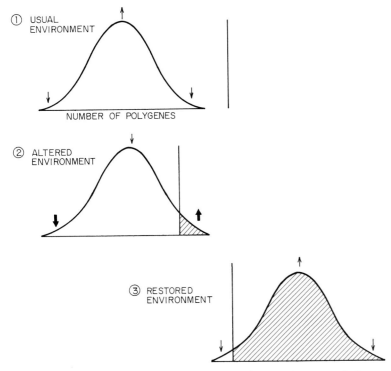

Figure 4-4 Genetic assimilation can occur if an environmental change causes the previously hidden genetic potential of some extreme individual to be exposed. (1) The ordinary environment never permits the development of the potential, but (2) a few individuals attain it when the environment changes. If the trait thus "unveiled" provides increased fitness, those genotypes with the potential will increase in the altered environment; and the population may then evolve to a point where most individuals develop the trait spontaneously even if the environment returns to its original state, as indicated in the bottom diagram (3).

developmental plasticity of any category of phenotypic traits, it is also theoretically the most subject to evolution by genetic assimilation. Behavioral scales, such as those that range within one species from territorial behavior to dominance hierarchies, could be created by the appearance of a few individuals capable of shifting their behavior in one direction or another when the environment is altered for the first time. If the environment remains changed in a way so as to strongly favor these genotypes, the species as a whole may shift further by dropping one end of the scale previously occupied. Most species of chaetodontid butterfly fishes, for example, are exclusively territorial in their habits. *Chelmon rostratus* and *Heniochus acuminatus*, however, form schools organized into dominance hierarchies (Zumpe, 1965). Other species in related groups display scales connecting the two behaviors. As different as the two extremes appear

superficially, it is not difficult to imagine how one could evolve into the other, especially if differing developmental capacities were subjected to natural selection with the aid of genetic assimilation. The process would be further intensified if members of the same species mimicked one another to any appreciable extent. Cultural innovation of the sort recorded in birds and primates could be the first step, provided that the creativity has a genetic basis. Finally, it is even possible for the most plastic species, including man, to pass through repeated assimilative episodes in the development of higher mental faculties.

Inbreeding and Kinship

Most kinds of social behavior, including perhaps all of the most complex forms, are based in one way or another on kinship. As a rule, the closer the genetic relationship of the members of a group, the more stable and intricate the social bonds of its members. Reciprocally, the more stable and closed the group, and the smaller its size, the greater its degree of inbreeding, which by definition produces closer genetic relationships. Inbreeding thus promotes social evolution, but it also decreases heterozygosity in the population and the greater adaptability and performance generally associated with heterozygosity. It is thus important in the analysis of any society to take as precise a measure as possible of the degrees of inbreeding and relationship.

Three measures of relationship, originally devised by Sewall Wright, are used routinely in population genetics:

Inbreeding coefficient. Symbolized by f or F, the inbreeding coefficient is the probability that both alleles on one locus in a given individual are identical by virtue of identical descent. Any value of f above zero implies that the individual is inbred to some degree, in the sense that both of its parents share an ancestor in the relatively recent past. (In defining "recent," we must recognize that virtually all members of a Mendelian population share a common ancestor if their pedigree is traced far enough back.) If the two alleles in question are identical (because they are descended from a single allele possessed by an ancestor), they are said to be *autozygous*; if not identical, they are called *allozygous*.

Coefficient of kinship. Also called the *coefficient of consanguinity*, the coefficient of kinship is the probability that a pair of alleles drawn at random from the same locus in two individuals will be autozygous. The coefficient of kinship is numerically the same as the inbreeding coefficient; it refers to two alleles drawn from the parents in one generation, whereas the inbreeding coefficient refers to the alleles after they have been combined in an offspring. The coefficient of kinship is ordinarily symbolized as f_{IJ} (or F_{IJ}), where I and J (or any other subscripts) refer to the two individuals compared.

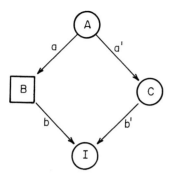

Figure 4-5 Pedigree of an organism (*I*) produced by the mating of two half sibs (*B* and *C*). The computation of the inbreeding coefficient of *I* is explained in the text.

Coefficient of relationship. Designated by *r*, the coefficient of relationship is the fraction of genes in two individuals that are identical by descent, averaged over all the loci. It can be derived from the previous two coefficients in a straightforward way that will be explained shortly.

Let us next examine the intuitive basis of the first two measures. Figure 4-5 presents a derivation of the inbreeding coefficient of an offspring (*I*) produced by a mating between half sibs (*B* and *C*), individuals related to each other by the joint possession of one parent. Females are enclosed in circles and males in squares, while the alleles are symbolized by lower-case letters (*a*, *a'*, *b*, *b'*). The inbreeding coefficient of *I* is computed as follows. Only the individuals descended from the common ancestor (*A*) are shown. The probability that *a* and *b* are the same is ½, since *a* makes up half the alleles in *B* at that locus and therefore half the gametes that *B* might contribute to *I*. The probability that *a* and *a'* are the same is also ½, because once one allele is chosen at random (label it *a*), the chance that the second allele chosen at random (label this *a'*) is the same as the first is ½, provided that *A* itself is not inbred and therefore is unlikely to have two identical alleles to start with. The probability that *a'* and *b'* are identical is ½, since *a'* makes up half the alleles at the locus and therefore half the gametes that *C* might contribute to *I*. The probability that *b* and *b'* are identical is the coefficient of kinship of *B* and *C*, as well as the inbreeding coefficient of *I*. Because *b* = *b'* if and only if *b* = *a* = *a'* = *b'*, the coefficient is the product of the three probabilities just as indicated:

$$f_{BC} = f_I = \frac{1}{2} \times \frac{1}{2} \times \frac{1}{2} = \frac{1}{8}$$

Notice that if we count the steps in the path leading from one parent to the common ancestor back to the second parent (*BAC*,

where the common ancestor is underlined), we obtain the number of times (three) by which the probability ½ must be multiplied against itself. This simple procedure is the basis of *path analysis*, by which coefficients in even the more complex pedigrees can be readily computed. Each possible path leading to every common ancestor is traced separately. The inbreeding coefficient is the sum of the probabilities obtained from every separate path. The technique is shown in the three somewhat more involved cases analyzed in Figure 4-6.

The analysis must be modified if the common ancestor is itself inbred. If its inbreeding coefficient is indicated as f_A, then the probability that two alleles drawn randomly from it will be autozygous is $\frac{1}{2}(1 + f_A)$, and the inbreeding coefficient of the ultimate descendant

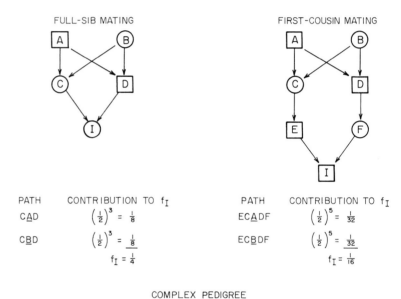

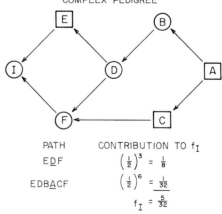

Figure 4-6 Path analysis and calculation of inbreeding coefficients in three pedigrees. The procedure is explained in the text.

(or at least of one separate path contributing to its coefficient) is

$$f_I = \left(\frac{1}{2}\right)^n (1 + f_A)$$

where n is the number of individuals in the path, as before.

The meaning of the coefficient of relationship can now be made clearer. It is related in the following way to the coefficient of kinship (f_{IJ}) and the inbreeding coefficients (f_I and f_J) of the two organisms compared:

$$r_{IJ} = \frac{2f_{IJ}}{\sqrt{(f_I + 1)(f_J + 1)}}$$

in the absence of dominance or epistasis. If neither individual is inbred to any extent, that is, $f_I = f_J = 0$, the fraction of shared genes (r_{IJ}) is twice their coefficient of kinship. But if each individual is completely inbred, that is, $f_I = f_J = 1$, r_{IJ} is the same as the coefficient of kinship. Suppose that r_{IJ} of two outbred individuals is known to be 0.5. This means that $\frac{1}{2}$ of the genes in I are identical by common descent with $\frac{1}{2}$ of the genes in J, when all the loci (or at least a large sample of them) are considered. Then if we consider one locus, the probability of drawing an allele from I and one from J which are identical by common descent (this probability is f_{IJ}, the coefficient of kinship) is the following: the probability of drawing the correct allele from I ($\frac{1}{2}$) times the probability of drawing the correct one from J ($\frac{1}{2}$), or $\frac{1}{4}$. In other words $r_{IJ} = 2f_{IJ}$. Suppose, in contrast, that I and J were totally inbred. In this unlikely circumstance all allele pairs in I and J are autozygous. As a result the fraction of alleles shared by I and J is the same as the fraction of loci shared by them. If 50 percent of the alleles in I are identical to 50 percent of the alleles in J, 50 percent of the loci are also shared in toto and 50 percent are not shared at all, because all the loci are autozygous.

The coefficients of kinship and relationship can be estimated indirectly, in the absence of pedigree information, by recourse to data on the similarity of blood types and other phenotypic traits among individuals, as well as information on migration (Morton, 1969; Morton et al., 1971; Cavalli-Sforza and Bodmer, 1971). In 1948 G. Malécot showed that in systems of populations with uniform rates of gene flow, the mean coefficient of kinship between individuals selected from different populations can be expected to decline exponentially as the distance (d) separating them increases:

$$f(d) = ae^{-bd}$$

where a and b are fitted constants. This result has been extended and further generalized by Imaizumi et al. (1970) and Morton et al.

(1971). The migration index (b) reflects the rate of gene flow within and between populations. To use an alternative expression, it decreases with the *viscosity*, or slowness of dispersal, of populations. Examples of Malécot's law from human populations are given in Figure 4-7.

As populations are fragmented and viscosity increases, the degree of kinship among immediately adjacent individuals grows larger. Consequently, the prospect for social evolution increases, since cooperative and even altruistic acts will pay off more in terms of the perpetuation of genes shared by common descent. Yet side effects also arise that can progressively reduce the fitnesses of both individuals and local groups when viscosity is increased, and hence bring social evolution to a standstill. As inbreeding increases, homozygous recombinants increase in frequency more than heterozygous ones, spreading the variation more evenly over the possible diploid types. More precisely, in the case of no dominance, the genetic variance within a local population is related to the inbreeding coefficient as

$$V_f = V_o(1 + f)$$

where V_o is the variance in the absence of inbreeding. In the case of dominance, the explicit relation is more complex (see Crow and Kimura, 1970). We can view inbreeding as having the effect of con-

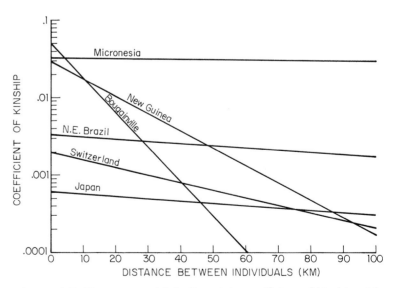

Figure 4-7 The exponential decline of the coefficient of kinship with distance in various human populations. The decline is steepest in the relatively isolated, immobile peoples of New Guinea and Bougainville and least marked in the highly migratory peoples of Micronesia (From Friedlaender, 1971; based on data from Imaizumi and Morton, 1969.)

gealing a population into an ensemble of little, semiisolated groups. If we measure the genetic variance of each group over a long period of time, taking into account genotypes that come and go by immigration and extinction, we find the variance of each group is less than if it were just one focus in a freely interbreeding population. But the little groups differ from one another enough so that if we measure the variance for the entire population (as in the formula just given) there will be a detectable increase over the variance found in an otherwise comparable freely interbreeding population. Populations can be subdivided this way, and they can also be subdivided by external forces such as physical barriers that prevent the exchange of genes. When subpopulations are isolated in this second way (we can now refer to them either as subpopulations of a larger population or as full populations belonging to a larger "metapopulation"), they will tend to diverge in gene frequencies both because of genetic drift, which is potentially most effective in populations of roughly a hundred members or less, and because of selection due to the inevitable differences in the environments occupied by the isolates. The result is an increase in the genetic variance measured over all the subpopulations. The precise relationship between the divergence of the subpopulations and the variance in genotypes was first expressed by the German biologist S. Wahlund in 1928 and is often called Wahlund's principle:

$$\bar{p}^2 = \overline{p^2} - V_p$$

Here V_p is the variance of the frequency (p) of a given allele for all the subpopulations; $\bar{p}^2$ is the proportion of the homozygotes one would expect from the Hardy-Weinberg formula after one generation if all the subpopulations were pooled and their members allowed to breed randomly; and $\overline{p^2}$ is the true mean proportion of the homozygotes in the divided state, defined as

$$\frac{n_1 p_1^2 + n_2 p_2^2 + \cdots + n_k p_k^2}{n_1 + n_2 + \cdots + n_k} = \overline{p^2}$$

where n_1, n_2, ... are the numbers of individuals in each of the subpopulations and p_1^2, p_2^2, ... are the frequencies of the homozygotes in the same subpopulations expected from the Hardy-Weinberg equilibrium. These relations hold, of course, only in the case of random breeding within each subpopulation and virtual total isolation between the subpopulations. When inbreeding is added, the frequency of the homozygotes goes up still farther, as previously indicated. If gene flow between subpopulations is permitted, the differences between them are reduced, their joint variance in gene frequencies decreases, and the total frequency of the homozygotes declines in accordance with Wahlund's principle:

$$\bar{p}^2 = \overline{p^2} - V_p$$

This relationship is especially worth noting in the analysis of the structure of social populations. Closed social groups form semiisolated subpopulations whose gene frequencies come to differ both by random deviation and by adaptation to local environments.

In any given generation inbreeding diminishes the proportion of heterozygotes (H_f) from that found in a comparable outbreeding population (H_o) by an amount equal to the inbreeding coefficient:

$$H_f = H_o(1 - f)$$

For example, if breeding in populations were limited to first cousins ($f = \frac{1}{16}$), the frequency of heterozygotes would be $\frac{15}{16}$ that predicted by the Hardy-Weinberg formula. A second mode of reduction of heterozygosity is due to genetic drift. The decrement of heterozygosity in time turns out to be a quite simple function, which can be derived in the following way (Crow and Kimura, 1970). We ask what the probability is that two gametes will have autozygous alleles on a given locus if they are drawn at random from a population of N individuals in generation $t - 1$ and used to constitute one of the N individuals in the next (t) generation. This is the sum of the two following probabilities. The first is that any two of the $2N$ gametes produced will represent the same locus on the same homologous chromosome of one individual (these two need not be combined into the same zygote); this probability is $1/(2N)$. The second is the probability that of the remaining fraction $1 - 1/(2N)$ of gametes, two drawn at random will be identical because they are of common descent. By definition, this latter probability is f_{t-1}, the inbreeding coefficient of the $t - 1$ generation. The sum of the two probabilities is by definition f_t, the inbreeding coefficient of the t generation:

$$f_t = \frac{1}{2N} + \left(1 - \frac{1}{2N}\right) f_{t-1}$$

Since H_t, the heterozygosity at any selected time t, is $H_o(1 - f_t)$, and since H_{t-1}, the heterozygosity at $t - 1$, is $H_o(1 - f_{t-1})$, we can rewrite the equation above as

$$H_t = \left(1 - \frac{1}{2N}\right) H_{t-1} = \left(1 - \frac{1}{2N}\right)^t H_o$$

In other words the amount of heterozygosity decreases each generation by a fraction equal to the reciprocal of the population size. This elementary result holds for completely random mating, including the possibility of self-fertilization. Removing the latter condition necessitates more complex formulas, but the qualitative result remains approximately the same. In many closed social groups, containing on the order of a hundred or fewer individuals over several generations, loss of heterozygosity by genetic drift must be a significant factor.

The combination of autozygous genes by pure chance due to the finiteness of population size can be viewed as a form of inbreeding. The preferential mating within the population of related individuals—inbreeding in the conventional sense—can be viewed as defining weakly divided subunits of the populations in a descending hierarchy of population organization. An estimation of the total degree of inbreeding can be made from a knowledge of autozygosity in the following way. The probability of getting two allozygous alleles when they are selected at random in the first generation is $1 - 1/2N$; if the population has been fixed at N individuals for t generations, the probability of allozygosity is $(1 - 1/2N)^t$. Quite independently, the probability of obtaining an allozygous pair in the face of inbreeding is $1 - f$, where f is the inbreeding coefficient. Then the total probability of obtaining allozygous alleles in a single draw is the product

$$1 - f_s = \left(1 - \frac{1}{2N}\right)^t \left(1 - f\right)$$

where f_s is the summed inbreeding coefficient. Suppose that a closed social group were newly composed from five individuals selected at random from a large population. After five generations, what would be the inbreeding coefficient of an offspring whose parents are first cousins? Recalling that the f of progeny of first cousins is $\frac{1}{16}$, we see that

$$1 - f_s = \left(1 - \frac{1}{10}\right)^5 \left(1 - \frac{1}{16}\right)$$
$$= 0.4982$$
$$f_s = 0.5018$$

In this case the role of genetic drift considerably outweighs the effect of the consanguineous mating; its strong influence can be demonstrated for populations of up to a hundred individuals or more. This surprising result may prove widely true for real social populations. Crow and Mange (1965), for example, found that in the Hutterite population the inbreeding coefficient due to genetic drift is quite important, about 0.04, but that the inbreeding coefficient due to consanguineous marriages is negligible.

In short, a crucial parameter in short-term social evolution is the size and degree of closure of the group. So far we have spoken of N, the number of individuals in the group (or in the population embracing it), as though it were composed of equal numbers of each

sex all equally likely to contribute progeny. This ideal state is seldom realized. Instead it is necessary to define the *effective population number*: the number of individuals in an ideal, randomly breeding population with 1:1 sex ratio which would have the same rate of heterozygosity decrease as the real population under consideration. Typically, the effective population number is well below the real population number. By measuring it, we obtain a truer picture of the likely course of microevolutionary events within the population. The formula for the effective population number (N_e) is the following:

$$\frac{1}{N_e} = \frac{1}{4N_m} + \frac{1}{4N_f}$$
$$N_e = \frac{4N_m N_f}{N_m + N_f}$$

where N_m and N_f are the number of males and females, respectively, that contribute to reproduction in the real population. The fraction $1/N_e$ is the probability that in an ideal panmictic population of N_e individuals any two alleles picked at random will come from the same individual. (Note that one allele is picked; then the chance that the next allele picked will come from the same individual is $1/N_e$.) In the real population, with N_m active males, the probability that a second gene comes from a male (not necessarily part of the same mating) is also $\frac{1}{2}$. The probability that both genes come from a *particular* male is

$$\frac{1}{2} \times \frac{1}{2} \times \frac{1}{N_m} = \frac{1}{4N_m}$$

Symmetrically, the probability that both genes come from a particular female is $1/(4N_f)$. Then the probability that both genes come from one individual regardless of sex (defined in the ideal equivalent population as $1/N_e$) is the sum $1/(4N_m) + 1/(4N_f)$. A more thorough explanation of the basic theory, taking into account the effect not only of inbreeding but also of variations in the fertility schedules, has been presented by Giesel (1971).

The effective population numbers of the few real populations measured so far have generally turned out to be low. In the house mouse *Mus musculus* they are on the order of 10 or less, with male dominance exerting a strong depressing influence (Lewontin and Dunn, 1960; DeFries and McClearn, 1970). Deer mice (*Peromyscus maniculatus*) form relatively stable territorial populations, in spite of the ebb and flow of emigrating juveniles, and the effective numbers range from 10 to 75 (Rasmussen, 1964; Healey, 1967). Leopard frogs (*Rana pipiens*) studied by Merrell (1968) had N_e values ranging from 48 to 102, which because of the strongly unequal sex ratios favoring males are well below the actual numbers of adult frogs inhabiting natural habitats. Tinkle (1965) studied side-blotched lizards (*Uta stansburiana*) with unusual care by marking and tracking young individuals until they reached reproductive age. He found that N_e ranged from 16 to 90 in six local populations, with a mean of 30; these figures did not depart far from the actual census numbers. From uncorrected census data, social vertebrates in general appear to have effective population sizes on the order of 100 or less. Social insects seem to be highly variable in this regard. Populations of rare ant species, including social parasites and inhabitants of caves and bogs, sometimes contain fewer than 1000 colonies, and the effective number of colonies is probably much lower (Wilson, 1963). Populations comprised of wasp colonies are relatively viscous, with founding queens sometimes returning to the neighborhood of their birth and even associating with sisters in the early stages of colony growth. My impression of bumblebees and stingless bees is that neither the males nor the females travel great distances, and the N_e of populations of colonies is likely to be low. In the case of most ants and termites the matter is more complicated. Nuptial swarms often contain immense numbers of individuals from hundreds or thousands of nests, and some travel for distances of hundreds or thousands of meters before mating. My guess is that N_e is often well above 100 and may be orders of magnitude higher.

The general occurrence of small effective deme sizes in social animals brings them into the range envisaged in Sewall Wright's original "island model" (1943): a population divided into many very small demes and affected by genetic drift that restricts genetic variation within individual demes but increases it between them. Such a population would conceivably be more adaptable than an undivided population of equal size because of its greater overall genetic variation. Where the genotypes of one deme fail, those of the next might succeed, with the end result of preserving the species. As a corollary result, such a population will also evolve more quickly.

We now ask, specifically what is the risk encountered by increased inbreeding and decreased heterozygosity by these social populations? Heterozygosity per se generally raises the viability and reproductive performance of organisms. The extreme case of the relation is heterosis, the temporary improvement in fitness that results from a massive increase in the frequency of heterozygotes over many loci from the outcrossing of two inbred strains. Wallace (1958, 1968) obtained essentially the same effect by irradiating populations of *Drosophila melanogaster* continuously. Instead of the expected decline in the population from accumulated lethal and subvital mutations, he got the opposite trend as sufficient numbers of these mutations expressed beneficial effects in the heterozygous state. Of course if a heterotic stock is then inbred, its performance declines precipitously because of the quick reversal from heterozygous to homozygous states created in a large fraction of the population through elementary Mendelian

recombination. Even so, ordinary populations sustain high levels of heterozygous loci, and any increase in inbreeding will result in a decrease in average population performance, part of which will be due to a raising of average mortality by the production of more lethal homozygotes. The formal theory of this decline has been considered at length by Crow and Kimura (1970) and Cavalli-Sforza and Bodmer (1971). The essential relation can be stated as follows. If some trait, such as size, intelligence, motor skill, sociability, or whatever, possesses a degree of heritability, and if some of the loci display either dominance or superior heterozygote performance, or both, inbreeding will cause a decline of the trait within the population. The decline will affect not only the trait averaged over the population as a whole, but also the performance of an increasing number of individuals. Suppose that in the case of a two-allele system (a_1 and a_2), the phenotypes consist of a quantity Y of a trait plus some other quantity (A, $-A$, or D) dependent on which alleles are represented in the three possible diploid combinations. In the case of inbreeding of the amount f, the combinations yield

GENOTYPE	a_1a_1	a_1a_2	a_2a_2
FREQUENCY	$p_1^2(1-f) + pf$	$2p_1p_2(1-f)$	$p_2^2(1-f) + p_2f$
PHENOTYPE	$Y - A$	$Y + D$	$Y + A$

The mean value of the trait ($\overline{Y}$) is the sum of the products of the phenotype values and phenotype frequencies:

$$\overline{Y} = Y + A(p_2 - p_1) + 2p_1p_2D - 2p_1p_2Df$$

The value of the trait thus diminishes as a linear function of dominance (A and D), of heterozygote superiority (D), and of the degree of inbreeding (f). The relationship holds only where there is no epistasis (interaction of alleles on different loci). When epistasis occurs, the function is nonlinear but still decreasing (Figure 4-8). A case of inbreeding depression of a human trait (chest circumference in males) is given in Figure 4-9. Further studies by Schull and Neel (1965) and others have demonstrated depression effects in overall size, neuromuscular ability, and academic performance. A recent study of children of incest in Czechoslovakia confirms the dangers of extreme inbreeding in human beings. A sample of 161 children born to women who had had sexual relations with their fathers, brothers, or sons were afflicted to an unusual degree: 15 were stillborn or died within the first year of life, and more than 40 percent suffered from various physical and mental defects, including severe mental retardation, dwarfism, heart and brain deformities, deaf-mutism, enlargement of the colon, and urinary-tract abnormalities. In contrast, a group of 95 children born to the same women through nonincestuous relations conformed closely to the population at large. Five died

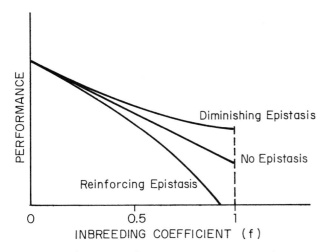

Figure 4-8   Decline in performance (or any trait of interest) as a function of the degree of inbreeding, in the absence or presence of epistasis. In diminishing epistasis, joint homozygosity on separate loci together reduces the effect less than the sum of the reductions when the loci are homozygous separately; in reinforcing epistasis the homozygous loci amplify one another's effects. (From Crow and Kimura, 1970.)

during the first year of life, none had serious mental deficiencies, and only 4.5 percent had physical abnormalities (Seemanova, 1972).

In addition to a straightforward decline in competence, the loss of heterozygosity reduces the ability to buffer the development of structures against fluctuations in the environment. Hence less heter-

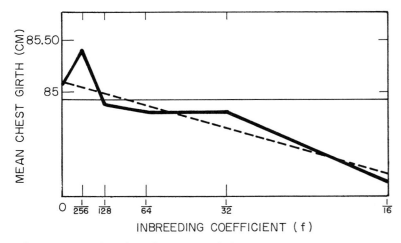

Figure 4-9   Inbreeding depression of chest size in men born in the Parma Province of northern Italy between 1892 and 1911. (From Cavalli-Sforza and Bodmer, 1971; after Barrai, Cavalli-Sforza, and Mainardi, 1964.)

ozygosity increases the chance of producing less adaptive variants such as phenodeviants. It further reduces the genetic diversity of offspring, a loss that can result in the loss of entire blood lines, or even social groups, when the environment changes.

In view of the clear dangers of excessive homozygosity, we should not be surprised to find social groups displaying behavioral mechanisms that avoid incest. These strictures should be most marked in small, relatively closed societies. Incest is in fact generally avoided in such cases. Virtually all young lions, for example, leave the pride of their birth and wander as nomads before joining the lionesses of another pride. A few of the young lionesses also transfer in this fashion (Schaller, 1972). A closely similar pattern is followed by many Old World monkeys and apes (Itani, 1972). Even when the young males remain with their troops they seldom mate with their mothers, possibly because of the lower rank they occupy with respect to both their mothers and older males for long periods of time. In the small territorial family groups of the white-handed gibbon *Hylobates lar*, the father drives sons from the group when they attain sexual maturity, and the mother drives away her daughters (Carpenter, 1940). Young female mice (*Mus musculus*) reared with both female and male parents later prefer to mate with males of a different strain, thus rejecting males most similar to the father. The discrimination is based at least in part on odor. Males do not make such choices (Mainardi, 1964; Mainardi et al., 1965; Kennedy and Brown, 1970). Similar effects have been demonstrated more recently in rats and guinea pigs (Marr and Gardner, 1965; Eisenberg and Kleiman, 1972). Despite such strong anecdotal evidence, however, we are not yet able to say whether incest avoidance in these animals is a primary adaptation in response to inbreeding depression or merely a felicitous by-product of dominance behavior that confers other advantages on the individual conforming to it. It is necessary to turn to human beings to find behavior patterns uniquely associated with incest taboos. The most basic process appears to be what Tiger and Fox (1971) have called the precluding of bonds. Teachers and students find it difficult to become equal colleagues even after the students equal or surpass their mentors; mothers and daughters seldom change the tone of their original relationship. More to the point, fathers and daughters, mothers and sons, and brothers and sisters find their primary bonds to be all-exclusive, and incest taboos are virtually universal in human cultures. Studies in Israeli kibbutzim, the latest by Joseph Shepher (1972), have shown that bond exclusion among age peers is not dependent on sibship. Among 2,769 marriages recorded, none was between members of the same kibbutz peer group who had been together since birth. There was not even a single recorded instance of heterosexual activity, despite the fact that no formal or informal pressures were exerted to prevent it.

In summary, small group size and the inbreeding that accompanies it favor social evolution, because they ally the group members by kinship and make altruism profitable through the promotion of auto-zygous genes (hence, one's own genes) among the recipients of the altruism. But inbreeding lowers individual fitness and imperils group survival by the depression of performance and loss of genetic adapt-ability. Presumably, then, the degree of sociality is to some extent the evolutionary outcome of these two opposed selection tendencies. How are the forces to be translated into components of fitness and then traded off in the same selection models? This logical next step does not seem feasible at the present time, and it stands as one of the more important challenges of theoretical population genetics. A few of the elements necessary for the solution will be given in the analysis of group selection to be provided in Chapter 5.

Assortative and Disassortative Mating

Assortative mating, or *homogamy*, is the nonrandom pairing of indi-viduals who resemble each other in one or more phenotypic traits. Human couples, for example, tend to pair off according to similarity in size and intelligence. Sternopleural bristle number, which may simply reflect total size, and certain combinations of chromosome inversions have been found to be associated with assortative mating in *Drosophila* fruit flies (Parsons 1967; Wallace, 1968). In domestic chickens and deer mice (*Peromyscus maniculatus*), color varieties prefer their own kind (Blair and Howard, 1944; Lill, 1968). Assortative mating can be based upon kin recognition, in which case its conse-quences are identical to those of inbreeding. Or it can be based strictly on the matching of like phenotypes, either without reference to kinship or in conjunction with the avoidance of incest, as in the case of human beings. "Pure" assortative mating of the latter type has effects similar to those of inbreeding, but it results in a less rapid passage to homozygosity, affects only those loci concerned with the homogamous trait or closely linked to it (whereas inbreeding affects all loci), and, in the case of polygenic inheritance, causes an increase in variance.

Experiments with *Drosophila* have established that when homo-gamy is imposed artificially on laboratory populations for several generations the resulting inbred strains tend to maintain it on their own thereafter. The basis of the discrimination is unknown but it could be the demonstrated ability of members of different inbred strains to recognize their own kind (Parsons, 1967; Hay, 1972). Thus disruptive selection, conceivably originating from a selective disad-vantage of intermediate phenotypes, can lead to assortative mating and an acceleration of the divergence of the evolving strains. The extreme end result might be the sympatric origin of two or more new species. Homogamy can also reinforce the divergence of isolated populations in the course of conventional geographic speciation. A

suggestive example was revealed by Godfrey's experiments with bank voles (*Clethrionomys glareolus*). Individuals taken from the main-land of Great Britain and three offshore islands preferred members of their own populations when allowed a choice, and they were able to discriminate on the basis of odor alone. When given no choice they mated with members of other populations, producing fertile offspring (Godfrey, 1958).

Disassortative mating has been documented fewer times in nature than assortative mating, and in a disproportionate number of in-stances it has involved chromosomal and genic polymorphs in insects (Wallace, 1968). The effects of disassortative mating are of course generally the reverse of those caused by assortative mating. In additive polygenic systems there is a tendency to "collapse" variation toward the mean. However, in the case of genetic polymorphism, diversity is preserved and even stabilized, since scarcer phenotypes are the beneficiaries of preferential mating and the underlying genotypes will therefore tend to increase until the advantage of scarcity is lost.

Because of its mathematical tractability and potential applications, nonrandom mating has been a perennially favorite subject of popula-tion geneticists. Successively more detailed and advanced accounts are to be found in the monographs by Crow and Kimura (1970), Wright (1969), and Karlin (1969), in that order.

Population Growth

Natural selection can be viewed simply as the differential increase of alleles within a population. It does not matter whether the popula-tion as a whole is increasing, decreasing, or holding steady. So long as one allele is increasing relative to another, the population is evolv-ing. In fact, a population can be evolving rapidly, responding to natu-ral selection and hence "adapting," at the same time that it is going extinct. The conceptualization and measurement of growth, then, is the meeting place of population genetics and ecology.

The rate of increase of a population is the difference between the rate of addition of individuals due to birth and immigration and the rate of subtraction due to death and emigration:

$$\frac{dN}{dt} = B + I - D - E$$

where N is the population size, and B, I, D, E are the rates at which individuals are born, immigrate, die, and emigrate. A society, even if nearly closed, comprises a population in which all four of these rates are significant. In larger populations, however, including the set of all conspecific societies that make up a given population, a realistic modeling effort can be started by setting $I = E = 0$ (no individuals enter the population or leave it) and varying B and D, the birth and death rates. In the simplest model of exponential growth, it is assumed that there exist some average fertilities and probabilities of death over all the individuals in the population. This means that B and D are each proportional to the number of individuals (N). In other words, $B = bN$ and $D = dN$, where b and d are the average birth and death rates per individual per unit time. Then

$$\frac{dN}{dt} = bN - dN$$

$$= (b - d)N = rN$$

where r $(= b - d)$ is called the intrinsic rate of increase (or "Malthusian parameter") of the population for that place and time. The solution of the equation is

$$N = N_o e^{rt}$$

where N_o is the number of organisms in the population at the moment we begin our observations (this can be any point in time chosen for convenience), and t is the amount of time elapsed after the observations begin. The units of time chosen (hours, days, years, or whatever) determine the value of r. (The symbol r is not to be confused with the same symbol used to denote the coefficient of relationship. The fact that the same letter has been used for two major parameters is one of the inconveniences resulting from the largely independent histories of ecology and genetics.)

Theoretically, each population has an optimum environment—physically ideal, with abundant space and resources, free of predators and competitors, and so forth—where its r would reach the maximum possible value. This value is sometimes referred to formally as r_{max}, the maximum intrinsic rate of increase. Obviously, the rates of increase actually achieved in the great majority of the less-than-perfect environments are well below r_{max}. For example, although the realized values of r of most human populations are very high, enough to create the current population explosion, they are still several times smaller than r_{max}, the value of r that would be obtained if human beings made a maximum reproductive effort in a very favorable environment. The values of r vary enormously among species. Almost all human populations increase at a rate of 3 percent or less per year ($r = 0.03$ per year). The value of r in unrestricted rhesus populations is about 0.16 per year, while in the prolific Norway rat it is 0.015 *per day.*

Since any value of r above zero will eventually produce more individuals of the species than there are atoms in the visible universe, the exponential growth model is obviously incomplete. The problem lies in the implicit assumption that b and d are constants, with values

independent of N. A new and more realistic postulate is that b and d are functions of N, say linear functions for simplicity:

$$b = b_o - k_b N$$
$$d = d_o + k_d N$$

In this case, b_o and d_o are the values approached as the population size becomes very small, k_b is the slope of the decrease for the birth rate, and k_d is the slope of the increase for the death rate. The equations state that the birth rate decreases and the death rate increases as the population increases, both of which are plausible assertions that have been documented in some species in nature. We substitute the new values of b and d into the model to find:

$$\frac{dN}{dt} = [(b_o - k_b N) - (d_o + k_d N)]N$$

This is one form of the basic equation for logistic population growth. Note that when b becomes equal to d, the population reaches a stable size. That is, the population can maintain itself at the value of N such that

$$b_o - k_b N = d_o + k_d N$$

$$N = \frac{b_o - d_o}{k_b + k_d}$$

This particular value of N is called the carrying capacity of the environment and usually is given the shorthand symbol K. For any value of N less than K the population will grow, and for any value greater than K it will decline; and the change will occur until K is reached (Figure 4-10). Taking the two shorthand notations

$$K = \frac{(b_o - d_o)}{(k_o + k_d)}$$

and

$$r = b_o - d_o$$

and substituting them into the logistic differential equation just derived, we obtain

$$\frac{dN}{dt} = rN\left(\frac{K - N}{K}\right)$$

This is the familiar form of the logistic equation for the growth and regulation of animal populations. Usually the equation is stated flatly

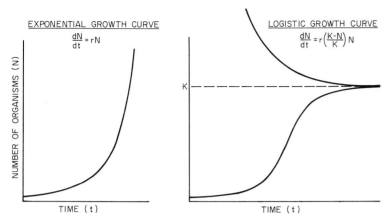

Figure 4-10 Two basic equations for the growth and regulation of populations (written as differential equations) and the solutions to the equations (drawn as curves). Two logistic curves are shown, one starting above K and descending toward this asymptote and the other starting from near zero and ascending toward the same asymptote.

in this way, then the constants are defined and discussed with reference to their possible biological meaning. The derivation given here reveals the intuitive basis for the model.

For all values of N less than or equal to K, the solution of the logistic equation gives a symmetric S-shaped curve of rising N as time passes, with the maximum population growth rate (the "optimum yield") occurring at $K/2$. Some laboratory populations conform well to the pure logistic, while a few natural populations can at least be fitted to it empirically. Schoener (1973) has recently shown that in at least one circumstance—the limitation of population growth by competition of individuals scrambling for resources as opposed to competition by direct interference—the growth curve cannot be expected to be S-shaped. Instead it will turn evenly upward and over to approach the asymptote. Other refinements designed to make the basic model more realistic have been added by Wiegert (1974).

Density Dependence

Why should populations be expected to attain particular values of the carrying capacity K, asymptotically or otherwise, and remain there? Ecologists often distinguish density-independent effects from density-dependent effects in the environment. A density-independent effect alters birth, death, or migration rates, or all three, without having its impact influenced by population density. As a result it does not regulate population size in the sense of tending to hold it close to K. Imagine an island whose southern half is suddenly blanketed by ash from a volcanic eruption. All of the organisms on this part

of the island, roughly 50 percent of the total from each population, are destroyed. Beyond doubt the volcanic eruption was a potent controlling factor, but its effect was density-independent. It reduced all of the populations by 50 percent no matter what their densities at the time of the eruption, and hence could not serve in a regulatory capacity. Most density-independent depressions in population size may be due to sudden, severe changes in weather. Journals devoted to birdlore, natural history, and wildlife management are filled with anecdotes of hail storms killing most of the young of local wading bird populations, late hard freezes causing a crash in the small mammal populations, fire destroying most of a saw grass prairie, and so forth. An important theoretical consideration is that populations whose growth is governed exclusively by density-independent effects probably are destined for relatively early extinction. The reason is that unless there are density-dependent controls always acting to guide the population size toward K, the population size will randomly drift up and down. It may reach very high levels for a while, but eventually it will head down again. And if it has no density-dependent controls to speed up its growth at lower levels while it is down, it will eventually hit zero. The density-independent population is like a gambler playing against an infinitely powerful opponent, which in this case is the environment. The environment can never be beaten, at least not in such a way that the population insures its own immortality. But the population, being composed of a finite number of organisms, will itself eventually be beaten, that is, reduced to extinction. For this reason biologists believe that most existing populations have some form of density-dependent controls that ward off extinction.

What are these density-dependent controls? First, consider the various forms of the quantitative effect they exert. The curve labeled A in Figure 4-11 is one that we intuitively expect to be associated with a fine degree of control in nonsocial populations. At excessively low population numbers, mating might be difficult and the per-individual growth rate correspondingly low. With a small increase in N this difficulty is remedied and the population, blessed with temporarily unlimited resources and light controls of other kinds, achieves its highest growth rate. As N goes up, however, the density-dependent controls begin to exert their effects, with the result that the population continuously decelerates as it approaches K. This is the form of density dependence suggested in the elementary logistic model. Curve B is a population with a less sensitive control. The population grows until it is close to or at K, then the control asserts itself abruptly. This effect is produced by many territorial systems and by shortages of certain types of nesting sites and food supplies. Curve C is what might be expected from a highly social species, in which a critical mass of individuals (N_{crit}) must be assembled if the population is to survive at all. Subsequent increase in population induces

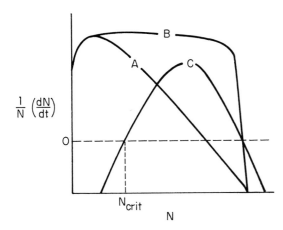

Figure 4-11 Three forms of density-dependence relations in the per-individual growth rate (dN/Ndt) of populations. *A* is the curve associated with a relatively fine degree of control, *B* is expected when the control is coarse or at least abrupt near equilibrium ($dN/dt = 0$), and *C* is a special form expected in highly social species.

a rise in the growth rate, perhaps for a substantial interval of N, before the inevitable decline in the growth rate sets in.

An astonishing diversity of biological responses have been identified as density-dependent controls. Most of them are implicated in one way or another in social behavior and, indeed, much social behavior is comprehensible only by reference to the role it plays in population control. This generalization will be borne out in the following briefly annotated catalog of the principal classes of controls.

Emigration

The single most widespread response to increased population density throughout the animal kingdom is restlessness and emigration. Hydras produce a bubble beneath their pedal disk and float away (Łomnicki and Slobodkin, 1966). Pharaoh's ants (*Monomorium pharaonis*) remove their brood from the nest cells, swarm feverishly over the nest surface, and depart for other sites located by worker scouts (Peacock and Baxter, 1950). Mice (*Mus, Peromyscus*) sharply increase their level of locomotor activity and begin to explore away from their accustomed retreats. Every overpopulated range of songbirds and rodents contains floater populations, consisting of individuals without territories who live a perilous vagabond's existence along the margins of the preferred habitats. Sometimes the movements become directed and unusually persistent, a trend that reaches an evolutionary extreme in the "marches" of the lemmings. As Christian (1970), Calhoun (1971), and others have repeatedly emphasized, the wanderers are the juveniles, the subordinates, and the sickly—the "losers" in the territorial contests for the optimum living places.

However, these meek of the earth are not necessarily doomed; their circumstances have simply forced them into the next available strategy, which is to "get out while the getting is good," to search with the possibility of finding a less crowded environment. In fact, many individuals do succeed in this endeavor, and as a result they play a key role in enlarging the total population size, in extending the range of the species, and possibly even in pioneering in the genetic adaptation to new habitats (Lidicker, 1962; Christian, 1970; G. C. Clough, in Archer, 1970).

Some insects respond to crowding by undergoing phase changes over one or two generations. The phenomenon is widespread in the noctuid moths, where it is associated with the rapid build-up and outbreak of opportunistic species. When caterpillars of the cotton leaf worm *Spodoptera littoralis* are crowded, they become darker, more active, and produce smaller adults (Hodjat, 1970). Crowded adults of many aphid species develop wings, turn from parthenogenesis to sexual reproduction, and fly away to new host plants. However, the most spectacular phase changes and emigrations occur in the "plague" locusts, which consist of many species of short-horned grasshoppers found in arid regions around the world. When these insects are crowded during periods of peak population growth, they undergo a phase change that takes three generations, from the *solitaria* phase that is first crowded through the intermediate *transiens* phase in the second generation to the *gregaria* phase in the third generation. The final adult products are darker in color, more slender, have longer wings, possess more body fat and less water, and are more active. In short, they are superior flying machines. Also, their chromosomes develop more chiasmata during meiosis, resulting in a higher recombination rate and, presumably, greater genetic adaptability. Finally, both the nymphs and adults are strongly gregarious, readily banding together until they create the immense plague swarms. Once in motion, the adult locusts persist for long distances. Swarms often fly intact from Eritrea to the island of Socotra, covering 220 kilometers of open water. When aided by wind, a few individuals leave the west African coast and land on the Azores, a distance of at least 1,900 kilometers from their point of departure. Most aspects of the biology of locust swarming are covered in the reviews of Waloff (1966), Norris (1968), Nolte et al. (1969, 1970), and Haskell (1970).

Stress and Endocrine Exhaustion

In 1939 R. G. Green, C. L. Larson, and J. F. Bell observed a population crash of the snowshoe hare (*Lepus americanus*) in Minnesota and drew a remarkable conclusion about it. They deduced the primary cause to be shock disease, a hormone-mediated idiopathic hypoglycemia that can be identified by liver damage and disturbances in several aspects of carbohydrate metabolism. The implication was that when conditions are persistently crowded the hares suffer an excessive endocrine response from which they cannot recover. Even when individuals collected during the population decline were placed in favorable laboratory conditions, they lived for only a short time. Many vertebrate physiologists and ecologists have subsequently explored the effect of crowding and aggressive interaction on the endocrine system. And conversely, they have speculated on the multifarious ways in which endocrine-mediated physiological responses can serve as density-dependent controls by increasing mortality and emigration, diminishing natality, and slowing growth. Among the best syntheses of this subject are those by Christian (1961, 1968), Etkin, ed. (1964), Esser, ed. (1971), Turner and Bagnara (1971), and von Holst (1972a). In general, raising the population density increases the rate of individual interactions, and this effect triggers a complex sequence of physiological changes: increased adrenocortical activity, depression of reproductive function, inhibition of growth, inhibition of sexual maturation, decreased resistance to disease, and inhibition of growth of nursing young apparently caused by deficient lactation. Stress-induced death has even been hypothesized to occur in cockroaches. Males of *Naupheta cinerea* that lose aggressive encounters with other males, and are forced into subordinate status, tend to die early even in the absence of visible injury or starvation (Ewing, 1967). The precise physiological basis and the form of endocrine mediation, if any, are unknown. The possible existence of a vertebratelike stress syndrome has seldom been considered in insects and other invertebrates and remains a promising subject for experimentation.

Reduced Fertility

An inverse relation between population density and birth rate has been demonstrated in laboratory and free-living populations of many species of insects, birds, and mammals (Lack, 1966; Clark et al., 1967; Solomon, 1969). Fission rates of protistans and lower invertebrates invariably decline in laboratory cultures if other restraining factors are removed and the organisms are allowed to multiply at will. Best et al. (1969) traced the control in the planarian *Dugesia dorotocephala* to a secretion released by the animals themselves into the surrounding water. In the house mouse (*Mus musculus*), a species probably typical of rodents in its population dynamics, the decline in birth rate in laboratory populations was found to be due to decreased fertility in mature females, inhibition of maturation, and increased intrauterine mortality (Christian, 1961). In fact, almost unlimited means exist by which crowding can reduce fertility. Pigeon fanciers are aware that when the birds are too crowded, the males interfere with one another's attempts to copulate, and female fertility declines. A similar effect has been reported by Adler and Zoloth (1970) in female rats. The mechanical stimulation caused by repeated copulation inhibits sperm transport and reduces the percentage of pregnancies.

Inhibition of Development

Parental care and the development of the young are both complex, fragile processes subject to density-dependent interference at any stage. John B. Calhoun's famous Norway rat colonies stopped reproducing when the population reached abnormally high densities largely because the females failed to build complete nests, causing the pups to leave the shelters prematurely. As a result, infant mortality reached 80 and 96 percent in two series of experiments (Calhoun, 1962a,b). The growth of the young was also retarded in the crowded rat colonies, a phenomenon that is one of the most widespread density-dependent controls in other kinds of animals. In *Animal Aggregations* (1931) Allee reviewed many such cases among the invertebrates and cold-blooded vertebrates. He hypothesized the existence of specific factors for each species that could be separated by appropriate experimentation, but the subject has not been pursued with any avidity by more recent investigators. One exception was Richards (1958) who, noting that the inhibition of growth of *Rana pipiens* tadpoles in excessively crowded cultures is due to the fouling of the water, traced the inhibitory agent to a peculiar type of cell passed in the feces. Some kinds of plants release toxic substances that inhibit the growth of smaller members of their own species (Whittaker and Feeney, 1971).

Infanticide and Cannibalism

Guppies (*Lebistes reticulatus*) are well known for the stabilization of their populations in aquaria by the consumption of their excess young. In one experiment Breder and Coates (1932) started two colonies, one below and one above the carrying capacity, by introducing a single gravid female in one aquarium and 50 mixed individuals in a second, similar aquarium. Both populations converged to 9 individuals and stabilized there, because all excess young were eaten by the residents. Cannibalism is commonplace in the social insects, where it serves as a means of conserving nutrients as well as a precise mechanism for regulating colony size. The colonies of all termite species so far investigated promptly eat their own dead and injured. Cannibalism is in fact so pervasive in termites that it can be said to be a way of life in these insects. When supernumerary reproductives of *Kalotermes flavicollis* are produced in laboratory colonies, they are soon pulled apart and eaten by the workers (Lüscher, 1952). Winged reproductives of *Coptotermes lacteus* prevented from leaving the nest on a normal nuptial flight are eventually killed and eaten by the workers (Ratcliffe et al., 1952). In general, when alien workers chance into a nest belonging to a colony of the same species, they are first disabled, typically by a mandibular strike from a soldier, and then consumed. Cook and Scott (1933) found that cannibalism became intense in colonies of *Zootermopsis angusticollis* when they were kept on a diet of pure cellulose and hence deprived of protein. When sufficient quantities of casein were added to their diet, cannibalism dropped almost to zero. The eating of immature stages is common in the social Hymenoptera. In ant colonies all injured eggs, larvae, and pupae are quickly consumed. When colonies are starved, workers begin attacking healthy brood as well. In fact, there exists a direct relation between colony hunger and the amount of brood cannibalism that is precise enough to warrant the suggestion that the brood functions normally as a last-ditch food supply to keep the queen and workers alive. In the army ants of the genus *Eciton*, cannibalism has apparently been further adapted to the purposes of caste determination. According to Schneirla (1971), most of the female larvae in the sexual generation (the generation destined to transform into males and queens) are consumed by workers. The protein is converted into hundreds or thousands of males and several of the very large virgin queens. It seems to follow, but is far from proved, that female larvae are determined as queens by this special protein-rich diet. Other groups of ants, bees, and wasps show equally intricate patterns of specialized cannibalism, a subject reviewed in detail by Wilson (1971a).

Nomadic male lions of the Serengeti plains frequently invade the territories of prides and drive away or kill the resident males. The cubs are also sometimes killed and eaten during territorial disputes (Schaller, 1972). High-density populations of langurs (*Presbytis entellus*, *P. senex*) display a closely similar pattern of male aggression. The single males and their harems are subject to harassment by peripheral male groups, who sometimes succeed in putting one of their own in the resident male's position. Infant mortality is much higher as a direct result of the disturbances. In the case of *P. entellus*, the young are actually murdered by the usurper (Sugiyama, 1967; Mohnot, 1971; Eisenberg et al., 1972).

It is also true that the young of a few vertebrates kill and eat one another. Crowding in ambystomid salamanders induces cannibalism among the aquatic larvae. The winners grow at increased rates by consuming smaller larvae that would otherwise die from starvation or from other ill effects of overcrowding. Consequently, at metamorphosis some individuals are larger and therefore better adapted to the land environment they enter, because larger size provides a higher volume/surface ratio and greater resistance to desiccation (Gehlbach, 1971). A closely similar process occurs in ponds overstocked with small-mouth bass, *Micropterus dolomieu* (Langlois, 1936).

Competition

Competition is defined by ecologists as the active demand by two or more organisms for a common resource. When the resource is not sufficient to meet the requirements of all the organisms seeking it, it becomes a limiting factor in population growth. When, in addition, the shortage of the resource limits growth with increasing severity as the organisms become more numerous, then competition is by definition one of the density-dependent factors. Competition can occur between members of the same species (intraspecific competition) or between individuals belonging to different species (interspecific competition). Either process can serve as a density-dependent control for a given species, although the more precise regulation of population size is likely to occur when the competition is primarily intraspecific. The techniques of competition are extremely diverse, and will be explored more fully in a later chapter on territory and aggression. An animal that aggressively challenges another over a piece of food is obviously competing. So is another animal that marks its territory with a scent, even when other animals avoid the territory solely because of the odor and without ever seeing the territory owner. Competition also includes the using up of resources to the detriment of other organisms, whether or not any aggressive behavioral interaction also occurs. A plant, to take an extreme case, may absorb phosphates through its root system at the expense of its neighbors, or cut off its neighbors from sunlight by shading them with its leaves.

For the moment, it is useful to classify competition into two broad modes, scramble and contest (Nicholson, 1954). Scramble competition is exploitative. The winner is the one who uses up the resource first, without specific behavioral responses to other competitors who may be in the same area. It is the struggle of small boys scrambling for coins tossed on the ground before them. If the boys stood up and fought, with the winner appropriating all the coins within a certain radius, the process would be contest competition. Examples of this latter, more fully animallike behavior are territoriality and dominance hierarchies. Competition theory is a relatively advanced field in ecological research; important recent reviews include those by Levins (1968), Pielou (1969), May (1973), and Schoener (1973).

Predation and Disease

Because their numbers can be counted, predators and parasites exert the most easily quantifiable density-dependent effects (see Figure 4-12). As local populations of the host species increase in numbers, its enemies are able to encounter and to strike individuals at a higher frequency. This "functional response," as it is called by ecologists (Holling, 1959), is enhanced in cases where the parasites and predators migrate to the foci of greatest density. Alternatively or concurrently, the parasites and predators can exert their influence on their victims by a long-term "numerical response," in which their own populations build up over two or more generations because of

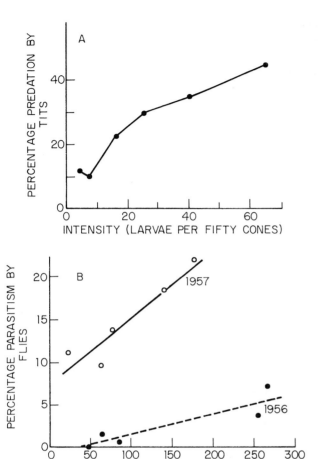

Figure 4-12 Density-dependent predation and disease in insects. *A*: The intensity of predation by the blue tit (*Parus caeruleus*) on the eucosmid moth *Ernarmonia conicolana* increases with the population density of the moth; the percentage predation data given refer to the populations of individual trees (Gibb, 1966). *B*: The intensity of fatal parasitism by the tachinid fly *Cyzenis albicans* on the winter moth *Operophtera bruceata* increases with the density of the moth; the data shown cover two years. (From Hassell, 1966.)

the increased survivorship and fecundity afforded by the improved food supply.

This governing relationship is potentially reciprocal. As the populations of the victims grow dense, the responses of their enemies become more efficient, and the growth rate of the victims is brought to zero or even reversed. With their food supply thus restricted, the parasites and predators ultimately halt their own growth. In simple ecosystems, predator-prey cycling can sometimes be observed across as many as three trophic levels. A simple and instructive example

of the balance between predator and prey is that of the wolves and moose of Isle Royale. Isle Royale is a 540-square-kilometer island located in Lake Superior near the Canadian shore. It is kept in its primitive condition by the U.S. National Park Service. Early in this century moose colonized Isle Royale, probably by walking over the 24-kilometer stretch of ice from Canada during the winter. In the absence of timber wolves and other predators, the moose increased rapidly. By the mid 1930's the herd had increased to between 1,000 and 3,000 animals. At this point the moose population far exceeded the carrying capacity of the island for moose, and the low vegetation on which they depend for existence was soon consumed. A population crash ensued, reducing the herd to well below the carrying capacity. As the vegetation grew back, the herd expanded rapidly again—and crashed again in the late 1940's. In 1949 timber wolves crossed the ice from Canada to Isle Royale. Their appearance had a marked and beneficial effect on the Isle Royale environment. The wolves reduced the number of moose to between 600 and 1,000, somewhat below the carrying capacity that would be determined by food alone. The browse vegetation has returned in abundance, and the moose now have plenty to eat. Their numbers are controlled by predation rather than by starvation. The timber wolf population has remained steady at between 20 and 25 individuals.

What controls the number of timber wolves? Why don't they just keep eating moose until none of these prey is left, then suffer a population crash of their own? The answer is very simple. The wolves catch all of the moose they possibly can, and their effort keeps the moose population down to 600 to 1,000 individuals. It is very hard work to trap and kill a moose. The wolves travel an average of 20 to 30 kilometers a day during the winter. Whenever they detect a moose they try to capture it. Most of their efforts fail. During one study conducted by L. David Mech, the wolf pack was observed to hunt various moose on 131 separate occasions. Fifty-four of the moose escaped before the wolves could even get close. Of the remaining 77 that the wolves were able to confront, only 6 were overcome. All this effort yielded a "crop" of about one moose every three days. That was enough to provide each of the wolves with an average of about 4 kilograms of meat per day. Apparently the wolves simply cannot increase the yield beyond this point, and their number has consequently stabilized. The moose, by unwillingly supplying the wolves with one of their members about every three days, have stabilized their own population. The predator-prey system is in balance. As a curious side effect, the moose herd is kept in good physical condition, since the wolves catch mostly the very young, the old, and the sickly individuals. And, finally, because the moose population is not permitted to increase to excessive levels, the vegetation on which they feed remains in healthy condition.

Like competition, predator-prey interaction lies at the heart of community ecology and has been the object of intensive theoretical and experimental research. Among the most significant recent reviews are those by Le Cren and Holdgate, eds. (1962), Leigh (1971), Krebs (1972), MacArthur (1972), and May (1973). A brief elementary introduction to the basic theory is provided by Wilson and Bossert (1971).

Genetic Change

Models of population dynamics conventionally assume that populations are genetically uniform with reference to density-dependent factors and do not evolve significantly during short-term fluctuations in numbers. If this restriction is removed, population control can be influenced in some interesting ways. Different genotypes can be subject to various density-dependent controls, with the result that the population fluctuates in size as one genotype replaces another. Suppose that when allele a predominates, the population equilibrates at a high level under the control of density-dependent effect A. However, selection favors allele b over a at this higher density. As b comes to predominate, the population shifts to a lower equilibrial density, mostly under the control of a new density-dependent effect, B. But at the level dictated by B, allele a is favored by selection, and the stage is set for the move back up to the higher level. Thus genetic polymorphism and the corresponding differences in density-dependent control can be coupled in a reciprocally oscillating system to create a population cycle. A system actually corresponding to this model has been worked out for the larch budmoth (*Zeirapheira griseana*) during many years of research by G. Benz, D. Bassand, and other Swiss entomologists (review in Clark et al., 1967). In the populations of Switzerland's Engadin Valley, a "strong" form gains the advantage at high densities by virtue of higher reproductive capacity and a greater tendency to disperse. Then, as peak densities are reached, the "weak" form is favored because of its greater resistance to granulosis virus. As the weak form begins to replace the strong form, the former is differentially attacked by hymenopterous parasites, thus starting the population on the downward arc of the cycle again.

C. J. Krebs (1964) and Dennis Chitty (1967a,b) have hypothesized a similar mechanism to explain population cycles in small mammal populations. They proposed that as density increases, selection favors genotypes that do not readily emigrate but more than hold their own by superiority in aggressive interactions. A sufficient frequency of aggressive encounters, enhanced by the selection process, helps bring the population into decline. At lower densities, aggressive genotypes are at a disadvantage, and the population as a whole evolves back toward gentler behavior. Krebs and his associates (1973) have demonstrated strong changes in certain transferrin alleles during various phases of the population cycles in voles (*Microtus*) and significant differences in frequencies of the same alleles between emigrating and

resident females. These data are consistent with the model but do not prove it. In particular, the direct connection between transferrin polymorphism and variation in aggressive and dispersal behavior has not been established. It will in any case be difficult to separate cause and effect in these systems. Does the genetic change really force the population cycles by aggressive overshoot, or does it merely track changes in the population density forced by other density-dependent effects? Krebs, Keller, and Tamarin (1969) have identified the true factors in *M. ochrogaster* and *M. pennsylvanicus* as emigration and food shortage, in that order of importance. Behavioral microevolution may function, if I have interpreted the rather intricate accounts correctly, to help pass populations back and forth between these two controls, creating cycles as a by-product. In short, the mechanism may not differ basically from that of the larch budmoth.

Social Convention and Epideictic Displays

Suppose that animals *voluntarily* agreed to curtail reproduction when they became aware of rising population density. For instance, males could compete with other males in a narrowly restricted manner for access to females, as in a territorial display, with the loser simply withdrawing from the contest short of bloodshed or exhaustion on either side. This technique of slowing population growth by ritualized means has been called conventional behavior by Wynne-Edwards (1962). Its most refined form might be the epideictic display, a conspicuous message "to whom it may concern" by which members of a population reveal themselves and allow all to assess the density of the population. The correct response to evidence of an overly dense population would be voluntary birth control or removal of one's self from the area. This idea, with strong roots going back to W. C. Allee (in Allee et al., 1949), was developed in full by Olavi Kalela (1954) and V. C. Wynne-Edwards. It is fundamentally different from the remainder of the conception of density dependence, because it implies altruism of individuals. And altruism of individuals directed at entire groups can evolve only by natural selection at the group level. Few ecologists believe that social conventions play a significant role in population control, and many doubt if such a role exists at all. The reason for scepticism is twofold. First, the intensity of group extinction required to fix an altruistic gene must be high, and the problem becomes acute when the altruism is directed at entire Mendelian populations. Because the formal theory of group selection is complex and has many ramifications in sociobiology, it will be left to a chapter by itself (Chapter 5). At that time the feasibility of population control by social conventions will be examined. The second reason for doubt is the difficulty of demonstrating the phenomenon in nature. To prove a functional social convention, and hence population-level selection, is to accomplish the onerous feat of proving (as opposed to disproving) a null hypothesis: the other density-

Table 4-1 The identity of density-dependent controls in representative animal species.

Species	Nature of the controls	Conditions of the study	Authority
COELENTERATA Hydra (*Hydra littoralis*)	Emigration by flotation	Laboratory; probably also primary control in nature	Lomnicki and Slobodkin (1966)
PLATYHELMINTHES Planarian (*Dugesia dorotocephala*)	Suppression of fissioning by secretions released by population into the water	Laboratory; highly effective, but other controls could intervene in nature	Best et al. (1969)
MOLLUSCA Pond snail (*Lymnaea elodes*)	Food supply	Experimental manipulation of natural populations	R. M. Eisenberg (1966)
ARTHROPODA Psyllid (*Cardiaspina albitextura*)	At low densities (low K), predation by birds and insect parasites; at high densities (high K), competition for food	Field studies	Clark et al. (1967)
Cabbage aphid (*Brevicoryne brassicae*)	Primarily emigration (by special alate forms); also reduced fecundity	Field studies	Clark et al. (1967)
Sheep blowfly (*Lucilia cuprina*)	Competition for food among adults, resulting in reduced fecundity	Experiments on laboratory populations	Clark et al. (1967)
Codling moth (*Cydia pomonella*)	Competition for feeding space among larvae and for cocooning sites	Introduced free populations in Australia	Clark et al. (1967)
Larch budmoth (*Zeiraphera griseana*)	Hymenopterous parasites and granulosis virus, which alternate in prevalence	Prolonged studies of natural populations	Clark et al. (1967)

Table 4-1 (*continued*)

Species	Nature of the controls	Conditions of the study	Authority
Sawfly (*Perga affinis*)	In some regions, emigration coupled with competition; in other regions, parasitism by other insects	Field studies	Clark et al. (1967)
European spruce sawfly (*Diprion hercyniae*), introduced populations in Canada	Erratic occurrence of disease and insect parasitism; weather plays a major role in these unstable populations	Field studies	Clark et al. (1967)
Ants (*Lasius flavus, L. niger, Myrmica ruginodis*)	Competition for nesting space, direct interference	Manipulation of natural populations	Brian (1956a,b), Pontin (1961)
Bumblebees (*Bombus* spp.)	At high densities (high K), competition for nest sites; at low densities (low K), probably competition for food	Manipulation of natural populations	Medler (1957)
REPTILIA Gecko (*Gehyra variegata*)	Territoriality; dispersal and differential mortality of excess	Field studies, laboratory experimentation	Bustard (1970)
AVES Wood pigeon (*Columba palumbus*)	Food supply	Field studies	Lack (1966)
Red grouse (*Lagopus lagopus*)	Territoriality, with surplus "floaters" being removed mostly by starvation, disease, and predation	Prolonged field studies and experimental manipulation of natural populations	Jenkins et al. (1963), Watson (1967)
Dioch (*Quelea quelea*)	Food supply	Field studies	Lack (1966)

Table 4-1 (*continued*).

Species	Nature of the controls	Conditions of the study	Authority
Great tit (*Parus major*)	Territoriality; floaters produce few young; food is probably limiting in at least some localities	Inferred from numerous field studies	Lack (1966), J. R. Krebs (1971)
Plain titmouse (*Parus inornatus*)	Territoriality, emigration of young; fate of floaters not established	Inferred from field studies	Dixon (1956)
Dickcissel (*Spiza americana*)	Territoriality, with elastic territory size; fate of floaters not established	Inferred from field studies	Zimmerman (1971)
MAMMALIA			
Wolf (*Canis lupus*)	Food supply, probably buffered to some degree by territoriality	Inferred from field studies	Murie (1944), Mech (1970)
Arctic ground squirrel (*Spermophilus undulatus*)	Territoriality, with surplus floaters emigrating and removed by predation	Inferred from field studies	Carl (1971)
Mice (*Mus, Peromyscus*), voles (*Microtus*), and lemmings (*Lemmus*)	Emigration is a primary control in many populations; see discussion in text	Extended field studies and experimental manipulation of populations	Anderson (1961), Caldwell and Gentry (1965), Frank (1957), Houlihan (1963), Krebs et al. (1969), Clough (1971)
Woodchuck (*Marmota monax*)	Territoriality, emigration of young; endocrine-mediated reduction of natality	Field studies	Snyder (1961)

dependent controls, based upon individual as opposed to group selection, must all be eliminated one by one.

In Table 4-1 are listed the density-dependent controls that have been documented in studies of laboratory and free-living populations of a wide diversity of animal species. The basis of selection of this sample was the thoroughness and reliability of the studies, rather than the balance of taxonomic representation. Several important generalizations emerge from these results, not the least of which is the great diversity of the operational factors. It is clearly quite useless to search for a single governing factor or set of factors. The closest approach to uniformity is to be found in the birds and mammals, where the combination of territoriality in adults and emigration by subordinate and young individuals appears to be widespread. But even here there are strong exceptions. For example, colonial birds such as the vegetarian pigeons and queleas are limited by food supply. An excellent experiment by Lidicker (1962) has revealed considerable variability in secondary controls in rodent species. Lidicker confined populations of four species (*Mus musculus, Peromyscus maniculatus, P. truei, Oryzomys palustris*) in similar enclosures and fed them ad libitum, thus removing the two cardinal controls of rodent populations, emigration and starvation. Growth of the *Mus* population and one of the *P. maniculatus* populations was halted by inhibition of reproduction in all the females. Growth of a second *P. maniculatus* population and two populations of *P. truei* was stopped by a combination of infant mortality, seasonal reproductive inhibition (which affects even individuals kept indoors), and nonseasonal reproductive inhibition in some females. Growth of the population of *O. palustris* was halted entirely by infant mortality.

Vertebrate populations have proved markedly more difficult to analyze than invertebrate populations. Much of the basic theory has therefore been constructed with reference to invertebrates, especially insects. The reason is evidently the greater complexity and flexibility of vertebrate systems, as well as the much greater practical problems encountered in studying large, slow-breeding animals. This difficulty of vertebrate ecology has had an important impact on the study of social systems by contributing confusion to many of the most basic concepts.

Intercompensation

A great deal of the variation in density-dependent controls between species, between laboratory and free-ranging populations of the same species, and even among free-ranging populations of the same species, is due to the property of intercompensation. This means that if the environment changes to relieve the population of pressure from a previously sovereign effect, the population will increase until it reaches a second equilibrium level where another effect halts it. For example, if the predators that normally keep a certain herbivore population in balance are removed, the population may increase to a point where food becomes critically short. If a superabundance of food is then supplied, the population may increase still further—until intense overcrowding triggers an epizootic disease or a severe stress syndrome. The rodent experiments of Calhoun, Christian, Krebs, Lidicker, and others have been instructive in revealing the sequences of intercompensating controls in a variety of species. Calhoun's "behavioral sink"—in which most individuals behaved abnormally and failed to reproduce—can be viewed as a rat population that was allowed to rise above nearly all the controls the species encounters in nature. Sociopathology, if caused by crowding, can be viewed as controls that are nonadaptive in the sense that they lie beyond the limit of a species' repertory and therefore do not contribute to either individual or group fitness.

Population Cycles of Mammals

The population cycles of mammals, and especially of rodents, have loomed large—too large—in the central literature of sociobiology. This is a doubly unfortunate circumstance because of the confusing, often bitter controversies that have risen around the cycles. The real problem, aside from the practical difficulties in obtaining data, is the fact that population cycles have traditionally been subjected to the advocacy method of doing science. Each of several density-dependent controls has had its own theory, school of thought, and set of champions: emigration (Frank, 1957; Caldwell and Gentry, 1965; Anderson, 1970; Krebs et al., 1973); stress and endocrine exhaustion (Christian, 1961; Davis, 1964; Christian and Davis, 1964); cyclical selection for aggressive genotypes (Krebs, 1964; Chitty, 1967a,b); predation (Pearson, 1966, 1971); nutrient depletion (Pitelka, 1957; Batzli and Pitelka, 1971). A plausible model and supporting data have been marshaled behind each process to advance it as the premier factor in nature. To express the matter in such a way is not to denigrate the work of these authors, which is of high quality and imaginative. And paradoxically, all could be at least partly correct. But inconsistencies have arisen from the tendency to generalize from restricted laboratory experiments and field observations of only one to several populations, together with a failure by a few key authors to perceive the possible role of intercompensation. It does seem plausible that intercompensation could be responsible for much of the great variation in operating controls from population to population and from one environment to another. If any rule can be drawn from the existing data, it is perhaps that in free-living rodent populations the principal density-dependent control is most often territoriality combined with emigration, followed by depletion of food supply and predation, in that order. Endocrine-induced changes are difficult to evaluate, but they appear to fall in the secondary ranks of the controls. When they occur they may affect female fertility primarily. Endocrine exhaustion, as easy as it is to induce in laboratory populations by the lifting of other controls, is perhaps rare or absent in most free-living populations. Genetic changes in aggressive behavior, already described in an earlier section, are also hard to evaluate. It seems probable that they amplify cycles but are nevertheless subordinate to territorial aggression and emigration as density-dependent controls.

Life Tables

The vital demographic information of a closed population is summarized in two separate schedules: the *survivorship schedule*, which gives the number of individuals surviving to each particular age, and the *fertility schedule*, which gives the average number of daughters that will be produced by one female at each particular age. First consider survivorship. Let age be represented by x. The number surviving to a particular age x is recorded as the proportion or frequency (l_x) of organisms that survive from birth to age x, where the frequency ranges from 1.0 to 0. Thus, if we measure time in years, and find that only 50 percent of the members of a certain population survive to the age of one year, then $l_1 = 0.5$. If only 10 percent survive to an age of 7 years, $l_7 = 0.1$; and so on. The process can be conveniently represented in survivorship curves. Figure 4-13A shows the three basic forms such curves can take. The curve for type I, which is approached by human beings in advanced civilizations and by carefully nurtured populations of plants and animals in the garden and laboratory, is generated when accidental mortality is kept to a minimum. Death comes to most members only when they reach the age of senescence. In survivorship of type II, the probability of death remains the same at every age. That is, a fixed fraction of each age group is removed—by predators, or accidents, or whatever—in each unit of time. The annual adult mortality of the white stork, for example, is steady around 21 percent, while that of the yellow-eyed penguin is 13 percent. Type II survivorship, therefore, takes the form of negative exponential decay. When plotted on a semilog scale (l_x on logarithmic scale, x on normal scale), the curve is a straight line. Type III is the most common of all in nature. It occurs when large numbers of offspring, usually in the form of spores, seeds, or eggs, are produced and broadcast into the environment. The vast majority quickly perish; in other words, the survivorship curve plummets at an early age. Those organisms that do survive by taking root or by finding a safe place to colonize have a good chance of reaching maturity. The shape of the survivorship curve depends on the condition of the environment,

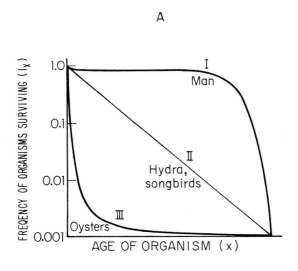

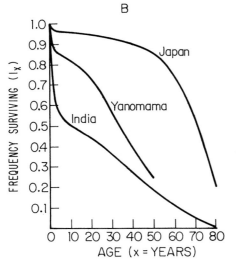

Figure 4-13 Survivorship curves. *A*: the three basic types. *B*: variation in the survivorship curves among human populations, from type I to type III (modified from Neel, 1970). The vertical axis of *A* is on a logarithmic scale.

with the result that it can vary widely from one population to another within the same species. In man himself, the variation ranges all the way from type I to type III (see Figure 4-13B).

The fertility schedule consists of the age-specific birth rates; during each period of life the average number of female offspring born to each female is specified. To see how such a schedule is recorded, consider the following imaginary example: at birth no female has yet

given birth ($m_0 = 0$); during the first year of her life still no birth occurs ($m_1 = 0$); during the second year of her life the female gives birth on the average to 2 female offspring ($m_2 = 2$); during the third year of her life she gives birth on the average to 4.5 female offspring ($m_3 = 4.5$); and so on through the entire life span. The fertility schedule can be represented even more precisely by a continuous fertility curve, an example of which is shown in Figure 4-14.

From the survivorship and fertility schedules we can obtain the *net reproductive rate*, symbolized by R_0, and defined as the average number of female offspring produced by each female during her entire lifetime. It is a useful figure for computing population growth rates. In the case of species with discrete, nonoverlapping generations, R_0 is in fact the exact amount by which the population increases each generation. The formula for the net reproductive rate is

$$R_0 = \sum_{x=0}^{\infty} l_x m_x$$

To see more explicitly how R_0 is computed, consider the following simple imaginary example. At birth all females survive ($l_0 = 1.0$) but of course have no offspring ($m_0 = 0$); hence $l_0 m_0 = 1 \times 0 = 0$. At the end of the first year 50 percent of the females still survive ($l_1 = 0.5$) and each gives birth on the average to 2 female offspring

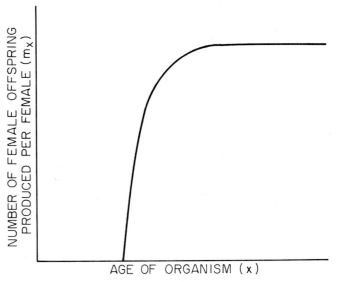

Figure 4-14 Fertility curve for the human louse. This example is typical of organisms that reach sexual maturity at a fixed age and remain fecund until death.

$(m_1 = 2)$; hence $l_1m_1 = 0.5 \times 2 = 1.0$. At the end of the second year 20 percent of the original females still survive $(l_2 = 0.2)$, and each gives birth on the average at that time to 4 female offspring $(m_2 = 4)$; hence $l_2m_2 = 0.2 \times 4 = 0.8$. No female lives into the third year $(l_3 = 0; l_3m_3 = 0)$. The net reproductive rate is the sum of all the l_xm_x values just obtained:

$$R_0 = \sum_{x=0}^{\infty} l_xm_x$$

l_xm_x at birth $(x = 0)$	l_xm_x 1st year $(x = 1)$	l_xm_x 2nd year $(x = 2)$	l_xm_x 3rd year $(x = 3)$
= 0	+ 1.0	+ 0.8	+ 0

$$= 1.8$$

We can proceed to the method whereby r, the intrinsic rate of increase, can be computed precisely from the survivorship and fertility schedules. We start with the solution of the exponential growth equation

$$N_t = N_0e^{rt}$$

Let t = maximum age that a female can reach, and N_0 be only one female. Thus we have set out to find the number of descendants a single female will produce, including her own offspring, the offspring they produce, et seq., during the maximum life span one female can enjoy. Since $N_0 = 1$

$$N_{\text{max age}} = e^{r(\text{max age})}$$

$$= \sum_{x=0}^{\text{max age}} l_xm_xe^{r(\text{max age} - x)}$$

In words, the total number of individuals stemming from a single female is the sum of the expected number of offspring produced by that female at each age x of the female (l_xm_x) times the number of offspring that each of these sets of offspring will produce from the time of their birth to the maximum age of the original female (max age − x). Substituting and rearranging, we obtain

$$e^{r(\text{max age})} = e^{r(\text{max age})} \sum_{x=0}^{\text{max age}} l_xm_xe^{-rx}$$

$$\sum_{x=0}^{\text{max age}} l_xm_xe^{-rx} = 1$$

Or, in continuous distributions of l_x and m_x,

$$\int_0^{\text{max age}} l_xm_xe^{-rx}dx = 1$$

For "max age" we can further substitute ∞, since the two are biologically equivalent. This formulation can be referred to as the Euler equation or the Euler-Lotka equation, after the eighteenth-century mathematician Leonhard Euler, who first derived it, and A. J. Lotka, who first applied it to modern ecology. Since we know the values of m_x and l_x, the Euler-Lotka equation permits us to solve for the intrinsic rate of increase, r. This process is often computationally tedious and thus usually requires the aid of a computer, but in principle it is straightforward.

The Stable Age Distribution

An important principle of ecology is that any population allowed to reproduce itself in a constant environment will attain a stable age distribution. (The only exception occurs in those species that reproduce synchronously at a single age.) This means that the proportions of individuals belonging to different age groups will maintain constant values for generation after generation. Suppose that upon making a census of a certain population, we found 60 percent of the individuals to be 0-1 year old, 30 percent to be 1-2 years old, and 10 percent to be 2 years old or older. If the population had existed for a long time previously in a steady environment, this is likely to be a stable age distribution. Future censuses will therefore yield about the same proportions. Stable age distributions are approached by any population in a steady environment, regardless of whether the population is increasing in size, decreasing, or holding steady. Each population has its own particular distribution for a given set of environmental conditions.

Stable age distributions, along with the resulting intrinsic rate of increase (r), can be computed with the aid of matrix algebra. Suppose that we represent the starting age distribution (at time $t = 0$) by the column vector

$$\begin{pmatrix} N_{00} \\ N_{10} \\ N_{20} \\ \vdots \\ N_{n0} \end{pmatrix}$$

where the $n + 1$ elements in the vector represent the proportion of females in each of $n + 1$ age groups into which we have divided the population. The first subscript denotes age of the organism; the second, the time the population is counted. This initial distribution can be the ultimate stable distribution or any deviation from it. We now transform the distribution by multiplying it against a projection matrix (or "Leslie matrix," after its inventor, P. H. Leslie), containing the survivorship and fertility schedules:

$$
\begin{pmatrix}
m_0 & m_1 & \cdots & m_{n-1} & m_n \\
P_0 & 0 & \cdots & 0 & 0 \\
0 & P_1 & \cdots & 0 & 0 \\
\cdot & \cdot & \cdot & \cdot & \cdot \\
0 & 0 & \cdots & P_{n-1} & 0
\end{pmatrix}
\begin{pmatrix}
N_{00} \\
N_{10} \\
N_{20} \\
\vdots \\
N_{n0}
\end{pmatrix}
$$

where m_i is the number of female offspring produced in each age interval ($i = 0, 1, \ldots, n - 1, n$), and P_i is the probability of survival within the interval t to $t + 1$ (P_i is distinguished from l_i, which is the probability of survival all the way from birth to age i). The product of the demographic matrix times the age distribution vector gives the age distribution (still a column vector) in the next interval of time. The population will converge to a stable age distribution if there exists a positive eigenvalue (λ) whose absolute value is greater than the other eigenvalues. At stability the absolute size of each size class, and therefore of the population as a whole, increases by a multiple of λ in each time interval. At $\lambda = 1$, the population is stationary ($dN/dt = 0$), but growth can also be negative ($\lambda < 1$) or positive ($\lambda > 1$) and still be associated with a stable age distribution. The eigenvector associated with λ is the stable age distribution. Full descriptions of matrix techniques in demography, with many special cases and applications of use in sociobiology, are given by Keyfitz (1968) and Pielou (1969).

Reproductive Value

Reflection on the properties of life tables leads to the following question: How much is an individual worth, in terms of the number of offspring it is destined to contribute to the next generation? Another way of putting the question is: If we remove one individual, in particular one female, how many fewer individuals will there be in the next generation? The answer depends very much on the age of the individual. If we destroy an old animal, past its reproductive period, the loss will not be felt in the next generation unless the animal has been contributing labor to a social group. But if we remove a young female just at the time she is ready to commence breeding,

the effect on the next generation will probably be considerable. The standard measure of the contribution of an individual to the next generation is called the *reproductive value*, symbolized by v_x, where the x in the subscript represents the age of the individual. The reproductive value is the relative number of female offspring that remain to be born to each female of age x. It can be expressed in words as follows:

$$
v_x = \frac{\begin{array}{c} \text{population growth due to a female of} \\ \text{age x through the remainder of her life} \end{array}}{\begin{array}{c} \text{population growth due to an average female,} \\ \text{regardless of age, for the period of the} \\ \text{remainder of the numerator female's life} \end{array}}
$$

The numerator female (age x) has the potential of reaching the maximum age for the species (max age). For each age y that is equal to or greater than x, the age at which we start observing, there is a probability of survival equal to l_y/l_x, in other words the conditional probability of a female reaching age y given that it has reached age x. At each age y of the numerator female a certain number of female offspring (m_y) will be produced; each of these sets of offspring will proceed to contribute to colony growth for the remainder of the numerator female's life, covering a period of time equal to max age $- y$, and each female born at age y of the numerator female will therefore contribute $e^{r(\text{max age}-y)}$ offspring during this time. The expected population growth due to a female x for the remainder of her life is therefore

$$
\sum_{y=x}^{\text{max age}} \frac{l_y}{l_x} m_y e^{r(\text{max age}-y)}
$$

Meanwhile an average female picked at random from the remainder of the population when the numerator female is at age x can be expected to contribute

$$
e^{r(\text{max age}-x)}
$$

female offspring by the time the numerator female has reached the maximum age. The reproductive value can now be restated as follows:

$$
v_x = \frac{\displaystyle\sum_{y=x}^{\text{max age}} \frac{l_y}{l_x} m_y e^{r(\text{max age}-y)}}{e^{r(\text{max age}-x)}}
$$

$$
= \frac{e^{rx}}{l_x} \sum_{y=x}^{\text{max age}} l_y m_y e^{-ry}
$$

or, in the more precise continuous form,

$$v_x = \frac{e^{rx}}{l_x} \int_x^{\text{max age}} e^{-ry} l_y m_y dy$$

where, again, max age and ∞ are biologically interchangeable.

The reproductive value typically is low at birth, because of the depressing effects of infant or larval mortality (low values of l_x for x near zero), then rises to a peak near the normal age of beginning reproductive effort, and finally falls off with increasing age because of the cumulative effects of mortality and diminishing fertility (see Figure 4-15). The reproductive value has several important implications for ecology and sociobiology. Consider first its relevance to the concept of optimum yield. A predator, or a human farmer or hunter, would want to do more than just try to keep the prey population at about the level that provides the greatest growth rate. Such a crude technique works only if the prey organisms all have about the same reproductive value. A truly skillful predator, or "prudent" predator, as some ecologists like to call it, would want to concentrate on the age groups with the lowest reproductive values. By this means it would obtain the largest amount of protein with the least subtraction from the growth of the exploited population. To take one example, poultry farms make use of the low reproductive value of eggs produced by continuously laying hens by removing them from the hens and selling them for profit. To butcher the hens themselves would be economically disastrous. At the opposite extreme is the case of migratory salmon. They die shortly after returning to freshwater streams to spawn. In the few days between spawning and death their reproductive value is zero, and their large bodies form a rich source of energy for predators and parasites, which can exploit them without subtracting from the growth of the salmon population. Is it possible that predators and parasites really evolve so as to select the age groups with the least reproductive value? Wolf packs prey most heavily on animals that are very young, or very old, or ill—in other words, animals with the smallest reproductive values. But this may be just coincidence; the same individuals are also the easiest to catch. The relation between predation and reproductive value is one that ecologists are just beginning to explore in a systematic fashion, and we cannot make any generalizations except the basic theoretical one already cited.

A second ecological process in which reproductive value is a major factor is colonization. New populations, especially those that colonize islands and other remote habitats, often are started by a very few individuals. The fate of such a founder population is clearly dependent on the reproductive value of its members. If the colonists are all old individuals past the reproductive period, the population is

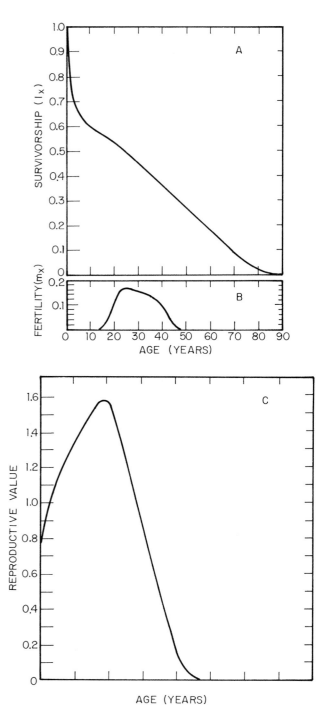

Figure 4-15 Survivorship curve (*A*), fertility curve (*B*), and reproductive value curve (*C*) in Taiwanese women in 1906. From *A* and *B* it is possible to compute curve *C*, as well as the value of the intrinsic rate of increase (*r*), which in this case is 0.017 per year. (Modified from W. D. Hamilton, 1966.)

doomed because $m_x = 0$, and $v_x = 0$. If the propagules are all very young individuals unable to survive by themselves in the new environment, the population is still doomed; this time $l_x = 0$, and $v_x = 0$. Obviously the best colonists are individuals with the highest v_x. Is it possible that species that regularly colonize new habitats have dispersal stages with both high mobility and high reproductive values? The evidence seems to favor this inference, although the relation of reproductive value to colonizing ability is still in an early stage of study (Baker and Stebbins, ed., 1965; MacArthur and Wilson, 1967; C. G. Johnson, 1969).

Finally, reproductive value plays an important role in evolution by natural selection. If a genetically less fit individual is removed from the population when it possesses a high reproductive value, its departure will have a relatively substantial influence on the evolution of the population. It is also true that genes which regularly cause mortality in individuals with high reproductive values will tend to be removed from the population more quickly than those that come into play at another age. It is in fact possible to account for the evolution of senescence by means of this concept. The prevailing theory was anticipated by August Weismann and put in successively more modern form by P. B. Medawar (1952), G. C. Williams (1957), W. D. Hamilton (1966), and J. M. Emlen (1970). Senescence, the increase in debility and mortality due to spontaneous physiological deterioration, is considered to be due to the fixation of genes that confer high fitness earlier in life but cause senescent degeneration later in life. If most of the members of a population are eliminated by predators, disease, and other "accidental" causes prior to reaching the age at which the genes bring senescence, the genes will be fixed because of the increased fitness they confer prior to senescence. In other words, genes that add to fitness when the reproductive value is high, and subtract from it later when the reproductive value is low, will tend to be fixed. When they are fixed, of course, they will in turn influence the l_x and m_x curves, and through them, the curve of reproductive values.

There exist circumstances in which the reproductive value of organisms can be sustained well above zero even when they cease reproduction. Aged members of lion prides, human societies, and presumably other highly organized societies can raise the survivorship of their own descendants by contributing to the effectiveness of the family. Not even social behavior is strictly required. If the species is distasteful or dangerous, and potential predators are capable of learning this fact well enough to avoid the species after a few initial contacts, it will pay older individuals to stay in circulation even if they have ceased reproduction. The reason is that by teaching the predators themselves, the parents better protect the offspring at no cost to the family's overall fitness, with the result that the reproductive value of the older organism is enhanced. Blest (1963) has cited as consistent with this conception an inverse relationship that he observed between palatability to predators and longevity in New World tropical saturniid moths.

Reproductive Effort

In the fundamental equations of population biology, effort expended on reproduction is not to be measured directly in time or calories. What matters is benefit and cost in future fitness. Suppose that the female of a certain kind of fish spawns heavily in the first year of her maturity, with the result that enough eggs are released to produce 20 surviving fry. However, the expenditure of effort and energy invariably costs the female her life. Imagine next a second kind of fish, the female of which makes a lesser effort, resulting in only 5 surviving fry but entailing a negligible risk to life, with the result that she can expect to make five or ten such efforts in one breeding season. The reproductive effort of the second fish, measured in units of future fitness sacrificed at each spawning, is far less than that of the first fish, but in this particular case we can expect populations of the second fish to increase faster. The general question is: In order to attain a given m_i at age i, what will be the reduction in future l_i and m_i? The problem has been the object of a series of theoretical investigations by G. C. Williams (1966a), Tinkle (1969), Gadgil and Bossert (1970), and Fagen (1972), who have used variations on the Euler-Lotka equation (or its intuitive equivalent) to investigate the effect on fitness of various relations between l_i and m_i over all ages. It makes sense to describe reproductive effort in terms of its physiological and behavioral enabling devices, such as proportion of somatic tissue converted to gonads and the amount of time spent in courtship and parental care. However, the performance of these devices must be converted into units in the life tables before their effects on genetic evolution can be computed.

Only fragmentary data exist that can be related to the reproductive effort models. The wildlife literature contains many anecdotes of male animals that lose their lives because of a momentary preoccupation with territorial contest or courtship. Schaller (1972), for example, observed that "when two warthog boars fought, a lioness immediately tried to catch one; a courting reedbuck lost his life because he ignored some lions nearby." When barnacles spawn, their growth rate is substantially reduced (Barnes, 1962), with the result that they are able to produce fewer gametes in the next breeding season and are more subject to elimination by other barnacles growing next to them. Murdoch (1966) demonstrated that the survival of females of the carabid beetle *Agonum fuliginosum* from one breeding season to the next is inversely proportional to the amount of reproduction in the first. In general, the smaller and shorter-lived the organism, the greater its reproductive effort as measured by the amount of fertility per season. A striking example from the lizards is given in Figure 4-16. The expected negative correlation between life span and fertility is based on the assumption, probably true for many kinds of organisms in addition to lizards, that there exists an inverse relation between the time an animal puts into reproduction and its chance of survival. However, in social animals this simple trade-off is easily averted. A dominant male, for example, may invest large amounts of its time in activities related more or less directly to reproduction, and still enjoy higher survivorship by virtue of its secure position within a territory or at the head of a social group.

The Evolution of Life Histories

The Euler-Lotka equation has potentially powerful applications throughout sociobiology. Each l_x and m_x value can have underlying social components. Conversely the adaptive value, r, of each genotype is determined in part by the way its social responses affect each l_x and m_x. Heritability in the $l_x m_x$ schedules has been documented in *Drosophila* (Dobzhansky et al., 1964; Ohba, 1967), *Aedes* mosquitoes (Crovello and Hacker, 1972), lizards (Tinkle, 1967), and human beings (Keyfitz, 1968); and it is surely a universal quality of organisms. Therefore the fine details of life history, meaning the survivor-

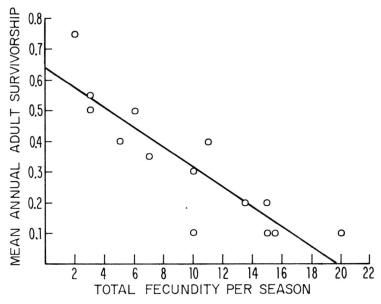

Figure 4-16 A rule of reproductive effort exemplified: the inverse relationship between the rate at which individual lizard females reproduce and the length of their lives, as measured by annual survivorship. Each point represents a different species. (From Tinkle, 1969.)

ship-fertility schedules and their determinants, can be expected to respond to natural selection. In fact, the entire evolutionary strategy of a species can be described abstractly by these schedules.

A basic and unusually flexible model of life history evolution has been provided by Gadgil and Bossert (1970). They accept, in agreement with most previous theory, that the optimum life history is the one whose set of l_x and m_x values provides the maximum r in the Euler-Lotka equation

$$\sum_{x=0}^{\text{max age}} l_x m_x e^{-rx} = 1$$

A population consists of genotypes, each of which possesses a particular $l_x m_x$ schedule in a particular environment. The one whose $l_x m_x$ schedule yields the highest value of r will rise in frequency, provided the environment holds steady. Suppose that the population is under the control of a density-dependent effect (other than predation or parasitism, which will be considered separately). Then the *degree of satisfaction*, ψ, is the index of the extent to which the effect is limiting. At lowest densities, when the control is negligible, ψ is equal to one (satisfaction with this aspect of the environment is total). As the population grows dense, and the control becomes severe, ψ approaches its minimum value of zero. Other parameters are:

α_i, the probability of survival from age i to age $i + 1$ for an individual making no reproductive effort at age i, in a nonlimiting, predator-free environment;

w_i, the size of the individual at age i;

δ_i, the increment in size from age i to $i + 1$ for an individual making no reproductive effort in a nonlimiting environment;

θ_i, the reproductive effort of the individual at age i;

η_i, the probability of escaping death by predation at age i.

The l_x and m_x values can then be computed by a stepwise accumulation of probabilities and increments:

$\alpha_i \cdot f_1(\theta_i)$, the probability of survival from age i to $i + 1$ in a nonlimiting, predator-free environment for an individual exerting reproductive effort θ_i at age i; the function f_1 will usually be assumed to be monotonically decreasing and taking values between zero and one.

$\delta_i \cdot f_2(\theta_i)$, the increment in size from age i to $i + 1$ in a nonlimiting, predator-free environment by an individual exerting reproductive effort θ_i at age i; the function f_2 will usually be assumed to be monotonically decreasing and taking values between zero and one.

$w_i \cdot f_3(\theta_i)$, the number of offspring produced at age i in a nonlimiting environment by an individual exerting reproductive effort θ_i; thus size of the organism is a determining influence; the function f_3 will usually be assumed to be monotonically increasing and taking values between zero and one.

$\alpha_i \cdot f_1(\theta_i) \cdot g_1(\psi_i)$, the probability of survival in a predator-free environment when the degree of satisfaction at age i equals ψ_i; g_1 is usually assumed to be a monotonically increasing function with values between zero and one.

$\delta_i \cdot f_2(\theta_i) \cdot g_2(\psi_i)$, the increment in size from age i to $i + 1$; g_2 is usually a monotonically increasing function with values between zero and one.

$w_i \cdot f_3(\theta_i) \cdot g_3(\psi_i)$, the number of offspring produced at age i; g_3 is usually a monotonically increasing function taking values between zero and one.

The system can now be completely defined:

$$l_x = \prod_0^{x-1} \alpha_i \cdot f_1(\theta_i) \cdot g_1(\psi_i) \cdot \eta_i \quad \text{(probability of survival to age x)}$$

$$w_x = w_0 + \sum_0^{x-1} \delta_i \cdot f_2(\theta_i) \cdot g_2(\psi_i) \quad \text{(size at age x)}$$

$$m_x = w_x \cdot f_3(\theta_x) \cdot g_3(\theta_x) \quad \text{(fertility at age x)}$$

These functions are substituted into the Euler-Lotka equation to determine which of the permissible parameter values yield the highest r. The parameters α_i, δ_i, and w_0 are the biological constraints on the life history; their values among the various genotypes are determined by the history of the species in ways that are external to the Gadgil-Bossert model. Similarly, the values of the parameters ψ_i and η_i define the environment according to circumstances also external to the model.

The Gadgil-Bossert model has produced serveral general results that are important in sociobiology. As illustrated in Figure 4-17, if the profit function of the reproductive effort is convex, or if the cost function is concave, the optimal strategy is probably to breed repeatedly (the condition called iteroparity). Otherwise, the optimal strategy is to breed in one suicidal burst (semelparity). The latter method, referred to by Gadgil and Bossert as "big bang" reproduction, is the kind found in migratory salmon, which spawn at the end of their long journey from the sea and then die, and bamboos, corypha palms, and century plants, which bloom in one massive burst at the end of their lives. For a given reproductive effort θ_j made at any age j, there is a profit to be measured in the number of offspring produced. There is also a cost to be measured in the lowered survival probability at age j and subsequent ages. The cost consists of the investment in energy and time, together with the reduced reproductive potential

at later ages, due to the slowed growth in turn caused by the effort θ_j. How would a profit function form a concave curve and thus favor semelparity? If a female salmon laid only one or two eggs, the reproductive effort, consisting principally of the long swim upstream, would be very high. To lay hundreds more eggs entails only a small amount of additional reproductive effort. For the opposite case, namely a convex profit curve favoring iteroparity, consider reproduction by a nidicolous bird. To produce a brood of several nestlings, the bird must expend a great deal of reproductive effort. To go beyond a normal brood size requires additional reproductive effort, and the pay-off in living young remains the same or is even lowered, because the parent birds cannot care for excess young.

A second result of the Gadgil-Bossert formulation, anticipated by

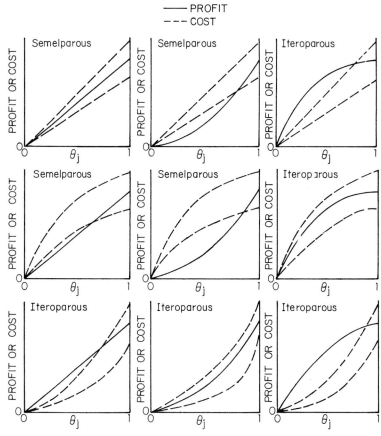

Figure 4-17 The two strategies of reproduction. Iteroparity (repeated reproduction) is the optimal strategy when either the profit function is convex or the cost function is concave. In other situations the optimal strategy is semelparity, or a single reproductive effort before death. (Modified from Gadgil and Bossert, 1970.)

Williams (1966a,b), is that the value of reproductive effort in itero-parous species should increase steadily with age. Consequently, an optimal strategy is one in which the amount of reproductive effort is stepped up with age. Fagen (1972), using the same model, found that the result depends on monotonicity of the functions (f_1, f_2, f_3) relating reproductive effort (θ) to survivorship, growth, and number of offspring. Suppose that α_i is not monotonic, that is, the survival rate in a nonlimiting environment does not move steadily up or down. If it starts high, drops, then rises again, the optimal reproductive effort can be bimodal through time, rising, dropping, and then rising again. Such an oscillation might occur if individuals were protected when young, then put in jeopardy when forced to become independent, only to find security again when attaining a territory or dominance in a social hierarchy. Conversely, special schedules of parameters can be arranged in which the optimal pattern of growth is to slow down at middle age, then to increase the growth rate once more. Such a sequence has actually been recorded in male elephants, certain seals, and toothed whales.

J. M. Emlen (1970), following on W. D. Hamilton's analysis of senescence (1966), used the Euler-Lotka equation to explore the effects of changes in the environment on the evolution of the survivorship and fertility schedules—those disasters or strokes of good fortune that alter conditions for certain of the age groups. How would the optimal schedules be changed, for example, if a new predator entered the range of the species and proved especially destructive to infants? To make such estimates, Emlen introduced measures of *selection intensity*, $I_q'(x)$ and $I_m'(x)$, for age-specific mortality and fertility, respectively. The selection intensity for age-specific mortality is defined as

$$I_q'(x) = \left| \frac{1}{R} \frac{\partial R}{\partial q_x} \right|$$

where R is N_t/N_{t-1}, the proportional change in the population size during the time interval $t - 1$ to t, and q_x is the mortality that occurs from age $x - 1$ to x. The selection intensity for age-specific mortality, then, is the degree to which a change in mortality at any age (x) causes a change in the overall growth of the population. We would expect genes with high selection intensities to change in frequency more rapidly than genes with low selection intensities. The greater intensities of certain genes could be due to the fact that they cause greater mortality, or act at a time when the reproductive value is higher, or both. Examples of $I_q'(x)$ curves are given in Figure 4-18A. In general, age-specific mortality in optimal life cycles should be high at or near conception, fall to a minimum during later prereproductive life, and then, after the age of first reproduction, rise steadily with age. The reasons for this inference are:

1. The $I_q'(x)$ curve descends monotonically throughout life; thus selection against mortality factors, including senescence, steadily weakens.

2. However, the mortality near birth is likely to be "precessive," moved by natural selection back toward the zygote, to minimize loss of parental investment. This effect will be enhanced in cases of prolonged parental care with heavy investment in a few offspring. The same result was obtained by Hamilton (1966).

3. Improvements in fitness measured by lowered mortality, insofar as they can be programmed by the fixation of modifier genes, will tend to be moved forward in prereproductive life to fall as close as possible to the onset of reproductive maturity, where they will have the greatest impact, that is, maximize $\partial R/\partial q(x)$. In fact, mortality curves do show the expected form where the requisite data exist, in barnacles, daphnia, fish, and birds. Human populations also conform, as illustrated in Figure 4-18B.

The selection intensity curve for fertility is defined in a parallel fashion as

$$I_m'(x) = \left| \frac{1}{R} \frac{\partial R}{\partial m_x} \right|$$

and its expected generalized form is represented in Figure 4-18C. Because the values of $I_m'(x)$ decrease monotonically with age, natural selection should act to move traits increasing fertility to an earlier and earlier age—until stopped by opposing selective forces. What these forces are is an interesting point for conjecture, because many of them certainly involve social development. Competing males, for example, need physical size to gain dominance, while social vertebrates of all kind need developmental time to learn their environment and to form bonds with other members of the group.

Emlen's model predicts that an increase in mortality at a certain age will, if sustained, encourage natural selection to raise relative mortality at the ages immediately preceding and following the afflicted age. This result accords with an earlier intuitive conjecture by L. B. Slobodkin that "the causes of mortality attract each other." The new mortality would also favor lowered fecundity immediately following the afflicted age. A sustained increase in fertility at a given age, occasioned, say, by improved nutritional status, will result in natural selection inducing higher mortality in early life as well as in the period immediately following the favored age. There would also be a tendency to reduce fertility in middle and late life. An essentially similar relation between enhanced fertility, shortened reproductive maturation time, and shortened longevity was deduced by Lewontin (1965) through a different modeling effort.

Longevity and low fertility are compensatory traits favored by natural selection under either one of two opposite environmental condi-

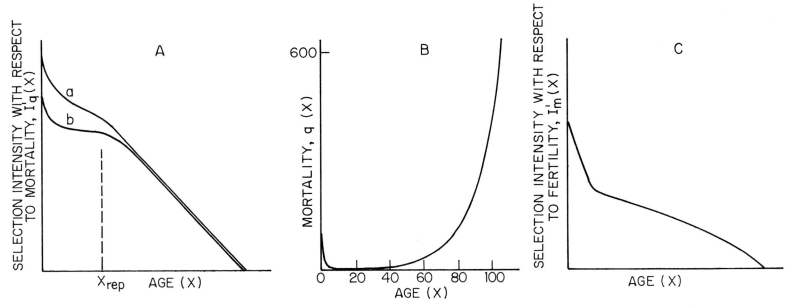

Figure 4-18 Age-specific selection intensity and mortality. *A*: The general predicted form of the selection-intensity curve with respect to mortality. A higher curve (*a*) is expected in species with strong parental care, as opposed to those (*b*) with little or none. The age of onset of reproduction is labeled x_{rep}. *B*: Mortality in man as a function of age, a curve of the kind expected from the generalized selection intensity curve for mortality. *C*: The general predicted form of the selection intensity curve with respect to fertility. (Based on J. M. Emlen, 1970.)

tions. If the environment is very stable and predictable, survivorship and hence longevity are improved for species that can appropriate part of the habitat, key their activities to its rhythms, or otherwise take advantage of the stability. Such organisms will not find it a good strategy to seed their homes with large numbers of offspring, who become potential competitors. At the other extreme, a harsh, unpredictable environment will cause some (but not all) species to evolve a tough, durable mature stage that utilizes its energies more successfully for survival than in reproductive effort. It can be shown that the best strategy for such organisms is to engage in highly irregular reproduction keyed to the occasional good times (Holgate, 1967). Longevity is further improved when the survival of progeny is not only low but unpredictable in time (Murphy, 1968).

Investigations of the evolution of life histories, together with the applications to sociobiology, include those by Cole (1954) and Anderson and King (1970) on general theory, Wilson (1966, 1971a) on applications to social insects, Istock (1967) on complex life cycles, and King and Anderson (1971) on the effects of population fluctuation.

r and *K* Selection

The demographic parameters *r* and *K* are determined ultimately by the genetic composition of the population. As a consequence they are subject to evolution, in ways that have only recently begun to be carefully examined by biologists. Suppose that a species is adapted for life in a short-lived, unpredictable habitat, such as the weedy cover of new clearings in forests, the mud surfaces of new river bars, or the bottoms of nutrient-rich rain pools. Such a species will succeed best if it can do three things well: (1) discover the habitat quickly, (2) reproduce rapidly to use up the resources before other, competing species exploit the habitat, or the habitat disappears altogether, and (3) disperse in search of other new habitats as the existing one becomes inhospitable. Such a species, relying upon a high *r* to make use of a fluctuating environment and ephemeral resources, is known as an "*r* strategist," or "opportunistic species" (MacArthur and Wilson, 1967). One extreme case of an *r* strategist is the fugitive species, which is consistently wiped out of the places it colonizes, and survives only by its ability to disperse and fill new places at a high rate (Hutchinson, 1951). The *r* strategy is to make full use of habitats that, because of their temporary nature, keep many of the populations at any given moment on the lower, ascending parts of the growth curve. Under such extreme circumstances, genotypes in the population with high r will be consistently favored (see Figure 4-19). Less advantage will accrue to genotypes that substitute an ability to compete in crowded circumstances (when $N = K$ or close to it) for the precious high r. The process is referred to as r selection.

A "K strategist," or "stable species," characteristically lives in a longer-lived habitat—an old climax forest, for example, a cave wall, or the interior of a coral reef. Its populations, and those of the species with which it interacts, are consequently at or near their saturation level K. No longer is it very advantageous for a species to have a high r. It is more important for genotypes to confer competitive ability, in particular the capacity to seize and to hold a piece of the environment and to extract the energy produced by it. In higher plants this K selection may result in larger individuals, such as shrubs or trees,

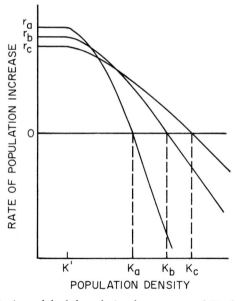

Figure 4-19 A model of the relation between r and K selection. Three genotypes (a, b, c) are envisaged as competing in natural selection. At low population levels, say below a critical value K', the populations grow at their unaltered intrinsic rates of increase (r_a, r_b, r_c). At equilibrium, when the growth rates are zero, each population is by definition at its carrying capacity (K_a, K_b, K_c). If the increase rate curves cross, as in this example, genotype a will prevail when the environments fluctuate enough to keep populations constantly growing (r selection); but genotype c will win in environments stable enough to permit the populations to remain at or near equilibration (K selection). (Modified from Gadgil and Bossert, 1970.)

with a capacity to crowd out the root systems of and to deny sunlight to other plants that germinate close by. In animals K selection could result in increased specialization (to avoid interference with competitors) or an increased tendency to stake out and to defend territories against members of the same species. All else being equal, those genotypes of K strategists will be favored that are able to maintain the densest populations at equilibrium. Genotypes less able to survive and to reproduce under these long-term conditions of crowding will be eliminated. The classical theorems of natural selection were mostly constructed with r selection implicitly in mind. It was MacArthur (1962) who first devised parallel theorems with explicit reference to K selection.

Of course the two forms of selection cannot be mutually exclusive. As suggested in the scheme of Figure 4-19, r is subject in all cases to at least some evolutionary modification, upward or downward, while few species are so consistently prevented from approaching K that they are not subject to some degree of K selection. King and Anderson (1971) and Roughgarden (1971) have, in fact, independently defined sets of conditions in which competing r and K alleles can coexist in balanced polymorphism. But in many instances where extreme K selection occurs, resulting in a stable population of long-lived individuals, the result must be an evolutionary decrease in r. For a genotype or a species that lives in a stable habitat, there is no Darwinian advantage to making a heavy commitment to reproduction if the effort reduces the chance of individual survival. At the opposite extreme, it does pay to make a heavy reproductive effort, even at the cost of life, if the temporary availability of empty habitats guarantees that at least a few of one's offspring will find the resources they need in order to survive and to reproduce. Most of the r strategists' offspring will perish during the dispersal phase, but a few are likely to find an empty habitat in which to renew the life cycle.

The degree of fluctuation of a population is not all that determines the fate of the r and K genes. The pattern of change itself can make a crucial difference (Mertz, 1971a,b). If a population fluctuates in a way that permits it to increase most of the time, as suggested in Figure 4-20, it will tend to evolve as an r selectionist in the usual manner. But if it fluctuates in a way that causes it to decline most of the time, genes will be favored that defer reproduction, maximize longevity, and slow the rate of decrease. An example of a chronically decreasing population may well be the California condor (*Gymnogyps californianus*), which has gradually retreated from its range of 10,000 years ago, extending from Florida to Mexico, to its present tiny refuge in central California. The condor is one of the longest lived and slowest breeding of all birds. Whether these demographic traits evolved in response to the retreat, or caused it, cannot be established with our present knowledge.

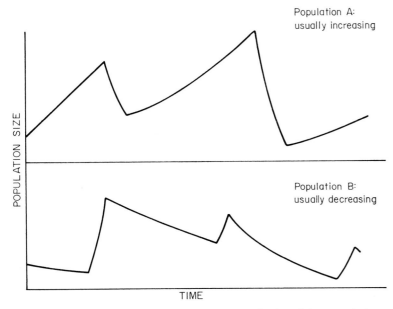

POPULATION SIZE

Population A:
usually increasing

Population B:
usually decreasing

TIME

Figure 4-20 Two opposite growth patterns displayed by populations with equal degrees of fluctuation. The resulting demographic evolution of population *A* is expected to differ in many details from that of population *B*. (From Mertz, 1971a.)

The expected correlates of *r* and *K* selection in ecology and behavior are numerous and complex (Table 4-2 and Figure 4-21). In general, higher forms of social evolution should be favored by *K* selection. The reason is that population stability tends to reduce gene flow and thus to increase inbreeding, while at the same time promoting land tenure and the multifarious social bonds that require longer life in more predictable environments.

The rodents are one of many groups of animals containing both *r*-selected and *K*-selected species. Judging from the account by Christian (1970), *Microtus pennsylvanicus* stands at the *r* extreme of the spectrum. In pre-Columbian times this abundant vole species may have been restricted to temporary wet grasslands, such as "beaver meadows" created by the abandonment of beaver dams. These temporary meadows give way rapidly to seral stages of reforestation, so that species dependent on them must adopt a strategy of rapid population growth and efficient dispersal. *M. pennsylvanicus* goes through marked population fluctuations that produce large numbers of "floaters," nonterritorial animals that emigrate long distances. Christian observed the invasion of one beaver meadow by these voles in less than a week after its creation, during a year when the *M. pennsylvanicus* population was very high. The voles had to cross inhospitable forest tracts to reach the newly opened habitat. The *r*

Table 4-2 Some of the correlates of *r* selection and *K* selection (modified from Pianka, 1970).

Correlate	*r* selection	*K* selection
Climate	Variable and/or unpredictable: uncertain	Fairly constant and/or predictable: more certain
Mortality	Often catastrophic, non-directed, density-independent	More directed, density-dependent
Survivorship	Often type III	Usually types I and II
Population size	Variable in time, non-equilibrium; usually well below carrying capacity of environment; unsaturated communities or portions thereof; ecological vacuums; recolonization each year	Fairly constant in time, equilibrium; at or near carrying capacity of the environment; saturated communities; no recolonization necessary
Intraspecific and interspecific competition	Variable, often lax	Usually keen
Relative abundance	Often does not fit MacArthur's broken-stick model (C. E. King, 1964)	Frequently fits the MacArthur broken-stick model (C. E. King, 1964)
Attributes favored by selection	1. Rapid development 2. High r_{max} 2. Early reproduction 4. Small body size 5. Semelparity: single reproduction	1. Slower development, greater competitive ability 2. Lower resource thresholds 3. Delayed reproduction 4. Larger body size 5. Iteroparity: repeated reproductions
Length of life	Short, usually less than 1 year	Longer, usually more than 1 year
Emphasis in energy utilization	Productivity	Efficiency
Colonizing ability	Large	Small
Social behavior	Weak, mostly schools, herds, aggregations	Frequently well developed

strategy preadapted *M. pennsylvanicus* to life in the rapidly changing, meadowlike environments of agricultural land, where today it is a dominant species over a large part of North America. Other North American microtine rodents, particularly the deer mice of the genus *Peromyscus*, are closer to the *K* end of the scale. They originally inhabited the continuous habitats of North America, particularly the

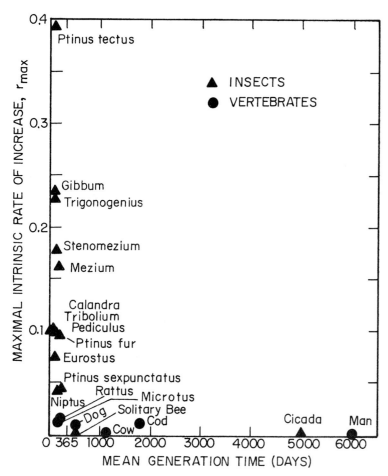

Figure 4-21 The threshold between *r* and *K* selection coincides in many groups of organisms with an increase in generation time from annual to perennial. Annual insects, with high r_{max}, tend to show the expected traits of *r* selectionists, but 13-year cicadas and social insects, including the honeybee, are more stable. Similarly, many rodents with less than annual breeding time, such as *Microtus* and *Rattus*, are among the vertebrate *r* selectionists. (From Pianka, 1970.)

eastern deciduous forests and the central plains. Their populations are more stable, and they seldom irrupt in the spectacular fashion of the voles and lemmings. The beaver (*Castor canadensis*) is close to what we can designate as a true *K* selectionist. To a large extent this mammal designs and stabilizes its own habitats with the dams and ponds it creates. Protected from predators by its large size and secure aquatic lodges, and provided with a rich food source, the beaver has both low mortality and low birth rates. The young disperse away from the parental lodges only after a couple of years of resi-

dence. As a result, beaver populations are much more stable than those of microtine rodents.

There has begun to emerge from the rodent studies a principle that may be applicable in other groups of animals as well: *the social tolerance of a species has evolved to fit the optimal population density and optimal population structure.* This generalization, first explicitly stated by Lidicker (1965) and Eisenberg (1967) and later developed independently from a different point of view by Christian (1970), can be broken down into three specifications. First, the lower the equilibrium density of the species in nature, the sooner its members begin to show some form of density-dependent social response, such as territoriality and emigration. Second, the thresholds of such responses are higher in opportunistic (*r*-selected) species than in more stable (*K*-selected) ones. Third, the thresholds for various social responses are highest *within* societies of the most social species, although the tolerance *between* such groups may be low in accordance with the first two relations, which pertain to populations as a whole rather than to societies. The Lidicker-Eisenberg principle has been documented by artificially increasing densities of rodents in laboratory enclosures to observe the onset of social responses, both the normal ones likely to serve in the density control of free-living populations and the pathological behavior that ensues when the ordinary density thresholds are transgressed.

Opportunistic as opposed to stable population strategies are expressed in diverse ways in other kinds of social organisms. In ectoproct bryozoans the three most common types of colony growth are (1) linear, in which the colony advances like a vine; (2) encrusting, in which the colony spreads over the surface in the manner of a lichen; and (3) three-dimensional, in which the colony grows in all directions like a miniature bush. Geometric models developed by K. W. Kaufmann (1970) show that linear forms produce the most larvae over a short period of time and are therefore best adapted for microhabitats with a short life. They are the *r* strategists. Three-dimensional forms have the greatest productivity over long spans of time, and presumably they also have some advantages in the crowding competition that determines the composition of stable communities of fouling organisms. Hence they are probably the *K* strategists. The encrusting ectoprocts occupy an intermediate position.

Among the birds, and particularly among the seabirds and certain other groups including the large birds of prey and carrion feeders, it is possible to discern ascending grades of *K*-selected demographic traits and population stability (Amadon, 1964; Ashmole, 1963). In competition for nest sites the tricolored blackbird (*Agelaius tricolor*) prevails over the closely related red-winged blackbird (*A. phoeniceus*) by virtue of its tolerance for much higher population densities. The tri-coloreds occupy smaller territories and are highly colonial, forming concentrations of up to 100,000 nests (Orians, 1961). One of the ultimate *K* selectionists must be the smaller adjutant stork (*Leptotilos javanicus*). According to Baker (1929), one colony had been known by the hill tribes af Assam since the beginnings of their surviving traditions. In 1885 the population was in virgin rain forest and consisted of 15 nests. By 1929 the forest had been cleared, and the colony was surrounded by cultivated land, but it still consisted of exactly 15 nests. Great stability seems to be a characteristic of many colonial bird species. The winter roosts of common crows (*Corvus brachyrhynchos*) in New York State and California have persisted for as long as 50 years despite radical changes in the surrounding vegetation (J. T. Emlen, 1938, 1940). The grounds on which male sharp-tailed grouse (*Pedioecetes phasianellus*) display to females have persisted since beyond the tribal memories of local Indians (Armstrong, 1947). Gannets (*Sula bassana*) have bred continuously on Bass Rock, in Scotland's Firth of Forth, since as far back as the fifteenth century, while a colony of grey herons (*Ardea cinerea*) has persisted on the castle park grounds at Chilham in Kent, England, from at least the thirteenth century (Gurney, 1913; Nicholson, 1929). These facts are of potentially great importance to the theory of the evolution of altruistic population control, since they indicate that in many of the more social species the rates of population extinction are far too low to generate the intensity of interpopulation selection necessary to favor genes that are altruistic with reference to the populations as a whole (see Chapter 5).

An unusual and interesting case of convergent *K* adaptation is to be found among the several independently evolved groups of mammalian anteaters. These animals occur in low densities, but they enjoy a relatively stable, evenly dispersed food source in the ant and termite colonies on which they have specialized. Aardvarks (*Orycteropus afer*, order Tubulidentata), scaly anteaters (*Manis* spp., order Pholidota), and great anteaters (*Myrmecophaga jubata*, order Edentata) are known for their solitary habits, low reproductive rates, persistent attachment of the young to the mother, and lack of aggressive behavior. It is likely that the same traits are shared by the lesser known aardwolf (*Proteles cristatus*, a hyaenid), the sloth bear (*Melursus ursinus*, a true bear), and the numbat (*Myrmecobius fasciatus*, a marsupial), all of which feed primarily on termites.

In contrast, the ultimate *r* selectionists are probably found among the arthropods. Many mite species, for example, are highly fugitive in their strategy. They depend on the discovery of bonanzas, such as large pieces of decaying food or large but short-lived insects that can be parasitized. As Mitchell (1970) has stressed in his recent analysis, the key to success for these organisms is the maximal dispersal of inseminated females. The enabling mechanisms include dispersal at a very young stage, reduction of the male/female ratio to maximize the absolute number of females, and decrease in the biomass of the dispersers to permit them to travel the greatest possible

distances as aerial plankton and as "hitch-hikers" on other organisms. There is also a tendency for the females to mate before dispersing, with the result that a single individual can found an entire population.

The Evolution of Gene Flow

The distance which organisms move from their place of birth is a constraining force in evolution. Slight movement results in a small effective population size, greater inbreeding, and a steady loss of genetic variability. A great deal of movement results in the genetic swamping of local adaptation and the rupture of social bonds. The fine details of this gene flow also have repercussions. A tendency for genotypes to migrate at different rates can result in geographic variation and balanced genetic polymorphism within species. A tendency for different sexes and age groups to migrate differentially can exert a profound influence on social structure.

Emigration is often strongly biased with respect to sex and age. The evidence also shows that young adults generally travel the farthest (Figure 4-22). These data are consistent with the theoretical inference drawn earlier that organisms evolve so as to travel at the time of their maximum reproductive value. Programmed dispersal is particularly stereotyped in insects (Johnson, 1969; Dingle, 1972a). It occurs not through local exploratory movements but through real migrations, during which insects travel in a hard, persistent manner and cannot easily be distracted by the stimuli that in other circumstances govern their lives. The process is highly adaptive, having evolved in response to the shortness of the life cycles of insects and the usually transitory

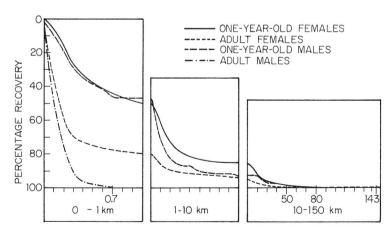

Figure 4-22 Age and sex differences in the dispersal of pied flycatchers (*Ficedula hypoleuca*) in Germany. The vertical axis consists of the cumulative recoveries of 4000 banded birds. (Redrawn from Berndt and Sternberg, 1969.)

nature of their breeding sites. As a rule, the intensity of the programmed migratory activity of an individual species is inversely
related to the stability of its preferred habitat. Migratory flight in
particular is the prime locomotory act of many if not most winged
insects. The flights follow patterns tailored to the individual needs
of the species. The members of some species, such as the plague
locusts and the migratory white butterfly *Ascia monuste*, conduct
lengthy powered flights in a single direction. A majority, however,
use their wings to work their way up into the wind and to maintain
themselves there while being carried along. The migration periods
are tightly scheduled. Flights are usually conducted by young adults,
especially females, who reduce ovarian development at the period of
maximum likelihood of flight. The migration of an insect is usually
triggered by token stimuli that herald the approach of favorable flight
conditions or otherwise inform the insect that its physiological state
is conducive to flight. The rigidity of many of the systems is exemplified by the bizarre case of the bark beetle *Trypodendron lineatum*.
When adults first leave their burrows they are positively phototactic
and attempt to fly. During the flight they swallow air until a bubble
forms in the proventriculus, the "gizzard" located at the rear of the
fore gut; the bubble causes them to revert to negative phototaxis and
settling. If the experimenter inflates the proventriculus of a previously
flying beetle, it will cease flight; but if he punctures the bubble it
starts flying again.

While vertebrates are not quite so mechanical as insects in their
responses, their dispersal patterns are still highly predictable. The
dispersants of rodent populations, from mice to beavers, are almost
invariably young adults, and their movements are precipitated by
aggressive interactions with the more secure, generally older territorial
residents. Sadleir (1965) proposed, and Healey (1967) proved experimentally, that adult aggressiveness of deer mice (*Peromyscus maniculatus*) peaks in the breeding season, at which time the juveniles are
maximally excluded, disperse the farthest, and suffer their highest
mortality. The rule of juvenile mobility is not invariable, however.
In the most social of all rodents, the black-tail prairie dog, it is the
adults who initiate new burrow systems and thus extend the limits
of the communities.

In semiclosed mammalian societies, such as baboon troops and lion
prides, the young males are the main dispersants. The pattern of gene
flow in these cases is quite consistent: the young animal leaves the
parental society, enters a nomadic period alone or with other members of the same sex, and finally joins a new group. In open societies
as well as otherwise nonsocial territorial systems, there appears to be
no overall strong sex bias in dispersal. In some species emigration
is undertaken principally by the males, in others by the females, and
in still others by both sexes equally.

Underlying the evolution of gene flow is the process of *migrant
selection*, the differential fitness of genotypes caused by variation in
their tendency to emigrate. A genotype more prone to move might
perish sooner, but in taking the gamble it has two potential payoffs.
First, it is more likely to colonize empty habitats and, as we noted
with respect to r selection, this advantage becomes overriding if the
preferred habitat of the species is very transient in nature. Second,
there may exist the "minority effect" discovered in *Drosophila* and
probably existing in at least some other animals. As males become
rarer relative to males of other genotypes, their mating success increases. Thus, immigrants arriving in a population genetically different from their own enjoy an initial advantage. Migrant selection can
parallel individual and group selection, in which case the three basic
forms of selection simply reinforce one another. Migrant genotypes
may, however, find themselves at a disadvantage in competition with
nonmigrant genotypes within established populations, or their presence may increase the probability of extinction of populations as a
whole. Under these conditions the selection pressures are counteracting, and a state of genetic polymorphism is likely to arise within
the species (Maynard Smith, 1964; Levins, 1970; Van Valen, 1971).
Migrant selection has been documented among the transferrin and
leucine aminopeptidase polymorphs of the voles *Microtus ochrogaster* and *M. pennsylvanicus* (Myers and Krebs, 1971); and its existence has been inferred in laboratory studies of house mice and
Drosophila (Thiessen, 1964; Narise, 1968), as well as in free populations of the butterfly *Euphydryas editha* (Gilbert and Singer, 1973).
In both the *Peromyscus* and *Drosophila* studies, polymorphism was
maintained by counteracting individual and migrant selection.

The properties of dispersal curves and the probabilities of successful
colonization under various environmental conditions have been formally investigated by MacArthur and Wilson (1967) and MacArthur
(1972). A significant distinction can be made between the exponential and normal decline of dispersing organisms through space. An
exponential distribution will result if the propagule moves in a constant direction with a constant probability that it will cease moving.
Such might be the case for passive terrestrial propagules carried over
the sea in a steady wind or in a steadily moving cyclone until, one
by one, the propagules hit the water. The number of propagules still
in motion after traveling a distance d would be $e^{-d/\Lambda}$, where Λ is
the mean dispersal distance for all the propagules. Exponential dispersal might prove to be common if not universal in plants and insects
that disseminate propagules passively through the air. A normal distribution, in contrast, can be expected when the animals move on
a randomly changing course, searching over the ground or in the air
without long-range orientation. It might also result from travel on a
sea-going "raft," such as a floating log, which has a normally distrib-

uted persistence time, or from movement along a set course for a period of time that is normally distributed for physiological reasons. The fraction of individuals still in motion at distance x falls off at the rate of e^{-x^2} rather than e^{-x}, the term in exponential dispersal. More precisely, a fraction

$$\sqrt{\frac{2}{\pi}} \frac{e^{-d^2/2}}{d}$$

reaches each distance d or beyond. These two types of curves can be expected to generate strong differences in the patterns of gene flow and colonization.

Analysis of the adaptive value of dispersal has been confused by disagreement over the level at which selection operates. W. L. Brown (1958) and Howard (1960) were thinking at least in part of group selection when they postulated the roles of dispersal to be reduction of inbreeding, extension of the range of the species, spread of new genes, and reinvasion of disturbed areas. Brown further hypothesized that population fluctuations speed these processes by serving as a kind of motor that drives general adaptation through entire species. Such "functions," if they exist as first-order Darwinian adaptations, would in many circumstances subordinate the welfare of the individual to that of the population. This explicit view was adopted by Wynne-Edwards (1962), who interpreted emigration to be one of the altruistic conventions used in the regulation of population density. Levins (1965) and Leigh (1971) have gone so far as to calculate the optimum rate of gene flow into a population in terms of its costs and benefits to the population as a whole. Leigh's reasoning is as follows. Suppose that in a changing environment one allele is substituted for another on the average of every n generations. Each substitution will reduce the population size by the fraction $(1/n) \log (s/u)$, where s is the selection coefficient and u is the proportion of the population that consisted, *prior* to the time the environment changed and the genotype gained the upper hand, of newly immigrated individuals belonging to the genotype. What is the generation-by-generation immigration rate (u) that will result in the least amount of loss to the population as a whole if the environment changes every n generations? Leigh showed that this optimum level is $u = 1/n$. If the effect exists in nature, we would expect a species living in a strongly fluctuating environment (high $1/n$) to adjust its rate and distance of dispersal upward.

It is indeed tempting to view species as homeostatic devices tinkering with their own population parameters, such as the dispersal and mutation rates. But there is an alternative hypothesis, developed by Lidicker (1962), Murray (1967), Johnson (1969), Gilbert and Singer (1973), and others, and formalized in mathematical models by D. Cohen (1967) and Gadgil (1971). It holds that dispersal behavior is shaped by natural selection at the individual level. Emigration is programmed in such a way as to take an individual from one locality when the odds favor (however slightly) that greater success will come from the attempt to settle in another locality. The population consequences of emigration are viewed as second-order effects. The reader will recognize that the evolution of dispersal is one more subject, like altruism and territorial behavior, in which the choice between hypotheses must turn on a precise assessment of the intensity of group selection. We are at last ready for a full review of this important but complex subject, which will be provided in the next chapter.

Chapter 5 **Group Selection and Altruism**

Reporter: *When you ran Finland onto the map of the world, did you feel you were doing it to bring fame to a nation unknown by others?*

Nurmi: *No. I ran for myself, not for Finland.*

Reporter: *Not even in the Olympics?*

Nurmi: *Not even then. Above all, not then. At the Olympics, Paavo Nurmi mattered more than ever.*

Who does not feel at least a tinge of admiration for Paavo Nurmi, the ultimate individual selectionist? At the opposite extreme, we shared a different form of approval, warmer in tone but uneasily loose in texture, for the Apollo 11 astronauts who left their message on the moon, "We came in peace for all mankind." This chapter is about natural selection at the levels of selection in between the individual and the species. Its pivot will be the question of altruism, the surrender of personal genetic fitness for the enhancement of personal genetic fitness in others.

Group Selection

Selection can be said to operate at the group level, and deserves to be called group selection, when it affects two or more members of a lineage group as a unit. Just above the level of the individual we can delimit various of these lineage groups: a set of sibs, parents, and their offspring; a close-knit tribe of families related by at least the degree of third cousin; and so on. If selection operates on any of the groups as a unit, or operates on an individual in any way that affects the frequency of genes shared by common descent in relatives, the process is referred to as kin selection. At a higher level, an entire breeding population may be the unit, so that populations (that is, demes) possessing different genotypes are extinguished differentially, or disseminate different numbers of colonists, in which case we speak of interdemic (or interpopulation) selection. The ascending levels of selection are visualized in Figure 5-1. The concept of group selection was introduced by Darwin in *The Origin of Species* to account for the evolution of sterile castes in social insects. The term intergroup selection, in the sense of interpopulation selection defined here, was used by Sewall Wright in 1945. Essentially the same expression (*Gruppenauslese*) was used independently and with the same meaning by Olavi Kalela (1954, 1957), while the phrase kin selection was coined by J. Maynard Smith (1964). The classification adopted here is approximately that recommended by J. L. Brown (1966). Selection can also operate at the level of species or entire clusters of related species. The process, well known to paleontologists and biogeographers, is responsible for the familiar patterns of dynastic succession of major groups such as ammonites, sharks, graptolites, and dinosaurs through geologic time (Simpson, 1953; P. J. Darlington, 1971). It is even possible to conceive of the differential extinction of entire ecosystems, involving all trophic levels (Dunbar, 1960, 1972). However, selection at these highest levels is not likely to be important in the evolution of altruism, for the following simple reason. In order to counteract individual selection, it is necessary to have population extinction rates of comparable magnitude. New species are not created at a sufficiently fast pace to be tested in this manner, at least not when the species are so genetically divergent as those ordinarily

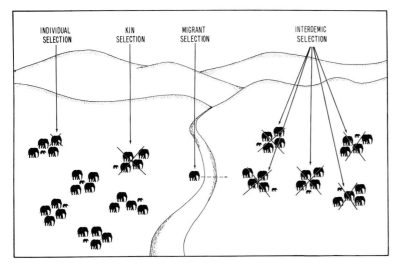

Figure 5-1 Ascending levels of selection. Group selection consists of either kin selection, in which the unit is a set of related individuals, or interdemic selection (also called interpopulation selection), in which entire populations are diminished or extinguished at different rates. The differential tendency to disperse is referred to as migrant selection.

studied by the biogeographers. The same restriction applies a fortiori to ecosystems.

Pure kin and pure interdemic selection are the two poles at the ends of a gradient of selection on ever enlarging nested sets of related individuals. They are sufficiently different to require different forms of mathematical models, and their outcomes are qualitatively different. Depending on the behavior of the individual organisms and their rate of dispersal between societies, the transition zone between kin selection and interdemic selection for most species probably occurs when the group is large enough to contain somewhere on the order of 10 to 100 individuals. At that range one reaches the upper limit of family size and passes to groups of families. One also finds the upper bound in the number of group members one animal can remember and with whom it can therefore establish personal bonds. Finally, 10 to 100 is the range in which the effective population numbers (N_e) of a great many vertebrate species fall. Thus, aggregations of more than 100 are genetically fragmented, and the geometry of their distribution is important to their microevolution.

Interdemic (Interpopulation) Selection

A cluster of populations belonging to the same species may be called a metapopulation. The metapopulation is most fruitfully conceived as an amebalike entity spread over a fixed number of patches (Levins,

1970). At any moment of time a given patch may contain a population or not; empty patches are occasionally colonized by immigrants that form new populations, while old populations occasionally become extinct, leaving an empty patch. If $P(t)$ is the proportion of patches which support populations at time t, m is the proportion receiving migrants in an instant of time (whether already occupied or not), and $\overline{E}$ is the proportion of populations becoming extinct in an instant of time,

$$\frac{dP}{dt} = mg(P) - \overline{E}P$$

The function $g(P)$ must decrease with the proportion of sites already occupied, a relation that can exist in the simple logistic form

$$\frac{dP}{dt} = mP(1 - P) - \overline{E}P$$

At equilibrium the proportion of occupied patches is

$$P = 1 - \frac{\overline{E}}{m}$$

where the metapopulation as a whole can persist only if $\overline{E} < m$. Thus the system is metaphorically viewed through evolutionary time as a nexus of patches, each patch winking into life as a population colonizes it and winking out again as extinction occurs. At equilibrium the rate of winking and the number of occupied sites are constant, despite the fact that the pattern of occupancy is constantly shifting. The imagery can be translated into reality only when the observer is able to delimit real Mendelian populations in the system. The complications that arise from this problem are illustrated in Figure 5-2.

In considering interdemic selection, it is important to distinguish the timing of the extinction event in the history of the population (Figure 5-3). There are two moments at which extinction is most likely: at the very beginning, when the colonists are struggling to establish a hold on the site, and soon after the population has reached (or exceeded) the carrying capacity of the site, and is in most danger of crashing from starvation or destruction of the habitat. The former event can be called r extinction and the latter K extinction, in appreciation of the close parallel this dichotomy makes with r and K selection. When populations are more subject to r extinction, altruist traits favored by group selection are likely to be of the "pioneer" variety. They will lead to clustering of the little population, mutual defense against enemies, and cooperative foraging and nest building. The ruling principle will be the maximum *average* survival and fertility of the group as a whole; in other words, the maximization of r.

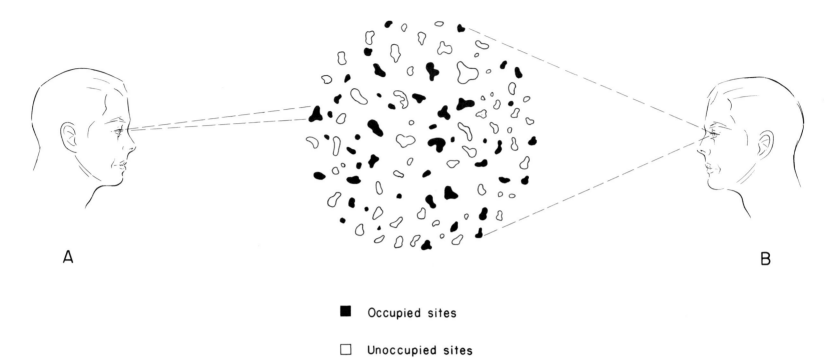

A   B

■   Occupied sites

□   Unoccupied sites

Figure 5-2   The metapopulation is a set of populations occupying a cluster of habitable sites. Because of constantly recurring extinction not totally canceled by new immigration, some percentage of the sites are always unfilled, although different ones are empty at different times.

Observer A precisely distinguishes each population and can correctly estimate extinction and immigration rates. Observer B incorrectly sees the entire metapopulation as one population and will underestimate the extinction and immigration rates.

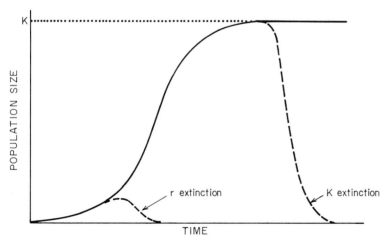

Figure 5-3   Extinction of a population probably most commonly occurs at an early stage of its growth, particularly when the first colonists are trying to establish a foothold (r extinction), or after the capacity of the environment has been reached or exceeded and a crash occurs (K extinction). The consequences in evolution are potentially radically different. (From Wilson, 1973.)

In K extinction the opposite is true. The premium is now on "urban qualities" that keep population size below dangerous levels. Extreme pressure from density-dependent controls of an external nature is avoided. Mutual aid is minimized, and personal restraint in the forms of underutilization of the habitat and birth control comes to the fore.

These two levels of extinction can be distinguished in the populations of the aphid *Pterocomma populifoliae* as described by Sanders and Knight (1968). The species is highly opportunistic, colonizing sucker stands of bigtooth aspen and multiplying rapidly to create small, isolated populations. Extinction rates are very high. The earliest colonies, composed of first-generation colonists, are wiped out by errant predators, including spiders and adult ladybird beetles. Older, more established colonies acquire resident predators, such as syrphid and chamaemyid flies and some ladybird beetles, who breed along with them. These predators, aided by the emigration of many of the surviving aphids themselves, often eliminate entire colonies.

Very young, growing populations are likely to consist of individuals who are closely related. Interdemic selection by r extinction is therefore intrinsically difficult to separate from kin selection, and in extreme cases it is probably identical with it. A second feature that makes the process difficult to analyze is the change in gene frequencies due to genetic drift. In populations of ten or so individuals drift can completely swamp out the overall effect of differential extinction within the metapopulation. For these reasons analysis has been concentrated on larger populations, and the most general results obtained are more easily applicable to interdemic selection by *K* extinction.

Our current understanding of counteracting interdemic selection can be most clearly understood if approached through its historical development. In 1932 Haldane constructed a few elements of a general theory that are equally applicable to kin and interdemic selection. He thought he could dimly see how altruistic traits increase in populations. "A study of these traits involves the consideration of small groups. For a character of this type can only spread through the population if the genes determining it are borne by a group of related individuals whose chances of leaving offspring are increased by the presence of these genes in an individual member of the group whose own private viability they lower." Haldane went on to prove that the process is feasible if the groups are small enough for altruists to confer a quick advantage. He saw that the altruism could be stable in a metapopulation if the genes were fixed in individual groups by drift, made possible by the small size of the groups or at least of the new populations founded by some of their emigrants. For some reason Haldane overlooked the role of differential population extinction, which might have led him to the next logical step in developing a full theory.

A separate thread of thought winds from Wahlund's principle (1926) to the development, in the 1930's and 1940's, of the "island model" of population genetics by Sewall Wright. For a formal, comprehensive review of the subject the reader is referred to the second volume of Wright's recent treatise (1969). The island model was related explicitly to the evolution of altruistic behavior by Wright in his 1945 essay review of G. G. Simpson's *Tempo and Mode in Evolution*. The formulation was nearly identical to that of Haldane, although made independently of it. Wright conceived of a set of populations diverging by genetic drift and adaptation to local environments but exchanging genes with one another. The pattern is that which Wright has persistently argued to be "the greatest creative factor of all" in evolution. In the special case considered here, the disadvantageous (for example, altruistic) genes can prevail over all the metapopulation if the populations they aid are small enough to allow them to drift to high values, and if the aided populations thereby send out a disproportionate number of emigrants. Like Haldane, Wright did not consider the effect of differential extinction on the equilibrial metapopulation. Nor did the model come any closer to a full theory of altruistic evolution. It is a curious twist that when W. D. Hamilton reinitiated group selection theory twenty years later, he was inspired not by the island model but by Wright's studies of relationship and inbreeding, which led to the topic of kin selection.

The next step in the study of interdemic selection was taken by ecologists largely unaware of genetic theory. Kalela (1954, 1957) postulated group selection as the mechanism responsible for reproductive restraint in subarctic vole populations. He saw food shortages as the ultimate controlling factor but believed that self-control of the populations during times of food plenty prevented starvation during food shortages. Kalela correctly deduced that self-control in matters of individual fitness can only be evolved if the groups not possessing the genes for self-control are periodically decimated or extinguished as a direct consequence of their lack of self-control. Kalela added one more feature to his scheme that substantially increased its plausibility. He suggested that rodent populations in many cases really consist of expanded family groups, so that self-restraint is the way for genetically allied tribes to hold their ground while other tribes of the same species eat themselves into extinction. In other words, the most forceful mode of interdemic selection is one that approaches a special form of kin selection. Kalela believed that the same kind of population structure and group selection might characterize many other rodents, ungulates, and primates. Independent but similar views were briefly expressed by Snyder (1961) and by Brereton (1962).

It remained for Wynne-Edwards, in his book *Animal Dispersion in Relation to Social Behaviour* (1962), to bring the subject to the attention of a wide biological audience. Wynne-Edwards' contribution was to carry the theory of self-control by group selection to its extreme—some of his critics would say the *reductio ad absurdum*—thereby forcing an evaluation of its strengths and weaknesses.

Food may be the *ultimate* factor, but it cannot be invoked as the *proximate* agent in chopping the numbers, without disastrous consequences. By analogy with human experience we should therefore look to see whether there is not some natural counterpart of the limitation-agreements that provide man with his only known remedy against overfishing—some kind of density-dependent convention, it would have to be, based on the quantity of food available but "artificially" preventing the intensity of exploitation from rising above the optimum level. Such a convention, if it existed, would have not only to be closely linked with the food situation, and highly (or better still perfectly) density-dependent in its operation, but, thirdly, also capable of eliminating the direct contest in hunting which has proved so destructive and extravagant in human experience.

The governing phrases in this scheme are "limitation-agreements" and "conventions." Social conventions are devices by which individual animals curtail their own individual fitness, that is, their survivorship, or fertility, or both, for the good of group survival. The density-dependent effects cited by Wynne-Edwards as involving social conventions run virtually the entire gamut: lowered fertility, reduced status in hierarchies, abandonment or direct killing of offspring, endocrine stress, deferment of growth and maturity. Sacrifice in each of these categories is viewed as an individual contribution to maintain populations below crash levels. Much of social behavior was reinterpreted by Wynne-Edwards to be epideictic displays, modes of communication by which members of populations inform each other of the density of the population as a whole and therefore the degree to which each member should decrease its own individual fitness. Examples of epideictic displays (which are distinguished from the epigamic displays that function purely in courtship) include the formation of mating swarms by insects, flocking in birds, and even vertical migration of zooplankton. The displays, then, are the most evolved communicative part of social conventions.

There has been a good deal of confusion, especially among non-biologists, over just what Wynne-Edwards had said that was different. He himself later stated (1971), "Seven years ago I put forward the hypothesis that social behavior plays an essential part in the natural regulation of animal numbers." That is not correct. The role of social behavior in population regulation is an old one and was never in dispute. What Wynne-Edwards proposed was the specific hypothesis that animals voluntarily sacrifice personal survival and fertility in order to help control population growth. He also postulated that this is a very widespread phenomenon among all kinds of animals. Furthermore, he did not stop at kin groups, as had Kalela, but suggested that the mechanism operates in Mendelian populations of all sizes, representing all breeding structures. Alternative hypotheses explaining social phenomena, such as nuptial synchronization, antipredation, and increased feeding efficiency, were either summarily dismissed or altogether ignored.

Wynne-Edwards' book had considerable value as the stalking-horse that drew forth large numbers of biologists, including theoreticians, who addressed themselves at last to the serious issues of group selection and genetic social evolution. It is also fair to say that in the long series of reviews and fresh studies that followed, culminating in G. C. Williams' *Adaptation and Natural Selection* (1966), one after another of Wynne-Edwards' propositions about specific "conventions" and epideictic displays were knocked down on evidential grounds or at least matched with competing hypotheses of equal plausibility drawn from models of individual selection. But for a long time neither critics nor sympathizers could answer the main theoretical question raised by this controversy: What are the deme sizes, interdemic migration rates, and differential deme survival probabilities necessary to counter the effects of individual selection? Only when population genetics was extended this far could we hope to evaluate the significance of extinction rates and to rule out one or the other of the various competing hypotheses in particular cases. Although some of the conceptual basis was independently formulated by Eshel (1972), who defined the crucial importance of migration rates in the evolution of altruism, the first strong efforts toward the construction

of a thorough, dynamic theory were made by Richard Levins and by Boorman and Levitt. Their models will now be summarized in turn.

The Levins Model

As we have seen, Levins (1970) conceived of a metapopulation occupying various fractions of a fixed number of habitable sites. Each population is subject to extinction but also has the opportunity to send forth N propagules that colonize previously empty sites. Now suppose there is an altruist gene occurring at a variable frequency x in each of the occupied sites. The proportion of populations containing exactly x altruist genes at time t will be denoted as $F(x, t)$, the overall gene frequency for the metapopulation as $\bar{x}$, the extinction rate of a population with x altruist genes as $E(x)$, and the mean extinction rate for all the populations as $\bar{E}$. Also, the frequency of the altruist gene in a founding group of N individuals is indicated as $N(x, \bar{x})$, and the rate at which individual selection reduces the gene frequency within a population as $M(x)$. The rate at which the proportion of populations with x genes changes through time is

$$\frac{dF(x, t)}{dt} = -E(x)F(x, t) + \bar{E}N(x, \bar{x}) + \frac{d}{dx}[M(x)F(x, t)]$$

This equation says that the proportion of populations in the metapopulation containing x altruist genes is declining because of extinction of such populations at the rate $-E(x)F(x, t)$, where $E(x)$ will be a generally declining function of x, that is, the more altruist genes there are, the lower the extinction rate. The equation also states that $F(x, t)$ is simultaneously changing because of new sites being colonized by groups of propagules with gene frequency x. When the proportion of sites occupied is at equilibrium, the proportion being newly occupied in each instant of time is $\bar{E}$, the proportion becoming extinct. Each population is founded by N individuals; the frequency of the altruist gene in these founder populations, which we designate $N(x, \bar{x})$, varies at random according to a binomial distribution around the metapopulation mean $\bar{x}$. In other words, the metapopulation is the source of the N migrants who found each new colony, and x, the frequency of the altruist genes among these founders, is a random variable dependent on N and $\bar{x}$. $N(x, \bar{x})$ is the binomial distribution (which can be approximated by the normal) of the gene frequencies in all founding populations, and the rate at which the altruist gene is changing because of colony foundation is therefore $\bar{E}N(x, \bar{x})$. $F(x, t)$ is also decreasing because of individual selection. By itself, each population has its gene frequency reduced toward zero by individual selection. The probability that a population with gene frequency x will be transformed into one with gene frequency $x - dx$

in an increment of time dt is $dtM(x)$. Then by individual selection alone

$$\frac{dF(x)}{dt} = -M(x)F(x) + M(x + dx)F(x + dx)$$

$$= \frac{d}{dx}[M(x)F(x,t)]$$

The rate at which populations in the metapopulation are slipping from x to $x - dx$, then, is dependent on the difference between the rate at which each passes from $x + dx$ to x and the rate at which it passes from x to $x - dx$.

The rate of change of the frequency of the altruist gene through the entire metapopulation is the mean of rates of change in all the constituent populations:

$$\frac{d\bar{x}}{dt} = \int_0^1 x \frac{dF(x, t)}{dt} dx$$

Levins' approach to the problem was to write parallel equations for the variance and higher central moments of the populations with reference to the gene frequency. Then $E(x)$ was expanded in Taylor series to obtain $E(0)$, the extinction rate of populations containing no altruists, and $E'(0)$, the rate at which the extinction rate declines as the first altruist genes are added. The easiest procedure was next to analyze the set of simultaneous equations for stability, where $x = 0$ and $E(x) = E(0)$. If a set of values for the individual selection intensity and other parameters yields instability in the ensuing matrix analysis, the implication is that x will move away from zero. In other words, the altruist gene will increase in frequency.

Suppose that selection is additive, following the relation

$$M(x) = s(\bar{x})(1 - \bar{x}) + s(1 - 2\bar{x})(x - \bar{x}) - s(x - \bar{x})^2$$

where s is the selection coefficient. The system is stable near $\bar{x} = 0$ if

$$-E'(0) < (N - 1)s + \frac{2Ns^2}{E(0)}$$

Analysis of this inequality shows that even if group selection, measured by $E'(0)$, is stronger than individual selection, the best it can do is to establish the altruist gene in a polymorphic state within the metapopulation. Prospects are better if the altruist gene is dominant.

In this case

$$M(x) = s\bar{x}(1 - \bar{x})^2 + s(1 - \bar{x})(1 - 3\bar{x})(x - \bar{x})4$$
$$- s(2 - 3\bar{x})(x - \bar{x})^2 + s(x - \bar{x})^2$$

When the altruist gene is fixed to start with ($x = 1$), then stability is achieved, and the gene remains fixed, provided that

$$-E'(1) > s$$

in other words, if the rate at which the altruist gene improves group survival as x approaches fixation is greater than the selection coefficient (see Figure 5-4). When $x = 0$, stability is abolished, and the altruist gene begins to increase in frequency, provided that the following inequality exists:

$$E'(0) > (N - 2)s + \frac{2Ns^2}{E(\bar{x})}$$

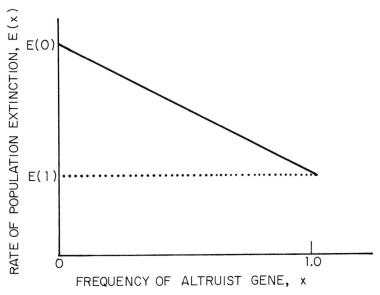

Figure 5-4 Group selection favoring an altruist gene. In this simplest possible model the rate of population extinction declines linearly as the frequency of the altruist gene in each population increases. The intensity of group selection is measured in two ways: first, by the extinction rates at various values of x, for example $E(x) = E(0)$, or $E(x) = E(\bar{x})$, the average extinction rate for all values of x; second, by the rate at which an increase in x lowers $E(x)$. In the elementary case depicted here, $E'(0) = E'(1)$. (From Wilson, 1973.)

In general, if $E' < s$ for any initial value of x, individual selection will prevail, and the altruist gene will be reduced toward zero or at least toward the mutational equilibrium. It is also necessary, in both the additive and dominance cases, to have a sufficiently high overall population extinction rate, measured by $E(0)$ or $E(\bar{x})$, to compensate for $2Ns^2$ in the righthand term of the inequality.

The Levins model advanced theory fundamentally by identifying and formalizing the parameters of extinction, relating them to migrant and individual selection, and introducing the technique of stability analysis to provide broad qualitative results. The shortcomings of the model include the uncertainty of the stability analysis (see Boorman and Levitt, 1973a), the failure to consider variation in the structure of the metapopulation, and the failure to analyze the enhancing effects of kin selection in the small founding groups postulated. More important, the results consist entirely of inequalities based on the stability analysis and are therefore not very heuristic. They do not provide a prescription for phenomenological models that can be applied to actual field studies. Levins showed us that evolution of altruistic traits by interpopulation selection is indeed feasible, and demonstrated that the conditions for its occurrence are stringent. But the model lacks sufficient structure to generate particular measurements and tests that might lead to an assessment of the places and times in which individual selection can be counteracted by interdemic selection.

Recently, B. R. Levin and W. L. Kilmer (personal communication) have overcome many of the technical difficulties in Richard Levins' model by studying similar island-model metapopulations with computer simulations. They realized that only by specifying the actual frequency distributions $F(x, t)$ through time would it be possible to design studies of real populations. Their experimental runs are stochastic processes in which fixed values are assigned to the individual selection coefficients, the extinction rates of the populations, and the rates at which individuals migrate between populations. The populations were either fixed in size or allowed to grow. The results so far are at least qualitatively consistent with the inequalities produced by Richard Levins' model. The advantage of the simulation technique is its potential realism—it is rather easily modified to accommodate special properties encountered in actual populations. The disadvantage, as in most simulation procedures, is the difficulty in defining the boundary conditions within which the phenomenon of interest can occur.

The Boorman-Levitt Model

S. A. Boorman and P. R. Levitt (1972, 1973a) made a second study with the same goal of predicting the full course of evolution by group selection. In order to characterize analytically the full course of evolution they envisaged a particular metapopulation structure different from that of Levins, consisting of a large, enduring central population and a set of marginal populations more liable to extinction (Figure 5-5). The altruist genes present in the marginal populations do not come to affect the population extinction rates until the populations have reached their demographically stable size, and individual selection does not operate in the marginal populations. Hence the Boorman-Levitt system allows for K extinction, whereas the Levins system more closely approximates the conditions that promote r extinction. Although Boorman and Levitt chose this particular structure in part for

BOORMAN-LEVITT METAPOPULATION

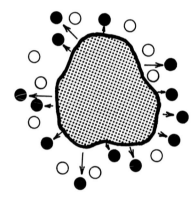

LEVINS METAPOPULATION

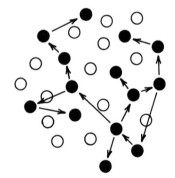

Figure 5-5 The metapopulations conceived in the Levins and Boorman-Levitt models. In the Levins system, small populations found other small populations in the habitable sites, and the altruist genes can decrease their probability of extinction from the moment of foundation (that is, they help avert r extinction). In the Boorman-Levitt system, marginal populations are derived from one large, stable population, and altruist genes do not influence extinction rates until the marginal populations have reached demographic carrying capacity (that is, they help avert K extinction). (From Wilson, 1973.)

its analytic tractability, it was a biologically happy choice as well. Many real metapopulations do in fact consist of large, stable "source" populations occupying the ecologically favorable portion of the range together with groups of smaller, semiisolated populations near the periphery of the range. The peripheral populations are more liable to extinction not only because of their smaller size but also because they more often exist in less favorable habitats.

The marginal populations have a gene frequency distribution evolving through time according to the following equation:

$$\frac{\partial}{\partial t}\phi(x, t) = -E(x)\phi(x, t) + \left\{\frac{\partial^2}{\partial_{x^2}}[V_{\delta_x}\phi(x, t)] - \frac{\partial}{\partial x}[M_{\delta_x}\phi(x, t)]\right\}$$

where $\phi(x, t)$ is the joint probability that a population will exist and have a gene frequency x at time t; $E(x)$ is the extinction rate as in the Levins equation; M_{δ_x} is the mean amount of change in gene frequency per generation, due principally to individual selection; and V_{δ_x} is the variance of the change in gene frequency per generation. After writing this equation, Boorman and Levitt gambled on an assumption that greatly simplified the analysis. They conjectured that if group selection is going to operate at all, it will probably require such high extinction rates that individual selection can be momentarily ignored: group selection and individual selection were thus "decoupled." Individual selection takes place in the central source population. In conjunction with genetic drift, it determines the initial low frequency of the altruist gene in the boundary population, which is founded at the level of carrying capacity. Extinction then proceeds in the boundary populations at a pace sufficiently fast to prevent significant further progress by individual selection within them. As a corollary, the populations are not replaced after extinction. The process of reduction in their numbers is allowed to run out in time until nearly all are gone.

The Boorman-Levitt model can be regarded as the mode of pure interdemic selection by means of K extinction that is the most likely to counteract individual selection. Its principal result is the demonstration that extinction of a severe and peculiar form is required to elevate the frequencies of altruist genes significantly—or of any kind of genes favored by group selection and opposed by individual selection. In particular, the extinction operator $E(x)$ must approach a step function, of the kind illustrated in Figure 5-6, in order to work. When it does work, the achievement comes after a close race between the rise of the frequency of the altruist gene in the metapopulation and the total extinction of the metapopulation. In order for the altruist gene to approach a frequency of 20 or 30 percent, most of the constituent populations must become extinct. Also, as suggested by Levins' model, the best the metapopulation can attain when starting from

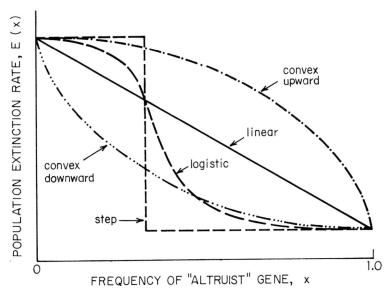

Figure 5-6 Various extinction rate functions that were applied in the Boorman-Levitt model. Only a steep logistic function or step function can produce a significant increase in the frequency of the altruist gene within the entire metapopulation as a result of pure interdemic selection.

low frequencies is polymorphism between the altruist and nonaltruist genes. An example of the process is given in Figure 5-7.

In summary, deductions from the two models agree that evolution of an altruist gene by means of pure interdemic selection, based on differential population extinction, is an improbable event. The metapopulation must pass through a very narrow "window" framed by strict parameter values: steeply descending extinction functions, preferably approaching a step function with a threshold value of the frequency of the altruist gene; high extinction rates comparable in magnitude (in populations per generation) to the opposing individual selection (in individuals per population per generation); and the existence of moderately large metapopulations broken into many semiisolated populations. Even after achieving all these conditions, the metapopulation is likely to be no more than polymorphic for the gene.

What this means in practice is that most of the wide array of "social conventions" hypothesized by Wynne-Edwards and other authors are probably not true. Moreover, self-restraint on behalf of the entire population is *least* likely in the largest, most stable populations, where social behavior is the most highly developed. Examples include the breeding colonies of seabirds, the communal roosts of starlings,

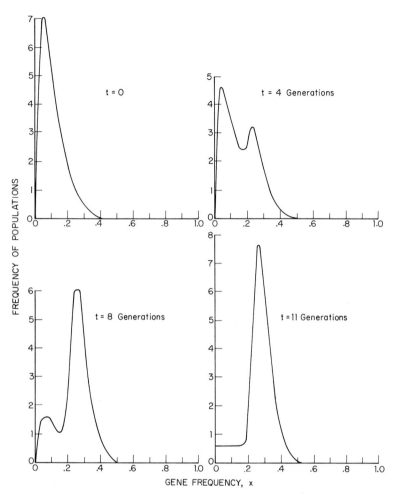

FREQUENCY OF POPULATIONS

t = 0

t = 4 Generations

t = 8 Generations

t = 11 Generations

GENE FREQUENCY, x

Figure 5-7 The rise of an altruist gene in a particular Boorman-Levitt metapopulation. The marginal populations have an effective population size of 200. They are derived from a large source population in which the frequency of the altruist gene is 0.1, sustained at equilibrium by a mutation rate of 10^{-4} per generation, which is opposed by individual selection at an intensity of 0.01 per generation. Extinction proceeds at the average rate of 0.1 populations per generation ($\int_0^1 E(x)dx = 0.1$). The extinction function is steeply logistic, with the altruist genes conferring little or no advantage to the populations below $x = 0.2$ and most or all of their advantage at all values above $x = 0.2$. By the time the population-frequency curve has reached the new modality (and the populations are polymorphic for the gene), most of the metapopulation has been extinguished. (Modified from Boorman and Levitt, 1973a.)

the leks of grouse, the warrens of rabbits, and many of the other societal forms cited by Wynne-Edwards as the best examples of altruistic population control. In these cases one must favor alternative hypotheses that involve either kin selection or individual selection. Even so, a mechanism for the evolution of population-wide cooperation has been validated, and the hypothesis of social conventions must either be excluded or kept alive for each species considered in turn. One should also bear in mind that the real population is the unit whose members are freely interbreeding. Such a unit can exist firmly circumscribed in the midst of a seemingly vast population—which is really a metapopulation in evolutionary time. Consider a population of rodents in which tens of thousands of adults hold small territories over a continuous habitat of hundreds of square kilometers. The aggregation seems vast, yet each ridge of earth, each row of trees, and each streamlet could cut migration sufficiently to delimit a true population. The effective population size might be 10 or 100, despite the fact that a hawk's-eye view of the entire metapopulation makes it seem continuous. Not even the little habitat barriers are required. If the rodents move about very little, or return faithfully to the site of their birth to breed, the population neighborhood will be small, and the effective population size low. The delimitation of such local populations could be sharpened by the development of cultural idiosyncrasies such as the learned dialects of birds (Nottebohm, 1970) or the inherited burrow systems of social rodents. With increasing delimitation and reduction in population size, the selection involved also slides toward the kin-selection end of the scale. To evaluate definitively the potential intensity of interdemic selection it is necessary to estimate the size of the neighborhood, the effective size of the populations, and the rate at which the true populations become extinct (see Figure 5-1).

The chief role of interdemic selection may turn out to lie not in forcing the evolution of altruistic density-dependent controls but rather in serving as a springboard from which other forms of altruistic evolution are launched. Suppose that the altruists also have a tendency to cooperate with one another in a way that ultimately benefits each altruist at the expense of the nonaltruists. Cliques and communes may require personal sacrifice, but if they are bonded by possession of one inherited trait, the trait can evolve as the groups triumph over otherwise comparable units of noncooperating groups. The bonding need not even require prolonged sacrifice, only the trade-offs of reciprocal altruism. The formation of such networks requires either a forbiddingly high starting gene frequency or a large number of random contacts with other individuals in which the opportunity for trade-offs exists (Trivers, 1971). These frequency thresholds might be reached by interdemic selection that initially favors other aspects of the behavior that are not altruistic.

There do exist special conditions under which interdemic selection can proceed without differential deme extinction, and in a way that might spread altruist genes rapidly in a population. Maynard Smith (1964) suggested a model in which local populations are first segregated and allowed to grow or to decline for a while in ways influenced by their genetic composition. Then individuals from different populations mix and interbreed to some extent before going on to form new populations. Suppose that the populations were mice in haystacks, with each haystack being colonized by a single fertilized female. If a/a are altruist individuals, and A/A and A/a selfish individuals, the a allele would be eliminated in all haystacks where A-bearing individuals were present. But if pure a/a populations contributed more progeny in the mixing and colonizing phases, and if there were also a considerable amount of inbreeding (so that pure a/a populations were more numerous than expected by chance alone), the altruist gene would spread through the population. D. S. Wilson (manuscript in preparation) argues that many species in nature go through regular cycles of segregation and mixing and that altruist genes can be spread under a wide range of realistic conditions beyond the narrow one conceived by Maynard Smith. All that is required is that the absolute rate of increase of the altruists be greater. It does not matter that their rate of increase relative to the nonaltruists in the same population is (by definition) less during the period of isolation. Provided the rate of increase of the population as a whole is enhanced enough by the presence of altruists, they will increase in frequency through the entire metapopulation.

What specific traits would interdemic selection be expected to produce? Under some circumstances the altruism would oppose r selection. There is a fundamental tendency for genotypes that have the highest r to win in individual selection, and their advantage is enhanced in species that are opportunistic or otherwise undergo regular fluctuations in population size. But the greater the fluctuation, the higher the extinction rate. Thus interdemic selection would tend to damp population cycles by a lower fertility and an early, altruistic sensitivity to density-dependent controls. There is also a fundamental tendency for genotypes that can sustain the highest density to prevail (K selection; see Chapter 4). But high density contaminates the environment, attracts predators, and promotes the spread of disease, all of which increase the extinction rates of entire populations. Altruism promoted by these effects might include a higher physiological sensitivity to crowding and a greater tendency to disperse even at the cost of lowered fitness. Levins (1970) has pointed out that mixtures of genotypes in populations of fruit flies and crop plants often attain a higher equilibrium density than pure strains, but under a variety of conditions one strain excludes the others competitively. If higher densities result in the production of more propagules without incur-

ring a greater risk of extinction to the mother population, an antagonism between group and individual selection will result. Also, genetic resistance to disease or predation often results in lowered fitness in another component, as exemplified by sickle-cell anemia. In the temporary absence of this pressure, individual selection "softens" the population as a whole, which will be disfavored in interdemic selection when the pressure is exerted again.

It is also true, as Madhav Gadgil has pointed out to me, that pure interdemic selection, acting apart from kin selection, can lead to exceptionally selfish and even spiteful behavior. Suppose, for example, that the particular circumstances of interdemic selection within a given species dictate a reduction in population growth. Then the "altruist" who curtails its personal reproduction might just as well spend its spare time cannibalizing other members of the population—also to the benefit of the deme as a whole. Another seemingly spiteful behavior that could be favored by K extinction is the maintenance of excessively large territories.

The evidence for interdemic selection is fragmentary and somewhat peculiar in nature. As a corollary result of their theory of island biogeography, MacArthur and Wilson (1967) showed that moderate to high colonization rates of empty environments implies correspondingly high population extinction rates. In particular, if $\hat{S}$ is the equilibrium number of species on an island or any other isolated habitat, $t_{0.9}$ is the time required to go from zero species to 90 percent of the equilibrium species number during the colonization process, and X_S is the turnover (that is, extinction) rate at equilibrium,

$$X_S = \frac{1.15\hat{S}}{t_{0.9}}$$

in the case where the extinction rate rises and the immigration rate declines in a linear manner. Applied to real cases of colonization, where $t_{0.9}$ and $\hat{S}$ can be approximated, this model predicts surprisingly high population extinction rates. For example, the birds of the island of Krakatoa, which was devastated by a volcanic eruption in 1883, reattained the expected faunistic equilibrium of 30 species in approximately 30 years, and this led to the prediction that species should be becoming extinct at the rate of approximately one a year. Incomplete census data taken in 1908, 1919–1921, and 1932–1934 indicate that the true extinction rate was at least 0.2 species per year, still a surprisingly high figure. More recent colonization experiments have also produced very high extinction rates within one order of magnitude of those predicted by the turnover equation. After small mangrove islands in the Florida Keys were fumigated to remove all animal life, the arthropods reattained the previous species equilibria in less than a year. With equilibrial species numbers between 20 and 40, the arthropod species were becoming extinct (and being replaced) at the rate of approximately 0.1 species per day or, given a month's generation time, 3 species per generation (Simberloff and Wilson, 1969; Wilson, 1969). Freshwater benthic protozoans studied by Cairns et al. (1969) reached equilibria of 30–40 species on artificial surfaces, at which time the extinction rate was one species per day. These rates are easily within the range required to power counteracting interdemic selection. MacArthur and Wilson further demonstrated the existence of a threshold equilibrium population number below which populations can be expected to become extinct at a high rate and above which they are relatively safe (Figure 5-8). Thus metapopulations broken into very small genetic neighborhoods can be expected to have high population extinction rates. This result has been extended and refined by Richter-Dyn and Goel (1972).

In spite of the frequently permissible conditions that exist in nature, actual cases of interdemic selection have only rarely been reported in the literature. One of the most promising circumstances in which to search for voluntary population control is the evolutionary reduction of virulence in parasites. Virulence often (but not always) comes from the capacity to multiply rapidly. Thus the condition is likely to evolve by individual selection. But too high a level of virulence kills off the hosts, perhaps before infection of other hosts is achieved, so that virulence will be opposed by interdemic selection. It may stretch credulity to think of an altruistic bacterium or self-sacrificing blood fluke, but in the sense that feeding ability or reproduction is curtailed in spite of competition from other genotypes, a parasite can be altruistic. This is precisely the course followed by the myxoma virus after it was introduced into Australia in 1950 to control rabbits. Early strains were too rapidly lethal to allow ready transmission by mosquitoes from one rabbit to another. Within less than ten years the virulence decreased dramatically, while simultaneously the resistance of the rabbits to all strains increased (Fenner, 1965).

Wild populations of the house mouse in the United States are polymorphic for mutant alleles at the T locus. The t alleles, which in certain combinations cause a tailless condition, are lethal or sterile in homozygous condition. At the same time they are strongly favored at gametogenesis; 95 percent of the sperm of heterozygous males contains the t allele. Deterministic models predict that when recessively lethal or sterile alleles have such a high segregation ratio, their heterozygotes should constitute between 60 and 95 percent of the population. But in real mouse populations the frequency of the heterozygotes is much lower, ranging between 35 and 50 percent. These lower frequencies can be explained by the fact that the populations are small enough, with effective sizes on the order of ten, for the t alleles to be fixed (that is, reach 100 percent) by genetic drift. When that happens, the population becomes extinct. As a result of relatively

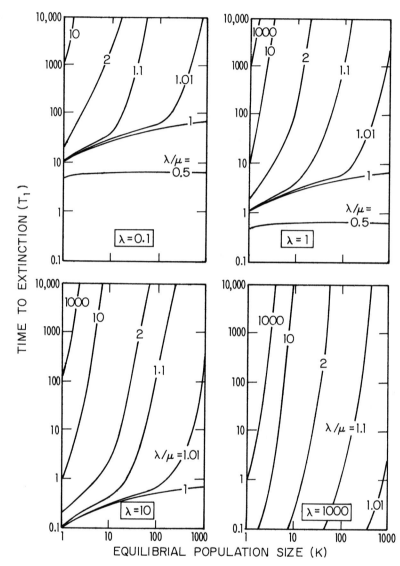

Figure 5-8 The threshold effect in extinction rate of populations. For a given individual birth rate (λ) and individual death rate (μ), there exists a narrow band of equilibrium population size below which the extinction rate is very high and above which it is very low. (From MacArthur and Wilson, 1967.)

frequent extinctions and hence reduction of t alleles in the metapopulation, the average frequencies would be expected to fall below the equilibrium frequencies predicted by the deterministic model. Lewontin and Dunn (1960), by simulating stochastic changes in populations with effective sizes of six and eight, demonstrated that the average frequencies really can equilibrate at the lower levels observed in nature. More recently, however, B. R. Levin et al. (1969) found that in at least some cases the true migration rates and effective size of the mice populations are too great for genetic drift to be effective. They have raised three alternative hypotheses that also account for low frequencies of *t* alleles, including less segregation distortion, selection against *t* heterozygotes, and systematic assortative mating. The four hypotheses are not mutually exclusive, and only further studies can assign them relative weights.

Circumstantial evidence of group selection, which may or may not favor altruistic behavior, is provided by the phenomenon of minimal population size of social species. When less than 10 males of the African village weaverbird (*Ploceus cucullatus*) form breeding colonies, they attract a much lower proportion of females than do the males in nearby colonies of larger, "normal" size (Collias and Collias, 1969). A captive group of blue-crowned hanging parrots (*Loriculus galgulus*) observed by Francine Buckley (1967) did not start emitting a full repertory of vocalizations or function as a synchronized flock until the number of members was increased from 3 to 12. It is easy to see that if a metapopulation is fragmented by a deterioration of the environment or dispersal into a new area, any population that manages to stay above the threshold size will hold a decisive advantage. Insofar as the tendency to form groups of varying size is heritable, the mean size of groups will then increase by evolution. If strong enough, the group selection can override individual selection favoring more solitary traits. Threshold population sizes above the level of the mated pair have also been documented in a few mammals and social insects. It must be added, however, that even in such colonial species no evidence exists that interdemic selection prevails over kin selection or even counteracts it. It is even possible that minimum population sizes are decided indirectly by some as yet unknown form of individual selection.

Kin Selection

Imagine a network of individuals linked by kinship within a population. These blood relatives cooperate or bestow altruistic favors on one another in a way that increases the average genetic fitness of the members of the network as a whole, even when this behavior reduces the individual fitnesses of certain members of the group. The members may live together or be scattered throughout the population. The essential condition is that they jointly behave in a way that benefits the group as a whole, while remaining in relatively close contact with the remainder of the population. This enhancement of kin-network welfare in the midst of a population is called *kin selection.*

Kin selection can merge into interdemic selection by an appropriate spatial rearrangement. As the kin network settles into one physical location and becomes physically more isolated from the rest of the

species, it approaches the status of a true population. A closed society, or one so nearly closed that it exchanges only a small fraction of its members with other societies each generation, is a true Mendelian population. If in addition the members all treat one another without reference to genetic relationship, kin selection and interdemic selection are the same process. If the closed society is small, say with 10 members or less, we can analyze group selection by the theory of kin selection. If it is large, containing an effective breeding size of 100 or more, or if the selection proceeds by the extinction of entire demes of any size, the theory of interdemic selection is probably more appropriate.

The personal actions of one member toward another can be conveniently classified into three categories in a way that makes the analysis of kin selection more feasible. When a person (or animal) increases the fitness of another at the expense of his own fitness, he can be said to have performed an act of *altruism*. Self-sacrifice for the benefit of offspring is altruism in the conventional but not in the strict genetic sense, because individual fitness is measured by the number of surviving offspring. But self-sacrifice on behalf of second cousins is true altruism at both levels; and when directed at total strangers such abnegating behavior is so surprising (that is, "noble") as to demand some kind of theoretical explanation. In contrast, a person who raises his own fitness by lowering that of others is engaged in *selfishness*. While we cannot publicly approve the selfish act we do understand it thoroughly and may even sympathize. Finally, a person who gains nothing or even reduces his own fitness in order to diminish that of another has committed an act of *spite*. The action may be sane, and the perpetrator may seem gratified, but we find it difficult to imagine his rational motivation. We refer to the commitment of a spiteful act as "all too human"—and then wonder what we meant.

The concept of kin selection to explain such behavior was originated by Charles Darwin in *The Origin of Species*. Darwin had encountered in the social insects the "one special difficulty, which at first appeared to me insuperable, and actually fatal to my whole theory." How, he asked, could the worker castes of insect societies have evolved if they are sterile and leave no offspring? This paradox proved truly fatal to Lamarck's theory of evolution by the inheritance of acquired characters, for Darwin was quick to point out that the Lamarckian hypothesis requires characters to be developed by use or disuse of the organs of individual organisms and then to be passed directly to the next generation, an impossibility when the organisms are sterile. To save his own theory, Darwin introduced the idea of natural selection operating at the level of the family rather than of the single organism. In retrospect, his logic seems impeccable. If some of the individuals of the family are sterile and yet important to the welfare of fertile relatives, as in the case of insect colonies, selection at the family level is inevitable. With the entire family serving as the unit of selection, it is the capacity to generate sterile but altruistic relatives that becomes subject to genetic evolution. To quote Darwin, "Thus, a well-flavoured vegetable is cooked, and the individual is destroyed; but the horticulturist sows seeds of the same stock, and confidently expects to get nearly the same variety; breeders of cattle wish the flesh and fat to be well marbled together; the animal has been slaughtered, but the breeder goes with confidence to the same family" (*The Origin of Species*, 1859: 237). Employing his familiar style of argumentation, Darwin noted that intermediate stages found in some living species of social insects connect at least some of the extreme sterile castes, making it possible to trace the route along which they evolved. As he wrote, "With these facts before me, I believe that natural selection, by acting on the fertile parents, could form a species which regularly produce neuters, either all of a large size with one form of jaw, or all of small size with jaws having a widely different structure; or lastly, and this is the climax of our difficulty, one set of workers of one size and structure, and simultaneously another set of workers of a different size and structure" (*The Origin of Species*, 1859: 24). Darwin was speaking here about the soldiers and minor workers of ants.

Family-level selection is of practical concern to plant and animal breeders, and the subject of kin selection was at first pursued from this narrow point of view. One of the principal contributions to theory was provided by Jay L. Lush (1947), a geneticist who wished to devise a prescription for the choice of boars and gilts for use in breeding. It was necessary to give each pig "sib credits" determined by the average merit of its littermates. A quite reliable set of formulas was developed which incorporated the size of the family and the phenotypic correlations between and within families. This research provided a useful background but was not addressed directly to the evolution of social behavior in the manner envisaged by Darwin.

The modern genetic theory of altruism, selfishness, and spite was launched instead by William D. Hamilton in a series of important articles (1964, 1970, 1971a,b, 1972). Hamilton's pivotal concept is *inclusive fitness:* the sum of an individual's own fitness plus the sum of all the effects it causes to the related parts of the fitnesses of all its relatives. When an animal performs an altruistic act toward a brother, for example, the inclusive fitness is the animal's fitness (which has been lowered by performance of the act) plus the increment in fitness enjoyed by that portion of the brother's hereditary constitution that is shared with the altruistic animal. The portion of shared heredity is the fraction of genes held by common descent by the two animals and is measured by the coefficient of relationship, r (see Chapter 4). Thus, in the absence of inbreeding, the animal and its brother have $r = \frac{1}{2}$ of their genes identical by common descent. Hamilton's key result can be stated very simply as follows.

A genetically based act of altruism, selfishness, or spite will evolve if the average inclusive fitness of individuals within networks displaying it is greater than the inclusive fitness of individuals in otherwise comparable networks that do not display it.

Consider, for example, a simplified network consisting solely of an individual and his brother (Figure 5-9). If the individual is altruistic he will perform some sacrifice for the benefit of the brother. He may surrender needed food or shelter, or defer in the choice of a mate, or place himself between his brother and danger. The important result, from a purely evolutionary point of view, is loss of genetic fitness—a reduced mean life span, or fewer offspring, or both—which leads to less representation of the altruist's personal genes in the next generation. But at least half of the brother's genes are identical to those of the altruist by virtue of common descent. Suppose, in the extreme case, that the altruist leaves no offspring. If his altruistic act more than doubles the brother's personal representation in the next generation, it will ipso facto increase the one-half of the genes identical to those in the altruist, and the altruist will actually have gained representation in the next generation. Many of the genes shared by such brothers will be the ones that encode the tendency toward altruistic behavior. The inclusive fitness, in this case determined solely by the brother's contribution, will be great enough to cause the spread of the altruistic genes through the population, and hence the evolution of altruistic behavior.

The model can now be extended to include all relatives affected by the altruism. If only first cousins were benefited ($r = \frac{1}{8}$), the altruist who leaves no offspring would have to multiply a cousin's fitness eightfold; an uncle ($r = \frac{1}{4}$) would have to be advanced fourfold; and so on. If combinations of relatives are benefited, the genetic effect of the altruism is simply weighted by the number of relatives of each kind who are affected and their coefficients of relationship. In general, k, the ratio of gain in fitness to loss in fitness, must exceed the reciprocal of the average coefficient of relationship ($\bar{r}$) to the ensemble of relatives:

$$k > \frac{1}{\bar{r}}$$

Thus in the extreme brother-to-brother case, $1/\bar{r} = 2$; and the loss in fitness for the altruist who leaves no offspring was said to be total (that is = 1.0). Therefore in order for the shared altruistic genes to increase, k, the gain-to-loss ratio, must exceed 2. In other words, the brother's fitness must be more than doubled.

The evolution of selfishness can be treated by the same model. Intuitively it might seem that selfishness in any degree pays off so long as the result is the increase of one's personal genes in the next generation. But this is not the case if relatives are being harmed to

Figure 5-9 The basic conditions required for the evolution of altruism, selfishness, and spite by means of kin selection. The family has been reduced to an individual and his brother; the fraction of genes in the brother shared by common descent ($r = \frac{1}{2}$) is indicated by the shaded half of the body. A requisite of the environment (food, shelter, access to mate, and so on) is indicated by a vessel, and harmful behavior to another by an axe. *Altruism*: the altruist diminishes his own genetic fitness but raises his brother's fitness to the extent that the shared genes are actually increased in the next generation. *Selfishness*: the selfish individual reduces his brother's fitness but enlarges his own to an extent that more than compensates. *Spite*: the spiteful individual lowers the fitness of an unrelated competitor (the unshaded figure) while reducing that of his own or at least not improving it; however, the act increases the fitness of the brother to a degree that more than compensates.

the extent of losing too many of their genes shared with the selfish individual by common descent. Again, the inclusive fitness must exceed 1, but this time the result of exceeding that threshold is the spread of the selfish genes.

Finally, the evolution of spite is possible if it, too, raises inclusive fitness. The perpetrator must be able to discriminate relatives from nonrelatives, or close relatives from distant ones. If the spiteful behavior causes a relative to prosper to a compensatory degree, the genes favoring spite will increase in the population at large. True spite is a commonplace in human societies, undoubtedly because human beings are keenly aware of their own blood lines and have the intelligence to plot intrigue. Human beings are unique in the degree of their capacity to lie to other members of their own species. They typically do so in a way that deliberately diminishes outsiders while promoting relatives, even at the risk of their own personal welfare (Wallace, 1973). Examples of spite in animals may be rare and difficult to distinguish from purely selfish behavior. This is particularly true in the realm of false communication. As Hamilton drily put it, "By our lofty standards, animals are poor liars." Chimpanzees and gorillas, the brightest of the nonhuman primates, sometimes lie to one another (and to zookeepers) to obtain food or to attract company (Hediger,

1955: 150; van Lawick-Goodall, 1971). The mental capacity exists for spite, but if these animals lie for spiteful reasons this fact has not yet been established. Even the simplest physical techniques of spite are ambiguous in animals. Male bowerbirds sometimes wreck the bowers of the neighbors, an act that appears spiteful at first (Marshall, 1954). But bowerbirds are polygynous, and the probability exists that the destructive bird is able to attract more females to his own bower. Hamilton (1970) has cited cannibalism in the corn ear worm (*Heliothis zea*) as a possible example of spite. The first caterpillar that penetrates an ear of corn eats all subsequent rivals, even though enough food exists to see two or more of the caterpillars through to maturity. Yet even here, as Hamilton concedes, the trait might have evolved as pure selfishness at a time when the *Heliothis* fed on smaller flowerheads or small corn ears of the ancestral type. Many other examples of the killing of conspecifics have been demonstrated in insects, but almost invariably in circumstances where the food supply is limited and the aggressiveness is clearly selfish as opposed to spiteful (Wilson, 1971b).

The Hamilton models are beguiling in part because of their transparency and heuristic value. The coefficient of relationship, *r*, translates easily into "blood," and the human mind, already sophisticated in the intuitive calculus of blood ties and proportionate altruism, races to apply the concept of inclusive fitness to a reevaluation of its own social impulses. But the Hamilton viewpoint is also unstructured. The conventional parameters of population genetics, allele frequencies, mutation rates, epistasis, migration, group size, and so forth, are mostly omitted from the equations. As a result, Hamilton's mode of reasoning can be only loosely coupled with the remainder of genetic theory, and the number of predictions it can make is unnecessarily limited.

Reciprocal Altruism

The theory of group selection has taken most of the good will out of altruism. When altruism is conceived as the mechanism by which DNA multiplies itself through a network of relatives, spirituality becomes just one more Darwinian enabling device. The theory of natural selection can be extended still further into the complex set of relationships that Robert L. Trivers (1971) has called *reciprocal altruism*. The paradigm offered by Trivers is good samaritan behavior in human beings. A man is drowning, let us say, and another man jumps in to save him, even though the two are not related and may not even have met previously. The reaction is typical of what human beings regard as "pure" altruism. However, upon reflection one can see that the good samaritan has much to gain by his act. Suppose that the drowning man has a one-half chance of drowning if he is not assisted, whereas the rescuer has a one-in-twenty chance of dying. Imagine further that when the rescuer drowns the victim also drowns, but when the rescuer lives the victim is always saved. If such episodes were extremely rare, the Darwinist calculus would predict little or no gain to the fitness of the rescuer for his attempt. But if the drowning man reciprocates at a future time, and the risks of drowning stay the same, it will have benefited both individuals to have played the role of rescuer. Each man will have traded a one-half chance of dying for about a one-tenth chance. A population at large that enters into a series of such moral obligations, that is, reciprocally altruistic acts, will be a population of individuals with generally increased genetic fitness. The trade-off actually enhances personal fitness and is less purely altruistic than acts evolving out of interdemic and kin selection.

In its elementary form the good samaritan model still contains an inconsistency. Why should the rescued individual bother to reciprocate? Why not cheat? The answer is that in an advanced, personalized society, where individuals are identified and the record of their acts is weighed by others, it does not pay to cheat even in the purely Darwinist sense. Selection will discriminate against the individual if cheating has later adverse affects on his life and reproduction that outweigh the momentary advantage gained. Iago stated the essence in *Othello*: "Good name in man and woman, dear my lord, is the immediate jewel of their souls."

Trivers has skillfully related his genetic model to a wide range of the most subtle human behaviors. Aggressively moralistic behavior, for example, keeps would-be cheaters in line—no less than hortatory sermons to the believers. Self-righteousness, gratitude, and sympathy enhance the probability of receiving an altruistic act by virtue of implying reciprocation. The all-important quality of sincerity is a metacommunication about the significance of these messages. The emotion of guilt may be favored in natural selection because it motivates the cheater to compensate for his misdeed and to provide convincing evidence that he does not plan to cheat again. So strong is the impulse to behave altruistically that persons in experimental psychological tests will learn an instrumental conditioned response without advance explanation and when the only reward is to see another person relieved of discomfort (Weiss et al., 1971).

Human behavior abounds with reciprocal altruism consistent with genetic theory, but animal behavior seems to be almost devoid of it. Perhaps the reason is that in animals relationships are not sufficiently enduring, or memories of personal behavior reliable enough, to permit the highly personal contracts associated with the more human forms of reciprocal altruism. Almost the only exceptions I know occur just where one would most expect to find them—in the more intelligent monkeys, such as rhesus macaques and baboons, and in the anthropoid apes. Members of troops are known to form coalitions or cliques and to aid one another reciprocally in disputes with other troop members. Chimpanzees, gibbons, African wild dogs, and wolves also beg food from one another in a reciprocal manner.

Granted a mechanism for sustaining reciprocal altruism, we are still left with the theoretical problem of how the evolution of the behavior gets started. Imagine a population in which a Good Samaritan appears for the first time as a rare mutant. He rescues but is not rescued in turn by any of the nonaltruists who surround him. Thus the genotype has low fitness and is maintained at no more than mutational equilibrium. Boorman and Levitt (1973b) have formally investigated the conditions necessary for the emergence of a genetically mediated cooperation network. They found that for each population size, for each component of fitness added by membership in a network as opposed to the reduced fitness of cooperators outside networks, and for each average number of individuals contacted in the network, there exists a critical frequency of the altruist gene above which the gene will spread explosively through the population and below which it will slowly recede to the mutational equilibrium (Figure 5-10). How critical frequencies are attained from scratch remains unknown. Cooperative individuals must play a version of the game of Prisoner's Dilemma (Hamilton, 1971b; Trivers, 1971). If they chance cooperation with a nonaltruist, they lose some fitness and the nonaltruist gains.

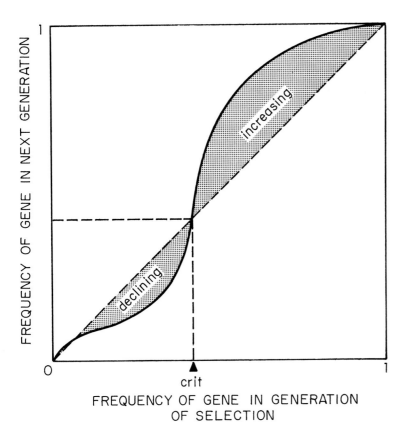

Figure 5-10 The condition for the genetic fixation of reciprocal altruism in a population. Above a critical frequency defined by the population size and the size and effectiveness of the cooperating network, the altruist gene increases explosively toward fixation. Below the critical frequency the gene recedes slowly toward mutational equilibrium. (Modified from Boorman and Levitt, 1973b.)

If they are lucky and contact a fellow cooperator, both gain. The critical gene frequency is simply that in which playing the game pays by virtue of a high enough probability of contacting another cooperator. The machinery for bringing the gene frequency up to the critical value must lie outside the game itself. It could be genetic drift in small populations, which is entirely feasible in semiclosed societies (Chapter 4), or a concomitant of interdemic or kin selection favoring other aspects of altruism displayed by the cooperator genotypes.

Altruistic Behavior

Armed with existing theory, let us now reevaluate the reported cases of altruism among animals. In the review to follow each class of be-

havior will insofar as possible be examined in the light of two or more competing hypotheses that counterpoise altruism and selfishness.

Thwarting Predators

The social insects contain many striking examples of altruistic behavior evolved by family-level selection. The altruistic responses are directed not only at offspring and parents but also at sibs and even nieces, nephews, and cousins (Wilson, 1971a). The soldier caste of most species of termites and ants is mostly limited in function to colony defense. Soldiers are often slow to respond to stimuli that arouse the rest of the colony, but when they do react, they normally place themselves in the position of maximum danger. When nest walls of higher termites such as *Nasutitermes* are broken open, for example, the white, defenseless nymphs and workers rush inward toward the concealed depths of the nests, while the soldiers press outward and mill aggressively on the outside of the nest. W. L. Nutting (personal communication) witnessed soldiers of *Amitermes emersoni* in Arizona emerge well in advance of the nuptial flights, wander widely around the nest vicinity, and effectively engage in combat all foraging ants that might have endangered the emerging winged reproductives. I have observed that injured workers of the fire ant *Solenopsis invicta* leave the nest more readily and are more aggressive on the average than their uninjured sisters. Dying workers of the harvesting ant *Pogonomyrmex badius* tend to leave the nest altogether. Both effects may be no more than nonadaptive epiphenomena, but it is also likely that the responses are altruistic. To be specific, injured workers are useless for more functions other than defense, while dying workers pose a sanitary problem. Honeybee workers possess barbed stings that tend to remain embedded in their victims when the insects pull away, causing part of their viscera to be torn out and the bees to be fatally injured (Sakagami and Akahira, 1960). The suicide seems to be a device specifically adapted to repel human beings and other vertebrates, since the workers can sting intruding bees from other hives without suffering the effect (Butler and Free, 1952). A similar defensive maneuver occurs in the ant *P. badius* and in many polybiine wasps, including *Synoeca surinama* and at least some species of *Polybia* and *Stelopolybia* (Rau, 1933; W. D. Hamilton, personal communication). The fearsome reputation of social bees and wasps is due to their general readiness to throw their lives away upon slight provocation.

Although vertebrates are seldom suicidal in the manner of the social insects, many place themselves in harm's way to defend relatives. The dominant males of chacma baboon troops (*Papio ursinus*) position themselves in exposed locations in order to scan the environment while the other troop members forage. If predators or rival troops approach, the dominant males warn the others by barking and may move toward the intruders in a threatening manner, perhaps accompanied by other males. As the troop retreats, the dominant males cover the rear (Hall, 1960). Essentially the same behavior has been observed in the yellow baboon (*P. cynocephalus*) by the Altmanns (1970). When troops of hamadryas baboons, rhesus macaques, or vervets meet and fight, the adult males lead the combat (Struhsaker, 1967a,b; Kummer, 1968). The adults of many ungulates living in family groups, such as musk oxen, moose, zebras, and kudus, interpose themselves between predators and the young. When males are in charge of harems, they usually assume the role; otherwise the females are the defenders. This behavior can be rather easily explained by kin selection. Dominant males are likely to be the fathers or at least close relatives of the weaker individuals they defend. Something of a control experiment exists in the large migratory herds of ungulates such as wildebeest and bachelor herds of gelada monkeys. In these loose societies the males will threaten sexual rivals but will not defend other members of their species against predators. A few cases do exist, however, that might be open to another explanation. Adult members of one African wild dog pack were observed to attack a cheetah and a hyena, at considerable risk to their own lives, in order to save a pup that could not have been a closer relation than a cousin or a nephew. Unattached Adélie penguins help defend nests and crèches of chicks belonging to other birds against the attacks of skuas. The breeding colonies of penguins are so strikingly large

and the defending behavior sufficiently broadcast to make it unlikely that the defenders are discriminating closely in favor of relatives. However, the possibility has not to my knowledge been wholly excluded.

Parental sacrifice in the face of predators attains its clearest expression in the distraction displays of birds (Armstrong, 1947; R. G. Brown, 1962; Gramza, 1967). A distraction display is any distinctive behavior used to attract the attention of an enemy and to draw it away from an object that the animal is trying to protect. In the great majority of instances the display directs a predator away from the eggs or young. Bird species belonging to many different families have evolved their own particular bag of tricks. The commonest is injury feigning, which varies according to the species from simple interruptions of normal movements to the exact imitation of injury or illness. The female nighthawk (*Chordeiles minor*) deserts her nest when approached, flies conspicuously at low levels, and finally settles on the ground in front of the intruder (and away from her nest) with wings drooping or outstretched (Figure 5-11). Wood ducks (*Aix sponsa*) and black-throated divers (*Gavia arctica*) spread one wing as if broken and paddle around in circles as if they were crippled. The prairie warbler (*Dendroica discolor*) plummets from the nest to the ground and grovels frantically in front of the observer. These performances can be quite affecting. New Zealand pied stilts (*Himanto-*

A   **B**

Figure 5-11 Distraction display of the female nighthawk. In order to draw intruders away from her nest, the bird often alights and either droops her wings (*A*) or holds them outstretched (*B*). (Original drawing by J. B. Clark; based on Gramza, 1967.)

pus picatus) are among the great actors of the animal world. Guthrie-Smith (1925) has described their response to intrusion in the vicinity of the nest as follows:

Dancing, prancing, galumphing over one spot of ground, the stricken bird seems simultaneously to jerk both legs and wings, as strange toy beasts can be agitated by elastic wires, the extreme length of the bird's legs producing extraordinary effects. It gradually becomes less and less able to maintain an upright attitude. Lassitude, fatigue, weariness, faint-ings—lackadaisical and fine ladyish—supervene. The end comes slowly, surely, a miserable flurry and scraping, the dying Stilt, however, even in *articulo mortis*, contriving to avoid inconvenient stones and to select a pleasant sandy spot upon which decently to expire. When on some shingle bank well removed from eggs and nests half a dozen Stilts—for they often die in companies—go through their performances, agonizing and fainting, the sight is quaint indeed.

Other behavior patterns besides injury feigning are utilized as distraction displays. Oystercatchers (*Haematopus ostralegus*) and dunlins (*Calidris alpina*) perform display flights of the kind usually limited to courtship. Many kinds of shore birds alternate injury feigning with squatting on the ground as though they were brooding eggs. Short-eared owls (*Asio flammeus*) and Australian splendid blue wrens (*Malurus splendens*) even pretend to *be* young birds, quivering their wings as though begging for food. Anecdotes in the literature indicate that predators are indeed attracted by the various kinds of distraction displays, and there can be little doubt that the adults engaging in the displays endanger their own lives while reducing the risk for their young.

There are other ways a defender can risk its life besides simply confronting the enemy. If the defender just attempts to alarm other members of its species, it attracts attention to itself and runs a greater risk. Alarm communication in social insects, described already in Chapter 3, is altruistic in a very straightforward way. In most species it is closely coupled with suicidal attack behavior. Even when the insect flees while releasing an alarm pheromone or stridulating, the signal cannot help but attract the intruder to it. Alarm communication in vertebrates is much more ambiguous. When small birds of many species discover a hawk, owl, or other potential enemy resting in the vicinity of their territories, they mob it while uttering characteristic clicking sounds that attract other birds to the vicinity. This behavior is relatively safe, because the predator is not in a position to attack. The aggression of the small birds often drives the predator from the neighborhood. Thus inciting or joining a mob would appear to increase personal fitness. However, the warning calls of the same small birds are very different in content and significance from mobbing calls. They are uttered by such diverse species as blackbirds, robins, thrushes, reed buntings, and titmice. When a hawk is seen flying overhead, the birds crouch low and emit a thin, reedy whistle.

In contrast to the mobbing call, the warning call is acoustically designed to make it difficult to locate in space. The continuity of the sound, extending over a half-second or more, eliminates time cues that reveal the direction. A pure tone of about 7 kiloherz is used, just above the frequency required for phase difference location but below the optimum for generating biaural intensity differences (Marler, 1957; Marler and Hamilton, 1966). The bird is evidently "trying" to avoid the great danger posed by the hawk. Then why does it bother at all? Why warn others if it has already perceived the danger itself? Warning calls seem prima facie to be altruistic. Maynard Smith (1965) hypothesized that they originate by kin selection; not just the mate and offspring but also more distant relatives are benefited. He devised a model proving that genes for such an altruistic trait can be maintained in a balanced polymorphic state. Next, G. C. Williams (1966a) and Trivers (1971) devised between them the following set of competing hypotheses that cover not only kin selection but also selection at the levels of the individual and the population.

Hypothesis 1. Warning calls function in the breeding season to protect the mate and young and are simply extended into the off season because it burdens the DNA to encode a seasonal adjustment (Williams, 1966a). Trivers has pointed out that this is an explanation of last resort. The hypothesis is made even less attractive by the fact that the same birds make intricate seasonal adjustments in almost every other aspect of their biology.

Hypothesis 2. Warning calls are fixed by interdemic selection (Wynne-Edwards, 1962). The considerable theoretical difficulties of such evolution have been discussed earlier in this chapter.

Hypothesis 3. Warning calls are fixed by kin selection and sustained outside the breeding season in evolutionary time owing to the probability that close kin are near enough to be helped (Maynard Smith, 1965).

Hypothesis 4. Warning calls evolve by individual selection because, in spite of first appearances, they actually help the bird giving the call. Such would be the case if predators are more likely to eat the caller when they succeed in eating a neighbor first (Trivers, 1971). Feasting on a neighbor can sustain the predator long enough for him to continue hunting, encouraging him to remain in the neighborhood. It can teach him how to catch members of the species, and give him a preference for that species. Thus warning calls may function as mobbing calls do after all, in the sense that they discourage predators from staying in the neighborhood.

At the present time no test has been devised to choose among these hypotheses. On the basis of plausibility alone, hypotheses 3 and 4 seem at least temporarily favored.

A parallel set of competing hypotheses must be devised to account for warning behavior in mammals. Colonies of black-tail prairie dogs and arctic ground squirrels set up waves of alarm calls when they sight a predator (King, 1955; Carl, 1971). Since the calling animals remain at or above the burrow entrance when they could scurry to safety, the action may be altruistic. However, a fully alerted colony cannot easily be exploited by a predator, which is thereby encouraged to postpone its attempt or to move out of the neighborhood altogether. Red deer and axis deer bark when an intruder approaches, and the herd moves off as a result. The warning might be altruistic, but, as Fraser Darling (1937) has suggested, the action could equally well be selfish, since with the entire herd alerted and moving away as a unit, the individual stands a better chance. When packs of wild dogs are sighted by Thomson's or Grant's gazelles these little antelopes run away in a conspicuous stiff-legged, bounding gait, with tails raised and white rumps flashing, a display called "stotting" or "pronking." The following observations were made by Estes and Goddard (1967) in the Ngorongoro Crater (see also Figure 5-12):

Undoubtedly a warning signal, it spread wavelike in advance of the pack. Apparently in response to the Stotting, practically every gazelle in sight fled the immediate vicinity. Adaptive as the warning display may seem, it nonetheless appears to have its drawbacks; for even after being singled out by the pack, every gazelle began the run for its life by Stotting, and appeared to lose precious ground in the process. Many have argued that the Stotting gait is nearly or quite as fast as a gallop, at any rate de-

Figure 5-12 Stotting in a female Thomson's gazelle: left, ordinary stotting; center, paddling with hind legs during extreme high stotting; right, landing from high stotting. (Modified from Walther, 1969.)

ceptively slow. But time and again we have watched the lead dog closing the gap until the quarry settled to its full running gait, when it was capable of making slightly better speed than its pursuer for the first half mile or so. It is therefore hard to see any advantage to the individual in Stotting when chased, since individuals that made no display at all might be thought to have a better chance of surviving and reproducing.

Other cursorial mammals, including other true African antelopes, the pronghorn antelope of North America, some species of deer, and the antelopelike caviid rodent *Dolichotis patagonum*, also stott or at least display a white rump flash in the face of predators. In many instances rump flashing appears to be used as a submissive signal directed toward other members of the same species (Guthrie, 1971). This behavior might have been extended along with the stotting gait to serve as an altruistic alarm system, as implied by the observations of Estes and Goddard. The extension could occur by kin selection. However, predators less relentless than the wild dog are likely to be confused and thwarted by the movement of an entire herd, and in this case the advantage would go to any individual that alarms the herd. A third hypothesis, suggested by Smythe (1970b), is that rump flashing and the stotting gait function as "pursuit invitations." When an animal sees a predator at a distance, it has begun a dangerous episode that may last a long time. Unless the predator pursues it immediately, the animal must spend time following the movements of the enemy or risk being surprised at close range. However, if the animal can induce pursuit by displaying at the time when it has the advantage in both distance and initiative, it stands a good chance of discouraging the predator and causing it to hunt for less alert prey. Just as in the case of the alarm calls of birds, no method has been developed to eliminate decisively any of the several available explanations.

Strong but not conclusive circumstantial evidence for kin selection is provided by the evolution of unpalatability in Lepidoptera. Suppose that mutants appeared in a moth or butterfly that made the individual repellent to predators. Each predator learns to avoid the mutant by eating one. The problem remains: How can the gene increase in frequency from low mutational levels if only a few predators in the neighborhood can be taught each generation, and if the price of teaching even this small fraction is the further reduction of the mutant frequency. There are three reinforcing processes by which the frequency might be increased. If mutants look, sound, or smell different from nonmutants, in other words if they are also aposematic, the predator can avoid them differentially, and the "lesson" will be taught to the predator population that much more easily. If attacks on the insects sometimes result in injury rather than death, the aposematic mutant can live to enjoy the fruits of its experience in terms of increased individual fitness. Finally, if the insect is surrounded by its relatives, even its death can result in a rise in inclusive fitness, because the unpalatability caused by genes held through common descent will spread. When kin selection is effective, we should expect that unpalatable species of insects will be those in which, by and large, relatives are most closely associated in populations. This does appear to be the case. Among the species of heliconiine butterflies, unpalatability of adults is associated with a tendency to form roosting aggregations at specific sites to which the insects return repeatedly. It is also associated with greater geographic subspeciation, generally a sign of lower gene flow within populations (Benson, 1971). Among the hemileucine saturniid moths, aposematic species are also the ones with the longest adult postreproductive life (Blest, 1963). The implication seems to be that it pays to stay around as long as possible after reproduction is finished in order to teach predators not to eat one's offspring. In contrast, cryptic saturniids have a short postreproductive life: it does *not* pay to teach predators that one's relatives are good to eat. The evolution of unpalatability by kin selection does not create altruism in the conventional sense. Heliconiine butterflies that reduce dispersal rates are not necessarily self-sacrificing. However, the process is fundamentally the same, in the sense that the genes of an individual shared with relatives by common descent are promoted by the individual's death.

Cooperative Breeding

The reduction of personal reproduction in order to favor the reproduction of others is widespread among organisms and offers some of the strongest indirect evidence of kin selection. The social insects, as usual, are clear-cut in this respect. The very definition of higher sociality ("eusociality") in termites, ants, bees, and wasps entails the existence of sterile castes whose basic functions are to increase the oviposition rate of the queen, ordinarily their mother, and to rear the queen's offspring, ordinarily their brothers and sisters. The case of "helpers" among birds is also strongly suggestive (Skutch, 1961; Lack, 1968; Woolfenden, 1974). Among the many cases of helpers assisting other birds to rear their young, including moorhens, Australian blue wrens, thornbills, anis, and others, the assistance is typically rendered by young adults to their parents. Consequently, just as in the social insects, the cooperators are rearing their own brothers and sisters (see Chapter 22).

In some respects "aunt" and "uncle" behavior in monkeys and apes superficially resembles the cooperative brood care of social insects and birds. Childless adults take over the infants of others for short periods during which they carry the young about, groom them, and play with them. The baby-sitting may seem to be altruistic, but there are other explanations. Adult males of the Barbary macaque use infants in ritual presentations to conciliate other adult males. The "aunts" of rhesus and Japanese macaques also use baby-sitting to form alliances with mothers of superior rank. Furthermore, the possibility

cannot be excluded that aunting behavior provides training in the manipulation of infants that improves the performance of young females when they bear their first young (see Chapter 16).

Outright adoption of infants and juveniles has also been recorded in a few mammal species. Jane van Lawick-Goodall (1971) recorded three cases of adult chimpanzees adopting young orphaned siblings at the Gombe Stream Reserve. As she noted, it is strange (but significant for the theory of kin selection) that the infants were adopted by siblings rather than by an experienced female with a child of her own, who could supply the orphan with milk as well as with more adequate social protection. During studies of African wild dogs in the Ngorongoro Crater conducted by Estes and Goddard (1967), a mother died when her nine pups were only five weeks old. The adult males of the pack continued to care for them, returning to the den each day with food until the pups were able to join the pack on hunting trips. The small size of wild dog packs makes it probable that the males were fathers, uncles, cousins, or other similarly close relatives. Males of the hamadryas baboon normally adopt juvenile females (Kummer, 1968). This unusual adaptation is clearly selfish in nature, since in hamadryas society adoption is useful for the accumulation of a harem.

Assistance in the reproductive effort of others can take even stranger forms. In the southeastern Texas population of the wild turkey (*Meleagris gallopavo*), brothers assist each other in the fierce competition for mates (Watts and Stokes, 1971). The union of brothers begins in the late fall, when the birds are six to seven months old. At that time the young males break away from their brood flock together. They maintain their bond as a sibling group for the rest of their lives, so that even when all its brothers die, a male does not attempt to join another sibling group. In the winter the brotherhood joins flocks of other juvenile birds. At this time their status is determined by a series of combats, in which the young males wrestle in fighting-cock style for as long as two hours, pecking at each other's head and neck and striking with the wings. The winner of the match becomes the dominant member of the pair for life. Such contests are conducted at three levels. First, the brothers struggle with each other until one emerges as the unchallenged dominant. Next, brotherhoods meet in contest until one group, usually the largest, achieves ascendancy over all the others in the winter flock. Finally, different flocks contend with one another whenever they meet, again settling dominance at the group level. The final result of this elaborate tournament is that one male comes to hold the dominant position in the entire local population of turkeys. When the males and females gather on the mating grounds in February, each brotherhood struts and fantails in competition with the others (Figure 5-13). The brothers display in synchrony with each other in the direction of the watching females. When a female is ready to mate, the subordinate brothers yield

Figure 5-13 Kin selection among males of the wild turkey. On the display grounds the brotherhoods, represented here by two pairs of brothers and one solitary cock, display to watching females by stereotyped strutting with their tails fanned and wings drooping. The brothers in each set display in synchrony. In the subsequent mating, subordinate brothers defer to the dominant male, and subordinate brotherhoods defer to the dominant brotherhood, usually the largest group. (From "The Social Order of Turkeys," by C. R. Watts and A. W. Stokes, 1971.

to their dominant sibling, and the subordinate brotherhoods yield to the dominant one. As a result only a small fraction of the mature males inseminate the females. Of 170 males belonging to four display groups watched by Watts and Stokes, no more than 6 cocks accounted for all the mating.

The Tasmanian hen *Tribonyx mortierii*, a flightless rail endemic to Tasmania, provides an equally strong case of kin selection among brothers (Maynard Smith and Ridpath, 1972). There is an excess of males among the juveniles, and males compete for females on the perennial territories. The territories are occupied either by mated pairs or by trios. Remarkably, most of the trios consist of a female and two brothers. The sibling cooperation pays off in inclusive fitness: the trios produce larger clutches, and successfully rear a higher percentage of chicks, than do the mated pairs. Such arrangements may be more widespread in the animal kingdom than previously suspected. Coe (1967) reported a case in the African rhacophorid frog *Chiromantis rufescens* of three males cooperating with a female to help build an egg nest. The nest was constructed from a fluid secreted by the female, which all four frogs beat into a thick white foam with a swimming motion of their hind legs (Figure 5-14). Although only one male was in the amplexus position, it was not determined whether he alone fertilized the eggs. Nor could the kinship of the males, if any, be estimated.

The cellular slime molds provide evidence of what seems to be altruistic cooperation at the single-cell level. Their life cycle, as exemplified by *Dictyostelium discoideum*, begins with the emergence of amebas from scattered spores (Bonner, 1967). In the beginning the amebas live independently from one another, moving sluggishly through the watery medium of their soil habitats, feeding on bacteria, and multiplying by fission. When food grows scarce and the popula-

Figure 5-14 Cooperative nest building in the African frog *Chiromantis rufescens.* The large female (far right) is assisted by three males in whipping her secretion into a froth. (From Coe, 1967.)

tion of cells dense, the amebas aggregate into much larger sluglike organisms called pseudoplasmodia. After migrating for a while, each pseudoplasmodium reshapes itself into a spore-producing structure consisting of a spherical mass supported by a thin stalk. The amebas that make it into the sphere generate the spores that start a new life cycle. Those that form the stalk do not reproduce. Virtually nothing is known about the kinship of the cooperating amebas in nature. It is probable that stalk and sphere cells are often closely related, perhaps even genetically identical, but such is not likely to be true all the time. The case of the altruistic amebas presents a theoretical problem no less challenging than that raised by the altruistic vertebrates and insects.

Food Sharing

Other than suicide, no behavior is more cleanly altruistic than the surrender of food. The social insects have carried food sharing to a high art. In the higher ants, the "communal stomach," or distensible crop, together with a specially modified gizzard, forms a complex storage and pumping system that functions in the exchange of liquid food between members of the same colony (Eisner, 1957). In both ants and honeybees, newly fed workers often press offerings of regurgitated food on nestmates without being begged, and they may go so far as to expend their supply to a level below the colony average (Gösswald and Kloft, 1960; Lindauer, 1961; Wallis, 1961; Lange, 1967). The regurgitation results in at least two consequences of importance to social organization beyond the mere feeding of the hungry. First, because workers tend to hold a uniform quantity and quality of food in their crops at any given moment, each individual is continuously apprised of the condition of the colony as a whole. Its personal hunger and thirst are approximately those of the entire colony, and in a literal sense what is good for one worker is good for the colony. Second, the regurgitated food contains pheromones, as well as special nutriments manufactured by exocrine glands and other substances of social importance. Besides contributing to colony organization, mutual feeding can be genuinely self-sacrificing. When honeybees are fed exclusively on sugar water, they can still raise larvae—but only by metabolizing and donating their own tissue proteins (Haydak, 1935). That this donation to their sisters actually shortens their own lives is suggested indirectly by the finding of de Groot (1953) that longevity in workers is a function of protein intake.

Altruistic food sharing among adults is also known among African wild dogs, where it permits some individuals to remain at the dens with the cubs while others hunt (Kühme, 1965; H. and Jane van Lawick-Goodall, 1971). The donors carry fresh meat directly to the recipients or else regurgitate it in front of them. Occasionally a mother dog will allow other adults to suckle milk. A bizarre case of regurgitation among adults has been observed in a captive colony of cattle egrets (*Bubulcus ibis*) by Koenig (1962). Young adult egrets continued to beg from their parents even after they started breeding. Part of the food they obtained in this way was passed on to their own offspring—the grandchildren of the original donors. However, the phenomenon may be abnormal. Crowded conditions in the cages led to unusual circumstances: nests being constructed on top of one another, proximity of parents and offspring prolonged into maturity, and close inbreeding.

Altruistic food sharing has been reported on several occasions in the higher anthropoids. In captive gibbons the exchange is initiated by one animal trying to take food from another, either by grasping the food or by holding the partner's hand while taking the food. The partner usually lets some of the food go without protest. Under some circumstances it will resist by keeping it out of reach or, rarely, by threatening or fighting. The offering of food without solicitation does not appear to occur (Berkson and Schusterman, 1964). Chimpanzees also successfully beg food from one another, especially part of the small mammals that the apes occasionally kill as prey. This benevolent behavior is in sharp contrast to that of baboons: when they kill and eat small antelopes, the dominant males appropriate the meat, and fighting is frequent (Kummer, 1971). Chimpanzees also communicate the location of foods to one another. Adults can remember the position of previous finds and lead others to the location by walking toward it in a characteristic fashion. If no one follows, the leader beckons with a wave of the hand or head, or else taps another chimpanzee on the shoulder, wraps an arm around its waist, and tries to induce it to walk in tandem (Menzel, 1971). Even more impressive, entire parties of chimpanzees often set up a loud booming clamor when they discover a fruit tree. Other groups within earshot (up to a full kilometer) respond boisterously and in many instances join the first group. The communication thus leads to a cooperative sharing of the food (Reynolds and Reynolds, 1965; Sugiyama, 1969).

Ritualized Combat, Surrender, Amnesty

The mere forebearance of an enemy can be a form of altruism. Fighting between animals of the same species is typically ritualized. By precise signaling, a beaten combatant can immediately disclose when it is ready to leave the field, and the winner will normally permit it to do so without harm. African wild dogs display submission by an open-mouth grimace, a lowering and turning of the head and neck, and a belly-up groveling motion of the body. The loser thus exposes itself even more to the bites of its needle-toothed opponent. But at this point the attack either moderates or stops altogether. Male mantis shrimps fight with explosive extensions of their second maxillipeds. One strike from these hammer-shaped appendages is enough to tear another animal apart. But fatalities seldom occur, because each shrimp is careful to aim at the heavily armored tail segment of its opponent (Dingle and Caldwell, 1969). Other examples of ritualized aggression can be multiplied almost endlessly from the literature, and indeed they form a principal theme of Konrad Lorenz's celebrated book *On Aggression.* They also pose a considerable theoretical difficulty: Why not always try to kill or maim the enemy outright? And when an opponent is beaten in a ritual encounter, why not go ahead and kill him then? Allowed to run away, to paraphrase the childhood rhyme, the opponent may live to fight another day—and win next time. So in a sense the kindness shown an enemy seems altruistic, an unnecessary risk of personal fitness. One explanation is that mercy is "good for the species," since it allows the greatest number of individuals to remain healthy and uninjured. That hypothesis requires interdemic selection of a high intensity, because at the level of individual selection the greatest fitness in such encounters would always seem to accrue to the genotype that "plays dirty." A second hypothesis is that ritualization arises from kin selection: the need to win fights without eliminating the genes shared with others by common descent. The explanation could well hold in many particular cases, for example the wrestling matches between the brother turkeys in Texas. But in other species the highly ritualized encounters are held between individuals that are at best distantly related. A third hypothesis, suggested by Maynard Smith and G. R. Price (1973; see also Price in Maynard Smith and Ridpath, 1972), explains ritualized fighting as the outcome of purely individual selection. It recognizes that a great many animal species actually display two forms of combat, ritualized fighting and escalated fighting. The escalated form is invoked when an animal is hurt by an opponent. This particular form of behavioral scaling will be stabilized in evolution because it is disadvantageous either to engage in escalated fighting too readily or never to use it at all.

The Field of Righteousness

In conclusion, although the theory of group selection is still rudimentary, it has already provided insights into some of the least understood and most disturbing qualities of social behavior. Above all, it predicts ambivalence as a way of life in social creatures. Like Arjuna faltering on the Field of Righteousness, the individual is forced to make imperfect choices based on irreconcilable loyalties—between the "rights" and "duties" of self and those of family, tribe, and other units of selection, each of which evolves its own code of honor. No wonder the human spirit is in constant turmoil. Arjuna agonized, "Restless is the mind, O Krishna, turbulent, forceful, and stubborn; I think it no more easily to be controlled than is the wind." And Krishna replied, "For one who is uncontrolled, I agree the Rule is hard to attain; but by the obedient spirits who will strive for it, it may be won by following the proper way." In the opening chapter of this book, I suggested that a science of sociobiology, if coupled with neurophysiology, might transform the insights of ancient religions into a precise account of the evolutionary origin of ethics and hence explain the reasons why we make certain moral choices instead of others at particular times. Whether such understanding will then produce the Rule remains to be seen. For the moment, perhaps it is enough to establish that a single strong thread does indeed run from the conduct of termite colonies and turkey brotherhoods to the social behavior of man.

Part II **Social Mechanisms**

Chapter 6

Group Size, Reproduction, and Time-Energy Budgets

Natural selection extended long enough always leads to compromise. Each selection pressure guiding genetic change in a population is opposed by other selection pressures. As the population evolves, the stronger pressure eventually weakens while opposing ones intensify. When these forces finally strike a balance, the population phenotypes can be said to be at their evolutionary optimum; and evolution has passed from the dynamic to the stabilizing state. A convenient way of visualizing the process with special reference to social evolution is shown in Figure 6-1. The axes of the graphs measure variation of two social traits in some quality, say, degree of complexity or intensity. The organisms in the population are represented by points on the plane, the position of each being determined by the phenotype it possesses in the two social traits. The cluster is densest near its center. This by definition constitutes the statistical mode of the population. For each environment there exists only one or a set of very few positions at which the statistical mode is favored by selection over less common phenotypes. If the population is not centered on one of these positions, the resulting dynamic selection will tend to move it there. Thus selection superimposes a kind of force field upon the plane of phenotypes. The position at which the population comes temporarily to rest is the ensemble of phenotypes around which selection forces are balanced, and it constitutes the evolutionary compromise.

If this equilibrium case is generally true in social systems, weak and intermediate stages in phylogenetic successions among living species represent earlier compromises rather than evolution in progress. The population phenotypes have simply been balanced by selection forces at some early point such as the lower lefthand area in Figure 6-1, rather than continuing to move away, as shown in the example

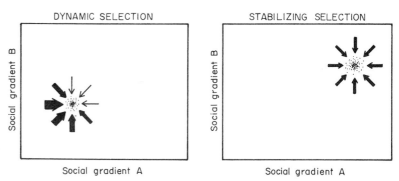

Figure 6-1 Evolution in two social traits is viewed here as the movement of an entire population of organisms on a plane of phenotypes. The rate and direction of movement is determined by the force field of opposing selection pressures (left figure). The stable states of social traits are reached when the selection pressures balance, the condition called stabilizing selection (right figure).

depicted. It is reasonable to postulate, as a working hypothesis, that most social species are at least temporarily stabilized. Some have halted well down on the scale, to remain "primitive" species, while others have moved farther along before stabilizing (as seen in the righthand graph of Figure 6-1), to become the "advanced" species.

Examples of counteracting selection forces are easy to find in nature. The intensity of aggressive behavior is undoubtedly limited by a destructiveness to self and relatives that causes a loss in genetic fitness comparable to the gains accrued from the defeat of the enemy. Destructive behavior is easy to document in nature. Male hamadryas baboons, for example, sometimes injure the females over which they are fighting, and bull elephant seals trample pups to death while flopping around in their spectacular territorial battles. Similarly, evidence of evolutionary compromise that limits destructiveness is readily found. In general, fighting among animals of the same species seldom passes from the ritual to the escalated stage, that is, to the point where serious injury is mutually inflicted (see Chapter 11). An obverse form of compromise is fashioned during the evolution of submissive behavior. Animals belonging to dominance systems submit to their superiors, signaling their state of mind with displays that are sometimes very specialized and elaborate, but they cannot be pushed beyond certain clear limits. At some level of harassment, the persecuted animal turns on its attacker with escalated fighting or deserts the group altogether. A more precise measure of the level of compromise can be obtained from the amount of time spent grooming other animals. In many dominance systems the subordinate individual grooms its superiors as a conciliation device. Rhesus monkeys are so punctilious in this matter that the rank of the animal can be ascertained simply by observing which group members it grooms and by whom it is groomed. How much time, from the animal's point of view, should be devoted to grooming others? Just enough to consolidate and advance its position. This cynical hypothesis is at least consistent with direct observations of the shifting dominance relations within rhesus troops.

Compromise is also manifest throughout the evolution of sexual behavior. The males of polygamous birds tend to evolve greater size, brighter plumage, and more conspicuous displays as devices for acquiring multiple mates. The trend is opposed by the greater ease with which predators are able to locate and capture the more dramatic males. As a result the sex ratio is progressively unbalanced with advancing age. The sex ratio of newly hatched great-tailed grackles (*Cassidix mexicanus*) is balanced, but within two months after the breeding season the ratio among first-year and adult birds is 1 male to 1.34 females, while five months later, in the following spring, the ratio has fallen to 1:2.42. Selander (1965), who discovered this case, believes that the higher mortality rate of males is partly a result of their greater vulnerability to predators, which in turn is due to their conspicuous coloration and loss of flight maneuverability caused by the long tails used in display. A second handicap appears to be their larger size, which reduces efficiency at foraging.

The Determinants of Group Size

The number of members in a society is an example of one of those elusive social phenotypes that can be wholly understood only by recourse to the concept of evolutionary compromise. We will approach this subject by considering first the purely functional parameters that influence group size, or more precisely those that determine the frequency distributions of groups of variable size. Then we will proceed to a consideration of the selection pressures that have led to particular values of the functional parameters. The total analysis must answer questions at two levels. First, what forces add individuals to groups and subtract them? And what magnitude of these forces must operate to create the observed frequency distributions? Second, to what extent has natural selection shaped responses to the forces, or even moderated the forces themselves? In proceeding from the first level to the next, the analysis will shift from phenomenological to fundamental theory.

The phenomenological theory has been largely developed by Joel E. Cohen (1969a, 1971). Cohen took his inspiration from earlier, partially successful attempts by sociologists, notably John James, J. S. Coleman, and Harrison White, to fit the size-frequency distributions of human groups to a Poisson distribution with the zero term (that is, frequency of groups with no members) eliminated. Groups were defined as clusters of people laughing, smiling, talking, working, or engaging in other activities indicating face-to-face interaction. The populations studied included pedestrians on city streets, masses of shoppers in department stores, and elementary school children at play. When the numbers of little clusters containing one person, two persons, and upward are counted, they fit in most instances a Poisson distribution with a truncated zero term. Cohen extended this approach to Old World monkeys. But he went deeper into the basic problem by deriving frequency distributions from a stochastic model in which groups of varying size and composition possess varying abilities to attract and to hold temporary members. Three parameters were entered: a is the rate (per unit of time) at which a single individual in a system of freely forming groups joins a group solely because of the attractions of group membership, independently of the size of any particular group; b is the rate at which the lone individual joins a group because of the attractiveness of individuals in the group, where the degree of attractiveness of the group can therefore be expected to change with the number of members in it; and d is the rate at which an individual member already in a group departs because of some personal decision of its own that is independent of the size of the group. Consider a closed population of individuals that are freely forming into casual groups containing variable numbers of individuals. The number of groups containing a certain number of members is designated as n_i, where $i(= 1, 2, 3 \ldots)$ represents the number of members. In the simplest kind of social system the rate of change in the number of casual groups of a certain size is conjectured to be

$$\frac{dn_i(t)}{dt} = an_{i-1}(t) + b(i - 1)n_{i-1}(t) - an_i(t) - b(i)n_i(t)$$
$$- d(i)n_i(t) + d(i + 1)n_{i+1}(t)$$

This formula states that in a short interval of time the number of groups of size i is being *increased* at time t by:

1. The number of groups one member less in size $(i - 1)$ times the rate (a) at which individuals join groups independently of their membership; this increases a given group of size $i - 1$ to size i. *Plus*

2. The number of groups one member less in size $(i - 1)$ times the rate $[b(i - 1)]$ at which individuals join groups of that size owing to the attractiveness of individuals in it. *Plus*

3. The number of groups one member more in size $(i + 1)$ times the rate $[d(i + 1)]$ at which individuals spontaneously leave groups of that size; this decreases a given group of size $i + 1$ to size i.

The number of groups of size i is being simultaneously *decreased* by:

1. The number of groups already at size i times the rate (a) at which individuals are attracted to groups regardless of membership; this increases the number of members from i to $i + 1$. *Plus*

2. The number of groups already at size i times the rate $[b(i)]$ at which individuals are attracted to groups of size i owing to membership. *Plus*

3. The number of groups already at size i times the rate of spontaneous departure from groups of that size $[d(i)]$.

A second equation is simultaneously entered for the rate of increase of solitary individuals in the system. The basic model can be made more complex, as Cohen (1971) has shown, by adding terms that include other increments of attraction and expulsion as functions of group size and composition.

The single most important result of Cohen's basic model is the demonstration that at equilibrium ($dn_i = 0$ for all i), the frequency distribution of casual group size in a closed population should be a zero-truncated Poisson distribution when $b = 0$, in other words when the specific membership or size of a group does not influence its attractiveness, and a zero-truncated binomial distribution when b is a positive number. Existing data from several primate species, including man, conform reasonably well to one or the other of the two

distributions. Cohen has further demonstrated that estimated values of the ratios a/d and b/d are a species characteristic. As shown in Table 6-1, a general decrease in the role of individual attractiveness is apparent when one passes from the more elementary to more advanced social groups. Whether this rule will hold in larger samples remains to be seen. The important point is that a great many previously jumbled numerical data have been put into preliminary order in a surprisingly simple way. Hope has thus been engendered that at least one of the coarser qualities of sociality, group size, can be fully derived from models that specify as their first principles the forms and magnitude of individual interactions.

In addition to the *casual societies*, or casual groups, just considered, there exist *demographic societies*. The difference between the two is only a matter of duration in time, but its consequences are fundamental. The casual group forms and dissipates too quickly for birth and death rates to affect its statistical properties; immigration and emigration into and out of the population as a whole are also insignificant. The demographic society, in contrast, is far more nearly closed than the casual group, and it persists for long enough periods of time for birth, death, and migration between demes to play leading roles. One way in which a population can exist at both levels is for a more or less closed society to exist demographically while the membership of casual groups within it changes kaleidoscopically on a shorter time scale. In a separate modeling effort, Cohen (1969b) showed that when members of demographic societies are born, die, and migrate from one society to another at positive rates not dependent on the size of the group to which they belong, the frequency distribution of societies with varying numbers of members can be expected to approximate the negative binomial distribution with the zero term truncated. If, on the other hand, the individual birth rate is temporarily zero, or the number of offspring born in each group per unit of time is constant regardless of the size of the group, the frequency distribution should approach a zero-truncated Poisson distribution. These predictions appear to be well borne out by existing data from primate field studies. Langur and baboon troops, in which the demographic parameters are more or less independent of group

Table 6-1   Values of the ratios of attraction rates (a, b) to the spontaneous departure rate (d) in groups of two monkey species and man, estimated from the basic Cohen model of casual group formation.

Group	a/d	b/d
Vervets (*Cercopithecus aethiops*)	1.15	0.66
Baboons (*Papio cynocephalus*)	0.12	0.16
Nursery children	0.33	0.10
Miscellaneous human groups	0.86	0

size, conform to a negative binomial distribution. Gibbon troops are societies in which only one infant is born at a time regardless of group size, which causes the individual birth rate to be a decreasing function of the number of members; group size in this case is Poisson distributed (see Table 6-2). During healthy periods, howler monkey troops fit negative binomial distributions, but following an epidemic in which young were temporarily eliminated, their size-frequency distribution shifted to the Poisson—as anticipated. It is a curious fact that although the form of the frequency distributions is correctly predicted by the most elementary stochastic model that incorporates demography, the dynamics of the model are not faithful to the single detailed set of demographic data (from yellow baboons) that were available to Cohen. In other words, the internal structure of the model must be made more complex in some way that cannot yet be guessed.

We can now turn to the evolutionary origins of group size by treating the entire subject in terms of the following argument. The immediately determining parameters are themselves adaptations on the part of individual organisms. The attractiveness of a group to a solitary animal is ultimately determined by the relative advantage of joining the group, measured by the gain in inclusive genetic fitness. Whether the organism attempts to migrate from one semiclosed demographic society to another is also under the direct sovereignty of natural selection. The birth rate, as shown earlier (Chapter 4), is another parameter very sensitive to selection, because it is not only a key component of reproductive fitness but also contributes—negatively—to the survival rate of the parents. Of all the parameters determining group size, only the death rate can be said to escape classification as a direct adaptation to the environment.

We can postulate that the modal size of groups will be simply the outcome of the interaction of the parameter values that confer maximum inclusive fitness. In all social species the modal group size will therefore represent a compromise. The size must be greater than one because of the advantages of group foraging, or group defense, or any one of the combination of the "prime movers" of social evolution reviewed in Chapter 3. But it cannot be indefinitely large, since beyond a certain number the food runs out, or the defense can no longer be coordinated effectively, and so forth. The upper limits of group size are unfortunately much more difficult to discern in field

Table 6-2 Size distribution of primate troops. (From Cohen, 1969b.)

Gibbon (*Hylobates lar*)			Baboon (*Papio*)		
Number of members in each troop	Number of troops observed	Number of troops predicted by Poisson distribution	Number of members in each troop	Number of troops observed	Number of troops predicted by negative binomial distribution
2	8	10.9	1–10	2	10.2
3	15	12.5	10–20	19	21.0
4	12	10.7	20–30	36	24.1
5	9	7.3	30–40	24	22.6
≥6	5	7.6	40–50	18	19.1
			50–60	16	15.2
			60–70	9	11.6
			70–80	10	8.6
			80–90	9	6.3
			90–100	2	4.5
			100–110	4	3.1
			110–120	. . .	2.2
			120–130	. . .	1.5
			130–140	. . .	1.0
			140–150	. . .	
			150–160	1	
			160–170	1	
			170–180	1	2.1
			180–190	. . .	
			190–	1	

studies than are the initial advantages favoring sociality at lower numbers. We can only speculate, for example, about the disadvantages of excessive size in fish schools. The food supply must ultimately be limiting, of course. As the schools grow larger their energy demands increase directly with the volume occupied by the fish, but the rate of energy acquisition increases with the outer surface of the fish school. Energy requirements, in other words, increase with the cube of the school's diameter, and energy input with its square—a disparity analogous to the weight-surface law of organismic growth. There are other potential disadvantages of large size. The Brock-Riffenburgh model (see Chapter 3) makes it plausible that by clumping into schools, fish are found less often by predators. If schools become very large, however, there is a strong incentive for predators to track them continuously and to develop special orientation and other behavioral devices for staying close. Goss-Custard (1970) has developed essentially the same argument with respect to feeding and defense in flocks of wading birds. Wildebeest feed socially; during the dry season in the Serengeti plains they migrate to new feeding grounds in vast herds. Groups of wildebeest also appear to have greater alertness to predators than do the solitary bulls, although the difference is not so striking as in gazelles and impalas. And belonging to large herds has its own clear dangers. According to Schaller (1972), "Wildebeest sometimes stampede toward a river from as much as 1 km away. The long column of animals hits the river at a run, and if the embankment is steep and the water deep the lead animals are slowed down while those behind continue to press forward until the river turns into a lowing, churning mass of animals some of which are trampled and drowned. One such herd I observed at Seronera left seven dead behind; several hundred may drown in such circumstances." Jarman (1974) has argued on the basis of an impressive amount of documentation that the upper limits of antelope herds, including those of wildebeest, are stringently set by the food habits of the species. For example, small-bodied browsers such as duikers and dik-diks remain in one small home range throughout the year, where they feed on such relatively dependable and densely concentrated items as flowers, twig tips, and bark; and they are consequently nearly solitary in habit. In contrast, the largest antelopes, including the wildebeest and eland, feed unselectively on a wide range of grasses and move seasonally within a very large home area. Partly in response to predators and partly as an adaptation to the patchiness and fluctuating quantity of their food supply, they roam in large herds. But their population density, and with it the upper limit of herd size, is restricted by the poor nutritive quality of the vegetation on which they feed.

Within closed human societies, similar principles are at work, although the role of Nature is not nearly so direct or harsh. In the 1800's and early part of the present century Mennonite communities of the rural United States needed about 50 families to achieve stabil-

ity. At this size the basic functions of commerce and services such as medicine and barbering could be assured. When only 40 families were present, the communities could still survive but were more vulnerable. With less than 40 families, inbreeding and disruption from more frequent marriages with outsiders became serious problems. When communities became very large, other kinds of disruption emerged: intracolonial rivalries developed, and the lay ministry became less effective. In more recent years the minimum viable group size dropped to 20 or 25 families as travel and communication with coreligionists in other parts of the country became easier (Allee et al., 1949).

The ultimate control of group size, being the result of evolutionary compromise, is most efficiently analyzed by optimization theory. In Figures 6-2 and 6-3 are shown two graphical models to illustrate this approach; their curves hypothesize the general form of the functions. The first, inspired by a proposition about group territoriality by Crook (1972), assumes an exclusive or at least overwhelming role for the energy budget. In this extreme case, foraging by small groups is more effective in energy yielded per individual animal per unit of time than is solitary foraging within populations of equal density. Crook

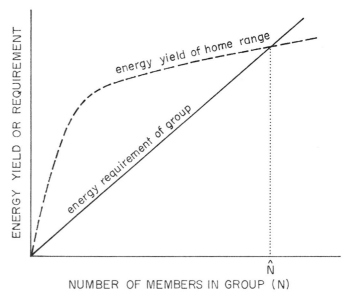

Figure 6-2 The optimization of group size is represented in this extreme model as an exclusive function of the energy budget. As group size increases, the energy requirement increases at the same rate, but the energy yield decelerates after an initial rapid rise. If unopposed by other selection pressures, the modal group size $\hat{N}$ should change in evolution to a point where the energy yield of the home range is fully utilized.

argued, correctly I believe, that although the energy requirements of the society increase linearly with the number of its members, the amount of territory that the group can effectively defend decelerates after a certain point. If defensible territory is translated into energy yield, it becomes clear that a maximum group size exists, above which demand exceeds the yield and either mortality or emigration must redress the balance. When the population as a whole is limited by energy, as opposed to some other density-dependent factor such as predation, group size will tend to evolve toward the maximum. There will also be a tendency for the group to become territorial in behavior. The more stable the environment, and the more evenly distributed the food in time and space, the more nearly the group size will approach the theoretical maximum. In a capricious environment, however, the optimal group size will normally be less than the theoretical maximum. The reason is that the energy yield of a home range measured over a long period of time is based on intervals of both superabundance and scarcity. The group must be small enough to survive the more prolonged periods of scarcity.

The energy-budget model takes into account only the components of fitness added by group superiority in territorial defense and foraging techniques. A more general modeling effort, which can encompass all of the components of genetic fitness, is presented in elementary form in Figure 6-3. Three ideas are incorporated into this more complicated graph: all components of fitness enhanced by group activity must inevitably decline beyond a certain group size; the increment curves, that is, contributions to fitness as a function of group size, usually differ from one component to another; and the optimum group size is that at which the sum of group-related increments in fitness is largest. This figure is no more than a representation of postulates; the data for drawing the fitness curves represented in it do not yet exist.

Among the social insects, group size is sometimes ultimately limited at least in part by the choice of nest site. Survival of a founding queen, swarm, or other colonizing unit often depends as much on the securing of an appropriate nest site as it does on the ability to forage for food. Often the nest site is clearly more important: because these insects can carry food reserves sufficient for days or weeks in their distensible crops or degenerating wing muscles, they do not have to forage very often; but they require constant protection from the enemies, including ants and a wide variety of predators, that threaten their existence every minute, and never so intensely as during the first few days. Social insects are typically specialized in their choice of nest sites. A great many tropical ant species nest only in cavities within the trunks or branches of trees; certain forms of *Azteca*, *Pseudomyrmex*, and *Tetraponera* are each limited to one particular tree species. Others require such havens as epiphytes, abandoned termite nests, the bark of living trees, and the subcortical spaces of

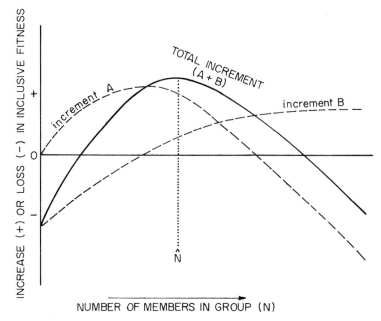

Figure 6-3 In this more general model, the optimum group size is given as a function of the maximum summed components of genetic fitness. Two social contributors to fitness are indicated as *A* and *B*; they could be, for example, increments due to superior group foraging and superior group defense against predators.

large logs at particular stages of decomposition. Still others construct fortresses out of excavated soil or carton made from chewed vegetable fibers. Substantial diversity also occurs in the social bees, wasps, and termites.

The evolutionary choice of a nest site sets an upper limit on the size of the mature colony. Species of meliponine bees that live in the narrow branches of forest trees have smaller colonies than those utilizing hollow tree trunks or cavities in the soil. The largest colonies of ants belong to species that nest in the soil or construct carton nests in trees. The smallest are specialized for life in small pieces of rotting wood embedded in leaf litter and humus. In the taxon cycle of the ants of New Guinea and other archipelagoes of the western Pacific, expanding species typically live in grassland, forest borders, and other ecologically marginal and variable habitats. Here they are required to nest mostly in the soil, and consequently are characterized by larger colony size, a greater tendency to utilize odor trails, and a more frequently expressed physical caste system. As species penetrate the inner forest habitats, their geographic ranges become more restricted, and many evolve into endemics limited to one or a few islands. A large percentage also become specialized for nesting in small pieces of rotting wood on the ground, the single most favored nest site of forest-dwelling ants. This nest-site specialization brings with it reduced mature colony size, less use of odor trails, and reduced physical caste systems (Wilson, 1959b, 1961). Brown and Wilson (1959) noted a similar trend within the ant tribe Dacetini and were able to link it with a shift in food habits. The morphologically most primitive species have large workers, nest in trees or in the soil, develop colonies containing hundreds or thousands of workers, and prey on a variety of small arthropods. The most advanced species, in *Strumigenys, Smithistruma*, and other genera, are characterized by small body size, a preference for small pieces of rotting wood or small cavities in rotting logs as nest sites, and a mature colony size of a few hundreds or less. They also prey exclusively on a limited number of kinds of springtails (Collembola) and other minute, soft-bodied arthropods that are most accessible to small ants living within the leaf litter.

Adjustable Group Size

The significance of the casual society, as opposed to the demographic, lies in its adjustable size. The number of animals can be fitted to the needs and opportunities of the moment. The total breeding population consists of a constantly moving set of nuclear units—the individual animals, the family, the colony, or whatever—that band together temporarily to form larger aggregates of variable size. The aggregates can be passive, consisting of nuclear units that relax their mutual aversion temporarily to utilize a common resource, or they can actively cooperate to achieve some common goal. The goal of the casual society may be any activity that promotes inclusive fitness, from feeding to defense or hibernation.

Excellent examples of casual groups are provided by what Kummer (1971) has called the fusion-fission societies of the higher primates. The nuclear unit of the hamadryas baboon society is the harem, consisting of a dominant male, his females and offspring, and often an apprentice male with a developing harem. In the evening the harems aggregate at the sleeping cliffs, where they are relatively well protected from leopards and other predators and where, in fact, the baboons are in a position to cooperate by constituting a more efficient alarm and defense system than individual harems could provide. In the mornings the sleeping groups separate into smaller foraging bands or individual harems that proceed separately to the feeding grounds. There exists a clear relation between the amount of resource and the size of such foraging groups. At Danakil, Ethiopia, isolated acacia trees, whose flowers and beans are the baboons' main food source, are gleaned by individual harems. The density of baboons in each tree is consequently such that each harem member is able to keep the usual feeding distance of several meters between it and other members. As a result the movements of subordinate animals are unimpeded by the dominants. In groves of ten or more acacias, the feeding group is the entire band, consisting of at least several harems whose mutual tolerance and attraction are unusually high. Again, density is low enough to preserve individual distance, and the feeding efficiency of each animal remains adequate. During the dry season water becomes the critical resource. River ponds are kilometers apart, and each is visited by aggregations of as many as a hundred or more baboons.

Chimpanzees are organized into still more flexible societies. Casual groups of very variable size form, break up, and re-form with ease, apparently in direct response to the availability of food (Reynolds, 1965). Chimpanzees must search for locally abundant food that is irregular in space and time. The ability to disperse and to assemble rapidly is clearly adaptive; chimps even use special calls to recruit others to rich food finds. This strategy may be contrasted with that of the gorilla. These great apes feed to a large extent on the leaves and shoots of plants, a relatively evenly distributed food resource. Gorilla societies are semiclosed and demographic in composition, and they patrol regular but broadly overlapping home ranges.

Some primitive human societies that depend on hunting and gathering of patchily distributed resources form casual societies not unlike the chimpanzee model (Lee, 1968; Turnbull, 1968; Sugiyama, 1972). The nuclear unit of the !Kung people of the Kalahari is the family or a small number of families. Units band together in camps for periods of two to several weeks, during which time most of the game captured by the men and the crop of nuts and other vegetable food gathered by the women and children are shared equally. The Mbuti pygmies of the Congo Forest have a much less organized society. The nuclear unit is more nearly the individual, rather than the family. Pygmies move about the forest according to the distribution of game, honey, fruit, and other kinds of patchily distributed foods. Groups form and divide in a very loose manner to make the fullest use of food discoveries.

Passive feeding aggregations are commonplace among the plains-dwelling ungulates. The herd size of axis deer in India, for example, is influenced by the food supply. In November and December 1964, when preferred forage was sparse and scattered, Schaller (1967) found the average herd size in Kanha Park to be a little under 5. In February, when green shoots of grass appeared in local patches, the herds increased to an average of 10.5 individuals. The nuclear unit of zebra societies is the stallion and his harem. Although stallions are aggressive toward one another, the harems readily join in herds of indefinitely large size to take advantage of favorable feeding areas (Klingel, 1967). A similar form of population organization has been described by Brereton (1971) in the open-country parrot species of Australia.

Group size is cooperatively adjusted in the social carnivores. Hunting groups of wolves, hyenas, and lions vary in size according to the difficulty of catching the prey pursued at the moment (Kruuk, 1972). When wolves hunt mountain sheep and caribou, only one or two pack members pursue a single animal, but when a moose is the target, ten or more individuals commonly join the chase. Lions chase gazelles and other small prey singly or in small groups. The formidable buffalo, however, often requires the effort of most or all the adult members of the pride.

The size of foraging parties of ants, honeybees, and other social insects is adjusted according to the richness and extent of each food find. Fire ant workers (*Solenopsis invicta*) are typical in laying odor trails back to the nest only when they perceive uncollected food. Nestmates then leave the nest and follow these trails; when they discover food anywhere along the way, they deposit trails of their own. The number of workers working together thus builds up until the food discovery is exhausted, then declines as the trails evaporate and decline in potency. This and other cases of "mass communication," an advanced phenomenon basic to the organization of many insect societies, will be described in greater detail in Chapter 8.

At least two ant species vary their group size over periods of demographic time in response to the alternating demands of foraging and hibernation. Colonies of the slave-maker *Leptothorax duloticus* tend to split up and disperse to multiple nest sites during the summer, when raids on surrounding nests of *L. curvispinosus* are being conducted. In the fall, they draw together in a smaller number of nuclear nests (Talbot, 1957). The Argentine ant *Iridomyrmex humilis* is an example of a "unicolonial" species: no clear lines can be drawn on the basis of aggressive responses between colonies, and the entire local breeding population therefore represents one immense colony. In warm weather the populations disperse outward to multiple nesting sites, where foraging is more even and efficient. As winter approaches, the population congeals into a much smaller number of hibernating units in the most protected nesting sites (M. R. Smith, 1936; Wilson, 1971a).

The Multiplication and Reconstitution of Societies

Relatively few observations have been made of the division and internal changes in animal societies through demographic time. The taxa which are best understood are the mammals, particularly the primates, and the social insects. A variety of multiplication procedures is used by both, some of which are similar and represent convergent evolution. In general, the societies of both taxa are matrifocal, and as a consequence societal division depends on the willingness of breeding females to form fresh associations with males and to move to new locations. At the same time, mammalian societies differ from insect societies in three basic details with respect to their multi-

plication and internal construction: they are genetically less uniform, they usually if not invariably divide as a result of aggressive interactions among the members or with invading outsiders, and the timing and behavioral responses of their emigrations are far less rigidly programmed.

In Old World monkey societies, which have been studied with exceptional care by Japanese and Americans during the past 20 years, male aggression is the initial impetus that leads to group reorganization. Disruption is caused by one or the other of three forms of interactions: the rise of young males within the hierarchy, the attractive power of solitary bachelors outside the troop, or invasion by bands of bachelors. During Sugiyama's (1967) field study of the langur *Presbytis entellus* at Dharwar, India, troops underwent an important reorganization on the average of once every 27 months. *P. entellus* troops consist either of groups of adult females and juveniles ruled over by a single male, or bands of bachelors. In one instance a group of 7 males attacked and displaced the resident male. Fighting then erupted among the usurpers, until 6 were ousted and only one remained in control. Two other changes directly observed by Sugiyama ended in troop division. Once a solitary male attacked and defeated the resident male, then decamped with all the members of the troop except one adult female, who remained behind with her old consort. In another instance, a large band of 60 or more bachelor males repeatedly attacked several bisexual troops, forcing the resident males to retreat temporarily. During the fighting, small groups of females joined the bachelor band, which eventually moved into a new territory. Finally, true to the despotic nature of langur society, all of the males except the dominant individual deserted, leaving him in control of the females. A common feature of the various divisions and reorganizations was the intolerance of the new leaders toward the offspring of the former resident males, leading in some cases to infanticide by biting. In general, the juvenile populations soon came to consist entirely of the offspring of the new tyrants.

Macaques have more stable communities. Changes occur less frequently than in the langur troops, and they are normally precipitated by events within the societies rather than by the invasion of outsiders. Troops of Japanese monkeys (*Macaca fuscata*) divide when subgroups of females and their offspring gradually drift away from the main troop, visiting the feeding areas at different periods and staying outside the influence of the dominant male. Under such circumstances they become associated with subordinate adult males who have also left the troop and live in solitude or in association with other expatriate males from the same original troop. The new troops then organize themselves into the mild dominance system that characterizes the Japanese species (Sugiyama, 1960; Furuya, 1963; Mizuhara, 1964). Warfare does not appear to occur between the bachelor groups and the established bands.

A feral population of rhesus monkeys (*Macaca mulatta*) on Cayo Santiago, a tiny island off the coast of Puerto Rico, has been closely monitored since Stuart Altmann began demographic studies there in 1955. The populations grew rapidly with the aid of ad libitum feeding, and by 1967 the original two troops had split in chain fashion to create a total of seven units (Koford, 1967). The basic process, as observed in the Japanese macaque, consists of the emigration of subgroups of females with offspring and relatives. Males frequently move from one group to another, often by first joining the all-male subgroup on the periphery of the main band and then moving into the band itself. Membership in the male subgroup is obtained by affiliating with a "sponsor," usually a brother or some other relative who made the move earlier (A. P. Wilson, 1968; cited by Crook, 1970).

The details of group fission differ strikingly from one mammalian species to the next. Mech (1970) has marshaled persuasive indirect evidence from his own data and those of Adolf Murie to suggest that new wolf packs are founded by an adult breeding pair who mate and leave the mother group. The new pack is soon enlarged by the first litter of about six pups. The young wolves then remain with the parents through at least the following winter while growing in size and acquiring competence in hunting. New packs of African wild dogs may also be founded by the departure of mated pairs. At least one instance of this kind has been observed by Hugo van Lawick. The females have very large litters, consisting of ten or more pups, and they are fiercely aggressive toward other bitches and their pups. Subordinate females are sometimes driven from the vicinity of the dens. If they depart permanently with a consort male, they constitute the potential nucleus of a new pack.

The black-tail prairie dog, a colonial rodent, has a radically different process of group fission (King, 1955). Burrows are occupied communally by coteries consisting of as many as two males and five females with their offspring. During the breeding season the females possessing young pups close off part of the burrow and defend it from other coterie members. The other adults, together with the yearlings, then construct new burrow systems nearby. The prairie dog "towns" are also extended by adults, who appear to be repelled by the incessant grooming demands of the juveniles. This pattern is the opposite of that of other mammals, including other rodents, in which it is the juveniles who form the bulk of the emigrants.

Mammallike group division occurs in a few species of termites, such as the members of *Anoplotermes* and *Trinervitermes*, in a scattering of ant taxa, including the army ants and the polygynous species of *Monomorium*, *Iridomyrmex*, and *Formica*, and in the stingless bees and honeybees. The process, which entomologists in the past have referred to variously as budding, hesmosis, and sociotomy, consists of the departure of functional reproductives with an attendant group of sterile workers sufficiently large to sustain them. Most of the species of ants and termites that use budding follow procedures that are relatively casual and depend on the discovery of new, additional nest sites. In contrast, army ants, particularly the genus *Eciton*, utilize a complex and stereotyped program. Their full life cycle was first elucidated by T. C. Schneirla and R. Z. Brown (1952). Through most of the year the mother queen is the paramount center of attraction for the huge population of workers. By serving as the focal point of the aggregating workers, she literally holds the colony together. The situation changes markedly, however, when the annual sexual brood is produced early in the dry season. This kind of brood contains no workers, but, in *E. hamatum* at least, it consists of about 1,500 males and 6 new queens. Even when the sexual larvae are still very young, a large fraction of the worker force becomes affiliated with the brood as opposed to the mother queen. By the time the larvae are nearly mature, the bivouac consists of two approximately equal zones: a brood-free zone containing the queen and her affiliated workers, and a zone in which the rest of the workers hold the sexual brood. The colony has not yet split in any overt manner, but important behavioral differences between the two sections do exist. For example, if the queen is removed for a few hours at a time, she is readily accepted back into the brood-free zone from which she originated, but she is also rejected by workers belonging to the other zone. Also, there is evidence that workers from the queen zone cannibalize brood from the other zone when they contact them.

The young queens are the first members of the sexual brood to emerge from the cocoons. The workers cluster excitedly over them, paying closest attention to the first one or two to appear (see Figure 6-4). Several days later the new adult males emerge from their cocoons. This event energizes the colony, sets off a series of maximum raids followed by emigration to a new bivouac site, and at last splits the colony. The raids are conducted along two radial odor trails from the old site. As they intensify during the day, the young queens and their nuclei of workers move out along one of the trails, while the old queen with her nucleus proceeds along the other. When the derivative swarm begins to cluster at the new bivouac site, only one of the virgin queens is able to make the journey to it. The others are held back by the clinging and clustering of small groups of workers. They are, to use Schneirla's expression, "sealed off" from the rest of the daughter colony. Like polar bodies created at the cellular level by oogenesis, they are useless rudiments, and are eventually abandoned and left to die. Now there exist two colonies: one containing the old queen; the other, the successful virgin, daughter queen. In a minority of cases the old queen is also superseded. That is, the old queen herself falls victim to the sealing-off operation, leaving both of the two daughter colonies with new virgin queens. This presumably happens most often when the health and attractive power of the old queen begin to fail before colony fission. The

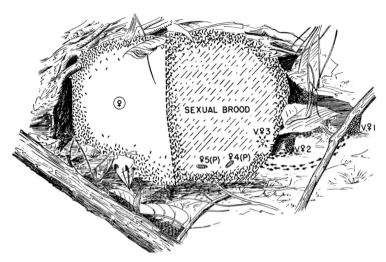

Figure 6-4 Colony division in army ants. The diagram shows a biv-ouacked mass of over a hundred thousand workers, queens, and males of *Eciton hamatum*, all constituting a single colony and the offspring of one queen. The left portion of the mass contains the mother queen but no immature stages, while the right portion contains the newly developing queens and males. Two of the virgin queens (v. ♀ 1 and v. ♀ 2) have emerged from their cocoons and moved to one side of the bivouac, to be attended by clusters of workers who still run back and forth to the bivouac along odor trails. A third virgin queen (v. ♀ 3) has emerged more recently and is still confined by a knot of workers at the edge of the mass, while two others remain within cocoons (*P*). The males are also still in the pupal stage. (From Schneirla, 1956.)

maximum age of the *Eciton* queen is not known, but is believed to be relatively great for an insect; a marked queen of *E. burchelli*, for example, has been recovered after a period of four and a half years. The males, in contrast, enjoy only one to three weeks of adult exist-ence. Within a few days of their emergence, at least some of them depart on flights away from the home bivouac in search of other colonies. It is also possible that a few remain behind to mate with their sisters; the matter has simply not been documented either way. In any case the new queens are fecundated within a few days of their emergence, and almost all of the males disappear within three weeks after that.

An equally elaborate but very different program is followed by the honeybee. Just before division, which occurs predominantly in the late spring, the colony contains a single mother queen and 20,000 to 80,000 workers. The first event is the construction by the workers of a small number of royal cells, which are large, ellipsoidal chambers usually placed along the lower margins of the combs. We know that

these cells will not be built so long as the mother queen is producing "queen substance" (*trans*-9-keto-2-decenoic acid) from her mandibu-lar glands in sufficient quantity for each worker to receive on the average of at least 0.1 microgram per day. But with the onset of the swarming season in late spring, the queen's production of this sub-stance falls off, and construction of royal cells ensues. The queen lays one egg in each royal cell, and the hatching larvae are fed special foods by the workers, which insure their development into queens. The growth of a new queen is astonishingly quick, requiring only 16 days from the laying of the egg to the eclosion of the adult bee, as opposed to 21 and 24 days for the worker and drone, respectively. While all of this is going on, the status of the mother queen changes. She still lays a few eggs, but her abdomen is reduced in size, and she begins to behave in an agitated fashion. The workers feed her less and even show mild hostility, pummeling and jumping on top of her. Eventually she is pushed out of the hive and flies off in the company of a large group of workers. Several such swarms may emerge around this time. The "prime" swarm, containing the old queen, usually leaves soon after the first royal cell has been capped, just prior to the pupation of the queen larva inside. The first "after-swarm," containing the first of the new queens, occurs around eight days later, very soon after the new queen emerges from the royal cell and mates (see Figure 6-5). The occurrence of afterswarms de-pends on the size and health of the colony, and the number of these events varies greatly. Eventually, however, about two thirds of all the workers leave the nest.

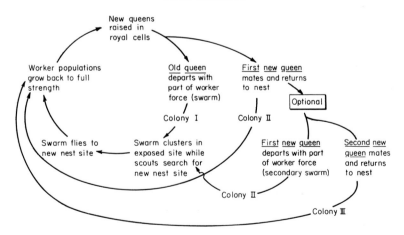

Figure 6-5 Colony division in the honeybee (*Apis mellifera*). Al-though only a single afterswarm is shown in this particular scheme, two or more occur in extreme cases in nature. (From Wilson, 1971a.)

The swarming bees fly en masse for a short distance from the old hive and settle onto an aerial perch, such as the trunk or branch of a tree or the side of a building, where they cluster tightly to form a solid mass of bodies. It is known that a second pheromone produced in the queen's mandibular glands, *trans*-9-hydroxy-2-decenoic acid, is necessary for this grouping behavior to be consummated. Scout bees now fly out from the bivouac in all directions in the search for a new permanent nest site. When a suitable site is found—a hollow tree, the enclosed eave of a building, an unoccupied commercial hive—the scouts return and signal the direction and distance of the find. This is accomplished by means of waggle dances performed on the sides of the swarm. Different scouts may announce different sites simultaneously, and a contest ensues. Finally the site being advertised most vigorously by the largest number of workers wins, and the entire swarm flies off to it. Now there are two colonies: the fraction back at the old nest which is about to acquire a newly fecundated daughter queen, and the fraction at the new nest which contains the old mother queen.

For a brief time, the workers at the parental nest are queenless. But the events that ensure requeening have long since been set in motion. Even before the construction of the queen cells that preceded swarming, the workers have built a group of drone cells, which look just like worker cells except that they are on the average slightly larger. Into these the mother queen lays unfertilized eggs, which, true to the haplodiploid mode of sex determination prevalent in most Hymenoptera, develop into males. When they are four days or more into adult life, the males begin leaving on mating flights, traveling short distances from the nests to special areas where they join loose swarms of males from other nests in the vicinity. Here, in sustained flight, they await the approach of the virgin queens.

The first virgin queen to emerge from a royal cell is the only adult member of her caste in the nest. Her mother has already departed, and her sister queens are still in their cells. She now searches through the colony for her rival sisters, exchanging with them special sound signals descriptively labeled as "piping" and "quacking." If her sisters emerge from their own cells while she is present, fighting ensues and is continued until her sisters are eliminated, either through swarming or through sequential killing, and only the original virgin queen is left. She is urged out of the nest and on to her nuptial flight by mildly aggressive behavior on the part of the workers. As she approaches the male congregations, she releases small quantities of 9-keto-decenoic acid from her mandibular glands. As this scent disperses downwind, it attracts males from distances of 10 meters or more. The mating is quick and violent; the male literally explodes his internal genitalia into the genital chamber of the queen and quickly dies. The queen makes as many as 3 flights a day for a total of up to 12 flights

or more, and on each flight she mates with a different male. Finally, she obtains enough sperm to last her lifetime. Then, she either participates in an afterswarm, making way for the next virgin queen to emerge and mate, or else she destroys the other young queens and takes over the nest. In either case, if conditions are favorable, her own daughter workers will cause the worker population under her control to double within a year, and the colony can divide again.

The great majority of ant and termite species accomplish colony multiplication by means of nuptial flights. The males and virgin queens of ants depart from the nests at certain hours of the day set by circadian rhythms. The timing varies among the species: mid-morning for some species, late afternoon for others, midnight or the predawn hours for still others, and so on around the clock. The queens are able to attract the males quickly. In at least one species, *Xenomyrmex floridanus*, they release a sex pheromone from the Dufour's gland (Hölldobler, 1971b). The sexual forms of the more abundant species mingle in nuptial swarms that form over conspicuous environmental features such as treetops and open fields. After being inseminated by one to several males, each queen drops to the ground, sheds her wings, and runs about in search of a suitable nest site. If she is one of the tiny minority that escapes death from predators and hostile colonies of her own species, she constructs a brood chamber and lays a batch of eggs. The first adults to emerge in the brood are small workers, all sterile females, who immediately assist their mother in rearing sisters and putting the colony on a firmer footing. The males play no part in this effort. Whether they have participated in mating or not, they soon separate from the nuptial groups and wander about alone, destined to die within hours from accidents or the attacks of predators. During the early stages of their growth, the new colonies produce only workers. After a period normally requiring one to several years, males and virgin queens begin to appear just before the season of nuptial flights. Colonies that generate new queens are said to be "mature," in the sense that they are now able to reproduce themselves directly.

Termite colonies multiply by means of nuptial flights similar in detail to those of ants, a fact remarkable in itself because the resemblance is due entirely to convergent evolution. *Incisitermes minor*, a member of the primitive family Kalotermitidae found in the southern and western United States, follows a program typical of most kinds of termites. After the winged males and queens leave the nest, their flight is aimless and wavering. The majority, however, manage to ascend 70 meters or more and to fly for distances of at least 100 meters and perhaps as much as a kilometer from the parental nest. As soon as the alate alights, it breaks off its wings by quickly spreading and lowering them until their tips touch the ground, then pivoting back and forth to bring pressure on the wings at the basal sutures.

Now the "dealate" runs excitedly in apparently random directions until it encounters a member of the opposite sex. The two individuals stop abruptly, turn face to face, and play their antennae over each other's heads.

The king makes advances toward the queen, the queen striking at the king with her head. After four or five such overtures, each of which is followed by a pause during which the termites stand facing each other with their antennae fanning slowly, the king is accepted or rejected. If he is rejected the queen turns and runs quickly away and the king goes in the opposite direction. If, however, the king is accepted, the queen turns quickly and speeds away, with the king in close pursuit. . . . Although the queen runs rapidly, the king keeps close to her and, when they become separated as occasionally happens, the king rapidly regains contact with her. . . . After pairing has been accomplished, separation seldom occurs. It is usually difficult to frighten members of a pair away from each other, and it appears that they seldom, if ever, leave one another for other mates, even though a number of unpaired termites are near. (Harvey, 1934)

This sequence of dealation, pairing, and tandem running is universal in the termites. In some groups, the tropical genus *Nasutitermes*, for example, the queens stand still for the most part and "call" males by means of sex pheromones released from intersegmental glands located on the abdominal dorsum. After pairing, the royal couple of *Incisitermes minor* undergo a radical change in behavior. During the nuptial flight and search for mates, the termites are attracted to light. As soon as they pair, however, they are repelled by light and become strongly attracted to wood. When they find a suitable spot, they begin excavating in the wood, alternating shifts, until they complete an entrance tunnel about a centimeter deep. The entrance hole is then sealed off with a puttylike mixture of chewed wood and cementlike secretion. Finally, the pair constructs its first royal cell, a small pear-shaped chamber at the bottom of the entrance tunnel.

When the royal cell has been finished, the queen lays from two to five eggs. Soon after hatching from these eggs, the fragile, chalk-white nymphs are fed by regurgitation and, after one or more molts, set about feeding themselves and enlarging the nest. The two activities are, in fact, the same thing! The royal pair stays in the advanced part of the main passage, while the nymphs dig out enlarged feeding chambers and side tunnels. By the end of the second year the young colony has consumed about 3 cubic centimeters of wood. The colony now consists of the royal pair, a single soldier, and ten or more pseudergates and nymphs. About a year is required for each soldier to develop from an egg to the mature form. The queen's abdomen begins to swell in two years, but the king looks about the same or, if anything, somewhat shrunken in size. After several more years winged sexual forms are produced, and the colony, like the ant colony which generates winged queens, is now referred to as being in a "mature" condition.

Time-Energy Budgets

The amount of time that an animal devotes to each activity and the energy expended on it differ markedly from one taxon to the next. Honeybees and *Pogonomyrmex* harvester ants devote roughly one-third of their time to various forms of work, one-third to resting, and one-third to patrolling through the nest (Wilson, 1971a). Male orang-utans spend about 55 percent of their time feeding, 35 percent resting, and 10 percent moving from one position in the tree canopy to another; the comparable figures for the female are 50, 35, and 15 percent, respectively (Peter S. Rodman, personal communication). Hummingbirds (*Calypte, Eulampis*) devote 76-88 percent of their time to sitting, 5-21 percent to foraging for nectar, 0.5-1.8 percent to flycatching, 0.3-6.4 percent to chasing other hummingbirds from their territories, and so on, the breakdown varying slightly according to the species of tree occupied (Wolf and Hainsworth, 1971).

Because the forms and priorities of social behavior are constrained to a large extent by the time-energy budgets of the species, it is important to establish the general principles of budget programming and to define the ecological forces that shape particular programs in evolution. The study of time-energy budgets is in a very early stage. It consists of three phases that must be fitted together to provide the total picture for a given species: bioenergetics, in which the caloric requirements of the animal are related to its size and activity patterns, and then compared with the energy harvested as a result of the activity; budget writing, in which a behavioral catalog, relatively finely divided into behavioral categories in the manner of descriptive ethology, is prepared and the time and energy costs are broken down according to it; and the ecological analysis, in which the natural environment of the species is analyzed to provide an evolutionary raison d'être for the details of the budget. These phases, which can be conducted together or in sequence, range from the purely physiological to the genetic and evolutionary, from the relatively simple to the difficult. Bioenergetics, the easiest of the three to pursue, is also the best documented at the present time; some of the generalizations resulting from it will be presented in a later review of territorial behavior (Chapter 12). The phase of most direct sociobiological interest, however, is the ecological analysis, which will be considered now.

Our knowledge is limited to mere fragments of data. Two preliminary generalizations can be made, both admittedly strongly conjectural in nature. The first can be called the *principle of stringency:* time-energy budgets evolve so as to fit the times of greatest stringency. Zoologists have often puzzled over the fact that animals in the midst of plenty spend a good deal of their time doing nothing. Lions resting next to zebra herds, barracudas hovering idle in front of passing schools of minnows, and birds perching for hours near fruit-laden bushes are disquieting to the thoughtful evolutionist. Why, he feels compelled to ask, haven't these species evolved so as to keep the members constantly foraging, consuming, growing, and reproducing? Shouldn't the most active genotypes have the greatest fitness? The answer is that animals and societies do not always live in the midst of plenty. Their time-energy budgets are adjusted to see them through periods of food shortage. Genotypes committed to the most rapid body growth and reproduction—the maximum consumers—will enjoy an advantage during the brief periods of resource surplus but will experience a severe setback, leading possibly to extinction, when times become hard. Among *K*-selected species, the more stable the environment and the less mobile the individual animals, the more prudent must be the investment in growth and reproduction, and hence the more idle and constrained animals will seem to be at any randomly chosen moment.

Periodic food shortages are not the only force favoring the evolution of idleness. A large percentage of the worker population of colonies of social insects (ants, social bees and wasps, and termites) are to be found resting throughout the day and night except during those rare episodes when the entire nest has been activated by an invasion or mechanical disturbance. Lindauer (1961) and Michener (1964a) have observed that this outwardly nonproductive activity, together with the seemingly aimless patrolling through the nest, actually enhances the capacity of the colony as a whole to respond to capricious changes in the environment. Patrolling workers assess the needs of the colony from moment to moment and are thus able to respond to local requirements with less delay. Resting workers constitute a reserve force, available for major emergencies, such as overheating of the nest or invasion by a predator, that require the simultaneous employment of many individuals. The idle force conforms to the principle of stringency, in the sense that its size is determined by the most severe requirements periodically imposed on the colony as a whole.

The second speculative proposition that can be made about the ecology of time-energy budgets is the *principle of allocation*. This states that the major requirements of animals differ greatly in the amounts of time and energy that it is profitable to devote to them in the currency of genetic fitness. Furthermore, as a rule these requirements descend in importance as follows: food, antipredation, and reproduction. Finally, to the extent that one priority is easily satisfied by a temporarily generous environment, more time and energy are devoted to the activities of the other priorities. Social insects, filter feeders, zooplankton, whales, elephants, and top carnivores such as wolves and hawks are food-limited. A very large proportion of their daily activity is devoted to securing food. Much of the aggressive behavior of such organisms is territorial and connected with the maintenance of a dependable food supply. Those that construct shelters, such as the social insects, use them as much to defend their territories against intruders as to ward off predators.

Antipredatory responses and reproductive behavior are effective and often elaborate, but they consume relatively little time and energy.

In sharp contrast, elephant seals on their hauling grounds have no serious food problems; females, in fact, have built up such great stores of fat that they can go without feeding throughout the nursing period. The islands on which the seals breed are also free of predators. As a result, the animals concentrate almost wholly on reproduction. The males have evolved spectacular reproductive adaptations, including great size, control of harems, and extremely aggressive behavior in maintaining dominance over unmated males in the vicinity. Most of their time is expended on fighting, mating, and resting. Mayflies devote virtually their entire adult lives to reproduction. They have eliminated the energy problem by shortening the adult life span to hours, and they thwart predators by emerging simultaneously in such large numbers that only a small fraction can be consumed.

The principle of allocation presumes nothing about cause and effect, except that the effort put into feeding, or antipredation, or reproduction tends to expand in evolution so as to fill the time made available to it. The expansion is halted by the dangers presented by the environment during the most difficult periods, as noted already with respect to episodic stringency. Furthermore, the compensation is more complex than any simple arithmetical trade-off. There are, for example, two extreme strategies that food-limited species can follow, which can result in very different time-energy budgets and social organizations. At one extreme there exist species that Schoener (1971) has called the "time minimizers," for which a predictable, reliable amount of energy is available so long as the energy source is protected. The species evolve in a way that minimizes the amount of time required to harvest the available energy; the remaining time can be devoted to other activities, including defending the food supply from intruders. Examples of this adaptive type are provided by a great diversity of kinds of insects, fish, and birds that maintain feeding territories (see Chapter 12). At the other extreme are "energy maximizers," species that consume all of the energy available regardless of the cost in time. Examples include the most opportunistic species, which grow and breed rapidly whenever they encounter conditions suitable for doing so. They appear able to circumvent the principle of stringency only by dispersing widely as the food supply dwindles, escaping extinction by hopping from one temporarily suitable patch of the environment to another.

The two kinds of strategists may or may not devote the same proportions of time ultimately to food gathering, if we put the defense of feeding territories into that broad category, but the enabling behavior patterns are vastly different, with significant consequences for the evolution of social behavior. As a rule, time minimizers will be territorial. And they can also defend territories in groups, which will then tend to be stable and well organized. Energy maximizers are more likely to be nonsocial or else travel in poorly organized herds.

Chapter 7   **The Development and Modification of Social Behavior**

Social behavior, like all other forms of biological response, is a set of devices for tracking changes in the environment. No organism is ever perfectly adapted. Nearly all the relevant parameters of its environment shift constantly. Some of the changes are periodic and predictable, such as the light-dark cycles and the seasons. But most are episodic and capricious, including fluctuations in the number of food items, nest sites, and predators, random alterations of temperature and rainfall within seasons, and others. The organism must track these parts of its environment with some precision, yet it can never hope to respond correctly to every one of the multifactorial twists and turns—only to come close enough to survive for a little while and to reproduce as well as most. The difficulty is exacerbated by the fact that the parameters change at different rates and often according to independent patterns. In each season, for example, a plant contends with irregularities in humidity on a daily basis, while over decades or centuries its species as a whole must adapt to a steadily increasing or decreasing average annual rainfall. An aphid has to thwart predators that vary widely in abundance from day to day, while over many years, the aphid species faces change not only in the abundance but also in the species composition of its enemies.

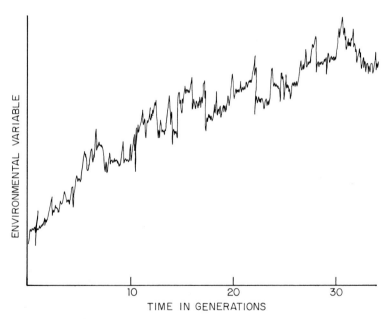

Figure 7-1 Environmental parameters fluctuate through time on a short-term basis while their mean values shift more gradually. Individual organisms must track the short-term changes with physiological and behavioral responses, while the species as a whole must undergo evolution (a genetic response) to cope with the long-term changes. The example shown here is imaginary.

Organisms solve the problem with an immensely complex multiple-level tracking system. At the cellular level, perturbations are damped and homeostasis maintained by biochemical reactions that commonly take place in less than a second. Processes of cell growth and division, some of them developmental and some merely stabilizing in effect, require up to several orders of magnitude more time. Higher organismic tracking devices, including social behavior, require anywhere from a fraction of a second to a generation or slightly more for completion. Figures 7-1 and 7-2 suggest how organismic responses can be classified according to the time required. All the responses together form an ascending hierarchy. That is, slower changes reset the schedules of the faster responses. For instance, a shift into a more advanced stage of the life cycle brings with it new programs of behavioral and physiological responses, and the release of a hormone alters the readiness to react to a given stimulus with learned or instinctive behavior. In both cases, the slower response alters the potential of the faster one. Even more profound changes occur at the level of entire populations during periods longer than a generation. In ecological time populations wax or wane, and their age structures shift, in reaction to environmental conditions. These are the demographic responses indicated by the middle curve of Figure 7-2. Ecological time is so slow that large sequences of organismic responses occur within it, few of which affect the outcome single-handedly. But ecological time is also generally too quick to bracket extensive evolutionary change. When the observation period is prolonged still further to about ten or more generations, the population begins to respond perceptibly by evolution. Long-term shifts in the environment permit certain genotypes to prevail over others, and the genetic composition of the population moves perceptibly to a better adapted statistical mode. The hierarchical nature of the tracking system is preserved, since the newly prevailing genotypes are likely to have different demographic parameters from those prevailing earlier, as well as different physiological and behavioral response curves. The time intervals are now spoken of as being evolutionary in scale—long enough to encompass many demographic episodes, so long, in fact, that separate events at the organismic level are reduced to insignificance.

The concept of the multiple-level, hierarchically designed tracking system has been developed in several contexts and to varying degrees of penetration by Pringle (1951), Bateson (1963), Skinner (1966), Manning (1967), Levins (1968), Kummer (1971), and Slobodkin and Rapoport (1974). It will be expanded in the remainder of this chapter to provide a clearer perspective of social behavior as a form of adaptation. The account begins at the evolutionary time scale and works downward through the hierarchy to learning, play, and socialization. The important point to keep in mind is that such phenomena as the hormonal mediation of behavior, the ontogenetic development of behavior, and motivation, although sometimes treated in virtual isola-

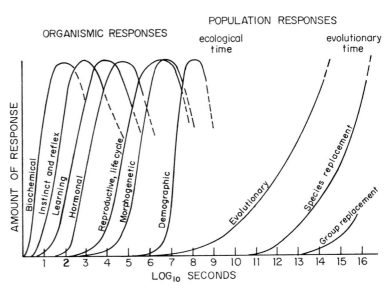

Figure 7-2   The full hierarchy of biological responses. Organismic responses are evoked by changes in the environment detectable within a life span, population responses to long-term trends. The hierarchy ascends with an increase in the response time; that is, any given response tends to alter the pattern of the faster responses. Beyond evolutionary responses are replacements of one species by another or even entire groups of related species by other such groups. The particular response curves shown here are imaginary.

tion as the proper objects of entire disciplines, or else loosely connected under the rubric of "developmental aspects of behavior," are really only sets of adaptations keyed to environmental changes of different durations. They are not fundamental properties of organisms around which the species must shape its biology, in the sense that the chemistry of histone or the geometry of the cell membrane can be so described. The phenomena cannot be generally explained by searching for limiting features in the adrenal cortex, vertebrate midbrain, or other controlling organs, for the reason that these organs have themselves evolved to serve the requirements of special multiple tracking systems possessed by particular species.

Tracking the Environment with Evolutionary Change

All social traits of all species are capable of a significant amount of rapid evolution beginning at any time. This statement may seem exaggerated at first, but as a tentative generalization it is fully justified by the facts. All that the potential for immediate evolution requires is heritability within populations. Moderate degrees of heritability have been demonstrated in the widest conceivable array of characteristics, including crowing and dominance ability in chickens, visual courtship displays in doves, size and dispersion of mouse groups, degree of closure of dog packs, dispersal tendency in milkweed bugs, and many other parameters in vertebrates and insects (see Chapter 4). One extensive research program devoted to the genetics of social behavior of dogs uncovered significant amounts of heritability in virtually every trait subjected to analysis (Scott and Fuller, 1965).

The speed with which a trait evolves in a population increases as does the product of its heritability and the intensity of the selection process. More precisely, $R = h_N^2 S$, where R is the response to selection, h_N^2 is heritability in the narrow sense, and S is a parameter determined by the proportion of the population included in the selection process and the standard deviation of the trait. Few persons, including even biologists, appreciate the speed with which evolution can proceed at the level of the gene. Consider first the theoretical possibilities. Let the frequency of a given gene in a population be represented by q (so that when $q = 0$ the gene is absent and when $q = 1$ it is the only gene of its kind at its chromosome locus); and let selection pressure against homozygotes of the gene be represented by s. When $s = 0$, individuals possessing nothing but the gene survive and reproduce as well as individuals with other kinds of genes. When $s = 1$, no such individuals contribute offspring to the next generation. In nature most values of s fall somewhere between 0 and 1. The rate of change in each generation in a large population will be

$$\frac{-sq^2(1-q)}{1-sq^2},$$

for which the simpler expression $-sq^2(1-q)$ is a good approximation, since sq^2 is usually a negligible quantity. The rate of change is greatest when $q = 0.67$ but falls off steeply as the gene becomes either rare or very abundant.

Figure 7-3 illustrates an actual case of microevolution of a character involving behavior in *Drosophila melanogaster*. Here $s = 0.5$, because the homozygote males (possessing two "raspberry" genes, which affect eye color and behavior) are about half as successful in mating as those males which possess one raspberry gene or none at all. The experimental curve of evolutionary change can be seen to be nicely consistent with the theoretical curve. In only ten generations the frequency of the gene declines from 50 percent to approximately 10 percent. Other eye mutants of *Drosophila* often show this degree of reduction in reproductive performance. The exact behavioral basis in the *yellow* mutant of *D. melanogaster* was elucidated by Bastock and Manning (1955) and Bastock (1956). They found that successful courtship by males of the species entails the following rigid sequence of maneuvers: (1) "orientation," in which the male stands close to or follows the female; (2) "vibration," in which he rapidly vibrates

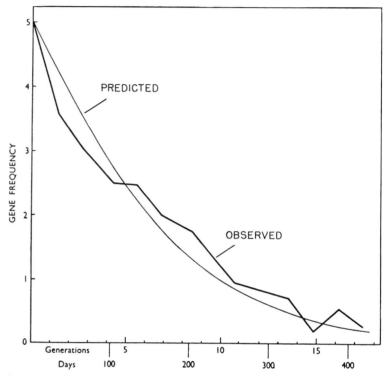

Figure 7-3 Rapid evolution in a behavioral trait under moderate selection pressure: the decline in percentage of a gene in a population when the homozygotes reproduce at 50 percent the rate of other genotypes. The smooth curve is the one predicted by theory. The irregular one, which fits the theoretical curve closely, shows the actual decline of the "raspberry" gene, which affects both eye color and behavior, in a laboratory population of the fruitfly *Drosophila melanogaster*. The frequency of the gene declines from 50 to about 10 percent in only ten generations. (From Falconer, 1960; experimental curve based on data from Merrell, 1953.)

his wings close to her head; (3) "licking," wherein the male extends his proboscis and licks the female's ovipositor; and (4) attempted copulation. The *yellow* homozygous males are wholly normal in the movements and sequence of these maneuvers, but they are less active at vibrating (movement number 2) and licking (movement number 3) than normal males. Hence they are less effective in achieving copulation. Such behavioral components are commonplace in the phenotypes of rapidly evolving *Drosophila* populations.

There exist numerous other examples in which significant evolution in major behavioral characters has been obtained in laboratory populations in ten generations or less—in accordance with the generous limits predicted by theory. Starting with behaviorally neutral stocks, Dobzhansky and Spassky (1962), Dobzhansky et al. (1972), and Hirsch (1963) have created lines of *Drosophila* whose adult flies orient toward or away from gravity and light. In only ten generations Ayala (1968) was able to achieve a doubling of the equilibrial adult population size within *Drosophila serrata* confined to overcrowded bottles. He then discovered that the result had come about at least in part because of a quick shift to strains in which the adults are quieter in disposition and thus less easily knocked over and trapped in the sticky culture medium (Ayala, personal communication).

Gibson and Thoday (1962), while practicing disruptive selection on a population of *Drosophila melanogaster* in order to create co-existing strains with high and low numbers of thoracic bristles, found that their two lines stopped crossbreeding in only about ten generations. In this brief span of time they had created what were, in effect, two species. The explanation, elucidated in later experiments by Thoday (1964), appears to be that linked genes favoring "homogamy"— the tendency for like to mate with like—were simultaneously and accidentally selected along with bristle number.

Evolution leading to rapid species formation can occur even without the application of such intense selection pressure. In the late 1950's, M. Vetukhiv set up six populations from a single highly heterozygous stock of *Drosophila pseudoobscura* derived from hybrids of populations from widely separate localities. Fifty-three months later the males were observed to prefer females from their own laboratory populations over those originating from the other five (Ehrman, 1964). After a strain of *D. paulistorum* spontaneously lost its interfertility with other laboratory strains in 1958–1963, thus creating an incipient species, Dobzhansky and Pavlovsky (1971) duplicated the result by taking parallel, interbreeding strains and deliberately selecting against the genotypes that hybridized. In this experiment, incipient species were created within ten generations. Quite a few cases of rapid microevolution of natural insect populations, some of them entailing behavioral traits, have been reported by Ford (1971) and his colleagues in England. In these examples selection coefficients (*s* in our earlier formula) typically exceeded 0.1.

Equally rapid behavioral evolution has been achieved in rodents through artificial selection, although the genetic basis is still unknown owing to the greater technical difficulties involved in mammalian genetics. Examples of the traits affected include running behavior in mazes, defecation and urination rates under stress, fighting ability, tendency of rats to kill mice, and tameness toward human observers (Parsons, 1967). Behavior can also evolve in laboratory rodent populations in the absence of artificial selection. When Harris (1952) provided laboratory-raised prairie deer mice (*Peromyscus maniculatus bairdii*), a form whose natural habitat is grassland, with both simulated grassland and forested habitats in the laboratory, the mice chose the grassland. This response indicated the presence of a genetic com-

ponent of habitat choice inherited from their immediate ancestors. Ten years and 12-20 generations later, however, the laboratory descendants of Harris' mice had lost this unaided tendency to choose the habitat (Wecker, 1963). If first exposed to grassland habitat, they later chose it, as expected. However, if they were exposed to woodland first, they later failed to show preference for either habitat. These results indicate that a predisposition remained to select the ancestral habitat, although the short period of evolution had weakened it considerably.

To summarize, there is every justification from both genetic theory and experiments on animal species to postulate that rapid behavioral evolution is at least a possibility, and that it can ramify to transform every aspect of social organization. The crucial independent parameters nevertheless remain the intensity and persistence of natural selection. If one or the other is very low, or if the selection is stabilizing, significant evolutionary change will consume far more time than the theoretical minimum of ten generations. Conceivably millions of generations might be needed. Thus while theory and laboratory experimentation have established the maximum possible rate of behavioral evolution, we must now as always return to nature to find out how far reality falls below it.

In order to make estimates of evolutionary roles in free-living populations, it is necessary to fall back on taxonomic measures. This method, which is exemplified in Table 7-1, consists of identifying the rank of the lowest taxa within a given phyletic group that display variation in the trait of interest. If different societies belonging to the same population differ to a significant degree from one another, and the variation has a strong genetic basis, the degree of heritability is high by definition and the trait is evolutionarily very labile. In cases where geographically discrete populations (demes) also differ markedly, the trait can now be hypothesized to evolve rapidly. If we must go to the species level to find variation in our social characteristic within a larger phylogenetic assemblage, the trait has evidently been evolving more slowly in that assemblage. Where the lowest taxa displaying variation are families or orders, evolution has been relatively very slow.

The reasoning behind this nomographic mapping is quite simple. Taxonomic categories (subspecies, species, genus, family, and so on) are based on increasing degrees of difference between populations. The larger the number of characteristics involved in the difference, and the greater the magnitude of the individual differences, the higher the populations are ranked in the ascending series of categories. In other words, the higher each ranks as a taxon. Although quantitative measures have been devised that can assign a single number to the total amount of difference (Jardine and Sibson, 1971), their magnitude depends arbitrarily on the sample of traits measured and the statistical technique employed. Furthermore, the breaking

Table 7-1 Rates of evolution of social traits, indicated by the rank of the lowest taxa between which significant amounts of variation occur within the phylogenetic line indicated.

Variable Trait	Taxon	Authority
VARIATION BETWEEN POPULATIONS OF SAME SPECIES		
Group size	African buffalo (*Syncerus caffer*)	Jarman (1974)
Degree of group cohesiveness	Baboon (*Papio anubis*)	Rowell (1966a)
Unimale or multimale society	Howler monkey (*Alouatta villosa*)	Chivers (1969)
Sex ratio	Baboon (*Papio anubis*)	Rowell (1966a)
	Vervet (*Cercopithecus aethiops*)	J. S. Gartlan, in Crook (1970)
	Langur (*Presbytis entellus*)	Ripley (1967), Sugiyama (1967)
Relative involvement of adults of both sexes and of juveniles in group in aggressive interactions	Rhesus monkey (*Macaca mulatta*)	Southwick (1969)
Presence or absence of territory and size of territory	Many examples—see Chapter 12	
Guarding of eggs by females or not	Marine iguana (*Amblyrhynchus cristatus*)	Eibl-Eibesfeldt (1966)
Optional carrying of infant on back (in addition to its hanging from belly) or not	Patas monkey (*Erythrocebus patas*)	Kummer (1971)
Dialects in vocal communication: differences are sometimes based in part on learning and "tradition drift"	Birds: species of *Certhia, Richmondena, Trachyphonus, Zonotrichia*, and many other genera	Lemon (1971a), Marler and Hamilton (1966), Nottebohm (1970), Thielcke (1965), Wickler and Uhrig (1969b)
	Mammals: elephant seals (*Mirounga angustirostris*)	LeBoeuf and Peterson (1969)
Dialects in waggle dance	Honeybee (*Apis mellifera*)	von Frisch (1967)
VARIATION BETWEEN SPECIES OF SAME GENUS		
Group structure: harems in *Papio hamadryas*, male hierarchies in other *Papio* species	Baboons (*Papio*)	Kummer (1971)
Group size	Monkeys: *Macaca, Presbytis*	Eisenberg et al. (1972)
	Antelopes: kobs (*Kobus*), reedbucks (*Redunca*), hartebeest (*Alcelaphus*), gazelles (*Gazella*), and others	Estes (1967), Jarman (1974)
Percentage of time spent in social behavior	Macaques (*Macaca*)	Davis et al. (1968)
Colonial vs. territorial nesting	Blackbirds (*Agelaius*)	Orians and Collier (1963)
Solitary vs. schooling behavior	Freshwater cichlids (*Tilapia*), reef fish (*Dascyllus*)	Dambach (1963), Fishelson (1964)
Mouthbrooding vs. substrate brooding	Freshwater cichlids (*Tilapia*)	Dambach (1963)
Dominance systems vs. territoriality	Numerous reptiles, birds, and mammals—see Chapter 13	
Presence or absence of sterile castes	Primitively social bees and wasps	Wilson (1971a)
VARIATION BETWEEN GENERA		
Percentage of time spent in grooming	Apes (Pongidae)	Schaller (1965b)
	Ants (Formicidae)	Wilson (1971a)
Solitary vs. communal nesting	Storks (Ciconiidae)	Kahl (1971)
Solitary vs. schooling behavior	Butterfly fish (Chaetodontidae)	Zumpe (1965)
VARIATION BETWEEN HIGHER TAXA		
Worker caste all-female or bisexual	Insects: Hymenoptera (bees, wasps, ants) vs. Isoptera (termites)	Wilson (1971a)
Presence or absence of sterile castes in advanced societies	Metazoans: arthropod superphylum (including insects) vs. chordate superphylum (including mammals)	Wilson (1971a)

points, such as the amount of difference required to place two species not only in different genera but also in different families, are wholly intuitive and vary in practice from one major group of organisms to another. Classifiers of mammals and birds, for example, tend to split species of a given degree of difference into higher ranking taxa than do entomologists or protozoologists. All this uncertainty notwithstanding, the taxonomic scale provides the soundest basis for estimating the amount of evolution that has taken place throughout all of the genome with reference to all characteristics. Social phenotypes make up a tiny fraction of the total pool of variable characteristics and are correlated only weakly if at all with most of the other characteristics. Hence total phenotypic difference between taxa—measured by the rank to which they have been separated by taxonomists—is a fair measure of the overall genetic divergence that has occurred between the taxa and, most important for our purposes, the relative amount of time that has passed since their initial divergence at the population level. Social divergence can be separated as the more or less dependent variable, and the remaining phenotypic difference used as a rough index of the time required to produce the social divergence.

To illustrate the method, we can do no better than refer to the communication systems of animals. The songs of courtship and of territorial advertisement often vary among bird and frog populations, creating the much-studied "dialect" phenomenon. Much of the variation is based on tradition drift and is mostly or wholly phenotypic, as in the white-crowned sparrow. But where genetic differences exist, they probably indicate a generally rapid rate of evolution. Although absolute time scales are lacking, the differences in some cases must have originated in no more than a few thousand years and possibly much less—perhaps in extreme cases even approaching the theoretical minimum. The lizard genus *Anolis*, a rapidly speciating assemblage of species belonging to the family Iguanidae, also varies at the population and species level in dewlap color and vertical bobbing patterns, which are components of courtship and territorial displays (Williams, 1972). But the basic movements, such as the bobbing itself and lateral body compression, are far more conservative. They are preserved in genera of iguanid lizards that diverged as far back as the Paleocene or upper Cretaceous, more than 50 million years ago. The same is true of the basic displays of the older bird taxa, such as the pelicans, doves, and ducks. The most slowly evolving of all groups, as Moynihan (1974) recently noted, may be the cephalopods. Three of the living major assemblages, the Teuthoidea (squids), the Sepioidea (cuttlefishes and their relatives), and the Octopoda (octopuses and argonauts), still share some basic displays despite the fact that they diverged at least as far back as the early Jurassic, or roughly 180 million years ago.

Table 7-1 presents some of the best-documented cases of variation in social phenotypes arrayed in upward sequence along the taxonomic scale. The reader will search in vain for any clear pattern. Few major categories of behavior can yet be characterized as either rapidly evolving or slowly evolving across all of the principal animal taxa. Territoriality and courtship displays tend to change easily, but as we have just seen there are striking exceptions. Only a small number of social traits, such as the presence of sterile castes and the existence of all-female societies in the insects, are very conservative. These last distinctions, incidentally, have endured since at least the middle of the Cretaceous Period, or 100 million years. The weakness of patterning is less surprising, however, when one considers the great diversity of ecological prime movers that have shaped the many social systems listed here and the opportunistic nature of evolution generally.

Additional insight into the relative rates of social evolution can be obtained by performing the obverse operation from that in Table 7-1. A comparison is made of the "sociograms" of populations of various rank, which list all of the known social behaviors and the amount of time devoted to each. In other words, instead of recording the lowest taxonomic level at which particular social behaviors diverge, one determines the overall differences in social behavior for particular taxonomic levels. The literature of comparative ethology is already filled with such information. Unusually thorough but otherwise typical paradigms have been provided by Poirier for the langur species *Presbytis entellus* and *P. johnii* (see Table 7-2), by Kummer (1971) for baboons, and by Struhsaker (1969) for cercopithecoids generally. If classificatory schemes such as these could be standardized for primates and exact sociograms prepared for species representing several different levels of taxonomic divergence, a much clearer picture would emerge of the rates of evolution of various categories of social behavior. New correlates with ecological adaptation would probably also appear in abundance. The same opportunity exists, of course, in the study of every other group of social species. A closely parallel effort has already been started on the camels and their relatives (Camelidae) by Pilters (1954).

One last parameter exists that must eventually be entered into the evolutionary equations: the complexity of the genetic change. We have seen that the substitution of single genes can be mostly completed in ten generations. Although the physiological effects of such a shift are likely to be relatively simple, they can cut deeply. The most important consequence is usually that some trait, say, aggressiveness or the ability to respond to an odor, might be reduced or lost. This would be due to the fact that new alleles most often act by diminishing or abolishing certain metabolic capabilities through the blocking of a single biochemical step; the effect on social behavior, if any, will most probably be impairment. Sometimes the need for a given social response is eliminated by a change in the environment. Day-flying species, to take one of many examples, can no longer use visual displays if they become nocturnal or cavernicolous. In such circumstances, the negating genes will be favored by natural selection

Table 7-2 Comparison of *Presbytis entellus* and *Presbytis johnii* communication systems; D, dominant; S, subordinate; E, equal. (From Poirier, 1970a.)

Patterns	*Presbytis johnii* D	S	E	*Presbytis entellus* D	S	E	Comments
Anal inspection	X	X	X	—	—	—	—
Biting air	X	—	X	X	—	X	—
Crouch, then stand suddenly	—	—	—	—	X	X	—
Displacement	X	—	X	X	—	—	—
Dominance pause	—	—	—	X	—	X	—
Embrace	X	X	X	X	X	X	—
Embrace/groom	X	—	X	—	—	—	—
Grin	X	—	—	X	—	X	*P. entellus* grimace?
Head bob	X	—	—	X	X	X	—
Looking away	—	X	X	—	X	—	—
Lunge in place	X	—	X	X	—	X	—
Mount	X	X	X	—	—	—	—
Observe dominance distance	—	X	X	—	X	—	—
Open mouth	X	—	X	—	—	—	—
Present	X	X	X	—	X	X	—
Stare threat	X	—	X	X	—	X	—
Shaking head	—	X	X	—	—	—	Only estrous *P. entellus* female utilizes it
Slap at	X	X	X	X	X	X	—
Slap ground	—	—	—	X	—	X	—
Sniffing mouth	X	—	X	—	—	—	—
Touch	X	X	X	X	—	—	—
Tongue in and out	—	—	—	—	X	X	—
Turn back on	X	X	X	X	X	X	—
VOCAL CUES							
Alarm call	X	X	X	X	X	X	—
Canine grind	X	—	X	X	X	X	—
Cough	—	—	—	X	X	X	—
Chuckle	—	—	—	—	—	—	*P. johnii* mothers use it
Growl	X	—	—	—	—	—	—
Gruff bark	X	X	X	X	X	X	*P. entellus* male and females both utilize it

Table 7-2 (*continued*).

Patterns	*Presbytis johnii* D	S	E	*Presbytis entellus* D	S	E	Comments
Grunt	X	X	X	X	X	X	—
Hiccup	X	X	X	X	X	X	*P. entellus* belch?
Hoho	X	X	X	—	—	—	—
Hollow subordinate sound	—	X	X	—	—	—	—
Pant	X	—	X	—	—	—	—
Roar	X	—	—	—	—	—	—
Scream	—	X	X	—	X	X	Only *P. johnii* infants use it
Screech	—	X	X	—	—	—	—
Squeal	—	—	X	X	X	X	Only *P. johnii* females use it
Subordinate segmented sound	X	X	X	—	—	—	—
Whooping	X	X	X	X	X	X	—
Warble	—	—	—	—	—	—	Only *P. johnii* mothers use it

through the principle of metabolic conservation, meaning that energy formerly devoted to the development and maintenance of useless structures adds genetic fitness when shunted to useful structures. Computer simulations, supported by laboratory experiments on *Drosophila* and other organisms, have established that traits controlled by a small number of polygenes can be altered with comparable alacrity, especially if they are dispersed over enough chromosomes to circumvent linkage disequilibrium. Quantitative characters under additive polygenic control can be readily changed in intensity or sign. A taxis, for instance, can be made stronger or reversed from positive to negative. Or a response to an odor can be changed from a mild attraction to either a strong attraction or a mild aversion.

A case of the next higher degree of complexity in genetic change is a shift in function. The ritualization of a preening movement of a bird to add a display in courtship, the alteration of biosynthesis in an exocrine gland to produce a new pheromone, the modification of olfactory receptors to detect the same pheromone, the involvement of a male in parental care, all such changes are more complicated than the loss of function or a mere shift in the intensity of its expression. It is reasonable to speculate that most changes of this third degree of complexity involve moderate to large numbers of polygenes and require at least hundreds or thousands of generations to pass from an early state of dynamic selection to stabilizing selection, during which time the adaptation is more or less perfected.

At the highest level of evolution are the origins of entirely new patterns or structures. The shift to stabilizing selection can be expected to require at least on the order of a thousand generations. Examples probably include the origins of the waggle dance of honeybees, in particular the familiar advanced version used by *Apis mellifera*; the unique social exocrine glands such as Nasanov's gland of honeybees and the postpharyngeal gland of ants; and human speech. With regard to the last example, Gottesman (1968) has gone so far as to estimate that during the approximately 35,000 generations required for the hypertrophy of the human brain, IQ increased by an average of 0.002 points per generation. Evolution of this magnitude need not proceed smoothly through time. It can advance in pulses, and perhaps stagnate altogether on "plateaus" during the intervening periods. The important point is that so many genes are recruited, and required coadaptation by other structures and functions becomes so extensive, that progress is likely to be detectable in no fewer than thousands of generations.

The Hierarchy of Organismic Responses

Scanning downward through the hierarchy of tracking devices, from evolutionary and morphogenetic change to the increasingly sophisticated degrees of learning, one encounters a steady rise in the specificity and precision of the response. A genetic or pervasive anatomical alteration is scarcely perfectible at all. It is final in the sense that the organism has made its choice and must live or die by it, with further change being left to later generations. Short-term learning, in contrast, can be shaped to very fine particularities in the environment—the direction of a flash of light or the strength of a gust of wind—and augmented or discarded quickly as new circumstances dictate. At this level of maximum precision in behavioral adaptation, the organism can remake itself many times over during its lifetime.

A clear trend in the evolution of the organismic hierarchies is the increasingly fine adjustments made by larger organisms. Above a certain size, multicellular animals can assemble enough neurons to program a complex repertory of instinctive responses. They can also engage in more advanced forms of learning and add an endocrine system of sufficient complexity to regulate the onset and intensity of many of the behavioral acts.

Species from euglena to man can be classified into evolutionary grades according to the length of the response hierarchy and the degree of concentration of power in the lower, more finely tuned responses. It would be both premature and out of place to attempt a formal classification here. Yet to support my main argument, let me at least suggest three roughly defined grades, one at the very bottom, one at the top, and one approximately midway between. The paradigms used are species, although the particular traits characterizing them define the grades in accordance with the usual practice of phylogenetic analysis.

Lowest grade: the complete instinct-reflex machine. The representative organism is so simply constructed that it must depend largely or wholly on token stimuli from the environment to guide it. Perhaps a negative phototaxis keeps it always in the darkness, a circadian rhythm makes it most active just before dawn, a shrinking photoperiod causes it to encyst in the fall, the odor of a certain polypeptide attracts it to prey and induces engorgement, an epoxy terpenoid identifies the presence of a mate and causes it to shed gametes, and so forth—in fact, with this short list we have come close to exhausting its repertory. Endowed with no more than a nerve net or simple central nerve cord containing on the order of hundreds or thousands of neurons, our organism has virtually no leeway in the responses it can make. It is like a cheaply constructed servomechanism; all its components are committed to the performance of the minimal set of essential responses. Possibly no real species exactly fits such a description, but the type is at least approached by sponges, coelenterates, acoel flatworms, and many other of the most simply constructed lower invertebrates (see Jennings, 1906; Corning et al., eds., 1973).

Middle grade: the directed learner. The organism has a fully elaborated central nervous system with a brain of moderate size, containing on the order of 10^5 to 10^8 neurons. As in the organisms belonging to the lowest grade, some of the behavior is stereotyped, wholly programmed, dependent on unconditioned sign stimuli, and species-specific. A moderate amount of learning occurs, but it is typically narrow in scope and limited in responsiveness to a narrow range of stimuli. It results in behavior as stereotyped as the most neurally programmed "instinct." The level of responsiveness may be strongly influenced by the hormone titers, which themselves are adjusted to a sparse set of cues received from the environment. The true advance that defines this intermediate evolutionary grade is the capacity to handle *particularity* in the environment. Depending on the species, the organism may be able to identify not just a female of the species, but also its mother; it not only can gravitate to the kind of habitat for which its species is adapted, but it can remember particular places as well, and will regard one area as its personal home range; it not only can hide but may retreat to a refugium, the location of which it has memorized; and so forth. Examples of this evolutionary grade are found among the more intelligent arthropods, such as lobsters and honeybee workers, the cephalopods, the cold-blooded vertebrates, and the birds.

Highest grade: the generalized learner. The organism has a brain large enough to carry a wide range of memories, some of which possess only a low probability of ever proving useful. Insight learning may be performed, yielding the capacity to generalize from one pattern to another and to juxtapose patterns in ways that are adaptively useful. Few if any complex behaviors are wholly programmed morphogenetically at the neural level. Among vertebrates at least, the endocrine system still affects response thresholds, but since most behaviors have been shaped by complex episodes of learning and are strongly dependent on the context in which stimuli are received, the role of hormones varies greatly from moment to moment and from individual to individual. The process of socialization in this highest grade of organism is prolonged and complex. Its details vary greatly among individuals. The key social feature of the grade, which is represented by man, the chimpanzee, baboons, macaques, and perhaps some other Old World primates and social Canidae (see Chapters 25 and 26), is a *perception of history*. The organism's knowledge is not limited to particular individuals and places with attractive or aversive associations. It also remembers relationships and incidents through time, and it can engineer improvements in its social status by relatively sophisticated choices of threat, conciliation, and formations of alliance. It seems to be able to project mentally into the future, and in a few, extreme cases deliberate deception is practiced.

The remainder of this chapter will complete the examination of the hierarchy of environmental tracking devices, commencing with morphogenesis and caste formation and proceeding downward to the most precise forms of learning and cultural transmission.

Tracking the Environment with Morphogenetic Change

The most drastic response to fluctuations in the environment short of genetic change itself is the modification of body form. Many phyletic lines of invertebrates have adopted this strategy. In principle, the genome is altered to increase its plasticity of expression. Two or more morphological types, which also normally differ in physiological and behavioral traits, are available to the developing organism. Acting on token stimuli that indicate the overall condition of the environment, the organism "chooses" the type into which it will transform itself. Thus, developing *Brachionus* rotifers grow long spines when they detect the odor of predaceous rotifers belonging to the genus *Asplancha.* The new armament prevents them from being consumed. For their part, *Asplancha* (specifically, *A. sieboldi*) can respond to the stimulus of cannibalism and supplementary vitamin E by growing into a gigantic form capable of consuming larger prey. The giant is only one of three distinct morphological types in which the species exists (Gilbert, 1966, 1973). Aphids of many species develop wings when the onset of crowded conditions is signaled by increased tactile stim-

ulation from neighbors. Given the power of flight, these insects are free to depart in search of uncrowded host plants. As populations of plague locusts grow dense, making contacts among the individual hoppers more frequent, they pass from the solitary to the gregarious phase. The transformation takes place over three generations (see Chapter 4). Locusts of the third generation belong to the fully gregarious form and are so different from their solitary grandparents that they can easily pass for a different species—and did, until the full life cycle was worked out by entomologists. The stimuli that trigger the phase transformation happen to be ones that provide reliable information on the degree of crowding. They include the sight of other small, moving objects, which draws the hoppers together, and the light touch of other bodies and appendages. Also important is the chemical "locustol," a phermone released in the feces of immature locusts. The substance has recently been identified by Nolte et al. (1973) as 2-methoxy-5-ethylphenol, an apparent degradation product from the metabolism of plant lignin.

The most elaborate forms of morphogenetic response are the caste systems of the social insects and the colonial invertebrates. With rare exceptions the caste into which an immature animal develops is based not on possession of a different set of genes but solely on receipt of such environmental stimuli as the presence or absence of pheromones from other colony members, the amount and quality of food received at critical growth periods, the ambient temperature, and the photoperiod prevailing during critical growth periods. The proportions of individuals shunted into the various castes are adaptive with respect to the survival and reproduction of the colony as a whole. Caste systems will be discussed in greater detail in Chapter 14.

Nongenetic Transmission of Maternal Experience

When mother rats are psychologically stressed in certain ways, the emotional development of their descendants is altered for up to two generations. In other words, the future of an individual can indeed be influenced in the womb. The first to lift this phenomenon from the realm of folklore was W. R. Thompson (1957). In order to determine the effect of pure "anxiety" of mother rats on the "emotionality" of the young, Thompson performed the following experiment. Anxiety was induced by conditioning female rats before pregnancy to associate the sound of a buzzer with the pain of an electric shock. Then during pregnancy the females were exposed to the sound of the buzzer alone, inducing stress of a mostly psychological nature. As measured solely by Thompson's tests, the offspring of stressed mothers displayed greater emotionality. Specifically, they took longer to leave their cage and to reach food when given the opportunity, and they traveled shorter distances away from the cage during individual forays. Ader and Conklin (1963) subsequently found that the litters of rats handled by the human experimenter during pregnancy were less emotional than those not handled. The pups of mothers that were handled not only crossed open spaces more readily, but they defecated less often while doing so. In order to eliminate postnatal influences, Ader and Conklin put half of the litters in both the experimental and control groups under the care of foster mothers with an experience opposite to that of the natural mothers.

Finally, Denenberg and Rosenberg (1967) established that the experiences of females can bias the behavior of even their grandoffspring. In the first step of their experiment, the future grandmother rats were either handled or not while they were still pups. The daughters of these females, destined to be the mothers of the experimental generation, were then either confined during their infancy to a small maternity cage or else allowed to live in a larger "free environment" cage that contained wooden boxes, a running disk, and other "toys." The interaction of these two classes of experience produced significant differences in the third generation. For example, descendants of nonhandled grandmothers whose mothers had been reared in a maternity cage were more active than descendants of nonhandled grandmothers whose mothers had been raised in a free environment. In other words, the maternal influence was shifted up or down in direction according to the experience of the grandmothers.

The mechanisms of the transgenerational effects remain unknown. Experience involving stress of any kind is known to invoke responses in the pituitary-adrenal complex, which in turn can influence the uterine development of fetuses in ways not understood at present. At the same time the possibility cannot be ruled out that the transmission is at least partly behavioral. Even in the Ader-Conklin experiment, which utilized foster mothers and presumably eliminated most postparturient contact of the natural mothers with their offspring, the offspring were not separated from the natural mothers until 48 hours after birth—perhaps time enough for some formative behavioral interactions to occur. Nevertheless, barring the future discovery of some wholly new biological system, this distinction is not really the main point of the experimental results. Their significance is the demonstration that in a mammal no more complex than a rat the histories of parents and grandparents can bias the behavioral development of individuals strongly, and with it their future status within societies and even the likelihood that they will survive and reproduce. What is true of rodents is almost certain to be true of more complexly social species such as the higher primates. Indeed, it is already known that the social status of male Japanese and rhesus macaques is determined to a large degree by the rank of their mothers. The early social interactions of the monkeys and the way they respond generally to other troop members are influenced by this single circumstance. A lineage of success and failure might easily result, reaching over three generations or more and incorporating experiential and endocrine factors that remain to be fully analyzed.

Hormones and Behavior

Elaborate endocrine systems have evolved in two principal groups of animals, the phylum Arthropoda, including particularly the insects, and the phylum Chordata, including particularly the vertebrates. Since these two taxonomic groups also represent end-points in the two great branches of animal phylogeny, namely the arthropod and echinoderm-chordate superphyla, their endocrine systems can safely be said to have evolved wholly independently. There are basic differences not only in structure and biochemistry but also in function. Arthropod hormones serve to mediate the events of growth, metamorphosis, and ovarian development. Their role in behavior appears to be limited to the stimulation of the production of pheromones and the indirect regulation of reproductive behavior through their influence on gonadal development. Vertebrate hormones have a much wider repertory. They help to regulate numerous purely physiological events, including growth, development, metabolism, and ionic balance. They also exercise profound effects on sexual and aggressive behavior, subjects that will be considered later in Chapters 9 and 11.

At this point there is a need only to draw two broad generalizations about the relation between hormones and behavior in vertebrates. The first is that the function of hormones is to "prime" the animal. Hormones affect the intensity of its drives, or to use a more neutral and professionally approved expression, the level of its motivational states. In addition, they directly alter other physiological processes and large sectors of the behavioral repertory of animals. However, they are relatively crude as controls. Their effects cannot be quickly turned on or off. They track medium-range fluctuations in the environment, such as the seasonal changes made predictable by steady increases or decreases in the daily photoperiod, the stress of extreme cold or threat by a predator, and the presence of a potential mate as signaled by releaser sounds, odors, or other stimuli. An animal cannot guide its actions or make second-by-second decisions through the employment of hormones. It must rely on quicker, more direct cues to provide a finer tuning of motivational states and to trigger specific actions. The second generalization is the intimate relationship, revealed by new techniques in microsurgery and histochemistry during the past twenty years, that exists between the behaviorally potent hormones and specific blocks of cells in the central nervous system.

Both of these features of hormone-behavioral interaction are well illustrated by the role of estrogen in the sexual behavior of female cats. An estrous female responds to the approach of a male by crouching, elevating her rump, deflecting her tail sidewise to expose the vulva, and pawing the ground with treading movements of her hind legs. She readily submits to being mounted. If not in estrus, she instead reacts aggressively to the close approach of a male. It is well known that estrus is initiated by the rise of the estrogen titer of the blood. But in what way does estrogen prime the animal for sexual behavior? Not, it turns out, by the estrogen-mediated growth of the reproductive tract. When castrated females are injected repeatedly with small doses of estrogen over a long period of time, the reproductive tract develops completely, yet sexual behavior is still not induced (Michael and Scott, 1964). The female sexual response depends on a more direct action of the hormone. When needles tipped in slowly dissolving estrogen are inserted into certain parts of the hypothalamus, the castrated cats display typical estrous behavior, even though their reproductive tracts remain undeveloped (Harris and Michael, 1964). Michael (1966) also discovered that radioactively labeled estrogen injected into the bloodstream is preferentially absorbed by neurons in just those areas of the hypothalamus most sensitive to direct applications of the hormones by needle.

The targeting of neurons by behaviorally active hormones is probably widespread among mammals. Fisher (1964) found that minute quantities of testosterone injected into the hypothalamus of rats evoked sexual and parental behavior. However, the results were not nearly as clear-cut as in the cats. Only a minority of individuals responded, and then in an often aberrant fashion: parental behavior was represented by attempts to carry other animals, including adults, back to the nest; and both sexes assumed the male position in attempts to copulate. It is nevertheless significant that Fisher got his results only with testosterone. Other chemicals and the use of electrical stimulation failed to produce even aberrant sexual behavior.

Corticosterones are released from the adrenal gland when mammals are stressed and play a key role in the general physiological adaptation of the body to the new circumstances (for details see Chapter 11). Zarrow et al. (1968) found that radioactive corticosterone is concentrated in the hypothalamus. Since infant rats as well as adults secrete the hormone when stressed, the possibility exists that the corticosterones and similar adrenocortical products act upon the developing brain to change many of the physiological and behavioral responses in an adaptive manner. Such a mechanism might even contribute to the transgenerational influence of maternal experience described previously (Denenberg, 1972). Still one more example of hormone targeting can be cited. Testosterone heightens general aggressivity in male animals and improves their performance during disputes over territory and status. When castrated male gerbils are injected with the hormone, they develop a larger ventral scent gland and commence marking their territories with the secretions. The same behavioral response is elicited by the injection of slowly dissipating testosterone directly into the preoptic area, which is located just anterior to the hypothalamus (Thiessen and Yahr, 1970).

To about the same degree that hormones control some aspects of behavior, behavior controls the release of hormones. The signals exchanged by members of the same species frequently act not only to induce overt behavioral responses in others but also to prime their physiology. Once modified in this way, the recipient animal responds to further signals with an altered behavioral repertory. The courtship of ring doves depends on an exact marching order of hormones timed by the perception of external signals. When a pair is placed together in a cage, the male begins to court immediately. He is the initiator because his testes are active and probably secreting testosterone. He faces the female and repeatedly bows and coos. The sight of the displaying male activates mechanisms in the female's brain, which in turn instruct the pituitary gland to release gonadotropins. These hormones stimulate the growth of the female's ovaries, which begin to manufacture eggs and to release estrogen into the bloodstream. The essential steps are thus concluded for the successful initiation of nest building and mating (Lehrman, 1964, 1965).

The release of reproductive hormones into the bloodstream of female mice is also sensitive to signals from other members of the same species (Whitten and Bronson, 1970; Bronson, 1971). In the manner of the medical sciences, the different kinds of physiological change are often called after their discoverers:

1. *Bruce Effect.* Exposure of a recently impregnated mouse female to a male with an odor sufficiently different from that of her stud results in failure of the implantation and rapid return to estrus. The adaptive advantage to the new male is obvious, but it is less easy to see why it is advantageous to the female and therefore how the response could have been evolved by direct natural selection.

2. *Lee-Boot Effect.* When about four or more female mice are grouped together in the absence of a male, estrus is suppressed and pseudopregnancies develop in as many as 61 percent of the individuals. The adaptive significance of the phenomenon is unclear, but it is evidently one of the devices responsible for the well-known reduction of population growth under conditions of high population density.

3. *Ropartz Effect.* The odor of other mice alone causes the adrenal glands of individual mice to grow heavier and to increase their production of corticosteroids; the result is a decrease in reproductive capacity of the animal. Here we have part but surely not all of the explanation of the well-known stress syndrome. Some ecologists have invoked the syndrome as the explanation of population fluctuation, including the occasional "crashes" of overly dense populations described in Chapter 4.

4. *Whitten Effect.* An odorant found in the urine of male mice induces and accelerates the estrous cycle of the female. The effect is most readily observed in females whose cycles have been suppressed by grouping; the introduction of a male then initiates their cycles more or less simultaneously, and estrus follows in three or four days.

Until the pheromones are identified chemically, the number of signals involved in the various effects cannot be known with certainty. Bronson (1971) believes that as few as three substances can account for all the observed physiological changes: an estrus-inducer, an estrus-inhibitor, and an adrenocortical activator. Martha McClintock (1971) reported a tendency toward synchronization in the menstrual timing of young women living in the same college dormitory, an effect not unlike that seen in the rodents. Whether odor is involved remains unknown.

Stress has an important influence on the mammalian endocrine system, a fact known to medical science since 1825, when Parry observed the onset of hyperthyroidism in a human being following a severely frightening experience. What has not been fully appreciated until recently, however, is the depth and scope of this influence. Systematic studies on the rhesus monkey have implicated at least the pituitary, thyroid, and adrenal glands, as well as the glandular elements of the reproductive organs of both sexes. The principal technique for identifying the effects, utilized largely by John W. Mason and his associates, is a special application of the Sidman avoidance procedure. The monkey is restrained in a chair within a soundproof room, a treatment which by itself is said not to create unusual stress. Then an electric shock is applied at 20-second intervals with no warning signal other than a red light left on for the full duration of the avoidance session. When the light flashes on, the monkey must press a hand lever that operates a microswitch, causing it to reset a 20-second timer. If the animal fails to press the lever during the subsequent 20 seconds, a circuit closes and a mild electric shock is administered to its feet. The shock intensity is adjusted to the minimal level required to maintain avoidance behavior. The obvious effect of such sessions is a sustained, generalized stress. The endocrine responses resulting from the procedure are documented in Figure 7-4. As Mason has argued, they could represent only a fraction of the total phenomena. Many of the responses further interact with one another, and these interactions ultimately result in complex changes in the physiology and behavior of the animal that are difficult to assess. The least of these changes go far beyond the simple conditioned response by which the monkey protects itself from electric shock. They involve behavioral parameters such as aggressiveness, proclivity to mate, willingness to explore, urination volume and frequency, and others.

It seems likely but has not been adequately proved that the same effects result from social stress within normally constituted societies. Rowell (1970) observed that subordinate female baboons (*Papio anubis*) beaten up by other females in the troop had longer menstrual cycles. When they were isolated from their adversaries, their perineal swelling increased in size, while their sexual skin changed from bright pink to pale greyish pink in color. There is no reason to expect that the hormonal changes induced by such stresses in the social environ-

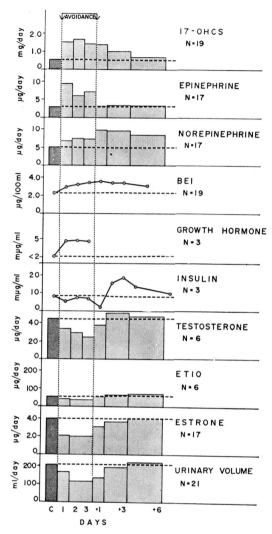

Figure 7-4 Endocrine responses to stress in the rhesus monkey. During a three-day "avoidance period" under very restricted laboratory conditions, the monkeys were required to press a bar at frequent intervals in order to avoid a mild electric shock. The hormone levels shown are the amounts in plasma or urine. The urinary volume indicated serves as an indirect measure of the antidiuretic effect of hormones. Following the three days of stress, the monkeys were monitored for an additional six days, during which time the hormone concentrations began to change back to the prestress levels. 17-OHCS means urinary 17-hydroxycorticosteroid; BEI, butanol extractable iodine, a measure of thyroid activity; and ETIO, etiocholanolone. (From Mason, 1968.)

ment are any less profound than those induced by experimental psychologists with electric shocks and other contrived stimuli.

Learning

The Directedness of Learning

Viewed in a certain way, the phenomenon of learning creates a major paradox. It seems to be a negating force in evolution. How can learning evolve? Unless some Lamarckist process is at work, individual acts of learning cannot be transmitted to offspring. If learning is a generalized process whereby each brain is stamped afresh by experience, the role of natural selection must be solely to keep the tabula rasa of the brain clean and malleable. To the degree that learning is paramount in the repertory of a species, behavior cannot evolve. This paradox has been resolved in the writings of Niko Tinbergen, Peter Marler, Sherwood Washburn, Hans Kummer, and others. What evolves is the directedness of learning—the relative ease with which certain associations are made and acts are learned, and others bypassed even in the face of strong reinforcement. Pavlov was simply wrong when he postulated that "any natural phenomena chosen at will may be converted into conditioned stimuli." Only small parts of the brain resemble a tabula rasa; this is true even for human beings. The remainder is more like an exposed negative waiting to be dipped into developer fluid. This being the case, learning also serves as a pacemaker of evolution. When exploratory behavior leads one or a few animals to a breakthrough enhancing survival and reproduction, the capacity for that kind of exploratory behavior and the imitation of the successful act are favored by natural selection. The enabling portions of the anatomy, particularly the brain, will then be perfected by evolution. The process can lead to greater stereotypy—"instinct" formation—of the successful new behavior. A caterpillar accidentally captured by a fly-eating sphecid wasp might be the first step toward the evolution of a species whose searching behavior is directed preferentially at caterpillars. Or, more rarely, the learned act can produce higher intelligence. As Washburn has said, a human mind can easily guide a chimpanzee to a level of performance that lies well beyond the normal behavior of the species. In both species, the wasp and man, the structure of the brain has been biased in special ways to exploit opportunities in the environment.

The documentation of the directed quality of learning has been extensive. Consider the laboratory rat, often treated by experimental psychologists in the past as if it were a tabula rasa. Garcia et al. (1968) found that when rats are made ill from x-rays at the time they eat food pellets (and not given any other unpleasant stimulus), they subsequently remember the flavor but not the size of the pellets. If they are negatively reinforced by a painful electric shock while eating

(and not treated with x-rays), they remember the size of the pellet associated with the unpleasant stimulus but not the flavor. These results are not so surprising when considered in the context of the adaptiveness of rat behavior. Since flavor is a result of the chemical composition of the food, it is advantageous for the rat to associate flavor with the after-effects of ingestion. Garcia and his coworkers point to the fact that the brain is evidently wired to this end: both the gustatory and the visceral receptors send fibers that converge in the nucleus of the fasciculus solitarius. Other sensory systems do not feed fibers directly into this nucleus. The tendency to associate size with immediate pain is equally plausible. The cues are visual, and they permit the rat to avoid such dangerous objects as a poisonous insect or the seed pod of a nettle before contact is made.

Very young animals display an especially sharp mosaic of learning abilities. The newborn kitten is blind, barely able to crawl on its stomach, and generally helpless. Nevertheless, in the several narrow categories in which it must perform in order to survive, it shows an advanced ability to learn and perform. Using olfactory cues, it learns in less than one day to crawl short distances to the spot where it can expect to find the nursing mother. With the aid of either olfactory or tactile stimuli it memorizes the route along the mother's belly to its own preferred nipple. In laboratory tests it quickly comes to tell one artificial nipple from another by only moderate differences in texture (Rosenblatt, 1972). Still other examples of constraints on learning are reviewed by Shettleworth (1972).

The process of learning is not a basic trait that gradually emerges with the evolution of larger brain size. Rather, it is a diverse array of peculiar behavioral adaptations, many of which have been evolved repeatedly and independently in different major animal taxa. In attempting to classify these phenomena, comparative psychologists have conceived categories that range from the most simple to the most complex. They have coincidentally provided a rank ordering of phenomena according to the qualities of flexibility in behavior, its precision, and its capacity for tracking increasingly more detailed changes in the environment. Excellent recent reviews of this rapidly growing branch of science have been provided by Hinde (1970), P. P. G. Bateson (1966), and Immelmann (1972).

The Ontogeny of Bird Song

The songs by which male birds advertise their territories and court females are particularly favorable for learning and other aspects of developmental analysis. The songs are typically complex in structure and differ strongly at the level of the species. Considerable variation among individual birds also exists, some of it subject to easy modification by laboratory manipulation. Following the pioneering work of W. H. Thorpe, who began his studies in the early 1950's, biologists have investigated every aspect of the phenomenon from its neurological and endocrine basis to its role in speciation. This advance has been made possible by a single technical breakthrough—the sound spectrograph, by which songs can be recorded, dissected into their components, and analyzed quantitatively. Perhaps the single most important result has been the demonstration of the programmed nature of learning in the ontogeny of song, a lock-step relation that exists between particular stimuli, particular acts of learning, and the short sensitive periods in which they can be linked to produce normal communication. Complete reviews have been provided by Hinde and his coauthors (Hinde, ed., 1969; Hinde, 1970) and by Marler and Mundinger (1971).

One of the more discerning studies has been conducted on the white-crowned sparrow *Zonotrichia leucophrys* of North America (Marler and Tamura, 1964; Konishi, 1965). The male song consists of a plaintive whistle pitched at 3 to 4 kiloherz, followed by a series of trills or *chillip* notes. Many variations occur, particularly in the form of "dialects" that distinguish geographic populations. Under normal circumstances full song develops when the birds are 200 to 250 days old, but Marler and Tamura showed that this capacity is present much earlier. Young birds captured at an age of one to three months in the area where they were born and kept in acoustical isolation later sang the song in the dialect of their region. Others removed from the nest at 3 to 14 days of age and raised by hand in isolation also developed a song; it possessed some though not all of the basic simplified structure characteristic of the species as a whole and had none of the distinctive features of the regional dialect. Evidently, then, the dialect is learned from the adult birds during rearing and before the young birds themselves attempt any form of song. Hand-raised sparrows will sing the dialect of their region or another region if taped songs of wild birds are played to them from the age of about two weeks to two months. Thus the species-specific skeleton of the song seems fully innate in the looser usage of that term, while the population-specific overlay is acquired by tradition. It turns out, however, that even the skeleton requires some elements of learning, albeit highly directed in character and virtually unalterable under normal conditions. Konishi found that when birds are taken from the nest and deafened by removal of the cochlea, they can produce only a series of unconnected notes when they attempt to sing. This remains true when the birds have been trained by exposure to adult calls. In order to put together a normal call, even the skeletal arrangement of the species, it is essential for the white-crowned sparrows to hear themselves as they sing the elements previously learned. The essential steps of development are summarized in Figure 7-5.

A closely parallel study was conducted on the chaffinch *Fringilla coelebs* of Europe by Thorpe (1954, 1961), Nottebohm (1967), and Stevenson (1969). Thorpe introduced the technique of playing syn-

thetic songs to the young birds in their sensitive periods to find which elements can be learned and which cannot. He found that "songs" constructed of pure tones had no effect, but real chaffinch songs chopped up and rearranged in various ways were learned in the modified form. Thus the young finches could be made to sing the song backward or with the end notes placed in the middle. Other details of the learning process, including the need for auditory feedback by the songster, were found to be essentially the same as in the white-crowned sparrow.

The infiltration of learning into the evolution of bird song introduces a closer fit of the individual's repertory to its particular environment. As Lemon and Herzog (1969) have said, learning permits the immediate satisfaction of communicative needs without recourse to the tedious process of selection over several generations. An individual bird achieves its vocal niche quickly in a complex environment of sound. As a result it can distinguish a potential mate of the same species from among a confusing array of related species. Where regional dialects and their recognition are based to some extent on adult learning, the bird can utilize familiarity with old neighbors to eliminate unnecessary hostile behavior. In the case of convergent duet singing, the bird can perfect communication with its mate and reduce the chance of being distracted by other members of the species.

The Relative Importance of Learning

The slow phylogenetic ascent from highly programmed to flexible behavior is nowhere more clearly delineated than in the evolution of sexual behavior. The center of copulatory control in male insects is in the ganglia of the abdomen. The role of the brain is primarily inhibitory, with the input of sexual pheromones and other signals serving to disinhibit the male and guide him to the female. The total removal of the brain of a male insect—chopping off the head will sometimes do—triggers copulatory movements by the abdomen. Thus a male mantis continues to mate even after his cannibalistic mate has eaten away his head. Entomologists have used the principle to force matings of butterflies and ants in the laboratory. The female is lightly anesthetized to keep her calm, the male is beheaded, and the abdominal tips of the two are touched together until the rhythmically moving male genitalia achieve copulation. A similar control over oviposition is invested in the abdominal ganglia of female insects. The severed abdomens of gravid female dragonflies and moths can expel their eggs in a nearly normal fashion.

The sexual behavior of vertebrates differs from that of insects in being controlled almost wholly by the brain, particularly regions of the cerebral neocortex. Furthermore, there exists within the vertebrates as a whole a correlation between the relative size of the brain—a crude indicator of general intelligence (Rumbaugh, 1970)—and the dependence of male sexual behavior on the cerebral neocortex

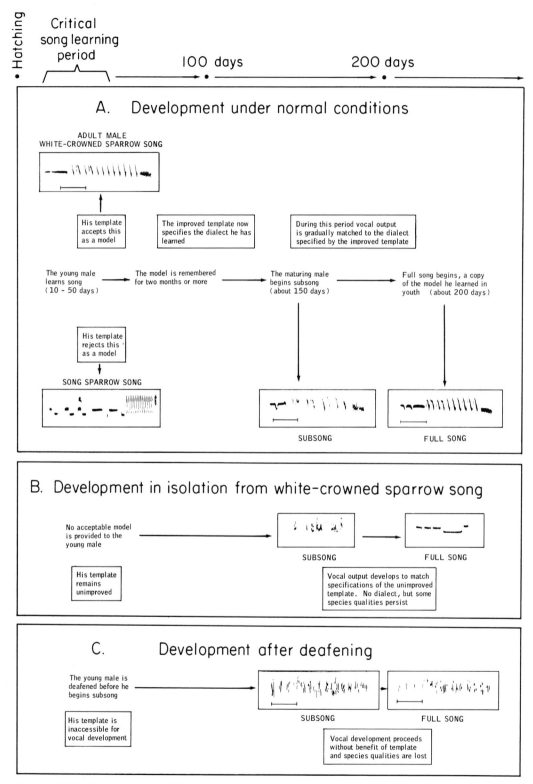

Figure 7-5 A case history of directed learning. The essential events in the development of song in the male white-crowned sparrow are expressed as a summary of the experiments by Marler, Tamura, and Konishi. (Courtesy of P. R. Marler.)

and social experience. As much as 20 percent of the cortex of male laboratory rats can be removed without visible impairment of their sexual performance. When 50 percent is removed, more than one-fifth of the animals still mate normally. In male cats, however, extensive bilateral injury to the frontal cortex alone causes gross abnormalities of sensorimotor adjustment. The animals display signs of intense sexual excitement in the presence of estrous females, but they are usually unable to make the body movements necessary for successful intromission. Higher primates, particularly chimpanzees and man, have prolonged, personalized sexual behavior which is even more vulnerable to cortical injury (Beach, 1940, 1964). The importance of social experience in sexual practice also increases with brain size, while the effectiveness of hormones in initiating or preventing it declines.

The cerebralization of sexual behavior is merely one facet of the increasing role of undirected learning that permits an ever tighter fit of behavior to short-term changes in the environment. The closer this fit, the more pliable the behavior patterns employed to create it. Both circumstances dictate a prolonged training period in young animals. Washburn and Hamburg (1965) have graphically expressed the argument with reference to primates:

In evaluating the importance of learning and skills in the behavior of free-ranging primates, it must be remembered that the criterion of success is survival in crises and not necessarily merely successful day-to-day behavior. Over a period of months the mother has only to make one mistake to kill the infant. When the play fighting of the older juvenile males changes to real fighting, skill means the difference between victory and a serious wound. It does not surprise us that athletes must practice constantly to be in top form, but it is easy to forget that the survival of animals under conditions of crisis may be just as demanding. Haddow (1952) describes seeing a group of feeding *Colobus* monkeys when a monkey-eating eagle suddenly came around the tree. These eagles fly below the level of the tops of the trees and appear with no warning. All the monkeys dropped down out of the high branches, except for one adult male which climbed *up* at the eagle. The precipitous downward flight of frightened monkeys is dramatic, but the point that should be stressed is that in the brief duration of such a crisis infants are retrieved and carried down, and leaps are taken that are much longer than those used in normal locomotion. This sudden flight requires the highest skills of climbing, and any error results in injury or death. The high incidence of healed fractures in monkeys and apes gives clear evidence that the selection for skill is important (Schultz, 1958), and such statistics are based on the animals which *survived*. The actual rates of injury must be higher, much higher in our opinion, but even so healed fractures are present in 50 percent of old gibbons. Many severe injuries do not result in fractures, so total injuries must far exceed this percentage. As Shultz points out, many of the injuries may be due to fighting. However that may be, our point here is that the criterion of successful learning through a prolonged youth is survival in crises and that such survival depends on knowledge and skill.

The principle is not confined to the mammals. The oystercatcher *Haematopus ostralegus* is unique among European shore birds in that the young are not self-supporting until they are fully fledged, or even later. The explanation appears to lie in the specialized and difficult feeding habits of the species, which are acquired over long periods of practice by the fledglings. The parents accompany their offspring out to the feeding grounds, where they search for small, hard-shelled bivalves together. The young birds then learn to open the mollusks by hammering them or by inserting the bill in just the right position (Norton-Griffiths, 1969).

Socialization

Socialization is the sum total of all social experiences that alter the development of an individual. It consists of processes that encompass most levels of organismic responses. The term and the set of diffuse ideas that enshroud it originated in the social sciences (Clausen, 1968; Williams, 1972) and have begun gradually to penetrate biology. In psychology socialization ordinarily means the acquisition of basic social traits, in anthropology the transmission of culture, and in sociology the training of infants and children for future social performance. Margaret Mead (1963), recognizing the different levels of organismic response implicit in the phenomenon, suggested that a distinction be made between true socialization, the development of those patterns of social behavior basic to every normal human being, and enculturation, the act of learning one culture in all its uniqueness and particularity. Vertebrate behaviorists who use the word socialization usually have in mind only the learning process (Poirier, 1972), but if comparative studies are to be made in all groups of animals its definition will have to embrace all the range of socially induced responses that occur within the lifetime of one individual. If that proposition is accepted, the following three categories can be recognized:

1. Morphogenetic socialization, for example caste determination.
2. Learning of species-characteristic behavior.
3. Enculturation.

Socialization has resisted deep analysis because of two imposing difficulties encountered by both social scientists and zoologists. The first is the considerable technical problem of distinguishing behavioral elements and combinations that emerge by maturation, that is, unfold gradually by neuromuscular development independently of learning, and those that are shaped at least to some extent by learning. Where both processes contribute, their relative importance under natural conditions is extraordinarily difficult to estimate. The second major problem is of course the complexity and fragility of the social environment itself.

In spite of this, experimental research has now been pursued to the point that a few interesting generalizations are beginning to emerge. As expected, the form of socialization is roughly correlated among species with the size and complexity of the brain and the degree of involvement of learning. Members of colonies of lower invertebrates and social insects are socialized principally by the physiological and behavioral events that determine their caste during early development. The specialized zooids of colonial coelenterates and bryozoans may be established solely by morphogenetic change imposed by their physical location among other zooids. Although the development of "social behavior" has not been analyzed in these animals, the visible responses are so elementary and stereotyped that learning seems unlikely to play any important role. Caste determination in social insects is achieved mainly through the physiological influence of adult colony members on the developing individual. Often, as in some ant species, it is a matter of the amount (and perhaps quality) of food given to the larva. In the honeybees, the quality of the food is paramount, depending on the presence or absence of certain unidentified elements in the royal jelly, which is fed to a few larvae sequestered in royal cells. In termites, inhibitory substances (pheromones) produced by the kings and queens force the development of the great majority of nymphs into one of the sterile worker castes. Only in the most primitively social of the insects do direct behavioral interactions play a key role. Among paper wasps of the genus *Polistes*, most of the females that found colonies together have been inseminated and possess similar reproductive capacities, but only one individual assumes a dominant role, becoming the egg-laying queen and forcing the others into subservient labor as the de facto worker caste. The ovarian development of the subordinates is reduced, an inhibition due at least in part to the greater amounts of energy the wasps must expend in foraging, nest building, and brood care. The subordinates are also placed at a disadvantage by being forced to surrender some of the booty collected on foraging trips to the dominant queen. A similar arrangement exists in the bumblebees, in which the larger-bodied queen arrogates to herself exclusively the privilege of laying eggs. When a queen paper wasp or bumblebee dies, some of the surviving females threaten and fight one another until a new functional queen emerges. The status of these females, and hence their socialization, is probably based to some extent on learning. They appear capable of recognizing one another as individuals and apparently are influenced in new encounters during the struggle for power by their past record of victories and defeats (Wilson, 1971a).

Once a social insect has matured into a particular caste, it launches into the complex repertory of behaviors peculiar to that form. In the 10 days after a typical honeybee worker emerges as an adult winged insect from the pupa, she engages expertly in a wide variety of tasks that includes at least some of the following: polishing and cleaning cells in the honeycomb and brood area, constructing new hexagonal cells out of wax to a precision of a tenth of a millimeter or better, attending the queen, flying outside the hive, ripening nectar into honey and storing it, feeding and grooming larvae, fanning on the comb to aid in thermoregulation, conducting and following waggle dances, and regurgitating with other workers. At 30 days the worker is old, her repertory largely behind her, and she has only a few more days of service as a forager left.

The role of learning in this brief but remarkable career has never been investigated, but it must be narrowly directed and stereotyped at best. We know that honeybees learn the odor of nestmates and the location of their hive and food sources. Tasks can be memorized and performed in a sequence, including the often complicated schedules of visits to flowers at specific times of the day. Isolated worker bees can be trained to walk through relatively complex mazes, taking as many as five turns in sequence in response to such clues as the distance between two spots, the color of a marker, and the angle of a turn in the maze. After associating a given color once with a reward of 2-molar sucrose solution, they are able to remember the color for at least two weeks. The location of a food site in the field can be remembered for a period of six to eight days; on one occasion bees were observed dancing out the location of a site following two months of winter confinement (Lindauer, 1961; Menzel, 1968). Nevertheless, these feats become less impressive when it is realized how narrowly and immediately functional they really are. Like the minor song dialects of finches, honeybee learning represents lesser variation that overlays basic behavior patterns that either develop regardless of experience or else are learned along strict channels during brief sensitive periods (Lindauer, 1970). The bee quickly learns the location of a distant site, for example. But the waggle dance by which she expresses this information is more complex and rigidly programmed. Do any learned components and autosensory feedback go into the early development of the bee's ability to dance? Until experiments are performed similar to those by Konishi and Nottebohm on the neurosensory basis of the development of bird song, we are not likely to know the answer.

Socialization has been much more intensively studied in primates than in other kinds of animals. The circumstance is fortunate for two reasons: the phylogenetic affinity of the Old World monkeys and apes to man plus the fact that socialization by learning appears to be deepest and most elaborate in these animals. Before describing the actual process as it is now understood, I believe it would be useful to outline the techniques of experimentation in a way that attempts to reflect the philosophy of the biologists who have conducted it. Their approach can be understood more readily if we draw an analogy between socialization and the biology of vitamins. In the evolution of a given species, a nutrient compound becomes a vitamin when it is so readily available in the normal diet that members of the species no longer need to synthesize it from simple components. In obedience to the principle of metabolic conservation, the species then tends to eliminate the biochemical steps required for the synthesis of the substance, thus permitting enzymatic protein and energy to be diverted for other, more urgent functions. At this point the molecule becomes "essential"—that is, a vitamin—in the sense that it must be included in the diet thereafter in order for the organism to thrive. Vitamin D, which regulates the absorption of calcium from the intestine by influencing membrane permeability or active transport, is a sterol produced from other sterols by irradiation with ultraviolet light. The human body obtains it easily, either in the diet or by transformation of dietary sterols. The existence of such vitamins can be discovered by systematically withholding suspected compounds from completely defined synthetic diets. Their role can be ascertained by a thorough study of the physiology of vitamin-starved animals. Important additional effects, sometimes harmful and sometimes beneficial, can be induced by enriching the diet with abnormally large amounts.

Through an analogous form of evolutionary decay, behavioral elements involved in socialization become increasingly dependent on experience for normal development. These elements are most easily identified by noting their reduction or disappearance when various forms of normal social experience are withheld. This is the method of *environmental deprivation*. Sometimes the same or additional elements can be discovered and characterized in part by increasing the amount of stimuli above the normal laboratory level and observing modifications in the opposite direction. This is the method of *environmental enrichment*.

Studies of socialization in primates, particularly in the rhesus monkey as the most favored species, have relied heavily on the technique of environmental deprivation. The rather involved results of these studies can be better understood if we order the experiments according to the amount of deprivation imposed, and hence the number and degree of perturbations they usually produce. The following list of experiences proceeds from the most to the least drastic.

Descending Degrees of Social Deprivation

1. The young monkey is raised by an artificial mother made of cloth and denied its real mother, peers, and all other social partners until maturity (Harlow and Zimmerman, 1959; Harlow, 1959; Harlow et al., 1966).
2. The young monkey is kept with its natural mother for part or all of its development but is not allowed contact with other monkeys until maturity (Mason, 1960, 1965; Hinde and Spencer-Booth, 1969).
3. The young animal is separated from its mother but allowed to associate with other monkeys of the same age (Sackett, 1970).

4. The young animal is raised by its natural mother in the midst of a troop but is temporarily separated as an infant for short periods of time (Spencer-Booth and Hinde, 1967, 1971; I. C. Kaufman and Rosenblum, 1967).
5. The young animal is raised with a normal social group but in a restricted laboratory environment as opposed to natural or semi-natural habitats (Mason, 1965).
6. The young animal is reared with its natural mother in as normal and complete a social setting as possible. The developmental schedule provides a control for the deprivation experiments. But differences among individuals inevitably arise owing to variation in heredity, social rank of the mother, illnesses and other events during development, and other uncontrolled circumstances. By careful clinical studies of individual case histories, considerable insight can be gained into the relative importance and interaction of the factors, although the system is too complex to permit quantitative assessments such as parameter estimates in multiple regression analyses. (N. R. Chalmers, Irven DeVore, R. A. Hinde, Jane B. Lancaster, Jane van Lawick-Goodall, G. D. Mitchell, F. E. Poirier, Timothy W. Ransom, Thelma E. Rowell, and others: excellent reviews have been provided by Alison Jolly, 1972a; Poirier et al., 1972; and Rowell, 1972.)

The story of socialization in monkeys and apes revealed by these studies is one of a gradual release of the young animal from the bosom of its mother into the increasingly chancy social milieu of the surrounding troop. Day by day, week by week, the infant reduces the amount of time it spends asleep or attached to its mother's nipple while lengthening the duration of its tentative explorations away from her and increasing the number of contacts made with other members of the troop. The amount of time apportioned to each activity changes linearly or in a logarithmic fashion with age, and the origin and slope of these proportions plotted as a function of age differ markedly between species (Figures 7-6 and 7-7).

Poirier (1972b), while conceding the continuous nature of social development, has suggested that it can be conveniently divided into four arbitrarily defined periods. In the first, the *neonatal period*, the animal is a helpless infant, limited to the ingestion of milk and forms of locomotion that hold it close to the mother's body. Contact with the mother is continuous and close. In the *transition period* the infantile movements are supplemented by adult locomotor and feeding patterns. The animal is still closely associated with the mother but leaves her for increasingly long periods to play and to feed on its own. For most primate species the transition period lasts for several months, ending when the company of the mother is no longer frequently sought. The monkey now passes into the time of *peer socialization*, when much of the contact is with group members other than

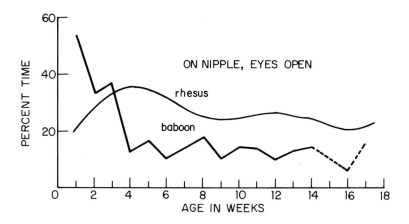

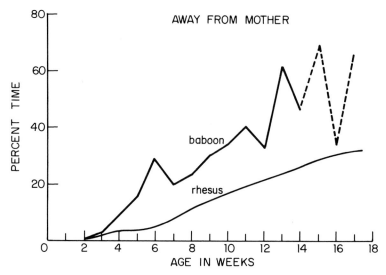

Figure 7-6 Early social development in the rhesus monkey (*Macaca mulatta*) and olive baboon (*Papio anubis*) is measured by time curves of two of the key activities of infants. The values shown are the percent of half-minute intervals of observation time spent respectively by six rhesus and four baboon infants. (Modified from Rowell, Din, and Omar, 1968.)

the mother. The most frequently sought peers are the mother's prior offspring, older females, and other youngsters of about the same age. The key events of this period are the completion of weaning and the gradual subsidence of the infantile behavior patterns. Finally, the animal enters the *juvenile-subadult period*, during which the infantile patterns disappear entirely and adult patterns, including sexual behavior, are first practiced. Females reach full adult status sooner

than males, and both sexes mature more quickly in short-lived species than in long-lived species.

The spreading nexus of relationships that characterizes the later periods of socialization has been analyzed by van Lawick-Goodall (1968), Burton (1972), Rowell (1972), and Hinde (1974). Only a few species have been studied in sufficient detail for long enough periods to draw firm conclusions. These include anubis baboons, macaques (Barbary, bonnet, Japanese, rhesus, and pigtail), vervets, and chimpanzees. The first contacts made by the youngster beyond its mother are normally with the maternal siblings. Even exceptionally restrictive, fearful mothers allow their infants to be approached by her older children, and sisters and half-sisters are the most favored of all. In the olive baboon, only the young females concern themselves with infants, while juvenile and subadult males restrict themselves to other young males in these age classes. Sibling relationships often endure into adulthood and form a principal basis for grooming partnerships and cliques. When rhesus males migrate to new troops, they some-

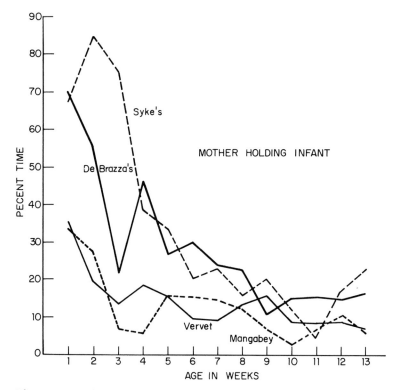

Figure 7-7 The time curves of four related species of African monkeys show the marked variation that exists in one aspect of social development. The species are Sykes' monkey (*Cercopithecus mitis*), DeBrazza's monkey (*C. neglectus*), the vervet (*C. aethiops*), and the gray-cheeked mangabey (*Cercocebus albigena*). (Redrawn from Chalmers, 1972.)

times penetrate these host societies by joining forces with brothers who have preceded them.

Young baboons and macaques begin to interact with peers when they are about six weeks old. They wander away from their mothers for long intervals and spend most of their waking time at play with other infants and juveniles. Both siblings and nonsiblings are now included in the widening circle of acquaintanceship. Virtually all of the components of aggressive and sexual behavior make their first appearance and are then strengthened and perfected by frequent practice. Initially the behavior patterns are almost always expressed in the context of play, and as such they consist of improperly connected, nonfunctional fragments. Later, during the juvenile-subadult period, these fragments are linked together to form the full repertories of serious aggressive and sexual behavior. The choice of playmates and the relationships forged with them during play often persist into adulthood and hence are crucial to the future social status of the individual. As Rowell has pointed out, the tense, formal relationships between the adult males of macaque troops do not arise simply in vacuo by random interactions at puberty. They grow gradually from the existing relationships of the juvenile males that seem at first to be relaxed and playful.

The nature of contacts between young and adults other than the mother varies strikingly among the primate species, sometimes in ways that have an apparently adaptive basis. "Aunts" are adult and older juvenile females, not necessarily close relatives who handle the infants of other females, carrying them about, grooming them, and inspecting their genitals. In certain species, such as Sykes' monkeys and vervets, the relationship is casual and apparently limited to older siblings. In rhesuses and baboons, aunting is much more pronounced. It often strongly affects the behavior of the mother and infant, and almost certainly influences the psychological development of the young animal. Infants distinguish between the aunts and their true mothers within a few days of their birth. They are less attracted to the former, and older individuals try to break away as soon as they lose sight of their mothers. Ordinarily they solicit the attentions of the aunts only if they are denied access to the mother. Thus it is the adult females who seek the role of aunt rather than the other way around. The benefit they receive is not immediately clear; it could be the establishment of a useful alliance with the mother to further status or the practice received in maternal behavior, or both (see Chapter 16).

Most adult males of some species, such as the patas monkey, Sykes' monkey, and the rhesus, ignore infants almost totally. At the other extreme, male Barbary macaques, who belong to the same genus as the rhesus, carry the young animals extensively, using them as devices to conciliate rival males. The relationship of male hamadryas baboons to young females is fundamental to the peculiar social organization of the species. Young subadult and adult males affiliate with the females while the latter are still infants and juveniles. As these consorts mature they seek the protection of the male and form the nucleus of his harem.

The younger the animal, the more traumatic is the effect of a given type of social deprivation. Thus isolation for six months will irreparably damage the social capacity of an infant monkey or ape, but it will have only a minor, temporary effect on a mature male. Furthermore, the greater the deprivation, the deeper and more enduring the result. Total deprivation is almost wholly crippling to the infant's development; it can be partly erased by permitting the infant access to peers for short but frequent intervals. Provided it lives with a normally constituted social group, a monkey raised in a restricted laboratory environment differs from a feral animal in only quantitative ways, such as the time required to achieve normal sexual behavior. Both of these principles have been well documented by the results of the many deprivation experiments that have been performed on rhesus monkeys. The results are briefly summarized in Figure 7-8.

The trauma of extreme deprivation was first clearly revealed in Harry F. Harlow's famous experiments on "mother love" and other aspects of socialization in the rhesus monkey. Infants were removed from their mothers and given a choice of two crude substitute models, one constructed of wire and the other of terry cloth. The young animals strongly preferred the cloth model, which they hugged and clutched much of the time. The softness of the material proved crucial; the models were accepted even when they were supplied with huge round eyes made of bicycle lamps and bizarre faces that made them seem more like toys or gargoyles than real monkeys. The young rhesus monkeys physically thrived when they were supplied with milk in ordinary baby bottles attached to the front of the dummies. In fact, it seemed at first that a superior substitute had been found for real mothers. The cloth models never moved, never rebuked, and were an absolutely dependable supply of food. But as the monkeys grew up and were permitted to join other monkeys, their social behavior proved abnormal, to such an extent that comparable deviations in human societies would be ruled psychotic. They were sometimes hyperaggressive and sometimes autistic; in the latter state they sat withdrawn while rocking silently back and forth. They also cried a great deal and sucked their own fingers and toes. As Harlow and W. A. Mason showed later, the males also grew up sexually incompetent. They tried to copulate with estrous females but were unable to assume the normal position, sometimes mounting the female from the side and sometimes thrusting against her back above the tail. Females proved equally abnormal. In estrus they refused to be mounted. When "raped" by experienced males, they bore infants but badly mistreated them, stepping on them and rejecting their

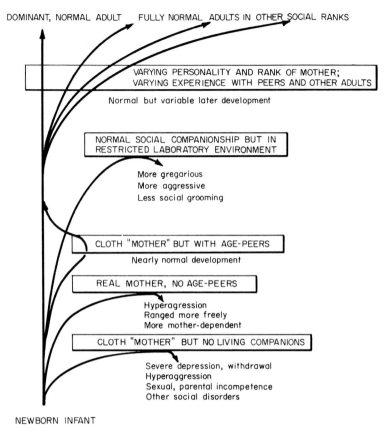

DOMINANT, NORMAL ADULT   FULLY NORMAL ADULTS IN OTHER SOCIAL RANKS

VARYING PERSONALITY AND RANK OF MOTHER;
VARYING EXPERIENCE WITH PEERS AND OTHER ADULTS

Normal but variable later development

NORMAL SOCIAL COMPANIONSHIP BUT IN
RESTRICTED LABORATORY ENVIRONMENT

More gregarious
More aggressive
Less social grooming

CLOTH "MOTHER" BUT WITH AGE-PEERS

Nearly normal development

REAL MOTHER, NO AGE-PEERS

Hyperagression
Ranged more freely
More mother-dependent

CLOTH "MOTHER" BUT NO LIVING COMPANIONS

Severe depression, withdrawal
Hyperaggression
Sexual, parental incompetence
Other social disorders

NEWBORN INFANT

Figure 7-8 The effects of social deprivation on behavior in the rhesus monkey.

attempts to nurse. Some of the infants managed to survive by persistence and ingenuity, but others had to be removed to save their lives. The degree of abnormality increased with the duration of the deprivation. Isolation for three months was followed by depression, but the monkeys recovered with no marked deviations. Isolation for six months resulted in extensive permanent damage, and isolation for a year caused virtually total impairment.

Although man-years of effort have gone into the rhesus socialization studies, the process is understood only to a limited degree. In particular, the interaction of factors has not been satisfactorily analyzed. We know that peers can be substituted for mothers to produce partly normal development of the rhesus infant, and vice versa, but the data give only a poor idea of the ways these two social factors influence each other in a normally constituted society. Infants without peers turn more to their mothers for play in the later stages of socialization. But mothers are not wholly adequate substitutes. Even their primary role is affected, since they tend to reject the advances

of their offspring more frequently (Hinde and Spencer-Booth, 1969). In a symmetrical fashion, peers may substitute for mothers but have obvious inadequacies, and their behavior as peers is modified by the greater demands of a motherless companion. The crucial stimuli for achieving socialization at various stages are also problematical. Harlow found that the sight of other monkeys is not by itself enough to avoid the effects of total isolation, and he concluded that physical contact and the involved sensorimotor processes of play with live companions are essential. Meier (1965), however, obtained results which indicate that visual contact *is* adequate. Perhaps the conditions of the two experimental groups differed in some unrecognized way. The complexities arising from the interaction of factors will be ultimately solved by sufficient experimental design and analysis of variance. But the road to a full understanding of such advanced social animals will be long. Those who have been most involved are keenly aware of this circumstance. After 2, 935 hours of observation in Tanzania and Uganda and six years of laboratory research on the olive baboon, Ransom and Rowell (1972) have chastely summarized the state of the art in primate studies as follows: "The task of discovering the factors and the intricate combinations that direct the form and development of social behavior has been begun in a number of field and cage studies of primates, including those of baboons reported here, but so far one of their main results has been to emphasize that a great deal more long-term observation, experimental manipulation, and analysis of primate behavior and group social structure in both field and cage situations will be well worthwhile."

More or less comparable studies have been undertaken on a few other vertebrate species, notably domestic fowl (Guhl, 1958; Guiton, 1959; Bateson, 1966; McBride et al., 1969); rodents, especially mice (Williams and Scott, 1953; Noirot, 1972), rats (Beach, 1964; Morrison and Hill, 1967), and squirrels (Horwich, 1972); dogs (Scott and Fuller, 1965); and wolves (Woolpy, 1968a,b; Woolpy and Ginsburg, 1967; Fox, 1971; Bekoff, 1972). Further aspects of the socialization process will be presented in later discussions of sexual behavior (Chapter 15) and parental care (Chapter 16).

Play

Play, virtually all zoologists agree, serves an important role in the socialization of mammals. Furthermore, the more intelligent and social the species, the more elaborate the play. These two key propositions having been stated, we must now face the question: What is play? No behavioral concept has proved more ill-defined, elusive, controversial, and even unfashionable. Largely from our personal experience, we know intuitively that play is a set of pleasurable activities, frequently but not always social in nature, that imitate the serious activities of life withcut consummating serious goals. Vince

Lombardi, the great coach of the Green Bay Packers, was once dismissed by a critic of football as someone who taught men how to play a boy's game. That was unfair. Human beings are in fact so devoted to play that they professionalize it, permitting the lionized few who turn it into serious business to grow rich.

The question before us then is to what extent animal play is also serious business. In other words, how can we define it biologically? Robert M. Fagen (1974) has pointed out that most of the confusion about play stems from the existence of two wholly different orientations in general writings on the subject. On the one hand are the *structuralists*, who are concerned only with the form, appearance, and physiology of play. Structuralists, such as Fraser Darling (1937), Caroline Loizos (1966,1967), and Corinne Hutt (1966), define play as any activity that is exaggerated or discrepant, divertive, oriented, marked by novel motor patterns or combinations of such patterns, and that appears to the observer to have no immediate function. *Functionalists*, on the other hand, define play as any behavior that involves probing, manipulation, experimentation, learning, and the control of one's own body as well as the behavior of others, and that also, essentially, serves the function of developing and perfecting future adaptive responses to the physical and social environment. For the functionalist, the wars of England were indeed won on the playing fields of Eton.

The functionalist concept dates to Karl Groos, who in *The Play of Animals* (1898) argued that although play carries no immediate risks or responsibilities, it serves to prepare the individual for the serious tasks of adult life. Konrad Lorenz (1950, 1956) adopted a similar viewpoint and added the hypothesis of a play "drive" that compels animals to learn in advance which instinctive acts are appropriate in a given contingency. In a more sophisticated recent refinement of the idea, Peter Klopfer (1970) postulates play to consist of "the tentative explorations by which the organism 'tests' different proprioceptive patterns for their goodness of fit." Learning proceeds in a directed, more or less programmed manner until the right combinations of stimuli and responses are achieved. Klopfer believes that essentially the same function is pursued in human play. In his view, creative thought and abstraction are forms of play, esthetics is the pleasure that results from biologically appropriate activity, and esthetic preference is a choice of objects or activities that induce the correct preprogrammed neural inputs or emotional states, independent of overt reinforcers.

Whether the functionalist hypothesis is incorporated into the definition or not, play can be usefully distinguished from pure *exploration*. To explore is to learn about a new object or a strange part of the environment. To play is to move the body and to manipulate portions of a known object or environment in novel ways. As Hutt says, the goal of exploration is getting to know the properties of the new object, and the particular responses of investigation are determined by its properties. True play proceeds only within a known environment, and it is largely manipulative in quality. In passing from exploration to play, the animal or child changes its emphasis from "What does this object do?" to "What can *I* do with this object?" Play can also be separated from pure problem solving, especially when the latter has a simply functional goal and does not entail the pleasurable learning of rules and variations. Jerome Bruner (1968) has attempted to capture this distinction in the following epigram: play means altering the goal to suit the means at hand, whereas problem solving means altering the means to meet the requirements of a fixed goal.

Robert Fagen (personal communication) has pursued the functionalist interpretation with a modification of the Gadgil-Bossert life history model. In this formulation play commonly entails an immediate cost in fitness due to such functions as the useless expenditure of energy, the increased vulnerability to watching predators, the risking of dangerous episodes with adults, and so forth. But fitness at *later* life stages is enhanced by the experience and the improved status that play confers. In more exact form, the proposition says that the $l_y m_y$ values are lowered at age $y = x$, when the play takes place, but they are raised at some age beyond x as a result. The play schedule—the intensities of play programmed for each age y—will evolve in such a way that the summed gains in $l_y m_y$ will exceed the summed losses over all ages y, that is, throughout the potential life span. By experimentally inserting a priori values of losses and gains into a numerical model, Fagen proved that play can be eliminated entirely from the behavioral repertory by natural selection. The additional constraints on its evolution are as follows. The amount of play can decrease monotonically from birth, peak unimodally at some later age, or even peak bimodally at two later ages. But if play exists at all, it is present at age 0. In more realistic terms we can translate this last result to read that animals belonging to a playful species can be expected to start playing as soon as they have developed coordinated movements of their bodies and limbs. Also, under a wide range of conditions, play will be most prominent at a relatively early age.

Play appears to be strictly limited to the higher vertebrates. No case has been documented in the social insects (Wilson, 1971a), and the phenomenon must be very scarce or altogether absent throughout the remainder of the invertebrates. To my knowledge no example has been documented in the cold-blooded vertebrates, including the fishes, amphibians, and reptiles. The sole dubious exception is the Komodo dragon *Varanus komodoensis*, the world's largest lizard. Craven Hill (1946) reported that a large individual in the London Zoo "played" repeatedly with a shovel by pushing it noisily over the stony floor of its cage. This behavior might just as well be interpreted as a redirection of foraging movements, in which logs or other objects are pushed aside in the search for prey. On the basis of Hill's single anecdote play cannot be said to have been conclusively demonstrated in reptiles. Unequivocal play behavior has been reported in a few species of birds, and appears to be especially well developed in crows, ravens, jackdaws, and other members of the family Corvidae, which are noted for their relatively high intelligence and unspecialized behavior. Hand-reared ravens (*Corvus corax*) display several patterns that the structuralists would classify as play, including repeated episodes of hanging from horizontal ropes by one leg while performing acrobatic stances with the head and free leg (Gwinner, 1966). Play, directed at both social companions and inanimate objects, occurs virtually throughout the mammals; it has been described in fruit bats (Neuweiler, 1969); the wombat, a large burrowing marsupial (Wünschmann, 1966); ground squirrels of the genus *Spermophilus* (Steiner, 1971) and tree squirrels of the genus *Sciurus* (Horwich, 1972); deer (Darling, 1937; Müller-Schwarze, 1968); antelopes of many species (Walther, 1964); pigs and other Suidae (Frädrich, 1965); goats (Chepko, 1971); the Indian rhinoceros (Inhelder, 1955); the European polecat (Poole, 1966); mongooses (Ewer, 1963, 1968); the European badger (Eibl-Eibesfeldt, 1950); sea lions (Farentinos, 1971); hyenas (Hugo and Jane van Lawick-Goodall, 1971); lions (Schaller, 1972); wolves and other canids (Mech, 1970; Bekoff, 1972); lemurs (Jolly, 1966); and other, higher primates generally (Jane van Lawick-Goodall, 1968a; Fady, 1969).

As the above phylogenetic distribution alone suggests, play is associated with a large brain complex, generalized behavior, and, most especially, a large role for learning in the development of behavior. The play activities of a kitten, an animal not very high on this scale, are direct and to the point. Much of it consists of mock-aggressive rushes and rough-and-tumble play with the mother and other kittens, clearly portending the territorial and dominance aggression of adult life. The more prolonged and elaborate patterns—the ones that make kittens so fascinating to watch—are the forerunners of the three basic hunting maneuvers of the adult cat. When the kitten spots the trailing end of a string, it slithers along the floor, tail twitching lightly, and suddenly pounces to press the string down to the ground with its claws. These are close to the exact motions by which an adult catches a mouse or some other small ground-dwelling animal. When a string is dangled in the air—sometimes even a mote of dust dancing in a beam of sunlight will do—the kitten chases it as an adult cat does a bird taking flight. Springing upward, it spreads its paws out and claps them together to seize the object in midair. Kittens also stand over objects and scoop them upward and to one side with a sweep of the paw. This last maneuver is quite possibly a rehearsal of the technique used later to capture small fish.

Play in the gray squirrel (*Sciurus carolinensis*) is also stereotyped and related to future function (Horwich, 1972). Soon after young squirrels are able to run in a smoothly coordinated fashion, they begin to scamper around by themselves, running quickly along tree branches or over the ground while executing sudden right-angle and face-about turns. Young males mount their female littermates in precocious sexual movements. The sisters respond by elevating the rump and pulling the tail inward or laying it over the back. Aggressive play might occur in the gray squirrel, but it is difficult to distinguish from real aggression, because at an early age the juveniles begin to quarrel over food and to establish dominance relations.

Play in the red deer of Europe is surprisingly sophisticated. Although some of the actions by the hinds are sexual in connotation, most of the play of both sexes is devoted to solitary running, chasing, and mock aggression. Darling (1937) described a "game" he called King-o'-the-Castle as follows:

A hillock is used as an objective, and each member of a group of deer calves tries to attain and occupy the summit. Rivalry is certainly strong in this type of play, but there seems to be no hint of mock combat in the actions of running up the hillock and shoving away the holder of the summit . . . No form of play continued for more than five minutes at a time, and the mock fights were little more than momentary. King-o'-the-Castle would start by one calf mounting the hillock and occasionally rising on its hind legs. This would seem to serve as invitation, for others would look up, leave their mothers, and run towards the hillock. The hillock was worn by the impress of many tiny feet, and it was obvious that this had become a traditional playing-place. When I say "traditional" I admit that association of the hillock with previous fun may influence their behaviour towards a repetition of the experience when they pass near it again. But I have seen deer calves come from a distance of fifty yards to their chosen hillock to begin playing, as if their play were premeditated.

Other red deer games include mock fighting, racing, and a form of tag in which individuals chase and flee from one another in rapid alternation.

Not unexpectedly, the animal species indulging in the most elaborate and free-ranging forms of play is the chimpanzee, the most intelligent of the anthropoid apes and man's closest living relative (van Lawick-Goodall, 1968a). Sessions are initiated by one or the other of two special invitation signals: the play-walk, in which the chimpanzee hunches its back into a rounded form, pulls its head slightly down and back between the shoulders, and takes small, stilted steps; and the play-face, a distinctive expression assumed by opening the mouth in a form neither aggressive nor fearful while partially or wholly exposing the teeth. In addition to conventional forms of chasing, mounting, and rough-and-tumble play, young chimps improvise an extraordinary variety of games.

During a chase one or more of the participants sometimes broke off and carried (in one hand, mouth, groin or between shoulder and neck) a leafy twig, spray of fruits or the like. On occasions the pursuing animal made repeated attempts to grab such "toys" from the other. One infant chasing round and round a clump of vegetation after his mature brother kept trying to catch the end of a palm frond which the latter was trailing behind him: each time the infant made a grab the elder male, looking back over his shoulder, deftly jerked the frond out of the infant's reach. One "toy" was a round hard-shelled fruit, and the playmates tried to grab it from each other.   (Jane van Lawick-Goodall, 1968a)

Finger wrestling and tickling were the commonest forms of play among the adults. When grown males and females played with infants and juveniles, they tickled them, sparred with light blows and kicks, and chased them around trees. Tickling often induced laughter. One young chimpanzee was seen to swing a clod of earth attached to a bunch of grass around its body and then to strike a companion repeatedly with it.

As Frank Beach (1945) first emphasized, animal play is not simply a melange of infantile behaviors. It progresses as part of the ontogeny, with specific patterns waxing and waning at different ages. The principle has been well documented by Horwich (1972), who compared the scheduling of sexual play across seven species of rodents. Each species can be characterized by the brief period during which the observed onset of the play takes place. In animals generally, the frequency of play rises quickly to a peak after the onset, then declines slowly through the juvenile and subadult stages, reaching its lowest level and often, particularly in the more primitive species, disappearing entirely at full sexual maturity (see Figure 7-9).

Another characteristic of play is the freedom with which behavioral elements are concatenated. The elements themselves can be well defined and are more or less consistent in form; they may even be closely similar to the serious adult behaviors they foreshadow. But the sequence in which they are put together is very variable and idiosyncratic—one might even say whimsical. It is possible that this trait of looseness is vital to the very process of environmental tracking itself. Play is the means by which the most appropriate combinations are identified, reinforced, and hence established as the future adult repertory. Fagen (1974) has analogized play with the process of chromosome mechanics, which have the same effect of multiplying diversity:

1. *Recombination.* In playful behavior the adult sequences of behavior break down. Behavioral elements, for example, threats, grooming, scampering, sexual posturing, are performed in novel and rapidly changing sequences that would be nonadaptive and perhaps even fatally dangerous within the serious contexts of adult life.

2. *Fragmentation.* Behavioral sequences are interrupted or discontinued; the normal adult beginnings and endings may be omitted.

3. *Translocation.* In play, the behavioral elements of different adaptive categories, for example of reproduction, feeding, and exploration, can be combined casually.

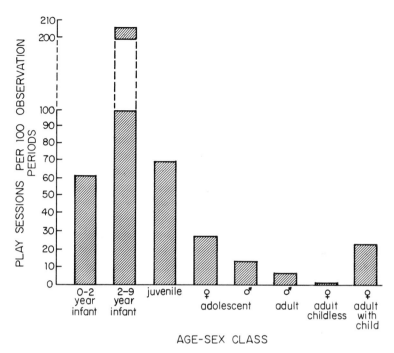

Figure 7-9 The frequency of play in various age-sex classes of wild chimpanzees. (Modified from Jane van Lawick-Goodall, 1968a.)

4. *Duplication.* Playful episodes can be extended indefinitely. Elements that are performed only rarely during serious adult life may be repeated frequently and even rhythmically.

Among the higher mammals, where it is most free-ranging, play loosens the behavioral repertory in each generation and provides the individual with opportunities to depart from the traditions of its family and society. Like sexual reproduction and learning generally, it is evidently one of the very broad adaptive devices sustained by second-order natural selection. At its most potent, in human beings and in a select group of other higher primates that includes the Japanese macaques and chimpanzees, playful behavior has led to invention and cultural transmission of novel methods of exploiting the environment. It is a fact worrisome to moralists that Americans and other culturally advanced peoples continue to devote large amounts of their time to coarse forms of entertainment. They delight in mounting giant inedible fish on their living room walls, idolize boxing champions, and sometimes attain ecstasy at football games. Such behavior is probably not decadent. It could be as psychologically needed and genetically adaptive as work and sexual reproduction, and may even stem from the same emotional processes that impel our highest impulses toward scientific, literary, and artistic creation.

Tradition, Culture, and Invention

The ultimate refinement in environmental tracking is tradition, the creation of specific forms of behavior that are passed from generation to generation by learning. Tradition possesses a unique combination of qualities that accelerates its effectiveness as it grows in richness. It can be initiated, or altered, by a single successful individual; it can spread quickly, sometimes in less than a generation, through an entire society or population; and it is cumulative. True tradition is precise in application and often pertains to specific places and even successions of individuals. Consequently families, societies, and populations can quickly diverge from one another in their traditions, the phenomenon described in Chapter 2 as "tradition drift." The highest form of tradition, by whatever criterion we choose to judge it, is of course human culture. But culture, aside from its involvement with language, which is truly unique, differs from animal tradition only in degree.

Some dialects in animal communication are learned, and to that extent they represent an elementary form of tradition. Local populations become differentiated through tradition drift, with several potentially adaptive consequences previously touched on in different contexts: the greater compatibility of mated pairs and enhanced efficiency of communication between the male and female, the implementation of the dear enemy effect, and the partial reproductive isolation of each deme distinguished by a dialect, which preserves the closeness of fit of its gene pool to the particular conditions of its local environment. All other factors being equal or negligible, the average geographic range of the dialects of a species diminishes as the behavioral plasticity of the species increases. The dialect ranges of the Indian hill mynah (*Gracula religiosa*), which is strongly imitative and has an unusually plastic call, do not extend beyond 17 kilometers (Bertram, 1970). Dialect formation based at least in part on learning between generations is widespread among species of songbirds (Thielcke, 1969). Geographic variation in vocalizations has been reported in several kinds of mammals, including pikas, pothead whales, and squirrel monkeys. Its significance is unknown, since the variation might be based on genetic differences and thus not constitute true tradition. LeBoeuf and Peterson (1969a) have suggested that the differences in vocalizations between island populations of elephant seals along the California coast are based at least in part on learning. Some of the populations have been founded by very small numbers of individuals during rapid overall population expansion over the past several decades. The first bull on Año Nuevo Island, for example, had a call distinguished by an unusually rapid burst of notes. It is possible, but not yet proved, that the sound could be imitated by younger males arriving later. Geographic variation in the waggle dance of the honeybee *Apis mellifera* is extensive, and has

been referred to as dialect formation by von Frisch (1967) and others. But in this case genetic analysis and experiments employing cross-hive adoption have shown that the differences are inherited rather than learned.

Most tradition in the best investigated animals is concerned with *Ortstreue*—fidelity to place—a German phrase for which there is no precise English word equivalent. *Ortstreue* is the tendency of individuals to return to the places used by their ancestors in order to reproduce, to feed, or simply to rest. Its most striking manifestation is in the fixed migration routes of birds and mammals. Each year ducks, geese, and swans migrate hundreds or thousands of kilometers, following the same traditional flyways, stopping at the same resting places, and ending at the same breeding and overwintering sites. Since these birds fly in flocks of mixed ages, there is abundant opportunity for the young to learn the travel route from their elders. The greater the fidelity shown by the birds to the flyways, the less the gene flow between local breeding populations, and consequently the stronger the geographic variation within the species (Hochbaum, 1955). Reindeer are perhaps the most birdlike of mammals in migratory behavior, showing comparable fidelity to their annual migratory routes and calving grounds (Lent, 1966). Migratory fish, including herrings (*Clupea*), eels (*Anguilla*), and salmon (*Oncorhynchus, Salmo*), return great distances to their spawning grounds. In at least the case of the salmon, the fish appear to be guided by odors of the streams to which they had become imprinted in the first weeks of their lives (Hasler, 1966, 1971). Consequently, tradition in the pure sense may not exist; *Ortstreue* is possible without tradition. Monarch butterflies (*Danaus plexippus*) are perhaps the long-distance champions among migratory insects; each spring they fly north and each fall south for distances of up to 1500 kilometers each way (Urquhart, 1960). One terminus on the west coast of North America is California. In some localities the overwintering monarchs settle in the same trees year after year. The famous "butterfly trees" of Pacific Grove have been used for at least 70 years. Monarchs are sufficiently long-lived for two generations to overlap, and it is possible that older individuals inadvertently guide the inexperienced ones to the winter sites. If that is indeed the case, these insects can be said to utilize a rudimentary form of tradition.

On a different scale, the game trails used by mammals are certainly traditional. Those of deer persist for generations, and where they follow paths smoothed into rocks, perhaps for centuries. Galápagos tortoises (*Geochelone elephantopus*) follow established trails during their annual migrations. At the beginning of the rainy season they descend from the moist habitats and waterholes of the highlands to lower elevations to feed and to lay eggs. Later they climb back to the highland refuges. In some places the tortoise trails run for kilometers and require days to follow, and they have almost certainly lasted for generations (Van Denburgh, 1914). The breeding grounds of colonial birds are also traditional. Wynne-Edwards (1962) has called attention to the Viking names of certain of the British Isles based on the kinds of birds that nested there in the eighth to tenth centuries. The names are still appropriate: for example, Lundy, "isle of puffins," and Sulisgeir, "gannets' rock." Even nests and roosts can be passed from one generation to the next. Osprey nests and the mud workings of swallows sometimes persist for decades. The lodges of muskrats and beavers last for at least several generations, while a few earthen dens of the European badger are said to be centuries old (Neal, 1948).

The display grounds of lek birds, including grouse, ruffs, manakins, pheasants, and birds of paradise, are usually fixed in location by strong traditions. Armstrong (1947) described how one population of ruffs in Britain insisted on returning to their ancestral arena even when a road was built right through it. "As I passed on my bicycle they ran from almost underneath the wheels and returned immediately to resume their antics." The lone surviving heath hen continued to visit the ancestral "booming field" on Martha's Vineyard, Massachusetts, for year after year until her death. It has not been determined how long the display grounds persist, but some probably last for decades or centuries. Beebe (1922), for example, found a Dyak tribe on Borneo that had been trapping argus pheasants from the same arena for many human generations. The home ranges and territories of closed societies are also bequeathed to descendants, their boundaries being taught inadvertently to the young animals by actions of experienced groupmates. Well-documented examples include the Australian magpie *Gymnorhina tibicen* (Carrick, 1963), feral domestic fowl (Collias et al., 1966), and a large array of lemuroids, monkeys, and apes (Alison Jolly, 1966, 1972a).

The greater the degrees of closure and philopatry of the society, and the more complex and prolonged the socialization of the young, the more important a role tradition assumes in social organization. Geist (1971a) has noted the confluence of all of these factors in the creation of strong traditions in bands of mountain sheep:

The manner in which mountain sheep establish home ranges is closely related to their social system. They inherit home ranges from their elders, by acquiring the movement habits of the latter; individual exploration plays a subordinate part. Females in general adopt the home ranges of the female group that raised them, but a few inherit the home ranges of another female group. Females have a critical period between one and two years of age in which they may switch to another female band if they happen to meet one, or follow a ram to another area and join females there. Young sheep may only follow individuals of their maternal band because they meet no one else.

Young rams desert the maternal "home range group" of females some time after their second year of life and join ram bands. They establish individual patterns of seasonal home ranges by acquiring the home ranges of various older rams they happen to follow . . .

Sheep society has all elements essential for smooth passage of home range knowledge, while minimizing dispersal by young sheep. Lambs are not driven off by the females after weaning or prior to bearing another lamb; rather, the juveniles desert their mothers and follow adults of their own choice. No social bonds are broken suddenly; separation of female and child is gradual and the juvenile is never forced from the band to wander on his own. The result is that young sheep are rarely alone. They are tolerated by whomever they follow—adult females, subadults, or mature rams.

While new lambs are being born, the yearlings tend to attach themselves to barren females. Female mountain sheep retain the juvenile trait of following others throughout their lives. Rams, in contrast, gradually separate from their companions over a period of seven to nine years. There is another difference: whereas females follow older females, especially those with lambs, males ordinarily follow the rams with the largest horns. When the rams become more independent, they are in turn followed by younger males and thus passively transmit the regional tradition to them. By the age of four and a half years the rams appear to be fixated on a home range pattern.

Among the higher primates, traditions sometimes shift in a qualitative manner. Poirier (1969a) observed changes in diet and foraging behavior in langurs (*Presbytis johnii*) of southern India after the monkeys' environment had been changed by human activity. One troop was forced into a new area when the habitat of its original home range was destroyed. Subsequently it altered its diet from *Acacia* to *Litasae* and *Loranthus*. Other troops have begun to shift to *Eucalyptus globulus*, an Australian tree that is being deliberately planted in place of the natural woodlands favored by the langurs. Although the adults are reluctant to eat anything but the leaf petioles of these aromatic trees, the infants sometimes consume the entire leaf. Poirier predicts that ultimately entire troops will incorporate eucalyptus as a principal food plant. Elsewhere, langurs are in the process of accommodating the encroachment of agriculture. In the Nilgiri area of India, potatoes and cauliflower were introduced not more than 100 years ago and are gradually replacing the natural woodland. The langurs come out of their refuges in the remaining forest patches to raid the crops. Not only do they feed on the vegetables in quantity, but they have learned to dig into the soil with their hands and to pull up the entire plants—a behavior pattern not yet seen in other langur troops.

It is probable that adaptations at this level of difficulty also occur in primate populations undisturbed by man. Desert-dwelling baboons, particularly *Papio hamadryas* but also some populations of *P. anubis*, eat dehydrated food for long periods of the year and must find drinking water on a daily basis. During the dry season the rivers shrink to scattered waterholes that become tepid and filled with algae. At this time the baboons use their hands to dig holes in the sand of the riverbeds. The locations are expertly selected, and the animals seldom have to go more than a foot down before they strike cool, clean water (Kummer, 1971).

The most carefully documented case histories of invention and tradition in primates have come from studies of the Japanese macaque *Macaca fuscata*. Since 1950 biologists of the Japanese Monkey Center have kept careful records of the histories of individuals in wild troops located at several places: Takasakiyama, near the northern end of Kyushu; Koshima, a small island off the east coast of Kyushu; and Minoo and Ohirayama on Honshu. More casual studies have been made at still other localities. At an early stage the Japanese scientists encountered differences between troops in the traditions of food-gathering behavior. The monkeys at Minoo Ravine had learned how to dig out roots of plants with their hands, while those at Takasaki-yama, although living in a similar habitat, apparently never used the technique. The population at Syodosima regularly invaded rice paddies to feed on the plants, but the troops at Takagoyama were never observed to do so, despite the fact that they had lived for many years in hills surrounded by paddies and occasionally passed through them during their nomadic wanderings (Kawamura, 1963).

When the biologists offered new foods to the monkeys, they directly observed both dietary extensions and the means by which these changes are transmitted through imitation. At Takasakiyama caramels were accepted readily by monkeys under three years of age, and candy eating then spread rapidly through this age class. Mothers picked up the habit from the juveniles and passed it on to their own infants. A few adult males most closely associated with infants and juveniles eventually accepted caramels also. Propagation of the habit was most rapid among young animals, and slowest among the sub-adult males who were farthest removed socially from the young and their parents. After eighteen months, 51.2 percent of the troop had been converted to candy eating (Itani, 1958). At Minoo, another troop added wheat to its diet when the grain was artificially supplied, but at a much faster rate and in a different pattern. The adult males were first to feed, and adult females and the younger animals quickly followed. Within only four hours the entire troop had adopted the habit (Yamada, 1958).

The scientists who summarized the early findings, including Kinji Imanishi (1958, 1963) and Syunzo Kawamura (1963), spoke of the macaque society as a "subhuman culture" or "preculture" and the dietary shifts as enculturation. If these terms were at all justified, they became much more so by the remarkable series of events witnessed about this time in a single troop on Koshima Island. Starting in 1952, the biologists began to scatter sweet potatoes on the beach in an attempt to supplement the diet of the monkeys. The troop then ventured out of the forest to accept the gift, and in so doing it extended its activities to an entirely new habitat. The following year

Kawamura (1954) observed the beginnings of a new behavior pattern associated with this habitat shift: some of the monkeys were washing sand off the potatoes by employing one hand to brush the sand away and the other to dip the potato into water. This and other subsequent behavioral changes were followed in detail during the ensuing ten years by Masao Kawai, who summarized the history of the population in 1965.

Potato washing was invented by a 2-year old female named Imo. Within ten years the habit had been acquired by 90 percent of the troop members in all age classes, except for infants a year old or less and adults older than 12 years. During the same period, the washing was transferred from the fresh water of the brook to the salt water of the sea. The behavior was most readily learned by juveniles between 1 and $2\frac{1}{2}$ years old, Imo's own age class. By 1958, five years after Imo invented it, potato washing was practiced by 80 percent of monkeys from 2 to 7 years in age. Older monkeys remained conservative; only 18 percent, all of them females, learned the behavior. Part of this conservatism is intrinsic to age and sex. Menzel (1966) subsequently tested Japanese monkeys by placing strange objects in their paths. Juveniles, for example, reacted to the sight of a yellow plastic rope much more strongly than adults. Up to the age of 3 years males responded as frequently as females. The response of adult males, however, fell to 18 percent, whereas nearly half of adult females still reacted. This is not to say that older animals were unaware of the rope, only that they were less inclined to explore it. Adult males, on seeing the object, typically deviated at a slight angle in their line of travel while glancing sidewise at it. Some of the conservatism was also a side product of the tendency of monkeys to learn from their closest companions. When the tradition of potato washing first spread, mothers learned from their children and juveniles from their siblings. Later, infants routinely picked up the habit from their mothers. Older monkeys, and especially the subadult and adult males who stayed near the periphery of the group, had fewer opportunities to learn in this way.

In 1955 Imo, the monkey genius, invented another food-gathering technique. The biologists had originally given wheat to the Koshima troop simply by scattering it onto the beach. The monkeys were then required to pick out the grains singly from among the particles of sand. Imo, now four years old, somehow learned to scoop handfuls of the mixed sand and wheat, carry them to the edge of the sea, and cast the mixture onto the water surface. When the sand sank, the lighter wheat grains were skimmed off the surface and eaten. The pattern by which this new tradition spread through the troop resembled that for sweet-potato washing. Juveniles passively taught their mothers and age-peers, and mothers their infants, but adult males largely resisted learning the technique. One important difference emerged, however: unlike potato washing, which spread most rapidly among monkeys one to two and a half years in age, wheat flotation was picked up most efficiently by members of the two-to four-year-old class, to which Imo herself belonged. The explanation of this difference in performance may lie in the relative complexity of the two tasks. Potato washing is only a slight modification of the procedure the macaques routinely follow when they pick up tubers and fruits from the ground with one hand and brush off dirt with the other. But the "placer mining" of wheat involves a qualitatively new element: throwing the food temporarily away and waiting a short period before retrieving it. It may well be that young animals are normally the inventors of new behavior patterns, but only those with several years of experience can manage the most complex tasks. This notion has received support from experiments by Atsuo Tsumori and his associates on the troops at Koshima, Ohirayama, and Takasakiyama (Tsumori et al., 1965; Tsumori, 1967). Peanuts were buried in the sand to a depth of 6-7 centimeters in full sight of the troop. At each place, a minority of the individuals succeeded at the first try in the moderately difficult task of digging up the peanuts. Thereafter, the habit spread through most of the remainder of each troop. The most innovative animals were young, with the best performance coming from those four to six years of age (see Figure 7-10).

The innovations of the Koshima troop have also provided a graphic illustration of the potential role of learned behavior as an evolutionary pacemaker. The food presented to the monkeys on the beach attracted them to a new habitat and presented them with opportunities for further change never envisioned by the Japanese biologists. Young monkeys began to enter the water to bathe and splash, especially during hot weather. The juveniles learned to swim, and a few even began to dive and to bring seaweed up from the bottom. One left Koshima and swam to a neighboring island. By a small extension in dietary opportunity, the Koshima troop had adopted a new way of life, or more accurately, grafted an additional way onto the ancestral mode. It is not too much to characterize such populations as poised on the edge of evolutionary breakthroughs, even though probably very few ever complete the process. An interesting parallel case in which the change has been carried to completion is provided by lizards. The species of the genus *Uta* are specialized for life in the deserts of western North America. They are in every respect among the most terrestrial of vertebrates, but contain one notable exception: on San Pedro Martir, a small desert island in the Gulf of Mexico, there exists an endemic species (*Uta palmeri*) which has assumed a partially marine existence. Individuals of *palmeri* are large, and they live in dense populations that probably could not be sustained by foraging on the land alone. Instead, they obtain a large part of their energy by entering the intertidal zone of the island at low tide to feed on a variety of the marine invertebrates living there. A further step in

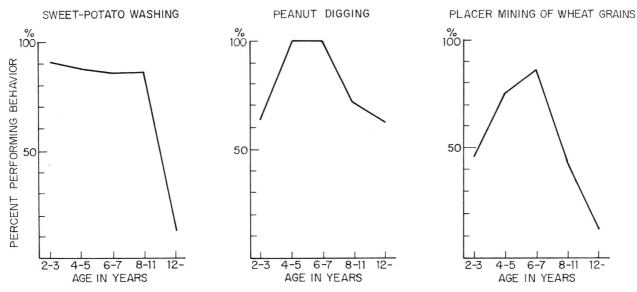

Figure 7-10 Innovation and tradition in Japanese monkeys as a function of age of the animal. The data, taken on the Koshima Island troop during August 1962, are the percentage of monkeys in various age groups that had acquired the potato-washing and placer-mining techniques up to that time and the percentage that successfully dug for peanuts after several trials. Potato washing is a relatively simple behavioral modification, while digging and placer mining are successively more difficult and complex tasks, a difference that may be reflected in the increasingly poor performance by the youngest monkeys in the latter two tasks. (Redrawn from Kawai, 1965a, and Tsumori et al., 1965.)

this general evolutionary progression has been taken by the marine iguana of the Galápagos Islands, *Amblyrhynchus cristatus*, which lives on lava outcrops on the edge of the sea and swims underwater to browse on algae.

Tool Using

Tools provide the means for quantum jumps in the process of invention. However, in the idiosyncratic world of animal behavior the phrase *tool using* must be carefully defined. John Alcock (1972) has characterized it as well as possible as the manipulation of an inanimate object, not manufactured internally by the organism, which is used in a way that improves the organism's efficiency in altering the position or form of some other object. Thus spider webs and wasp nests, although ranking among the wonders of inanimate nature, are not tools. Nor is string pulling by tits and raccoons tool using, because other inanimate objects are not the goal of the manipulation.

Even when restricted by definition in this way, the use of tools is an extraordinarily diversified and widespread phenomenon among insects, birds, and mammals. The following list has been assembled from reviews by Millikan and Bowman (1967), Evans and Eberhard (1970), van Lawick-Goodall (1970), Struhsaker and Hunkeler (1971), Alcock (1972), Jones and Kamil (1973), and R. E. Silberglied (personal communication); and it is probably nearly complete for all groups other than the primates.

——— Solitary wasps of the genus *Ammophila* pound shut their nest entrances with a small pebble held in the mandibles.

——— Ant lions and worm lions, which are larvae of the neuropterous insect genus *Myrmeleon* and fly genera *Lampromyia* and *Vermilio*, respectively, knock insect prey down into their pits by hurling sand at them with tosses of the head.

——— The archer fish *Toxotes jaculatrix* spits drops of water at insects and spiders, knocking them into the water where they can be caught in the fish's mouth.

——— On the Galápagos Islands, at least four species of Darwin's finches belonging to three genera use twigs, cactus spines, and leaf petioles to dig insects out of crevices in tree bark. The tool is held in the beak and used essentially as an extension of the beak. Only one of the species, the woodpecker finch *Cactospiza pallida*, employs the behavior routinely.

——— While searching for insects on tree trunks, the brown-headed nuthatch *Sitta pusilla* of the southern United States occasionally holds a fragment of bark in its bill and uses it to pry loose other pieces still in place.

——— The black-breasted buzzard *Hamirostra melanosterna* of Australia, which in American parlance would be called a broad-winged hawk, carries rocks and lumps of soil into the air and drops them onto the eggs of birds, especially those of the ground-nesting emu. The buzzard then feeds on the contents.

——— The Egyptian vulture *Neophron percnopterus* (a member of a group of highly modified carrion-eating hawks) picks rocks up in its beak and hurls them at ostrich eggs in order to break the shells open.

——— The black cockatoo *Probosciger aterrimus* of the Aru Islands grasps nuts in its beak with the aid of a leaf while cracking them open, a technique rather like our holding a jar in a towel to gain better traction while the lid is twisted off. Alfred Russell Wallace's account of this behavior in *The Malay Archipelago* (1869) may be the first published record of tool using in animals below the primates.

——— Captive northern blue jays (*Cyanocitta cristata*) have been observed tearing out strips of newspaper and using them to rake in food pellets placed out of the bill's reach beyond the mesh wall of the cage (Jones and Kamil, 1973).

——— The sea otter *Enhydra lutris* collects stones and shells from the ocean bottom, places them on its stomach while floating on its back at the surface, and uses them as anvils against which it pounds and cracks open mussels and other hard-shelled mollusks.

Early observers were inclined to treat tool-using behavior as evidence of hidden resources of intelligence and insight learning. A close examination of the best-studied examples does not support this optimistic conclusion. In nearly every case, as Alcock pointed out, the patterns of behavior are relatively stereotyped and might easily have arisen by a redirection or some other elementary modification of preexisting behavior patterns. For example, sand throwing by ant lions and worm lions is closely similar to the motions by which those insects excavate their pits in the soil. Stone bombing by the Egyptian vulture and black-breasted buzzard could have arisen fortuitously by the redirection of the carrying response by which the birds transport prey. Such a transference is made more likely when the bird has been frustrated by large, otherwise unbreakable eggs. Even the highly specialized spitting behavior of the archer fish is more plausibly explained as the end product of a series of small evolutionary steps than as some extraordinary piece of reasoning. It is interesting that two of the more dramatic examples of tool using are associated with a major shift in adaptive behavior in response to an unusual ecological opportunity. The woodpecker finch lives on an archipelago that harbors no true woodpeckers (members of the family Picidae) or any other species specialized for the gleaning of insects from tree crevices and burrows. Cactus spines and twigs are poor substitutes for the long, chisellike bill and coiled tongue of a picid, but they are nevertheless adequate in the absence of competition. The sea otter has opened a new habitat for mammals of its type. It needed only the addition of a crude tool to its natural swimming powers and manual dexterity in order to exploit a whole new food source. Perhaps early man provides a third example of tool using in the service of an adaptive shift. When australopithecines turned increasingly to hunting, crude stone and bone tools replaced the claws and carnivorous dentition that had been lost far back in the prehominid ancestors of the early Tertiary Period. Although advanced intelligence subsequently evolved in concert with tool using, it was not a prerequisite for its inception.

Tool using occurs sporadically among the species of higher primates, mostly to a degree no greater than in other vertebrate groups. However, the chimpanzee has a repertory so rich and sophisticated that the species stands qualitatively above all other animals and well up the scale toward man. The details of chimpanzee tool using have been uncovered over many years by Savage and Wyman (1843–44), who described a chimpanzee in the wild using a rock to crack open a small fruit, and by Köhler (1927), Beatty (1951), Merfield and Miller (1956), Kortlandt and Kooij (1963), Struhsaker and Hunkeler (1971), and McGrew and Tutin (1973); but most especially by Jane van Lawick-Goodall (1968a,b; 1970, 1971), whose lengthy studies of the population at the Gombe Stream Park, Tanzania, added the greatest number of acts to the repertory, established the prevalence of tool using under natural conditions, and called the phenomenon to the attention of a wide audience of biologists and laymen. The known categories of tool using are listed below.

1. *Using saplings and sticks as whips and clubs.* The behavior was first observed in chimpanzees attacking a leopard (experiment by Kortlandt and Kooij). Van Lawick-Goodall saw it occur in highly variable form during several kinds of aggressive and playful encounters between the Gombe Stream chimpanzees.

2. *Aimed throwing.* Kortlandt and Kooij observed wild chimpanzees throwing sticks at a stuffed leopard. Van Lawick-Goodall saw adolescents throwing sticks at one another in apparent play. She also frequently witnessed chimpanzees throwing sticks, stones, and handfuls of vegetation at antagonists with clearly hostile intent. The targets included other apes during bouts of chasing and bluff charging, human beings who blocked their access to bananas, and baboons during encounters at the feeding areas. The objects hurled were often large enough to intimidate the baboons and human beings. But the effectiveness was otherwise not impressive; of 44 objects thrown, only 5 hit their objectives, and these targets were all within 2 meters. The aim was generally good but the object usually fell short.

3. *Use of sticks, twigs, and grasses for capturing ants and termites.* At the Gombe Stream Park, these objects were poked into holes in nests and withdrawn to obtain the ants or termites clinging to them. The behavior is thus a kind of "fishing" for insects that would otherwise remain inaccessible beneath the ground. Sometimes the tools were carefully prepared before use. Stems or twigs were stripped of their leaves with the hand or lips to make them fit into the holes, while grass blades were sometimes split apart to make them narrower.

4. *Use of sticks, twigs, and grasses as olfactory aids.* The objects were pushed down into holes in ant and termite nests, withdrawn, and sniffed. Apparently the results of this test helped the chimpanzee to decide whether to continue fishing further into the nest in an exploratory fashion or to search elsewhere.

5. *Use of sticks as levers.* At the Gombe Stream Park the chimpanzees tried to open boxes containing bananas by inserting sticks beneath the lids and into other crevices (see Figure 7-11). Although these efforts were clumsy by human standards, they occasionally succeeded, and the habit gradually spread through the troop.

6. *Use of sticks and stones to open fruits and nuts.* Following the initial observation of Savage and Wyman some 125 years previously, Struhsaker and Hunkeler witnessed numerous incidents in which chimpanzees broke open nuts by pounding them with sticks and stones. In one case a stone implement weighed about 16 kilograms. The chimps placed the nuts on depressions in exposed tree roots before cracking them. These observations took place in Ivory Coast, West Africa. Van Lawick-Goodall's animals, living far away in Tanzania, did not exhibit the behavior. The difference between the populations may represent another example of tradition drift.

7. *Use of sticks in dental grooming.* A captive adult female at the Delta Regional Primate Center in Louisiana consistently groomed the teeth of a young male with the aid of small sticks. She concentrated on the location of two fresh cavities and a loose, movable tooth among the molars (McGrew and Tutin).

8. *Use of leaves as a drinking and feeding tool.* When Gombe Stream chimpanzees were unable to reach drinking water at the bottom of tree hollows, they dipped leaves to retrieve it. The animals first chewed the leaves briefly to crumple them. Then they used them like sponges: holding the leaves between the index and second fingers, they dipped these objects into the hollows, pulled them out, and finally sucked water from their surfaces. Teleki (1973) observed chimpanzees use leaves to wipe brains from the skulls of freshly killed baboons.

9. *Use of leaves for body wiping.* The Gombe Stream chimpanzees commonly used leaves to wipe their bodies free of feces, blood, urine, semen, and various forms of sticky foreign material such as overripe bananas. "A 3-year-old, dangling above a visiting scientist, Professor R. A. Hinde, wiped her foot vigorously with leaves after stamping on his hair" (van Lawick-Goodall, 1968a).

The richness and variety of these observations provide an unusual opportunity to learn how tool using is acquired and passed along to social companions. In 1963 K. R. L. Hall proposed that the employment of tools by primates generally represents an extension of aggressive behavior under conditions that inhibit direct attack. Frustrated by the inability to carry through overt aggression, the animal turns to inanimate objects on which displacement activity or redirected aggression can be performed. Stick throwing, for example, might result when a chimpanzee seizes an object in redirected hostility and then accidentally flings it away while making an incomplete attack movement with its arm in the direction of the live opponent. Although possibly true for aimed throwing, this theory is patently inadequate

Figure 7-11 A chimpanzee uses a stick in an unsuccessful attempt to pry open a bunker containing bananas. The scene is in the Gombe Stream National Park of Tanzania. (Photograph by Peter Marler.)

for most other cases of tool using in chimpanzees. For years experimental psychologists have observed that captive chimpanzees possess strong exploratory tendencies. New objects are routinely inspected and handled with no reward other than the performance of the activity (Schiller, 1957; Butler, 1965). Van Lawick-Goodall found such behavior to be normal in wild troops. In the course of nesting and feeding, the chimpanzees of the Gombe Stream Park idly broke off branches and twigs and stripped leaves and peeled bark from stems. While traveling through trees they used their hands to snap loose dead branches and drop them to the ground. Most of the known techniques of tool using might easily have originated from such generalized investigative and play behavior. It is easy to visualize chimpanzees scratching and prodding the surface of the ground with sticks in a playful manner until they accidentally catch insects. Thus reinforced, they could perfect the "fishing" by seeking new places and practicing movements that yielded the largest number of insects. The usefulness of leaves as sponges might be perceived by intelligent animals that habitually handle and chew them. Leaves dangling in depressions and hollows will yield more water; it is not too difficult a step for chimpanzees to place the leaves in such places and then retrieve them.

Learning and play are indisputably vital to the acquisition of tool-using skills by chimpanzees. Schiller (1952, 1957) found that when two-year-old infants are deprived for one year of all opportunity of playing with sticks, their subsequent ability to solve problems with the aid of sticks is significantly reduced. Given access to play objects, young animals in captivity undergo a slow, relatively inflexible maturation in skills. Under two years of age they simply touch or hold objects without attempting to manipulate them. As they grow older they increasingly employ one object to hit or prod another, while simultaneously improving in the solution of problems that require the use of tools. Jane van Lawick-Goodall observed a similar progression in the wild chimpanzee troop. Infants as young as six weeks reached out to leaves and branches. Older infants constantly inspected their environment with their eyes, lips, tongues, noses, and hands, while frequently plucking leaves and sticks and waving them about. They then advanced to tool-using behavior in small steps. For example, one eight-month-old infant added grass stems to his other "toys," but for the special purpose of wiping them against other objects such as stones and his mother, the behavior pattern uniquely associated with ant and termite fishing. During play, other infants "prepared" grass stalks as fishing tools by shredding the edges off wide blades and chewing the ends off long stems.

Of equal importance, van Lawick-Goodall acquired direct evidence of imitative behavior in the transmission of these traditions. On many occasions she saw infants watch adults as they used tools, then pick the tools up and use them after the adults had moved on. On two occasions a three-year-old youngster observed his mother intently as she wiped dung from her bottom with leaves; then he picked up leaves and imitated the movements, although his bottom was not dirty. Chimpanzees almost certainly invent and propagate traditions in a manner similar to that observed directly in the Japanese monkeys. The use of sticks to pry open banana boxes is a case in point. The behavior spread gradually through the Gombe Stream troop, evidently aided by imitation. One female new to the area remained hidden in the bushes while watching others trying to open the boxes. On her fourth visit she walked into the open, immediately picked up a stick, and began to poke it at the boxes. It is extremely unlikely that a chimpanzee would have responded this quickly had it simply stumbled on the boxes without prior experience.

Because chimpanzees are unique among animals in their level of intelligence and phylogenetic proximity to man, it is of surpassing interest to know all of the many ways they use tools and form traditions. Each scrap of information on this subject obtained in future field and laboratory studies, however loosely connected to previous information, should be regarded as potentially important.

Chapter 8   **Communication: Basic Principles**

What is communication? Let me try to cut through the Gordian knot of philosophical discussion that surrounds this word in biology by defining it with a simple declarative sentence. Biological communication is the action on the part of one organism (or cell) that alters the probability pattern of behavior in another organism (or cell) in a fashion adaptive to either one or both of the participants. By adaptive I mean that the signaling, or the response, or both, have been genetically programmed to some extent by natural selection. Communication is neither the signal by itself nor the response; it is instead the relation between the two. Even if one animal signals and the other responds, there still has been no communication unless the probability of response was altered from what it would have been in the absence of the signal. We know that in human beings communication can occur without an outward change of behavior on the part of the recipient. Trivial or otherwise useless information can be received, mentally noted, and never used. But in the study of animal behavior no operational criterion has yet been developed other than the change in patterns of overt behavior, and it would be a retreat into mysticism to try to add mental criteria. At the same time there exist certain probability-altering actions which common sense forbids us from labeling as communication. An attack by a predator certainly alters the behavior patterns of the intended victim, but there is no communicating in any sense in which we would care to use the word. Communication must also be consequential to some reasonable degree. If one animal simply pauses to watch as another moves by unknowingly at a distance, the passing animal has altered the behavior pattern of the first. But the passing animal was not really communicating in any way that could alter its own behavior or affect its relationship to the observing animal in the future. Perception occurred in this case, but not communication.

J. B. S. Haldane once said that a general property of communication is the pronounced energetic efficiency of signaling: a small effort put into the signal typically elicits an energetically greater response. This cannot be a universal prescription, but it is faithful enough to our intuition to permit the explicit exclusion of certain kinds of interactions. Two animals goring each other during an escalated territorial bout can be said to have ceased communicating and to have commenced fighting. But to lift a friend from the ground in an abrazo is true communication that surely violates Haldane's principle.

To finish drawing the boundaries of our definition, consider the following two unusual examples that involve microorganisms. When bioluminescent bacteria of the genus *Photobacterium* are inoculated into a fresh medium, they are unable to produce a sufficient quantity of luciferase to generate light. After a while the growing bacteria secrete an activator substance of low molecular weight that promotes the synthesis of luciferase in bacteria of the same strain (Eberhard, 1972). Is this chemical synergism a form of communication? It can be designated as such or not, according to convenience. Lower organisms such as *Photobacterium*, the interactions of which tend to be strictly physiological rather than behavioral, often create a gray zone of phenomena in which communication cannot be sharply demarcated. The second example includes three links in the communicative chain rather than the usual two. Hoyt et al. (1971) discovered that the sex attractant used by females of the grass grub beetle *Costelytra zealandica* is manufactured by symbiotic bacteria. These organisms live in the beetle's collaterial glands, which are located beneath the vagina and serve the primary function of secreting a protective coating for the eggs. In this example, who is communicating with whom? Of course the question is basically frivolous: the beetles have simply added an entire organism to their biosynthetic machinery. The serious point to be made is that communication is an adaptive relation between the organism that signals and the one that receives, regardless of the complexity and length of the communication channel.

Human versus Animal Communication

The great dividing line in the evolution of communication lies between man and all of the remaining ten million or so species of organisms. The most instructive way to view the less advanced systems is to compare them with human language. With our own unique verbal system as a standard of reference we can define the limits of animal communication in terms of the properties it rarely—or never—displays. Consider the way I address you now. Each word I use has been assigned a specific meaning by a particular culture and transmitted to us down through generations by learning. What is truly unique is the very large number of such words and the potential for creating new ones to denote any number of additional objects and concepts. This potential is quite literally infinite. To take an example from mathematics, we can coin a nonsense word for any number we choose (as in the case of the "googol," which designates a 1 followed by 100 zeros). Human beings utter their words sequentially in phrases and sentences that generate, according to complex rules also determined at least partly by the culture, a vastly larger array of messages than is provided by the mere summed meanings of the words themselves. With these messages it is possible to talk about the language itself, an achievement we are utilizing here. It is also possible to project an endless number of unreal images: fiction or lies, speculation or fraud, idealism or demagoguery, the definition depending on whether or not the communicator informs the listener of his intention to speak falsely.

Now contrast this with one of the most sophisticated of all animal communication systems, the celebrated waggle dance of the honeybee (*Apis mellifera*), first decoded in 1945 by the German biologist Karl von Frisch. When a foraging worker bee returns from the field after discovering a food source (or, in the course of swarming, a desirable new nest site) at some distance from the hive, she indicates the location of this target to her fellow workers by performing the waggle dance. The pattern of her movement is a figure eight repeated over and over again in the midst of crowds of sister workers. The most distinctive and informative element of the dance is the straight run (the middle of the figure eight), which is given a particular emphasis by a rapid lateral vibration of the body (the waggle) that is greatest at the tip of the abdomen and least marked at the head.

The complete back-and-forth shake of the body is performed 13 to 15 times per second. At the same time the bee emits an audible buzzing sound by vibrating her wings. The straight run represents, quite simply, a miniaturized version of the flight from the hive to the target. It points directly at the target if the bee is dancing outside the hive on a horizontal surface. (The position of the sun with respect to the straight run provides the required orientation.) If the bee is on a vertical surface inside the darkened hive, the straight run points at the appropriate angle away from the vertical, so that gravity temporarily replaces the sun as the orientation cue. (See Figure 8-1.)

The straight run also provides information on the distance of the target from the hive, by means of the following additional parameter: the farther away the goal lies, the longer the straight run lasts. In the Carniolan race of the honeybee a straight run lasting a second indicates a target about 500 meters away, and a run lasting two seconds indicates a target 2 kilometers away. During the dance the follower bees extend their antennae and touch the dancer repeatedly. Within minutes some begin to leave the nest and fly to the target. Their searching is respectably accurate: the great majority come down to search close to the ground within 20 percent of the correct distance.

Superficially the waggle dance of the honeybee may seem to possess some of the more advanced properties of human language. Symbolism occurs in the form of the ritualized straight run, and the communicator can generate new messages at will by means of the symbolism. Furthermore, the target is "spoken of" abstractly: it is an object removed in time and space. Nevertheless, the waggle dance, like all other forms of nonhuman communication studied so far, is severely limited in comparison with the verbal language of human beings. The straight run is after all just a reenactment of the flight the bees will take, complete with wing buzzing to represent the actual motor activity required. The separate messages are not devised arbitrarily. The rules they follow are genetically fixed and always designate, with a one-to-one correspondence, a certain direction and distance.

In other words, the messages cannot be manipulated to provide new classes of information. Moreover, within this rigid context the messages are far from being infinitely divisible. Because of errors both in the dance and in the subsequent searches by the followers, only

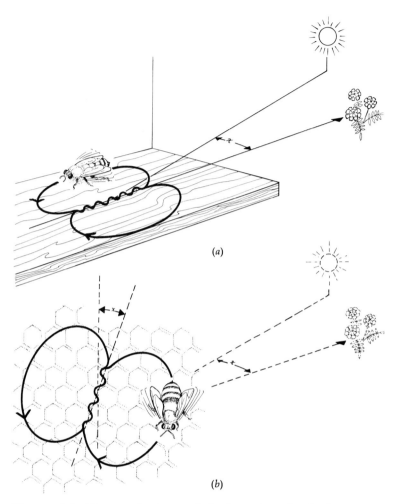

Figure 8-1 The waggle dance of the honeybee. As the bee passes through the straight run she vibrates ("waggles") her body laterally, with the greatest movement occurring in the tip of the abdomen and the least in the head. At the conclusion of the straight run, she circles back to about the starting position, as a rule alternately to the left and right. The follower bees acquire the information about the food find during the straight run. In the example shown here the run indicates a food find 20° to the right of the sun as the bee leaves the nest. If the bee performs the dance outside the hive (*a*), the straight run of the dance points directly toward the food source. If she performs a dance inside the hive (*b*), she orients herself by gravity, and the point directly overhead takes the place of the sun. The angle x (= 20°) is the same for both dances. (From Curtis, 1968a; based on von Frisch.)

about three bits of information are transmitted with respect to distance and four bits with respect to direction. This is the equivalent of a human communication system in which distance would be gauged on a scale with eight divisions and direction would be defined in terms of a compass with 16 points. In reading single messages, northeast could be distinguished from east by northeast, or west from west by southwest, but no more refined indication of direction would be possible. A thorough account of the work of von Frisch and his students is given in his master work *Tanzsprache und Orientierung der Bienen* (1965) or its English translation by L. E. Chadwick (1967). A briefer review, including critiques and more recent studies, is provided by Wilson (1971a). The design features of human language as opposed to communication in animals, particularly honeybees, were first systematically analyzed by Hockett (1960) and Altmann (1962b) and have been more recently reevaluated by the same authors (Altmann, 1967b,c; Hockett and Altmann, 1968). The main points of their formal system are included in the looser, more flexible account that follows.

Discrete versus Graded Signals

Animal signals can be partitioned roughly into two structural categories: discrete and graded, or, as Sebeok (1962) designated them, digital and analog. Discrete signals are those that can be presented in a simple off-or-on manner, signifying yes or no, present or absent, here or there, and similar dichotomies. They are most perfectly represented in the act of simple recognition, particularly during courtship. The steel-blue back and red belly of the male three-spined stickleback (*Gasterosteus aculeatus*) is an example of a discrete signal. Another is the ritualized preening of the male Mandarin duck (*Aix galericulata*), who whips his head back in a striking movement to point at the bright orange speculum of his wing. Still other examples are provided by the bioluminescent flashing sequences of fireflies (Figure 8-2). Discreteness of form also characterizes the communion signals by which members of a group identify one another and stay in contact, such as the duetting of birds and certain grunting calls of ungulates. Discrete signals become discrete through the evolution of "typical intensity" (Morris, 1957). That is, the intensity and duration of a behavior becomes less variable, so that no matter how weak or strong the stimulus evoking it, the behavior always stays about the same.

In contrast, graded (analog) signals have evolved in a way that increases variability. As a rule the greater the motivation of the animal or the action about to be performed, the more intense and prolonged the signal given. The straight run of the honeybee waggle dance denotes rather precisely the distance from the hive to the target. The "liveliness" or "vivacity" of the dance and its overall

Figure 8-2 Discrete signals in the sexual communication of fireflies. The flashing and flight paths of the males belonging to nine species of *Photinus* are shown here as they would look in a time-lapse photograph. Each species has a distinct, relatively invariable (hence discrete) flashing pattern. When a female on the ground observes the pattern of her own species, she flashes in response, attracting the male down to her. Fireflies are actually lampyrid beetles. (From Lloyd, 1966.)

displays are increasingly obvious in their meaning. The new components are added one by one or in combination: the mouth opens, the head bobs up and down, characteristic sounds are uttered, and the hands slap the ground. By the time the monkey combines all these components, and perhaps begins to make little forward lunges as well, it is likely to carry through with an actual attack (Figure 8-3). Its opponent responds by retreating or by escalating its own displays. These hostile exchanges play a key role in maintaining dominance relationships in the rhesus society.

Squirrels reveal gradually rising hostility by tail movements that increase from a slow waving back and forth to violent twitching. Birds often indicate aggressive tendencies by ruffling their feathers or spreading their wings, movements which create the temporary illusion that they are larger than they really are. Many kinds of fish achieve the same deception by spreading their fins or extending their gill covers. Lizards raise their crests, lower their dewlaps, or flatten the sides of their bodies to give an impression of greater depth. In short, the more hostile the animal, the more likely it is to attack and the bigger it seems to become. Such exhibitions are often accompanied by graded changes in both color and vocalization, and even by the scaled release of characteristic odors. (See Figure 8-4.)

Gradation in one form or another characterizes most of the major categories of communication in animal societies. Birds and mammals transmit a rich array of messages, some of which are qualitatively different in meaning, by gradually varying postures and sounds (Andrew, 1972). Ants, to cite a very different kind of organism, release quantities of alarm substances in approximate relation to the degree to which they have been stimulated. Fire ants deposit trail scent in amounts that reflect both the hunger of the colony and the richness of the food find (Hangartner, 1969a; Wilson, 1971a). The amplification of a signal can be accomplished simply by the gradual increase of the power output, movement, melanophore contraction, or whatever component contains the information. Or it can be achieved by adding wholly new components. A striking example of the second method is found in the mobbing calls of certain birds (see Figure 8-5).

The Principle of Antithesis

One of the most general principles of animal communication was first recognized by Charles Darwin in *The Expression of the Emotions in Man and Animals* (1872). Labeled by him the Principle of Antithesis, it can be expressed in an oversimplified manner as the following duality: When an animal reverses its intentions, it reverses the signal. The signal antitheses are most sharply defined in aggressive interactions. An animal that approaches another in a conciliatory mood, or else has lost a fight and is trying to appease the victor, uses postures

duration increase with the quality of the food find and the favorableness of the weather outside the hive. Graded communication is also strikingly developed in aggressive displays among animals. In rhesus monkeys, for example, a low-intensity aggressive display is a simple stare. The hard look a human being receives when he approaches a caged rhesus is not so much a sign of curiosity as it is a cautious display of hostility. Rhesus monkeys in the wild frequently threaten one another not only with stares but also with additional displays on an ascending scale of intensity. To the human observer these

Figure 8-3 Graded signals in the aggressive displays of a rhesus monkey (top) and a green heron (bottom). In the rhesus what begins as a display of low intensity, a hard stare (left), is gradually escalated as the monkey rises to a standing position (middle) and then, with an open mouth, bobs its head up and down (right) and slaps the ground with its hands. If the opponent has not retreated by now, the monkey may actually attack. A similarly graduated aggressive display is characteristic of the green heron. At first (middle) the heron raises the feathers that form its crest and twitches the feathers of its tail. If the opponent does not retreat, the bird opens its beak, erects its crest fully, ruffles all its plumage to give the illusion of increased size, and violently twitches its tail (right). Thus in both animals the more probable the attack, the more intense the aggressive display. (Based on Altmann, 1962a, and Meyer-riecks, 1960; from Wilson, 1972b. From "Animal Communication" by E. O. Wilson. © 1972 by Scientific American, Inc. All rights reserved.)

and movements that are the opposite of aggressive displays. Darwin's own description of antithetic signaling in dogs (see Figure 8-6) is graphic and precise:

When a dog approaches a strange dog or man in a savage or hostile frame of mind he walks upright and very stiffly; his head is slightly raised, or not much lowered; the tail is held erect and quite rigid; the hairs bristle, especially along the neck and back; the pricked ears are directed forwards, and the eyes have a fixed stare. These actions, as will hereafter be explained, follow from the dog's intention to attack his enemy, and are thus to a large extent intelligible. As he prepares to spring with a savage growl on his enemy, the canine teeth are uncovered, and the ears are pressed close backwards on the head; but with these latter actions we are not here concerned. Let us now suppose that the dog suddenly discovers that the man he is approaching, is not a stranger, but his master; and let it be observed how completely and instantaneously his whole bearing is reversed. Instead of walking upright, the body sinks downwards or even crouches, and is thrown into flexuous movements; his tail, instead of being stiff and upright, is lowered and wagged from side to side; his hair instantly becomes smooth; his ears are depressed and drawn backwards, but not closely to the head; and his lips hang loosely. From the drawing back of the ears, the eyelids become elongated, and the eyes no longer appear round and staring.

When displaying aggressively, a gull stretches its head forward, the ritualized intention movement by which the bird indicates it is ready

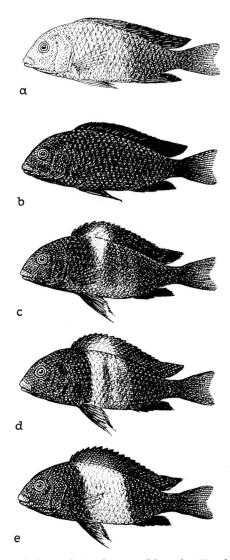

Figure 8-4 Graded signals in the mouthbrooder *Tropheus moorei*, a cichlid fish endemic to Lake Tanganyika: *a,* coloration assumed when the fish is frightened and strongly submissive; *b,* neutral coloration; *c-e,* the increasing expression of a yellow band around the middle of the fish accompanies the "quiver dance," used both in courtship and as an appeasement signal. (From Wickler, 1969a.)

to peck at its enemy. But in order to appease an opponent, a gull turns its head 90° to the side. Two gulls attempting to conciliate reciprocally will stand side by side, or face each other with their bodies, but they will be momentarily gazing in opposite directions (N. Tinbergen, 1960). Dominant male rhesus monkeys raise their tails and

heads and display their testicles by lowering them; subordinates lower their tails and heads and raise their testicles. The dominant males also mount their subordinates in ritual pseudocopulation; the subordinates present themselves in a pseudofemale posture to be mounted. Although such examples can be multiplied at length, not all displays opposite in meaning are also antithetical in appearance to the human observer. Even appeasement displays sometimes incorporate wholly new elements unrelated to hostile signaling. Hyenas, for example, rely heavily on penis displays to conciliate one another; even the females are equipped with pseudopenes which they use with convincing skill (Kruuk, 1972). Rodents and primates routinely utilize grooming, while some birds and mammals revert to begging and other juvenile postures (Wickler, 1972a).

Signal Specificity

The communication systems of insects, of other invertebrates, and of the lower vertebrates (such as fishes and amphibians) are characteristically stereotyped. This means that for each signal there is only one response or very few responses, that each response can be evoked by only a very limited number of signals, and that the signaling behavior and the responses are nearly constant throughout entire

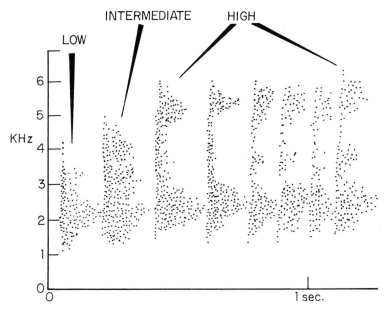

Figure 8-5 Intensification of the meaning of a sound signal by the addition of components. As shown in this spectrogram, the urgency of the mobbing call of the European blackbird (*Turdus merula*) is increased gradually by adding higher frequencies. (Modified from Andrew, 1961.)

Figure 8-6 The principle of antithesis is exemplified in Darwin's 1872 figure of the aggressive postures of dogs. In the upper figure a dog approaches another animal in a fully aggressive posture. The lower figure shows the same dog in a conciliatory stance, in which virtually all of the signals of the aggressive display have been reversed.

populations of the same species. An extreme example of this rule is seen in the phenomenon of chemical sex attraction in moths. The female silkworm moth draws males to her by emitting minute quantities of a complex alcohol from glands at the tip of her abdomen. The secretion is called bombykol (from the name of the moth, *Bombyx mori*), and its chemical structure is *trans*-10-*cis*-12-hexadecadienol.

Bombykol is a remarkably powerful biological agent. According to estimates made by Dietrich Schneider and his coworkers at the Max Planck Institute for Comparative Physiology at Seewiesen in Germany, the male silkworm moths start searching for the females when they are immersed in as few as 14,000 molecules of bombykol per

cubic centimeter of air. The male catches the molecules on some 10,000 distinctive sensory hairs on each of his two feathery antennae. Each hair is innervated by one or two receptor cells that lead inward to the main antennal nerve and ultimately through connecting nerve cells to centers in the brain. The extraordinary fact that emerged from the study by the Seewiesen group is that only a single molecule of bombykol is required to activate a receptor cell. Furthermore, the cell will respond to virtually no stimulus other than molecules of bombykol. When about 200 cells in each antenna are activated per second, the male moth starts its motor response (Schneider, 1969). Tightly bound by this extreme signal specificity, the male performs as little more than a sexual guided missile, programmed to home on an increasing gradient of bombykol centered on the tip of the female's abdomen—the principal goal of the male's adult life.

Such highly stereotyped communication systems are particularly important in evolutionary theory because of the possible role they play in the origin of new species. Conceivably one small change in the sex-attractant molecule induced by a genetic mutation, together with a corresponding change in the antennal receptor cell, could result in the creation of a population of individuals that would be reproductively isolated from the parental stock. Persuasive evidence for the feasibility of such a mutational change has been adduced by Roelofs and Comeau (1969). They found two closely related species of moths (members of the genus *Bryotopha* in the family Gelechiidae) whose females' sex attractants differ by only the configuration of a single carbon atom adjacent to a double bond. In other words, the attractants are simply different geometric isomers. Field tests showed not only that a *Bryotopha* male responds solely to the isomer of his own species but also that his response is inhibited if some of the other species' isomer is also present. An even more extreme case—the ultimate possible, in fact—has been reported by Minks et al. (1973). The two tortricid moth species *Adoxophyes orana* and *Clepsis spectrana* utilize the same two isomers, *cis*-9- and *cis*-11-tetradecenyl acetate, as their female sex attractant. However, they manufacture and release them in different proportions, and the different blends are sufficient to affect the male responses and hence isolate the species from each other.

For each such case of extreme specificity there exist others in which signals are shared by more than one kind of animal. Among moths of the families Saturniidae and Tortricidae, specificity of the sex pheromone often exists at the species-group level, meaning that the males respond to the pheromones emitted by females of both their own and closely related species (Priesner, 1968; Sanders, 1971). Under natural conditions the species depend on other kinds of prezygotic isolating mechanisms to avoid hybridization, particularly differences in preferred habitats, seasons of emergence, and times of peak mating activity.

Other kinds of signals are known which are clearly not designed to impart specificity. The alarm substances of ants, termites, and social bees consist of an astonishing diversity of terpenes, hydrocarbons, and esters, most of which have low molecular weights. In spite of the fact that they differ in composition and proportionality from one species to another, they are generally active across broad taxonomic groups. When an agitated honeybee worker discharges isoamyl acetate or 2-heptanone, it alarms not only her nestmates but also any ant or termite that happens to be in the near vicinity. This phenomenon is precisely what the evolutionist would expect. Privacy is not a requirement of alarm communication, and when the communication is coupled with interspecific aggressive behavior, signals should be expected to affect enemies as well as nestmates. The same differences in breadth of activity are found among the communication systems of birds and are subject to the same explanation. Territorial and courtship displays, including advertising songs, are characteristically elaborate and species-distinct. Their exceptional complexity and repetitive patterning are in fact the reasons why human beings consider them beautiful. But esthetics are not the primary consideration for birds. The displays are sufficient, in the great majority of cases, to isolate the members of each species of bird from all other species breeding in the same area. Where "mistakes" occur, resulting in interspecific territorial combat or hybrids, they are usually limited to closely related species and most often to those that have come into close contact only in the most recent geologic time. In contrast, the mobbing calls of small birds, which assemble other birds for cooperation in driving a predator from the neighborhood, are very similar from one species to another and are understood by all. The gulls (family Laridae) provide an excellent capsular illustration of the specificity rule. The sequence of courtship displays of each sympatric species is distinctive: in one species the long call is followed by mewing, in another the choking display is followed by the long call, and so forth. The precise forms of the separate components also vary. During the lengthy exchanges leading to pair bonding, these signals make it unlikely that any gull will choose a partner of the wrong species. In contrast, the displays of aggression and appeasement are simple in execution and uniform across species. In a closely parallel manner, the intergroup spacing and within-group rallying calls of *Cercopithecus* monkeys are species-specific, but their alarm calls are relatively constant and can be understood from one species to the next (Marler, 1973).

Even though much convergence of signals exists in aggressive interactions, there is no universal code to which all species of a group subscribe. In the mammals, for example, we find appeasement behavior following much the same form in species after species: the animal tends to crouch, often rolling over to expose its flank or belly. Konrad Lorenz suggested that in some mammals such as the dog the exposure

of these most vulnerable parts cancels the aggressive impulse of the opponent. However, the belly-up posture does not invariably mean submission. Among shrews it signifies hostility and dominance—with good reason, since the shrew's best fighting position is on its back (Ewer, 1968). Two points should be stressed: first, that evolution is entirely opportunistic and not bounded by any goal-directed rules, however intuitively appealing they may seem; and second, that display behaviors are among the most evolutionarily labile of all phenotypic traits.

Signal Economy

When evaluated by human standards, the number of signals employed by each species of animal seems severely limited. In order to establish some quantitative measure pertaining to this intuitive generalization, let us define a signal as any behavior that conveys information from one individual to another, regardless of whether it serves other functions as well. Most communication in animals is mediated by displays, which are behavior patterns that have been specialized in the course of evolution to convey information. In other words, a display is a signal that has been changed in ways that uniquely enhance its performance as a signal. The hawk warning call of a songbird, the hostile eyelid flashing of a male baboon, the zigzag dance of a courting male stickleback, and the release of sex attractant by a female moth are all examples of displays. By confining our attention for the moment to displays, we can delimit the set of the most important, easily diagnosed signals. Recent field studies have established the curious fact that even the most highly social vertebrates have no more than 30 or 40 separate displays in their entire repertory. Data compiled by Martin H. Moynihan (see Table 8-1) indicate that throughout the vertebrates the number of displays varies from species to species by a factor of only three or four. More precisely the number ranges from a minimum of 10 in certain fishes to a maximum of 37 in the rhesus monkey, one of the primates closest to man in complexity of social organization. The full significance of this rule of relative inflexibility is not yet clear. It may be that the maximum number of displays any animal needs in order to be fully adaptive in any ordinary environment, even a social one, is 10 to 40. Or it may be, as Moynihan has suggested, that each number represents the largest amount of signal diversity the particular animal's brain can handle efficiently in quickly changing social interactions. Moynihan's hypothesis includes an ingenious model of evolutionary turnover in signals, such that the number employed by a species at any given point in evolutionary time represents a dynamic equilibrium. As old displays decline, perhaps in competition with new displays that are more efficient in conveying the same information, they can be expected to become linked as a component with

Table 8-1 The number of displays of vertebrate species. (From Moynihan, 1970a.)

Species	Number of displays
FISHES	
Guppy (*Poecilia reticulata*)	15
Ten-spined stickleback (*Pygosteus pungitius*)	11
River bullhead (*Cottus gobio*)	10
Badis (*Badis badis*)	26
Mouthbrooder (*Tilapia natalensis*)	21
Sunfish (*Lepomis gibbosus*)	15
BIRDS	
Mallard duck (*Anas platyrhynchos*)	19
Skua (*Catharacta skua*)	18
White-hooded gull (*Larus modestus*)	28
Green heron (*Butorides virescens*)	26
American coot (*Fulica americana*)	25
Great tit (*Parus major*)	17
Chaffinch (*Fringilla coelebs*)	25
English sparrow (*Passer domesticus*)	15
Green-backed sparrow (*Arremenops conirostris*)	21
Eastern kingbird (*Tyrannus tyrannus*)	18
MAMMALS	
Deer mouse (*Peromyscus maniculatus*)	16
Black-tailed prairie dog (*Cynomys ludovicianus*)	18
Elk (*Cervus canadensis*)	26
Grant's gazelle (*Gazella granti*)	25
European polecat (*Mustela putorius*)	25
Coati (*Nasua narica*)	17
Ring-tailed lemur (*Lemur catta*)	34
Night monkey (*Aotus trivirgatus*)	16
Rufous-naped tamarin (*Saguinus geoffroyi*)	32
Dusky titi (*Callicebus moloch*)	27
Rhesus monkey (*Macaca mulatta*)	37

another display or else become increasingly rare and exaggerated in form. It is also generally true, as noted by Marler (1965), that the most stereotyped and complex displays of primates are the rarest in occurrence. Examples include hoot drumming in chimpanzees and chest beating in gorillas. There is a curious analogy in this principle to Zipf's law of human linguistics: the longer the word, the less frequently it is used.

In the extent of their signal diversity the vertebrates are closely approached by social insects, particularly honeybees and ants (Butler, 1969; Wilson, 1971a). The number of known signal categories within single species of these insects falls between 10 and 20. The honeybee

has been the most thoroughly studied of all the social insects. Apart from the waggle dance, its known communicative acts are mediated primarily by pheromones, the glandular sources of which have now been largely established. Other signals include tactile cues involved in food exchange and several dances that are different in form and function from the waggle dance. The fire ant *Solenopsis invicta*, another relatively well-analyzed species, has a comparable mix of chemical and tactile displays (Table 8-2). It is further true that nonsocial insects possessing the most complicated communication systems, for example the crickets (Alexander, 1961), have nearly as many displays as the social insects, although the systems serve fewer functions.

The relative simplicity of the signals in some display categories has resulted in striking examples of parallel and convergent evolution. True chameleons, constituting the family Chamaeleontidae, and false chameleons, composed of *Anolis* and related genera in the family Iguanidae, have converged in many respects as part of a mutual adaptation to a fully diurnal, arboreal existence. In particular, they share a distinctive form of visual aggressive display: the body is flattened and presented sidewise, the gular sac is spread, the mouth is opened,

Table 8-2 Known categories of communication among workers of the imported fire ant *Solenopsis invicta*. (Modified from Wilson, 1962a.)

Stimulus	Transmission	Response
Nest odor	Chemical	Nil, if odor is undisturbed
Casual antennal or bodily contact	Tactile	Turning-toward movement or increased undirected movement
Abdominal vibration	Sound from stridulation	Function unknown
Body surface attractants	Chemical	Oral grooming
Carbon dioxide	Chemical	Clustering and digging
Liquid food begging	Tactile	Regurgitation
Regurgitation	Chemical, at least in part	Feeding
Emission of Dufour's gland substance as trail	Chemical	Attraction, followed by movement along trail; used in mass foraging and colony emigration
Emission of Dufour's gland substance during attack	Chemical	Attraction to disturbed worker
Emission of cephalic substance	Chemical	Alarm behavior

the entire body sways, and the head is bobbed up and down (Kästle, 1967). Hyenas and social canids (wolves, wild dogs) have converged strongly in social structure, despite the fact that their common ancestry dates all the way back to the early Tertiary and hyenas are more closely related to cats and viverrids than to canids. The tail signals and body postures used in communication are also remarkably similar (Kruuk, 1972).

The general paucity of signal diversity in animal communication contrasts sharply with the seemingly endless productivity of human language. Yet certain intriguing parallels exist between man and other organisms. The paralinguistic signals of each human culture, including hand gestures and eyebrow raises, are roughly comparable in number to the displays of animals; the average person uses about 150–200 of these "typical" nonverbal gestures while communicating. The sound structure of language is based upon 20 to 60 phonemes, the precise number again varying according to culture. Perhaps 60 phonemes represent the maximum number of simple discrete vocalizations that the ear can distinguish, just as 30 or 40 displays are perhaps the most that an animal can distinguish efficiently. Human language is created by the sequencing of these sounds into an ascending hierarchy of morphemes, words, and sentences, which contain sufficient redundancy to make them easily distinguishable.

The Increase of Information

Although the number of displays catalogued by ethologists is 50 per species or less, the actual number of messages may be far greater. In the simplest systems a display may have only a single meaning, with no nuances permitted. Sexual communication in insects and other invertebrates is often of this kind. The whine of a female mosquito in flight, the ultraviolet flash of a male sulfur butterfly's wings, the release of cis-11-tetradecenyl acetate by a female leaf roller moth, each occurs in only one context and conveys the single unalterable message of sexual advertisement. However, in some invertebrate and the great majority of vertebrate systems, the number of messages that can be conveyed by one signal is increased by enrichment devices. The signal can be graded; it can be combined with other signals either simultaneously or in various sequences to provide new meaning; or it can be varied in meaning according to the environmental context. The extremes of enrichment are found, as might be expected, in the higher primates. Furthermore, the elementary concept of the social releaser, first developed in studies of sexual and aggressive behavior of birds and insects, has tended to break down most dramatically in mammals and particularly the higher primates. A full understanding of animal communication therefore depends on a systematic account of the enrichment devices, a brief version of which follows.

Adjustment of Fading Time

Any signal that is limited in space and time potentially provides information about both of these parameters. A predator-warning call designed to conceal the position of the signaler transmits only the information that a disturbing object has been sighted. In contrast, a territorial advertisement call, which can be easily localized, conveys both the challenge and the location of a part of the territory. This general mode of enrichment is especially clear in cases of chemical communication. The interval between discharge of the pheromone and the total fade-out of its active space can be adjusted in the course of evolution by altering the Q/K ratio, that is, the ratio of the amount of pheromone emitted (Q) to the threshold concentration at which the receiving animal responds (K). Q is measured in number of molecules released in a burst, or in number of molecules emitted per unit of time, while K is measured in molecules per unit of volume (Bossert and Wilson, 1963). Where location of the signaler is relevant, the rate of information transfer can be increased by lowering the emission rate (Q) or raising the threshold concentration (K), or both. This adjustment achieves a shorter fade-out time and permits signals to be more sharply pinpointed in time and space. A lower Q/K ratio characterizes both alarm and trail systems.

In the case of ingested pheromones, the duration of the signal can be shortened by enzymatic deactivation of the molecules. When Johnston et al. (1965) traced the metabolism of radioactive trans-9-keto-2-decenoic acid fed to worker honeybees, they found that within 72 hours more than 95 percent of the pheromone had been converted into inactive substances consisting principally of 9-ketodecanoic acid, 9-hydroxydecanoic acid, and 9-hydroxy-2-decenoic acid.

Increase in Signal Distance

If part of the message is the location of the signaler, the information in each signal increases as the logarithm of the square of the distance over which the signal travels. In chemical systems it is the active space, or the space within which the concentration of the pheromone is at or above threshold concentration, that must be expanded. An increase in active space can be achieved either by increasing Q or decreasing K. The latter is more efficient, since K can be altered over many orders of magnitude by changes in the sensitivity of the chemoreceptors, while a comparable change in Q requires enormous increases or decreases in pheromone production and capacity of the glandular reservoirs. The reduction of K has been especially prevalent in the evolution of airborne insect sex pheromones, where threshold concentrations are sometimes on the order of only hundreds of molecules per cubic centimeter. When the pheromone is expelled downwind, a relatively small amount is required to create very long active spaces, because orientation can be achieved by anemotaxis, or movement against or with the wind, rather than by a more laborious movement up or down the odor gradient. As a consequence Q can be kept small. The rate of information transfer is kept down, in the sense that signals cannot be turned on or off as rapidly. But the total amount of information eventually transmitted is increased, since a very small target can be pinpointed within a very large space.

The signaler may also identify the location of an object in space. When a sentinel male baboon barks to alert his troop, the troop members first look at him, then follow the direction of his gaze in an attempt to locate the disturbing stimulus (Hall and DeVore, 1965). The straight run of the waggle dance of the honeybee conveys detailed information on the location of targets and hence is more efficient than the round dance, which merely alerts bees to the existence of food somewhere near the hive (von Frisch, 1967).

Increase in Signal Duration

When a signal is broadcast continuously, the potential amount of information transferred increases evenly with time. Anatomical structures used in courtship are examples of more or less effortless sustained signals. The antlers of a male deer, the swollen buttocks of an estrous chimpanzee female, and the brightly colored legs of a sexually active male heron all continuously broadcast the reproductive status of their owners. Scent posts left by territorial mammals also signal continuously, and in addition, as the scent weakens by diffusion or changes chemically, its concentration provides further information on the age of the signal and the probability that the signaler is still in the vicinity.

Structures built by animals can provide the most durable signal source of all. Such communication can be labeled *sematectonic* from the Greek *sema* (sign, token) and *tekton* (craftsman, builder). This term is recommended as a substitute for stigmergy ("incite to work") or stigmergic communication, coined by Grassé to refer specifically to the guidance of work performed by social insects through the evidences of work previously accomplished. There is a need for a more general, somewhat less clumsy expression to denote the evocation of any form of behavior or physiological change by the evidences of work performed by other animals, including the special case of the guidance of additional work.

Students of social insects have been aware of sematectonic communication for nearly two centuries. Pierre Huber (1810) said of nest building in the ant *Formica fusca*, "From these observations, and a thousand like them, I am convinced that each ant acts independently of its companions. The first that hits upon an easy plan of execution immediately produces the outline of it; others only have to continue

along those same lines, guided by an inspection of the first efforts." A modern, more explicit example of Huber's principle is provided by the cooperative labor in weaver ants of the genus *Oecophylla*. These insects are wholly arboreal, and they construct their nests of green leaves held together by sticky larval silk. In order to make a nest wall, it is necessary for groups of workers to pull leaves together simultaneously while others move the larvae back and forth like animated shuttles. How is this cooperation achieved? The solution, discovered by Sudd (1963), involves a simple form of sematectonic communication. As shown in Figure 8-7, workers work independently in their first attempts to pull down or roll up leaves. When success is achieved by one or more of them at any part of a leaf, other workers in the vicinity abandon their own efforts and join in. Other examples of sematectonic communication, with a discussion of their role in the organization of social insects generally, are provided by Wilson (1971a).

Sematectonic communication is by no means limited to social insects. When the larvae of stem-dwelling eumenine wasps pupate, their bodies must be aligned so that the insects face the outer, open end of the hollow stem. If the pupae are accidentally pointed in the opposite direction, the newly eclosing adult wasps try to dig down through the pith of the stem toward the trunk of the bush or tree—and they die in the process. How does a larva know the correct direction to face when it is about to turn into a pupa? Kenneth Cooper (1957) found that the mother wasp provides the necessary information for her offspring when she first constructs the cell in the hollow stem. She makes the texture and concavity (versus convexity) of the terminal wall leading to the outside different from the texture and concavity of the wall leading inward toward the tree trunk. The larva

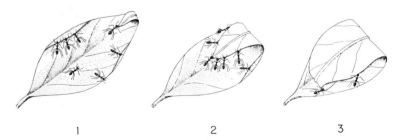

1   2   3

Figure 8-7   The initiation of cooperative nest building in the weaver ant. When workers first attempt to fold a leaf (left) they spread over its surface and pull up on the edge wherever they can get a grip. One part (in this case, the tip) is turned more easily than the others, and the initial success draws other workers who add their effort, abandoning the rest of the leaf margin (center). The result (right) is a rolled leaf of a kind frequently encountered in *Oecophylla* nests. (From Sudd, 1963; from *Science Journal*, London, incorporating *Discovery*.)

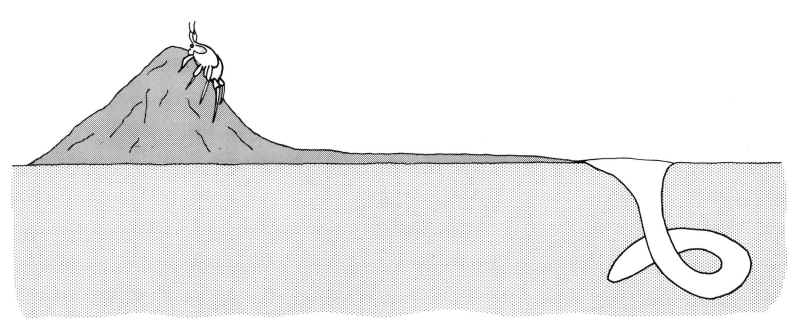

Figure 8-8 Sematectonic communication in ghost crabs (*Ocypode saratan*) is based on an elaborate structure built in the sand by adult male ghost crabs. The sand pyramid on the left is connected by a path-way to a shallow vestibule and spiral burrow on the right. The mere sight of these complexes repulses other males and attracts females. (Redrawn from Linsenmair, 1967.)

instinctively uses this information to orient its body at pupation, even though it has never had direct contact with the mother or access of any kind to the outside world. Cooper was able to change the orientation of larvae at will by placing them in artificial stems containing cells of deliberately varied construction.

The mere sight of nests constructed by other animals also can acquire a communicative function. Adult male ghost crabs (*Ocypode saratan*) build peculiar architectural complexes in their sandy habitats, each unit consisting of a pyramid of excavated sand, a pathway, a vestibule, and a spiral burrow (Figure 8-8). Burrows dug by other members of the same species are nonspiral and lack pyramids. Linsenmair's experiments (1967) showed that the complexes represent "petrified display signals" that force other males to construct their own burrows at a minimum distance of 134 centimeters. Adult females are attracted by the pyramids and use them to find the spiral burrows, which serve as the mating sites. The structures appear to be employed strictly for communication. The males occupy them for only four to eight days, during which time they do not feed.

Gradation

All other circumstances being equal, graded messages convey more information than equivalent discrete messages. Consider the simplest possible case, in which one discrete signal is compared with a single signal selected from along a point in a gradient. The discrete signal can only exist or not exist. In the absence of alternate signals in the same message category, it conveys at most one bit of information. The graded signal, in contrast, exists or does not exist, and when it exists it further designates a point on the gradient. The additional number of bits yielded is a function of the logarithm of the total number of points on the gradient that can be discriminated. Now suppose that the two systems being compared are (1) a set of discrete signals arrayed along a gradient, labeled, say, 1 to 10 along a scale of rising intensity, and (2) a continuously varying signal arrayed along the same scale. Let the precision of emission and reception be the same in both cases. It can be shown that the continuously varying system will always carry more information than the one divided into discrete steps. This principle can be seen more clearly by comparing the honeybee waggle dance, which is a continuously varying system, with an imaginary equivalent system divided into discrete messages. We know that because of errors in both the dance and the execution of the outward flights directed by the dance, the amount of information conveyed about direction is rather limited, consisting of about four bits. This is the equivalent of pinpointing any target within one or other of 2^4, or 16 equiprobable compass sectors. Suppose that the imaginary competing system has 16 discrete signals representing the same compass sectors. If the precision of transmission is the same as in the real waggle dance, the information transmitted will be less than four bits per dance. This is because some of the bees will inevitably end up in sectors other than the one designated by the signal, and the probability and degree of their errors would then have to be translated into bits and deducted from the four bits that represent the maximum in a perfect discrete system.

It is possible for the messages of graded signals to shift not only in intensity but also in qualitative meaning. Workers of the harvester ant *Pogonomyrmex badius* react in strongly varying ways to their principal alarm substance, 4-methyl-3-heptanone, which is released from the mandibular gland reservoir when the ants are disturbed. Workers respond to threshold concentrations averaging 10^{10} molecules per cubic centimeter by simply moving toward the odor source. When a zone of concentration one or more orders of magnitude greater than this amount is reached, the ants switch into an alarm frenzy. If the ants are then exposed to high concentrations for more than a minute or two, many change from alarm to digging behavior.

Composite Signals

By combining signals it is possible to give them new meanings. The theoretical upper limit of a combinatorial message is the "power set" of all of its components, or the set of all possible combinations of subsets. Thus, if *A*, *B*, and *C* are three discrete signals, each with a different meaning, and each combination produces still one more message, the total ensemble of messages possible is the power set consisting of seven elements: *A*, *B*, *C*, *AB*, *AC*, *BC*, and *ABC*. No animal species communicates in just this way, but many impressive examples have been found in which conspicuous signals are used effectively in different combinations to provide different meanings. A case from the horse family (Equidae) embracing both discrete and graded signals is shown in Figure 8-9. A zebra or other equid shows hostility by flattening its ears back and friendliness by pointing them upward (discrete signals). In both postures the intensity is indicated by the degree to which the mouth is opened simultaneously (a graded signal). The mare is able to produce a third message by adding two more components: when ready to mate, she presents the stallion with the threat face but at the same time raises her hindquarters and moves her tail aside.

Chemical communication, like visual communication, lends itself easily to the production of composite messages. Many species of insects and mammals possess multiple exocrine glands, each of which produces pheromones with a different meaning. Kullenberg (1956), for example, found that females of certain aculeate wasps release simple attractants from the head that act in concert with sexual excitants released from the abdomen. Different substances with different meanings can also be generated by the same gland. A minimum of 32 compounds have been detected in the heads of honeybee

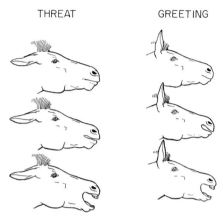

THREAT   GREETING

Figure 8-9 Composite facial communication in zebras (*Equus burchelli*). Threat is indicated by laying the ears back (a discrete signal) and opening the mouth ever wider to indicate increasing amounts of hostility (a graded signal). When making a friendly greeting, the zebra opens its mouth variably in the same way, but now points the ears upward. (Modified from Trumler, 1959.)

queens, including methyl 9-ketodecanoate, methyl 9-keto-2-decenoate, nonoic acid, decanoic acid, 2-decenoic acid, 9-keto-decanoic acid, 9-hydroxy-2-decenoic acid, 10-hydroxy-2-decenoic acid, 9-keto-2-decenoic acid, and others (Callow et al., 1964). Most or all are present in the mandibular gland secretion. The biological significance of most of these substances is still unknown. Some are undoubtedly precursors to pheromones, but at least 2 are known pheromones with contrasting effects. The first, 9-keto-2-decenoic acid, is basically an inhibitor. Operating in conjunction with additional scents produced elsewhere in the body, it reduces the tendency of the worker bees to construct royal cells and to rear new queens, who would then be rivals of the mother queen. It also inhibits ovarian development in the workers, in effect preventing them from entering into rivalry with the queen. The second mandibular gland pheromone, 9-hydroxy-2-decenoic acid, causes clustering and stabilization of worker swarms and helps to guide the swarms from one nest site to another (Butler et al., 1964). A rich mixture of chemicals is also found in the castoreum of the beaver. About 45 substances have been identified, including a surprisingly diverse array of alcohols, phenols, ketones, organic acids, and esters, as well as salicylaldehyde and castoramine ($C_{15}H_{23}O_2N$) (Lederer, 1950). Although no behavioral function has yet been demonstrated, it is likely that a deliberate testing of the idea will reveal some of the substances to be pheromones. In fact, the American naturalist Ernest Thompson Seton once speculated that castoreum-impregnated scent posts serve as "mudpie telegrams" by which the beavers communicate.

There are a few examples of pheromones acquiring additional or even different meanings when presented in combination. When released near fire ant workers, cephalic and Dufour's gland secretions cause alarm behavior and attraction, respectively; when expelled simultaneously by a highly excited worker, they cause oriented alarm behavior. Honeybee workers confined closely with a queen for hours acquire scents from her that, possibly in combination with their own worker-recognition scent, cause them to be attacked by nestmates (Morse and Gary, 1961).

Among vertebrates especially, signals transmitted through different sensory channels are often combined in ways that increase information. In some instances the signals are simply redundant: the simultaneous hissing and body jerking of a chameleon, for example, and the stretch display of a male snowy egret delivered with its courtship call; in both cases the precision of the message is increased, although the combinations add no new meaning not already present in the separate elements. Components in different modalities can be added as part of the graded intensification of a signal. In closely grouped societies of primates, such as the dense troops of macaques and baboons, the threats of lowest intensity are typically visual in nature. When these visual signs are intensified, characteristic sounds are added for reinforcement. Workers of the ant *Camponotus socius* employ odor trails for either one of two purposes, to recruit nestmates to newly discovered food sources or to lead them to new nest sites (Hölldobler, 1971a). Recruitment is specified by adding a waggling motion of the head while nest transfer is specified by a back-and-forth jerking motion of the entire body (see Figure 8-10). Other animals use more or less orthogonal gradients of display, so that a different message is identified with each point on the intersection of the two gradients. To visualize the structure of such an intersecting system, consider the extreme theoretical case in which there are m possible signals belonging to one message category, whether in the form of that number of discrete signals or as m distinguishable points on a continuous spectrum. Suppose that the species further uses a second, related message category containing n possible signals. Then the two categories in combination can yield up to mn messages. For example, during the early stages of courtship, males of some species of fishes and birds present combinations of pure threat and pure courtship signals, indicating varying degrees of readiness to admit particular females into their territories. The females respond with appropriate conciliatory displays. In time the displays of the males shift to a predominantly sexual cast, and the pair bond is achieved (Baerends and Baerends-van Roon, 1950; Tinbergen, 1952, 1959; Meyerriecks, 1960).

It is also possible for hostile and submissive displays to be combined orthogonally to generate new messages. In other words, the displays do not form a simple spectrum ranging from most hostile at one end to most submissive at the other, but rather constitute two sets of signals that can be presented either separately or in combination. When combined, the signals create a message containing a high level of ambiguity. In the domestic cat a high-intensity threat combined with a high-intensity fear display produces the "halloween cat" posture: body raised on fully extended limbs and mouth closed (threat); also, body arched and ears flattened (fear). This mosaic of postures is ambiguous with reference to the basic signals but provides new information in a different category. The cat can be interpreted by the human observer, and presumably by other cats as well, as being in a highly excited state, ready to be tipped into either a violent fight or precipitous flight. This message is distinct from the high-intensity states of the purely aggressive or purely submissive postures; it is also different in meaning from the more relaxed mosaic posture of a cat displaying low-intensity aggression and fear (Leyhausen, 1956). This rather involved interpretation has received some neurophysiological support. Using implanted electrodes, J. L. Brown et al. (1969) elicited composite aggressive and flight behavior in cats by simultaneously stimulating hypothalamic centers previously shown to control the responses independently. Similar combinations of signals, more or less orthogonal in nature, have been described in the wolf and dog (Schenkel, 1947).

Syntax

True syntax in the sense of human linguistics, wherein the meaning of combinations of signals depends on the order of appearance, has not yet been demonstrated in animals. The one possible exception is play invitation, to be described later in a discussion of metacom-

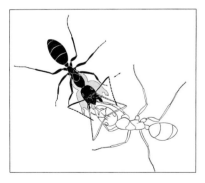 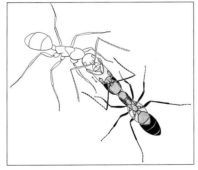

RECRUITMENT TO FOOD   TRANSPORT TO NEST SITE

Figure 8-10 Composite signaling in ant workers (*Camponotus socius*). Odor trails laid from the hindgut are used to guide nestmates either to food or to new nest sites; as shown here, the former is indicated by a waggling of the head and the latter by a back-and-forth jerking movement. (From Hölldobler, 1971a.)

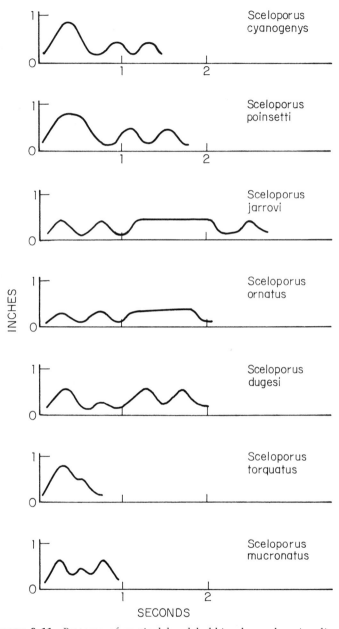

INCHES

SECONDS

Figure 8-11 Patterns of vertical head bobbing by male spiny lizards (genus *Sceloporus*) are sufficiently distinctive to permit females to select mates belonging to their own species. The depth of the bob is given on the ordinate. Although the sequences of movements are specific, they do not constitute true syntax. (Modified from Hunsaker, 1962.)

munication, and even this is at best a marginal case. True syntax occurs when separate signals, say, *A*, *B*, and *C*, that have distinct meanings when alone create new messages when presented in various orders: *AB*, *CBA*, *CAB*, and so forth. In human speech, each of the three permutations "George hunts," "George hunts the bear," and "The bear hunts George" has a very different meaning. No comparable process of message formation is known to exist in animal communication.

Even so, the distinctiveness of a *single* message often depends on the ordering of the elements that compose it. The head-bobbing movements of spiny lizards (*Sceloporus*) are sequenced through time in ways that permit females to recognize males of their own species (Figure 8-11). Experimental models made to bob according to various of the species patterns attract watching females of the species being imitated (Hunsaker, 1962). But the sequence of bobs is not a sentence that individual lizards break down and rearrange; it is a solitary, unalterable signal. Similarly, Brémond (1968) found that the sequence of notes sung by males of the European robin (*Erithacus rubecula*) is important for recognition by other members of the same species. In contrast, sequence of notes is not important in the songs of the wood lark (*Lullula arborea*) and indigo bunting (*Passerina cyanea*) (Tretzel, 1966; S. T. Emlen, 1972). Like lizard nodding, robin song falls short of syntactic organization. It is merely a case of temporal cues being added to pitch, duration, and other components that impart distinctiveness to unit signals.

It is further true that the performance of one act or the emission of one signal influences the probability of occurrence of the next one to follow. In other words, separate displays are presented not as an independent trials process, but as a Markov process in which the probability of occurrence of one display is influenced to some degree by the nature of the displays just performed. This is the general impression of virtually all ethologists who have considered the subject, and it has been documented by the excellent statistical study of the goldeneye duck *Bucephala clangula* by Dane and Van der Kloot (1964). Because of the very large number of observations required to estimate transition probabilities of second order or greater, the maximum length of the stochastic chains in communication systems is laborious to measure, sometimes requiring thousands of observations. Nevertheless, in their analysis of aggressive interactions of hermit crabs (Paguridae), Hazlett and Bossert (1965) were able to establish the existence of at least second-order probabilities. This means that for a sequence of three acts, say *A-B-C*, the probability of occurrence of *C* is affected not only by the prior occurrence of *B* but also by that of *A*. The sequences involve series of behaviors in single individuals and also the probability that one individual will respond when presented with one or two signals from another individual. The same result was obtained by Altmann (1965) in his study of the rhesus monkey, a species with a far more complicated repertory than the hermit crabs. Yet in neither kind of animals are constrained transition probabilities the equivalent of syntactical language. Hazlett, Bossert, and Altmann have made it clear that these constraints do not create packages of sequenced signals with any special meaning beyond the mere sum of the messages implicit in the separate signals.

Metacommunication

A peculiar form of composite signaling is metacommunication, or communication about the meaning of other acts of communication (Bateson, 1955). An animal engaged in metacommunication alters the meaning of signals belonging to categories other than the original signals that are being transmitted either simultaneously or immediately afterward. Altmann (1962a,b), who first applied this concept extensively to the behavior of nonhuman primates, recognized two circumstances in which metacommunication occurs. The first is status signaling. A dominant male rhesus monkey can be recognized by his brisk, striding gait; his lowered, conspicuous testicles; the posture of his tail, which is held erect and curled back at the tip; and his calm "major-domo" posture, during which he gazes in a confident, unhurried manner at any other monkey catching his attention. A subordinate male displays the opposite set of signals (Figure 8-12). Similar signaling has been recorded in other species of macaques and baboons. Altmann's hypothesis is that the displaying animal communicates its own knowledge of its status and therefore the likelihood that it will attack or retreat if confronted. Since the individual troop members know one another personally, they can judge for themselves whether particular rivals are prepared to alter the dominance order. They evaluate the general "attitude" of the other members of the society. This explanation is eminently plausible but has not yet been subjected to any convincing test.

The second form of primate metacommunication is play invitation. The play of rhesus monkeys, like that of most other mammals, is devoted largely to mutual chasing and mock fighting. The invitation signals consist of gamboling and gazing at playmates from between or beside their own legs with their heads upside down. In the play that ensues, the monkeys wrestle and mouth one another vigorously. Although easily capable of hurting one another, the monkeys seldom do. Real damage will result later from escalated versions of the same behavior during bouts of intense aggression. Play signaling says approximately the same thing as this simple human message: "What I am doing, or about to do, is for fun; don't take it seriously. In fact—join me!"

Metacommunicative play signaling in dogs was first described by Charles Darwin, in 1872: "When my terrier bites my hand in play, often snarling at the same time, if he bites and I say *gently, gently*, he goes on biting, but answers me by a few wags of the tail, which

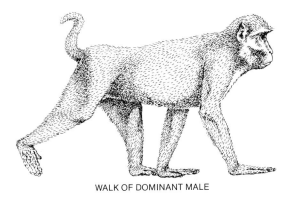

WALK OF DOMINANT MALE

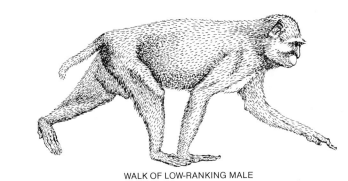

WALK OF LOW-RANKING MALE

Figure 8-12 Metacommunication in rhesus monkeys includes status signals. The postures and movements of individuals indicate the rank they occupy in the dominance order. (From Wilson et al., 1973; based on S. A. Altmann.)

seems to say 'Never mind, it is all fun.'" Dogs initiate play with one another by abruptly crouching down with their forelegs extended stiffly forward and by barking. Both signals appear to be ritualized aggression intention movements (Loizos, 1967; Bekoff, 1972). At the same time the dogs keep their eyes wide open and their ears pulled forward. Domestic cats and lions use a similar posture but omit the vocalization (see Figure 8-13). Juvenile male squirrels make a high bounding leap onto their female litter mates, followed by movements that resemble mating and grooming by adults (Horwich, 1972). Chimpanzees, baboons, and Old World monkeys present the "play-face" or "relaxed open-mouth display" (Andrew, 1963b; van Hooff, 1972): the mouth is opened widely with the lips still covering most or all of the teeth, and the mouth corners are not pulled forward as in the overtly aggressive bared-teeth display (see Figure 8-14). The body and eyes continue to move in a relaxed manner, while breathing becomes quick and shallow. In chimpanzees the accelerated breathing is vocalized as a series of hoots that sound like "ahh ahh ahh." The play-face

Figure 8-13 Play initiation in lions exemplifies one of the two principal forms of metacommunication used by mammals. Above: an adult male invites a cub to play by lowering his forequarters. Below: he clubs it lightly on the head. (From Schaller, 1972.)

in fact may be homologous to the relaxed grin of human beings. A man who taps a friend on the arm or punches him lightly on the chest is unlikely to receive a hostile response if he remembers to grin broadly at the same time. In Western cultures the combined gestures are routinely an invitation to friendly banter.

Context

Even though an animal is limited to a small signal repertory, it can greatly increase the information transmitted by presenting each signal in different contexts. The meaning of the signal then depends on the other stimuli impinging simultaneously on the receiver. Consider an imaginary extreme case in which an animal is limited to one signal that generally alerts other members of the same species. Particularity is added as follows: when presented in the face of danger, the signal serves to alarm; given while the animal is in its own territory, it is a threat to sexual rivals or an invitation to potential mates; when presented to offspring, it means that food is about to be offered—and so on.

W. J. Smith (1963, 1969a,b) has stressed the importance of contextual change for the enrichment of communication in birds. The male of the eastern kingbird *Tyrannus tyrannus*, for example, emits a

Figure 8-14 The relaxed open-mouth display of the crab-eating monkey (*Macaca fascicularis*) on the right is a play invitation signal of a kind widely used by Old World monkeys and apes, and may be homologous to the grin of human beings. The two monkeys shown here are engaged in play-fighting. (Redrawn from van Hooff, 1972.)

general purpose call that sounds roughly like "Kitter!" The Kitter is evoked in a variety of contexts when the bird experiences indecisiveness or interference in its attempt to approach some object—a perch, a mate, or another bird. When a lone male flies from perch to perch in his newly delimited territory, the Kitter is employed to attract a female and to warn off potential rivals. Later, the same signal is evidently used as an appeasement signal by the male when approaching his mate.

Mammals employ contextual information extensively. Stotting is shown by Grant's and Thomson's gazelles during flight from predators such as wild dogs, and it is also employed by adults during intraspecific chases (Estes, 1967). The roaring of lions serves at least four functions according to context: it helps individuals belonging to the same pride to find one another during separations; it is a bond-reinforcing signal for members of the pride while the animals are in contact; it serves as a spacing device for neighboring prides; and it functions as the most spectacular vocal display during close aggressive interactions (Schaller, 1972). The meaning of the greeting ceremony of wolves, during which one animal attempts to lick the muzzle of another, also depends on social circumstances. It is sometimes used as a submission signal, especially toward the most dominant male. Pups employ a closely similar or even identical version to beg food, while the adults participate in an excited version of the greeting ritual when prey is scented (Mech, 1970). The African wild dog uses an apparently homologous behavior just before the pack begins to run after prey.

The social insects have developed forms of contextual enrichment even more extreme than those of the vertebrates. The honeybee queen substance, 9-keto-2-decenoic acid, functions as a caste-inhibitory pheromone inside the hive, as the primary female sex attractant during the nuptial flight, and as an assembly scent for the colony during swarming. The waggle dance of the honeybee guides workers to new food finds; it also directs the swarm to new nest sites. The Dufour's gland secretion of the fire ant *Solenopsis invicta* is an attractant that is effective on members of all castes during most of their adult lives. Under different circumstances it serves variously to recruit workers to new food sources, to organize colony emigration, and—in conjunction with a volatile cephalic secretion—to evoke oriented alarm behavior.

Mass Communication

Many of the most highly organized communication systems of social insects contain components of information that cannot be passed from one individual to another, but only from one group to another. This is the phenomenon which I have called mass communication (Wilson, 1962a, 1971a). The number of fire ant workers leaving the

nest is controlled by the amount of trail substance emitted by nestmates already in the field. Tests involving the use of enriched trail pheromone have shown that the number of individuals attracted outside the nest is a linear function of the amount of substance presented to the colony as a whole. Under natural conditions this quantitative relation results in the adjustment of the outflow of workers to the level needed at the food source. Equilibration is then achieved in the following manner. The initial build-up of workers at a newly discovered food source is exponential, and it decelerates toward a limit as workers become crowded on the food mass because workers unable to reach the mass turn back without laying trails and because trail deposits made by single workers decline to below threshold concentrations within a few minutes. As a result, the number of workers at food masses tends to stabilize at a level that is a linear function of the area of the food mass. Sometimes, for example when the food find is of poor quality or far away, or when the colony is already well fed, the workers do not cover the find entirely, but equilibrate at a lower density. This additional mass communication of quality is achieved by means of an "electorate" response, in which individuals choose whether to lay trails after inspecting the food find. If they do lay trails, they adjust the quantity of pheromone according to circumstances (Hangartner, 1969a). The more desirable the food find, the higher the percentage of positive responses, the greater the trail-laying effort by individuals, the greater the amount of trail pheromone presented to the colony, and hence the greater the number of newcomer ants that emerge from the nest. Consequently the trail pheromone, through the mass effect, provides a control that is more complex than could have been assumed from knowledge of the relatively elementary forms of individual behavior alone.

The waggle dance of the honeybee regulates the number of workers at finds by means of mass communication closely paralleling that in the fire ant odor trail. A second example of mass communication in honeybees is displayed in hive-cooling behavior (see Chapter 3). The air-conditioning system is tuned by the willingness of nest workers to receive loads of water from the foragers that bring it in from the field. When enough droplets are distributed, and the temperature falls, the nest workers seek water less actively from incoming foragers, who must search longer for a willing recipient to whom they can regurgitate their load. As a consequence the flow of water into the hive slows and finally ceases (Lindauer, 1961). The encouragement or discouragement of water carriers, controlling water intake of the entire colony, is therefore a form of mass communication analogous in many respects to odor-trail recruitment in ants. In both systems, the quantitative needs of the colony as a whole can be measured and filled only by the summation of large numbers of actions by individual workers.

The Measurement of Communication

Communication has been defined as the process by which behavior of one individual alters the probability of behavioral acts in other individuals. This concept has the advantage of being directly translatable into a mathematical statement. Our formalism recognizes the following minimal set of six entities:

Individuals	A	B
Acts	X_1	X_2
Probabilities of acts occurring	$p(X_1)$	$p(X_2)$

Communication occurs when $p(X_2|X_1) \neq p(X_2)$. In words, the conditional probability that act X_2 will be performed by individual B given that A performed X_1 is not equal to the probability that B will perform X_2 in the absence of X_1.

Given that some amount of information is transmitted, how can it be measured? The basic quantitative unit is the bit, which is short-hand for binary digit. One bit is the amount of information required to control, without error, which of two equiprobable alternates is to be chosen by the receiver. Imagine an ultrasimple social system consisting of a territorial bird facing a series of intruders. Each invader pays no attention to the resident until it is presented with one or the other of two equally likely signals: if the resident raises its wings the intruder invariably leaves; if it lowers its wings the intruder invariably moves forward. Each presentation of a signal therefore transmits one bit of information. If four equiprobable messages can be sent, each signal contains two bits of information; a system of eight equiprobable messages contains three bits of information per signal, and so on. In short, the number of bits is the power to which the number 2 must be raised to yield the number of equiprobable messages. Where H is the number of bits and N the number of messages,

$$N = 2^H$$

$$H = \log_2 N$$

The preference for the binary system is due to its convenience and familiarity in many other branches of science and in engineering. The binary vocabulary, it will be recalled, grows exponentially with the number of digits used: two messages, 0 and 1, from one binary digit; four messages, 00, 01, 10, and 11, from two binary digits; and so on. With equal validity we could use trinary digits (0, 1, 2), in which case the number of messages would go up with the number of digits as the power of 3, and the unit of information would be called a "trit." Or we could employ the full decimal array of numbers

(0, 1, 2, ..., 8, 9), have the number of possible messages increase as the power of 10, and call the basic unit a "dit."

Suppose next that the separate messages are not equiprobable. In this case the amount of information transmitted is invariably less than $\log_2 N$. The meaning of the loss of information is easy to grasp intuitively. When all signals are equiprobable, the uncertainty connected with the identity of each future signal is at its maximum. We say that when the signal is delivered, it reduces uncertainty by the greatest possible amount (that is, by $\log_2 N$). But when one signal is more frequent than the others, we recognize that there is less uncertainty about each undelivered message. When identified, it is more likely to be the common message than one of the less common ones. Suppose that the imaginary bird just cited delivered one of its two messages almost all the time, and the second message only very rarely. There is correspondingly little uncertainty about which message will come next and hence little information per message. The amount of potential information in each message is, for any such message system, calculated by the Shannon-Wiener formula:

$$H(X) = -\sum p(i) \log_2 p(i)$$

where $p(i)$ is the probability of each signal X_i. The negative of the sum of terms is taken because the logarithms of all $p(i) > 0$ are negative, and $H(x)$ would be negative. A simple example of a computation is presented in Table 8-3. The value of $H(X)$ is 0.948, slightly less than one bit. Note that the information content is below that of a system with two equiprobable signals, and far below the content of a system containing four equiprobable signals, $H(X) = 2$.

The Shannon-Wiener measure has several strong a priori mathematical advantages. (1) It is independent of the scale used; one can compare systems measured in angstroms and meters, compass degrees and color divisions, and so forth. (2) It can be computed for both continuous and discrete variates. (3) It is a continuous function of $p(i)$. (4) Because of the logarithmic transformation, rare messages

Table 8-3 The computation of information in an imaginary four-signal system by means of the Shannon-Wiener formula.

Message—X_i	Frequency of Message—$p(i)$	$p(i) \log_2 p(i)$
X_1	0.80	−0.257
X_2	0.13	−0.382
X_3	0.06	−0.243
X_4	0.01	−0.066
	$\sum_i = 1.0$	$-H(X) = \sum_i = -0.948$

contribute very little to the measure; it is possible to miss a great many rare messages in the course of compiling a behavioral catalog and still underestimate $H(X)$ by only a small amount. (5) It is additive; that is, if two signal systems (say, X and Y) are employed independently, the total information in both is simply the sum of their separate information contents. This last property will quickly become clear by noting that if m equiprobable signals exist in X and n in Y, there exist mn equiprobable combinations of two signals; $H(X + Y) = \log_2 mn = \log_2 m + \log_2 n = H(X) + H(Y)$.

The information in the signal is called the signal entropy. In a noiseless system, where each kind of signal evokes one and only one kind of response, without error, the information transmitted between the sender and the receiver is exactly the source entropy. But few communication systems, and almost certainly none employed by animals, contain such a perfect design. Noise permeates most systems in the form of the potential triggering of more than one response by one signal (*ambiguity* on the part of the receiver) and the potential effectiveness of more than one signal in evoking a given response (*equivocation* on the part of the signaler). To make this notion clearer, suppose that an animal species is discovered that has a very rich repertory of signals. We might conclude, as popular writers are always doing for the porpoise for example, that the repertory reflects high intelligence and a complicated code of communication. But then it is discovered that all the signals evoke only one kind of response. The equivocation is thus so great that the communication code is comparable to a one-signal–one-response system. This noise must be subtracted from the total information in the signals (or in the responses) in order to measure the amount of constraint between the signals and the responses. That constraint is the true information transmitted by the signal.

The full procedure for measuring information in a two-animal system is given in Figure 8-15 and Table 8-4. The essential data are the probabilities of each pairwise combination of X_i and Y_j. For example, in Table 8-4 note that the probability that any given signal and response will be X_4 and Y_2, respectively, is 0.042; in other words 4.2 percent of all signal-response combinations observed are X_4 followed by Y_2. Signal X_4, to make the model somewhat more realistic, could be crest raising; response Y_2 might be a subsequent retreat from the territory.

The amount of information transmitted in each signal is equal to the signal entropy minus the equivocation. The method for measuring signal entropy is provided in Table 8-4. The equivocation is obtained by taking each Y_j in turn and noting the conditional probabilities for each X_i that evokes it. Thus when Y_1 is evoked, X_2 is responsible $0.001/0.01 = 0.1$ (10 percent) of the time; X_3 is responsible 0.5 of the time; and X_4 is responsible the remaining 0.4 of the time. We compute the entropy of these three values for the category Y_1:

$$H_j(i) = -\sum_i p_j(i)\log_2 p_j(i) = -\sum_i p_1(i)\log_2 p_1(i)$$

Then we weight the entropy with the frequency with which the response Y_1 occurs. This is $p(j) = p(1) = 0.001 + 0.005 + 0.004 = 0.010$. The weghted $H_1(i)$ is $p(1) \cdot H_1(i) = 0.01 H_1(i)$. The same procedure is followed for each of the five remaining Y_j's. The sum of all six values of $p(j) \cdot H_j(i)$ is the equivocation.

The amount of information transmitted can also be obtained in symmetric fashion by subtracting the amount of ambiguity from the response entropy. The relationships of the essential components are indicated in Figure 8-16.

Few attempts have been made to measure the amount of information transferred in animal communication systems. Hazlett and Bossert (1965) characterized the full aggressive communication system of hermit crabs (Paguridae) and measured transition probabilities for chains of up to three behavioral acts. They found the average information transmitted per signal to vary among eight species from 0.35 to 0.52 bits, with an overall average of 0.41 bits, a remarkable degree of taxonomic consistency. Transmission rates varied overall from a range of 0.4-1.0 bits per second in the slowest transmitter, *Paguristes grayi*, to 0.9-4.4 bits per second in the fastest species, *Pagurus bonairensis*. A similar study of mantis shrimps (Stomatopoda) conducted by Dingle (1972b) yielded somewhat higher values: 0.64-0.79 bits transmitted per signal in *Gonodactylus spinulosus* and 0.63-1.03 bits transmitted per signal in *G. bredini*. *G. spinulosus* transmitted at the rate of 0.021-8.58 bits per second and *G. bredini* at 0.014-6.27 bits per second. The upper rate values are surprisingly high, extending into the estimated lower range of information transfer in human speech, which is 6-12 bits per second.

Suppose that information comes from a continuously varying signal source, such as a sweep in pitch, amplitude, or color, and that the frequency of signals within this gradient fits a normal distribution. Shannon (in Shannon and Weaver, 1949) showed the signal entropy in such cases to be

$$H = \log_2 \sqrt{2\pi e}\,\sigma$$

where e is the base of natural logarithms and σ is the standard deviation. Haldane and Spurway (1954) applied Shannon's formula to the angular scattering of newcomer honeybees around target baits following waggle dances, the dispersion of which is assumed to fit a normal distribution. In the absence of any information at all, the uncertainty with respect to direction of the target is

$$H_1 = \log_2 360°$$

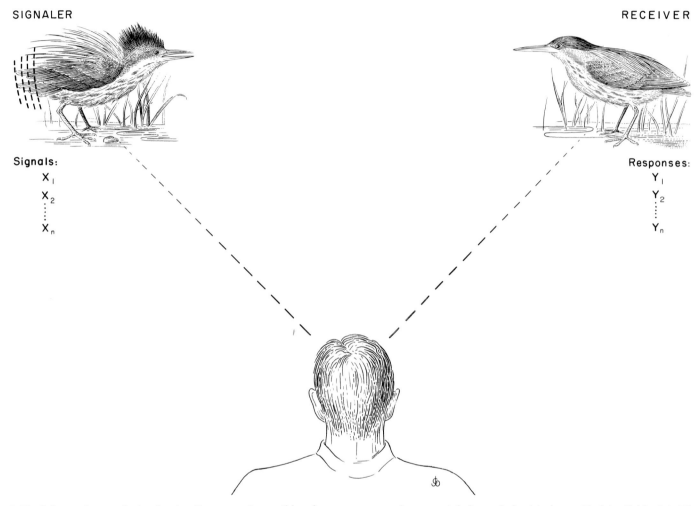

SIGNALER

RECEIVER

Signals:

X_1
X_2
$\vdots$
X_n

Responses:

Y_1
Y_2
$\vdots$
Y_n

Figure 8-15 Information analysis of a dyadic system is possible when the observer can distinguish a set of signals (X_i) emitted by one animal and a set of responses (Y_j) given by the second animal. The probabilities of the combinations of each pair of X_i and Y_j are estimated and consti-tute the essential data of the kind provided in Table 8-4. The animal species arbitrarily selected here for purposes of illustration is the green heron.

where H_1 is the number of bits required to reduce the uncertainty to an interval of one degree. If H_2 is the uncertainty remaining after the message is received by the newcomer bees, then $H_1 - H_2$ is the amount of information transmitted per waggle dance. Hence,

$$H_\theta = H_1 - H_2 = \log_2 360° - H_2$$

If it is accepted that the dispersion of the searching newcomer bees around the target has a one-dimensional normal distribution, then

$$H_\theta = \log_2 360° - \log_2 \sqrt{2\pi e}\sigma_\theta$$

$$= \frac{\log_{10} \dfrac{360°}{\sigma_\theta}}{\log_{10} 2} - 2.0471$$

Wilson (1962a) adapted the method to distance communication as well and used it to make estimates from the fire ant data and the data of von Frisch and Jander (1957) on honeybees. The essential results are shown in Figure 8-17. Notice that the two systems transmit

Table 8-4 Computation of signal entropy, receiver entropy, equivocation, and ambiguity in an imaginary communication system. (Modified from Quastler, 1958.)

				Response				$p(i)$	$-p(i)\log_2 p(i)$
	$Y_j = Y_1$	Y_2	Y_3	Y_4	Y_5	Y_6			
$X_i = X_1$	—	.001	—	—	—	—	.001	.01
X_2	.001	.007	.006	.001	—	—	.015	.09
X_3	.005	.022	.060	.027	.005	—	.119	.37
X_4	.004	.042	.156	.152	.039	.001	.394	.53
X_5	—	.009	.075	.175	.095	.010	.364	.53
X_6	—	.001	.011	.035	.039	.010	.096	.32
X_7	—	—	—	.003	.006	.002	.011	.07
$p(j)$	.010	.082	.308	.393	.184	.023	1.000	$\sum = 1.92$
$-p(j)\log_2 p(j)$	.07	.30	.52	.53	.45	.13	$\sum = 2.00$	

Action (signal)

$$H(X) = -\sum_i p(i)\log_2 p(i) = 1.92 \text{ bits} \quad \text{Source entropy}$$

$$H(Y) = -\sum_j p(j)\log_2 p(j) = 2.00 \text{ bits} \quad \text{Receiver entropy}$$

$$H_Y(X) = -\sum_j p(j) \cdot H_j(i) = 1.70 \text{ bits} \quad \text{Equivocation (of source)}$$

$$H_X(Y) = -\sum_i p(i) \cdot H_i(j) = 1.78 \text{ bits} \quad \text{Ambiguity (of receiver)}$$

$$T(X, Y) = H(X) - H_Y(X) \quad \text{Information transmitted}$$
$$= H(Y) - H_X(Y)$$
$$= 0.22 \text{ bits}$$

$$H_j(i) = -\sum_i p_j(i)\log_2 p_j(i)$$

$$H_i(j) = -\sum_j p_i(j)\log_2 p_i(j)$$

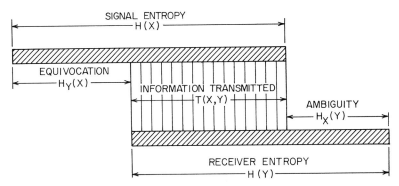

Figure 8-16 The relations between information functions represented graphically. (Redrawn from Quastler, 1958.)

roughly comparable amounts of information with reference to both direction and distance. The amount of directional information in the ant odor trail, however, increases with the length of the trail. This is because the width of the active space remains constant, and the follower ants stay about as close to the true path of the trail layer all along the length of the trail. Consequently the angular deviation away from the true path, with reference to the nest, decreases as the trail lengthens away from the nest. Thus the directional errors committed by the followers decrease, and the amount of directional information in the trail itself increases as the trail lengthens.

The estimates of distance information in the waggle dance contain a relatively small error resulting from a mistake in the original von Frisch–Jander statistics. R. Boch (personal communication) has

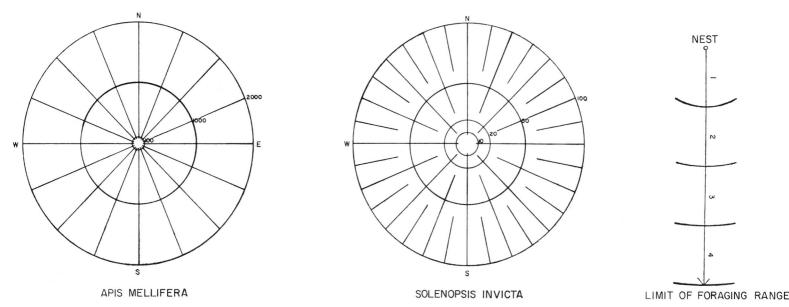

APIS MELLIFERA   SOLENOPSIS INVICTA   LIMIT OF FORAGING RANGE

Figure 8-17 Information analysis of communication in social insects. These abstract figures represent the amounts of information transmitted by the honeybee (race *carnica*) around the time it performs the waggle dance and by the fire ant when it lays an odor trail. Left: The "bee compass" indicates that the worker honeybee receives up to four bits of information with respect to direction, or the equivalent of acquiring information necessary to allow it to pinpoint a target within one of 16 equiprobable angular sectors. The compass lines are represented arbitrarily as bisecting the sectors. The amount of direction information remains independent of distance, given here in meters. This last estimate is probably subject to revision, as explained in the text. Center: The "fire ant compass" shows approximately how direction information increases with distance, given here in millimeters. Right: The "distance scale" of both bee and fire ant communication shows that approximately two bits are transmitted, providing sufficient information for the worker to pinpoint a target within one of four equal concentric divisions between the nest and the maximum distance over which a single message can apply. (From Wilson, 1962a.)

pointed out that, with the exception of one experiment conducted in 1949, von Frisch and his associates always captured the bees landing on the target dishes without counting them. In most cases, therefore, the performances of the "errorless" bees were not entered into the calculations of the standard deviations, and my 1962 estimates of H are too low. A reasonable adjustment can be made. From the 1949 data, together with comparisons made with the proportion of erring versus nonerring bees in direction experiments, it seems very unlikely that the errorless bees were more than three times as numerous as the bees in error. A likely upper limit for transmitted distance information is therefore 2.3-4.3 bits, with 3 bits (instead of 2 bits as shown) being a "typical" intermediate value. Only new data can establish the true value with confidence. Also, there is at present no way to assess the contribution of Nasanov gland substances and other pheromones to the transmitted information in the honeybee. But even with these adjustments it can still be said that the information combined from the waggle dance and odor is comparable to that transmitted by the fire ant odor trail.

The information analysis of invertebrate systems seems a priori to be practicable, especially when we can separate distinctive categories such as aggressive exchange between pairs and recruitment. Vertebrate behavior, however, offers new levels of difficulty. Altmann (1965a) has nevertheless made a heroic attempt to perform a total analysis of the rhesus monkey, one of the most complex of all animals. After defining a repertory of 120 behavior patterns, some communicative in nature and some not, he estimated the source entropy of the repertory to be 6.9 bits per act. The constraint on this behavior due to immediately antecedent social interaction, and hence the information received from communication with the monkey troop, was approximately 1.9 bits per act. This latter statistic is undoubtedly a lower bound, since some additional constraint on behavior could be demonstrated as far back as two social interactions. Although Altmann's effort fell short of its original goal of completely specifying the rhesus social system, it did produce an unusually thorough catalog of rhesus behavior and the first clear picture of transition probabilities in behavioral sequences of an advanced nonhuman primate.

The Pitfalls of Information Analysis

Now, having established the plausibility of measuring communication, we must consider a series of technical and conceptual difficulties that in many instances make the technique inapplicable. Perhaps the most serious difficulty is simply how to recognize all of the signals and to perceive when other organisms respond. The long-enduring status signals of monkeys and wolves, for example, are perfectly apparent to human observers. But just when do the animals respond to them? Do they continuously modify their behavior, or do they only take note of the signals when other relevant messages are added, such as aggression, appeasement, or sexual advertisement? Other kinds of signals have a primer effect, rather than a releaser effect. That is, they act through the neuroendocrine system to modify the physiological state of the receiver animal (Wilson and Bossert, 1963). The animal affected is thus "primed" for a new behavioral repertory, which will be evoked by new kinds of messages in the future. An example of a primer signal is the bowing of a male ring dove in the initial stages of courtship, which activates centers in the brain of the female dove that in turn induce secretion of pituitary gonadotropins. The gonadotropins induce growth of the ovaries and the release of estrogen, which readies the female for sexual and nest-building behavior (Lehrman, 1965). How can this complex chain of physiological events, which extends over two days, be quantified in bits? Primer pheromones create a double problem, being not only prolonged in effect but also unusually cryptic and elusive for the human observer. Urinary products of mice alter the reproductive physiology of female mice in various ways, causing such ultimate effects as synchrony of estrus, pseudopregnancy, and abortion. The queen substances of honeybees, wasps, and ants repress ovarian development in workers, while those of termites inhibit development of nymphs into queens. Because the action of these primer substances seldom includes overt behavioral responses, information transfer must be measured in terms of physiological change and altered behavioral potential.

Part of the solution of decoding cryptic messages may come from the development of new methods of monitoring nervous and other physiological activity directly, rather than continued dependence on inference from overt behavior. Otto von Frisch (1966a,b) used electrocardiograms to follow the heart activity of stressed animals. In blackbirds (*Turdus merula*) and lapwings (*Vanellus vanellus*), he found that young birds decreased the frequency of their heartbeats when they heard the alarm calls of their parents. Adults increased their heartbeat rates upon seeing a man or dog but failed to react when they viewed a chipmunk or heard the chipmunk's warning call. Young hares (*Lepus europaeus*) cowered and their heartbeats fell to as low as 50 percent of the normal rate when they spotted predators. Electrocardiograms have also been used to study the emotional re-

sponses of guinea pigs under social stress by Fara and Catlett (1971). Dietrich von Holst (1969) developed another, more visual technique to measure "emotional" response in tree shrews (*Tupaia glis belangeri*). Hairs on the tails of these animals are raised by musculi arrectores pilorum, which are under the control of the sympathetic nervous system. After a tree shrew has habituated to the environment of its cage, tail ruffling is elicited almost exclusively by social stimuli. In von Holst's experiments, particular forms of social interaction, including dominance, sexual activity, and parent-young relations, were each associated with a predictable degree and duration of tail ruffling.

Even with such refinements in estimating physiological change, serious technical problems can be expected to persist. Some animals, especially primates and other behaviorally advanced mammals, may be more sophisticated in communication then even the most careful observers have appreciated. Wolfgang Köhler (1927) provided a striking example when he characterized the subtlety of chimpanzee communication:

Chimpanzees understand "between themselves," not only the expression of *subjective moods* and emotional states, but also of definitive desires and urges, whether directed towards another of the same species, or towards other creatures or objects. I have mentioned the manner in which some of them used the "language of the eyes" when in a state of sexual excitement. A considerable proportion of all desires is naturally expressed by slight initiation of the actions which are desired. Thus, one chimpanzee, who wishes to be accompanied by another, gives the latter a nudge, or pulls his hand, looking at him and making the movements of "walking" in the direction desired . . . The summoning of another animal from a considerable distance is often accompanied by a beckoning very human in character. The chimpanzee also has a way of "beckoning with the foot," by thrusting it forwards a little sideways, and scratching it on the ground.

Menzel (1971) has shown how chimpanzees use these and other postures and gestures to lead troop mates to food.

The problem is exacerbated when groups of animals display simultaneously. To what extent, we need to ask, is the behavior of an arbitrarily selected group member being influenced (in bits per second) by its nearest neighbor, as opposed to its second, third, or even seventh nearest neighbor?

If the observer goes so far as to limit information analysis to dyads and overt signals and responses, he will encounter still another residue of problems. Graded signals, for example, are hard to dissect into messages. In his study of the red colobus monkey (*Colobus badius*) Marler (1970) found that "the whole vocabulary recorded seems to comprise one continuously graded system," one that is singularly difficult to separate into messages that can be rendered meaningful by human standards. Another common difficulty is that the signal entropies of messages sometimes change with experience, as when one animal becomes dominant over another (Dingle, 1972b), with age, and with context. The male of the black-throated green warbler (*Dendroica virens*) uses two song types. Type *A* is given most frequently in the presence of conspecific males. Song *B*, which functions more in general advertisement, is automatically emitted by sufficiently primed males even in the absence of rival males. The relative frequencies of the two songs shift along a continuum according to the circumstances in which the male finds himself: the proportion of *A* tends to increase at the expense of *B* even in the absence of rivals when the male is near the edge of his territory and during the first and last bouts of singing during the day. Other males can presumably "read" a great deal from the proportions of *A* and *B* , but the system defies conventional information analysis (Lein, 1972).

In summary, an array of technical difficulties, some hidden in nature, create pitfalls for most attempts at information analysis and make it unlikely that entire behavioral repertories can be analyzed in this way. Even so, such quantification remains quite feasible and desirable in a few exceptional cases. The comparison of recruitment efficiencies of two radically different systems can be accomplished, as it has been for the honeybee waggle dance and the fire ant odor trail. Furthermore, the complexity and rate of information transfer in narrow, easily perceived categories such as aggressive interactions can be measured. One also needs to have a reason for acquiring these estimates other than the mere collapse of data into numbers of bits for purposes of simplification.

Redundancy

If a zoologist were required to select just one word that characterizes animal communication systems, he might well settle on "redundancy." Animal displays as they really occur in nature tend to be very repetitious, in extreme cases approaching the point of what seems like inanity to the human observer. The trait is most exaggerated in sexual displays and territorial advertisement. The courtship of the goldeneye duck *Bucephala clangula* is a typical example: head-throw, bowsprit, head-up, nodding, head-throw-kick, masthead, ticking—the displays follow one another in a jumble of unpredictable combinations, the duck repeating them, pausing a while, and repeating again, hour upon hour for days on end (Dane et al., 1959).

Why must animals be so tedious? One could reasonably expect them to use one or two displays adequate to the purpose and then simply cease. Nevertheless, it is possible to conceive of several circumstances in which redundancy is at a premium. Suppose that the signals are graded and the relationship between individuals finely calibrated and varying through time. This will be the case when two adversaries are jockeying for position, for example, or when a male

and a female are establishing a pair bond. Under these conditions, signals need to be constantly repeated in order to reassess the relationship at each point of change.

Redundancy can also serve to sustain a state of arousal. The members of many animal species accomplish more during courtship than the establishment of a pair bond and psychological preparation for mating. By means of hormonal mediation they also alter each other's reproductive physiology and future behavioral repertory. Since these priming effects take relatively long periods of time, arousal must be sustained. Hartshorne (1958), Moynihan (1966), and Barlow (1968) suggested that the arousal is enhanced by the employment of multiple signals that are different in form but redundant in meaning. Two techniques can be envisaged by which such novelty is attained. One is to keep some components stereotyped, to make identification more nearly certain, while varying others in unpredictable ways. During courtship the orange chromide *Etroplus maculatus*, a cichlid fish studied by Barlow, holds the amplitude and basic form of the quivering display constant while varying such parameters as the duration of the display, the phasing of the pelvic fin movements with the head quivering, the tilt of the body, and the orientation toward the mate. The second technique, so conspicuously developed in the goldeneye, is simply to vary the sequencing of the multiple displays.

Another circumstance favoring redundancy is the existence of a substantial risk that solitary signals will be missed or misinterpreted. As a result, reinforcing signals with identical meaning are needed for insurance. Rand and Williams (1970) have noted that the potential informational content of the combined dewlap color and display movements of *Anolis* lizards far exceeds that needed to separate the ten or so species that coexist on the largest islands of the West Indies. They hypothesize that the redundancy serves to insure precision in transmission under the prevalent conditions of poor visibility, which are caused by weak light and dense foliage in the forest habitats.

Finally, redundancy in certain cases may prove on closer examination to be more apparent than real. Lein (1973) recently discovered that the five song types of male chestnut-sided warblers (*Dendroica pensylvanica*) vary along a series according to morphology and volume, from those that have strong end-phrases and carry long distances to others that lack end-phrases and are not noticeable at long distances. The relative frequency of usage of each type depends on the location of the bird within his territory and his distance from the nearest singing male. Thus the song types might constitute different messages, serving to transmit differences in the degrees of arousal and insecurity on the part of the songster. At the very least Lein's analysis should provide a warning that signals can be classified as redundant only when the contexts of their transmission are identical or at least meaningless to both participants in the communication.

Chapter 9 **Communication: Functions and Complex Systems**

The total analysis of a communication system is a relatively simple concept to create—and a forbiddingly difficult task to accomplish. The analysis falls into three parts: the identification of the *function* of the message, that is, what it means to the communicants and therefore ultimately what role it plays in altering genetic fitness; the inference of the *evolutionary* or *cultural derivation* of the message; and the full specification of the *channel*, from the neurophysiological events that initiate the signaling behavior to the processes by which the signal is emitted, conducted, received, and interpreted. Philosophers have not overlooked the fact that human thought is just a special case of communication. Some have viewed the study of communication as coextensive with logic, mathematics, and linguistics. C. S. Pierce, Charles Morris, Rudolf Carnap, and Margaret Mead, for example, have used the word *semiotic* (or semiotics) to designate the analysis of communication in the broadest sense. One of the rather few useful insights originating from these attempts at synthesis is the recognition that even in human language a word or phrase conveys only a minute fraction of the stimuli associated with the referent. "A tree," for example, alludes to a short list of properties, including certain general attributes of plants, woodiness, canopy atop trunk, and relatively large size. It does not specify details of molecular structure, principles of forest ecology, or any other of the expanding range of qualities of "treeness" that dendrologists have only begun to delineate. In short, even human language is concerned with what the ethologists designate as sign stimuli. T. A. Sebeok (1963, 1965), reflecting on zoology from the viewpoint of a linguist, recognized that animal communication deals far more explicitly than human language with signs and on that basis alone can provide useful guides for the deeper analysis of linguistics. He suggested that the study of animal communication be called "zoosemiotics," in recognition of the fact that it is compounded of two elements: first, the strongly evolutionary emphasis of ethology, which describes whole patterns of behavior under natural conditions and deduces their adaptive significance in the genetic sense; and second, the logical and analytic techniques associated with human-oriented semiotics. There have been other efforts to adapt the principles of human semiotics to the description of animal communication. Hockett and Altmann have systematically listed the design features of human speech and used them to reclassify certain phenomena in animal behavior. Other investigators, whose work was reviewed in Chapter 8, have carried the mathematical techniques of information theory, developed originally to study human communication, into the study of animal systems. Marler (1961, 1967) has adapted the objective linguistic classification systems of Morris (1946) and Cherry (1957) in an attempt to further deanthropomorphize descriptions of animal behavior.

But there is danger in forcing too early a marriage of animal behavior studies and human linguistics. Human language has unique properties facilitated by an extraordinary and still largely unexplained growth of the forebrain. The deep grammars hypothesized by Chomsky and Postal, if they really exist, are likely to be as diagnostic a trait of *Homo sapiens* as man's bipedal stride and peculiar glottolaryngeal anatomy—and consequently constitute a *de novo* adaptation that cannot be homologized. The introduction of linguistic terminology into zoology, and the reverse, should be attempted only in an exploratory and heuristic manner, with no congruence between zoological and linguistic classifications being forced all the way. It is in this spirit that I wish to take a strongly phenomenological approach to animal communication, beginning with observed facts and classifying them tentatively by induction.

The Functions of Communication

Social behavior comprises the set of phenotypes farthest removed from DNA. As such it is an evolutionarily very labile phenomenon, the one most subject to amplification in the transcription of information from the genes to the phenotype of individuals. It is also the class of phenotypes most easily altered by addition or subtraction of otherwise unrelated components. Hence when communication is viewed over all groups of organisms, the included behaviors become so eclectic in nature as to be beyond all hope of homology, and so divergent in function as to defy simple classification.

By focusing on this matter of lability and heterogeneity in social behavior we come quickly to the crux of the problem. The key idea, seldom recognized explicitly in the past, is that the art of classification is central to the study of function in social behavior. In fact, zoosemiotics presents just one more case of the two classical problems of taxonomic theory: how to define the ultimate unit to be classified (in this case, the function), and how to cluster these units into a hierarchy of categories that serve as a useful shorthand while at the same time staying reasonably close to phylogeny. In the taxonomy of organisms the basic unit is the species. Groups of species judged by largely subjective criteria to most closely resemble one another owing to common descent are grouped into genera. Similar and related genera are clustered into families, families into orders, and so on upward to the phyla and kingdoms. In creating classifications of functions, animal behaviorists employ the message as the basic unit. Although not taxonomists working with species, they more or less consciously perform the same mental operations. A set of messages labeled a "message category" is the semiotic equivalent of a genus or family, and it is no more intuitively sound than the definition of the individual messages clustered within it. No a priori formula can give the message category crisper definition or deeper meaning. For this reason the best way to consider the significance of animal

communication is to start with a simple, relatively finely defined catalog of functional categories, that is, the "species" of our semiotic classification, and then to proceed with clustering. The categories discussed below provide a relatively complete coverage of existing knowledge, but they are not as finely divided as possible. Thus sexual signals could be broken down into at least six subcategories, many of which overlap broadly, while caste inhibition signals in social insects could be more precisely matched to many of the distinguishable castes, and so forth.

Facilitation and Imitation

Both the induction of behavior by the mere presence of another member of the species and the close imitation of another's behavior patterns (see Chapter 3) can be construed as acts of communication in the broadest sense. It can be reasonably argued that the action of the model animal is not "intended" to modify the behavior of the follower animal. To use MacKay's (1972) expression, the model does not possess an evaluator of the effect of its own actions, and hence the transmission is a perception rather than a communication. This semantic distinction can be sidestepped for the moment by noting that in many instances there probably does exist some degree of appraisal. The actions of members of family groups and closely knit societies are often highly coordinated, and it is advantageous to the leaders as well as to the followers that the group act as a unit. Components of locomotion have often been modified to serve as signals for inducing locomotion in groupmates: the swing step of hamadryas baboons, for example, or ritualized wing flicking by birds in flocks. Social insects have carried facilitation to an extreme in the coordination of group activity. Wasps departing on foraging trips tend to activate nearby wasps into flight also. Ants and termites initiating soil excavation and other nest construction activities attract nestmates, who join in the labor. The mere sight of another individual in rapid motion excites and attracts workers of large-eyed ant species. This form of communication, called kinopsis, aids in the capturing and subduing of prey. The result of such facilitation is the concentration of group effort in space and time, a form of coordination which is manifestly to the advantage of both the signaler and the responder.

Monitoring

A complementary function of facilitation and imitation is the persistent observation of the activities of other animals. The presence of food or water, the intrusion of a territorial rival, and the appearance of a predator can all be "read" from the actions of neighbors. Whether monitoring is true communication even in the most liberal sense can be debated.

Contact

Social animals use signals that serve in some circumstances just to keep members of a group in touch. The habit is particularly well developed in species that move about under conditions of poor visibility. South American tapirs (*Tapirus terrestris*) use a short "sliding squeal" to stay in touch in the dense vegetation of their rain forest habitat (Hunsaker and Hahn, 1965). The lemurlike sifaka (*Propithecus verreauxi*) uses a cooing sound for the same purpose (Alison Jolly, 1966). Duetting, during which pairs of animals exchange notes in rapid succession, functions as a contact-maintaining signal. The phenomenon occurs widely in frogs, in birds, and in at least two species of primates, the tree shrew *Tupaia palawanensis* and the siamang *Symphalangus syndactylus* (Hooker and Hooker, 1969; Williams et al., 1969; Lamprecht, 1970; Lemon, 1971b). The extraordinary songs of the humpback whale (*Megaptera novaeangliae*), recently analyzed by Payne and McVay (1971), possibly serve to keep members of families or herds in touch, particularly during their long transoceanic migrations. Duetting of birds and the whale songs will be discussed in fuller detail later in this chapter.

Individual and Class Recognition

The capacity to recognize different castes is widespread in the social insects. Nest queens are treated in a preferential way, and workers easily distinguish them from virgin members of the same caste. Within the largest colonies these fecund individuals are as a rule heavily attended by nurse workers, who constantly lick their bodies and offer them regurgitated food and trophic eggs. Among honeybees at least, three unique pheromones have been implicated in this special treatment: *trans*-9-keto-2-decenoic acid and *trans*-9-hydroxy-2-decenoic acid from the mandibular glands and an unidentified volatile attractant from Koschevnikov's gland, located at the base of the sting. Workers of stingless bees (Meliponini) engaged in brood cell construction give way at the approach of the nest queen and permit her to eat the regurgitated nectar and pollen placed in the cell, as well as any of their own eggs laid on top. The several castes of the termite genus *Kalotermes* manufacture widely differing quantities of the principal volatile attractant 2-hexenol, and hence vary in their capacity to serve as foci in clustering. Within colonies of the social Hymenoptera, males are usually discriminated against as a group. They are offered less food by the workers (all of whom are female), and in times of starvation they are frequently driven from the nest or killed.

In addition to these instances of caste identification by workers, there exists evidence of an even finer ability to detect differences among life stages. In the relatively primitive myrmicine ant genus *Myrmica*, workers are evidently not capable of distinguishing the tiny first instar larvae from eggs, so that when eggs hatch the larvae are left for a time in the midst of the egg pile. As soon as they molt and enter the second instar, however, the larvae are removed by the workers and placed in a separate pile (Weir, 1959). The tendency to segregate eggs, larvae, and pupae into separate piles is a nearly universal trait in ants. An identification substance can be extracted from larvae of the fire ant *Solenopsis invicta* and transferred to previously inert dummies, causing workers to carry the dummies to the larval piles (Glancey et al., 1970). Furthermore, workers of most ant species are able to distinguish larvae of two or more size classes (LeMasne, 1953). The same capacity is possessed to an extreme degree by the primitively social allodapine bees (Sakagami, 1960). In *Monomorium pharaonis* the workers are further able to tell male eggs from female eggs (Peacock et al., 1954).

In most species of social insects, caste and life stage identification appears to be accomplished by antennal contact. This fact in itself suggests chemoreception, although Brian (1968) has speculated that several age classes of *Myrmica* larvae might also be distinguished by certain differences in hairiness that are quite apparent to the human observer under low magnification. Two cases are known in which the communication appears to be by means of odors transmitted over a distance. When workers of *Pogonomyrmex badius* lay trophic eggs (specialized eggs used only for feeding other individuals), they search for hungry larvae while sweeping their antennae through the air. When they come within about a centimeter of the head of the larva they move directly to it and unerringly place the egg onto its mouthparts. Free (1969) has demonstrated that the smell of the larvae alone causes the honeybee workers to forage for pollen. The effect is enhanced if the workers are given direct access to the larvae.

The ability to distinguish infants, juveniles, and adults is a universal trait of the vertebrates. Several sensory modalities are routinely employed, including particularly sound, vision, and smell. Often the response is quite specialized and insectlike in its stereotyped quality. The cichlid fish *Haplochromis bimaculatus* distinguishes larvae from fry by odor alone. The characteristic responses of the adults can be obtained by placing them either in "fry water" or "larva water" from which the immature stages have previously been removed. Parents of altricial birds recognize the nestlings at least in part by the distinctive appearance of their gaping maws. In a few species, such as the estrildid finches, the effect is enhanced by a strikingly colored mouth lining, which may be further embellished by special paired markings (Nicolai, 1964; Eibl-Eibesfeldt, 1970). However, contexual stimuli are frequently required. Young robins, for example, must be within the nest perimeter to be recognized by their parents. Those placed only a few centimeters outside are in danger of starving.

Among species of the higher vertebrates it is commonplace for individuals to be able to distinguish one another by the particular way they deliver signals. Indigo buntings, American robins, and certain other songbirds learn to discriminate the territorial calls of their neighbors from those of strangers that occupy territories farther away. When a recording of a song of a neighbor is played near them, they show no unusual reaction, but a recording of a stranger's song elicits an agitated aggressive response. This is the dear enemy phenomenon, to be described in greater detail in a later discussion of territoriality (Chapter 12). Analyses by Falls (1969), Thielcke and Thielcke (1970), and Emlen (1972) have revealed the particular components of songs, such as absolute frequency (in the white-throated sparrow) and detailed phrase morphology (in the indigo bunting), that vary from individual to individual and are evidently used by the birds to make identifications.

Families of seabirds depend upon a similar personalization of sig-

nals to keep together as a unit in the dense, clamorous breeding colonies. A sleeping herring gull (*Larus argentatus*) is awakened by the call of its mate but is undisturbed by similar calls from other gulls around its nest (Tinbergen, 1953). Occasionally, gannets (*Sula bassana*) are seen to turn in the direction of their mates before these birds fly into view. It is possible that the landing calls, which White and White (1970) found to vary markedly from one gannet to another, serve as the identification cues. The young of the common murre (*Uria aalge*), a large auk, learn to react selectively to the call of their parents in the first few days of their lives, and the parents also quickly learn to distinguish their own young (Tschanz, 1968). Adults of royal terns (*Sterna maxima*) recognize their own chicks by their calls and occasionally by their visual appearance alone (see Figure 9-1). Still more remarkable, they recognize their own eggs when these are removed by the experimenter and placed in an adjacent nest (Buckley and Buckley, 1972).

Figure 9-1 Royal tern adults and chicks mingle in a loose crèche. Under such circumstances the adults can readily pick out their own offspring through individual traits in the voice and appearance of the young birds. In the center a parent bird shields its chick with an outspread wing. (From Buckley and Buckley, 1972.)

We can expect to find personalized elements in pair bond and contact signals in some species. Mated pairs of the boubou shrike *Laniarius aethiopicus* learn to sing duets with each other. In so doing they work out combinations of phrases that are sufficiently individual to enable them to recognize each other even though both remain hidden much of the time in dense vegetation (Thorpe and North, 1965, 1966).

Mammals are at least equally adept at discriminating among individuals of their own kind. A wide variety of cues are employed by different species to distinguish mates and offspring from outsiders. The faces of gorillas, chimpanzees, and red-tailed monkeys (*Cercopithecus nictitans*) are so variable that human observers can tell individuals apart at a glance. It is plausible that the equally visual nonhuman primates can do as well (Marler, 1965; van Lawick-Goodall, 1971). Some mammal species use secretions to impart a personal odor signature to their environment or to other members in the social group. As all dog owners know, their pet urinates at regular locations within its territory at a rate that seems to exceed physiological needs. What is less well appreciated is the communicative function this compulsive behavior serves: a scent included in the urine identifies the animal and announces its presence to potential intruders of the same species. Scent marking probably serves as a repulsion device in the ancestral wolf to keep the pack territory free of intruders. As Heimburger (1959) has shown, this behavior is widespread, if not universal, among other species of Canidae. Tigers and domestic cats establish scent posts and partial territories in approximately the same fashion. Brown lemurs (*Lemur fulvus*), which are among the most olfactory of the primates, can recognize individuals of the same species on the basis of the perineal gland scent (Harrington, 1971). Rodents utilize scent heavily in their social interactions; gerbils, for example, can discriminate individuals on the basis of urinary odors even when the urine is diluted a thousandfold (Dagg and Windsor, 1971).

Males of the sugar glider (*Petaurus papuanus*), a New Guinea marsupial with a striking but superficial resemblance to the flying squirrel, go even further. They are able to discriminate odors at the specific, group, and individual levels. The male marks his mate with a secretion from a gland on the front of his head. He uses other secretions, originating on his feet, on his chest and near his arms, together with his saliva, to mark his territory. In both instances the odors are specific enough for the male to distinguish them from those of other sugar gliders (Schultze-Westrum, 1965; see Table 9-1). In a closely parallel manner, males of the European rabbit *Oryctolagus cuniculus* use anal gland secretions to mark the territory of the warrens they dominate. The secretions, which are mixed with urine, are specific in odor to individuals. The anal glands, along with the submandibular glands that serve a similar function, are developed in correlation with the animal's rank, so that only the leading males

Table 9-1 Sources of secretions used by male phalangers (*Petaurus breviceps*) to mark territories and partners. (From Schultze-Westrum, 1965.)

Source of odor	Anal	Oral	Pedal	Flank	Sternal	Partner marking
Primary odors from						
Frontal gland						+
Sternal gland					+	
Glands of anal regions	+		+			
Plantar glands			+			
Secondary odor mixtures from						
Saliva		+				
Fur				+		+

Table 9-1 header spanning columns Anal, Oral, Pedal, Flank, Sternal: "Territory marking"

are able to impart their own odor signature (Mykytowycz, 1965, 1968).

Mammalian scent appears to acquire its individuality through subtle variations in the blend of complex mixtures. The tarsal scent of the black-tailed deer (*Odocoileus hemionus columbianus*), for example, contains dozens of components that vary in proportion from animal to animal. The deer sniff and lick each other's tarsal organs and are capable of discriminating individuals on this chemoreceptive basis alone. Dietland Müller-Schwarze, R. M. Silverstein, and their coworkers have found that at least four substances elicit responses qualitatively identical to that produced by the total tarsal scent. They are unsaturated lactones with about 12 carbon atoms; the principal component has been identified as *cis*-4-hydroxydodec-6-enoic acid lactone (Müller-Schwarze, 1969). The use of personal odor signals in recognition is widespread in still other groups of mammals, having been documented in groups as different as mice (Hahn and Tumolo, 1971) and lions (Schaller, 1972), but the chemical nature of the identification pheromones remains largely unexplored.

Individual recognition has also been discovered in two nonsocial invertebrate species whose adults form long-lasting sexual pair bonds: the starfish-eating shrimp *Hymenocera picta* (Wickler and Seibt, 1970) and the desert sowbug *Hemilepistus reaumuri* (Linsenmair and Linsenmair, 1971). The sowbugs also distinguish their brood from the broods of other parents, utilizing individuals' secretions that are exchanged back and forth between the young animals (Linsenmair, 1972). The employment of body surface odors for the recognition of fellow members of a colony is a nearly universal trait in the fully social arthropods, which is to say the social insects. In most but not all species, workers instantly recognize aliens and drive them from the nest or kill them. Nixon and Ribbands (1952) showed experimentally that recognition scents in the honeybee are derived at least in part from diet, while Lange (1960) proved that both the diet of the workers and the chemical nature of the nest wall can contribute to the colony odor of the ant *Formica polyctena*. In addition to such extrinsic elements of the epicuticular odor, there are genetically determined components that allow workers to discriminate members of alien species and probably, to some extent, alien insects as well. Some evidence has been presented to suggest that the colony odor of carpenter ants (*Camponotus*) at least partially originates in the queen (Hölldobler, 1962). The literature on colony odors, which is extensive but still embryonic, has recently been reviewed by Wilson (1971a). Of crucial importance, nothing is known about the chemistry of the odors. Until such information is forthcoming, it will be futile to speculate on the secretory origins of the odors, their transmission, or the relative contributions to colony discrimination of the genetic and phenotypic variances within and among species.

Purists may argue that identification of an individual does not constitute true communication. Nevertheless, all the rest of the social repertory is dependent on the constant input of such information. Slight alterations of this input cause a prompt change in the interactions of group members. If experimenters remove an ant larva from the brood pile and place it in an adjacent, less suitable chamber, the workers promptly pick it up and carry it back. If experimenters wash its cuticle lightly with a solvent to disturb the larva's odor, it is killed and eaten instead. When experimenters give a goat kid to a mother not imprinted on its personal odor, it is driven away and allowed to starve. In each of these cases other behavioral patterns are activated, but they are dependent in their timing and orientation on the constant reception of identification signals.

Status Signaling

Various peculiarities in appearance and signaling, often metacommunicative in nature, serve to identify the rank of individuals within dominance hierarchies. This subject was discussed in another context in Chapter 8.

Begging and Offering of Food

Elaborate systems of begging and feeding have evolved repeatedly in the birds and mammals. Nestling birds recognize returning parents by landing calls, the sight of the movement of the parents' bodies above the nest rim, the jarring of the nest as the adults alight, or combinations of such signals. They then respond by gaping. The visual releasers in the maw of the young bird induce the parent to drop pieces of food into it or to regurgitate to it. Other, more specific signals may accompany the exchanges. The conspicuous red dot on the lower beak of the adult herring gull guides the young to the exact spot of the parent's anatomy where they are most likely to receive regurgitated food (Tinbergen, 1951). As they grow older and more agile, the offspring commonly use conspicuous wing movements while begging. The bald ibis (*Geronticus eremita*) and Australian wood swallows (*Artamus*) spread their wings and wave them slowly, while songbirds quiver the wings (Immelmann, 1966; Wickler, 1972a). Among precocial birds, begging and feeding are absent or else replaced by a special form of feeding enticement. When the hen of a domestic fowl discovers food she lures her chicks to her side by clucking. She may also peck conspicuously at the ground, pick up bits of food and let them fall to the ground again (Wickler, 1972a).

Mammals that feed their young primarily through lactation use relatively simple begging and feeding signals. Among deer, antelopes, and related ungulates, mothers that bear single young or twins stand in an open stiff-legged pose and let their young approach them from beneath for suckling. Those giving birth to multiple young, such as the pigs and their relatives (Suidae), lie on their side (Fraser, 1968). In both kinds of ungulates the young are strongly precocial; in the extreme case of the wildebeest and pigs, they are able to walk and follow the mother within an hour after birth. During infrequent visits to feed her young, the mother tree shrew (*Tupaia glis belangeri*) simply straddles them in a stiff-legged pose (Martin, 1968). The parents of some mammal species use special techniques to shift their offspring to more adult forms of food. As the young of squirrels, rats, and other rodents grow older, they learn to take food directly from the mouth of the mother when she brings it to the nest area, thus gaining experience about the preferred items they will encounter when they start to forage on their own. Meerkats (*Suricata suricatta*), relatives of the mongoose, have added feeding incitement to this procedure. When the mother brings food home she first offers it to her young while holding it in her mouth. If they do not respond adequately, she leaps around in front of them until they take the food directly from her (Ewer, 1963). Jackals, African wild dogs, and wolves regurgitate to their young like birds, and the young have evolved an appropriate form of communication to initiate the behavior: they vigorously nuzzle the lips of the adults in an attempt to induce the regurgitation, sometimes forcing their heads inside the open jaws to take food directly from the parent's mouth (Mech, 1970; H. and J. van Lawick-Goodall, 1971).

An aberrant form of food exchange is employed by the koala (*Phascolarctos cinereus*), a specialized Australian marsupial that feeds exclusively on eucalyptus leaves. For about a month prior to the time that the infant koala begins to eat leaves on its own, the mother supplements its milk diet with a special form of feces which is unlike her normal droppings. This material, which consists of a soft paste of half-digested leaves, is licked directly from the mother's anus by the young koala (Minchin, 1937). The behavior is remarkably similar to anal trophallaxis of termites, to be described shortly, and it may serve the same purpose of transferring symbiotic digestive microorganisms from one individual and one generation to the next.

Begging and food exchange among adults, as opposed to exchange between adults and young, is rare in vertebrates. African wild dogs just returning from a successful hunt regurgitate to others who remained behind (Kühme, 1965a,b). Adult macaques, baboons, gibbons, and chimpanzees occasionally beg, either by gentle attempts to take food from others in accompaniment with conciliatory postures or by presenting their hands with palms up. Among baboons and chimpanzees, begging and sharing are most prominently displayed on those rare occasions when one of the animals captures a small antelope, monkey, or some other prey and has control of fresh meat.

The exchange of food reaches its extreme development in social insects, and in fact it is fundamental to the organization of their colonies. When the food is in liquid form, delivered by regurgitation from the crop or as secretions from special glands associated with the alimentary tract, the exchange is referred to as *trophallaxis*. Trophallaxis is very widespread but not universal among the higher social insects. It occurs generally through the eusocial wasps, including the socially rather primitive *Polistes*. It appears in a highly irregular pattern among the bees, reflecting both the phylogenetic position of the species concerned and the constraints placed on it by the food habits and nest forms of these insects. In the bumblebees, a primitively social group, the workers simply place pollen on the eggs or larvae, and very little direct contact occurs between adults and larvae. Furthermore, the exchange of liquid food is extremely rare (Free, 1955b). The halictine bees seal their brood cells, but the females of at least some of the lower social species open the cells to add provisions at frequent intervals, while those of the higher social species keep the cells open all the time and attend the larvae regularly (Batra, 1964; Plateaux-Quénu, 1972). Even so, there is as yet no evidence that the adults regurgitate to the larvae or even to one another, and deliberate efforts to induce such exchange in laboratory colonies of *Lasioglossum* (incorporating *Dialictus* and *Evylaeus*) have failed (Sakagami and Hayashida, 1968; Michener et al., 1971). Sealed brood cells prevent the adults of stingless bees (Meliponini) from feeding the larvae, but regurgitation among adults is a common event (Sakagami and Oniki, 1963). Although the brood cells of honeybee colonies are kept open and workers provision them continuously, they do not feed the larvae by direct regurgitation onto the mouthparts. Adult honeybees, by contrast, regurgitate to one another at a very high rate. Workers regurgitate water, nectar, and honey to one another out of their crops, but larvae and queens receive most of their protein from royal jelly or brood food secreted by the hypopharyngeal glands (Free, 1961b). Allodapine bees regurgitate to their larvae, which are kept

exposed in the central nest cavity, but not to one another (Sakagami, 1960).

Trophallaxis in ants also reflects phylogeny. The workers of all species of the myrmecioid complex so far studied engage in liquid food exchange. In the primitive bulldog ants, comprising the subfamily Myrmeciinae, the habit is either rare or else frequent but poorly executed (Freeland, 1958). In the higher myrmecioid subfamilies (Aneuretinae, Dolichoderinae, Formicinae) the exchange is frequent, and in the last two subfamilies it is prevalent enough to result in a fairly even distribution of liquid food throughout the worker force of the colony. Among the major groups of the poneroid complex, trophallaxis is much more variable and shows fewer extremes of development than among the myrmecioids. It is apparently absent altogether in *Amblyopone*, one of the most primitive living ponerines, but it has been noted to occur to a limited extent in other ponerines wherever a special search has been made for it (Haskins and Whelden, 1954). Some ant species, for example certain species of *Myrmecia*, *Pogonomyrmex*, *Leptothorax*, *Dolichoderus*, *Iridomyrmex*, and *Formica*, supplement trophallaxis with the laying and donation of special alimentary eggs (trophic eggs); and at least one species of *Pogonomyrmex* (*P. badius*) has supplanted trophallaxis entirely with this peculiar form of food exchange.

Trophallaxis is also general in the termites (Alibert, 1968; Noirot, 1969a). In all the lower termite species examined to date, belonging to the families Kalotermitidae and Rhinotermitidae, the members of the colony feed one another with both "stomodeal food," which originates in the salivary glands and crop, and "proctodeal food," which originates in the hindgut. The stomodeal material is a principal source of nutriment for the royal pair and the larvae. It is a clear liquid, apparently mostly secretory in origin but with an occasional admixture of woody fragments. The proctodeal material is emitted from the anus. It is quite different from ordinary feces since it contains symbiotic flagellates that feces completely lack, and it has a more watery consistency. Evidently the principal function of proctodeal trophallaxis is the donation of flagellates to nestmates that lose them while molting. Termites have the typically insectan trait of shedding the chitinous linings of both their foreguts and hindguts with each molt. The lining of both parts of the intestine are eliminated through the anus one or two days after the molt, and they carry with them the vital symbiotic protozoans from the hindgut. The newly molted termite, now deprived of its only means of digesting cellulose, must obtain a new protozoan fauna from its nestmates. The proctodeal fluid elicited in anal trophallaxis almost certainly serves as a secondary source of nutriment as well, but its importance in this regard has not been analyzed. The higher termites (Termitidae) do not depend on symbiotic flagellates for digestion of cellulose, and they have also lost the habit of anal trophallaxis. At the same time, the immature stages have become completely dependent on stomodeal liquid. Unlike the larvae of the lower termites, termitid larvae are morphologically very distinct from older individuals, which undergo radical transformation either in the second or third molt. Until this occurs they are entirely white, with soft exoskeletons and nonfunctional mandibles. Noirot has suggested that the liquid fed to them consists of pure saliva. The older nymphs of the Termitidae also receive stomodeal liquid, but are able to feed on woody material and fragments of fungi as well.

Complex forms of food exchange are also practiced by some of the presocial arthropods. Female burrowing crickets (*Anurogryllus muticus*) feed their nymphs with trophic eggs (West and Alexander, 1963). The female of burying beetles (*Necrophorus*) interacts with her larvae in very much the same way a mother bird interacts with her nestlings. As she approaches them, they lift the forepart of their bodies into the air and make grasping motions with their legs in what appear to be begging movements. The female then opens her mandibles and regurgitates liquid to each larva in turn (Pukowski, 1933). Even more surprising, the females of a few species of spiders, members of the family Theridiidae, regularly regurgitate to their young (Kaston, 1965; Kullmann, 1968).

Trophallaxis, insofar as it has been analyzed, is regulated by combinations of chemical and tactile signals. In general, potential donors recognize and approach potential recipients primarily by chemical cues, perhaps abetted by touch; but begging is achieved by specialized tactile signals. Highly motivated ant donors approach nestmates head-on, opening their mandibles wide and regurgitating droplets of liquid as offerings. In contrast, begging consists in good part of rapid but light drumming of the antennae or forelegs on the potential donor's labium, the hinged mouthpart located just beneath the oral opening. This action causes a reflexive regurgitation of the crop contents (Hölldobler, 1970). In a like manner, a termite initiates anal trophallaxis by caressing the terminal abdominal segments of another individual with its antennae and mouthparts, causing the extrusion of a proctodeal droplet.

Free (1956) used a series of ingenious experiments to analyze the releasers of trophallactic behavior in the honeybee. More soliciting and offering is directed at the head than at any other part of the body, and a freshly severed head is sufficient to elicit either reaction. Free noted that heads belonging to nestmates were favored over those belonging to aliens. So important is odor in fact that he even obtained occasional responses with small balls of cotton that had been rubbed against bees' heads. The antennae are also potent stimuli. Heads lacking antennae are less effective than those that possess them, and the loss can be restored by inserting imitation wire antennae of the right length and diameter into heads lacking antennae. Apparently the antennae serve not only as releasers but also as guides for the bees when they touch one another with their lower mouthparts.

Montagner's study (1966, 1967) of social wasps belonging to the genus *Vespula* indicates that trophallactic communication is both subtle and prolonged. When Montagner repeated Free's experiments, using the wasps instead of honeybees, the results were mostly negative. It is clear that the odor of severed worker heads attracts other workers, who seem prepared to engage in food exchange, but the inert head is not sufficient by itself to induce begging or offering. Artificial wire antennae fixed to severed heads and vibrated at 20 to 100 cycles per second induced some regurgitation, but the live workers broke off contact in seven seconds or less. Montagner has shown that trophallaxis is sustained only when the pair engage in continuous reciprocal antennal signaling according to the specific pattern illustrated in Figure 9-2.

Trophallaxis in social insects is a complex subject with ramifications extending into the physiology of caste determination, dominance behavior, division of labor, the spread of pheromones, and many other aspects of colony organization. The large literature dealing with it has recently been summarized and interpreted by Wilson (1971a).

Grooming and Grooming Invitation

Grooming is an eclectic set of behaviors evolved in various combinations by many different phylogenetic lines of animals. Although the behaviors superficially resemble one another, they differ in many mechanical details and serve a diversity of functions. Therefore the clumping of all kinds of grooming into a single functional category is frankly an artifice taken for convenience and partly as a concession to our imperfect knowledge of the adaptive significance of most of its individual variants.

One generalization can nevertheless be drawn at this time about the meaning of grooming in both the vertebrates and social insects. Vertebrates use allogrooming (the grooming of other individuals) to some extent as a cooperative hygienic device, and this is likely to be its primitive function. However, allogrooming is one of the most easily ritualized of all social behaviors, and it has been repeatedly and consistently transformed into conciliatory and bonding signals. Often these social functions completely overshadow the hygienic function, which in extreme cases may be entirely absent. In social insects allogrooming is still largely a mysterious process. It could be basically hygienic, although direct evidence on this point is lacking. In some cases it distributes pheromones and may also serve to spread and imprint the colony odor. Therefore, in social insects as in vertebrates, allogrooming appears to have evolved at least to some extent into a group bonding device.

Allogrooming in birds, more precisely referred to as allopreening, is preeminently if not exclusively devoted to communication (Sparks, 1965, 1969; Harrison, 1965). The behavior has a scattered phylogenetic distribution within this group and occurs in only a minority of the

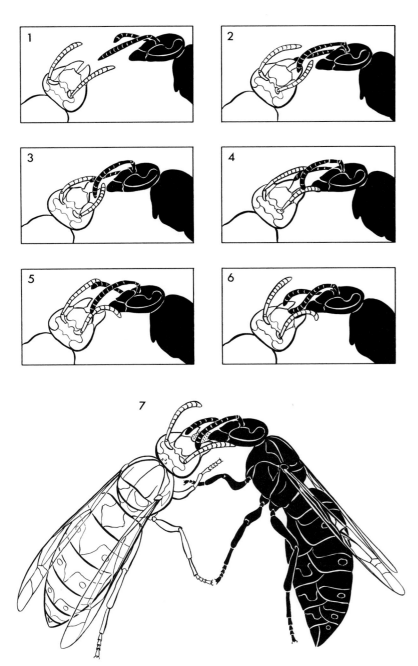

Figure 9-2 Trophallaxis between workers of the social wasp *Vespula germanica*. The solicitor on the right approaches the donor and places the tips of her flexible antennae on the donor's lower mouthparts (*1, 2*); the donor responds by closing her antennae onto those of the solicitor (*3*), who then begins gently to stroke her antennae up and down over the lower mouthparts (*4–7*). If this interaction continues, the donor will begin to regurgitate, and the solicitor is able to feed. (From Wilson, 1971a, based on Montagner.)

species. It is limited almost wholly to species in which there is a great deal of bodily contact, such as waxbills (Estrildidae), babblers (Timaliidae), white-eyes (Zosteropidae), and parrots (Psittacidae). Advanced social behavior may be as important a correlate as bodily contact, because some crows and related birds (Corvidae) allopreen while maintaining individual distances. Allopreening usually functions as an appeasement display: when birds respond to threats or attacks as if they were about to be preened, the attack is typically inhibited. In at least one species, the cowbird *Molothrus ater*, the behavior occurs as a redirected activity when the bird's attempts to attack are frustrated. Besides being associated with bodily contact and sociality, allopreening is most frequent in species that are sexually dimorphic, form persistent pair bonds, or both; and it reaches its highest intensity within species when birds are first brought together or reunited after a prolonged absence. Birds use distinctive invitation postures, such as fluffing of the feathers and withdrawing of the head as though to protect the eyes. In most instances these postures are plausibly interpreted as modified appeasement or retreat movements. The grooming movements are directed mostly at the head, one of the few areas of the body which the groomee is unable to reach itself. This circumstance may indicate that allopreening in birds also has a purely functional, cleaning component that has been largely overshadowed in evolution by its ritualization into a signal.

Allogrooming is widespread in rodents, where it consists of a gentle nibbling of the fur (see Figure 9-3). The behavior is most frequently observed in conflict situations, although it also occurs under other circumstances where neither participating animal appears to be aggressive or tense. Allogrooming is found sporadically in other mammals. Perhaps the most ritualized form is displayed by the mouflon

Figure 9-3 Grooming in wild rats consists of a gentle nibbling of the fur. As in other mammals, its function in communication is primarily to conciliate. (From Barnett, 1958.)

(*Ovis ammon*), a mountain sheep of the southern Mediterranean region. Soon after two males have fought for dominance, the loser performs an appeasement ceremony in which he licks the winner on the neck and shoulders. Often the dominant animal kneels forward on his carpal joints, that is, on his "wrists," in order to assist the process (Pfeffer, 1967).

Among the primates, allogrooming is a way of life. By passing from phylogenetically lower to higher groups, one can detect a marked shift from a dependence on the mouth and teeth to a nearly exclusive use of the hands. Tree shrews groom with their teeth and tongues, employing the procumbent lower incisors as a "tooth comb." Lemurs, the most primitive living animals that are indisputably primates, employ the teeth, tongue, and hands in close coordination (Buettner-Janusch and Andrew, 1962). In higher primates the hands are the principal grooming instruments. The basic movements consist of drawing hair through the thumb and forefinger, rubbing the thumb in a variable rotary pattern against the direction of the hair tracts, and lightly scratching and raking the hair and skin with the nails. Objects loosened by these actions are conveyed to the mouth to be tasted and sometimes eaten (Sparks, 1969). Unusual movements have been recorded. The gelada, a long-haired ground monkey, presses the hair down with wide sweeps of one hand while continuing to pick with the other. Surprisingly, chimpanzees and gorillas manipulate hair a good deal with their very mobile lips, a behavior that probably has been secondarily evolved.

Allogrooming in higher primates serves at least in part to clean fellow troop members. Parasites are systematically removed by hand movements, while wounds are cleaned by hand and sometimes even licked (Carpenter, 1940; Simonds, 1965). In his study of the Nilgiri langur (*Presbytis johnii*), Poirier (1970a) found that 62 percent of the time spent in allogrooming was directed at areas of the body the recipient could not reach itself.

At the same time, most primate species employ allogrooming in a strongly social role. During moment-to-moment encounters between troop members, grooming is reciprocally related to aggression: as one interaction goes up in frequency and intensity, the other goes down. In tense situations animals either offer to groom or present their bodies for grooming. These gestures are seldom followed by serious further threat or fighting, and in fact they seem to avert aggression. In most higher primate species, the dominant animals seem to be groomed disproportionately by their subordinates. The relationship, if borne out by future studies, is wholly in accord with our idea of the primitive role of the behavior, since it is the recipient who receives the greatest benefit. Yet even this straightforward generalization has at least one puzzling exception: in the spider monkey *Ateles geoffroyi* the high-ranking troop members perform most of the grooming (Eisenberg and Kuehn, 1966). Mothers concentrate mostly on

their infants, members of cliques on one another, and breeding adults on their sexual consorts. Saayman (1971b) showed how the pattern of grooming of female chacma baboons changes in a way that makes the social significance indisputable. Flat and lactating females groom less and interact chiefly with one another. Females in the estrous cycle are the class most involved in grooming bouts. In their follicular phase they consort most with juvenile and subadult males; at mid-cycle they shift to adult males while grooming to a lesser extent with subadult males. A closely similar pattern has been reported in rhesus monkeys by Kaufmann (1967). Of equal importance, allogrooming is most prominently developed and has its most clearly apparent social role in the aggressively organized species (Marler, 1965; Sparks, 1969). It is the most time-consuming mode of social interaction in macaques and baboons, but is infrequent and more limited to a cleaning function in the relatively pacific gorilla and red-tailed monkey. Chimpanzees provide ambiguous evidence with reference to this rule. Jane van Lawick-Goodall has referred to allogrooming as the principal social activity at the Gombe Stream Park. But Reynolds and Reynolds (1965) witnessed only 57 clear-cut examples of the behavior during 300 hours of observation in the Budongo Forest, a datum more consistent with the relatively nonaggressive nature of this species.

Although grooming invitation behavior in higher primates is generally unspecialized, it varies markedly from species to species. A gorilla first presents its buttocks in ritual appeasement and then offers the part it wishes groomed to the partner (Schaller, 1965a,b). Rhesus monkeys block the movements of a prospective groomer, but in a relaxed unaggressive manner. They also lie down on their sides with their backs to the animal being solicited, or else present the neck or chest. Baboons typically offer a hip or shoulder and Sykes' monkeys (*Cercopithecus mitis*) the top of the head, while talapoins (*C. talapoin*) lie down facing away from the groomer in order to present the back of the neck (Rowell, 1972). While the groomer concentrates intently on the spot it is cleaning, the recipient relaxes its body and may even close its eyes and appear to sleep. Both partners periodically smack their lips, a general primate appeasement signal.

Workers of a majority of social insects groom nestmates with their glossae (tongues) and, much less frequently, their mandibles, which are the functional equivalents of the primate hands. At least some of the social wasps, for example *Polistes* (Eberhard, 1969), engage in grooming. But the phenomenon is only occasional in the meliponines (Sakagami and Oniki, 1963) and evidently very rare or absent in the bumblebees (Free and Butler, 1959) and the primitively social halictine bee *Dialictus zephyrus* (Batra, 1966). The significance of allogrooming in social insects is not really understood. We can only guess that its cleansing action is in some way beneficial. Allogrooming probably plays some role in the transfer of colony odors and pheromones. For example, it is known that the queen substance of the honeybee, 9-ketodecenoic acid, is initially transmitted in this fashion from the queen to the worker (Butler, 1954). Bumblebees, in significant contrast, neither groom nestmates nor employ a queen substance. Ants probably also spread phenylacetic acid and other biocidal metapleural gland secretions in this fashion, thus protecting the colony against the growth of fungi and bacteria, one of the principal hazards of subterranean social life (Maschwitz et al., 1970). According to Haydak (1945) and Milum (1955), honeybee workers employ a special invitation display that these authors call the grooming dance or shaking dance. The inviting worker shakes her body rapidly back and forth and from side to side, while attempting to comb her thoracic hairs with her middle legs. Often, but not always, this behavior induces a nearby worker to approach and employ her mandibles to groom the hairy coat on the petiole and base of the wings. These are the parts which a bee is unable to clean herself.

Alarm

To alarm a groupmate is to alert it to any form of danger. As a rule, the danger is an approaching predator or territorial invader. But it can be anything else: in termites, for example, alerting trail substances are released in the presence of a breach in the nest wall. Although most alarm signals are general in the scope of their designation, a few are narrowly specific. According to Eberhard (1969), paper wasps (*Polistes*) respond to certain parasites of their own brood in a unique way. In particular, when an ichneumonid wasp of the genus *Pachysomoides* is detected on or near the nest, the paper wasps launch into an intense bout of short runs and wing flipping, which quickly spreads through the entire colony. Mammalian alarm calls are mostly nonspecific, but the vervet (*Cercopithecus aethiops*), an arboreal African monkey, uses a lexicon of at least four or five sounds to identify enemies. A snake evokes a special chutter call and a minor bird or mammalian predator an abrupt *uh!* or *nyow!* As soon as a major bird predator is seen, the vervets emit a call that sounds like *rraup;* when either the bird or a major mammal predator is close, the monkeys chirp and produce the threat-alarm bark. The responses of the vervets vary according to the different signals. The snake chutter and minor-predator calls direct the attention of the monkeys to the danger. The *rraup* call, indicating the presence somewhere of a large bird, causes the monkeys to scatter out of the open areas and treetops and into closed vegetation. The other alarm calls cause the monkeys to look at the predator while retreating to cover (Struhsaker, 1967c).

It is possible that an alarm pheromone exists in rodents. When house mice and rats are stressed, either by electric shock or (in the case of rats) aggressive encounters, an odor is produced that causes avoidance in conspecific animals (Müller-Velten, 1966; Carr et al.,

1970; Eisenberg and Kleiman, 1972). Müller-Velten found that the odor is released with the urine and remains potent for between 7 and 24 hours.

Responses to alarm signals differ markedly among species and according to circumstances in which individuals find themselves. In their studies of formicine ants, Wilson and Regnier (1971) classified species roughly into those that display predominantly aggressive alarm, in which workers orient aggressively toward the center of disturbance, and those that react with panic alarm, in which the workers scatter in all directions while attempting to rescue larvae and other immature stages. There is good evidence that aggressive alarm is the more general form and has evolved as part of an "alarm-defense system," in which the spraying of defensive chemicals and other forms of attack on enemies come to serve as increasingly effective alerting signals for nestmates as well. Certain other aspects of alarm communication, with special reference to the evolutionary origin of sociality, have been reviewed in Chapter 3.

Distress

The young of a diversity of bird and mammal species utilize special distress calls to attract adults to their sides. The chicks of precocial birds, such as domestic fowl, ducks, and geese, pipe in a way that is indistinguishable from the call emitted when they are cold or hungry (Lorenz, 1970). The pups of African wild dogs give a special "lamenting call" (*Klage*) when deserted (Kühme, 1965b). The young of collared lemmings (*Dicrostonyx groenlandicus*) emit unique ultrasonic chirps when subjected to cold stress or sudden, nonpainful tactile stimuli; these sounds attract the adult females (Brooks and Banks, 1973). When young primates are threatened, they call the adults with shrieks or screams. Vervet infants separated from their mothers use a scale of several vocal signals, culminating in a high-intensity squeal or scream (Struhsaker, 1967c). Stridulation in leaf-cutter ants, a squeaking sound created by scraping a ridge on the third abdominal segment against a row of finer ridges on the fourth segment, appears to serve primarily or exclusively as a distress signal. The ants begin squeaking when trapped in close quarters, particularly when they are pinned down by a predator or caught in a cave-in. The sound alone brings nestmates to their aid (Markl, 1968).

Assembly and Recruitment

No firm line exists between these two functions. Assembly can be defined crudely as the calling together of members of a society for any general communal activity. Recruitment is merely a special form of assembly, by which groupmates are directed to some point in space where work is required.

Assembling signals serve above all to draw societies into tighter physical configurations. The bright spotting and banding of coral fishes, called "poster coloration" by zoologists, is a case in point. Experiments by Franzisket (1960) showed that *Dascyllus aruanus* are attracted by the black-and-white banding pattern characteristic of their species, and that the response helps to hold individuals together in schools. The striking, individualistic colors of many species of coral fish may be required for quick, precise assembly and coordination of schools among the large numbers of other kinds of fish that crowd their habitat (W. J. Hamilton, III, personal communication). It is probably not coincidence that poster coloration is also well developed in the rich freshwater fish faunas of the tropics. Keenleyside (1955) found that the conspicuous dark spot on the dorsal fin of the water goldfinch *Pristella riddlei*, a schooling fish of the Amazon, is an aggregation stimulus. *Nannacara, Geophagus,* and a few other tropical cichlid fishes actually "summon" their young with a short lateral head movement, which appears to be a ritualized form of departure swimming. Among adult zebra cichlids (*Cichlasoma nigrofasciatum*), a high-intensity form of this display further serves as a warning signal; the young fish seeing it not only swim to the mother but also cluster beneath her belly. In *Neetroplus carpintis* the head movement is very exaggerated and serves only as an alarm signal. Hence the assembly function has evidently been lost in evolution (Lorenz, 1971).

Armstrong (1971) suggested that the white plumage of some species of herons, terns, pelicans, and other marine birds aids in assembling flock members near newly discovered fish shoals. The howling of wolves collects members of the pack that have scattered over the large territory routinely patrolled by these animals (Mech, 1970). Chimpanzees have a functionally similar, booming call that alerts distant troop members when a food tree is discovered (Sugiyama, 1972).

The known techniques of assembly in social insects are almost entirely chemoreceptive in nature. Termites are attracted to one another over distances of at least several centimeters by the odor of 3-hexen-1-ol emitted from the hindgut, while fire ants find one another by moving up carbon dioxide gradients (Wilson, 1962a; Verron, 1963; Hangartner, 1969b). Somewhat more complex forms of pheromone-meditated attraction and assembly have been found in the honeybee. When workers have discovered a new food source or have been separated from their companions for a long period of time, they elevate their abdomens, expose their Nasanov glands, and release a strong scent consisting of a mixture of geraniol, nerolic acid, geranic acid, and citral (von Frisch and Rösch, 1926; Butler and Calam, 1969). These pheromones draw other workers over considerable distances. Citral constitutes only 3 percent of the total volume of the fresh secretion, but it is by far the most potent attractant. It has also been demonstrated (Velthuis and van Es, 1964; Mautz et al., 1972) that swarming bees expel the Nasanov gland scent when they first encounter the queen, thus attracting other workers to the vicinity. The substances therefore function as true assembly pheromones. Evidently

the discovery of food lowers the threshold of the response and turns the pheromones secondarily into recruitment signals.

A second, more nearly continuously emitted assembly scent is found in the hive odor. This pheromone, which assists honeybees in finding their hives, appears not to be specific to colonies and may be identical with the "footprint substance" which is laid down continuously around the nest and food sites by worker bees. The latter scent is sometimes paid out in the form of a trail around the hive and serves as a rudimentary guide to honeybees seeking the nest entrance (Butler et al., 1969). Soil from around the nest entrance is highly attractive to workers of the harvester ant *Pogonomyrmex badius.* Members of each colony are able to recognize their own nest material (Hangartner et al., 1970). Bert Hölldobler and I have also established that the odor from around *P. badius* workers is attractive to other workers of the same species, even when separated from the nest. It is possible, but as yet unproven, that this substance constitutes part of the nest odor.

By far the most dramatic form of assembly in social insects is exercised by the mother queens of colonies. Except in the most primitively social species, any well-nourished, fertilized queen attracts a retinue of workers who tend to press close in with their heads facing her. When the pheromones are extracted and transferred to olfactorily inert dummies, the dummies serve as an attraction center temporarily as potent as the natural queen. Some of the pheromones have been chemically identified in the honeybee queen. One is the queen substance, *trans*-9-keto-2-decenoic acid. The second is a fat-soluble substance of unknown identity produced by Koschevnikov's gland, a tiny cluster of cells located in the sting chamber and whose principal duct opens between the overlapping spiracle and quadrate plates. These two pheromones are responsible at least in part for the formation of the retinue that surrounds the queen at all times. When the colony divides by the process of swarming, the attractive power of the 9-ketodecenoic acid comes to play a new role. Workers are drawn to the queen in midair by flying upwind when they smell the pheromone. As the queen flies to the swarm site, and later to the new nest (in both cases guided by scout bees), the bulk of the worker force follows along in the wake of her evaporating 9-ketodecenoic acid. Once a new destination is reached, however, this substance is not adequate to settle the flying workers. Now a second mandibular gland pheromone produced by the queen, *trans*-9-hydroxy-2-decenoic acid, comes into play. Workers can smell this substance only over a short distance. Those that do smell it begin to call in other workers by dispersing their own Nasanov gland scent, and, in a short time, the entire colony forms a quiet, stable cluster. There may be still more to the story. The substituted acids are but 2 of at least 32 components present in the mandibular glands of the queen. Other substances identified include methyl-9-ketodecanoate, nonanoic acid, and a variety of other esters and acids (Callow et al., 1964). The possibility that these and other as yet unidentified secretions manufactured by other glands in the queen's body (Renner and Margot Baumann, 1964) have some communicative function remains largely unexplored.

True recruitment is a form of communication that is apparently limited to the social insects. Ants, bees, wasps, and termites have evolved a multitude of ingenious signaling devices to assemble workers for joint efforts in food retrieval, nest construction, nest defense, and migration (see Chapter 3 and, for greater detail, Wilson, 1971a).

Leadership

A few vertebrate and insect species use signals that seem explicitly designed to initiate and to direct the movement of groups. Parents and young of precocial birds use an elaborate system of signals to coordinate their travels. The mother mallard (*Anas platyrhynchos*), for example, walks ahead at a pace just slow enough for her ducklings to stay close behind, all the while emitting a special guiding call. When a duckling falls too far to the rear, it begins piping in distress. The response of the mother mallard is immediate and automatic. She comes to a halt, extends her body, flattens her feathers, and calls more loudly. If the stragglers do not find their way to her in a short time, she runs back to them, momentarily forgetting the ducklings close to her. When she reaches the stragglers, they all exchange greeting and "conversation" calls. Meanwhile, the leading ducklings begin to pipe, causing the mother to run ahead again to attend to them. This time the laggard ducklings run forward a few meters before becoming lost a second time. By dashing back and forth, greeting and guiding, the mother eventually brings the two groups together so that the entire family can once again be on its way (Lorenz, 1970).

As Lorenz has also shown, larger flocking birds that cannot take flight easily have evolved special signals to induce simultaneous departures by members of the flock. Mallards "talk" back and forth with rising intensity while moving their beaks in what appears to be a ritualized flight intention movement. Geese perform a similar ceremony, but the head movements, which consist of brief lateral head shakings, are not so easily associated with locomotion. Other kinds of birds use auditory as well as visual signals. Cockatoos emit a loud shriek. Domestic pigeons and their wild rock dove ancestors (*Columba livia*) clap their wings loudly, with the duration of the signal indicating approximately how long the bird intends to fly. For short flights, no signal at all is given. A long journey is encoded by a prolonged bout of clapping before take-off. The reader will note the remarkable similarity that exists between this graded signal and the straight run of the honeybee waggle dance, which increases in duration with the distance from the hive to the target. Once in flight,

birds can present "poster signals" on their wings or tails to induce following by stragglers still on the ground. Oskar and Magdalena Heinroth (1928) showed that the patterns of wing coloration differ from species to species in ways that permit human observers to identify them at a glance. The principle is exactly the same as that used in the design of maritime flags.

A comparable signal is contained in the "swing step" of dominant male hamadryas baboons. These animals control the movements of their group to an unusual degree for primates. When they wish to depart, they take large, rapid steps while lifting their tails and swaying their buttocks rhythmically from side to side. These movements appear to induce subordinates to fall in behind (Kummer, 1968).

The honeybees have evolved two spectacular forms of leadership signal that exceed anything known in the nonhuman vertebrates. The first, of course, is the waggle dance. The second is the much less famous buzzing run, also called the breaking dance or *Schwirrlauf*, which is used by honeybee colonies to initiate swarming. Just before the swarm occurs, most of the bees are still sitting idly in the hive or outside in front of the entrance. As midday approaches and the air temperature rises, one or several bees begin to force their way through the throngs with great excitement, running in a zigzag pattern, butting into other workers, and vibrating their abdomens and wings in a fashion similar to that observed during the straight run of the waggle dance. The sound produced is very different from that produced during the straight run, however, raising the possibility that it is an important part of the signal to swarm (Esch, 1967a). The *Schwirrlauf* is swiftly contagious, and, within a minute or two, a dozen or more workers are engaged. As Lindauer (1955) describes it, "Like an avalanche the number of buzzing runners grows, many of them rush to the hive entrance, arousing similarly those slothful ones who had gathered together like a tuft before the flight opening, others hover briefly about the hive but return once again to continue their buzzing runs. In about 10 minutes the moment for departure has arrived . . . then the bees nearest the hive entrance rush forth and in a dense stream all follow. The queen too has been aroused, and if she does not follow the swarming bees out at once she is badgered without interruption by bees buzzing and running until she has found the hive entrance and hurls herself into the swarm cloud" (translation by L. E. Chadwick in von Frisch, 1967). The phenomenon is remarkable in that it is the only clear-cut example I know of an autocatalytic reaction in an animal communication system. The signal itself produces the same signal in others, with the result that a chain reaction and a behavioral "explosion" occur. Of course this is just the effect that is needed to insure a simultaneous action by the ten thousand or more individuals who fly from the hive. The buzzing run is also practiced when the queen is accidentally displaced from the rest of the swarm. In this circumstance it serves to get the workers airborne and actively searching for the queen (Mautz et al., 1972). Other possible cases of autocatalytic communication are found in the preflight behavior of flocking birds. The head tossing of Canada geese, for example, builds among the family members until finally the gander is involved and the flight begins.

Incitement To Hunt

The greeting ceremony, a display widely distributed in the dog family (Canidae), has been broadened in function by the African wild dog to include the initiation of mass hunting. Here is how Hugo van Lawick-Goodall (in H. and J. Lawick-Goodall, 1971) recorded the start of one hunt in the Ngorongoro Crater:

Just as the sun was setting old Genghis rose to his feet and yawned as he stretched himself. He trotted over to where Havoc, Swift and Baskerville lay together. At his approach they jumped up and all four began nosing and licking each other's lips, their tails up and wagging, their squeaks gradually changing to frenzied twittering. In a moment all the adult dogs had joined them and soon the pack was swirling round and round in the greeting ceremony. Amidst the confusion of legs and tails and lean lithe bodies I caught a glimpse of Havoc and Swift, their wide open mouths touching, their tongues curled back in their mouths; a momentary flash of old Yellow Peril piddling all over his toes in excitement; a sudden picture of Juno, her forelimbs flat on the ground and her rump up in the air as she twisted round to lick Genghis on the lips. And then, as suddenly as it had begun, the wild flurry of activity subsided and the pack started to trot away from the den on its evening hunt.

Van Lawick-Goodall felt intuitively that the ceremony expresses the unity of the pack for the purposes of the hunt. "I submerge my identity," the signals might say if somehow translated into human speech. "I will do my share of the hunting, I will share in the feeding. Let's go! Let's go!"

The grand masters of group hunting in the social insects, the invertebrate equivalents of the African wild dog, are the legionary ants. The workers of the large colonies of driver ants (*Dorylus*), army ants (*Eciton*, *Labidus*, *Neivamyrmex*, and other genera), and other members of the Dorylinae and Ponerinae organize their hunts by mutual tactile stimulation and a constant, cooperative form of exploratory trail laying. T. C. Schneirla (1940) described the procedure in *Eciton burchelli* as follows:

Upon arriving in new terrain the worker slows up and meanders noticeably in her course, now with a jerky movement of the anterior body. Within the limited advance made before she withdraws, the worker's body is held closer to the ground than before in a characteristic sprawly posture, legs extended and moving somewhat stiffly. Together with the wavering of the anterior body there is a rapid wasplike semirotatory vibration of the antennal funiculi. The extended antennae are bent downward and in their rapid beating tap the ground at frequent intervals. After having advanced hesitantly a few centimeters in this manner, the worker leans forward in an abrupt pause which may be repeated very

rapidly or followed by another short advance, then she quickly turns and runs back into the swarm.

During her brief advance into new territory ahead of the swarm the pioneer worker lays a small amount of trail pheromone from the tip of her abdomen, which draws other workers in the same direction. Meanwhile most of the swarm is moving forward in a chaotic manner, seizing prey and passing the victims back through the feeder columns to the central bivouac area holding the queen and immature stages:

It is important not to understate the great variability of individual behavior in the swarm, in describing constant trends. When *Eciton* workers cross paths in the swarm there occur all degrees of contact from momentary brushing of antennae or legs to a forcible collision. Ants that collide head-on draw back more or less abruptly and both may turn away or (if running slowly) slip past each other; those running against each other sidewise usually change their courses somewhat according to the force of the contact; or when a worker is overtaken from the rear her pace is accelerated by the bump if she is not actually overrun.

Yet out of all this disorder the characteristic swarm of *Eciton burchelli* emerges: a roughly elliptical mass of workers, 10–15 meters or more across and 1–2 meters in depth, connected by two or more thick feeder columns of workers leading back to the point of origin at the bivouac site, with the forward edge growing at the speed of 30 centimeters a minute. How is it created? Schneirla noted that two antagonistic forces continuously work on the individual ants in the swarms. The first is pressure—the tendency of ants to move away from places where crowding becomes too tight. As newcomers press in, mostly from the direction of the bivouac, they stimulate workers already present to turn and move away from them. This activity in turn induces workers still farther away to move outward, which generates a centrifugal wave of excitation and movement. The second force is drainage: as places are vacated by workers, other workers in adjacent crowded areas tend to fill them again. Drainage is thus the simple opposite of pressure, and it, too, exerts its influence by wavelike propagations through the swarm. As pressure builds from the rear by the constant influx of newly arriving workers, the ants already constituting the swarm attempt to move forward and to the side. However, the slow progress of the pioneer ants at the edge of the swarm impedes the movement of other ants at the heads of the columns and causes them to fan out into the terminal swarm formation. For some unknown reason, the impedance is greater at the front than along the sides, so that the swarm flattens into an elliptical shape.

Synchronization of Hatching (Embryonic Communication)

The young of precocial birds belonging to the same clutch have a strong incentive to hatch as close together in time as possible. The brooding mother and the first-hatched young will be on the move within hours; chicks left behind in the egg will perish. Synchronized hatching of entire broods, requiring at most one or two hours, is a general trait of precocial birds, including particularly pheasants, partridges, grouse, ducks, and rheas. When the eggs of these species are incubated separately, the hatching times are spread over a period of days; but when they are kept together, hatching is synchronous. Margaret Vince (1969) has obtained strong experimental evidence that the coordination is achieved by sound signals exchanged by the chicks while they are still in the eggs. The vocalizations become loudest and most persistent just prior to hatching. The most characteristic sound is a regular loud click, audible when the egg is held to the ear. It is not caused by a tapping against the shell, as biologists once widely believed, but is a true vocalization associated with breathing movements.

Initiation of Physical Transport

Ant workers routinely pick up nestmates and carry them from one place to another. The behavior is most consistently displayed by colonies migrating from one nest site to another. Upon discovering a superior nest site, scout workers of many species lay odor trails back to the old nest. The trail pheromone alone is enough to induce some workers to move out and investigate the site, and in the fire ant *Solenopsis invicta* it serves as the nearly exclusive basis of colony emigration. In many other ants, more primitive in this regard, the most important method of initiating migration is adult transport. The scouts simply pick up other colony members and carry them to the new site. When the colony occupies multiple nest sites, adult transport sometimes occurs continuously from one site to another. Økland (1934), while studying the European wood ant *Formica rufa*, was the first to realize that the phenomenon can be an important means of colony integration. In the closely related species *F. polyctena*, adult transport between multiple nest sites is seasonal, reaching its maximum during spring and autumn. In one colony of approximately a million workers studied by Kneitz (1964) in Germany, between 200,000 and 300,000 transportations occurred in the course of a year. Most of the workers doing the carrying were older foragers, and most of the workers being carried were younger individuals, of the kind that engage principally in nursing and the storage of food within their crops.

The value of simple emigration from a bad nest site to a good one is clear enough, and the function can be said to be basic and primitive in the ants. In the higher ants, adult transport has evolved into an elaborate, stereotyped form of communication. Among the Formicinae, the transporter approaches the transportee face to face, antennates it rapidly on the surface of the head, and attempts to seize it by the mandibles while jerking its own body rapidly backward. If the transportee is receptive it folds its antennae and legs in against the body in the pupal position and allows itself to be lifted from the ground.

As it is pulled up, it curls its abdomen forward. The transporter then swiftly carries it to the destination. In contrast, the transporting worker in most Myrmicinae seizes the transportee just beneath the mandibles or by the neck, and the transportee curls its body over the head of the transporter with its abdomen pointing upward or to the rear. Other taxonomic groups display their own characteristic variations on transport communication (Wilson, 1971a).

Basic transporting behavior has been adapted to new ends by a few ant species. In *Manica rubida* and *Leptothorax acervorum* it is used to remove alien workers from the colony territory (Le Masne, 1965; Dobrzański, 1966). Interestingly enough, the subdued aliens respond with the same submissive behavior as nestmates. Slave-making ants of the *Formica sanguinea* group routinely carry nestmates back and forth between the home nest and nests of other ants they are raiding. The tendency is carried to an extreme in the phylogenetically related *Rossomyrmex proformicarum*. The workers travel to the target nest of *Proformica* in pairs, one ant carrying the other in typical formicine fashion (Arnoldi, 1932).

Among termites the transport of nestmates other than eggs is a rare event. It does occur in at least a few higher termite species, for example members of *Anoplotermes* and *Trinervitermes*, on the infrequent occasions when colonies or fragments of colonies emigrate from one nest site to another. Young larvae are then carried in the mandibles of the adult workers, but most older larvae are required to walk. Adult and brood transport is unknown in the social bees and wasps, evidently because of the difficulty in carrying such heavy burdens in flight. When a colony of honeybees, stingless bees, or polistine wasps emigrates, in the course of either absconding or colony multiplication, the brood is left behind, and the new nest is peopled entirely by adult queens and workers who travel under their own power.

Nothing wholly comparable to the elaborate transport behavior of ants is known in the vertebrates. However, the carrying of young by mammals is sometimes stereotyped. Mothers of the dog and cat families (Canidae, Felidae) carry their young by the soft and ample nape of the neck. When picked up, the cub usually hangs limply, a posture that aids the mother in her efforts. Shrews seize the young almost at random. Most rodents favor the dorsal surface of the young, although muskrats, squirrels, and the murine *Apodemus* favor a ventral grip, with young squirrels then curling around to take hold of the mother's head (Ewer, 1968). The females of small rodents sometimes carry their young still attached to the nipples. The incisors of young wood rats (*Neotoma fuscipes*) are especially modified to serve as gripping organs and hence can be said to have evolved as part of the regular transport mechanism of this species (Gander, 1929).

Play Invitation

The specialized signals used by mammals to initiate play with groupmates have been reviewed in Chapter 8.

Work Initiation

Social insects routinely use sematectonic communication, that is, the evidence of work already completed, to initiate and guide specific forms of nest construction (see Chapter 8).

Threat, Submission, and Appeasement

The complex, often graded system of signals that mediate agonistic behavior have already been introduced in Chapter 8. They will be considered in more detail in later accounts of aggression, territoriality, and dominance (Chapters 11–13).

Nest-Relief Ceremony

In bird species where both parents care for helpless young, one typically remains at the nest while the other forages. When the forager returns it then relieves the mate from nest duty. The changing of the guard is a delicate operation in which recognition of the mates is first established by personalized sounds and other signals, then mutual agreement to change is reached by ceremonies special to the occasion (Armstrong, 1947; Lorenz, 1971). In some species the ceremony is obviously related phylogenetically to appeasement behavior used in agonistic encounters, including the tense give and take of the original formation of the pair bond. The male grey heron (*Ardea cinerea*) relieves his mate with a series of typically conspicuous reciprocal communications. He first alights on the rim of the nest with a vigorous flapping of wings, to which the female responds by stretching her neck upward and crying out several times. Now the pair stand back to back while calling loudly. Finally, the male bends his head down with the crest raised, snaps his beak several times, and settles on the nest, after which the female departs. Sometimes the routine is varied as follows: the male stretches his neck and head upward, raises his crest, and flaps his wings, while the female performs a muted version of the same display. The male nightjar (*Caprimulgus europaeus*) flies in to the nest uttering a characteristic churring sound, and the female responds with the same note. He next settles close to her and, as they sway gently from side to side, eases her off the nest and takes her place. The female then flies off. Mated pairs of some kinds of birds occasionally flair up in aggression toward each other when the placatory function of the nest-relief ceremony fails. In fact, gentoo penguins (*Pygoscelis papua*) routinely threaten their approaching mates and will actually peck them if they close in too quickly. Fighting is circumvented by an elaborate bowing and hissing.

Sexual Behavior

The full course of sexual activity is a tightly orchestrated sequence of behaviors that differ radically in form and function while remaining channeled toward the single act of fertilization. At least five such classes of acts can be distinguished: sexual advertisement, courtship, sexual bonding, copulatory behavior, and postcopulatory displays. In addition a few signals are known with the explicit function of inhibiting reproduction. These categories will be treated later in a special chapter on sexual behavior (Chapter 15).

Caste Inhibition

The queens of the most advanced social insects secrete pheromones that inhibit the development of immature stages into new queens. The result is the production of a high proportion of infertile workers that protect and feed the mother queen. The consort males of termites also produce a substance that inhibits male nymphs from developing into their own caste. In honeybees the female substance has been identified as the ubiquitous *trans*-9-keto-2-decenoic acid, which is secreted by the hypertrophied mandibular glands of the hive queen. The odor of the pheromone prevents workers from constructing the enlarged royal cells in which new queens can be reared from early larval stages. Each spring the ketodecenoic acid production of the hive queen is lowered, permitting the production of a few new queens and the subsequent multiplication of the colony by fission. The queen substances of ants, which have not been chemically identified, appear to influence the treatment of the larvae by the workers in a way that slows the development of individuals who show too much promise of turning into queens. In contrast, termite royal pheromones act directly on the developmental physiology of the growing nymphs. The full story of the inhibitory pheromones of social insects cannot be separated from the many other morphological and physiological processes of caste determination as a whole. The reader is referred to the recent review of this complex subject by Wilson (1971a).

The Higher Classification of Signal Function

The grail of zoosemiotics is the perception of deep structure in animal communication systems. Zoologists would be gratified if they could list those broad categories of messages whose identity somehow reveals the mind of the animal and what it is really trying to communicate. Hope would exist if by a combination of logical analysis and reorganization of data we proved that message categories are not endlessly proliferated in evolution, and that animals are able to say only a few things to one another.

This goal, I believe, can never be reached. Worse still, the harder different zoologists try to attain it, the more discordant will be their results—and the greater the overall amount of confusion in the literature. The primary difficulty is the one already indicated at the beginning of this chapter, that higher classifications of communicative acts (or deeper ones, in a psychological sense) are a straightforward taxonomic exercise limited by a built-in arbitrariness in the definition of unit categories and clustering procedures. The difficulty is exacerbated by the fact that social behavior is very far removed from the genotype and is unusually genetically labile. As investigators expand the classification above the level of the family (above, for example, the Felidae, Canidae, Hyaenidae, and other families) to the level of orders or greater to embrace all such units, similarities in behavior are increasingly likely to be convergent. Thus to collect behaviors of different species in a single category is increasingly a matter of judging analogy rather than homology, a largely subjective procedure. Zoosemiotics is closely similar in this regard to phytosociology, the classification of plant communities, and descriptive biogeography, which seeks to classify the world into regions, biome-types, and lesser units. In both of these disciplines, competing pyramids of units and higher categories have been painstakingly built-up, only to collapse in a bewildering debris of contradictory definitions and arcane terminology.

Nevertheless, if the construction of categories is hopeless, it is also profitable. Loose classifications, when not taken too seriously, can provide new insights into old phenomena and they can suggest new avenues for future research. This is the spirit in which we should review previous (and conflicting) systems by previous authors. Sebeok (1962), for example, suggested that all communication serves six basic functions. Two occur in many animals: emotive, or the induction of emotional response, and phatic, the establishment and maintenance of contact. The third and fourth functions, which are used by at least some animal species, are cognitive, which imparts objective information unrelated to emotion, and conative, which simply commands and directs activity. Sebeok considered the fifth function, metacommunication, to be exclusively human; we now believe that it occurs in many other mammal species. The sixth function, the poetic, was also posited to be strictly human. It is the evocation of complex, personal emotional images, allusory in nature, triggering memories and impulses based upon past associations that can be spelled out only with great difficulty when messages are kept exclusively cognitive in nature. Marler (1961), following Morris' 1946 system for human linguistics, recognized four functions of signals orthogonal in nature to those defined by Sebeok. Any signal can contain components that serve variously as *identifiors*, which specify a certain place and time; as *designators*, which identify the nature of the object toward which the attention of the responder is directed; as *prescriptors*, which designate the appropriate action for the responder to follow; and as *appraisors*, which allow the responder to react more to one object (or signaler) than to another. Consider a male bird singing on his territory. For a female bird passing nearby, the male's advertisement song identifies his position and that of at least part of his territory; it designates that he belongs to the correct species and is an appropriate sexual partner; it prescribes that the female should approach and adopt certain postures, following which the next stages of courtship can be activated; and, finally the song contains measures of volume, precision, and persistence that may allow the female to appraise the male as a partner in competition with other singing males.

W. J. Smith (1969a) recognized, independently of Moynihan, that the number of displays used by each individual vertebrate species falls within a narrow range, from about 10 to perhaps as many as 40 or 50 in the most social species. Smith grouped these displays into 12 clusters, or "messages", which he perceived to cover all kinds of vertebrates studied carefully to date, from songbirds to prairie dogs. The diagnostic traits of the messages are intuitive, a posteriori and cut across the systems proposed earlier by Sebeok and Morris. This kind of discordance, it should be added, is a common result of independent revisions in pure taxonomy. Smith's messages can be briefly characterized as follows:

1. *Identification:* the same as Morris' identifior.
2. *Probability:* the likelihood that the signaler will follow through with the act to which the signal refers; thus, in graded signals, a higher intensity usually means a higher probability of action.
3. *General set:* components or separate messages that have no independent meaning but indicate that the animal is very likely to take action of some unspecified nature.
4. *Locomotion:* messages associated with the onset or termination of locomotion, or those that are emitted solely while the animal is in motion.
5. *Attack:* any hostile act or display.
6. *Escape:* messages emitted when the animal is retreating from an aggressive interaction or any other aversive stimulus.
7. *Nonagonistic subset:* any signal that indicates the animal will not attack.
8. *Association:* special messages that are given when an animal is trying to approach and stay close to another without attempting hostile or sexual behavior.
9. *Bond-limited subset:* messages connected with the maintenance of tighter, more persistent bonds, as between a mated pair or between parents and offspring.
10. *Social play:* specifically, play initiation.
11. *Copulation:* messages used only just before or during attempts to copulate.
12. *Frustration:* behavior that occurs only when the animal is thwarted in the execution of other kinds of acts, such as copulation or aggression, for which it has been primed by physiological change or prior signaling.

Smith (personal communication) has since modified this list to some extent but still recognizes only about ten behavioral categories. His program is to utilize Cherry's distinction, originally applied to human language, between the study of the "message" of signs (semantics) and the study of the significance the signs have for the communicants (pragmatics). Cherry's third major division is syntactics, the study of signals as physical phenomena, a discipline with unambiguous goals. For the zoologist, a purely semantic approach for determining what information is encoded about behavior would be simply to connect signals with what is actually being done when the signal is given, for example copulation, locomotion, excretion, or with that behavior to which the animal is predisposed. A more pragmatic approach is to consider the ultimate function of the message, in other words its long-term adaptive significance for both communicants. Smith's 1969 classification is clearly intended to be semantic and hence more "objective." While objectivity is a desirable goal, any attempt to separate meaning from function in animal communication seems to create more ambiguity than it removes. Moreover, the clustering of many nonhomologous phenomena into classes is a departure from real objectivity and basically an arid procedure, difficult to grasp and to utilize in ordinary practice. True to its nature as a taxonomic exercise, it is very much like listing genera or families with diagnostic characteristics appended, but in the absence of a catalog of the constituent species. Every good taxonomist knows that such a list can be confidently used only by experts who already know the species and who are prepared to consult their own knowledge to evaluate the reviser's opinion of the best way to cluster the species. We should continue to make and to revise such lists, but not to take any one of them very seriously.

Complex Systems

It is a common misconception, held even by zoologists, that most animal communication consists of simple signals that reciprocate as stimuli and responses. Such a digital simplicity does indeed occur among microorganisms and many of the lower metazoan invertebrates. But where animals possess brains containing, say, on the order of ten thousand neurons or better, their social behavior tends to be much more devious and subtle. This generalization can best be made with examples. We will start with two well-analyzed "ordinary" communication systems, aggression in hamsters and courtship in doves, to show how intricate such behavioral exchanges really can be. Then we will move to several of the most advanced animal systems so far discovered, in order to gain a sense of the upper limit reached by animal communication systems as a whole.

Aggression in the Hamster

Female hamsters (*Mesocricetus auratus*) are intensely aggressive when not in estrus, being able to dominate even males. When two strange females are placed together, they fight until one gains clear power over the other. The contest is by no means just rough-and-tumble fighting. It follows a series of maneuvers as precise and orderly as a Greco-Roman wrestling match. After approaching nose to nose, the rodents perform one or the other of three movements: circling, following, or standing upright to face each other (see Figure 9-4). These maneuvers can alternate for an indefinite period, and any one

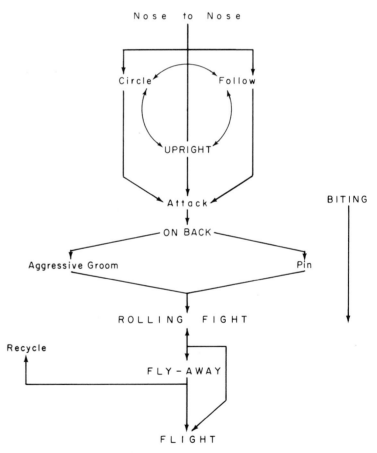

Figure 9-4 Stereotyped aggressive communication in hamsters. An encounter between two strange females follows a predictable series of maneuvers and fighting techniques. In extreme cases the sequence leads to the explosive "fly-away" escape and retreat of the loser. The example shows the organized nature of what might seem at first to be a very simple social interaction. (From Floody and Pfaff, 1974.)

can serve as the immediate prelude to an attack. The contest proceeds from intermediate forms of exchange, including pinning and aggressive grooming, to an escalated, rolling fight. Either hamster can terminate the fight by the ''fly-away'' maneuver, in which the animal disengages itself with an explosive extension of the hind limbs. Eventually the loser retreats from the scene or else accepts a fully subordinate status in the presence of the winner.

Reproduction in the Ring Dove

The reproductive behavior of ring doves (*Streptopelia risoria*) appears on casual observation to be mediated by a relatively few simple signals exchanged between the mated pair over a period of several weeks. In fact, as the careful researches of D. S. Lehrman and his associates have shown, it is a physiological drama that unfolds through the precise orchestration of communication, external stimuli, and hormone action (Lehrman. 1964, 1965). The full cycle runs six to seven weeks (Figure 9.5). As soon as an adult male and female are placed together in a cage containing nesting materials, the male begins to court by bowing and cooing. After a few hours the birds select a concave nesting site (a bowl works well in the laboratory) and crouch in it, uttering a characteristic cooing sound. Soon afterward the two birds carry material over to the site and build a loose nest with it. After several days of building activity, the female becomes closely attached to the nest, and soon afterward lays two eggs. Thereafter the two birds take turns incubating. Experiments by Lehrman and his coworkers indicate that the sight and sound of the mate alone stimulates the pituitary gland to secrete gonadotropins. These substances induce an increase in estrogen, which triggers nest-building behavior, and progesterone, which initiates incubation behavior. Another pituitary hormone, prolactin, causes growth in the epithelium of the crop. The sloughed-off epithelium functions as a kind of ''milk'' which is regurgitated to the squabs. Prolactin also sustains incubation behavior. When the squabs reach two to three weeks of age, the parents begin to neglect them, and soon the parents initiate a new endocrine-behavioral cycle. In the laboratory the process recycles continuously around the year.

Extreme Courtship Displays in Insects and Vertebrates

Although the brains of insects are orders of magnitude smaller than those of vertebrates, their most elaborate displays are at least equally complex. This generalization is illustrated by the waggle dance of the honeybee and the combined odor trails and tactile displays of certain ants. It is further exemplified by the courtship displays of many kinds of insects. Probably the most complex pattern known is that of the acridid grasshoppers belonging to the genus *Syrbula* (see Figure 9-6).

COURTSHIP

NEST CONSTRUCTION

INCUBATION

"NURSING"

Figure 9-5 Programmed reproductive communication in the ring dove. The reproductive cycle takes six to seven weeks and is mediated by interacting stimuli from the mate, the nest materials, and several hormones secreted in sequence. (From Wilson et al., 1973; based on D. S. Lehrman.)

As described by Otte (1972), the displays used in the sequences are mostly composed of one or the other of several kinds of sounds made by stridulation, combined with special caresses with the antennae and wings. Perhaps the most elaborate courtship process known in vertebrates is that of the ruff *Philomachus pugnax*. Males perform on leks in which they are positioned according to their status in a dominance hierarchy. A total of at least 22 visual displays are employed, with males of different ranks distinguished by the subsets of signals they

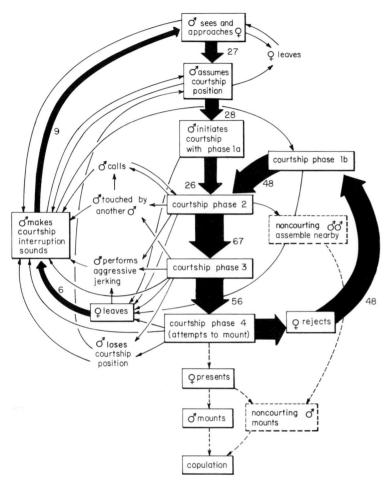

Figure 9-6 The most complex courtship procedure known in insects is that used by grasshoppers of the genus *Syrbula*, in this case *S. admirabilis*. The number of observations of each transition between steps is given next to the arrows and is further indicated by the thickness of the arrows. The separate signals, including those labeled phases, consist of combinations of vocalizations and movements of body parts. (From Otte, 1972.)

employ (Hogan-Warburg, 1966; Rhijn, 1973). My subjective impression is that the courtship repertories of the insect *Syrbula* and the bird *Philomachus* are roughly comparable in complexity.

Whale Songs

The most elaborate *single* display known in any animal species may be the song of the humpback whale *Megaptera novaeangliae*. First recognized by W. E. Schevill and later analyzed in some detail by Payne and McVay (1971), the song lasts for intervals of 7 to more

than 30 minutes duration. The really extraordinary fact established by Payne and McVay is that each whale sings its own particular variation of the song, consisting of a very long series of notes, and it is able to repeat the performance indefinitely (Figure 9-7). Few human singers can sustain a solo of this length and intricacy. The songs are very loud, generating enough volume to be heard clearly through the bottoms of small boats at close range and by hydrophones over distances of kilometers. The notes are eerie yet beautiful to the human ear. Deep basso groans and almost inaudibly high soprano squeaks alternate with repetitive squeals that suddenly rise or fall in pitch. The functions of the humpback whale song are still unknown. There is no evidence that special information is encoded in the particular sequence of notes. In other words, the song evidently does not contain sentences or paragraphs but consists of just one very lengthy display. The most plausible hypothesis is that it serves to identify individuals and to hold small groups together during the long annual transoceanic migrations. But the truth is not known, and the phenomenon may yet hold some real surprises. Other whale species use vocalizations, some of them similar to a few of the components of the humpback whale song (Schevill and Watkins, 1962; Schevill, 1964), but none is known to approach the complexity of this one species.

Gorilla and Chimpanzee Displays

Among land animals the most complicated single displays are probably those performed by the two evolutionarily most advanced and intelligent species of anthropoid apes, the gorilla and the chimpanzee. The famous chest-beating display of the gorilla occurs infrequently and is given only by the dominant silver-backed males. According to Schaller (1965a), the entire display consists of nine acts, which may be presented singly or in any combination of two or more. When in combination there is a tendency for the behaviors to appear in the following predictable sequence:

1. To start, the gorilla sits or stands while emitting from 2 to 40 clear hoots, at first distinct but then becoming slurred as their tempo increases.

2. The hoots are sometimes interrupted as the gorilla plucks a leaf or branch from the surrounding vegetation and places it between his lips in what appears to be a ritualized form of feeding.

3. Just before reaching the climax of the display, the animal rises on his hind legs and remains bipedal for several seconds.

4. While rising, he often grabs a handful of vegetation and throws it upward, sideways, or downward.

5. The climax of the display is chest beating, in which the standing gorilla raises his bent arms laterally and slaps his chest alternately with open, slightly cupped hands from 2 to 20 times. The beats are very fast, about 10 a second. Gorillas sometimes beat their abdomens and thighs, as well as branches and tree trunks.

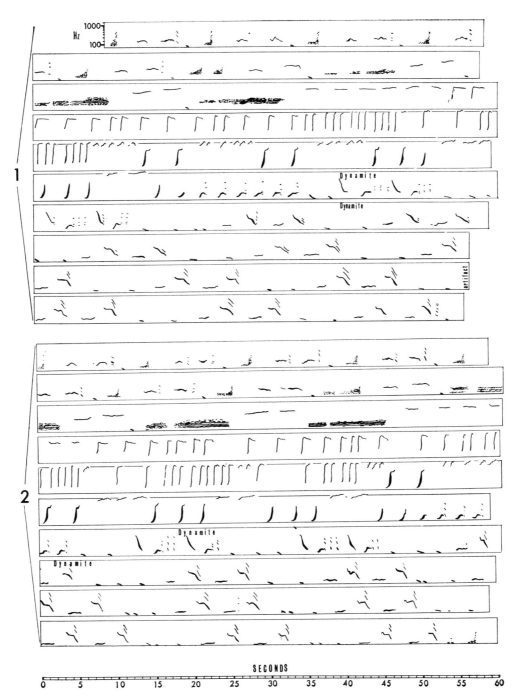

Figure 9-7 The song of the humpback whale. Lasting for as long as 30 minutes or more, it is perhaps the most complex single display thus far discovered in animals. The spectrographic tracings labeled 1 and 2 in this diagram represent two repetitions of the same song given by a single animal near Bermuda. The remarkable consistency in note sequence can be confirmed by comparing the two records step by step. (From Payne and McVay, 1971. Copyright © 1971 by the American Association for the Advancement of Science.)

6. A leg is sometimes kicked into the air while the chest is being drummed.

7. During or immediately following chest beating, the gorilla runs sideways, first a few steps bipedally and then quadrupedally, for 3 to 20 or more meters.

8. While running, the gorilla sweeps one arm through the vegetation, swats the undergrowth, shakes branches, and breaks off trees in its path.

9. The final act of a full display consists of thumping the ground with one or both palms.

The display appears to function very generally in advertisement and threat. It is seen most frequently when the male encounters a man or another gorilla troop, or when some other member of the troop begins to display. But it also occurs during play and sometimes even without any outside stimulus evident to the observer.

Even stranger are the "carnivals" of chimpanzee troops. From time to time, at unpredictable periods of the day or night, groups of the apes unleash a deafening outburst of noise—shouting at maximum volume, drumming trunks and buttresses of trees with their hands, and shaking branches, all the while running rapidly over the ground or brachiating from branch to branch (Sugiyama, 1972). The awed human observer feels he is in the presence of pandemonium. Reynolds and Reynolds (1965) described their experience in the Budongo Forest of Uganda as follows: "Inside the forest we were attempting to locate the chimpanzees to observe, if possible, the behavior associated with the tremendous uproar. Unfortunately this proved impossible. Calls were coming from all directions at once and all groups concerned seemed to be moving about rapidly. As we oriented toward the source of one outburst, another came from another direction. Stamping and fast-running feet were heard sometimes behind, sometimes in front, and howling outbursts and prolonged rolls of drums (as many as 13 rapid beats) shaking the ground surprised us every few yards." Unlike gorilla chest beating, the chimpanzee choruses are communal in nature. Far from serving to intimidate and disperse animals, they appear to keep scattered troops in touch and even to bring them closer together. The frenzies occur most often when the apes are on the move or have gathered for the first time in a feeding area. Sugiyama and the Reynolds believe that they serve in part to recruit other chimpanzees to newly discovered fruit trees, but the evidence is thin. The possibility remains open that the display serves other, perhaps wholly unexpected functions.

Duetting

For the ultimate in precision and coordination in displays, as opposed to mere complexity, we must turn to duetting in birds. An extreme manifestation is found in the communication systems of African shrikes (*Laniarius*), which have been studied at length by Thorpe

(1963b), Wickler (1972b,c), and their associates. Mated pairs of these birds keep in contact by calling antiphonally back and forth, the first vocalizing one or more notes and its mate instantly responding with a variation of the first call. So fast is the exchange, sometimes taking no more than a fraction of a second, that unless an observer stands between the birds or uses sophisticated recording equipment he does not realize that more than one bird is singing (see Figure 9-8). In at least one of the species, the boubou shrike (*L. aethiopicus*), the members of the pair learn to sing duets with each other. They work out combinations that are sufficiently individual to enable them to recognize each other even when out of sight.

Duetting of one form or another has been evolved, probably independently, by a wide diversity of birds—cranes, sea eagles, geese, quail, grebes, woodpeckers, barbets, megapode scrub hens, kingfishers, cuckoo-shrikes, *Melidectes* honey-eaters, and many others. The ex-

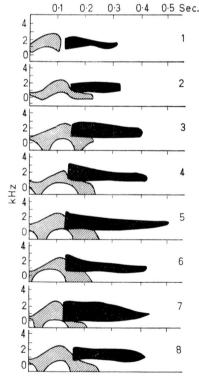

Figure 9-8 Duetting in African shrikes. Eight reciprocal calls of a mated pair of black-headed gonoleks (*Laniarius erythrogaster*) are represented here by sound spectrograms. The call of one bird is shown as the cross-hatched area, and its partner's almost instant response as the black area. The apparent subzero frequencies are due to distortion and interference below 50 cycles. (From Hooker and Hooker, 1969, after Thorpe, 1963b.)

changes vary greatly in form from one group to the next. In general however, they do show some broad ecological correlates of probable adaptive significance. Duetting species are typically monogamous. The two sexes usually resemble each other, and the mated pairs live in environments where it seems to be clearly advantageous to remain in touch over long periods of time. In an analysis of duetting in the New Guinea bird fauna, Diamond and Terborgh (1968) noted, as had previous ornithologists in other parts of the world, that many of the species live in thick vegetation, where the birds often cannot see each other and frequent vocal exchanges are necessary for continuous contact. But in some other species visual contact is maintained most of the time, and still duetting occurs. For these cases Diamond and Terborgh hypothesized that the exchanges are part of an adaptation for breeding in a capriciously fluctuating environment, where the partners must stay closely bonded and be prepared to start breeding on short notice when conditions become favorable. A third, competing hypothesis raised by the same authors is that duetting reduces the chances of hybridization with closely related species. Other information on the evolution, ecological significance, and ontogenetic development of duetting in birds of Africa has been supplied by Wickler and his associates (Wickler, 1972b,c; Wickler and Uhrig, 1969; von Helversen and Wickler, 1971) and by Todt (1970).

Chapter 10 **Communication: Origins and Evolution**

Where do the communication codes of animals come from in the first place? By comparing the signaling behavior of closely related species, zoologists can sometimes link together the evolutionary steps that lead to even the most bizarre communication systems. Any evolutionary change that adds to the communicative function has been called "semanticization" by Wickler (1967a). At one conceivable extreme of the semanticizing process, only the response evolves. Thus the sensory apparatus and behavior of the species is altered in such a way as to provide a more adaptive response to some odor, movement, or anatomical feature that already exists and that itself does not change. Male lobsters and decapod crabs, for example, respond to the molting hormone (crustecdysone) of the female as if it were a sex attractant. It is possible, although not yet proven, that crustecdysone has assumed a signaling function entirely through an evolved change in male behavior. The vast majority of known cases of semantic alteration, however, involve *ritualization*, the evolutionary process by which a behavior pattern changes to become increasingly effective as a signal. Commonly and perhaps invariably, the process begins when some movement, anatomical feature, or physiological trait that is functional in quite another context acquires a secondary value as a signal. For example, members of a species can begin by recognizing an open mouth as a threat or by interpreting the turning away of an opponent's body in the midst of conflict as an intention to flee. During ritualization such movements are altered in a way that makes their communicative function still more effective. Typically, they acquire morphological support in the form of additional anatomical structures that enhance the conspicuousness of the movement. They also tend to become simplified, stereotyped, and exaggerated in form. In extreme cases the behavior pattern is so modified from its ancestral state that its evolutionary history is all but impossible to decipher. Like the epaulets, shako plumes, and piping that garnish military dress uniforms, the practical functions that originally existed have long since been obliterated in order to maximize efficiency in communication.

Ritualized biological traits are referred to as displays. A special form of display recognized by zoologists is the *ceremony*, a highly evolved set of behaviors used to conciliate and to establish and maintain social bonds. We are all familiar with ceremonies in our own social life. Although the American culture is still too young to have many rituals that are truly indigenous, an interesting set can be seen in the yearly commencement at Harvard University. During this seventeenth-century affair the governor of Massachusetts is escorted by mounted lancers, the sheriffs of Middlesex and Essex counties appear in formal dress to represent civil authority, and a student gives a Latin oration. Each of the performances has lost its original function and is perpetuated only as ceremony in the truest sense. In a closely parallel fashion, animals use ceremonies to reestablish sexual bonds, to change position at the nest, and to avoid or reduce aggression during close interactions. Ceremony, to use Edward Armstrong's phrase, is the evolved antidote to clumsiness, disorder, and misunderstanding.

The ritualization of vertebrate behavior often begins in circumstances of conflict, particularly when an animal is undecided whether to complete an act. Hesitation in behavior communicates to onlooking members of the same species the animal's state of mind or, to be more precise, its probable future course of action. The advertisement may begin its evolutionary transformation as a simple intention movement. Birds intending to fly typically crouch, raise their tails, and spread their wings slightly just before taking off. Many species have independently ritualized one or more of these components into effective signals (Daanje, 1950; Andrew, 1956). In some species white rump feathers produce a conspicuous flash when the tail is raised. In others the wing tips are flicked repeatedly downward to uncover conspicuous areas on the primary feathers of the wings. In their more elementary forms the signals serve to coordinate the movement of flock members and perhaps also to warn of approaching predators. When hostile components are added, such as thrusting the head forward or spreading the wings as the bird faces its opponent, flight intention movements become ritualized into threat signals. But the more elaborate and extreme manifestations of this form of ritualization occur where the basic movements are incorporated into courtship displays (see Figure 10-1).

Signals also evolve from the ambivalence created by the conflict between two or more behavioral tendencies (Tinbergen, 1952). When

Figure 10-1 Flight intention movements have been ritualized to serve as a courtship display in the male European cormorant *Phalacrocorax carbo*. In the presence of the female, the male performs a conspicuous but nonfunctional modification of the take-off leap. (From Hinde, 1970, after Kortlandt, 1940. From *Animal Behaviour*, by R. A. Hinde. Copyright © 1970 by McGraw-Hill Book Company. Used with permission.)

a male faces an opponent, undecided whether to attack or to flee, or approaches a potential mate with strong tendencies both to intimidate and to court, he may at first choose neither course of action. Instead he performs a third, seemingly irrelevant act. He redirects his aggression at some object nearby such as a pebble, a blade of grass, or a bystander, who then serves as a scapegoat. Or the animal may abruptly switch to a *displacement activity:* a behavior pattern with no relevance whatever to the circumstance in which the animal finds itself. The animal preens itself, for example, or launches into ineffectual nest-building movements, or pantomimes feeding and drinking. Such redirected and displacement activities have often been ritualized into strikingly clear signals used in courtship. As Tinbergen expressed the matter, these new signals are derived from preexisting motor patterns—they have been "emancipated" in evolution from the old functional context.

The concept of ritualization was originated by Julian Huxley in his 1914 study of the great crested grebe *Podiceps cristatus*, and developed still more explicitly in a later monograph on the red-throated diver *Gavia stellata* (1923). In his initial work Huxley was struck by the apparently symbolic nature of the simple movements by which a grebe climbs out of the water onto the nest platform. The bird's approach in this manner indicates its willingness to mate, and its movements and postures on the platform have been modified further to lead to copulation. Although the great crested grebe is phylogenetically a primitive bird, it employs some of the most elaborate courtship and pair-bonding displays to be found anywhere in the vertebrates. Much later, Huxley's observations were extended and strengthened by K. E. L. Simmons (1955) and reinterpreted according to the concepts of modern ethological theory by Huxley himself (1966). The displays, three of which are illustrated in Figure 10-2, not only are of historical interest but also provide an excellent paradigm of hypothesis formation on the precise course of ritualization. Each of the grebe's remarkable ceremonies is performed with maximum intensity when the mated birds come together after a period of separation. Each is comprised of postures and movements that are among the most conspicuous in the bird's repertory: the raising of the crest during the head-shaking ceremony, for example, the diving motions prior to the penguin dance, and the wing spreading of the cat display. Finally, several of the components can be reasonably homologized with more basic threat and appeasement movements used by the birds under other circumstances. This is notably true of head shaking, in which the grebes approach as though hostile but then flick their bills downward and sideways out of the attack posture.

For a decade or so after the appearance of Tinbergen's 1952 article, most ethologists found it easy to interpret communication systems like that of the great crested grebe by what came to be known as

Figure 10-2 Three bond-forming ceremonies of the great crested grebe. Left: mutual head-shaking ceremony, apparently ritualized from turning-away movements in which a bird switches from aggression to appeasement. Center: mutual penguin-dance ceremony, during which the two birds present each other with waterweeds of the kind used in the nest; this ceremony is hypothesized to have originated as a ritualized form of displacement nest building. Right: the reciprocal discovery ceremony; one partner rises slowly from the water, while the other spreads its wings in the cat display, a movement that combines elements of defense and courtship. (From Simmons, 1955.)

the conflict theory of the origin of displays. The explanations were seemingly undergirded by Tinbergen's neurophysiological model. A displacement activity was simply "an activity belonging to the executive motor pattern of an instinct other than the instinct(s) activated." The irrelevant discharge of activity was forced by a "surplus of drive." In evolutionary time the adoptive executive center took over the instinct in its new context, molded it into a signal by the usual forms of ritualization, and emancipated it from the old executive center. The newly created display was then free to evolve solely with reference to the communication system it served. Perhaps the most thoroughgoing application of the conflict theory was made by Moynihan (1958) in his studies of hostile behavior of North American gulls. On the basis of subjective impressions in the field, Moynihan attempted to map the position of the various agonistic gull displays on a two-dimensional field defined along one axis by the graduated shift from a predominantly attack drive to a predominantly escape drive, and along the second axis by the intensity of hostile motivation. For example, the choking display was interpreted as resulting when a highly excited animal is balanced between drive and escape, while the aggressive upright display was seen to be the response of a bird with a weak but mostly aggressive tendency.

Subsequent neurophysiological experimentation on birds and mammals failed to provide confirmation of the existence of the executive centers, innate releasing mechanisms, and other key elements of the primitive Lorenz-Tinbergen models; and the conflict theory has been accordingly modified well away from its original, provocative form. As developed by Andrew (1963, 1972), Wickler (1969b) and others, particularly with reference to mammals, the newer view is roughly as follows. Many signals do evolve from ritualized intention and displacement activities, much as conceived by Daanje and Tinbergen. But ritualization is a pervasive, highly opportunistic evolutionary process that can be launched from almost any convenient behavior pattern, anatomical structure, or physiological change—not just from displacement activities. As Andrew has stressed, signals must be closely analyzed with respect to the immediate biological context in which they occur and without reference to preconceived notions of conflicting drives and the like. When we do that, it becomes clear that all manner of biological processes, from blushing and sweating to mucus secretion and defecation, some of them under the control of the autonomic nervous system, have been appropriated by one species or another. The following examples illustrate the almost protean nature of the process.

Ritualized predation. As part of his courtship ceremonies, the male grey heron (*Ardea cinerea*) routinely performs what is clearly a modified fishing movement. With his crest and certain body feathers erected, he points his head down as though striking at an object in front of him and snaps the mandibles with a loud clash (Verwey, 1930).

Ritualized food exchange. Billing in birds serves multiple functions centered on the establishment and maintenance of bonds (Wickler, 1972a). In some species, such as the masked lovebird *Agapornis personata*, it is used by mated pairs as a greeting ceremony or to end quarrels. In others, for example the Canada jay *Perisoreus canadensis*, billing is an appeasement signal employed by subordinate birds in flocks. The display has evidently originated as a ritualized variant of food exchange between young and adults. When a subordinate bird employs it in appeasement it is usually similar or identical to the begging motions of a young bird, which include a squatting body posture and wing quivering. The billing of mated pairs is often accompanied by actual feeding of one bird by the other. The male masked lovebird regularly feeds the female, who remains at the nest to care for the brood. Male terns of the genus *Sterna* feed their partners just before or during copulation through motions that appear identical to the feeding of young (Nisbet, 1973).

The greeting ceremony of wolves and African wild dogs is roughly analogous to billing in birds. Subordinate individuals approach higher-ranked pack members in a groveling posture and enthusiastically nip and lick the mouth area. Packs of African wild dogs also use the behavior to incite and perhaps to coordinate chases. The greeting ceremony seems to have been derived from the begging motions of pups, an elementary motion that induces the adults to regurgitate pieces of meat to them. An intermediate behavioral variant is "snuffling" among adult wolves, in which one animal probes its nose and mouth around the lip area of another in an apparent attempt to learn whether it has eaten recently (Mech, 1970).

The *ne plus ultra* of ceremonial food exchange is to be found in the central act in the courtship ceremonies of certain dance flies belonging to the family Empididae (Kessel, 1955). Primitive empidids, or at least those species that are primitive in reproductive behavior, engage in a form of courtship behavior basically similar to that of other flies. But because empidids are predaceous, the female occasionally seizes and eats the male. The males of a few species, such as certain members of *Empis*, *Empimorpha*, and *Rhamphomyia*, avoid this fate by first catching another kind of fly and presenting it as a wedding gift to the female. While she feeds on the victim, the male copulates with her in safety. The second step in the ritualization sequence is exhibited by some *Hilara* and *Rhamphomyia*. The male catches the prey, but instead of searching for a female, he joins other males in an aerial dance. The swarm of males is now the attractant to the female, who flies into it and finds a mate. In later stages, which have been meticulously traced by Kessel and others through the mazelike taxonomy of the empidid species, the dancing males begin to add threads or globules of silk to the gift prey to make the nuptial swarms more conspicuous. Then (in certain forms of *Empis*) the entire prey is covered by a sheet of silk, producing the

first balloon. Ritualization is quite advanced at this point, but it has still further to go. In some *Empis* and *Empimorpha*, the size of the prey is reduced, so that the gift consists mostly of a balloon. In fact, the prey insect is so small, and so crushed and dry, apparently because of prior feeding on it by the male, that it can no longer serve as a significant meal for the female empidid. The final stage in the evolutionary sequence can now be guessed. In *H. granditarsus* and *H. sartor* the male does not obtain a prey at all but only spins a balloon, which is nevertheless willingly accepted by the female. It is a curious twist of history that this last stage was also the first to be discovered, by Baron Osten-Sacken in 1875. Osten-Sacken's bafflement was of course complete. No doubt we would still be speculating about the evolution of this behavior if the remarkable series of intermediate species had not subsequently been brought to light by the combined efforts of generations of entomologists.

Lip smacking. The higher primates, typified by the yellow baboon *Papio cynocephalus*, use lip smacking as an all-purpose conciliatory greeting. They employ it most noticeably during sexual encounters or in response to sexual objects. The signal appears to consist of rapidly repeated sucking movements. Anthoney (1968) traced its ontogenetic development in young baboons from elementary nursing directed at the mother to a separate greeting behavior directed at other members of the troop. Several anatomical features are especially effective at inducing lip smacking; all are pink like the mother's nipple and several resemble it in shape. They include the nipples and sexual skin of the estrous female, the penis of the male, and the face and perineum of the young infant.

Smiling and laughing. Van Hooff (1972) believes that smiling and laughter in human beings can be homologized in a straightforward way with similar and equally complex displays used by the other higher primates. Smiling, according to van Hooff's hypothesis, was derived in evolution from the "bared-teeth display," one of the phylogenetically most primitive social signals. The members of most primate species assume this expression when they are confronted with an aversive stimulus and have a moderate to strong tendency to flee. The display intensifies when escape is thwarted. In higher primates the bared-teeth display is commonly silent in expression. Among chimpanzees it is furthermore graded in intensity and is used flexibly to establish friendly contacts within the troop. The "relaxed open-mouth display," often accompanied by a short expirated vocalization, is a signal ordinarily associated with play. In man these two signals, the silent bared-teeth display and the relaxed open-mouth display, appear to have converged to form two poles in a new, graded series ranging from a general friendly response (smile) to play (laughter). A third kind of signal that developed from the archaic facial expressions is the bared-teeth scream display. This behavior, which is widespread in primates but missing in man, indicates extreme fear and submission, as well as readiness to attack if the animal is pressed further. (See Figure 10-3.)

Ritualized flight. The male courtship of some bird species features a labored, conspicuous form of flight during which special plumage patterns are revealed to maximum advantage. An example is given in Figure 10-4. The males of many species or oedipodine grasshoppers perform display flights that appear to attract females watching from the ground. During the performances they fly upward while flashing their brightly colored hindwings or rapidly snapping their hindwings to create a peculiar vocalization called "crepitation" by entomologists (Otte, 1970).

Ritualized respiration. African chameleons display on their territories by pumping the sides of their bodies in and out in an exaggerated respiratory movement. This behavior is accompanied by head wagging and jerking, which appear to be ritualized defensive head thrusts (Kästle, 1967).

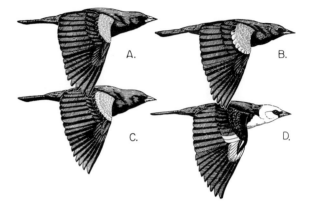

Figure 10-4 The ritualization of the flight in male blackbirds. Normal flight is shown in three races of the red-winged blackbird *Agelaius phoeniceus* (*A–C*) and yellow-headed blackbird *Xanthocephalus xanthocephalus* (*D*). *E* and *F* present two views of a male red-winged blackbird in ritual flight, and *G* a side view of a male yellow-headed blackbird in the same form of display. (From Orians and Christman, 1968. Originally published by the University of California Press; reprinted by permission of the Regents of the University of California.)

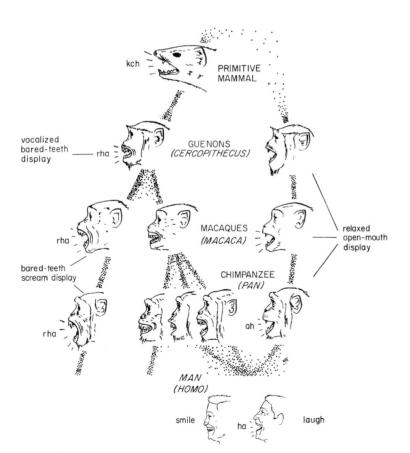

Figure 10-3 Phylogeny of facial signals in primates. (Modified from van Hooff, 1972.)

Ritualized excretion and secretion. The early development of the concept of signal evolution was based almost exclusively on visual and auditory signals, which are the easiest for human beings to perceive. Now that studies of chemical communication have attained equal prominence, examples have come to light of the parallel modification of secretory and excretory products. In order to mark their scent posts, various mammals use the metabolic breakdown products released in urination and defecation, as well as special glandular products released by glands associated with the urethra and anus. Some of the species, such as the giant rat of Africa (*Cricetomys gambianus*), the mongoose, and other viverrids, employ hand stands and other movements distinct from basic urination and defecation in order to deposit the scent on tree trunks and other objects above ground level. Odorous components of urine in mice have taken on a regulatory function in reproduction, serving to block or to coordinate estrus and pregnancy according to circumstances. Domestic swine boars release a substance with the urine that induces lordosis in the sow. In a parallel fashion, army ants and formicine ants lay odor trails from material in the hindgut. Although trail marking is a wholly distinctive behavior, it is reasonably hypothesized to have originated as a ritualized form of defecation.

Ritualization of waste products need not be confined to feces and urine. The sex attractant of female rhesus monkeys emanates from the vagina. It has recently been found to consist of a mixture of at least five short-chain fatty acids (Curtis et al., 1971). These substances are ordinary products of lipid metabolism and might well have been appropriated in evolution by a partial ritualization of such materials excreted at low concentrations through the epidermis.

The body surface of any vertebrate contains at least trace amounts of hundreds of secretory and excretory substances that are available for the semanticizing process. The epidermal mucus of fish might serve as an example. Recent experiments (Rosen and Cornford, 1971) show that the hydrodynamic properties of fish slime permit fish to attain a considerably higher speed than would be otherwise possible, and we can assume that the increase of velocity is a primary function of the material. But fish slime is also a veritable chemist's shop of water-soluble odorants, a property that preadapts it for communicative functions. Nordeng (1971) discovered that young char (*Salmo alpinus*) are attracted to the streams occupied by their parents. He hypothesized that the anadromous migration of these fish is guided by the specific odors of the parents' skin mucus.

Easily the most bizarre case of chemical ritualization demonstrated to date is the exploitation of cyclic AMP for communicative purposes by the cellular slime molds. This substance, cyclic-3'5'-adenosine monophosphate, serves as an intracellular messenger in all organisms. It stimulates certain forms of genetic expression, and, in vertebrates at least, mediates between hormones arriving at the cell membrane and selected target enzymes inside the membrane (Pastan, 1972). The life history of the cellular slime molds consists of an alternation of an ameba stage with a multicellular pseudoplasmodium stage, which travels about in a sluglike manner before finally coming to rest and producing spores from elevated fruiting bodies. The pseudoplasmodium is created by an aggregation of the single-celled amebas, and the aggregation is guided by minute amounts of a substance called acrasin. Recently acrasin has been identified as cyclic AMP (Konijn

et al., 1967). Why this particular compound, out of the many generated by amebas, was selected in evolution to serve as the amebic pheromone is still very much a mystery.

Automimicry. Ritualization in some of its most extreme and elegant forms occurs when one sex or life stage evolves to imitate communication in another class of individuals belonging to the same species. By exploiting the responses of the model, the mimic increases its own fitness. Because automimicry is employed in social behavior of one form or another, the model also benefits—or at least is not seriously harmed. The concept of automimicry has been developed principally by Wolfgang Wickler (1962, 1967, 1969). One of his more striking examples is illustrated in Figure 10-5. The males of certain species of mouthbrooder fishes in the freshwater genus *Haplochromis* have conspicuous spots aligned in a row on their anal fins. These marks resemble, in varying degrees of precision, the eggs carried by the females in their mouths for protection. Females have a strong tendency to pick up eggs that they accidentally drop from their mouths. The males exploit this behavior by displaying their anal fin spots close to the lake bottom. When a female approaches and attempts to pick up the "eggs," she receives a mouthful of sperm instead, thus inadvertently fertilizing the true eggs carried in her mouth.

A similar mutually beneficial form of trickery is performed by the female hyena, who possesses an extraordinarily realistic pseudopenis that she uses as part of appeasement signaling. Folklore has held since at least the time of Aristotle that the laugh of the hyena represents mischievous delight over its ability to change sex. In fact, penile displays are important appeasement signals in the aggressively organized hyena societies. Wickler has also stressed the automimetic nature of such shifts from sexual to social behavior as a significant event in the evolution of social life of the primates. Around the time of estrus the females of many Old World monkey species develop a large red swelling of the naked portions of the skin around the genital orifice. In extreme cases the growth becomes so excessive that the animal has difficulty sitting down. The female presents sexually to males by crouching and lifting her hindquarters to give maximum exposure to the genital area. Although the color and unique form of the sexual skin have not been proved by experiments to serve as visual releasers of copulatory behavior, it is entirely reasonable to suppose that they do. Males inspect the genital area, including the skin, and they also sometimes sniff at it, a behavior perhaps indicating the presence of sexual pheromones of the kind that have been identified in the rhesus monkey. Males of the hamadryas baboon and some other species possess a permanently colored rump, which they present when greeting and appeasing other males. This sexual charade is often carried so far that the male receiving the presentation mounts briefly in an imitation of copulation—an ex-

Figure 10-5 Automimicry in the mouthbrooder fish *Haplochromis burtoni*. Above: The female is attracted to spots on the anal fin of the larger male, because these marks resemble eggs that she carries in her mouth. Below: When she attempts to pick up the "eggs" she receives sperm released by the male instead, causing her real eggs to be fertilized. (From Wickler, 1967a.)

change that one writer has compared to a military salute. That such homosexual encounters entail true automimicry is suggested by the fact that males possess colored rumps only in those species in which the female sexual skin is transformed during estrus. The case is especially strong where the females of only a few species in a large group have estrous swellings, and the males of the same species, and only the same species, possess similarly colored rumps. Among the many kinds of Old World leaf-eating monkeys, including the langurs (*Presbytis*), proboscis monkeys (*Nasalis*), doucs (*Pygathrix*), and guerezas and colobus monkeys (*Colobus*), only the red colobus (*C.*

badius) and olive colobus (*C. verus*) have red female estrous swellings. And in only these two species do the males have altered rumps, which resemble the female swellings not only in color but also in shape. A more recent review of sexual displays in primates, including an evaluation of Wickler's hypothesis of automimicry, has been provided by Crook (1972).

Ab initio signals. Although behaviorists have correctly concentrated their attention on the shift of function in the process of ritualization, it is also possible that signaling organs and behaviors can arise *de novo* in the primary service of communication. Certain glands of social insects appear to fall in this category, including the sternal gland of termites and Pavan's gland of dolichoderine ants used in trail laying, the anal gland of dolichoderine ants employed in alarm and defense, the postpharyngeal gland of all ant groups employed in larval feeding, and Nasanov's gland of honeybees used in attraction and assembly (Wilson, 1971a). None of these structures appears to have any precursor in the nonsocial insects. Of course, it can be argued that the glands unavoidably arose from preexisting undifferentiated epidermal cells, but this form of evolution is not ritualization in anything close to the sense exemplified by the classical vertebrate studies.

The Sensory Channels

The concept of ritualization and its aftermath have left us with a picture of extreme opportunism in the evolution of communication systems, in which signals are molded from almost any biological process convenient to the species. It is therefore legitimate to analyze the advantages and disadvantages of the several sensory modalities as though they were competing in an open marketplace for the privilege of carrying messages. Put another, more familiar way, we can reasonably hypothesize that species evolve toward the mix of sensory cues that maximizes either energetic or informational efficiency, or both. Let us now examine each of the sensory modalities with special reference to their competitive ability, meaning the relative advantages and disadvantages of their physical properties.

Chemical Communication

Pheromones, or substances used in communication between members of the same species, were probably the first signals put to service in the evolution of life. Whatever communication occurred between the ancestral cells of blue-green algae, bacteria, and other procaryotes was certainly chemical, and this mode must have been continued among the eucaryotic protozoans descended from them. At this state of our knowledge it is still reasonable to speculate with J. B. S. Haldane that pheromones are the lineal ancestors of hormones. When the metazoan soma was organized in evolution, hormones appeared simply

as the intercellular equivalent of the pheromones that mediate behavior among the single-celled organisms. With the emergence of well-formed organ systems among the platyhelminths, coelenterates, and other metazoan phyla, it was possible to create more sophisticated auditory and visual receptor systems equipped to handle as much information as the chemoreceptors of the single-celled organisms. Occasionally these new forms of communication have overridden the original chemical systems, but pheromones remain the fundamental signals for most kinds of organisms. This important fact was not fully appreciated in the early days of ethology, when attention was naturally drawn to the visual and auditory systems of birds and other large vertebrates whose sensory physiology most closely resembles our own. But now chemical systems have been discovered in many microorganisms and lower plants and in most of the principal phyla. They continue to turn up with great dependability in species whenever a deliberate search is made for them, to an extent that makes it reasonable to conjecture that chemical communication is virtually universal among living organisms. Table 10-1 contains a progress report of the ongoing phylogenetic survey being conducted by many investigators. Not only are chemical systems widespread, but they are at least as equally diverse in function as visual and auditory systems.

Chemical communication of a high degree of sophistication also occurs in exchanges between species that are closely adapted to each other, particularly between symbionts and predators and their prey. The term *allomone* has been coined by W. L. Brown and Thomas Eisner (in Brown, 1968) for interspecific chemical signals. Later Brown et al. (1970) muddied the nomenclatural water a bit by recommending a distinction between allomones, which are adaptive to the sender, and "kairomones," which are adaptive to the receiver. This is a difficult and occasionally impossible choice to make in practice, and the prudent course would seem to be to drop the latter term and continue to use "allomone" in the broader sense.

Chemical signals possess several outstanding advantages. They transmit through darkness and around obstacles. They have potentially great energetic efficiency. Less than a microgram of a moderately simple compound can produce a signal that lasts for hours or even days. Pheromones are energetically cheap to biosynthesize, and they can be broadcast by an operation as simple as opening a gland reservoir or everting a glandular skin surface. They have the greatest potential range in transmission of any kind of signal used by animals. At one extreme pheromones are conveyed by contact chemoreception or over distances of millimeters or less, which makes them ideal for communication among microorganisms. At the other extreme, and without radical alteration of design in biosynthesis and reception, they can generate active spaces as much as several kilometers in length. The potential life of chemical signals is very great, rivaled in

Table 10-1   The phylogenetic distribution of chemical communication systems.

Taxa	Activity of pheromone	Chemical nature of pheromone	Authority
PROTISTA			
Volvox sp.	Female substance induces gonidia to develop into sperm packets	High molecular weight, over 200,000; probably a protein	Starr (1968)
Paramecium bursaria	Mate recognition, by cilial contact	Apparently a protein	Siegel and Cohen (1962)
ALGAE			Müller et al. (1971)
Ectocarpus siliculosus (brown alga)	Female gamete attracts male gametes	Allo-*cis*-1-(cyclo-heptadien-2′,5′-yl)-butene-1	
FUNGI			
Allomyces sp. (water mold)	Sperm attractant produced by female gametes; active at 10^{-10} M	Sirenin: oxygenated sesquiterpene with cyclohexane center; $C_{15}H_{24}O_2$	Machlis et al. (1968)
Achlya bisexualis (water mold)	Induction of antheridial hyphae on the male plant; active at 2×10^{-10} gm/ml	Antheridiol: a steroid $C_{29}H_{42}O_5$	Barksdale (1969)
Mucor mucedo	Induction of sexual hyphae in opposite sex	A "gamone": $C_{20}H_{25}O_5$	Plempel (1963)
Dictyostelium discoideum (slime mold)	Attraction and aggregation of ameboid cells	Acrasin: cyclic 3′,5′-adenosine monophosphate	Konijn et al. (1967), Bonner (1974)
TRACHEOPHYTA			
Pteridium and other ferns	Future female gametophytes secrete antheridogen, which induces nearby gametophytes to develop antheridia (male organs)	Unknown	Voeller (1971)
ASCHELMINTHES			
Brachionus spp. (rotifer)	Recognition of females by males, followed by breeding	Not a protein; otherwise unknown	Gilbert (1963)
ANNELIDA			
Lumbricus terrestris (earthworm)	Alarm and evasion; secreted in mucus	Unknown	Ressler et al. (1968)
MOLLUSCA			
Helisoma spp. and some other aquatic snails	Alarm: self-burying or escape from water	Polypeptides from tissue; mol. wt. about 10,000	Snyder (1967)
ARTHROPODA			
Amphipoda (*Gammarus duebeni*)	Female sex attractant	Unknown	Dahl et al. (1970)
Decapoda			
Portunus (crab)	Female sex attractant	Unknown	Ryan (1966)
Cancer, Pachygrapsus	Female sex attractant	Probably crustecdysone	Kittredge et al. (1971)
Cirripedia			
Balanus balanoides and *Elminius modestus* (barnacles)	Aggregation and settlement of larvae, by contact with pheromone on substratum	Protein	Crisp and Meadows (1962)
Arachnida			
Lycosidae (wolf spiders)	Female sex attractant	Unknown	Kaston (1936)
Salticidae (jumping spiders)	Female sex attractant	Unknown	Crane (1949)

Table 10-1 (*continued*).

Taxa	Activity of pheromone	Chemical nature of pheromone	Authority
Insecta	*Sex attractants.* Female attractants are common and very widespread, having been demonstrated in the following orders: Dictyoptera including Isoptera, Lepidoptera, Coleoptera, Hymenoptera, Diptera. Male attractants and "aphrodisiac" agents are also common and widespread, having been reported from Dictyoptera (Blattaria only), Hemiptera, Mecoptera, Neuroptera, Lepidoptera, Coleoptera, Diptera, Hymenoptera. See reviews by Jacobson (1972), Butler (1967), Wilson (1968, 1970), Shorey (1970), Silverstein (1970), and Roelofs and Comeau (1971).		
	Alarm substances, trail substances, recognition odors, etc. These occur in most social insects. Review by Wilson (1971a).		
CHORDATA Vertebrata	*Sex attractants, both male and female.* These are widespread in fish, amphibians, reptiles, and mammals, although they are still poorly documented in most groups. See reviews by Bardach and Todd (1970), Burghardt (1970), Ralls (1971), Bronson (1971), and Eisenberg and Kleiman (1972). These pheromones are now known to be common in primates, including even the female rhesus (Rowell, 1971). The strong possibility of their occurrence in human beings is discussed by Comfort (1971).		
	Dominance odors and territorial and home-range markers. These are common in mammals (see, e.g., Mykytowycz, 1964; Schultze-Westrum, 1965; Thiessen et al., 1968; Thiessen, 1973; Eisenberg and Kleiman, 1972). Individual odor involved in territorial defense has been reported in fish (Todd, 1971).		

animal systems only by the sematectonic visual uses present in nest architecture. When put down as scent posts or odor trails, pheromones also have a strange capacity for transmitting into the future. Even the animal that created the signal has the opportunity to come back and make use of it at a later time.

The outstanding disadvantages of chemical communication are slowness of transmission and fade-out. Because pheromones must be diffused or carried in a current, the animal cannot convey a message quickly over long distances, nor can it abruptly switch from one message to another. Although rats are able to distinguish the odors of dominant animals from the odors of submissive ones (Krames et al., 1969), no evidence exists of pheromones that transmit rapid changes in aggressiveness and status in the manner that is routine in auditory and visual communication. Furthermore, no case of information transfer by frequency and amplitude modulation of chemical emissions has been reported in any kind of animal, although this possibility has scarcely begun to be considered by biologists. As Bossert (1968) showed, the amount of potential information that might be encoded in this manner is surprisingly high. Under two special circumstances, when transmission occurs in still air over a distance of the order of a centimeter or less, or when it is accomplished in a steady, moderate wind, modulation is not only practicable but highly efficient. Under extremely favorable conditions, a perfectly designed system could transmit on the order of 10,000 bits of information a second, an astonishingly high figure considering that only one substance is involved. Under more realistic circumstances,

say for example in a steady 400-centimeters-per-second wind over a distance of 10 meters, the maximum potential rate of information transfer is still quite high—over 100 bits a second, or enough to transfer the equivalent of 20 words of English text per second at 5.5 bits per word. For every pheromone released independently, the same amount of capacity could be added to the channel capacity. We can hardly expect any animal species to achieve more than a minute fraction of the theoretical capacity calculated by Bossert. To do so would require the evolution of a symbolical and syntactical language, something no animal species has achieved in any other sensory modality. But it is conceivable that modulation has been added somewhere to pheromone communication in order to increase signal specificity, just as a great many visual and acoustical systems have acquired signal modulation in some animal species. To doubt it on the grounds that no examples are yet known is not enough, for human observers are incapable of detecting odor waves, especially under the environmental circumstances Bossert shows to be optimal for the evolution of odor modulation.

Even so, there is abundant evidence that animals have not in general relied on modulation of single chemical signals but have resorted to the only other course left open to them—the multiplication of glands or other principal biosynthetic sites to permit the independent discharge of pheromones with different meanings. The most olfactory mammals are covered with such signal sources. The black-tailed deer *Odocoileus hemionus*, for example, produces pheromones in at least seven sites: feces, urine, tarsal glands, metatarsal glands,

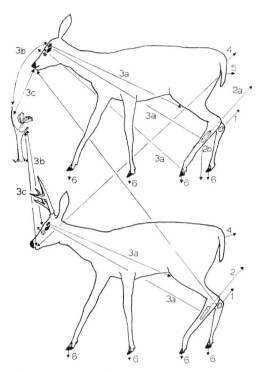

Figure 10-6 Sources and pathways of transmission of pheromones in the black-tailed deer. The scents coming from the tarsal organ (*1*), metatarsal gland (*2a*), tail (*4*) and urine (*5*) are all transmitted directly through the air. While the deer reclines the metatarsal gland also touches the ground (*2b*). The deer rubs its hindleg over its forehead (*3a*) and the forehead is rubbed against dry twigs (*3b*), which are sniffed and licked (*3c*). Finally, interdigital glands (*6*) leave scent directly on the ground. (From Müller-Schwarze, 1971.)

preorbital glands, forehead "gland," and interdigital glands. Insofar as they have been analyzed experimentally, the substances from each source have a different function (Müller-Schwarze, 1971; see Figure 10-6). Additional pheromone-producing glands occur elsewhere in other kinds of mammals: the body flanks, the chin, the perineum, the pouches of female marsupials, and so on. The social insects have carried this method of informational enrichment to still greater lengths. The workers and queens of the most advanced social Hymenoptera are walking batteries of exocrine glands (see Figure 10-7).

The size of pheromone molecules that are transmitted through air can be expected to conform to certain physical rules (Wilson and Bossert, 1963). In general, they should possess a carbon number between 5 and 20 and a molecular weight between 80 and 300. The a priori arguments that led to this prediction are essentially as follows. Below the lower limit, only a relatively small number of kinds of

molecules can be readily manufactured and stored by glandular tissue. Above it, molecular diversity increases very rapidly. In at least some insects, and for some homologous series of compounds, olfactory efficiency also increases steeply. As the upper limit is approached, molecular diversity becomes astronomical, so that further increase in molecular size confers no further advantage in this regard. The same consideration holds for intrinsic increases in stimulative efficiency, insofar as they are known to exist. On the debit side, large molecules are energetically more expensive to make and to transport, and they tend to be far less volatile. However, differences in the diffusion coefficient due to reasonable variation in molecular weight do not cause much change in the properties of the active space, contrary to what one might intuitively expect. Wilson and Bossert further predicted that the molecular size of sex pheromones, which

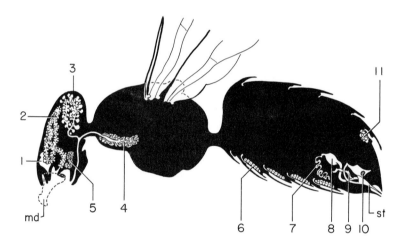

1.	Mandibular gland	2-heptanone	Alarm; queen substances (multiple functions in sex and colony control)
2.	Hypopharyngeal gland	Royal jelly	Larval food
3.	Head labial gland	?	Cleaning, dissolving, digestion (?)
4.	Thorax labial gland	?	Cleaning, dissolving, digestion (?)
5.	Postgenal gland	?	?
6.	Wax gland	Beeswax	Nest construction
7.	Poison gland	Venom	Defense
8.	Poison gland reservoir	Venom	Defense
9.	Dufour's gland	?	?
10.	Koschevnikov's gland	?	In queen, attraction of workers
11.	Nasanov's gland	Geraniol; citral, nerolic acid	Assembly, orientation to swarm

Figure 10-7 The numerous exocrine glands of the honeybee that are devoted to social organization. The honeybee is an example of a species that has enlarged its chemical "vocabulary" by the involvement of additional glands in the production of pheromones. Not indicated in this diagram is a site at the base of the worker sting (*st*) that produces isoamyl acetate, an alarm substance. The outline of the mandible (*md*) is also indicated by a dashed line.

generally require a high degree of specificity as well as stimulative efficiency, would prove higher than that of most other classes of pheromones, including, for example, the alarm substances. The empirical rule displayed by insects is that most sex attractants have molecular weights that are between 200 and 300, while most alarm substances fall between 100 and 200. Some of the evidence for this last statement, together with a discussion of the exceptions, has been reviewed by Wilson (1968b).

When we come to pheromones transmitted in water, however, a very different situation exists. The rules concerning diversity of molecular species are, of course, the same; but the rates at which given substances are passed into the medium from films or droplets, as well as the diffusion coefficients, are drastically altered. What kind of molecules might be expected in the aqueous pheromones? Only within the last several years have enough chemical characterizations been made to permit some generalizations. So far as molecular size is concerned, the substances fall into two distinct classes. One is exemplified by the sex pheromones of fungi and *Lebistes*, along with acrasin, the aggregating attractant of the slime mold. These substances are comparable in size to the airborne sex attractants of terrestrial animals. The diffusion coefficients of most water-soluble substances in this range of molecular weights is of the order of 10^{-5} in water and between 10^{-1} and 10^{-2} in air. A thousandfold or more decrease in diffusivity makes a great deal of difference in the properties of the active space. At least in the case of discontinuous pheromone release, the maximum radius of the space is the same in water as in air. But the time required to reach the maximum radius and the interval between release of the pheromone and the disappearance of the active space (that is, the fade-out time) are approximately 10,000 times greater in water than in air. How then can aquatic and marine organisms use molecules of this size? A better question is: How can organisms transmit pheromones through water at all? There are in fact two ways in which the same substance can be employed as efficiently in water as in the air: (1) the Q/K ratio (the ratio of molecules emitted to minimum density of molecules causing the response) can be adjusted appropriately; and (2) the pheromone can be spread more quickly by placing it in natural currents or creating artificial currents.

By extending the diffusion theory of Bossert and Wilson (1963), I have examined the possibilities of adjusting the Q/K ratios in aqueous systems with the following results. In order for the same substance to generate about the same intervals to maximum radius and fade-out in water as on land, it would be necessary for the Q/K ratio to be about a million times greater in water. In other words, the aquatic or marine species (in the single-puff model, where all the molecules are released at once) would have to increase the amount of pheromone solute a millionfold, or lower its response threshold to a millionth, or achieve some equivalent combined alteration of the two

parameters in order to achieve the same signal times as a terrestrial species using the same pheromone in air (Wilson, 1970). This adjustment, incidentally, would result in a hundredfold *increase* in the maximum radius of the active space.

Such a huge increment in Q/K has not been as difficult to attain as it might first seem. The most promising parameter is the emission rate, Q. When a pheromone is emitted as a film, spray, or droplet in air, the emission rate is largely a function of the vapor pressure. Within most homologous series, vapor pressure falls off steeply with an increase in molecular weight. In the alkane series, for example, the emission rate—measured in molecules per second from a surface of fixed area—declines somewhat less than one order of magnitude with every CH_2 group added. Proteins and other macromolecules have for practical purposes zero vapor pressure and cannot be transmitted by air unless they are somehow adsorbed onto bubbles or dust particles or absorbed into droplets of mist. But the same is not true for water transport. The solubility of large polar molecules is moderately high and can conceivably provide the requisite increase in Q in water as opposed to air.

Proteins in fact make up a large fraction of the known waterborne pheromones. In the case of the protistan pheromones and aggregation substance of barnacles, transport of these substances raises no problems since communication is by contact chemoreception or transmission over short distances. In the case of the snail alarm substances, the species have evidently made use of the fact that injured individuals release large quantities of their blood and tissue proteins into turbulent water, involuntarily of course. The ability of the liberated proteins to diffuse is limited but still adequate to generate a large active space. The diffusion coefficients of proteins in water at 20°C range from 0.34×10^{-7} to 1.6×10^{-8}. The long duration of the signal would be in accord with the behavior of the responding snails, who bury themselves or leave the water altogether.

Although the transmission rate to a fixed distance can be increased by enlarging the Q/K ratio, such an adjustment will also increase the time to fade-out. Consequently, in cases where a reasonable short fade-out time is required, we can expect to find additional devices, such as unstable molecular structure or enzymatic deactivation, that cancel the signals. These devices should be more prominently developed in waterborne systems than in airborne systems of similar function.

The foregoing discussion should suffice to point out some of the aspects of analysis that can be undertaken to advance this largely unexplored subject in the behavioral ecology and sociobiology of aquatic and marine organisms. For most of these organisms pheromones and allomones are the cardinal or even exclusive means of communication. One of the more sinister potential effects of chemical pollution is interference with such biological systems.

Auditory Communication

Like pheromones, sonic signals flow around obstacles and can be broadcast day and night in all weather conditions. They are intermediate in energetic efficiency between pheromones, which require little effort to transmit, and visual displays, many of which require extensive movement of the entire body. Sounds have considerable potential reach, exceeding the capacity of pheromones and light under a wide range of real conditions. Fraser Darling (1938) noted that the calls of gulls and other colonial seabirds are heard by other birds in the breeding colonies for distances of up to 200 meters. This is also approximately the reach of a great many species of songbirds and calling insects living in various habitats. Under the best of conditions, the most vocal of vertebrates can be heard for much greater distances. The roaring of male colobus and howler monkeys can be heard by human observers for over one kilometer. The champions among birds, and possibly terrestrial animals as a whole, are the colonially breeding grouse (Tetraonidae). The booming calls of the males reach for over a kilometer in the open country around the display grounds; and in a few species, such as the black grouse *Lyrurus tetrix* of Europe and the greater prairie chicken *Tympanuchus cupido* of North America, they can be heard for 3 to 5 kilometers. The visual displays of the same species, in contrast, can be seen for distances of no more than one kilometer and usually for much less (Hjorth, 1970).

This is not to say that animals evolve so as to shout over the greatest possible distances. On the contrary, the volume and frequency of animal calls seem designed to reach just those individuals of concern to the signaler and no others. To broadcast beyond them is to provide an unnecessary and dangerous homing beacon for predators. In some cases, of course, it is to the animal's advantage to project signals as far as possible. Males displaying on their leks, infants lost and in distress, and social animals calling in alarm while fleeing from a predator all seek maximum volume and transmission distance. The mobbing calls of birds provide an excellent example. They appear to be designed to carry long distances and to permit easy localization of the predatory birds that the mobs are attacking. In contrast, mothers calling young to their sides, group members maintaining contact in thick vegetation, and mated pairs performing nest-changing ceremonies use more modest, private signals that seldom reach past the ears of the intended receivers. Moynihan (1969) has utilized this principle to explain the difference in pitch found in the calls of various species of New World monkeys. The distance champions among these animals, the howler monkeys (*Alouatta*), use low-pitched roars to signal to rival groups far out of sight in the dense canopy of the rain forest. Other species, including the tamarins and night monkeys, emit high-pitched calls, which dissipate energy in the air more rapidly than sounds of lower frequency and hence die out

in shorter time. These high-frequency calls are also restricted by their greater tendency to scatter when they strike the numerous leaves and branches surrounding the signaler animal. The calls are used in a variety of circumstances, including short-range contact between neighboring troops. Moynihan has argued from several lines of evidence that the higher pitch of the signals is not an automatic outcome of the smaller size of the monkeys but is a specially evolved trait that enhances privacy and hence relief from the more intense predation generally suffered by small animals.

By far the most favorable design feature of vocal communication, the one that surely led to its adoption in the evolution of human linguistics, is its flexibility. Where pheromones must be deployed among multiple gland reservoirs to increase the rate of information transfer to any appreciable degree, all of the requisite sound signals can be generated from a single organ. Simple mechanical adjustments of the organ permit it to vary the volume, pitch, harmonic structure, and sequencing of notes that in combination create a vast array of distinguishable signals. The rapidity of the transmission of sounds and the equal quickness of their fade-out provide the basis for a very high rate of information transfer.

Bird song represents one of the pinnacles of auditory communication. It has been subjected to every level of analysis, from neurophysiological to evolutionary, by a large group of able investigators (see especially Thorpe, 1961, 1972b; Konishi, 1965; Hinde, ed., 1969; and Chapter 7 in this book). A basic dichotomy has emerged from this work between call notes and songs. Call notes are much the simpler in structure, consisting of one or a few short bursts of sound. Their functions are among the most direct and elementary in the repertory of the species: alarm, mobbing, distress, contact-maintenance, flight intention, and the like. They are also efficient in design. Distress and mobbing calls, for example, typically consist of loud, short notes covering a wide range of frequencies (see Chapter 3). Each of these features promotes localization over moderate to long distances. Fear trills and predator warning calls, by contrast, are longer and cover fewer frequencies, properties that cause them to be audible but difficult to locate (Marler, 1957). Call notes are given by a majority or all of the members of the species during most or all of the year. Bird songs, in contrast, are given most commonly by males during the breeding season. Typically they are elaborate in structure, lengthy in duration, and simply broadcast into the environment without serving any obvious immediate purpose. Functions do exist, however, and most can be characterized by a single word: identification. The male uses his song to announce that he is a member of such and such a species, a sexually mature male on a territory, and prone to some measurable degree to undertake the actions of territorial defense and courtship. A second male belonging to the same species recognizes that the singer will defend his territory, and—by the arcane rules of instinctive behavior—probably win. The conspecific female, on the other hand, is informed that she will receive courtship displays if she ventures close enough.

Why are bird songs so complex? It has long been recognized that the vocalizations of males are important premating isolating mechanisms. This means that they collaborate with other kinds of genetically based differences to prevent species from interbreeding. In fact, as W. H. Thorpe has said, "it is virtually impossible to think of two closely related species of birds which, possessing full song, are not thereby specifically distinguishable." Bird watchers know that many complexes of very similar species, such as the *Empidonax* flycatchers of North America, are best identified in the field by their songs, the same cues that the birds themselves use during the breeding season. According to current speciation theory, most or all bird species begin the multiplication process when a single, ancestral species is broken into two or more geographically isolated populations. The barrier causing the fragmentation can be any impassable feature of the environment—a dry valley separating mountain forests, a mountain ridge separating dry valleys, a sea strait separating two islands, or whatever. As these daughter populations subsequently evolve, they inevitably diverge from one another in many genetically determined traits, representing the multiple differences in the environments they inhabit. Given sufficient time, the populations become so different that hybridization is made difficult when and if the geographic barrier is removed. If different enough, they may be wholly segregated into separate preferred habitats, or breed at different seasons, or simply not respond to one another's courtship displays. The genetically determined differences, which block mating attempts between the newly formed species, are the premating isolating mechanisms. Suppose that the populations rejoined before the premating devices became perfect, so that substantial hybridization occurred. Hybrids of genetically very divergent populations, especially those in the F_2 generation and beyond, tend to be sterile or inviable. As a consequence, a selective premium is put on genotypes that are so different from the opposing species that interspecific mating is avoided and gametes are not wasted on making hybrids. The theoretically expected result, which can take place in as little as ten generations of intense interaction, is character displacement, in this case the reinforcement of premating isolating mechanisms. It is to be expected that among newly formed bird species, the male song will be frequently implicated in displacement; and among related species occupying the same geographic range, the song will evolve so as to be among the most clearly distinguishing features.

A correlate of the theory is that the larger the number of species occupying a given area, the more elaborate (hence, distinctive) the male songs and other courtship displays. Although the evidence for this predicted phenomenon is patchy and equivocal (Thielcke, 1969;

Grant, 1972), it is consistent with the theory and in certain cases strongly suggestive. Most notably, bird species native to islands where they are in contact with few or no related species tend to have either more variable songs, which overlap those of similar species on the mainland, or else songs that are simpler in structure. The blue titmouse *Parus caeruleus*, an endemic of Teneriffe and the only member of its genus on the island, uses an extraordinary range of songs, some peculiar to itself and others resembling the songs of various *Parus* species found on the European mainland. A similarly variable repertory is employed by the Canary Islands chiffchaff *Phylloscopus collybita*. But the chaffinches endemic to the Islands, *Fringilla teydea* and *F. coelebs tintillon*, have simpler songs than their European counterparts (Marler, 1960). More detailed and persuasive evidence exists for the evolutionary displacement of courtship calls of certain frogs (Littlejohn and Loftus-Hills, 1968), but overall the data are too thin to permit application of the theory to animal species generally.

Speciation is not the only force that injects complexity into bird song. Somehow the birds recognize intensity and mood and, in at least a few species, the individual identity of the songster. These functions require that additional, special properties be built into the song. In 1960 Peter Marler speculated that such components of information are encoded into different parts of the song, perhaps into separate segments of the individual notes themselves. Whether true or false, this hypothesis is at least heuristic, because it suggests that much of the analysis of bird song is really a problem of decomposing the song and decoding its information content according to functional categories. Some documentation of the idea has recently come from S. T. Emlen's study of the indigo bunting *Passerina cyanea*. By observing the responses of males to taped songs of similar species as well as to indigo bunting songs experimentally modified on the recordings, Emlen was able to estimate the role of major components of structure, sequence, and timing (see Figure 10-8). His most interesting finding was that the several major categories of identification are indeed deployed into separate features of the songs. Components of species recognition are those that are generally constant within populations, individual recognition components vary from one male's song to the next, and motivational cues reside in the components that vary markedly within the repertories of individual birds. Most of the segments contribute some form of information, in some cases redundantly with others. However, at least one of the most conspicuous features, the sequence of the notes, conveys no apparent message to other male buntings.

Bird species vary greatly in the mode of development of their songs. In some the male song is transmitted from generation to generation entirely by heredity, with no learning required. Members of other species, including the chaffinch, must hear the singing of other conspecific individuals in order to develop part or all of the normal song.

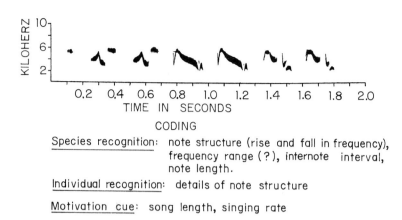

CODING

Species recognition: note structure (rise and fall in frequency),
frequency range (?), internote interval,
note length.

Individual recognition: details of note structure

Motivation cue: song length, singing rate

Figure 10-8 The information content of the song of the male indigo bunting. The figure shows the sound spectrogram of a typical song. Components that are inferred but not proven to have a stated function are indicated by a question mark. (Based on Emlen, 1972.)

The learning process has the effect of permitting territorial males to imitate one another, one mechanism that leads to the formation of local regional dialects. Familiarity with neighbors results in the dear enemy phenomenon (see Chapter 8) and a reduction of unnecessary territorial strife. It can also hasten speciation by congealing overall variation around certain genetic forms that correspond to genetically semiisolated species. And, finally, it has been demonstrated that mated pairs of some species work out individual duets that tighten their bonding and improve their vocal contact.

The singing of crickets, cicadas, and other insects is much simpler than that of birds. Insects are tone-deaf; they cannot perceive differences in pitch. Identification cues are added principally by modifying the intensity of the sounds and the rapidity with which they are produced. Examples of three songs that might be distinguished by insects are: "CheeeCHEEEcheeeCHEEE...," "Cheee
CHEEE   cheee   CHEEE...," and "Cheee
cheee   cheee   cheee." This is the reason that insect sounds seem so unmelodic and monotonous to human ears. Yet (as illustrated in Figure 10-9) a large number of messages can be generated without the benefit of pitch and harmony.

Surface-Wave Communication

Water striders (family Gerridae) are stilt-legged insects that live entirely on the surface film of quiet bodies of water. Although moderate in size, they are supported by surface tension. It has long been known that water striders are sensitive to water waves, which they detect with proprioceptors in the legs. They dart toward and seize insects that fall into the water, but they flee from more severe perturbations

of the sort caused by fish and other potential vertebrate enemies. Recently Wilcox (1972) discovered that at least one species, a *Rhagadotarsus* living in northeastern Australia, conducts most of its courtship by means of the propagation of patterned surface waves. The signals passed back and forth between the sexes at various stages of the courtship differ in the frequency and pacing of the ripples. The sequence begins when a male grasps a floating or fixed object on the water surface and vibrates in a way that sends out waves at the rate of 17-29 per second. Females nearby respond by moving toward the source. When one approaches to within 5-10 centimeters of the male, he switches to "courtship calling" and finally to pure courtship signals. At 2-3 centimeters the female responds with courtship signals of her own, followed by a series of tactile signals that lead finally to copulation. During and immediately after pairing, the male sets off still another kind of courtship rippling. During a brief interval following copulation, the female excavates a hole in the object to which she

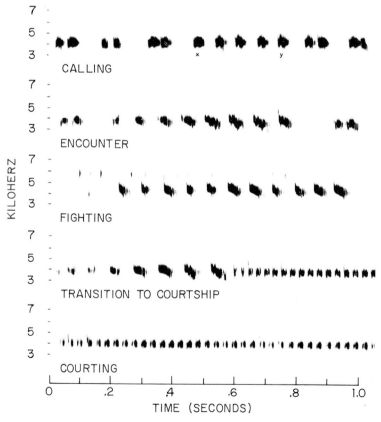

Figure 10-9 The acoustic repertoire of the cricket *Teleogryllus commodus*. Since insects are tone-deaf, the signals are differentiated on the basis of volume and rate of emission. (Redrawn from Alexander, 1962.)

and the male have been attached, and lays her eggs. When she leaves, the male begins a new round of calling.

A basically similar mode of communication is used by a few kinds of spiders with the aid of their webs. The females of some species of cobweb spiders in the genus *Theridion* feed their young by regurgitation and allow them to share the prey. Nørgaard (1956) has described two specialized forms of web communication between the mother of the European *T. saxatile* and her offspring. The young remain in the mother's web for about a month after hatching and feed simultaneously on the prey items captured, almost 90 percent of which are ants. While still very young, they remain inside a retreat located in the center of the tangled web and let the mother kill all of the entangled prey. Whenever they venture too close to a struggling ant, the female turns toward them and thrums her forelegs on the web strands rather like a musician plucking the strings of his instrument. The young respond to the thrumming by running back into the retreat. As the young grow larger they become able to participate in the capture, and now, instead of warning them off, the mother summons them with a different, sweeping motion of the forelegs.

Tactile Communication

Communication by touch is maximally developed just where we would expect to find it, in those intimate sequences of aggregation, conciliation, courtship, and parent-offspring relations that bring animals into the closest bodily contact. For tightly aggregating species, such as hibernating beetles and pods of moving fish (see Chapter 3), body contact is both the evident goal and the signal that terminates the searching behavior. But in some instances it triggers other physiological and behavioral changes that lead animals into new modes of existence. Tactile stimulation in aphids is the dominant cue that transforms these insects from wingless into winged forms. The alates reproduce sexually and disperse much more easily, thus alleviating population pressure in the mother colony while founding new colonies on additional host plants (Lees, 1966). Locust nymphs reared in groups learn to distinguish their fellow hoppers from other dark objects of equal size, and they greet them with the typical social responses of the species—kicking their own hind legs, twirling their antennae, and inspecting the bodies of the other locusts with their palps and antennae. Peggy Ellis (1959) simulated the socialization process by rearing nymphs in isolation but in constant contact with fine, constantly moving wires. The locusts achieved a normal level of response by this form of tactile stimulation alone. Equally profound effects occur in the vertebrates. Experiments by N. T. Adler and his associates (Adler, 1969; Adler et al., 1970) revealed that the multiple intromissions of male rats, which ordinarily precede ejaculation, induce two adaptive physiological changes within the female. First,

these presumably tactile stimuli increase the rate of transport of the sperm to the uterus. Second, through a neuroendocrine reflex not yet fully elucidated, the stimuli raise the amount of progesterone and 20α-OH-pregn-4-ene-3-one in the blood and hence increase the percentage of successful implantations by fertilized ova in the uterine wall.

Visual Communication

Directionality is the paramount feature of systems of visual communication. Visual images are instantly pinpointed in space: the honeybee, a typical large-eyed insect, can distinguish two points that subtend an angle of approximately 1°, while the human eye, which is typically mammalian in construction, has an angle of resolution of 0.01°. Light signals lend themselves to either one or the other of two opposite strategies of signal duration. At one extreme, patterns of shading and coloration can be grown more or less permanently into the surface, or else added temporarily by special pigment deposition, chromatophore expansion and contraction, and so forth, providing signals of long duration at minimal energy cost. Hence whenever vision is possible, optic signals are found to be paramount in the identification of species as individuals, as well as the status of individuals within dominance systems. At the opposite extreme, visual signals can be designed in such a way as to provide rapid fade-out and turnover. Consequently they are routinely coupled in evolution with acoustic signals to transmit the most rapidly fluctuating moods of courtship and aggressive encounters.

But the distinctive features of light signals are advantageous only under limited conditions. In the absence of light, visual communication fails unless the animals can generate their own signals by bioluminescence. Visual communication further works only when the signals are directed at the photic receptors. In order to communicate with any precision, two animals must not only perform the appropriate actions but orient themselves correctly for each transmission. This probably explains the fact that although many animal species are known whose systems are wholly chemical, and many others whose systems are almost exclusively auditory, there are few if any that depend to a comparable degree on vision.

Electrical Communication

Sharks and rays, catfish, common eels (Anguillidae), and electric fish (Gymnotidae, Mormyridae, Gymnarchidae) are capable of sensing and orienting to low-frequency, feeble voltage gradients (Kalmijn, 1971; Bullock, 1973). Electroreception is widely used as a prey-seeking device. By means of feeble, steady electric fields that leak out of flatfish, sharks are able to locate these prey even when they are buried in sand. Furthermore, electric fish generate their own fields by means of electric organs consisting of highly modified muscle tissue. When prey or other objects in the water disturb the field, their presence is betrayed to the fish even when all other sensory cues are lacking (Lissmann, 1958). In view of this degree of sophistication, it is perhaps not surprising to find that at least some of the electric fish also use their fields to communicate with one another (Möhres, 1957; Valone, 1970; Black-Cleworth, 1970). Black-Cleworth showed that individuals of *Gymnotus carapo* recognize and tend to avoid the normal electrolocating pulses of members of their own species. Attacks are preceded and accompanied by sudden increases in the discharge frequencies, a pattern similar to the acceleration of pulses triggered when prey are located. Attacking fish also suddenly cease discharges for periods of less than 1.5 seconds. Both the sharp increase in discharge frequency and the discharge break are followed by the retreat of the receiving animals. The two behaviors can therefore be interpreted as threat signals.

We do not know whether electrocommunication occurs in animals other than the electric fish because the phenomenon can only be revealed by special techniques. The advantages of this sensory channel are considerable. Like sound, electric fields can be detected in the

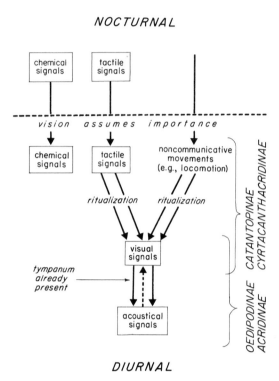

Figure 10-11 The evolution of communication in grasshoppers. The ancestral forms are hypothesized to have been nocturnal, relying heavily on pheromones and tactile signals. The more primitive Catantopinae and Cyrtacanthacridinae use a more or less even mixture of chemical, tactile, and visual signals. The Oedipodinae and Acridinae have added acoustical signals and use them to a prevailing degree with visual signals; at the same time pheromones and tactile signals have receded to a lesser role. (From Otte, 1970.)

dark, and they flow around ordinary obstacles. They are also strongly directional, and, insofar as they prove to be used by a relatively few species, they provide a high degree of privacy. At the same time, they can be used only in relatively quiet water and can be employed only over a short range.

Evolutionary Competition among Sensory Channels

If the theory of natural selection is really correct, an evolving species can be metaphorized as a communications engineer who tries to assemble as perfect a transmission device as the materials at hand permit. Microorganisms, sponges, fungi, and the lowest metazoan invertebrates are all but stuck with chemoreception and tactile responses. Visual and auditory systems require multicellular receptor

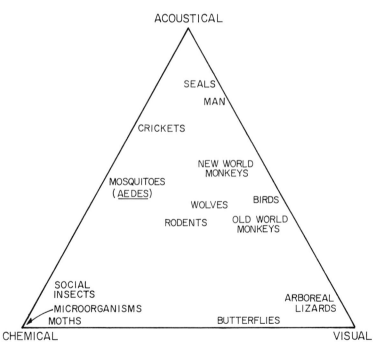

Figure 10-10 The relative importance of sensory channels in selected groups of organisms. The nearness of the group to each apex indicates, by wholly subjective and intuitive criteria, the proportionate usage of the channel in the species signal repertory. Tactile, surface-wave, and electrical channels are not included.

organs and, in the case of auditory signals, special sound-producing organs as well. Electric and surface-wave systems also depend on multicellular signaling and receiving devices. In general, the more primitive the organism and the simpler its body plan, the more it depends on chemical communication.

The effects of phylogenetic constraint on the selection of sensory channels are apparent to a lesser degree throughout the higher invertebrates and vertebrates. For example, consider why butterflies are colorful and silent. They seem bright and cheerful to us in large part because we are vertebrates heavily dependent on vision, and butterflies have tended to develop poisonous and distasteful substances to repel vertebrate predators while simultaneously evolving audacious color patterns to provide warnings about their unpalatable condition (Brower, 1969). They have also evolved distinct ultraviolet wing and body patterns, visible to one another but not to vertebrates, which serve as the medium of much of their private communication (Silberglied and Taylor, 1973). Why haven't they also evolved elaborate acoustical signals like birds? Butterflies and birds live in the same environments, fly at approximately the same heights, and communicate over comparable distances. The answer appears to be that the bodies of adult butterflies, unlike those of birds, are too small and

constructed on too delicate a plan to allow the development of the noisy sound-producing machinery required to transmit effectively over long distances.

Within their personal phylogenetic constraints, species have chosen and molded sensory channels in astonishingly diverse combinations (see Figure 10-10). They have also attained an efficiency of design that should impress any human engineer. In the case of butterflies again, it can be noted that like moths they use sex pheromones extensively but, unlike moths, they transmit the pheromones principally by contact or through air over distances of no more than a few centimeters. The reason for this curtailment may well be that the thermal updrafts and turbulence of the daytime atmosphere preclude the formation of long active spaces. Ethologists have had no difficulty making such correlations between environment and sensory modes across widely separated phylogenetic groups. Some of the best evolutionary reconstructions have traced shifts from one modality to another at the species level and are based on enough detail to be fully persuasive. An example is Otte's analysis of communicative evolution in grasshoppers, summarized in Figure 10-11. Many of the better-documented vertebrate cases have been ably reviewed by Wickler (1967, 1972).

Chapter 11 **Aggression**

What is aggression? In ordinary English usage it means an abridgment of the rights of another, forcing him to surrender something he owns or might otherwise have attained, either by a physical act or by the threat of action. Biologists cannot improve on this definition, even in the narrow context of animal behavior, except to specify that in the long term a loss to a victim is a real loss only to the extent that it lowers genetic fitness. In an attempt to be more precise, many writers have turned to the word "agonistic," coined by Scott and Fredericson (1951) to refer to any activity related to fighting, whether aggression or conciliation and retreat. However, agonistic behavior cannot be defined any more precisely than aggressive behavior or fighting behavior in particular cases, and the term is ordinarily useful only in pointing to the close physiological interrelatedness of aggressive and submissive responses. But we should not worry too much about terminology. The essential fact to bear in mind about aggression is that it is a mixture of very different behavior patterns, serving very different functions. Here are its principal recognized forms:

1. *Territorial aggression.* The territorial defender utilizes the most dramatic signaling behavior at its disposal to repulse intruders. Escalated fighting is usually employed as a last resort in case of a stand-off during mutual displays. The losing contender has submission signals that help it to leave the field without further physical damage, but they are ordinarily not so complex as those employed by subordinate members of dominance orders. By contrast, females of bird species entering the territories of males often use elaborate appeasement signals to transmute the aggressive displays of the males into conciliation and courtship.

2. *Dominance aggression.* The aggressive displays and attacks mounted by dominant animals against fellow group members are similar in many respects to those of the territorial defenders. However, the object is less to remove the subordinates from the area than to exclude them from desired objects and to prevent them from performing actions for which the dominant animal claims priority. In some mammalian species, dominance aggression is further characterized by special signals that designate high rank, such as the strutting walk of lemmings, the leisurely "major-domo" stroll with head and tail up of rhesus macaques, and the particular facial expressions and tail postures of wolves. Subordinates respond with an equally distinctive repertory of appeasement signals.

3. *Sexual aggression.* Males may threaten or attack females for the sole purpose of mating with them or forcing them into a more prolonged sexual alliance. Perhaps the ultimate development in higher vertebrates is the behavior of male hamadryas baboons, who recruit young females to build a harem and continue to threaten and harass these consorts throughout their lives in order to prevent them from straying.

4. *Parental disciplinary aggression.* Parents of many kinds of mammals direct mild forms of parental aggression at their offspring to keep them close at hand, to urge them into motion, to break up fighting, to terminate unwelcome suckling, and so forth. In most but not all cases the action serves to enhance the personal genetic fitness of the offspring.

5. *Weaning aggression.* The parents of some mammal species threaten and even gently attack their own offspring at the weaning time, when the young continue to beg for food beyond the age when it is necessary for them to do so. Recent theory (see Chapter 16) suggests that under a wide range of conditions there exists an interval in the life of a young animal during which its genetic fitness is raised by continued dependence on the mother, while the mother's fitness is simultaneously lowered. This conflict of interests is likely to bring about the evolution of a programmed episode of weaning aggression.

6. *Moralistic aggression.* The evolution of advanced forms of reciprocal altruism carries with it a high probability of the simultaneous emergence of a system of moral sanctions to enforce reciprocation (see Chapter 5). Human moralistic aggression is manifested in countless forms of religious and ideological evangelism, enforced conformity to group standards, and codes of punishment for transgressors.

7. *Predatory aggression.* There has been some question about whether predation can be properly classified as a form of aggression (for example, Davis, 1964). Yet if one considers that cannibalism is practiced by many animal species, sometimes accompanied by territoriality and other forms of aggression, and sometimes not, it is hard to regard predation as an entirely different process.

8. *Antipredatory aggression.* A purely defensive maneuver can be escalated into a full-fledged attack on the predator. In the case of mobbing the potential prey launches the attack before the predator can make a move. The intent of mobbing is often deadly and in rare instances brings injury or death to the predator.

Previous authors, particularly Tinbergen (1971), Barlow (1968), Moyer (1969, 1971), and J. L. Brown (1970b), have stressed the eclectic nature of "aggression." Aggressive behavior serves very diverse functions in different species, and different functional categories evolve independently in more than one control center of the brain. Moyer constructed the following classification of seven categories based both on animal and human behavior: predatory, intermale, fear-induced, irritable, territorial, maternal, and instrumental. The eight provisional categories I have recommended here are similar but somewhat less introspective, and they conform more closely to the true adaptive categories observed in the natural behavior of the mass of animal species. Barlow cited an illuminating example of multiple forms of aggression coexisting in the behavior of rattlesnakes. When two males

compete, they intertwine their necks and wrestle as though testing each other's strength, but they do not bite. In contrast, the snake stalks or ambushes prey—it strikes from any of a number of positions. Also, it does not give warning with its rattle. When confronted by an animal large enough to threaten its safety, the rattlesnake coils, pulls its head forward to the center of the coil in striking position, and raises and shakes its rattle. It may also rear the head and neck into a high S-shaped posture. However, if the intruder is a king snake, a species specialized for feeding on other snakes, the rattlesnake switches to a wholly different maneuver: it coils, hides its head under its body, and slaps at the king snake with one of the raised coils.

Aggression and Competition

The largest part of aggression among members of the same species can be viewed as a set of behaviors that serve as competitive techniques. Competition, as most ecologists employ the word (Miller, 1967), means the active demand by two or more individuals of the same species (intraspecific competition) or members of two or more species at the same trophic level (interspecific competition) for a common resource or requirement that is actually or potentially limiting. This definition is consistent with the assumptions of the Lotka-Volterra equations, which still form the basis of the mathematical theory of competition (Levins, 1968). The theory of population biology suggests that competitive phenomena are meaningfully divided into two large classes: *sexual competition* and *resource competition.* The former is exemplified by the violent *machismo* of males in the breeding season and especially upon the communal display grounds: the horn fighting of male sheep, deer, and antelopes, the spectacular displays and fighting among males of grouse and other lek birds, the heavyweight battles of elephant seals for the possession of harems, and others. The struggle for possession of multiple females is competition for a very special kind of resource. It becomes a significant part of the repertory when r selection is paramount or when other environmental pressures are relaxed to the extent that males can afford to invest the large amounts of time and energy required to be a polygamist. The theory of this subject will be developed in the discussion of the evolution of sexual behavior in Chapter 15.

Nonsexual aggression practiced within species serves primarily as a form of competition for environmental resources, including especially food and shelter. It can evolve when shortages of such resources become density-dependent factors (see Table 11-1 and the introductory discussion of density dependence in Chapter 4). However, even in this circumstance aggression is only one competitive technique among many that can emerge. For reasons that we are only beginning to understand, species may elect to compete by means of scrambling methods that do not include aggressive encounters. The following

Table 11-1 A simplified classification of the density-dependent factors that reduce population growth rates. The factors grouped under contest competition are asterisked to stress that aggressive behavior constitutes only one alternate outcome in the evolution of density-dependent controls.

A. Competition
  * 1. Contest competition
    * a. Fighting and cannibalism
    * b. Territoriality
    * c. Dominance orders
  　2. Scrambling competition
B. Predation and disease
C. Emigration
D. Noncompetitive modification of the environment

---

generalizations about competition in animals also pertain to the evolution of aggressive behavior (Wilson, 1971b).

1. The mechanisms of competition between individuals of the same species are qualitatively similar to those between individuals of different species.

2. There is nevertheless a difference in intensity. Where competition occurs at all, it is generally more intense within species than between species.

3. Several theoretical circumstances can be conceived under which competition is perpetually sidestepped (Hutchinson, 1948, 1961). Most involve the intervention of other density-dependent factors of the kind just outlined or fluctuations in the environment that regularly halt population growth just prior to saturation.

4. Field studies, although still very fragmentary in nature, have tended to verify the theoretical predictions just mentioned. Competition has been found to be widespread but not universal in animal species. It is more common in vertebrates than in invertebrates, in predators than in herbivores and omnivores, and in species belonging to stable ecosystems than in those belonging to unstable ecosystems. It is often forestalled by the prior operation of other density-dependent controls, the most common of which are emigration, predation, and disease.

5. Even where competition occurs, it is frequently suspended for long periods of time by the intervention of density-independent factors, especially unfavorable weather and the frequent availability of newly created empty habitats.

6. Whatever the competitive technique used—whether direct aggression, territoriality, nonaggressive "scrambling," or something else—the ultimate limiting resource is usually food. Although the documentation for this statement (Lack, 1966; Schoener, 1968a) is still thin enough to be authoritatively disputed (Chitty, 1967b), there still seem to be enough well-established cases to justify its provisional acceptance as a statistical inference. It is also true, however, that a minority of examples involve other limiting resources: growing space in barnacles and other sessile marine invertebrates (Connell, 1961; Paine, 1966); nesting sites in the pied flycatcher (von Haartman, 1956) and Scottish ants (Brian, 1952a,b); resting places of high moisture in salamanders (Dumas, 1965) and of shade in the mourning chat in African deserts (Hartley, 1949); nest materials in rooks (C. J. F. Coombs in Crook, 1965) and herons (A. J. Meyerriecks, University of South Florida, Tampa, personal communication).

The Mechanisms of Competition

If aggressive behavior is only one form of competitive technique, consider now a series of cases that illustrate the wide variation in this technique actually recorded among animal species. We will start with aggression in its direct and most explicit form and then, by passing from species to species, examine the increasingly more subtle and indirect forms.

Direct Aggression

When the barnacles of the species *Balanus balanoides* invade rock surfaces occupied by the barnacle species *Chthamalus stellatus*, they eliminate these competitors by direct physical seizure of the attachment sites. In one case studied by Connell (1961) in Scotland, 10 percent of the individuals in a colony of *Chthamalus* were overgrown by *Balanus* within a month, and another 3 percent were undercut and lifted off in the same period. A few others were crushed laterally by the expanding shells of the dominant species. By the end of the second month 20 percent of the *Chthamalus* had been eliminated, and eventually all disappeared. Individuals of *Balanus* also destroy one another but at a slower rate than they do members of the competitor species.

Ant colonies are notoriously aggressive toward one another, and colony "warfare" both within and between species has been witnessed by many entomologists (for example, Talbot, 1943; Haskins and Haskins, 1965; Yasuno, 1965). Pontin (1961, 1963) found that the majority of the queens of *Lasius flavus* and *L. niger* attempting to start new colonies in solitude are destroyed by workers of their own species. Colonies of the common pavement ant *Tetramorium caespitum* defend their territories with pitched battles conducted by large masses of workers. The adaptive significance of the fighting has been made clear by the recent discovery that the average size of the worker and the production of winged sexual forms at the end of the season, both of which are good indicators of the nutritional status of the colony, increase with an increase in territory size (Brian, Elmes, and Kelly, 1967). The following description by Brian (1955) of fighting among workers belonging to different colonies of *Myrmica ruginodis* is typical of a great many territorial ant species. The dispute in this particular case was brought about when workers from one colony approached those of another colony at a sugar bait.

If its approach is incautious, the feeder turns round . . . and grapples, and the pair fall to the ground and break. On the other hand, the incomer may approach slowly, and examine the abdomen of the feeder carefully without disturbing it; then it grips it by the pedicel (with the mandibles) and lifts it up. In this grip the lifted ant invariably remains quiescent, and is carried right back to the nest of the incomer. Sometimes under circumstances when a perfect grip is not obtained, other ants may become involved, and a group of three or four workers, composed of individuals from both nests, may struggle backwards and forwards along a line between the nest and the source (no perceptible track is formed). Mortality does not occur in the field, but those ants that are successfully dragged into the opposing nest will probably be dismembered. Hence the outcome of these struggles should favor the colony that brings the most workers to the site; that is, it will be related to colony size, proximity and recruitment ability.

One of the more dramatic spectacles of insect biology is provided by the large-headed soldiers of certain species belonging to the genus *Pheidole*. These individuals have mandibles shaped approximately like the blades of wire clippers, and their heads are largely filled by massive adductor muscles. When clashes occur between colonies the soldiers rush in, attack blindly, and leave the field littered with the severed antennae, legs, and abdomens of their defeated enemies. Brian (1956) has provided evidence that interference among colonies leads to replacement and "dominance hierarchies" among Scottish ant species that place the winners in the warmest nest sites. He identified the three following competitive techniques: (1) gradual encroachment on the competitor's nest; (2) occupation of nest sites abandoned by competitor colonies following adverse microclimatic change (for example, the nest chambers becoming temporarily too wet or cold), the occupation being accomplished when conditions improve but before the competitor can return; (3) siege, involving continuous harassment and fighting, until the competitor evacuates the nest site. Interference at the colony level sometimes leads to the total extirpation of one species by another from a local area. This extreme result occurs most frequently in unstable environments, such as agricultural land, or when newly introduced species invade native habitats. An example is given in Figure 11-1.

There can be no question that fighting and even cannibalism are normal among the members of some insect species. In the life cycle of certain species of parasitic Hymenoptera belonging to the families Ichneumonidae, Trigonalidae, Platygasteridae, Diapriidae, and Serphidae the larvae undergo a temporary transformation into a bizarre fighting form that kills and eats other conspecific larvae occupying the same host insect. This reduces the number of parasites to a

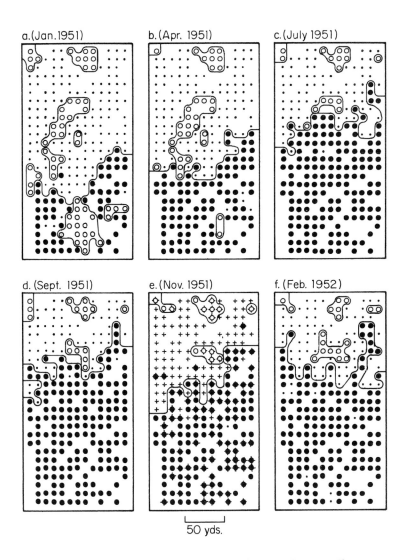

a. (Jan. 1951)   b. (Apr. 1951)   c. (July 1951)

d. (Sept. 1951)   e. (Nov. 1951)   f. (Feb. 1952)

50 yds.

● = Palm occupied by A. longipes
○ = Palm occupied by O. longinoda
· = Palm in which A. longipes and O. longinoda were absent
— Approx. boundary of territories occupied by A. longipes and O. longinoda

+ P. punctulata nesting at base of palm
✦ P. punctulata and A. longipes present on palm
◇ P. punctulata and O. longinoda present on palm

Figure 11-1 The exclusion of the ant *Oecophylla longinoda* by its competitor *Anoplolepis longipes* in a coconut plantation in Tanzania. The exclusion occurs through fighting at the colony level. In areas of sandy soil with sparse vegetation, *Anoplolepis* replaces *Oecophylla*, but where the vegetation is thicker and the soil less open and sandy, the reverse often occurs. A third species, *Pheidole punctulata*, is occasionally abundant but plays a minor role. (From Way, 1953.)

number that more easily grow to the adult stage on the limited host tissue available. Two of the cannibalistic species are illustrated in Figure 11-2.

Murder and cannibalism are also commonplace in the vertebrates. Lions, for example, sometimes kill other lions. In his study of the Serengeti prides, Schaller (1972) observed several fights between males that ended fatally. He also recorded a case of the killing and can-

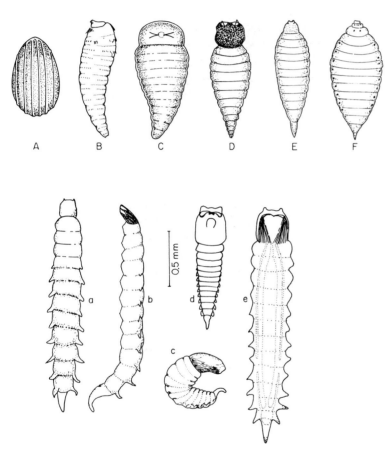

Figure 11-2 In certain species of parasitic wasps, the larvae undergo a temporary transformation into a bizarre fighting form, equipped with a sclerotized head and large mandibles. While in this instar the larvae inhabiting a single host insect fight together until only one is left alive. In the upper row are shown the egg and successive larval instars of the trigonalid *Poecilogonalos thwaitesii*; the fighting stage is the fourth instar (*D*). In the lower row are the first (*a, b*) and second (*c–e*) larval instars of the ichneumonid *Collyria calcitrator*; the fighting stage in this case is the second instar. (From *Entomophagous Insects* by C. P. Clausen. Copyright © 1940 by McGraw-Hill Book Company. Used with permission.)

nibalism of cubs after one of the protector males died and the territory was invaded by several other prides. Less severe fighting is more frequent, and it results in injuries and infections that ultimately shorten the lives of many individuals. Hyenas are truly murderous by human standards (Kruuk, 1972). They are also habitual cannibals. Mothers must stand guard while their cubs are feeding on a carcass in order to prevent them from being eaten by other members of the clan. Neighboring clans sometimes engage in pitched battles over carcasses of prey that one or the other of the groups has killed. The following account is taken from Kruuk's protocols:

The two groups mixed with an uproar of calls, but within seconds the sides parted again and the Mungi hyenas ran away, briefly pursued by the Scratching Rocks hyenas, who then returned to the carcass. About a dozen of the Scratching Rock hyenas, though, grabbed one of the Mungi males and bit him wherever they could—especially in the belly, the feet, and the ears. The victim was completely covered by his attackers, who proceeded to maul him for about 10 min. while their clan fellows were eating the wildebeest. The Mungi male was literally pulled apart, and when I later studied the injuries more closely, it appeared that his ears were bitten off and so were his feet and testicles, he was paralyzed by a spinal injury, had large gashes in the hind legs and belly, and subcutaneous hemorrhages all over. . . . The next morning I found a hyena eating from the carcass and saw evidence that more had been there; about one-third of the internal organs and muscles had been eaten. Cannibals!

The annals of lethal violence among vertebrate species are beginning to lengthen. Male Japanese and pig-tailed macaques have been seen to kill one another under seminatural and captive conditions when fighting for supremacy (Kawamura, 1967; Bernstein, 1969). When a new group of Barbary macaques was introduced into the Gibraltar population, severe fighting broke out that resulted in some deaths (Keith, 1949). In central India, roaming langur males sometimes invade established troops, oust the dominant male, and kill all of the infants (Sugiyama, 1967). Young black-headed gulls that wander from the parental nest territory are attacked and sometimes killed by other gulls (Armstrong, 1947), while in the brown booby (*Sula leucogaster*) of Ascension Island, the first-hatched young routinely thrusts its second-hatched sibling from the nest (Simmons, 1970). A case of cannibalism of an infant by an adult has even been reported in the chimpanzee, but the event does appear to be truly rare (Suzuki, 1971).

The evidence of murder and cannibalism in mammals and other vertebrates has now accumulated to the point that we must completely reverse the conclusion advanced by Konrad Lorenz in his book *On Aggression*, which subsequent popular writers have proceeded to consolidate as part of the conventional wisdom. Lorenz wrote, "Though occasionally, in the territorial or rival fights, by some mishap a horn may penetrate an eye or a tooth an artery, we have never found that the aim of aggression was the extermination of fellow members of the species concerned." On the contrary, murder is far more common and hence "normal" in many vertebrate species than in man. I have been impressed by how often such behavior becomes apparent only when the observation time devoted to a species passes the thousand-hour mark. But only one murder per thousand hours per observer is still a great deal of violence by human standards. In fact, if some imaginary Martian zoologist visiting Earth were to observe man as simply one more species over a very long period of time, he might conclude that we are among the more pacific mammals as measured by serious assaults or murders per individual per unit time, even when our episodic wars are averaged in. If the visitor were to be confined to George Schaller's 2900 hours and one randomly picked human population comparable in size to the Serengeti lion population, to take one of the more exhaustive field studies published to date, he would probably see nothing more than some play-fighting—almost completely limited to juveniles—and an angry verbal exchange or two between adults. Incidentally, another cherished notion of our wickedness starting to crumble is that man alone kills more prey than he needs to eat. The Serengeti lions, like the hyenas described by Hans Kruuk, sometimes kill wantonly if it is convenient for them to do so. As Schaller concludes, "the lion's hunting and killing patterns may function independently of hunger."

There is no universal "rule of conduct" in competitive and predatory behavior, any more than there is a universal aggressive instinct—and for the same reason. Species are entirely opportunistic. Their behavior patterns do not conform to any general innate restrictions but are guided, like all other biological traits, solely by what happens to be advantageous over a period of time sufficient for evolution to occur. Thus, if it is of even temporary selective advantage for individuals of a given species to be cannibals, at least a moderate probability exists that the entire species will evolve toward cannibalism.

Mutual Repulsion

When workers of the ants *Pheidole megacephala* and *Solenopsis globularia* meet at a feeding site, some fighting occurs, but the issue is not settled in this way. Instead, dominance is based on organizational ability. Workers of both species are excitable and run off the odor trails and away from the food site when they encounter an alien. The *Pheidole* calm down, relocate the odor trails, and assemble again at the feeding site more quickly than the *Solenopsis*. Consequently they build up their forces more quickly during the clashes and are usually able to control the feeding sites. *S. globularia* colonies are nevertheless able to survive by occupying nest sites and foraging areas in more open, sandy habitats not penetrated by *P. megacephala* (Wilson, 1971b). Pharaoh's ant (*Monomorium pharaonis*) is unusually effective at competing with other species at food

sites. It repels them with the odor of a substance released from the poison gland (Hölldobler, 1973).

Other examples are known in which competition for resources is conducted by indirect forms of repulsion. Females of the tiny wasp *Trichogramma evanescens* parasitize the eggs of a wide variety of insect host species by penetrating the chorion with their ovipositors and inserting their own eggs inside. Other females of the same species are able to distinguish eggs that have already been parasitized, evidently through the detection of some scent left behind in trace amounts by the first female; thus alerted, they invariably move on to search for other eggs (Salt, 1936).

Chemical aggression and interference can be both insidious and unpredictable in their effects. If a female mouse recently inseminated by one male is placed with a second male belonging to a different genetic strain, she will usually abort and quickly become available for a new insemination. The aborting stimulus is an as yet unidentified pheromone produced in male urine that is smelled by the female and activates the pituitary gland and corpora lutea (Bruce, 1966; Bronson, 1969). An equally significant effect has recently been demonstrated by Ropartz (1966, 1968). Investigating the causes of reduced fertility in crowded populations, he found that the odor of other mice alone causes the adrenal glands of individual mice to grow heavier and to increase their production of corticosteroids, resulting eventually in a decrease in reproductive capacity and even death of the animal.

The Limits of Aggression

Why do animals prefer pacifism and bluff to escalated fighting? Even if we discount the very large number of species in which density-dependent controls are sufficiently intense to prevent the populations from reaching competitive levels, it still remains to be explained why overt aggression is lacking among most of the rest of the species that do compete. The answer is probably that for each species, depending on the details of its life cycle, its food preferences, and its courtship rituals, there exists some optimal level of aggressiveness above which individual fitness is lowered. For some species this level must be zero, in other words the animals should be wholly nonaggressive. For all others an intermediate level is optimal. There are at least two kinds of constraints on the evolutionary increase of aggressiveness. First, a danger exists that the aggressor's hostility will be directed against unrecognized relatives. If the rates of survival and reproduction among relatives are thereby lowered, then the replacement rate of genes held in common between the aggressor and its relatives will also be lowered. Since these genes will include the ones responsible for aggressive behavior, such a reduction in inclusive fitness will work against aggressive behavior as well. This process will continue until the difference between the advantage and disadvantage, measured in units of inclusive fitness, is maximized.

Second, an aggressor spends time in aggression that could be invested in courtship, nest building, and the feeding and rearing of young. Dominant white leghorn hens, for example, have greater access to food and roosting space than subordinates, but they present less to cocks and hence are mated fewer times (Guhl, Collias, and Allee, 1945). It is plausible that the average level of aggression in these hens represents the optimum balance struck to obtain the greatest difference between the advantages and disadvantages of aggression generally. However, the case cannot be made, because the experiments were not extended long enough to determine whether the dominant hens actually laid fewer eggs as a result of their reduced sexual activity (Guhl, 1950). Adverse effects of such "aggressive neglect" have been more convincingly documented in pigeons (Castoro and Guhl, 1958), gannets (Nelson, 1965), and in sunbirds and honeyeaters (Ripley, 1959, 1961). The particularities of the environments of different species can sometimes be related directly to the forms and intensities of aggressive behavior that characterize them. Species of chipmunks (*Tamias* and *Eutamias*), for example, vary notably in the amount of territorial defense they display. According to Heller (1971), territorial intensity is determined in evolution by the interaction of the magnitude of the need to gain absolute control over the territory and the cost of defending the territory in terms of energy loss and risk from predators. These factors differ greatly from one habitat to another, sufficiently, according to Heller, to account for

the fact that some *Eutamias* species are strongly territorial while others are apparently nonterritorial. As Table 11-2 shows, territorial defense has evidently evolved when the food supply is limited enough to be worth defending, but only if there is also no overriding cost entailed in the defense.

The evolutionary compromise can even extend to the fine details of aggressive behavior. The kittiwake (*Rissa tridactyla*) is a gull with the unique habit of nesting on tiny cliff ledges next to the sea. The birds are capable of only limited movements after landing. They have accordingly restricted their aggressive behavior. The upright threat posture employed by all other gull species has been abandoned, and the birds do nothing more than seize and twist one another's beaks. Because immature kittiwakes that fall off the ledges are invariably doomed, their behavior is uniquely modified to prevent accidents: instead of running when attacked, they turn their heads and completely hide their beaks in an extreme appeasement display (Esther Cullen, 1957).

The Proximate Causes of Aggression

Aggression evolves not as a continuous biological process as the beat of the heart, but as a contingency plan. It is a set of complex responses of the animal's endocrine and nervous system, programmed to be summoned up in times of stress. Aggression is genetic in the sense, defined earlier (Chapter 4), that its components have proved to have a high degree of heritability and are therefore subject to continuing

Table 11-2 Presence or absence of territorial behavior in species of chipmunks as an evolutionary compromise between opposing ecological forces; + indicates a condition favoring territoriality, − indicates a condition opposing it. (Based on Heller, 1971.)

Species	Food supply	Energetic cost of territorial defense	Danger incurred from predators due to territorial defense	Presence or absence of territory
Alpine chipmunk (*Eutamias alpinus*)	On limited area, in short season +	Relatively low +	Relatively low: many hiding places among alpine rocks +	Present
Yellow pine chipmunk (*Eutamias amoenus*)	On limited area, in short season +	Moderate or low +	Relatively low: many hiding places on forest floor +	Present
Least chipmunk (*Eutamias minimus*)	On limited area, probably seasonal +	High because of hot, dry environment −	?	Absent
Lodgepole chipmunk (*Eutamias speciosus*)	Widespread, abundant, diverse and year-round −	Probably moderate or low +	Probably high −	Absent

evolution. The documentation for this statement is substantial and has been reviewed by Scott and Fuller (1965) and McClearn (1970). Aggression is also genetic in a second, looser sense, meaning that aggressive and submissive responses of some species are specialized, stereotyped, and highly predictable in the presence of certain very general stimuli. The adaptive significance of aggression, its ultimate causation and the environmental pressures that guide the natural selection of its genotypic variation, should be an object of analysis whenever aggressive or submissive components are discerned in any form of social behavior.

The proximate causes of the variation will now be examined. They are most easily understood when classified into two sets of factors. The first is the array of external environmental contingencies to which the animal must be prepared to respond, including encounters with strangers from outside the social group, competition for resources with other members of its own group, and daily and seasonal changes in the physical environment. All of these exigencies provide stimuli to which the animal's aggressive scale must be correctly adjusted. The second set of stimuli is the internal adjustments through learning and endocrine change by which the animal's aggressive responses to the external environment are made more precise.

External Environmental Contingencies

Encounters outside the group. The strongest evoker of aggressive response in animals is the sight of a stranger, especially a territorial intruder. This xenophobic principle has been documented in virtually every group of animals displaying higher forms of social organization. Male lions, normally the more lethargic adults of the prides, are jerked to attention and commence savage rounds of roaring when strange males come into view. Nothing in the day-to-day social life of an ant colony, no matter how stressful, activates the group like the introduction of a few alien workers. The principle extends to the primates. Southwick (1967, 1969) conducted a series of controlled experiments on confined rhesus monkeys in order to weigh the relative importance of several major factors in the evocation of aggression. Food shortages actually caused a decrease in aggressive-submissive interactions, since the animals reduced all social exchanges and began to devote more time to slow, tedious explorations of the enclosure. Crowding of the monkeys induced a somewhat less than twofold increase in aggressive interactions. The introduction of strange rhesus monkeys, however, caused a fourfold to tenfold increase in such interactions. The experiment put a more precise measure on what is observed commonly in the wild. The rate of aggression displayed when two rhesus groups meet, or a stranger attempts to enter the groups, far exceeds that seen within the troops as they pass through the stressful episodes of their everyday life.

Food. The relation of aggressive behavior to the supply and distribution of food is generally complex in animals and difficult to predict for any particular species. In general, aggressive-submissive exchanges increase sharply when food is clumped instead of scattered and domination of one piece of the food or of a small area of ground on which food is concentrated becomes profitable. Baboons ordinarily forage like flocks of birds, fanning out in a search for small vegetable items that are picked off the ground and eaten quickly. The troop members seldom challenge one another under these circumstances. But when a clump of grass shoots is discovered in elephant dung, or a small animal is killed, the baboons threaten one another and may even fight over the food. The quickest way for an observer to witness aggression and the dominance order is to feed the baboons pieces of bread or some other rich items of food. N. R. Chalmers (cited by Rowell, 1972) observed that when white cheeked mangabeys (*Cercocebus albigena*) feed on jackfruit, which are very large fruits growing directly on the trunk of the tree, they interact aggressively about ten times more frequently than when feeding on other kinds

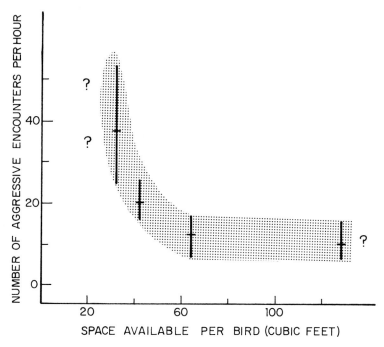

Figure 11-3 As crowding is increased in caged house finches (*Carpodacus mexicanus*), and the amount of space per bird decreased, the rate of aggressive interactions goes up exponentially. The means and ranges of variation of the rates are given for several quantities of space. (Redrawn from Thompson, 1960.)

of small fruit scattered through the forest canopy. Food shortages can exacerbate aggression, but only if the food is distributed in a defensible pattern. Pure hunger can even lower the rate of interactions by producing listlessness and causing animals to scatter away from one another in search of food. This rather surprising response was observed by Hall (1963a) in chacma baboons marooned on an island by the rising waters of the Kariba Dam. It was also observed in rhesus monkeys on Cayo Santiago when the regular supply of monkey chow for this semiwild population failed to arrive (Loy, 1970).

Crowding. As animals move into ever closer proximity, the rate at which they encounter one another goes up exponentially. All other things being equal, the frequency of aggressive interactions goes up at the same exponential rate (see Figure 11-3). The space-aggression curves of some species, however, are more complex. At intermediate densities crayfish (*Orconectes virilis*) form territories, but at extremely high densities they collapse into peaceful aggregations (Bovbjerg and Stephen, 1971). When individuals of *Dascyllus aruanus,* an Australian reef fish, are crowded around pieces of synthetic coral in increasing densities, the rate of their aggressive encounters first rises, then drops (Figure 11-4). Aggression also rises as a function of group size quite independently of density (Sale, 1972). Experiments on the European rabbit by Myers et al. (1971) have also demonstrated a rise in aggressive interactions with increased density. However, a second, more surprising effect was observed: if density is kept constant, but the total space occupied by the group as a whole is lowered (by reducing the number of rabbits in the group to keep density constant), the rate of aggression still rises. Thus rabbits are

sensitive not only to the proximity of other rabbits but also to the absolute amount of room within which the crowding must be accommodated.

Seasonal change. The aggressive interactions of most animal species peak in the breeding season. Fighting among tigers, for example, is limited to the contest between males for estrous females. Baikov (1925) described his experience with the Manchurian tiger as follows: "I have spent many nights in the taiga alone with my fellow hunters, sitting by the fire and listening to tigers challenging their rivals— resounding through the frost-bound forests; but though the battle ground is invariably drenched with blood, such encounters never end in death." The sifaka (*Propithecus verreauxi*), a Madagascan lemur, is placid through most of the year but erupts in savage fighting during the breeding season (Alison Jolly, 1966). The female reindeer is a passive animal during most of her life. But just before and after giving birth she becomes aggressive toward other herd members, especially toward the yearlings. Rhesus macaques are exceptionally aggressive animals, even for Old World monkeys, and their societies are based to a large degree on dominance orders maintained by virtually continuous aggressive confrontations. Even so, hostility among males reaches a peak during the mating season, and females are involved in the greatest amount of fighting during both the mating and birth seasons. Injuries and deaths are also most common at these times (Wilson and Boelkins, 1970). Other seasonal patterns can be cited at length from the literature on the life histories of both the vertebrates and invertebrates.

Learning and Endocrine Change

Previous experience. A variety of experiences in the life of an animal can influence the form and intensity of its aggressive behavior. Aggression in laboratory rats can be increased by straightforward instrumental training. The behavior amplified in these studies is the "pain-aggression" response: when two rats are presented with certain painful stimuli, such as an electric shock, they attack each other by standing face-to-face on their hind legs, thrusting their heads forward with mouths open, and vigorously thrusting and biting at each other. Neal E. Miller (1948) trained rats to fight in the absence of an electric shock by terminating the shock just as the animals assumed the fighting stance. More recently, Vernon and Ulrich (1966) succeeded in inducing the pain-aggression response in the absence of pain by means of classical associational training. A previously neutral sound, consisting of an electrically generated 1.32 kiloherz tone of 60 decibels, was played simultaneously with electric shock during repeated trials. After a time the rats came to assume the stereotyped fighting posture when stimulated by the sound alone.

Instrumental amplification of aggressive behavior is to be regarded

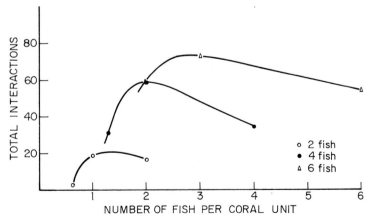

Figure 11-4 The rate of aggressive interactions in the reef fish *Dascyllus aruanus* rises steeply as density is increased, then drops. Aggressive behavior is also a function of total group size, as shown by the positions of the three space-aggression curves. (Redrawn from Sale, 1972.)

as a laboratory manifestation of the socialization by which animals learn their place in territorial and dominance relationships under natural conditions. As animals move up in rank, their readiness to attack increases, particularly when they encounter rivals who have been defeated previously. Animals that are defeated consistently in encounters with one set of opponents become psychologically "down," display timidity when they encounter new sets of opponents, and thus are more likely to retain their low rank than others who have known early triumphs (Ginsburg and Allee, 1942; McDonald et al., 1968). Although the effect is generally thought of as a vertebrate trait, it also occurs in insects. Free (1961a) discovered its existence when conducting routine experiments in which the dominant worker of one queenless bumblebee colony was introduced into the nest of another. Characteristically the intruder bee was challenged by the resident dominant worker, and the two then grappled and fought. Eventually one signaled its submission by hiding in the corner of the nest box. When the introduced bee was returned to its own colony its status then depended on whether it had won or lost in the strange colony. If it had won, it invariably regained its former dominant position in the mother colony. But if it had been beaten, it assumed a subordinate status. Similarly, Alexander (1961) was able to reverse the dominance order of male crickets by repeatedly "defeating" the dominant male between encounters by means of artificial stimuli.

The normal aggressive responses of mammals are also influenced by the socialization process. Male house mice (*Mus musculus*) reared in isolation after weaning are less aggressive than those reared in social groups. The longer the mice are exposed to others, the greater their aggressiveness toward strangers at later times. The critical period is relatively long; in experiments by King (1957) isolation as late as 20 days still had a depressing effect on subsequent responses. Male deer mice (*Peromyscus maniculatus*) develop within even more stringent conditions. In order to show aggressiveness toward other males in the absence of females, they must have had extensive sexual experience (Bronson and Eleftheriou, 1963).

Hormones and aggression. The endocrine system of vertebrates acts as a relatively coarse tuning device for the adjustment of aggressive behavior. The interactions of the several hormones involved in this control are complicated (see Figure 11-5). However, they can be understood readily if the entire system is viewed as comprising three levels of controls: the first determines the state of preparedness (androgen, estrogen, and luteinizing hormone), the second the capacity for a quick response to stress (epinephrine), and the third the capacity for a slower, more sustained response to stress (adrenal corticoids).

The level of preparedness to fight is what we usually refer to as aggressiveness, in order to contrast it with the act of aggression. Aggressiveness, as Rothballer (1967) has said, is a threshold. It can be measured either by the amount of the provoking stimulus required to elicit the act or by the intensity and prolongation of the act in the face of a given stimulus. The class of hormones most consistently associated in the vertebrates with heightened aggressiveness is the androgens, which are 19-carbon steroids, with methyl groups at C-10 and C-13, secreted by the Leydig cells of the testes. The behaviorally most potent androgen is evidently testosterone. It has been known since the experiments of Arnold Berthold in 1849 that roosters stop crowing and fighting when they are castrated, but retain these behaviors if testes from other roosters are implanted in their abdominal cavities. In recent years it has been demonstrated that the behaviors can be restored by injection of appropriate amounts of testosterone proprionate. A similar effect has been demonstrated in a wide range of species, including swordtail fish, gobies (*Bathygobius*), anolis lizards, fence lizards (*Sceloporus*), painted turtles (*Chrysemys*), night herons, doves, songbirds, quail, grouse, deer, mice, rats, and chimpanzees (Scott and Fredericson, 1951; Davis, 1964; Andrew, 1969; Floody and Pfaff, 1974). Immature males, including boys, can be brought into maturity more quickly by injections of testosterone proprionate, and in some species even the behavior of females can be strongly masculinized. The effects of the androgens extend deeply into physiological and social traits that are coupled to aggressiveness. When Mongolian gerbils (*Meriones unguiculatus*) of either sex are castrated, they resorb the ventral sebaceous glands with which territories are marked. The glands are regenerated and territorial behavior resumed when testosterone proprionate is injected (Thiessen, Owen, and Lindzey, 1971). As Allee, Collias, and Lutherman first showed in 1939, hens given small doses of testosterone become more aggressive and move up in rank within the dominance hierarchies of the flocks. Watson and Moss (1971) reported that red grouse cocks (*Lagopus lagopus*) implanted with androgen became more aggressive, nearly doubled their territory size, devoted more time to courtship, and mated with two hens instead of the usual one. Two nonterritorial cocks that had been in poor physical condition regained good condition and drove back territorial cocks to establish new territories on their own. Although they remained unmated that year, they survived the winter and set up territories the following year. One of the cocks was able to acquire a hen the following summer. Without the implant both cocks would almost certainly have perished during the winter.

Among the vertebrates, a seasonal rise in the androgen titer of males is generally associated with an increase in aggressiveness, an establishment or enlargement of territory in those species that are territorial, and the onset of sexual behavior. In short, the androgens initiate the breeding season. Also, dominance among males is correlated with their androgen level. The relation between the hormone and behavior is nevertheless much more complicated than a simple chemical reaction. In higher vertebrates dominance depends to a large extent on experience and on the deference shown by other members of the group on

ENDOCRINE GLAND

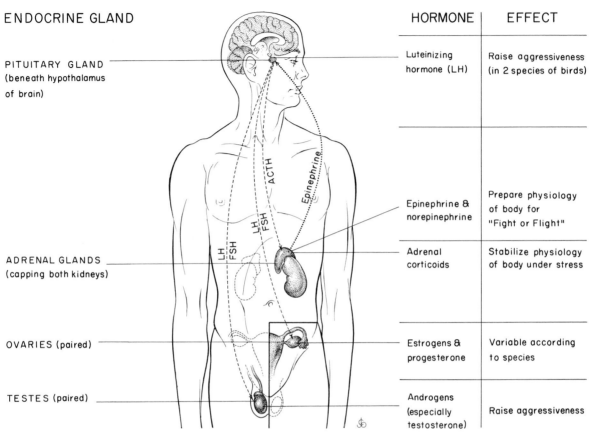

HORMONE	EFFECT
Luteinizing hormone (LH)	Raise aggressiveness (in 2 species of birds)
Epinephrine & norepinephrine	Prepare physiology of body for "Fight or Flight"
Adrenal corticoids	Stabilize physiology of body under stress
Estrogens & progesterone	Variable according to species
Androgens (especially testosterone)	Raise aggressiveness

PITUITARY GLAND (beneath hypothalamus of brain)

ADRENAL GLANDS (capping both kidneys)

OVARIES (paired)

TESTES (paired)

Figure 11-5 The principal hormones that affect aggressive behavior in mammals. The pituitary gland, stimulated by impulses from the hypothalamus and to a lesser degree by epinephrine ("adrenalin"), releases corticotrophin (ACTH), which enlarges the cortex of the adrenal gland and raises the output of the adrenal corticoids. The pituitary gland also releases the luteinizing hormone (LH), which stimulates the production of androgens in the testes of the male. In the female, LH acts synergistically with the follicle-stimulating hormone (FSH) to promote the secretion of estrogen by the follicles of the ovary. Stimulation of certain neurons of the autonomic nervous system, controlled mainly by the hypothalamus, causes the medulla of the adrenal gland to release epinephrine. This scheme is based on results obtained piecemeal from a variety of vertebrate species, and the use here of the human system is for convenience only.

the basis of past performance. Birds are notably inconsistent in their reactions. Davis (1957), for example, showed that testosterone does not affect the rank of starlings in the roost hierarchies. Males of blackbirds (*Turdus merula*) and many other bird species continue to defend territories in the fall and winter, when the gonads are very small (Snow, 1958). The explanation of these inconsistencies could lie in the continued role of even low levels of androgen or in the overriding effects of other hormones. A new range of possibilities was opened by Mathewson's discovery (1961) that injections of the luteinizing hormone (LH) increase aggressiveness and dominance rank in starlings where testosterone fails. One function of LH is to stimulate the production of testosterone. Davis (1964) has suggested that LH

has the more fundamental role in controlling aggression and that the function has been shared with testosterone only as a later evolutionary development. However, the data are not yet adequate to test this hypothesis.

Among higher primates the relationship of androgens to aggressiveness may be more complex still. Rose et al. (1971) found that plasma testosterone levels are correlated with aggressiveness in male rhesus monkeys. However, the correlation with rank in the dominance order was less precise, since high-ranking males had lower levels. Equally surprising, the testosterone titers of the lowest-ranking males were higher than those of solitary caged animals. The possibility exists that aggressiveness induces testosterone secretion, via the brain-pituitary-testis route, rather than the other way around. Or, equally likely, hormone production and the behavior pattern are both enhanced by other, as yet unidentified stimuli such as experience and input from other hormones that influence the pituitary-testis axis.

The estrogens induce a confusing variety of effects on aggressive behavior. The great bulk of these substances is produced by the ovaries, but small quantities are also found in the adrenals, placentae, testes, and even spermatozoans. Thus although the estrogens are primarily female hormones, they can also play some role in male physiology. In general, high estrogen levels promote feminization, female sexual responses, and hence less aggressiveness except when the individual is defending young or, less frequently, competing with other adults. Males vertebrates injected with estrogen typically become less aggressive. A red grouse cock implanted with estrogen by Adam Watson (Watson and Moss, 1971) lost his hen and eventually his territory. In contrast, castrated males of the golden hamster (*Mesocricetus auratus*) regain aggressive traits when injected with estrogen, while females remain unaffected (Vandenbergh, 1971). When female hamsters are given both estrogen and progesterone, which duplicate the conditions for a normal estrus, aggressiveness is strongly suppressed (Kislak and Beach, 1955). Female chimpanzees become more aggressive with estrogen treatment (Birch and Clark, 1946). In humans, either the effects are negligible or sufficiently subtle that they can be defined only by psychoanalytic study (Gottschalk et al., 1961). There does seem to be some diminution of anxiety and aggressiveness in women in the middle of the menstrual cycle, when ovulation is scheduled to occur (Ivey and Bardwick, 1968). At this time receptivity should be greatest, and both estrogen and progesterone levels are at their peak. In sum, we can infer from the fragmentary evidence that estrogen influences vertebrate aggressivity in ways that are highly conditional. When it is adaptive to be sexually submissive, particularly at estrus when progesterone levels are high, aggressiveness is suppressed. At other times estrogen may actually raise aggressiveness in a way that helps a female to maintain status and to defend offspring. The inhibiting effect on male aggressiveness could well be a meaningless artifact.

Given that LH and the gonadal hormones maintain a vertebrate in a state of readiness appropriate to its rank and reproductive status, epinephrine is the hormone by which it makes a fine adjustment to emergencies that arise moment by moment. Epinephrine is a catecholamine, a derivative of tyrosine secreted primarily by the medullas of the adrenal glands. Release of epinephrine into the bloodstream is stimulated by sympathetic nerves and thus ultimately falls under the command of the hypothalamus. The substance is complementary to the other principal catecholamine, norepinephrine, which is coupled with the parasympathetic nervous system and has generally different, sometimes opposite physiological effects. Epinephrine acts quickly in conjunction with the sympathetic system to prepare the entire body for "fight or flight." The heart rate and systolic blood pressure go up. Vasodilation occurs over the body, and the eosinophil count rises. The blood flow through the skeletal muscle, brain, and liver increases by as much as 100 percent. Blood sugar rises. Digestion and reproductive functions are inhibited. In man at least there is also an onset of a feeling of anxiety. Epinephrine is released whenever the vertebrate is placed in a stressful situation, whether cold, a "narrow escape," or hostility from some other member of the species. The hormone does not itself cause the animal to be aggressive but instead prepares it to be more efficient during aggressive encounters. Under certain conditions epinephrine also acts to promote the release of corticotrophin (ACTH) from the anterior lobe of the pituitary, with a consequent release of adrenal corticoids and a gearing of the body for more prolonged adjustment to stress.

Norepinephrine is also released in response to general stress but independently of epinephrine. Where epinephrine triggers a massive general response of the body, mobilizing glycogen as blood glucose and redistributing blood to the action centers, norepinephrine acts mainly to sustain blood pressure. It promotes heart action and vasodilation while having relatively little effect on the rate of blood flow or metabolism. Thus epinephrine conforms closely to Walter Cannon's original emergency theory of medullary adrenal action with norepinephrine playing a secondary, principally regulatory role. A curious effect discovered in human beings is that violent participation in aggressive encounters induces the release of relatively large quantities of norepinephrine together with only moderate amounts of epinephrine, while the *anticipation* of aggressive interaction, in the form of anger or fear, favors only the release of epinephrine. Professional hockey players on the bench, for example, secrete only epinephrine at the same time their teammates playing on the floor are secreting mostly norepinephrine.

Under conditions of stress, ACTH from the anterior lobe of the pituitary induces an outpouring of steroids from the cortex of the adrenal glands. When the stress is prolonged, the adrenals increase in weight and sustain a high production of these corticosteroid hormones. The secretion contains a variety of active substances, including cortisone, cortisol, corticosterone, and others. Their functions vary from one group of vertebrates to the next, but in general one class of substances helps to preserve ionic balance in the blood and tissue fluids, while another controls the body's reaction to infection by reducing inflammation, lowering the eosinophil count, and killing lymphocytes in lymph nodes. Certain of the hormones also promote the deposition of glycogen in liver. Thus some of the adrenal corticoids are opposite in effect to the catecholamines. They serve as a braking device on the body's emergency mobilization system. Some quantity of adrenal corticoids is required at all times, even when the animal is not under any particular stress. Adrenalectomized animals produce symptoms identical to those of patients with Addison's disease: hypoglycemia, gastrointestinal disturbances, reduced blood pressure and body temperature, kidney failure, and the inability to stand stress of any kind. Without adrenal corticoids the animal's (or human being's) condition deteriorates in the face of temperature extremes, prolonged activity, infections, intoxication, and so forth. When subjected to prolonged stress, a normal animal undergoes what Hans Selye (1956) has called the "General Adaptation Syndrome." The G.A.S. is envisaged as proceeding in the following sequence of three stages:

1. *Stage of alarm.* The pituitary gland, activated by the brain, releases ACTH, which in turn induces the release of adrenal corticosteroids into the blood. The corticosteroids mediate their various effects, abetting and controlling the animal's fast response to the emergency and helping to stabilize its physiology. If the stress continues, the animal enters the second stage.

2. *Stage of resistance.* The greater demand for corticosteroids induces growth of the adrenal gland. Aggressive interactions are among the most potent of stressors. When laboratory mice made hyperexcitable by long-term isolation are exposed to a trained fighter for only 15 minutes, their plasma corticosterone level is greatly elevated and remains high for over 24 hours (Bronson, 1967). Fighting for as little as 5 minutes per day for five days results in the enlargement of the adrenal glands by as much as 38 percent (Welch and Welch, 1969; see also Figure 11-6). The system remains stabilized but, if Selye's hypothesis is correct, continued stress brings the animal to the third, pathological stage.

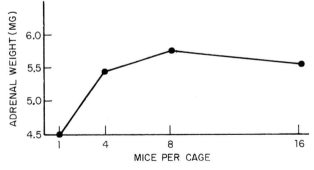

Figure 11-6 The increase in adrenal weight of mice with increase in crowding in laboratory cages. This response occurs generally when mammals are stressed for prolonged periods and is associated with a rise in the production of steroids from the adrenal cortex. (Redrawn from Davis, 1964.)

3. *Stage of exhaustion.* The body is not able to stand the exposure to increased corticosteroid loads. Even though the hormones protect it in certain ways, they weaken it in others. Thus, large quantities of the antiinflammatory corticosteroids may enable the animal to survive an emergency by avoiding excessive inflammation, but in the long run the same action increases the chances of infection. Proinflammatory and antiinflammatory steroids might cancel each other's effects for a time, but high titers of the two in combination can cause liver damage. The evidence for such an "adaptation disease" comes chiefly from pathological effects induced by implantation of large amounts of corticosteroids into laboratory animals. Whether the degenerative phase of the General Adaptation Syndrome occurs commonly in nature, or at all, is still a matter of controversy (Turner and Bagnara, 1971). Correlations have been variously demonstrated between corticosteroid levels and lowered fertility, decreased antibody production, renal failure, and decreased resistance to trypanosomiasis and other diseases (Christian, 1955; Noble, 1962; Jackson and Farmer, 1970; D. von Holst, 1972a,b). These results are consistent with Selye's hypothesis but, in the absence of a demonstration of direct causation, they do not prove it. As a consequence, the sequential chain of causation postulated by J. J. Christian and others, which runs from an increase in population density to aggressive interaction to increased adrenocorticoid output to population control, must also be regarded as speculative. It is possible to forge such a chain in laboratory populations, but this accomplishment is far from proof that the phenomenon exists in nature (see Chapter 4).

The subject of behavioral endocrinology has been dominated almost exclusively by vertebrate studies. It will perhaps surprise the reader to learn that this circumstance may be wholly justified. No evidence exists, at least to my knowledge, of any hormonal system that regulates aggressive behavior in invertebrates, including insects (see also Barth, 1970, and Truman and Riddiford, 1974). Ewing (1967) has reported death in *Naupheta* cockroaches associated with aggression-induced stress, but this in no way implicates the endocrine system. Furthermore, although the development of sex differences in insects is mediated by hormones, the direct physiological effects are not known to include the alteration of aggressiveness.

Human Aggression

Is aggression in man adaptive? From the biologist's point of view it certainly seems to be. It is hard to believe that any characteristic so widespread and easily invoked in a species as aggressive behavior is in man could be neutral or negative in its effects on individual survival and reproduction. To be sure, overt aggressiveness is not a trait in all or even a majority of human cultures. But in order to be adaptive it is enough that aggressive patterns be evoked only under certain conditions of stress such as those that might arise during food shortages and periodic high population densities. It also does not matter whether the aggression is wholly innate or is acquired part or wholly by learning. We are now sophisticated enough to know that the capacity to learn certain behaviors is itself a genetically controlled and therefore evolved trait.

Such an interpretation, which follows from our information on patterned aggression in other animal species, is at the same time very far removed from the sanguinary view of innate aggressiveness which was expressed by Raymond Dart (1953) and had so much influence on subsequent authors:

The blood-bespattered, slaughter-gutted archives of human history from the earliest Egyptian and Sumerian records to the most recent atrocities of the Second World War accord with early universal cannibalism, with animal and human sacrificial practices or their substitutes in formalized religions and with the worldwide scalping, head-hunting, body-mutilating and necrophiliac practices of mankind in proclaiming this common blood lust differentiator, this mark of Cain that separates man dietetically from his anthropoidal relatives and allies him rather with the deadliest of Carnivora.

This is very dubious anthropology, ethology, and genetics. It is equally wrong, however, to accept cheerfully the extreme opposite view, espoused by many anthropologists and psychologists (for example, Montagu, 1968) that aggressiveness is only a neurosis brought out by abnormal circumstances and hence, by implication, nonadaptive for the individual. When T. W. Adorno, for example, demonstrated (in *The Authoritarian Personality*) that bullies tend to come from families in which the father was a tyrant and the mother a submerged personality, he identified only one of the environmental factors affecting expression of certain human genes. Adorno's finding says nothing about the adaptiveness of the trait. Bullying behavior, together with other forms of aggressive response to stress and unusual social environments, may well be adaptive—that is, programmed to increase the survival and reproductive performance of individuals thrown into stressful situations. A revealing parallel can be seen in the behavior of rhesus monkeys. Individuals reared in isolation display uncontrolled aggressiveness leading frequently to injury. Surely this manifestation is neurosis and nonadaptive for the individuals whose behavioral development has been thus misdirected. But it does not lessen the importance of the well-known fact that aggression is a way of life and an important stabilizing device in free-ranging rhesus societies.

This brings us to the subject of the crowding syndrome and social pathology. Leyhausen (1965) has graphically described what happens to the behavior of cats when they are subjected to unnatural crowding: "The more crowded the cage is, the less relative hierarchy there is. Eventually a despot emerges, 'pariahs' appear, driven to frenzy and all kinds of neurotic behaviour by continuous and pitiless attack by all others; the community turns into a spiteful mob. They all seldom relax, they never look at ease, and there is a continuous hissing, growling, and even fighting. Play stops altogether and locomotion and exercises are reduced to a minimum." Still more bizarre effects were observed by Calhoun (1962) in his experimentally overcrowded laboratory populations of Norway rats. In addition to the hypertensive behavior seen in Leyhausen's cats, some of the rats displayed hypersexuality and homosexuality and engaged in cannibalism. Nest construction was commonly atypical and nonfunctional, and infant mortality among the more disturbed mothers ran as high as 96 percent.

Such behavior is obviously abnormal. It has its close parallels in certain of the more dreadful aspects of human behavior. There are some clear similarities, for example, between the social life of Calhoun's rats and that of people in concentration and prisoner-of-war camps, dramatized so remorselessly, for example, in the novels *Andersonville* and *King Rat*. We must not be misled, however, into thinking that because aggression is twisted into bizarre forms under conditions of abnormally high density, it is therefore nonadaptive. A much more likely circumstance for any given aggressive species, and one that I suspect is true for man, is that the aggressive responses vary according to the situation in a genetically programmed manner. It is the total *pattern* of responses that is adaptive and has been selected for in the course of evolution.

The lesson for man is that personal happiness has very little to do with all this. It is possible to be unhappy and very adaptive. If we wish to reduce our own aggressive behavior, and lower our catecholamine and corticosteroid titers to levels that make us all happier, we should design our population densities and social systems in such a way as to make aggression inappropriate in most conceivable daily circumstances and, hence, less adaptive.

Chapter 12 **Social Spacing, Including Territory**

Some animals, planktonic invertebrates for example, drift through life without fixed reference points in space. They contact other members of the species only fleetingly as sexual partners and serve briefly, if at all, as parents. Other animals, including nearly all vertebrates and a large number of the behaviorally most advanced invertebrates, conduct their lives according to precise rules of land tenure, spacing, and dominance. These rules mediate the struggle for competitive superiority. They are enabling devices that raise personal or inclusive genetic fitness. In order to understand the rules it is necessary to begin with an elementary classification of special social relationships in which they are involved:

Total range: the entire area covered by an individual animal in its lifetime (Goin and Goin, 1962).

Home range: the area that an animal learns thoroughly and habitually patrols (Seton, 1909; Burt, 1943). In some cases the home range may be identical with the total range; that is, the animal familiarizes itself with one area and never leaves it. Many times the home range and the territory are identical, meaning that the animal excludes other members of the same species from all of its home range. In the great majority of species, however, the home range is larger than the territory, and the total range is much larger than both. Ordinarily the home range is patrolled for food, but in addition it may contain familiar look-out positions, scent posts, and emergency retreats. It can also be shared jointly by the members of an integrated social group.

Core area: the area of heaviest regular use within the home range (Kaufmann, 1962; see Figure 12-1). Core areas can be confidently delimited in such species as coatis and baboons, in which they are associated with sleeping sites located in a more or less central position in the home range. Fundamentally, however, the precise limits of both the home range and the core area are arbitrary, depending on the time the observer spends in the field and the minimum number of times he requires an animal to visit a given locality in order for the locality to be included. The solution to the problem, as Jennrich and Turner (1969) have pointed out, is simply to define a home range as the area of the smallest subregion of the total range that accounts for a specified proportion of the summed utilization. A smaller proportion can be selected to circumscribe the core area as any subregion in which the visitation is strongly disproportionate. The two specifications should prove most useful when comparing societies or the social systems of different species.

Territory: an area occupied more or less exclusively by an animal or group of animals by means of repulsion through overt defense or advertisement (Noble, 1939; J. L. Brown, 1964; Wilson, 1971b). As I will show in a later discussion, this definition is but one nuance of several that have been advanced during the past twenty years. It has been selected here because it fits the intuitive concept of most

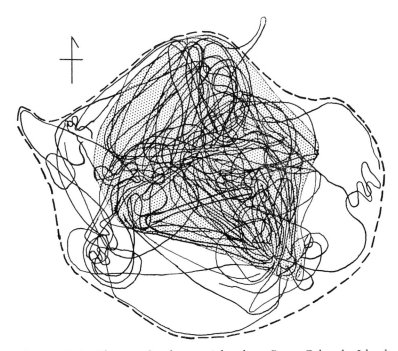

Figure 12-1 The travels of a coati band on Barro Colorado Island during a nine-month period. The home range is enclosed by a dashed line, while the core area is stippled. The core area is simply the arbitrarily delimited portion of the home range subject to heaviest use. The maximum (east-west) diameter of this particular home range was 700 meters; the single loop shown extending beyond the northern edge of the home range represents a single incident in which the band appeared to be lost. (From Kaufmann, 1962. Originally published by the University of California Press; reprinted by permission of the Regents of the University of California.)

investigators and, more important, is the most practical in application. The territory need not be a fixed piece of geography. It can be "floating" or *spatiotemporal* in nature, meaning that the animal defends only the area it happens to be in at the moment, or during a certain time of day or season, or both.

Individual distance (*social distance*): the minimum distance that an animal routinely keeps between itself and other members of the same species (Hediger, 1941, 1955; Conder, 1949). Each species has a characteristic minimum distance that can be measured when animals are not on their territories. The measurement becomes meaningless on territories, since the minimum distance to the nearest neighbor then changes continuously as the animal moves about inside its territory. Outside the territory, individual distance varies from zero in aggregating species to a meter or more in some large birds and mammals. When this distance is greater than zero, the animal en-

forces the spacing by either retreating from the encroaching neighbor or threatening it away. Individual distance is not to be confused with *flight distance*, the minimum distance an animal will allow a predator to approach before moving away (Hediger, 1950).

Dominance: the assertion of one member of a group over another in acquiring access to a piece of food, a mate, a place to display, a sleeping site or any other requisite that adds to the genetic fitness of the dominant individual (see Chapter 13).

When the behaviors of many animal species are compared, their separate manifestations of home range, territory, individual distance, and dominance are seen to form a continuously graded series. Each species occupies its own position along the gradients. Some encompass a large segment of one or more of the gradients, utilizing them as behavioral scales to provide variable responses to a changing environment. Others are fixed at one point.

Let us now use the classification as a framework on which to arrange phenomena and to analyze their adaptiveness, starting with individual distance as the simplest form and proceeding eventually to dominance. In the end, in Chapter 13, we will come back to the concepts of gradients and scaling.

Individual Distance

Paul Leyhausen used the following German fable to clarify the significance of individual distance: "One very cold night a group of porcupines were huddled together for warmth. However, their spines made proximity uncomfortable, so they moved apart again and got cold. After shuffling repeatedly in and out, they eventually found a distance at which they could still be comfortably warm without getting pricked. This distance they henceforth called decency and good manners."

Individual distance is the compromise struck by animals that are both attracted to other members of their own species and repelled by them at short distances. A few social animals do not observe any distance at all. Striped mullet (*Mugil cephalus*), for example, swim in tight pods with their bodies repeatedly touching, while members of many insect and snake species form hibernating aggregations by simply piling on top of one another. But most kinds of animals observe a more or less precise distance that is a species characteristic (see Figure 12-2). The swallow *Hirundo rustica* maintains 15 centimeters, the black-headed gull *Larus ridibundus* 30 centimeters, the greater flamingo *Phoenicopterus ruber* 60 centimeters and the sandhill crane *Grus canadensis* 175 centimeters (Hediger, 1955; Miller and Stephen, 1966). When animals are thrown together by an experimenter, they quickly spread out until they reattain the correct distance for their species. Bovbjerg (1960) found that the rate of dispersal of the Pacific shore crab *Pachygrapsus crassipes* is a direct function

Figure 12-2 Individual distance in domestic pigeons. (Photograph by Stanley Baumer.)

of the initial density, reaching zero when the normal spacing is attained. When caddisfly larvae are forced together, they fight among one another while they disperse, stopping when each is just far enough from all of its neighbors to spin a funnel-shaped, insect-catching net without interference from the nets constructed by the others (Glass and Bovbjerg, 1969). If forced into abnormal proximity in cages, many kinds of mammals, including rhesus monkeys, tend to compensate by spending long hours hiding from others behind whatever objects are available, or else gazing at the ground or out windows.

Individual distance can also be maintained by purely chemical cues. Adults of the flour beetle *Tribolium confusum* aggregate at low population density, apparently distribute themselves randomly at intermediate density, and distribute uniformly at high density (Naylor, 1959). The last effect is evidently due to the secretion of quinones, which act as repellents above a certain concentration. Loconti and Roth (1953) have shown that these substances are produced by thoracic and abdominal glands and, in the case of *T. castaneum*, consist primarily of two quinones. A similar effect may be responsible for the dispersal of young *Zinaria* millipedes, which secrete a substance, apparently hydrocyanic gas, during dispersal from their initial aggregations. However, the evidence is anecdotal.

Edward Hall (1966), grasping the implications of this zoological principle for human beings, suggested the need for a discipline of "proxemics," the systematic study of the use of space as a specialized component of culture. Civilized man, Hall argued, uses walls to provide a sense of adequate space in his "unnaturally" dense habitations. Cultures differ greatly in their individual distances. Mediterranean peoples, including the French, tolerate closer packing in restaurants and other meeting places and stand closer to one another when speaking than do northern Europeans. As a result an Englishman is likely to consider an Italian crude and forward, while the Italian views the Englishman as cool and impolite. The German concept of private occupancy differs from the idea of space of most other cultures and permeates the thought process of the German's daily existence.

The German's ego is extraordinarily exposed, and he will go to almost any length to preserve his "private sphere." This was observed during World War II when American soldiers were offered opportunities to observe German prisoners under a variety of circumstances. In one instance in the Midwest, German P.W.s were housed four to a small hut. As soon as materials were available, each prisoner built a partition so that he could have *his own space* . . .

Public and private buildings in Germany often have double doors for soundproofing, as do many hotel rooms. In addition, the door is taken very seriously by Germans. Those Germans who come to America feel that our doors are flimsy and light. The meanings of the open door and the closed door are quite different in the two countries. In offices, Americans keep doors open; Germans keep doors closed. In Germany, the closed door does not mean that the man behind it wants to be alone or undisturbed, or that he is doing something he doesn't want someone else to see. It's simply that Germans think that open doors are sloppy and disorderly.   (Hall, 1966)

Hediger's studies (1950, 1955) of domestic and zoo animals have shown how complex and finely calibrated individual distance can be in other mammals. A sheep grazes with its head down and its body held at 60° from its nearest neighbor while shifting about to maintain the characteristic individual distance. The animal's attention is constantly divided between its food and its neighbor. When forage grows scarce, the bond is broken and the flock scatters out to form new, random patterns. In order to handle lions and tigers, animal trainers in circuses exploit the delicate mental balance between flight and aggression. A lion in a cage moves away from the trainer when the flight distance is encroached, and it continues to retreat until backed against the wall. When the trainer now closes the gap to the individual distance, the lion begins to stalk the trainer. If the trainer eases backward, placing a stool in front of him, the lion will climb the obstacle. The act is based on a judgment to within centimeters of the distances the big cats will advance or retreat. Whips and blank-loaded pistols serve as little more than stage props.

A "Typical" Territorial Species

Figure 12-3 shows a mutual threat display between two male pike blennies, illustrating what zoologists consider to be some of the most general characteristics of aggressive behavior, and particularly of territorial behavior. For the reader unfamiliar with the natural history of territoriality, these fish will provide an interesting introduction. The behavior displayed is "typical" in most respects: (1) it is most fully developed in adult males; (2) there is a clearly delimited area within which each male begins to display to intruders of the same

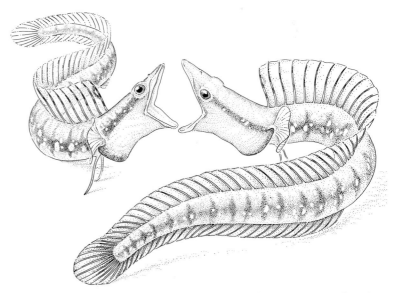

Figure 12-3   Territorial display and fighting between two male pike blennies (*Chaenopsis ocellata*). (From Wilson et al., 1973. Based on Robins, Phillips, and Phillips, 1959.)

species, especially other adult males; (3) the resident male—or, as in these blennies, the larger male—usually wins the contest; (4) some of the most conspicuous and elaborate behaviors of the entire species' repertory occur during these particular exchanges; (5) the posturings make the animal appear larger and more dangerous; (6) the exchanges are mostly limited to bluffing, and even if fighting occurs it does not ordinarily result in injury or death. As we shall see shortly, almost all of these generalizations are violated by one species or another. For the moment, however, let us examine *Chaenopsis* a little more closely as a baseline example.

Chaenopsis ocellata is a small fish, 7 centimeters in length, that dwells on the bottom of shallow water inshore from southeastern Florida to Cuba. The males occupy the abandoned burrows of annelid worms. The approach of any animal to within 25 centimeters excites the interest of the male, who lifts his head and erects his dorsal fin. If the intruder is another male pike blenny, the alert posture is escalated into a full-scale threat display, marked by a rapid increase in the respiratory rate, an intense darkening of the spinous portions of the dorsal fin and of the head, spreading of the pectoral fins, and finally a wide gaping of the mouth and spreading of the azure branchiostegal membranes. In most cases this dramatic transformation is enough to turn the intruder back. If the stranger persists in its onward course, however, the resident male carries through an attack. The two blennies meet snout to snout and then raise the anterior two-thirds of their bodies off the substrate, curling their tails on the bottom for support. The mouths are gaped widely and placed in contact with each other, the branchiostegal membranes are kept fully extended, and the pectoral fins are fanned rapidly to maintain position. If the two fish are nearly equal in size they may rise and fall in this ritualized combat several times without losing oral contact. If one male is smaller, he usually concedes after the completion of the first rising contact. The winning male is the one that suddenly shifts its mouth sideways and clamps down hard. The loser then abruptly folds in its dorsal fin and branchiostegal membranes, and contact is broken as both males drop to the bottom. Uninjured, the beaten male leaves the scene. Female pike blennies are not challenged by resident males. Very probably the tolerance toward them is the prelude to courtship during the breeding season (Robins et al., 1959).

The History of the Territory Concept

Aristotle and Pliny noted the demarcation and defense of territories by male birds, and the phenomenon was then sporadically rediscovered through the first centuries of modern science. In Rome in 1622, G. P. Olina commented upon the nightingale's "freehold." John Ray, after reading Olina, wrote in 1678: "It is proper to this bird at its first coming to occupy and seize upon one place as its freehold, into

which it will not admit any other nightingale but its mate." Gilbert White was perhaps the first to perceive the effect of territory on population density. In February 1774, he wrote to Daines Barrington that "during the amorous season, such a jealousy prevails between the male birds that they can hardly bear to be together in the same hedge or field. Most of the singing and elation of spirits at that time seem to me to be the effect of rivalry and emulation: and it is to this spirit of jealousy that I chiefly attribute the equal dispersion of birds in the spring over the face of the country."

The modern study of territory can be said to have begun with the publication of *Der Vogel und sein Leben* in 1868 by Johann Bernard Theodor Altum, a professor at Münster and later at the forestry college at Eberswalde. A translation and commentary on the pertinent parts have been provided by Ernst Mayr (1935). Bernard Altum's book was written in good part to serve as an answer to Alfred Brehm's *Das Leben der Vögel* (1867), which described birds as though they felt and thought like human beings. Altum insisted on the dictum *Animal non agit, agitur* (animals do not act, they are acted upon; that is, they respond to stimuli and to drives, including the territorial drive). He clearly perceived not only the population consequences but also the individual adaptive value of territoriality: "If a locality produces a great deal of food, the result of favorable soil, vegetation, and climatic conditions, the size of territories may be reduced to some extent. We call such localities excellent Warbler, Nightingale, etc., terrain, but even here territorial boundaries cannot be absent. It is not at all remarkable that for each species of bird the size of these necessary territories is adjusted to its exact ecological requirements and its specific food. While, for example, the Sea-eagle has a territory an hour's walk in diameter, a small wood lot is sufficient for the Woodpecker, and a single acre of brush for the Warbler. All this is well-balanced and well-contrived."

In 1903 C. B. Moffat, writing about the behavior of robins, introduced the word "territory" into the English scientific literature. But it was H. Eliot Howard who, in his celebrated *Territory in Bird Life* (1920) and subsequent works, finally took up where Altum had left off. Here is how he expressed the intricate interplay of aggressive and sexual responses in waterhens:

This then is the problem—to make persistent attack with reluctant response on the pond agree with free response and no attack in the meadow. I shall say that he remembers his mate, and therefore has no interest in other hens. Then if she, in memory, excludes other hens as objects of interest, she excludes them as objects of attack; if she excludes them as objects of sexual interest only, that does not agree with what he does in the meadow, does not tell why he attacks; if she, and nothing else, excites him to attack the stranger, there is no accounting for the way he limits attack to a region; and if the region, nothing else, excites him to attack hens, then the only purpose the region serves is to damn his chances of pairing. But turn him into a bachelor, and no hen suffers from him. So the cock is guided by his perception of three things—the strange hen, his mate, and the pond; but neither one nor another by itself provokes his attack. (*A Waterhen's Worlds.* 1940)

Howard can be said to have made three principal contributions to the study of territoriality in birds. First, he subjected the behavior to a systematic inquiry, revealing a surprising richness of detail and variation among the bird species. This detail he attributed to the separate adaptations of species to their environment. In the course of his review Howard noted that aggression also occurs between species, with hostility at a maximum between the most closely related species. Second, Howard articulated the close connections that exist between territorial aggressive display and courtship. He anticipated Fraser Darling in postulating that the displays between the members of a mated pair synchronize their reproductive conditions. Finally, Howard extended and strengthened the idea that territoriality sets an upper limit to the density of bird populations.

Since 1920 the number of studies devoted to territoriality has risen exponentially, until today it ranks with aggression and dominance as one of the several most intensely studied topics of sociobiology. Territorial behavior has been documented in all major groups of vertebrates and in several of the invertebrate phyla. To complete this brief historical survey, it is appropriate to cite five other more recent contributions as being especially notable for their originality and impact.

1. Margaret M. Nice (1937, 1941, 1943). Nice's studies of the life history and behavior of the song sparrow (*Melospiza melodia*) were unusual at the time for their thoroughness and objectivity. They helped to set new standards for field research on sociobiology, including the description of territorial and reproductive behavior.

2. C. R. Carpenter (1934, 1940). In his studies of howling monkeys in Panama and gibbons in Thailand, Carpenter established the importance of territoriality in the social life of nonhuman primates. He further recognized group territoriality as a phenomenon distinct from individual or pair territoriality.

3. W. H. Burt (1943). In this short article Burt explicitly distinguished home range from territory and subsequently helped to sort out a great deal of confusing data on the behavior of mammals.

4. G. A. Bartholomew and J. B. Birdsell (1953). This essay, one of the first efforts to reconstruct the ecology of early man, included speculation on territorial behavior in *Australopithecus*. The authors postulated that territory served as a principal regulating device when australopithecine populations were in approximate demographic equilibrium.

5. R. A. Hinde (1956). In reviewing the evidence on birds accumulated to the mid-1950's, Hinde stressed the heterogeneous quality of the behaviors that constitute territoriality and the multiple functions they are likely to serve. He also demonstrated that much of the existing evidence on function was ambiguous, thus challenging ornithologists and others to devise more rigorous tests in their field studies.

The Multiple Forms of Territory

Territoriality, like other forms of aggression, has taken protean shapes in different evolutionary lines to serve a variety of functions. And like general aggression, it has proved difficult to define in a way that comfortably embraces all of its manifestations. The problem becomes simpler, however, when we notice that previous authors have tended to speak at cross purposes. A few have defined territory in terms of economic function: the territory is said to be the area which the animal uses exclusively, regardless of the means by which it manages its privacy. Pitelka (1959), for example, argued that "the fundamental importance of territory lies not in the mechanism (overt defense or any other action) by which the territory becomes identified with its occupant, but the degree to which it in fact is used exclusively by its occupant." A majority of biologists, in contrast, define territory by the mechanism through which the exclusiveness is maintained, without reference to its function. They follow G. K. Noble's (1939) simplification of Eliot Howard's concept by defining territory as any defended area. Or, to use D. E. Davis' alternate phrase, territorial behavior is simply social rank without subordinates.

I am convinced that this time the majority is right for practical reasons, that defense must be the diagnostic feature of territoriality. More precisely, territory should be defined as an area occupied more or less exclusively by animals or groups of animals by means of repulsion through overt aggression or advertisement. We know that the defense varies gradually among species from immediate aggressive exclusion of intruders to the subtler use of chemical signposts unaccompanied by threats or attacks.

Maintenance of territories by aggressive behavior has been well documented in a great many kinds of animals. Dragonflies of the species *Anax imperator*, for example, patrol the ponds in which their eggs are laid and drive out other dragonflies of their species as well as those of the similar-appearing *Aeschna juncea* by darting attacks on the wing (Moore, 1964). Orians (1961b) found that the tri-colored blackbird *Agelaius tricolor* is excluded by the red-winged blackbird *A. phoeniceus* in the western United States by a different kind of interaction. Colonies of the former species do not defend territories and are consequently interspersed in the seemingly less favorable nesting sites not preempted by the aggressive *A. phoeniceus* males.

A somewhat less direct device of territorial maintenance consists of repetitious vocal signaling. Familiar examples include some of the more monotonous songs of crickets and other orthopteran insects (Alexander, 1968), frogs (Blair, 1968), and birds (Hooker, 1968). Such vocalizing is not directed at individual intruders but is broadcast as a territorial advertisement. An even more circumspect form of advertisement is seen in the odor "signposts" laid down at strategic spots within the home range of mammals. Leyhausen (1965) has pointed out that the hunting ranges of individual house cats overlap considerably, and that more than one individual often contributes to the same signpost at different times. By smelling the deposits of previous passersby and judging the duration of the fading odor signals, the foraging cat is able to make a rough estimate of the whereabouts of its rivals. From this information it judges whether to leave the vicinity, to proceed cautiously, or to pass on freely. Comparable advertisement has been reported in the insects. After females of the apple maggot fly (*Rhagoletis pomonella*) oviposit beneath the skin of an apple, they drag their extended abdomens over the surface of the fruit for about 30 seconds while laying down a pheromone. The scent is sufficient to deter other females from ovipositing on the same apple for as long as four days, giving the larvae of the first female a decisive head start (Prokopy, 1972).

We do not have the information needed to decide whether occupied land is generally denied at certain times to other members of the species by means of chemical advertisement. Animal behaviorists have naturally focused their attention on the more spectacular forms of aggressive behavior that arise during confrontations. When such behavior is lacking, one is tempted to postulate that exclusion is achieved by advertisement of one form or other. It remains to be pointed out that the exclusive use of terrain must be due to one or the other of the following five phenomena: (1) overt defense, (2) repulsion by advertisement, (3) the selection of different kinds of living quarters by different life forms or genetic morphs, (4) the sufficiently diffuse scattering of individuals through random effects of dispersal, or (5) some combination of these effects. Where interaction among animals occurs, specifically in the first two listed conditions, we can say that the occupied area is a territory.

In animals that are long-lived and endowed with a good memory, territorial exclusion can be based on episodes that occurred long before the human observer appeared on the scene. The mammalogical literature abounds with accounts of "nonterritorial" social systems that might well have passed through periods of more overt exclusion unknown to the investigator. Fraser Darling's herds of red deer, for example, showed no open displays during the time he watched them, although each occupied an exclusive range; but the ranges were fixed before the study period began. On St. Kilda, feral ewe groups have been observed to occupy exclusive core areas for five years without apparent aggressive encounters (Grubb and Jewell, 1966). Schenkel (1966a) likewise concluded that black rhinoceroses are nonterritorial. These great animals mark scent posts with feces and urine and show "excitement" when they encounter one another or smell the scent posts. Schenkel conceded that there is an occasional component of aggression in the encounters but believed that the communication really expresses "an atmosphere of familiarity or solidarity." The last wild populations of brown bears in Europe, located in the Tyrolean Alps of northern Italy, have also been considered to be nonterritorial, but on even slimmer evidence (Krott and Krott, 1963).

All of these negative examples must be regarded as inconclusive. This is not to say that private tenancy cannot be maintained without active exclusion or warning advertisement, only that the negative evidence is not decisive. The episodes establishing the territorial boundaries and dominance relations might have taken place years ago. Where regions are occupied by whole families and groups who pass them to descendants by tradition, exclusion might occur only once in many generations. Moreover, when populations are below the density permitted by the carrying capacity of the environment, territorial defense may be muted or temporarily suspended altogether. Fortunately, the mammalogist need not wait a lifetime to test these hypotheses. If one or the other of the ideas is correct, territorial conflict should be easily observed when animals move into new regions, or the turnover of groups is hastened artificially, or population densities are increased experimentally.

Territorial behavior is widespread in animals and serves to defend any of several kinds of resources. In Table 12-1 are listed examples of studies that have established the primary function of the territory in particular species with a reasonable degree of confidence. This list is short, because circumstances must be unusually favorable for the observer to identify certain resources as the ones defended while striking others from consideration. In the case of bird and ungulate leks, the territories are set up by males and used almost exclusively for breeding. They do not supply protection from predators. Indeed, males of ungulates such as the wildebeest are subject to the heaviest danger from lions and other predators while on their leks (Estes, 1969). Furthermore, the dense concentrations of animals around the display grounds make the land less favorable for feeding. Lek birds such as prairie chickens and turkeys even go elsewhere to feed. So the resource guarded on the lek territories is simply the space for sexual display, plus the females that respond to the male within that space.

Analyses of other forms of territories require different modes of inference but can be made just as positive. Female marine iguanas (*Amblyrhynchus cristatus*) in the Galápagos Islands are normally casual about egg laying, merely placing the eggs in loose soil and departing. On Hood Island, however, nest sites are scarce. The females compete for the limited space by assuming a bright coloration similar to that of the males and fighting in a tournamentlike fashion. The winners get to lay their eggs in the favored spots. Afterward, they survey the sites from lookout positions on nearby rocks, descending

Table 12-1 Examples of territorial behavior in which the primary function has been reasonably well established.

Species	Resource protected	Authority
MOLLUSCA		
Owl limpets (*Lottia gigantea*)	Food supply	Stimson (1970)
ANNELIDA		
Polychaete worms (*Nereis caudata*)	Retreat (constructed tube)	Evans (1973)
CRUSTACEA		
Amphipods (*Erichthonius*)	Shelter and feeding area	Connell (1963)
Spiny lobsters (*Jasus lalandei*)	Shelter	Fielder (1965)
ARANEA		
Sheet-web spider males (*Linyphia triangularis*)	Access to females	Rovner (1968)
ODONATA		
Damselfly and dragonfly males	Space for sexual display	Johnson (1964), Bick and Bick (1965)
HYMENOPTERA		
Males of solitary bees (*Anthidium*)	Space for sexual display	Haas (1960)
Cicada-killer wasp males (*Sphecius speciosus*)	Emergence holes of females, thus priority in mating	Lin (1963)
Ant colonies	Nest sites and food, according to species	Wilson (1971a)
Termite colonies	Nest sites and food, according to species	Wilson (1971a)
Hawaiian drosophilid fly males	Space for sexual display (on leks)	Spieth (1968)
PISCES		
Garden eels (*Gorgasia sillneri*)	Food	Clark (1972)
Reef fish (*Pomacentrus flavicauda*)	Food, shelter	Low (1971)
Reef fish males (*Abudefduf zonatus*)	Spawning sites	Keenleyside (1972)
Garibaldi (*Hypsypops rubicunda*)	Food and reproductive nest site	Clarke (1970)

Table 12-1   (*continued*).

Species	Resource protected	Authority
AMPHIBIA		
Newt males (*Triturus*)	Space for sexual display	Gauss (1961)
Dendrobatid frog females (*Phyllobates trinitatis*)	Shelter site; possibly also surrounding feeding area	Test (1954)
Pipid frog males (*Hymenochirus boettigeri*)	Space for sexual display	Rabb and Rabb (1963)
Bullfrog males (*Rana catesbeiana*)	Space for sexual display	Capranica (1968), Emlen (1968)
Green frog males (*Rana clamitans*)	Space for sexual display	Martof (1953)
Plethodontid (lungless) salamanders (*Desmognathus, Eurycea, Hemidactylium*)	Shelter site; possibly also surrounding feeding area	Grant (1955), Brandon and Huheey (1971), D. B. Means (personal communication)
REPTILIA		
Galápagos marine iguanas (*Amblyrhynchus cristatus*)	Males: space for resting and sexual display; females: egg-nest site	Eibl-Eibesfeldt (1966)
Land iguanas (*Iguana iguana*)	Females: egg-nest site	Rand (1967)
"False chameleons" (*Anolis lineatopus*)	Males: females and food; females: food	Rand (1967)
AVES		
Red grouse (*Lagopus lagopus*)	Food supply; and for males, access to healthiest females	Jenkins et al. (1963), Watson and Moss (1971)
Dunlins (*Calidris alpina*)	Food supply	Holmes (1970)
Hummingbirds (*Amazilia, Archilochus, Panterpe, Phaeochroa*)	Food supply	Pitelka (1942), Stiles and Wolf (1970), Wolf and Stiles (1970)
Ovenbirds (*Seiurus aurocapillus*)	Food supply	Stenger (1958)
Song sparrows (*Melospiza melodia*)	Food supply	Yeaton and Cody (1974)
Mockingbirds (*Mimus polyglottos*)	Food supply	Hailman (1960)

Table 12-1 (*continued*).

Species	Resource protected	Authority
Long-billed marsh wrens (*Telmatodytes palustris*)	Food supply	Verner and Engelson (1970)
Lek birds: cocks of the rock, manakins, prairie chickens, turkeys, oropendolas, etc.	Space for courtship display	Gilliard (1962), Drury (1962), Ellison (1971)
Colonial ground nesting birds: albatrosses, gulls, terns, etc.	Space for courtship and nesting; spacing may be further increased to reduce intensity of predation on eggs, hence eggs are protected	Hinde (1956), Rice and Kenyon (1962), Tinbergen (1967), Tinbergen et al. (1967)
MAMMALIA		
Uganda kob males (*Kobus kob*)	On the leks, space for sexual display	Buechner (1961), Leuthold (1966)
Wildebeest (*Connochaetes taurinus*)	Space for sexual display	Estes (1969)
Waterbucks (*Kobus ellipsiprymnus*)	Food supply	Kiley-Worthington (1965)
Vicuñas (*Vicugna vicugna*)	Food supply	Koford (1957)
Tree squirrels (*Tamiasciurus*)	Food supply	C. C. Smith (1968), Kemp and Keith (1970)
Yellow-winged bats (*Lavia frons*)	Food supply	Wickler and Uhrig (1969a)
Rhesus monkeys (*Macaca mulatta*)	Food supply	Neville (1968)

occasionally to sniff and taste the sites and to scratch a little more earth on top (Eibl-Eibesfeldt, 1966).

Still other forms of evidence may be less direct but equally strong. In his study of *Pomacentrus flavicauda*, a fish of the Great Barrier Reef, Low (1971) noted that each territory covers a particular kind of interface of sand and coral in which sheltering crevices and an adequate supply of algae are located. The fish apparently never leave this spot. They challenge not just rival *P. flavicauda* but any intruder belonging to an alga-feeding species. Nonherbivores are ignored. When resident *P. flavicauda* were removed by Low, alga-feeders of several species moved into the vacated territories. Theoretically, the existence of feeding competitors of other species should reduce the density of food and force individuals to expand their territory size

in order to harvest the same quantity of energy. The higher the number of competing species, the larger we should expect to find the average territory size—insofar as territory size is plastic and other factors, such as habitat differences, are accounted for. Precisely this result, in impressive detail, has recently been obtained in field studies of song sparrows by Yeaton and Cody (1974; see Figure 12-4). Not only did these investigators find the expected close positive correlation, but they were able to predict the approximate mean territory sizes at different localities from a knowledge of the competing species and the estimated competition coefficients between them and the song sparrow.

The functions served by territorial defense, like those of most other components of social behavior, are idiosyncratic and difficult to classify. We can nevertheless distinguish several major categories in

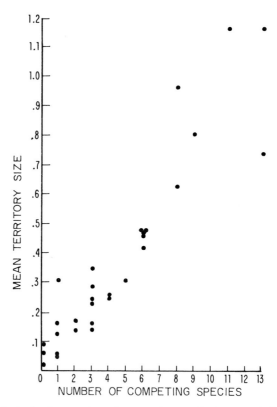

Figure 12-4 As the number of competing species increases, the average territory size of the song sparrow *Melospiza melodia* also increases. Each point represents a different locality in the Pacific Northwest or Wyoming. The localities with the lowest numbers of competing species are on islands. The correlation is consistent with the interpretation that competition reduces food density and forces an enlargement of territory size to satisfy energy requirements. (From Yeaton and Cody, 1974.)

which the known or probable function matches the size and location of the defended space. The following classification is an extension of one developed for birds in sequential contributions by Mayr (1935), Nice (1941), Armstrong (1947), and Hinde (1956). I have modified it slightly in order to cover other groups of animals as well.

Type A: a large defended area within which sheltering, courtship, mating, nesting, and most food gathering occur. This type of territory exists in especially high frequency among species of benthic fishes, arboreal lizards, insectivorous birds, and small mammals.

Type B: a large defended area within which all breeding activities occur but which is not the primary source of food. Examples of species utilizing this less common type include the nightjar *Caprimulgus europaeus* and reed warbler *Acrocephalus scirpaceus*.

Type C: a small defended area around the nest. Most colonial birds utilize such a restricted form of territory, including a majority of seabirds, herons, ibises, flamingos, weaver finches, and oropendolas. Examples among the insects include sphecid wasps ("mud daubers") and bees that nest in aggregations.

Type D: pairing and/or mating territories. Examples include the leks of certain insects, including male damselflies and dragonflies, as well as those of birds and ungulates.

Type E: roosting positions and shelters. Many species of bats, from flying foxes to the cave-dwelling forms of *Myotis* and *Tadarida*, gather in roosting aggregations within which personal sleeping positions are defended. The same is true of such socially roosting birds as starlings, English sparrows, and domestic pigeons. A wide variety of invertebrates, fishes, amphibians, and reptiles defend personal retreats from which they venture periodically to feed. They either breed in or around the retreat, or else leave it temporarily for the sole purpose of breeding. A less familiar but typical example is provided by the spadefoot toads (*Scaphiopus*), which hide in private burrows during the day and emerge to feed on damp or rainy nights. At irregular intervals, usually after heavy rains, they briefly congregate in shallow pools in order to breed.

It is useful to recognize two additional classifications orthogonal to the one just given. For territories of types A and B, territorial defense can be absolute or spatiotemporal. That is, the resident can guard its entire territory all of the time, or it can defend only those portions of the territory within which it happens to encounter an intruder at close range. Spatiotemporal feeding territories are quite common in mammals, especially those that are frugivores or carnivores, because the area that must be covered to secure enough food is typically too large to be either monitored or advertised continuously. Absolute feeding territories are more frequently encountered in birds, which have excellent vision, access to vantage points, and the flight speed to scan relatively large foraging areas. This difference between the two major vertebrate groups is the reason why territo-

riality was originally elucidated in birds, and why its general significance in that group has never been in doubt, while in mammals the subject has always been plagued by seemingly inadequate data and semantic confusion.

The third classification is the following: the territory can be either fixed in space or floating. Animals are committed to floating territories when the substratum on which they depend is mobile. An example is the bitterling (*Rhodeus amarus*), a fish that lays its eggs in the mantle cavity of *Anodonta* mussels and other freshwater bivalves. Each bitterling limits sexual fighting to the vicinity of its mussel, and when the mussel moves around, the fish's territory moves with it (Tinbergen, 1951). A few animals shift their territories about on a fixed substratum. Male damselflies (*Argia apicalis*) fly back and forth daily between their nocturnal roosts and the ponds where breeding occurs. When they arrive over the pond surface they space themselves at 2-meter intervals, exclude other males from the areas they have staked out, and attempt to mate with any females that enter. Each day the locations of the mating territories shift as the males disperse into new positions (Bick and Bick, 1965). Floating territories occur even in a few kinds of birds. The territory of the ovenbird (*Seiurus aurocapillus*) consists of a well-defined but fluctuating area that changes day by day and even hour by hour (Stenger and Falls, 1959).

Functions have occasionally been ascribed to territories other than the primary one indicated by the evidence. In particular, the resident animal is said to become familiar with its domain and as a result more expert at finding food and evading predators there (Hinde, 1956). This is no doubt true, but the same benefit will accrue to any animal that stays in one place, whether it defends a territory or not. The diagnostic quality of territory is defense, and its function is the resource defended. Ordinarily the particular resources defended are the ones that affect genetic fitness most crucially, but familiarity with the resources is a prerequisite of territoriality and not a function. It has also been argued, by Wynne-Edwards (1962) and others, that territories "function" to limit populations. To be sure, they often have that effect, but they are unlikely to serve as an adaptive device in population control. To consider the evidence behind this generalization, the reader is referred back to the theory of interpopulation selection reviewed in Chapter 5. Finally, it has been suggested that territoriality serves to prevent epizootics (Tavistock, 1931; additional review in Hinde, 1956). Again, such an effect, if it occurs at all, may be no more than a felicitous outcome of territorial behavior, and not the principal selective force shaping the particularities of territorial behavior.

The results of 30 years of field research reveal territoriality to have a patchy phylogenetic distribution. It occurs widely through the vertebrates and is common, but far less general, among the arthropods, including especially the crustaceans and the insects. True territorial behavior has also been reported in one species of mollusk, the owl limpet *Lottia gigantea*, by Stimson (1970) and in a few nereid polychaete worms (Evans, 1973). Among the most comprehensive reviews written on a taxonomic basis are the following: insects generally (Johnson, 1964; MacKinnon, 1970), social insects (Wilson, 1971a), crustaceans (Connell, 1963; Dingle and Caldwell, 1969; Bovbjerg and Stephen, 1971; Linsenmair and Linsenmair, 1971), spiders (Rovner, 1968), fishes (Gerking, 1953; Clarke, 1970; Low, 1971), frogs (Duellman, 1966, 1967; Lemon, 1971b), lizards (Kästle, 1967; Rand, 1967), birds (Hinde, 1956; J. L. Brown, 1964, 1969; Lack, 1966, 1968), mammals generally (Ewer, 1968), and primates (Bates, 1970; Alison Jolly, 1972a).

The Theory of Territorial Evolution

"If you shoot a gibbon, you leave seven lonely rivers."
—Saying of the Skaw Karen, North Thailand

The home range of an animal, whether defended as a territory or not, must be large enough to yield an adequate supply of energy. At the same time it should ideally be not much greater than this lower limit, because the animal will unnecessarily expose itself to predators by traversing excess terrain. This optimal-area hypothesis seems to be borne out by what little data we have bearing directly on the energy yield of home ranges. C. C. Smith (1968), for example, found that the territory size of *Tamiasciurus* tree squirrels appears to be adjusted to provide just enough energy to sustain an animal on a year-round basis. In 26 territories Smith measured, the ratio of energy available to energy consumed during one-year periods varied from just under 1 to 2.8, with a mean of 1.3. The poorer the energy yield per square unit of a given habitat, the larger the territory each squirrel occupied to compensate.

The same basic principle emerged in a more intuitive way from the Altmanns' study of the yellow baboons of Amboseli, the most thorough analysis of its kind conducted on a primate species. Each day the baboon troops move out from the sleeping trees along paths that lead them to waterholes and feeding grounds. Their direction, their pace, and the periods of time through the day they spend in each sector appear to be based on the memory and judgment of the troop leaders. The track of a troop's movement on any single day makes little sense by itself. But when many combined track segments are plotted against different times of the day, the pattern of troop activity is seen to follow a strong diurnal rhythm (Figure 12-5). The tracks spread amebalike out from the sleeping trees, pause at the waterholes, diffuse to the maximum area at midday, and contract back to the sleeping trees at dusk. The home range appears to be just large enough

to sustain the troop, and the frequency with which the baboons head toward given locations is roughly proportional to the expected yield of food from each. In studying these data one is reminded of R. J. Herrnstein's principle of quantitative hedonism. Herrnstein found that pigeons trained at two disks, one located to the left and the other to the right, will try one as opposed to the other in precise proportion to the percentage of times each disk reinforces the pigeon with food when it is pecked (Herrnstein, 1971a). In other words, where P stands for the number of pecks, R for the number of reinforcements, and the subscripts l and r indicate the left disk and right disk, respectively,

$$\frac{P_l}{P_l + P_r} = \frac{R_l}{R_l + R_r}$$

If different feeding sites and temporary waterholes reward animals with varying degrees of satisfaction, we can hypothesize that the pattern of movement through the home range will reflect this heterogeneity in a fashion consistent with Herrnstein's principle or some modification of it.

The optimum-yield hypothesis is further supported by data revealing a general correlation among terrestrial vertebrates between the size of the animal and the size of the home range it occupies. This relation, which is surprisingly consistent, was first demonstrated by McNab (1963) in the mammals and extended by later authors to other vertebrate groups. The relationship obtained by comparing many species fits roughly the following logarithmic function:

$$A = aW^b$$

where A is the home range area of a given species, W the weight of an animal belonging to it, and a and b fitted constants. It is also approximately true that the rate of energy utilization (E) is a linear function of the metabolic rate (M), that is,

$$E = cM$$

where c is another fitted constant. Finally, the metabolic rate, M, increases as a logarithmic function of the animal's weight, W:

$$M = \alpha W^\beta$$

where α and β are two more fitted constants. It follows that the area of the home range is a logarithmic function of energy needs. The values of a, b, α, and β for three taxonomic groups of vertebrates are given in Table 12-2. It can be seen that each group has a distinctive set of values, reflecting peculiarities in locomotion and efficiency in the harvesting of energy. Schoener (1968a) has further demonstrated

1KM

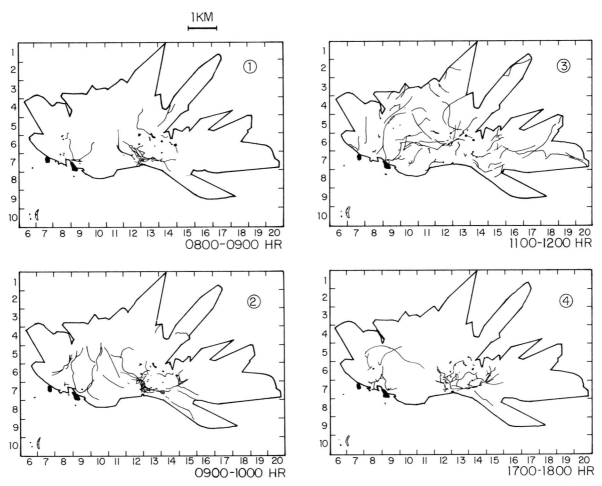

Figure 12-5 The diurnal cycle of utilization of home range by a troop of yellow baboons. The limit of the home range is indicated as a solid dark line, while the observed tracks of the troop at various intervals during the day are drawn as solid thin lines. The small solid patches in the center of the home range and at the southwest corner are the permanent waterholes. (Modified from Altmann and Altmann, 1970.)

that the slope of the curve relating home range (or territory) to body weight in birds depends on their diet. As shown in Figure 12-6, the slope is greatest for predators, least for herbivores, and intermediate for species with mixed diets. This relationship convincingly supports the optimum-yield hypothesis. The hypothesis now reads that as predators grow larger, prey of suitable size grow scarcer and the predators must search over disproportionately larger areas to secure the minimum ration of energy. But why should the suitable prey grow scarcer? There are two reasons. Within any trophic level, say the herbivores or first-level carnivores, most organisms are concentrated at the small end of the size scale; hence disproportionately fewer items will be suitable for the bigger carnivore. Also, as a predator grows larger, it is more likely to feed on other predators, which are

scarcer by virtue of the ecological efficiency rule. Schoener has provided evidence that the mammalian data can be decomposed in the same way, again yielding higher slopes for the predators. The data for lizards are not yet adequate to test the hypothesis in this third group (Turner et al., 1969).

One should keep in mind that these quantitative relationships pertain only to undefended home ranges and to feeding territories, which are a special kind of home range. Other forms of territories, for example those deployed around shelters or display positions, are subject to wholly different sets of controls and are likely to be related to physiological properties of the animals in diverse ways. Even the feeding territories are sometimes demarcated by factors less complex or more complex than energy yield. Garden eels (*Gorgasia*), being

Table 12-2 Regressions of metabolic rate (M) and area of home range (A) on body weight (W) in three groups of vertebrates. (M in kcal/day for mammals and birds, cm³ O₂/hr for lizards. A in acres for mammals and birds, m² for lizards. W in kg for mammals and birds, g for lizards.)

Group	Relationship	Function	Authority
Mammals	Basal metabolism and body weight	$M = 70W^{0.75}$	Kleiber (1961)
	Home range and body weight	$A = 6.76W^{0.63}$	McNab (1963)
Birds	Basal metabolism and body weight	$M = kW^{0.69}$	Lasiewski and Dawson (1967)
	Home range and body weight	$A = kW^{1.16}$	Schoener (1968a)
Lizards	Standard (30°C) metabolism and body weight	$M = 0.82W^{0.62}$	Bartholomew and Tucker (1964)
	Home range and body weight	$A = 171.4W^{0.95}$	Turner et al. (1969)

plankton feeders, live in a superabundance of food. But being sedentary bottom dwellers evidently subject to heavy predation, they also do not leave their burrows. Therefore the radius of the feeding territory of each eel, which it defends from adjacent eels, is the exact length of its body (Clark, 1972). Territories of tube-dwelling amphipods (*Erichthonius braziliensis*) are governed in an identical fashion. The crustaceans, which graze on algae, utilize and defend all of the feeding area that they can reach without losing contact with their tubes (Connell, 1963). In various species of animals, closed groups rather than individuals or mated pairs occupy the territories. In these cases the weight-area law may still hold, but there is likely to be a further correlation between the size of the group and the quality of the habitat it can hold, causing a scatter in the regression. Troops of vervet monkeys (*Cercopithecus aethiops*), to take one example, are highly variable in size. The largest troops dominate the smallest, which are forced into less favorable terrain and must defend larger areas in order to satisfy their energy needs (Struhsaker, 1967a). Such complexities arising from higher social organization are probably responsible for the extreme variation in home range area among the primates generally, as revealed in the data recently collated by Bates (1970).

But why should animals bother to defend any part of their home range? MacArthur (1972) proved that pure contest competition for food is energetically less efficient than pure scramble competition. This is a paradox easily resolved. Territoriality is a very special form of contest competition, in which the animal need win only once or a relatively few times. Consequently, the resident expends far less

energy than would be the case if it were forced into a confrontation each time it attempted to eat in the presence of a conspecific animal. Its energetic balance sheet is improved still more if it comes to recognize and to ignore neighboring territorial holders—the dear enemy phenomenon to be examined later in this chapter.

Clearly, then, a territory can be made energetically more efficient than a home range in which competition is of the pure contest or pure scramble form. But if this is the case, why are not all species with fixed home ranges also strictly territorial? The answer lies in what J. L. Brown (1964) has called economic defendability. Natural-selection theory predicts that an animal should protect only the amount of terrain for which the defense gains more energy than it

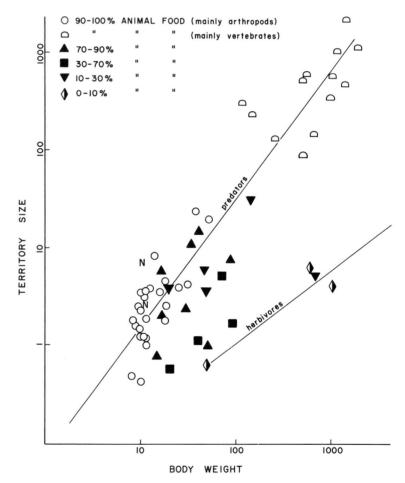

Figure 12-6 The relationship of territory size (in acres) to body weight (grams) for birds of various feeding categories. Each point represents a different species. Omnivores (10–90 percent animal food) are shaded, herbivores half-shaded, and predators clear. N = nuthatch species. (From Schoener, 1968a.)

expends. In other words, if an animal, a carnivore for example, occupies a much larger territory than it can monitor in one quick survey, it may find trotting from one end of its domain to the next just to oust intruders an energetically wasteful activity. Consequently, natural selection should favor the evolution of a spatiotemporal territory rather than an absolute territory. The carnivore will devote most of its energies to hunting prey, challenging only those intruders it encounters at close range. Or else it will deposit scent at strategic positions through the territory in an effort to warn off intruders. Arboreal lizards, such as iguanids, agamids, and chamaeleontids, which are able to scan large areas visually, also tend to maintain absolute territories. Terrestrial forms, such as many scincids, teiids, and varanids, generally have spatiotemporal territories or broadly overlapping home ranges (Judy Stamps, personal communication).

Horn (1968) utilized this same concept when investigating the conditions favoring colonial nesting in blackbirds. He proved that when resources are uniformly distributed and continuously renewing, there is an advantage to maintaining a complete defense of whatever portions of the area can be patrolled in reasonably short periods. But when food is patchily distributed and occurs unpredictably in time, it does not pay to defend fixed areas. The optimum strategy is then to nest colonially and to forage in groups. By this means the individual is able to utilize the knowledge of the entire group. Economic defensibility is actually only one important component of fitness that determines the evolution of territorial behavior. As Heller's work on chipmunks showed (Heller, 1971), territorial defense is curtailed if it exposes animals to too much predation. There is also the phenomenon of aggressive neglect: defense of a territory results in less time devoted to courtship, fewer copulations, and neglected and less fit offspring. In short, the territorial strategy evolved is the one that maximizes the increment of fitness due to extraction of energy from the defended area as compared with the loss of fitness due to the effort and perils of defense.

Schoener (1971) has taken the first step toward parameterizing this theory of territorial evolution. It is possible to estimate the permeability of a territory, measured by the density of intruders tolerated at any given moment of time, if we view the permeability as the balance struck when the rate of invasion by intruders becomes equal to the rate at which they are being expelled (Figure 12-7). In the simplest case, the invasion rate decreases linearly with increased density of invaders already on the territory. It may simply equal the following product:

Invasion Rate

The perimeter of the defended area	×	A constant determined by the probability an outsider will invade	×	$\left(1 - \dfrac{(N/A)}{H}\right)$

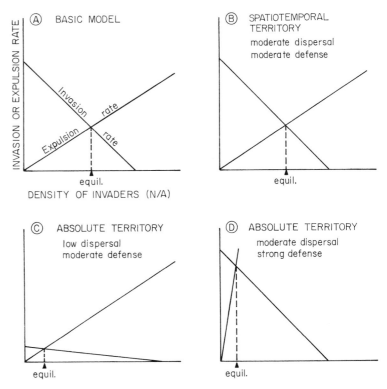

Figure 12-7 Schoener's model of territorial defense and three extensions that can be made from it.

where N/A is the density of invaders on the territory (number of invaders divided by the area of the territory), and H is the maximum density that can occur under any condition. The rate at which the territory holder chases out or destroys invaders might also be linear:

Expulsion Rate

Rate in area/time at which defender searches	×	Probability of expelling an invader when encountered	×	The density of invaders	×	$\dfrac{1}{\text{Time spent by each invader in the territory}}$

In Figure 12-7B-D are given three elementary extensions of Schoener's model. We note that a moderate to high invasion rate relative to the perimeter of the home range, combined with a moderate to low defensive response, produces a spatiotemporal territory. An absolute territory, in which all of a fixed area is defended all of the time, results when the dispersal rate of invaders is low, defense is strong, or both. The parameters envisaged by Schoener are probably correct, but we have no present way of estimating the form of the invasion and expulsion curves. Nor does the model incorporate the energy balance, which is all-important in natural selection, or other components of fitness. A somewhat similar model, less precisely parameterized but incorporating notions of energy gain and loss, has been independently developed by Crook (1972). Crook's reasoning with special reference to the optimization of group size was presented earlier, in Chapter 6.

Special Properties of Territory

Territorial behavior involves much more than the mere expulsion of intruders. And territories are more than defended areas: they possess both structure and dynamism and can be described as fields of variable intensity. Territories change in size and shape through the seasons and as the animal matures and ages. Field studies have revealed the following rich set of phenomena, some very general and others restricted to one or a small number of species.

The Elastic Disk

Territorial size in most animal species varies to a greater or lesser degree with population density. Julian Huxley (1934) compared the variable territory to an elastic disk with the resident animal at its center. When overall population density increases and pressure builds along its perimeter, the territory contracts. But there is a limit beyond which the animal cannot be pushed. It then stands and fights, or else the entire territorial system begins to disintegrate. When by contrast, the surrounding population decreases, the territory expands. But, again, there is a limit beyond which the animal does not try to extent its control. In very sparse populations, either territories are not contiguous, or their boundaries simply become too vague to define

Figure 12-8 shows an example of the elastic territories of the dunlin (*Calidris alpina*), a kind of sandpiper. In the subarctic locality of Kolomak, at 61°N in Alaska, food is relatively abundant and reliable, and populations of the sandpipers attain a density of 30 pairs per hectare. Farther north at 71°N, at the arctic site of Barrow, the food supply is unpredictable and the summers shorter. Here the birds live at densities one-fifth those at Kolomak, or about 6 pairs per hectare. Since the territorial boundaries are contiguous at both localities, the average territorial size at Barrow is five times that at Kolomak.

The pattern of usage of the territory sometimes changes as it is compressed or relaxed. Judith Stenger Weeden (1965) found that tree sparrows (*Spizella arborea*) intensively utilize all of their space when compressed, but in sparser populations, where each bird has more room, the territory is divided into a core area of intensive usage and an outer cortex which is visited less frequently. Compressible home

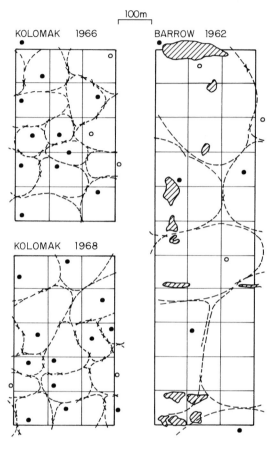

Figure 12-8 The elastic disk phenomenon is shown by territories of the dunlin, a species of sandpiper that breeds in Alaska. At Barrow the population density is one-fifth that at Kolomak. Since the territorial boundaries are contiguous at both localities, the Barrow territories are therefore five times as large. A solid dot represents the location of a nest, an open dot the probable vicinity of a nest not uncovered. (Redrawn from Holmes, 1970.)

ranges, sometimes visibly defended and sometimes not, have been reported in lizards of the genus *Uta* (Tinkle, 1967), many bird groups (Nice, 1941; Kluijver and L. Tinbergen, 1953; Stenger Weeden, 1965), hyraxes (J. B. Sale, in Jewell, 1966), *Suncus* house shrews (Myers et al., 1971), *Microtus* voles (Frank, 1957), *Peromyscus* deer mice (White, 1964), and the European rabbit *Oryctolagus cuniculus* (Myers et al., 1971).

The "Invincible Center"

Unless an adult male bird is grossly overmatched or ill, he is usually undefeatable by conspecific birds at the center of his territory (N.

Tinbergen, 1939; Nice, 1941). This circumstance is only the extreme manifestation of the more general principle that the aggressive tendencies of animals, and the probability that any given intensity of aggression or display will result in the domination of rivals, increase toward the center (J. R. Krebs, 1971). What exactly is the "center"? In uniform terrain it is customarily the geometric center, but in a heterogeneous environment the behavioral center is more likely to be the location of either the animal's shelter or the most concentrated food supply within the territory, whichever is the more vital to the welfare of the animal. The territory holder spends most of its time near the center, routinely performing courtship displays and constructing shelters there. Male tree sparrows begin their day with a reaffirmation of the core area through song followed by more lengthy trips into the cortex (Stenger Weeden, 1965).

Let us conjecture that each species is characterized by a particular gradient of aggressiveness and dominance measured from the territorial perimeter to the center. In some cases the gradient will be near zero; that is, the cortex will be defended as vigorously as the core. In others it will be steep and perhaps shift in value with increasing distance, showing a pattern of slow inclines or plateaus separated by sudden upward steps. This hypothesis is consistent with observed peculiarities in the behavior of certain species. Steller's jays (*Cyanocitta stelleri*) and bicolored antbirds (*Gymnopithys bicolor*) do not defend clear territorial boundaries, just concentric zones of outwardly decreasing dominance, so that in certain intermediate areas the birds are either neutral or maintain a feeble balance with their neighbors (J. L. Brown, 1963; Willis, 1967). In such cases a description of the dynamics of the system will of course be more precise than any map of static territorial boundaries.

Polygonal Boundaries

When circular disks made of a very flexible substance are pressed against one another along their edges, they deform into hexagons, which provide the maximum congruence of boundaries. Perfectly fitted hexagons of equal size leave no spaces between them. The waxen cells of a honeycomb, for example, are constructed as hexagonal columns by honeybees. When territories are absolute, sharply bounded, and maintained in areas of high population density, they press against one another in a manner analogous to plastic disks. It would be to the advantage of the territory holders to utilize space maximally by defending perimeters that are polygonal in form, tending toward the optimum hexagonal shape. Precisely this phenomenon has been described in dunlins by Grant (1968), who reanalyzed R. T. Holmes's Alaskan data. Polygons are not evident in the maps by Holmes reproduced in Figure 12-8, but they are evident in a few territories in which the boundaries were traced with special care. At Berkeley in 1972, George W. Barlow showed me remarkable clusters

of polygonal territories that had been formed by male mouthbrooder fish (*Tilapia mossambica*) kept in shallow outdoor tanks. Most of the figures were six-sided, and a few appeared to be five-sided. We were unable to find any that were clearly four- or seven-sided (see Figure 12-9).

Changes with Season and Life History Stages

The values of the parameters that define the optimum territorial size shift during the life cycles of most kinds of animals. When a male bird begins to court, he needs to defend fewer positions than later, when his nestlings demand large amounts of food. But he needs to defend the positions more often, since the population as a whole is more mobile and floater males challenge more frequently. Such changes have been documented by Hinde (1952) in the great tit *Parus major* and by Marler (1956) in the chaffinch *Fringilla coelebs*. Males of these small European birds begin the breeding season by singing and fighting around selected display sites. Only later do they extend their defense to the entire territory. The boundaries of the territories of black-capped chickadees (*P. atricapillus*) strongly fluctuate through the breeding season, first expanding slightly as nests are built, then contracting drastically at the egg and nestling stages, and finally expanding again when the young reach the fledgling stage (Figure 12-10). In fact, the patterns of change vary greatly from species to species throughout the birds. The male of the mockingbird *Mimus polyglottos* stays on his territory the year round, expanding its size at the beginning of the breeding season in the spring (Hailman, 1960). The green heron *Butorides virescens*, in contrast, arrives at the breeding ground in the spring and immediately sets up a full-sized territory about 40 meters in greatest diameter. Thereafter, the defended area steadily shrinks until it is limited to the immediate vicinity of the nest, at which time the mated pair cooperate in defense (Meyerriecks, 1960). Some of the mystery presented by these differences vanishes when we consider the natural history of the species. The male mockingbird defends a feeding territory, which must be expanded and maintained while the young are growing. The male green heron, however, first defends a courtship display area. Later, he and his mate are no longer courting. They feed in shallow water outside the breeding area, and need only to defend their nest and young.

Seasonal variation of home range and territory is no less complicated and idiosyncratic in the mammals. Red squirrels (*Tamiasciurus hudsonicus*) maintain two forms of territories in the mixed forests of Alberta. "Prime territories" are defended the year round by successful adults in mature coniferous stands, which provide a continuous supply of seeds that serve as food. Other habitats, especially those with a high proportion of deciduous trees, yield seeds only during the growing season. During the winter the resident red squirrels, consisting mostly of juveniles, defend caches of seeds gathered in the warmer months. Wildebeest males, to take another, radically different example, defend display areas during the breeding season and travel with the nomadic herds the rest of the year (Estes, 1969).

In some mammals the home range and territory vary not only by season but also through the life of the animal. Young gray squirrels (*Sciurus carolinensis*) simply expand their home range outward in all directions from their birthplace, starting at the age of about two months and completing the process about six months later (Horwich,

1972). On the Scottish isle of Rhum, the ontogeny of the red deer follows a more leisurely and complex course (Lowe, 1966). During the first three years of its life the young deer gradually dissolves its ties with its family and moves up to the watershed in the island's center, which it proceeds to explore during the summer months. Its home range is next established in the watershed and gradually extended below as it starts to breed and to move back and forth with the seasons. Finally, in old age, the deer abandons the upper,

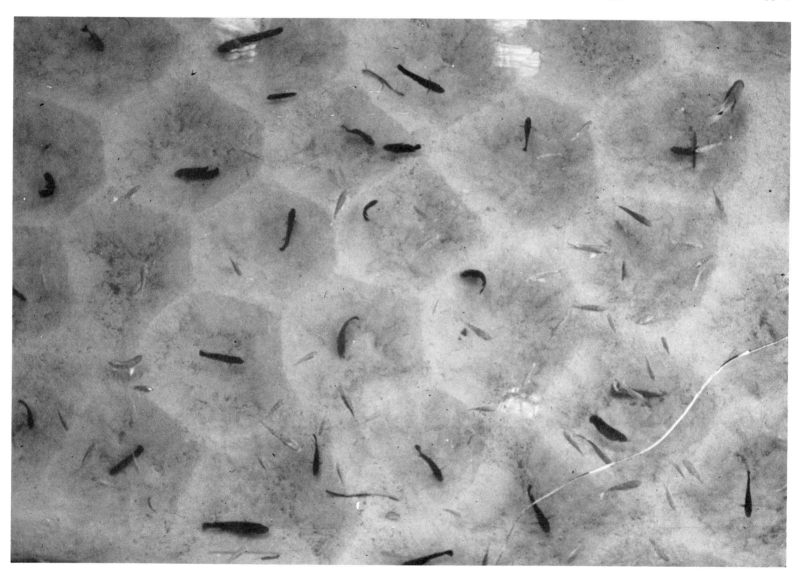

Figure 12-9 Hexagonal territorial boundaries of male mouthbrooder fish (*Tilapia mossambica*) can be clearly seen in this photograph as ridges of heaped-up sand around the depressions scooped out by the fish. Each territory is occupied by a single male, which is distinguishable from the other fish by his dark breeding coloration and generally larger size. (From Barlow, 1974b.)

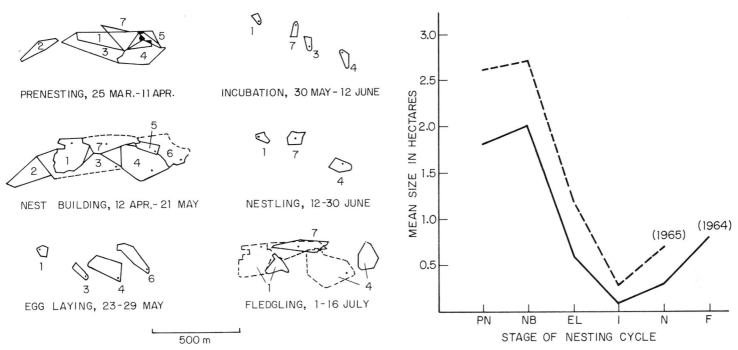

Figure 12-10 The territories of adult black-capped chickadees change in size and shape through different phases of the breeding season. In the left figure, territorial boundaries are shown as lines and nest sites as dots. The figure on the right shows the change in average territorial size through the six phases of the season. (Redrawn from Stefanski, 1967.)

summer portion of the range and spends all of its time in the lower reaches.

Other Determinants of Territory Size

Empirical studies of territorial behavior occasionally disclose previously unexpected parameters. Van den Assem (1967), for example, discovered that twice as many territories of the three-spined stickleback (*Gasterosteus aculeatus*) are established when males are introduced simultaneously into an aquarium as when they are introduced one by one. The existence of a similar phenomenon is indicated by less direct evidence in white wagtails (*Motacilla alba*), blackbirds (*Turdus merula*), and pig-tailed macaques (*Macaca nemestrina*). The tighter packing appears to result at least in part from the reduced hostility of animals toward one another when they enter a strange environment at the same time (Bernstein, 1969; J. R. Krebs, 1971). Tradition and the "personality" of the territory holder can also exert an influence, especially in the longer-lived, more intelligent mammals. Southwick and Siddiqi (1967) have provided a suggestive anecdote from their observations of wild rhesus monkeys in India. The home range of a certain troop covered 16 hectares while the dominant male was healthy but fell to less than 4 hectares when he was injured.

As soon as the leader died from the injury, a formerly subordinate male within the group took over his role—yet the home range stayed at the reduced size.

Nested Territories

Social organizations are known in which an overlord male maintains a territory subdivided in turn by females. In the dwarf cichlid *Apistogramma trifasciatum* (Burchard, 1965) and some species of the lizard genus *Anolis* (Rand, 1967), the females defend their domains against one another but not against the male. Such a nested territory is actually a combined territorial system and dominance order.

The Dear Enemy Phenomenon

A territorial neighbor is not ordinarily a threat. It should pay to recognize him as an individual, to agree mutually upon the joint boundary, and to waste as little energy as possible in hostile exchanges thereafter. In cases where the Fraser Darling effect is also operative, the boundary becomes an important source of social stimulation. James Fisher (1954) recognized most of these principles in birds when he stated, "The effect is to create 'neighborhoods' of individuals who while masters on their own definite and limited properties are bound firmly and *socially* to their next-door neighbors by what in human terms would be described as a dear-enemy or rival-friend situation, but which in bird terms should more safely be described as mutual stimulation." The ability of birds to distinguish the songs of neighbors from those of strangers has since been proved by playing recordings of both kinds of individuals to the territorial males and observing their responses. The species tested include cardinals, *Richmondena cardinalis* (Lemon, 1967); ovenbirds, *Seiurus aurocapillus* (Weeden and Falls, 1959); white-throated sparrows, *Zonotrichia albicollis* (Falls, 1969); indigo buntings, *Passerina cyanea* (S. T. Emlen, 1971); great tits, *Parus major* (J. R. Krebs, 1971); and Indian hill mynahs *Gracula religiosa* (Bertram, 1970). In general, when a recording of a neighbor is played near a male, he shows no unusual reactions, but a recording of a stranger's song elicits an agitated aggressive response. But if the stranger is from such a remote area that its song belongs to a different dialect, the response is weaker.

The adaptive significance of the dear enemy phenomenon is problematical. We cannot say whether it is energy conservation, social stimulation, or both. But given that the phenomenon is adaptive, what mechanisms make it possible? There appear to be three. The first is simply habituation, the form of learning in which responses to a stimulus decrease in time as the animal becomes more familiar with it. In an almost literal sense, the neighbors "tame" one another. Habituation may be reinforced by a second effect discovered in human beings and laboratory rats by R. B. Zajonc (1971): merely repeating exposure of an individual to a given stimulus object is enough to increase its attraction to the object. In other words, fixation can occur without reinforcement. The more an animal or person is exposed to an initially neutral stimulus, the more attractive the stimulus becomes. Rats exposed by Zajonc to a certain kind of music, namely, Mozart or Schoenberg, later chose to listen to the one with which they were familiar. (Classicists will be comforted to learn that unconditioned rats chose Mozart over Schoenberg.) Human beings given nonsense words or unfamiliar Chinese ideograms later judged them good as opposed to bad in comparison with other words and ideograms presented for the first time. The more the faces of particular strangers are shown in photographs, the more experimental subjects are inclined to think well of them. In short, familiarity induces warm feelings. Or, as Zajonc has put it: "Familiarity does not breed contempt. Familiarity breeds!" The probable adaptiveness of the effect can be easily guessed. The more something stays around without causing harm, the more likely it is to be part of the favorable environment. In the primitive lexicon of the emotive centers, strange means dangerous. It is perhaps adaptive to become homesick in foreign places, or even to suffer culture shock. And for animals, it would seem prudent to treat a familiar and relatively harmless enemy as dear.

The third mechanism favoring dear enemy recognition is convergence in the dialect used. When Indian hill mynahs set up territories next to one another, their songs change so as to converge in dialect (Bertram, 1970). The same behavior is exhibited by the American cardinal, the pyrrhuloxia (*Pyrrhuloxia sinuata*), the chaffinch, and the great tit (Hinde, 1958; Gompertz, 1961; Lemon, 1968). The tendency appears to be greatest in species that remain in territories for the longest periods of time each year.

Tolerance and even mutually beneficial exchanges with familiar neighbors appear to occur in at least some mammals, and we should not be surprised to find that it is a general phenomenon. Estes (1969) detected what he considered to be a cooperative greeting step in the highly ritualized challenge displays of neighboring bull wildebeest. The bulls appear to know one another as individuals, and the establishment of a territory by a newcomer has many of the outward signs of joining a club. Deer mice (*Peromyscus maniculatus*) are much more hostile at their territorial borders to strangers than to old neighbors. Healey (1967) has proposed that a cluster of compatible neighbors is in fact the true social unit among these little rodents.

Territories and Population Regulation

In a seminal paper on the population dynamics of titmice in Holland, H. N. Kluijver and L. Tinbergen (1953) concluded that territoriality plays a precise role in the regulation of populations. They recognized that the habitable environment is divided into areas that are optimal for breeding and others that are suboptimal. Kluijver and Tinbergen postulated that the optimal habitats support the densest, most stable populations—the endemic core of the species. The suboptimal habitats support sparser, less stable breeding populations. In the spring the birds that arrive first settle in the optimal habitats, spacing themselves out by territorial exclusion until the area is filled. Territoriality prevents overpopulation and thus guards the population from excessive fluctuation. Kluijver and Tinbergen referred to the stabilization as buffering. Late arrivals spill over into the suboptimal habitats, where they exist in more scattered territories or wander as floaters. These marginal populations are not buffered. They breed far less, their mortality is higher, especially in the fall and winter, and their numbers fluctuate more widely. The optimal habitat of the great tit *Parus major*, for example, is comprised of narrow strips of broad leaved woodland, while the surrounding zones of pine woods serve as the suboptimal habitat.

In a later, detailed review of the subject, J. L. Brown (1969) envisaged the buffer effect as a three-stage development in the build-up of bird populations:

Level 1. At the lowest population density, territories are not circumscribed by competition. No individual is prevented from settling in the best habitat.

Level 2. As the population density rises, some individuals are excluded from the optimal habitats and are forced to establish territories in the poorer habitable areas.

Level 3. At the highest densities, some individuals are prevented from establishing territories altogether. They exist as a floating population that drifts back and forth, in and around the established territories. Brown inferred the floaters to be part of the buffering process for the optimal habitat and, to a lesser degree, for the less favorable habitats in which nesting to any extent occurs. When birds die on their territories, floaters move in to take their place and thereby maintain an approximately constant density in the habitable areas.

The literature is filled—indeed, it overflows—with tortuous discussions about the role of territory in the regulation of populations. Boiled down, the argument turns on whether exclusion regulates the populations, or whether food supply ultimately plays this role. The question is often raised whether natality and mortality, conditioned principally by food supply, fluctuate enough to override the buffering effect of territoriality. However, the problem presents no great conceptual difficulty if we state it as an evolutionary hypothesis, consistent with the theory of population biology and phrased so as to make it subject to testing and modification. Food supply, the conjecture goes, is very likely to be the ultimate limiting factor. In some species, such as the pied flycatcher *Ficedula hypoleuca* (von Haartman, 1956), specialized nest sites might be the ultimate limiting factor. Whatever the resource, however, territorial behavior is the mechanism for defending it when it is in short supply. The buffer effect causing population stability is the by-product of territorial behavior. This completes the statement of the hypothesis that territoriality evolves by selection at the level of the individual.

A second, competing hypothesis is that territoriality evolves by group selection, particularly interpopulation selection. This model also assumes that food, or less probably some other resource, is the ultimate limiting factor. Territoriality is a device evolved by the entire population, including the unfortunate floaters, to hold population densities at or below the densities that can be sustained by the environment. The regulation is achieved at least in part by altruistic restraint and even self-sacrifice, especially on the part of the floaters.

Existing evidence strongly favors the first, individual-selection hypothesis. Consider the Levins and Boorman-Levitt models presented in Chapter 5. From the large amounts of data on birds, we can obtain a rough idea of the magnitude of individual mortality associated with territoriality as opposed to the rates of extinction in territorial populations (see, for example, the reviews by Lack, 1966; Brown, 1969; and J. R. Krebs, 1971). These data seem decisive. The differential mortality associated with territorial exclusion is heavy, on the order

of 10 percent or more per generation, enough to drive the evolution of territorial behavior with even a small amount of heritability in innate components of the behavior. The rates of population extinction, by contrast, must be very low, even if we assume restricted genetic neighborhoods and small effective population sizes. The group-selection hypothesis therefore appears to be excluded in at least the bulk of the better-analyzed cases.

Turning to a closer scrutiny of the individual-selection hypothesis, we should in the spirit of strong inference try to exclude it also and hence force a reexamination of existing evolutionary theory. This hypothesis might be in trouble if we find that the energy yield of feeding territories is normally well in excess of the requirements of the residents. In the face of intense competition and risk, territory holders should be expected to limit their defense to something close to the area of minimally sufficient energy yield. If species generally do so, the fact is consistent with the individual-selection hypothesis. If they do not, an additional explanation is required. The existing data indicate that the hypothesis is still safe on this basis. A great deal of variation in home range and territory size within the best-analyzed species of birds and mammals (see Table 12-1) can be inversely correlated with the quality of the environment and hence its energy yield. However, few quantitative balance sheets of energy requirements and energy yield have been prepared of the kind needed for a truly rigorous testing of the hypothesis. C. C. Smith (1968) found that the yield of tree squirrel territories averaged 1.3 times the requirements and, even more favorable to the individual-selection hypothesis, 5 out of the 26 territories measured had a yield/demand ratio of less than 1. In other words, the population as a whole is crowded up against its lower energy limit, and individual selection must be strong even among the successful territorial holders.

Some writers have been troubled by the free use of the expressions "regulation" and "density dependence" with reference to territorial exclusion. They seem to believe that in order to qualify as regulating devices, territories of a particular species must be elastic, so that as density increases territory size decreases, and consequently population growth decelerates gradually as a result of territorial behavior. By this strict conception, true regulation is not involved if the environment simply fills up with inelastic territories until there is suddenly no more room. But elasticity is not really crucial. Deceleration can operate according to either a continuous function or a step function. Although the continuous function alone yields the classic logistic growth curve, both relations are truly regulatory and density-dependent.

The existence of abundant suboptimal territories and floaters has now been documented in so many species of birds and mammals, belonging to so many genera and higher taxa, as to suggest that they are a very general if not universal phenomenon. The readiness of floaters to fill vacated territories has been amply demonstrated by experiments in which territory holders were simply trapped or shot. The first such removal experiment was performed by Stewart and Aldrich (1951) and Hensley and Cope (1951). These investigators censused the territorial males of 50 species of birds in a 16-hectare plot of spruce-fir woodland, then shot as many birds as they could during a three-week period. The results were startling: the total number of territorial birds removed during the experiment was three times the number originally estimated to be present. A similar effect was subsequently obtained for other kinds of birds, including grouse and ptarmigan (Bendell and Elliot, 1967; Watson, 1967; Watson and Jenkins, 1968), oystercatchers (Harris, 1970), sandpipers (Holmes, 1966), *Agelaius* blackbirds (Orians, 1961b), gulls (Patterson, 1965), *Zonotrichia* sparrows (Mewaldt, 1964), *Tadorna* ducks (Young, 1964), and titmice (Krebs, 1971); deer mice (Healey, 1967), voles (Smyth, 1968), woodchucks (Lloyd et al., 1964) and other rodents among the mammals (see also the review by Archer, 1970); fish (Gerking, 1953; Clarke, 1970); and dragonflies (Moore, 1964). The evidence is generally strong that floaters provide the bulk of replacements, as opposed to late arrivals who would obtain territories of their own in any case. Where territories are persistent, enduring either year-round or at least through more than one cycle, it is invariably the juveniles who are forced into the marginal habitats and floater populations. Particularly

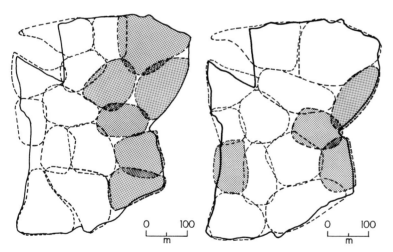

Figure 12-11 When birds are removed from their territories, they are quickly replaced by nonterritorial individuals belonging to the same species. In this experiment, John R. Krebs shot the six pairs of great tits occupying the territories shown as the stippled areas in the map on the left. During the following three days, the surviving residents expanded and shifted their holdings while four new pairs moved in (the map on the right). The final result was a restoration of a complete mosaic over the wood. (Redrawn from Krebs, 1971.)

detailed data on the role of floaters have been provided by the work of J. R. Krebs on the great tit (see also Figure 12-11) and by Watson, Jenkins, and their coworkers on the red grouse. In each instance the replacements damp fluctuation in the size of the territorial populations.

Interspecific Territoriality

Interspecific competition is one of the prime movers of social evolution. When two ecologically similar species first meet, either they coexist stably or one eliminates the other from the zone of overlap. The conditions for coexistence are curiously inverted in nature. One species "tolerates" the existence of the other if its own density-dependent controls stabilize the population before the other species is crowded out. Symmetrically, its competitor must possess stringent enough density-dependent controls to permit the first species to survive. Ideally, each species has its own set of density-dependent controls, and we speak of them as resulting from differences in the niches of the species. Suppose that the ecological niches of two competitors differ in the kind of food preferred and that food shortage is the primary density-dependent factor. The two species will coexist if shortages of the preferred food of each species bring each to zero population growth before one crowds out the other. Essentially the same result is obtained when other requisites are limiting, or even when the principal control is predation. If predator a stops the population growth of prey species A before A eliminates B, and predator b stops species B before it eliminates A, then A and B will coexist. Notice that when the two competitors first meet, the niche differences that guarantee their coexistence are simply accidental outcomes of their divergent evolution in the period prior to contact.

Although two competing species may prove basically compatible, by definition they reduce each other's niche space and biomass. Current ecological theory teaches that a reduction in niche is more likely to take the form of surrender of some of the habitats to the competitor than of surrender of some of the preferred food items: if species A occupies habitats 1 and 2, and species B occupies 2 and 3, we may find that species A yields 2 to B. It is also possible, but less likely, that both A and B will remain in habitat 2, but A will no longer be able to utilize certain food items found there (see Figure 12-12). The addition of only a single competitor can thus quickly and drastically reduce the realized niche of a species. In the initial stages of species packing leading to the formation of a plant and animal community, competitors are required to settle for realized niches that are more or less imperfect subsets of their fundamental (that is, complete potential) niches. Profound effects on social evolution can then ensue, as stressed earlier in Chapter 3.

Of all the possible competitive devices, none is more dramatic in

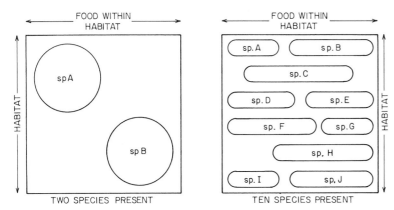

Figure 12-12 The compression hypothesis of interspecific competition. As more species are packed into a community (left to right), the habitats occupied by particular species shrink, but acceptable food items within the occupied habitat are not changed. The actual diet may become more restricted in space, but the range of items is not as likely to be reduced. Conversely, as species invade a species-poor area from a species-packed source (right to left), it is primarily the utilized habitat that expands. The model applies only to short-term, nonevolutionary changes. (Redrawn from MacArthur and Wilson, 1967.)

its initial effect than interspecific territoriality. The more closely two territorial species resemble each other, the more likely they are to defend their territories against each other. The reason is simple: the releasers for intraspecific recognition and aggression are more likely to be sufficiently similar to trigger territorial behavior. As a consequence, the circumstance most favorable to interspecific territoriality is the first contact between two cognate species that have just evolved from a single parental species.

Interspecific territoriality is relatively common in bird species and has been the subject of a number of careful studies by ornithologists writing from different points of view (Simmons, 1951; Lanyon, 1956; T. H. Hamilton, 1962; Johnson, 1964; Orians and Willson, 1964; Grant, 1966; Cody, 1969; Cody and Brown, 1970; Murray, 1971). It has also been discovered in ants (Wilson, 1971a), crayfish (Bovbjerg, 1970), *Anolis* lizards (Rand, 1967), squirrels (Ackerman and Weigl, 1970), chipmunks (J. H. Brown, 1971), pocket gophers (Miller, 1964), and gibbons (Berkson et al., 1971).

Species that contend with one another on territories can be expected to evolve in such a way as ultimately to reduce the interference and hence minimize the loss of genetic fitness. The several pathways evolution can follow are presented in Figure 12-13. This scheme has been inferred principally from general speciation theory combined with field studies on birds. The potential impact on the evolution of territoriality is twofold. The dominant species, the one

that wins in most or all of the contests, may evolve in such a way as to resemble the subordinate species more closely. The Darwinian advantage that its members gain is a more efficient exclusion of the competitor and larger amounts of resources per unit area defended. Thus interspecific territoriality can be one of the causes of *character convergence*, a puzzling phenomenon reported in the zones of overlap of a few birds (Moynihan, 1968; Cody, 1969). By contrast, the subordinate species, and under some circumstances the dominant species as well, is likely to undergo *character displacement*, an evolutionary divergence from the competitor in the zone of overlap. Murray (1971) has suggested the following three alternative ways that displacement can proceed in birds, a scheme which should apply equally well to other kinds of animals. (1) The subordinate species evolves so that

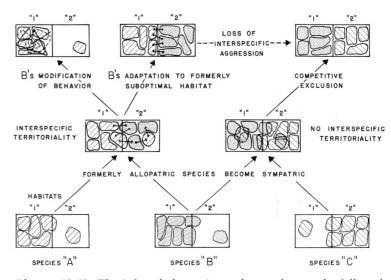

Figure 12-13 The inferred alternative pathways that can be followed when two territorial species become sympatric (overlap in range) after evolving in separate (allopatric) ranges. Species *A* and species *B* are best adapted to habitat *1*, while species *C* is best adapted to habitat *2*. If species *B* and *C* become sympatric, compete for some resource, and are not interspecifically territorial, their cooccupancy of habitats *1* and *2* (center right) would not persist indefinitely before competitive exclusion resulted in habitat segregation (upper right). If species *A* and *B* become sympatric, do not compete for resources (for example, food, nesting sites) other than territorial space, and are interspecifically territorial, then species *B*, if subordinate to *A*, will be forced out of its optimal habitat (center left). Species *B* may either modify its territorial behavior (upper left) or become adapted to its suboptimal habitat (upper center). If species *B* subsequently loses its interspecific territoriality through divergence resulting from intraspecific selection for different colors and other recognition signals, then habitat segregation not distinguishably different from that resulting from competitive exclusion could occur. (From Murray, 1971.)

it no longer fights when attacked by the dominant species. Provided it is still able to acquire enough resources it can coexist in the optimal habitats with the dominant species. This is evidently the route followed by the tri-colored blackbird *Agelaius tricolor*, sharp-tailed sparrow *Ammospiza caudacuta*, and reed warbler *Acrocephalus scirpaceus*, which live in close association with dominant congeners. (2) One or both of the species diverge sufficiently in appearance so that interspecific aggression is no longer provoked by either species, with the result that the formerly subordinate species is able to expand its ecological range and to reenter the optimal habitats. (3) The subordinate species "gives up" and adapts to the suboptimal habitats, which now become its favored habitat. The relations between territoriality, evolutionary convergence or divergence, and population stability are indicated in Figure 12-14.

The vertebrate literature is rich in documentation of each of the various possible outcomes of interspecific territorial aggression. In western North America, for example, the yellow-headed blackbird *Xanthocephalus xanthocephalus* is dominant over the red-winged blackbird *Agelaius phoeniceus*. When they nest together in the marshy habitats preferred by both, the yellowheads force the redwings out of the most favored nesting sites (see Figure 12-15). The yellowheads are also dominant when they meet at feeding grounds away from the breeding territories (Orians and Willson, 1964).

Alison Jolly (1966) discovered that the ring-tailed lemur *Lemur catta* and Verreaux's sifaka *Propithecus verreauxi*, two sympatric prosimians of Madagascar, engage in a form of aggressive "play" that lies somewhere between tolerance and the full-scale territorial exclusion which the two animals impose within their own species. The following incident was typical:

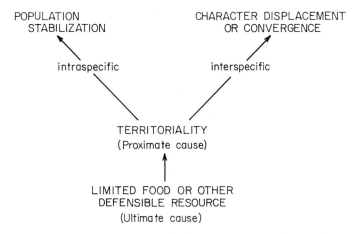

Figure 12-14 The mediating role played by territoriality in character convergence and displacement and the evolution of population stability.

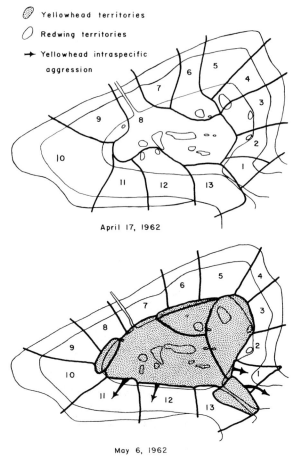

April 17, 1962

May 6, 1962

Figure 12-15 Territorial exclusion in two species of blackbirds. The late-arriving yellowheads forced the red-winged blackbirds from the favored central locations in the marsh. The arrows indicate places where yellowhead aggression was directed against other yellowheads instead of redwings. (From Orians and Willson, 1964.)

On August 16 and August 24, 1963, and, in more leisurely fashion, on March 23, 1964, a whole troop of *L. catta* barred the *Propithecus'* way, while the *Propithecus* returned their teasing. Again, the animals leaped toward each other, stared, feinted approach, but never came into contact. All the game lay in leaps and counterleaps, the *Propithecus* trying to pass through the *L. catta* troop, the *L. catta* attempting to keep in front of them, facing the other direction. Since there are about twenty *L. catta* to five *Propithecus*, the *L. catta* had an advantage: if one animal does not outguess the *Propithecus'* next move, another can do so.

There is no reason to believe that this degree of interference seriously affects the population stability of either species. It seems to represent either a permanent accommodation or a transition point in character displacement.

Chapter 13 **Dominance Systems**

Dominance behavior is the analog of territorial behavior, differing in that the members of an aggressively organized group of animals coexist within one territory. The dominance order, sometimes also called the dominance hierarchy or social hierarchy, is the set of sustained aggressive-submissive relations among these animals. The simplest possible version of a hierarchy is a *despotism:* the rule of one individual over all other members of the group, with no rank distinctions being made among the subordinates (C. C. Carpenter, 1971). More commonly, hierarchies contain multiple ranks in a more or less linear sequence: an alpha individual dominates all others, a beta individual dominates all but the alpha, and so on down to the omega individual at the bottom, whose existence may depend simply on staying out of the way of its superiors. The networks are sometimes complicated by triangular or other circular elements (Figure 13-1), but such arrangements seem a priori to be less stable than despotisms or linear orders. In fact, Tordoff (1954) found that triangular loops first established by a captive flock of red crossbills (*Loxia curvirostra*) were disruptive, so that changes in the order increasingly replaced them with straight chains. The dominance order of a flock of roosters assembled by Murchison (1935) was at first unstable and contained triangular elements, but it later settled into a slowly changing linear order (Figure 13-2). Ivan Chase (personal communication) obtained direct evidence that straight-chain hierarchies can result in higher group efficiency. When triads of hens formed a linear dominance order, a certain amount of food was eaten quickly by the alpha bird, sometimes assisted by the beta individual. When the dominance

Figure 13-1 Three elementary forms of networks found in dominance orders. More complex networks are built up of combinations of such elements. (From Wilson et al., 1973.)

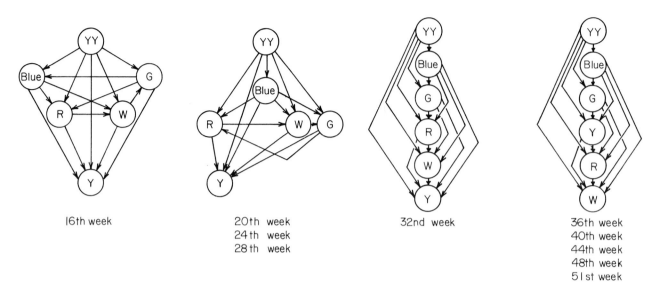

16th week   20th week   32nd week   36th week
                 24th week   40th week
                 28th week   44th week
                                                         48th week
                                                         51st week

Figure 13-2 Shifts in the hierarchy of a newly formed flock of roosters. Triangular subnets give way to a more stable, linear order. The letters and the name "Blue" designate the individual roosters. (Redrawn from Murchison, 1935.)

orders were circular, however, the hens fed warily, individuals frequently displaced one another, and the food was consumed more slowly.

Hierarchies are formed in the course of the initial encounters between animals by means of repeated threats and fighting. But after the issue has been settled, each individual gives way to its superiors with a minimum of hostile exchange. The life of the group may eventually become so pacific as to hide the existence of such ranking from the observer—until some minor crisis happens to force a confrontation. Troops of baboons, for example, often go for hours without displaying enough hostile exchanges to reveal their hierarchy. Then in a moment of tension—a quarrel over an item of food is sufficient—the ranking is suddenly revealed, appearing in graphic detail rather like an image on photographic paper dipped in developer fluid.

The societies of some species are organized into *absolute dominance hierarchies*, in which the rank order is the same wherever the group goes and whatever the circumstance. An absolute hierarchy changes only when individuals move up or down the ranks through further interactions with their rivals. Other societies, for example crowded groups of domestic cats, are arrayed in *relative dominance hierarchies*, in which even the highest-ranking individuals yield to subordinates when the latter are close to their personal sleeping places (Leyhausen, 1956, 1971). Relative hierarchies with a spatial bias are intermediate in character between absolute hierarchies and territoriality.

In its stable, more pacific state the hierarchy is sometimes supported by "status" signs. The identity of the leading male in a wolf pack is unmistakable from the way he holds his head, ears, and tail, and the confident, face-forward manner in which he approaches other members of his group. He controls his subordinates in the great majority of encounters without any display of overt hostility (Schenkel, 1947). Similarly, the dominant rhesus male maintains an elaborate posture signifying his rank: head and tail up, testicles lowered, body movements slow and deliberate and accompanied by unhesitating but measured scrutiny of other monkeys that cross his field of view (Altmann, 1962).

Finally, dominance behavior is mediated not only by visual signals but also by acoustic and chemical signals. Mykytowycz (1962) has found that in male European rabbits (*Oryctolagus cuniculus*) the degree of development of the submandibular gland increases with the rank of the individual. By means of "chinning," in which the lower surface of the head is rubbed against objects on the ground, dominant males mark the territory occupied by the group with their own submandibular gland secretions. Recent studies of similar behavior in flying phalangers and black-tailed deer indicate that territorial and other agonistic pheromones of these species are complex mixtures that vary greatly among members of the same population (Schultze-Westrum, 1965; Wilson, 1970). As a result individuals are able to distinguish their own scents from those of others.

History of the Dominance Concept

The development of the dominance concept, which covers a period of over 170 years, has been a progression from an elementary idea based on simple animals to complex and shifting theory based on the most complex animals. Dominance orders were first explicitly recognized by the Swiss entomologist Pierre Huber (1802) in his pioneering study of bumblebees. He noted that when a queen lays her eggs, some of the workers try to steal and eat them, but the queen usually repels these intruders with great fury. Similar observations were made by a long line of subsequent investigators. The Austrian entomologist Eduard Hoffer (1882–1883) discovered that the dominance relationship among bumblebees is orderly and predictable. As he described it: "Punishment is almost always meted out with the legs and mandibles, and the guilty individual never even attempts to defend herself, all her efforts being directed toward making the quickest possible escape. The punishment is sometimes so rough that the poor creature is seriously wounded or even killed." Once egg stealing has been rebuffed for a few hours, the attempts by the workers "become less and less frequent, and finally cease altogether; and these same little insects which previously tried their very best to destroy the newly-laid eggs, now become attentive guardians and nurses of their embryo brothers and sisters; they keep them warm and tenderly provide them with nourishment continuously thereafter."

The early observers of bumblebees were puzzled by what seemed to them an impulse toward restraint and conservation. Huber offered a surprisingly modern hypothesis that invokes density-dependent control of the kind envisioned as a "social convention" by Wynne-Edwards. He stated, "The bumblebees are the largest insects that feed on honey; and if their number trebled or quadrupled, other insects would not find any nourishment, and perhaps their species would be destroyed." Pérez (1899), however, viewed egg eating as simply evidence of selfish behavior on the part of the workers and therefore an imperfection in the social order. There is still another, perhaps more straightforward explanation, one that involves a function adaptive to the colony as a whole. Lindhard (1912), in confirming and extending Hoffer's observations, recorded an instance in which a queen of *Bombus lapidarius* extracted the egg of a rival worker and fed it to a queen larva. Also noting that very few if any worker-laid eggs ever survive, he hypothesized that these eggs are not meant to develop but instead serve as a kind of "royal food" for prospective queens. If there is any truth in this conclusion, it is consistent with the general occurrence of trophic eggs laid by workers in other kinds of social Hymenoptera (Wilson, 1971a).

The general significance of the bumblebee studies was not appreciated at the time and did not enter the mainstream of the behavioral literature. It remained for the Norwegian biologist Thorleif Schjelderup-Ebbe (1922, 1923, 1935) to start afresh with the vertebrates. Experimenting with domestic fowl, he showed that the members of the flock recognize one another on the basis of memories that last as long as two or three weeks. In the course of aggressive encounters they establish the "peck order" by which access to roosts and food is rigidly determined. During the 1930's and 1940's Carl Murchison, Warder C. Allee, Nicholas E. Collias, and their associates greatly extended knowledge of the domestic fowl by charting the development of hierarchies in flocks, statistically analyzing the factors that determine rank in individuals, and experimenting with the effects of androgens on aggression and dominance. Other investigators, particularly C. R. Carpenter, J. T. Emlen, D. W. Jenkins, Bernard Greenberg, E. P. Odum, and J. P. Scott, described hierarchies in both free-living and confined groups of other kinds of birds and vertebrates. By 1949, when Allee and Alfred E. Emerson wrote the first truly modern sociobiological synthesis as part of their contribution to *Principles of Animal Ecology,* dominance hierarchies of the elementary peck-order form were universally regarded as a basic mechanism of social organization in animals, and investigators were searching avidly in all directions to uncover new examples. Dominance orders were rediscovered in social insects by G. Heldmann (1936a,b) and L. Pardi (1940), both of whom studied the European paper wasp *Polistes gallicus.* It is ironic, and a commentary on the episodic quality of the history of science generally, that Pardi, whose work really brought the dominance concept back to invertebrate zoology, was influenced not by the earlier studies on bumblebees but by the later experiments on domestic fowl.

Serious difficulties in the dominance concept appeared as soon as the idea was extended to the more complex social life of primates by Zuckerman (1932) and Maslow (1936, 1940). Sexual attraction and selection were strongly implicated in the aggressive interactions; in fact Zuckerman postulated them to be the binding force of primate sociality. Maslow discovered that baboons and macaques use homosexual mounting as a ritualized form of aggression, with subordinates presenting their rumps in the female receptive posture as a sign of submission and conciliation. Some writers (see especially Schenkel, 1947; Altmann, 1962) then recognized that in both primates and wolves a rich repertory of signals is used to denote status in a manner not directly coupled with aggressive interactions. Status signs were seen to be metacommunicative, indicating to other animals the past history of the displaying individual and its expectation of the outcome of any future confrontations. DeVore and Washburn (1960) modified altogether the Hobbesian interpretation of primate dominance hierarchies by pointing out that subordinates as well as dominants live peaceably most of the time—first because the evolved rules of hierarchical behavior produce stable social systems, and second because the advantages of belonging to a group far outweigh the disadvantages of being a subordinate. Moreover, the existence of peer cliques and alliances between certain dominant and subordinate animals serves to mitigate the effects of subordination still more.

As investigations deepened, students of primate behavior found the concept of dominance order increasingly unsatisfactory both as an explanatory scheme and as a device for analyzing individual behavior. In the 1960's K. R. L. Hall, I. S. Bernstein, Thelma Rowell, J. S. Gartlan and others started a new approach to the study of status, in which roles of classes of individuals were identified and classified. For example, a male can serve as the "control" animal (Bernstein, 1966), maintaining vigilance, interposing himself between an intruder and the group, and terminating fights between group members—and yet still not exercise aggressive dominance toward other troop members in the narrow, traditional sense. Control animals exist, in fact, in troops of capuchin monkeys (*Cebus albifrons*), even though no dominance orders have been detected. Michael Chance and Clifford Jolly (1970) stressed the role of certain animals as attraction centers that determine the geometry and orientation of primate troops. (Roles will be given special attention in the following chapter.)

Still another, wholly different kind of complication has been encountered in the social insects, one that arises from the subtle interplay between selfish and altruistic behavior in these highly group-selected animals. Montagner (1966) found, for example, that a peculiar form of dominance hierarchy exists in the higher social wasps of the genus *Vespula*. Liquid food exchange is the medium of the hierarchy, and workers contend, often aggressively, for the privilege of receiving the crop contents of other workers, who regurgitate to them. This display of apparently selfish behavior among members of the worker caste is not typical of social insects generally and cannot easily be explained by the theory of natural selection. It is true that the *Vespula* workers lay some eggs of their own which develop into males, and it is at least conceivable that dominant individuals perpetuate the "selfish" genes underlying dominance behavior by contributing more than their share of the eggs. To be sure, dominance hierarchies appear early in the growth of the colony, long before male eggs are laid, and there is only a loose positive correlation between ovarian development and dominance rank. But dominance can still conceivably be interpreted as selfish behavior based on genes favored in the worker-male hereditary lines. However, a second explanation of the phenomenon, clearly altruistic in content, emerges when the organization of the colony as a whole is examined. Montagner has shown that the dominance hierarchies are the basis of an efficient division of labor among the workers. The "low-ranking" workers are the foragers, who gather the food and nest-construction

materials and turn them over to higher-ranking workers on entering the nest. The highest-ranking workers remain in the nest, attending the larvae and building and repairing the brood cells. Thus the dominance behavior serves as a mechanism that apportions the colony labor, and one can reasonably suppose that it contributes to the fitness of the colony as a whole. A similar hierarchical organization, based on liquid food exchange and associated with ovarian development, has been discovered in the ant *Formica polyctena* by Lange (1967). Unlike the *Vespula*, the *Formica* do not display overt aggression in their interactions.

The *Formica polyctena* case forms a transition between that of *Vespula* and a more subtle but interesting situation in the honeybee *Apis mellifera*. In the latter species there is also a kind of "dominance hierarchy" of food exchange by which food flows from the foragers to the nurse bees. There is no overt aggression, and most of the bees change their status, as they grow older, from "dominant" nurses to "subordinate" foragers. Most important, workers do not normally contribute to drone production, so we can discount the selfish-gene hypothesis. In short, what appears to be selfish behavior when viewed in a few individuals over a short period of time is more evidently altruistic behavior when interpreted at the level of the colony over a longer period of time. Other cases of aggression within insect colonies, particularly those associated with the initiation of nuptial flights, can be similarly explained as mechanisms of social integration evolved by colony-level selection (Wilson, 1971a, 1974c).

In the subsequent discussion we will temporarily avoid such complications, confining our attention to the forms of dominance behavior mediated by aggression and inferentially based upon natural selection at the level of the individual.

Examples of Dominance Orders

Dominance hierarchies, like territories, are distributed in a highly irregular fashion through the animal kingdom. Among the invertebrates the hierarchies appear to be limited principally to evolutionarily more advanced forms characterized by large body size (Allee and Dickinson, 1954; Lowe, 1956). Among the insects, hierarchies are most clearly developed in species that are fully social yet still primitively organized, such as the bumblebees and paper wasps (Wilson, 1971a; Evans and Eberhard, 1970). Crane (1957) has reported nonterritorial aggressive interactions among male heliconiine butterflies, but interprets them as "simply a fragment of the appetitive portion of the courtship pattern." Dominance occurs in a few spiders under peculiar circumstances. Males of some species, such as the crab spider *Diaea dorsata*, fight for access to the female (Braun, 1958). When males of the sheet-web spider *Linyphia triangularis* encounter one another on the webs of females, they use their enlarged chelicerae and fangs to fight. While residing on the web for a day or two, the winning male also dominates the female (Rovner, 1968). Pagurid crabs contend for the possession of mollusk shells to use as shelters, a form of aggression that could be classified as either territoriality or dominance, and the largest individuals usually win (Hazlett, 1966, 1970). Certain crayfish (*Cambarellus, Procambarus*) are ordinarily territorial, but when forced together they form neat, stable linear dominance hierarchies (Lowe, 1956; Bovbjerg, 1956).

Many kinds of fish show a similarly easy transition between territorial defense and dominance orders. However, those that normally live in schools do not organize themselves into hierarchies, at least not of a stable nature based upon individual recognition (Thines and Heuts, 1968; McDonald et al., 1968; McKay, 1971). When certain anurans, including the leopard frog *Rana pipiens* and African clawed frog *Xenopus laevis*, are crowded together, they form dominance hierarchies (Haubrich, 1961; Boice and Witter, 1969). The question remains as to whether such species ever aggregate in the free state. T. R. Alexander (1964) has observed behavior close to a natural dominance order in the giant toad *Bufo marinus*. When individuals in a free-living population gathered at dishes of food set out for them, they fought and displaced one another according to a consistent order. Other members of the frog genus *Rana*, such as *R. catesbeiana*, sometimes feed in close quarters, but none to my knowledge has been seen to be organized by aggressive interactions.

Dominance orders, relatively stable in nature and based at least in part on memory, have been documented in virtually all bird groups that forage in flocks or roost communally (Armstrong, 1947; Crook, 1961; Crook and Butterfield, 1970). Similarly, the vast majority of mammal species forming groups with any degree of social complexity also display dominance (Tembrock, 1968; Crook et al., 1970). Hierarchies are well developed in kangaroos and wallabies (La Follette, 1971), rodents (Calhoun, 1962; Barnett, 1963; Archer, 1970), pinnipeds (Peterson and Bartholomew, 1967; Le Boeuf and Peterson, 1969a), and ungulates (de Vos et al., 1967; Schaller, 1967; Estes and Goddard, 1967; Geist, 1971). The literature on primate dominance is vast, but good reviews of most aspects of it have been prepared by Alison Jolly (1966, 1972a), Gartlan (1968), Poirier (1970), Chance and C. J. Jolly (1970), and Baldwin (1971). In general, prosimians (particularly lemurs) and Old World monkeys have moderate to strong hierarchies, anthropoid apes have weakly developed hierarchies, and species of New World monkeys vary greatly, some lacking hierarchies altogether and others showing weak to moderate dominance orders.

Let us now examine a small sample of case histories of particular species, selected for both phylogenetic diversity and the representation they provide of some of the extreme variants of dominance relations.

Domestic Fowl (*Gallus gallus*)

The common domestic fowl, sometimes referred to as *Gallus domesticus*, is descended from populations of the red jungle fowl (*G. gallus*), a smaller, ground-dwelling bird that ranges in the free state from north central India and Indochina south to Sumatra. The domesticated form was the first vertebrate species in which dominance relations were systematically investigated (Schjelderup-Ebbe, 1922), and it has been the most intensively studied of all the animal species since that time. During the past 30 years A. M. Guhl and his associates at Kansas State University have concentrated on nearly every conceivable aspect of the subject; most of their key results and reviews are given in Guhl (1950–1968), Guhl and Fischer (1969), Craig et al. (1965), Craig and Guhl (1969), and Wood-Gush (1955). The social behavior of chickens is relatively simple and is based to a large extent on the dominance order. As soon as a new flock is created by the experimenter, the power struggle begins. The hierarchy that quickly forms is in a literal sense a peck order: the chickens maintain their status by pecking or by a threatening movement toward an opponent with the evident intention of attacking in this manner. High-ranking birds are clearly rewarded with superior genetic fitness. They gain priority of access to food, nest sites, and roosting places, and they enjoy more freedom of movement. Dominant cocks mate far more frequently than subordinates. But dominant hens mate somewhat less frequently than others, because they display submissive and receptive postures to the cocks less consistently. Nevertheless, the fitness of dominant hens is probably greater because of the more than compensating advantages gained in access to food and nests. Cocks form a separate hierarchy above that of the hens. The adaptive explanation for this disjunction is that it facilitates mating: cocks subordinate to hens are unable to copulate with them. The heritability of fighting ability has been well documented. Strong genetic variation in this trait exists both between and within the various breeds of fowl. It is an interesting commentary on the evolution of dominance that when poultry breeders select for strains that lay more eggs, they also produce more aggressive chickens. In other words, the breeders have simply picked out the dominant birds, which incidentally have the most access to nests (McBride, 1958).

The critical size of a flock of hens is ten. In flocks below approximately that number, as Schjelderup-Ebbe demonstrated in his original study, triangular and square elements straighten out and the resulting linear orders are stable for periods of months. In flocks above that number, looped elements stay common and the hierarchy continues to shift at a relatively high rate. However, revolts and rapid shifting can occur even in the small groups when one or more of the subordinate hens are injected with testosterone propionate. It is to the advantage of a chicken to live in a stable hierarchy. Members of flocks kept in disorder by experimental replacements eat less food, lose more weight when their diet is restricted, and lay fewer eggs. Chickens remember one another well enough to maintain the hierarchy for periods of only up to two or three weeks. If separated for much longer periods, they reestablish dominance orders as if they were strangers. However, a chicken can be removed repeatedly for shorter intervals and reinserted without changing its rank.

To what extent do peck orders of the domestic fowl reflect their ancestral social condition? Collias et al. (1966) observed flocks of red jungle fowl allowed to range freely over the exhibit areas of the San Diego Zoo. They found that the population as a whole was divided into several flocks which maintained extremely stable territories centered on the flock roosting sites. Although the flocks were large and changed rapidly in composition owing to mortality, the flock territories remained the same. Most specific roosts contained between 5 and 20 birds at any given time. But the truly basic social unit, also observed in the wild in India and Ceylon by Collias and Collias (1967), is a dominant cock accompanied by one to several hens and often one or more subordinate cocks, who follow at a distance. Thus the natural groups of these birds, insofar as contention for mating and roosting places are concerned, fall easily within Schjelderup-Ebbe's range for stable dominance hierarchies.

Leopard Frogs (*Rana pipiens*)

Territorial animals forced into close proximity in laboratory pens or as an outcome of unusual environmental conditions typically shift to a primitive form of dominance order. Boice and Witter (1969) recorded a typical example of the phenomenon in the leopard frog. The animals enclosed together show little awareness of one another except when food is presented. Then the dominant animals, ordinarily the largest, push into the most favorable positions and maintain a clear space around themselves by nudging away competitors. This is accomplished by what appears to be purposeful shoving with the forelegs. When earthworms are fed to the group, the dominant frogs respond more quickly and with a higher frequency of success. If a high-ranking frog lands past the worm when it leaps, it very quickly returns; whereas a low-ranking frog usually quits for long periods of time. Frogs of approximately equal rank occasionally nip at one another when contending for the same worm. *Rana pipiens*, like other frogs and toads observed in temporary aggregations, does not utilize special aggressive or status displays during the dominance interactions.

Paper Wasps (*Polistes*)

In the primitive insect societies strife and competition prevail, leading to the emergence of a single female, the queen, who physically dominates the rest of the adult members of the colony. In the more ad-

vanced insect societies, particularly those in which strong differences between the queen and worker castes exist, the queen also exercises control over the workers but usually in a more subtle manner devoid of overt aggression. In many cases, as in the honeybee, hornets, and many ants and termites, the dominance is achieved by means of special pheromones that inhibit reproductive behavior and the development of immature members of the colony into the royal caste. The gradual change from what might be regarded as brutish dominance to the more refined modes of queen control is one of the few clear-cut evolutionary trends that extend across the social insects as a whole (Wilson, 1971a).

The existence of a primitive dominance system in the European paper wasp *Polistes gallicus* was intimated by the studies of Heldmann (1936a,b), who found that when two or more females start a nest together in the spring, one comes to function as the egg layer, while the others take up the role of workers. The functional queen feeds more on the eggs laid by her partners than the reverse, thus exercising a form of reproductive dominance. Pardi (1940, 1948) discovered that the queen establishes her position and controls the other wasps by direct aggressive behavior, and he went on to analyze this form of social organization in detail. Dominance behavior has since been documented in *P. chinensis* by Morimoto (1961a,b), *P. fadwigae* by Yoshikawa (1963), and *P. canadensis* and *P. fuscatus* by Mary Jane West Eberhard (1969). The relationship among the adult members of the *Polistes* colony is somewhat more elaborate and stereotyped in form than that in bumblebee colonies. Instead of simple despotism by the queen, there exists in most cases a linear order of ranking involving a principal egg layer, the queen, and the remainder of the associated females, called auxiliaries, who form a graded series in the relative frequencies of egg laying, foraging, and comb building (Table 13-1). Dominant individuals receive more food whenever food is exchanged, they lay more eggs, and they contribute less work. They establish and maintain their rank by a series of frequently repeated aggressive encounters. At lowest intensity the exchange is simply a matter of posture; the dominant individual rises on its legs to a level higher than the subordinate, while the subordinate crouches and lowers its antennae. At higher intensities leg biting occurs, and at highest intensity the wasps grapple and attempt to sting each other. During the brief fights, the contestants sometimes lose their hold on the nest and fall to the ground. Injuries are rare, although Eberhard once saw a female of *P. fuscatus* killed in a fight. The severest conflicts occur between wasps of nearly equal rank and during the early days of the association when the nest is first being constructed. As time passes, the wasps fit more easily into their roles, and aggressive exchanges become more subdued, less frequent, and eventually purely postural in nature. When the first adult workers eclose from the pupal instar during later stages of colony develop-

Table 13-1 Division of labor according to rank within a group of colony-founding females of the paper wasp *Polistes fuscatus* observed for 26 daylight hours between May 18 and June 14, 1965. (From Eberhard, 1969.)

Identification number	Dominance rank	Number of eggs laid	Number of eggs of others eaten	Number of new cells started	Foraging rate (loads per hour observed)	Number of loads received from associates
13	1	9	4	0	0.08	25
34	2	5	2	1	0.50	20
35	3	0	0	1	1.41	0
28	4	0	0	2	1.56	5
15	5	0	0	8	1.80	3
6	6	0	0	0	1.22	1
18	7	0	0	0	1.50	0

ment, their relations to the foundresses are invariably subordinate and nonviolent. The workers also form a hierarchy among themselves; those that emerge early tend to dominate their younger nestmates. If the alpha female is removed, the next highest auxiliaries intensify their hostility to one another until one of their number becomes unequivocally dominant; then the exchanges subside to the previous low level. Eberhard (1969) found that the tropical species *P. canadensis* differs from all the temperate species studied to date in that its contests are more violent and result in the losers leaving the nest to attempt nest founding on their own elsewhere. The queens are also less dependent on the indirect technique of egg eating to maintain reproductive control. Thus *P. canadensis* approaches a condition of true despotism, whereas the temperate species so far studied possess a colony organization that can be characterized as an uneasy oligarchy.

The dominant females of *Polistes* colonies maintain their superior reproductive status by three means: they demand and receive the greatest share of food whenever it becomes scarce; they lay the greatest number of eggs in newly constructed brood cells; and they remove and eat the eggs of subordinates when these rivals succeed in laying in empty cells. The size and degree of development of the ovaries is loosely correlated with the rank of the wasp (Pardi, 1948; Deleurance, 1952; Gervet, 1962). When a female slips in rank, her ovaries also decrease in size. It is tempting to think that subordination leads to decrease in ovarian development simply because the individual receives less food; in other words, she is subjected to "nutritional castration" of the sort originally conceived by Marchal (1896, 1897). The same effect might be achieved by the closely related phenomenon of "work castration," in which the subordinates are forced to expend more of their energy reserves in foraging and nest building (Plateaux-Quénu, 1961; Spradbery, 1965). But the matter is far more complex than this. Energy deprivation and ovarian development very likely play some kind of a role in establishing the rank of a wasp, but other factors are at least equally important. When Gervet (1956) chilled the queens of *P. gallicus* overnight, to a degree that inhibited their egg production but not their daily activity, they retained their dominant status. They left more brood cells unfilled, however, and this in turn stimulated ovarian development in the subordinates. Deleurance (1948) was able to remove all of the ovaries of dominant *P. gallicus* by surgery, yet even this treatment did not affect their rank. Thus rank seems to determine ovarian development, while the reverse is not the case. Moreover, Gervet's result precludes the possibility that ovarian development is controlled in a straightforward manner by nutrition. What evidently matters most is behavioral control over the empty brood cells. But what determines this more purely psychological phenomenon in turn is not at all clear. Experience seems to have something to do with it, since the first females arriving at new nest sites tend to dominate later arrivals. It is also true, as noted already, that older workers tend to dominate their younger nestmates.

The despotisms and dominance hierarchies of *Bombus* and *Polistes* have many qualities in common with those of aggressively organized vertebrate societies. But are they also similar in being based on recognition of individuals and memory of past experiences? This is at least a theoretical possibility in the smaller *Bombus* and *Polistes* colonies, which contain from several to a few tens of individuals. It is still a possible factor in insect colonies of a few hundred individuals, where the queens and high-ranking auxiliaries constitute a small company of elites that stay close to the active brood cells and are therefore in frequent contact. However, it becomes very difficult to conceive in the largest insect societies. Sakagami (1954) observed mild hostility among worker honeybees when the queen was absent, but he could see no evidence of a regular, *Polistes*-like dominance hierarchy. He argued that since honeybee colonies contain tens of thousands of workers, few of whom live for more than a month, such a complex system of personal relationships is an impossibility. Sakagami's reasoning is correct to a point, but it does not preclude a looser dominance system based on variation in individual aggressiveness—as opposed to learned, pairwise relationships.

Spider Monkeys (*Ateles geoffroyi*)

The New World monkeys have not attained the degrees of social complexity found in the most advanced Old World monkeys and anthropoid apes. Spider monkeys can be taken as typical for the diffuse, muted form of dominance relations they display (Eisenberg and Kuehn, 1966). Aggression occurs: members of a group threaten one another by shaking branches, grinding their teeth, coughing, hissing, and even roaring. The animals slap and kick one another, and they sometimes slash at one another with their canines or nip with hard bites of the incisors. Dominants sometimes chase subordinates. However, such overt aggressive behavior occurs rather rarely. Males tend to be dominant over females and adults over juveniles, but the order is not linear and is difficult to define from the infrequent, often unpredictable exchanges of pairs. *Ateles geoffroyi* does not employ aggressive presenting and mounting, status posturing, or any of the other highly ritualized threat and conciliatory signals so prominently used by macaques and baboons among the Old World monkeys. Grooming is uncommon, and high-ranking animals groom more than they are groomed, the reverse of the grooming trend in most other monkeys. The adult male sometimes halts fighting between others, but he does not otherwise play the role of a control animal. In short, the relatively simple social organization of the spider monkey is reflected in its primitive and infrequently used dominance system.

Thick-Tailed Galagos (*Galago crassicaudatus*)

By proceeding to the prosimians, the most primitive of living primates, we encounter a dominance system even less ordered than that of the spider monkey. Within a group of eight galagos observed by Pamela Roberts (1971) at the Duke Primate Facility, the males performed no grooming, and one individual was dominant in despotic fashion over the other three. Females groomed often and displayed strong dominance behavior, but the relations were shifting. In one instance a female dominated another who was larger and generally more aggressive. Instead of a true dominance system, Roberts interpreted the basis of the galago society to be "individual preferences and aversions."

Minnows (*Poeciliopsis*)

This short list of examples is appropriately closed with a truly aberrant case, chosen to suggest how dominance behavior can be integrated by unexpected means into sexual and social behavior. *Poeciliopsis* is one of the groups of vertebrates (others include *Ambystoma* salamanders and certain lizards of the genera *Cnemidophorus* and *Lacerta*) that posses unisexual-bisexual species complexes. This means that in addition to the normal, ancestral species, which remains bisexual, there exists a parthenogenetic female strain which is gynogenetic, that is, produces other females without fertilization. The *Poeciliopsis* unisexuals, however, still require insemination by a male of the bisexual species in order to produce eggs, even though the sperm serve strictly as a stimulus and do not succeed in fertilizing the eggs. In the bisexual schools, the dominant males show an almost absolute preference for normal females. Subordinate males, ordinarily the less mature and experienced individuals, inseminate the unisexual females. Thus the gynogenetic "species" is maintained in a parasitic fashion by the exploitation of the dominance system of the parental species (McKay, 1971; Moore and McKay, 1971).

Special Properties of Dominance Orders

The Xenophobia Principle

The relative calm of a stable dominance hierarchy conceals a potentially violent united front against strangers. The newcomer is a threat to the status of every animal in the group, and he is treated accordingly. Cooperative behavior reaches a peak among the insiders when repelling such an intruder. The sight of an alien bird, for example, energizes a flock of Canada geese, evoking the full panoply of threat displays accompanied by repeated mass approaches and retreats (Klopman, 1968). Chicken farmers are well aware of the practical implications of xenophobia. A new bird introduced into an organized flock will, unless it is unusually vigorous, suffer attacks for days on end while being forced down to the lowest status. In many cases it will simply expire with little show of resistance. Southwick's experiment (1969), cited in Chapter 11, demonstrated that the appearance of a newcomer is the single most effective means of increasing aggressive behavior in a troop of rhesus monkeys, most of the hostility being directed against the stranger. Human behavior provides some of the best exemplification of the xenophobia principle. Outsiders are almost always a source of tension. If they pose a physical threat, especially to territorial integrity, they loom in our vision as an evil, monolithic force. Efforts are then made to reduce them to subhuman status, so that they can be treated without conscience. They are the gooks, the wogs, the krauts, the commies—not like us, another subspecies surely, a force remorselessly dedicated to our destruction who must be met with equal ruthlessness if we are to survive. Even the gentle Bushmen distinguish themselves as the !Kung—the human beings. At this level of "gut feeling," the mental processes of a human being and of a rhesus monkey may well be neurophysiologically homologous.

The Peace of Strong Leadership

Dominant animals of some primate societies utilize their power to terminate fighting among subordinates. The phenomenon has been described explicitly in rhesus and pig-tailed macaques (Bernstein and Sharpe, 1966; Tokuda and Jensen, 1968) and in spider monkeys (Eisenberg and Kuehn, 1966). In squirrel monkeys this control function appears to operate in the absence of dominance behavior (Baldwin, 1971). Species organized by despotisms, such as bumblebees, paper wasps, hornets, and artificially crowded territorial fish and lizards, also live in relative peace owing to the generally acknowledged power of the tyrant. If the dominant animal is removed, aggression sharply increases as the previously equally-ranked subordinates contend for the top position.

The Will to Power

In a wide range of aggressively organized mammal species, from elephant seals, harem-keeping ungulates, and lions to langurs, macaques, and baboons, the young males are routinely excluded by their dominant elders. They leave the group and either wander as solitary nomads or join bachelor herds. At most they are tolerated uneasily around the fringes of the group. And, predictably, it is the young males who are also the most enterprising, aggressive, and troublesome elements. They contend among one another for in-group dominance and sometimes form separate bands and cliques that cooperate in reducing the power of the dominant males. Even the personalities of males in the two categories differ. The "establishment" males of a Japanese macaque troop remain calm and detached when shown a novel object, and thus do not risk the loss of their status. It is the females and young animals who explore new areas and experiment with new objects. The obvious parallels to human behavior have been noted by several writers, but most explicitly and persuasively by Tiger (1969) and Tiger and Fox (1971).

Social Inertia

When strange animals are thrown together, aggressive interactions are at first very frequent. As time passes hostilities decrease in frequency, and at a steadily decreasing rate, until the number of interactions per unit time is approximately constant. The gradual mitigation of aggression is due to the sorting out of individuals in rank and to habituation to the increasingly familiar signals provided by these individuals. Guhl (1968) has referred to the viscosity of such

a stabilized system as social inertia. An animal that attempts to change its position in a fixed dominance hierarchy is less likely to succeed than if it made the exertion during the early, fluid stages of the formation of the hierarchy.

Nested Hierarchies

Societies that are partitioned into units can exhibit dominance both within and between the components. Thus flocks of white-fronted geese (*Anser albifrons*) develop a rank order of the several subgroups (parents, mated pairs without goslings, free juveniles) superimposed over rank ordering within each one of the subgroups (Boyd, 1953). Brotherhoods of wild turkeys contend for dominance, especially on the display grounds, and within each brotherhood the brothers establish a rank order (Watts and Stokes, 1971). Team play and competition between human tribes, businesses, and institutions are also based upon nested hierarchies, sometimes tightly organized through several more or less autonomous levels.

The Advantages of Being Dominant

In the language of sociobiology, to dominate is to possess priority of access to the necessities of life and reproduction. This is not a circular definition; it is a statement of a strong correlation observed in nature. With rare exceptions, the aggressively superior animal displaces the subordinate from food, from mates, and from nest sites. It only remains to be established that this power actually raises the genetic fitness of the animals possessing it. On this point the evidence is completely clear.

Consider, to start, the simple matter of getting food. Wood pigeons (*Columba palumbus*) are typical flock feeders. Solitary birds are attracted by the sight of a group feeding on the ground, and no doubt there is great advantage to following the lead of others in locating food. Dominant birds place themselves at the center of the flock. Murton et al. (1966) noted that these individuals feed more quickly than those on the edge of the flock, and especially those on the forward edge, who constantly interrupt their pecking to look back at the advancing center. By shooting pigeons at dusk just before they flew to the roosts, Murton and his coworkers established that the subordinate birds accumulate less food. In fact, they have only enough to last the night, and they are in danger of perishing if the temperature drops sharply during the night or bad weather prevents foraging the next day.

Without systematic studies that include an evaluation of this question, it is impossible to guess whether the relation between status and food-gathering ability is a crucial one. Studies of maternal care in sheep and reindeer have revealed that low-ranking females are among the most poorly fed animals and also among the poorest mothers (Fraser, 1968). The teat order of piglets is a feeding dominance hierarchy in microcosm with an apparently direct adaptive basis. During the first hour of their lives the piglets compete for teat positions that, once established, are maintained until weaning. The piglets struggle strenuously, using temporary incisors and tusks to scratch one another (McBride, 1963). Preference is for the anterior teats, which provide more milk than the posterior teats and keep the piglets attached to them farther away from the trampling of the hind legs of the mother. The more milk a young pig receives, the more it weighs at weaning. The gradient of milk yield in the teats is probably great enough to provide a selective pressure for the competition to evolve. Gill and Thomson (1956) found that the four anteriormost piglets studied in each of a series of eight litters obtained an average of 15.3 percent more milk than the four posteriormost piglets. Those who occupied the three anteriormost pairs of teats got 83.8 percent more milk than the small group relegated to the posteriormost three or four pairs. Not surprisingly, piglets able to shift teat preference during early lactation moved their position forward. The orienting stimulus by which piglets find their correct positions quickly, even when the teats are partly hidden from view and smeared with mud, has not been established with certainty, but by process of elimination it would seem to involve smell. Piglets are often seen to rub their noses on the udder around the teat, and McBride has made the intriguing suggestion that they are depositing a personal scent.

Teat orders have also been reported in cats, and dominance of some degree may be involved: hungry kittens challenge and scratch trespassers in the vicinity of their personal teats. Ewer (1959), who made a special study of the phenomenon, believes that the function of teat fixation is feeding efficiency—an orderly assembly that minimizes time and effort. Also, fixation insures that there will always be one functioning nipple for each kitten, since a nipple left unused for several days ceases to produce milk. However true this may be, it is also the case that the posteriormost four pairs of nipples of the mother cat are richest in milk. Whether the gradient is sufficient to make competition for these nipples adaptive is not known. Teat orders have a less than complete phylogenetic distribution. They have been searched for without success, for example, in dogs (Rheingold, 1963), the viverrid *Suricata suricatta* (Ewer, 1963), the African giant rat *Cricetomys gambianus* (Ewer, 1967), and the tree shrew *Tupaia glis* (Martin, 1968).

The evidence favoring the hypothesis of dominance advantage in reproductive competition is even more persuasive. A recent experiment by DeFries and McClearn (1970) on laboratory mice deserves to be cited for the cleanness of its design. Groups were assembled of three males, distinguishable by genetic markers, and three females. In each of the replications the males fought for a day or two and established rigid hierarchies. The relationship between dominance and genetic fitness, as detected by the genetic markers in the offspring, was striking. In 18 of 22 groups established, the dominant male sired all of the litters. In 3 of the triads a subordinate male sired one litter, and in only one case did a subordinate male succeed in siring two litters. Dominant males, constituting one-third of the population, were the fathers of 92 percent of the offspring. Similar correlations, some weak and others strong, have been reported in the dominance systems of domestic fowl (Guhl et al., 1945), Norway rats (Calhoun, 1962), rabbits (Myers et al., 1971), elephant seals and other pinnipeds (Le Boeuf, 1972), and deer, mountain sheep, and other ungulates (Schaller, 1967; Geist, 1971).

The reproductive advantages conferred by dominance are preserved even in the most complex societies. The females of anubis baboons copulate with juvenile and subordinate males during the time of partial swelling of their sexual skins. But during the five to ten days of maximum swelling, when ovulation occurs, only the most dominant males of the troop copulate with the females (DeVore, 1971). Polygyny characterizes many primitive human cultures and is probably generally associated with other forms of behavioral dominance. Supplementary wives are traditionally the reward for male achievement, usually judged by material standards, and for longevity. Among the Yanomama Indians of Brazil, studied by James Van Neel and associates (Neel, 1970; MacCluer et al., 1971), the politically dominant males father a strongly disproportionate number of the children. Because polygyny is accompanied by a substantial amount of female infanticide, women are in short supply, and many men are forced either to remain bachelors or to raid other villages (which, of course, increases the shortage in the raided villages). The Yanomama feel obligated to trade wives, and they attempt to do so with the most powerful lineages within each village. Familial fertility is thus reinforced by the fact that a young man belonging to a politically strong family has many sisters and half-sisters who can be traded to insure his own polygyny. Can such a system really promote the evolution of behavioral dominance? Neel makes the following provocative comment: "Even with allowance for the happy accident of a large sibship, the open competition for leadership in an Indian community probably results in leadership being based far less on accidents of birth and far more on innate characteristics than in our culture. Our field impression is that the polygynous Indians, especially the headmen, tend to be more intelligent than the nonpolygynous. They also tend to have more surviving offspring. Polygyny in these tribes thus appears to provide an effective device for certain types of natural selection. Would that we had quantitative results to support that statement!"

The adaptive value of priority of access to nesting sites and shelters is a hypothesis less easily tested. Convincing evidence, however, has been produced in the case of Canada geese (*Branta canadensis*) by

Collias and Jahn (1959). The female selects the nest site, and she is escorted by the most aggressive male available to join her. Pairs low in dominance ranking are repeatedly evicted from nest sites by other birds, and their breeding attempts are significantly delayed. Dominant stream trout (*Salmo trutta* and *S. gairdneri*) studied by Jenkins (1969) enjoyed a freer choice of current and shelters than subordinates. Since the number of hierarchically organized groups is a simple function of the suitable living areas present along the stream channels, it seems probable that the dominant fish enjoy the highest survival rates.

The top ranking animal in a hierarchy is also under less general stress. It therefore expends less energy coping with conflict, and it is less likely to suffer from endocrine hyperfunction. Erickson (1967) found, for example, that subordinate pumpkinseed sunfish (*Lepomis gibbosus*) initiated fewer aggressive acts than did the dominant fish; and to the extent that they were the target of aggression, they developed larger interrenal glands, the source of corticosteroids in fish. Two classes of males in rhesus troops engage in the least aggressive behavior: the lowest-ranking individuals, who remain on the periphery of the troop and are systematically excluded from the best food and resting places, and the dominant males, who enjoy their privileges with a minimum of effort. Tension and antagonism are greatest in the middle-ranking males, who are continuously striving to move up in the dominance hierarchies (Kaufmann, 1967).

Finally, rank sometimes carries perquisites that might further enhance survival value. Among macaques, dominant females are the

Figure 13-3 The rank order of male rhesus monkeys is revealed by the direction of the grooming. The monkeys performing the grooming in this trio are of ascending rank from right to left. (From Kaufmann, 1967.)

beneficiaries of aunting behavior. Grooming, which serves as a basic cleaning operation as well as a social signal, is received by dominant animals of both sexes more than it is given. In rhesus monkeys, rank order can be reliably measured by the directionality of the grooming (see Figure 13-3).

The Compensations of Being Subordinate

Defeat does not leave an animal with a hopeless future. The behavioral ontogenies of species seem designed to give each loser a second chance, and in some of the more social forms the subordinate need only wait its turn to rise in the hierarchy. The most frequent recourse, from insects to monkeys, is emigration. A common principle running throughout the vertebrates is that juveniles and young adults are the ones most likely to be excluded from territories, most probable to start at the bottom of the dominance orders, and therefore most likely to be found wandering as floaters and subordinates on the fringes of the group. In the more nearly closed societies such wanderers are preponderantly males. Emigration is a common form of density-dependent control of populations. Natural-selection theory teaches that where the emigration behavior is programmed to occur at a certain life stage and at a certain population density, and involves a determined outward journey as opposed to mere aimless drifting, the chances of success on the part of the migrant at least equal those of otherwise equivalent animals who remain at home. Quite coincidentally, the migrants play a key role in dispersing genes between populations and extending the boundaries of the species. They contribute, as it were, the biogeographic turgor, by which the species as a whole maintains its maximum spread and overall density. Certain authors, notably Christian (1970) and Calhoun (1971), have imputed even greater potential to subordinates and migrants. The wanderers are the ones most likely to pioneer in new habitats, to experiment with new forms of adaptation, to learn more quickly and to adjust the cultural capacity of the species by genetic assimilation. Outcasts, to put the idea in its starkest form, are the cutting edge of evolution. This is an attractive hypothesis but still just speculation. It is equally easy to build a model in which the "establishment" center of the population accounts for most of the evolution. It is in the center that we find the greatest amount of genetic diversity. There the habitable areas are most extensive and ecologically diverse, while the dense, relatively stable populations inhabiting them are subject to the maximum variety of social interaction among individuals. From such ingredients evolution can lead to each of the qualities identified with outcasts in the alternative model. To refine and test these and still other, competing hypotheses remains an important task of sociobiology.

Other functions of subordinates can be defined that are probably adaptive to the dominants but not to the subordinates. The omega individuals can serve as an "aggression sink." Bernard Greenberg (1946) found that when subordinate, nonterritorial green sunfish (*Lepomis cyanellus*) are removed from aquaria, the remaining territorial residents increase aggressive interactions with one another. When a strange fish is then introduced, it becomes the new target for attacks. This omega effect is somewhat artificial in that subordinate fish in free-ranging populations can be expected to move away and to try to establish a territory in a less optimal habitat.

Kin selection might provide the means by which subordination pays out a genetic benefit. If an animal that has little chance of succeeding on its own chooses instead to serve a close relative, this strategy may raise its inclusive fitness. A concrete example is provided by the social insects. When fertile females of the paper wasp genus *Polistes* emerge from hibernation and begin searching for a nest site, they tend to settle in the neighborhood of the nest in which they were born the previous summer. Groups of these wasps, many of whom are sisters, commonly cooperate in founding a new nest, with one assuming the dominant, egg-laying role and the others turning into functional workers. This voluntary subordination is not easy to explain, for even if the associated females were full sisters, the subordinate female would be taking care of nieces with a coefficient of relationship of $\frac{3}{8}$, whereas she could choose to care for her own daughters and share a bond of $\frac{1}{2}$. The missing piece of the theory has been supplied by what might be termed the "spinster hypothesis," invented by Mary Jane West (1967). West points out that nest-founding females of *Polistes* vary greatly in ovarian development and that rank in the dominance hierarchy varies directly with the development. It is further true that most new *Polistes* nests fail. Consequently, the probability of a female with low fertility establishing and bringing a nest through to maturity may simply be so low that it is more profitable, as measured by inclusive fitness, for these high-risk individuals to subordinate themselves to female relatives in foundress associations.

In still other societies we encounter direct incentives for subordinates to stay with their group. Individual macaques and baboons cannot survive for very long on their own, especially away from the sleeping areas, and they have almost no chance at all to breed. As Stuart Altmann and others have shown, even a low-ranking male still eats well if he belongs to a troop, and he gets an occasional chance to copulate with estrous females. Furthermore, patience can turn half a loaf into a full one, because the dominant animals will eventually grow old and die. The European black grouse *Lyrurus tetrix* even observes a kind of seniority system on the display grounds: the yearling cocks occupy peripheral territories, which attract few females; at two years of age they move into second-ranking positions near the center; and at three years of age they have the chance to become dominant cocks (Johnsgard, 1967). The turnover of dominant males may be a general phenomenon. Fraser Darling observed that red deer stags do not eat while herding a harem. After about two weeks they are easily defeated by a fresher, often younger stag. They then retire and wander to higher ground to feed, to regain their strength, and perhaps to try again. Dominant male impalas also wear themselves out quickly, yielding to fresher rivals or falling victim to predators.

Many kinds of monkeys and apes possess what Eisenberg and his coworkers (1972) have called the age-graded-male system, which is essentially the same seniority sequence that exists in the black grouse. In this system a single older, dominant male tolerates younger males and may even cooperate with them in foraging and group defense. When the alpha male weakens or expires from age or injury, one of the older lieutenants takes his position. The age-graded-male organization is apparently intermediate in evolution between the unimale society, in which the ruling male tolerates no subordinates, and the multimale society, in which multiple adult males enjoy approximately equal rank. Most of the known examples are found among the macaques, drills, and guenons. The gorilla troop, with its dominant but highly tolerant silver-backed male, is a noteworthy example among the anthropoid apes.

An age-graded system has also been reported in the primitively social wasp *Belonogaster junceus* of Africa (Roubaud, 1916). All of the colony members are approximately the same size, with well-developed ovaries, and all or nearly all are inseminated within about a week following eclosion. Prior to insemination and for some time afterward the young females serve as workers. According to the hypothesis suggested by Roubaud, they are kept sterile by a combination of hard work and lack of nourishment. However, as they grow older they somehow assume the role of egg layers. Thus no permanent caste division exists, and all the females have essentially the same rank when status is averaged over a lifetime. One searches in vain through Roubaud's account for evidence of dominance hierarchies, but of course in 1916 Roubaud was not aware of the concept and could easily have neglected to record the pertinent observations. Similar age-graded societies may exist in the primitively social bees and in the stenogastrine wasp genus *Parischnogaster* (Yoshikawa et al., 1969).

The Determinants of Dominance

What qualities determine the status of an individual? Surprisingly little critical work has been directed to this important question, and investigators with useful data often present the results tangentially while discussing other topics. Much of the most substantial information is presented as a phylogenetic catalog in Table 13-2. Our current knowledge can be summarized in the form of the following loose principles:

1. Adults are dominant over juveniles, and males are usually dominant over females. In multimale societies, it is typical for the rank ordering of the males to lie entirely above that of the females, or at most to overlap it slightly. In such cases juvenile males sometimes work their way up through the female hierarchy before achieving greater than omega status with reference to the males. Exceptional species in which females are dominant over males include the brown booby *Sula leucogaster* (Simmons, 1970), the hyena (Kruuk, 1972), the vervet (*Cercopithecus aethiops*), and Sykes' monkey (*C. mitis*) (Rowell, 1971).

2. The greater the size of the brain and the more flexible the behavior, the more numerous are the determinants of rank and the more nearly equal they are in influence. Also, the more complex and orderly are the dominance chains. These correlations are very loose, and they become apparent at our present state of knowledge only when species are compared over the greatest phylogenetic distances. Arthropods, including social insects, display relatively simple types of aggressive behavior that result in despotisms, short-chain hierarchies of elementary structure, or chaotic systems in which dominance is established anew with each contact (as in the wasp *Vespula*). Fish, amphibians, and reptiles also form despotisms and short-chain hierarchies. Birds and mammals commonly form long-chain hierarchies, the members of which defend territories communally. In some of the higher monkeys and apes, we see the emergence of coalitions of peers, protectorates by dominant individuals, and strong maternal influence in the early establishment of rank.

3. The greater the cohesiveness and durability of the social group, the more numerous and nearly coequal the correlates of rank and the more complex the dominance order. The male rank orders of antelopes, sheep, and other ungulates, especially those formed temporarily during the breeding season, are predominantly based on size, with age perhaps being a second, closely associated factor (see Figure 13-4). In the more aggressively organized Old World monkeys, particularly the baboons and macaques, status is based more on childhood history as it relates to the mother's rank, to membership in coalitions, and to "luck"—whether the animal is a member of an old family, for example, or has just immigrated from a neighboring troop, or has been fortunate enough to catch a stronger opponent in a weak moment when it could be defeated. When a group is newly constituted, such as a group of hens or rhesus monkeys thrown together in an enclosure, the initial dominance orders tend to be established on the basis of size, strength, and aggressivenss. But later the other more personal and experiential factors assert themselves as well.

Few studies have been conducted with the explicit goal of assigning weights to the determinants of rank. The most instructive to date is N. E. Collias' (1943) analysis of domestic fowl. Collias measured

Table 13-2   The correlates of dominance rank order.

Species	Factor	Degree of correlation	Resource controlled	Authority
INSECTS				
Hercules beetle (*Dynastes hercules*): males	Size	High	Females	Beebe (1947)
Paper wasps (*Polistes* spp.): females	Sequence of arrival at nest-founding site (queens only)	High	Food, oviposition rights	Pardi (1948), Eberhard (1969), Wilson (1971a)
	Age, i.e., sequence of emergence in nest (workers only)	High	Food	Pardi (1948), Eberhard (1969), Wilson (1971a)
parasitic queens	Queens of parasitic species dominate those of host species by greater strength and persistence	High	Food, oviposition rights	Scheven (1958); see Chapter 17
Bumblebees (*Bombus* spp.): females	Size, caste, and ovary development (all strongly correlated)	High	Food, oviposition rights	Free (1955, 1961)
	Victory or defeat in previous encounters	Low	Food, oviposition rights	Free (1955, 1961)
parasitic queens	Queens of parasitic species dominate those of host species by greater defensive ability and persistence	High	Food, oviposition rights	Plath (1922), Free and Butler (1959), Wilson (1971a); see Chapter 17
Stingless bees (*Melipona quadrifasciata*): females	Caste	High, probably total	Food, oviposition rights	Sakagami, Montenegro, and Kerr (1965)
Ants: queens (females), between species of *Formica* and *Myrmica*	Size	High	Nest sites	Brian (1952)
CRUSTACEANS				
Spiny lobsters (*Jasus lalandei*)	Size	High	Shelter	Fielder (1965)
Crayfish (*Cambarellus, Procambarus*)	Size	Moderate to high	Space, converted from territoriality	Lowe (1956), Bovbjerg (1956)
FISH				
Green sunfish (*Lepomis cyanellus*)	Victory or defeat in previous encounters	Moderate and partial: other factors experimentally excluded	Space, converted from territoriality	McDonald et al. (1968)
Swordtails (*Xiphophorus*)	Size, and victory or defeat in previous encounters	Moderate	?	Thines and Heuts (1968)
Minnows (*Poeciliopsis*): males	Age, experience	Moderate	Females	McKay (1971)
REPTILES				
Galápagos tortoises (*Geochelone elephantopus*): males	Size, particularly height to which head can be raised	Strong ("invariable")	Females	MacFarland (1972)

Table 13-2 (*continued*).

Species	Factor	Degree of correlation	Resource controlled	Authority
BIRDS				
Black grouse (*Lyrurus tetrix*): males	Age	Moderate or strong	Females	Johnsgard (1967)
Brown boobies (*Sula leucogaster*): nestlings	Time of emergence	Strong	Probably food	Simmons (1970)
RODENTS				
Voles, deer mice (*Clethrionomys, Microtus, Peromyscus*)	Size	Strong	Space, at least partly converted from territories	Grant (1970)
UNGULATES				
New world camelids (alpacas, guanacos, llamas): females	Age primarily, size secondarily	Moderate or strong	?	Pilters (1954)
Mountain sheep (*Ovis canadensis, O. dalli*): males	Size, especially horn size	Moderate or strong	Females	Geist (1971)
Dairy cattle: females	Size	Strong	?	Schein and Fohrman (1955)
Gaur (*Bos gaurus*): males	Size	Moderate	Females	Schaller (1967)
Axis deer (*Axis axis*): males	Size, especially antler length	Moderate	Females	Schaller (1967)
PRIMATES				
Japanese and rhesus macaques (*Macaca fuscata, M. mulatta*)	Complex social factors: primarily mother's rank; also health, recency of arrival in troop, "personality," history of previous wins and losses, (in female) time of reproductive cycle	?	Food, resting places	B. K. Alexander and Hughes (1971), Bartlett and Meier (1971), A. Jolly (1972a), Rowell (1963), Southwick and Siddiqi (1967), Sade (1967)
Mangabeys (*Cercocebus*): females	Age and time of reproductive cycle	?	?	Chalmers and Rowell (1971)

the following intuitively promising qualities in a series of White Leghorn hens: their general health as indicated by weight and general vigor of movement, their age, their stage of molting, their level of androgen as indicated by the size of the comb, and their rank in the home flock from which they were drawn. The hens were then matched pairwise on neutral ground and the outcome of their aggressive interactions recorded. Winning in these encounters was found to depend most on the absence of molt, followed in order by comb size, earlier social rank, and weight. Age did not seem to matter. All of these factors in combination accounted for only about half the variance. Collias suggested that the additional contributors to rank included differences in fighting skill, luck in landing blows, degree of wildness and aggressivity, slight differences in handling, and the physical resemblance of particular opponents to past despots. Of

course, most or all of these components are heritable, so it is correct to say that to a degree not yet measured the status of hens is determined genetically.

A similar multiplicity of factors has been discovered in the more social mammals. Hormone levels are deeply implicated. An increase in androgen titer, and hence masculinization of anatomical and behavioral traits, tends to move individuals upward in the hierarchy. The adrenal hormones also appear to have a role. Candland and Leshner (1971) found that dominant males in a laboratory troop of squirrel monkeys had the highest levels of 17-hydroxycorticosteroids and the lowest levels of catecholamines (epinephrine plus norepinephrine). The 17-ketosteroids, however, were related to dominance by a J-shaped function: dominant males had medium titers, middle-ranking males low titers, and low-ranking males high titers with levels

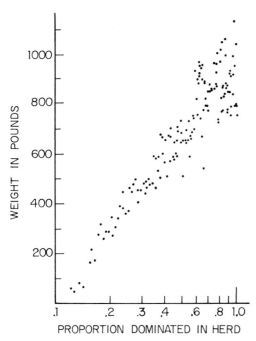

Figure 13-4 Dominance orders of cattle are typical of ungulates in being based primarily on the size of the animal. (Redrawn from Schein and Fohrman, 1955.)

rising as rank fell. Candland and Leshner then turned the procedure around to see if the dominance order could be predicted from the hormone levels. Prior to forming a laboratory troop of five squirrel monkeys, they obtained baseline measures of urinary steroids and catecholamines in the separate animals. The concentration of 17-hydroxycorticosteroids was sufficient to predict the subsequent rank order, while the catecholamine titer was a J-shaped function of declining rank. These results, while very suggestive, do not constitute proof. The mere existence of the correlation between rank order and hormone level does not establish causation in either direction. Moreover, both could stem from other, prior determinants such as age, health, and experiences unrelated to dominance interactions.

The status of parents can also matter. Among Japanese macaques the sons of high-ranking females are able ipso facto to spend more time in the center of the troop and to associate more closely with dominant males during their childhood. They tend to receive the cooperation of the leaders and to succeed them in position when they die. Sons of low-ranking mothers, in contrast, remain near the periphery of the group and are the first to emigrate. The existence of such a hereditary aristocracy results in greater group stability (Kawai, 1958; Kawamura, 1958, 1967; Imanishi, 1963). Kawai has made the useful distinction between *basic rank*, the outcome of the interaction of

two monkeys unaffected by the influence of kin, and *dependent rank*, in which kinship plays a biasing role. Similar dependent succession has been studied in rhesus macaques by Koford (1963), Sade (1967), Marsden (1968), and Missakian (1972). Young rhesus monkeys of both sexes begin play-fighting with older infants or yearlings. The outcome is rather unfair: each animal defeats age peers whose mothers rank below its own, and each is defeated by age peers with higher-ranking mothers. As the monkeys mature they extend their dominance position into the existing hierarchy of adults, thus coming to rank just below their mothers. Females remain at approximately this level. Males, however, tend to change in rank upward or downward, possibly as a result of physiological variation.

It is not wholly imprecise to speak of much of the residual variance in dominance behavior as being due to "personality." The dominance system of the Nilgiri langur *Presbytis johnii* is weakly developed and highly variable from troop to troop. Alliances are present or absent, there is a single adult male or else several animals coexist uneasily, and the patterns of interaction differ from one troop to another. Much of this variation depends on idiosyncratic behavioral traits of individuals, especially of the dominant males (Poirier, 1970).

At this stage of our analysis it may seem that dominance orders can be fully characterized if we are given knowledge of the finite set of characteristics that determine individual competence: size, age, hormone-mediated aggressiveness, and so forth, up to and including the subtle components of personality. But this turns out not to be the case. Mathematical analysis has revealed that the observed degrees of orderliness and stability of many of the hierarchies in chicken flocks and other animal groups cannot be easily explained even with a full knowledge of the determinants and their correlations with fighting ability. The basis for this surprising result was established by Hyman Landau (1951-1965), who pioneered in the mathematical analysis of animal social behavior. The key of Landau's analysis is the following index, which he devised for the measurement of a strength of a hierarchy (h):

$$h = \frac{12}{n^3 - n} \sum_{a=1}^{n} \left(v_a - \frac{n-1}{2} \right)^2$$

where n is the number of animals in the group and v_a is the number of group members that the ath animal dominates. The term $12/(n^3 - n)$ normalizes h such that its value ranges from 0 to 1. A low Landau index value indicates a weak hierarchy: a value of zero means that each animal dominates an equal number of group members. A high index value means a strong hierarchy. A "perfect" score of 1 is received by a completely linear order, while an order with a score of 0.9 would still be intuitively judged to be strong when seen in graphical form. Landau utilized the index to prove several useful theorems. He derived the mean value, $E(h)$, of a system in which (1) the outcome of each pairwise encounter, p, is a probability such that $p_{jk} + p_{kj} = 1$ where j and k are the animals contending for dominance, and (2) the probabilities are determined by components of ability, such as size or aggressiveness, that are distributed in a set fashion. Landau then showed that as the number of uncorrelated ability components and the size of the group pass to infinity, the index approaches zero. In more realistic terms this result means that as the number of uncorrelated components of ability is increased in moderate to large groups, the strength of the hierarchy declines sharply. In short, the more complex the society, the more likely it is to be egalitarian. Landau also pointed to a paradox. The data on chickens have revealed very strong hierarchies. In fact, $h = 1$ routinely when group size is less than about ten. Yet the largest correlation obtained by Collias (1943) between a single ability factor and dominance ($r = 0.593$) yields a Landau index of only 0.34. In fact, extremely high correlations are required to produce strong hierarchies, higher than those that appear to exist in some hierarchical societies.

Landau's paradox was the point of departure for a new effort by Ivan Chase (1973, 1974). Recognizing that the best step in such cases is to construct hypotheses and pit them against one another, Chase formalized biological thinking on dominance orders into two basic models. The first model envisages a round-robin tournament in which each animal fights or simply compares itself with every other member of the group, and thereafter dominates all the other animals beaten in the initial encounters. The probability of winning or losing is set for each pairwise encounter without reference to any particular biological property. Under these conditions a strong hierarchy cannot be established. Chase proved the result using an extension of the Landau formula, but the essential argument can be intuitively grasped as follows. Some animals will have a high probability of success in competitions, and some a low probability, but most will be just moderately successful. Thus a majority will win an intermediate number of contests, and the probabilistic nature of the outcome will prohibit the pattern of successes from falling into a simple linear order.

Chase's second hypothesis postulates a high statistical correlation between components of ability and position in the hierarchy. This hypothesis cannot be eliminated altogether, yet, as intimated by Landau's earlier result, it demands almost impossibly stringent conditions. For a perfectly linear hierarchy (a commonly observed condition) the correlation coefficient must equal unity. To produce a moderately strong hierarchy, correlation coefficients greater than 0.9 are required, accounting for more than 80 percent of the variance.

Thus it is difficult, perhaps sometimes impossible in practice, for strong hierarchies to be generated by a simple pairwise matching of attributes by group members. But how else can a dominance order be formed? Chase views the formation as a magnification process, in which combinations of ability and luck increasingly drive some animals downward in rank while lifting others upward. Aggressive animals will seek out others, while more timid ones will consistently avoid confrontations. Repeatedly successful encounters increase the probability of success in later encounters, and make a contest with a timid animal still more of a mismatch. Accidental events, such as fatigue on a certain day or a chance blow, will start an animal upward or downward. The dominance order will stabilize as all of the pairwise encounters become strongly asymmetric, with one contestant clearly dominating another, and the order approaches one of the few available stable states at or near linearity. Chase's hypothesis will be difficult to prove or ·disprove. However, its plausibility is enhanced by the independent experimental demonstration of the magnification process. Warren and Maroney (1958) found that among rhesus monkeys the differentiation between winners and losers in pairwise contests increased with time. The overall scores of the initially successful animals rose as the scores of the initial losers fell. If at the beginning of the Warren-Maroney experiment the monkeys had been joined in a single group, the hierarchy would have been weak. But in later stages of the experiment such a combination would have produced a much stronger hierarchy, essentially as predicted by the Chase model.

Intergroup Dominance

Sometimes groups dominate groups in much the same way that group members dominate one another. Intergroup dominance is not often seen in nature, because contact between well-organized societies regularly occurs along territorial boundaries where power is more or less balanced. However, if the territories are spatiotemporal, dominance orders can appear when groups meet in overlapping portions of the home ranges. Phyllis Jay (1965) observed such a pattern in low-density populations of the common langur (*Presbytis entellus*) at Kaukori and Orcha in northern India. Because the langur troops possessed distinct core areas and followed their own routes while foraging, they seldom encountered one another. When contacts did occur, the larger group took precedence, with the smaller group simply remaining at a distance until the larger group moved away.

Intergroup hierarchies can also be created by confining societies in spaces smaller than the average territory occupied by a single group. When this is done to colonies of social insects, the result is almost invariably fatal for the weaker unit (Wilson, 1971a). While studying the phenomenon systematically in rhesus monkeys, Marsden (1971) discovered an interesting secondary effect. As the subordinate troops retreated into a smaller space, their members fought less among one another. But within the dominant group, which was in the process of acquiring new space, aggressive interactions increased. Marsden's effect, if it occurs at all generally, has important implications for the evolution of cooperative behavior.

Interspecific Dominance

Dominance orders have often been encountered among species that belong to the same taxonomic group. As a rule, the more closely related and ecologically similar the species, the more pronounced the dominance by members of one over members of the other. Species with large individuals dominate those with small individuals, except where one or more species is social, in which case the one forming the largest, best-organized groups dominates the others. MacMillan (1964) found that among seven rodent species living in the semidesert of southern California, the largest routinely dominate the smaller. Encounters seldom lead to fighting, because the subordinate species flees at the sight of the larger. In Yellowstone National Park, the large mammals advance or retreat according to the following descending dominance order: adult human beings, bison, elk, mule deer, pronghorn antelope, and moose or white-tailed deer (McHugh, 1958).

When certain species of birds, including nuthatches, warblers, chickadees, and others, flock together in foraging groups, they form interspecific dominance hierarchies. One common result is a displacement of species into narrower feeding niches than the ones enjoyed when the same species feed alone. In such cases the dominant species have access to the most predictable portion of the food supply (Morse, 1967, 1970). Interspecific dominance has also been reported in mixed schools of three species of the freshwater fish genus *Cichlasoma* in Nicaragua (Barlow, 1974a).

Scaling in Aggressive Behavior

The general pattern of scaling in aggressive behavior among animals is summarized in Figure 13-5. This scheme is the culmination of a long history of investigation by many zoologists. Perhaps the first explicit description of scaling was that of H. H. Shoemaker (1939), who found that canaries forced together in small spaces become organized into dominance orders. Given more space, they establish territories (the natural condition for *Serinus canaria* in the wild), even though low-ranking individuals continue to be dominated around bath bowls, feeding areas, and other nonterritorial public space. The phenomenon has been subsequently documented in other birds (review in Armstrong, 1947), sunfishes and char (Greenberg, 1947; Fabricius and Gustafson, 1953), iguanid lizards (L. T. Evans, 1951, 1953), house mice (Davis, 1958), Norway rats (Barnett, 1958;

Calhoun, 1962), *Neotoma* wood rats (Kinsey, 1971), woodchucks (Bronson, 1963), and cats (Leyhausen, 1956). Kummer (1971) developed the concept with special reference to the social evolution of primates.

The existing data permit several generalizations about aggressive scaling. The clearest cases are found in species, such as certain lizards and rodents, in which the normal state is for solitary individuals or pairs to occupy territories. When forced together, groups of these individuals shift dramatically to despotisms or somewhat more complex dominance orders (see Figure 13-6). In most such cases the shift from territoriality to a dominance system is really superficial in nature. In the case of despotism, one individual in effect retains its territory while merely tolerating the existence of the others. Such transitions are not limited to laboratory experiments. In Mexico Evans (1951) found a crowded colony of the large black lizard *Ctenosaura pectinata* living on a cemetery wall, which provided shelter from which the lizards ventured to feed in a nearby cultivated field. At least eight adult males made up a dominance hierarchy, with one individual playing the role of a strong tyrant.

Although some species display phenotypic variability that covers a substantial portion of the gradients, in other words utilize true behavioral scaling, many others are fixed at a single point. The males of sea lions, elephant seals, and other harem-forming pinnipeds maintain territories with about the same intensity regardless of population density. The adaptive significance of such rigidity is clear. Aggressive behavior in these animals serves the single function of acquiring

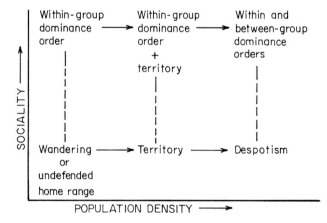

Figure 13-5 The prevailing patterns of scaling in aggressive behavior. The solid lines indicate true scaling: the transitions commonly observed to be part of the phenotypic variation of individuals. The dashed lines represent shifts unlikely to occur except by genetic evolution, which permits a species to substitute one pattern of scaling for another in the course of social evolution.

Figure 13-6 Despotism in the iguanid lizard *Leiocephalus carinatus*. When groups of this normally territorial West Indian species are forced together, one individual (*foreground*) dominates all of the others by tail curling and other threat signals, as well as by fighting. (From L. T. Evans, 1953.)

harems. The means by which this goal is reached and its value to genetic fitness are unaffected by changes in the density of animals on the hauling grounds. In such cases, shifts from one point to another on the behavioral gradients occur by evolution, but probably only when changed environmental circumstances alter the optimum social strategy.

Finally, the patterns outlined in Figure 13-5 occur very generally among the vertebrates. The only exceptions I know are displayed by certain fishes. When salmon (*Salmo salar*) and trout (*S. trutta*) are crowded in hatcheries enough to disrupt territorial behavior, they shift not to hierarchies but to schools (Kalleberg, 1958). The same transition occurs in crowded natural populations of the ayu (*Plecoglossus altivelis*), a Japanese salmonid (Kawanabe, 1958). The banded knife-fish (*Gymnotus carapo*), one of the species that orient and communicate by electric discharges, displays the reverse behavioral scaling from all other known vertebrates: dominance hierarchies at low densities and territories at higher densities (Black-Cleworth, 1970). Invertebrates, including insects, have not been systematically studied with respect to the plasticity of aggressive behavior and the possible existence of behavioral scaling. When they are, a good chance exists that new kinds of transitions will be found that deviate far from the standard vertebrate pattern.

Chapter 14 **Roles and Castes**

Society, in the original, quasi-mystical vision created by Durkheim and Wheeler, is a superorganism that evolves to greater complexity through the complementary processes of differentiation and integration. As the society becomes increasingly efficient, larger, and geometrically more structured, its members become specialized into roles or castes and their relationships become more precisely defined through superior communication. Whole new ways of life—the practice of agriculture, industrialization, the storage of vast amounts of information, travel over fantastic distances, and more—await the society that can correctly engineer the division of labor of its members. Even the lowly ants have invented agriculture and slavery.

Although castes in social insects have been clearly understood since at least as far back as Charles Butler's *The Feminine Monarchie* (1609), zoologists have been slow to recognize even the rudiments of such differentiation in the nonhuman vertebrates. The vertebrate society has been traditionally viewed as a congeries of individuals distinguished from one another by age, sex, and sometimes status. Each member was considered to be endowed with the total repertory of its sex and to occupy a social position that can be largely defined by only two parameters: the position in the dominance order and the tendency to assume leadership during group movement or defense. But a comparison with human social organization leads to the question of whether more subtle roles exist. Is there an underlying differentiation of behavior in higher vertebrate societies that foreshadows the extreme division of labor in human societies? This important question began to be addressed only about ten years ago, when several students of primate behavior, notably Hall (1965), Bernstein and Sharpe (1966), Rowell (1966b), and Gartlan (1968) became dissatisfied with the effectiveness of the dominance concept as an analytic tool in the description of societies. They borrowed the concept of *role* from sociology, a second case of biology taking an idea from the social sciences (the other example is socialization, described in Chapter 7). And almost immediately confusion and doubt arose concerning the meaning and usefulness of the word (see Hinde, 1974). It is therefore necessary to begin with an attempt to define this and allied terms in a way that represents a consensus of most authors.

Role: a pattern of behavior that appears repeatedly in different societies belonging to the same species. The behavior has an effect on other members of the society, consisting either of communication or of activities that influence other individuals indirectly—or both. An animal, like a human being, can fill more than one role. For example, it might function as a control animal in terminating disputes and also as a leader when the group is on the move. Ideally, the full description of all roles together, insofar as they can be meaningfully distinguished, will fully define the society. In the broadest sense, male behavior during copulation constitutes a role, as does maternal care, despite the fact that primatologists have not yet found it useful to speak of such behaviors in just that way. Idiosyncratic actions of individuals do not constitute roles; only regularly repeated categories fulfill the criterion. For example, the animal or set of animals that regularly watches for predators near the periphery of the group is playing a role, but a particular male who prefers to watch from a certain tree is not. Thus when Saayman (1971a) spoke of the "roles" of three male chacma baboons in one particular troop as coincident with detailed differences in their behavior, he stretched the definition too far.

Caste: a set of individuals, smaller than the society itself, which is limited more or less strictly to one or more roles. Where the role is defined as a pattern of behaviors, which particular individuals may or may not display, the caste is defined obversely as a set of individuals characterized by their limitation to certain roles. In human societies a caste is a hereditary group, endogamously breeding, occupied by persons belonging to the same rank, economic position, or occupation, and defined by mores that differ from those of other castes. In social insects a caste is any set of individuals of a particular morphological type, or age group, or both, that performs specialized labor in the colony. It is often more narrowly defined as any set of individuals that are both morphologically distinct and specialized in behavior. A caste system may or may not be based in part on genetic differences. In the stingless bee genus *Melipona*, queens are determined as complete heterozygotes in multiple-locus systems, but in most or all other social insects the caste of individuals is fixed by purely environmental influences.

Polyethism: the differentiation of behavior among categories of individuals within the society, especially age and sex classes and castes. Both role playing and caste formation lead automatically to polyethism. In the social insects polyethism refers particularly to division of labor. A distinction is sometimes made, in compliance with the narrower usage of the word "caste," between caste polyethism, in which morphological castes are specialized to serve different functions, and age polyethism, in which the same individual passes through different forms of specialization as it grows older (Wilson, 1971a).

The Adaptive Significance of Roles

The differentiation of behavior within a society can best be measured by the judicious selection of sets of individuals and the comparison of their behavior patterns. Tables 14-1 and 14-2 present the results of such analyses in a colony of ants and in troops of primates, respectively. Notice that the two matrices closely resemble each other. The idea can be fleetingly entertained that they provide the means of comparing such different societies in a quantitative manner. The

independent category of ants is the caste and the dependent one division of labor, whereas the monkey troop is partitioned into age-sex classes and "role profiles." But the distinction at this level is trivial. The castes of insects are based on age and sex in addition to size, while their places in the division of labor could equally well be labeled role profiles.

The deeper difference between the two patterns lies in the nature of the adaptiveness of the differentiation of behavior. We ask: At what level has natural selection acted to shape these varying profiles? The reader will recognize one more version of the central problem of group selection in social evolution. The problem must be solved with reference to polyethism before the full significance of behavioral differentiation can be disclosed. For social insects the problem appears to be essentially solved. Selection is largely at the level of the colony. Castes are generated altruistically—they perform for the good of the colony. Caste systems and division of labor can therefore be treated by optimization theory. Vertebrates, however, are as usual mired in ambiguity. Kin selection is undoubtedly strong in small, closed societies such as primate troops. Hence an adult male of a single-male group can play the role of sentinel and defender in an altruistic

Table 14-1 Division of labor among workers of the ant *Daceton armigerum* by head width. (From Wilson, 1962b.)

Type of labor	Head width (mm)				Total number of observations
	1	2	3	4	
Surinam colony					
Total population (in artificial nest, April 5)[a]	13	60	20	9	102
Disposing of corpses and refuse[b]	0	19	12	2	33
Dismembering and feeding on fresh prey in nest[b]	0	14	25	5	44
Feeding larvae by regurgitation[b]	8	15[c]	3[c]	1[c]	27
Attending egg-microlarva pile[b]	24	3	0	0	27
Foraging in the field[a]	0	0	4	10	14
Trinidad colony					
Foraging in the field[a]	1	91	77	12	181
Resting in way-station[a]	0	8	19	10	37
Carrying prey[a]	0	1	1	10	12

[a]Numbers refer to separate, individual workers.
[b]Numbers refer to separate behavioral acts, without regard to the number of workers engaged.
[c]Consisting mostly of callows.

Table 14-2 Differentiation of behavior in troops of the vervet monkey *Cercopithecus aethiops*, given as frequencies of contributions by age-sex classes to several categories of behavior. (Based on Gartlan, 1968.)

Behavior ("roles")	Age-sex class				
	Adult males	Adult females	Juvenile males	Subadult females	Infants
Territorial display	.66	0	.33	0	0
Vigilance; look-out behavior	.35	.38	.03	.12	.12
Receiving friendly approaches	.12	.46	.04	.27	.12
Friendly approach to others	.03	.32	0	.47	.15
Chasing of territorial intruders	.66	0	.33	0	0
Punishing intragroup aggression	1.00	0	0	0	0
Leading in group movement	.32	.49	0	.16	0

manner. He risks injury or death for the good of the society. But his role is not quite the same as the role of defender in the insect society, for the male vertebrate is defending his own offspring. Much of role playing in vertebrate societies is patently selfish. The forager who discovers food gets the first share; the male who visits new troops improves his chances of rising in status by finding weaker opponents. Each behavior must be interpreted unto itself. Only when the contribution of the behavior to individual as opposed to group fitness is assayed will it become feasible to distinguish roles that are the secondary outcome of indiviual adaptations from those that are "designed" with reference to the optimum organization of the society. Meanwhile the concept of the vertebrate role must be regarded as loose and even potentially misleading. We shall explore this matter further, but first it will be useful to examine the less ambiguous paradigm of castes in insect societies and lower invertebrates. Here the basic theory has been initiated to which vertebrate societies can eventually be referred.

The Optimization of Caste Systems

Caste in the social insects is a large and complicated subject that I recently reviewed in *The Insect Societies*. For the reader interested in such topics as the physiology of caste determination or a detailed comparison of termite and ant systems, the book will serve as an introduction and guide to the literature. Here we shall consider only two topics of general interest: the defensive castes of ants and termites, which illustrate the extremes of specialization and altruism found in the social insects as a whole, and the theory of caste ergonomics, through which the problem of optimization can be approached.

In the case of advanced polymorphism in ant colonies, especially complete dimorphism where intermediates have dropped out and the two remaining size classes are strikingly different in morphology,

members of the larger class usually serve as soldiers. Often they play other roles as well. Soldiers of some species of *Camponotus* and *Pheidole* assist in food collection, and their abdomens swell with liquid food. Recent work has revealed that their per-gram capacity is much greater than that of their smaller nestmates, and they therefore serve as living storage casks (Wilson, 1974a). But it is apparent that the extensive changes in the head and mandibles that make the soldiers so deviant are directed primarily toward a defensive function. One of three fighting techniques is employed, depending on the form of the soldier. In one form the soldier may use the mandibles as shears or pliers: the mandibles are large but otherwise typical, the head is massive and cordate, and the soldiers are adept at cutting or tearing the integument and clipping off the appendages of enemy arthropods. Examples are found in *Solenopsis, Oligomyrmex, Pheidole, Atta* (Figure 14-1), *Camponotus, Zatapinoma,* and other genera of diverse taxonomic relationships. W. M. Wheeler, in his essay "The Physiognomy of Insects" (1927), pointed out that the peculiar head shape of this kind of soldier is due simply to an enlargement of the adductor muscles of the mandibles, which imparts to the mandibles greater cutting or crushing power. A second form of soldier has pointed, sickle-shaped or hook-shaped mandibles that are used to pierce the bodies of enemies. Some formidable examples are the major workers of the army ants (*Eciton*) and driver ants (*Dorylus*), which are able to drive off large vertebrates with their simultaneous bites and stings. The third basic type of soldier is less aggressive, using its head instead to block the nest entrance—thus serving literally as a living door. The head may be shield-shaped (many members of the tribe Cephalotini) or plug-shaped (*Pheidole lamia* and several subgenera of *Camponotus*). The colonies possessing such forms usually nest in cavities in dead and living plants and cut nest entrances with diameters just a little greater than the width of the head of an individual soldier. In the case of the soil-dwelling *Camponotus ulcerosus*, a

Figure 14-1 A soldier of the leaf-cutting ant *Atta cephalotes* is surrounded by smaller nestmates. The middle-sized workers shown here are most active in foraging outside the nest, while the smallest individuals specialize more in the care of the brood. Soldiers weigh as much as 90 milligrams and the smallest workers as little as 0.42 milligrams. (Photograph by courtesy of C. W. Rettenmeyer.)

carton shield is constructed at the ground surface with a single aperture that closely approximates the head of the soldier in size and shape (Creighton, 1953).

The behavior of the ant soldiers is often extremely specialized and simplified. An efficient form of colony defense is achieved by the integration of the responses of such specialists with those of other castes. This principle is nicely exemplified by the blocking-type soldiers of the North American cephalotine *Paracryptocerus texanus.* The entrance hole to the arboreal nest is somewhat larger than the head of the soldier and is blocked by the combined mass of its head and expanded prothorax, both of which structures are heavily armored and pitted. The head is held obliquely, rather like the blade of a miniature bulldozer. This posture, combined with the thrust and pull of the short, powerful legs, allows the soldier to press intruders right out of the nest. When a minor worker returns to the entrance after a trip to forage for food, it palpates the soldier with its antennae, causing it to crouch down and make just enough room for the smaller ant to squeeze into the nest (Creighton and Gregg, 1954).

The soldier is also the most specialized caste found in the termites. The soldier castes of ants and termites display many remarkable convergences in anatomy and behavior. The three basic forms found in ants—the shearer-crusher, the piercer, and the blocker—also occur in various termite species. In addition there are bizarre "snapping" soldiers in *Capritermes, Neocapritermes,* and *Pericapritermes* (Kaiser, 1954; Deligne, 1965). Their mandibles are asymmetrical and so arranged that the flat inner surfaces press against each other as the adductor muscles contract. When the muscles pull strongly enough, the mandibles slip past each other with a convulsive snap, in the same way that we snap our fingers by pulling the middle finger past the thumb with just enough pressure to make it slide off with sudden force. If the mandibles strike a hard surface, the force is enough to throw the soldier backward through the air. If they strike another insect, which seems to be the primary purpose of the adaptation, a stunning blow is delivered. Even vertebrates receive a painful flick. The mandibles of *Pericapritermes* in particular are modified in such a way that the left mandible alone strikes out, so that the target can be hit only if it is located on the right side of the soldier's head.

The premier combat specialists are the soldiers that employ chemical defense. The mandibulate soldiers of the very primitive Australian termite *Mastotermes darwiniensis* produce almost pure *p*-benzoquinone from glands that open into the mouth cavity (Moore, 1968). When a soldier bites an adversary, the quinone is mixed with amino acids and protein in the saliva, soon producing a dark, rubberlike material that entangles the victim. Excess quinone probably acts as an irritant. The mandibulate soldiers of the Termitidae, the largest and phylogenetically most advanced of the termite families, have independently modified their salivary glands to the same end. When

Protermes soldiers attack, they emit a drop of pure white saliva that spreads between the opened mandibles. When they bite, the liquid spreads over the opponent. In general, the salivary glands of termitid soldiers are better developed than those of their worker nestmates, and they sometimes reach a huge size in proportion to the remainder of the body. The salivary reservoirs of *Odontotermes magdalenae* swell out posteriorly to fill most of the anterior segments of the abdomen. Those of *Pseudacanthotermes spiniger* fill nine-tenths of the abdomen. The soldiers of *Globitermes sulfureus* are quite literally walking chemical bombs. Their reservoirs fill the anterior half of the abdomen. When attacking, they eject a large amount of yellow liquid through their mouths, which congeals in the air and often fatally entangles both the termites and their victims. The spray is evidently powered by contractions of the abdominal wall. Occasionally these contractions become so violent that the wall bursts, spraying defensive fluid in all directions.

In still another, independent evolutionary development, members of the termitid subfamily Nasutitermitinae have carried chemical defense to a separate, equally bizarre extreme. In the advanced nasutitermitine species the frontal gland of the head has been enlarged and the surrounding portion of the head capsule drawn out into a conical organ that roughly resembles a great nose on the front of the soldier's head—hence the expressions "nasus" to describe the organ and "nasute soldier" to describe the caste (Figures 14-2 and 14-3). The most primitive nasutitermitine genera, namely, *Syntermes, Cornitermes, Procornitermes, Paracornitermes,* and *Labiotermes,* have typically mandibulate soldiers. Certain phylogenetically intermediate genera, such as *Rhynchotermes* and *Armitermes,* are characterized by soldiers that possess both hooked mandibles and nasute head capsules. These individuals are therefore "double threats" in their defensive roles. According to Sands (1957), the nasus has evolved twice through such an intermediate step. The mandibles have been subsequently reduced in size within several independent phyletic lines. The extreme form of the nasute soldier, in which the mandibles have become small, nonfunctional lobes, originated independently in at least nine instances within eight genera. This remarkable flurry of convergent evolution, together with the outstanding diversity and abundance of the higher Nasutitermitinae in the tropics, is evidence that the nasute technique of chemical defense is highly successful. With the aid of its fontanellar "gun," fired by a contraction of powerful mandibular muscles, a nasute soldier is able to eject the frontal gland material over a distance of many centimeters. The soldier's aim is quite accurate in spite of the fact that it is completely blind. The nature of the nasute soldier's orientation device has yet to be studied, although, by process of elimination, it seems almost certainly to be olfactory or auditory in nature. After firing, the soldier wipes its nasus on the ground and retreats into the nest, apparently lacking enough

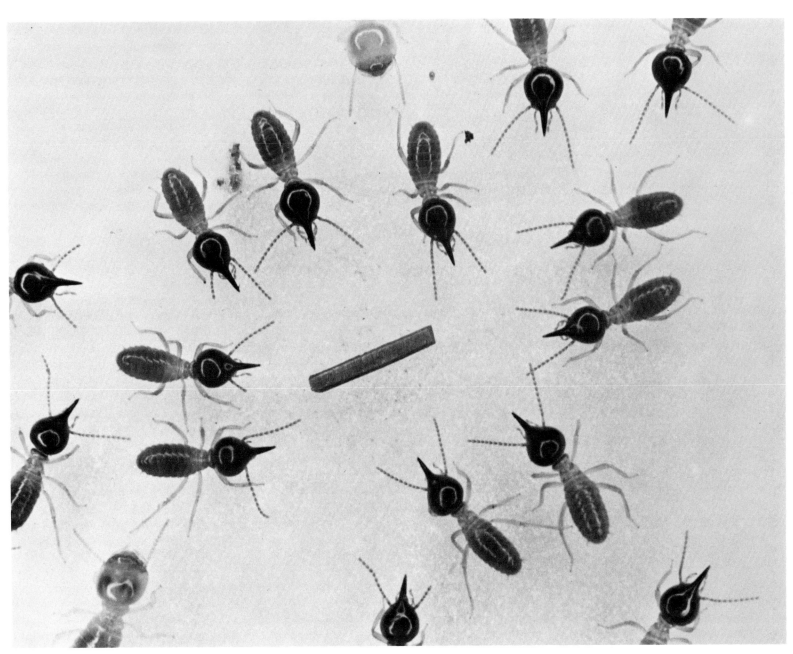

Figure 14-2 Communal defense systems reach their highest expression in the nasutitermitine termites. In the experiment shown here, performed by Thomas Eisner and Irmgard Kriston, the nasute soldiers have been distracted from their foraging columns by the metal bar, which is rotated by a spinning magnet located beneath the platform. Some have already attempted to entangle the bar with sticky chemical secretions sprayed from the spoutlike "nasus" on the head. Two workers, distinguished by the absence of the nasus, stand back from the ring of soldiers at the lower left and top of this photograph. The termites shown belong to the Australian species *Nasutitermes exitiosus*. (Photograph by courtesy of Thomas Eisner.)

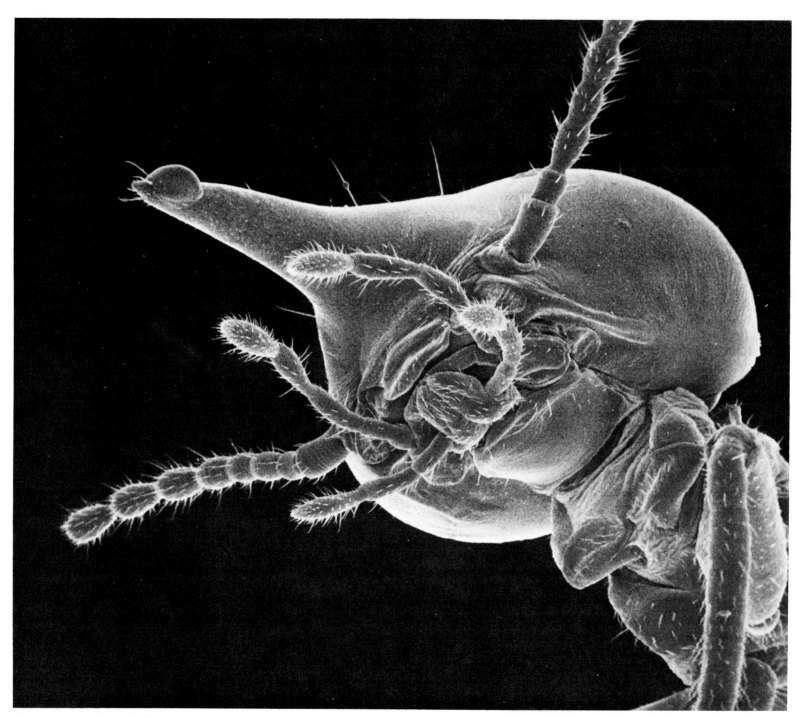

Figure 14-3 The head of a nasute termite soldier *Nasutitermes exitiosus*) seen from below. In this scanning electron micrograph, which magnifies the head 90 times, a droplet of defensive secretion can be seen adhering to the tip of the nasus. (Photograph by courtesy of Thomas Eisner.)

material to make a rapid series of shots. Because nasute soldiers are able to strike and disable an adversary at a considerable distance, they seldom become fatally entangled in their own secretions. They therefore have an advantage over the soldiers of many other termitid species who are forced to apply mandibular gland secretions at short range. According to Ernst (1959), the frontal-gland secretion of *Nasutitermes* is nontoxic and functions solely as a mechanical entrapment device. Moore (1964, 1969), who has studied the chemistry of the Australian species, reports that the defensive secretion consists primarily or wholly of terpenoids. The volatile fraction contains α-pinene as the principal component and β-pinene, limonene, and monocyclic isomers as minor components. The "resinous" fraction consists of a number of closely related polyacetoxy diterpenoids, which become increasingly viscous and sticky when exposed to the air. As the volatile components evaporate, they also serve as alarm substances, so that when one soldier fires at a given target others are likely to attack it also.

The extreme soldier castes of some ant and termite species are so specialized that they function as scarcely more than organs in the body of the colony superorganism. Their existence supports the soundness of the procedure of accepting colony selection as a mechanism, and it seems correct to press further with optimization theory based on the assumption that the mechanism operates generally. For, if selection is mostly at the colony level, and workers are mostly or wholly altruistic with respect to the remainder of the colony, their numbers and behavior can be closely regulated through evolution to approach maximum colony fitness. In the ergonomics theory developed earlier (Wilson, 1968a), I postulated that the mature colony, on reaching its predetermined size, can be expected to contain caste ratios that approximate the *optimal mix*. This mix is simply the ratio of castes that can achieve the maximum rate of production of virgin queens and males while the colony is at or near its maximum size. It is helpful to think of a colony of social insects as operating in somewhat the same way as would a factory constructed inside a fortress. Entrenched in the nest site and harassed by enemies and capricious changes in the physical environment, the colony must send foragers out to gather food while converting the secured food inside the nest into virgin queens and males as rapidly and as efficiently as possible. The rate of production of the sexual forms is an important, but not an exclusive component of colony fitness. Suppose we are comparing two genotypes belonging to the same species. The relative fitness of the genotypes could be calculated if we had the following complete information: the survival rates of queens and males belonging to the two genotypes from the moment they leave the nest on the nuptial flights; their mating success; the survival rate of the fecundated queens; and the growth rates and survivorship of

the colonies founded by the queens. Such complete data would, of course, be extremely difficult to obtain. In order to develop an initial theory of ergonomics, however, it is possible to get away with restricting the comparisons to the mature colonies. In order to do this and still retain precision, it would be necessary to take the difference in survivorship between the two genotypes outside the period of colony maturity and reduce it to a single weighting factor. But we can sacrifice precision without losing the potential for general qualitative results by taking the difference as zero. Now we are concerned only with the mature colony, and, given the artificiality of our convention, the production of sexual forms becomes the exact measure of colony fitness. The role of colony-level selection in shaping population characteristics within the colony can now be clearly visualized. If, for example, colonies belonging to one genotype contain, on the average, 1,000 sterile workers and produce 10 new virgin queens in their entire life span, and colonies belonging to the second genotype contain, on the average, only 100 workers but produce 20 new virgin queens in their life span, the second genotype has twice the fitness of the first, despite its smaller colony size. As a result, selection would reduce colony size. The lower fitness of the first genotype could be due to a lower survival rate of mature colonies, or to a smaller average production of sexual forms for each surviving mature colony, or to both. The important point is that the rate of production can be expected to shape mature colony size and organization to maximize this rate.

The production of sexual forms is determined in large part by the number of "mistakes" made by the mature colony as a whole in the course of its fortress-factory operations. A mistake is made when some potentially harmful contingency is not successfully met—a predator invades the nest interior, a breach in the nest wall is tolerated long enough to desiccate a brood chamber, a hungry larva is left unattended, and so forth. The cost of the mistakes for a given category of contingencies is the product of the number of times a mistake is made times the reduction in queen production per mistake. With this formal definition, it is possible to derive in a straightforward way a set of basic theorems on caste. In the special model, the average output of queens is viewed as the difference between the ideal number made possible by the productivity of the foraging area of the colony and the number lost by failure to meet some of the contingencies. (The model can be modified to incorporate other components of fitness without altering the results.) The evolutionary problem which I postulate to have been faced by social insects can be solved as follows: the colony produces the mixture of castes that maximizes the output of queens. In order to describe the solution in terms of simple linear programming, it is necessary to restate the solution in terms of the dual of the first statement: the colony evolves the mixture of castes that allows it to produce a given number of queens with a minimum quantity of workers. In other words, the objective is to minimize the energy cost.*

The simplest case involves two contingencies whose costs would exceed a postulated "tolerable cost" (above which, selection takes place), together with two castes whose efficiencies at dealing with the two contingencies differ. The inferences to be made from this simplest situation can be extended to any number of contingencies and castes.

The most important step is to relate the total weights, W_1 and W_2, of the two castes in a colony at a given instant to the frequency and importance of the two contingencies and the relative efficiencies of the castes at performing the necessary tasks. By stating the problem as the minimization of energy cost (see Wilson, 1968a), the relation can be given in linear form as follows:

Contingency Curve 1

$$W_1 = \frac{\ln F_1 - \ln k_1 x_1}{\alpha_{11}\ln(1 - q_{11})} - \frac{\alpha_{12}\ln(1 - q_{12})}{\alpha_{11}\ln(1 - q_{11})} W_2$$

Contingency Curve 2

$$W_1 = \frac{\ln F_2 - \ln k_2 x_2}{\alpha_{21}\ln(1 - q_{21})} - \frac{\alpha_{22}\ln(1 - q_{22})}{\alpha_{21}\ln(1 - q_{21})} W_2$$

W_1 is the weight of all members belonging to caste 1 in an average colony.

W_2 is the weight of all members belonging to caste 2 in an average colony.

F_1 and F_2 are the highest tolerable costs due to contingencies 1 and 2.

α_{11} is a constant such that $\alpha_{11}W_1$ gives the average number of individual contacts with a contingency of type 1 by members of caste 1 during the existence of the contingency.

α_{12} is a constant such that $\alpha_{12}W_2$ gives the average number of individual contacts with a contingency of type 1 by members of caste 2 during the existence of the contingency.

α_{21} and α_{22} are constants similar to the above two but with reference to contingencies of type 2.

q_{11} is the probability that, on encountering contingency 1, a worker of caste 1 responds successfully.

q_{12} is the probability that, on encountering contingency 1, a worker of caste 2 responds successfully.

*Levins (1968) has rederived the same theorems in terms of the opposite dual in order to align them with his general theory of fitness sets. His method has pedagogic advantages, but it is more difficult to relate to the underlying behavioral phenomena.

q_{21} and q_{22} are the probabilities of the above two but with reference to contingency 2.

x_1 and x_2 are the average costs (in this case, measured in nonproduction of virgin queens) per failure to meet contingencies 1 and 2, respectively.

k_1 and k_2 are the frequencies of contingencies 1 and 2, respectively, for a given period of time.

I have presented this amount of detail to illustrate one particular form that contingency curves might take, using conventions that relate to intuitively simple ideas concerning behavior. In fact, no contingency curves of actual species have been drawn. At present, the required steps of defining contingencies and measuring their effects in natural populations are technically formidable. The important point is that under a very wide range of conceivable conditions the contingency curves would be linear or almost linear, or at least could be rendered graphically in linear form.

The optimal mix of castes is the one that gives the minimum summed weights of the different castes while keeping the combined cost of the contingencies at the maximum tolerable level. The manner in which the optimal mix is approached in evolution is envisaged as follows. Any new genotype that produces a mix falling closer to the optimum is also one that can increase its average net output of queens and males. In terms of energetics, the average number of queens and males produced per unit of energy expended by the colony is increased. Even though colonies bearing the new genotype will contain about the same adult biomass as other colonies, their average net output will be greater. Consequently, the new genotype will be favored in colony-level selection, and the species as a whole will evolve closer to the optimal mix.

The general form of the solution to the optimal-mix problem is given in Figure 14-4. It has been postulated that behavior can be classified into sets of responses in a one-to-one correspondence to a set of kinds of contingencies. Even if this conception only roughly fits the truth, it is enough to develop a first theory of ergonomics. For example, the graphical presentation in Figures 14-5 and 14-6 shows that so long as the contingencies occur with relatively constant frequencies, it is an advantage for the species to evolve so that in each mature colony there is one caste specialized to respond to each kind of contingency. In other words, one caste should come into being that perfects the appropriate response, even at the expense of losing proficiency in other tasks.

A curious possible effect in the evolution of castes is illustrated in Figure 14-7. This theorem was derived as an answer to the following question: If proliferation and divergence of castes are the expected consequences of selection at the colony level, why have they not reached greater heights throughout the social insects? In fact, these

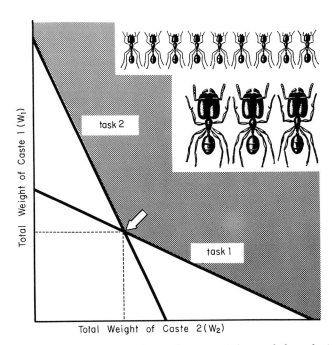

Figure 14-4 This diagram shows the general form of the solution to the optimal-mix problem in evolution. In this simplest possible case, two kinds of contingencies ("tasks") are dealt with by two castes. The optimal mix for the colony, measured in terms of the respective total weights of all the individuals in each caste, is given by the intersection of the two curves. Contingency curve 1, labeled "task 1," gives the combination of weights (W_1 and W_2) of the two castes required to hold losses in queen production to the threshold level due to contingencies of type 1; contingency curve 2, labeled "task 2," gives the combination with reference to contingencies of type 2. The intersection of the two contingency curves determines the minimum value of $W_1 + W_2$ that can hold the losses due to both kinds of contingencies to the threshold level. The basic model can now be modified to make predictions about the effects on the evolution of caste ratios of various kinds of environmental changes. (From Wilson, 1968a.)

qualities vary greatly from genus to genus and even from species to species. The only answer consistent with the theory is that, as in most evolving systems, the various levels reached by individual species are compromises between opposing selection pressures. The obvious pressure that must oppose proliferation and divergence is fluctuation of the environment. From Figure 14-7 we can see that a long-term change can eliminate a caste if the caste that supersedes it (by taking over its tasks through superior numbers) is not very specialized. In this example, contingency 2 has increased in frequency (or importance) enough to shift the contingency curve to the right of the contingency 1 curve intercept of the W_2 axis. Consequently, the number of caste 2 workers required to take care of contingency 2 is also more than

enough to take care of contingency 1. The presence of caste 1 now reduces colony fitness, and if the environmental change is of long duration, caste 1 will tend to be eliminated by colony-level selection. In this case the species tracks the environment to acquire a new optimal mix that just happens to eliminate the superseded caste. Thus, if the critical features of the environment are changing at a rate slow enough to be tracked by the species but too fast to permit much specialization of individual castes, both the number and the degree of specialization of castes will be kept low.

At another level, the critical features of the environment may be changing too fast to be tracked genetically, yet too slowly to provide each colony with a consistent average for the duration of its life. In this case, a mix of specialized castes would be inferior to a few generalized forms able to adapt to new circumstances.

This form of ergonomic theory also reveals two ways in which the consequences of colony-level selection can be the exact opposite of those stemming from individual-level selection. In Figures 14-8 and 14-9 a relation is shown to exist between the prior degree of caste specialization and the magnitude of change in the optimal mix that is invoked by a given change in the environment. The castes repre-

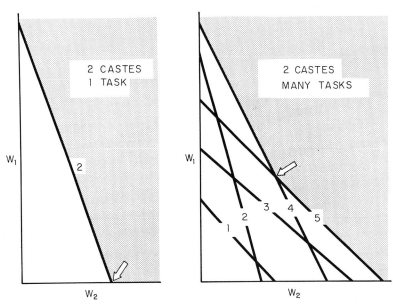

Figure 14-5 The diagram on the left shows that, when there are more castes than tasks, the number of castes will be reduced in evolution to equal the number of tasks. The surplus castes removed will be the least efficient ones (in this case, caste 1). The diagram on the right shows that if there are more tasks than castes, the optimal mix of castes will be determined entirely by those tasks, equal or less in number to the number of castes, which deal with the contingencies of greatest importance to the colony (in this case, tasks 4 and 5). (From Wilson, 1968a.)

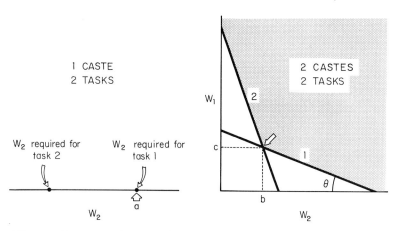

1 CASTE
2 TASKS

W_2 required for task 2

W_2 required for task 1

W_2

a

W_1

2

2 CASTES
2 TASKS

c

1

θ

b

W_2

Figure 14-6 It is always to the advantage of the species to evolve new castes until there are as many castes as contingencies, and each caste is specialized uniquely on a single contingency. This theorem can be substantiated readily by comparing the two graphs in this figure. With the addition of caste 1 in the righthand figure, the total weight of workers is changed from a to $b + c$. Since caste 1 specializes in task 1, θ is acute; therefore, $a - b > c$ and $a > b + c$ for all a, b, and c. (From Wilson, 1968a.)

sented in Figure 14-8 are relatively unspecialized. Task 2 is shown to become somewhat less common (or less important); this results in a shift of the contingency curve toward the origin without a change in slope. As a consequence, the optimal mix changes from one comprised predominantly of caste 2 to one comprised predominantly of caste 1. In contrast, the castes represented in Figure 14-9 are highly specialized, and a shift in the contingency curve results in little change in caste ratios. These models lead to the conclusion that species with initially unspecialized castes will have on the average fewer castes and more variable caste ratios, and this effect will be enhanced in fluctuating environments. The more specialized the castes become in evolution, the more entrenched they become, in the sense that they are more likely to be represented in the optimal mix regardless of long-term fluctuations in the environment. Here we have a peculiar theoretical result of colony-level selection, the opposite of individual-level selection. For in classical population genetic theory, which is based on individual selection, it is the generalized genotypes and species, and not the specialized ones, that are most likely to survive in the face of long-term fluctuation in the environment.

The second peculiar result of colony-level selection, illustrated in Figure 14-10, involves the relation between the efficiency and the numerical representation of a given caste. If, in the course of evolution, one caste increases in efficiency and the others do not, the proportionate total weight of the improving caste will decrease. In

other words, the expected result of colony-level selection is precisely the opposite of that of individual selection, which would be an increase in the more efficient form.

Ergonomic theory will not be easy to test. The required steps of defining contingencies and measuring their effects in natural populations will require closer attention to the biology of insect colonies than has been attempted in the past. Yet I can see no way of probing very deeply into the evolution of castes except by this means, or at least by comparable studies guided by some other, more clever form of ergonomic theory.

There exists a small amount of indirect empirical evidence relevant to the ergonomic theorems just presented. It is the case, for example, that some phyletic ant lines have lost a caste (the soldier caste) secondarily. Although the theory allows for this possibility, it is not proved by its realization. A second, more suggestive piece of evidence is the fact that physical castes are more frequent in tropical ant faunas than in temperate ant faunas. This rule is consistent with the postulate that castes always tend to proliferate in evolution but are simulta-

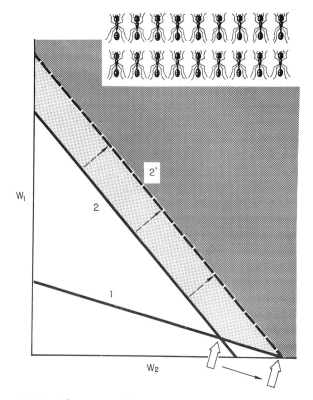

Figure 14-7 A long-term change in the environment can cause the evolutionary loss of a caste, even when the task to which the caste is specialized remains as frequent and important as ever. (From Wilson, 1968a.)

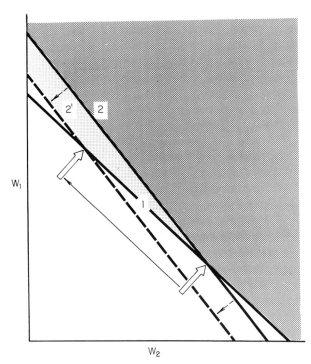

Figure 14-8 If the castes are relatively unspecialized, small but long-term changes in the environment will ultimately result in large evolutionary changes in the optimal caste ratios. (From Wilson, 1968a.)

neously being reduced in response to fluctuations in the environment, the degree of response being proportionate to the degree of fluctuation. Third, it is a fact consistent with the theory, but still far from proving it, that the most specialized castes are found primarily in tropical genera and species. The bizarre soldiers of ant genera such as *Paracryptocerus*, *Pheidole* (*Elasmopheidole*), *Acanthomyrmex*, *Zatapinoma*, and *Camponotus* (*Colobopsis*) and of termite genera such as *Nasutitermes*, *Mirotermes*, *Anacanthotermes*, and *Capritermes* are all but limited to the tropics and subtropics. Polymorphism in temperate ant species, representing the less extreme members of *Pheidole*, *Solenopsis*, *Monomorium*, *Myrmecocystus*, and *Camponotus*, is predominantly of the simpler forms produced by elementary allometry. This climatic correlation is predictable from the theorem that specialization in castes already in existence should increase indefinitely until countered by opposite selective pressures imposed by fluctuations in the environment.

Schopf (1973) has pointed out that ergonomic theory also applies to zooid differentiation and division of labor in colonies of invertebrates. In the ectoprocts, constituting the major element of the old phylum Bryozoa, individual zooids often resemble the beaks of birds (avicularia), whips (vibracularia), and other strange forms (see Chap-

ter 19). Preliminary studies indicate that each type has a distinctive behavior shaped to a particular function. Zooid polymorphism is maximally developed in the most stable environments. The condition was found to be present in 75 percent of the species sampled from the tropics, the Arctic, and the deep sea. By contrast, polymorphism was absent in faunas collected from estuaries, the least stable of the environments studied.

Roles in Vertebrate Societies

We can now consider the key question about roles in vertebrate societies, which is the following: To what extent are age-sex classes and other categories of individuals defined by behavioral profiles comparable to the castes of invertebrates? In other words, can there be an ergonomics of vertebrate societies? The answer, as suggested earlier, lies in the intensity of group selection with reference to behavioral differentiation.

The best way to attack the problem may be to partition the behavioral differences into *direct* and *indirect roles*. A direct role is a particular behavior or set of behaviors displayed by a subgroup that benefits other subgroups and therefore the group as a whole. An

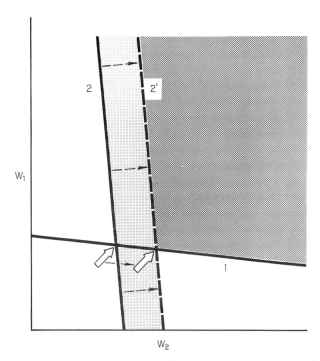

Figure 14-9 The more specialized the castes are in aggregate, the less evolutionary change there will be in the optimal mix in the face of long-term environmental change. (From Wilson, 1968a.)

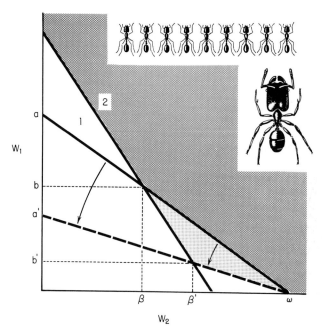

Figure 14-10 If one caste increases in efficiency during the course of evolution and the others do not, the proportionate total weight of the improving caste will decrease. This theoretical result of colony-level selection is the opposite of what might be expected from individual-level selection, which tends to increase improving phenotypes. (From Wilson, 1968a.)

indirect role is behavior that benefits only the individuals that display it and is neutral or even destructive to other subgroups. The direct role is favored by group selection or at least does not run counter to it. It can be detrimental to the individual and the individual's progeny, as in the actions of castes of ants and zooids of invertebrate colonies. In this case favorable group selection almost certainly occurs. Or the direct role can add to individual fitness while at the same time reinforcing group survival or at least not diminishing it. Direct roles favored by group selection are subject to ergonomic optimization with respect to the numbers of individuals playing the role and the intensity with which it is expressed. This is evidently what Gartlan (1968) had in mind when describing primate societies in terms of roles: "The group is an adaptive unit, the actual form of which is determined by ecological pressures. Different roles of relevance to particular ecological conditions are performed by different animals." An indirect role, in contrast, is simply the outcome of selfish behavior that can be manifested by some but not all members of the society. If the magnitude of individual genetic selection is at least comparable to the rate of group extinction opposing it, the role will be maintained in a balanced polymorphic state (see Chapter 5).

But in no sense can the indirect role be ergonomically optimized with reference to the society as a whole.

Most roles so far defined in nonhuman vertebrates are apparently indirect in nature. Consider the "leaders" in flocks of European wood pigeons. They constitute the advancing front of the feeding assemblies, but they are in that position only because they are displaced by the dominant birds who control the center. Because they constantly glance backward toward the advancing group, they eat less and are more prone to starvation in hard times (Murton et al., 1966). Some authors have spoken of the role of young birds and mammals as dispersants of the species, colonizing new terrain and exchanging genes between populations. While it is true that individuals journey farther while they are young, the difference is generally the outcome of their subordinate position in their place of birth. Its adaptive basis is the greater chance it gives young animals to gain territory or to rise to dominance in new places. The role the young play with respect to population dynamics and gene flow is probably wholly indirect. Fruit bats (*Pteropus giganteus*) form large daytime resting aggregations in certain trees in the Asiatic forests. Each male has his own resting position, with subordinate individuals occupying the lower limbs and hence suffering the most exposure to ground-dwelling predators. The subordinate males usually see danger first and alert the remainder of the colony by their excited movements. They serve as very effective sentinels for the group as a whole (Neuweiler, 1969), but their role is clearly indirect in nature. Female hamadryas baboons are half the size of the males and subordinate to them. When troops feed in acacia groves, the males take precedence in gathering flowers and seeds, while the females are able to glean from the smaller branches that cannot support the weight of the males (Kummer, 1971). The "roles" of the two sexes in making full use of the food resource are thus clear-cut, but they are indirect in the sense just defined. Similar examples can be multipled indefinitely.

Cases of direct roles are much harder to find among the vertebrates. Adults of the African wild dog appear to divide labor in a way that benefits the pack as a whole. Adults, including the mother bitch, remain behind with the pups during a chase, and the successful hunters regurgitate meat to them upon returning to the den. Adult male olive baboons cooperate to an impressive degree in policing the area around the troop. When juveniles register alarm or excitement, the nearest adult male investigates the cause. If his own reaction is strong enough, the other adult males rush to his assistance (Rowell, 1967). Silver-backed males of the mountain gorilla play multiple roles in the troops they lead. When a group loses the leader-male, it appears to search for a new one (Schaller, 1963). Perhaps no evidence more strongly suggests the direct nature of a role than the effort by the group to recruit another animal to fill it.

Do castes occur in vertebrate societies in addition to direct roles?

If group selection is strong enough, there is no reason why caste systems cannot have evolved. They might even have a purely physiological basis, as in most social insects. In that case individuals at inception would possess equal potential for development into any caste. Once an animal crossed a certain threshold in growth and differentiation, its caste would be fixed for some period of time. Although physiological caste systems seem intuitively to be the most easily generated and optimized in evolution, we cannot discount the possibility that genes influencing some aspects of behavior exist in a state of balanced polymorphism, for the reason that their carriers benefit noncarriers and therefore the group as a whole. Such genes might be altruistic or nonaltruistic, that is, either counteracting or reinforcing individual-level selection. To put it another way, relatively strong group selection might tend to favor the evolution of genetic diversity within as opposed to between societies. A society with genes near the ergonomic mix would have higher fitness than those away from the mix, including those possessing less diversity. However, specific gene *frequencies* are harder to maintain by selection than specific *genes*, and such genes can individually program physiological caste systems.

Castes in vertebrates, if such exist, should take the form of distinctive physiological or psychological types that recur repeatedly at predictable frequencies within societies. Some would probably be altruistic in behavior—homosexuals who perform distinctive services, celibate "maiden aunts" who substitute as nurses, self-sacrificing and reproductively less efficient soldiers, and the like. The most direct and practicable test of the caste hypothesis is whether phenotypic variance within societies exceeds that of comparable samples from closely related nonsocial or at least less social species. If it does not, the hypothesis is negated; if it does, the hypothesis is supported but still not proved. In cases where greater variance is further associated with higher genetic diversity the possibility of a genetic caste system is indicated. Vertebrate zoologists appear not to have consciously investigated these possibilities, although Trivers' recent theoretical work on parent-offspring conflict envisages celibate and other self-sacrificing behavioral types as one possible outcome of vertebrate kin selection (see Chapter 16). The evidence is sparse and equivocal. Jolicoeur (1959) reports that populations of wolves are highly polymorphic in size and color, while Fox (1972) has detected strong differences among members of the same litter in reactivity, exploratory behavior, and prey-killing ability. Other authors have commented on the existence of striking variation within packs of African wild dogs and of differences in facial features of baboons and chimpanzees that permit human observers to recognize individuals at a glance. Such variation is subject to several competing explanations, but at least it is consistent with the hypothesis of hereditary castes in advanced mammalian societies.

The existence of direct roles and castes in vertebrate societies is thus indicated only by marginal evidence limited to the most social of the mammals. This limitation puts the utility of the role as a scientific concept in considerable doubt. The word can be used in a metaphorical sense, intuitively and changeably defined, but it is not likely to acquire a firm operational definition in the immediate future. The classification of indirect roles remains a formidably difficult—and perhaps useless—task. After a brave start, the primate literature has foundered in a simple listing of categories. Some authors, for example Bernstein and Sharpe (1966) and Crook (1971), virtually equated roles with role profiles. The category was defined by sex, age, and perhaps also some diagnostic social trait, and then its other statistically distinctive qualities were explored. Thus reference was made to the "roles" of the control male, the secondary male, the isolate male, the central female, the peripheral female, and others. Gartlan (1968), in contrast, equated roles with acts of social behavior: territorial vigilance, approaching other troop members in a friendly manner, and so forth. Such orthogonal classifications are multiplicative when taken together and hence increase confusion at an exponential rate. Moreover, in the absence of operational definitions based on ergonomics, categories within one classification can be subdivided right down to the level of the group member, a procedure already approached by Saayman (1971a) and Fedigan (1972).

To make this criticism is not to doubt the value of lists of social behaviors and analyses of behavioral profiles. All that is suggested here is that they be called by their correct names and not obscured by unnecessary reference to roles. If this is true, can it be said that the concept of the role has any useful place at all in vertebrate sociobiology? The answer is a qualified yes. There exist a few patterns of social behavior that can be conveniently labeled roles and treated as separate elements in the analysis of certain vertebrate societies. Two of them, leadership and control, will now be briefly reviewed. It is only necessary to bear in mind that each is a heterogeneous collection of behaviors defined loosely by function across species, and referred to as a role because of its employment by a subgroup of the society in affecting the behavior and welfare of the group as a whole.

Leadership

When zoologists speak of leadership, they usually mean the simple act of leading other group members during movement from one place to another. In many instances such a role is filled casually, even accidentally. Schooling fish, such as mullet and silversides, are "led" from moment to moment by whatever fish happen to be brought to the forward edge by the movement of the school as a whole. Individuals frequently try to turn inward toward the center of the the group, so that the second ranks are brought to the front. When the school encounters a predator or impassable object, the members turn away individually. As a result, the entire school wheels to the side or reverses direction, and fish along the flank or rear become the new leaders (Shaw, 1962). The least-organized bird flocks, for example the feeding groups of starlings, move in a similar fashion, with the leaders often being simply the fastest fliers (Allee, 1942). In flocks of ring doves and jackdaws the most experienced birds in a particular situation take the initiative, and others follow (Lorenz, 1935; Collias, 1950). Leadership in large ungulate groups is also casual and shifting in character. The vanguard of reindeer herds consists chiefly of the most timid and restless individuals, who are first to stop eating, first to rest and chew their cud, and first to get up again (V. M. Sdobnikov, in Allee et al., 1949). Within herds of chital (*Axis axis*) and gaur (*Bos gaurus*) in India, either adult females or males lead, with females predominating. In times of danger the first individual chital who breaks away is followed by the rest (Schaller, 1967).

A few mammalian species possess stronger forms of leadership, more nearly consonant with the role as played by human beings. When members of a wolf pack travel in single file, any one of several individuals can take the lead. But during chases the dominant male assumes command. He directs the attack on prey and sometimes the pursuit after others give up. In one instance observed by Mech (1970), a male brought his pack to a halt by turning and lunging at his followers. Dominant males also take the lead in challenging intruding packs. Herds of red deer are much better organized than those of most other ungulates. A fertile hind consistently leads the group, and her followers sometimes include even young stags. A second fertile hind normally brings up the rear (Darling, 1937). Herds of the African elephant are organized in a nearly identical fashion. Clans of mountain sheep are also highly structured. The adult males and females usually stay apart except during the rutting season, and leadership is assumed in each group by the largest and oldest individuals (Geist, 1971a).

In a few species leadership shifts moment by moment from one individual or subgroup to another in a predictable manner according to changing circumstances. Zebra family groups are small and organized into strong dominance hierarchies, with the alpha male in the topmost position. When the group goes to water holes, the stallion leads; but when it departs, the dominant mare takes over the lead and the stallion brings up the rear (Klingel, 1968). The shift seems to be adaptive, since it always places the stallion between his family and the water hole, where the largest number of predators are concentrated. Yearling cattle go through a patterned shift less easily explained. When the herd is moving casually and freely, the middle- and high-ranking individuals are near the front, with the former usually taking the actual lead. But when the cattle are forced into movement, the low-ranking individuals go first (Beilharz and Mylrea, 1963).

Control

Since the role concept was first introduced into the primate literature, the key paradigm has been the control animal. The point stressed by Bernstein, Crook, Gartlan, and other writers is that dominance and the control function are separable. The two forms of interaction are generally correlated but nevertheless distinct. In some animals, for example the squirrel monkey *Saimiri sciureus*, a control animal exists but there is no overt dominance order. This is an important generalization, but unfortunately it has been semantically obscured. There has been a failure to distinguish between one or more control behaviors, which can be defined if need be down to the neuromuscular mechanisms, and the behavior profile of control animals. The elemental behavior pattern constituting the role is the intervention in aggressive episodes with the result of reducing or halting them. In monkey groups control is almost always achieved by threat or punishment. Kawamura (1967) describes it in the Japanese macaque as follows: "When one monkey of the troop is being attacked by another and emits an exaggerated cry for help, the leader males quickly rush in to attack and punish the aggressor. When the leaders arrive on the scene, many other monkeys flatter them while the aggressor attacks still another monkey as a new enemy, thereby adding to the confusion. Because the monkeys create such a furor, observers wonder at times whether the real purpose of the display is to punish the aggressor. Usually, however, the leaders do eventually find the original offender and punish it, even though it appears as though they are no longer angry with it." If an animal performing control behavior is characterized further, he is usually found to be prominent in leading the group in defense against intruders and to serve as an attention focus for other members of the group. But it must be admitted that these are additional roles and not part of control behavior per se, unless we care to broaden the definition of control to the point of uselessness. The correct way to analyze roles is to define them as discrete behavior patterns in particular species, to establish their degree of correlation within the group members, and finally to identify categories of individuals according to the roles usually invested in them. The correspondence of role profiles to certain age and sex groups, or even to castes, is an important but separable issue.

Roles in Human Societies

The very poverty and vagueness of roles in nonhuman primate societies underscores their richness and importance in human behavior. Human existence, as Erving Goffman and his fellow microsociologists have argued, is to a large extent an elaborate performance of roles in the presence of others. Each occupation—the physician, the judge, the waiter, and so forth—is played just so, regardless of the true workings of the mind behind the persona. Significant deviations are interpreted by others as signs of mental incapacity and unreliability.

Role playing in human beings differs from that in other primates, including even the chimpanzee, in several ways intimately connected to high intelligence and language. The roles are self-conscious: the actor knows that he is performing for the sake of others to some degree, and he continuously reassesses his persona and the impact his behavior is having on others. Models from his own social class and occupation are chosen and imitated. Role playing is thorough. The individual may change his clothing and personality and even his manner of speech while off the job, but while on it his performance must be consistent or others will suspect him of insincerity or incompetence. Human roles are very numerous. In advanced societies each individual is familiar with the behavioral norms of scores or hundreds of occupations and social positions. Division of labor is based on these memorized distinctions, in a fashion analogous to the physiological determination of castes in social insects. But whereas social organization in the insect colonies depends on programmed, altruistic behavior by an ergonomically optimal mix of castes, the welfare of human societies is based on trade-offs among individuals playing roles. When too many human beings enter one occupation,

their personal cost-to-benefit ratios rise, and some individuals transfer to less crowded fields for selfish reasons. When too many members of an insect colony belong to one caste, various forms of physiological inhibition arise, for example the underproduction or overproduction of pheromones, which shunt developing individuals into other castes.

Nonhuman vertebrates lack the basic machinery to achieve advanced division of labor by either the insect or the human methods. Human societies are therefore unique in a qualitative sense. They have equaled and in many cultures far exceeded insect societies in the amount of division of labor they contain. We can speculate that if the evolutionary trajectory of higher nonhuman primates were now to be continued beyond the chimpanzee, it would reach a role system similar to the human model. With an increase in intelligence would come the capacity for language, the consciousness of personae, the long memories of personal relationships, and the explicit recognition of "reciprocal altruism" through equal, long-term trade-offs. Did in fact such qualities emerge as a consequence of higher intelligence during human evolution? Or was it the other way around—intelligence constructed piece by piece as an enabling device to create the qualities? This distinction, which is not trivial, will be explored further in the more extended discussion of man in Chapter 27.

Chapter 15 **Sex and Society**

Sex is an antisocial force in evolution. Bonds are formed between individuals in spite of sex and not because of it. Perfect societies, if we can be so bold as to define them as societies that lack conflict and possess the highest degrees of altruism and coordination, are most likely to evolve where all of the members are genetically identical. When sexual reproduction is introduced, members of the group become genetically dissimilar. Parents and offspring are separated by at least a one-half reduction of the genes shared through common descent and mates by even more. The inevitable result is a conflict of interest. The male will profit more if he can inseminate additional females, even at the risk of losing that portion of inclusive fitness invested in the offspring of his first mate. Conversely, the female will profit if she can retain the full-time aid of the male, regardless of the genetic cost imposed on him by denying him extra mates. The offspring may increase their personal genetic fitness by continuing to demand the services of the parents when raising a second brood would be more profitable for the parents. The adults will oppose these demands by enforcing the weaning process, using aggression if necessary. The outcomes of these conflicts of interest are tension and strict limits on the extent of altruism and division of labor.

The strong tendency of polygamous species to evolve toward sexual dimorphism reinforces this canonical genetic constraint. When sexual selection operates among males, adults become larger and showier, and their behavior patterns and ecological requirements tend to diverge from those of the females. One consequence is a partitioning of the incipient society, not into castes designed to promote the efficiency of the society but into secondary sex roles that enhance individual as opposed to group genetic fitness. The antagonism between sex and sociality is most strikingly displayed in the social insects. Only in some of the higher termites is caste determination based on sex differences. Specifically, in the primitive nasutitermitine genus *Syntermes* and the fungus-growing Macrotermitinae the large workers are males and the small ones females; in the amitermitine *Microcerotermes* and in the higher Nasutitermitinae the reverse is true. Also, among a majority of the nasutitermitine species the soldiers are normally all males, while in the Macrotermitinae and Termitinae they are normally all females. In contrast, caste determination does not appear to be linked to sex throughout the remainder of the higher termites or in the lower termite families, the Mastotermitidae, Kalotermitidae, Hodotermitidae, and Rhinotermitidae. In the social Hymenoptera, comprised of the ants and social bees and wasps, members of the sterile castes are invariably females. Males cannot in any reasonable sense be considered castes. They are highly specialized for the single act of insemination, which normally takes place outside the nest. While still within the nest prior to mating, males live a mostly parasitic existence, cared for by the female members of the colony.

The inverse relation between sex and social evolution becomes still clearer when a phylogenetic survey of the entire animal kingdom is conducted. The vertebrates are all but universally sexual in their mode of reproduction. Judging from Uzzell's recent review (1970), the relatively few cases of parthenogenetic populations recorded in fishes, amphibians, and lizards are local derivatives that do not evolve far on their own. And with the exception of man, vertebrates have assembled societies that are only crudely and loosely organized in comparison with those of insects and other invertebrates. Sex is a constraint overcome only with difficulty within the vertebrates. Sexual bonds are formed by a courtship process typically marked in its early stages by a mixture of aggression and attraction. Monogamy, and especially monogamy outside the breeding season, is the rare exception. Parent-offspring bonds usually last only to the weaning period and are then often terminated by a period of conflict. Social ties beyond the immediate family are mostly limited to a few mammal groups, such as canids and higher primates, that have sufficient intelligence to remember detailed relationships and thereby to form alliances and cliques. Even these are relatively unstable and in most species mixed with elements of aggression and overt self-serving.

The highest forms of invertebrate sociality are based on nonsexual reproduction. The phylogenetic groups possessing the highest degrees of caste differentiation, namely the sponges, coelenterates, ectoprocts, and tunicates, are also the ones that create new colony members by simple budding. The social insects reproduce primarily by sexual means, and the limited amounts of conflict that occur within the colonies can be traced to genetic differentiation based on sexual reproduction. The Hymenoptera, the order in which advanced social life has most frequently originated, is also characterized by haplodiploidy, a mode of sex determination that causes sisters to be more closely related genetically to each other than parents and offspring. According to prevailing theory (see Chapter 20), this peculiarity accounts for the fact that the worker castes of ants, bees, and wasps are exclusively female. Thus increased sociality in insects appears to be based on a moderation of the shearing force of sexuality. In the invertebrates as a whole, sociality is also loosely associated with hermaphroditism. Groups in which the two conditions coexist include the sponges (Porifera), corals (Anthozoa), ectoprocts, and sessile tunicates. However, a few colonial groups are not hermaphroditic, while many hermaphrodites are noncolonial. (A thorough general review of hermaphroditism, with an investigation of its adaptive significance, is provided by Ghiselin, 1969.)

In short, social evolution is constrained and shaped by the necessities of sexual reproduction and not promoted by it. Courtship and sexual bonding are devices for overriding the antagonism that arises automatically from genetic differences induced by sexual reproduction. Because an antagonistic force is just as important as a promotional one, the remainder of this chapter will present a systematic review of the current theory of the evolution of sex and its multifaceted relationships to social behavior.

The Meaning of Sex

Sexual reproduction is in every sense a consuming biological activity. Reproductive organs tend to be elaborate in structure, courtship activities lengthy and energetically expensive, and genetic sex-determination mechanisms finely tuned and easily disturbed. Furthermore, an organism reproducing by sex cuts its genetic investment in each gamete by one-half. If an egg develops parthenogenetically, all of the genes in the resulting offspring will be identical with those of the parent. In sexual reproduction only half are identical; the organism, in other words, has thrown away half its investment. There is no intrinsic reason why gametes cannot develop into organisms parthenogenetically instead of sexually and save all of the investment. Why, then, has sex evolved?

It has always been accepted by biologists that the advantage of sexual reproduction lies in the much greater speed with which new genotypes are assembled. During the first meiotic division, homologous chromosomes typically engage in crossover, during which segments of DNA are exchanged and new genotypic combinations created. The division is concluded by the separation of the homologous chromosomes into different haploid cells, creating still more genetic diversification. When the resulting gamete is fused with a sex cell from another organism, the result is a new diploid organism even more different than the gamete from the original gametic precursor. Each step peculiar to the process of gametogenesis and syngamy serves to increase genetic diversity. To diversify is to adapt; sexually reproducing populations are more likely than asexual ones to create new genetic combinations better adjusted to changed conditions in the environment. Asexual forms are permanently committed to their particular combinations and are more likely to become extinct when the environment fluctuates. Their departure leaves the field clear for their sexual counterparts, so that sexual reproduction becomes increasingly the mode.

The precise means by which this adaptability is rewarded is less certain. Two hypotheses have been proposed, called by Maynard Smith the long-term and the short-term explanations, respectively. The long-term explanation first took form in the writings of August Weismann, R. A. Fisher, and H. J. Muller, and was given quantitative expression by Crow and Kimura (1965). In essence it says that entire populations evolve faster when they reproduce by sex, and as a result they will prevail over otherwise comparable asexual populations. Suppose that two favorable mutations, $a \rightarrow a'$ and $b \rightarrow b'$, occur at very low frequencies on different loci. In asexual populations the frequency with which the most favored combination, a'/b', is assembled is the product of the two mutation rates. Because the rates are very low, this event might never occur. However, in sexual populations the rate of combination is much higher, because a'/b' can be generated not only by coincident mutations but also by the mating of an a'-bearing individual with a b'-bearing individual. Maynard Smith (1971) refined the Crow-Kimura model to show that if N is the size of the population, l the number of loci at which favorable mutations are possible but have not yet occurred, and μ the mutation rate per locus, sexual reproduction will accelerate evolution provided that $N > 1/10\mu$. Also, in the case of very large populations, say, on the order of 10^7 or greater, evolution in the sexual population will be accelerated by approximately l. The process is further speeded when two populations invade a new environment simultaneously and interbreed to combine notably different sets of genes. The plausibility of the argument is improved by this latter, more dynamic version of the model. It is well known that the ranges of all but the most conservative, K-selected species are constantly expanding and contracting. During periods of expansion, propagules from neighboring populations can be expected to mingle repeatedly. If they interbreed, their offspring will constitute the cutting edge in the evolution of the species as a whole.

The alternative, "immediate-explanation" argument has been developed most cogently by G. C. Williams (1966a) as part of his overall critique of group-selection theory. According to this hypothesis, sexual reproduction evolves because it permits an individual parent to diversify its own offspring and thus meet unpredictable changes in the environment encountered from one generation to the next. Consider an asexual organism that is heterozygous at a particular locus, say, one possessing the genotype a/b. It is capable of producing only a/b offspring, and its fitness is thus dependent on the environment being favorable for this one genotype. In contrast, an a/b sexual organism mating with another a/b organism can generate three genotypes among its offspring: a/a, a/b, and b/b. The sexual strain has a better chance of meeting contingencies than does the asexual strain. If, for example, the environment changes so as to permit only b/b to survive, the sexual strain will persist and the asexual one will become extinct. This hypothesis is in accord with peculiarities in the life cycle of organisms that undergo alternation of generations. There are many kinds of animals, such as the freshwater hydras and aphids, which breed asexually when times are favorable for rapid local population growth. This is the part of the life cycle in which social organization is most likely to appear. But as the environment deteriorates, or a change in photoperiod portends the approach of winter, the animals shift to sexual reproduction followed by dispersal and encystment or some other form of dormancy. In other words, the sexual phase of the life cycle spreads the organisms out, diversifies them

genetically, and prepares them physiologically for hard times (see Bonner, 1965).

Maynard Smith has weakened the credibility of the immediate-explanation hypothesis somewhat by demonstrating that in order to favor the evolution of sexual as opposed to asexual reproduction, the environment must be unpredictable on a generation-to-generation basis. This means that biologically potent variables, such as temperature, humidity, insolation, and so forth, must change signs frequently. Only under this condition will new combinations of genes, and the sexual process that creates them, be favored strongly enough to add genetic fitness at the individual as opposed to the population level. Maynard Smith interprets such rapid fluctuations in sign to be an extreme, improbable case. It is indeed extreme but may not be improbable. Features of the environment of major importance in adaptation are numerous enough, fluctuate rapidly enough, and may well be poorly correlated enough to create the needed conditions in a majority of species. At this early juncture in the development of the theory it should be stressed that the long-term and immediate explanations of the origin of sexuality are not incompatible. The relative weight of their influence is likely to vary according to the predictability of the environment and certain population characteristics of the evolving species. Sexuality will be favored by a lowered autocorrelation in environmental conditions, by the more intense action of natural selection, and by lower mutation rates and higher dispersal rates within the population. Clearly, the biologies of most kinds of organisms, from bacteria to elephants, lie within the envelope of these variables that favors sexual reproduction (Williams and Mitton, 1973). What varies, and is relevant to sociobiology, is the intensity of the process, as measured by the degree of outbreeding, the amount of dispersal before and after reproduction, and the amount of time devoted to sexual reproduction. Each parameter can be viewed as an adaptation in itself, never far removed from the direct influence of the environment.

Evolution of the Sex Ratio

Why are there usually just two sexes? The answer seems to be that two are enough to generate the maximum potential genetic recombination, because virtually every healthy individual is assured of mating with a member of another (that is, the "opposite") sex. And why are these two sexes anatomically different? Of course in many microorganisms, fungi, and algae, they are not; gametes identical in appearance are produced (isogamy). But in the majority of organisms, including virtually all animals, anisogamy is the rule. Moreover, the difference is usually strong: one gamete, the egg, is relatively very large and sessile; the other, the sperm, is small and motile. The adaptive basis of the differentiation is division of labor enhancing individual fitness. The egg possesses the yolk required to launch the embryo into an advanced state of development. Because it represents a considerable energetic investment on the part of the mother the embryo is often sequestered and protected, and sometimes its care is extended into the postnatal period. This is the reason why parental care is normally provided by the female, and why most animal societies are matrifocal. The spermatozoan is specialized for searching out the egg, and to this end it is stripped down to the minimal DNA-protein package powered by a locomotory flagellum. Scudo (1967), entirely on the basis of an analysis of the searching role of the sperm, concluded that anisogamy must be developed to a high degree before its advantages outweigh those of the ancestral state of isogamy.

It is also generally profitable for parents to produce equal numbers of offspring belonging to each sex. Such mechanisms as XY and XO sex determination (where X and Y represent sex chromosomes and O denotes the absence of a chromosome) are not to be viewed as some inevitable result of chromosome mechanics but rather as specialized devices favored by natural selection because they generate 50/50 sex ratios with a minimum of complication. The evolutionary process thought to underlie the 50/50 ratio was first modeled by R. A. Fisher (1930). In barest outline "Fisher's principle" can be stated as follows. If male births in a population are less frequent than females, each male has a better chance to mate than each female. All other things being equal, the male is more likely to find multiple partners. It follows that parents genetically predisposed to produce a higher proportion of males will ultimately have more grandchildren. But the tendency is self-negating in the population as a whole, since the advantage will be lost as the male-producing gene spreads and males become commoner. As a result the sex ratio will converge toward 50/50. An exactly symmetric argument holds with reference to the production of females. Subsequent authors, notably MacArthur (1965), Hamilton (1967), and Leigh (1970), have refined and extended this model to the point where the following more precise statement can be made. Ideally a parent will not produce equal *numbers* of each sex; it should instead make equal *investments* in them. If one sex costs more than the other, the parent should produce a correspondingly smaller proportion of offspring belonging to it. Ordinarily, cost can be assessed in amounts of energy expended. Thus if a newborn female weighs twice as much on the average as a male, and no further parental investment is made after birth, the optimal sex ratio at birth should be in the vicinity of 2 males/1 female. Probably an even more precise assessment than energy expenditure is reproductive effort, the decrement in future reproductive potential as a consequence of the present effort (see Chapter 4). When parental care is added, differences in the amount of care devoted to the two sexes must be added to the deficit side of the ledger. If a daughter, for example, proves twice as costly to raise to independence as a son,

the optimum representation of females among the offspring is cut by one-half. Once parental care ends, differential mortality between the sexes has no effect on the optimum sex ratio.

Other selection pressures can intervene to shift the ratio away from numerical parity. Parasitic species that found populations with small numbers of inseminated females are not bound by Fisher's principle (Hamilton, 1967). Because a large percentage of matings are between sibs, many of the males seeking mates will be in competition with other males who share sex-determining genes by common descent. In the parasitic life style it is advantageous to produce as many inseminated females as possible, even at the expense of unbalancing the initial sex ratio in favor of females. This advantage will override the selection working to restore male representation to parity, since the Fisher effect is weakened by inbreeding. Hamilton proved that under such conditions the "unbeatable" sex ratio will be $(n - 1)/kn$, where k is either 1 or 2, depending on the mode of sex inheritance, and n is the number of females founding the population. (Sex ratios are conventionally given as male-to-female.) When $n = 1$, the ideal ensemble is all-female, but the practical solution is either gynandromorphism or the production of a single male capable of fertilizing all of his sisters. The parasitic Hymenoptera appear to have solved this problem by haplodiploidy, in which males originate from unfertilized eggs and females from fertilized ones. A female has the capacity to control the sex of each offspring simply by "choosing" whether to release sperm from her spermatheca, the sperm-storage organ, just before the egg is laid. This control is used by some hymenopterous species to yield other sex ratios appropriate to special circumstances. The social bees, wasps, and ants ordinarily produce males only prior to the breeding season, reverting to all-female broods during the remainder of the year. A common pattern seen in parasitic wasps is the production of all-male broods on small or young hosts and an increasing proportion of females on hosts capable of supporting a larger biomass (Flanders, 1956; van den Assem, 1971).

With physiological control of sex determination so prominently developed in the insects, the possibility should not be overlooked that it also occurs at least to a limited extent in the vertebrates. Trivers and Willard (1973) have constructed an ingenious argument to reveal which peculiarities can be expected to result from such an adaptive distortion of the sex ratio. Their reasoning proceeds syllogistically as follows:

1. In many vertebrate species, large, healthy males mate at a disproportionately high frequency, while many smaller, weaker males do not mate at all. Yet nearly all females mate successfully.

2. Females in the best physical condition produce the healthiest infants, and these offspring tend to grow up to be the largest, healthiest adults.

3. Therefore, females should produce a higher proportion of males

when they are healthiest, because these offspring will mate most successfully and produce the maximum number of grandchildren. As the females' condition declines, they should shift increasingly to the production of daughters, since female offspring will now represent the safer investment.

The first two propositions have been documented in rats, sheep, and human beings. The rather surprising conclusion of the argument (no. 3) is also consistent with the evidence. It provides a novel explanation for some previously unexplained data from mink, pigs, sheep, deer, seals, and human beings. For example, in deer and human beings environmental conditions adverse for pregnant females are associated with a reduced sex ratio, favoring the birth of daughters. The most likely mechanism is differential mortality of the young in utero. It is known that stress induces higher male fetal mortality in some mammals, especially during the early stages of pregnancy. The ultimate cause of the mortality could be natural selection in accordance with the Trivers-Willard principle.

In a few animal species ratios are stabilized by the capacity of individuals to change sex as a response to either the sex or social status of others. Fishes of the families Labridae, Scaridae, and Serranidae are capable of rapid sex reversals in either direction, in some cases switching back and forth in concert with the reverse changes in single partners. Thus the partners literally exchange sex with each other. As strange as this seems, it is by no means the most bizarre adaptation. Social groups of the tropical Pacific labrid *Labroides dimidiatus* consist of one male and a harem of females that occupy a common territory. The male suppresses the tendency of the females to change sex by aggressively dominating them. When he dies, the dominant female in the group immediately changes sex and becomes the new harem master (Robertson, 1972).

In considering the subtleties of parental investment, one must not overlook the equally important role of other demographic processes in the determination of sex ratios. The adult sex ratio is in fact a product of three quantities: the ratio at birth, the difference in maturation times between males and females, and differential mortality. All three, and not just the initial sex ratio, can be expected to be functions of sexual selection (Trivers, 1972). Differential maturation and mortality should be counted among the results of the social system rather than as independent variables affecting it.

Sexual Selection

The final question in our basic series about the nature of sex is: Why do the sexes differ so much? The traits of interest are the secondary sexual characteristics, which occur in addition to the purely functional differences in the gonads and reproductive organs. The males of many species are larger, showier in appearance, and more aggressive

than the females. Often the two sexes differ so much as to seem to belong to different species. Among the ants and members of such aculeate wasp families as the Mutillidae, Rhopalosomatidae, and Thynnidae, males and females are so strikingly distinct in appearance that they can be matched with certainty to species only by discovering them *in copula*. Otherwise experienced taxonomists have erred to the point of placing them in separate genera or even families. The ultimate vertebrate case is encountered in four families of deep-sea angler fishes (Ceratiidae, Caulophrynidae, Linophrynidae, Neoceratiidae) in which the males are reduced to parasitic appendages attached to the bodies of the females.

Part of the solution to the mystery of sexual divergence was supplied by Charles Darwin in his concept of *sexual selection*, first developed at length in *The Descent of Man and Selection in Relation to Sex* (1871). According to Darwin, sexual selection is a special process that shapes the anatomical, physiological, and behavioral mechanisms that function shortly before or at the time of mating and serve in the process of obtaining mates. He excluded selection that leads to the evolution of such primary reproductive traits as the form of the male gonads or the egg-laying behavior of females. Darwin reasoned that competition for mates among the members of one sex leads to the evolution of traits peculiar to that sex. Two distinct processes were judged to be of about equal importance in the competition. They are, in Julian Huxley's (1938) phraseology, *epigamic selection*, which consists of the choices made between males and females, and *intrasexual selection*, which comprises the interactions between males or, less commonly, between females. To use Darwin's own words, the distinction is between "the power to charm the females" and "the power to conquer other males in battle." As early as 1859, when he first used the expression "sexual selection," Darwin envisaged it as basically different from most forms of natural selection in that the outcome is not life or death but the production or nonproduction of offspring.

Pure epigamic selection is not easy to document in the field. The displays of male birds are ordinarily directed at both males and females, and sexual selection is based as much on the territorial exclusion of rival males as on competition for the attention of potential mates. Epigamic selection unalloyed by intermale aggression can be seen during part of the courtship rituals of the ruff *Philomachus pugnax*, a European shore bird. The males are highly variable in color and display frenetically on individual territories that are grouped tightly together in a communal arena (see Figure 15-1). The rivals scuttle about with their ruffs expanded and wings spread and quivering. Sometimes they pause to touch their bills to the ground or shudder their entire bodies. Females wander singly or in groups from territory to territory, expressing their willingness to mate by crouching. The possession of a territory is not essential in all cases. Females

Figure 15-1 The competitive epigamic display of the ruff *Philomachus pugnax*. The males occupy small territories and display to females wandering through them. This is the only species of bird in which such marked individual variation in plumage occurs. (From Lack, 1968.)

have been observed to follow individual satellite males as they wandered from the territory of one dominant male to the territory of another (Hogan-Warburg, 1966). True epigamic selection also occurs in *Drosophila*. The *yellow* mutant of *D. melanogaster* is characterized not only by the altered body color from which it draws its name but also by subtle alterations in male courtship activity. One step in the display is wing vibration, a ritualized flight movement which is perceived by the female's antennae. The vibration bouts of *yellow* males are shorter in duration and spaced further apart than those of the normal genotypes, and they are less successful in obtaining the appropriate response from the female (Bastock, 1956). Maynard Smith (1956) obtained a similar result when he compared the performances of *D. subobscura* males from inbred and outbred lines. A typical outbred male displayed greater "athletic ability" in maintaining contact with the female and attempting to elicit the appropriate responses from her. The movements required during this exchange are difficult to perform. The male first taps the female on the head with his front legs, then moves around to approach her head-on while extending his proboscis. The female sidesteps rapidly back and forth and the male must shift to maintain his position facing her. Outbred males simply show greater vigor and skill in executing these maneuvers. Heritability in virtually every component of mating behavior has been documented in *Drosophila*, and it is fairly easy to demonstrate differences in mating performances between strains (Kessler, 1966; Petit and Ehrman, 1969).

Epigamic competition can be based on criteria other than showiness and athletic prowess during courtship. Disassortative mating can have the same effect. As the frequencies of certain genes decline, their bearers become increasingly favored. If a frequency exists below

which the rarer genotypes are more successful at mating, a state of balanced genetic polymorphism is reached. The phenomenon has been extensively documented in *Drosophila* (Petit and Ehrman, 1969), to the point that it now appears to be a general although not universal phenomenon within the genus. One especially interesting example from *D. melanogaster* is given in Figure 15-2. The U-shaped selection curve of the white mutant, dipping below parity near the center, establishes two equilibrium frequencies. When the starting frequency of *white* males is less than 80 percent, the frequency will tend to move toward a stable equilibrium of 40 percent. If the starting frequency is about 80 percent, the *white* males will tend to increase still more on the road to fixation. In the laboratory populations, other selective pressures besides epigamic competition work on the *white* mutants, and fixation is not reached.

Sexual selection need not be based on polygamy. Darwin (1871) imagined one process that can equally well be based on monogamous matings:

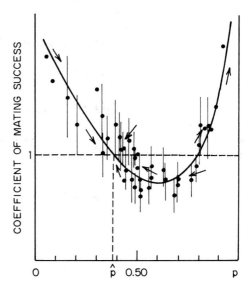

Figure 15-2 Frequency-dependent sexual selection in the fruit fly *Drosophila melanogaster*. The mating success of the *white* mutant (causing white eye color) is plotted as a function of the frequency of *white* males in the male population. When the coefficient of mating success is 1, *white* and normal males are equally successful at mating; when the coefficient is above 1, *white* males are more successful, and when it is below 1 they are less successful. When the frequency of *white* males begins below 80 percent, it tends to seek an equilibrium value of 40 percent. When the starting point is greater than 80 percent, the *white* gene is increased still further. (Modified from Petit and Ehrman, 1969.)

Let us take any species, a bird for instance, and divide the females inhabiting a district into two equal bodies, the one consisting of the more vigorous and better nourished individuals, and the other of the less vigorous and healthy. The former, there can be little doubt, would be ready to breed in the spring before the others. There can also be no doubt that the most vigorous, best nourished and earliest breeders would on an average succeed in rearing the largest number of fine offspring. The males, as we have seen, are generally ready to breed before the females; the strongest, and with some species the best armed of the males, drive away the weaker; and the former would then unite with the more vigorous and better nourished females, because they are the first to breed. Such vigorous pairs would surely rear a larger number of offspring than the retarded females, which would be compelled to unite with the conquered and less powerful males, supposing the sexes to be numerically equal; and this is all that is wanted to add, in the course of successive generations, to the size, strength and courage of the males, or to improve their weapons.

In short, females will tend to select males on the basis of breeding time, with an earlier time being correlated with greater fitness in both sexes. Lack (1968) criticized the model with a counterargument that breeding time is ultimately determined by the availability of food. But the criticism is not relevant. By means of a formal model, O'Donald (1972) proved that so long as a correlation exists between breeding time and fitness, females will tend to evolve to breed at the time the most fit males are active. Under some circumstances evolution will proceed rapidly even when the intrinsic fitness of the females themselves does not vary with breeding time. The selection intensity is notably frequency-dependent. When a sufficiently small fraction of females exercise a choice at mating time, the superior males at first enjoy a strong advantage, but the edge soon dwindles to near zero. When a large number of females make choices, the superior genotype spreads rapidly, as intuition suggests. In either case the mean breeding time of both sexes will eventually evolve toward the environmental optimum, of which Darwin's early session is a special case.

Pure epigamic display can be envisioned as a contest between salesmanship and sales resistance. The sex that courts, ordinarily the male, plans to invest less reproductive effort in the offspring. What it offers to the female is chiefly evidence that it is fully normal and physiologically fit. But this warranty consists of only a brief performance, so that strong selective pressures exist for less fit individuals to present a false image. The courted sex, usually the female, will therefore find it strongly advantageous to distinguish the really fit from the pretended fit. Consequently, there will be a strong tendency for the courted sex to develop coyness. That is, its responses will be hesitant and cautious in a way that evokes still more displays and makes correct discrimination easier.

In intrasexual selection, which is based on aggressive exclusion among members of the courting sex, the matter is settled in a more direct way. A member of the passive sex simply chooses the winner or, to put the matter more realistically, it chooses from among a group of winners who represent a small subset of the potential mates. By picking a winner, the individual not only acquires a more vigorous partner but shares in the resources guarded by it. The latter consideration can be overriding. Males of the long-billed marsh wren (*Telmatodytes palustris*) attempt to stake out territories in stands of cattail, which provide the richest harvest of the aquatic invertebrates on which the birds feed. There they build as many nests and attract as many females as they can. Strong indirect evidence compiled by Verner and Engelsen (1970) suggests that the females choose territories according to the richness of the food and that they somehow are able to assess this quality without reference to the displays of the males. The richer the territory, the easier it is for the male to secure food, and the more time it has to build and maintain nests. Verner and Engelsen believe that the number of visible nests serves as a primary visual index.

When resources are not part of the bargain, intrasexual conflict often evolves onward to acquire a style and intensity impressive even to the most hardened human observer. Grouse, capercaillie, and most other tetraonid birds are highly polygamous. The males compete on communal display grounds, where only a tiny fraction succeed in inseminating the females. The young are raised exclusively by the females in areas well removed from the display grounds. Consequently, for the male everything turns on prowess at the display grounds. Here is John W. Scott's description of male fighting in the prairie sharp-tailed grouse, *Pedioecetes phasianellus* (Scott, 1950).

Dominance is primarily determined by vicious fighting between cocks. Both wings and beaks are used in a rapid interchange of blows, too rapid for the eye to see. Feathers are frequently pulled out. The battles begin suddenly and continue for some seconds without pause. Sometimes, after a brief pause, the fighting continues as viciously as ever. After two or three rounds like this, one cock gives up, turns, and runs to escape, as rapidly as he can, pursued by the victor, who continues to peck at the vanquished even in pursuit. I have seen this pursuit extend 100 feet or more. By this time, other cocks join in the pursuit and the victim may be chased out of the area, the victor in the meantime returning to his accepted location. A kind of gang fighting also occurs when the master cock attempts mating. As the master cock mounts the hen, he is always attacked by one, two, or three other cocks. So rapid is their action that, before a quick copulation can take place, the master cock is hit hard and sometimes displaced. A short fight ensues which usually results in the withdrawal of the attacking cock to a safe distance.

Rampant *machismo* has also evolved in some insects with similar mating patterns. The horns of male rhinoceros beetles and their relatives and the mandibles of male stag beetles are among the weapons used. Beebe (1947) has described fighting in the hercules beetle (*Dynastes hercules*), a gigantic member of the Scarabaeidae from

South America (see Figure 15.3). The battle follows a highly predict-able sequence from the moment it is joined:

The projecting horns touch and click, spread wide and close, the whole object of this opening phase being to get a grip outside the opponent's horns. When the four horns are closed together, there is a dead-lock. All force is now given over to pinching, with the apparent desire to crush and injure some part of head or thorax . . . Again and again, both opponents back away, freeing their weapons, and then rush in for a fresh grip. When a favorable hold is secured outside the other's horns, a new effort, exercised with all possible force, is initiated. This is a series of lateral jerks, either to right or left, with intent to shift the pincer grip farther along the thorax as far as the abdomen and if possible on to mid-elytra. In addition, if the hold is at first confined to the incurving horn tips, the shift must be ahead, so that the final grasp brings into play the two opposing sets of teeth on the horns. Once this hold is attained and a firm grip secured the beetle rears up and up to an un-believably vertical stance. At the zenith of this pose it rests upon the tip of the abdomen and the tarsi of the hind legs, the remaining four legs outstretched in mid-air, and the opponent held sideways, kicking impotently. This posture is sustained for from two to as many as eight seconds, when the victim is either slammed down, or is carried away in some indefinite direction to some indeterminate distance, at the end of which the banging to earth will take place. After this climax, if the fallen beetle is neither injured nor helpless on its back, it may either renew the battle, or more usually make its escape.

The displays of species showing extreme intrasexual selection function both to attract females and to intimidate other males. Precopulatory displays are short or absent. The male hercules beetle, for example, evidently engages in none whatever. Occasionally he picks a female up and carries her aimlessly about for a short while, but the significance of the behavior is unknown. During both transportation and copulation the female remains outwardly passive.

Preoccupation with the more dramatic vertebrate examples leads to the impression that intrasexual competition is exclusively precopulatory in timing, ending with the act of insemination. But, as the classification of modes of sexual selection in Table 15-1 suggests, numerous postcopulatory devices exist, some of which are refined and ingenious in nature. Male mice are capable of inducing the Bruce effect: when they are introduced to a pregnant female, their odor alone is enough to cause her to abort and to become available for reinsemination. Nomadic male langurs routinely kill all of the infants of a troop after they drive off the resident males; the usurpers then quickly inseminate the females. A similar form of infanticide is perpetrated by male lions. By far the greatest diversity of postcopulatory techniques occurs in the insects (Parker, 1970a). The reason for this phylogenetic peculiarity appears simple. Female insects generally need to fertilize a great many eggs, often during a prolonged period; at the same time they must economize on the weight of sperm carried in their spermatheca. In the extreme case, exemplified by parasitic wasps, honeybees, ants, and at least some *Drosophila*, the spermato-

zoans are paid out one to an egg. As a consequence, males still find some profit in trying to inseminate females that have already mated. Their sperm can displace at least some of those inserted by their predecessors. In grouse locusts (*Paratettix texanus*), flour beetles (*Tribolium castaneum*), *Drosophila*, and a few other insects, the last male to mate successfully is often the one that fathers most of the offspring, because his spermatozoans are concentrated at the entrance to the female's sperm receptacle.

The threat posed by sperm displacement has provoked the evolution of a series of countermeasures making up much of the list of devices in Table 15-1. Mating plugs, commonly added to the female's genital tract by the coagulation of secretions from the male accessory gland, occur through a very wide array of insect groups. Some authors have concluded that the plugs serve chiefly to prevent sperm leakage, but in at least some of the Lepidoptera and in water beetles of the genera *Dytiscus* and *Cybister* the principal function appears to be prevention of subsequent matings. Also, competitive sperm blockage has not been ruled out as at least a secondary function in the great majority of remaining cases. In one extraordinary species, the cerato-pogonid fly *Johannseniella nitida*, the body of the male itself serves as the plug. Following copulation the female eats her mate, leaving his genitalia still attached. Copulatory plugs also occur in some mammals, including marsupials, bats, hedgehogs, and rats. The coagulation of seminal fluid is induced by the enzyme vesiculase, which in rodents is secreted by a "coagulating gland" adjacent to the seminal vesicle (Mann, 1964). Again, the role of the plug is traditionally considered to be the prevention of sperm leakage, but the prevention of additional inseminations remains an equally viable hypothesis.

During copulation the male may transmit substances that reduce the receptivity of the female. Craig (1967) has postulated that such a pheromone, which he calls "matrone," is secreted by the accessory gland in mosquitoes of the genus *Aedes*. A similar agent is produced by male house flies (*Musca domestica*) from secretory cells lining the male ejaculatory duct (Riemann et al., 1967). An even more effective means of thwarting sperm displacement is prolonged copulation. Male house flies remain *in copula* about an hour in spite of the fact that virtually all of the sperm are transferred during the first 15 minutes. The male's steadfastness works to his disadvantage in another way, however, because he loses valuable time in which other females could be inseminated. Mating is a highly competitive activity in this and most other kinds of flies. Even more extreme cases of prolonged copulation have been reported in other insects. Flies of the genus *Cylindrotoma* and moths commonly copulate for a full day. Male insects generally thwart attempts to dislodge them by remaining tightly attached. The external genitalia of such groups as moths, wasps, and flies are complex structures fitted with hooks, spines, and claspers. They provide some of the most reliable characteristics used

Table 15-1 The modes of sexual selection.

I. Epigamic Selection
    A. *Based on choices made among courting partners*
        1. The choice among the different types of suitors is dependent on their relative frequencies
        2. The choice is not frequency-dependent
    B. *Based on differences in breeding time: superior suitors offer to breed more at certain times than at others*
II. Intrasexual Selection
    C. *Precopulatory competition*
        1. Differential ability in finding mates
        2. Territorial exclusion
        3. Dominance within permanent social groups
        4. Dominance during group courtship displays
    D. *Postcopulatory competition*
        1. Sperm displacement
        2. Induced abortion and reinsemination by the winning suitor
        3. Infanticide of loser's offspring and reinsemination by the winning suitor
        4. Mating plugs and repellents
        5. Prolonged copulation
        6. In "passive phase" of courtship, suitor remains attached to partner during a period before or after copulation
        7. Suitor guards partner but without physical contact
        8. Mated pair leaves vicinity of competing suitors

by taxonomists to distinguish species. As O. W. Richards (1927b) first suggested, these devices may have evolved through intrasexual selection to prevent takeover by rival males during copulation.

Parker (1970a,b) has distinguished a "passive phase" in the courtship of many insect species, during which the male attaches himself physically to the female for more or less prolonged intervals without sexual contact. The attachment, which according to species occurs before or after copulation has taken place, prevents rival males from mounting the female. The tandem position of dragonflies is the most familiar example. The male holds on to the abdomen of the female, and the two fly about together while the female lays eggs on the water surface. The strategem is not of exclusive benefit to the male. The female must oviposit in areas normally dominated by territorial males who attempt to seize her on sight; if she were not flying in tandem

Figure 15-3 On the floor of a Venezuelan rainforest, two males of the hercules beetle fight for dominance and access to a nearby female. The struggle consists to a large degree of grappling and lifting with the huge horns that sprout from the head and prothorax. The orchid shown in this illustration is *Teuscheria venezuela;* mosses, liverworts, and lichens cover other parts of the ground and litter. (Original drawing by Sarah Landry.)

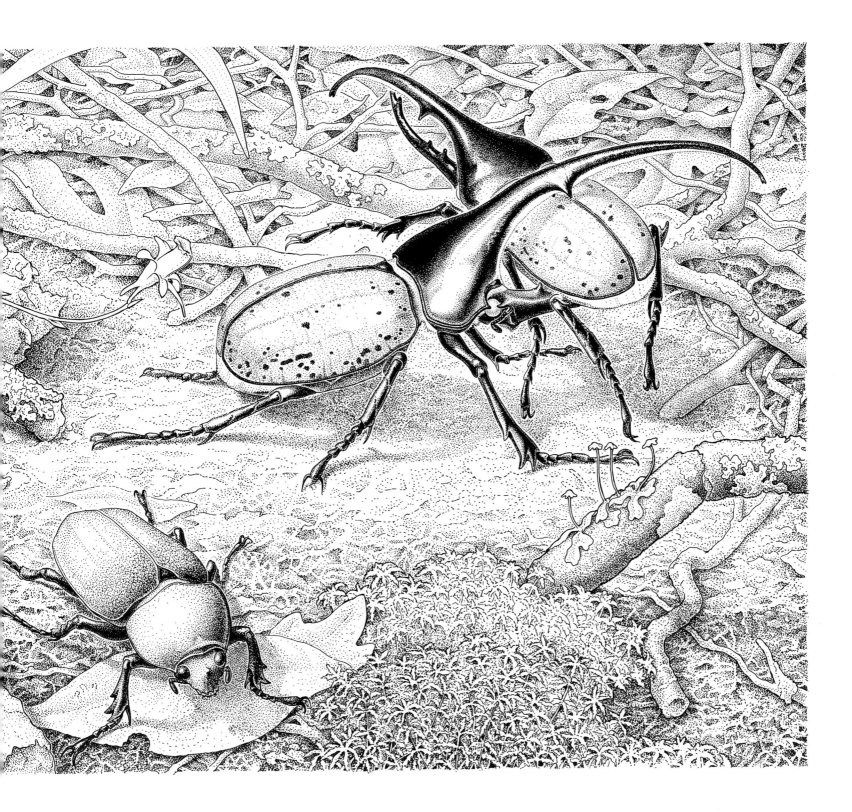

her attempts to oviposit would be repeatedly interrupted by useless sexual approaches. Males of the yellow dung fly *Scatophaga stercoraria* not only stand over the females during the passive phase but fight off intruding rivals. Their maneuvering is as stereotyped and as skilled as a jujitsu exercise (see Figure 15-4). Similar forms of active defense may be employed by males even when they are not in direct contact with their partners. For example, not all dragonflies utilize a lengthy tandem phase. After mating and flying a short time in tandem, females of *Calopteryx maculatum* begin to lay eggs on submerged vegetation. The male perches on a nearby support and flies out to attack any other male that comes close to his mate.

Finally, the mated pair may simply remove themselves from the presence of other suitors. Males of the ant *Pheidole sitarches* form

Figure 15-4 Fighting between males of the yellow dung fly *Scatophaga stercoraria* during the passive phase of courtship. The female is at the bottom. The male that inseminated her earlier is in the middle and is seen in the act of thrusting aside a rival who attacked him a moment before. The attack came from his left, causing him to lift his left middle leg in order to prevent penetration from that side. He then raised his entire body to push the intruder over and away in the opposite direction. (From Parker, 1970b.)

conspicuous mating swarms, into which the virgin queens fly to be mated. As soon as a male attaches himself to a queen, the couple cease flying and drop to the ground, where mating is completed. The queen then detaches her own wings and crawls off to start a new colony, thus effectively preventing further insemination (Wilson, 1957). When reeves make their choice among competing male ruffs, they often fly well away from the arena before assuming the crouching, invitatory posture that makes coition possible (Selous, 1927).

It must be kept in mind that the aggression displayed during intrasexual competition is of a special kind. In Chapter 11 I argued that most forms of animal aggression are techniques that evolve when shortages of resources chronically limit population growth. The aggressive behavior thus becomes part of the density-dependent controls. In the case of intrasexual selection there is also competition for a limiting resource. But the shortage, usually of females available for insemination but sometimes of males available to care for the females' offspring, does not limit population growth, and the aggression does not contribute to the density-dependent controls. Indeed, intrasexual selection is likely to become most intense when other resources, such as land and food, are in greatest supply and population growth most rapid. At that time females are able to reproduce at higher rates, which places a premium on fertility per se, and the abundance of other resources frees the males for pursuit of the females. The principle of allocation comes into play (see Chapter 6): the male behavior evolves so as to carry intrasexual competition to its greatest heights. The most elaborate forms of courtship display and intrasexual aggression develop under conditions in which males have the fewest problems with food and predators. The lek systems of insects, birds, and African grassland antelopes such as the Uganda kob are located away from the feeding grounds. The violent dominance hierarchies of the elephant seal and some other pinnipeds have evolved on island hauling grounds where both time devoted to feeding and mortality from predators are minimal. Thus not only do ordinary competition and intrasexual selection differ basically from each other, but they are in conflict. Insofar as social behavior evolves as a response to resource shortages and predation, the principle of allocation reinforces the antagonism between sexual reproduction and social evolution.

The Theory of Parental Investment

The ultimate basis of sexual selection is greater variance in mating success within one sex. Because of anisogamy, females—defined as the sex producing the larger gametes—are virtually assured of finding a mate. The eggs are the limiting resource. Females therefore have more to offer in terms of energetic investment with each act of mating, and they are correspondingly more likely to find a mate.

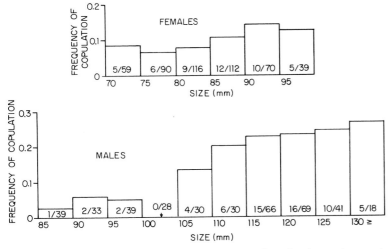

Figure 15-5 Bateman's principle illustrated in lizards: the variance of reproductive success is greater in males than in females in the Jamaican species *Anolis garmani*. Reproductive success is measured by the number of copulations observed per number of individuals (male or female) in each overlapping 5-millimeter size category. The data also show that male success increases with size. (Redrawn from Trivers, 1972.)

Males invest relatively little with each mating effort, and it is to their advantage to tie up as many of the female investments as they can. This circumstance is reversed only in the exceptional cases where males devote more effort to rearing offspring after birth. Then the females compete for males, in spite of the initial advantage accruing from anisogamy. Active competition for a limiting resource tends to increase the variance in the apportionment of the resource. Some individuals are likely to get multiple shares, others none at all. The resulting differential in reproductive success leads to evolution in secondary sexual characteristics within the more competitive sex.

This difference in variance was documented by Bateman's classic experiment (1948) on *Drosophila melanogaster*. The technique consisted simply of introducing five males to five virgin females, so that each female could choose among five males and each male had to compete with four other males. The flies carried chromosomal markers allowing Bateman to distinguish them as individuals. Only 4 percent of the females failed to mate, and even this small minority were vigorously courted. Most of those that mated did so only once or twice, by which time they had received more than sufficient sperm. In contrast, 21 percent of the males failed to mate, and the most successful individuals produced almost three times as many offspring as the most fertile females. Furthermore, most of the males repeatedly attempted to mate, and, in contrast to the females, their reproductive success increased in a linear proportion to the number of times they copulated.

Data on reproductive success in wild populations are few. One set comparable to those of Bateman have been obtained from the Jamaican lizard *Anolis garmani* by Trivers (see Figure 15-5). Other animals in which the variance in reproductive success of males exceeds that of females include dragonflies, the dung fly *Scatophaga stercoraria*, the common frog *Rana temporaria* of Europe, prairie chickens and other lek-forming grouse, elephant seals, and baboons. Indirect evidence suggests the widespread occurrence of this difference in variance in other vertebrates. Building on this principle, Trivers (1972) has constructed the outlines of a general theory of parental investment intended to account for a wide range of differing patterns of sexual and parental behavior. His arguments are based on the graphical analysis of *parental investment*, which is defined as any behavior toward offspring that increases the chances of the offspring's survival at the cost of the parent's ability to invest in other offspring. A second variable analyzed is *reproductive success*, measured by the numbers of surviving offspring. The central principle of sexual selection is reformulated in the graph presented in Figure 15-6. Here we see that one sex, usually the female, invests more heavily in each offspring. An egg costs more than a spermatozoan in the sense that it more drastically reduces the number of additional eggs that can be pro-

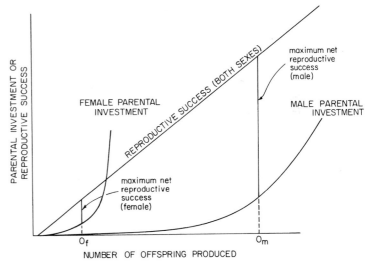

Figure 15-6 A conflict of the sexes arises when the optimum number of offspring (O_f for the female, O_m for the male) differs between them. In the imaginary case depicted here, the usual situation is exemplified: the female must expend a greater effort to create offspring, and her greatest net production of offspring comes at a lower number than in the case of the male. Under these conditions the male is likely to turn to polygamy in order to attain his optimum number. In a few species the situation is reversed, and the female is polygamous. (Modified slightly from Trivers, 1972.)

duced at that time or later. The parent that commits itself to the greater part of parental care, again usually the female, will find it difficult or impossible to begin reproducing again until the first off-spring are fledged. Hence for that parent investment rises more quickly as a function of the number of offspring produced per reproductive episode. The reproductive success curve, however, will be the same for both sexes: a one-to-one linear increase with the number of offspring produced, by which it is in fact defined. Consequently, the parent making the greater investment per offspring will want to stop at fewer offspring than its mate. From this disparity flows Bateman's principle, that variance in net reproductive success will be greater in the sex with the smaller per-offspring investment. Furthermore, this sex will experience the more intense degree of sexual selection and be prone to evolve the more extreme epigamic displays and techniques of intrasexual selection.

Although the basic theory is built on parameters difficult to measure in practice, there is an indirect way it can be decisively tested. In the exceptional cases where males have taken on more than their share of parental care, we should also find the exceptional circumstance that females are the competitive sex, using the more conspicuous displays and perhaps contending directly for possession of the males. This prediction is easy to confirm in full. Species in which such a reversal of sex role exists and is associated with extended male parental care include the following: pipefishes and seahorses of the family Syngnathidae (Fiedler, 1954); Neotropical "poison-arrow" frogs of the family Dendrobatidae (Dunn, 1941; Sexton, 1960); jacanas, which are gallinulelike wading birds (Mathew, 1964); four species of tinamous in the genera *Crypturellus* and *Nothocercus* (Lancaster, 1964); phalaropes (Höhn, 1969; Hildén and Vuolanto, 1972); the painted snipe *Rostratula benghalensis* (Lowe, 1963); the button quail *Turnix sylvatica* (Hoesch, 1960); and the Tasmanian native hen *Tribonyx mortierii* (Ridpath, 1972).

The Trivers mode of analysis can be extended to a consideration of parental investment through time in order to interpret patterned changes in sexual interaction. Figure 15-7 presents the cumulative investment curves of the male and female of an imaginary bird species. The principles it illustrates can be broadened with little effort to include any kind of animal as well as human beings. At each point in time there will be a temptation for the partner with the least accumulated investment to desert the other. This is particularly true of the male immediately following insemination. The female's investment has surged upward, while that of the male remains small. As parental care by each sex accumulates, the tendency to desert will depend not only on the difference in the amount of investment but on the ability of the partner to rear the offspring alone. If one partner is deserted it will no doubt try to finish the job, since so much has been committed already. But if a substantial risk exists that a solitary parent will fail because the task is overwhelming, desertion carries

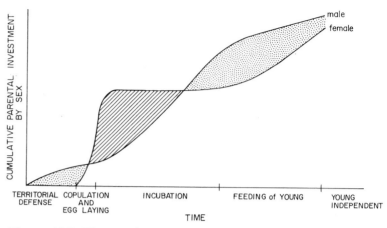

Figure 15-7 The cumulative parental investment of two mated animals can change through time, causing shifts in their attitudes and relationships. In this imaginary example modeled on the bird life cycle, the male has more to lose at certain stages (stippled) and the female at others (cross-hatched). *Territorial defense:* the male defends the area to protect food supply and nest sites. *Copulation and egg laying:* the female commits her eggs to the male while the male commits his defended nest to the female. *Incubation:* the male incubates eggs while the female does nothing relevant to the offspring; consequently, the cumulative female investment remains constant while that of the male rises, and the two investments attain parity a second time. *Feeding of young:* each parent feeds the young but the female does so at a more rapid rate, causing the investments to converge a third time. (Modified slightly from Trivers, 1972.)

the risk to the potential deserter of a loss in genetic fitness. When the expected loss in fitness is not likely to be compensated by success in future matings with other partners, desertion is likely to be rare at this stage of the cycle. As Trivers has pointed out, there may come a time when the investments of both partners are so great that natural selection will favor desertion by either partner even if the investment of one is proportionately less. This is so because desertion places the faithful mate in a cruel bind: it has invested so much that it cannot abandon the investment regardless of the difficulties that lie ahead. Under these circumstances the relationship between the partners may develop into a game as to which can desert first. The outcome might be determined not so much by wiliness as by the opportunities that exist for the possession of a second mate. Rowley (1965) described a parallel episode in the Australian superb blue wren *Malurus cyaneus*. Two neighboring pairs happened to fledge their young simultaneously and could not tell them apart, so that all were fed indiscriminately as in a crèche. One pair then deserted in order to start another brood. The remaining couple continued to care for all of the young, even though they had been cheated.

Out of this unsentimental calculus of marital conflict and deceit can be drawn a new perception of cuckoldry. When fertilization occurs by insemination, the universal mode of reptiles, birds, and mammals, the male cannot always be completely sure that his mate's eggs have been fertilized by his own sperm. To the degree that he invests in the care of the offspring, it is genetically advantageous to him to make sure that he has exclusive access to the female's unfertilized eggs. Frequently the preemption comes about as a bonus resulting from other kinds of behavior. Males that exclude others from territories or control them within dominance systems are avoiding sperm competition. The same effect is achieved through the time lag that normally occurs between bonding and copulation in monogamous birds (Nero, 1956), an interval that serves as a de facto quarantine period for the detection of alien sperm. The theory suggests that a particularly severe form of aggressiveness should be reserved for actual or suspected adultery. In many human societies, where sexual bonding is close and personal knowledge of the behavior of others detailed, adulterers are harshly treated. The sin is regarded to be even worse when offspring are produced. Although fighting is uncommon in hunter-gatherer peoples such as the Eskimos, Australian aborigines, and !Kung Bushmen, murder or fatal fighting appears to be frequent in these groups in comparison with other societies and is usually a result of retaliation for actual or suspected adultery.

Until recently biologists and social scientists have viewed courtship in a limited way, regarding it as a device for choosing the correct species and sex and for overcoming aggression while arousing sexual responsiveness in the partner. The principal significance of Trivers' analysis lies in the demonstration that many details of courtship can also be interpreted with reference to the several possibilities of maltreatment at the hands of the mate. The assessment made by an individual is based on rules and strategies designed by natural selection. Social scientists might find such an interpretation rather too genetic or even amoral for their tastes, yet the implications for the study of human behavior are potentially very great.

The Origins of Polygamy

Because of the Bateman effect, animals are fundamentally polygamous. At the start, virtually all are anisogamous. In those species in which parental care is also lacking, variance in reproductive success is most likely to be greater in males than in females. Under many circumstances the addition of parental care will reinforce this inequality, because parental investment in postnatal care is seldom quantitatively the same in both sexes. Monogamy is generally an evolutionarily derived condition. It occurs when exceptional selection pressures operate to equalize total parental investment and literally force pairs to establish sexual bonds. This principle is not compromised by the fact that the great majority of bird species are monogamous. Although polygamy in birds is in most cases a phylogenetically

derived condition, the condition represents a tertiary shift back to the primitive vertebrate state. Monogamy in modern birds was almost certainly derived from polygamy in some distant avian or reptilian ancestor.

Before examining the evidence behind these generalizations, let us define the essential terminology pertaining to mating systems. *Monogamy* is the condition in which one male and one female join to rear at least a single brood. It lasts for a season and sometimes, in a small minority of species, extends for a lifetime. *Polygamy* in the broad sense covers any form of multiple mating. The special case in which a single male mates with more than one female is called polygyny, while the mating of one female with more than one male is called polyandry. Polygamy can be simultaneous, in which case the matings take place more or less at the same time, or it can be serial. Simultaneous polygyny is sometimes referred to as harem polygyny. In the narrower sense preferred by zoologists, polygamy also implies the formation of at least a temporary *pair bond*. Otherwise, multiple matings are commonly defined as *promiscuous*. But the word *promiscuity*, as Selander (1972) has pointed out, carries the incorrect connotation that the matings are random, even though in fact they are usually highly selective in a way that leads to the evolution of secondary sexual characteristics. Selander has proposed the alternative expression *polybrachygamy*. Although the term is technically and etymologically correct, it may prove too cumbersome to gain wide usage.

Entirely by itself, anisogamy favors polygamy as broadly defined. There also exist five general conditions that promote polygamy still further. They are (1) local or seasonal superabundance of food; (2) risk of heavy predation; (3) precocial young; (4) sexual bimaturism and extended longevity; and (5) nested territories due to niche division between the sexes. All but the last were discovered in birds, where polygamous and monogamous species commonly coexist and provide the opportunity for evolutionary comparisons; but the same biases probably operate with equal force on other, less well studied groups.

Local or Seasonal Superabundance of Food

Armstrong (1955), on the basis of his study of the common wren, *Troglodytes troglodytes*, of Europe, hypothesized that monogamy in birds evolves when food is limiting, the population is at or near its maximum, and it is therefore of advantage for the male to stay with the female and to help her rear the young. Polygamy evolves in species that enjoy a superabundance of food during the breeding season, a circumstance permitting the female to raise the offspring alone while the male goes off to search for new mates. Crook (1964) used essentially the same argument to account for polygamy among the more than 100 species of ploceine weaver finches of Africa and Asia. He noted that the species inhabiting humid environments, particularly forests, are primarily monogamous and display few secondary sexual differences in anatomy. Polygamy and sexual dimorphism, by contrast, are common although not universal traits in the species that occupy grasslands and other arid environments. The distinction, Crook suggested, stems from a difference in diet. Forest dwellers are mostly insectivores and can depend on a relatively steady yield from the same places during lengthy breeding seasons. As a result, the adult birds tend to form monogamous pair bonds and to defend territories in pairs. Most species inhabiting arid country feed on seeds and other plant materials that are present in superabundance for short annual seasons. The males are therefore freed from the necessity of parental care during the breeding season and can spend their time trying to inseminate additional females.

Lack (1968) identified a flaw in Crook's evolutionary argument by noting that nearly all other seed-eating birds are monogamous. Yet the correlation *within* the Ploceinae holds (see also Moreau, 1960), and the prospect remained for development of a stronger and more general theory to account for these and similar facts in other bird groups. An important effort to do so was made by Orians (1969), who drew on his own work on blackbirds (*Agelaius, Xanthocephalus*), as well as that of Verner (1965) on wrens, to support the following argument. When the female bird chooses a courting male, she need not depend entirely on her assessment of the male's physical readiness. In many species it will be to her advantage also to consider the quality of the habitat in which the territory is located. The site should be rich in resources and provide protection against predators and inclement weather. If the environments of different territories vary sufficiently in quality, a female will gain more in genetic fitness by joining other females in the single rich territory of a polygynous male than by becoming the sole partner of a monogynous male on poor land. This conception is parameterized and put in simple graphical form in the Orians-Verner model shown in Figure 15-8. Given rising curves of female success in monogynous and polygynous groups as a function of increasing environmental quality, there exists some minimal difference between the richest and the poorest territories such that it is better for a female to join a harem in the richest territory than to become the sole partner of a male in a very poor territory. This minimal difference has been called the "polygyny threshold" by Verner and Willson (1966).

Several results flow from the Orians-Verner model that encompass the special cases identified by Armstrong and Crook. Habitats that are highly variable in productivity will be more likely to contain territories that vary in excess of the polygyny threshold. Marshes are notably variable in this way; differences in energy yield between the surrounding aquatic environments are often tenfold or greater. According to Verner and Willson, 8 of the 15 known polygynous species of North American passerine birds nest in marshes, despite the fact

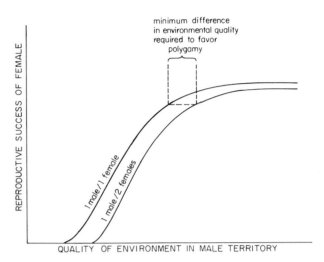

Figure 15-8 The Orians-Verner model of the conditions required for the evolutionary origin of polygyny. If reproductive success rises steeply enough as a function of the environmental quality of the male territory, and if individual male territories vary sufficiently in quality—attaining the "polygyny threshold"—it will be of advantage for some females to join a harem instead of becoming the sole partner of a male in a poor territory. The same model can be applied to polyandry by simply reversing the sexes. (Modified from Orians, 1969.)

that marsh-nesting birds constitute only about 5 percent of the total fauna. In Africa polygyny is also prevalent among the marsh-nesting weaver finches (Crook, 1964). Early successional stages in the growth of vegetation also provide highly variable environments, and the opportunistic bird species that exploit them are as a rule wide-ranging and unspecialized. Five of the 15 polygynous species of North American passerines breed in prairie or savanna habitats, while 2, the dickcissel *Spiza americana* and bobolink *Dolichonyx oryzivorus*, are restricted to the earliest successional stages of grassland vegetation. Finally, when nest sites limit population density but food does not, the quality of the male territory can be expected to vary strongly according to the suitability of the nest sites it contains. The polygynous wrens *Troglodytes aedon* and *T. troglodytes* nest in preformed cavities but are unable to excavate their own, and the same is true of the polygynous pied flycatcher (*Ficedula hypoleuca*) of Europe (von Haartman, 1954). It is probably significant that the polygynous weaver finches of Africa and Asia not only enjoy abundant food resources during their breeding season but also nest in trees, which are in short supply in the most favored habitats.

When the implications of this formulation are pressed somewhat further, we again encounter the basic antagonism between sex and social behavior. If it is advantageous for females to join territorial harems on the better side of the polygyny threshold, it will be still more advantageous for each to be the sole member of a harem. Consequently, we should expect to find conflict among the harem females. Furthermore, a conflict of interest might develop between the two sexes; the females want as few companions as possible, but the interests of the male are best served by maintaining the number of females who together can rear the largest number of offspring within the limits of the territory. This is in fact the condition encountered by Downhower and Armitage (1971) in their study of the yellow-bellied marmot (*Marmota flaviventris*), a polygynous territorial rodent of the Rocky Mountains. The reproductive success of individual female marmots declines with an increase in the size of the harem. The decrease is strongly marked in the number of litters per female, the average number of offspring each female cares for, and, above all, the number of yearlings per female present in the harem. From the yearling data, which give the final reproductive success of the females, it is easy to calculate that the average optimum number of females per male is between two and three (see Figure 15-9).

Risk of Heavy Predation

Heavy predation on territorial animals will tend to favor monogamy, provided the parents are capable of warding off the predators and the presence of both adults gives extended protection. Von Haartman (1969) noted that polygyny in European birds is not associated with habitat, as in the North American fauna, but rather with the preferred nest site. Many of the polygynous species build domed nests or utilize holes. It is von Haartman's belief that these sites give extra protection

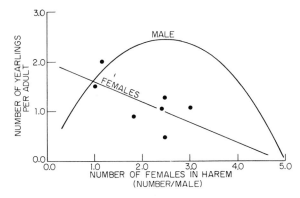

Figure 15-9 The reproductive success of yellow-bellied marmots, measured by the number of yearlings produced, is affected by harem size in different ways for females and males. It decreases steadily for females, so that the optimum harem size for this sex is one; but reproductive success rises and then falls for males, reaching its peak between two and three females per male. The data points refer to observed yearlings per female. (Modified from Downhower and Armitage, 1971.)

from predators, allowing the male to spend more time in defense of the territory and courtship of additional females. It is also possible that closed nests provide superior insulation for the young, which reduces their maintenance cost. Von Haartman's hypothesis, it will be noted, is not inconsistent with the Orians-Verner model. Yet it does add the independent element of an improved energy balance for all of the territories that embrace the polygyny threshold, whether they are relatively poor or rich in nest sites.

Another way to protect young animals from predators is to remove them as far as possible from the courtship performances of the parents. Such an adaptive response could be part of the explanation for polygyny in grouse, birds of paradise, and other birds that mate on communal display arenas. Once inseminated, females are free to withdraw and to devote themselves to the rearing of their young in quieter, food-rich areas.

Precocial Young

When females can guide the young to the best feeding areas and keep them out of sight of predators, the need for male participation is sharply reduced. The males are then freer to devote themselves to epigamic displays and fighting for mates, activities that often occur on special, communal mating grounds. As Orians (1969) has pointed out, the expected correlation occurs in nature but it is surprisingly loose. Polygynous species occur frequently among the Phasianidae (pheasants, partridges, peafowl, chickens, quail) and Tetraonidae (grouse), and a few occur among the shore birds, for example the buff-breasted sandpiper *Tryngites subruficollis*, pectoral sandpiper *Erolia melanotos*, and great snipe *Capella media*. On the other hand, monogyny is the rule in the Charadriidae (plovers) and Scolopacidae (sandpipers), as well as in the Anatidae (swans, geese, and ducks). The reasons for these numerous exceptions remain unknown.

Sexual Bimaturism and Extended Longevity

When the courting sex is long-lived, its members can defer reproduction until they are large and mature enough to gain a dominant position. Reproductive dominance can result in the insemination of enough females to more than compensate for the loss due to the earlier deferment. This condition by itself will favor polygyny. According to Wiley (1974), longevity and sexual bimaturism are the conditions prevailing in the polygynous species of grouse. During their first year, males do not mate, or at least they mate very seldom, whereas females breed readily in the first year. The same basic condition, called sexual bimaturism, is widespread in other polygynous birds and mammals. It has been documented for example in red-winged blackbirds (Peek, 1971), manakins (Snow, 1963), the southern elephant seal *Mirounga leonina* (Carrick et al., 1962), and mountain sheep (Geist, 1971a).

Nested Territories Due to
Niche Division between the Sexes

Nested territories due to intersexual differences in ecology can occur only in species that breed within the confines of a feeding territory. If the female is smaller than the male or for some other reason requires less space, and if she cares for the offspring on her own, polygyny will be favored because more than one female can afford to live within the territory of a single male. This ecological divergence has evidently been a factor in the evolution of polygyny in the lizard genus *Anolis* (Rand, 1967; Schoener, 1967; Schoener and Schoener, 1971a, b), and it may also be implicated in polygyny in the gecko *Gehyra variegata* (Bustard, 1970).

The Origins of Monogamy and Pair Bonding

In the animal world, fidelity is a special condition that evolves when the Darwinian advantage of cooperation in rearing offspring outweighs the advantage to either partner of seeking extra mates. Three biasing ecological conditions are known that seem to account for all of the known cases of monogamy: (1) the territory contains such a scarce and valuable resource that two adults are required to defend it against other animals; (2) the physical environment is so difficult that two adults are needed to cope with it; and (3) early breeding is so advantageous that the head start allowed by monogamous pairing is decisive.

Defense of a Scarce and Valuable Resource

An estimated 91 percent of all bird species are monogamous during at least the breeding season. This adaptation provides superior defense for scarce nest sites, or territories containing scattered renewable food sources, or both (Lack, 1966, 1968). A few species even form life-long bonds. One well-analyzed example is the oilbird (*Steatornis caripensis*) of Trinidad and northern South America. According to Snow (1961), the permanence of the bond stems ultimately from the combined circumstances that the birds nest in caves, are long-lived, and breed very slowly. Appropriate nesting sites along the cave walls are extremely few, and their scarcity appears to be the principal factor limiting the size of the population. Cooperative defense of the sites is needed to maintain them during the long reproductive lives of the mated couple. Yellow-winged bats (*Lavia frons*) are unusual among the insectivorous chiropterans in the way they feed. Like the flycatchers among birds, they rest in trees until a prey comes near, then fly off after it. The bats use echolocation for orientation and rely on their unusually large wings to maneuver through the treetops. As soon as a prey is captured the bats return to their roosts. *L. frons* is also unusual in forming monogamous sex bonds. Wickler and Uhrig

(1969a) hypothesize that the novel feeding strategy of the species requires maintenance of a fixed territory, and that monogamy has evolved as a device to aid its defense. In short, the species has converged in this respect with the insectivorous forest-dwelling birds, of which the great majority are monogamous.

Monogamy is a much rarer event in the invertebrates than in the vertebrates, characterizing perhaps less than one invertebrate species out of ten thousand. Where it occurs, improved defense of a resource often seems to be its principal function. Pairs of beetles (*Necrophorus*) appropriate the corpses of small animals and defend them against other pairs. Only by complete control of a cadaver can the female hope to rear a single brood through to maturity. The starfish-killing shrimp *Hymenocera picta* is unusual among crustaceans in forming long-lasting sex bonds. It is also exceptional in the large size of its prey—a single large starfish can nourish a pair for a week. Defense of the prey and of the area surrounding it is far more important for *H. picta* than for more conventional crustaceans that feed on plankton, algae, and detritus (Wickler and Seibt, 1970).

Adaptation to a Difficult Physical Environment

I know of no case in which monogamy serves exclusively as a device for overcoming challenges offered by the physical environment. However, at least one species exists in which generally severe conditions have intensified the need for defense of the territory to the point of favoring permanent monogamy. The sowbug *Hemilepistus reaumuri* is an isopod crustacean that lives in the dry steppes of Arabia and North Africa. During the hottest, driest time of the year the crustaceans are forced to retreat deep into burrows in order to survive. Linsenmair and Linsenmair (1971) discovered that each burrow is occupied by an adult pair that remain mated for life. The highly territorial behavior of the mated sowbugs prevents overcrowding of the burrows and depletion of the scarce, unpredictable food supply in the surrounding area. This example may be regarded as a special case of the defense of a limiting resource. It deserves particular attention because it identifies the possibility that parameters in the physical environment affect the entire population simultaneously. This is the opposite phenomenon from the Orians-Verner effect, in which the environment varies individual success to a degree that promotes polygamy.

An Early Start in Breeding

When the timing of breeding is important, cooperation between the mated pair can provide a decisive edge. In his excellent study of the kittiwake gull (*Rissa tridactyla*) in Britain, Coulson (1966) found that about 64 percent of the breeding birds retained their mate from the previous season. Females in this category began to lay eggs three to seven days earlier, produced a higher seasonal total, and reared a larger number of chicks than did comparable individuals that had taken a new mate. The difference stemmed from the ability of "married" pairs to cooperate more smoothly from the beginning of courtship and nesting. Yet divorces were common. Over the 12-year period of the study, Coulson found that two-thirds of the birds that changed partners did so while the first mate was still living in the colony. Also, birds that failed to hatch eggs the previous season were three times more likely to change mates than those that had enjoyed success. The latter correlation suggests that "divorce" is adaptively advantageous to birds originally bound to a reproductively incompatible partner. Although data on breeding success are lacking, a similar explanation may hold for the monogamy of the least sandpiper (*Calidris minutilla*) and stilt sandpiper (*Micropalama himantopus*), members of a family of birds, the Scolopacidae, noted for its conspicuously polygamous species. The two species breed in northern Canada, where a rapid start in breeding is required to take sufficient advantage of the lush but very short spring and summer (Jehl, 1970).

Communal Displays

Communal sexual displays provide some of the great spectacles of the living world. In southeast Asia thousands of male fireflies sit in certain trees in the forest and flash synchronously and rhythmically through the night. The light is so strong and the location of the trees so consistent that mangrove trees inhabited close to the shore line are used as navigational aids (Buck, 1938; Lloyd, 1966, 1973). Millions of 13-year and 17-year locusts gather to mate in one or another forest locality in the eastern United States; the singing of the males can be literally deafening to the human ear (Alexander and Moore, 1962). These insect performances are rivaled by arena mating in birds, the battles royal of male mountain sheep and elk, and others among the more dramatic vertebrate displays.

The primary role of communal displaying appears to be enhancement of attractiveness through the increase in volume and reach of the signal. In simplest terms, a group of males is more likely to attract a single female than is a solitary male, and a male is more likely to encounter a receptive female when he is in a group. The effect can be strengthened further when the display grounds are situated in an open space, atop a prominent high point, or in some other distinctive location that makes orientation easy. Such features characterize the swarm areas of parasitic wasps, ants, nematoceran flies, and other communally breeding insects. Birds also commonly rely on landmarks. Moreover, the display grounds of many species are traditional in nature, remembered by older individuals from one season to the next.

Data on the role of predation in the evolution of communal displays are ambiguous. Under certain circumstances, admittedly neither documented adequately nor understood in theory, predation might increase in the dense aggregations enough to serve as a counteracting selection force that ultimately limits the size and conspicuousness of the displaying groups. But under other conditions the opposite might occur. If the aggregations are short-lived and spaced far enough apart in time, they could saturate local predator populations and reduce individual mortality due to predation. This does indeed appear to be the primary strategy of the periodic cicadas (see Chapter 3). Lloyd (1966) has put a similar construction on the extreme periodicity of some firefly species. *Lampyris knulli* is active for only about half an hour each evening, while *Photinus collustrans* flashes for less than 25 minutes. In most communally displaying birds, the mating grounds are far removed from the nests, with the possible result that predators are distracted away from the inconspicuously colored females and their young. But the advantage of this disjunction might easily be turned around if suitable nest sites are scarce and difficult to locate. Then the individuals that display on ecologically desirable ground might gain enough fitness to override the loss of some of their offspring by predation. When males and females of the West African firefly *Luciola discollis* are located at the site of a previously existing population, and hence a proven breeding ground, the males emit special flashes that attract females ready to oviposit. The males are also able to call in unmated females to the same areas (Kaufmann, 1965). In short, the only general adaptive significance of communal displays that can be inferred at this time is enhanced signaling. Other environmental factors such as predation and the patchy distribution of breeding areas can affect the character and location of the displays, but by themselves they do not appear adequate to explain why the displays are communal as opposed to solitary in nature.

When an area is consistently used for communal displays, it is referred to as a *lek* or *arena*. The animals are said to be engaged in lek displays or arena displays, and the entire breeding system as a lek or arena system. In the original, ornithological usage (see Mayr, 1935; Armstrong, 1947; Lack, 1968) the true lek is also defined as being removed from the nesting and feeding areas, and this seems to be a useful qualification to retain for animals generally. Less useful restrictions which cannot be applied universally to other groups that have evolved otherwise similar behavior are (1) polygamy and (2) the special circumstance that pairs meet only for the purpose of mating (Gilliard, 1962). The males may display on little separate territories or not; in the ornithological literature each territory is sometimes referred to as a court.

The most complicated and spectacular lek systems occur in birds. The phenomenon has arisen independently in lines belonging to ten families: the ruff *Philomachus pugnax* and great snipe *Capella media*

(Scolopacidae, which also includes sandpipers and curlews); many grouse species, including capercaillie and blackcock (Tetraonidae); a few hummingbirds (Trochilidae); the argus pheasant *Argusianus argus* (Phasianidae); most manakins (Pipridae); the cock of the rock *Rupicola rupicola* (Cotingidae, a large New World family that also includes cotingas, fruit-eaters, tityras, and becards); the bustard *Otis tarda* (Otidae); some bowerbirds (Ptilonorhynchidae); two species of birds of paradise (Paradisaeidae); and Jackson's dancing whydah *Drepanoplectes jacksoni* (Ploceidae, a large Old World family that also includes the weaver finches and *Passer* sparrows). The males belonging to species on this list are among the most colorful of the bird world. The brilliant red cock of the rock, for example, is easily the most spectacular cotingid, and the birds of paradise are justly considered the most beautiful of all birds. The basis of the correlation is that lek systems in birds are universally associated with extreme polygyny and sexual dimorphism, both of which promote secondary sexual evolution in males. Good reviews of various aspects of the subject have been provided by Armstrong (1947), Gilliard (1962), Snow (1963), Lack (1968), Selander (1972), and Wiley (1974).

For an especially instructive example, let us consider the lek system of the most advanced tetraonid, the sage grouse *Centrocercus urophasianus*. The behavior of this bird, an inhabitant of the northern sagebrush plains, has been studied in depth by Scott (1942), Dalke et al. (1963), and most especially and recently by R. Haven Wiley (1973). Each lek contains a mating center, toward which the breeding adults crowd centripetally. The population at such places is often very large. As many as 400 males spread out over a lek area of a hectare or more, and a comparable number of females visit for short intervals to be mated. A large fraction of the males occupy little territories, each of which covers from 10 to 100 square meters. But only the males whose territories overlap the mating center are accepted by the females. As a result, in each season a group of less than 10 percent of the males achieve more than 75 percent of the copulations. The territories are relatively stable within each season and from one year to the next. As long as he is able, each male comes back to the same spot at the start of the breeding season in February and March. Neighbors sometimes interrupt attempts at copulations but usually only in the vicinity of the territorial boundary. The original system postulated by Scott, of a master cock assisted by a limited number of subcocks organized into a dominance hierarchy, has been disproved by the more detailed studies of Wiley. Instead, a male becomes successful at breeding by acquiring a territory at the breeding center. In effect, there is a waiting list for this prime real estate. Yearling males establish territories around the periphery of the lek, and with luck and maturity they gradually move inward toward the center as vacancies occur.

The displays of the territorial sage grouse males toward one another and the females are elaborate and dramatic. The extreme display is the strut, illustrated in Figure 15-10. The male inflates his chest sac, an elastic extension of the esophagus that has a capacity of 4 to 5.5 liters. The strutting posture is assumed, in which the body is tilted upward with the head held high. The male erects the white feathers and thin plumes on the side of his neck and expands the yellowish combs over his eyes. Then suddenly and twice in quick succession he raises his chest sac as high as possible and drops it. In the instant before the sac is elevated he extends his wings forward, and as the sac rises he pulls the wings backward over the stiff, specialized feathers lining the sides of his chest. This motion produces a swishing sound. As the sac is raised and dropped it expands with air to reveal two bare yellowish-olive patches of skin on the chest. When the sac is dropped the second time, it is compressed by the contraction of the muscles in the skin of the chest. This releases the air into the pockets of bare skin, which abruptly balloon forward and then collapse, the movement producing two sharp snapping noises 0.1 second apart. The snaps are preceded by several brief, soft cooing sounds. Thus the entire acoustic part of the strutting display, which lasts only a little more than 2 seconds, is an arresting "swish-swish-coo-oo-poink!" According to Wiley the sound can be heard by human ears up to several hundred meters from the lek, and the bobbing white chests can be seen over a kilometer away.

Birdlike lek systems occur in many of the open-country antelopes of Africa, including the common and defassa waterbucks (*Kobus ellipsiprymnus, K. defassa*), Uganda kob (*K. kob thomasi*), puku (*K. vardoni*), springbuck (*Antidorcas marsupialis*), Grant's and Thomson's gazelles (*Gazella granti, G. thomsoni*), wildebeest (*Connochaetes taurinus*), and others classified by Jarman (1974) on behavioral and ecological grounds as Class C and Class D species (see Chapter 24). In the Uganda kob, an antelope that has carried this trend unusually far, successful males cram small territories next to one another in sites well removed from the feeding and watering areas. Receptive females wander through the networks as part of the nursery herds and are mated by those territorial males able to detain them. Bachelor herds roam the periphery of the lek, sometimes joining the nursery herds there, but their members are seldom if ever able to copulate (Buechner, 1963; Buechner and Roth, 1974; Leuthold, 1966). The antelope systems are more diffuse than those of the lek-forming birds. In fact, they are geometrically intermediate between the full feeding territories that characterize the forest antelopes and the extreme leks found in birds.

Full-scale leks are formed by the fruit-eating bat *Hypsignathus monstrosus* of Africa (Bradbury, 1975). The adults display the greatest sexual dimorphism of all the 875 bat species of the world. The males

Figure 15-10 The lek of a sage grouse in Montana. Each of the three displaying cocks occupies a small territory at the mating center of the lek. The less showy hens crowd in from all directions to be mated by these favored few. The other cocks in the lek area, who occupy more peripheral territories out of sight in this photograph, seldom have a chance to mate. The lek system of the sage grouse is the evolutionarily most advanced known within the Tetraonidae. (From Wiley, 1973.)

have grotesque muzzles and enlarged larynxes. During the breeding season they gather in nocturnal aggregations at traditional sites in the forest canopies. Each male stakes out a small territory that it defends from rivals with harsh cries and gasps. From the moment of arrival at his post the male sings, emitting metallic notes at 80–120 times a minute while beating his partially unfolded wings at twice this rate. Females visit the lek and fly along its axis, causing sudden increases in the rates of display as they pass by. A similar lek system occurs in the related species *Epomophorus gambianus* (Booth, 1960).

The endemic fruit flies of Hawaii are distinguished within the large family Drosophilidae and perhaps among flies generally by their possession of a true lek system (Spieth, 1968). Males congregate on the stems of tree ferns and other arboreal sites that are both exposed and well removed from the flowers on which the flies feed and oviposit. The males differ strikingly in appearance from the females and are characterized by patterns of banding and spotting on the wings. Spieth hypothesizes that the segregation of the mating arenas, with the concomitant evolution of sexual dimorphism, has been caused ultimately by predation. Drosophilids are among the dominant insects of the Hawaiian forests and a prime target of such common native insectivorous birds as the elepaio (*Chasiempis sandwichensis*), which collect the flies at their feeding sites. By transferring courtship to special arenas, the males appear to have reduced the magnitude of this threat.

According to Bert Hölldobler (personal communication) the harvesting ant *Pogonomyrmex rugosus* has altered the basic nuptial flight pattern of ants into an arena system. At one locality the males were observed to leave the nests earlier in the day than the females and to gather in a dense congregation on the ground in a restricted area measuring 60 by 80 meters. As the females flew in and landed, each was instantly smothered by crowds of 10–30 males who fought in an attempt to mate with her. At night the surviving males retreated into crevices in the soil. The next day they emerged again to be joined by additional males and females in new rounds of frenzied mating. The evolutionary origin of the *Pogonomyrmex* arena system seems clear. It is simply a grounded nuptial flight that is renewed daily.

Other Ultimate Causes of Sexual Dimorphism

The reader will recognize the following thread of reasoning that runs strongly through the theory of sexual evolution: polygamy, promoted by one or more forces in the environment such as the very unequal apportionment of territorially defended resources, leads to increased sexual selection, which leads in turn to increased sexual dimorphism.

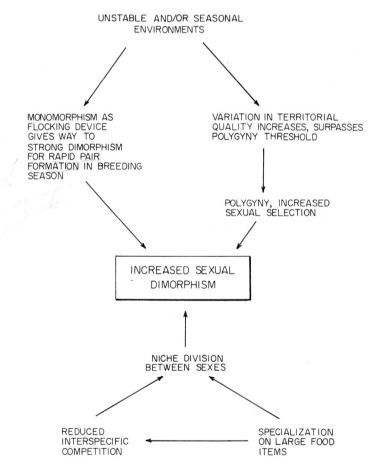

Figure 15-11 The alternate chains of events leading to increased sexual dimorphism in birds.

But as we have seen exemplified in other sociobiological phenomena, the final effect—in this case the enhancement of sexual dimorphism—can be approached along other evolutionary pathways. These alternate chains of causation, as understood principally from studies on birds, are represented diagrammatically in Figure 15-11.

The relation between strong sexual dimorphism and unstable environments was first examined in depth by Moreau (1960) in the ploceine weaver finches and by Hamilton (in Hamilton and Barth, 1962) in the parulid warblers and other passerine birds of the New World. Most of the species of *Ploceus* that breed in the dry habitats of Africa move in large, itinerant flocks and wear dull plumage when out of the breeding season. In a parallel fashion, the passerine birds that migrate the farthest in the New World assume the dullest plumage and often associate in feeding flocks during the off season. Hamilton and Barth, following Moynihan (1960), concluded that the convergence toward dull plumage is a device for reducing hostile interaction during flock formation. The strongest seasonal dimorphism is assumed by the species that have the longest migration routes and the shortest periods in which to breed. It seems to follow that dimorphism has been heightened in such cases by the need to form pair bonds as quickly as possible, a requirement that amplifies the usual process of sexual selection. Jehl (1970) has reached a similar conclusion in his study of dimorphism and breeding strategy among the arctic sandpipers. In the tropics, species of parulid warblers and many other passerine species are generally monomorphic but the plumage of both sexes is as brilliantly colored as the seasonal plumage of the males in migratory species. The explanation for this separate trend appears to be essentially the same as for duetting: pair bonds are permanent, and a premium is placed on continuous communication in these birds' complex tropical environments, where it is often difficult for them to see.

Selander (1966, 1972) has observed that sexual dimorphism is commonly based on a difference in food preference in species of birds that specialize on large food items. The relationship has been documented in woodpeckers, hawks and their allies, owls, frigate birds, jaegers and skuas, and the extinct huia (*Heteralocha acutirostris*) of New Zealand. Its basis appears to be the relative scarcity of the items, which places a premium on a division of the niche between mated birds that must cooperate in utilizing the same resources in order to rear young. The hypothesis gains strength from the discovery by Schoener (1965) that one category of birds showing interspecific character displacement in bill size is comprised of species that feed on scarce food items, especially those unusually large in size. A third, independent line of evidence comes from the data collected by Schoener (1967, 1968b) on West Indian *Anolis*. When small and medium-sized species of this insectivorous lizard occur alone on small islands, the sexes diverge in size. To a startling degree, the head length of the male approaches a mean of 17 millimeters, and the head length of the female approaches a mean of 13 millimeters, regardless of the species. The implication appears to be that there exists an optimum division of labor between the males and the females living inside the territories of the males. Such ecological partitioning does not require group selection; it can stem entirely from selection at the individual level. Specifically, that female best survives who is able to eat well within the territory of a male, while that male breeds the most who is best able to accommodate females.

Sexual dimorphism need not be entirely anatomical in nature. Males and females can also divide food niches through differences in behavioral responses. Among some species of *Dendrocopos* woodpeckers, the two members of a mated pair forage close together but maintain a spatial separation by virtue of the dominance of the male over the female. As they work over the same area, they utilize different tree strata—living as opposed to dead trees and branches, branches of different sizes—or use different foraging techniques to glean the same spots (Ligon, 1968; Jackson, 1970; Kilham, 1970).

Chapter 16   **Parental Care**

The pattern of parental care, being a biological trait like any other, is genetically programmed and varies from one species to the next. Whether any care is given in the first place, and of what kind and for how long, are details that can distinguish species as surely as the diagnostic anatomical traits used by taxonomists. The females of most hemipterous bugs, for example, simply deposit their eggs on the host plant and depart. In a few cases one parent—whether the female or the male depends on the species—stands guard over the egg mass until the nymphs emerge. Adults of a subset of these species protect the nymphs as well, standing near or over them and warding off predatory insects. In a still smaller group, which includes the tingid *Gargaphia solani* and scutellerid *Pachycoris fabricii,* the young orient to the mother and follow it from place to place. Some evidence exists that the females of one species, the Brazilian pentatomid *Phloeophana longirostris,* also provide nourishment to the nymphs (Bequaert, 1935). Some arachnids abandon their eggs or guard them to the time of hatching; others carry their newly hatched young around in brood pouches on the abdomen (see Figure 16-1). Parental care in vertebrate species is even more diversified. Birds are constrained by their own warm-bloodedness, which requires that the eggs and young be kept within a narrow range of temperature. But the 8,700 living bird species use virtually every conceivable device for accomplishing this (Kendeigh, 1952). Many species, from ostriches to pheasants, have precocial young that are able to run and feed within hours after emergence from the egg. The megapodes of Australia and southeastern Asia not only possess precocial young but have given up nearly every trace of postnatal parental care. The female simply buries the eggs in sand, volcanic ash, or mounds of rotting vegetation and allows the sun and heat of decomposition to provide the heat for incubation. At the opposite extreme are species in which one of the parents sits on the eggs without food until the young birds hatch; these spartan types include the emus, the eider duck, the argus pheasant, and the golden pheasant. Altricial bird species, those with helpless young that require protection and nursing within a nest, also vary greatly in the amount and kind of aid they provide. Lesser but still impressive amounts of diversity exist within the fishes (Sterba, 1962; Wickler, 1963; Barlow, 1974a), amphibians (Noble, 1931; Goin and Goin, 1962), reptiles (Tinkle, 1969; Greer, 1971; Neill, 1971), and mammals (Rheingold, ed., 1963; Fraser, 1968; A. Jolly, 1972a). Such variation is evidently due to the sensitivity of parental behavior to natural selection.

The Ecology of Parental Care

By bits and pieces a true theory of parental care has begun to take form (Cole, 1954; Williams, 1957, 1966a, b; Hamilton, 1966; Wilson, 1966, 1971a; Tinkle, 1969; Gadgil and Bossert, 1970; Emlen, 1970;

Figure 16-1 Parental care in arachnids. A female of the whip scorpion *Mastigoproctus giganteus* carries her newly hatched prenymphs in and around the brood chamber of her abdomen. (From Weygoldt, 1972.)

Trivers, 1974). Expressed in the language of population biology, it postulates a web of causation leading from a limited set of primary environmental adaptations through alterations in the demographic parameters to the evolution of parental care as a set of enabling devices. The reader can gain the essential idea by studying the diagram in Figure 16-2. The proposition states that when species adapt to stable, predictable environments, K selection tends to prevail over r selection, with the following series of demographic consequences that favor the evolution of parental care: the animal will tend to live longer, grow larger, and reproduce at intervals instead of all at once (iteroparity). Further, if the habitat is structured, say, a coral reef as opposed to the open sea, the animal will tend to occupy a home range or territory, or at least return to particular places for feeding and refuge (philopatry). Each of these modifications is best served by the production of a relatively small number of offspring whose survivorship is improved by special attention during their early devel-

opment. At the opposite extreme, species sometimes penetrate new, physically stressful environments by developing idiosyncratic protective devices that include care of offspring through the most vulnerable period of their development. Specialization on food sources that are difficult to find, to exploit, or to hold against competitors is occasionally augmented by territorial behavior and the strengthened defense of the food sources when offspring are present. A few species of vertebrates even train their offspring in foraging techniques. Finally, the activity of predators can prolong parental investment to protect the lives of the offspring. All four of these environmental prime movers—stable, structured environments leading to K selection, physical environments that are unusually difficult, opportunities for certain types of food specialization, and predator pressure—can act singly or in combination to generate the evolution of parental care. Let us now examine some of the logic and evidence behind the theory.

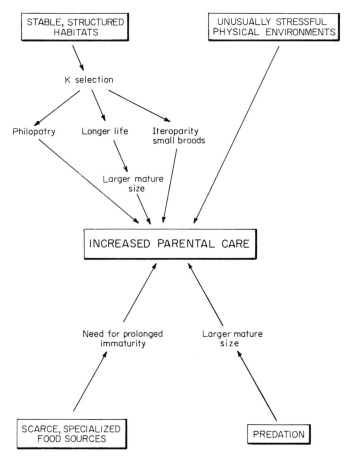

Figure 16-2 The prime environmental movers and intermediate biological adaptations that lead to increased parental care.

Iteroparity and Reduced Brood Size

As explained in Chapter 4, life cycles can be expected to be molded in a way that maximizes the sum of the products of survival and reproduction in each interval through time. This sum cannot be expanded indefinitely, because survivorship (ℓ_x) at each age x is generally an inverse function of reproduction (m_x) at the same age. In most instances reproductive effort diminishes not only present and future survivorship but also future fertility. Thus to add a unit of personal genetic fitness by "cashing in" on reproduction at any age is to subtract some quantity of personal genetic fitness by alteration of the life cycle. It follows that the life cycle of each population will evolve in a way that makes the best compromise in terms of units of genetic fitness summed over the lifetimes of individuals. The

Gadgil-Bossert model confirms that iteroparity is the optimum strategy if the cost in fitness due to reproduction gradually increases with age, or if the benefit gradually decreases, or both. If neither condition holds, the best pattern is semelparity—reproduction in one large, usually suicidal burst. Iteroparity is the rule in the vertebrates and among the solitary members of the Orthoptera and Hymenoptera, the orders of insects that have given rise to the advanced social species. If iteroparity did not exist and constitute a preadaptation in these particular groups, then sociality would be a rare and for the most part weakly developed phenomenon.

This reasoning leads to the theory of brood size. The smaller the brood, the more likely the iteroparous adult is to care for it. Also, the more effort the parent puts into rearing young, the more precisely controlled will be the brood size. The idea was first developed by David Lack (1954, 1966), who showed that clutches of songbirds deviating by as few as one or two eggs from the average number of the species produce fewer fledglings than do those at the average size. Lack argued that fewer eggs fell below the parents' potential to raise them, while too many eggs resulted in undernourishment and high mortality for the growing brood as a whole. Wynne-Edwards (1962) offered a competing hypothesis, that clutch size is adjusted altruistically by parents to prevent overpopulation (see Chapter 5). Both logic and evidence have subsequently favored Lack's view. In particular, Cody (1966) extended the theory of clutch size in a way that permits an independent test of Lack's hypothesis. Cody recognized three adaptive "goals" between which some compromise is essential: large clutch size, efficient food search, and effective predator escape. He postulated that large clutch size increases r, efficient food search increases K, and predator avoidance increases both, so that the absence of predators from a particular area will not alter the balance between clutch size and feeding efficiency. Cody's argument yields several nonobvious predictions. On the seasonal north temperate mainland, where r selection is generally more important than K selection, clutch size will be larger and feeding efficiency somewhat less. This effect should be lessened on offshore islands at the same latitudes, which enjoy a generally milder and less fluctuating climate. Toward the mainland tropics, where predation and K selection are considered to be more important, the compromise should lean toward feeding efficiency and predator avoidance, and clutch size should be diminished accordingly. On tropical islands yet another trend can be expected to appear: predators are less important and selection already leans toward feeding efficiency, so that the reduction in clutch size should be less than on the nearby mainland. All of these predictions are consistent with the evidence (see also MacArthur, 1972). Similar theory, appropriately modified to take into account special biological properties, will apply to the brood size of other kinds of animals that supply postnatal care.

Longevity and Delayed Maturity

The original formulations of the evolution of senescence by Peter Medawar and G. C. Williams predicted that selection for genes postponing mortality will be most intense at the ages of greatest reproductive value. Hence, senescence will increasingly intrude after the organism has attained reproductive age. Hamilton and Emlen subsequently deduced that mortality will be concentrated in the earliest ages in those species with the most substantial parental care. The reason is that when embryos are defective, or the newly born young are ailing, it is often more profitable for the parent to jettison them, "cut its losses," and begin again with new offspring. The more the parental investment measured by the length of the gestation period, the larger the size of the offspring at birth, and the greater the amount of care devoted to the neonatal infant, the earlier the mortality will be programmed. When a heavy early investment is made, prolonged postnatal care, extended immaturity, and long life are likely to emerge as coadaptations. Furthermore, the older the parent, the more personal risks it is likely to take on behalf of the offspring. Emlen has called attention to the possibility of the evolution of spite as a complementary adaptation. The parent that invests heavily in offspring is apt to behave destructively toward the offspring of unrelated individuals, and this hostility will reach its maximum at the age of their greatest reproductive value. It could be significant in the context of this theory that human beings tend to respond with the most unreasoning fear and hostility not to the small children and elders of strange groups but to their late adolescents and young adults.

Long life and low fertility can be mutually reinforcing in still another way. Suppose that long life has been favored by circumstances wholly independent of reproductive effort—say, a rich, stable environment relatively free of predators. Suppose further that conditions are not favorable for the emigration of offspring. Then progeny of this *K*-selected species are likely to become direct competitors of the parents. If the parents have lived only part of their life span, each unit of genetic fitness gained by an offspring at their expense is compensated for by as little as one-half a unit of inclusive fitness. On this basis alone it pays not to reproduce frequently. In practice, competition from offspring offers no separate theoretical problem, since it can be computed as part of the reproductive effort.

The positive correlations between low reproductive effort, lateness of maturity, and large investment in individual offspring generally hold through the major vertebrate groups. An example from the lizards, compiled by Tinkle, is given in Table 16-1. In these animals greater reproductive effort is reflected in the proportionately larger weight and size of the clutches and in the greater effort devoted to courtship, which in turn is measured by increases in the degrees of sexual dimorphism and elaborateness of courtship behavior.

Large Size

Longer-lived animals not only mature later, but also are generally larger. The expected correlation between size and parental care holds in the birds and mammals, but it is weak or absent in the fishes and reptiles. Williams (1966a) reviewed the evidence in the latter group in some depth and concluded that the lack of correlation is due to a compromise with extraneous factors. A small fish may have a greater need for defending its eggs in nests, while its size makes it less effective in providing protection against predators. Oral incubation and viviparity are alternatives available to it, but they impose a reduction in fertility that make them less valuable. The social insects appear to conform to the a priori size rule. The cryptocercid cockroaches, which are closely related to the primitive termites and show strongly developed parental care, are large in size, long-lived, and breed very slowly. The nonsocial tiphiid wasps considered closest among living Hymenoptera to the ancestors of the ants display similar traits in comparison with most of the remainder of the Hymenoptera. The same is true of the solitary vespid wasps that are most closely related to the social species. No clear relation either way exists in the bees.

Table 16-1 The number of early-maturing and late-maturing lizard species that display more or less reproductive effort and parental investment. N = number of species in sample. (From Tinkle, 1969.)

Variable	Early-maturing ($N = 35$)	Late-maturing ($N = 23$)
Clutch weight/body weight	0.2–0.3 (mean 0.25), $N = 5$	0.1–0.4 (mean 0.30), $N = 9$
Number of clutches	1–6 (mean 3), $N = 17$	1–2 (mean 1.1), $N = 19$
Sexual dimorphism	Strong in 16 of 19 species	Weak or absent in 11 of 17 species
Courtship type	Elaborate in 5 of 9 species	Elaborate in 1 of 6 species
Territoriality	All of 10 species	In 10 of 14 species
Viviparity	In 1 of 35 species	In 7 of 23 species
Parental care	In 2 of 35 species	In 5 of 14 species

Philopatry

Parental care is facilitated by the existence of nest sites where the young can be left for intervals while the parents forage for food and ward off predators. The most primitive living members of the ants, social wasps, and social bees prepare secured nest sites to which they home efficiently. Entomologists generally agree with the view of Wheeler (1923, 1933), later strengthened by Evans (1958) for wasps, that the elaboration of nesting behavior has been a crucial factor in the repeated emergence of social behavior in the Hymenoptera. Colonies of cryptocercid cockroaches, as well as of the primitive termites closely related to them, occupy rotting logs and other cellulose sources for most or all of their lives. Fish species evincing the greatest amounts of postnatal parental care also typically occupy territories on mudflats, coral reefs, and other bottom habitats. They are to be contrasted with other species that spend their lives wandering through open water (Barlow, 1974a). Among the mammals, nests and territories are not essential for parental care. The females of some migratory ungulates, such as wildebeest and reindeer, rear their young with the aid of neither. But in the great majority of species the females, occasionally aided by males, keep their young in defended nest sites. Finally, the dependence of parental care by birds on nesting and homing is nearly universal and does not need further comment with reference to the present argument.

Unusually Stressful Physical Environments

Next to a stable, predictable environment, the condition most likely to promote the evolution of parental care is very nearly the direct opposite. When species penetrate new habitats in which one or more physical parameters are exceptionally stressful, they sometimes add parental care as the only means of advancing the young to a developmental stage in which they are able to cope with the new conditions. *Bledius spectabilis* is a beetle that lives in an extreme environment for an insect—the intertidal mud of the northern coast of Europe, where it constantly faces the hazards of high salinity and oxygen shortage. The species is also exceptional within the large taxonomic group to which it belongs (family Staphylinidae) in the amount of care given by the female to her brood. She keeps the larvae in a burrow, protects them from intruders, and brings them fresh algae at frequent intervals (Bro Larsen, 1952). The plethodontid salamanders are unusual in the degree to which they have penetrated the land environment. Their eggs are laid in the soil, pieces of wood, or equivalent sites, and are often protected by the mother until they hatch. Instead of passing through an aquatic larval stage, the young hatch directly into a form resembling a miniature adult. Highton and Savage (1961) found that in *Plethodon cinereus* the presence of the mother is important for normal development of the eggs. The yolk

is utilized more fully, the embryos grow to a larger size, and over twice the number of young survive than in comparable groups of eggs deprived of maternal protection. The mothers also actively defend their eggs from other females. Frog species that lay their eggs on land, including those that provide some degree of parental care, are almost without exception dwellers in humid mountainous areas. The phenomenon is especially common in the tropical highlands. There the environment is still stressful but evidently much less so than in the drier, more seasonal lowlands. Goin and Goin (1962) view this behavioral change as a repetition of the first step in evolution that led to the origin of the reptiles from amphibians during the late Paleozoic, a time of extensive mountain formation.

Scarce or Difficult Food Sources

The slowest breeding of all birds are the eagles, condors, and albatrosses. Only one young is fledged at a time, and the full breeding cycle occupies more than a year. Maturity requires at least several years; condors and royal albatrosses do not begin to breed until they are about nine years old. The ecological trait that all these species have in common is dependence on food that is sparse and difficult to obtain (Amadon, 1964). Foraging consists of long, skillful searches. Homing occurs over long distances, and resourcefulness in transport is often required. Some eagles, for example, wander over thousands of square kilometers in search of prey. During the breeding season, however, movement must be severely restricted. The male normally does all of the hunting for himself, his mate, and the single chick. When the young bird is nearly grown, the female begins to hunt also. The prey of the largest eagles are moderate-sized mammals such as tree sloths, monkeys, and small antelopes. Bringing them to the nest after the kill has been made often requires exceptional strength and skill. Little wonder, then, that the young must attain a large size themselves before attempting an independent existence. The crowned eagle (*Stephanoaetus coronatus*) perhaps represents the ultimate case. Mated pairs breed in alternate years and require at least 17 months to fledge a single offspring. The young bird sometimes kills for itself long before it leaves its parents. A similar phenomenon may occur in certain smaller seabirds that search for food over large areas of the open sea. Royal terns (*Thalasseus maximus*) and frigate birds (*Fregata*) continue to feed their offspring after the latter have left the nest. The same kind of birds have been observed practicing "play" activities that appear to contribute to hunting skills. They snatch objects from one another's beaks in mid-flight, break branches from trees, and follow one another in close formation while swooping low over the water (Ashmole and Tovar S., 1968). A specialized example of protracted parental investment occurs in bee-eaters (Meropidae), in which the offspring are fed nonvenemous insects in the nests and receive additional care outside the nest while they learn the difficult art of devouring bees, their primary prey (Fry, 1972).

A comparable degree of prolonged immaturity characterizes large mammalian carnivores such as wolves, African wild dogs, and the great cats. Lions engage in training sessions during which the adult females initiate their young to the hunting of prey. According to Schenkel (1966b) these exercises resemble real stalking up to a point but are not carried through to the kill. The following is a typical example:

At dawn the two mothers approached the six cubs, A_1 and B_1, and were immediately greeted by them. After a little play all the lions settled down on a minute elevation on the otherwise flat terrain, covered sparsely with gall acacias, and watched the surroundings. When two wildebeeste bulls passed the group at a distance of 50–60 yd, one of the mothers rose immediately followed by the other one. Both stepped forward in a stalking gait with a distance of about 15 yd between them, in order to approach the bulls transversely from behind. Without hesitation the young lions joined in, forming an irregular front line and taking advantage of the acacias as cover. As the bulls walked at fair speed, the lions did not get much closer and one after the other including the mothers gave up. Only two young lions continued to stalk until they had to cross a nearly completely bare flat. Here the wildebeeste bulls detected them and ran away in an easy gallop.

When the female lions leave on "real" hunting trips, they walk off from the cubs in a determined gait and the youngsters do not even try to follow. The invitation presented to the cubs by the mothers appears to be the start of an elaborate play session. Schenkel's wild cubs began to hunt on their own when they were about 20 months old, while they were still under the care of the females. Their first victims were warthogs, but they also stalked wildebeest and zebras frequently. When some of the young lions busied themselves at these activities, the other cubs watched intently from a short distance.

Parent-Offspring Conflict

The traditional view of the relationship between parent and offspring has always assumed unilateral parental investment. The offspring was considered to represent so many units of genetic fitness to the parent, a more or less passive vessel into which a certain amount of care is poured to enlarge the investment. Until recently, behaviorists have not come to grips with the phenomenon of parent-offspring conflict during the weaning period. As the juvenile grows older, the mother discourages it with increasing firmness. For example, the female macaque removes the juvenile's lips from her nipple by pushing its head with the back of her hand; she holds its head beneath her arm, or strips the infant away from her body altogether and deposits it on the ground. The juvenile, sometimes screaming in protest, struggles to get back into a favorable clinging position (Rosenblum, 1971a).

Among ungulates the discouragement often shades into open hostility. A young moose passes through two crises during the period of declining dependence on its mother. The first is in the spring, when it is one year old and its mother has just given birth to a new calf. The dam suddenly turns hostile and drives the yearling from her territory. The young moose lingers in the immediate vicinity and repeatedly attempts to return to the dam. In the fall, at the onset of the rutting season, the territorial barriers relax and the yearling is able to draw close to its mother again. But this new proximity precipitates the second crisis. Dams now treat their daughters as rivals, while bulls chase away the young males as if they were adults. At this stage the young animal finally becomes independent of its mother (Margaret Altmann, 1958).

Mammalogists have commonly dealt with conflict as if it were a nonadaptive consequence of the rupture of the parent-offspring bond. Or, in the case of macaques, it has been interpreted as a *mechanism* by which the female forces the offspring into independence, a step designed ultimately to benefit both generations. Hansen (1966), writing on the rhesus, expressed this second hypothesis as follows: "One of the primary functions that the mother monkey served was seen in her contribution to the gradual, but definite, emancipation of her infant. Although this process was aided and abetted by the developing curiosity of the infants to the outer world, this release from maternal bondage was achieved in considerable part by responses of punishment and rejection." A similar explanation was advanced by Hinde and Spencer-Booth (1967). But the data are equivocal. Kaufmann (1966) concluded from his own study of free-ranging bands that young rhesus monkeys are drawn away from their mothers more by attraction to other monkeys than by maternal rejection. The data of Hinde and Spencer-Booth on captive animals do not contradict this view. They can be interpreted as indicating that maternal rejection tends to increase attempts by the infants to stay with their mothers, independently of the attractiveness of other troop members.

A wholly different approach to the subject has been taken by Trivers (1974). Rather than viewing conflict as the rupture of a relationship, or a device promoting the independence of the young animal, Trivers interprets it as the outcome of natural selection operating in opposite directions on the two generations. How is it possible for a mother and her child to be in conflict and both remain adaptive? We must remember that the two share only one-half of their genes by common descent. There comes a time when it is more profitable for the mother to send the older juvenile on its way and to devote her efforts exclusively to the production of a new one. To the extent that the first offspring stands a chance to achieve an independent life, the mother is likely to increase (and at most, double) her genetic representation in the next breeding generation by such an act. But the youngster cannot be expected to view the matter in this way at all. So long as the continued protection of its mother increases its own inclusive genetic fitness, the young animal should try to remain dependent.

If the mother's inclusive fitness suffers first from the relationship, conflict will ensue. More precisely, selection will favor rejection behavior on the part of the mother when the cost in units of fitness to herself exceeds the benefit in the same units, while the offspring will try to hang on until the cost to its mother exceeds twice the benefits to itself. At that point the offspring's inclusive fitness is reduced and independence becomes profitable. We can expect that when the offspring is very small in size, the ratio of cost-to-mother/benefit-to-offspring is also very small and the mother and offspring will "agree" to continue the dependent relationship. As the youngster grows it becomes increasingly more expensive to maintain in units of inclusive fitness, so that the following two thresholds are crossed in sequence:

Cost-to-mother/benefit-to-offspring exceeds 1: the conflict begins as the mother's fitness declines but the inclusive fitness of the offspring is not yet diminished by the relationship.

Cost-to-mother/benefit-to-offspring exceeds 2: the conflict ends and the offspring willingly leaves, because the inclusive fitness of both participants is now diminished.

The hypothetical time course of the relationships is represented in Figure 16-3. The principal conflict can be expected to begin during the period of weaning, when the young animal becomes independent of milk or other food provided directly by the parent. Weaning conflict has been documented in a variety of mammals, including rats, dogs, cats, langurs, vervets, baboons, and the rhesus and other species of macaque. Among birds it has been recorded in the herring gull,

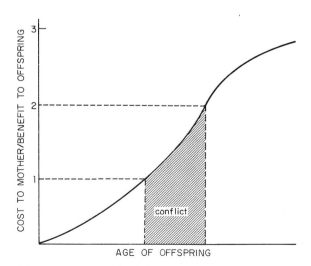

Figure 16-3 Trivers' model of the timing of parent-offspring conflict.

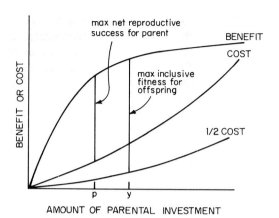

Figure 16-4 Parent-offspring conflict of varying degrees throughout the period of parental care is possible under the conditions envisaged here. The benefit, cost, and half the cost of a parental act toward an offspring at one moment in time are shown as functions of the amount of the parental investment in the act. An example of an investment in mammals would be the quantity of milk provided during one day of nursing. At *p* the parent's inclusive fitness (benefit minus cost) is maximized; at *y* the offspring's inclusive fitness (benefit minus ½ cost) is maximized. The parent and offspring are consequently selected to disagree over whether *p* or *y* should be invested. (Modified from Trivers, 1974.)

red warbler, Verreaux's eagle, and white pelican, and is perhaps widespread in altricial birds generally. It even appears to occur in mouth-brooding fish (Reid and Atz, 1958). Another form of parent-offspring conflict, known in rodents and ungulates, is territorial exclusion.

The period of conflict is in fact one extreme case, entailing disagreement over whether any aid at all will be given the offspring. It is equally easy to conceive of circumstances earlier in the life of the young animal when the interests of both individuals are served when the parent helps, but disagreement will exist over *how much* help is to be given. This lesser conflict results from the fact that whereas the adult is selected to bestow the amount of investment that maximizes the difference between benefit and cost, the offspring is selected to try to secure an amount of investment that maximizes the difference between the benefit to itself and the cost to its mother devalued by the relevant coefficient of relationship, which is normally ½. The two functions are visualized graphically in Figure 16-4.

Trivers' hypothesis is consistent with the time course of conflict observed in cats, dogs, sheep, and rhesus monkeys. In each of these species the conflict begins well before the onset of weaning and tends to increase progressively thereafter. In dogs (Rheingold, 1963) and cats (Schneirla et al., 1963) the period of maternal care has been explicitly divided into three successive intervals characterized by increasing conflict. In the first stage, most nursing is initiated by the mother, and she seldom if ever resists the infant's advances. During the second interval, approaches are initiated by both individuals with about equal frequency; the mother occasionally rejects the offspring and may even treat it with hostility. The third interval is the period of weaning. The young animal initiates most or all nursing episodes and is usually rejected. Paradoxically, but in a manner consistent with the theory, the more wide-ranging and independent the juvenile becomes, the more frequently it seeks to renew contact with the mother.

Trivers' model can be fitted closely in nonobvious ways to the detailed results of experiments conducted on rhesus infant development by Robert Hinde and his associates (Hinde and Spencer-Booth, 1971; Hinde and Davies, 1972a,b). When an infant was separated from its mother for a few days and then reunited with her, it sought contact more frequently than it did previous to the separation. In contrast, control infants left with their mothers decreased their rate of contact during the same interval. Separated infants also displayed more signs of distress, such as calls and immobility, after they were reunited with their mothers. The more the mothers rejected the infants prior to the separation, the more distress they displayed afterward. Even more significantly, infants separated by having their mothers removed were more distressed than those who were taken from their mothers. All of these results are consistent with the view that the infant strives to increase the amount of maternal investment and is alert to signs that the mother is reducing the investment. The data are not consistent with the two principal competing hypotheses, namely, that maternal rejection has been selected as a device for promoting independence in the young, or that independence is attained primarily by attraction to other members of the society.

Trivers has elaborated his model to account for inclusive selection through a wide array of relatives and nonrelatives. However, the refinement may eventually prove applicable only to human beings and a few of the other most intelligent vertebrates. Consider the offspring that behaves altruistically toward a full sibling. If it were the only active agent, its behavior would be selected when the benefit to the sibling exceeded 2 times the cost to itself. From the mother's point of view, however, inclusive fitness is gained whenever the benefit to the sibling simply exceeds the cost to the altruist. Consequently, there is likely to evolve a conflict between parents and offspring in the attitudes toward siblings: the parent will encourage more altruism than the youngster is prepared to give. The converse argument also holds: the parent will tolerate less selfishness and spite among siblings than they have a tendency to display, since its inclusive fitness will begin to suffer at a lower intensity. The same inequalities hold through an indefinitely widening circle of relatives and nonrelatives. Altruistic acts toward a first cousin are ordinarily selected if the benefit to the cousin exceeds 8 times the cost to the altruist, since

the coefficient of relationship of first cousins is $\frac{1}{8}$. However, the parent is related to its nieces and nephews by $r = \frac{1}{4}$, and it should prefer to see altruistic acts by its children toward their cousins whenever the benefit-to-cost ratio exceeds 2. Parental conscientiousness will also extend to interactions with unrelated individuals. From a child's point of view, an act of selfishness or spite can provide a gain so long as its own inclusive fitness is enhanced. The exploited individual (or society as a whole) may retaliate against the individual and one or more members of its family. But the act will be favored in selection if the benefit it brings is greater than the loss inflicted by the retaliation, where the loss is summed over the individual and its relatives and devalued by the appropriate coefficients of relationship. The parents can be expected to view the matter according to approximately the same calculus. However, since they lose more inclusive fitness by costs inflicted on the offender's siblings and other relatives, they are likely to be less tolerant of the act. In human terms, the asymmetries in relationship and the differences in responses they imply will lead in evolution to an array of conflicts between parents and their children. In general, offspring will try to push their own socialization in a more egoistic fashion, while the parents will repeatedly attempt to discipline the children back to a higher level of altruism. There is a limit to the amount of altruism the parents wish to see; the difference is in the levels that selection causes the two generations to view as optimum. Trivers has summarized the argument as follows: "Conflict during socialization need not be viewed solely as conflict between the culture of the parent and the biology of the child; it can also be viewed as conflict between the biology of the parent and the biology of the child." Above all, the young individual is not simply a malleable organism being molded by its parents, as psychologists have conventionally viewed it. On the contrary, the youngster can be expected to be receptive to some of the actions of its parents, neutral to others, and hostile to still others.

The implications of the conflict theory do not stop here. Under some circumstances parents can be expected to influence the offspring's behavior into its adult life. Altruistic behavior might be induced when it results in an increase in inclusive fitness through benefits bestowed on the parents and other relatives. The celibate monk, the maiden aunt, or the homosexual need not suffer genetically. In certain societies their behavior can redound to improved fitness of parents, siblings, and other relatives to an extent that selects for the genes that predisposed them to enter their way of life. Moreover, their relatives, and especially their parents, will respond in a way that reinforces the choice. The social pressure need not be conscious, at least not to the extent of explicitly promoting the welfare of the family. Instead, it is likely to be couched in the sanctions of custom and religion. In case the benefit-to-cost ratio is less than 1 for the individual but greater than 1 for members of the family, a conflict will arise that makes selection for the trait far less likely. Even so, such an evolutionary trend can be sustained where the greater experience and initial advantage of the relatives prove overwhelming.

Parental Care and Social Evolution in the Insects

One of the few really fundamental differences between insect and vertebrate societies exists in the realm of parental care. The species of social insects display a great diversity in the form and intensity of attention paid the young, and this variation is only weakly correlated with the degree of complexity of social organization. In some of the most advanced species virtually no contact occurs between the adults and immature forms. Moreover, it is probable that parent-offspring interaction has little effect on the development of social behavior. Vertebrate species, in contrast, display a strong correlation between the amount of parental care and the complexity of social organization. Furthermore, the behavior of the parent strongly influences the social development of the offspring. Both of these relations are especially well marked within the mammals.

Consider first one of the groups of social insects with intimate parental care. The social wasps of the genera *Vespa* and *Vespula*, called hornets or yellowjackets in the United States, house their growing larvae in individual hexagonal paper cells that are packed together on horizontal brood combs. When hungry, the larvae attract the attention of the workers by rhythmically scraping the sides of the brood cells with their heads, producing a sound not unlike the crunching of lettuce (Ishay and Landau, 1972). The workers feed their charges with finely masticated but undigested pieces of insect prey, which are placed directly on the mandibles. From time to time the larvae exude droplets of salivary secretion, which the workers quickly lap up (Figure 16-5). The paired salivary glands of the wasp larvae are relatively huge, each divided posteriorly into ventral and dorsal rami that wind through the body cavity. The first solid clue to the significance of the larval secretion was obtained by Montagner (1963), who found that males of *Vespula* obtain nutrition from larval saliva. At the beginning of queen rearing, the males are rebuffed by the workers when they beg, and they then turn to larval secretions as their principal source of food. In biochemical studies of the same species of *Vespula*, Maschwitz (1966b) confirmed that the larval salivary secretion is both attractive and nutritious. It contains an average of 9 percent trehalose and glucose, about four times the concentration in the larval hemolymph. Maschwitz believes that the sugars alone are enough to induce adult feeding, and there is no evidence of additional attractants. The secretion contains mostly sugars; amino acids and proteins are present in the saliva but at only one-fifth the

Figure 16-5 A worker of the social wasp *Vespula vulgaris* feeding on the salivary secretion of a larva. (From Maschwitz, 1966b.)

concentration in the hemolymph. Montagner and Maschwitz both consider the larval secretions to constitute a colony food reserve, serving the same function as the crop liquid passed among adults. The salivary glands are thus the functional analog of the crop of adult workers. Maschwitz calculated that a microliter of larval saliva provides the energy to keep a worker *Vespula* alive for up to 1.8 hours, and the sugar released by a large larva in a single "milking" suffices a worker for half a day. Subsequently, Ishay and Ikan (1969) added a fascinating new twist to the story. They discovered that not only do the larvae of the wasp species they studied in Israel, *Vespa orientalis*, supply the adults with salivary carbohydrates, but also that the larvae are the only colony members capable of converting proteins to carbohydrates in the first place. Only these individuals possess chymotrypsin and carboxypeptidase A and B. There is no evidence that the adults can engage in protein digestion at all. The larvae manufacture glucose, fructose, and sucrose, as well as unidentified trisaccharides and tetrasaccharides and feed them back to the worker nurses. The inability of the adult wasps to engage in gluconeogenesis is unusual in insects and it was, of course, quite unexpected as a social mechanism. The principal significance of the findings on the vespine wasps is that they demonstrate for the first time that larvae can behave altruistically toward adults and that they therefore contribute, by virtue of their behavior patterns, to the homeostatic machinery of the colony. Their monopoly of gluconeogenesis shows that advanced wasp societies have gone very far in arranging a biochemical division of labor between adults and young.

The case in the wasps thus seems strong at first for regarding the complexity of adult-offspring relations as a true measure of social evolution. If we pass next to the honeybee *Apis mellifera*, generally interpreted as occupying one of the summits of insect social evolution, this intuitive impression is sustained. Although direct trophallaxis does not occur between adults and larvae, the larval cells are repeatedly visited by nurse workers, who clean them and place fresh food next to the larvae. Further, a weak correlation exists in the ants between the intimacy of brood care and other parameters of social organization. Queens and workers of the genus *Myrmecia*, the most primitive living ants of the large myrmecioid complex, scatter the eggs over the floor of the nest. The larvae feed directly on fresh pieces of insect prey and are able to crawl short distances under their own power. Regurgitation between the adults and larvae rarely if ever occurs (Haskins and Haskins, 1950; Freeland, 1958). A comparable low level of brood care is maintained by species of *Amblyopone*, the most primitive living members of the second major division of ants, the poneroid complex (Haskins and Haskins, 1951; R. W. Taylor, personal communication). Within the anatomically and socially more advanced lines of the myrmecioid and poneroid complexes, the brood is more closely attended by the workers and the kinds of interactions are more numerous. Eggs are characteristically bunched with the first instar larvae. Larvae are typically immobile except for the mouthparts, and they donate salivary secretions to the workers in return for fragments of solid food and nutritive liquid regurgitated from the workers' crops (Le Masne, 1953, Wilson, 1971a). The secretions of at least one species, *Monomorium pharaonis*, have been shown to prolong the survival of otherwise unfed workers (Wüst, 1973). This saliva contains substantial quantities of amino acids and proteins, including proteases, but few or no fats or carbohydrates. The watery fraction of the secretions also assists workers to survive prolonged dry periods. Thus the adult-larva symbiosis in these ants is reciprocally beneficial, although it does not appear to be quite as specialized as in *Vespa orientalis*. A parallel evolutionary trend has been followed by the termites. The immature stages, or nymphs, of the primitive

families Kalotermitidae and Rhinotermitidae are self-reliant. Indeed, they individually contribute more to the labor of the colony than the adults. Upon attaining nearly full growth they remain indefinitely in a nymphlike stage that entomologists call a pseudergate. Only a small fraction of the pseudergates transform into the fully developed, irreversible castes of soldiers, reproductive females, and reproductive males. Social organization of the more primitive termites is therefore based to a large extent on "child labor" (Lüscher, 1961b; E. M. Miller, 1969). In the phylogenetically more advanced Termitidae, which are generally referred to as the higher termites and constitute 75 percent of all of the known 1,900 termite species of the world, the young are much more dependent on the adults. In its first two or three instars the developing termitid is usually classified as a larva, despite the fact that it is essentially a nymph in form, because it is helpless, does not participate in the colony labor, and is fed salivary secretions by the older nymphs and the true worker caste (Noirot, 1969b).

The overall phylogenetic correlation that seems to exist between the intimacy of brood care and other parameters of social organization is sharply disrupted by the stingless bees. These insects, constituting the tribe Meliponini of the family Apidae, are phylogenetically close to the honeybees. In most regards they are as socially advanced as *Apis mellifera*. In some species the colonies are as large and the anatomical and behavioral differences between the queen and worker castes as well marked as in the honeybee. Colony division is achieved by a type of swarming which is different from that in the honeybees but equally complicated. The queen repeatedly performs a bizarre ritual during which she consumes stored food and a worker-laid egg before laying an egg of her own in each cell of the brood comb. Some of the species deposit odor trails to food finds. The workers of one form, *Melipona quadrifasciata*, lead nestmates to the target with buzzing zigzag runs that are nearly as sophisticated as the waggle dance of the honeybee (Esch, 1967 a,b). Yet in spite of all these social adaptations there is no contact at all between the adult meliponine bees and their larvae. In the classic manner of bees, the workers fully provision each brood cell at the outset with pollen and nectar. As soon as the queen lays an egg in the cell, it is tightly capped by the workers and essentially abandoned. When the larva inside hatches, it feeds on the provisions stored around it, grows through each molt, and pupates, all entirely on its own. Its first contact with other members of the colony comes as it emerges from the cell as a fully developed, winged worker. Within hours the young bee embarks on a complicated round of tasks that, in the two to three months it has to live, includes nest construction, foraging, provisioning of brood cells, and sometimes emigration to new nest sites. These acts appear to be shaped little if any by socialization. In fact, when Nogueira-Neto (1950) introduced meliponine bees as pupae into the nests of other species and allowed them to eclose as adults, they proceeded to construct brood combs and other nest structures in the form characteristic of their own species rather than that of their hosts.

The exact reverse of the situation in stingless bees is found in many subsocial insects. Here there exists complex, intimate parental care but no additional social organization. The negative correlation seen in the bees is thus preserved. For example, the adults of the scolytid beetle genera *Gnathotrichus*, *Monarthrum*, and *Xyloberus* keep their young in "cradles," which are short diverticula from main galleries carved through dead wood, and they feed them on a special fungus that is cultured for the purpose. In *Monarthrum* the mated pair work together to excavate the nest and to care for the brood. The mother beetle lays eggs singly in circular pits carved into the walls of the main gallery on opposite sides parallel to the grain of the wood. According to Hubbard (1897), the eggs are loosely packed in the pits with chips and with mycelia taken from nearby fungus beds. As soon as the larvae hatch, they begin to eat the fungus from the chips and to eject the refuse from the cradles. As they grow they enlarge the cradles by chewing at the walls with their mandibles and swallowing the wooden chips. The fragments are passed through the intestines undigested and voided with the feces, cemented together with yellowish excrement. The pellets are pushed from the cradles by the larvae and picked up by their mother, who carries them to the fungus garden bed. The mother guards the young constantly throughout their development. As soon as the fungus wads at the cradle entrances are consumed, she replaces them with fresh material. Yet the association ends as soon as the larvae pupate; the mother is gone when the young adults emerge from the chambers.

The most advanced form of parental care in the subsocial insects is perhaps that displayed by the burying beetles of the genus *Necrophorus* (Pukowski, 1933; Niemitz and Krampe, 1972). In May the overwintered adults begin to search for the dead bodies of small vertebrates, such as birds, mice, and shrews. If a male encounters a corpse, he takes the "calling" posture, lifting the tip of his abdomen into the air and releasing a pheromone. The substance appears to attract only females belonging to the same species. If more than a single pair find a corpse—and sometimes as many as ten do—fighting ensues, male against male and female against female, until only a single pair is left. The winners then excavate the soil beneath or around the body until their prize is partly buried. At the same time they chew and manipulate the putrefying mass until it is roughly spherical in shape and can be rolled downward into a burrow excavated beneath it. Then the beetles seal off the burrow from below, entombing themselves with the rotting ball. The female proceeds to eat out a crater-shaped depression on the top of the ball and to spread her feces over the surface. When the larvae hatch, they sit in the crater like so many baby birds in the nest. Pukowski's observations show that they also interact with the parents much like the young of altricial birds. "As the female approaches the crater, the larvae lift the fore part of their bodies in unison, so that their legs are grasping air. The beetle stands directly over the larvae and strikes the food ball or the larvae with trembling motions of her fore legs. Now the female opens her mandibles, and one of the larvae swiftly inserts its head between the mandibles and tightly into the mouth opening. If circumstances are right one can see a brown liquid pass from the mother's mouth to the larvae. After a few seconds the beetle pulls back and attempts to put her mouth down on the head of another larva. Without doubt the brood is being fed by the female." Niemitz and Krampe have further demonstrated that the adult alerts the larvae with a distinctive chirping sound.

When the *Necrophorus* larvae are only five to six hours old they begin to feed on the putrefying ball directly. However, they continue to receive regurgitated food from the mother at irregular intervals, and for a time after each molt they are wholly dependent on this source of nutriment. If the mother is removed while the larvae are immature, they start to pupate, but are unable to complete the transformation. In two of the six European species that have been studied (*N. germanicus* and *N. vespilloides*) the male assists his mate in feeding the young, although he is less active than she. In spite of the close interactions of the adults with the larvae, they do not remain to see their offspring emerge from the pupae. So far as we know, the strong development of parental care in this and in many other groups of beetles has never led to the origin of a level of social organization comparable to that found in the most primitive termites and eusocial Hymenoptera, in other words, to close cooperation between members of two or more generations (von Lengerken, 1954; Wilson, 1971a).

Parental Care and Social Evolution in the Primates

To illustrate the more orderly relation that exists between brood care and social organization in the mammals, we can do no better than examine the extremes within the primates. The tree shrews, constituting the family Tupaiidae, are so primitive that their placement as primates has been disputed, some authors preferring to place them within the Insectivora close to the elephant shrews. However, the weight of evidence, chiefly osteological and serological in nature, indicates that they fit close to if not directly in line with the origin of the primates. Tree shrews turn out to have a simple but quite peculiar form of parental care, characterized by absenteeism on the part of the mother. Martin (1968) was able to follow the complete breeding cycle in six pairs of *Tupaia glis belangeri* under seminatural conditions. Sexual pair bonding is strong, and the male marks both the cage and his mate with scent. The male and female sleep together in a nest that is constructed mostly by the male. When the female becomes pregnant she builds a second nest, in which she later bears her young. The litter size is typically two. The mother soons leaves the helpless infants and returns to the first nest to rejoin the male. Thereafter, she visits the nursery nest only every 48 hours. The infants move from one of her six nipples to another in an unsystematic way. After a few minutes the mother shakes them off and runs away. The young are left to groom themselves and, presumably, to clean out their own feces. The adults do not retrieve them when they are left outside the nest, and the infants do not utter distress calls. Even when picked up, they fail to assume the curled-up posture taken by many mammal infants to aid their parents during transport. The parent tree shrews are repelled by the odor of the infants' urine; in one case when the young were born in the parental nest, the adults moved out the following day. Parents forced to remain too close to their young often kill and eat them. When they are about 33 days old, the young tree shrews start emerging from their nest for short foraging trips, and the rhythm of maternal visits begins gradually to break down. At first the young return to their own nest at night and whenever they are frightened, but after 3 days they shift to the parental nest. Sexual maturity is reached at about 90 days of age. Afterward the young evidently scatter to find mates and territories of their own. At least some variation in maternal behavior exists within the genus. Sorenson (1970) found that females of *T. (Lyonogale) tana* visit and nurse their young twice a day.

Although a low level of parental care was undoubtedly primitive in the early mammals, some question remains as to whether the remarkable absentee system of *Tupaia* is really a baseline or instead represents a special, secondary adaptation. Martin believes that absenteeism is primitive, that the nursery nest is the first, crude step in the elaboration of parental care. According to this hypothesis, grooming, nest cleaning, infant transport, and close protection of the young were steps added in later primate lineages. The alternative hypothesis would be that *Tupaia* has discarded some or all of these elements as a secondary adaptation to dissociate the parents as much as possible from the young. What would the basis for such segregation be? The chief advantage is that most of the mother's activities will not lead predators to the young, and might even lead them away. But as Martin has argued, there are potent disadvantages in absenteeism. The young are deprived of the parents' warmth and hour-by-hour protection from those predators that the adults can repel. Furthermore, a predator that locates its prey by their odor will have less trouble locating an uncleaned nest.

If absenteeism is the primitive pattern, it should appear regularly in other groups of mammals judged to be primitive on anatomical grounds. So far, a pattern closely similar to that of *Tupaia* has been described only in rabbits, particularly the European *Oryctolagus cuniculus* (Deutsch, 1957; Mykytowycz, 1959). Absenteeism of a sort is displayed by mothers of certain ungulates, such as Grant's gazelle and the elk (Ewer, 1968; M. Altmann, 1963); the Steller sea lion,

Alaskan fur seal, and possibly other pinnipeds (Bartholomew, 1959; Peterson and Bartholomew, 1967); and many kinds of bats (Novick, 1969). In each case, however, the pattern is different from that of *Tupaia* and *Oryctolagus*, and the behavior is clearly a secondary adaptation to special environmental circumstances. Turning to the Insectivora, which contain many of the most primitive of the living eutherian mammals, we find that with the exception of the elephant shrews (Macroscelididae) most or all of the known species have altricial young that are closely tended by the mother in a single nest (Crowcroft, 1957; Eisenberg, 1966; Eisenberg and Gould, 1970). Macroscelidid infants, unlike those of tupaiids, are precocial. Thus the hypothesis that absentee parental behavior is primitive is not supported by the phylogenetic evidence. But it nevertheless remains true that the tree shrews provide a minimum opportunity for the socialization of their young, and this condition is associated with nearly solitary existence on the part of the adults. Approximately the same level of solitary existence occurs in conjunction with simple maternal care in some of the other anatomically primitive primates, including tarsiers, dwarf lemurs (*Cheirogaleus*), and probably the aye-aye *Daubentonia madagascariensis* (Napier and Napier, 1967; Petter and Petter, 1967).

Above this lowest level, higher evolutionary grades exist in parental care that can be correlated with other intuitively suitable parameters of social organization. It would be premature to try to define these grades with any precision. Ontogenetic studies of parent-offspring relations are now being pursued in many species of primates, and the data will in time lend themselves to an appropriate and illuminating synthesis by experts. The following brief generalization is perhaps the best that can be made. Cope's Rule holds more or less within the primates: there has been an overall gradual increase in body size through geological time, with a few phyletic lines reversing the process and becoming smaller. As size increases, longevity and the periods of gestation and immaturity lengthen (see Table 16-2). Closely related to this trend are three ethoclines that can be discerned by comparison of the better-studied species: (1) the degree of socialization increases, and in particular the young become more dependent on learning for the acquisition and perfecting of social acts (see also Chapter 7); (2) the behaviors involved in parent-offspring interactions become more numerous, and the interactions more frequent; (3) the circle of group members involved in the socialization of the infant widens, and care bestowed on it by group members other than the parents becomes more extensive and complex. Evolutionary grades can be defined as horizontal lines drawn between more or less correlated points on the different ethoclines.

The highest grade below man is occupied by the chimpanzee. The development of the young chimpanzee has been studied in captive groups by many investigators, including R. M. Yerkes and his associates (Yerkes, 1943), Mason and Berkson (1962)), and others. But our

Table 16-2 The duration of life periods in seven species of primates, illustrating the overall tendency for increase within the order. (From Napier and Napier, 1967.)

Species	Gestation period (days)	Infantile phase (years)	Juvenile phase (years)	Adult phase (years)	Life span (years)
Lemur	126	$\frac{3}{4}$	$1\frac{3}{4}$	11+	14
Rhesus macaque	168	$1\frac{1}{2}$	6	20	27–28
Gibbon	210	2(?)	$6\frac{1}{2}$	20+	30+
Orang-utan	275	$3\frac{1}{2}$	7	20+	30+
Chimpanzee	225	3	7	30	40
Gorilla	265	3+	7+	25	35(?)
Man	266	6	14	50+	70–75

chief understanding of the process comes from the naturalistic studies conducted by Jane van Lawick-Goodall (1967; 1968a,b; 1971) on the wild population of the Gombe Stream National Park in Tanzania. The main significance of van Lawick-Goodall's research has been to show how exceptionally subtle, complicated, and even manlike is the social development of the young chimpanzee. The process unfolds over a period of a little more than ten years. It consists in essence of the slow gaining of locomotory competence, during which the infant leaves the mother for lengthening intervals to explore the environment, manipulate objects, and play with other members of the troop. At first the youngster is broadly tolerated by the adults it encounters, and their friendly responses contribute an important part of its socialization. But as it approaches maturity, it begins to receive rebuffs from adults. Aggressive interactions intensify as the chimpanzee next finds its way into the adult hierarchy. Although the mother mildly rejects her youngster's attempts to nurse during the weaning period, she remains an ally and solace throughout its adolescence.

The newborn chimp is about as helpless as a human infant. In the first few days it is supported almost continuously by the mother. Its eyes seem not to be focused, and its only movements consist of directing the head in search of the nipple. By the end of the second week the baby chimp can grip objects with its hands as well as push and pull. By 7–10 weeks it evidently sees objects clearly, because it starts to reach its hands toward nearby leaves and its mother's face. Subsequently it begins to crawl about on the mother's body, and attempts to pull away from her by grasping twigs and other objects and pulling itself toward them. The mother frequently plays with the infant. At 16-24 weeks the infant breaks its total physical dependence on the mother. It licks and sucks a little solid food, takes its first quadrupedal steps, and climbs small branches. By this time the motor development of the chimpanzee infant has accelerated well beyond that of the human infant.

During the remainder of the first year of its life, the young chimp rapidly perfects its locomotory movements and its manipulation of objects. Its social milieu expands as it leaves its mother to be patted or groomed by other adults and to play with older infants, juveniles, and adolescents. The infant also "greets" other troop members who approach it. During the second year of life the first adult postures and ritualized gestures are displayed. In the final stage of infancy, ranging from two and a half to three years after birth, the young chimpanzee is weaned. With increasing frequency the mother now rejects its attempts to nurse, although she continues to protect it from other chimpanzees. For the first time adults are seen to occasionally rebuff the youngster's approaches, and it grows more cautious.

The juvenile stage, which begins at about the end of the third year, is defined by van Lawick-Goodall as that in which the chimpanzee no longer suckles or rides on its mother's back but is not yet sexually mature. The juvenile makes its own nest but continues to move around with its mother during most of its waking hours. The rebuffs from older individuals become increasingly severe, forcing major social readjustments. With the attainment of sexual maturity—"adolescence"—the chimpanzee begins a long initiation into the full adult society. Its relationships become more stable, and its actions become more deliberate and cautious. As in human beings, the transition to complete maturity occurs gradually over years.

Other Animal Ontogenies

Ontogeny is currently one of the several most actively pursued subjects of animal behavior. Its study is still in an early descriptive and nomothetic stage—shifting, creative, and diffuse. Lehrman and Rosenblatt (1971) captured the mood in the following statement:

In the study of behavioral development, as in the study of other aspects of behavioral biology, it is neither possible nor necessary to agree about a single formulation of *the* major problems, for the purpose of defining and delimiting the paths to be followed by scientific investigation. The diversity of conceptual and methodological approaches (and of investigative techniques) is limited only by the ability of investigators to perceive new relationships and ask new questions about them.

Of necessity much of this effort centers on parent-offspring interactions. In addition to the insect and primate examples described in the previous sections, the following species deserve to be cited as emerging paradigms of the parent-offspring relationship: the ring dove *Streptopelia risoria* (Lehrman, 1965; Wortis, 1969), the Asiatic jungle fowl and its derivative the domestic fowl *Gallus gallus* (McBride et al., 1969), the laboratory rat *Rattus norvegicus* (Rosenblatt and Lehrman, 1963; Rosenblatt, 1965), the moose *Alces alces* and elk *Cervus canadensis* (Margaret Altmann, 1960, 1963), the reindeer *Rangifer tarandus* (Espmark, 1971), the domestic cat *Felis domestica* (Schneirla et al., 1963; Rosenblatt, 1972), the lion (Schaller, 1972), the wolf *Canis lupus* (Woolpy, 1968b), the African wild dog *Lycaon pictus* (H. and Jane van Lawick-Goodall, 1971), the langurs *Presbytis entellus* and *P. johnii* (Jay, 1963; Poirier, 1970a, 1972), the vervet *Cercopithecus aethiops* (Struhsaker, 1967a,b; Lancaster, 1971), baboons (DeVore, 1963a; Kummer, 1968; Ransom and Rowell, 1972), the gorilla (Schaller, 1963), and the rhesus monkey *Macaca mulatta* (Kaufmann, 1966; Rosenblum, 1971a; Rowell, 1972; Hinde, 1974; see also the review in Chapter 7). General reviews of the ontogeny of mammalian relationships have been provided by Moltz (1971) and Poirier (1972). Comparable studies of the social insects remain to be conducted.

Although vertebrate studies are marked by eclecticism, as Lehrman and Rosenblatt said, much of the work seems motivated by a very few strong, albeit implicit themes. One is environmentalism. The background of a majority of the researchers is in anthropology or experimental psychology, in which there exists a bias to assign as much of the measured intraspecific variance of behavioral traits as possible to environmental influences. There is nothing wrong with this attitude; it can be quite heuristic so long as it is kept explicit. The bias results in a determined probe to catalog and weigh all possible environmental factors, both those manifest in naturalistic studies of free-ranging populations and others that become apparent only when their effects are magnified through experimental manipulation. Behavioral genetics is still at a relatively elementary level of analysis, especially with reference to parental care (see McClearn and DeFries, 1973). Evolutionary studies are also quite elementary, consisting mostly of the deduction of dendrograms of particular sets of behavior but lacking reference to the modern techniques of phylogenetic analysis used by systematists.

The environmentalist theme is derived from another, still more defensible theme, that of the relevance of comparative studies to developmental social psychology in human beings. The hope exists that behavioral homologues and analogues are being found that can shed light on human behavior. For this reason, many of the best studies of parent-offspring ontogenies are quite literally clinical in

detail. They meticulously pursue fine details of variation among individuals, including experimentally induced abnormalities. The evolutionary biologist is tempted to regard some of this variation as developmental noise, and he is likely to oversimplify by incorporating it in single quantitative measures of statistical dispersion. The developmental psychologists cannot be too far off the correct path; it is better to have too much information than too little, especially when a discipline has only weakly defined its questions. Meanwhile, the environmentalists and evolutionists should agree on one important point: when parent-offspring relationships regularly affect the structure of societies, they deserve special attention as group-level mechanisms. Whether the mechanisms are true adaptations at the level of the group or the accidental outcome of individual adaptation is just one more version of the central question of the level of natural selection, already discussed in Chapters 5 and 14.

Alloparental Care

When other members of the society assist the parent in the care of offspring, the potential for social evolution is enormously enlarged. The socialization of individuals can be shaped in new ways, dominance systems can be altered, and alliances can be forged. The term "aunt" was used by Rowell et al. (1964) to denote any female primate other than the mother who cares for a young animal. No genetic relationship was implied; the inspiration is the British "auntie," or close woman friend of the family. The parallel term "uncle" was suggested by Itani (1959) for the equivalent male associates in macaque societies. Since the use of both expressions must always be accompanied by a disavowal of any necessary hereditary relationship, it seems more useful to employ the neutral terms *alloparent* (or helper) and alloparental care. "Allomaternal" and "allopaternal" can then be used as adjectives to distinguish the sex of the helper.

Alloparental care is mostly limited to advanced animal societies. It is, for example, the essential behavioral trait of the higher social insects, the form of altruism displayed by the sterile worker castes. In at least 60 species of birds, including certain babblers, jays, wrens, and others, young adults assist the parents in raising their younger siblings (see Chapter 22). When eider ducks (*Somateria mollissima*) migrate, mothers and their offspring are often joined by barren females whose behavior closely resembles that of the mothers. Fraser Darling (1938), who discovered the phenomenon, even called these individuals "aunties" well before the term was introduced to primate studies. Among mammals, the phenomenon has been reported in porpoises (*Tursiops*) and both the African and Asiatic elephants (Eisenberg, 1966). But alloparental care is most richly expressed in the primates. It has been recorded in lemurs (*Lemur catta, Propithecus verreauxi*), New World monkeys (*Lagothrix lagothricha, Saimiri sciureus*), Old World monkeys (species of *Cercopithecus, Colobus,*

Macaca, Papio, and *Presbytis*), and chimpanzees (see DuMond, 1968; Spencer-Booth, 1970; Lancaster, 1971; A. Jolly, 1972a; Rowell, 1972; and especially the general review by Hrdy, 1974).

Within individual primate species, important differences exist between female and male alloparents. Females generally restrict their efforts to the fondling of infants, play, and "baby sitting"; males not only perform these roles but under various conditions also affiliate with infants as future sexual consorts and exploit them as conciliatory objects during encounters with other males. The evidence suggests that the function of the behavior generally differs in males and females, and in the case of male helpers it varies greatly from species to species. Thus alloparental care, like so many patterns of social behavior, is a heterogeneous category that can be understood only with reference to the natural history of individual species.

Allomaternal behavior has been particularly well studied in the rhesus monkey. Rowell et al. (1964) found that adult females are attracted to the newborn infants of others, examine them closely and reach out in an effort to touch them. This exploration is at first cautious, and the mother typically repels it with aggressive displays. The females then use apparent subterfuge to make the approach. They sidle up to the mother while pretending to forage, or groom the mother until she is sufficiently distracted, then shift their attention to the infant. When the youngster becomes independent enough to crawl away for short periods, the allomothers respond with a curious mixture of maternal and sexual attentions. They crouch over the youngster, wrap their arms around it as though to pick it up, and sometimes touch their muzzles to its head. They also pick it up and hold it to their bellies, or seem to try to mount it sexually while making pelvic thrusts. In time the allomothers commence playing with the infant in a gentle manner, crouching down with their mouths open, cuddling the baby, and tugging gently at its arms. They seldom initiate the kind of rough-and-tumble exchange that characterizes play between rhesus juveniles. The females observed by Rowell and her associates also assumed a protective role:

As the infants grew, aunts sometimes watched them when they tried new physical feats and hovered anxiously nearby, going to the rescue if necessary. They seemed to be aware of dangers to young infants—for instance showing care when using the heavy swing door connecting the two parts of the pen if babies were near, and occasionally holding it open for an infant to scramble through. When a baby approached the observer an aunt would sometimes threaten, with the result that the baby went away, and on a few occasions an aunt punished a mother female who had been aggressive to a baby.

Eventually the mother comes to trust the females and to use them as baby sitters while she conducts foraging trips. Most of the time, however, the helpers serve only as alert sentinels.

The details of allomaternal care vary greatly among the species of primates, even within the cercopithecoid monkeys. A mother langur (*Presbytis entellus*) allows other females to handle her infant within several hours of its birth, in fact as soon as it is dry. As many as eight individuals may trade it back and forth on the first day (Jay, 1965). Baboons are more restrictive. Females contend with mothers for access to the infants, and some high-ranking females repeatedly harass the mothers in efforts to obtain this privilege. A few even turn into compulsive "baby thieves." In other species, such as the vervet and other guenons (constituting the genus *Cercopithecus*), allomaternal care is limited to adolescent and nulliparous females. Experienced mothers pay little attention to the babies of others (Rowell, 1972). In contrast, mothers of *Lemur catta* allow only other mothers to handle their infants (Jolly, 1966). Macaques and chimpanzees are matrifocal in the choice of helpers; strong preference is given to the older female siblings of the infant. Other Old World primates are relatively flexible. Lancaster (1971) found that in vervets the amount of access is determined more by the general permissiveness of the mother than by the degree of genetic relationship.

Why should females care for the infants of others, and why should mothers tolerate such behavior? Wholly plausible explanations exist for the actions of both participants. First, young females who handle infants gain experience as mothers before committing themselves to motherhood. Gartlan (1969) and Lancaster (1971) have argued that although maternal care may possess innate basic components, it is a sufficiently complex and physically difficult activity to require practice. In this view, play-mothering is one of the final episodes of the socialization process. The relevant evidence is somewhat equivocal but as a whole seems to support the hypothesis with reference to at least a few primate species. Of the seven langur females that Phyllis Jay observed to drop infants through awkwardness, all were very young and in four cases known to be nulliparous. Similarly, female vervets seen by Gartlan to be carrying infants upside down or in other unusual positions were all less than fully grown. The crucial experience appears to be contact with other animals, including infants, rather than just primiparity. When female monkeys and chimpanzees are reared in the wild, their competence at caring for their firstborn is evidently as high as it will ever be. In the rhesus, for example, primiparous mothers are more anxious in manner and reject their infants less firmly than multiparous mothers but are equivalent in the way they retrieve, constrain, cradle, and nurse them (Seay, 1966). When females are reared in varying degrees of social isolation, however, their initial responses are far less adequate and only at later births do they approach normalcy. The rhesus monkeys reared with artificial mothers by Harlow and his coworkers rejected and mistreated their firstborn in a way that would have been fatal to many. But of six such individuals, five gave adequate care to their second infants (Harlow et al., 1966). A captive gorilla who had killed her first infant cared for her second (Schaller, 1963). Similar improvement with experience has been reported in chimpanzees (van Lawick-Goodall, 1969).

A second line of evidence supporting the learning-to-mother hypothesis is the fact that in most species in which allomaternal care is prominent, and in at least some where it occurs infrequently, the behavior is displayed principally by juvenile and subadult females. Of 347 allomaternal contacts recorded in vervets by Lancaster, 295 were initiated by nulliparous females between one and three years of age; the remaining 52 involved females three years or older who were experienced mothers. In caged rhesus populations, two-year-old females are the most active allomaternal class. Nulliparous individuals are more hesitant than experienced mothers in approaching infants, but they nevertheless initiate contacts at a higher rate (Spencer-Booth, 1968).

Thus the evidence conforms to the learning-to-mother hypothesis, although it cannot yet be said to prove it. But even if we grant for the moment that allomothers are benefiting in this way, why should the real mothers tolerate it? The mothers can be expected to lose fitness by turning their children over to the ministrations of incompetents. One reason for taking the risk could be kin selection. By permitting daughters, nieces, and other close female relatives to practice with their children, mothers can improve their inclusive fitness through the proliferation of additional kin when the relatives bear their own first offspring. Such selection will not work if the infants used in practice are damaged as much as the first babies of females who lack practice, for in this case the mother will be trading damage to an infant with $r = \frac{1}{2}$ for potential benefit to an infant (yet to be born) with $r = \frac{1}{4}$ or less. In practice, however, there is no reason to expect an even trade. In most species of primates, mothers do not release their infants to allomaternal care until the behavior of the babies has developed somewhat. Even then the mother is alert, sometimes to the point of aggressiveness, and she permits the infant to be kept only for short periods of time. In other words, the allomaternal helpers can practice with the babies without endangering them nearly as much as if they were given sole responsibility. The kin-selection hypothesis might be tested by establishing the degrees of relationship of mothers and helpers, but so far the data are insufficient (Sarah Blaffer Hrdy, personal communication).

Other advantages can accrue to the mother. Circumstances are conceivable in which an infant will be saved by allomaternal care when the mother is ill, injured, or temporarily lost from the troop. Thus the helpers might serve as emergency nurses—seldom used but vitally important on rare occasions. Even under normal circumstances the use of the helpers as baby sitters frees the mother for foraging. Among Nilgiri langurs (Poirier, 1968), vervets (Lancaster, 1971), patas, and rhesus monkeys (Hrdy, 1974), the mother often deposits her infant near another female before departing to feed some distance away.

Finally, allomaternal care can lead to alliances between females that are useful to one or both of them. Rhesus mothers generally permit only subordinate females to handle their infants. Consequently any advantage will belong to the helpers, which might be the reason they are persistent in this species in the face of early resistance from the mothers. However, whether such relationships regularly lead to elevations in rank in this and other primate species, either unilaterally or reciprocally, is not known.

Because primate societies are usually matrifocal, the paternity of infants cannot be determined by casual observation of wild populations. Consequently true paternal care is difficult to separate from allopaternal care, and in most instances there is no reason to expect the males themselves to know the difference. Yet variation in the form of male care among the primate species strongly suggests that a clear distinction between paternal and allopaternal behavior exists. In species characterized by the presence of a single male in the troop or at least one or a very few dominant males likely to be fathers, the males tend to show an almost maternal solicitude toward infants. At one extreme, the male marmoset (*Callithrix*) carries his twin offspring until their combined weight equals his own, turning them over to his mate only for feeding. The male siamang (*Symphalangus syndactylus*) sleeps with the juvenile while the female sleeps with the infant. During the day a switch is made, and the male carries the infant while the family forages (A. Jolly, 1972a). In some troops of the Japanese macaque, dominant males routinely associate with year-old juveniles while the mothers are occupied with newborn infants (Itani, 1959). In their long-term study of olive baboons (*Papio anubis*) at Ishasha, Ransom and Ransom (1971) found that the most intensive male care of infants occurred when an adult male formed a close consort relationship with a multiparous female and was probably the father of her youngest offspring. Of six such males observed by the Ransoms, one was quite old and past his prime while another was barely mature and still a peripheral member of the troop. The remaining four were ranking males at or near their prime. Even where quasimaternal care by males is rare or absent, the single or dominant male typically protects infants in times of danger. This is true of male patas, who perform distraction displays toward predators away from the troops, and of male langurs and squirrel monkeys, who have been observed to rush to the defense of distressed infants (McCann, 1934; DuMond, 1968).

In situations where the males are less likely to be the fathers, the forms of interactions are different and appear to serve other ends. In the olive baboon troops studied by the Ransoms, young males took an intense interest in the female offspring of low-ranking females. The possibility exists that this kind of affiliation leads to consort formation and more successful breeding on the part of subordinate males. The same explanation could hold for a peculiar pattern observed in the Japanese macaque by researchers of the Japanese Monkey Center. Whereas males caring for year-old infants showed no sex preference, those caring for two-year-old juveniles preferred females. The trend is carried to an extreme in the case of male hamadryas baboons, who adopt juvenile females to add to their harem (Kummer, 1968).

A second form of undoubted allopaternal behavior is what Deag and Crook (1971) have called "agonistic buffering." It has often been observed that the presence of infants inhibits aggression among the adult members of primate troops. For example, van Lawick-Goodall (1967) noticed that chimpanzee mothers are attacked far less often and with less intensity when they carry infants on their backs than when the infants are carried in a ventral, less visible position. Among the aggressively organized cercopithecoid monkeys, mothers carrying infants are generally the least likely to be attacked by other adults. This tendency has been utilized by the males of a few species, who pick up and carry infants as a safeguard when approaching higher-ranking males who would ordinarily rebuff them. An extreme form of agonistic buffering is practiced by the Barbary macaque *Macaca sylvanus*. As Deag and Crook observed, often a subordinate male picks up an infant, holds it in one hand, and runs as far as 40 meters straight to another male, to whom the infant is then presented. The subordinate monkey next assumes a pseudofemale sexual posture, and the dominant animal mounts him while mouthing the infant or pulling it off to one side. Sometimes the infant is simply picked up by one of the monkeys and placed in an intermediate position on the ground. Itani (1959) reported a closely similar usage of infants as "passports" by male Japanese macaques. According to the Ransoms, anubis baboon males sometimes place themselves next to infants in moments of stress, and in a few cases they carry them on their belly or back. Males of the langur *Presbytis johnii* have been observed to insinuate themselves into alien troops by first associating with infants and juveniles. On one occasion observed by Poirier (1968), a band of three males played almost exlusively with one older infant. Once acceptance was gained, the dominant member of the trio abandoned the youngster. According to Struhsaker (cited by Mitchell, 1969), male vervets use a similar stratagem to penetrate alien troops.

Adoption

Although allomaternal care is widespread in the primates, only under special circumstances does it lead to the full adoption of strange infants. In particular, mothers who are nursing young of their own (and are therefore best able to rear orphans) are typically hostile to strange infants who try to approach them. A possible exception is provided by the langur *Presbytis johnii*. Lactating females were seen by Poirier (1968) to respond permissively toward other infants. When more than one infant struggled to gain access to a nipple, mothers showed no preference for their own young. Even in species with aggressive females, orphans are probably rarely left to starve. Female macaques who have lost their own babies readily accept other infants, and they may even go so far as to kidnap them (Itani, 1959; Rowell, 1963). Since mothers are more likely to lose their infants than the reverse, it is probable that orphans will find a willing foster mother. Even if none is available, other females might be able to assume the role fully. When caged rhesus females were induced to adopt infants under experimental conditions, they began to produce apparently normal milk (Hansen, 1966). The foster mothers observed in captive groups of rhesus monkeys usually served first as helpers. This circumstance makes it more likely that in nature a relative will adopt an orphaned infant. Van Lawick-Goodall (1968a) recorded three instances of adoption in wild chimpanzees of the Gombe Stream National Park, two by older juvenile sisters and one by an older brother. Similarly, foster mothers observed by Sade (1965) in the feral rhesus population of Cayo Santiago were older sisters.

Adoption can occur in ants as an occasional outcome of territorial aggression. When I placed colonies of *Leptothorax curvispinosus* close together in the laboratory, the larger ones raided the smaller, killed or drove away the adults, and carried the brood into their own nests. The pupae were treated normally, and the adult workers eclosing from them were fully accepted by their captors. When such behavior is extended across species, an important preadaptation exists for slave making. During experiments similar to that just described, *L. ambiguus* raided *L. curvispinosus* and carried off the brood to their own nest, where they licked and cared for the pupae in a normal manner. The adults were assisted during eclosion from the pupae but then executed by the host workers within a few hours. This pattern of behavior is close to the primitive slave-making behavior of the obligatorily parasitic *L. duloticus* (Wilson, 1974b). *Formica naefi* is another species with behavior that preadapts it for slave making. When Kutter (1957) placed colonies of this member of the *exsecta* group near colonies of species in the *fusca* group, the *naefi* attacked their neighbors and carried away both the brood and the adult workers. Although the behavior has not been seen directly in nature, Kutter noted that all larger *naefi* colonies excavated in the field contain a few *fusca*-group workers. This evolutionary stage is but a short step from the elementary slave-making behavior shown by *F. sanguinea* and other species in the genus.

Certain other aspects of the formation of parent-offspring bonds and the basis of adoption in animals were discussed in Chapter 7.

Chapter 17 **Social Symbioses**

Symbiosis, defined as the prolonged and intimate relationship of organisms belonging to different species, is conventionally illustrated in the biological literature by interactions of pairs of organisms. But many other cases are known of individuals that enter symbiosis with societies, and even of symbioses between entire societies. The adaptations at the social level are no less diverse than those at the organismic level.

In every regard the insects excel the vertebrates in the development of social symbiosis. Although the reasons for this difference are not yet fully clear, the following observations seem collectively to constitute a plausible hypothesis. To begin with the organization of insect societies is based to a much higher degree than that of vertebrates on altruism and the frequent repetition of altruistic acts. Social insects regurgitate, allogroom, recruit, and perform other services in a manner unrelated to either dominance or the peculiarities of personal recognition and kinship within the limits of the colony. This indiscriminate generosity opens multiple lines of entry into the energy flow of the colony. Once the would-be symbiont has gained at least partial acceptance, it can tap liquid food during regurgitation, lick nutritive secretions from the bodies of its hosts, consume the immature stages, and permit itself to be recruited to food sources outside the nest. Insect societies are typically organized into castes, each of which performs a limited set of roles and communicates in a narrow and specialized manner with other castes. The individual insect lacks a broad awareness of the roles of other colony members, and this makes it easy for social symbionts to insert themselves into the colony as pseudocastes.

The connection between impersonality and social symbiosis is illustrated still further within the vertebrates. Birds having altricial young are especially vulnerable to brood parasitism, in which females of other species insert their eggs into the nests and trick the hosts into raising their young. The switch is possible first because eggs are relatively anonymous objects, recognized by the adult birds through a relatively small set of vaguely characterized releasing stimuli. These stimuli are easily supplanted during experiments in which supernormal stimuli such as larger size and altered surface patterns are preferred over the normal traits of the egg. But even more important, the nestlings are usually anonymous, being recognized by a few cues such as the position they occupy in or near the nest, the general appearance of their maw during gaping movements, and specialized begging sounds. Like ant larvae, they are little more than helpless eating machines. In contrast, the young of precocial birds form close bonds with the parents in the first hours after hatching, and they are required to follow them on trips through particular habitats and in the performance of specialized feeding maneuvers. The intrusion of parasitic precocial young into such a regimen intuitively seems difficult, and in fact only one example is known—the black-headed

354   Part II   Social Mechanisms

duck *Heteronetta atricapilla* of South America. Even this species is revealing in its own way. The parasitic duckling remains in the host nest for only one to one and a half days, then departs to grow up in solitude (Weller, 1968). Social parasitism is virtually unknown in the mammals, perhaps because of the intimate and highly personalized relationships that result from live birth. There is no egg stage to serve as an entrée for the would-be nest parasite.

Social symbiosis is a bewilderingly rich and complex subject, which I reviewed not long ago in *The Insect Societies* (1971a). Many of the details are of interest only to entomologists. The remainder of this chapter will be devoted to the principles of the subject, with some exemplification selected with particular reference to the broader issues of sociobiology. It will be shown that each form of vertebrate social symbiosis has a counterpart in the insects, with the vertebrate cases forming a tiny subset of the universe of possible symbioses more nearly expressed to completeness by the insects. The phenomena will be classified according to the terminology used by most American biologists (see Table 17-1). *Symbiosis* includes all categories of close, protracted interaction. When the symbiosis benefits one participating species while neither benefiting nor harming the other, it is referred to as *commensalism.* An interaction that benefits both partners is *mutualism,* the special case that European biologists commonly refer to as "true" symbiosis. Finally, when one species benefits at the expense of the other, the relationship is called *parasitism.* In its consequences on population growth parasitism does not differ fundamentally from predation.

Social Commensalism

Entomologists make a distinction between *compound nests* of social insects, in which two or more species live very close to one another but keep their brood separated, and *mixed colonies,* in which the brood are placed together and tended communally. Many pairs of ant species are found in compound nests. When the relationship is obligatory for one species, the relationship is ordinarily parasitic rather than neutral or mutualistic. But in many other instances the association is facultative, probably even accidental. In the simplest situation, sometimes labeled plesiobiosis, different ant species nest very close to one another, but engage in little or no direct communication—except when their nest chambers are broken open, in which case fighting and brood theft typically ensue. The less similar the species are morphologically and behaviorally, the more likely they are to cluster together in a plesiobiotic relationship. To express it another way, the most closely related species of ants are the least likely to tolerate one another's presence. Do some plesiobiotic ants benefit from the association? Like many presumptive cases of commensalism, this question has never been examined closely in field studies. The answer is probably yes; we know that some forms of

Table 17-1   The modes of social symbiosis.

Social Commensalism: one species benefits, the other is unaffected.
Plesiobiosis: the close association of the nests of two or more species without mixing and with little or no benefit to either participant. Example: ant species that habitually nest in close association.
Nest commensalism: one species lives in the nests of the other, scavenging on refuse or preying on the scavengers, in either case without harming or benefiting the hosts. Examples: specialized millipedes, beetles, and other arthropods that live with army ants.
Mixed flocks, herds, and schools: in vertebrates, members of one or more species join foraging groups of others; in some cases the passive nuclear (attractive) species are probably not affected significantly by the presence of the others. Examples: mixed flocks of chickadees, sparrows, and other birds forming winter foraging groups in North America.

Social Mutualism: both species benefit.
Mixed flocks, herds, and schools: in some of the mixed vertebrate groups, the passive nuclear species as well as the followers are probably benefited.
Trophobiosis: one species yields food to another, ordinarily in a form of little use to the donor, in exchange for protection from parasites, predators, and inclement weather. Examples: aphids and other homopterous insects kept as "cattle" by ants.
Parabiosis: species nest in close association, defend the nest jointly, forage together, and may even share food, but they do not rear the offspring together. Examples: certain ant species in South and Central America.

Social Parasitism: one species benefits, the other suffers.
Trophic parasitism: one species steals food from societies of the other. Examples: nest robbing in stingless bees, nest eating in termites, thievery of prey by hyenas.
Xenobiosis: one species nests close to or within the nests of another and begs food from it. Example: the "shampoo ant" *Leptothorax provancheri,* which lives on regurgitated food from *Myrmica brevinodis.*
Temporary social parasitism: one species spends part of its life cycle as a parasitic member in societies of another species and the remainder of the time in free life. Example: certain ants, bees, and wasps, the single queens of which invade colonies of other species and preempt the position of the original queen; brood parasitism in birds.
Slave making (dulosis):   one species raids the nests of another, captures its immature forms, and allows them to eclose as adult slaves in its own nest. Examples: ants belonging to six north temperate genera.
Inquilinism: one species spends its entire life cycle as a parasite within the societies of another. Examples: certain ants, social bees, and social wasps, especially in the cold temperate zones; also, numerous other arthropods living as "guests" in the nests of all kinds of social insects, in both the tropics and temperate zones.

social parasitism are derived from the intimate cohabitation of dissimilar species, a trend to be described shortly, and it is likely that ant species exist that are preadapted for this change because they benefit from the close association without straining the resources of their plesiobiotic partners. This subject is likely to become a rich ground for field research.

A legion of nonsocial arthropods have been modified for a commensalistic existence within the nests of social insects, living as scavengers on the refuse piles or preying on the scavengers. They include squamiferid isopods, gamasid and uropodid mites, entomobryid and isotomid collembolans, nitidulid and endomychid beetles, and many others. Some of the symbionts, for example the entomobryid collembolans, avoid their hosts by swiftness of foot. Others, such as the sluglike larvae of the syrphid fly *Microdon*, rely on slow movement combined with a neutral body odor (Wheeler, 1910; Wilson, 1971a). Still others, including nicoletiid silverfish and white scavenger millipedes of the family Stylodesmidae, even go so far as to run with army ants and are virtually accepted as nestmates by their hosts (Rettenmeyer, 1962, 1963a; see Figure 17-1).

True social commensalism is rare in the vertebrates. By this I mean

Figure 17-1 Social commensalism in insects. A thysanuran ("silverfish") belonging to the species *Trichatelura manni* runs in the middle of a raiding column of the tropical American army ant *Eciton hamatum*. The little insect follows the odor trails of the ants, licks their body surfaces, and shares their prey. The principal worker castes of the ants are also illustrated in this photograph; the *Trichatelura* is preceded and followed by minor workers, while two large, light-headed soldiers flank it on the left. (Photograph by C. W. Rettenmeyer.)

that few cases are known with certainty of individuals or societies that insert themselves into the midst of other societies in an entirely unobtrusive manner. Schooling fish probably mix in this fashion on occasion, an association analogous to the plesiobiotic nesting of social insects. Undoubted social commensals are represented by trumpet fishes of the genus *Aulostomus*, found in tropical American waters. Individuals have been observed by Eibl-Eibesfeldt (1955) to ride the backs of parrot fish or to join schools of surgeon fish from which they dart periodically to seize smaller fish as prey (see Figure 17-2). The behavior appears to be an extension of the tendency of *Aulostomus* to hide among coral branches roughly shaped like the bodies of fish. Probably the members of some mixed-species flocks of birds behave in a commensalistic manner toward one another. Since other interactions in the flocks are either mutualistic or unknown in nature, this subject will be deferred to a later, special section.

Social Mutualism

An extreme development of mutualistic symbiosis is represented by the associations between homopterous insects such as aphids and their ant hosts. The ants provide protection from predators and parasites, and the homopterans "repay" them with honeydew expended as excrement. The system, called trophobiosis, is based on the remarkable food habits of the symbionts. When aphids feed on the phloem sap of plants, they pass a sugar-rich liquid through their gut and back out through the anus in only slightly altered form. During the passage of this honeydew, as much as one-half of the free amino acids are absorbed by the gut, sugars are partly absorbed and converted into glucosucrose, melezitose, and higher oligosaccharides,

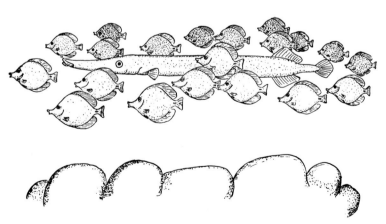

Figure 17-2 Social commensalism in fish. The trumpet fish (*Aulostomus*) uses a school of yellow surgeon fish (*Zebrasoma flavescens*) for camouflage. (From Eibl-Eibesfeldt, 1955.)

while organic acids, B-vitamins, and minerals are probably partially taken up as well. To process a large volume of phloem sap and discard the excess as honeydew evidently costs the aphid less in calories than a more nearly total extraction from smaller quantities of sap. The ants simply extract some of the nutrient residue to their own profit. In order to evoke the flow of honeydew the ants stroke the aphids with their antennae, a behavior not basically different from the antennal and tarsal strokings by which ants induce nestmates to regurgitate food.

Honeydew is produced by some other kinds of homopterans, namely scale insects (Coccidae), mealybugs (Pseudococcidae), jumping plant lice (Chermidae = Psyllidae), treehoppers (Jassidae, Membracidae), leafhoppers (Cicadellidae), froghoppers or spittle insects (Cercopidae), and members of the "lantern-fly" family (Fulgoridae). A few species in all of these families, with the possible exception of the Cercopidae and Fulgoridae, have entered into mutualisms with ants. Both the homopterans and their ant hosts have undergone anatomical and behavioral changes in the service of the symbiosis (Wheeler, 1910; Auclair, 1963; Way, 1963). The homopterans ease out the honeydew droplets when solicited by the ants, rather than ejecting them at a distance in the manner of nonsymbiotic species (see Figure 17-3). Individuals of the black bean aphid (*Aphis fabae*) show the following sequence of specialized responses in the presence of ants: the abdomen is raised slightly, the hind legs are kept down instead of being lifted and waved as in unattended aphids, and the honeydew droplet is emitted slowly and held on the tip of the abdomen while it is being consumed by the ants. In at least some species of aphids and scale insects, only a light touch on the back is required to induce the extrusion of the droplet.

The extreme myrmecophilous aphids have evolved to the status of little more than domestic cattle. They have reduced or lost the usual defensive structures found in free species, including the defensive abdominal spouts called cornicles that secrete a quickly hardening wax, the dense shrouds of flocculent wax filaments secreted by special epidermal glands, the sclerotized exoskeletons, and the modifications of the legs for jumping. But they have acquired a new organ that appears to serve trophobiosis exclusively, a circlet of hairs surrounding the anus that holds the honeydew droplet in place while it is being eaten by the ant. Long anal hairs borne by certain mealybugs seem to have the same function. The life cycles of trophobiotic homopterans, documented at length by Zwölfer (1958) and others, have been modified in ways that promote synchronization with the activities of the host ants.

For their part, the homopteran-tending ants have acquired behavior that is clearly specialized to serve the symbiosis. Some species care for their cattle within their nests. The early, classic studies of S. A. Forbes and F. M. Webster revealed that eggs of the corn-root aphid

Figure 17-3 Trophobiosis, a form of social mutualism, is exemplified by the relationship of the ant *Formica polyctena* to scale insects (Coccidae). Here three workers attend one of these "cattle"; the scale insect has extruded a droplet of honeydew that is about to be licked up by the ant on top. (Photograph by Bert and Turid Hölldobler.)

(*Aphis maidiracis*) are kept by colonies of the north temperate ant *Lasius neoniger* in their nests throughout the winter. The following spring the workers carry the newly hatched nymphs to the roots of nearby food plants. When the corn plants are uprooted, the ants transport the aphids to new, undisturbed root systems. During the late spring and summer some of the aphids develop wings and fly away from the host nests to seek new plants. If they settle on roots within the territory of another *Lasius* colony, they are adopted; otherwise they begin an independent existence not basically different from that of nonsymbiotic species. The behavior of host ants has been modified even in small details to raise the efficiency of the trophobiosis. It has been established beyond doubt that the workers carry their homopterans to the appropriate part of the food plant and at the correct stage of the trophobionts' development. Such behavior has been documented, for example, in the case of the subterreanean ant *Acropyga* and its root coccids, in the weaver ant *Oecophylla* and its scale insects, and in *Lasius* and its aphids. Even more impressive is the fact that the queens of certain species of *Acropyga* and *Cladomyrma* carry coccids in their mandibles during the nuptial flights. In a real sense the homopterans have been integrated into the colonies of the host ants.

Ants are not the only organisms that attend homopterans. Stingless bees of the genus *Trigona* collect honeydew directly from membracids in Brazil, and at least one species palpates the treehoppers to induce flow. In *The Naturalist in Nicaragua* (1874) Thomas Belt reported that polybiine wasps of the genus *Brachygastra* attend membracids in closely similar fashion. The soliciting signals are not difficult to produce: I have "milked" coccids myself with one of my own hairs. They have been evolved repeatedly in other, nonsocial insects, including silvanid beetles, lycaenid butterflies, and flies of the genus *Revellia*. Trophobionts also occur outside the order Homoptera. One species of plataspidid hemipteran is attended by ants of the genus *Crematogaster* in Ceylon. Larvae of many species of lycaenid butterflies are kept by ants, rewarding their hosts with a sugary liquid secreted by an unpaired gland on the dorsum of the seventh abdominal segment.

Parabiosis

In 1898 Auguste Forel labeled as *parabiosis* a novel form of symbiosis that he had discovered in South American ants. Colonies of the tree-dwelling *Crematogaster limata parabiotica* and *Monacis debilis* commonly nest in close association, with the nest chambers kept separate but connected by passable openings. Also, workers of the two species run together along common odor trails. Wheeler (1921) found the same phenomenon in Guyana and established that the two species collect honeydew together from membracid treehoppers. He

discovered a similar association between the *Crematogaster* and the large formicine ant *Camponotus femoratus*. Both species were observed utilizing common trails and gathering honeydew from jassids and membracids on the same plants, as well as nectar from the same extrafloral nectaries of *Inga*. Not only were the ants tolerant of each other in this competitive situation, they were on friendly terms. They "greeted" each other on the trails by calm mutual strokings of the antennae, and on three occasions Wheeler observed *Camponotus* workers regurgitate to individuals of *Crematogaster*.

It is not known whether the ant parabioses are mutualistic or parasitic in nature. At best the distinction must be a subtle one in such a complex relationship. The *parabiotica* form of *Crematogaster limata* is evidently always associated with other ants, and it may prove to be a distinct sibling species. Either way, the prima facie case for mutualism is strong at this time. The broods are never mixed, and as Weber (1943) has pointed out on the basis of his own studies, all of the parabiotic species participate vigorously in defending the nest against invaders. There is no evidence that the *Crematogaster* harms the other species. On the contrary, *Camponotus femoratus* maintains flourishing populations in localities where virtually every colony lives in parabiosis with *Crematogaster*.

Mixed Species Groups in Vertebrates

Throughout the world, small insectivorous birds gather in flocks of two or more species to forage together. These groupings are true flocks in the sense that at least some of the birds seek one another out and stay together while flying from one place to another. They are to be distinguished from mere aggregations, which are groups that gather passively around a localized food or water source. Thus a group of titmice and woodpeckers moving as a unit through the canopy of a deciduous forest is a flock, but a band of formicariid thrushes following in the wake of an army ant march is an aggregation (see Hinde, 1952; Rand, 1954; Willis, 1966).

Mixed-species bird flocks are loosely organized and constantly shift in composition. Members may remain together for hours or all of a day, sometimes regrouping each morning. Turnover is increased when species are left behind as a result of slower horizontal speeds or when individuals become members only during the time the flock passes through their territories. It is especially strong when seasonal migrants travel through the area and join for short feeding bouts (Morse, 1970). The species composition of a flock changes accordingly, but certain species are more consistently present and indeed serve as the *primum mobile* of the associations. Moynihan (1962), building on the work of Winterbottom (1943, 1949) and Davis (1946) on tropical bird faunas, proposed the following loose classification, which can be usefully applied to mixed flocks in general.

Nuclear species.   These are the bird species that contribute significantly to the formation and cohesion of the flocks. They may or may not actually lead the other birds; the important fact is that the flocks are not likely to persist without them. Some forms, called the active nuclear species, seek other birds and follow them. Others, the passive nuclear species, form the attractive elements.

Attendant species.   These are regular members of flocks, but they are not as attractive to other birds as passive nuclear species. Their membership in flocks is less consistent than that of the nuclear species.

Accidental species.   These are the birds that join flocks only on rare occasions.

Moynihan's categories grade into one another. The species composition and relative abundance of the separate categories change kaleidoscopically from flock to flock and from time to time within the same flock. In Panama, for example, flocks composed chiefly of tanagers and honeycreepers spend most of their time in treetops close to the edge of the forest. Upon entering localities where tall trees are scarce, they sometimes descend for short periods of time to the tops of low scrub. There they are joined by the green-backed sparrow (*Arremonops conirostris*) and dusky-tailed ant-tanager (*Habia fuscicauda*), species limited to this type of vegetation. When the flocks return to the trees their guests drop out.

Each species displays traits that fit it to particular roles according to the companions it meets in the mixed flocks. The plain-colored tanager (*Tangara inornata*) of Panama is an example of a powerful passive nuclear element. It forms large, cohesive flocks on its own and attracts other common nuclear species such as the blue tanager (*Thraupis episcopus*) and the green honeycreeper (*Chlorophanes spiza*), to a degree indicating that it is signaling in a specialized manner. In fact, some of the social behavior patterns of the plain-colored tanager appear to be adapted particularly for the assembling of flocks within the species. Wing flicking and tail flicking are ritualized flight-intention movements used by many songbird species to coordinate group movement. In the plain-colored tanager they are exaggerated and more frequent than in related species. Call notes of the kind used in flock organization are exchanged at a higher rate and supplant song altogether. Hostile interactions are reduced. The vigorous movements and repeated calling of this tanager make it more conspicuous than other species and evidently provide much or perhaps all of its attractive power. The summer tanager (*Piranga rubra*) is an example of an extreme attendant species. It joins the tanager-honeycreeper associations only during the winter migratory sojourns and is most common along the edges of forests. Summer tanagers do not form flocks on their own. They join the mixed assemblages exclusively as individuals, so that only seldom is more than one seen in the same flock. Because they fly in silence at the edge of the flocks, they do not ordinarily attract other birds themselves.

These two examples from the Panama forests illustrate the strong component of preadaptation in the formation of mixed-species flocks. It is possible that some of the behavior has evolved to promote interspecific association. If so, this postadaptation is most likely to be encountered in older, more complex faunas such as those inhabiting tropical rain forests. But, clearly, strong preadaptation and not postadaptation is the key to the origin of this particular form of symbiosis. Moynihan has reasoned that as few as two species are enough to create a well-integrated flock, provided they possess the correct, previously tailored behavioral profiles. There must be a passive nuclear species with strong tendencies to form conspicuous monospecific flocks, and an attendant species with little or no tendency to group independently. The initial advantages can be expected to accrue to the attendant species and that is where we should look for postadaptation. No one has yet devised a method for separating evolutionary progress made before and after the flock formation, but Vuilleumier (1967) has at least identified highly simplified flocks on which such an analysis can eventually be conducted. In the *Nothofagus* forests of Patagonia flocks are made up of at most four species and are more loosely organized than in Central America. The passive nuclear species is the small ovenbird *Aphrastura spinicauda*, a restless, highly vocal insectivore that forms flocks of 4 to 15 individuals. The sociality of *A. spinicauda* is wholly self-contained. The flocks move in tight formations while searching for insects over the trunks and branches of the trees. Cohesion is maintained by frequently repeated contact calls. At any given time about 60 percent of the flocks contain a second, larger ovenbird species, *Pygarrhichas albogularis*. The *Aphrastura* appear indifferent to these guests, but the *Pygarrhichas* actively seek and follow them, and they are indeed seldom found alone. Of equal significance, the *Pygarrhichas* do not form flocks independently. Thus the Patagonian mixed flocks consist of Moynihan's two essential elements, each represented by a single species. The symbiosis appears to be commensalistic, with *Aphrastura* the host and *Pygarrhichas* the benefited guest. Two other attendant species, the woodpecker *Dendrocopos ligniarius* and tyrant flycatcher *Xolmis pyrope*, may also be commensals but are relatively insignificant, since they occur in less than 10 percent of the *Aphrastura* flocks.

Over the years various authors have postulated three adaptive advantages of joining mixed-species flocks: increased avoidance of predation, increased foraging efficiency, and the use of grouping as an epideictic display to control population growth in the manner envisaged by Wynne-Edwards. The first two hypotheses, which are not mutually exclusive, have received strong supporting evidence from field work on flocks in the United States and Europe. It is well known and was established earlier in this book (Chapter 3) that individuals in some flocks of birds and other animal groups are subject to less predation because the total state of alertness of the group exceeds that of solitary individuals. In the pinelands of Louisiana this is evidently the benefit enjoyed by three attendant species, the eastern bluebird (*Sialia sialis*), slate-colored junco (*Junco hyemalis*), and chipping sparrow (*Spizella passerina*). They are to a large extent ground foragers, whereas the chickadees, warblers, and other nuclear elements of the flocks are arboreal. Thus not only do the two elements have divergent food niches, but the differences are so great as to make it cumbersome for the attendant species to remain with the flock. Why do they do so? These small birds are especially vulnerable to predators while foraging through the sparse ground cover of the pine forests, and they take advantage of the early warning systems provided by the birds foraging above them. All three attendant species, for example, respond to the warning call of one of the nuclear species, the Carolina chickadee *Parus carolinensis*, by scattering simultaneously and alighting in the lower limbs of the pines (Morse, 1970).

The hypothesis of improved foraging efficiency is favored by even more persuasive evidence. Again, we know from observations of single-species flocks that groups are often able to find food more quickly than individuals, especially when the resources are sparse and scattered. Also, flock formation appears to be a device brought into action by individuals of some species to cope with periods of food shortage. When the food supply is ample, European titmice remain territorial throughout the winter. When the supply is poor, the birds join in flocks and forage together (Hinde, 1952). According to Morse (1967), brown-headed nuthatches (*Sitta pusilla*) in Louisiana markedly decreased their participation in flocks when pine seeds became temporarily superabundant. If this form of behavioral scaling is generally employed by the members of mixed-species flocks, we should expect to find an inverse relationship between the population density of the birds, reflecting food availability, and the percentage of individuals participating in flocks. Just this relationship has been documented in considerable detail by the studies of Morse (1967, 1970). Earlier, in Chapter 3, it was shown that the commonality of the flock can draw on the experience or the luck of a few leaders in progressing from one patch of food to another. The same principle appears to be at work in Morse's mixed flocks: "The mixed forest was the area of lowest population density in Maryland. Upon 15 occasions flocks that had been foraging for one hour or more in deciduous forests adjacent to mixed forests were observed to fly almost directly through several hundred meters of mixed forest, scarcely stopping to forage in transit. Never was the opposite tendency (to forage in mixed forests and to fly directly through deciduous forests) noted." Morse also found that the larger the flock, the more rapidly it moves from place to place. The advantages in discovering and utilizing new food sources must be great enough to overcome the unavoidable disadvantage of competing for food with other flock members in close quarters. Cody

(1971), in his combined field and theoretical study of mixed-species flocks of finches in the Mohave Desert, was dealing with an environment in which the food is richer and more evenly distributed. In this special case, it was evident that the resource can be harvested by groups more efficiently if they move in formation than if they mill chaotically along separate paths. Hence for a given population density it is of advantage for the individual bird to be a member of a group.

Superiority at foraging, particularly in food-poor areas, could also be a significant factor in the tropics. Moynihan (1962) believes that antipredation is the main selective force operating on tropical flocks, but he also notes that mixed flocks occur more commonly in relatively unfavorable or partly isolated habitats. By foraging in groups, certain species and the commensals that attend them are more likely to succeed at invading these ecologically marginal areas.

Competition in mixed-species flocks is diminished somewhat by division of the food niche among the member species. Some kinds of birds dominate others, pushing them into special corners of the foraging space. In the eastern United States, the most abundant nuclear elements, including the chickadees, titmice, and kinglets, are also the behavioral dominants. Pine warblers (*Dendroica pinus*), to take another example, displace brown-headed nuthatches (*Sitta pusilla*) through aggressive interactions; when the two species travel together, the nuthatches stay more on twigs and the distal parts of limbs, while the warblers concentrate on tree trunks and the proximal parts of limbs. When the nuthatches are alone, they expand their operations to concentrate more heavily on the warbler's preferred niche (Morse, 1967). The reverse process, ecological convergence, has been recorded by Moynihan (1962) in two of the Panama species. When foraging alone the silver-billed tanager (*Ramphocelus carbo*) works through moderately to very low scrub, while the black-throated tanager (*R. nigrogularis*) remains at a somewhat higher level in low trees. When the two flock together, however, the black-throated tanager usually moves to the lower vegetation preferred by its partner, and there both appear to eat the same food. Whether any given two species displace each other or converge will depend to a large extent on the initial difference between their preferred niches, as well as on the intensity of the hostile response each shows to alien forms generally. Moynihan (1968) has postulated the existence of "social mimicry," a convergence of conciliatory and contact signaling among species that serves to diminish hostility among species forming the mixed flocks.

Mixed-species schools of marine fishes have been reported occasionally in the literature (see Eibl-Eibesfeldt, 1955; Breder, 1959; Shaw, 1970; Ehrlich and Ehrlich, 1973), but their ecological significance remains to be carefully investigated. The smaller cetaceans also form compound groups. In the Mediterranean, Atlantic dolphins (*Del-*

phinus delphis) often swim with either blue-white dolphins (*Stenella caeruleoalba*) or pilot whales (*Globicephala melaena*), while mixed schools consisting of Risso's dolphins (*Grampus griseus*), right whale dolphins (*Lissodelphis borealis*), and pilot whales have been sighted off the California coast (Fiscus and Niggol, 1965; Pilleri and Knuckey, 1969). Mixed groups of bats occur commonly among species that roost in aggregations (Bradbury, 1975). Interspecific groups of herbivorous mammals are commonplace on the African plains, consisting of various combinations of impala, wildebeest, Coke's hartebeest, gazelles, zebra, giraffes, warthogs, and baboons. Each species is sensitive to the alarm responses of at least some of the others (Washburn and DeVore, 1961; Altmann and Altmann, 1970; Elder and Elder, 1970), so that large groups of any species composition are more alert to the approach of predators than are small groups and solitary animals. A few records of mixed-species troops of primates have also been published. Species of guenons (*Cercopithecus*) and guerezas (*Colobus*) frequently mingle in foraging parties in the African forests; in particular, the lesser spot-nosed guenon (*Cercopithecus petaurista*) occasionally combines with no less than three other members of the genus (Marler, 1965). In Malaya Bernstein (1967) watched a mated pair of gibbons that closely affiliated with a troop of banded leaf monkeys (*Presbytis melalophos*). The male was particularly well integrated, feeding, resting, and traveling regularly in the midst of the group. At the Gombe Stream National Park, van Lawick-Goodall (1971) often saw immature chimpanzees and baboons play together. This seems strange, for the adults are aggressive toward one another and chimpanzee males sometimes kill baboon infants for food. Altmann and Altmann (1970) also observed young baboons playing with young vervet monkeys at Amboseli. Mixed-species groups occur in the platyrrhine monkeys of the New World: both spider monkeys and squirrel monkeys frequently join bands of capuchins (Bernstein, 1964b; Moynihan, personal communication).

Mixed-species interactions and social mimicry may occur across still wider taxonomic gulfs. Moynihan (1968, 1970b) has presented fascinating evidence of an indirect nature that suggests the existence of a loose commensal relation between monkeys and birds. The rufous-naped tamarin is a small monkey found in scrub and forest along the Pacific coast of Central America. The same habitats contain dense populations of several species of tyrannid flycatchers. The tamarins and the birds feed on the same kinds of fruits and insects. This resource is distributed in a patchy and irregular manner that requires constant searching by both animals. Some of the calls of the monkeys consist of rattles and plaintive whistles remarkably similar to the assembly calls of the birds. Although direct proof is lacking, Moynihan believes that the monkeys probably use the flycatchers as guides, moving in the direction of feeding grounds when the birds announce their presence.

Trophic Parasitism

Perhaps the simplest form of social parasitism consists of the intrusion of one species into the social system of another just deeply enough to steal food. German writers have coined exactly the right word, *Futterparasitismus*, to describe this symbiosis. Hyena packs, to take the only mammalian example of which I am aware, parasitize wild dogs. They try to appropriate newly slain zebra and other large animals that the dogs are more skillful at hunting, and they even go so far as to run closely behind the dog packs during the chases (Estes and Goddard, 1967). Basically similar behavior is displayed by some ants. R. C. Wroughton (quoted by Wheeler, 1910) discovered an Indian species of *Crematogaster* that ambushes workers of *Monomorium* as they return home along the foraging trails. The little highwaymen take away the seeds that the *Monomorium* have collected for their own use. Other small ant species, including *Solenopsis* and related myrmicine genera, live in the walls of large nests built by other ants and termites and enter the living quarters to steal food and to prey on the inhabitants. The best-known examples are the tiny "thief ants" belonging to the subgenus *Diplorhoptrum* of *Solenopsis*, which burrow next to the nests of much larger ant species, stealthily enter their chambers, and prey on the brood. Species of *Carebara* in Africa and tropical Asia frequently construct their nests in the walls of termite mounds and are believed to prey on the inhabitants (Wheeler, 1936).

Stingless bees of the genus *Lestrimelitta* are specialized for a different method of thievery. *L. limao*, a species common from Mexico to Argentina, makes its living by invading the nests of the free-living stingless bees *Melipona* and *Trigona* and seizing their stored supplies (Sakagami and Laroca, 1963). The *Lestrimelitta* lack a scopa (pollen basket composed of long hairs) on the hind legs, a structure evidently lost in evolution as part of their parasitic adaptation. Instead, they carry the pilfered supplies in their crops and later place them in their own storage pots in the form of honey-pollen mixtures. While invading nests the bees release a mandibular gland substance with a strong lemonlike odor, the principal component of which is citral (Blum, 1966). Sometimes the raiders occupy the plundered nest and thereby multiply their own colonies. The evolutionary origin of the *Lestrimelitta* behavior is easy to imagine. Both honeybees and nonparasitic stingless bees occasionally engage in robbing, both within and between species. For *Lestrimelitta* the pattern has simply become an obligatory way of life. Precursor behavior is also exhibited by primitively social sweat bees of the family Halictidae. In the spring the hibernating assemblages of fertile young *Halictus scabiosae* females break up, and some of the auxiliaries disperse away from the home nests. Many of these individuals construct new nests of their own. Others, however, invade the newly founded nests of other halictid species, most frequently *Evylaeus nigripes*, and drive out or kill the rightful occupants (Knerer and Plateaux-Quénu, 1967).

A peculiar variation of trophic parasitism is practiced by certain kinds of termites. Members of three genera, *Ahamitermes*, *Incolitermes*, and *Termes*, specialize on living in cavities in the nest walls of other termite species and feeding on the supporting carton material. In other words, some termites have termites in their houses! The winged reproductive forms of two of the species, *Incolitermes pumilis* and *Termes insitivus*, even go so far as to enter the host chambers occasionally and to mingle briefly with the host colonies (Calaby, 1956; Gay, 1966). Mound-building termites are especially vulnerable to nest parasitism. The mounds are solidly constructed, often the most durable features in the ground environment, and they present conspicuous landmarks to the flying reproductive forms in search of a nest site. They also provide unusually favorable microenvironments, so long as the colonizing reproductives are able to remain hidden from their hosts in the mound walls. The precursor patterns of behavior are to be found in territorial competition between species. Numerous cases have been reported of colonies of two or more termite species living in close association, and these commonly represent different genera and even different families. Often the relationship is exploitative in nature, one species appropriating part of the nest of another. Of 150 species studied by Ernst (1960) in Africa, 70 percent were at least occasionally disturbed by other species encroaching on their nests.

Xenobiosis

By a subtle shift in behavior, nest robbers can become tolerated guests. An intermediate evolutionary stage, called xenobiosis, exists in nature that falls just short of the complete mixing of the two participating species. Xenobionts live in the walls or nest chambers of their hosts and move freely among them, but the immature stages are still kept separate. The classic case of xenobiosis is the relationship of the little "shampoo ant" *Leptothorax provancheri* to its host *Myrmica brevinodis*, studied by Wheeler (1910) at his summer home in Connecticut. Species of *Leptothorax* characteristically nest in tight little spaces, inside hollow twigs lying on the ground, rotted acorns, and abandoned beetle galleries in dead trees. The workers forage singly, and when they encounter other ants they usually move away in a quiet, unobtrusive way. Because of these traits, colonies of *Leptothorax* are often found close to the nests of larger ants, and their workers are able to move easily among their neighbors. The trend has been extrapolated into xenobiotic parasitism by *L. provancheri*. The shampoo ant has been found living only in close association with colonies of *Myrmica brevinodis*. Both species occur widely through the northern United States and southern Canada. Colonies of *M. brevinodis* con-

struct their nests in the soil, in clumps of moss, and under logs and stones. The smaller *L. provancheri* colonies excavate their nests near the surface of the soil and join them to the host nests by means of short galleries open at both ends. They keep their brood strictly apart. The *Myrmica* are too large to enter the narrow *Leptothorax* galleries, but the *Leptothorax* move freely through the nests of the hosts. Rather than foraging for their own food, the *Leptothorax* workers depend almost entirely on liquid regurgitated by the host workers. They also mount the *Myrmica* adults and lick them in what Wheeler has described as "a kind of feverish excitement," to which the hosts respond with "the greatest consideration and affection." Although Wheeler at first believed that the *Leptothorax* were providing a beneficial "shampoo," he later conceded that they are probably no more than parasites. Yet they are far from being helpless. When isolated in artificial nests in the laboratory they construct their own nests and rear their own brood, and they are also able to feed themselves, albeit awkwardly.

Similar xenobiotic behavior has been reported in *Leptothorax diversipilosus* of the western United States (Alpert and Akre, 1973) and *Formicoxenus nitidulus*, a close relative of *Leptothorax* in Europe (Stumper, 1950; Wilson, 1971a). In Central America the small myrmicine *Megalomyrmex symmetochus* lives xenobiotically with the fungus-growing ant *Sericomyrmex amibilis*. The *Sericomyrmex* form modest-sized colonies, comprised of 100-300 workers and a queen, that nest in the wet soil of the forest clearings. They subsist entirely on a special fungus raised on beds of dead vegetable material. On Barro Colorado Island, Panama, where Wheeler (1925) discovered them, the *Megalomyrmex* form smaller colonies, consisting of 75 adults or less, that live directly among the fungus gardens of the host. Since the *Sericomyrmex* also place their brood in the gardens, the young of both species become mixed to a limited extent. However, the *Megalomyrmex* tend to segregate their brood in little clumps, each closely attended by a few workers, and neither feeds or licks the brood of the other. The most remarkable fact is that the *Megalomyrmex* appear to subsist exclusively on the fungus. This represents a major dietary shift that must have occurred relatively recently in the evolution of the genus. Because liquid food exchange is uncommon or lacking in fungus-growing ants, the *Megalomyrmex* xenobionts do not secure nutriment from the *Sericomyrmex* in this way. They do, however, lick the body surfaces of their hosts.

Temporary Social Parasitism in Insects

Life cycles of ants that include periods of temporary social parasitism were first elucidated by Wheeler (1904) in the course of his studies of *Formica microgyna* and related species. Closely parallel symbioses have since been discovered in a diversity of genera belonging to the subfamilies Myrmicinae, Dolichoderinae, and Formicinae. The newly inseminated queen finds a host colony belonging to a different species and secures adoption, either by forcibly subduing the workers or by conciliating them in some fashion. The original host queen is then assassinated by the intruder or by the original queen's own workers, who somehow come to favor the parasite. When the first parasite brood matures, the worker force turns into a mixture of hosts and parasites. Finally, since the host queen no longer is present to replace them, the host workers gradually die out over the following months, and the colony comes to consist entirely of the parasite queen and her offspring.

Some members of the mound-building *Formica exsecta* group are facultative temporary parasites. The majority of new colonies are founded by the adoption of queens by their own colonies after they have been inseminated during the nuptial flights. But a few individuals wander further afield and attempt entrance into nests of *Formica fusca* and related species. They stalk the host colonies and either penetrate by stealth or else permit themselves to be carried in by host workers. Those approached by workers lie down and "play dead" by pulling their appendages into the body in the pupal posture. In this position they are picked up by the host workers and carried down into the nests without any outward show of hostility. Later they somehow manage to eliminate the host queen and take over the reproductive role (Kutter, 1956, 1957).

Further subtleties have been perfected by the related formicine genus *Lasius*. Apparently all of the species of the subgenera *Austrolasius* and *Chthonolasius* are temporary parasites on the much more abundant, free-living members of the subgenus *Lasius*. At least some forms of *Dendrolasius*, a fourth subgenus, are temporary hyperparasites, taking over the nests of *Chthonolasius* after the *Chthonolasius* colonies have grown to the free-living state. The relationship between these various species is obligatory, not optional as in *Formica exsecta* and its relatives. Homospecific adoption is not practiced. When newly mated queens of *L. umbratus* are searching for a host colony they first seize a worker in their mandibles, kill it, and run around with it for a while before attempting to invade a host nest (K. Hölldobler, 1953). Apparently all of the parasitic *Lasius* get rid of the host queens, but the exact method employed is still unknown in most cases. The tiny queens of *L. reginae*, a species discovered in Austria by Faber (1967), eliminate their rivals by rolling them over and throttling them (Figure 17-4). Assassination is also the technique practiced by the queens of the aptly named dolichoderine parasites *Bothriomyrmex decapitans* and *B. regicidus* in gaining control of colonies of *Tapinoma* (Santschi, 1920).

The European species of the myrmicine ant genus *Epimyrma* comprise among themselves a remarkably clear evolutionary progression

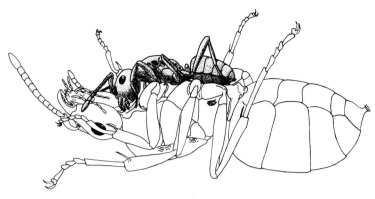

Figure 17-4 Temporary social parasitism in ants. A newly mated queen of *Lasius reginae* has entered a nest of the host species *L. alienus* and is strangling the queen. The *alienus* workers will then care for the offspring of the parasite, and when these workers eventually die from old age and other causes, the colony will consist of pure *reginae*. (From Faber, 1967.)

leading from temporary social parasitism to full inquilinism, in which the parasitic form spends its entire life cycle in the nests of the host species (Gösswald, 1933; Kutter, 1969). In at least five of the eight known species, a worker caste still exists, but it is relatively scarce and rather similar morphologically to the queen, although still a discrete apterous phase. It apparently never aids the host workers in foraging, nest labor, or brood care. All of the species of *Epimyrma* are parasitic on *Leptothorax*. The queen mates in her home nest, then leaves the nest, sheds her wings, and searches for a new host colony. The mode of entry and subsequent behavior vary greatly among the various species. The queen of the French species *E. vandeli*, upon approaching a *L. unifasciata* colony, makes repeated hostile approaches to the host workers and "intimidates" them, to use Kutter's expression. If she succeeds in entering the nest, she kills the host queen and secures complete adoption by the rest of the colony. The queen of *E. goesswaldi*, on the other hand, calms the host workers (*L. unifasciata* in Germany) by stroking them with her antennae and lower mouthparts. Once inside the nest, she mounts the host queen from the rear, seizes her around the neck with her saber-shaped mandibles, and kills her. *E. stumperi*, studied in Switzerland by Kutter, uses still another variation to enter the nests of its host, *L. tuberum*. The queen first stalks the host colony with slow, deliberate movements. When approached by the *Leptothorax* workers, she "freezes," crouches down, and seems to feign death. After a time she begins to mount the workers from the rear, strokes their bodies with her foreleg combs, and grooms herself, perhaps thereby passing nest odors back and forth. With this display of sophistication in evidence, it is not surprising to find that queens of

E. stumperi are able to penetrate host colonies more quickly than the other *Epimyrma* species so far studied. Once inside the nest, the *E. stumperi* queen begins an implacable round of assassination directed at the host queens, of which there are usually at least several in the *L. tuberum* colonies. She mounts each queen in turn, forces her to roll over, then seizes her by the throat with her mandibles. The sharp tips of the mandibles pierce the soft intersegmental membrane of the neck of the victim. The *Epimyrma* maintains her grip for hours or even days, until the *Leptothorax* queen finally dies. Then she moves on to the next queen, and this procedure is repeated until none is left. It is a matter of more than ordinary interest that the *E. stumperi* workers also occasionally mount *Leptothorax* workers and go through an ineffectual rehearsal of the assassination behavior, but without harming their "victims" and with no visible benefit to the parasites. This seems best interpreted as a partial transfer of the queen's behavioral pattern to the vestigial worker caste where it has neither positive nor harmful effects.

But why does the *Epimyrma* queen go to all this trouble? Since all the *Epimyrma* species have already entered a permanently inquiline state, with total dependence on the host workers, it would seem an error to exterminate the host queens, which are, after all, the source of the labor force. However, the *Epimyrma* habit of reginicide cannot be written off simply as an unfortunate vestige from an earlier time when the *Epimyrma* ancestors were temporary parasites. It turns out that, when deprived of their own queens, some of the *Leptothorax* workers begin laying eggs, even in the presence of the *Epimyrma* queen. These develop into workers and thus insure an indefinite continuation of the worker force. Even so, there is one species, *E. ravouxi*, a parasite of *L. unifasciatus*, that has taken the final step of permitting the host queens to live. *E. ravouxi*, in other words, has moved on into advanced inquilinism, and in this respect it is indistinguishable from other inquiline species whose probable evolutionary history is not nearly so well displayed.

An equally clear evolutionary sequence leading from temporary social parasitism to inquilinism has been worked out in social wasps by Taylor (1939), Sakagami and Fukushima (1957), Beaumont (1958), Scheven (1958), and others. The essential steps are the following:

1. *Facultative, temporary parasitism within species.* In *Polistes* and *Vespa*, overwintered queens sometimes attack established colonies of their own species and displace the resident egg-laying queen.

2. *Facultative, temporary parasitism between species.* Queens of the Asiatic hornet *Vespa dybowskii* are able to found colonies on their own, but they prefer to enter small colonies of *V. crabro* or *V. xanthoptera* and to usurp the position of the mother queen. The parasitism is facilitated by the fact that *V. dybowskii* emerges from hibernation later than do the other species, so that vulnerable young host colonies are present in large numbers when the *dybowskii*

queens start searching for a nest site. By the end of the summer the last of the host workers have died of natural causes, and the colony then consists entirely of *dybowskii* workers and the newly emerged *dybowskii* males and virgin queens.

3. *Obligatory, temporary parasitism between species.* This stage, so common in the ants, has not yet been documented in the social wasps, even though it seems to be a probable step on the road to full inquilinism.

4. *Obligatory, permanent parasitism between species (inquilinism).* The three parasitic *Polistes* species of Europe, *atrimandibularis, semenowi*, and *sulcifer*, are workerless, and the queens have completely lost the ability to build nests or to care for the young. The queens force their way onto the paper combs of host colonies belonging to other species of *Polistes*. Relying on greater physical strength and staying power, they take over the dominant position from the resident egg-laying queens. The host queens conquered by *P. atrimandibularis* and *P. semenowi* are permitted to remain in the nest in the role of subordinate workers. Those displaced by *P. sulcifer* always disappear, however.

The bumblebees present still another, independent sequence running from temporary parasitism to inquilinism (Free and Butler, 1959; K. W. Richards, 1973). As in the social wasps, facultative temporary parasites are common but no case of obligatory temporary parasitism is yet known. Inquilinism is richly represented by *Bombus hyperboreus* of the Canadian Arctic together with 18 species of the derivative, wholly parasitic genus *Psithyrus*.

Brood Parasitism in Birds

The temporary social parasitism of the ants, bees, and wasps is closely paralleled by brood parasitism in birds. Obligate brood parasitism is practiced by about 80 species, and it has evolved independently seven times, in cowbirds *Molothrus* (Icteridae), the cuckoo weaver *Anomalospiza imperbis* (Ploceidae), the combassous and widow birds of the subfamily Viduinae (Ploceidae), the honeyguides (Indicatoridae), the Old World cuckoos constituting the subfamily Cuculinae (Cuculidae), the South American cuckoos *Tapera* and *Dromococcyx* in the subfamily Neomorphinae (Cuculidae), and the black-headed duck *Heteronetta atricapilla* (Anatidae). The subject has been extensively documented by F. Haverschmidt, F. C. R. Jourdain, Jürgens Nicolai, C. I. Vernon, and especially Herbert Friedmann among modern investigators. The following brief account is based largely on the excellent reviews by Lack (1968) and Meyerriecks (1972).

No less than 50 species of cuckoos are obligatory brood parasites, and virtually every phase of their reproductive biology bears the stamp of this adaptation. Their call is loud and simple. The vernacular name cuckoo is itself onomatopoeic for the European *Cuculus can-*

orus, which in addition to "*cuc-coo*" also emits a deep "*wow-wow-wow.*" The hawk cuckoo or brainfever bird of Asia (*C. varius*) has a piercing whistle, rising in intensity with each repetition. The accepted explanation for the loudness is that the birds are scarce and therefore require a loud call to communicate, while the simplicity is thought to stem from the fact that the young must acquire the song entirely by inheritance. Cuckoos generally exploit passerine birds smaller than themselves, and the hosts end up rearing the parasites to the exclusion of their own young. Three species often parasitize crows and other corvids, however, and when they do the young parasites are raised in the company of the host brood. The female of the European cuckoo ranges widely over a large defended territory in her search for nests in the process of construction. When host nests are discovered that are too advanced for successful parasitization, the female destroys them, forcing the birds to lay again.

Cuculids around the world have evolved a remarkable variety of devices that intimidate or trick the hosts into accepting their eggs. Two Indian hawk cuckoos, *Cuculus varius* and *C. sparverioides,* resemble the sparrowhawks *Accipiter badius* and *A. virgatus,* respectively, in plumage. *C. varius* also mimicks *A. badius* in flight and has been observed to lure its host from the nest. The European cuckoo resembles the sparrowhawk *A. nisus* and flies like it during the breeding season. The adaptive significance of the convergence may lie in the fact that songbirds avoid accipitrine hawks that are passing overhead and therefore are less likely to defend their nests. The Indian drongo-cuckoo *Surniculus lugubris* resembles the drongo *Dicrurus macrocercus* not only in its plumage but also in its forked tail and distinctive breeding call. The *Surniculus* have been hypothesized to take advantage of the intimidation of songbirds by drongos. An elegant ruse is employed by the Indian koel *Eudynamis scolopacea* to overcome its host, the crow *Corvus splendens.* The male approaches the host nest, calls loudly, and allows itself to be driven off. While the crow is occupied in this way, the female koel slips in quickly and lays her egg.

Cuckoos are well adapted for inserting their eggs safely into the host nests. The female has an unusually extrusible cloaca, which functions like an insect ovipositor by permitting her to drop eggs into crevices and holes that the smaller hosts occupy but that the parasite herself is too large to enter. The shells of the eggs are typically thicker than in the case of most birds, evidently to reduce the danger of breakage when they are dropped, as opposed to laid, into the nests.

Egg mimicry is the rule in the cuckoos and most other brood parasites. The eggs of cuckoos are close in size to those of the hosts, which necessitates that they are also unusually small in proportion to the female's body size. The reduction serves a dual function, since it further permits an increase in the number of eggs laid in a season. One female European cuckoo, for example, was observed to lay a total of 61 eggs over four breeding seasons, 58 of them in nests of a single host species, the meadow pipit *Anthus pratensis.* The eggs of brood parasites also tend to resemble those of the hosts in color. In the case of the European cuckoo the color mimicry exists in a form that still poses a first-class scientific mystery. Each female in a population belongs to what ornithologists call a gens (plural: gentes), all members of which lay their eggs principally in nests of the same single species of host. Even more remarkable, the eggs of a gens mimic those of the host in size and color. Thus local populations of the European cuckoo are divided into coexisting "host races" on the basis of both behavior and egg morphology. The main hosts of the three gentes found together in Finland, for example, are the European redstart *Phoenicurus phoenicurus,* which lays blue, unspotted eggs; the brambling *Fringilla montifringilla,* with pale blue eggs covered by heavy reddish spots; and the pied wagtail *Motacilla alba,* with white eggs flecked in gray. It would appear that the gentes are kept in partial genetic isolation by a preference on the part of individual females for hosts belonging to the species that reared them. Perhaps this choice is based on imprinting on the young bird while it is still in the nest. But the problem is complicated by the fact that a single male sometimes mates with females belonging to more than one gens. No genetic mechanism is known whereby egg mimicry can be kept consistent within female lines, unless the genes controlling egg color and size are located on the odd chromosome. In birds, unlike *Drosophila* and man, it is the female that has the unmatched pair of chromosomes.

Upon hatching, the young cuckoo usually eliminates the eggs and nestlings of its host, reserving the entire nest to itself. When one of these host eggs or nestlings presses against the back of a *Cuculus* nestling in the first day or two of its life, the little bird climbs backward up the side of the nest and heaves it outside. Young *Indicator* honeyguides are equipped with sharp hooks on the tips of their mandibles which they use to pierce and kill the host nestlings (see Figure 17-5). In two cuckoo species, *Clamator glandarius* and the koel *Eudynamys scolopacea,* the young also resemble those of the host. In both instances the host species are corvids, which are relatively large birds, and the parasitic young are raised in company with the host young. The significance of the mimicry seems to be that when the host parents have the opportunity to examine more than one young bird visually, they are likely to eject the parasite as the odd member of the group. Additional evidence in support of the mimicry hypothesis in cuckoos is the nature of geographic variation in the koel. In Asia young koels are covered by plumage resembling that of the immature crows with which they live. But in Australia, where koels parasitize honey-eaters and magpie-larks and eject the host young from the nest, the parasite young are not mimetic. They are instead colored like the adult koel female.

Figure 17-5 Brood parasitism in birds. These two drawings depict methods by which the young parasite nestlings dispose of the brood of their hosts. On the left a newly hatched European cuckoo ejects an egg of the reed warbler *Acrocephalus scirpaceus*. On the right a newly hatched honeyguide (*Indicator indicator*) uses its hooked bill to attack and kill the host nestlings. (From Lack, 1968.)

Undoubted mimicry of a high order of precision is displayed by the young of the combassous and widow birds of Africa. The nestlings possess the distinctive feeding guides of the host young, which are a particular pattern of colored spots on the mouth linings combined with two spherical tubercles located at the corners of the mouth. When the young bird gapes in the begging posture, the tubercles stand out so conspicuously that they were once thought to be luminescent. The host species build globular nests with dark interiors. As parents enter these structures to feed the nestlings, the tubercles are like dimly lit bulbs that guide their feeding efforts to the gaping mouths. The resemblance is probably not due to evolutionary convergence, however. The parasitic Viduinae are closely related to the host species, and it seems more likely that the feeding guides were possessed by their free-living ancestors and served as a preadaptation helping to point evolution toward parasitism.

The possible evolutionary origin of brood parasitism can be inferred by comparing the parasitic species with their closest free-living relatives. The cowbirds of the New World tropics are especially informative in this regard. Five species are parasites on other icterids and small passerine birds. A fifth, the bay-winged cowbird *Molothrus badius*, may form the connecting link with the nonparasitic ancestors. It normally uses the nests of other birds, although it still incubates its own eggs and rears the young to maturity. Sometimes bay-winged cowbirds try to build their own nests, but they have only partial

success. The seemingly probable next step in phylogeny beyond this species is facultative parasitism, in which the females lay in the nests of other species and allow the hosts to rear their young but also occasionally build nests of their own. Precisely this stage is represented by the shiny cowbird *Molothrus bonariensis*. The ultimate conceivable development in brood parasitism in birds would be total dependence on a single host species—the stage represented by the screaming cowbird *Molothrus rufo-axillaris*, whose status is rendered still more bizarre by the fact that it parasitizes the sole nonparasitic cowbird, *M. badius*.

Finally, the giant cowbird *Scaphidura oryzivora* of South and Central America has evolved along a path that has taken it out of parasitism and into mutualistic symbiosis with its hosts. The full *Scaphidura* story, worked out in meticulous detail by Neal G. Smith (1968), is the most complex example of social symbiosis known in the vertebrates. It is based on polymorphism among females combined with the presence or absence of protection provided to the hosts by social insects. *S. oryzivora* is dependent on oropendolas and caciques, which are colonially nesting members of the grackle family Icteridae. Five classes of *S. oryzivora* females can be distinguished on the basis of egg coloration and choice of hosts; namely, three mimics of the oropendola genera *Zarhychus*, *Psarocolius*, and *Gymnostinops*, a mimic of caciques (*Cacicus*), and "dumpers," which lay nonmimetic eggs of a generalized icterid form (see Figure 17-6). Females that lay mimetic eggs resembling those of a given icterid host constitute a gens comparable to the host-specific units described earlier for the European cuckoo. But the giant cowbirds go further than the cuckoos; the eggs are true not only to a particular host species but also to the local population with which the cowbirds live. Females that produce mimetic eggs are shy in behavior. They skulk around the host colonies and wait until a host female has left the nest before inserting a single egg. Dumpers, by contrast, are aggressive, drive off host females, and lay two to five eggs in each nest. In a symmetric manner, the oropendolas and caciques are polymorphic in their response to the giant cowbirds. Adults in "discriminator" populations reject any cowbird egg that is not closely mimetic, while those in "nondiscriminator" populations accept eggs that vary in color, pattern, and size.

In order to understand the meaning of this striking variation in both the parasite and host populations it is necessary to turn to a major enemy of both, the botflies of the genus *Philornis*. These insects infest many of the icterid nests, burrowing into the flesh of the nestlings and killing many of them. Oropendolas and caciques have "discovered" two ways of reducing botfly attacks. By building their nests near large colonies of social wasps (*Protopolybia* and *Stelopolybia*) and stingless bees (*Trigona*) the birds somehow receive protection for their young. These social insects repel the botflies in a way that has not been ascertained. The second mode of protection available to the birds comes from being "parasitized" by the cowbirds.

Figure 17-6 Brood parasitism by the giant cowbird *Scaphidura oryzi-vora*. In the photograph at left, a female parasite peers into the nest of an oropendola (*Zarhynchus wagleri*), while the host perches nearby. This particular host was a nondiscriminator, a genetic type that lives in a mutualistic relationship with its cowbird. The righthand photograph shows another cowbird female examining an oropendola nest. (Photographs courtesy of Neal G. Smith.)

Almost incredibly, oropendolas and caciques nesting away from the protection of wasps and bees and therefore exposed to heavy botfly attacks do better if their nests are parasitized by young cowbirds. The reason is that the guests preen their nestmates, removing the eggs and maggots of the botflies. They protect themselves by aggressively snapping at any moving object, including adult botflies invading the nest. The preening and snapping behaviors are unique within the altricial passerine birds.

Now the essential elements of the story can be fitted together. Host colonies that do not have the protection of wasps or bees are the nondiscriminators with respect to the cowbirds; they permit the eggs of dumper cowbirds to remain in the nest and therefore gain in genetic fitness. Host colonies that do enjoy the protection of wasps or bees discriminate against cowbirds. The parasites associated with them have evolved egg mimicry and shy behavior to overcome their resistance. The entire set of populations of each species of oropendolas and caciques remains in a polymorphic condition because of the uncertainty in local situations of whether protection against botflies will come from the insects or from the cowbirds. The cowbirds, in turn, maintain their own polymorphism as a mixed strategy that takes fullest advantage of their hosts. In populations where wasps and bees are present, the cowbirds are not needed and are therefore parasites; but where the insects are absent, the cowbirds live in mutualism with their hosts.

Slavery in Ants

Since the Swiss entomologist Pierre Huber first reported it in *Recherches sur les moeurs des fourmis indigènes* (1810), slavery in ants has been the object of close analysis by biologists. Dulosis, as the phenomenon is sometimes more technically labeled, has arisen six times independently. Within the Myrmicinae, it is primitively developed in *Leptothorax duloticus* and is the exclusive way of life in *Harpagoxenus* and *Strongylognathus*, which are phylogenetic derivatives of *Leptothorax* and *Tetramorium*, respectively. Within the Formicinae, slave making occurs throughout the *Formica sanguinea* complex of species, as well as in *Polyergus* and *Rossomyrmex*, which are derived from *Formica*. The slave raids of most of these forms are dramatic affairs in which the workers go out in columns, forcibly penetrate the nests of colonies belonging to other, related species, and bring back pupae to their own nests. The pupae are allowed to develop into workers, which become fully functional members of the colony. The slaves simply accept their captors as sister workers and proceed to perform as they would in their own nest. The workers of most slave-making species, by contrast, seldom if ever join in the ordinary chores of foraging, nest building, and rearing of the brood, all of which they leave to the slaves.

Charles Darwin was fascinated by the implications of ant slavery. In *The Origin of Species* he devised the first hypothesis of how the behavior originated in evolution. He proposed that the ancestral species began by raiding other kinds of ants in order to obtain their pupae for food. Some of the pupae survived in the storage chambers long enough to eclose as adult workers, whereupon they accepted their captors as nestmates. This fortuitous addition to the work force helped the colony as a whole, and consequently there was an increas-

ing tendency, propelled by natural selection, for subsequent generations to raid other colonies solely for the purpose of obtaining slaves. Recently I presented an alternative scheme based on studies of *Leptothorax* (Wilson, 1974b). Two free-living species of the genus, *L. ambiguus* and *L. curvispinosus*, readily raid other colonies of the same or other species, drive off and kill the adults, and capture the brood when the nests are placed too close together. The brood is tolerated and allowed to mature, perhaps because it has a less distinctive colony odor. Newly eclosed *curvispinosus* workers are permitted to live when their captors are themselves *curvispinosus*, but they are killed after a day or two if the captors are *ambiguus*. Thus the preadaptation to dulosis appears to be a combination of territorial behavior and the tolerance of brood. *Leptothorax* species (and probably many other ant species as well) might become obligatory slave-makers by merely extending their territorial limits and then coming to depend on the workers captured in this manner. Just such an early stage is represented by the rare slave-maker *L. duloticus* (see Figure 17-7). When I deprived a *duloticus* colony of its *curvispinosus* slaves, the workers regained most of the general behavioral repertory, grooming and cleaning the brood, handling nest materials, and foraging to a limited extent for food. However, they were less than competent at most of these tasks, and they showed a fatal inability to retrieve and feed on insect prey and other solid food. *L. duloticus*, in other words, is still not far removed from its closest free-living relatives in the genus, but behavioral decay has progressed enough to make it an obligatory parasite.

Approximately the same evolutionary grade is occupied by *Formica sanguinea*, the species in which dulosis was discovered by Huber. More recent studies have been summarized by Dobrzański (1961, 1965) and Wilson (1971a). The "sanguinary ants," so named for their blood-red thorax, head, and appendages, are very aggressive and territorial. They are not obligatory slave holders, since colonies are often found in which no slaves are present, and the ants are able to live indefinitely on their own in the laboratory. The commonest slaves taken belong to the *fusca* group of *Formica*, including *fusca*, *lemani*, and *rufibarbis*; less commonly exploited are *gagates*, *cunicularia*, *transkaukasica*, and *cinerea*, all of which are also members of the *fusca* group as conceived in the broadest sense. As a rule, *sanguinea* colonies raid colonies nearest their own nests, and the seeming preferences are merely a reflection of the local relative abundance of the slave species. Two or even three slave species are sometimes present in a given *sanguinea* nest simultaneously, and the composition of slaves may change from year to year. Each *F. sanguinea* colony conducts at most two or three raids a year, in July and August after the reproductive forms have left the nest on their nuptial flights. At any time of the day, but usually in the morning, a large detachment of workers leaves the nest and heads in a straight line for the target

Figure 17-7 Slavery in ants. Workers of the slave-making species *Leptothorax duloticus* are indicated by arrows; other workers in this photograph belong to the slave species *L. curvispinosus*. Also present are brood of both species in all stages of development. The similarity of the slave-maker to the enslaved ant is an example of the rule of phylogenetic closeness that applies to most kinds of social parasitism. (From Wilson, 1974b.)

nest of the slave species. The raiding party is actually a loose phalanx up to several meters across. It may travel for as far as 100 meters. Upon arriving at the target nest, the *sanguinea* workers wait a while around the entrance, then enter one after another. The resident workers usually try to flee, carrying their eggs, larvae, and pupae out of the nest to scramble away over the ground and up nearby grass blades. They are attacked by the *sanguinea* workers only when they offer hostile resistance. The raiders finally straggle back to their own nest carrying the captured pupae.

The communicative signals that trigger and orient the raids by colonies of the *sanguinea* group of slave-making ants have been identified at least in part by Regnier and Wilson (1971). We found that workers of the American species *Formica rubicunda* readily follow artificial odor trails made from whole body extracts of *rubicunda* workers and applied with a camel's hair brush over the ground in the vicinity of the nest. When the trails were laid away from the nest opening in the afternoon, at about the time raids are usually conducted, the *rubicunda* workers showed behavior that was indistinguishable from ordinary raiding sorties. They ran out of the nest and along the trail in an excited fashion, and, when presented with colony fragments of a slave species (*F. subsericea*), they proceeded to fight with the workers and to carry the pupae back to the nest. It seems probable that under normal circumstances lone *rubicunda* scouts lay odor trails from the target slave nests they discover to the home nest, and the raids result when nestmates follow the trails out of the home nest back to the source. This is evidently the general mode of communication among slave-making ants. Strong evidence has been adduced to indicate its existence in *Harpagoxenus* and *Leptothorax* (Wesson, 1939, 1940) as well as in *Polyergus* (Talbot, 1967). The tendency of *F. sanguinea* workers to fan out into "phalanxes" during their outward march does not conflict with this interpretation; several odor trails could be involved, around which orientation is less than perfect.

The general biology and raiding behavior of *Formica subintegra*, an American representative of the *sanguinea* group, have been studied by Wheeler (1910) and Talbot and Kennedy (1940). The latter investigators, by keeping a chronicle over many summers of a population on Gibraltar Island, in Lake Erie, found that raiding is much more frequent than in *sanguinea*. Some colonies raided almost daily for weeks at a time, striking out in any one of several directions on a given day. Occasionally the forays continued on into the night, in which case the *subintegra* workers remained in the looted nest overnight and returned home the following morning. Regnier and Wilson (1971) discovered that each *subintegra* worker possesses a grotesquely hypertrophied Dufour's gland, which carries approximately 700 micrograms of a mixture of decyl, dodecyl, and tetradecyl acetates (see Figure 17-8). These substances are sprayed at the defending colonies during the slave raids. They act at least in part as "propaganda substances" because they help to alarm and disperse the defending workers. The acetates are in fact ideally designed for this purpose in accordance with the "engineering rules" for the evolution of pheromones described in Chapter 10. Having a higher molecular weight than ordinary alarm substances, they evaporate at a slower rate and exert their effect for longer periods of time. The larger size of the molecules also gives the acetates the potential for lower response thresholds, although this possibility has not been tested experimentally. The *subintegra* workers themselves are not adversely affected by the odor of their own acetates. They are excited and attracted by these substances, exactly the responses needed to conduct successful slave raids. The discovery of the propaganda substances appears to solve a puzzle first noted by Huber in 1810, namely, the readiness with which colonies of slave species yield to the raiders. As Huber expressed it, "One of the principal features of the wars levied on the Ash-colored ants [*F. fusca*] seems to consist of exciting fear, and this effect is so strong that they never return to their besieged nest, even when the oppressors [*F. sanguinea*] have retired to

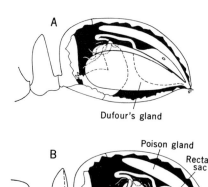

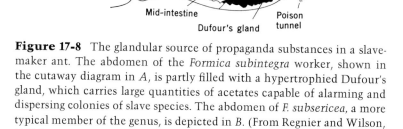

Figure 17-8   The glandular source of propaganda substances in a slave-maker ant. The abdomen of the *Formica subintegra* worker, shown in the cutaway diagram in *A*, is partly filled with a hypertrophied Dufour's gland, which carries large quantities of acetates capable of alarming and dispersing colonies of slave species. The abdomen of *F. subsericea*, a more typical member of the genus, is depicted in *B*. (From Regnier and Wilson, 1971.)

their own nest; perhaps they realize that they could never remain in safety, being continually liable to new attacks by their unwelcome visitors." The fear and the realization are evidently due to the perversion by the slave-makers of the normal chemical communication used by the slaves in their own nests.

The eight species of the Old World genus *Strongylognathus* present among themselves the full transition from slave making to full inquilinism. Since the first observations by Forel (1874), this wholly parasitic genus of ants has been intensively studied by many authors, among whom the most recent and thorough are Kutter (1923, 1969) and Pisarski (1966). *Strongylognathus* is closely related to *Tetramorium*, and its species enslave members of the latter genus. The most favored slave species is *T. caespitum*, one of the most abundant and widespread ant species of Europe. *S. alpinus* has a life cycle more or less typical of the majority of *Strongylognathus* species. It is at an evolutionary level somewhat less advanced than that of *Harpagoxenus* in the one special sense that the behavior of its workers is less degenerate. The workers, like those of most parasitic ant species, do not forage for food or care for the immature stages; nevertheless, they still feed themselves and assist in nest construction. The raids of *alpinus* are notoriously difficult to observe. They occur in the middle of the night and take place, for the most part, along underground galleries. The *alpinus* workers are accompanied by *T. caespitum* slaves, who, true to the aggressive nature of their species, join in every phase of the raid. Warfare against the target colony is total: the nest queen and winged reproductives are killed, and all of the brood and surviving workers are carried back and incorporated into the mixed colony. This union of adults should not be too surprising when it is recalled that *T. caespitum* colonies, even in the absence of *Strongylognathus*, frequently conduct pitched battles that sometimes terminate in colony fusion. The *S. alpinus* workers are well equipped for lethal fighting. Like many other dulotic and parasitic ant species, they possess saber-shaped mandibles adapted for piercing the heads of victims that resist them (see Figure 17-9). The mode of colony multiplication is not known, but it is at least clear that the host queen is somehow eliminated in the process.

One member of the genus, *Strongylognathus testaceus*, has completed the transition to complete inquilinism. The *Tetramorium* queen is tolerated and lives side by side with the *S. testaceus* queen. There are fewer *testaceus* than host workers, the usual situation found in advanced dulotic species. The *testaceus* workers do not engage in ordinary household tasks and are wholly dependent on the host workers for their upkeep. But the key fact is that they also do not engage in slave raids. Somehow the reproductive ability of the host queen is curtailed. She generates only workers and no reproductives. Only the *S. testaceus* queen is privileged to produce both castes. Nevertheless, the presence of the *Tetramorium* queens permits the mixed colonies to attain great size. Wasmann found one comprised of between 15,000 and 20,000 *Tetramorium* workers and several thousand *Strongylognathus* workers. The brood consisted primarily

of queen and male pupae of the inquiline species. It is evident that *S. testaceus* is in a stage of parasitic evolution just a step beyond that occupied by *S. alpinus*. The worker caste of *testaceus* has been retained, and it still has the murderous-looking mandibles dating from the species' dulotic past, but it has evidently lost all of its former functions and is in the process of being reduced in numbers. Probably *S. testaceus* is on the way to dropping the worker caste altogether, a final step that would take the species into the ranks of the extreme inquilines.

Inquilinism in Ants

We have seen that full inquilinism, in which the social parasite is dependent on its host throughout its life cycle, can be reached through several evolutionary approaches. This information is summarized diagrammatically in Figure 17-10. Once a species enters the final evolutionary sink of inquilinism, it seems to evolve quickly into a state of abject dependence on the host species. It acquires an increasing number of traits that together constitute the "inquiline syndrome." The worker caste is lost, and the queen tends to be replaced by fertile queen-worker intercastes called ergatogynes. If the queen persists, she and the male are reduced in size, often dramatically so; in some species, including *Teleutomyrmex schneideri*, *Aporomyrmex ampeloni*, and *Plagiolepis xene*, the queen is actually smaller than the worker caste of the host species. The male develops a pupalike form: its body is thickened, the articulations of the petiole and postpetiole are broadened, the genitalia become permanently exserted, the cuticle is thinned and depigmented, and the wings are reduced

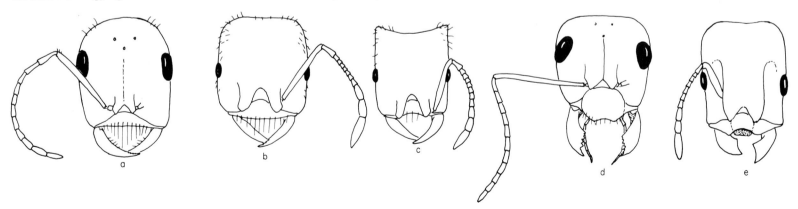

Figure 17-9 Heads of workers of five species of dulotic ants, showing varying degrees of modification of the mandibles for fighting during the slave raids: (*a*) *Polyergus rufescens*, (*b*) *Strongylognathus alpinus*, and (*c*) *S. testaceus* have saber-shaped mandibles used to pierce the exoskeletons of their victims; (*d*) *Formica sanguinea*, a facultative slave-maker whose workers still carry on normal work loads in their own nests, has unmodified mandibles with a full set of teeth on the gripping edge; (*e*) *Harpagoxenus sublaevis* has sharp, clipper-shaped mandibles used to nip and cut the appendages of opponents. (From Kutter, 1969.)

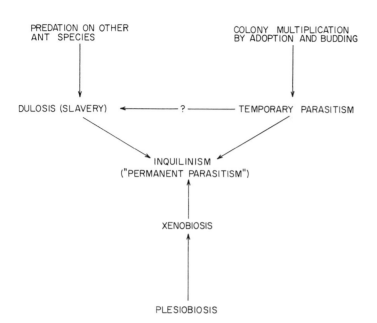

Figure 17-10 The evolutionary pathways of social parasitism in ants. (From Wilson, 1971a.)

or lost. Mating often takes place in the nest, and the dispersal distance of the fecundated queen afterward is limited. Probably as a result of the curtailment of the nuptial flight, or perhaps as its cause, the populations of inquilinous species are characteristically fragmented and very local in occurrence. Anatomical structures are reduced and simplified all over the body: wing venation is partially lost, the antennal segments are reduced in number by fusion, and the mouthparts are simplified and weakened. Many of the exocrine glands are diminished or lost, including some employed in chemical communication by other, free-living ant species. The central nervous system is reduced in size and complexity and the behavioral repertory is drastically narrowed. The parasites depend increasingly on their ability to attract the host workers and to trick them into donating liquid food by regurgitation.

Morphological and behavioral decay, evidently irreversible in nature, has been documented step by step during the comparison of species across many ant genera. The sequence proceeds from the beginnings of the inquilinous state, as in the myrmicine genera *Kyidris* (Wilson and Brown, 1956) and *Strongylognathus* (Kutter, 1969), through intermediate conditions in which the worker caste is in the process of being lost, as in the parasitic members of the formicine genus *Plagiolepis* (Le Masne, 1956a; Passera, 1968), to the most bizarre, degenerate species completely lacking workers. In the

last category can be placed *Teleutomyrmex schneideri*, which perhaps deserves the title of the "ultimate" social parasite. This remarkable species was discovered by Kutter (1950) in the Saas-Fee, an isolated valley of the Swiss Alps near Zermatt. Its behavior has been studied by Stumper (1950) and Kutter (1969), its neuroanatomy by Brun (1952), and its general anatomy and histology by Gösswald (1953). *T. schneideri* is a parasite of *Tetramorium caespitum*. Like many other inquilinous species, it is phylogenetically closer to its host than to any of the other members of the ant fauna to which it belongs. This tendency in social parasitism of ants is sometimes called "Emery's rule," in recognition of the first formulation by the Italian myrmecologist Carlo Emery in 1909. In fact, *Teleutomyrmex schneideri* may have been derived directly from a temporarily free-living offshoot of *Tetramorium caespitum*, since the latter form is the only nonparasitic member of the tribe Tetramoriini known to be native at the present time to central Europe. Figure 17-11 presents a hypothesis of the origin of this and other species exemplifying Emery's rule in terms of the modern theory of geographic speciation.

It is difficult to conceive of any stage of social parasitism more advanced than that actually attained by *Teleutomyrmex*. The species, which possesses no worker caste, lives exclusively within the nests of its hosts. The queens, which contribute nothing to the labor of the colony, are tiny in comparison with other ants, especially other tetramoriines, averaging only about 2.5 millimeters in length. They

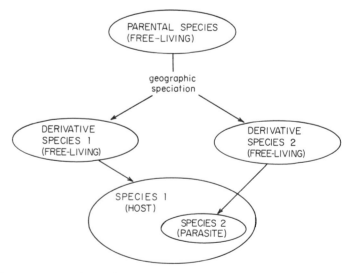

Figure 17-11 The evolutionary origin of socially parasitic species. This diagram depicts the steps through which a parasitic species can originate and come to live as a social parasite with its closest living relative, in accordance with Emery's rule. (From Wilson, 1971a.)

are unique among all known social insects in being ectoparasites, meaning that they spend much of their time riding on the backs of their hosts (see Figure 17-12). The anatomy of the *Teleutomyrmex* queen is strikingly modified to serve this peculiar habit. The ventral surface of the gaster (the large terminal part of the body) is strongly concave, permitting the parasites to press their bodies close to those of their hosts. The tarsal claws and arolia (foot pads) are unusually large, allowing the parasites to secure a strong grip on the smooth chitinous surface of the hosts. The queens have a marked tendency to grasp objects quite unlike that displayed by any other ants. Given a choice, they will position themselves atop the body of the host queen, either on the thorax or on the abdomen. Deprived of the nest queen, they will seize a virgin *Tetramorium* queen, or a worker, or a pupa, or even a dead queen or worker. Stumper observed a case in which eight *Teleutomyrmex* queens simultaneously grasped a *Tetramorium* queen, completely immobilizing her. The parasites evidently receive their nourishment from liquid materials regurgitated to them by the workers. Each queen lays eggs on the average of one every 30 seconds. The infested *Tetramorium* colonies are smaller than those free of *Teleutomyrmex* in the same localities, but they still contain thousands of workers and appear to function normally. The single host queen found in each nest continues to lay eggs but the emerging larvae are unable to develop into any caste but workers. This reproductive "castration" of the host colony has been reported in other parasitic ants that tolerate the host queens. It is clearly to the advantage of the parasite to keep the host queens alive so long as they produce only workers, which then prolong the survival and increase the reproductive rate of the parasites. The physiological mechanism of castration is still unknown.

The *Teleutomyrmex* queen has undergone extensive morphological deterioration correlated with her total dependence on the social system of *Tetramorium*. The labial and postpharyngeal glands are reduced, and the maxillary and metapleural glands are completely absent. The integument is thin and less pigmented and sculptured than that of *Tetramorium*. As a result of these reductions the queens are shining and brown, an appearance that contrasts with the opaque blackish-brown of their hosts. The sting and poison apparatus are reduced; the mandibles are so degenerate that the parasites almost certainly cannot secure food on their own; the brain is reduced in size and ganglia 9-13 are fused into a single piece; and so on. In its essentials the life cycle of *Teleutomyrmex schneideri* resembles that of other known extreme ant parasites. Mating takes place within the host nest. The fecundated queens then either shed their wings and join the small force of egg layers or else fly out in search of a new *Tetramorium* nest to invade. This reproductive restriction no doubt contributes to the fact that *Teleutomyrmex schneideri* is one of the rarest and most locally distributed insect species in the world.

The General Occurrence of Social Parasitism in Insects

Social parasitism in higher social insects is mostly confined to the temperate zones, in particular the cooler parts of the United States, Canada, Europe, and Asia, South Africa, and central Argentina. A few inquilinous ant species have been collected in the tropics, for example the extreme workerless parasite *Anergatides kohli* from the Congo and the strange postxenobiotic members of the genus *Kyidris* from New Guinea and Madagascar. From morphological evidence alone, numerous African and Asian species of *Crematogaster* in the subgenera *Atopogyne* and *Oxygyne* are likely to be temporary social parasites (Wheeler, 1925). The same is true of *Azteca aurita* and *A. fiebrigi* of the New World tropics and all of the species of *Rhoptromyrmex*, which are widespread in South Africa, Asia, New Guinea, and Australia (Brown, 1964). Also, inquilinous species of allodapine bees have been discovered in East Africa, Malaya, and Queensland (Michener, 1970). Even so, the vast majority of proven social parasites are found in cool climates. The disparity is too great to be attributable to differences in thoroughness of collecting. Furthermore, a disproportionate number of ant parasites occurs in mountainous and arid regions. Numerous species have been discovered in the Alps, no less than six in the little valley of Saas-Fee alone, while a majority of the described North American inquilines occupy limited ranges in the mountains of Texas, New Mexico, Colorado, and California. Not a single slave-making species has been discovered in either the tropical or south temperate zones.

Two hypotheses are available to account for the richness of social parasitism in cold climates. Richards (1927a), and later Hamilton (1972), argued that in bumblebees the crucial preadaptation is the existence of two closely related species, one northern in distribution and the other southern. When the southern species penetrates the range of the northern species, it will at first tend to emerge later in the spring than the northern species within the zone of overlap. A second precondition, which has been well documented in some species of *Bombus* as well as in social wasps of the genera *Polistes* and *Vespa*, is the tendency of queens to invade colonies of their own species. Given the availability of a closely related species that already has well-developed colonies and that in the initial stages of range invasion is more abundant as well, there might be a tendency for the invader to evolve in the direction of interspecific parasitism. As the parasitism advances from the facultative temporary state to complete inquilinism, the range of the southern invader would be wholly absorbed within that of the host species. The data on geographic ranges and stages of parasitism are still inadequate, or at least insufficiently analyzed, to test Richards' hypothesis. The second mechanism, perhaps compatible with the first, was suggested by Wilson

Figure 17-12 The "ultimate" social parasite may well be the little ant *Teleutomyrmex schneideri*, shown here with its host *Tetramorium caespitum.* The two *Teleutomyrmex* queens sitting on the thorax of the host queen have not yet undergone ovarian development, and their abdomens are consequently flat and unexpanded. One still bears her wings and is almost certainly not yet fecundated. The third *Teleutomyrmex* queen, which rides on the abdomen of the host queen, has an abdomen swollen with hyperdeveloped ovarioles. A host worker stands in the foreground. (Drawing based on a painting by Walter Linsenmaier, courtesy of Robert Stumper.)

(1971a). It is possible that cooler temperatures ease the introduction of parasitic queens by dulling the responses of the host colonies. Queens in laboratory culture can be more easily combined with groups of alien workers if all are first chilled to immobility and then allowed to warm up together. In nature parasite queens need not wait for winter to exploit this advantage. Some degree of chilling, say down to 10° or 15°C, occurs commonly during the cool summer nights in mountainous regions, right in the middle of the season of nuptial flights.

Breaking the Code

The deep penetration of alien insect societies by inquilines has been achieved with the aid of physiological and behavioral convergence toward the hosts. The inquilines have broken the code of the social insects. To varying degrees the individual parasite species track down their host colonies, gain acceptance as members, and persuade the workers to feed them. The entomologist who studies social symbiosis is presented with an unusual opportunity to identify the minimal set of signals that hold an insect society together. Because parasites commonly deal in supernormal stimuli, the investigator also has a better than ordinary chance of characterizing the signals physiologically.

A case in point is the elegant series of experiments conducted by Bert Hölldobler (1967–1971) on the behavior of the staphylinid beetle *Atemeles pubicollis*, an inquiline of ant colonies in Europe. *Atemeles* emigrating from one host colony to another are guided by the odor of the ants. Hölldobler found that the beetles pay no attention to the exits of laboratory nests containing colonies so long as the test arenas are kept in still air. But when a weak air current is drawn first over the colonies and then in the direction of the beetles, the *Atemeles* run upwind and congregate around the nest exits. This set of stimuli closely approximates the stimuli encountered in nature, where beetles must search for the scattered host nests over the surface of the ground. By moving upwind in odor-laden air currents, they are able to orient over much greater distances than by following odor gradients alone. Hölldobler also discovered that the odor preference of the *Atemeles* changes with age. As Erich Wasmann had learned many years previously, and Hölldobler confirmed, the beetles migrate from nests of *Formica* to those of *Myrmica* six to ten days after they eclose from the pupae. The following spring the beetles return to nests of *Formica*, where they reproduce. The physiological basis of the shift is quite elementary. Laboratory experiments revealed that the beetles are attracted to air laden with *Myrmica* odor in preference to that of *Formica* following eclosion, but when they emerge from hibernation their preference switches back to *Formica*. The adaptive significance seems to be the availability of ant larvae, which are the chief

prey of the beetles. *Myrmica* maintain larvae in their nests throughout the late fall, winter, and early spring, but *Formica* do not. However, during the summer the larger *Formica* colonies provide the richer source of larvae.

Admittance to the host colony is gained by intricate maneuvers in which two and sometimes three exocrine glands are brought into play. After reaching the nest entrance a beetle wanders around until it encounters a worker. Then it turns to present its "appeasement gland," located at the tip of the abdomen (see Figure 17-13). The secretion of the gland is at least partly proteinaceous and it contains no appreciable amounts of carbohydrates. The ant feeds on the material, seeming to grow calmer in the process. Then it moves over to the "adoption glands," which it also licks. After this second repast, the ant transports the beetle into the nest. If the ruse fails, and the *Atemeles* is attacked, it is able to use repugnatorial secretions from defensive glands to keep the ant away.

Once inside the nest, the beetle is easily able to induce its hosts to regurgitate food. Hölldobler demonstrated that the only signal required is the minimal tactile stimulus used by the ants themselves. The most susceptible worker ant is one that has just finished a meal and is searching for nestmates with which to share the contents of its elastic crop. In order to gain its attention, a nestmate (or social parasite such as *Atemeles*) has only to tap its body lightly with antennae or forelegs. This causes the donor to turn and to face the individual that gave the signal. If tapped lightly and repeatedly on the labium, it will regurgitate. Other ants ordinarily use their fore tarsi for this purpose, while the adult *Atemeles* tap with either their tarsi or antennae. The larvae of *Atemeles* lack appendages of sufficient length and must curve the front parts of their bodies upward and push the labia against those of the host ants. Even these clumsy imitations are enough if the donors are heavily laden with crop liquid.

A more sophisticated feat is for the symbionts to dupe the host workers into treating them as particular immature stages in the host brood development. The grub-shaped larvae of the *Atemeles* accomplish this with distinction. They are picked up by the *Formica* workers and placed among the host larvae, which they proceed to consume voraciously. Hölldobler discovered that a substance associated by the workers with their own brood can be separated from the bodies of the beetle larvae. When he extracted the parasites in acetone and soaked dummies in the mixture, the dummies became temporarily attractive to the workers and were treated as pieces of brood. The substances are evidently secreted by pairs of glands located on the upper surface of each segment of the parasite's body.

In summary, the *Atemeles* have penetrated the heart of the ant colony by the production of no more than two or three "pseudopheromones" and the imitation of two elementary tactile signals. They have taken advantage of the relative impersonality of insect

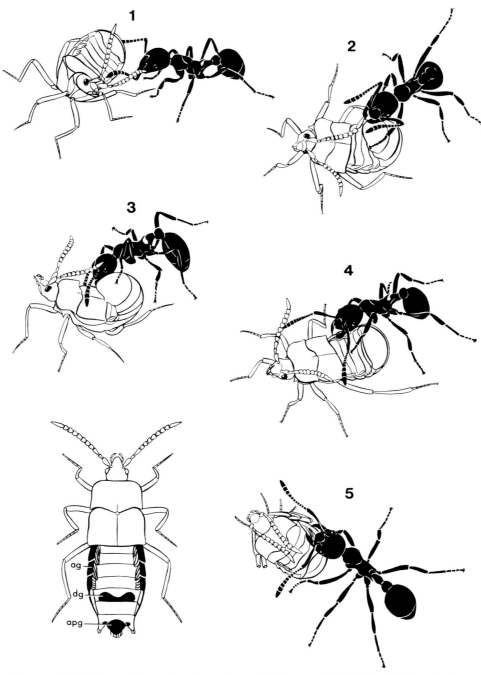

Figure 17-13 The social parasite *Atemeles pubicollis*, a staphylinid beetle, has "broken the code" of the ant *Myrmica* to gain partial membership in its colonies. These drawings depict the maneuvers by which one of the adult beetles induces a worker ant (shown in black) to carry it into the nest. The figure at the lower left indicates the location of the three principal abdominal glands of the parasite: (*ag*) adoption glands; (*dg*) defensive glands; (*apg*) appeasement gland. The beetle presents its appeasement gland to a worker of *Myrmica* that has just approached it (*1*). Upon licking the gland opening (*2*), the worker moves around to lick the adoption glands (*3, 4*), after which it carries the beetle into the nest (*5*). (After Hölldobler, 1970.)

societies and the narrow sensory *Umwelt* of their hosts. Noting the strikingly different appearance of such parasites and their distinctive behavior, we can only marvel at the simplicity of the codes by which such complex societies can be organized. As Wheeler said, "Were we to behave in an analogous manner we should live in a truly Alice-in-Wonderland society. We should delight in keeping porcupines, alligators, lobsters, etc., in our homes, insist on their sitting down to table with us and feed them so solicitously with spoon victuals that our children would either perish of neglect or grow up as hopeless rhachitics." The scientific exploration of the labyrinthine world of social symbiosis has just begun, and the discoveries yet to be made will sustain our sense of wonder for a long time to come.

Part III **The Social Species**

Chapter 18 **The Four Pinnacles of Social Evolution**

To visualize the main features of social behavior in all organisms at once, from colonial jellyfish to man, is to encounter a paradox. We should first note that social systems have originated repeatedly in one major group of organisms after another, achieving widely different degrees of specialization and complexity. Four groups occupy pinnacles high above the others: the colonial invertebrates, the social insects, the nonhuman mammals, and man. Each has basic qualities of social life unique to itself. Here, then, is the paradox. Although the sequence just given proceeds from unquestionably more primitive and older forms of life to more advanced and recent ones, the key properties of social existence, including cohesiveness, altruism, and cooperativeness, decline. It seems as though social evolution has slowed as the body plan of the individual organism became more elaborate.

The colonial invertebrates, including the corals, the jellyfishlike siphonophores, and the bryozoans, have come close to producing perfect societies. The individual members, or zooids as they are called, are in many cases fully subordinated to the colony as a whole—not just in function but more literally, through close and fully interdependent physical union. So extreme is the specialization of the members, and so thorough their assembly into physical wholes, that the colony can equally well be called an organism. It is possible to array the species of colonial invertebrates into an imperceptibly graded evolutionary series from those forming clusters of mostly free and self-sufficient zooids to others with colonies that are functionally indistinguishable from multicellular organisms.

The higher social insects, comprised of the ants, termites, and certain wasps and bees, form societies that are much less than perfect. To be sure, they are characterized by sterile castes that are self-sacrificing in the service of the mother queen. Also, the altruistic behavior is prominent and varied. It includes the regurgitation of stomach contents to hungry nestmates, suicidal weapons such as detachable stings and exploding abdomens used in defense of the colony, and other specialized responses. The castes are physically modified to perform particular functions and are bound to one another by tight, intricate forms of communication. Furthermore, individuals cannot live apart from the colony for more than short periods. They can recognize castes but not individual nestmates. In a word, the insect society is based upon impersonal intimacy. But these similarities to the colonies of lower invertebrates are balanced by some interesting qualities of independence. Social insects are physically separate entities. The secret of their success is in fact the ability of a colony to dispatch separately mobile foragers that return periodically to the home base. Also, the queens are not always the exclusive egg layers. Female workers sometimes insert eggs of their own into the brood cells. Because these eggs are unfertilized, they develop into males. The evidence is strong that in some species of ants, bees, and wasps a low-keyed struggle continually takes place between queens and workers for the opportunity to produce sons. Conflict is sometimes overt in more primitive forms. Groups of female wasps starting a colony together contend for dominance and the egg-laying rights that go with the alpha position. Losers perform as workers, once in a while stealthily inserting eggs of their own into empty brood cells. In this case there is evidence of individual recognition. Similarly, bumblebee queens control their daughters by aggression, attacking them whenever they attempt to lay eggs. If the queen is removed from the relatively simple wasp and bumblebee societies, certain of the workers fight among one another for the right to replace her.

Aggressiveness and discord are carried much further in vertebrate societies, including those of mammals. Selfishness rules the relationships between members. Sterile castes are unknown, and acts of altruism are infrequent and ordinarily directed only toward offspring. Each member of the society is a potentially independent, reproducing unit. Although an animal's chances of survival are reduced if it is forced into a solitary existence, group membership is not mandatory on a day-to-day basis as it is in colonial invertebrates and social insects. Each member of a society is on its own, exploiting the group to gain food and shelter for itself and to rear as many offspring as possible. Cooperation is usually rudimentary. It represents a concession whereby members are able to raise their personal survival and reproductive rates above those that would accrue from a solitary life. By human standards, life in a fish school or a baboon troop is tense and brutal. The sick and injured are ordinarily left where they fall, without so much as a pause in the routine business of feeding, resting, and mating. The death of a dominant male is usually followed by nothing more than a shift in the dominance hierarchy, perhaps accompanied, as in the case of langurs and lions, by the murder of the leader's youngest offspring.

Human beings remain essentially vertebrate in their social structure. But they have carried it to a level of complexity so high as to constitute a distinct, fourth pinnacle of social evolution. They have broken the old vertebrate restraints not by reducing selfishness, but rather by acquiring the intelligence to consult the past and to plan the future. Human beings establish long-remembered contracts and profitably engage in acts of reciprocal altruism that can be spaced over long periods of time, indeed over generations. Men intuitively introduce kin selection into the calculus of these relationships. They are preoccupied with kinship ties to a degree inconceivable in other social species. Their transactions are made still more efficient by a unique syntactical language. Human societies approach the insect societies in cooperativeness and far exceed them in powers of communication. They have reversed the downward trend in social evolution that prevailed over one billion years of the previous history of life. When placed in this perspective, it perhaps seems less surprising

that the human form of social organization has arisen only once, whereas the other three peaks of evolution have been scaled repeatedly by independently evolving lines of animals.

Why has the overall trend been downward? It must have something to do with the greater physical malleability of the lower invertebrates. Because their body plan is so elementary, such colonial animals as coral polyps and bryozoans can be grossly modified to permit an actual physical union with one another. Compared with insects and vertebrates, they require less "rewiring" of nerve cells, rerouting of circulatory systems, and other adjustments of organ systems needed to coordinate colonial physiology. The generally sedentary habits of zooids also make it easier for them to be fused. However, this advantage is not the decisive one; some of the most elaborate invertebrate colonies, including those of siphonophores and thaliaceans, are also the most motile. Their simple body construction also makes it possible for lower invertebrates to reproduce by directly budding off new individuals from old. Colonies therefore consist of genetically identical individuals. And this, in the final analysis, is the most important feature of all. Absolute genetic identity makes possible the evolution of unlimited altruism. It is already the basis for the extreme specialization and coordination of somatic cells and organs within the metazoan body. The most advanced colonial invertebrates have followed essentially the same road, leading to superorganisms whose organs are created by the extreme modification of zooids (see Chapter 19).

The social insects have none of the preadaptations of the lower invertebrates. Their bodies are in many ways as elaborately constructed as those of the vertebrates, and they are fully motile. Physical union is impossible. Yet they have produced altruistic castes and degrees of colony integration almost as extreme as it is possible to imagine. I would like to suggest that part of the explanation of this achievement is the sheer enormity of the sample size. Over 800,000 species of insects have been described, constituting more than three-quarters of all the known kinds of animals in the world. In the late Paleozoic Era the ancestors of the living insects were among the first invaders of the land, and they took full advantage of this major ecological opportunity. Whereas the ocean and fresh water were already crammed with the representatives of many animal phyla, most of which had originated in Precambrian times, the land was like a new planet, filling with plant life and nearly devoid of animal competitors. The result was an adaptive radiation of species unparalleled before or since. The purely statistical argument states that out of this immense array of new types, it was more probable for at least a few extreme social forms to arise—in comparison, say, with the mere 7000 species of annelid worms or 5300 species of starfish and other echinoderms. The argument can perhaps be made clearer by imagining a concrete example: if the rate of invention of advanced sociality were 10^{-12} per species per year for animals generally, 800,000 insect species would certainly have achieved it many times by chance alone, whereas 10,000 species belonging to another phylum might never do so.

The statistical argument gains strength if we calculate on the basis of numbers of genera, families, and higher taxa instead of species, because these higher taxonomic categories reflect stronger degrees of ecological difference. A wolf, for example, representing the family Canidae, differs more in ecology from a deer (family Cervidae) than it does from other canids such as foxes and wild dogs. The very extensive insect radiation has made it more probable that at least one entire group of species has arisen that is especially predisposed toward sociality. Such a group can in fact be identified; it is the order Hymenoptera, composed of the ants, bees, and wasps. Although constituting only about 12 percent of the living insect species, the hymenopterans hold a near monopoly on higher social existence. Eusociality, the condition marked by the possession of sterile castes, has originated within this order on at least eleven separate occasions, within the roachlike ancestors of the termites once, and in no other known insect group. This remarkable fact brings us back to the overriding factor of kinship. Because of the haplodiploid mode of sex inheritance (to be explained in Chapter 20), female hymenopterans are more closely related to their sisters than they are to their daughters. Thus, all other circumstances being equal, it is genetically more advantageous to join a sterile caste and to rear sisters than it is to function as an independent reproductive. Hymenopterans have other preadaptive traits that make the origin of social life easier, including a tendency to build nests, a long life, and the ability to home; but the haplodiploid bias remains the one feature that they possess uniquely. Even so, the maximum degree of relationship among hymenopteran sisters (measured by the coefficient of relationship, r) is $\frac{3}{4}$, which is substantially less than that in the colonial invertebrates ($r = 1$). The 25 percent or more of genetic difference is enough to explain the amount of discord observed within the hymenopteran societies.

In vertebrates the maximum r between siblings is $\frac{1}{2}$, meaning 50 percent identity of genes by common descent, the same degree that exists between parents and their offspring. As a result no special genetic advantage accrues to members of a sterile caste, and with the remotely possible exception of homosexuals in human beings (see Chapter 27) no sterile caste is known to have originated. The nonhuman vertebrates as a whole are more social than the insects in the one special sense that a larger percentage have achieved some level of sociality. But their most advanced societies are not nearly so extreme as those of the insects. In other words, a strong impelling force appears to generate social behavior in vertebrate evolution, but it is brought to a halt by the equally strong countervailing force of lower genetic relationship among closest relatives. Consequently it seems

best to dwell not on the matter of genetic relationship, which is simple and canonical, but on the nature of the impelling force. This force, I would like to suggest, is greater intelligence. The concomitants of intelligence are more complex and adaptable behavior and a refinement in social organization that are based on personalized individual relationships. Each member of the vertebrate society can continue to behave selfishly, as dictated by the lower degrees of kinship. But it can also afford to cooperate more, by deftly picking its way through the conflicts and hierarchies of the society with a minimum expenditure of personal genetic altruism. We must bear in mind that whereas the primary "goal" of individual colonial invertebrates and social insects is the optimization of group structure, the primary "goal" of a social vertebrate is the best arrangement it can make for itself and its closest kin within the society. The social behavior of the lower invertebrates and insects has been evolved mostly through group selection, whereas the social behavior of the vertebrates has been evolved mostly through individual selection. The requisite refinement and personalization in vertebrate relationships are achieved by (1) enriched communication systems; (2) more precise recognition of and tailored responses to groupmates as individuals; (3) a greater role for learning, idiosyncratic personal behavior, and tradition; and (4) the formation of bonds and cliques within the society. Let us examine each of these qualities briefly.

The majority of vertebrate species utilize at least two or three times more basic displays than do most insect species, including even the social insects. But the actual number of messages that can be transmitted is far greater, for two reasons. First, context is more important to the meaning of each vertebrate display. A distinct message can be associated with the place the display occurs, the time of year, or even the sex and rank of the animal. The signal is also more likely to be part of a composite display. For example, a movement of the head may accompany one or the other of several vocalizations, each of which lends it a different meaning. Scaling is also more prominently developed in vertebrates than in insects. Variations in the intensity of the signal, often very slight, are used to convey subtle changes in mood. All of these improvements together enlarge vertebrate repertories to such an extent that they are able to transmit perhaps an order of magnitude more bits of information per second than insect repertories. We cannot be sure of the exact amount, owing to the severe technical difficulties encountered in measuring the information content of more complex communication systems (see Chapter 8).

The recognition of individuals as such is mostly a vertebrate trait. Tunicate colonies "recognize" those of differing genotypes by failing to coalesce with them on contact (Burnet, 1971). *Drosophila* adults can identify the odors of different genetic strains when choosing mates (Hay, 1972), while social insects generally discriminate nestmates from all other members of the species through colony odors that adhere to the surface of the body (Wilson, 1971a). These responses are to classes of individuals and not to separate organisms, however. Only a few cases of truly personal recognition have been documented in the invertebrates. When females of the social wasp *Polistes* found colonies together they organize themselves into dominance hierarchies that appear to be based on knowledge of one another as individuals (Pardi, 1948; Mary Jane Eberhard, 1969). Sexual pair bonds are formed by individual starfish-eating shrimp *Hymenocera picta* (Wickler and Seibt, 1970) and desert sowbugs *Hemilepistus reaumuri* (K. E. and C. Linsenmair, 1971; K. E. Linsenmair, 1972). Both of these species use bonding as a device to cope with specialized ecological requirements. Other examples among invertebrates will no doubt be discovered, but they will certainly continue to constitute a very small minority. Vertebrates, in contrast, generally have the power of personal recognition. It is probably lacking in schooling fishes, in amphibians, and in at least the more solitary reptiles. But personal recognition is a widespread and possibly universal phenomenon in the birds and mammals, the two vertebrate groups containing the most advanced forms of social organization.

Vertebrates are also capable of quick forms of learning that fit them to the rapidly changing nexus of relationships within which they live. When an ant colony faces an emergency, its members need only respond to alarm pheromones and assess the general stimuli they encounter. But a rhesus monkey must judge whether the excitement is created by an internal fight, and if it is, learn who is involved, remember its own past relation to the participants, and judge its immediate actions according to whether it will personally benefit or lose by taking action of its own. The social vertebrate also has the advantage of being able to modify its behavior according to observations of success or failure on the part of the group as a whole. In this manner traditions are born that endure for generations within the same society. Play became increasingly important as vertebrate social evolution advanced, facilitating the invention and transmission of traditions and helping to establish the personalized relationships that endure into adulthood. Socialization, the process of acquiring these traits, is not the cause of social behavior in the ultimate, genetic sense. Rather, it is the set of devices by which social life can be personalized and genetic individual fitness enhanced in a social context (see Chapter 7).

Finally, the typically vertebrate qualities of improved communication, personal recognition, and increased behavioral modification make possible still another property of great importance: the formation of selfish subgroups within the society. It is possible for mated pairs, parent-offspring groups, clusters of siblings and other close kin, and even cliques of unrelated individuals to exist within societies without losing their separate identities. Each pursues its own ends, imposing severe limits on the degree to which the society as a whole can operate as a unit. The typical vertebrate society, in short, favors individual and in-group survival at the expense of societal integrity.

Man has intensified these vertebrate traits while adding unique qualities of his own. In so doing he has achieved an extraordinary degree of cooperation with little or no sacrifice of personal survival and reproduction. Exactly how he alone has been able to cross to this fourth pinnacle, reversing the downward trend of social evolution in general, is the culminating mystery of all biology. We will return to it at the end of the survey of social organisms composing the remainder of this book.

Chapter 19 **The Colonial Microorganisms and Invertebrates**

For years the study of colonial organization in microorganisms and the lower animals has been steeped in a dilemma. On the basis of several criteria many of the species can be considered to belong to the highest social grade ever attained in three billion years of evolution. The very term *colony* implies that the members are physically united, or differentiated into reproductive and sterile castes, or both. When the two conditions coexist in an advanced stage, the "society" can be viewed equally well as a superorganism or even as an organism. Many invertebrate zoologists have pondered and debated this philosophical distinction. The dilemma can be restated simply as follows: At what point does a society become so nearly perfect that it is no longer a society? On what basis do we distinguish the extremely modified zooid of an invertebrate colony from the organ of a metazoan animal?

These questions are not trivial. They address a theoretical issue seldom made explicit in biology: the conception of all possible ways by which complex metazoan organisms can be created in evolution. To make the issue wholly clear let us go directly to the *ne plus ultra* of invertebrate social forms, and of animal societies generally, the colonial hydrozoans of the order Siphonophora. Approximately 300 species of these bizarre creatures have been described. Vaguely resembling jellyfish, all live in the open ocean where they use their stinging tentacles to capture fish and other small prey. The most familiar genus is *Physalia*, the Portuguese man-of-war. Other examples are *Nanomia* and *Forskalia*, illustrated in Figures 19-1 and 19-2. These creatures resemble organisms. To the uninitiated they appear basically similar to scyphozoans, the "true" jellyfish of the ocean, which are unequivocally discrete organisms. Nevertheless, each siphonophoran is a colony. The zooids are extremely specialized. At the top of each *Nanomia* sits an individual modified into a gas-filled float, which gives buoyancy to the rest of the colony strung out below it. Nectophores act like little bellows, squirting out jets of water to propel the colony through the water. By altering the shape of their openings they are able to alter the direction of the jets and hence the path followed by the colony. Through their coordinated action the *Nanomia* colony is able to dart about vigorously, moving at any angle and in any plane, and even executing loop-the-loop curves. Lower on the stem sprout saclike zooids called palpons and gastrozooids, which are specialized for the ingestion and distribution of nutrients to the remainder of the colony. Long branched tentacles arise as organs from both the palpons and gastrozooids. They are used to capture prey and perhaps to defend the colony as well. The roster of specialists is completed by the sexual medusoids, which are responsible for the production of new colonies by conventional gamete formation and fertilization, and the bracts, which are inert, scalelike zooids that fit over the stem like shingles and evidently help protect it from physical damage. New zooids are generated by budding in one or the other of two growth zones located at each end of the nectophore region.

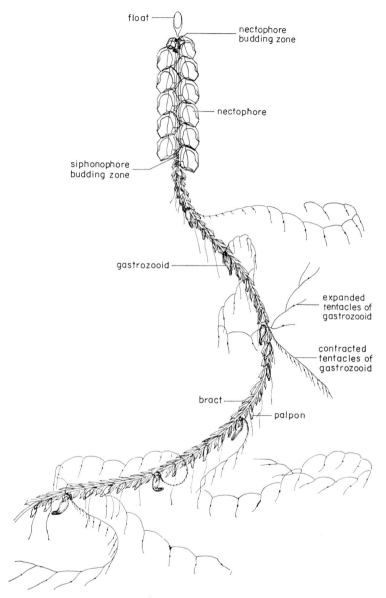

float
nectophore budding zone
nectophore
siphonophore budding zone
gastrozooid
expanded tentacles of gastrozooid
contracted tentacles of gastrozooid
bract
palpon

Figure 19-1 A colony of the siphonophoran *Nanomia cara*. The float that provides buoyancy, the nectophores that propel the entire ensemble, the gastrozooids that capture and digest prey, and other individual colony members, such as the bracts and palpons, are modified to such an extreme as to be comparable to the organs of single metazoan animals. (Modified from Mackie, 1964.)

The behavior and coordination of *Nanomia* colonies has been the object of excellent studies by Mackie (1964, 1973). The zooids behave to some degree as independent units but they are also subject to considerable control from the remaining colony members. Each nec-

tophore, for example, has its own nervous system, which determines the frequency of contraction and the direction in which the water is squirted. However, the nectophore remains quiescent except when aroused by excitation arriving from the rest of the colony. When the rear portion of the colony is touched the forward nectophores begin to contract and then others join in. Experiments have shown that the coordination is due to conduction through nerve tracts that connect the nectophores. When the float is touched the nectophores reverse the direction of their jet propulsion and drive the colony backward. The conduction that coordinates this second movement is not through nerves but through sensitive cells in the epithelium. The gastrozooids of *Nanomia* are more independent in action. Several gastrozooids may collaborate in the capture and ingestion of a prey, but their movements and nervous activity are wholly separate. They and the palpons, which are auxiliary digestive organs, pump digested food out along the stem to the rest of the colony. Even the empty zooids participate in the peristaltic movements, with the result that the digested material is flushed more quickly back and forth along the stem. In other respects, however, their behavior remains independent.

Much of the difficulty in conceptualizing *Nanomia* and other siphonophores as colonies rather than organisms stems from the fact that each of the entities originates from a single fertilized egg. This zygote undergoes multiple divisions to form a ciliated planula larva. Later, the ectoderm thickens and begins to bud off the rudiments of the float, the nectophores, and other zooids. Fundamentally the process, which specialists in colonial invertebrates call "astogeny," does not differ from the ontogeny of scyphozoan medusas or other true individual organisms among the Coelenterata. The resolution of the paradox is that siphonophores are both organisms *and* colonies. Structurally and embryonically they qualify as organisms. Phylogenetically they originated as colonies. Other hydrozoans, including the anthomedusans and leptomedusans as well as milleporine and stylasterine "corals," display every stage in the evolution of coloniality up to the vicinity of the siphonophore level. Some species form an elementary grouping of fully formed, independently generated polypoids that are nevertheless connected by a stolon that either runs along the substrate or rises stemlike up into the water. In others there is a physical union of the zooid body wall along with varying degrees of specialization. As castes differentiate among the zooids, as for example in the familiar leptomedusan genus *Obelia*, some lose their reproductive capacity, while the reproductive individuals (gonozooids) lose the capacity to feed and to defend themselves. In the end, among the evolutionarily most advanced species, the ensemble collapses into a highly integrated unit. It assumes a distinctive and relatively invariable form that subordinates the zooids to mere coverings, floats, tentacle bearers, and other organlike units. This last stage is the one attained by the order Siphonophora and to a less spectacu-

Figure 19-2 A complex siphonophore, *Forskalia tholoides.* This creature represents little more than an enlargement and elaboration of the *Nanomia* colony plan depicted in Figure 19-1. (From Haeckel, 1888.)

lar degree by *Velella* and other pelagic hydroids of the order Chondrophora.

The achievement of the siphonophores and chondrophores must be regarded as one of the greatest in the history of evolution (Mackie, 1963). They have created a complex metazoan body by making organs out of individual organisms. Other higher animal lines originated from ancestors that created organs from mesoderm, without passing through a colonial stage. The end result is essentially the same: both kinds of organisms escaped from the limitations of the diploblastic (two-layered) body plan and were free to invent large masses of complicated organ systems. But the evolutionary pathways they followed were fundamentally different.

Another unique feature of coloniality in the lower invertebrates is the capacity of some nonrelated colonies to fuse into single units under certain conditions. H. Oka (in Burnet, 1971) found that colonies of the tunicate *Botryllus* will combine if they possess at least one "recognition" gene in common. When a colony is divided in two and the fragments are juxtaposed, the parts fuse with no difficulty. This result is to be expected, since the colony is a clone of genetically identical animals. If two unrelated colonies are brought into contact, however, a zone of necrotic material develops between them. All colonies are heterozygous, so that the recognition genes postulated by Oka can be represented as AB, CD, and so forth. (They remain heterozygous because the follicular cells surrounding the ovum prevent the entrance of any sperm identical to the ovum.) If the colonies in contact have any recognition genes in common, for example, AB meets BD or BC meets AC, fusion occurs. If all the genes are different, as in AB versus CD, necrotic rejection ensues. It remains to be seen whether Oka's mechanism, or anything like it, occurs in other colonial microorganisms and invertebrates. As Theodor (1970) has pointed out, recognition of "self" as opposed to "not-self" appears to be widespread if not universal in the invertebrates, where it is based on ectodermal histo-incompatibility analogous to the immune responses of vertebrates. The conditions under which the incompatibility can be overridden have been investigated only in the case of Oka's tunicates.

The Adaptive Basis of Coloniality

The Darwinian advantage of membership in colonies is not immediately obvious. At the highest levels of integration, a majority of the zooids do not reproduce, and many are freely autotomized when injury or overgrowth causes them to encumber the colony as a whole. Even fully formed zooids may suffer some disadvantages from colony membership. Bishop and Bahr (1973) demonstrated that as colony size in the freshwater bryozoan *Lophopodella carteri* increases, the clearance rate of water per zooid (for example, microliter/zooid/minute) decreases. Hence larger colony size results in less food for

each of the member individuals. Arrayed against these negative factors are a variety of advantages that have been documented in one or more species of colonial organisms:

1. *Resistance to physical stress in the neritic benthos.* Coloniality is most common among the invertebrates that inhabit the bottom of the ocean in the shallow water along the seacoast. There wave action is strongest and sessile organisms are most likely to be choked by sedimentation. Coral reefs and the fouling communities on pilings and rocks are made up principally of colonial coelenterates, bryozoans, and tunicates. Careful studies of zoantharian corals have revealed that the massed calcareous skeletons of the zooids, when constructed in certain ways, anchor the colonies more securely to the sea floor and increase the survival time of the organisms they protect (Coates and Oliver, 1973). The colony raises individuals from the bottom, away from the densest concentrations of suspended soil particles. The orientation of the zooids in corals and other colonial forms allows them to generate faster currents than any attainable by isolated single organisms of similar construction (Hubbard, 1973; see Figure 19-3).

2. *Liberation of otherwise sessile forms for a free-swimming, pelagic existence.* The zooids of siphonophore and chondrophore colonies are polypoids, which are basically hydralike individuals adapted for attachment to the sea floor and a sedentary existence. By modifying some of the polypoids into floats and swimming bells, the colonies have been able to swim free in the open ocean. Some of the members, such as the gastrozooids, gonozooids, and bracts, are still polypoid in construction, but they are easily carried along by the swimming specialists (Phillips, 1973).

3. *Superior colonizing and competitive abilities.* As Bonner (1970) has emphasized, the clear advantage gained by aggregation in the

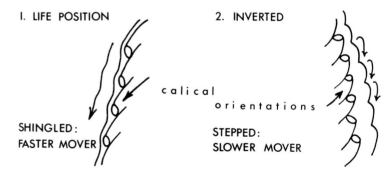

Figure 19-3 The orientation of the zooids in the coral *Montastrea* (left) allows them to create faster currents than single individuals or zooids arranged in other positions (right). (From Hubbard, 1973. Reprinted with permission from *Animal Colonies: Development and Function through Time*, ed. R. S. Boardman, A. H. Cheetham, and W. A. Oliver, Jr. Copyright © 1973 by Dowden, Hutchinson and Ross, Inc., Publishers, Stroudsburg, Pennsylvania.)

myxobacteria and cellular slime molds is the capacity to elevate fruiting bodies on stalks. The spores liberated from the fruiting bodies consequently travel farther than would have been the case had they been formed by the individual bacteria or myxamebas still in the soil. Dispersal is the "aim" of sporulation, because aggregation is induced when local environmental conditions deteriorate. Coloniality in the sessile invertebrates is associated not with the enhancement of dispersal but with the improvement of colony growth and survival following dispersal. Asexual budding is the fastest form of growth, especially when performed laterally to create an encrusting assemblage. It also enables a colony to overgrow and to choke out competing forms. Corals, for example, compete with one another like plants, by cutting out the light of those beneath or by covering and suffocating competitors occupying the same surface. In both instances the capacities to produce large masses and to continue growing at high densities are decisive.

Kaufmann (1973) has modeled growth in bryozoan colonies with reference to the distinction between colonizing and competitive abilities. He started with the reasonable assumption that the limiting factor of larva production is the rate of energy consumption, which is proportional in turn to the number of feeding zooids. The energy must be apportioned among budding new zooids, creating new nonfeeding heterozooids, adding calcification to the colony, and producing larvae. Encrusting and vinelike species have the highest r and could be expected to prevail under circumstances in which new space is frequently made available. In other words, they would be opportunistic, thriving in the most fluctuating and short-lived habitats. But like r selectionists generally, they would have less competitive ability under crowded conditions. The K selectionists are most likely to be the encrusting and bushy forms with heavy calcification and many heterozooids. Such species make their own space by eliminating smaller and more delicate competitors.

4. *Defense against predators.* Most of the heterozooids of the polymorphic Ectoprocta for which functions have been established specialize in the defense of the colony, either by adding to the strength of the colony wall or by actively repelling invaders (Kaufmann, 1971; Schopf, 1973, and personal communication).

General Evolutionary Trends in Coloniality

The most nearly complete systematic account of colonial life in the invertebrates is that by the Russian zoologist W. N. Beklemishev (1969). After surveying most of the colonial taxa and providing a simple morphological classification of the assemblage, Beklemishev formulated the major evolutionary rules that he believed apply broadly across the invertebrates. His thinking was influenced by two venerable ideas, the concept of the superorganism and the view that

biological complexity evolves by the dual processes of the differentiation and integration of individuals. He accordingly identified three complementary trends as the basis of increasing coloniality: (1) the weakening of the individuality of the zooids, by physical continuity, sharing of organs, and decrease in size and life span, as well as by specialization into simplified, highly dependent heterozooids; (2) the intensification of the individuality of the colony, by means of more elaborate, stereotyped body form and closer physiological and behavioral integration of the zooids; and (3) the development of cormidia, or "colonies within colonies." Within at least the cheilostome ectoprocts, Banta (1973) has concluded that coloniality first increased by division of labor, presumably in association with the delimitation of cormidia, then by physiological interdependence of the polymorphic zooids, and finally by structural interdependence of the zooids. In later stages of evolution all three processes proceeded concurrently.

The cormidia are particularly interesting, because they correspond to the organ systems and appendages of metazoan individuals. Examples of cormidia include the nectosome, or region of swimming bells (nectophores), in *Nanomia* and other siphonophores; the "leaves," "petals," and branches of sea pens and certain other octocorallian corals; the shoots or internodes in ectoproct colonies; and others. In *Muggiaea* and related calycophoran siphonophores, one kind of cormidium is so nearly independent as to exist on the borderline between the cormidium as an organizational unit and a full colony. It consists typically of a helmet-shaped bract, a gastrozooid with a tentacle, and one or more gonophores of one sex, which double as swimming bells. When fully developed these units break loose and lead a temporarily free existence. Known as eudoxomes, they were considered to be distinct species of Siphonophora until their true relationship to the larger colony unit was discovered (Hyman, 1940; see Figure 19-4).

In Table 19-1 are given the taxa within which colonial development has occurred, together with those features of the life cycle most affected by it. In the sections to follow several of the groups will be described in sufficient additional detail to serve as paradigms of the principal features of coloniality. They are presented in phylogenetic order, from the relatively primitive slime molds to the advanced triploblastic bryozoans. The point to be remembered, however, is that although these organisms vary greatly in phylogenetic position on the basis of their overall biology, each is distinguished by colonial specialization of a very high order.

Slime Molds and Colonial Bacteria

The remarkable life cycle of the slime mold *Dictyostelium*, the best-known member of the Acrasiales, is of general interest to biologists because it provides a model system of a developing multicellular organism that can be experimentally manipulated with relative ease.

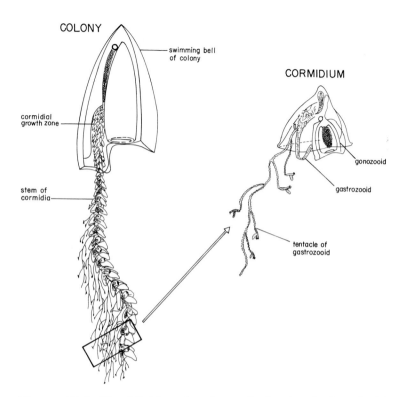

COLONY

swimming bell
of colony

cormidial
growth zone

stem of
cormidia

CORMIDIUM

gonozooid

gastrozooid

tentacle of
gastrozooid

Figure 19-4   The full hierarchy of organizational units in colonial invertebrates is displayed by the siphonophore *Muggiaea*. The cormidium comprises a group of zooids (individuals) that are coordinated among themselves strongly enough to be recognized as a distinct element within the colony. In the case of *Muggiaea*, the cormidium nearly ranks as a full colony, since it can break away and lead a separate existence for a while. (Modified from Hyman, 1940.)

For sociobiologists it has the more special attraction of displaying perhaps the most advanced social behavior of single-celled organisms—the aggregation of the myxamebas that initiates the multicellular half of the life cycle. The biology of *Dictyostelium* and related slime molds has been perceptively reviewed by Bonner (1967, 1970), one of the chief contributors to its study.

The *Dictyostelium* cycle can be conveniently marked as beginning with the settling of spores onto soil, leaf litter, or rotting wood. The emerging cells are single-celled and behave like "true" amebas; they creep through liquid films, engulfing bacteria and dividing at frequent intervals. The cells are completely independent of one another so long as a rich supply of food is available. When the food grows scarce, however, a dramatic change occurs. Certain amebas become attraction centers, and the remainder of the population streams toward them. Soon the random array is transfigured into rosettes of amebas, with a rising center and radiating arms composed of the amebas still migrating inward. As the aggregation congeals further it assumes a

sausage shape averaging $\frac{1}{2}$ to 2 millimeters in length. This new entity, called a pseudoplasmodium or grex, now performs like a multicellular organism. It has distinct front and hind ends, and moves slowly in the direction of heat and light. Up to one or two weeks later the pseudoplasmodium transforms into a fruiting body, with some of the former amebas contributing to the base and stalk and others to the spore-bearing spheres at the tip. Each species of cellular slime mold has a distinctive version of this final, most complex life stage (see Figure 19-5). The adaptive significance of the life cycle is not hard to decipher. Because of their small size, the amebas have the largest surface-to-volume ratio and therefore the greatest capacity to feed and

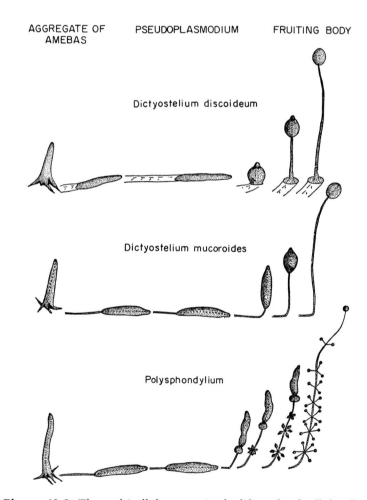

AGGREGATE OF   PSEUDOPLASMODIUM   FRUITING BODY
AMEBAS

Dictyostelium discoideum

Dictyostelium mucoroides

Polysphondylium

Figure 19-5   The multicellular stages in the life cycle of cellular slime molds. At the left is the raised aggregation of amebas. This turns over to become the migrating pseudoplasmodium (center), which eventually transforms into the fruiting body (right). Each species differs in details of the external morphology of these stages. (Modified from Bonner, 1958.)

Table 19-1 The phylogenetic distribution and characteristics of colonies in microorganisms and invertebrates.

Kind of organism	Colonial organization	Authority
KINGDOM MONERA PHYLUM SCHIZOPHYTA (BACTERIA) Class Myxobacteria (slime bacteria)	Individual bacteria glide on slime trails; when certain amino acids are deficient in the medium, the cells aggregate and form distinctive fruiting bodies.	Bonner (1955), Doetsch and Cook (1973)
KINGDOM PROTISTA PHYLUM MASTIGOPHORA (FLAGELLATES)	Simple motile colonies in *Gonium, Pandorina*, and other forms are clusters of cells embedded in clear mucilage; *Volvox* is composed of 500–50,000 cells in mucilage and connected by cytoplasmic strands, with complex asexual budding and differentiated sex cells.	Hyman (1940), Grassé (1952a), Curtis (1968b), Barnes (1969)
PHYLUM CILIOPHORA (CILIATES)	Sessile, stalked colonies of varying complexity; in extreme cases, e.g., *Zoothamnium*, numerous cells occur, some of which are sporelike colonizing forms.	Summers (1938), Hyman (1940), Corliss (1961)
PHYLUM MYXOMYCOTA (SLIME MOLDS) Order Labyrinthulales	Parasitic fungi that form networks of cells and occasionally dense aggregations. Only one genus, *Labyrinthula*, on plant hosts.	Bonner (1967)
Order Plasmodiophorales	Parasitic on plants. Single-cell swarmers penetrate host cells, multiply, and fuse into a mass lacking a cell wall (plasmodium) that continues invasion of host tissue.	Bonner (1967)
Order Myxomycetales (true slime molds)	Decomposer organisms, e.g., *Physarum*, that live in the soil and decaying wood. The plasmodium grows by nuclear division and accretion rather than cell aggregation; as the environment deteriorates, the plasmodium often transforms into spore-bearing fruiting structures. Occasionally separate zygotes or plasmodia fuse to start a new plasmodium, thus constituting a primitive aggregation.	Alexopoulos (1963), Bonner (1967, 1970)
Order Acrasiales (cellular slime molds)	Decomposer organisms, e.g., *Dictyostelium, Hartmanella, Polysphondylium*, that live in the soil and decaying wood. Individual ameboid cells (myxamebas) emerge from spores and feed independently. When the environment deteriorates the myxamebas aggregate into a motile pseudoplasmodium that migrates for a time and then produces a stalked, multicellular fruiting body that sheds spores.	Bonner (1967, 1970)
KINGDOM ANIMALIA PHYLUM PORIFERA (SPONGES)	A motile ciliated larva develops into the wholly sessile sponge that in many species, e.g., *Esperiopsis, Hircinia*, and *Microciona*, buds off semiindependent units organized around separate excurrent openings (oscula). These clusters have been variously called individuals or colonies and are clearly intermediate in level of organization. Neighboring sponges of the same species sometimes fuse and reorganize as a new individual (or colony).	Hyman (1940), Fry, ed. (1970), Hartman and Reiswig (1973), Simpson (1973)
PHYLUM COELENTERATA OR CNIDARIA (HYDRAS, JELLYFISH, AND RELATED FORMS) Class Hydrozoa (hydras and hydroids)	Most species possess a polypoid stage in the life cycle, and the vast majority of these are colonial. The colonies grow from single zygotes and vary greatly in form according to species. All have at least two types of individuals (zooids)—one serving for prey capture and digestion (gastrozooids) and the other for reproduction (gonozooids). Most colonies are sessile, arborescent forms. In a few species the colonies float in the open sea and resemble medu-	Hyman (1940), Garstang (1946), Barnes (1969), Beklemishev (1969), Mackie (1964, 1973), Phillips (1973)

Table 19-1 (continued).

Kind of organism	Colonial organization	Authority
Class Hydrozoa (*cont.*)	soids, or "jellyfish"; the extreme of this trend, and of invertebrate coloniality generally, occurs in the order Siphonophora. Some hydrozoan species have a true medusoid stage, but it is solitary.	
Class Anthozoa (corals, sea anemones, and related forms)	With the exception of the sea anemones and ceriantharians, these consist of colonial forms. Colonies grow from a planula larva that settles on the substrate; new zooids are budded asexually and are closely connected although usually either whole or nearly whole organisms, distinguishable as typical coelenterate polyps. An enormous variety of colony forms are produced by the ten living orders. Some secrete calcareous exoskeletons, forming the tropical coral reefs.	Hyman (1940), Barnes (1969), Beklemishev (1969), Boardman et al. (1973)
PHYLUM PLATYHELMINTHES (FLATWORMS)		
Class Turbellaria (planarians and other free-living flatworms)	In the rhabdocoel genera *Catenula, Microstomum,* and *Stenostomum,* individuals reproduce by transverse fission, then remain attached to form linear colonies of zooids.	Hyman (1951a), Beklemishev (1969)
Class Cestoda (tapeworms)	The larva of *Cysticercus* proliferates a system of numerous attached but semiindependent cysts; that of *Hymenolepis* produces branched colonies with multiple stolons and cysticercoid heads. The proglottids of mature tapeworms, with independent reproductive systems, can be loosely interpreted as zooids and the entire tapeworm body as a colony.	Hyman (1951a), Beklemishev (1969)
PHYLUM ROTIFERA (ROTIFERS)	A few species form elementary colonies in which individuals are attached at the base. In the sessile genus *Floscularia,* young rotifers attach to the tubes of older ones. *Conochilus* is a colonial pelagic form; the animals radiate in all directions from a common center and float through open water as a unit.	Edmondson (1945), Hyman (1951b), Barnes (1969)
PHYLUM ENTOPROCTA (MOSS ANIMALS OR ENTOPROCTS)	Most species are colonial, in most cases sending multiple stalks with one or more zooids up from a horizontal creeping stolon. The zooids are generalized and semiindependent, resembling primitive hydrozoans in level of organization.	Hyman (1951b), Beklemishev (1969)
PHYLUM ANNELIDA (SEGMENTED WORMS)		
Class Polychaeta (polychaete worms, all marine)	A few species in *Autolytus, Myrianida,* and other genera form linear chains by transverse fission of individuals. *Autolytus* is also dimorphic: the first maternal individual is asexual and the daughters sexual.	Beklemishev (1969)
PHYLUM ARTHROPODA (ARTHROPODS)		
Class Crustacea (crustaceans)	In *Thompsonia socialis,* a parasitic barnacle in the order Rhizocephala, the sacciform larva buds off progeny, and all remain connected by a common root system.	Beklemishev (1969)
Class Insecta (insects)	Eusocial colonies, characterized by the presence of sterile worker castes, occur in all species of ants and termites and in many species of bees and wasps. The colonies are founded by a fertilized queen (ants, bees, wasps) or a mated queen and king (termites); the reproductive individuals then produce workers and, when the colony attains a certain size, other reproductive forms. Unlike other invertebrate colonies, those of insects consist of members that are physically separate and capable of independent locomotion.	Wilson (1971a); see Chapter 20 of this book
PHYLUM PHORONIDA (PHORONIDS, WORMLIKE MARINE ANIMALS)	*Phoronis ovalis* reproduces by budding and transverse fission to form branching, sessile colonies.	Hyman (1959), Beklemishev (1969)

Table 19-1 *(continued)*.

Kind of organism	Colonial organization	Authority
PHYLUM ECTOPROCTA OR BRYOZOA (ECTOPROCTS, MOSS ANIMALS, BRYOZOANS)	Almost all of the 4000 living species are colonial and sessile. The colonies, which grow by budding, vary greatly in form according to species. Some are flat and encrusting, others erect and branching, and still others massed into lobes resembling corals. The colonies of most species of the class Gymnolaemata are polymorphic, with many of the zooids ("heterozooids") reduced or otherwise modified to serve in defense, cleaning, or reproduction.	Hyman(1959),Beklemishev (1969), Ryland (1970), Boardman et al. (1973), Larwood et al. (1973)
PHYLUM HEMICHORDATA (ACORN WORMS) Class Pterobranchia (pterobranchs)	*Cephalodiscus* and *Rhabdopleura* form sessile colonies by budding. In *Rhabdopleura* the zooids remain attached at their base to a common stolon that creeps over the substrate.	Barrington (1965), Beklemishev (1969)
PHYLUM CHORDATA (VERTEBRATES AND INVERTEBRATE CHORDATES), SUBPHYLUM UROCHORDATA (TUNICATES) Class Ascidiacea (ascidians or sea squirts)	The largest group of urochordates, entirely benthic marine and sessile. Colonial organization has arisen independently in at least several lines and varies greatly among species in form and integration. In the simplest colonies, such as those of *Perophora*, the individual zooids rise from a vinelike stolon. Intermediate degrees of cohesion involve uniting the basal part of the body and forming a common tunic. In the extreme cases of *Cyathocormus* and *Coelocormus* the colonies are organized like those of some advanced sponges: the buccal siphons ("mouths") of the member zooids open to the outside, but the atrial siphons empty into a common cloaca; also the tunics are shaped into a single formed structure.	Barrington (1965), Barnes (1969), Beklemishev (1969)
Class Thaliacea (thaliaceans)	All thaliaceans are free-swimming plankton feeders of the open sea. *Pyrosoma* forms elaborate, bilaterally symmetrical colonies with the zooids arranged as feeding units along the wall and their atria leading into a common cloaca, which empties in turn from a large opening at one end of the colony.	Barrington (1965), Barnes (1969), Beklemishev (1969), Griffin and Yaldwyn (1970), Baker (1971)

reproduce rapidly when conditions are favorable. When local conditions turn bad, aggregation and migration change the strategy to one of maximum dispersal.

The substance inducing the aggregation of the amebas is called acrasin and has been identified as adenosine-3',5'-cyclic-monophosphate (cyclic AMP) in the species *Dictyostelium discoideum* (Konijn et al., 1967). When the amebas are deprived of food they go through a period of differentiation, called the interphase, lasting for six to eight hours. Then the amount of cyclic AMP given off rises dramatically, from 10^{-12} moles at the outset to 10^{-10} moles at the peak six hours later, a hundredfold difference. There is a simultaneous hundredfold increase in the sensitivity of the amebas to cyclic AMP. For a time the basis for the orientation was a mystery. How could an ameba move up an acrasin gradient when functioning itself as a high point in a local gradient? The answer turns out to be that the amebas signal back and forth to one another during the process. Cyclic AMP is released in pulses, which are evidently sharpened in definition by the subsequent release of acrasinase, an enzyme that converts cyclic 3',5' AMP to 5' AMP. Amebas respond to these pulses by emitting a pulse of their own approximately 15 seconds later, then moving for about 100 seconds toward the original signal source. The period between pulses is approximately 300 seconds, and while moving the ameba does not respond to further signals. With each ameba acting as a local signal source, the population as a whole tends to travel toward its nearest neighbors, forming the early aggregation streams. Within the streams the flow continues toward the original signal source, which now becomes the overall aggregation center. Robertson et al. (1972) were able to induce this entire process by the appropriately paced electrophoretic release of cyclic AMP from a microelectrode. The amebas in the culture dish obediently clustered together at the needle point of the microelectrode.

The changes that occur within the pseudoplasmodium are less well understood and offer challenges of even greater magnitude. During the migration the amebas in the cell mass undergo differentiation:

those in the forward third become somewhat larger than the amebas in the posterior two-thirds. They also stain differently with several kinds of dyes. The division between the two parts is sharp and foreshadows the coming formation of the fruiting body. At the end of the migration the pseudoplasmodium rounds up into a ball. The larger tip cells grow still larger and plunge into the interior of the ball, where they begin to form the stalk of the fruiting body. As other tip cells pile on, the stalk elongates, lifting the smaller posterior cells into the air in the form of a rounded sac. Soon the posterior cells are transformed into spores. This division of labor is curious, since it means that certain cells perpetuate themselves as spores with the aid of self-sacrifice on the part of those imprisoned in the stalk. If the amebas are genetically identical no theoretical problem exists, because the process is not fundamentally different from tissue differentiation in a metazoan organism. But if the cells are genetically different, which is possible because of their origin from multiple spores at the beginning of the life cycle, reproductive subordination of the kind witnessed in *Dictyostelium* will be countered by individual selection, and its evolution must be explained by selection at a higher level.

One of the most notable of all cases of convergent evolution is the close resemblance of the life cycle of the myxobacteria to that of the cellular slime molds. The bacteria are procaryotes and the fungi are eucaryotes. These two kinds of microorganisms thus fall on opposite sides of the deepest chasm in all of evolution—one even greater than the separation between such unicellular eucaryotes as protozoans and the more primitive multicellular animals. Yet their life cycles resemble each other in many details. One of the most elaborate bacterial forms is *Chondromyces* (Figure 19-6). The "myxospores" of *Chondromyces* are actually small cysts 50 microns or greater in diameter that contain several thousand bacteria each. When they split open, the rod-shaped bacteria inside stream out like a flame from the mouth of a dragon, to use Bonner's felicitous expression. The cells then glide over the surface along slime trails, absorbing nutrients and multiplying by conventional fission. Large numbers move en masse, each bacterium tending to follow the trails of others. Like foraging colonies of army ants, they press first in one direction and then another, sometimes fanning out as though in search of new food. Occasionally groups contract into solid aggregates. Not only do the individual cells divide, but masses from different cysts combine, so that the moving sheet of bacteria soon attains impressive proportions. When food grows short, more precisely when certain amino acids run low in the surrounding medium, the bacteria congeal and form the characteristic fruiting bodies. The stems are supported by hardened slime, while the accumulation of carotenoid pigments impart beautiful shades of red, pink, violet, or yellow to the fruiting bodies.

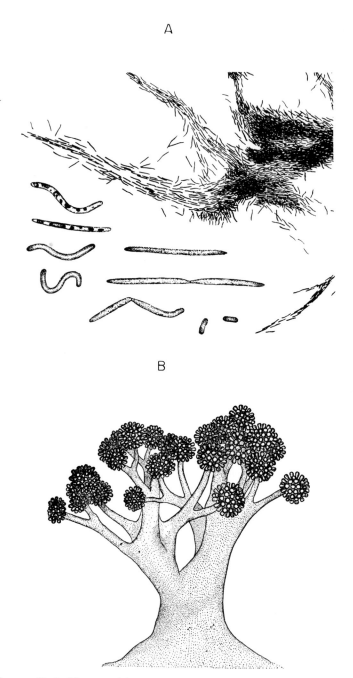

Figure 19-6 The social bacterium *Chondromyces*. *A:* the general appearance of vegetative cells and of the foraging cell swarms in *C. aurantiacus*. *B:* the fruiting body of *C. crocatus*, a large, complex structure measuring more than 0.5 millimeter across. (From Thaxter, 1892.)

The Coelenterates

The fullest demonstrable range of colonial evolution has occurred within the phylum Coelenterata. Although the individual organisms themselves all retain the basic diploblastic body plan, the associations show a huge amount of diversity, from the solitary scyphozoan jelly-fish, hydras, and sea anemones through virtually every conceivable gradation to colonies so fully integrated as to be nearly indistinguishable from organisms. Some of the colonies are sessile, mosslike forms, others complex motile assemblages resembling jellyfish.

The living species of corals present a tableau of the evolution of sessile colonies (Bayer, 1973). In the order Stolonifera are found colonies of the most basic type, consisting of nearly independent zooids connected by a living stolon. Growth is plantlike, with new zooids sprouting up from the stolon as it creeps over the substrate (see Figure 19-7). In other species of Stolonifera, new zooids originate from stolonic outgrowths on the body walls of older zooids. When they reach a certain height they generate still other new individuals in the same fashion. The result is a dendritic colony that increases in density from bottom to top. Within the order Alcyonacea further integration is achieved by the formation of a common jellylike meso-gloea in which the gastrovascular cavities are closely packed. The zooids of species producing the largest colonies are differentiated into two forms: the more elementary autozooids, which eat, digest, and distribute nutrients to the remainder of the colony; and the siphono-zooids, which possess the reproductive organs and circulate water by means of large ciliated grooves running down the side of the pharynx (see Figure 19-7). The apparent significance of the differentiation lies in this last function, which prevents water stagnation at high population densities. The ultimate colonial development among the corals has been attained by the alcyonacean *Bathyalcyon robustum*. Each mature colony consists of a single giant autozooid, in the body wall of which are embedded large numbers of daughter siphonozooids. In effect, the siphonozooids have become organs of the parental auto-zooid.

A wholly different colonial strategy appears in the order Gorgo-nacea. Here growth is treelike, and integration consists of little more than regularity in the patterns of branching. The colonies of some species resemble fans, others palm leaves, while still others rise in delicate exploding whorls of zooid-bearing branches. Among the Coralliidae, which are undoubted gorgonians, the zooids are as fully dimorphic as in the large alcyonaceans. But the similarity is probably due to convergence. In the coralliids the autozooids, not the siphono-zooids, contain the gonads.

Figure 19-7 Two grades in the colonial evolution of corals. Top: a simple form (*Clavularia hamra*), in which largely independent zooids grow from near the terminus of the ribbon-shaped stolon. Bottom: a more advanced species (*Heteroxenia fuscescens*) in which the zooids collabo-rate to construct a massive calcareous base. This specimen is cut in half to reveal the two types of zooids: the siphonozooids, which have a ciliated groove along their sides and gonads that penetrate deeply into the base, and the autozooids, which lack these organs. The siphonozooids specialize in reproduction and the creation of water currents, while the autozooids eat and digest the food. (From Bayer, 1973; after H. A. F. Gohar. Reprinted with permission from *Animal Colonies: Development and Function through Time*, ed. R. S. Boardman, A. H. Cheetham, and W. A. Oliver, Jr. Copyright © 1973 by Dowden, Hutchinson and Ross, Inc., Publishers, Stroudsburg, Pennsylvania.)

The Ectoprocts

The phylum Ectoprocta, or Bryozoa, containing the bulk of the "bryozoans" of older zoological classifications, displays the most advanced colonial organization of any of the coelomate groups. The specialization of the zoids is extreme, rivaling that found in the acoelomate siphonophores. The vast majority of ectoproct species are sessile, forming encrusting or arborescent colonies on almost any available firm surface in both marine and freshwater environments. To the naked eye some of the aggregations resemble sheets of fine lacework, others miniature moss or seaweed. The colonies of one genus, *Cristatella*, are ribbon-shaped bodies that creep over the substrate at rates not exceeding 3 centimeters a day (see Figure 19-8). Ectoprocts feed on plankton, which they capture with lophophores, hollow organs crowned with ciliated tentacles. All of the species are colonial. The zoids communicate by pores through the skeletal walls. The pores are usually plugged by epidermal cells; only in *Plumatella* and other freshwater members of the class Phylactolaemata do holes exist, permitting a free flow of coelomic fluid.

Polymorphism of zoids is limited to the primarily marine class Gymnolaemata. The diversity of specialists among these individuals is enormous, and bryozoologists have scarcely begun to study them systematically. Even the classification of the basic morphological types is in a state of flux. Autozoids, which are individuals with independent reproductive and feeding organs, are distinguished as a major category from heterozoids, the class of all specialists together (see Figure 19-9). One of the most distinctive types of specialists is the avicularium, found in some species of the order Cheilostomata. The operculum of this zoid has been modified into a sharp-edged lid that opens and snaps shut by means of opposing sets of muscles. The pedunculate avicularia of *Bugula* and *Synnotum* bear a remarkable resemblance to the head of a bird and are in fact capable of turning and biting at intruding objects. The difference between such an individual and the presumed ancestral autozoid is much greater than the difference between any two castes of the social insects and is matched elsewhere in the animal kingdom only by the diversification of the siphonophore zoids. Other ectoproct specialists include the vibraculum, in which the operculum is modified into a flexible bristle that can be lashed back and forth; spinozoids, characterized by spines that project from the body wall; gonozoids, which are specialized for sexual reproduction; and interzoids, highly reduced forms that serve as pore-plates or pore-chambers fitted between completely formed neighboring zoids. Kenozoids consist of a wide variety of supporting and anchoring elements, such as the rhizoids and other

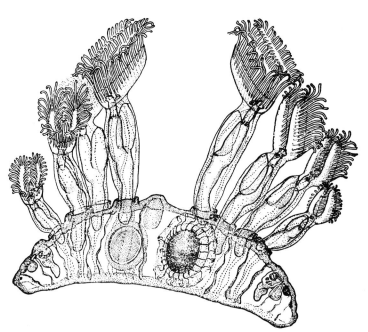

Figure 19-8 The motile colony of the freshwater ectoproct *Cristatella mucedo*. Left: view of an entire colony creeping over plant stems; individual zoids project their brushlike lophophores upward while beneath them, round statoblasts, which are asexually produced reproductive bodies, can be seen embedded in the gelatinous supporting structure of the colony (from J. Jullien, 1885). Right: transverse section of a *Cristatella* colony (from Brien, 1953).

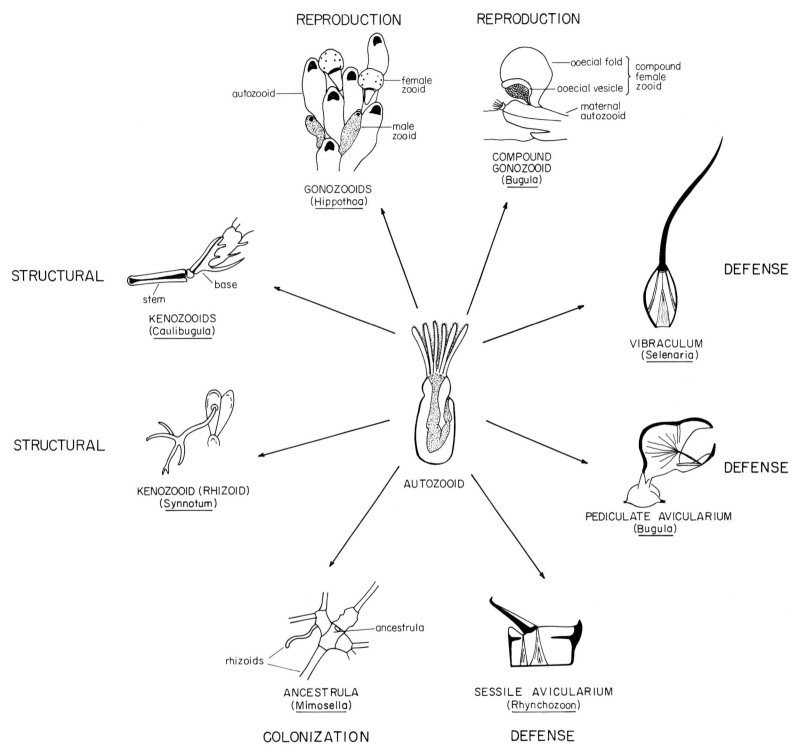

REPRODUCTION

autozooid

female zooid

male zooid

GONOZOOIDS
(Hippothoa)

REPRODUCTION

ooecial fold
ooecial vesicle } compound female zooid
maternal autozooid

COMPOUND
GONOZOOID
(Bugula)

STRUCTURAL

stem   base

KENOZOOIDS
(Caulibugula)

STRUCTURAL

KENOZOOID (RHIZOID)
(Synnotum)

AUTOZOOID

DEFENSE

VIBRACULUM
(Selenaria)

DEFENSE

PEDICULATE AVICULARIUM
(Bugula)

ancestrula

rhizoids

ANCESTRULA
(Mimosella)

COLONIZATION

SESSILE AVICULARIUM
(Rhynchozoon)

DEFENSE

Figure 19-9 The differentiation of individuals in colonies of bryozoans. Each of the basic types of specialists (heterozooids) have evolved from the more primitive autozooid, which is still capable of both reproduction and feeding. Some of the major categories of heterozooids are given in this diagram, along with the names of representative genera possessing the particular forms depicted. Only several kinds of heterozooids are encountered in any one species.

root attachments, the tubular elements of the stolons from which other zooids sprout, and attachment disks. Kenozooids, together with interzooids, are often so simplified in structure that their identity as individuals rather than organs can sometimes be established only by comparing them with more completely formed, intermediate evolutionary stages. Their existence led Silén (1942) to provide the ultimate structural definition of a zooid (that is, an individual organism) in the Ectoprocta as a cavity enclosed by walls. Still other morphological types exist. Nanozooids are dwarf replicas of autozooids that occur in colonies of *Diplosolen* and *Trypostega*. Their function, if any, is unknown. The founding member of the ectoproct colony, called the ancestrula, often takes the shape of the first segment of a stolon. Compound zooids also occur. As Woollacott and Zimmer (1972) showed, the rounded brood chamber or ooecium of *Bugula* and other members of the order Eurystomata is not just a modified zooid. It consists of two parts, between which the embryo is brooded: an inner fold, which is an evagination of the body wall of the maternal gonozooid, and an outer calcified wall, which is a distinct kenozooid derived from the autozooid next to the maternal individual. No more than four or five kinds of heterozooids, including reproductive specialists, are found in any one species.

The structure, embryonic origin, and evolution of heterozooids have been the subject of excellent reviews by Ryland (1970), Boardman and Cheetham (1973), Cheetham (1973), and Silén (1975). Relatively few studies of their behavior have been conducted. A notable exception is the work of Kaufmann (1971) on the pedunculate avicularia of *Bugula*. As their anatomy alone suggests, these bizarre creatures are defensive specialists, but their range of effectiveness is surprisingly narrow. They can seize and immobilize animals that are either 0.5-4 millimeters in length with numerous appendages or else have wormlike bodies less than 0.05 millimeters in diameter. In practice this means that their key role is to prevent tube-building gammarid crustaceans from settling on the colony. Most animals outside this range, including the larvae of fouling organisms and most potential predators, are inhibited very little by the avicularia. The general analysis of the ecology of bryozoan polymorphism has been explored in a pioneering article by Schopf (1973). In accordance with the ergonomics theory of caste (see Chapter 14), the highest frequency of polymorphic cheilostome species was found to occur in the most stable environments, in particular the tropical continental shelves and the deep sea. The most advanced forms of heterozooids also are concentrated in such places. Whether or not Schopf's correlations have been correctly interpreted in detail, they point the way to the next logical phase of investigation of bryozoan natural history.

Chapter 20 **The Social Insects**

The social insects challenge the mind by the sheer magnitude of their numbers and variety. There are more species of ants in a square kilometer of Brazilian forest than all the species of primates in the world, more workers in a single colony of driver ants than all the lions and elephants in Africa. The biomass and energy consumption of social insects exceed those of vertebrates in most terrestrial habitats. The ants in particular surpass birds and spiders as the chief predators of invertebrates. In the temperate zones ants and termites compete with earthworms as the chief movers of the soil and leaf litter; in the tropics they far surpass them.

Insects present the biologist with a rich array of social organizations for study and comparison. The full sweep of social evolution is displayed—repeatedly—by such groups as the halictid bees and sphecid and vespid wasps. There are so many species in each evolutionary grade that one can sample them as a statistician, measuring variance and partialing out correlates. And the subject is in its infancy. The great majority of social genera are unexplored behaviorally, so that one can only guess the outlines of their colonial organization. The ants are a prime example. Somewhat fewer than 8000 species have been described in the most limited sense of being assigned a name in the binomial system. W. L. Brown, the leading authority on ant classification, estimates that at least another 4000 remain to be discovered, and the rate at which new species appear in the literature seems more than ample to bear him out. Of the perhaps 12,000 living species, less than 100, or 1 percent, have been studied with any care, and less than 10 thoroughly and systematically. The species comprise about 270 genera (Brown, 1973), which represent the grosser units that can be compared with greatest profit when making comparative studies. I judge only 49 to have been subjected to careful sociobiological inquiry, and in most cases these studies have been rather narrowly conceived and executed.* The remaining 220 genera are known mostly from natural history notes having little to do with their social behavior. Perhaps half of these can be said to be virtually unknown in every respect except habitat and nest site. The other major groups of social insects, the termites, the social bees, and the social wasps, have been even less well explored.

Hence, despite the fact that there have been about 3,000 publications on social wasps, 12,000 on termites, 35,000 on ants, and 50,000

*The 49 best-known ant genera are *Myrmecia, Amblyopone, Onychomyrmex, Rhytidoponera, Cerapachys, Belonopelta, Leptogenys, Odontomachus, Eciton, Labidus, Neivamyrmex, Dorylus* (including *Anomma*), *Aenictus, Pseudomyrmex, Myrmica, Pogonomyrmex, Aphaenogaster, Messor, Pheidole, Melissotarsus, Leptothorax, Harpagoxenus, Tetramorium, Teleutomyrmex, Anergates. Strongylognathus, Monomorium, Solenopsis, Myrmicina, Cardiocondyla, Crematogaster, Daceton, Trichoscapa, Strumigenys, Smithistruma, Kyidris, Cyphomyrmex, Trachymyrmex, Acromyrmex, Atta, Aneuretus, Iridomyrmex, Tapinoma, Oecophylla, Plagiolepis, Lasius, Formica, Polyergus,* and *Camponotus.* References are given in Table 20-2.

on social bees (Spradbery, 1973), the development of insect socio-biology lies largely in the future. The number of investigators, the rate of publication, and the accretion of knowledge are all in what appears to be the early exponential phase of growth. Entomologists have also only begun to address the central questions of the subject, which can be stated in the following logical sequence:

——What are the unique qualities of social life in insects?

——How are insect societies organized?

——What were the evolutionary steps that led to the more advanced forms of social organization?

——What were the prime movers of social evolution?

These questions have been considered systematically in my earlier book *The Insect Societies*, to which the reader is referred for fuller explanation. Still more detailed and specialized accounts are available on social bees (Michener, 1974), honeybees in particular (Chauvin et al., 1968), social wasps (Kemper and Döhring, 1967; Richards, 1971; Spradbery, 1973), and termites (Howse, 1970; Krishna et al., 1969, 1970). The remainder of this chapter consists of a précis of insect sociobiology, which starts with partial answers for the questions just given and proceeds to a more systematic account of the behavior of key groups.

What Is a Social Insect?

The "truly" social insects, or eusocial insects, as they are often more technically labeled, include all of the ants, all of the termites, and the more highly organized bees and wasps. These insects can be distinguished as a group by their common possession of three traits:

(1) individuals of the same species cooperate in caring for the young; (2) there is a reproductive division of labor, with more or less sterile individuals working on behalf of fecund nestmates; (3) and there is an overlap of at least two generations in life stages capable of contributing to colony labor, so that offspring assist parents during some period of their life. These are the three qualities by which the majority of entomologists intuitively define eusociality. If we bear in mind that it is possible for the traits to occur independently of one another, we can proceed with a minimum of ambiguity to define *presocial* levels on the basis of one or two of the three traits. Presocial refers to the expression of any degree of social behavior beyond sexual behavior yet short of eusociality. Within this broad category there can be recognized a series of lower social stages, which are defined in matrix form in Table 20-1. The full logic of the reconstruction of social evolution, worked out over many years by such veteran entomologists as Wheeler (1923, 1928), Evans (1958), and Michener (1969), can be perceived by examining the matrices closely. In the *parasocial* sequence, adults belonging to the same generation assist one another to varying degrees. At the lowest level, they may be merely communal, which means that they cooperate in constructing a nest but rear their brood separately. At the next level of involvement, *quasisociality*, the brood are attended cooperatively, but each female still lays eggs at some time of her life. In the *semisocial* state, quasisocial cooperation is enhanced by the addition of a true worker caste; in other words, some members of the colony never attempt to reproduce. Finally, when semisocial colonies persist long enough for members of two or more generations to overlap and to cooperate, the list of three basic qualities is complete, and we refer to the species

Table 20-1 The degrees of sociality in the insects, showing intermediate parasocial and subsocial states that can lead to the highest (eusocial) form of organization.

	Qualities of sociality		
Degrees of sociality	Cooperative brood care	Reproductive castes	Overlap between generations
Parasocial sequence			
Solitary	−	−	−
Communal	−	−	−
Quasisocial	+	−	−
Semisocial	+	+	−
Eusocial	+	+	+
Subsocial sequence			
Solitary	−	−	−
Primitively subsocial	−	−	−
Intermediate subsocial I	−	−	+
Intermediate subsocial II	+	−	+
Eusocial	+	+	+

(or the colony) as being eusocial. Precisely this sequence has been envisioned by Michener and his coworkers as one possible evolutionary pathway taken by bees.

The alternate sequence is comprised of the *subsocial* states. In this case there is an increasingly close association between the mother and her offspring. At the most primitive level, the female provides direct care for a time but departs before the young eclose as adults. It is possible then for the care to be extended to the point where the mother is still present when her offspring mature, and they might then assist her in the rearing of additional brood. It remains only for some of the group to serve as permanent workers, and the last of the three qualities of eusociality has been attained. The subsocial route is the one believed by Wheeler and most subsequent investigators to have been followed by ants, termites, social wasps, and at least a few groups of the social bees.

The eusocial insects are listed and their habits very briefly summarized in Table 20-2. It is fair to say that as an ecological strategy eusociality has been overwhelmingly successful. It is useful to think of an insect colony as a diffuse organism, weighing anywhere from less than a gram to as much as a kilogram and possessing from about a hundred to a million or more tiny mouths. It is an animal that forages amebalike over fixed territories a few square meters in extent. A colony of the common pavement ant *Tetramorium caespitum*, for example, contains an average of about 10,000 workers that weigh 6.5 grams in the aggregate and control 40 square meters of ground. The average colony of the American harvester ant *Pogonomyrmex badius*, a larger species, contains 5,000 workers who together weigh 40 grams and patrol tens of square meters. The giant of all such "superorganisms" is a colony of the African driver ant *Dorylus wilverthi*, which may contain as many as 22 million workers weighing a total of over 20 kilograms. Its columns regularly patrol an area between 40,000 and 50,000 square meters in extent. The solitary tiphioid wasps, the nearest living relatives of ants, are by comparison no more than minuscule components of the insect fauna. Termites similarly outrank cockroaches. The cryptocercid cockroaches, the advanced subsocial forms closest to the ancestry of the termites, are in particular an insignificant cluster of species limited to several localities in North America and northern Asia. Only in the social bees and wasps have species in intermediate evolutionary grades held their own in competition with eusocial forms. They provide an array from which the full evolutionary pageant can now be deduced.

The Organization of Insect Societies

Once a species has crossed the threshold of eusociality there are two complementary means by which it can advance in colonial organization: through the increase in numbers and degree of specialization of the worker castes, and through the enlargement of the com-

munication code by which the colony members coordinate their activities. This statement is the insectan version of the venerable prescription that a society, like an organism and indeed like any cybernetic system, progresses through the differentiation and integration of its parts. In Chapter 14 I derived the less obvious theorem that castes tend to proliferate in evolution until there is one for each task. The theoretical limit has probably not been attained by many species of social insects, but it has been pursued by the most advanced forms to the extent that the number of discernible, functional types within the worker caste is often five or more and perhaps often exceeds ten. The reason for the vagueness of the estimate is simple. Castes can be physical, meaning that they are based upon permanent anatomical differences between individuals. Or they can be temporal, which means that individuals pass through developmental stages during which they serve the colony in various ways. The individual, in other words, belongs to more than one caste in its lifetime. Purists may hesitate to call a developmental stage a caste, but examination of ergonomic theory will show why it must be so defined.

Three basic physical castes are found in the ants, all members of the female sex: the worker, the soldier, and the queen. I refer to them as basic because they exist usually, but not always, as sharply distinctive forms unconnected to any other castes by intermediates. The males constitute an additional "caste" only in the loosest sense. No certain case of true caste polymorphism *within* the male sex has yet been discovered. Two forms of the male occur in some species of *Hypoponera*, but even in these cases they are not known to coexist in the same colony. Soldiers are often referred to as major workers, and the smaller coexisting worker forms as minor workers. Where soldiers exist in a species, minor workers are also invariably found. The latter caste is the more versatile of the two, typically attending to food gathering, nest excavation, brood care, and other quotidian tasks. In many species the soldiers assist to some degree, but in most cases they serve as defenders of the nest and living vessels for the storage of liquid food (see Chapter 14). The entire worker caste has been lost in many socially parasitic species, while in a few free-living species, especially in the primitive subfamily Ponerinae, the queen has been completely supplanted by workers or workerlike forms. In only a minority of species are all three female castes found together. All ant species, however, produce males in abundance as part of the normal colony life cycle.

In the course of evolution these castes have been elaborated in various, often striking ways. Sometimes the derived form bears little resemblance to the ancestral type, as for example the soldiers of *Acanthomyrmex*, whose tiny bodies are partially tucked beneath their massive heads, or the huge, bizarrely formed queens of the army ants. Also, intermediates sometimes connect the basic female castes: ergatogynes between workers and queens and media workers between minor and major workers.

Table 20-2 A synopsis of the social insects.

Taxonomic grouping	Natural history and social behavior	References
ORDER HYMENOPTERA (wasps, ants, and bees) SUPERFAMILY SPHECOIDEA (sphecoid wasps) FAMILY SPHECIDAE SUBFAMILY SPHECINAE *Trigonopsis* and others	Worldwide. Most sphecine species are solitary. *Trigonopsis cameronii* is communal: up to four females cooperate in constructing a large nest, but brood are still reared separately.	W. G. Eberhard (1972)
SUBFAMILY PEMPHREDONINAE *Microstigmus* and others	Worldwide. Most pemphredonine species are solitary. As many as 11 females of *Microstigmus comes*, a Central American species, cooperate to build and to guard a single communal nest. Preliminary evidence indicates that one individual serves as the egg layer, the others as workers. Hence the species is semisocial or primitively eusocial.	Matthews (1968a,b)
SUBFAMILY CRABRONINAE TRIBE CRABRONINI *Crabro, Moniaecera,* and others	Worldwide. Most crabronine species are solitary. Two to three females of *Moniaecera asperata* belonging to the same generation nest communally but do not cooperate in brood rearing. Similar behavior has been recorded in *Crossocerus dimidiatus*.	Evans (1964), Evans and Eberhard (1970), Peters (1973)
TRIBE CERCERINI *Cerceris* and others	Worldwide. Most cercerine species are solitary. In *Cerceris rubida* as many as five daughters remain with the mother, guarding the nest cooperatively but still rearing their own brood separately.	Grandi (1961)
SUPERFAMILY VESPOIDEA (vespoid wasps) FAMILY EUMENIDAE SUBFAMILY EUMENINAE *Eumenes, Ancistrocerus, Leptochilus, Odynerus, Pterocheilus,* and others: potter wasps and related forms	Nearly worldwide. Most species are solitary, but a few practice progressive provisioning, i.e., are subsocial. *Synagris cornuta* even macerates the prey and feeds them directly to the larvae.	Evans (1958), Evans and Eberhard (1970)
FAMILY VESPIDAE SUBFAMILY STENOGASTRINAE *Stenogaster, Eustenogaster, Liostenogaster, Parischnogaster*	India to New Guinea. Subsocial, with prey being macerated and fed directly to the larvae. In some *Parischnogaster* and *Stenogaster*, daughters remain with their mothers for at least a while, adding cells and rearing their own offspring. In at least one species of *Parischnogaster* there is also a division of labor, with unfertilized females serving as workers; the genetic relationship of the two castes is not yet known.	F. X. Williams (1919), Iwata (1967), Yoshikawa et al. (1969)

Table 20-2 (*continued*).

Taxonomic grouping	Natural history and social behavior	References
SUBFAMILY POLISTINAE TRIBE ROPALIDIINI *Ropalidia*	Africa to tropical Asia, New Guinea, and Australia. Colonies of females appear to be quasisocial, and there may exist at least a temporary division of labor between female reproductives and workers.	Yoshikawa (1964), Iwata (1969), Spradbery (1973)
TRIBE POLYBIINI *Polybia, Belonogaster, Polybioides, Apoica, Mischocyttarus, Brachygastra* (=*Nectarina*), *Chartergus, Synoeca*, and others: polybiine wasps	Tropical portions of Africa, Asia, and (especially) the New World. All species appear to be eusocial, with females divided into reproductive and worker castes that are at least functionally distinct. Many of the New World species have advanced social organizations, marked by castes that are physically distinct and large, long-lived colonies. Nests are often elaborately structured, evidently "designed" in good part to provide defense against ants and other predators. Colonies usually multiply by swarming, in which supernumerary queens depart for new nest sites accompanied by part of the worker force.	Richards and Richards (1951), Richards (1969), Evans and Eberhard (1970), Pardi and Piccioli (1970), Wilson (1971a), Jeanne (1972), Schremmer (1972), Spradbery (1973)
TRIBE POLISTINI *Polistes:* paper wasps	Worldwide, containing over 150 species. Primitively eusocial. Four species, sometimes separated as the genus "*Sulcopolistes*," are social parasites on other species of *Polistes*. Colonies usually founded by single queens or small groups of queens belonging to the same generation, but sometimes (in tropical species) generated by swarming. See account elsewhere in this chapter.	Pardi (1948), Deleurance (1957), Yoshikawa (1963), Eberhard (1969), Yamane (1971), Guiglia (1972), Spradbery (1973)
SUBFAMILY VESPINAE *Vespa, Provespa, Vespula:* hornets, yellowjackets	Eurasia, North Africa, and North America, with the greatest diversity in eastern Asia. All species are advanced eusocial, with physically distinct queen and worker castes, colonies containing hundreds or thousands of adults, elaborately constructed paper carton nests, and advanced forms of chemical and auditory communication. Six species of *Vespa* and *Vespula* are social parasites on members of the same genera, while *Vespa mandarinia* preys to some extent on the brood of other vespine species. See account elsewhere in this chapter.	Ishay et al. (1967), Kemper and Döhring (1967), Guiglia (1972), Ishay and Landau (1972), Spradbery (1973)
SUPERFAMILY FORMICOIDEA (ants) FAMILY FORMICIDAE (ants) THE "MYRMECIOID COMPLEX" SUBFAMILY SPHECOMYRMINAE *Sphecomyrma*	Fossil only; Cretaceous of United States, Canada, and Russia. The most primitive known ants, forming the link between modern members of the myrmecioid complex and solitary tiphioid wasps.	Wilson et al. (1967), G. Dlusski (personal communication)

Table 20-2 (*continued*).

Taxonomic grouping	Natural history and social behavior	References
SUBFAMILY MYRMECIINAE *Myrmecia* and others: bull ants or bulldog ants	Living genera are *Myrmecia* of Australia and New Caledonia and *Nothomyrmecia* of Australia. Other genera have been found in fossil form in the early Tertiary of Europe and South America. Morphologically the most primitive members of the myrmecioid complex. The living species are advanced eusocial insects, with well-developed physical differences between the queen and worker castes, colonies with hundreds or thousands of adults, and complex forms of chemical communication. But compared with other ants, myrmeciines have some primitive social traits: colony-founding queens still forage for food while raising the first brood, larvae are motile to some extent, workers do not recruit others to food, and so on. See account elsewhere in this chapter.	Wheeler (1933), Haskins and Haskins (1950), Haskins (1970), Gray (1971a,b), Wilson (1971a)
SUBFAMILY PSEUDOMYRMECINAE *Pseudomyrmex, Pachysima, Tetraponera, Viticicola*	Tropics around the world, including Australia. Slender, almost exclusively arboreal forms. Some of the species live in obligatory symbiosis with plants such as *Acacia, Barteria,* and *Vitex,* defending them from herbivorous insects and mammals. Larvae have special pouches on the venter of the body beneath the head in which workers place their food.	Janzen (1967, 1972), Wilson (1971a)
SUBFAMILY DOLICHODERINAE TRIBE ANEURETINI *Aneuretus* and others	The only known living species, *Aneuretus simoni,* is limited to Ceylon. Fossil species have been described from the Oligocene of United States and Europe. *A. simoni* is sociobiologically not basically different from the remainder of the Dolichoderinae, considered to be derived from older aneuretines. The worker caste is divided into minor and major subcastes, colonies are large, and chemical communication, which includes trail systems, is relatively advanced.	Wilson et al. (1956)
TRIBE DOLICHODERINI *Dolichoderus, Acanthoclinea, Hypoclinea, Monacis,* and others	*Hypoclinea* is worldwide; other genera are limited to Australia or the New World tropics. Colonies are very large, and often live in multiple nests connected by long odor trails. In the case of *Hypoclinea* at least, the odor trails are also used as part of trunk foraging routes to reach honeydew-producing homopterous insects and other food sources. *Dolichoderus* and *Monacis,* the endemic New World genera, are also exclusively arboreal. The sting is replaced by chemical substances, mostly terpenoids manufactured by the anal gland, that serve to repel enemies and to alarm nestmates. Large, occasionally hard-bodied ants.	Wheeler (1910), Kempf (1959), Wilson (1971a)

Table 20-2 (continued).

Taxonomic grouping	Natural history and social behavior	References
TRIBE LEPTOMYRMECINI *Leptomyrmex*	Australia and New Caledonia. Large slender ants noted for the presence of a wingless, ergatoid queen, as well as the existence of a replete caste among the larger workers. Sting reduced but still may be functional.	Wheeler (1910, 1934)
TRIBE TAPINOMINI *Tapinoma, Azteca, Bothriomyrmex, Conomyrma, Dorymyrmex, Forelius, Iridomyrmex, Liometopum, Technomyrmex,* and others; includes Argentine ants (*Iridomyrmex humilis*) and meat ants (*I. detectus* and other, similar Australian species)	Worldwide, especially diverse and abundant in Australia, New Guinea, and the New World. Large to very large colonies, often occupying multiple nest sites connected by long-lasting trunk trails. Trunk trails also lead to food sources, including honeydew-producing homopterous insects. As in the Dolichoderini, the sting has been replaced by a chemical alarm-defense system. The workers are small to medium in size, soft-bodied, and swiftly moving.	Wheeler (1910), M. R. Smith (1936), Blum and Wilson (1964), Markin (1970), Benois (1973)
SUBFAMILY FORMICINAE TRIBE MYRMECORHYNCHINI *Myrmecorhynchus*	Australia. A relatively primitive arboreal formicine that nests in hollow twigs and other arboreal sites. Workers are divided into major and minor castes. Natural history little known. In these and all other formicine ants, the sting has been replaced with formic acid, which is manufactured by a special poison gland separate from the sting apparatus.	Wilson (1971a)
TRIBE GESOMYRMECINI *Gesomyrmex*	Tropical Asia. Also known from the Baltic amber (Oligocene) of Europe. Relatively primitive, arboreal ants with major and minor worker subcastes. Natural history little known.	Wheeler (1910), Wilson (1971a)
TRIBE MYRMOTERATINI *Myrmoteras*	Tropical Asia. Small, predaceous ants with elongate mandibles that open more than 180° and close like a spring trap. Form small colonies that nest in soil of rainforest. Natural history otherwise little known.	Wilson (1971a)
TRIBE MELOPHORINI *Melophorus, Lasiophanes, Notoncus, Prolasius*	*Lasiophanes* occurs in temperate South America; other genera are Australian. Medium-sized, soil-dwelling ants that form moderately to very populous colonies. Some *Melophorus* species harvest seeds.	W. L. Brown (1955, 1973)
TRIBE PLAGIOLEPIDINI *Plagiolepis, Acantholepis, Acropyga, Anoplolepis*	*Acropyga*, which occurs from tropical Asia to Australia and in the New World tropics, is characterized by blind, subterranean workers that attend homopterous insects living on roots; in some places it is abundant enough to be an agricultural pest. The remaining genera are limited to the Old World; most forage above ground and are diverse in body form, size, and ecology.	Bünzli (1935), Weber (1944), Steyn (1954), E. S. Brown (1959)

Table 20-2 (continued).

Taxonomic grouping	Natural history and social behavior	References
TRIBE OECOPHYLLINI *Oecophylla:* weaver ants	Tropical Asia to northern Australia; and equatorial Africa. Also Oligocene of Europe and Miocene of Kenya. Large, aggressive, arboreal ants. The workers construct nests by fastening leaves together with larval silk; they employ the larvae as shuttles, holding them in their mandibles and weaving them back and forth along the edges of the leaves. The workers are divided into a major and a minor caste, the former comprising the foragers and the latter the nurses.	Ledoux (1950), Way (1954a,b), Sudd (1963), Wilson and Taylor (1964), Wilson (1971a)
TRIBE FORMICINI *Formica, Acanthomyops, Brachymyrmex, Lasius, Paratrechina, Polyergus, Pseudolasius,* and others	Worldwide. Rivaling the Camponotini as the dominant group of the Formicinae; also the largest and most diverse. Among the many adaptive types are the large, diurnal ants of the genus *Formica,* including species that parasitize or enslave others. The genera *Polyergus* and *Rossomyrmex* are exclusively slave-makers, exploiting *Formica. Acanthomyops* is evidently temporarily parasitic on *Lasius;* during its free-living phase the colony consists of large numbers of blind, subterranean workers that attend homopterous insects on roots. *Brachymyrmex* consists of very small ants; some species forage above ground and others are exclusively subterranean. *Pseudolasius* workers are divided into subcastes, but those of other species are monomorphic. Many other examples could be cited to illustrate the great diversity in physical form and social systems that often occur even within genera of the Formicini.	Wilson (1955a, 1971a), Sudd (1967), Wing (1968), W. L. Brown (1973), Francoeur (1973)
TRIBE CAMPONOTINI *Camponotus, Calomyrmex, Opisthopsis, Polyrhachis;* some *Camponotus* species are called carpenter ants	*Camponotus* is the largest and most widely distributed of all ant genera, occurring on every continent and from treeline to tropical forests. Its species show more ecological and behavioral diversity than most entire ant tribes. Nearly all *Camponotus* species are polymorphic, possessing well-defined major and minor subcastes. The workers are large and the colonies are relatively populous, in some cases containing hundreds or thousands of adults. *Polyrhachis* is a large and diverse genus of large monomorphic ants; its species range from Africa to Australia. *Calomyrmex* and *Opisthopsis* are monomorphic forms limited to Australia.	Sudd (1967), Sanders (1970), Lévieux (1971), Wilson (1971a), Benois (1972), Hölldobler et al. (1974)
THE "PONEROID COMPLEX" SUBFAMILY PONERINAE TRIBE AMBLYOPONINI *Amblyopone* and others	*Amblyopone,* a worldwide genus, is one of the most primitive groups of ants and the most primitive	Wheeler (1916, 1933), Wilson (1958a), W. L. Brown (1960, 1973),

Table 20-2 *(continued)*.

Taxonomic grouping	Natural history and social behavior	References
Amblyopone and others *(cont.)*	within the poneroid complex. The colonies are small, loosely organized, and construct poorly formed nests in the soil. The differences between the queen and worker castes are typical of ants but only weakly developed. Communication appears to be primitive. Colonies are founded by single queens that leave the nest periodically to search for prey. Workers sometimes transport the larvae to large prey rather than the reverse. *Myopopone*, found from tropical Asia to New Guinea, has similar habits. *Onychomyrmex*, an Australian genus, has independently evolved army-ant behavior, including nomadism and mass attacks on prey, to overcome large insects. The biology of *Mystrium* and *Prionopelta* is little known.	W. L. Brown et al. (1970), Haskins (1970), Gotwald and Lévieux (1972)
TRIBE ECTATOMMINI *Ectatomma* and others	*Ectatomma* (New World tropics), *Acanthoponera* (New World tropics), *Gnamptogenys* and *Heteroponera* (New World tropics and Indo-Australian region), and *Rhytidoponera* (tropical Asia to Australia) are medium to large ants, with moderately populous colonies, and have relatively advanced social organizations. *Discothyrea* and *Proceratium* are secretive ants that form small colonies and prey on spider eggs. The primitive ectatommines are considered to have given rise to the Myrmicinae, possibly in the late Mesozoic.	W. L. Brown (1957, 1973), Wilson (1971a)
TRIBE TYPHLOMYRMECINI *Typhlomyrmex*	New World tropics. Natural history little known. Form moderate-sized colonies, often forage in masses under bark.	Brown (1965, 1973)
TRIBE PONERINI *Ponera* and others	The most diversified and abundant tribe of the Ponerinae. *Ponera* and *Hypoponera*, which are worldwide in distribution, consist of diminutive, timid ants that live in small loosely organized colonies in rotting wood and the soil and feed on collembolans and other small arthropods. Although very abundant, their behavior is still little known. *Cryptopone* is a pantropical genus, even more secretive in behavior. In the Old World tropics occur an array of medium to large species in *Bothroponera*, *Brachyponera*, *Diacamma*, and *Myopias* which form larger, evidently well-organized colonies. *Myopias* is notable for its extreme prey specialization; certain species feed only on millipedes, others on beetles, or ants, and	Wheeler (1936), Wilson (1955b, 1958a,c,1971a),LeMasne(1956b), W. L. Brown (1965 and contained references, 1973, 1975), Gotwald and Brown (1966), Colombel (1970a,b), Haskins and Zahl (1971), Lévieux (1972)

Table 20-2   (*continued*).

Taxonomic grouping	Natural history and social behavior	References
Ponera and others (*cont.*)	so forth. *Leptogenys* is a pantropical genus of slender, fast-moving ants. Many species form small to moderate-sized colonies that prey on sowbugs (isopod crustaceans). Others, forming large to enormous colonies, behave like army ants, marching in columns to prey on termites and other arthropods. *Simopelta* also behaves like an army ant, raiding colonies of other ants; the same is true of *Termitopone* of the New World tropics and *Megaponera* and several other genera of large ants in Africa, which feed on termites. *Odontomachus* and *Anochetus* are pantropical in distribution; these ants are predaceous, using slender, traplike mandibles to seize prey.	
TRIBE PLATYTHYREINI *Platythyrea*	Pantropical. Slender, very fast-moving, often arboreal ants. Biology little known; often prey on termites.	W. L. Brown (1952a, 1975), Wilson (1971a)
TRIBE CYLINDROMYRMECINI *Cylindromyrmex*	New World tropics. Small colonies nest in rotting wood and plant cavities and appear to prey on termites. Natural history otherwise unknown.	W. L. Brown (1975)
TRIBE CERAPACHYINI *Cerapachys, Leptanilloides, Simopone, Sphinctomyrmex*	*Cerapachys* (pantropical) and *Sphinctomyrmex* (Old World tropics) conduct mass attacks on other ant species, which appear to be their exclusive prey.	Wheeler (1936), Wilson (1958a,b), W. L. Brown (1975 and personal communication)
TRIBE ACANTHOSTICHINI *Acanthostichus*	New World tropics. Colonies nest in soil and raid termites.	W. L. Brown (personal communication)
SUBFAMILY LEPTANILLINAE *Leptanilla*	Pantropical. Minute, highly secretive ants that apparently behave like army ants, but their biology is still almost entirely unknown.	Wilson (1971a)
SUBFAMILY DORYLINAE TRIBE DORYLINI *Dorylus:* driver ants or Old World army ants	Old World tropics. Advanced army ants. *Dorylus* attacks termites and a wide variety of other arthropods. Its colonies are the largest of any social insect, occasionally containing over 20 million workers.	Wheeler (1922), Raignier and van Boven (1955), Schneirla (1971), Wilson (1971a), Raignier (1972)
TRIBE AENICTINI *Aenictus:* army ants or Old World army ants	Old World tropics. Advanced army ants that prey mostly on wasps and other ant species.	Wilson (1964, 1971a), Schneirla (1971)
SUBFAMILY ECITONINAE TRIBE ECITONINI *Eciton, Labidus, Neivamyrmex:* army ants or New World army ants	New World tropics. Advanced army ants. The colonies feed by conducting massive raids on other arthropods, especially social wasps and ants. See account elsewhere in this chapter.	Borgmeier (1955), Schneirla (1971), Wilson (1971a), W. L. Brown (1973)

Table 20-2   (*continued*).

Taxonomic grouping	Natural history and social behavior	References
TRIBE CHELIOMYRMECINI *Cheliomyrmex:* army ants or New World army ants SUBFAMILY MYRMICINAE TRIBE MELISSOTARSINI	New World tropics. Army ants. Relatively primitive in morphology; behavior unknown.	Gotwald (1971), Schneirla (1971)
Melissotarsus, Rhopalomastix	*Melissotarsus,* which occurs in Africa and Madagascar, forms moderate-sized colonies that nest in the bark of standing trees and attend coccids. The biology of *Rhopalomastix,* a tropical Asiatic genus, is unknown.	Delage-Darchen (1972)
TRIBE MYRMICINI (broad sense) *Myrmica, Aphaenogaster, Cardiocondyla, Chelaner, Leptothorax, Lordomyrma, Messor, Monomorium, Myrmecina, Oligomyrmex, Pheidole, Pheidologeton, Pogonomyrmex, Pristomyrmex, Solenopsis, Tetramorium, Triglyphothrix, Veromessor,* and others.	Worldwide. The largest, most ecologically diversified tribe of ants. Most genera are monomorphic, but others show various degrees of worker caste differentiation and a few (e.g., *Oligomyrmex, Pheidole*) are strongly dimorphic. Most species are generalized insectivores, collecting honeydew as well. But *Messor, Pogonomyrmex, Veromessor,* and many members of *Monomorium* and a few other genera have become harvesting ants, depending to a large extent on seeds.	Ettershank (1966), Cole (1968), Wilson (1971a), W. L. Brown (1973)
TRIBE OCHETOMYRMECINI *Ochetomyrmex, Blepharidatta, Wasmannia*	New World tropics. *Wasmannia* forms large colonies with multiple queens; its small workers utilize trunk odor trails.	Wilson (1971a)
TRIBE ATTINI *Atta, Acromyrmex, Apterostigma, Cyphomyrmex, Trachymyrmex,* and others: fungus-growing ants; *Atta* and *Acromyrmex* are also known as leaf-cutting ants	Tropics and warm temperate zones of the New World. Workers rear and consume specialized symbiotic fungi. The substratum on which the fungus is cultured varies from genus to genus: *Cyphomyrmex* and *Trachymyrmex* utilize insect dung extensively, *Atta* and *Acromyrmex* cut fresh leaves and flowers, and so forth.	Weber (1966, 1972), Martin and Martin (1971), Wilson (1971a), Cherrett (1972), Martin et al. (1973)
TRIBE MERANOPLINI *Meranoplus, Calyptomyrmex, Mayriella*	Old World tropics, Australia. Small to medium colonies. *Meranoplus* harvest seeds; behavior otherwise little known.	Wilson (1971a)
TRIBE CATAULACINI *Cataulacus*	Old World tropics. Arboreal; biology mostly unknown.	Bolton (1974)
TRIBE CEPHALOTINI *Cephalotes, Paracryptocerus, Procryptocerus, Zacryptocerus*	New World tropics. Wholly arboreal. Most species have polymorphic worker castes; in some cases they are divided into extreme minor and major forms. The majors use their shield-shaped heads to guard the nest entrances. Omnivorous, relying extensively on the scavenging of dead insects and other small animals.	Kempf (1951, 1958), Creighton and Gregg (1954), Wilson (1971a)
TRIBE CREMATOGASTRINI *Crematogaster*	Worldwide. One of the largest genera, especially abundant and diverse in the Old World tropics. The moderate to very large colonies are either soil-dwell-	Soulié (1960a,b, 1964), Buren (1968), Leuthold (1968a,b), Hocking (1970)

Table 20-2 (continued).

Taxonomic grouping	Natural history and social behavior	References
Crematogaster (cont.)	ing or arboricolous, depending on the species. The abdomen is heart-shaped and can be folded forward over the head, permitting toxic secretions to be applied by the otherwise nonfunctional sting. The workers seek insects and honeydew along well-developed trail systems that come from a unique source—the tibial glands of the hind legs.	
TRIBE METAPONINI *Metapone*	Pantropical. Large, heavily armored ants that nest in small colonies in dead wood and evidently feed on termites. The biology of this relatively scarce group is otherwise unknown.	Wilson (1971a), W. L. Brown (personal communication)
TRIBE DACETINI *Daceton, Acanthognathus, Colobostruma, Epopostruma, Mesostruma, Orectognathus, Smithistruma, Strumigenys,* and others	Worldwide; especially diverse and abundant in the tropics. The more primitive genera, including *Daceton* of South America and *Orectognathus* of Australia, are medium-sized to large ants that forage in the open and prey on a variety of arthropods. Morphologically more advanced genera are comprised of small, secretive ants that prey on soft-bodied arthropods, especially springtails (collembolans).	W. L. Brown (1952b, 1973), W. L. Brown and Wilson (1959), Wilson (1962b), W. L. Brown and Kempf (1969)
TRIBE BASICEROTINI *Basiceros* and others	*Basiceros*, a genus limited to the New World tropics, contains large, sluggish ants that feed on termites. *Eurhopalothrix* and *Rhopalothrix*, which are pantropical, consist of diminutive, secretive forms, at least one of which preys on small, soft-bodied arthropods.	W. L. Brown and Kempf (1960)
SUPERFAMILY APOIDEA (bees) FAMILY HALICTIDAE SUBFAMILY NOMIINAE *Nomia:* sweat bees	Most species are evidently solitary. Some Old World species are communal and perhaps even quasisocial or semisocial.	Michener (1974)
SUBFAMILY HALICTINAE TRIBE AUGOCHLORINI *Augochlora, Augochloropsis, Neocorynura,* and others: sweat bees	Primarily New World, especially tropical. Most species are solitary, but others display varying degrees of communal, semisocial, or primitively eusocial behavior. Colonies small.	Michener (1974)
TRIBE HALICTINI *Halictus, Agapostemon, Dialictus, Evylaeus, Lasioglossum, Paralictus, Pseudagapostemon, Sphecodes,* and others: sweat bees	Worldwide, consisting of a very large number of species. The group tends to be replaced in New World tropics by Augochlorini. Most species are solitary, but a few are communal and many are primitively eusocial. Occasional eusocial species have well developed queen-worker differences and form perennial colo-	Sakagami and Michener (1962), Ordway (1965, 1966), Batra (1966, 1968), Knerer and Atwood (1966), Michener (1966a,b, 1974), Knerer and Plateaux-Quénu (1967a,b), Michener and Kerfoot (1967),

Table 20-2   (*continued*).

Taxonomic grouping	Natural history and social behavior	References
Halictus, etc. (*cont.*)	nies with hundreds of members; but the majority are monomorphic and form small, relatively short-lived colonies. *Paralictus* and *Sphecodes* are nonsocial nest parasites.	Sakagami and Hayashida (1968), Wille and Orozco (1970), Eickwort and Eickwort (1971, 1972, 1973a,b), Plateaux-Quénu (1972, 1973), Brothers and Michener (1974)

FAMILY ANDRENIDAE
  SUBFAMILY ANDRENINAE
    Andrena and a few other smaller genera
	Principally north temperate zone. Most species are solitary, a few parasocial (presumably communal).	Michener (1974)

SUBFAMILY PANURGINAE
    Panurgus, Calliopsis,
    Meliturga, Nomadopsis,
    Panurginus, Perdita, and
    others
	Worldwide. Most species are solitary but a few, as in *Panurgus* and *Perdita*, are communal.	Michener (1974)

FAMILY MEGACHILIDAE
  SUBFAMILY MEGACHILINAE
    TRIBE MEGACHILINI
      Megachile, Chalicodoma, Chelostoma,
      Hoplitis, Osmia, and others
	Worldwide. The great majority of species in this large assemblage are solitary, but a few members of *Chalicodoma* and *Osmia* are communal or possibly quasisocial.	Michener (1974)

TRIBE ANTHIDIINI *Anthidium* and others	Worldwide. Most species are solitary, but a few members of *Dianthidium, Heteranthidium*, and *Immanthidium* are communal.	Michener (1974)

FAMILY ANTHOPHORIDAE
  SUBFAMILY ANTHOPHORINAE
    TRIBE EXOMALOPSINI
      Exomalopsis, Paratetrapedia, and others
	New World, especially tropical. All *Exomalopsis* are colonial so far as known, and probably communal.	Michener (1974)

TRIBE EUCERINI
      Eucera, Melissodes, Peponapis, Svastra,
      Tetralonia, and others
	Worldwide. Mostly solitary but a few species of *Eucera, Melissodes*, and *Svastra* nest in colonies and are presumably communal.	Michener (1974)

SUBFAMILY XYLOCOPINAE
    TRIBE CERATININI
      Ceratina, Allodape, and others; including principally the allodapine bees
	Worldwide. All genera but *Ceratina* and *Manuelia* are restricted to the tropics and south temperate portions of the Old World and constitute a group called the allodapine bees. *Eucondylops, Inquilina*, and *Nasutapis* are social parasites on other allodapines. *Halterapis* and the two nonallodapines are solitary; the remainder of the allodapines are subsocial or primitively eusocial. See account elsewhere in this chapter.	Skaife (1953), Sakagami (1960), Michener (1961a, 1962, 1966d, 1971, 1974)

TRIBE XYLOCOPINI *Xylocopa* and others: carpenter bees and others	Throughout tropical and warm temperate regions of world. Solitary, with quasisocial behavior occasional in a few species.	Michener (1974)

Table 20-2 (*continued*).

Taxonomic grouping	Natural history and social behavior	References
FAMILY APIDAE SUBFAMILY BOMBINAE TRIBE EUGLOSSINI *Euglossa, Eulaema, Euplusia,* and others: orchid bees	New World tropics. Species, many of which are large and metallic in coloration, are variously solitary, communal, or quasisocial. In quasisocial forms, there are rarely more than 20 females in a nest and usually fewer than 10.	Dodson (1966), Roberts and Dodson (1967), Zucchi et al. (1969), Michener (1974)
TRIBE BOMBINI *Bombus, Psithyrus:* bumblebees	Primarily north temperate zone; mostly adapted to cold. Species of *Bombus* are almost all primitively eusocial; those of *Psithyrus* are social parasites on *Bombus.* See account elsewhere in this chapter.	Sladen (1912), Plath (1934), Free and Butler (1959), Sakagami and Zucchi (1965), Michener (1974)
SUBFAMILY APINAE TRIBE MELIPONINI *Melipona, Dactylurina, Lestrimelitta, Meliponula,* and *Trigona:* stingless bees or stingless honeybees	Tropics around the world; especially abundant and diverse in the New World. All of the species are perennial and highly social, with strong differences between the queen and worker castes, very large colonies (occasionally containing tens of thousands of adults), complex nests that vary in architecture from species to species, and elaborate systems of chemical and auditory communication. No social parasites are known, but species of *Lestrimelitta* rob the food stores of other meliponines.	Schwarz (1948), Michener (1961b, 1974), Kerr et al. (1967), Nogueira-Neto (1970a,b), Sakagami (1971), Wille and Michener (1973)
TRIBE APINI *Apis.* Four species occur: the eastern hive bee *A. cerana,* the giant honeybee *A. dorsata, A. florea,* and the common honeybee or western hive bee *A. mellifera:* true or stinging honeybees	Originally restricted to Europe, Asia, and Africa, but *A. mellifera* has been spread around the world by man. All four species are highly social, with strong queen-worker differences, colonies numbering into the tens of thousands, elaborate nest architecture, and advanced forms of communication, including the waggle dance. See account elsewhere in this chapter.	Von Frisch (1954, 1967), Lindauer (1961), Chauvin, ed. (1968), Morse and Laigo (1969), Michener (1973, 1974)
ORDER ISOPTERA (termites) FAMILY MASTOTERMITIDAE *Mastotermes*	Australia; also Eocene and later Tertiary fossils from Europe and North America. *M. darwiniensis,* the only surviving species, is by far the most primitive living termite. See account elsewhere in this chapter.	Gay and Calaby (1970)
FAMILY KALOTERMITIDAE *Kalotermes, Calcaritermes, Cryptotermes, Glyptotermes, Neotermes, Rugitermes,* and others: dry wood termites	Worldwide. Relatively primitive morphologically and behaviorally. Called dry wood termites because colonies usually nest primarily in wood, without soil connections. Nests consist of ill-defined excavation galleries. Nymphs active in colony labor; in later stages they become pseudergates, which are capable of transforming into soldiers or reproductive castes.	Emerson (1969), Bess (1970), Krishna (1970), Weesner (1970)

Table 20-2 (continued).

Taxonomic grouping	Natural history and social behavior	References
Kalotermes, etc. (cont.)	On at least 8 occasions soldiers have developed plug-shaped heads with which they block entrances to the nest. See account elsewhere in this chapter.	
FAMILY HODOTERMITIDAE SUBFAMILY HODOTERMITINAE *Hodotermes, Microhodotermes, Anacanthotermes:* harvester termites	Primarily African, but extending to Middle East and tropical Asia. Large, relatively primitive species that form populous ground-nesting colonies. The large-eyed workers forage above ground for grasses and seeds, which are stored in special nest chambers.	Bouillon (1970), Roonwal (1970), Lee and Wood (1971), Watson et al. (1972)
SUBFAMILY CRETATERMITINAE *Cretatermes*	Described from a mid-Cretaceous fossil found in Labrador, about 100 million years old. The oldest known termite, and with the fossil ants of the subfamily Sphecomyrminae one of the oldest known social insects.	Emerson (1967)
SUBFAMILY TERMOPSINAE *Termopsis, Archotermopsis, Hodotermopsis, Porotermes, Zootermopsis:* damp wood termites	Warm temperate zones of Europe, Asia, and North America. Immature forms consist mostly of pseudergates fully involved in colony labor and capable of transforming into soldiers and reproductive castes. Nests consist of poorly defined galleries excavated in moist, rotting wood.	Castle (1934), Krishna (1970), Stuart (1970), Weesner (1970), Lee and Wood (1971)
FAMILY RHINOTERMITIDAE *Rhinotermes, Coptotermes, Heterotermes, Psammotermes, Reticulitermes, Schedorhinotermes,* and others	Worldwide. A large assemblage of genera and species, roughly intermediate in morphology, caste development, and social behavior between the most primitive termites (Mastotermitidae, Kalotermitidae) and the "higher" termites (Termitidae). The species are ecologically very diverse. *Reticulitermes* ranges well north into North America and Europe, forming sometimes large, diffuse colonies in dead wood, sometimes with both primary and supplementary reproductive castes present. *Psammotermes* of Africa and Arabia penetrates farthest into arid deserts, feeding on animal dung and dry wood. At least some members of *Rhinotermes*, a neotropical genus, nest in a wide variety of sites in rain forests. In Australia the colonies of some species of *Coptotermes* contain over a million members, which build large mounds (the only rhinotermitids to do so) and are serious pests of timber, attacking even living trees. The Asiatic species *C. formosanus* has been accidentally spread by man to many parts of the world, where it is destructive to buildings, utility poles, and other wooden structures.	Araujo (1970), Bess (1970), Gay and Calaby (1970), Harris (1970), Weesner (1970), Emerson (1971)

Table 20-2 (*continued*).

Taxonomic grouping	Natural history and social behavior	References
FAMILY SERRITERMITIDAE *Serritermes*	The only known species is *S. serrifer*, which occurs in the nest walls of *Cornitermes* mounds in Brazil.	Araujo (1970)
FAMILY TERMITIDAE *Termes, Ahamitermes, Acanthotermes, Amitermes, Armitermes, Capritermes, Cornitermes, Cubitermes, Drepanotermes, Labiotermes, Macrotermes, Nasutitermes, Ophiotermes, Pericapritermes, Rhynchotermes, Syntermes,* and many others: higher termites	Worldwide. An extremely large, evolutionarily advanced assemblage constituting 75 percent of the known termite species. The morphology of the soldiers is exceptionally diversified, serving as the basis for most practical classification and phylogenetic studies by taxonomists. Extreme types include the nasute soldiers (e.g., in *Nasutitermes*), which shoot drops of repellent liquid from spout-shaped organs on the head; soldiers (e.g., in *Pericapritermes*) that pull their twisted mandibles apart with violent force like the snapping of fingers; and others. The termitid species are also notable for the extent of their ecological radiation. Besides more "conventional" adaptive types there are specialized fungus growers (*Macrotermes* and other members of the subfamily Macrotermitinae); species in *Ahamitermes, Incolitermes,* and *Termes* that live only in the nest walls of host termites; and many others. See further account elsewhere in this chapter.	Araujo (1970), Bouillon (1970), Krishna and Weesner, eds. (1969, 1970), Roonwal (1970), Ruelle (1970), Lee and Wood (1971), Maschwitz et al. (1972), Sands (1972)

Although the workers of only a minority of ant species are divided into physical subcastes, all studied thus far undergo complex physiological and behavioral changes with age. These changes constitute shifts from one temporal caste to another. The case of *Formica polyctena*, the European wood ant analyzed by Otto (1958), appears to be typical. Each worker, while completely bound to the colony, addresses its activities indiscriminately to all members of the colony or, at most, to all members of a given caste or life stage. About half the time of the worker is spent at rest and half engaged in some social activity or foraging. There is a tendency for workers to spend at least 50 days after their emergence from the cocoon in what the German investigators call the *Innendienst*—service inside the nest. The activities of the *Innendienst* include care of the brood, queens, and other adult workers, handling of dead prey in the nest chambers, and nest cleaning. Although a few workers specialize on one or two of these tasks, the majority perform most or all of the tasks at some time. After about 50 days, most workers shift permanently to the *Aussendienst*, during which they forage and work on nest construction. There is a further specialization possible in nest construction, in that some workers concentrate on excavating within the nest while others gather materials for roofing. Individual behavioral ontogenies vary greatly in both content and timing. For example, many workers pass through the *Innendienst* without attending the brood at all.

During the *Innendienst*, the ovaries of the workers contain eggs. Toward the end of this period, resorption of the eggs begins, and, by the onset of *Aussendienst*, the resorption is total. Other, suggestive changes occur in several exocrine glands. Workers concentrating on excavation within the nest, for example, have somewhat larger mandibular gland nuclei than other workers. When the many subtle changes of this kind are added up, age polyethism in ants is seen to be extremely complex. Virtually all categories of social behavior have proved to change to some degree, many forming discordant patterns when viewed in combination.

Even the larval stage can serve as a caste, despite the fact that larvae are virtually immobile in most kinds of ants. In many species the larvae present salivary gland secretions to the adults. It used to be thought that they were merely ejecting liquid waste material, but now the evidence is strong that the material has nutritive value and under certain circumstances plays an important role in the economy of the colony. For example, workers of *Monomorium pharaonis* withstand desiccation for longer periods of time when allowed access to the larval secretions, while the queens of *Leptothorax curvispinosus* feed constantly on these secretions, in a way that leaves little doubt that they are receiving food.

Although termites are phylogenetically unrelated to ants, they have evolved a caste system that is remarkably similar to the ant system in several major respects. Like the ants, they have produced a soldier caste that is highly specialized in both head structure and behavior for colony defense, and a minor worker caste that is numerically dominant in the colonies, morphologically similar from species to species, and behaviorally versatile. The number of physical castes in the phylogenetically most advanced termite species is somewhat greater than in the most highly evolved ant species, but the average degree of specialization of individual castes is about the same. Finally, the higher termites have developed temporal polyethism resembling that of the ants in broad outline.

Differences also exist. The neuter castes of termites consist of both sexes, rather than females alone as in the ants, and there are no termite "drones" that live solely for the act of mating and that are programmed for an early postreproductive death. Where ant larvae are grublike and incapable of contributing to the labor of the colony other than through the biosynthesis of nutrients, immature termites are active nymphs not radically different in form and behavior from the mature stages. In the more primitive termites the nymphs contribute to the work of the colony; in other words, there is an employment of "child labor." This is not the case in the higher termites (the family Termitidae), where the immature forms are wholly dependent on a well-differentiated worker caste. Finally, the termites generally have a wide array of "supplementary reproductives," fertile but wingless individuals of both sexes that develop in colonies whenever the primary reproductives are removed. The nearly universal occurrence of these substitute castes provides termite colonies with a degree of resiliency leading in extreme cases to a potential immortality seldom encountered in ants and other social hymenopterans.

Caste finds a wide range of expression among the species of social bees and wasps. In the primitively eusocial halictine bees, it emerges as a mere psychological difference among morphologically similar adults, but goes on to include, in a few species, several striking forms of queen-worker dimorphism. In the honeybee species *Apis mellifera* strong morphological and physiological differences exist between queens and workers, and the caste of individuals is determined by a complex interaction between pheromone-mediated behavior on the part of nurse workers and specialized diets fed to the larvae. Next, at least one group of species of stingless bees, the genus *Melipona*, has superimposed a genetic control of caste upon the conventional physiological device employed by related groups. Most of these phylo-

genetic advances, with the most conspicuous exception being the invention of genetic control, have been paralleled in the evolution of the social wasps. Together the social bees and wasps differ from the ants and termites in one major respect: for some reason none of them has fashioned well-defined worker subcastes. It is true that the species with very large colonies display a division of labor comparable to that of the most advanced ants and termites. But where the division in the latter insects is based in part on physical subcastes and in part on programmed, temporal polyethism, in most bees and wasps it is based almost entirely on temporal polyethism.

Certain other evolutionary trends can be recognized in bees and wasps. As colony size has grown in the course of evolution, the differences between the queen and worker castes have been exaggerated, intermediate forms have disappeared, and the behavior of the queen has become increasingly specialized and parasitic. The ultimate stage is attained in the honeybees and stingless bees, whose queens never attempt to start colonies on their own and are reduced to the status of little more than egg-laying machines. Correlated with this trend has been a subtle shift in the power structure of the colony. Among the primitively social groups, particularly the halictine bees, the bumblebees, and the primitive polistine wasps, the queen maintains a dominant position primarily by aggressive behavior toward her sisters, daughters, and nieces. In more complexly social species, reproductive control is exercised through inhibitory pheromones.

Although morphological subcastes are generally so weakly developed in bees and wasps as to be almost nonexistent when compared with those of ants and termites, size effects do occur. Larger members of a given colony tend to forage more, and smaller members tend to devote themselves to brood care and nest work. In honeybees, the larger an individual bee the more quickly it passes through the normal ontogenetic stages of behavior, terminating in a period devoted principally to foraging. Throughout the course of evolution to higher levels of eusociality, there has been a tendency to produce ever more elaborate patterns of temporal division of labor, the most extreme cases again being those of honeybees and stingless bees. This temporal polyethism, like that of ants and termites, is typically a sequence leading from nest work and brood care to foraging. To my knowledge only one exception has been reported, that of the Japanese paper wasp *Polistes fadwigae*, in which a very weak polyethism follows the opposite sequence.

The ways by which social insects communicate are impressively diverse. They include tappings, stridulations, strokings, graspings, antennations, tastings, and puffings and streakings of chemicals that evoke various responses from simple recognition to recruitment and alarm. We must add to this list other, often subtle and sometimes even bizarre, effects: the exchange of pheromones in liquid food that inhibit caste development, the soliciting and exchange of special "trophic" eggs that exist only to be eaten, the acceleration or inhibition of work performance by the presence of other colony members nearby, various forms of dominance and submission relationships, programmed execution and cannibalism, and still others.

Three generalizations are useful in gaining perspective on this subject. First, most communication systems in the social insects appear to be based on chemical signals. The known visual signals are sparse and simple. In some groups, particularly the termites and subterranean ants, they play no role in the day-to-day life of the colony. Airborne sound is only weakly perceived by social insects and has not been definitely implicated in any important communication system. Many species, however, are extremely sensitive to sound carried by the substrate, but they evidently employ it only in limited fashion, chiefly during aggressive encounters and alarm signaling. Modulated sound signals appear to play a role in recruitment in the advanced stingless bees of the genus *Melipona* and in the honeybees, which have incorporated them into the waggle dance. Touch is universally employed by insect colonies, but, with the possible exception of dominance and trophallaxis control in the vespine wasps, it has not been molded into a Morse-like system capable of transmitting higher loads of information.

In contrast, chemical signals, evoking the sensations of either odors or smells, have been implicated in almost every category of communication. In 1958 I suggested that the separation of these substances by the dissection of their glandular sources could provide the means of analyzing much social behavior that had previously seemed intractable: "The complex social behavior of ants appears to be mediated in large part by chemoreceptors. If it can be assumed that 'instinctive' behavior of these insects is organized in a fashion similar to that demonstrated for the better known invertebrates, a useful hypothesis would seem to be that there exists a series of behavioral 'releasers,' in this case chemical substances voided by individual ants that evoke specific responses in other members of the same species. It is further useful for purposes of investigation to suppose that the releasers are produced at least in part as glandular secretions and tend to be accumulated and stored in glandular reservoirs" (Wilson, 1958d). With each improvement in organic microanalysis permitting the separation and bioassay of secretory substances, new evidence has been added to support this conjecture. Pheromones, as the chemical releasers were first called by Karlson and Butenandt (1959), may be classified as olfactory or oral according to the site of their reception. Also, their various actions can be distinguished as releaser effects, comprising the classical stimulus-response mediated wholly by the nervous system (the stimulus being thus by definition a chemical "releaser" in the terminology of animal behaviorists), or primer effects, in which endocrine and reproductive systems are altered

physiologically. In the latter case, the body is in a sense "primed" for new biological activity, and it responds afterward with an altered behavioral repertory when presented with appropriate stimuli. Examples of releaser pheromones include the alarm and trail substances of workers and attractive scents of queens, while the best-understood primer pheromones include the substances secreted by queen and king termites that inhibit the development of nymphs into their own castes. A pheromone may have both releaser and primer effects: 9-keto-decenoic acid, the principal "queen substance" produced by honeybee queens, attracts males and inhibits the building of royal cells on the part of workers (releaser effect); it also inhibits the development of worker ovaries (primer effect). The sum of current evidence indicates that pheromones play the central role in the organization of insect societies.

The second generalization is that most of the communication systems have parallels in behavior patterns already present in some form or other in solitary and presocial insects. Nest building is a case in point. The primitive ants, termites, and social wasps build nests that are scarcely more complicated than those of many of their solitary relatives. The nests of primitively social bees are frequently simpler than those of their solitary relatives. Elaboration of nest structure occurred in certain phyletic lines after the eusocial state was attained, and its evolution can be easily traced. The dominance hierarchies that play a key role in bumblebee and wasp societies have a precedent in the territorial behavior of many solitary insect species, including at least a few hymenopterans. Elaborate brood care, a hallmark of higher sociality, has its precursor in progressive larval feeding in a multitude of subsocial species belonging to several insect orders. Alarm substances are in many cases simply modified defensive secretions, and trail substances have a parallel in the odor spots used to mark the nuptial flight paths of the males of some solitary Hymenoptera. Michener, Brothers, and Kamm (1971) have concluded that in the primitively social halictine bees, "Mechanisms of social integration (resulting in division of labor and differentiation of castes) mostly appear to involve behavioral features of the solitary ancestors and accidental results of joint occupancy of nests." Even the elements of the honeybee waggle dance, the distant apex of insect social evolution in the eyes of most biologists, have precursors: the modulated rocking behavior of saturniid moths, which varies in duration according to the length of the flight just completed and which thus resembles the straight run of the bee dance; the oriented "dances" of hungry *Phormia regina* flies after they have been given a small drop of sugar water; and the ability of some solitary insects to shift from light to gravity orientation when placed on dark vertical surfaces.

This brings us finally to the third generalization about communication in insect societies. The remarkable qualities of social life are mass phenomena that emerge from the integration of much simpler individual patterns by means of communication. If communication itself is first treated as a discrete phenomenon, the entire subject is much more readily analyzed. To date it has been found convenient to recognize about nine categories of responses in social insects, as given in the following list:

1. Alarm
2. Simple attraction (multiple attraction = "assembly")
3. Recruitment, as to a new food source or nest site
4. Grooming, including assistance at molting
5. Trophallaxis (the exchange of oral and anal liquid)
6. Exchange of solid food particles
7. Group effect: either increasing a given activity (facilitation) or inhibiting it
8. Recognition, of both nestmates and members of particular castes
9. Caste determination, either by inhibition or by stimulation

Most of these categories have been examined elsewhere in the present book (see especially Chapters 3 and 8–10), as well as in the monographs cited earlier.

The Prime Movers of Higher Social Evolution in Insects

The single most notable fact concerning eusociality in insects is its near monopoly by the single order Hymenoptera. Eusociality has arisen at least 11 times within the Hymenoptera: at least twice in the wasps, more precisely at least once each in the stenogastrine and vespine-polybiine vespids and probably a third time in the sphecid genus *Microstigmus*; 8 or more times in the bees; and at least once or perhaps twice in the ants. Yet throughout the entire remainder of the Arthropoda, true sociality is known to have originated in only one other living group, the termites. This dominance of the social condition by the Hymenoptera cannot be a coincidence. Throughout at least the Cenozoic Era less than 20 percent of all insect species have belonged to the order. Furthermore, eusociality is limited within the Hymenoptera to the aculeate wasps and to their immediate descendants, the ants and the bees, which, together, constitute no more than 50,000 estimated living species, or perhaps 6 percent of the total number of insect species in the world. This overwhelming phylogenetic bias is the most important clue we have to go on in searching for the prime movers of higher social evolution.

The tendency of aculeate Hymenoptera to evolve eusocial species can probably be ascribed in part to their mandibulate (chewing) mouthparts, which lend themselves so well to the manipulation of objects, or to the penchant of aculeate females for building nests to which they return repeatedly, or to the frequent close relationship between mother and young. These and perhaps some other biological features are prerequisites for the evolution of eusociality. But they

are shared in full by many other, species-rich groups of arthropods, including the spiders, earwigs, orthopterans, and beetles, none of which, with the exception of the cockroaches that gave rise to termites, achieved full sociality. Time and again phyletic lines have pressed most of the way to eusociality, in some cases to the very threshold, and then unaccountably stopped.

At the present time the key to hymenopteran success appears to be haplodiploidy, the mode of sex determination by which unfertilized eggs typically develop into males (hence, haploid) and fertilized eggs into females (hence, diploid). Haplodiploidy is a characteristic of the Hymenoptera shared by only a few other arthropod groups (certain mites, thrips, and whiteflies; the iceryine scale insects; and the beetle genera *Micromalthus*, *Xylosandrus*, and, perhaps, *Xyleborus*). Two authors have independently suggested a connection between haplodiploidy and the frequent occurrence of eusociality. Richards (1965) suggested that the control which haplodiploidy grants the female over the sex of her own offspring has eased the way to colonial organization. This is undoubtedly true. The postponement of male production until late in the season, by the simple expedient of passing sperm through the spermathecal duct to meet all eggs, is a characteristic of advanced sociality, for example, in the annual halictid bees (Knerer and Plateaux-Quénu, 1967b). At the same time, it is not a characteristic of many other Halictidae that are primitively eusocial—but eusocial nonetheless. In other words, sex control by the mother is a general feature of higher social evolution but not a prerequisite for the attainment of full sociality.

Hamilton (1964) created an audacious genetic theory of the origin of sociality that assigns a wholly different central role to haplodiploidy. Working from traditional axioms of population genetics, he first deduced the following principle that applies to any genotype: in order for an altruistic trait to evolve, the sacrifice of fitness by an individual must be compensated for by an increase in fitness in some group of relatives by a factor greater than the reciprocal of the coefficient of relationship (r) to that group. As explained in Chapter 4, the coefficient of relationship (also called the degree of relatedness) is the equivalent of the average fraction of genes shared by common descent; thus, in sisters r is $\frac{1}{2}$; in half-sisters, $\frac{1}{4}$; in first cousins, $\frac{1}{8}$; and so on. The following example should make the relation intuitively clearer: if an individual sacrifices its life or is sterilized by some inherited trait, in order for that trait to be fixed in evolution it must cause the reproductive rate of sisters to be more than doubled, or that of half-sisters to be more than quadrupled, and so on. The full effects of the individual on its own fitness and on the fitness of all its relatives, weighted by the degree of relationship to the relatives, is referred to as the "inclusive fitness." This measure can be treated as the equivalent of the classical measure of fitness, which takes no account of effects on relatives. Hamilton's theorem on altruism consists merely of a more general restatement of the basic axiom that genotypes increase in frequency if their relative fitness is greater.

Next Hamilton pointed out that owing to the haplodiploid mode of sex determination in Hymenoptera, the coefficient of relationship among sisters is $\frac{3}{4}$; whereas, between mother and daughter, it remains $\frac{1}{2}$. This is the case because sisters share all of the genes they receive from their father (since their father is homozygous), and they share on the average of $\frac{1}{2}$ of the genes they receive from their mother. Each sister receives $\frac{1}{2}$ of all her genes from the father and $\frac{1}{2}$ from the mother, so that the average fraction (r) of genes shared through common descent between two sisters is equal to

$$\left(1 \times \frac{1}{2}\right) + \left(\frac{1}{2} \times \frac{1}{2}\right) = \frac{3}{4}$$

Therefore, in cases where the mother lives as long as the eclosion of her female offspring, those offspring may increase their inclusive fitness more by care of their younger sisters than by an equal amount of care given to their own offspring. In other words, hymenopteran species should tend to become social, all other things being equal.

This strange calculus, when extended to other kin (see Table 20-3), leads to even stranger conclusions. Consider, for example, the prediction that males should be more consistently selfish than females toward everyone else in the colony. This is expected to be the case because under all conditions except complete queen domination of the workers, a male's expected reproductive success is greater than that of a similar sized female (see below). In order for selection to favor male altruism, such altruism would have to confer greater benefits than similar altruism by a female—an unlikely situation. Not only is this prediction met in nature; its fulfillment seems explicable only by this particular theory. The selfishness of male behavior is well known but has never before been adequately explained—in our language, the word "drone" has come to designate any lazy, parasitic person. Not only do hymenopteran males contribute virtually nothing to the labor of the colony, but they are also highly competitive in begging food from female members of the colony and become quite aggressive in contending with other males for access to females during the nuptial flights. Nature has even provided a control experiment: termites are not haplodiploid and yet have equaled the hymenopter-

ans in social evolution, for different reasons that will be discussed later. According to the theory termite males should not be drones. And they are not. Males constitute approximately half of the worker force, contribute an equal share of the labor, and are as altruistic to nestmates as are their sisters.

A second, nonobvious prediction of the theory is that workers of hymenopteran colonies should favor their own sons over their brothers. In other words, workers should lay unfertilized eggs and try to rear them to the exclusion of the queen's unfertilized eggs. This bias follows in part from the simple fact that females are related to their sons by a degree of $\frac{1}{2}$ but to their brothers by a degree of only $\frac{1}{4}$. It is enhanced by the relations between sister workers, in a manner to be explained shortly. Although the result seems odd, it can be reasonably well documented. Males are commonly derived from worker-laid eggs in nests of paper wasps (Yamanaka, 1928), bumblebees (Ronaldo Zucchi, personal communication), stingless bees of the genus *Trigona* (Bieg, 1972), and ants of the genera *Oecophylla* and *Myrmica* (Ledoux, 1950; Brian, 1968). The origin of males from workers appears to be a widespread phenomenon in the social Hymenoptera. But it is not universal; in the ant genera *Pheidole* and *Solenopsis*, for example, ovaries are completely lacking in the worker caste.

A still more detailed and rigorous test of the kin-selection hypothesis can be made by examining the asymmetries within the haplodiploid system (Trivers, 1975). The test can be made objective by challenging it with a competing hypothesis. In particular, Brothers and Michener (1974) and Michener and Brothers (1974) have proposed that eusocial behavior in halictid bees evolved by the successful domination and control of some female bees over others—as opposed to "voluntary" submission of the dominated bees due to kin selection. They have noted that queens of the primitively eusocial bee *Lasioglossum zephyrum* control other adult females by a pair of simple behaviors. Other adult females are systematically nudged, an action that appears to be aggressive in nature and may have the effect of inhibiting ovarian development. The individuals most frequently nudged are the ones with the largest ovaries, and hence the greatest potential as rivals to the queens. Nudging is followed by backing, in which the nudger retreats down the nest galleries, apparently attempting to draw the other bee after it. The effect is to maneuver

Table 20-3   The degrees of relatedness (r) among close kin in hymenopteran groups. (From Trivers, 1975; modified from Hamilton, 1964.)

	Mother	Father	Sister	Brother	Son	Daughter	Nephew or niece
Female	$\frac{1}{2}$	$\frac{1}{2}$	$\frac{3}{4}$	$\frac{1}{4}$	$\frac{1}{2}$	$\frac{1}{2}$	$\frac{3}{8}$
Male	1	0	$\frac{1}{2}$	$\frac{1}{2}$	0	1	$\frac{1}{4}$

the follower closer to the brood cells where it can assist in the construction and provisioning of the cells used by the queens. The bees that follow the most consistently are the ones with the smallest ovaries. It is not difficult to imagine, along with Michener and Brothers, that sterile castes can evolve if certain genotypes arise that are very powerful in controlling nestmates. Alexander (1974) has independently advocated the influence of exploitation, especially of parents over offspring, as a general factor in the social evolution of insects.

Trivers has shown how to discriminate between the kinship and exploitation hypotheses, by making use of the asymmetries in the haplodiploid system. According to the exploitation hypothesis, we expect a queen in full control of a colony to produce an equal dry weight of reproductive females (new, virgin queens) and males. This would be in accordance with the original Fisher model that predicts a maximum benefit/cost ratio when the energetic investments in the two sexes are equal, that is, when the dry weight of the queens produced is equal to the dry weight of the males produced (see Chapter 15). On the other hand, kin selection in haplodiploid systems will lead to strong deviations from the 1:1 ratio. Two circumstances involving kin selection are possible:

1. *Denying the queen the production of males.* If a worker is able to assist her mother in raising the queen's daughters (and her own sisters), but lays unfertilized eggs and succeeds in having the colony raise only her own sons, she trades an average r to her own offspring of $\frac{1}{2}$ for an r (to sisters and sons) of $\frac{5}{8}$ (average of $\frac{3}{4}$ and $\frac{1}{2}$). If the other workers collaborate with the egg-laying worker, they will raise sisters and nephews and thereby trade an r to their own offspring of $\frac{1}{2}$ for an average of $\frac{9}{16}$. Finally, the queen also gains by the arrangement, because she now has daughters and grandsons at an average $r = \frac{3}{8}$; whereas if the workers left and had their own offspring exclusively, the queen would have only granddaughters and grandsons at $r = \frac{1}{4}$. However, the arrangement is still inferior to the one in which the workers let her have all the daughters and sons. If workers do manage to produce the males, then most of the females in the colony—the queen and the nonlaying workers—will prefer to invest the same in new queens as in males. For example, the queen will be related by $r = \frac{1}{2}$ to the new queens (her daughters) and by $r = \frac{1}{4}$ to the males (grandsons via laying workers); but a male is in turn twice as valuable, per unit investment, as a new queen, because he will father females related by $r = 1$ and males (via laying workers) related by $r = \frac{1}{2}$, while a new queen will (like her mother) produce females related by $r = \frac{1}{2}$ and grandsons related by $r = \frac{1}{4}$. Nonlaying workers also prefer equal investment: they are related to new queens by $\frac{3}{4}$ and to males by $\frac{3}{8}$, but (as just shown) a male is twice as valuable, per unit investment, as a new queen. When laying workers only produce some of the males, the situation is complicated, but Trivers (1975) has shown that the queen still prefers nearly equal investment,

while nonlaying workers begin to prefer more investment in the females. The more males that come from the queen, the sharper the conflict over the ratio of investment.

2. *Allowing the queen to produce males but controlling the ratio in other ways.* Even if the queen is permitted to be the mother of all the males, the workers can still adjust the ratio to their optimum as opposed to the queen's optimum. The methods at their disposal are differential destruction according to sex of the eggs, larvae, and pupae. The evidence already exists that the rate of colony growth, in *Leptothorax* ants at least, is determined almost entirely by the workers and not the queens (see Wilson, 1974d). In the case where the queen lays all of the eggs, the workers trade $r = \frac{1}{4}$ for $r = \frac{3}{4}$ if they invest in a sister instead of a brother. The equilibrium ratio should be 3:1 in favor of queens (sisters) as opposed to males (brothers), since the expected reproductive success of the males will then be three times that of the queens on a per-gram basis, balancing the one-third initial investment.

In summary, the kin-selection hypothesis predicts that to the extent that workers control the reproduction of the colony—one might even say to the extent that they "exploit" the queen—the ratio of investment will fall between 1:1 and 3:1 in favor of queen production. If the mother queen is in control, that is "exploiting" the workers, the ratio should be the usual Fisherian 1:1. For various species of ants thus far measured, the ratio is significantly greater than 1:1, and in many cases it falls very close to 3:1 (Trivers, 1975).

Trivers' remarkable result appears to confirm the operation of kin selection in ants as the controlling force, as opposed to individual selection leading to domination and exploitation by the queen. Needless to say, both processes might conceivably operate, and in fact the existence of dominance systems in primitively social bees and wasps leaves open the possibility that individual-selected exploitation does play a role. But to what degree does "dominance" behavior in a species such as *Lasioglossum zephyrum* really represent control? The behavior could be simply part of the communication system by which individuals with different capacities accept the most appropriate roles, that is, the roles that maximize personal fitness. Lin and Michener (1972) in fact anticipate such an arrangement in the evolution of *Lasioglossum* and other social Hymenoptera. They see the early role of workers as being not necessarily altruistic or even based on kin selection. An auxiliary female can gain some amount of personal fitness by laying eggs surreptitiously; she can also be prepared to take over as the principal egg layer if the queen dies or leaves. In an environment where there are few opportunities to start new nests, such compromises can yield a higher average number of offspring than the attempt to proceed alone. This kind of cooperative behavior can conceivably evolve in the absence of kin selection.

Yet in the final analysis, even after the parameters of exploitation and compromise are added to the equation, nothing but kin selection seems to explain the statistical dominance of eusociality by the Hymenoptera. Kin selection still appears to be the force that guided one phyletic group after another across the threshold of eusociality and permitted colony-level selection to take command.

It remains to be pointed out that although termites are not haplo-diploid, they possess one remarkable feature that may provide the clue to their social beginnings: along with the closely related crypto-cercid cockroaches, they are the only wood-eating insects that depend on symbiotic intestinal protozoans. As first pointed out by L. R. Cleveland (in Cleveland et al., 1934), the protozoans are passed from old to young individuals by anal feeding, an arrangement that neces-sitates at least a low order of social behavior. Cleveland postulated that termite societies started as feeding communities bound by the necessity of exchanging protozoans and, in a sequence that is the reverse of hymenopteran social evolution, only later evolved social care of the brood. It is not theoretically necessary to the origin of eusociality for sibs to be unusually closely bound by kinship in the hymenopteran manner. Williams and Williams (1957), in an exten-sion of the Wright theory of group selection (1945), demonstrated that eusocial behavior, including the formation of sterile, altruistic castes, can evolve in insects if competition between groups of sibs is intense enough. The point is that the termites have gone this far. The achievement is remarkable, and biologists should continue to reflect on the conditions that made it possible.

The Social Wasps

Although only about 725 species of truly social wasps are known (see Richards, 1971), the study of their behavior has repeatedly yielded results of major interest. Four of the basic discoveries of insect socio-biology—nutritional control of caste (P. Marchal, 1897), the use of behavioral characters in studies of taxonomy and phylogeny (A. Ducke, 1910, 1914), trophallaxis (E. Roubaud, 1916), and domi-nance behavior (G. Heldman, 1936a,b; L. Pardi, 1940)—either origi-nated in wasp studies or were based primarily on them. Even more important, the living species of wasps exhibit in clearest detail the finely divided steps that lead from solitary life to the advanced euso-cial states (Wheeler, 1923; Evans, 1958; Evans and Eberhard, 1970).

Eusocial behavior in wasps is limited almost entirely to the family Vespidae. The only known exception is an apparently primitive euso-cial organization recently discovered in the sphecid *Microstigmus comes* (Matthews, 1968). In order to put these and other social hymenopterans in perspective consider the phylogenetic arrangement given in Figure 20-1 of the seven superfamilies of the aculeate Hymenoptera. The aculeates, as they are familiarly called by ento-mologists, include the insects referred to as "wasps" in the strict sense. Also placed in this phylogenetic category are the ants (Formi-coidea), which are considered to have been derived from the scoloid

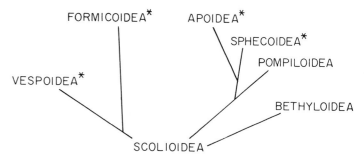

Figure 20-1 Evolution of the aculeate Hymenoptera, which include the "wasps" in the strict sense. An asterisk identifies the superfamilies in which eusocial behavior occurs, having evolved two or more times in certain cases. The Vespoidea and Sphecoidea are superfamilies of wasps; the Formicoidea are the ants and the Apoidea are the bees. (Modified from Evans, 1958.)

wasp family Tiphiidae, and bees (Apoidea), which are considered to have originated from the wasp superfamily Sphecoidea. The Ves-poidea is comprised of three families, the Masaridae, Eumenidae, and Vespidae. These wasps are often called the Diploptera because of the extraordinary ability of the adults to fold their wings longitudinally. The trait does not occur in the stenogastrine vespids or in the great majority of Masaridae, but its absence there may be a derived rather than a primitive characteristic. Vespoids are further distinguished from other wasps by the manner in which the combined median vein and radial sector slant obliquely upward and outward from the basal portion of the fore wing. Most can also be recognized at a glance by the presence of a notch on the inner margin of each eye.

Among the more primitively eusocial vespid wasps are the paper wasps of the genus *Polistes*. Various of the 150 species are found throughout the world with the exception of New Zealand and the polar regions, and in Europe and North America *Polistes* colonies outnumber those of all other social wasps combined. *P. fuscatus*, the familiar brown paper wasp of temperate North America, has been the subject of an excellent study by Mary Jane West Eberhard (1969). This species has an annual life cycle, with each colony lasting only for a single warm season. In the colder parts of the United States, the only individuals to overwinter are the queens. After being insemi-nated by the short-lived males in late summer and fall, they take refuge in protected places such as the spaces between the inner and outer walls of houses and beneath the loose bark of trees. In the spring the ovaries begin to develop several weeks before nest initiation, and during this time queens often aggregate in sunny places. Then, pre-sumably when their ovaries reach an advanced stage of development, the queens begin to sit alone on old nests and future nest sites, where they react aggressively to other females who come close.

Eberhard found that nests in Michigan are usually started by a single female. Of 38 nests observed during May when they contained only from one to ten cells, 37 were attended by a single female. A single nest had two foundresses when it was less than 24 hours old. However, by the time the first brood appears in late June, the majority of the foundresses have been joined by from two to six auxiliaries— overwintered queens who for some reason have not managed to start a nest of their own. These wasps are usually subordinate in status and reproductive capacity to the foundresses. Their subordinacy is expressed behaviorally in overt ways: the auxiliaries assume submissive postures, undertake food-gathering flights and regurgitate to the dominant foundress, and defer to the foundress in egg laying. The foundress not only attempts to prevent her associates from laying eggs; she also eats their eggs when they occasionally sneak them into unoccupied cells. In time the ovaries of the subordinates regress. Marking experiments have revealed that such auxiliaries prefer to associate with foundresses who are sisters. But they move rather readily from nest to nest during the period of colony founding, and a few even attempt to start their own nests while serving as subordinates in established nests.

Through the summer, and on to the onset of the colony's decline and dissolution in early fall, the adult population grows rapidly (Figure 20-2). The complete development from egg to adult takes an average of 48 days, so that roughly three widely overlapping, complete brood sequences can be completed in a season. By the end of summer as many as 200 or more adult individuals may have been reared in a single nest, but their mortality is consistently high, and only a fraction are to be found together at a given time. The first individuals to appear are all workers, that is, females whose wings are generally less than 14 millimeters in length and whose ovaries are undeveloped. Together with the foundress, and possibly the original auxiliaries, they make up the entire adult population until the end of July. They carry on all the work of the colony: foraging for insect prey, nectar, and wood pulp for nest construction, building new cells onto the edge of the nest, and caring for the brood and nonworking adults of the colony. In early August males and "queens" (larger females capable of overwintering) begin to emerge; these purely reproductive forms come to replace the workers entirely by fall. The reproductives are essentially parasites, and as they grow in number they exert an increasingly disruptive influence on the life of the colony. The males are treated aggressively by the workers, and during the peak of male abundance in mid-August the chasing of males is a conspicuous feature of behavior on the nest.

Around the middle of August the *Polistes fuscatus* males begin to leave the nests and to cluster in cracks and on old, abandoned nests. Later, females begin to join these groups. Mating takes place on or close to sunlit structures nearby or within the cavities destined to serve as hibernacula. With the onset of winter the males die off and

the inseminated females hibernate singly to await the coming of spring and the renewal of the colony life cycle.

The *Polistes* life cycle illustrates one of the most important generalizations concerning the sociobiology of wasps. Since the time of von Ihering (1896) it has been noted repeatedly that the nests of tropical species tend to be founded by multiple foundresses while those of temperate species tend to be founded by single females that overwinter in solitude. The extreme development of the first type is seen in some of the tropical Polybiini, in which new colonies are started by swarms of morphologically similar individuals who leave the old nest at about the same time. The extreme development of the second type is shown by the temperate species of Vespinae, in which new colonies are always begun in the spring by a single fecundated individual belonging to the morphologically very distinct queen caste. An extension of the generalization, but not an essential part of it, is that colonies of the tropical swarming species tend to have multiple functional queens but those of temperate species have only one queen.

Primary monogyny, in which single queens start their own colonies, is generally regarded as having evolved from primary polygyny, in which groups of queens cooperate in colony founding during the process of swarming. As Wheeler (1923) pointed out, such a transition is easily visualized: "We might, perhaps, say that our species of *Vespa* and *Polistes* each year produce a swarm of females and workers but that the advent of cold weather destroys the less resistant workers and permits only the dispersed queens to survive and hibernate till the following season." *Polistes* is of special interest because its species display the intermediate steps in this transition. The temperate species *P. fuscatus* is primarily monogynous, to be sure, but the founding queen is usually joined by others within days or even within hours after nest construction begins, so that the initial state is nearly polygynous. An even closer approach to swarming is practiced by *P. canadensis*, a species that ranges from the southern United States to Argentina and, in spite of its name, is tropical in origin (Rau, 1933; Eberhard, 1969). In Central and South America a new nest is started by a female who goes directly to the new nest site from the old nest still occupied by her sisters. Often such pioneers are provoked to leave when they fight over the dominant position, a contest which is more overt and evenly matched than in *P. fuscatus*. Just as in *fuscatus*, however, the *canadensis* foundress is quickly joined by other individuals. After quarreling, one female takes precedence, and the colony becomes functionally monogynous. Since the primitive species of *Polistes* are tropical, it seems clear that the cold temperate species have intercalated a hibernation episode in the colony life cycle without having changed social behavior in any important way.

In order to find a consistent alteration in the social organization of wasps that can be linked to climatic adaptation, it is necessary to turn to the Vespinae. This group of species, called hornets or

Figure 20-2 A colony of the paper wasp *Polistes fuscatus* in Michigan. The nest, which is viewed from below, consists of a single comb of brood cells fashioned from chewed vegetable fibers. Most of the adult wasps seen here are females and workers. Some bear paint marks used by the investigator as an aid in recognizing individuals. New cells are added on the periphery, with the result that the youngest members of the brood are located initially at this position. The heads and thoraces of mature larvae can be seen in the cells at the top of the photograph. Somewhat older pupae are located in the capped cells near the center of the comb. Finally, the center is occupied by uncapped cells from which fully adult worker wasps have emerged. Eggs have already been laid in some of the cells, initiating a new generation. (Photograph by courtesy of Mary Jane West Eberhard.)

yellowjackets in English-speaking countries and the "true wasps" in Germany (or *Hornisse* in the case of *Vespa crabro*), is concentrated in tropical Asia but has penetrated deep into the temperate zones of Eurasia and North America. All vespines are eusocial or else social parasites on their eusocial relatives. They are notable for the advanced state of this sociality relative to most of the Polistinae, even though in temperate species the colony life cycle is only annual in nature. The queen is, on the average, much larger in size than the worker caste and is the principal or sole egg layer (Figure 20-3).

The life cycle of the vespines is basically similar to that of *Polistes*, except that the queen is not joined by auxiliaries during nest founding in the spring. Little variation in details of the cycle occurs among the species. As a rule only the queens hibernate. A few workers have been found still alive in midwinter in warm climates, but it is doubtful that they play any role in nest founding. In the spring the queen selects the nest site, gathers fragments of dead wood and vegetable fibers, and chews them into a pulp to construct the first cells of the nest. One to three thin paper envelopes are added to enclose the first several cells. The queen next lays an egg in each cell and, when the

first brood of larvae hatches, feeds them with insects caught fresh each day and chewed into a pulp. Soon after the first workers eclose, they begin to forage for insects on their own and to add materials to the nest. Now the queen only rarely leaves the nest, and, as the season progresses, she gives up all activities except egg laying. Throughout the summer the workers continue to add new cells to the combs as well as new pillars and combs. The nest as a whole grows outward and downward, assuming an ever larger and more globular shape as the workers tear away old portions and add new material. The wasps capture a wide variety of soft-bodied insects to take back to the nest, favoring bees, flies, and both adult and larval lepidopterans. The giant workers of *Vespa mandarinia* prey extensively on other species of vespine wasps. As few as ten individuals can destroy an entire colony of honeybees within an hour, in the process crushing 5000 or more of the bees with their mandibles.

Toward the end of the summer vespine wasps belonging to temperate species construct larger cells on the brood combs, and in these they rear a crop of several dozen to several hundred queens and males. About this time the mother queen dies, and brood production ceases. The virgin queens and males leave the nest and mate, and, as cold weather approaches, the last few workers in the nest die or wander off. The males, after feeding in solitude on nectar at flowers for a few days or weeks, also perish. But the newly inseminated queens enter hibernacula in the form of spaces under bark of trees and between stacked pieces of cordwood, abandoned beetle burrows in decaying logs, and similar refuges, and prepare to wait out the winter.

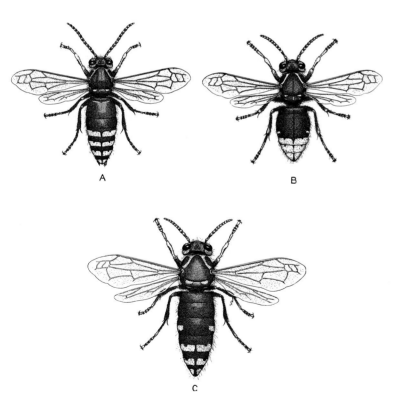

Figure 20-3 The three adult castes of the bald-faced hornet, *Vespula maculata*, a highly social North American vespine wasp: *A*, male; *B*, worker; *C*, queen. (From Betz, 1932.)

The Ants

Ants are in every sense of the word the dominant social insects. They are geographically the most widely distributed of the major eusocial groups, ranging over virtually all the land outside the polar regions. They are also numerically the most abundant. At any given moment there are at least 10^{15} living ants on the earth, if we assume that C. B. Williams (1964) is correct in estimating a total of 10^{18} individual insects—and take 0.1 percent as a conservative estimate of the proportion made up of ants. The ants contain a greater number of known genera and species than all eusocial groups combined.

The reason for the success of these insects is a matter for conjecture. Surely it has something to do with the innovation, as far back as the mid-Cretaceous period 100 million years ago, of a wingless worker caste able to forage deeply into soil and plant crevices. It must also stem partly from the fact that primitive ants began as predators on other arthropods and were not bound, as were the termites, to a cellulose diet and to the restricted nesting sites that place colonies within reach of sources of cellulose. Finally, the success of ants might be explained in part by the ability of all of the primitive species and most of their descendants to nest in the soil and leaf mold, a location that gave them an initial advantage in the exploitation of these most energy-rich terrestrial microhabitats. And perhaps this behavioral adaptation was made possible in turn by the origin of the metapleural gland, the acid secretion of which inhibits growth of microorganisms. It may be significant that the metapleural gland (or its vestige) is the one diagnostic anatomical trait that distinguishes all ants from the remainder of the Hymenoptera.

The "bulldog ants" of the genus *Myrmecia* are important in several respects for the study of sociobiology. They are among the largest ants, workers ranging in various species from 10 to 36 millimeters in length, but are nevertheless easy to culture in the laboratory. They are, next to *Nothomyrmecia* and *Amblyopone*, the most primitive of the living ants. The first encounter with foraging *Myrmecia* workers in the field in Australia is a memorable experience for an entomologist. One gains the strange impression of a wingless wasp just on its way to becoming an ant: "In their incessant restless activity, in their extreme agility and rapidity of motion, in their keen vision and predominant dependence on that sense, in their aggressiveness and proneness to use the powerful sting upon slight provocation, the workers of many species of *Myrmecia* and *Promyrmecia* show more striking superficial resemblance to certain of the Myrmosidae or Mutillidae than they do to higher ants" (Haskins and Haskins, 1950).

Yet so little do ants vary in the broad features of their societal life cycles that *Myrmecia* can be taken as an adequate paradigm for most of this group of insects. The colonies are moderate in size, containing from a few hundred to somewhat over a thousand workers. They capture a wide variety of living insect prey, which they cut up and feed directly to the larvae. The ants are formidable predators, being able to haul down and to paralyze honeybee workers. They also collect nectar from flowers and extrafloral nectaries, which appears to be the main article in their diets when the nest is without larvae. In most species the queens are winged when they emerge from the pupae, whereas the workers are smaller and wingless—the universal condition of ants. Intermediates between the two castes normally occur in some species, and occasionally the usual queen caste has been replaced either by ergatogynes with reduced thoraces and no wings or by mixtures of ergatogynes and short-winged queens. However, these exceptions represent secondary evolutionary derivations and not the primitive states left over from the ancestral wasps. In some of the larger species, such as *M. gulosa*, the worker caste is differentiated into two overlapping subcastes. The larger workers do most or all of the foraging, while the smaller ones devote themselves principally to brood care.

Many species of *Myrmecia* engage in a spectacular mass nuptial flight. The winged queens and males fly from the nests and gather in swarms on hilltops or other prominent landmarks. As the females fly within reach they are mobbed by males, who form solid balls around them in violent attempts to copulate. After being inseminated, the queen sheds her wings, excavates a well-formed cell in the soil beneath a log or stone, and commences rearing the first brood of workers. In 1925 John Clark made the discovery, later confirmed and extended by Wheeler (1933) and Haskins and Haskins (1950), that the queens do not follow the typical "claustral" pattern of colony founding seen in higher ants. That is, they do not remain in the initial cell and nourish the young entirely from their own metabolized fat bodies and alary muscle tissues. Instead, they periodically emerge from the cells through an easily opened exit shaft and forage in the open for insect prey. This "partially claustral" mode of colony foundation, which is now known to be shared with most of the Ponerinae, is regarded as a holdover from a more primitive form of progressive provisioning practiced by the nonsocial tiphiid wasp ancestors. More recently C. P. Haskins (1970) has shown that *Myrmecia* queens also use nutrients metabolized from their own tissues to help raise the brood. The important difference is that they do not depend upon it.

A typical colony of *Myrmecia* is depicted in Figure 20-4. As Haskins has stressed, bulldog ants display a mosaic of primitive and advanced traits in their social biology. In Table 20-4 I have classified many of the recorded traits according to this simple dichotomy. It must be added at once that this effort at a synthesis is no more than a set of phylogenetic hypotheses. The "higher ants" with which *Myrmecia* is compared are all the living subfamilies except the Myrmeciinae and Ponerinae. The last two subfamilies, which are the most primitive living subfamilies of the myrmecioid and poneroid complexes, respectively, share some (but not all) of the primitive traits listed for *Myrmecia*. In sum, the behavior of *Myrmecia* is well advanced into the eusocial level in most essential features, yet marked by a residue of primitive traits which gives us an indistinct and tantalizing view of what the ancestral Mesozoic ants must have been like.

The adaptive radiation of ants from something like the *Myrmecia* prototype has been extraordinarily full. Food specialization in many species is extreme, exemplified by the species of the ponerine genus *Leptogenys* that prey only on isopod crustaceans; by certain *Amblyopone* that feed exclusively on centipedes (Wilson, 1958a; Gotwald and Lévieux, 1972); by species of the ponerine genera *Discothyrea* and *Proceratium* that feed only on arthropod eggs, especially those of spiders (Brown, 1957); by certain members of the myrmicine tribe Dacetini that prey only on springtails (Brown and Wilson, 1959); and by ponerines in the genus *Simopelta* and in the tribe Cerapachyini, all of which, so far as we know, prey exclusively on other ants (Wilson, 1958b; Gotwald and Brown, 1966). The majority of ant groups exhibit a high degree of variability in prey choice, while a

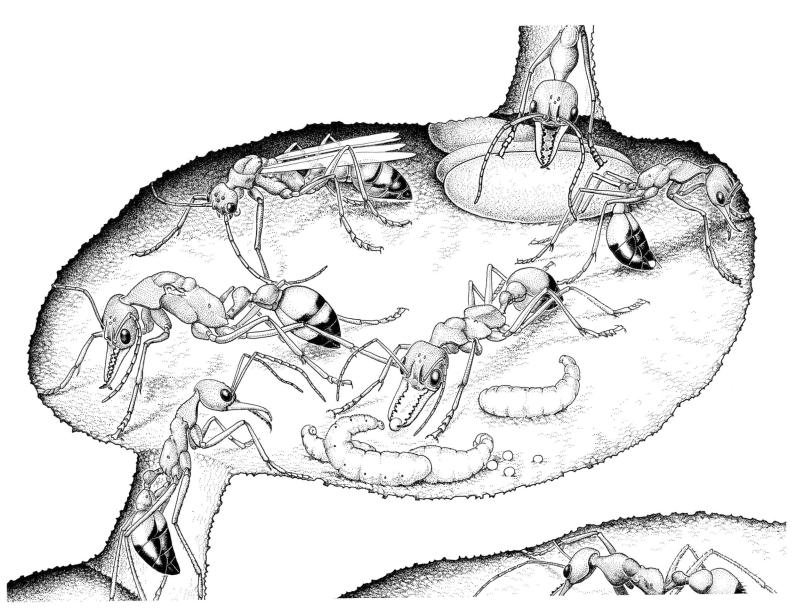

Figure 20-4 A view inside the earthen nest of a colony of primitive Australian bulldog ants (*Myrmecia gulosa*). To the extreme left is the mother queen, distinguished by her larger size, heavier thorax, and three ocelli in the center of her head. Behind her is a winged male, who is her son. The other adults are workers, all daughters of the queen. To the right, one lays a trophic egg while another offers one of its trophic eggs to a wormlike larva. Spherical queen-laid eggs, which will be permitted to hatch into larvae, are scattered singly over the nest floor. To the rear of the chamber are three cocoons containing pupae of the ants. (Drawing by Sarah Landry; from Wilson, 1971a.)

few have come to subsist on seeds. Still others rely primarily or exclusively on the anal "honeydew" excretions of aphids, mealybugs, and other homopterous insects reared in their nests. Unquestionably the most remarkable group of all are the fungus-growing ants of the myrmicine tribe Attini. The 11 genera and 200 attine species are limited entirely to the New World. They are extremely successful in the tropics—in Brazil *Atta* is the most destructive insect pest of agriculture—and a few species range as far north as New Jersey in

Table 20-4 Behavioral and other traits of *Myrmecia*. (Modified from Wilson, 1971a; based on data of C. P. Haskins.)

Primitive traits
1. Multiple queens occur in many nests
2. The eggs are spherical and lie apart from one another on the nest floor
3. The larvae are fed directly with fresh insect fragments
4. The larvae are able to crawl short distances unaided
5. The adults are highly nectarivorous and collect insects mainly as food for the larvae
6. Transport of one adult by another is rare, awkward in execution, and not accompanied by tonic immobility on the part of the transportee
7. There is neither recruitment among workers to food sources nor any other apparent form of cooperation during foraging
8. Alarm communication is slow and inefficient; the nature of the signal is still unknown
9. Colony founding is only partially claustral
10. When deprived of workers, nest queens can revert to colony-founding behavior, including foraging above ground

Advanced traits also found in higher ants
1. The queen and sterile worker castes are very distinct from each other, and intermediates are rare
2. Worker polymorphism occurs in many species, manifested as the coexistence of two well-defined worker subcastes
3. The colonies are moderately large and the nests regular and fairly elaborate in construction
4. Regurgitation occurs both among adults and between adults and larvae
5. Adults groom one another as well as the brood
6. Trophic eggs are laid by the workers and fed to other workers and the queen
7. The workers cover the larvae with soil just prior to pupation, thus aiding them in spinning cocoons; and they assist the newly eclosed adults in emerging from the cocoons
8. Nest odors exist and territorial behavior among colonies is well developed
9. Workers respond to the odor of oleic acid and possibly other decomposition products in dead nestmates by carrying the corpses out of the nest

the United States (Weber, 1972). These ants rear specialized symbiotic yeasts or fungi on organic material that they gather and carry into their nests. The substratum varies according to species: in *Cyphomyrmex rimosus*, for example, it is chiefly or entirely caterpillar feces; in *Myrmicocrypta buenzlii*, dead vegetable matter and insect corpses; and in the famous leaf-cutters ants in *Atta* and *Acromyrmex*, fresh leaves, stems, and flowers. The art of gardening has been highly

developed in these ants, and has even been extended to the "manuring" of the fungi with fecal droplets rich in chitinases and proteases (Martin and Martin, 1971; Martin et al., 1973).

As explained in Chapter 17, social parasitism attains its most advanced development in ants. A finely graded series of stages in the evolution of the phenomenon is displayed by various species up to and including degenerate forms of slavery in which the slave-maker workers are capable only of conducting raids and are totally dependent for minute-to-minute care on their slave workers. Other evolutionary lines lead to total inquilinism, in which the worker caste is lost.

Nesting habits have been no less diversified. A few ant species, such as members of the genus *Atta* and the extreme desert dwellers *Monomorium salomonis* and *Myrmecocystus melliger*, excavate deep galleries and shafts down into the soil, sometimes to depths of 6 meters or more (Jacoby, 1952; Creighton and Crandall, 1954). In contrast, some arboreal members of the subfamily Pseudomyrmecinae and dolichoderine genus *Azteca* are limited to cavities of one or a very few species of plants. Some of the host plants in turn are highly specialized to house and nourish ant colonies. Experiments have shown that these plants are probably unable to survive without their insect guests (Janzen, 1967, 1969, 1972). The tiny myrmicine *Cardiocondyla wroughtoni* sometimes nest in cavities left in dead leaves by leaf-mining caterpillars, while a few formicine species, *Oecophylla longinoda* and *O. smaragdina*, *Camponotus senex*, and certain species of *Polyrhachis*, have evolved the habit of using silk drawn from their own larvae to construct tentlike arboreal nests.

In certain respects the army ants constitute one of the most advanced grades of social evolution within the insects. A colony on the march presents one of the great spectacles of nature. Wheeler expressed it in the following way in *Ants: Their Structure, Development and Behavior* (1910): "The driver and legionary ants are the Huns and Tartars of the insect world. Their vast armies of blind but exquisitely coöperating and highly polymorphic workers, filled with an insatiable carnivorous appetite and a longing for perennial migrations, accompanied by a motley host of weird myrmecophilous campfollowers and concealing the nuptials of their strange, fertile castes, and the rearing of their young, in inaccessible penetralia of the soil— all suggest to the observer who first comes upon these insects in some tropical thicket, the existence of a subtle, relentless and uncanny agency, directing and permeating all their activities."

The years since Wheeler's characterization have seen the mystery largely solved. It was T. C. Schneirla (1933–1971) who, by conducting patient field and laboratory studies over virtually his entire career, first unraveled the complex behavior and life cycles of *Eciton*, *Neivamyrmex*, and other New World species. His results have been confirmed and greatly extended by others, especially Rettenmeyer

(1963b). Meanwhile, the essential features of the life cycle of the African driver ants (*Dorylus*) have been worked out by Raignier and van Boven (1955) and Raignier (1972).

Let us turn to *Eciton burchelli*, a big, conspicuous swarm raider found in humid lowland forests from southern Mexico to Brazil. A day in the life of an *E. burchelli* colony begins at dawn, as the first light suffuses the heavily shaded forest floor. At this moment the colony is in bivouac, meaning that it is temporarily camped in a more or less exposed position. The sites most favored for bivouacs are the spaces beneath the buttresses of forest trees and beneath fallen tree trunks, or any sheltered spot along the trunks and main branches of standing trees to a height of 20 meters or more above the ground.

Most of the shelter for the queen and immature forms is provided by the bodies of the workers themselves. As they gather to form the bivouac, they link their legs and bodies together with their strong tarsal claws, forming chains and nets of their own bodies that accumulate layer upon interlocking layer until finally the entire worker force constitutes a solid cylindrical or ellipsoidal mass up to a meter across. For this reason Schneirla and others have spoken of the ant swarm itself as the "bivouac." Between 150,000 and 700,000 workers are present. Toward the center of the mass are found thousands of immature forms, a single mother queen, and, for a brief interval in the dry season, a thousand or so males and several virgin queens. The dark-brown conglomerate exudes a musky, somewhat fetid odor.

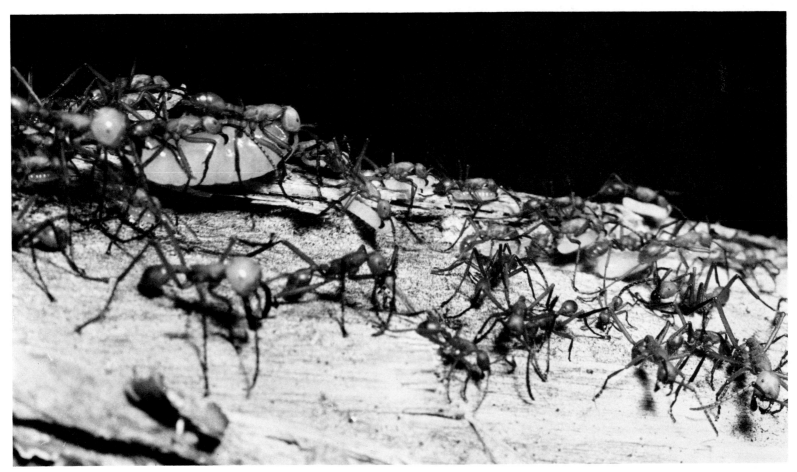

Figure 20-5 When army ants move from one bivouac to another during their nomadic phase, the workers transport the larvae by carrying them slung beneath their bodies. In this photograph of an emigrating *Eciton hamatum* colony, workers are also seen (upper left) carrying the large larva of a polybiine wasp, which had been captured as prey. Two soldiers, easily distinguished by their large, light-colored heads and long mandibles, can be seen on the left. Two media workers carry the wasp larva, while a third faces the observer on the extreme right. (Photograph courtesy of C. W. Rettenmeyer.)

When the light level around the ants exceeds 0.5 foot candle, the bivouac begins to dissolve. The chains and clusters break up and tumble down into a churning mass on the ground. As the pressure builds, the mass flows outward in all directions. Then a raiding column emerges along the path of least resistance and grows away from the bivouac at a rate of up to 20 meters an hour. No leaders take command of the raiding column. Instead, workers finding themselves in the van press forward for a few centimeters and then wheel back into the throng behind them, to be supplanted immediately by others who extend the march a little farther. As the workers run on to new ground, they lay down small quantities of chemical trail substance from the tips of their abdomens, guiding others forward. A loose organization emerges in the columns, based on behavioral differences among the castes. The smaller and medium-sized workers race along the chemical trails and extend them at the points, while the larger, clumsier soldiers, unable to keep a secure footing among their nest-mates, travel for the most part on either side. The location of the *Eciton* soldiers misled early observers into believing that they are the leaders. As Thomas Belt put it, "Here and there one of the light-colored officers moves backwards and forwards directing the columns." Actually the soldiers, with their large heads and exceptionally long, sickle-shaped mandibles, have relatively little control over their nestmates and serve instead almost exclusively as a defense force. The smaller workers, bearing shorter, clamp-shaped mandibles, are the generalists. They capture and transport the prey, choose the bivouac sites, and care for the brood and queen.

At the height of their raids the *Eciton burchelli* workers spread out into a fan-shaped swarm with a broad front. Dendritic columns, splitting up and recombining again like braided ropes, extend from the swarm back to the bivouac site where the queen and immature forms remain sequestered in safety. The moving front of workers flushes a great harvest of prey: tarantulas, scorpions, beetles, roaches, grasshoppers, wasps, ants, and many others. Most are pulled down, stung to death, cut into pieces, and quickly transported to the rear. Even some snakes, lizards, and nestling birds fall victim.

As one might expect, the *burchelli* colonies have a profound effect on the animal life of those particular parts of the forest over which the swarms pass. E. C. Williams (1941), for example, recorded a sharp depletion of the arthropods at those spots on the forest floor where a swarm had struck the previous day. But the total effect on the forest at large may not be very significant. On Barro Colorado Island, which has an area of approximately 16 square kilometers, there exist only about 50 *burchelli* colonies at any one time. Since each colony travels at most 100 to 200 meters in a day (and not at all on about half the days), the collective population of *burchelli* colonies raids only a minute fraction of the island's surface in the course of one day, or even in the course of a single week.

Even so, it is a fact that the food supply is quickly and drastically reduced in the immediate vicinity of each colony. Early writers jumped to the seemingly reasonable conclusion that army ant colonies change their bivouac sites whenever the food supply is exhausted. At an early stage in his work, Schneirla (1933, 1938) discovered that the emigrations are subject to an endogenous, precisely rhythmic control unconnected to the immediate food supply. He proceeded to demonstrate that each *Eciton* colony alternates between a *statary phase*, in which it remains at the bivouac site for as long as two to three weeks, and a *nomadic phase*, in which it moves to a new bivouac site at the close of each day, also for a period of two to three weeks. The basic *Eciton* cycle is summarized in Figure 20-6. Its key feature is the correlation between the *reproductive cycle*, in which broods of workers are reared in periodic batches, and the *behavior cycle*, consisting of the alternation of the statary and nomadic phases. The single most important feature of *Eciton* biology to bear in mind in trying to grasp this rather complex relation is the remarkable degree to which development is synchronized within each successive brood. The ovaries of the queen begin to develop rapidly when the colony enters the statary phase, and within a week her abdomen is greatly swollen by 55,000 to 66,000 eggs. Then, in a burst of prodigious labor lasting for several days in the middle of the statary period, the queen lays from 100,000 to 300,000 eggs. By the end of the third and final week of the statary period, larvae hatch, again all within a few days of one another. A few days later the "callow" workers (so called because they are at first weak and lightly pigmented) emerge from the cocoons. The sudden appearance of tens of thousands of new adult workers has a galvanic effect on their older sisters. The general level of activity increases, the size and intensity of the swarm raids grow, and the colony starts emigrating at the end of each day's marauding. In short, the colony enters the nomadic phase. The nomadic phase itself continues as long as the brood initiated during the previous statary period remains in the larval stage. As soon as the larvae pupate, however, the intensity of the raids diminishes, the emigrations cease, and the colony (by definition) passes into the next statary phase.

The activity cycle of *Eciton* colonies is truly endogenous. It is not linked to any known astronomical rhythm or weather event. It continues at an even tempo month after month, in both wet and dry seasons throughout the entire year. Propelled by the daily emigrations of the nomadic phase, the colony drifts perpetually back and forth over the forest floor. The results of experiments performed by Schneirla indicate that the phases of the activity cycle are determined by the stages of development of the brood and their effect on worker behavior. When he deprived *Eciton* colonies in the early nomadic phase of their callow workers, they lapsed into the relatively lethargic state characteristic of the statary phase, and emigrations ceased.

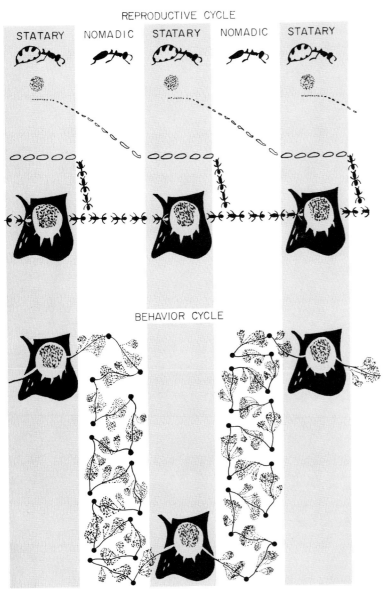

REPRODUCTIVE CYCLE

STATARY   NOMADIC   STATARY   NOMADIC   STATARY

BEHAVIOR CYCLE

Figure 20-6 The monthly colony cycle of the army ant *Eciton burchelli*. The alternation of the statary and nomadic phases consists of distinct but tightly synchronized reproductive and behavior cycles. During the statary phase the queen (top) lays a large batch of eggs in a brief span of time; the eggs hatch into larvae; the pupae derived from the previous batch of eggs develop into adults; and (below) the colony remains in one bivouac site. During the nomadic phase, the larvae complete their development; the new workers emerge from their cocoons; and the ants change their nest sites after the completion of each day's swarm raid. (Redrawn from "The Army Ant" by T. C. Schneirla and G. Piel. © by Scientific American, Inc. All rights reserved.)

Nomadic behavior was not resumed until the larvae present at the start of the experiments had grown much larger and more active. In order to test the role of larvae in the activation of the workers, Schneirla divided colony fragments into two parts of equal size, one part with larvae and the other without. Those workers left with larvae showed much greater continuous activity. The nature of the stimuli inducing the activity, whether chemical, tactile, or whatever, remains to be determined.

In his interpretive writings, Schneirla typically failed to distinguish between proximate and ultimate causation. After demonstrating the endogenous nature of the cycle, and its control by synchronous brood development, he dismissed the role of food depletion. The emigrations, he repeatedly asserted, are caused by the appearance of callow workers and older larvae; they are not caused by food shortage. He overlooked the fuller evolutionary explanation combining the two causations: that the adaptive significance of the emigrations is to take the huge colonies to new food supplies at regular intervals, and that in the course of evolution the emergence of callows has come to be employed as the timing signal. To state this another way, if there is a selective advantage for colonies to move frequently to new feeding sites (and all of the evidence from the *Eciton* studies suggests that this is so), then worker behavior would tend to evolve in such a way as to synchronize the emigrations precisely with the presence of the life stages that cause the greatest food shortage. The internalization of the proximate cause of emigration does not alter the nature of the ultimate cause of emigration, which seems almost certainly to be the chronic depletion of food sources.

In 1958 I traced the probable evolutionary steps leading to army-ant behavior by comparing the behavior of group-raiding ants in the subfamily Dorylinae (the advanced army ants) and in the related subfamily Ponerinae. It had been stated repeatedly by previous entomologists that compact armies of ants are more efficient at flushing and capturing prey than are assemblages of foragers acting independently. This observation is certainly correct, but it is not the whole story. Another, primary function of group raiding becomes clear when the prey preferences of the group-raiding ponerines and dorylines are compared with those of ponerines that forage in solitary fashion. Most nonlegionary ponerine species for which the food habits are known take living prey of approximately the same size as their worker caste or smaller. As a rule they must depend on proportionately small animals that can be captured and retrieved by lone foraging workers. Group-raiding ants, on the other hand, feed on large arthropods or the brood of other social insects, prey not normally accessible to ants foraging solitarily. Thus, the species of *Onychomyrmex* and the *Leptogenys diminuta* group specialize on large arthropods; those of *Eciton* and *Dorylus* prey on a wide variety of arthropods that include social wasps and other ants; species of *Simo-*

pelta and the Cerapachyini specialize on other ants; and *Megaponera foetans* and certain other large African and South American Ponerini prey on termites.

From this generalization and a close comparison of species it has been relatively easy to reconstruct the steps in evolution leading to the full-blown legionary behavior of the Dorylinae.

1. Group raiding was developed to allow specialized feeding on large arthropods and other social insects. Group raiding without frequent changing of nest sites might occur in *Cerapachys* and related genera, but if so, this probably represents a short-lived stage, which soon gives way to the next step.

2. Nomadism was either developed at the same time as group-raiding behavior, or it was added shortly afterward. The reason for this new adaptation was that large arthropods and social insects are more widely dispersed than other types of prey, and the group-predatory colony must constantly shift its foraging area to tap new food sources. With the acquisition of both group-raiding and nomadic behavior, the species is now truly "legionary," that is, an army ant in the functional sense. Most of the group-raiding ponerines have evidently reached this adaptive level. Colony size in these species is on the average larger than in related, nonlegionary species, but it does not approach that attained by *Eciton* and *Dorylus*.

3. As group raiding became more efficient, still larger colony size became possible. This stage has been attained by many of the Dorylinae, including the species of *Aenictus* and *Neivamyrmex* and at least a few members of *Eciton*.

4. The diet was expanded secondarily to include other smaller and nonsocial arthropods and even small vertebrates and vegetable matter; concurrently, the colony size became extremely large. This is the stage reached by the driver ants of Africa and tropical Asia (*Dorylus*), the species of *Labidus*, and *Eciton burchelli*, most or all of which also utilize the technique of swarm raiding as opposed to column raiding.

The Dorylinae, then, constitute either a phyletic group of species or a conglomerate of two or more convergent phyletic groups that have triumphed as legionary ants over all their competitors. They not only outnumber other kinds of legionary ants in both species and colonies, but they tend to exclude them altogether. Cerapachyines, for example, are relatively scarce throughout the continental tropics wherever dorylines abound, but they are much more common in remote places not yet reached by dorylines—for example, Madagascar, Fiji, New Caledonia, and most of Australia.

The Social Bees

All the bees together constitute the superfamily Apoidea. On morphological grounds they fall closest to the sphecoid wasps, although the lack of an adequate fossil record has made it impossible to pin-point the exact ancestral phyletic line. In a word, the Apoidea can be loosely characterized as sphecoid wasps that have specialized on collecting pollen instead of insect prey as larval food. The adults are still wasplike in that they eat nectar (and sometimes store it, in the form of honey), but, unlike the vast majority of true wasps, including all of the sphecoids, they feed their larvae on pollen or pollen-honey mixtures. Some of the eusocial species feed their larvae on specialized glandular products derived ultimately from pollen and nectar.

Eusociality has arisen at least eight times within the Apoidea by both the parasocial and subsocial routes, and presociality of nearly every conceivable degree has emerged on an uncounted number of other occasions. This prevalence and great variability of social behavior in bees provides an opportunity to study the evolution of social behavior paralleled only in the wasps, an opportunity that has only begun to be exploited.

Among the more primitively eusocial forms are the allodapine bees. These insects hold a particular interest for two additional reasons. First, in contrast to the larvae of other kinds of bees, those of allodapines are kept together and fed progressively with small meals (see Figure 20-7). Second, as a concomitant of this peculiar habit, allodapine species display among themselves the evolutionary transition from solitary to eusocial behavior by way of subsocial stages. The essential facts were discovered by H. Brauns (1926) in his work on the South African *Allodape* and have been greatly extended in recent years by field studies conducted in Asia, Australia, and Africa by K. Iwata, C. D. Michener, T. Rayment, S. F. Sakagami, and S. H. Skaife.

Allodape angulata, a South African species, is a good example of a eusocial allodapine (Skaife, 1953). The colonies nest in dead flower stalks and a variety of other kinds of plants whose stems have pithy centers. The colony life cycle begins when the adults of the new generation emerge in the middle of summer, a period extending from the end of December to early February. They remain together in a largely quiescent state through the remainder of the summer and the following fall, then disperse to new nest sites. Breeding takes place shortly afterward, in July and August. Now the solitary, mated females begin new colonies. In the typical sequence the female digs a short cavity in the pith of a stem and lays a large, white, and slightly curved egg at the bottom. During the four to six weeks required for the eggs to hatch, the mother remains on guard at the nest entrance, and she extrudes the hind portion of her abdomen outward whenever she is disturbed. As the young develop, she arranges them in order of size, with the pupae nearest the entrance, followed by the larger larvae and so on down to the eggs, which are always grouped at the bottom of the tube, much as shown in Figure 20-7. Newly hatched larvae are fed with a colorless liquid regurgitated by the mother. The older ones are given little balls of a paste made of pollen and nectar.

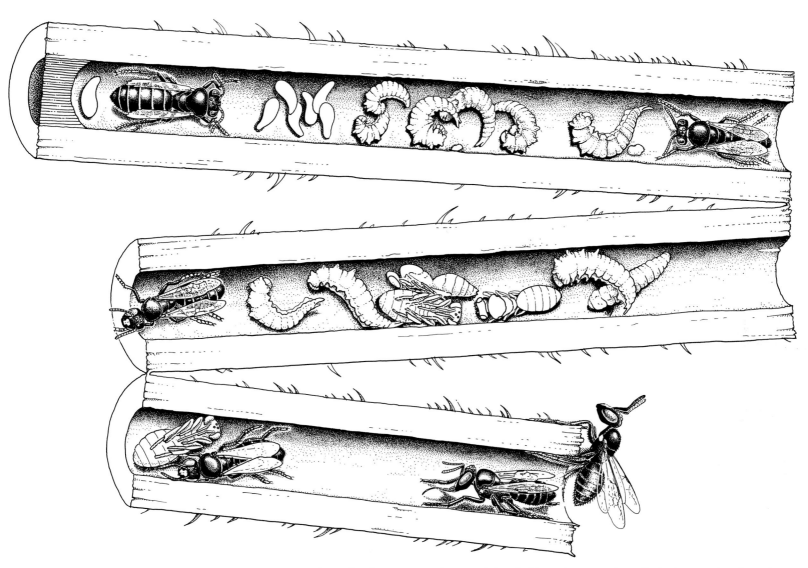

Figure 20-7 *Braunsapis sauteriella*, a primitively eusocial bee that has evidently reached this state by the subsocial route. This populous colony occupies a hollow stem of *Lantana camara* in Formosa. In typically allodapine fashion, the brood is freely arranged in a common chamber (rather than being spaced in individual brood cells), and the larvae have been fed progressively with small meals. The eggs, whose huge size is a characteristic of this and some other allodapine species, have been placed in a cluster at the bottom of the nest, while the mother queen rests nearby. Pollen is stored in small deposits on the nest wall. The larvae are fed at frequent intervals with little pollen balls. (Drawing by Sarah Landry; from Wilson, 1971a.)

After seven or eight weeks, with the coming of early summer in November, the first larvae pupate. By January all of the first brood have emerged as adults. Just about this time the mother *Allodape*, now a year old, may lay three or four more eggs. Then, after a few more days or weeks have elapsed, she dies. The members of the second brood, attended by their sisters, emerge as adults in late summer or early autumn. During this final episode the males of the first brood occasionally leave the nest to get food for themselves, but they never take part in rearing the later brood.

The bumblebees represent a somewhat farther advance into eusociality. Comprising about 200 species of the genus *Bombus*, they are notable as social insects primarily adapted to colder climates. Most are restricted to the temperate zones of North America and Eurasia, and several are found near the Arctic Circle and the treeless summits of high mountains. Two occur as far north as Ellesmere Island—and one of these is a social parasite of the other! A few species reach in the other direction as far as Tierra del Fuego and the mountains of Java, and a single species is even common in the Amazon rain forests.

In the North Temperate Zone the life cycle of *Bombus* is annual. Only the fertilized bumblebee queens hibernate. The history of a colony unfolds in the following way. In early spring the solitary queen leaves her hibernaculum and searches on wing until she finds an abandoned nest of a field mouse or some other similarly shaped cavity, in an open but relatively undisturbed habitat such as a fallow field or abandoned garden. She pushes her way into the nest and then modifies it for her own use by constructing an entrance tunnel and lining the inner cavity with fine material teased out of the nest walls. While in the nest the queen begins to secrete wax in the form of thin plates from intersegmental glands on the abdomen. From this material she fashions the first egg shell in the form of a shallow cup set onto the floor of the nest cavity. Next she places a pollen ball into the egg cell and lays 8–14 eggs onto the surface of the ball. Finally, she constructs a dome-shaped roof of wax and other materials over the cell, so that the entire brood cell is sealed and spherical in shape. About the time the first eggs are laid, the queen also constructs a wax honeypot just inside the entrance of the nest cavity and begins to fill it with some of the nectar gathered in the field. When the first workers emerge, they assist the queen in expanding the nest and caring for additional brood, as depicted in Figure 20-8.

Depending on the species of *Bombus* involved, the larvae are fed by one or the other of two very different techniques. In one group of species, the "pollen storers," the pollen is placed in abandoned cocoons, which may be extended with wax layers until they form cylinders as high as 6 or 7 centimeters. From time to time pollen is removed from this modified cocoon and fed into the brood cell in the form of a viscous liquid mixture of pollen and honey. The queen and workers of the pollen-storer species do not feed the larvae directly. Instead, they make a small breach in the larval cell and regurgitate the pollen-honey mixture next to the larvae. In the second group of species, the "pouch makers" or "pollen makers," the queens and workers build special wax pouches adjacent to groups of larvae and fill them with pollen. The larvae then feed as a group directly from the pollen mass. Occasionally, the pouch makers also feed larvae by regurgitation, and groups of larvae destined to become queens are fed exclusively in this manner.

By the end of summer the colony contains, again according to the species, from around 100 to 400 workers. As fall approaches the annual colonies produce males and queens and begin to break up. The demise of the bumblebee colonies seems to be controlled by

endogenous factors. In the mild climate of northern New Zealand, species of *Bombus* introduced from Europe fly at all times of the year, and solitary queens can start nests during at least nine months of the year. Colonies sometimes overwinter and attain unusual size. In spite of this opportunity for perennial growth, however, the New Zealand colonies never return to the production of workers after they have reared queens.

Mating behavior varies greatly among the species of *Bombus*. In some, the males hover around the nest entrances and wait for the young queens to emerge. In others, the male selects a prominent object, such as a flower or fence post, and alternately stands on it and hovers over it, ready to dart at any passing object that resembles a queen in flight. In a third group of species, the males establish flight paths that they mark at intervals by dabbing spots of scent from the mandibular gland onto objects along the route. The males fly around the paths hour after hour, day after day, waiting for the approach of the females. After mating, the queens hibernate in specially excavated chambers in the soil, and the following spring they initiate new colonies.

Queen bumblebees differ from workers only in their larger size and the greater extent of their ovarian development, and intermediates between the two castes are common. There is also great variation in size within the worker caste. The larger workers tend to forage more, and the smaller workers spend more time in nest work. In a few species, the smallest workers do not fly and are thus bound to the nest permanently. Nest guarding occurs in some species and is usually undertaken by workers who possess better-developed ovaries.

Within the Apidae, whose species constitute the *haut monde* of the social bees, *Bombus* occupies a relatively lowly position. Its solutions to the problems of social organization are as a rule crude, and it has not achieved many of the more spectacular control mechanisms that distinguish honeybees and the meliponine stingless bees from the primitively eusocial sweat bees of the family Halictidae. In Table 20-5 I have indicated the characteristics which, in my opinion, are more primitive, or at least simpler, in the context of the biology of the Apidae as a whole.

The common honeybee *Apis mellifera* can be taken as representative of the most advanced social bees. By the general intuitive criteria of social complexity—colony size, magnitude of queen-worker difference, altruistic behavior among colony members, periodicity of male production, complexity of chemical communication, regulation of the nest temperature and other evidences of homeostatic behavior— the honeybee is at about the level of the other highest eusocial insects, that is to say, the stingless bees, the ants, the higher polybiine and vespine wasps, and the higher termites. In one feature, the waggle dance, the species comes close to standing truly apart from all other insects. The really remarkable aspect of the waggle dance is that it is a ritualized reenactment of the outward flight to food or new nest

Figure 20-8 A colony of the European bumblebee *Bombus lapidarius.* The nest has been fashioned out of the center of an abandoned mouse nest in an old cultivated field. The large queen sits atop a cluster of cocoons inside which are worker pupae (one pupa has been exposed to show its position). At the upper and lower left are three communal larval cells: the waxen envelopes of the bottom two have been torn open to reveal the larvae inside. Large waxen honeypots occupy the left and center of the ensemble. At the lower right are clusters of abandoned cocoons, which are now used to store pollen. (Drawing by Sarah Landry; from Wilson, 1971a.)

sites; it is performed within the nest and somehow understood by other workers in the colony, who are then, and this must be counted the remarkable part, able to translate it back into an actual, unrehearsed flight of their own. A similar ability to interpret modulated symbols is evidently shared by certain meliponine bees, who transmit sound signals correlated in duration and frequency with the distance of food finds. But other cases of symbolical communication have yet to be demonstrated in the social insects.

At the risk of oversimplification, it can be said that the key to understanding the biology of the honeybee lies in its ultimately tropical origin. It seems very likely that the species originated somewhere in the African tropics or subtropics and penetrated colder climates

Table 20-5 Primitive (or at least relatively simple) social traits in *Bombus*, compared with the more advanced traits found in the highest social bees, the honeybees of the genus *Apis* and the stingless bees of the tribe Meliponini. (From Wilson, 1971a.)

Bombus	*Apis* and Meliponini
Queens and workers differ morphologically to a slight degree, and intermediates are common.	Queens and workers are morphologically very different from each other, and intercastes are normally absent.
The life cycle is annual, at least among the majority of species; new colonies are founded by single queens; and the mature colony size is small.	The life cycle is perennial; new colonies are started by swarming, and colony size is moderate to very large.
The queen maintains reproductive dominance by aggressive behavior, and the workers tend to behave toward one another in the same way. Workers occasionally steal eggs from one another and the queen.	The queen maintains reproductive dominance by pheromones, at least in *Apis*, and aggressive behavior is muted or absent. Egg stealing is unknown except as a ritual form of eating by the meliponine queens.
The larvae are often reared in groups and must compete with other larvae for food placed indiscriminately in their vicinity.	The larvae are reared in separate cells on the brood comb, which greatly increases the chances for individual attention on the part of the nurses and control of caste determination.
The larvae are fed with raw pollen and regurgitated mixtures of pollen and honey.	In *Apis*, larvae are fed at least in part with special food manufactured by the mandibular and pharyngeal glands.
The adults rarely regurgitate food directly to other adults or try to groom them.	Both grooming and direct transfer of food by regurgitation are very frequent and, in the case of *Apis* at least, known to play an important role in communication and regulation.
The queen regulates colony growth by building all of the egg cells herself and laying in them, following the same behavior patterns by which she initiates the colony.	The queen plays no direct role in colony growth or in the construction of the brood combs. The workers determine these matters and are subject to much more feedback from the environment outside the nest.
Temporal division of labor is weakly developed.	A temporal division of labor is strongly developed, in which the young adult worker first engages in brood care (or nest work), then nest work (or brood care), and finally in foraging. In *Apis*, at least, this progression is associated with orderly changes in the exocrine glands.
Chemical alarm communication is lacking.	Chemical alarm communication is well developed and involves pheromones apparently especially evolved for the purpose.
Recruitment among workers is lacking.	Recruitment is well developed and mediated by special assembling or trail pheromones; in *Apis* there is also a symbolic waggle dance.

prior to the time it came under human cultivation. Thus, unlike the vast majority of social bees endemic to the cold temperate zones, the honeybee is perennial, and, being perennial, it is able to grow and to sustain large colonies. Having large colonies, it must forage widely and exploit efficiently the flowers within the flight range of its nests;

the waggle dance and the release of scent from the Nasanov gland of the abdomen are clearly adaptations to this end. Also, being ultimately tropical in origin, its colonies multiply by swarming; there is no need to have a hibernation episode in the colony life cycle as in the temperate paper wasps and bumblebees. And finally, since the queen is relieved of the necessity to overwinter and initiate colonies in solitude, she has regressed in evolution toward the role of a simple egg-laying machine, with the result that the queen and worker castes differ drastically from each other in both morphology and physiology. Within the scope of these interlocking effects are to be found just about all of the phenomena that distinguish *Apis mellifera* from the exclusively cold temperate bee species (see Figure 20-9). When we turn to the tropical faunas and consider what else has evolved to eusocial levels within the Apoidea, the contrasts are not nearly so sharp. The prevailing group of tropical eusocial bees, the Meliponini, not only resemble *Apis* in their life cycle, but are comparable to it in complexity of social organization. Of course, a great many, perhaps most, of the primitively eusocial bees exist in the tropics, but this does not affect the important generalization that the most advanced bee societies are tropical in origin.

The Termites

Termites are almost literally social cockroaches. Detailed similarities exist in anatomy between the most primitive termite family, the Mastotermitidae, and the relatively primitive wood-eating cockroaches constituting the family Cryptocercidae. Even the intestinal microorganisms that digest cellulose are similar. Of the 25 species of hypermastigote and polymastigote flagellate protozoans found in the gut of the cockroach *Cryptocercus punctulatus*, all belong to families also found in the more primitive termites. Even one genus, *Trichonympha*, is shared. These intestinal protozoans can be successfully "transfaunated" from cockroach to termite and vice versa. It is of course too much to hope that any of the living cockroaches are really the ancestors of the termites. All known cockroaches have horny fore wings; the clear, membranous wings of the termites are more primitive. Other differences indicate that the two groups of insects originated from a common, cockroachlike ancestor. But they are not cardinal distinctions, and some entomologists have gone so far as to place termites in the same order (Dictyoptera) as the cockroaches and mantids.

Because the termites have climbed the heights of eusociality from a base extremely remote in evolution from the Hymenoptera, it is of great interest to know whether their social organization differs from hymenopteran organization in any fundamental way. Although value judgments of the degree of convergence of two radically differing stocks are difficult to make, much less to justify quantitatively, I believe the following assessment can reasonably be made. The termites have adopted mechanisms that are mostly but not entirely similar to those in the ants and other social Hymenoptera. Also, the level of complexity of termite societies is approximately the same as that in the more advanced hymenopteran societies. In Table 20-6, I

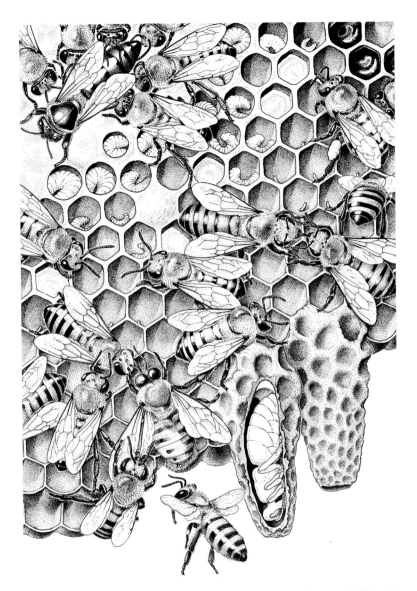

Figure 20-9 A portion of a colony of honeybees. In the upper lefthand corner the mother queen is surrounded by a typical retinue of attendants. She rests on a group of capped cells, each of which encloses a developing worker pupa. Many of the open cells contain eggs and larvae in various stages of development, while others are partly filled by pollen masses or honey (extreme upper right). Near the center a worker extrudes its tongue to sip regurgitated nectar and pollen from a sister. At the lower left another worker begins to drag a drone away by its wings; the drone will soon be killed or driven from the nest. At the lower margin of the comb are two royal cells, one of which has been cut open to reveal the queen pupa inside. (Drawing by Sarah Landry; from Wilson, 1971a.)

Table 20-6 Basic similarities and differences in social biology between termites and higher social Hymenoptera (wasps, ants, bees). Similarities are due to evolutionary convergence. (From Wilson, 1971a.)

Similarities	Differences Termites	Eusocial Hymenoptera
1. The castes are similar in number and kind, especially between termites and ants.	1. Caste determination in the lower termites is based primarily on pheromones; in some of the higher termites it involves sex, but the other factors remain unidentified.	1. Caste determination is based primarily on nutrition, although pheromones play a role in some cases.
2. Trophallaxis occurs and is an important mechanism in social regulation.	2. The worker castes consist of both females and males.	2. The worker castes consist of females only.
3. Chemical trails are used in recruitment as in the ants, and the behavior of trail laying and following is closely similar.	3. Larvae and nymphs contribute to colony labor, at least in later instars.	3. The immature stages (larvae and pupae) are helpless and almost never contribute to colony labor.
4. Inhibitory caste pheromones exist, similar in action to those found in honeybees and ants.	4. There are no dominance hierarchies among individuals in the same colonies.	4. Dominance hierarchies are commonplace, but not universal.
5. Grooming between individuals occurs frequently and functions at least partially in the transmission of pheromones.	5. Social parasitism between species is almost wholly absent.	5. Social parasitism between species is common and widespread.
6. Nest odor and territoriality are of general occurrence.	6. Exchange of liquid anal food occurs universally in the lower termites, and trophic eggs are unknown.	6. Anal trophallaxis is rare, but trophic eggs are exchanged in many species of bees and ants.
7. Nest structure is of comparable complexity and, in a few members of the Termitidae (e.g., *Apicotermes*, *Macrotermes*), of considerably greater complexity. Regulation of temperature and humidity within the nest operates at about the same level of precision.	7. The primary reproductive male (the "king") stays with the queen after the nuptial flight, helps her construct the first nest, and fertilizes her intermittently as the colony develops; fertilization does not occur during the nuptial flight.	7. The male fertilizes the queen during the nuptial flight and dies soon afterward without helping the queen in nest construction.
8. Cannibalism is widespread in both groups (but not universal, at least not in the Hymenoptera).		

have listed the principal known similarities and dissimilarities of the two kinds of societies. This simplified accounting does not overlook the fact, which was stressed earlier, that a great deal of important variation also occurs within the social Hymenoptera. Surely the similarities are remarkable in themselves. They seem to tell us that there are constraints in the machinery of the insect brain that limit not only the options of social organization but also the upper limit that the degree of organization can attain. These limits appear to have been reached between 50 and 100 million years ago in both the termites and the social Hymenoptera.

The most primitive living termite and sole surviving member of the Mastotermitidae, *Mastotermes darwiniensis*, is found over most of the northern half of Australia. In some ways it acts very strangely for a Mesozoic relic. It is the most destructive termite species in Australia and the most destructive insect of any kind in the northern part of the continent. The colonies, which nest in the soil, are immense, the largest containing over a million individuals. The diet of

M. darwiniensis is the most catholic of any known termite; one might even say it resembles that of the cockroach. Workers have been observed attacking poles, fences, wooden buildings, living trees, crop plants, wool, horn, ivory, vegetables, hay, leather, rubber, sugar, human and animal excrement, and the plastic lining of electric cables. Unattended homesteads in the outback have been reduced to dust in only two or three years—house, fences, and all. Colonies of *M. darwiniensis* occupy many kinds of nest sites through a wide range of habitats, and they are able to excavate rapidly in both soil and wood. Their subterranean nests, which are often fragmented and connected by covered passageways constructed on the surface of the ground, are difficult to detect. The galleries run outward for as much as 100 meters or more from the nest. Most are shallow, extending no more than 40 centimeters below the surface. One gallery system, however, was uncovered by quarrying operations at a depth of 4 meters.

Considering its phylogenetic position and economic importance, surprisingly little is known concerning the biology of *Mastotermes*, including the most basic facts of the life cycle. One curious fact is that the primary reproductives are rare. Multiple supplementary reproductives appear to be the rule, and colony multiplication often occurs by budding. When groups of nymphs are detached from the main colony, some are able to develop into reproductive castes. Eggs are laid in packets of about 20 each, in a form resembling the oothecae of cockroaches. Nuptial flights occur regularly, but their relative contribution to the formation of new colonies is unknown.

The Kalotermitidae, known as the dry wood termites, are anatomically relatively primitive although still considerably advanced over the Mastotermitidae. Their sociobiology is a mosaic of elementary and advanced traits. The colonies, which rarely contain more than a few hundred individuals, live in ill-defined galleries inside the wood on which they feed. The termites rely on an intestinal flagellate fauna to digest the wood and do not utilize symbiotic fungi or store food. When the primary queens and males are lost, they are quickly replaced by secondary "neoteinics" that transform in one molt from a labile, workerlike caste called pseudergates. When present, the primary reproductives prevent the transformation of pseudergates by means of inhibitory pheromones passed out of their anuses. Soldier inhibition also occurs, but the physiological basis is not yet known. The exchange of oral and anal liquids, as well as integumentary exudates, occurs very frequently among all members of the colony. Anal exchange is essential to the transmission of flagellates to young nymphs and newly molted individuals of all ages.

It is a curious fact that most kalotermitids, as well as most other relatively primitive termite groups, are concentrated in the temperate zones. The tropics, constituting the true headquarters of the world fauna, are dominated by the "higher" termites of the family Termitidae. The majority of the termitids are soil dwellers and are responsible for most of the elaborately structured mounds that are such a conspicuous feature of the tropical landscape. Various of their species have specialized on virtually every conceivable cellulose source. To reach this food, workers extend galleries through the soil, or construct covered trailways over the surface of the ground, or even march in columns along exposed odor trails.

As an example of a relatively unspecialized termitid, we can take *Amitermes hastatus*, which has been studied in detail by Skaife (1954a,b; 1955). The species occurs in South Africa, in the mountains of the southwest Cape at elevations from about 100 to 1,000 meters above sea level. It nests in the sandy soil of the natural veld, throwing up conspicuous hemispherical or conical mounds constructed of a black mixture of soil and excrement. In the late summer months of February and March large numbers of white nymphs with wing pads are to be found in the larger nests. By the end of March, or April at the latest, these individuals have transformed into winged repro-

ductives. For several weeks the alates wander slowly through the nest. Then, soon after the onset of the autumn rains, the nuptial flight occurs. One day between 11 o'clock in the morning and 4 o'clock in the afternoon, immediately after a ground-soaking rain and with the temperature rising, the exodus begins. The workers first excavate large numbers of tightly grouped exit holes, each about 2 millimeters in diameter, giving the apex of the mound the appearance of a coarse sieve. True to the pattern of most termite species, this is the only time the workers breach the walls of their nest and expose themselves to the outside air. Workers, soldiers, and alates boil out of the holes in a state of intense excitement, the alates fly off almost immediately, and within three or four minutes the termites retreat back down into the nest, plugging the exit holes after them. Most, but not all, of the alates leave in this first flight. A few remain behind to participate in later departures. The alates are feeble flyers; many do not travel more than 50 or 60 meters from the nest before alighting. As soon as they land they break off their wings at the basal fracture line by swiftly pressing the wing tips to the ground. The subsequent pairing and nest-founding behavior follows the same basic sequence as in *Kalotermes*. The construction of the initial nest chamber is conducted principally by the queen; sometimes the king does not assist at all. The pair remain in the incipient nest through the winter and apparently do not copulate until the arrival of warmer weather. In

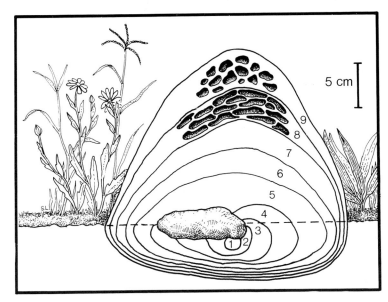

Figure 20-10 The growth of a typical mound of the South African termitid *Amitermes hastatus*, over a period of nine years. Each successive year's growth is indicated by a number. Representative outer and inner cells are shown at the top of the mound. There is no royal cell. (Based on Skaife, 1954a.)

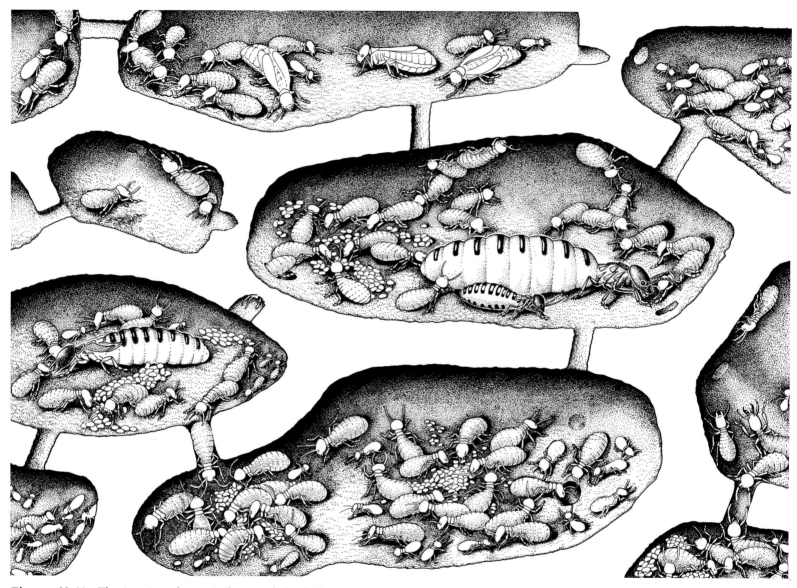

Figure 20-11 The interior of a typical nest of the higher termite *Amitermes hastatus* of South Africa. The primary queen and much smaller primary male sit side by side in the middle cell. To the lower left can be seen a secondary queen, who is also functional in this case. In the chamber at the top are reproductive nymphs, characterized by their partially developed wings. Workers attend the queens and are especially attracted by their heads, to which they offer regurgitated food at frequent intervals. Other workers care for the numerous eggs. A soldier and presoldier (nymphal soldier stage) are seen in the lower right chamber, while worker larvae in various stages of development are found scattered through most of the chambers. (Drawing by Sarah Landry; from Wilson, 1971a.)

the spring months of October and November the queen lays the first five or six eggs. The individuals of the first brood develop into stunted workers. Soldiers make their appearance in later broods, and finally after four years alate reproductives are produced. The growth of a typical nest is displayed in Figure 20-10. Skaife has estimated the age of some mounds of *Amitermes hastatus* to be greater than 15 years, but judging strictly from the size of the mounds, he did not believe any to be more than 25 years old. This mortal state of individual colonies, if true, is an unexpected feature, because presumably the colonies are capable of producing secondary reproductives when the queen dies. When the primary queen does fail, the workers put her to death, apparently by licking her abrasively. As Skaife describes it, "She is surrounded by a crowd of workers, all with their mouthparts applied to her skin, and this goes on for three or four days, her body slowly shrinking until no more than the shrivelled skin is left."

Secondary and tertiary queens do appear in the presence of the queen—at least sometimes (see Figure 20-11). Skaife, however, was unable to rear them in queenless colonies kept in artificial nests, and he found that only about 20 percent of the natural mounds contained them. Clearly, then, either the supplementary reproductives are rare, or appear only under special conditions, or the colonies that possess them are relatively short-lived.

Chapter 21 **The Cold-Blooded Vertebrates**

The fishes, amphibians, and reptiles are sophisticated in some of the elements of social organization but not in the ways the elements are assembled. In territoriality, courtship, and parental care, these cold-blooded vertebrates are the equal of mammals and birds, and various of their species have served as key paradigms in field and laboratory investigations. But for some reason, possibly lack of intelligence, they have not evolved cooperative nursery groups of the kind that constitute the building blocks of mammalian societies. For other reasons, possibly the lack of haplodiploid sex determination or the presence of the right ecological imperatives, they have not become altruistic enough to generate insectlike societies. Even so, the cold-blooded vertebrates offer special attractions in the study of sociobiology. As the remainder of this chapter will show, schooling in fishes has unique features which are just now beginning to be appreciated. In a sense schooling is sociality in a new physical medium, making three-dimensional geometry important for the first time in social organization (all other societies consist of individuals arrayed on a plane). The amphibians are no less interesting but for wholly different reasons. Recent research has shown that frogs possess well-developed, highly diversified social systems parallel with those possessed by birds. Since they are phylogenetically far removed from the birds, and the traits under consideration are labile at the level of the genus and species, frogs provide us with an independent evolutionary experiment just beginning to be examined. Much the same is true of the reptiles, particularly the territorial species of lizards.

Fish Schools

In 1927 Albert E. Parr published an article that opened the subject of schooling to objective biological research. Rejecting vague earlier notions of a "social instinct," he postulated that fish schools result from the balance struck between the programmed mutual attraction and repulsion of individual fish based on the visual perception of one another. Species differ in the degree to which they are committed to schooling and in the form of the groupings. Parr identified schooling by implication as an adaptive biological phenomenon, to be analyzed like any other at both the physiological and evolutionary levels. The past 50 years have seen the accumulation of a very large amount of information on the behavioral basis of schooling and its ecological significance that confirms the validity of Parr's approach. The best recent reviews are those of Shaw (1970), who covers the large English and German literature well, and Radakov (1973), who deals with the equally large Russian literature. The Soviet studies, hitherto mostly unknown to Western zoologists, have been well financed because of their potential application to the fisheries industry. They are notable in the attention they pay to the ecological significance of the schools, in line with the more modern aspects of sociobiological research being conducted in other countries.

A fish school, to cite Radakov, is "a temporary group of individuals, usually of the same species, all or most of which are in the same phase of the life cycle, actively maintain mutual contact, and manifest, or may manifest at any moment, organized actions which are as a rule biologically useful for all the members of the group." One can quarrel with this characterization, adding, deleting, or modifying the separate qualifications; but intuitive semantic argumentation has already clouded the "theory" of this subject for too long. Radakov's definition, which is close to the consensus, is more than adequate for a description of the current substantive issues.

At a distance a fish school resembles a large organism. Its members, numbering anywhere from two or three into the millions, swim in tight formations, wheeling and reversing in near unison. Either dominance systems do not exist or they are so weak as to have little or no influence on the dynamics of the school as a whole. There is, moreover, no consistent leadership. When the school turns to the right or left, individuals formerly on the flank assume the lead (see Figure 21-1). The average school size varies according to species, as does the spacing of its members, its average velocity, and its three-dimensional shape (Breder, 1959; Pitcher, 1973). Although the fish are usually aligned with military precision while the group is on the move, they assume a more nearly random orientation while resting or feeding. Their alignments also shift in particular ways when the fish are attacked by predators (see Figure 21-2). Spacing within the moving school is evidently determined to a large extent by hydrodynamic force. Individual fish tend to seek positions in which they can be as close as possible to their neighbors without suffering serious

loss of efficiency due to turbulence created by the other fish (Rosen, 1959; Breder, 1965; Shuleikin, 1968). Each individual generates a trail of dying vortices behind it. In most schools the side-to-side spacing is slightly more than twice the distance from the flank of one fish to the outer edge of the trail of vortices close to the zone of their production. It is even possible for fish to coast along the edges of vortices for short distances, utilizing the energy expended by the schoolmates in front of them. But energy expenditure is not the sole consideration. Schools sometimes condense into what Breder has called "pods," in which the bodies of the members actually touch. Under some circumstances such formations can help protect individual fish from predators. Young catfish in the genus *Plotosus*, for example, mass together in a solid ball when disturbed, their sharp pectoral fins projecting out in all directions like thorns on a cactus. In general, fish tend to form the most compact schools when well fed and to thin out and become less aligned when hungry. The shift can be interpreted as the surrender of some of the advantage of predator avoidance in exchange for an increased probability of finding food.

Extensive experimentation by Shaw and others has shown that the orientation of individual fish to their school is primarily visual. Minnows, in particular *Menidia menidia* and *Atherina mochon*, display the appropriate optomotor reactions in the first few days of life and achieve parallel alignment soon afterward. *Menidia* reared in isolation still form schools but far less smoothly than those raised in groups. Jack mackerel (*Trachurus symmetricus*) adjust their velocity to match that of their schoolmates, paying closest attention to

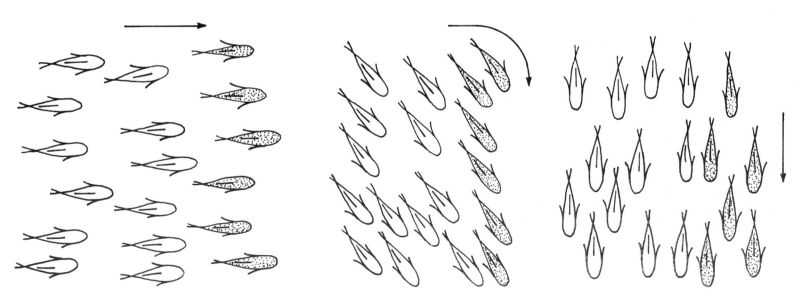

Figure 21-1 A school of fish changes its leadership when it changes direction. The leaders at the left (stippled) are shifted to the flank when the school makes a 90° turn, as shown in sequence in the center and at the right. (Modified from Shaw, 1962.)

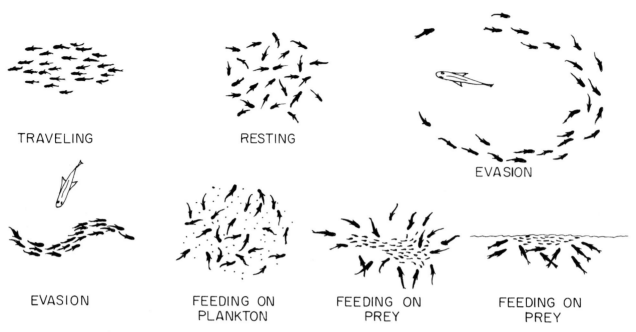

TRAVELING

RESTING

EVASION

EVASION

FEEDING ON
PLANKTON

FEEDING ON
PREY

FEEDING ON
PREY

Figure 21-2 In open water members of fish schools change their alignments according to conditions in the environment. In general, organiza-tion declines and behavior is individualized when the fish rest or feed. (Modified from Radakov, 1973.)

those directly off their flank (Hunter, 1969). The orientation is also rheotactic in part: fish tend to swim upstream, and they skirt around the edge of vortices. Occasionally schools show some degree of geometric structuring, with fish at different positions in the school differing somewhat in their behavior. Each fall striped mullet (*Mugil cephalus*) migrate from the bays of the Gulf Coast and eastern seaboard of the United States into the open sea in order to reproduce. Their dense schools constantly change shape, shifting with fluid ease into circles, discs, ellipses, triangles, crescents, and lines. Individual fish also constantly change their positions. The more densely packed ranks at the rear churn the water with random movements and break off frequently into smaller, diverging subgroups that may or may not rejoin the school. McFarland and Moss (1967) found that the concentration of environmental oxygen drops significantly from the front to the rear of large schools (see Figure 21-3). They concluded that this factor alone could account for the churning movements, for the break-up to the rear, and for much of the continuing change in the overall shape of the mullet school. In contrast, the ambient pH did not appear to vary enough to play a role.

The hydrodynamic constraints also dictate that the members of each school be about the same size. The range in size in fact seldom exceeds 1:0.6, where unity is the size of the largest school members (Breder, 1965). If small fish attempted to swim with large ones, they would find it difficult to hold the same velocity. They would also be unable to maintain the correct interindividual distances required to avoid the slowing effects of turbulence. If fish swam with school-mates of varying size, they would constantly have to readjust the spacing to fit the individuals closest to them at each moment, a maneuvering probably too complex to achieve.

Many obligatorily schooling fish engage in forms of communication with functions other than the coordination of movement. The characid *Pristella riddlei*, a South American freshwater species, possesses a conspicuous black patch on its dorsal fin, which is jerked rapidly up and down when the fish is alarmed (Keenleyside, 1955). The striking black-and-white banding on the body of *Dascyllus aruanus*, a Pacific coral fish, serves to attract school members together (Franzis-ket, 1960). A few species, especially those that school at night, use sounds as an apparent contact signal (Winn, 1964). An alarm substance (*Schreckstoff*) is present in the skin of cyprinid minnows, catfish, and other ostariophysan fishes. When a member of the school is injured, release of the material in the water causes the others to scatter (Pfeiffer, 1962; Tucker and Suzuki, 1972).

Why do fishes school in the first place? They are obviously able to do so only when they are not bound to a permanent territory. Species that spend part or all of their lives feeding in open water, moving opportunistically from one site to another, are the ones with the potential to evolve schooling behavior. It is possible to infer the ecological factors that "free" species from a territorial existence by comparative analyses of species varying strongly in territorial behavior. An excellent example is contained in the recent study of blennies of the genus *Hypsoblennius* by Stephens et al. (1970). Along the coast of southern California two dominant species occupy nearly exclusive zones: *H. jenkinsi* is limited to mussel beds, clam burrows, and worm tubes in the subtidal zone, while *H. gilberti* occurs above it in the intertidal zone, occupying "home" rock pools at low tide and wandering widely through the intertidal zone and adjacent subtidal cobble at high tide. It is a reasonable inference that the two species displace each other competitively. The more stable and predictable environment of *jenkinsi* allows the adults to stay within 1 meter of their retreats, and they defend this territory strenuously against other *Hypsoblennius*. *H. gilberti*, by contrast, is forced to wander widely for as much as 15 meters away from its home base in order to feed. It defends this much larger home range only weakly, if at all. Schooling behavior is very likely to evolve from such an opportunistic

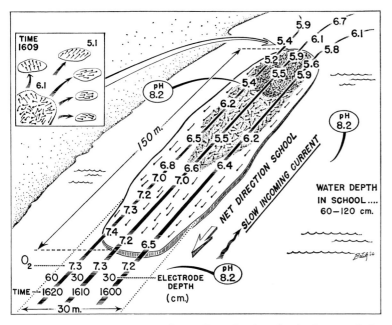

Figure 21-3 The structure of a mullet school. Individuals toward the rear of these large migratory formations are more densely packed. They roil the water by swimming in various directions away from the main course, and, as depicted in the inset, often break away as divergent subgroups. This activity causes a continuing change in the shape of the school and in the relative positions of its individual members. The behavior could be caused by the drop in concentration of environmental oxygen, which is documented in this particular example. The ambient pH appears to remain too invariable to play a significant role. (From McFarland and Moss, 1967; copyright © 1967 by the American Association for the Advancement of Science.)

strategy. What are required are conditions that sever wandering individuals from their home bases and make it more advantageous for the fish to migrate continuously from place to place. The reverse evolution, from nomadic schooling to the solitary occupancy of territory, is equally plausible. Some species, such as the sticklebacks, alternate the two behaviors in the same life cycle, departing from the feeding schools at the onset of the breeding season to set up territories.

Nomadism is a necessary condition for the evolution of schooling behavior without being a sufficient one. Nor can any other single ecological imperative be assigned as *the* prime factor. Schooling is a highly eclectic phenomenon that originated independently in numerous phylogenetically distinct groups (Shaw, 1962). Perhaps 2,000 marine species school. Most belong to three orders that include the most abundant fish of the sea: the Clupeiformes, or herrings; the Mugiliformes, comprising the mullets, atherinid "minnows," and related forms; and the Perciformes, which include the schooling jacks, pompanos, bluefishes, mackerels, tuna, and occasional schooling snappers and grunts. A single freshwater order, the Cypriniformes, contains another 2,000 schooling species. These include the freshwater minnows and characins. The evidence is now overwhelming that a variety of advantages accrue from the behavior, and that these apply singly or in various combinations according to the species:

1. *Protection from predators.* The strongest and most distinctive changes in schooling behavior occur when the fish are confronted by a predator. Some species, such as sticklebacks and catfishes, close ranks. Most spring away as a school, often taking off at a sharp angle from the original course. Still others, such as the sand eels of the genus *Ammodytes*, flee only a short distance before regrouping to form a circle around the predator. If the larger fish charges, the *Ammodytes* wheel away to either side, then close ranks to surround it once again (Kühlmann and Karst, 1967). Radakov observed that when schools of the Caribbean fish *Atherinomorus stipes* are presented with a frightening stimulus, a "wave of disturbance" passes through the school at a speed greater than the movement of individual fish. The intensity of the wave diminishes with distance, so that the response to a weaker stimulus can be localized within the school. These and similar observations have led to the suggestion by Parr and later investigators that the school behaves in such a way as to confuse the predator. The effect presumably decreases the rate of individual captures below that which would prevail if the fish were uncoordinated. It is also likely that the school as a whole can detect predators more quickly than lone fish and thus give individual school members a better chance to escape. Direct evidence on these points is meager, but recently S. R. Neil (cited by Pitcher, 1973) found that under laboratory conditions attacking pike and perch are less successful with schools than with solitary prey. Williams (1966a) has pointed

out that the tendency of fish to seek cover will promote the cohesiveness of schools. Because it is relatively dangerous to swim apart from the school or even along its edge, each fish exhibits a marked tendency to turn inward toward the center of the school. The result is that fish schools proceed by a constant rolling inward of the vanguard; a few fish press forward for short distances as others crowd them from behind, but they turn back to yield the leadership. Another way schooling might confound predators is to shrink the total populations to a smaller number of points in space. Unless the predator can then track schools for long intervals, its feeding rate is actually likely to decline (Brock and Riffenburgh, 1960). Under these circumstances the predator that develops a special ability at locating and tracking schools will enjoy a special advantage. Larger animals can also utilize prey that would otherwise be too small to serve as adequate food items. For example, Bullis (1960) saw a large white-tip shark biting mouthfuls of thread herrings (*Polydactylus*) from a dense school as though it were eating an apple. The same school was attacked by boobies, which floated on the water above it and reached down to take gulps of the little fish.

2. *Improved feeding ability.*   In theory at least, individual members of the school can profit from the discoveries and previous experience of all other members of the school during the search for food. This advantage was documented earlier with reference to bird flocks (see Chapters 3 and 17). It can become decisive, outweighing the disadvantages of competition for food items, whenever the resource is unpredictably distributed in patches. Thus larger fish that prey on schools of smaller fish or cephalopods might be expected to hunt in groups for this reason alone. In fact, many of the largest predators, which have the least reason to fear predation themselves, do run in schools. Increased searching efficiency of individuals as a benefit of school membership can be demonstrated in laboratory experiments. O'Connell (1960) conditioned a group of Pacific sardines (*Sardinops caerulea*) to search for food pellets in response to a 5-second light signal. The quickness and vigor of the response climbed steadily with repeated trials, and they were not diminished by the substitution of unconditioned fish for 41 percent of the school. The newcomers searched in apparent response to the activity of the others.

3. *Energy conservation.*   As mentioned earlier, schooling fish can ride the edges of vortices made by other school members in front of them, thus utilizing energy that would otherwise be lost while conserving their own. It is also possible that heat is retained by the crowding, an important consideration for cold-water species. Hergenrader and Hasler (1967) found that when winter temperatures fell to 0–5°C in Wisconsin's Lake Mendota, solitary individuals of the yellow perch *Perca flavescens* swam at only half the velocity achieved by members of schools.

4. *Reproductive facilitation.*   Fish species that range widely through open water exist in population densities far below the densities of species that remain in special habitats on the sea bottom. Membership in schools almost certainly makes it easier for individuals to find mates or to spawn near others, but whether this advantage has been sufficient by itself to cause the evolution of schooling cannot be decided on the basis of existing evidence.

The Social Behavior of Frogs

The popular image of frogs and other anurans, held even by many zoologists, is one of simple creatures that lead a monotonous, solitary existence, interrupted only by brief bouts of courtship and spawning. In fact the life histories of the hundreds of anuran species are enormously diverse. Although a great many do follow the basic aquatic egg-tadpole-adult sequence, the events often entail elaborate communication and even temporary social organization of breeding groups. Furthermore, profound changes in the life cycle have occurred, especially among tropical forms. Some species carry the tadpoles on the back or in the vocal pouch of the male, and others build nests in vegetation above streams so that the tadpoles can drop into the water when they hatch. Still others omit the tadpole stage altogether. Each adaptation is accompanied by modifications in sexual communication and the roles of the sexes.

Territoriality is the rule in the families Dendrobatidae, Hylidae, Leptodactylidae, Pipidae, and Ranidae (Sexton, 1962; Duellman, 1966; Bunnell, 1973). At dusk, male bullfrogs (*Rana catesbeiana*) leave their retreats and take up calling stations in open water, where they adopt a characteristic high floating position by inflating the lungs completely with air. This exposes the brilliant yellow gular area, which may serve as a supplementary visual signal when the frogs emit their deep-throated calls. If one male approaches another closer than by about 6 meters, the resident individual gives a sharp, staccato "hiccup" vocalization and advances a short distance toward the intruder. In most cases the intruder withdraws. If he does not, the two frogs join in battle. One may leap at or on top of the other, forcing him away. More commonly, however, the two males wrestle face to face with their arms locked around each other, kicking violently with their hind legs, until one is forced over onto his back (S. T. Emlen, 1968). Similar encounters occur in dendrobatid frogs, which defend territories on land (see Figure 21-4).

The evolution of social behavior of frogs and other amphibians has been played out during the transition from an aquatic to a terrestrial existence. A partial escape from the water has been achieved by numerous phyletic lines of frogs independently, to differing degrees and with the aid of a variety of alterations in the life history. Jameson (1957) has identified four parallel trends that appear to represent coadaptations with an increased terrestrial existence: (1) the transfer

Figure 21-4 Males of a tropical frog (*Dendrobates galindoi*) wrestle for possession of a territory. In most cases spacing is maintained by repetitious calling. (From Duellman, 1966.)

of much or all of the courtship and spawning behavior from aquatic to terrestrial sites; (2) the apposition of the cloacas during egg laying; (3) the increasing role of the female in courtship; and (4) the increasing care of the eggs by one sex or the other. The shifting roles of the two sexes in courtship is particularly interesting. The male of the primitive tailed frog *Ascaphus truei* has no voice and must seek out the passive female. He uses an intromittent organ to fertilize the eggs. In this case, however, basic morphology may not mean basic sexual behavior. The more primitive condition seems to be represented by forms that breed in purely aquatic habitats, such as *Bombina*, *Xenopus*, *Scaphiopus*, and most *Bufo*. The males, sometimes forming aggregations and sometimes spaced out in permanent territories, attract the females to the breeding sites with the use of distinctive calls. The males of some species of *Scaphiopus* are extremely active, pursuing any female as soon as she is spotted. The males of other forms, including *Bufo*, *Rana*, *Rhacophorus*, and *Syrrhophus*, pursue a potential mate only when she approaches closely. Some *Scaphiopus*, *Gastrophryne*, and *Hyla* must be touched by the female before they cease calling and initiate the next phase of courtship. In the final step of this sequence, females of *Dendrobates* pursue the males while they move about and continue to call. No ecological correlate of the evolutionary trend has been established. The current theory of sexual selection, presented in Chapter 15, would by itself

suggest that males are pursued when they provide enough parental care to make them a limiting resource to the females. It may be significant that *Dendrobates* males receive the eggs of the females at terrestrial sites and later carry the tadpoles to the water.

When males gather to call in choruses, they are in reality forming leks similar to those of birds. The sounds of the group carry much farther and can be sustained more continuously than those of a lone male. A member of a chorus presumably has a better change of mating than if he were singing elsewhere, alone and in competition with the group. Choruses are typically formed by species that breed in rain pools and bodies of fresh water temporarily swollen by rain. They produce some of the most spectacular sounds of nature. The wailing of thousands of spadefoot toads (*Scaphiopus*) in a Florida roadside ditch, in the pitch-black darkness of a hot summer night, brings to mind the lower levels of the Inferno. It might be counterpointed a short distance away by the soft trilling of *Hyla avivoca* or the sharp, metallic ringing of *Pseudacris ornata*. Choruses of South American frogs sometimes consist of bedlamlike mixtures of ten or more species.

In 1949 C. J. Goin made the surprising discovery that *Hyla crucifer* males call in trios, so that each chorus is made up of numerous trios. Since then duets, trios, and even quartets have been recognized in other species of *Hyla*, *Centrolenella*, *Engystomops*, *Gastrophryne*, *Pternohyla*, and *Smilisca*, representing independent lines of evolution in several frog families (Duellman, 1967). Males of the leptodactylid genus *Eleutherodactylus* remain within their home ranges while duetting with their neighbors (Jameson, 1954; Lemon, 1971b). Duets consist of the alternating of notes between individuals, often at precise intervals. Removal of one of the frogs causes a disruption of the singing by the other, although as Lemon showed it is possible to substitute a tape recording for the missing member. If a frog's partner ceases to call while he is highly stimulated, he may shift his position while emitting occasional calls, in an apparent attempt to find a new

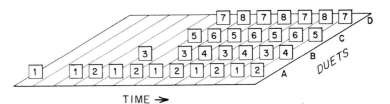

TIME →

Figure 21-5 The call sequence of four pairs of male *Smilisca baudini*, each of which sings in duets of rapidly alternating notes. The eight frogs are indicated by numbers and the pairs by letters along the edges of the plane. The leading pair (individuals *1* and *2*) usually initiated the choruses, which serve to attract females. (Modified from Duellman, 1967.)

partner (Duellman, 1967). There is some evidence of dominance within the little groups, a trait that also characterizes the lek systems of birds. When Duellman removed the loudest member from each of a series of trios of the Central American chorusing frog *Centrolenella fleischmanni*, the two survivors remained silent for a while and then called only sporadically. When "subordinate" members were removed, the leader continued to call at about the same rate. Brattstrom (1962) found that the leader of *Engystomops pustulosus* groups not only initiates most of the sequences but also has the greatest success at breeding. It is also possible for groups to lead other groups, as noted by Duellman in *Smilisca baudini*. A leader of the first duet gives one note (a distinctive "wonk!"), pauses, then gives another single note or a series of two or three notes. If his partner does not respond he waits for up to several minutes and repeats the invitation. When the second frog starts to call, the pair then exchange notes in precise and rapid alternation. Typically other pairs next join in until the chorus is in full swing (see Figure 21-5). Periodically the entire aggregation stops abruptly, only to be reactivated soon afterward by the leading pair.

The Social Behavior of Reptiles

The behavior of reptiles has been poorly explored in comparison with that of birds and mammals. Although part of the reason is the secretiveness of the animals, the major factor is that their behavior tends to be reduced markedly in captivity. Tinkle's experience with *Uta stansburiana* is typical. When transferred to the laboratory his lizards underwent a sharp curtailment of normal aggressive and sexual behavior, and homosexual matings, never observed in the wild, became frequent (Tinkle, 1967). It is commonly believed that reptiles lack complexity in all aspects of their behavior and are relatively unintelligent. But as Brattstrom (1974) and others have found, this conception is based on observations of captive animals kept in cages with cool, oversimplified interiors. When temperatures are carefully raised to the levels preferred by wild populations, which are often surprisingly high, the performances of the animals improve dramatically. In earlier studies, for example, some lizards took over 300 trials to learn a simple T maze. Placed in the normal temperatures determined by field measurements, other individuals performed comparable tasks in 15 trials or less. Lizards can even be trained to press a bar to obtain more heat for their cages. Full repertories of social behavior depend not only on adequate warmth but also on the placement of rocks, plants, and other objects in the cages to simulate the three-dimensional visual environment to which the species is adapted.

The picture that is at last beginning to emerge of reptilian social life is one of considerable diversity among species, with a few flashes of sophistication. The average complexity of social behavior is probably below that of the birds and mammals. That is, many more species are strictly solitary, while very few possess social systems even approaching the middle evolutionary grades of these two other vertebrate groups. Nevertheless, among the reptiles as a whole are to be found a surprising array of adaptations, some of them advanced even by mammalian standards.

Consider home range and territoriality. As in the remainder of the vertebrates, these phenomena are highly labile. Within the lizards a broad ecological basis underlying the form of land tenure can be detected. Most members of the families Agamidae, Chamaeleontidae, Gekkonidae, and Iguanidae sit and wait for their prey, often in exposed situations, and they rely heavily on optical cues. They also tend to be territorial, watching their domain constantly and warning off invaders of the same species with visual signals. In contrast, members of the Lacertidae, Scincidae, Teiidae, and Varanidae typically search for their food in places where vision is obstructed. Many root through the soil and leaf litter, depending strongly on olfactory cues. Probably as a consequence of this behavior, their home ranges overlap broadly. If territories exist, they are spatiotemporal. Considerable variation in land usage also exists within species. In both the land and marine iguanas of the Galápagos, territorial defense is limited to the breeding season. In *Uta stansburiana* it varies in form and intensity between localities. Many cases have been documented of a density-dependent shift between strict territoriality at one extreme and coexistence of adults organized into dominance hierarchies at the other. When black iguanas (*Ctenosaura pectinata*) occur in less disturbed habitats, so that individuals are able to spread out, each solitary adult male defends a well-defined territory. Evans (1951) found a population in Mexico which was compressed onto the rock wall of a cemetery. During the day the lizards went out into the adjacent cultivated fields to feed. At the rock wall retreat there was not enough space to permit multiple territories, even though the food supply in the fields was ample to support a sizable population. As a result the males were organized into a two-layered dominance hierarchy. The leading male was truly a tyrant. He regularly patrolled his domain, opening his jaws to threaten any rival who hesitated to retreat into a crevice. Each subordinate possessed a small space which he defended against all but the tyrant. During a study of a related species, *C. hemilopha*, Brattstrom (1974) was able to simulate this transition in the laboratory. When five males were placed in a large outdoor cage with four rock piles, the four largest individuals each took possession of a rock pile. When the four piles were then combined into one, the lizards formed a dominance hierarchy based on size. Scaling between territoriality and dominance hierarchies is not invariably dependent on changing density. In *Anolis aeneus* the main factor appears to be the thickness of cover, with hierarchies forming in dense vegetation (Stamps, 1973). The position of populations of *Uta stansburiana* on the scale is evidently the outcome of varying schedules of mortality and degrees of *r* selection (Tinkle, 1967).

Reptilian displays associated with aggression and courtship are intermediate in complexity between those of frogs and birds. On the basis of intensive studies Kästle (1963) distinguished four basic types in the grass anolis *Norops auratus*, while Rand (1967b) recorded seven in *Anolis lineatopus*. Submissive behavior is nearly as well developed as threat displays, and in some cases it permits the close coexistence of two or more animals. Subordinate males of the bearded dragon *Amphibolurus barbatus*, an Australian agamid lizard, halt the threats of their superiors by pressing their bodies to the ground and waving one or the other of their hands in a characteristic movement. By this means they are able to pass freely through the territory of the dominant animal. Males of the Lake Eyre dragon *Amphibolurus reticulatus* use an even more curious signal. They flip over on their backs and wait until the tyrant passes (Brattstrom, 1974). The desert tortoise *Gopherus agassizi* of the southwestern United States may have carried dominance systems one step further. Males fight strenuously, pausing only when one of the rivals retreats or is turned over on his back. To be upside down is a mortal threat to a tortoise; he cannot easily right himself and is in danger of being overheated by the sun. According to Patterson (1971), the loser emits a distinctive sound that induces the winner to turn him right-side-up.

Most reptilian dominance systems appear to be little more than transmuted forms of territorial hegemony, with a tyrant permitting a few subordinates to exist within his domain. The subordinates themselves are seldom organized. One exception exists in *Anolis aeneus*. Multiple females live within the territories of single males and are themselves arrayed into hierarchies consisting of at least three levels (Stamps, 1973).

It is commonplace for male lizards to tolerate multiple females within their territories. This form of polygyny has been reported in the gekkonid *Gehyra* and the iguanids *Anolis, Amblyrhynchus, Chalarodon,* and *Tropidurus*. Such associations, however, are not true harems in the strict sense applied to birds and mammals. The females are tolerated, but they are not specifically recruited or defended. The closest approach to a true harem is found in the chuckwalla *Sauromalus obesus*, a large herbivorous lizard of the southwestern United States (Berry, 1971). Tyrant males maintain large territories, within which subordinate males are permitted to hold restricted territories of their own in the vicinity of rock piles and basking sites. Females also have territories within the tyrant's domain, which are larger than those of the subordinate males. During the breeding season the tyrant visits each female daily, restricting the other males to their territories. Only he mates with the females.

Parental care is generally poorly developed in reptiles. It has been observed in both wild and captive king cobras (*Ophiophagus hannah*) by Oliver (1956). The females build nests and defend them against all intruders—making these large snakes especially dangerous to man. Since snakes are otherwise the least social of all the reptiles, this unique behavior pattern is quite remarkable and makes the king cobra one of the most promising reptile species for future field investigations. It may also be surprising to learn that the most advanced forms of parental care are practiced by the crocodilians—the alligators, crocodiles, caimans, and related forms. The females of all of the 21 living species lay their eggs in nests and defend them against intruders (Greer, 1971). The more primitive behavior is hole nesting, employed by the gharial and 7 species of crocodiles. The remaining crocodilians, including alligators, caimans, the tomistoma, and the remaining crocodile species, build mound nests of leaves, sticks, and other debris. The mounds serve to raise the eggs above rising water, and they probably also generate extra heat by decomposition. Just before they hatch, the young emit high-pitched croaks, particularly when disturbed by nearby movements. The response of the mother is to start tearing material off the top of the nest. Her assistance is probably essential for the escape of the young in many cases, since the outer shell of the nest is baked into a hard crust by the sun after the eggs are buried. In some species at least, the mother also leads the young to the edge of the water and protects them for varying periods afterward.

Crocodilians are archosaurs, the last surviving members of the group of ruling reptiles that dominated the land vertebrate fauna of Mesozoic times. Since they practice a relatively sophisticated form of maternal care, it is entirely reasonable to inquire whether dinosaurs, their distant relatives, lived in social groupings. A few scraps of evidence exist to indicate that this could have been the case for at least some of the species. The celebrated egg clutch of *Protoceratops* discovered by the 1922 American Museum of Natural History expedition to Mongolia appears to have been buried in a sand nest, perhaps not much different from the hole nest of modern crocodilians. More significant, however, are the dinosaur footprints and trackways that have been discovered in Texas and Massachusetts (Bakker, 1968; Ostrom, 1972). The animals that made them appear to have passed in groups, laying down tight rows of foot tracks. At Davenport Ranch, Texas, 30 brontosaurlike animals evidently progressed as an organized herd. The largest footprints occur only at the periphery of the trackway and the smallest near the center. Furthermore, the largest plant-eating dinosaurs may not have been the sluggish, stupid creatures envisioned in popular accounts of the past. Bakker (1968, 1971) has argued on the basis of very general physiological principles and new anatomical reconstructions that many of the species were erect in carriage, homoiothermal, and swiftly moving. Herds of brontosaurs and ornithiscians might have roamed the dry plains and open forests much like the antelopes, rhinoceroses, and elephants of the present time. In Figure 21-6 Sarah Landry and I have taken the maximum amount of liberty in reconstructing this scene. The animals shown are *Diplodocus*. Since they were among the largest of the dinosaurs, we have assigned them the same social organization as African elephants.

Figure 21-6 A speculation concerning the social life of dinosaurs. The reconstructed habitat is a dry flood plain in Wyoming in Late Jurassic times. The large sauropod dinosaurs are *Diplodocus.* Because they were the closest ecological equivalents of modern plains ungulates and elephants, they have been arbitrarily assigned the same social organization as elephants. A herd of females and young moves in from the left, led by an old matriarch. In the foreground two males fight for dominance, neck-wrestling like giraffes and clawing at one another with their elongated middle toenails. The *Diplodocus* were among the largest of all dinosaurs; adults reached 30 meters in length, stood about 4 meters at

the shoulder and could extend their heads 10 meters into the air when they reared up on their hind legs. Here they are represented as agile open-country animals and not sluggish aquatic forms of the kind popularized in the older literature. A "pack" of flesh-eating dinosaurs, *Allosaurus*, is seen in the right background. To the left a small "flock" of bipedal dinosaurs scurries through a stand of horsetails. Other characteristic plants are the cycadeoid *Williamsonia*, the palmlike plant to the right, a true cycad just in front of it, and araucarian pines in the background. (Drawing by Sarah Landry; based on Robert T. Bakker, 1968, 1971, and personal communication, and John H. Ostrum, 1972.)

Chapter 22   **The Birds**

Birds are the most insectlike of the vertebrates in the details of their social lives. A few species, including the African weaverlike birds *Bubalornis albirostris* and *Philetairus socius*, the wattled starling *Creatophora cinerea*, the West Indian palmchat *Dulus dominicus*, and the Argentinian parrot *Myiopsitta monachus*, build communal nests in which each pair occupies a private chamber and rears its own brood. The advantage of collaborating to this extent appears to be the improvement in defense against predators (Lack, 1968). In the language of entomology, such birds form communal groups. They are closely paralleled by certain bee species, including *Augochloropsis diversipennis*, *Lasioglossum ohei*, and *Pseudagapostemon divaricatus* (Michener, 1974). Insects in the communal stage are considered to be on the "parasocial" route of evolution that can eventually lead to full-fledged colonies with sterile castes. Communal nesting is distinguished from cooperative breeding, in which more than one pair of adults join at the same nest to rear young together. In many bird species certain of the individuals, known as helpers, assist in raising the young of others and do not lay eggs on their own. This, too, is notably insectlike. When helpers attach themselves to breeders at the very start, as in the long-tailed tit *Aegithalos caudatus*, the species resembles the "semisocial" bee and wasp species, which are also on the parasocial route. When helpers consist of offspring from former broods who remain with the parents at the nest site, a condition exemplified by the social jays, the entomologists would classify the species as "advanced subsocial," equally well along the alternate, subsocial route of evolution. Whether or not the distinction between the parasocial and subsocial states will prove to be as useful in the study of birds as it has been in entomology, it is undeniable that the presence of helpers is an advanced social trait by insectan standards. To attain the level of ants and termites all that would be needed is for a helper "caste" to evolve, whose members remain permanently in the role. So far as is known this final step has never been taken by any bird species. Bird helpers are potentially fully reproductive and ready to start their own nests whenever the opportunity arises.

The resemblance between birds and insects does not stop with the matching of stages of social evolution. Birds are also the only vertebrates with true social parasites. Moreover, the form of the behavior—brood parasitism—resembles the temporary social parasitism of ants in many details. The birds have not carried the trend to the extremes achieved by the social insects, but a few of the bird species occupy advanced intermediate positions by insectan standards. For further information on these phenomena the reader is referred to Chapter 17.

The reason for the resemblances, I believe, lies in the mode of parental care shared by both groups. Birds, like the presocial and social insects, provide extended parental care requiring repeated expeditions to gather food for the young. In the great majority of cooperatively breeding bird species, as well as in those that are hosts for brood parasites, the young are altricial—helpless at birth—and must be kept in specially constructed nests. These two factors together appear to be the basis for the widespread occurrence of bonding between the two parents, a condition that is relatively infrequent in other vertebrate groups. The stage is set, first, for older siblings and other kin to improve their inclusive genetic fitness by assisting their parents, and, second, for parasitic forms to exploit the process by inserting their eggs into the nests. Parasitism may be promoted further by the relative anonymity of altricial young and the stereotypy of the communication between them and the parents.

The reader is by this time aware that the *elements* of social behavior in birds have played a large role in the development of the general principles of sociobiology. In particular, the adaptive significance of aggregations has been analyzed with special reference to bird flocks (see Chapters 3 and 17), while the study of communication—and with it the larger discipline of ethology—has been based to a large extent on birds (Chapters 8–10). Birds provide much of the documentation of territoriality and dominance (Chapters 12–13), the endocrine control of reproductive and aggressive behavior (Chapters 7 and 11), sexual behavior with special reference to colonial nesting and polygamy (Chapter 15), parental care (Chapter 16), and brood parasitism and mixed-species foraging groups (Chapter 17). Most of these components are conventional in the sense that their properties are shared by most other vertebrates. What is needed now, and the remainder of this chapter will provide, is a closer examination of the most advanced *patterns* of avian social organization, particularly those based upon cooperative breeding.

Considering the large amount of effort that has gone into the field study of birds around the world, the analysis of cooperative breeding has been a surprisingly late development. In 1935 Alexander F. Skutch could report examples from less than 10 species, 3 of which he had discovered himself. In 1961, when he summarized the subject again, helpers of one kind or another had been reported in more than 130 species, ranging from flamingos to swallows, woodpeckers, wrens, and members of a startling array of other families. Fry reexamined the matter in 1972 and interpreted the phenomenon to be well developed in about 60 species belonging to perhaps 30 families. Either way, the list of examples continues to grow, and cooperative breeding can now be regarded as occurring as a regular feature in nearly 1 percent of the world fauna.

Ornithologists have gained some understanding of the ecological basis of cooperative breeding (Lack, 1968; J. L. Brown, 1968). Brown in particular has evaluated the demographic factors involved and thereby aligned this aspect of bird sociality for the first time with the theory of population biology. In Figure 22-1 I have presented a simple scheme that attempts to link together causal and intermediate

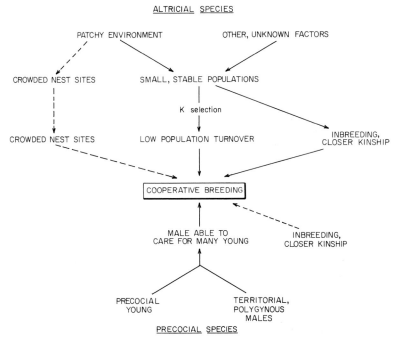

Figure 22-1 The hypothesized chain of causation leading to the evolution of cooperative breeding, the most advanced form of social behavior known in birds. The solid lines indicate relationships that are documented and considered to be of crucial importance; dashed lines suggest those that are still undocumented but may play at least auxiliary roles.

factors to account for all of the known cases of cooperative breeding. Note that there appear to be two major pathways leading to the phenomenon. One has been taken by species with precocial young (able to leave the nest soon after birth) and the other by species with altricial young (helpless at birth). The form of the communal nesting also differs in an important way. In the first group of species, which includes the ostrich *Struthio camelus*, the rhea *Rhea americana*, and several species of the primitive tropical American birds called tinamous, two to four hens lay in one nest guarded by a single male. The male generally takes exclusive charge of the nest, although in the ostrich he is sometimes assisted by the dominant female. Females of the tinamou *Nothocercus bonapartei* remain in the male's territory and are prepared to lay another clutch if the first is destroyed. But in *Crypturellus boucardi* and *Nothoprocta cinerascens* they move on to lay for other males. The environmental prime movers of this peculiar form of mutual tolerance among females are unknown, but certain conditions that predispose species to acquire it in evolution are clear enough. First, the precocial nature of the young means that a single parent can look after all their needs. It is then advantageous for a male to control a territory into which he can entice multiple females. This is basic polygyny, and it is conceivable that a large variance in the quality of the male territories exists, as predicted by the Orians-Verner model (see Chapter 15). What is peculiar, however, is the fact that individual females do not attempt to preempt access to the male and the single nest within each territory. One would expect them at least to follow the pattern common in other bird species of subdividing the male's territory and constructing private nests of their own. The reason they do not do so may well be that they are closely related. A little band of sisters could gain maximum inclusive fitness if it performed as a unit, especially if it could make use of more than one male, as in the case of the *Crypturellus* and *Nothoprocta* tinamous. The study of kinship within this small group of birds will no doubt prove instructive.

The second class of cooperatively nesting birds is much larger, containing over 90 percent of the known species. As suggested in Figure 22-1, there appear to be several causative factors, which are complexly linked with each other. These factors have been elucidated separately in studies by several authors. Pulliam et al. (1972), for example, suggested that gregariousness in the yellow-faced grassquit *Tiaris olivacea* intensifies as population size is reduced and inbreeding thereby increased. In the Jamaican population, which is nearly continuous and hence relatively large, individuals are strongly territorial. The populations of Costa Rica are small and semiisolated, and the members gather in relatively large flocks. On Cayman Brac, both the population and the flocks are intermediate in size. The implication of this finding is that the smaller the effective population size, the greater the degree of kinship among interacting individuals, and the less likely they are to respond aggressively. Several zoologists who have investigated cooperatively breeding species, including Davis (1942) in the case of crotophagines and Brown (1972, 1974) in the case of jays, have similarly commented on the small size and stability of the populations.

The division of species into small, semiisolated populations is itself an effect of other, more purely environmental factors. The factors have not been identified conclusively in birds, but their general nature can be guessed. First, it is evident that the species preadapted for sociality have become specialized on patchily distributed resources. The form of the patchiness has a profound influence on the kind of sociality evolved. Where the resources are fine-grained, meaning that individual birds search from one patch to another during the course of single foraging trips (Levins, 1968), the result is likely to be the formation of flocks. Food and water are the resources most likely to be fine-grained, whereas nesting and roosting sites tend to remain fixed and stable. As a result, the birds maintain individual territories where they breed, but form flocks to search for food and water. The more unpredictable these resources are in space and time, the more pronounced the optimum flocking behavior. This causal relationship appears to be the most plausible explanation for flocking

by terns and some other colonial seabirds (Ashmole, 1963), starlings (Hamilton and Gilbert, 1969), and Australian desert-dwelling parrots (Brereton, 1971). If the principal resources are more nearly coarse-grained (widely distributed or large enough to require careful exploration one by one), the result is likely to be radically different. Now individuals wander less widely. Populations are restricted to limited habitats and are more prone to be both genetically isolated from one another and smaller in size. The possible result is the entrainment of events represented in Figure 22-1. Small, isolated populations tend to be stable and *K*-selected. *K* selection favors longer life, less potential fecundity, and a more prolonged parent-offspring relationship (see Chapters 4 and 16). All of these alterations in the life cycle promote cooperation and altruism during the reproductive period.

In essence, then, the evolutionary origin of cooperative breeding is viewed as depending most upon small effective breeding size. Species that practice flocking behavior might also evolve cooperative breeding if the nest sites are so restricted that population size is reduced and kinship significantly increased. However, the two processes can be entirely decoupled. Many flocking species form large breeding colonies in which average kinship is low, and intrasexual competition and aggression at the breeding sites are consequently intense. Conversely, species that live and feed in special habitats—the coarse "grains" of the environment—may utilize habitat patches so extensive, or exist in such high densities, that their population size is relatively large. They too will evolve intrasexual competition and aggression at the breeding sites. Cooperative breeding, according to the current hypothesis, depends upon the existence of a limiting resource—food, nest site, or whatever—that keeps populations small, philopatric, and isolated.

Even if this reasoning proves correct, it will leave an important question unanswered: Why do species evolve one feeding and nesting strategy as opposed to another? A detailed answer, at least with reference to birds, is outside the scope of this book. The choices made by particular species are the result of adaptive radiation, followed by the formation of communities of species that displace one another into various ecological roles. Some of the basic theory has been presented in Chapters 3 and 4; more detailed expositions are given by MacArthur (1972a,b) and Cody (1974).

We will now turn to two examples in which the evolution of cooperative breeding has been relatively well worked out by a comparison of closely related species. Such phylogenetic studies provide the best means of establishing the adaptive basis of the phenomenon as well as of discovering new forms of social behavior.

The Crotophaginae

The Crotophaginae, consisting of the guira cuckoo *Guira guira* and the anis of the genus *Crotophaga*, constitute one of the six subfamilies of the cuckoo family Cuculidae. The crotophagines are entirely limited to the tropical and subtropical portions of the New World. Although there are only four species, the diversity in their social behavior is sufficiently great to permit a plausible reconstruction of social evolution. The study of the crotophagines by David E. Davis (1942) is notable in being one of the first to explore the ecological basis of social evolution in any vertebrate group, and it is still both modern and definitive. Additional information has since been added by Skutch (1959) and Lack (1968).

All four species live mostly in open habitats and are characterized by "noisy, ostentatious habits." They associate in conspicuous flocks of about a dozen individuals, foraging as a group and sleeping in the same tree at night. Each flock defends its territory from other members of the same species by aggressive displays and fighting. During the breeding season the birds build a communal nest in which up to several females lay eggs. The males contribute to the construction of nests and to the subsequent rearing of the young. At least some of the first fledged offspring assist with the rearing of subsequent broods, while a few participate in breeding in the following year. The crotophagine colony is a semiclosed group. A small percentage of individuals, as yet unmeasured, migrate from one group to another, but their entry into a new flock is achieved only after they have overcome threats and fighting.

Davis has distinguished three progressive stages in the evolution of cooperative behavior. Communal nesting is only facultative in *Guira guira*. Some of the mated pairs stake out a small territory of their own within the group territory, build separate nests, and rear their young apart from the others. Thus guira cuckoos still occasionally follow the basic avian pattern of pair bonding and territoriality, but they differ in always being allied with a particular flock in nonbreeding activities. An early stage of social evolution is further indicated by the fact that the group defends its territory only weakly. The greater ani *Crotophaga major* almost always nests communally, although the flock remains composed of mated pairs. Group territorial defense is still only weakly developed. Finally, in the smooth-billed ani *C. ani*, and evidently in the groove-billed ani *C. sulcirostris* as well, communal nesting is carried to an extreme. Polygamy or promiscuity is practiced consistently, several females contribute to the same clutch, and the entire flock defends its territory vigorously as a unit.

The ultimate causes of the crotophagine trend are not known. The sex ratio is in favor of males, a phenomenon often encountered in other cooperatively breeding bird species. This could predispose helper behavior to evolve, since unmated males gain fitness if they devote their energies to rearing siblings. However, the sex ratio is itself an evolutionary product, easily shifted by small genetic changes. An unbalanced ratio could be one of the coadaptations of cooperative breeding rather than a cause. The prime mover is more likely to be an environmental factor. Davis noted that crotophagines nest and

sleep in clumps of trees that are widely scattered through the tropical grasslands. He suggested that the birds are simply forced together by lack of space. It seems more likely that the significant effects of the patchiness are the smaller size of the local breeding populations and their genetic isolation from one another.

The Jays

The most recent and edifying studies of bird sociality have dealt with the New World jays (J. L. Brown, 1972, 1974; Woolfenden, 1974). With the possible exception of the piñon jay *Gymnorhinus cyanocephala*, the eight genera form a close phylogenetic grouping. Like other members of the Corvidae, including the crows, magpies, nutcrackers, and choughs, the jays are adaptable omnivores strongly disposed toward social behavior. Their social systems range from the basic avian pattern of pair bonding with defended territories to some of the more extreme forms of colonial nesting and cooperative breeding known in the birds.

Brown (1974) points out that social evolution within this group

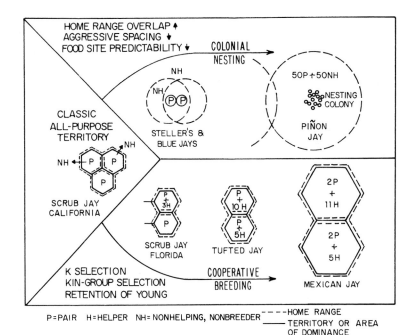

Figure 22-2 The two pathways to advanced sociality in the New World jays. The upper route terminates in the piñon jay, in which pairs of birds build their nests close together in "colonies." The members of each colony also forage together in tight flocks. The alternate route leads to cooperative breeding, in which helper birds assist in rearing the offspring of other adults. This second trend culminates in the Mexican jay. (Modified slightly from J. L. Brown, 1974.)

Figure 22-3 The helper phenomenon in the Florida scrub jay *Aphelo-coma coerulescens.* This drawing depicts a typical scene at the Archbold Biological Station in central Florida. At the nest the two parents and a yearling feed the nestlings, which are the siblings of the helpers. To the right two other helpers have spotted an indigo snake (*Drymarchon corais*), one of the dangerous predators of jay nestlings. One crouches on the ground in a threat posture. The other perches nearby in the "hiccup stance," an alarm signal that will soon alert the birds at the

nest. The habitat is the floristically peculiar Florida "scrub" to which *A. coerulescens* is restricted. The nest is constructed of dead twigs in a low myrtle oak (*Quercus myrtifolia*). Other typical plants include wire grass (*Aristida oligantha*), seen in the lower righthand corner, and saw palmetto (*Serenoa repens*) and sand pine (*Pinus clausa*) in the right background. (Drawing by Sarah Landry; based on G. E. Woolfenden, 1974 and personal communication.)

has followed two alternate pathways (see Figure 22-2). One culminates in the only colonially nesting species, the piñon jay. Up to several hundred adult pairs build nests in clusters and forage together in closely packed flocks that "roll" through the open woodland like groups of starlings or wood pigeons. Only the immediate vicinity of the nest is defended by the resident pair, and the colony as a whole does not protect its home range from other piñon jays. Some adults serve as helpers, but the phenomenon is not nearly so well developed as in the scrub jay and Mexican jay. A possible early intermediate stage is represented by Steller's jay, *Cyanocitta stelleri*. This species is not truly colonial, since the nests are evenly spaced owing to aggressive behavior on the part of the resident pairs. But the home ranges are left mostly undefended, and as a result they overlap widely. Steller's jay can be interpreted as a species whose territorial defense has begun to diminish, setting the stage for the clumping of the nests into a colonial system.

Cooperative breeding is well advanced in the Florida scrub jay (*Aphelocoma coerulescens*), the behavior of which has been painstakingly studied over a period of five years by G. E. Woolfenden (1973, 1974, and personal communication). This handsome blue-and-white bird is limited to the "scrub" of peninsular Florida, a highly discontinuous, sandy habitat with a distinctive flora. The eastern North American form of *A. coerulescens* is so attached to the scrub that it is the most distinctive of the Floridian birds, having never been recorded beyond the borders of the state. Its populations are very stable and bear the expected marks of prolonged *K* selection. Individuals are long-lived for wild birds, often surviving for eight years or more. They do not begin to breed before they are two years old. Pairs are bonded for life and occupy permanent territories. Approximately half of the breeding pairs studied by Woolfenden were assisted by helpers; the number actually fluctuated from year to year, varying from 36 to 71 percent. The helpers did not participate in nest construction or incubation, but they were active in every other activity, including defense of the territory and nest from other jays, attacks on predators, and feeding the young (see Figure 22-3).

By marking large numbers of jays and following them through the first several years of their lives, Woolfenden was able to determine the relationships and ultimate fates of the helpers. In 74 seasonal breedings (complete seasons of breeding by individual pairs), helpers assisted both parents 48 times, a father and stepmother 16 times, a mother and stepfather twice, a brother and his mate 7 times, and an unrelated pair only once. Thus the closest kin are strongly preferred—and a basis for the evolution of the altruistic trait by kin selection exists. Woolfenden was also able to demonstrate that the presence of the helpers actually increases the rate of reproduction of the breeders and hence their own inclusive fitness. Among 47 seasonal breedings by unassisted pairs observed over a period of sev-

eral years, the average number of fledglings produced per pair was 1.1, while the average number of offspring still alive three months after fledging was 0.5. In contrast, 59 seasonal breedings by pairs accompanied by helpers produced an average of 2.1 fledglings per pair, and 1.3 of these were still alive three months after fledging. Hence the presence of helpers increased the replacement rate of the jay family by a factor of two to three. Woolfenden was aware that breeding pairs lacking helpers are also the youngest and least experienced and that this factor alone might account for the difference. But when experience was partialed out, by eliminating inexperienced birds, the role of helpers remained equally strong. Finally, the analysis was made still more rigorous by comparing the success of the same pairs of birds during years in which they had helpers and years in which they were alone. Again, the advantage of being helped proved clear-cut.

Surprisingly, the enhancement of reproduction does not appear to be a result of the increased feeding rate of the young. The number of helpers had no influence on the number of offspring fledged, and the weight of the fledglings had no discernible effect on their subsequent survival rate. The most likely remaining hypothesis is that helpers increase survival rates by improving communal defense against predators, notably the large snakes that are especially dangerous to the nestlings. The helpers add to the vigilance system of the family, and they assist in the mobbing of snakes that approach too closely to the nest. But whether their presence actually reduces mortality of the young birds remains to be established.

The data on the Florida scrub jay are important because little other evidence exists to indicate whether cooperative breeding really improves reproductive success, in other words, whether helpers really help. In only one additional species, the superb blue wren *Malurus cyaneus*, has such an enhancement been documented (Rowley, 1965). Fry's data on the bee eater *Merops bulocki* also suggest enhancement, but they are not statistically significant. Gaston's study of the long-tailed tit *Aegithalos caudatus* in England indicates that helpers have no effect, while helpers of the Arabian babbler *Turdoides squamiceps* may even hinder reproduction (Amotz Zahavi, personal communication).

If in some species helpers do not help the breeders, the implication is that they themselves benefit in some way from the relationship. Woolfenden has found that this is probably the case even in the "altruistic" scrub jays. A strict dominance order exists among the nonbreeders in each family group, with males above females. If the breeder male dies or leaves, he is most likely to be replaced by the dominant helper male. It is also true that the presence of helpers results in some expansion of the territory, which may ultimately grow in area by one third or more. When this occurs, the dominant helper male sometimes sets up a personal territory within that of the group, pairs, and begins to breed on his own. In short, the population grows to some extent by budding, with helpers being the beneficiaries. Thus the helper phenomenon could be due at least in part to individual selection. The relative contributions of individual and kin selection to the evolution of cooperative breeding in this and other bird species remain to be measured.

The Mexican jay *Aphelocoma ultramarina* displays the farthest known extension of cooperative breeding within the New World jays (Brown, 1972, 1974). The *A. ultramarina* group is in fact an extended family of the *A. coerulescens* type. Each exclusive home area is ordinarily occupied by a flock of 8 to 20 individuals, which include 2 or more breeding pairs of birds. The nestlings are fed by all members of the group, with roughly half of the visits being made by the parents. Mexican jays do not attempt to pair and breed until they are three or more years old, and most probably spend their entire lives within the family territory. It seems probable that Mexican jays generate new flocks at least in part the way scrub jays do, by the budding off of subgroups into new, adjacent home areas. If so, adjacent groups are likely to be more closely related than is usually the case in birds.

Chapter 23 Evolutionary Trends within the Mammals

The key to the sociobiology of mammals is milk. Because young animals depend on their mothers during a substantial part of their early development, the mother-offspring group is the universal nuclear unit of mammalian societies. Even the so-called solitary species, which display no social behavior beyond courtship and maternal care, are characterized by elaborate and relatively prolonged interactions between the mother and offspring. From this single conservative feature flow the main general features of the more advanced societies, including such otherwise diverse assemblages as the prides of lions and the troops of chimpanzees:

——— When bonding occurs across generations beyond the time of weaning, it is usually matrilineal.

——— Since the adult females are committed to an expenditure of substantial amounts of time and energy, they are the limiting resource in sexual selection. Hence polygyny is the rule in mammalian systems, and harem formation is common. Monogamous bonding is relatively rare, having arisen in such scattered forms as beavers, foxes, marmosets, titis, gibbons, and nycterid bats. In this regard the mammals depart from the largely monogamous birds. They are also distinguished by the absence of any species that shows reversal of sex roles, wherein females court the males and then leave them to care for the young.

Although these very broad generalizations can be safely made, most of the sociobiology of mammals is in an early stage of exploration, well behind that of the insects and birds. Most accounts of natural history touch on the subject only in an anecdotal fashion, especially in the case of burrowing and nocturnal species. Authors often erroneously label dense populations and breeding aggregations as "colonies" and mothers accompanied by larger offspring as "bands." The sociobiology of the majority of the families and genera of two of the greatest mammalian orders, the bats and rodents, is virtually unknown. The same is true of the marsupials, which represent a remarkable experiment in social evolution comparable to that of the eutherians.

Table 23-1 presents much of our existing knowledge of mammalian social systems in a highly condensed, synoptic form. It is difficult if not impossible to put this information into one grand evolutionary scheme. In the first place the data are still too fragmentary. But more fundamentally, most social traits in mammals are very labile. Beyond the universal occurrence of maternal care and the most obvious immediate consequences just listed, particular features of social organization occur in a highly patchy manner within taxonomic units as small as the family and genus. The bats, documented in Table 23-2, are an interesting case in point. Various species within the same family and even within the same genus sometimes occupy three or more "grades" of social evolution. In a given taxon some may be solitary, others monogamous or harem-forming or living in permanent

Table 23-1 The families of living mammals (names ending in -idae), with representative genera, mode of social life, and selected key references containing sociobiological information. The classification follows Anderson and Jones, ed. (1967). See also Walker, ed. (1964), for a detailed bibliography of earlier literature on behavior and ecology, and "Recent Literature of Mammalogy," a bibliographic series published as a continuing supplement of the *Journal of Mammalogy.*

Kind of mammal	Sociobiology	References
ORDER MONOTREMATA		
TACHYGLOSSIDAE		
Echidnas, spiny ant-eaters (*Tachyglossus, Zaglossus*). Australia, New Guinea.	*Solitary.* Probably territorial when free-living; dominance hierarchies formed by groups in close captivity. Female lays egg directly into pouch, later forages for 1–2-day periods while young remain in safe retreat. No male cooperation.	M. Griffiths in Ride (1970), Brattstrom (1973)
ORNITHORHYNCHIDAE		
Duck-billed platypus (*Ornithorhynchus*). Australia.	*Solitary.* Female lays eggs in closed burrow; young remain in burrow for 17 weeks while female forages. No male cooperation.	Troughton (1966), Ride (1970)
ORDER MARSUPIALIA		
DIDELPHIDAE		
Opossums (*Didelphis, Chironectes, Philander*). New World, esp. tropics.	*Solitary.* In *Didelphis,* newborn young carried in pouch; they later travel with mother for a short time, usually clinging to fur of her back. No male cooperation.	Reynolds (1952), Llewellyn and Dale (1964), McManus (1970)
DASYURIDAE		
Marsupial cats (*Dasyurus* and other genera), marsupial "mice" and "rats" (*Antechinus, Sminthopsis*), marsupial anteaters (*Myrmecobius*), Tasmanian devils (*Sarcophilus*), Tasmanian wolves (*Thylacinus*). Australia.	*Solitary.* Females of at least some species utilize dens. Young carried in pouch, later accompany mother for brief periods. Young of *Dasyurus* and *Sarcophilus* play-fight. No male cooperation. In at least *Antechinus* and *Sarcophilus,* broadly overlapping home ranges.	Fleay (1935), Calaby (1960), Eisenberg (1966), Troughton (1966), Van Deusen and Jones (1967), Guiler (1970), Lidicker and Marlow (1970), Ride (1970), Wood (1970)
NOTORYCTIDAE		
Marsupial "moles" (*Notoryctes*). Australia.	*Probably solitary.*	Van Deusen and Jones (1967), Ride (1970)
PERAMELIDAE		
Bandicoots (*Perameles, Isoodon*). Australia.	*Solitary.* Probably territorial. Nests in mounds of vegetation. Young stay in pouch at first, then remain for short time with mother. No male cooperation.	Mackerras and Smith (1960), Troughton (1966), Van Deusen and Jones (1967), Ride (1970)
CAENOLESTIDAE		
Rat opossums (*Caenolestes, Lestoros*). South America.	*Probably solitary.*	Van Deusen and Jones (1967)

Table 23-1 (*continued*).

Kind of mammal	Sociobiology	References
PHALANGERIDAE Cuscus (*Phalanger*), ringtails (*Hemibelideus, Pseudocheirus*), gliders (*Petaurus*), koalas (*Phascolarctos*), honey possums (*Tarsipes*). Australia.	*Diverse.* Some species are solitary. *Petaurus* lives in family groups dominated by the male, and several generations may live together; *Pseudocheirus* has a similar but apparently looser organization. Species are evidently territorial, and in *Pseudocheirus* aggressiveness increases with population density. In *Phascolarctos* maternal care is prolonged for up to 1 year.	Schultze-Westrum (1965), Eisenberg (1966), Troughton (1966), Ride (1970)
PHASCOLOMYIDAE Wombats (*Lasiorhinus, Vombatus*). Australia.	*Solitary.* Females give birth to single young, which are kept in a pouch and then associate with the mother for up to several months. Individuals occupy complex tunnel systems.	Troughton (1966), Wünschmann (1966), Van Deusen and Jones (1967), Ride (1970)
MACROPODIDAE Kangaroos, quokkas, wallabies, and related forms (*Macropus, Dendrolagus, Hypsiprymnodon, Megaleia, Petrogale, Potorous, Setonyx*, and others). Australia, New Guinea.	*Diverse.* Some species are solitary or paired, e.g., in *Hypsiprymnodon* and *Potorous; Setonyx* forms unorganized aggregations with male dominance hierarchies; *Megaleia* and *Macropus* occur in bands with weak organization (see description elsewhere in this chapter).	Hughes (1962), Caughley (1964), Eisenberg (1966), Troughton (1966), Packer (1969), Ride (1970), Russell (1970), Kitchener (1972), Grant (1973), Kaufmann (1974a–c)
ORDER INSECTIVORA ERINACEIDAE Hedgehogs and moon rats (*Erinaceus, Echinosorex, Paraechinus*, and others). Old World.	*Solitary.* Mother cares for young, with no male participation.	Eisenberg (1966), Findley (1967), Matthews (1971)
TALPIDAE Moles and desmans (*Talpa, Condylura, Desmana*, and others). North America, Eurasia.	*Solitary.* Mother cares for young, with no male participation.	Eisenberg (1966), Findley (1967), Matthews (1971)
TENRECIDAE Tenrecs (*Tenrec, Dasogale, Echinops, Hemicentetes, Microgale, Potamogale, Setifer*, and others). Madagascar, West Africa.	*Diverse.* Mother cares for young in burrows; in *Hemicentetes* and *Tenrec* young follow mother on foraging trips. Small male groupings may occur outside breeding season in *Setifer. Hemicentetes* is the most social: same denning space is used by several females, possibly related, by their offspring, and by a single male.	Dubost (1965), Eisenberg and Gould (1970)
CHRYSOCHLORIDAE Golden moles (*Chrysochloris* and others). South Africa.	*Solitary.*	Findley (1967), Matthews (1971)

Table 23-1 (*continued*).

Kind of mammal	Sociobiology	References
SOLENODONTIDAE Solenodons (*Atopogale*, *Solenodon*). West Indies.	*Solitary or primitively social*. Extended families may occupy same burrows.	Eisenberg and Gould (1966), Findley (1967), Matthews (1971), Eisenberg (personal communication)
SORICIDAE Shrews (*Sorex, Blarina, Crocidura, Suncus,* and others). Worldwide.	*Solitary*. Young of *Crocidura* and *Suncus* seize tails and form chains behind their mothers when alarmed.	Crowcroft (1957), Shillito (1963), Quilliam et al. (1966)
MACROSCELIDIDAE Elephant shrews (*Macroscelides, Elephantulus,* and others). Africa.	*Solitary*.	J. C. Brown (1964), Findley (1967), Ewer (1968), Matthews (1971), Sauer and Sauer (1972)
ORDER DERMOPTERA CYNOCEPHALIDAE "Flying lemurs" or colugos (*Cynocephalus*). Tropical Asia.	*Solitary or aggregating*. Extreme gliding forms. One young born a year, clings to mother's abdomen. No nest is built. Occasionally adults form loose aggregations lacking internal organization.	Wharton (1950), Eisenberg (1966), Findley (1967), Matthews (1971)
ORDER CHIROPTERA 19 families. See Table 23-2.	*Very diverse*, both between and within families. Some species are solitary (e.g., *Epomops, Eptesicus, Lasiurus*); others are pair-forming (*Kerivoula, Lavia, Taphozous*), harem-forming (*Saccopteryx, Tadarida*), or live in permanent large male-female aggregations (*Pteropus, Saccopteryx*). About 50 percent of tropical and 20 percent of temperate species are social to some degree. See elsewhere in this chapter.	Eisenberg (1966), Koopman and Cockrum (1967), Davis et al. (1968), LaVal (1973), Bradbury (1975)
ORDER PRIMATES	*See Chapter 26*.	
ORDER EDENTATA MYRMECOPHAGIDAE Anteaters (*Myrmecophaga, Cyclopes, Tamandua*). Central and South America.	*Solitary*. Single offspring carried by mother on back, for up to a year in the case of *Myrmecophaga*.	Krieg (1939), Schmid (1939), Barlow (1967), Matthews (1971)
BRADYPODIDAE Tree sloths (*Bradypus, Choloepus*). Central and South America.	*Solitary*. Territorial defense by fighting, following a period of wandering. Females carry single young on chest or back for a month or more.	Beebe (1926), Barlow (1967), Montgomery and Sunquist (1974)
DASYPODIDAE Armadillos (*Dasypus* and others). New World, esp. tropics.	*Solitary*. Occupy definite home range. Up to 12 polyembryonically generated young born at a time; they are precocious, being capable of foraging with their mother within hours after birth.	Taber (1945), Talmadge and Buchanan (1954), Barlow (1967)
ORDER PHOLIDOTA MANIDAE Pangolins, scaly anteaters (*Manis*). Africa and tropical Asia.	*Solitary*. One or two young carried by mother on back and tail.	Rham (1961), Pagès (1965, 1970, 1972a,b), Barlow (1967)

Table 23-1 (*continued*).

Kind of mammal	Sociobiology	References
ORDER LAGOMORPHA OCHOTONIDAE Pikas (*Ochotona*). Asia and western North America.	*Solitary in "colonies."* Populations are dense and local, but individual animals maintain solitary territories within them.	Haga (1960), Broadbooks (1965), Layne (1967)
LEPORIDAE Rabbits and hares (*Lepus, Oryctolagus, Pronolagus, Sylvilagus,* and others). Worldwide; introduced into Australia.	*Diverse.* Territorial behavior is widespread. Some species are solitary, e.g., *Lepus.* In the European rabbit *Oryctolagus cuniculus*, some males contain multiple females in their territories, and the females are arranged in a loose dominance order. Young adult offspring are also tolerated for a time within the *Oryctolagus* warren. Some groups dominate others, occupying larger territories.	Southern (1948), Lechleitner (1958), Mykytowycz (1958–60, 1968), O'Farrell (1965), Ewer (1968), Mykytowycz and Dudziński (1972)
ORDER RODENTIA (43 living families) APLODONTIDAE Mountain beavers (*Aplodontia*). Western North America.	*Solitary in "colonies."* Populations are dense and local, but individual animals maintain small solitary territories around their burrows.	Anthony (1916), McLaughlin (1967)
SCIURIDAE Squirrels (*Sciurus, Aeretes, Cynomys, Eutamias, Petaurista, Spermophilus, Tamias, Tamiasciurus,* and many others), marmots, and woodchucks (*Marmota*). Worldwide.	*Diverse.* Territorial behavior is widespread if not universal. Some species are solitary (in *Sciurus, Tamiasciurus*), others form harems (*Marmota*) or temporary winter bands (*Glaucomys*). *Cynomys ludovicianus*, the black-tail prairie dog, forms complex "coteries" of adults of both sexes and all ages; see description later in this chapter.	Layne (1954), Robinson and Cowan (1954), King (1955), Bakko and Brown (1967), Broadbooks (1970), Dunford (1970), Waring (1970), Brown (1971), Carl (1971), Downhower and Armitage (1971), Heller (1971), Yeaton (1972), Barash (1973, 1974a), Drabek (1973), Smith et al. (1973)
GEOMYIDAE Pocket gophers (*Geomys, Thomomys,* and others). New World.	*Solitary.* Subterranean, defend burrow systems.	Eisenberg (1966), McLaughlin (1967)
HETEROMYIDAE Kangaroo rats, spiny rats, and allies (*Heteromys, Diplodomys,* and others). New World.	*Solitary.* Generally territorial, occupying exclusive burrow systems.	Eisenberg (1963, 1966, 1967), McLaughlin (1967), Rood and Test (1968)
CASTORIDAE Beavers (*Castor*). North America and Europe.	*Familial groups.* Mated pair, yearlings, and new young occupy lodge. Young disperse at two years of age. Lodge area defended against other families.	Tevis (1950), Eisenberg (1966), Wilsson (1971), Bartlett and Bartlett (1974)

Table 23-1 (*continued*).

Kind of mammal	Sociobiology	References
ANOMALURIDAE Scaly-tailed squirrels (*Anomalurus*). Tropical Africa.	*Familial groups.* Apparently occur in pairs.	McLaughlin (1967)
CRICETIDAE Hamsters (*Cricetus* and others), wood rats (*Neotoma*), rice rats (*Oryzomys* and others), tree mice (*Thomasomys* and others), deer mice (*Peromyscus*), pericots (*Phyllotis*), maned rats (*Lophiomys*), lemmings (*Lemmus* and others), voles (*Microtus* and many others), muskrats (*Ondatra* and others), gerbils (*Gerbillus* and others), and other forms. 97 genera. Worldwide.	*Diverse.* Most species are solitary, and possibly all are territorial. Some *Peromyscus* species show male-female bonding for varying periods, and in a few cases larger winter aggregations. Some *Microtus* live in extended mother-litter associations, especially at high population densities. *M. brandti,* a prairie dweller, forms coteries of mixed sexes similar to those of the prairie dog *Cynomys.*	Linsdale and Tevis (1951), Eibl-Eibesfeldt (1953), F. Petter (1961), Eisenberg (1962–1968), Errington (1963), Lidicker (1965), Arata (1967), Healy (1967), Dunaway (1968), King (1968), Linzey (1968), Packard (1968), Stones and Hayward (1968), Baker (1971), Matthews (1971), Getz (1972), Myton (1974)
SPALACIDAE Mole rats (*Spalax*). Middle East.	*Solitary,* territorial.	Arata (1967)
MURIDAE Old World rats and mice (*Mus, Aethomys, Apodemus, Dendromus, Rattus* and many others). 98 genera. Throughout Old World.	*Diverse.* Many of the species are solitary. Harems and loose family bands are formed in *Mus* and *Rattus.*	Calhoun (1962), Barnett (1963), Eisenberg (1966), Arata (1967), Saint Girons (1967), Ropartz (1968), Ewer (1971), Matthews (1971), Wood (1971), R. M. Davis (1972)
GLIRIDAE Dormice (*Glis* and others). Europe, Middle East, and Africa.	*Aggregative and familial.* Winter aggregations of both sexes occur; in captivity families stay together at least temporarily.	Koenig (1960), Eisenberg (1966), Arata (1967)
ZAPODIDAE Jumping mice (*Zapus* and others), birch mice (*Sicista*). North temperate zone.	*Solitary,* evidently territorial.	Quimby (1951), Whitaker (1963), Eisenberg (1966), Arata (1967)
DIPODIDAE Jerboas (*Dipus* and others). North Africa, Asia.	*Solitary,* evidently territorial.	Eisenberg (1966, 1967), Arata (1967)

Table 23-1   (continued).

Kind of mammal	Sociobiology	References
HYSTRICIDAE Old World porcupines (*Hystrix* and others). Africa to China.	*Diverse.* Some species evidently live in pairs, and a few in communal warrens.	Starrett (1967)
ERETHIZONTIDAE New World porcupines. (*Erethizon* and others). Alaska to South America.	*Solitary.* Overlapping home ranges which are marked by scent. Breeding is slow and maternal care prolonged.	Eisenberg (1966), Starrett (1967)
CAVIIDAE Guinea pigs and cavies (*Cavia, Microcavia*, and others). Patagonian hare (*Dolichotis*). South America.	*Solitary and territorial.* One male may accommodate multiple females within his area, while the females maintain dominance in smaller areas of their own. In *Microcavia*, females tolerate daughters until birth of next litter, while males congregate near receptive females and form dominance order. *Dolichotis* mated pairs appear to share territories through successive litters.	King (1956), Kunkel and Kunkel (1964), Rood (1970), Eisenberg (personal communication)
HYDROCHOERIDAE Capybaras (*Hydrochoerus*). South and Central America.	*Social.* Largest of all rodents. Form small herds of 3–30, composed of mixed sexes and ages and consisting at least in part of family groups.	Starrett (1967), Matthews (1971)
CAPROMYIDAE Hutias (*Capromys, Geocapromys*, and others).	*Solitary.* Some dispersion and mutual avoidance in the wild, but relatively tolerant and form groups in captivity.	Clough (1972)
MYOCASTORIDAE Nutrias, coypus (*Myocastor*). South America.	*Familial.*	Ehrlich (1966)
DASYPROCTIDAE Agoutis (*Dasyprocta* and *Myoprocta*), pacas (*Cuniculus* and others). South and Central America.	*Solitary or pair bonded.* Groups form in captivity but dispersion appears to be the rule in the wild. Male and female share territory, each defending against members of its own sex.	Starrett (1967), Kleiman (1971, 1972a), Eisenberg (personal communication)
CHINCHILLIDAE Chinchillas (*Chinchilla*), viscachas (*Lagidium, Lagostomus*). South America.	*Diverse.* The mountain viscacha (*Lagidium*) forms coteries of 2–5 individuals of both sexes and various ages, consisting at least in part of families. Coteries occur close together in "colonies" (dense populations) of up to 75 individuals. In breeding season females become hostile, males wander through burrow system and may den in groups.	Pearson (1948), Starrett (1967)
THRYONOMYIDAE Cane rats (*Thryonomys*). Africa.	*Harem.*	Ewer (1968)
OTHER RODENT FAMILIES 24 other living families, many small and rare, and generally poorly known.	*Diverse.* Poorly known.	Arata (1967), McLaughlin (1967), Packard (1967), Starrett (1967)

Table 23-1 (*continued*).

Kind of mammal	Sociobiology	References
ORDER MYSTICETI		
BALAENIDAE Right whales (*Balaena,* *Caperea*)	*Diverse. Caperea:* solitary or in pairs. *Balaena:* solitary or family pods of male, female, and young.	Slijper (1962), Norris (1966, 1967), Rice (1967), Mörzer Bruyns (1971)
ESCHRICHTIIDAE Gray whales (*Eschrichtius*)	*Seasonally varying.* The whales migrate singly or in pods of up to 12; on arctic feeding grounds they form large loose aggregations, with females and calves tending to stay apart from males.	Same as above
BALAENOPTERIDAE Rorquals: blue, sei, and fin whales (*Balaenoptera*); and humpback whales (*Megaptera*).	*Social.* Pods of variable size, coalescing into larger aggregations in rich feeding areas. *Megaptera* well known for its elaborate songs; this whale often forms family pods of male, female, and young.	Slijper (1962), Norris (1966, 1967), Rice (1967), Mörzer Bruyns (1971), Payne and McVay (1971)
ORDER ODONTOCETI		
ZIPHIIDAE Beaked whales, including bottle-nosed whales (*Ziphius* and others).	*Diverse.* Some species evidently solitary (*Mesoplodon, Ziphius*), but *Hyperoodon* forms tightly packed, well-coordinated pods of 10 or more individuals.	Norris (1966, 1967), Rice (1967), Mörzer Bruyns (1971)
MONODONTIDAE Belugas (*Delphinapterus*) and narwhals (*Monodon*).	*Social.* Form schools of variable size.	Same as above
PHYSETERIDAE Sperm whales (*Physeter, Kogia*).	*Social.* Females and calves travel in tight nursery schools, accompanied by one or more large adult bulls ("schoolmasters"); younger males often form loose bachelor schools. The schools sometimes coalesce in large temporary aggregations of as many as 1000 individuals.	Caldwell et al. (1966), Norris (1966, 1967), Rice (1967)
PLATANISTIDAE Long-snouted river dolphins (*Platanista, Inia,* and others).	*Social.* Travel in small schools of less than a dozen.	Layne (1958), Layne and Caldwell (1964), Rice (1967)
STENIDAE Rough-toothed dolphins (*Steno*) and ridge-backed dolphins (*Sotalia* and *Sousa*).	*Social.* Travel in schools of variable size; as many as 1000 individuals reported in schools of *Steno*, but membership of less than 10 is more common.	Rice (1967), Mörzer Bruyns (1971)
PHOCOENIDAE Porpoises (*Phocoena* and others).	*Social.* Schools usually contain 6 members or less; they sometimes chase fish in crescentic formations.	Rice (1967), Mörzer Bruyns (1971), Vaughan (1972)

Table 23-1 (*continued*).

Kind of mammal	Sociobiology	References
DELPHINIDAE Ocean dolphins, "porpoises," killer whales (*Delphinus, Orcaella, Orcinus, Lissodelphis, Grampus, Stenella, Tursiops,* and others).	*Social.* Schools highly variable in size; aggregations of up to 100,000 have been observed in saddle-backed dolphins (*Delphinus delphis*). Pilot whales (*Globicephala scammoni*) travel over broad front, often broken into subgroups of same age and sex, then "loaf" in mixed groups. Killer whales (*Orcinus orca*) hunt sea lions, whales, and other dolphins in well-coordinated packs.	Tavolga and Essapian (1957), Norris and Prescott (1961), Dreher and Evans (1964), Norris (1966, 1967), Rice (1967), Evans and Bastian (1969), Pilleri and Knuckey (1969), Martinez and Klinghammer (1970), Mörzer Bruyns (1971), Caldwell and Caldwell (1972), Saayman et al. (1973), Tayler and Saayman (1973)
ORDER CARNIVORA Dogs, cats, raccoons, bears, otters, weasels, skunks, civets, hyenas, and allies.	*See Chapter 25.*	
ORDER PINNIPEDIA OTARIIDAE Sea lions (*Otaria, Eumetopias, Neophoca, Zalophus*), fur seals (*Arctocephalus, Callorhinus*).	*Social.* In breeding season, otariids assemble in large herds on the beaches and other protected shoreline sites, where largest males guard territories containing harems of females and young.	McLaren (1967), Orr (1967), Peterson and Bartholomew (1967), Stains (1967), Peterson (1968), Schusterman and Dawson (1968), Farentinos (1971), Matthews (1971), Stirling (1971, 1972), Caldwell and Caldwell (1972), Nishiwaki (1972)
ODOBENIDAE Walrus (*Odobenus*)	*Social.* Male herds remain apart from those composed of females and young except during the breeding season, when they fight among each other. Harems do not appear to be formed. Mother-offspring relationships are maintained for 3 years.	Eisenberg (1966), Perry (1967), Stains (1967)
PHOCIDAE Earless seals, including common seals (*Phoca*), gray seals (*Halichoerus*), leopard seals (*Hydrurga*), elephant seals (*Mirounga*), bearded seals (*Erignathus*), monk seals (*Monachus*), hooded seals (*Cystophora*), and others.	*Highly diverse.* From nearly solitary (*Hydrurga*), to gregarious but with little organization and promiscuous sexual activity (*Erignathus, Monachus*), pair formation during breeding season with spacing between families (*Cystophora*), and otariidlike harem formation (*Halichoerus, Mirounga*).	Bartholomew (1952, 1970), Scheffer (1958), Bartholomew and Collias (1962), Carrick et al. (1962), Eisenberg (1966), Stains (1967), Peterson (1968), Ray et al. (1969), Nicholls (1970), Caldwell and Caldwell (1972), Le Boeuf et al. (1972), Nishiwaki (1972), Le Boeuf (1974)

Table 23-1 (*continued*).

Kind of mammal	Sociobiology	References
ORDER TUBULIDENTATA ORYCTEROPODIDAE Aardvarks (*Orycter-opus*). Africa.	*Solitary.* Female accompanied by 1 and sometimes 2 offspring.	Eisenberg (1966), Hoffmeister (1967), Pagès (1970)
ORDER HYRACOIDEA PROCAVIIDAE Hyraxes or dassies (*Pro-cavia, Dendrohyrax, Heterohyrax*). Africa, Arabian peninsula.	*Social. Dendrohyrax* in family groups consisting of male, female, and their offspring; *Procavia* in "colonies" within which the unit is a male and harem of females and offspring.	Coe (1962), Eisenberg (1966), Hoffmeister (1967), Rahm (1969), Matthews (1971)
ORDER SIRENIA DUGONGIDAE Dugongs (*Dugong*). East Africa to Solomon Islands.	*Social.* Occur in small groups, which in at least some cases are family groups.	Eisenberg (1966), Jones and Johnson (1967)
TRICHECHIDAE Manatees (*Trichechus*). Florida to South America, West Africa.	*Solitary or weakly social.* Basic unit is the mother and a single young, but loose aggregations formed under some circumstances.	Moore (1956), Eisenberg (1966), Bertram and Bertram (1964), Jones and Johnson (1967)
ORDER PERISSODACTYLA Horses, zebras, asses, tapirs, and rhinoceroses.	*See Chapter 24.*	
ORDER ARTIODACTYLA Pigs, peccaries, hippopotami, camels, chevrotains, deer, giraffes, antelopes, cattle, goats, sheep, and allies.	*See Chapter 24.*	
ORDER PROBOSCIDEA Elephants	*See Chapter 24.*	

Table 23-2 Phylogenetic distribution of social systems within the bats, showing the great diversity at the level of the genus and below. (Based on Bradbury 1975.)

Kind of bat	A Solitary except for copulation and mother-offspring association	B Sexes separate except for mating	C Sexes segregate at parturition; sexes together at other times	D Monogamous families	E Year-round harems	F Year-round multimale, multifemale groups
PTEROPODIDAE (fruit-eating bats, "flying foxes")						
Pteropus eotinus		X				
P. geddiei		X				
P. giganteus						X
P. poliocephalus		X				
P. scapulatus		X				
Epomops franqueti	X					
Megaloglossus woermanni	X					
Rousettus leschenaulti			X			
RHINOPOMATIDAE (mouse-tailed bats)						
Rhinopoma hardwickei			X			
EMBALLONURIDAE (sac-winged bats, ghost bats)						
Balantiopteryx plicata			X			
Diclidurus alba	X					
Rhynchonycteris naso						X
Saccopteryx bilineata					X	
S. leptura						X
Taphozous melanopogon			X			
T. nudiventris			X			
T. peli				X		
NYCTERIDAE (hispid bats)						
Nycteris (*arge, hispida, nana*)				X		
RHINOLOPHIDAE (horseshoe bats, Old World leaf-nosed bats)						
Hipposideros atratus		X				
H. beatus				X		
H. brachyotis				X		
H. commersoni			X			
H. diadema			X			
Rhinolophus rouxi		X				
R. clivosus			X			
R. lepidus			X			
PHYLLOSTOMATIDAE (American leaf-nosed bats)						
Macrotus waterhousii			X			
Mormoops megalophylla		X				
Phyllostomus discolor					X	
P. hastatus					X	

Table 23-2 (*continued*).

Kind of bat	A Solitary except for copulation and mother-offspring association	B Sexes separate except for mating	C Sexes segregate at parturition; sexes together at other times	D Monogamous families	E Year-round harems	F Year-round multimale, multifemale groups
VESPERTILIONIDAE (common bats)						
Antrozous pallidus			X			
Eptesicus fuscus			X			
E. minutus	X					
E. rendalli	X					
Kerivoula (*harrisoni, papillosa, picta*)				X		
Lasiurus borealis	X					
L. cinereus	X					
Myotis (*austroriparius* and others)			X			
Miniopterus australis			X			
M. schreibersii			X			
Plecotus auritus			X			
P. townsendii			X			
Pipistrellus (*pipistrellus* and others)			X			
MOLOSSIDAE (mastiff bats, free-tailed bats)						
Tadarida brasiliensis			X			
T. major			X			
T. midas					X	
T. pumila			X			

groups of mixed sexes. The combination of such systems displayed by related species varies from family to family and is not easily predicted from existing knowledge of other aspects of natural history. Bradbury (1975), whose excellent review is the basis for this conclusion, cites an example from the genus *Saccopteryx* to illustrate how subtle the environmental factors can be that control social evolution. On Trinidad, groups of *S. bilineata* rest principally on the buttresses of large trees. When disturbed by a bird or mammal the bats drop to safety in the dark recesses between the buttresses and remain motionless. This habit allows the formation of moderately large, stable aggregations and, from that, a more elaborate social system. The males keep year-round harems while competing with each other by means of complex singing, barking, gland shaking, and hovering. The related species *S. leptura* occurs in the same localities but forms groups of five individuals or less on the exposed boles of trees. When disturbed they immediately fly off to some other, usually well-known site. Evidently as a result of this escape strategy, and more particularly

the small group size it necessitates, the *S. leptura* males do not form harems, and their signal repertory is smaller than that of *S. bilineata*. The *Saccopteryx* case is strongly reminiscent of the two principal defense strategies of social wasps. Some species, in particular the numerous members of the neotropical genus *Mischocyttarus*, form small, rapidly maturing colonies that fly to new sites when attacked by army ants or some other formidable predator. Others, such as the members of *Chartergus*, *Polybia*, and *Vespa*, build fortresslike nests that can withstand almost any predator. The latter species are characterized by very large colonies, marked physical differences between queens and workers, and more elaborate communication systems (Jeanne, 1975).

A few other trends are visible within the Chiroptera as a whole. Smaller species of bats, which have the most difficulty with thermoregulation, tend to nest more in protected sites such as caves and the hollows of large trees. Consequently, they form larger aggregations and as a rule cluster while resting, traits that set the stage for the evolution of the more advanced forms of social organization. But the correlation is weak. One of the most spectacular lek systems occurs in the large, sexually dimorphic *Hypsignathus monstrosus*, an African bat that also rests in the open in the forest canopy. Huge permanent aggregations are formed in trees by some of the large fruit-eating bats of the family Pteropidae, evidently as a protective device against predators. Overall correlations between diet and social systems are still weaker and perhaps even nonexistent.

A relative intractability to quick evolutionary analysis also characterizes the other mammalian orders. This is very much the case in the largest and most interesting eutherian groups, including the rodents, artiodactyls, and primates. It is also true of the marsupials, which provide us with the one great evolutionary experiment outside the eutherians. In the case of artiodactyls and primates, the analysis has begun to reach sufficient depth to establish correlations at the level of the genus and species. These mammal groups will be the subjects of later special chapters. Also, it is now possible to assess to some extent the relative degree of evolutionary lability in individual social traits. In Chapter 27, the procedure will be used to help reconstruct the early evolution of man.

General Patterns

The details of mammalian social evolution are best summarized not by a general phylogenetic tree but by the Venn diagram displayed in Figure 23-1. This arrangement recognizes that the close mother-offspring relationship is universal and that the other social traits are added or subtracted at the genus or species level with relative ease. The square encloses the set of all mammalian species at a given moment in time. Evolutionary changes in individual species are depicted as tracks through time across boundaries of the subsets. Additional, smaller subsets can be delimited. Details vary, for example, in the mode of intrasexual cooperation, the degree of cohesion, and the openness of the societies. Also, most of the forms of interaction change seasonally in one species or another, and the patterns of these changes differ at the species level.

Yet despite the patchy distributions of particular social systems among the species, certain broad phylogenetic trends can be detected within the Mammalia as a whole and within a few of its larger orders (Eisenberg, 1966). Stem groups such as the more primitive living marsupials and insectivores tend, as expected, to be solitary. Species that forage nocturnally or underground are also predominantly solitary. As a rule the most complex social systems within each order occur in the physically largest members. This is true, for example, of the marsupials, rodents, ungulates, carnivores, and primates. Perhaps the trend partially reflects the simple fact that the largest animals forage above ground and during the day. But another significant correlate must be their increased intelligence. The biggest forms in

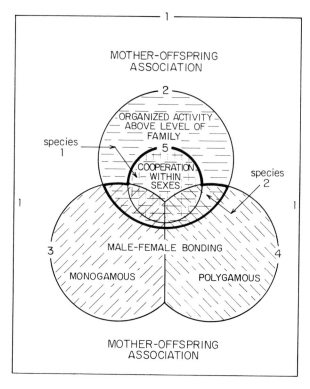

"SOLITARY"

　1 only: mother-offspring groups, with
　　　　males visiting only for purposes
　　　　of mating.

SOCIAL

　2 only: unorganized herds, schools, and
　　　　other motion groups.

　3 only: mated pairs, often territorial.

　4 only: harems, often territorial.

　5: packs, bands, troops

　2+3, 2+4, and 5 (heavy outline):
　　　　structured societies above the
　　　　level of the family.

Figure 23-1 The diversity of mammalian social systems is represented by a Venn diagram delimiting species that possess combinations of particular social traits. The square encloses all mammalian species at a given moment of time and the circles the subsets of species that possess individual social traits. The heavy line in the center encloses the mammal species that are considered to have the most advanced social organizations. Phylogenetic trees and evolutionary grades are not used because of the great complexity of pattern created by lability in most social traits at the level of the genus and species, making generalized diagrams of this nature impracticable. However, the inferred evolution of individual species can be represented as tracks through the subsets, as illustrated by the cases of the imaginary species *1* and *2*.

each taxonomic group, regardless of their way of life, ordinarily possess larger, more complexly structured brains and are capable of greater feats of learning. Finally, species adapted to life in open environments are more likely to be social. For example, the most social of all marsupials are the species of wallabies and kangaroos that graze in the grasslands and open woodlands of Australia. The few rodent species known to form coteries of mixed sexes are all inhabitants of grasslands. Among the ungulates, the great herds are formed predominantly by species limited to grasslands and savannas. Although the herds are very loosely structured in most cases, those of horses, mountain sheep, elephants, and a few other forms comprise cohesive, highly organized societies.

The remainder of this chapter will be devoted to three mammalian species possessing the most advanced form of social behavior in their own groups. The whiptail wallaby and the black-tail prairie dog are located at the apices of the marsupials and rodents, respectively. The bottle-nosed dolphin is a promising but still enigmatic species that will represent the cetaceans (orders Mysticeti and Odontoceti, including all the whales and dolphins), the least understood of all major groups of mammals. In the subsequent four chapters, which conclude the book, more nearly complete reviews will be presented of the ungulates, carnivores, and primates.

The Whiptail Wallaby (*Macropus parryi*)

Whiptail wallabies, which are probably the most social of all living marsupials, range from northern Queensland to northeastern New South Wales. Their preferred habitat is open *Eucalyptus* woodland with an abundance of grass. These attractive little macropods are diurnal grazers, feeding exclusively on grass and some other herbaceous plants, including ferns. A free-living population was studied by John H. Kaufmann (1974a) at Gorge Creek, in the Richmond Range of New South Wales, for a period of 13 months. The animals were found to be grouped into three loosely organized "mobs" which remained stable throughout the year. Each mob contained 30 to 50 members. The adult sex ratio differed greatly from one group to the next, and it is possible that the overall ratio was balanced in the population as a whole. Although the data are fragmentary, it is apparent that at least some of the subadult males wander from one mob to the next, while females rarely if ever do so.

The three mobs occupied nearly exclusive home ranges 71, 99, and 110 hectares in extent, respectively. The overlap was only about 10 hectares in the two cases where it could be reliably measured. Meetings between mobs were uncommon but amicable. They resulted in a temporary fusion of the groups into single aggregations that rested and fed together. On such occasions the wallabies treated individuals belonging to other groups much as they did members of their own group. Animals of all ages mingled easily, while the adult males fought

for dominance and courted females with no apparent particular reference to mob affiliation.

Within the home range of the mob as a whole, individual members utilized smaller areas of their own. Kaufmann was able to distinguish a relatively consistent daily pattern of movement by each mob, in which individuals aggregated during the night among the trees and broke into irregular smaller groups to forage over open ground during the day. The pattern varied somewhat in detail among the mobs. One, for example, regularly broke into large subgroups of 15 or more while progressing down off a wooded ridge early in the morning. During the middle of the day the members moved about in scattered groups which changed frequently in size and composition. By late afternoon some of the subgroups merged before moving back up the ridge. Sometimes virtually the entire mob reassembled before returning. The other two mobs, living in areas with different arrangements of vegetation, did not migrate up and down the ridges. However, they still tended to aggregate in open areas during the day.

This casual regime was reflected in a weak, individualistic mode of social organization. The wallaby mob was little more than a loosely structured aggregation, with individuals and small groups carrying on differing activities in close proximity to one another (see Figure 23-2). Dominance hierarchies existed among the subadults and adults. They were diffuse and infrequently expressed by the females but strongly marked, linear, and reinforced at frequent intervals in the case of the males. The aggressive behavior was highly ritualized. Its mildest form was physical displacement, in which one wallaby caused another to move aside. The first animal sometimes simply approached and sniffed the other or touched its nose, inducing it to step away. Sometimes it leaped at its opponent from behind and seized it around the middle. Displacement occurred most frequently when males contended for access to a female or when females were trying to ward off amorous males. In the case of male conflict, displacement often led to chasing and fighting. Kaufmann was impressed by the "gentlemanly" nature of fighting in the Gorge Creek population. One male usually challenged another by standing upright—the fighting position—and perhaps also by placing his paws gently on the opponent's neck or upper body. When the challenge was accepted the fight proceeded in a predictable manner. The combatants faced each other erect, rearing to the fullest possible height by standing on their toes. They then pawed with open hands at each other's head, shoulders, chest, and throat. More force was put in the return motion than in the extension. Sometimes pawing gave way to wrestling, in which the two males seized each other around the neck or shoulders and tried to throw each other over. In a small percentage of fights one animal kicked his opponent in the abdomen with his hind legs. This was done with far less than maximum force and usually indicated that the kicker was about to give up. Fighting clearly served to reinforce the dominance relationships among males. It was initiated in most

Figure 23-2 A mob of the whiptail wallaby (*Macropus parryi*), considered to be the most social of all the marsupials. The scene is Gorge Creek, New South Wales, in the early morning. The entire group is still assembled, but the wallabies will soon begin to break up into smaller subgroups that move into more open areas to feed. There is virtually no coordination of the mob as a whole. Individuals and small groups carry on diverse activities within close proximity of one another. In the foreground various females and joeys (young animals) rest, groom, and go through the first motions of feeding. To the left two females can be seen sniffing each other for identification. To the right of center two

males spar in the ritual combat that determines rank in the dominance hierarchy. A third male watches the encounter. To the rear of this group a male inspects the cloacal area of a female, a frequent procedure used to "check" whether females are in estrus. In the left background a courting male bends toward an estrous female while pawing up earth and grass. Three subordinate males surround the pair, each ready to commence courtship displays of his own should the dominant male leave the vicinity. The habitat is open woodland. The ground cover consists principally of grass and clover with a sprinkling of bracken ferns and thistles. (Drawing by Sarah Landry; based on J. H. Kaufmann, 1974.)

cases by the higher-ranking animal, and was most vigorous among males of nearly equal rank. It was never observed to result in visible injury.

Superior rank paid off in access to estrous females. In the few hours in which a female remained in this condition as many as a half dozen or more males trailed her. But usually only the alpha male copulated with her. When this individual was occupied with another estrous female, the second-ranking male took his place. The shortness of duration and the unpredictability of timing of estrus resulted in a great deal of sexual searching on the part of the males. In fact, the commonest overt social interaction seen among the whiptail wallabies was the sexual "checking" of females by males. Kaufmann believes that virtually every male checks most or all of the females in the mob every day. Since the females are out of estrous on all but a small fraction of days, and only dominant individuals have a reasonable chance of success, most of the effort must come to nothing. Nevertheless, it keeps each male in a state of readiness for the opportunity that may eventually come his way. At Gorge Creek the checking procedure was initially olfactory. Typically, the male approached the female from behind and quickly sniffed at her tail, perhaps going so far as to lift the tail and to paw and lick the female's cloaca. Occasionally the female responded by urinating into the male's mouth. Next the male stood in front of the female, pushing his head toward her or waving it back and forth and up and down. Sometimes he crossed his arms over his chest or placed them gently on her head or shoulders. When the female was not in estrus her usual response was to move away or to hit at him with her paws until he retreated. As the female entered estrus the approaches became more prolonged and persistent. At first low-ranking males trailed her, but at the peak of estrus they were invariably forced away by the highest-ranking male in the neighborhood. An exclusive consort relationship was then established that lasted from one to four days. Sometimes the female broke into a run and led her consort and the other males on a wild chase.

The Gorge Creek whiptails often sniffed one another in various nonsexual contexts, leading to the suspicion that olfactory communication is a strong supplement to the more obvious visual displays. Allogrooming was surprisingly infrequent. It consisted principally of licking and was mostly limited to interactions between mothers and their offspring and between juveniles, only rarely occuring between fighting males. Thus the behavior does not play the conciliatory role so conspicuous in primates and other eutherian orders.

Play among the wallabies was weakly developed in comparison with most eutherians. It was almost entirely limited to interactions between mothers and their offspring and consisted of mock sexual and aggressive movements. When fighting was begun by subadult males, it was "serious" in form and led directly to the formation of

dominance hierarchies. Thus in this context fighting was already functional and not true play.

In summary, the whiptail wallaby is of exceptional interest because it represents the limit of social evolution in a major group of mammals phylogenetically remote from all others that have been studied to date. Although complex, ritualized behaviors have emerged in the evolution of courtship and aggression, resulting in a well-developed dominance system among the males, the wallabies apparently have not produced other modes of internal organization Their aggregations are stable and the group home ranges are persistent and nearly exclusive. Tolerance between groups is remarkably high and may be facilitated by recognition of individuals across the groups. In this one, special way whiptails resemble chimpanzees. But in other aspects their behavior is strongly individualistic, and the total social pattern over short periods of time tends to be chaotic. Although aggression plays an important role in whiptail social life, allogrooming has not evolved to a compensatory level as it has in most eutherians. Finally, the relationships between the mother and her offspring are as complex as in the social eutherians, but relationships between young age peers remain rudimentary. Social play among the peers is virtually nonexistent despite the fact that the particular adult interactions to which play normally relates are as complex and personalized in the whiptails as in other mammals.

The Black-Tail Prairie Dog (*Cynomys ludovicianus*)

Rodent species that live in the most exposed habitats tend to form dense local populations. Within these "colonies" individuals or small social groups maintain their separate burrow systems and defend small territories around the burrow entrances. Examples include the arctic ground squirrel (*Spermophilus parryi*) on the open tundra, marmots (*Marmota*) and viscachas (*Lagidium*) in alpine meadows, the vole *Microtus brandti* in grasslands, and others. The culmination of this trend, which has appeared independently in several lines, is represented by the black-tail prairie dog (*Cynomys ludovicianus*) of the northern plains. This species has been studied intensively in the wild in the Black Hills of South Dakota by J. A. King (1955). His results have been confirmed and extended by W. J. Smith et al. (1973) with reference to both free-living populations and a captive colony at the Philadelphia Zoo. The communication system has been the subject of a meticulous study by Waring (1970).

In the Black Hills, local populations, sometimes referred to as towns, contain as many as 1000 individuals. The towns are physically divided by ridges, streams, or bands of vegetation into wards. The wards are comprised in turn of the coteries, the real social units, which are separated by behavioral rather than environmental features.

The average coterie composition in the population studied by King was 1.65 adult males, 2.45 adult females, 3.57 immature males, and 2.36 immature females. The largest group discovered contained 38 individuals—2 adult males, 5 adult females, 16 immature males, and 15 immature females. The larger coteries were usually soon reduced in size by fission and emigration of individuals.

The members of coteries share burrows and clearly recognize one another as associates. When any two prairie dogs meet they "kiss," touching lips with the mouths open and teeth exposed. This identification exchange perhaps originated as a ritualized threat display. When the kissing animals are members of the same coterie, they may simply brush on past each other. But just as often they proceed to groom each other. One lies down while the other nibbles through its fur with its teeth. Occasionally the kiss ends with the two animals lying side by side for a while and then moving off to feed in concert. When the two are strangers the kiss leads to a different sequel. The tails are raised, exposing the anal glands. The rodents take turns sniffing the glands until finally one gives up and leaves the vicinity.

The most extraordinary single feature of the social life of these animals is the fact that the coterie territorial limits are passed on by tradition. The population of each coterie constantly changes over a period of a few months or years, by death, birth, and emigration. But the coterie boundary remains about the same, being learned by each prairie dog born into it. The young animals evidently acquire this information through repeated episodes of grooming from other members of the coterie along with rejection by territorial neighbors. New coteries are formed by adult males who venture into adjacent empty land and commence burrowing there. They are followed by a few adult females. The juveniles and subadults are left behind in the old burrows. The coterie system partially breaks down each year in the late winter and early spring, when the females raising pups defend parts of the burrow system against all comers.

Allogrooming, in sharp contrast to the situation in the whiptail wallabies, is the most common form of social interaction in prairie dogs. Pups are especially fond of the activity, and they frequently pursue adults in order to present themselves for grooming. In addition, prairie dogs employ an exceptionally rich repertory of auditory and visual signals. When potential predators approach the towns, a wave of barking—actually a high-pitched nasal yipping—spreads from burrow to burrow. The call reaches its highest intensity when a hawk or eagle is sighted overhead. At this time it becomes so different in pitch, rate, and duration as to effectively constitute a distinct display. Another kind of bark, slow and intermittent, is given when an animal defends its territory. The vocalization may be accompanied by tooth chattering, especially when the animal is seriously threatening its opponents. Females defending their burrows give a distinctive muffled bark. When a prairie dog is chased after losing a fight,

it typically emits a churring sound, which may serve as a signal of submission that reduces hostility in the pursuer. Finally, the most dramatic of all the displays is the "confident" territorial call. The animal rears up on its hind legs, emitting a loud syllable by inspiration on the way up, then comes back down while delivering a second syllable through expiration. The double cry is sometimes given with such force that the prairie dog leaps off the ground. It may even topple over backward. King has compared the vocalization to the advertisement song of a male bird secure on its own territory. To a human observer the prairie dog seems to be saying, "This is my coterie's territory. Nothing can drive me away. Strangers keep out."

The association between life in open environments and advanced social organization in the black-tail prairie dog and other rodents is one of the strongest such correlations to be found in all of the mammals. What, if any, are the prime movers in the environment? One suggested by King and more or less accepted by the majority of other students of the subject (for example Carl, 1971, and Smith et al., 1973) is predation. When a rodent becomes specialized for life in the most exposed habitats, it substitutes dense aggregations and a communal alarm system for the cover of rocks and vegetation. At the same time the black-tail prairie dog has largely shifted its diet from the grasses of the undisturbed prairie to the forbs that flourish in the soil excavated from the burrow systems. This rodent has used its social life to modify the environment to its liking. Or should we say instead that the prairie dog has modified its liking to the socially altered environment? One is tempted to select the latter hypothesis, which implies that predation was indeed the prime mover and that other changes were postadaptations forced by the original change. But at this point there is no way of being sure. In either case it is clear that social life has permitted the development of denser rodent populations in certain sections of the prairie than would have been possible otherwise. The demographic concomitants of the security of coterie existence are low birth rates and long average life. The behavioral concomitants are a rich new repertory of signals specialized for groupmate recognition and varying forms of territorial defense.

Dolphins

Are bottle-nosed dolphins more intelligent than other animals and perhaps the equals of human beings? Do they communicate with one another by a highly sophisticated but alien language not yet decoded by human observers? These are notions widely held by the public and even among scientists, thanks largely to John C. Lilly's two books *Man and Dolphin* (1961) and *The Mind of the Dolphin: A Nonhuman Intelligence* (1967). In my opinion, Lilly's books are misleading to the point of bordering on irresponsibility. Lilly opens with an astonishing assertion: "Within a decade or two the human species will establish communication with another species: nonhuman, alien, possibly extraterrestrial, more probably marine; but definitely highly intelligent, perhaps even intellectual." This encounter will reveal "ideas, philosophies, ways, and means not previously conceived by the minds of men" (Lilly, 1961). It will quickly become the concern of governments, just as the atomic bomb brought nuclear physics into the realm of public policy. In order to support his thesis, Lilly turns to an account of the bottle-nosed dolphin. But having raised his readers' expectations so high, he chastely warns that he might be wrong about the dolphin. This way of discussing a subject he stoutly defends: Doesn't science progress by the negation of hypotheses?

Although Lilly never states flatly that the dolphin and other delphinids are the alien intelligences he seeks, he constantly implies it. "They may have a nomadic culture, they may herd their own fish—we do not know. These are facts yet to be determined." Anecdotes are used to launch sweeping speculations. A case of rapid retreat by killer whales from a whaling fleet leads to the conjecture that the whales might have been saying to one another, "There is a thing sticking out in front of some of these boats that can shoot a sharp thing that can go into our bodies and explode. There is a long line attached to it by which they can pull us in." This fantasy is then turned into a premise for even stronger discussion and speculation: "Now let us contrast this 'conversation' with that of the school of fish . . In the first place, a lot of information is transmitted about another object, not killer whales, and this object is differentiated from similar objects in the neighborhood. A particular aspect of the different objects is dangerous, and they say it is dangerous." This example fairly represents the overall quality of Lilly's documentation and logic. Objective studies of behavior under natural conditions are missing, while "experiments" purporting to demonstrate higher intelligence consist mostly of anecdotes lacking quantitative measures and controls. Lilly's writing differs from that of Herman Melville and Jules Verne not just in its more modest literary merit but more basically in its humorless and quite unjustified claim to be a valid scientific report.

I have dealt frankly with these two books because they are possibly the most widely read works on sociobiology and therefore have been extraordinarily misleading to both the general public and a wide audience of scientists. They have served as the source of innumerable popular articles, several similar books by other authors, and a successful motion picture. Most zoologists simply ignore them when writing reviews of social behavior, but this noncommital attitude only serves to perpetuate the myth that Lilly helped to create. It is important to emphasize that there is no evidence whatever that delphinids are more advanced in intelligence and social behavior than other animals. In intelligence the bottle-nosed dolphin probably lies somewhere between the dog and rhesus monkey (Andrew, 1962). The communication and social organization of delphinids generally appear to be of a conventional mammalian type.

The factual basis on which the alien culture myth was created is the undeniably large size of the dolphin's brain and the exceptional ability of the animal to imitate. As pointed out by McBride and Hebb (1948), the brain of the Atlantic bottle-nosed dolphin *Tursiops truncatus* is about as large as a human being's, weighing approximately 1600 to 1700 grams, and it is also comparable in the degree of cortical convolution. But the brain size and cortical area alone are not precise measures of intelligence. The mass tends to increase in relation to body size, so that the sperm whale, a gigantic distant relative of the dolphin, possesses a brain weighing as much as 9200 grams. Perhaps the sperm whale is really a genius in disguise; the possibility cannot be totally discounted. But consider the brain of the elephant, which weighs approximately 6000 grams, or four times as much as that of a human being. The behavior of this largest of land animals is now well enough known for us to be reasonably sure that its intelligence is far below the human level and probably comparable to that of the more intelligent cercopithecoid monkeys and apes. Furthermore, in signal repertory and social organization the elephant does not differ radically from other ungulates (see Chapter 24). Thus brain size, while being roughly correlated with intelligence, is not a precise measure of it.

The significant question remains, however, as to why the dolphin brain is so large. The answer may lie in the dolphin's truly remarkable imitative powers. These animals are not only as easily trained as seals and chimpanzees to perform circus tricks, they show a strong tendency to imitate the actions of other species in the absence of reinforcement. Lilly reported that some captive dolphins answered laughter, whistles, and Bronx cheers with similar sounds. Phrases such as "One, two, three," "TRR," and "It's six o'clock" were also mimicked, albeit poorly. When Brown et al. (1966) placed an Atlantic bottle-nosed dolphin in the same tank with a Pacific *Stenella* dolphin, it made a spinning leap like that of the *Stenella* after seeing this distinctive maneuver only once. In the wild, bottle-nosed dolphins do not leap in this manner and the Atlantic specimen had never previously had an opportunity to see a spinning leap. Tayler and Saayman (1973) have provided a remarkable series of additional examples involving captive Indian Ocean bottle-nosed dolphins (*Tursiops aduncus*). When placed in the same tank as Cape fur seals, they imitated the seals' sleeping postures and various swimming, comfort, and sexual movements. One dolphin observed a diver cleaning algae from an observation window, then proceeded to repeat the movements while making sounds similar to those made by the air-demand valve and emitting streams of bubbles resembling the diver's exhaust air. Another watched a diver remove algae from the flow of a tank with a mechanical scraper, then manipulated the tool itself well enough to loosen some of the algae, which it proceeded to eat. In this final case the dolphin displayed a capacity comparable to the learning of the use of tools by chimpanzees.

Why has the dolphin become such a superb imitator? Andrew offered a plausible hypothesis for the vocal mimicry. As in the mimicking birds and primates, the behavior might cause a convergence of signals among group members and permit individuals to recognize their own group at a distance. This faculty would seem to be especially valuable to animals that cruise the open sea at high speeds, repeatedly joining and breaking away from schools of their own species. This factor alone could account for the hypertrophy of the capacity for vocal mimicry and the enlargement of the brain. Moreover, the dependence of delphinids on echolocation for orientation and the detection of prey has preadapted them for a strongly developed system of auditory communication. The tendency to imitate movements is less easily explained. Our knowledge of the behavior of free-ranging dolphin schools is still fragmentary, although studies are currently under way (see Saayman et al., 1973). There is a possibility that the members of schools adapt quickly to special challenges from the environment, profiting from the maneuvers of the most successful individuals during escapes from predators or the pursuit of fish. Such flexibility could also lead to coordinated behavior under particular circumstances. Hoese (1971) witnessed two bottle-nosed dolphins cooperate to strand small fish by pushing waves onto the muddy shore of a salt marsh. The dolphins then rushed up onto the bank for a short distance, seized the fish, and slid back down into the water.

Another form of cooperative behavior occurs during the rescue of disabled animals. When a member of a delphinid school is harpooned or otherwise injured, the usual response of the remainder of the school is to desert the area, leaving the injured member to its fate. But occasionally the school clusters around the animal and lifts it to the surface of the water, where it can continue to breathe. The following incident was recorded by Pilleri and Knuckey (1969) in the Mediterranean.

A school of approximately 50 *Delphinus delphis* was sighted. As soon as the Zodiac approached, they increased speed, dived and changed direction under water. The school reassembled behind the Zodiac. The yacht took over the chase and an animal was wounded by the harpoon. We saw quite clearly how other dolphins came immediately to the help of the wounded animal on the starboard side of the yacht. They supported the wounded dolphin with their flippers and bodies and carried it to the surface. It blew 2-3 times and then dived. The whole incident lasted about 30 seconds and was repeated twice when the animal appeared unable to surface alone. All the animals including the wounded dolphin then dived and swam quickly out of sight.

This scene is depicted in Figure 23-3. Similar behavior has been observed in both free-ranging and captive bottle-nosed dolphins (Caldwell and Caldwell, 1966). It represents a form of altruistic behavior comparable to acts of rescue observed in wild dogs, African elephants, and baboons. However, it does not necessarily reflect a higher order of intelligence. By itself the behavior is not as complicated as say, nest building by weaver birds or the waggle dance of honeybees. It could well represent an innate, stereotyped response to the distress of companions. Drowning that results from an incapacitating injury must be one of the chief causes of mortality among cetaceans. The automatic rescue of offspring and other relatives contributes greatly to inclusive fitness and is likely to have been fixed in the innate behavioral repertoire of the species.

Allomaternal behavior is also well developed in *Tursiops truncatus* (Tavolga and Essapian, 1957). In captivity at least, older, nonpregnant females associate with pregnant females and help to tend the newborn calves by swimming next to them. They sometimes lift stillborn calves to the surface in what can be interpreted as a rescue attempt.

The schools of social delphinids are highly variable in size. Those of the Pacific bottle-nosed dolphin *Tursiops gilli* consist of both sexes and usually contain 20 members or fewer, although exceptional groups of up to a hundred have been sighted. The species almost always swims in association with the pilot whale *Globicephala scammoni* (Norris and Prescott, 1961). In the Mediterranean the group size of *Delphinus delphis* and *Stenella styx* usually ranges between 10 and 100 individuals, although occasionally schools containing hundreds or even thousands of members have been seen (Pilleri and Knuckey, 1969). Several geometric formations of the schools have been noted, each with an apparently different function (see Figure 23-4). By watching from an underwater vehicle Evans and Bastian (1969) were able to learn that free-swimming *S. attenuata* form three kinds of schools distinguishable on a demographic basis. The first consists of a lone male, sometimes accompanied by a female; the second, 4 to 8 subadult males; and the third, 5 to 9 adult females and young. This triple array is strongly reminiscent of the herd organization of many ungulate species, in which males remain apart from nursery groups except during the breeding season. The impression is strengthened by the fact that captive Atlantic bottle-nosed dolphins form dominance hierarchies, with a senior bull ruling over subordinate males and females. The bull is especially aggressive during the breeding season, when he bites and rakes other adults with his teeth. He controls juveniles by ramming them with his head, striking them with his flukes, and threatening them with loud percussive jaw claps. Adult females sometimes dominate both lower-ranking males and other females, although the relationships are loose and imprecise (Tavolga and Essapian, 1957; Tavolga, 1966). The resemblance of these features to ungulate social behavior may have a basis in ecology. Like

Figure 23-3 Altruistic, cooperative behavior in the dolphin *Delphinus delphis*. On the left a group assists an individual that has just been struck by an electroharpoon. As described in the text, the dolphin was harpooned from a research vessel in the western Mediterranean. With blood pouring from its side, the injured animal is unable to rise to the surface to breathe and would soon drown if others did not push it upward, as shown here. Other members of the school mill nearby; to the far right can be seen two youngsters crowding close to their mothers. (Drawing by Sarah Landry; based on a written account by Pilleri and Knuckey, 1969.)

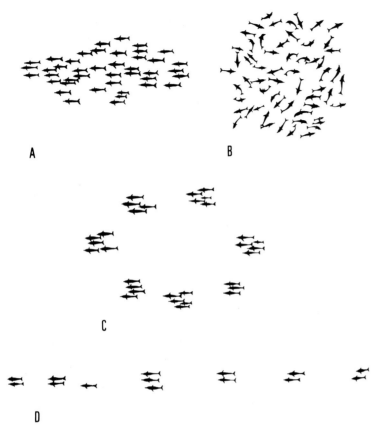

A

B

C

D

Figure 23-4 Principal formations of dolphin schools observed in the Mediterranean. *A:* navigating formation, during which the school swims in a constant direction, with cows accompanied by calves often being the most closely grouped (all delphinid species observed). *B:* feeding formation (*Delphinus delphis, Stenella styx*). *C:* hollow circle, apparently a "parade" formation for silent navigation through clear water (*Tursiops truncatus*). *D:* "parade" formation, used during silent navigation through clear water (*D. delphis*). (From Pilleri and Knuckey, 1969.)

the ungulates of the savannas and semideserts, delphinids "graze" and "browse" over wide areas. Their food consists of fish rather than vegetation, but the resource is similar in being patchily distributed in space and time. Under these circumstances it is generally advantageous to move in herds of variable size, with male and nursery groups capable of independent travel (see Chapter 3).

The communication systems of delphinids appear to be of approximately the same size and complexity as those of most other species of birds and mammals. Dreher and Evans (1964) were able to distinguish 16 distinct whistles in the Atlantic bottle-nosed dolphin, 16 in the Pacific bottle-nosed dolphin, and 19 in *Delphinus bairdi.* When any two of these three species are compared in detail, 60–70 percent of the signals are found to be held in common. To these can be added several percussive sounds produced by slapping the water with the flukes and snapping the jaws (Caldwell and Caldwell, 1972; see also Busnel and Dziedzic, 1966). Thus a reasonable first estimate of the total number of signals would lie between 20 and 30, well below the total systems of the rhesus monkey, the chimpanzee, and other higher nonhuman primates but comparable to those of most other vertebrates. However, this approximation could easily be too low. Because of the difficulty of studying free-living schools, investigations of the sociobiology of dolphins and other cetaceans are still in an early stage. It is extremely difficult to mark individual animals and to follow them during their lengthy travels in the open water of the sea. Moreover, the auditory signals employed in communication may be difficult to distinguish from the ultrasonic emissions used for purposes of echolocating prey and orienting under conditions of poor visibility. Finally, the challenge of communication in a featureless space may include unique problems that have been solved by cetaceans in ways unattainable by other marine animals. In particular, there is a strong prospect that these mammals have evolved long-distance contact signaling to hold the families and schools together. Such a function has already been suggested for the elaborate songs of the humpback whale.

Chapter 24 **The Ungulates and Elephants**

The ungulates, or hoofed mammals, are a heterogeneous assemblage once placed under the single order Ungulata but now recognized to comprise two phylogenetically distinct orders: the Perissodactyla, or odd-toed ungulates, including the horses, rhinoceroses, and tapirs; and the Artiodactyla, or even-toed ungulates, including the camels, pigs, deer, giraffes, antelopes, cattle, goats, sheep, and related forms. Ungulates are vegetarians whose limbs are largely specialized for running to escape big cats and other mammalian carnivores. Hooves in effect place the feet on points, permitting a faster striding rate and greater speed when the animals are running in the open. Elephants are called subungulates, an allusion to the fact that they originated from the same ancestral stock as the ungulates. They, too, are vegetarians but rely more on sheer bulk and strength to defeat predators.

Throughout most of the Cenozoic Era, for roughly 50 million years, the perissodactyls declined while the artiodactyls and elephants expanded. Since Pleistocene times, during the past 3 million years, artiodactyls and elephants have also declined. But the artiodactyls suffered the least of the three groups, so that today they are by a wide margin the dominant large herbivores throughout the world. And the premier artiodactyls are the ruminants, the suborder Ruminantia, comprised of the deer, antelopes, cattle, sheep, and their allies. Ruminants are distinguished by their peculiar mode of digestion. Food is swallowed with a minimum of mastication and later brought up from the four-chambered stomach as a cud which is then chewed and reswallowed. A huge population of symbiotic protozoans and bacteria living in the stomach breaks down the cellulose and is then itself partially digested and absorbed. The technique of rumination, combined with the use of microorganisms, allows the animals to feed on rough forage more efficiently and has undoubtedly contributed to their general ecological success.

Two characteristics of ungulates make them especially favorable for studies of social evolution: their strong tendency toward herd formation and the relatively large number of species (187 worldwide). In the past ten years there has been a dramatic upsurge of studies of both captive and free-ranging populations. Much of the information is summarized in condensed form in Table 24-1. The social systems considered together present a relatively simple pattern that can be transformed with minor distortion onto a single axis, or "sociocline." At one end is the primitive state shared with most other mammalian groups, including undoubtedly the Paleocene condylarths that gave rise to the ungulates and elephants: adults live alone except to pair, and the young animals remain closely associated with their mother until they are partially or fully grown. Some ungulates, for example the moose, retain this elementary organization while forming temporary aggregations at the best feeding grounds. Other species, such as the horses, pigs, and many antelopes, have taken a major additional step. Multiple female-offspring units are allied for prolonged periods

Table 24-1 Orders and lower taxa of ungulates, their sociobiological traits, and selected references.

Kind of ungulate	Sociobiology	References
ORDER PERISSODACTYLA		
FAMILY EQUIDAE (horses, asses, zebras). 1 genus, 7 species. Africa, Middle East to Central Asia; introduced worldwide.	*Harem-forming or male-territorial.* In Burchell's zebra (*Equus burchelli*) and the primitive horse (*E. caballus przewalskii*) groups of mares and their colts run together, with the mares typically forming dominance hierarchies. In most or all species this herd is controlled by a single male, who dominates and leads the females and aggressively excludes other males. A stallion of Burchell's zebra controls up to 6 mares, so that the herd may total 15 individuals if young are included. When the stallion disappears the herd stays together until another stallion takes over. The primitive horse has the basic harem system, but feral races diverge in various ways, showing the trait to be labile. In Grevy's zebra (*E. grevyi*) and the wild ass (*E. asinus*), males are territorial while mares and their colts roam in separate herds.	McKnight (1958), Klingel (1965, 1968, 1972), Eisenberg (1966), Tyler (1972), Estes (personal communication)
FAMILY TAPIRIDAE (tapirs). 1 genus, 4 species. Central and South America, Indochina to Sumatra.	*Solitary or (possibly) in pairs.* Social behavior in the wild is little known; adults appear to recognize and tolerate their own offspring even after separation into overlapping home ranges.	Hunsaker and Hahn (1965), Eisenberg (1966 and personal communication), Matthews (1971)
FAMILY RHINOCEROTIDAE (rhinoceroses). 4 genera, 5 species. Africa, tropical Asia.	*Diverse.* White rhinoceroses (*Ceratotherium simum*) of Africa form family groups of up to 5 members and at least temporary herds of up to 24 individuals. Other species are evidently solitary. Territoriality occurs in the black rhinoceros (*Diceros bicornis*) and white rhinoceros (*C. simum*); it may occur generally in other species, but data are few.	Ripley (1952, 1958), Hutchinson and Ripley (1954), Sody (1959), Lang (1961), Goddard (1967, 1973), Dorst (1970), Owen-Smith (1971, 1974), Mukinya (1973)
ORDER ARTIODACTYLA		
SUBORDER SUINA (pigs, peccaries, hippopotami)		
FAMILY SUIDAE (pigs, hogs). 5 genera, 8 species. Old World exclusive of Australia and Oceania.	*Diversely social.* In the European wild pig (*Sus scrofa*), several sows and their litters live together in bands; individual females break away from the group at the beginning of the birth season. African bush-pigs (*Potamochoerus porcus*) live in herds of 6–20 (occasionally up to 40) dominated by a master boar. Warthogs (*Phacochoerus aethiopicus*) live in family groups of male, female, and 1 or 2 successive litters. Several families sometimes join into larger herds. Warthog groups are territorial, and males fight for dominance.	Frädrich (1965, 1974), Gundlach (1968), Dorst (1970)
FAMILY TAYASSUIDAE (peccaries). 1 genus, 2 species. Southwestern U.S. to South America.	*Social.* The collared peccary *Pecari angulatus* forms herds of mixed sexes, averaging about 10 members but with up to 50. Reproduction occurs throughout the year. Home ranges of herds are mostly exclusive. Females are dominant over males and initiate most courtship and mating. No clear leadership exists within the herd.	Eisenberg (1966), Sowls (1974)
FAMILY HIPPOPOTAMIDAE (hippopotami). 2 genera, 2 species. Africa.	*Diverse.* Pygmy hippopotami (*Choeropsis liberiensis*) live singly or in pairs. Hippopotami (*Hippopotamus amphibius*) are highly social. Females and their young live in herds of 5–15, with males on the periphery. Dominant males apparently mate first. Herds move from water to feeding areas along regular trails marked by scent posts.	Verheyen (1954), Eisenberg (1966), Dorst (1970)
SUBORDER RUMINANTIA (ruminants)		
FAMILY CAMELIDAE (camels, llamas). 3 genera, 4 species. North Africa to temperate Asia, South America.	*Harem-forming.* Small herds of females and young are controlled by a single dominant male. The harems are seasonal in camels but permanent in the vicuña. See description of vicuña elsewhere in this chapter.	Koford (1957), Gauthier-Pilters (1959, 1974), Franklin (1973, 1974)

Table 24-1 *(continued).*

Kind of ungulate	Sociobiology	References
FAMILY TRAGULIDAE (chevrotains, mouse deer). 2 genera, 4 species. West Africa, tropical Asia.	*Solitary.* Live alone or in pairs during breeding season.	Davis (1965), Dubost (1965), Dorst (1970)
FAMILY CERVIDAE (deer and allies, including caribou, elk, and moose). 16 genera, 37 species. Worldwide except for sub-Saharan Africa, Australia, and Oceania.	*Diverse.* Some species are solitary, including the roe deer (*Capreolus capreolus*), white-tail deer (*Odocoileus virginianus*), and moose (*Alces americana*). Others form herds of females and young controlled by 1 or a few males during the rutting season. Caribou (*Rangifer*) form large migratory herds, with most males staying apart from females except during rutting season; a few males sometimes remain with the females throughout the year.	Darling (1937), Linsdale and Tomich (1953), Dasmann and Taber (1956), Geist (1963), Eisenberg (1966), Vos et al. (1967), Kelsall (1968), Prior (1968), Dorst (1970), Espmark (1971), Brown (1974), Houston (1974), Peek et al. (1974)
FAMILY GIRAFFIDAE (giraffes and okapis). 2 genera, 2 species. Africa.	*Diverse.* The forest-dwelling okapi (*Okapia johnstoni*) is solitary. The giraffe forms herds of 2 to 40 members, occasionally as many as 70. Some are exclusively male; others contain multiple females and young accompanied by a single bull or several bulls of which one is dominant. Mixed herds are usually led by a female.	Innis (1958), Dorst (1970), Matthews (1971), Foster and Dagg (1972)
FAMILY ANTILOCAPRIDAE (pronghorns). 1 genus, 1 species. Western North America.	*Harem-forming.* Males establish territories and sequester harems during the breeding season.	Buechner (1950), Eisenberg (1966), Bromley (1969)
FAMILY BOVIDAE (cattle, buffaloes, bison, antelopes). 44 genera, 111 species. North America, Eurasia, Africa.	*Very diverse.* See Table 24-2 and discussion elsewhere in this chapter.	Schloeth (1961), Tener (1965), Eisenberg (1966), Estes (1967, 1969, 1975a), Hanks et al. (1969), Pfeffer and Genest (1969), Dubost (1970), Leuthold (1970, 1974), Roe (1970), Geist (1971), Hendrichs and Hendrichs (1971), Kiley (1972), Shank (1972), Whitehead (1972), Jarman and Jarman (1973), Gosling (1974), Jarman (1974), Joubert (1974)
ORDER PROBOSCIDEA FAMILY ELEPHANTIDAE (elephants). 2 genera, 2 species. Africa, tropical Asia.	*Highly social.* Herds of females and young are attended by dominant males during the breeding season. Males usually live in bachelor herds and occasionally alone. See fuller description elsewhere in this chapter.	Kühme (1963), Hendrichs and Hendrichs (1971), Sikes (1971), Eisenberg (1972), Eisenberg and Lockhart (1972), Douglas-Hamilton (1972, 1973), McKay (1973), Laws (1974)

Table 24-2 Social traits of selected ungulate species, showing the full range of social organization. (Based principally on Eisenberg, 1966; additional data from Klingel, 1968; Tyler, 1972; Douglas-Hamilton, 1972; Owen-Smith, 1974.)

Kind of ungulate	Adults solitary except at pairing; territorial or not	Loose grouping of adults; female-offspring units not allied	Small groups of allied females with young; males solitary	Large herds. Also leks: males hold territories on traditional breeding grounds; herds otherwise unisexual	Large herds. Permanent harem by territorial male; female-offspring units allied to each other	Large herds. Seasonal harems by male; female-offspring units allied to each other	Large herds. Unisexual subgroupings. Several males associate with female herd at rut or permanently
ORDER PERISSODACTYLA							
FAMILY EQUIDAE							
Burchell's zebra (*Equus burchelli*)					X		
Wild horse (*Equus caballus*)					X		
FAMILY RHINOCEROTIDAE							
White rhinoceros (*Ceratotherium simum*)		X————varies————X(?)					
ORDER ARTIODACTYLA							
FAMILY SUIDAE							
Wild pig (*Sus scrofa*)			X				
FAMILY HIPPOPOTAMIDAE							
Hippopotamus (*Hippopotamus amphibius*)							X
FAMILY CAMELIDAE							
Vicuña (*Vicugna vicugna*)					X		
Camel (*Camelus bactrianus*)					X————varies————X		
FAMILY TRAGULIDAE							
Chevrotains (*Tragulus*)	X						
FAMILY CERVIDAE							
SUBFAMILY CERVINAE							
Red deer (*Cervus elaphus*)						X	
SUBFAMILY ODOCOILEINAE							
Mule deer (*Odocoileus hemionus*)			X				
Moose (*Alces alces*)		X					
Roe deer (*Capreolus capreolus*)			X				
FAMILY ANTILOCAPRIDAE							
Pronghorn (*Antilocapra americana*)						X	
FAMILY BOVIDAE							
SUBFAMILY CEPHALOPHINAE							
Blue duiker (*Cephalophus maxwelli*)	X						
SUBFAMILY BOVINAE							
Wild cattle (*Bos taurus*)							X
Bison (*B. bison*)							X

Table 24-2   (continued).

Kind of ungulate	Adults solitary except at pairing; territorial or not	Loose grouping of adults; female-offspring units not allied	Small groups of allied females with young; males solitary	Large herds. Also leks: males hold territories on traditional breeding grounds; herds otherwise unisexual	Large herds. Permanent harem by territorial male; female-offspring units allied to each other	Large herds. Seasonal harems by male; female-offspring units allied to each other	Large herds. Unisexual subgroupings Several males associate with female herd at rut or permanently
SUBFAMILY HIPPOTRAGINAE							
Kob (*Kobus kob*)				X			
Bubal hartebeest					X		
(*Alcelaphus buselaphus*)							
SUBFAMILY ANTELOPINAE							
Impala (*Aepyceros melampus*)						X	
Gerenuk (*Litocranius walleri*)						X	
SUBFAMILY CAPRINAE							
TRIBE SAIGINI							
Saiga (*Saiga tatarica*)						X	
TRIBE RUPICAPRINI							
Chamois (*Rupicapra rupicapra*)						X	
Mountain goat (*Oreamnos americanus*)			X				
TRIBE OVIBOVINI							
Musk ox (*Ovibos moschatus*)						X	
TRIBE CAPRINI							
Mountain sheep (*Ovis canadensis*)							X
Wild goat (*Capra hircus*)							X
ORDER PROBOSCIDEA							
FAMILY ELEPHANTIDAE							
Indian elephant (*Elephas maximus*)							X
African elephant (*Loxodonta africana*)							X

of time, during which the members recognize one another and may or may not exclude strangers. Finally, the elephants have carried this tendency to its extreme, with tight kinship groups persisting across generations. The adult cows assist others altruistically in times of stress, young are nursed indiscriminately by whichever members happen to be lactating, and a single old matriarch leads the group in every progression and formation.

In short, ungulate and elephant societies are matrifocal assemblages capable of considerable sophistication. The role of males varies greatly among species in a manner that can be viewed as orthogonal to the evolution of the female-offspring units. In all known species the males compete in some manner for access to the females. Some do so simply by territorial defense, servicing females whose home ranges overlap their own or, as in migratory populations of wildebeest, whichever females pass through their domains while in estrus. The males of at least one antelope species, the Uganda kob of Africa, concentrate their territories into leks, which are traditional sites visited by the females for the primary purpose of mating. The males of other species, in such diverse groups as the horses, camels, and bush-pigs, contend for dominance over nursery herds. The most successful individuals then enjoy unlimited access to the estrous females. In still other species, including the elk and pronghorn "antelope" of North America, males control harems only during the season of rut.

The full array of the social states is given in the column headings of Table 24-2. The catalog of species presented in the table shows that ungulates are about as labile in primary social traits as other major mammalian groups, including the marsupials, rodents, carnivores, and primates. In particular, individual social qualities other than the mother-offspring bond vary widely at the level of the family and genus. A distinctive feature of ungulate social life appears to be the rarity of prolonged male-female bonding. Where the home ranges of males and females of "solitary" species overlap, couples may occupy extensive areas exclusively, but in only a very few cases are they known to cooperate in territorial defense or the rearing of young in the manner so common in birds and carnivores. Yet the "solitary" tragulids and antelopes are still only poorly known in the wild, and pair bonding may well prove to be more general than previously thought (Estes, 1974).

The Ecological Basis of Social Evolution

The array of social states displayed by ungulate and elephant species can also be viewed as ensembles of points in three-dimensional space, the axes of which are herd size, intensity of alliance among the adult females, and the form of attachment of males to the female herds. The correlations among these variables are weak. The ecological imperatives that determine the position of each species have been considered in a preliminary manner by Eisenberg (1966) in his review of mammalian sociobiology and investigated at greater depth in special studies of sheep, deer, and bison by Geist (1971a,b), Asiatic ungulates and elephants by Eisenberg and Lockhart (1972), and African bovids by Estes (1974), Jarman (1974), and Leuthold (1974). What the combined data of these writers reveal is the elaboration of social behavior as a consequence of the shift from the cover of closed forests to more open habitats such as savannas, grasslands, and meadows. Estes has argued that the extraordinary speciation of the African antelopes, constituting 37 percent of the entire ungulate fauna of the world, was made possible by this change. A majority of the species remaining in the forests are small and solitary, while most of those on the plains are to some extent social. The great herds of the Serengeti and other savanna reserves are the visible evidence of this correlation. It is no coincidence that the herds are pursued by lions, the most social of the cats, and by wild dogs, the most social of the canids. In a word, the wildlife spectacles of Africa are to a large degree based on social organization.

Using quantitative data, Jarman has analyzed the fine structure of antelope sociobiology. The strength of his method comes from the detailed documentation of the increase in herd size and social complexity that accompanies the increase in body size among the antelope species, and a piecing together of all that has been learned about their diets and habitat preferences. The models advanced to explain this information are tightly argued, gaining additional strength from the fact that they cover most of the major aspects of ecology and behavior. Jarman's complete theory can be summarized in the form of a stepwise argument:

1. The open habitats of Africa—grassland, savanna, and light woodland—contain the highest biomass and species diversity of antelopes. These sites also have the highest but least uniform plant production, largely because of the synchronized emergence of grass early in the growing season. At the same time, grass plants tend to be more homogeneous in food value than do browse plants, which offer only scattered edible parts over their surfaces.

2. Small antelopes tend to be more selective feeders. They are able to bite off individual plant parts, whereas larger species must eat bunches of parts at a time. Furthermore, owing to the surface-to-mass law the smaller species have a higher per-gram metabolic requirement, and as a consequence they must eat food items of higher energetic value. Because such items, which exist in the form of particular plant species and special parts on plants, are scarcer and more dispersed, the biomass levels of the smaller species are less than those of larger species. This is particularly true in open habitats, where grass provides large quantities of lower-quality food of the kind more efficiently utilized by larger antelopes (see Bell, 1971).

3. Jarman's five principal categories of social organization, which correspond roughly to the array given in Table 24-2, are correlated closely with the feeding styles and average body sizes of the individual antelope species. These relationships are summarized in Table 24-3. The smallest species are forced by the nature of their diet to be more widely dispersed. They are solitary or at most live in small groups. The larger the animal, the more likely it is to occur in the open, where it can take fuller advantage of grass as food. Also, it is more likely to benefit from herd membership through the improved avoidance of predators. Smaller antelopes depend almost entirely on the communal alarm system to reduce the risk of being eaten, while the largest ones supplement the alarm system by relying on the confrontation of predators with solid defensive formations or even the launching of communal attacks. These two factors, dense biomass through the utilization of grass and the need for communal defense in exposed sites, have combined to promote herd formation. The larger the average size of the ungulate, the larger the stable groups formed.

The relationships revealed by Jarman's analysis and in other ungulate studies have been skillfully codified by Geist (1974). His formulations, some of which are empirical statements and others hypotheses from deduction, have brought the study of ungulate sociobiology to the edge of population biology. In this regard the study of ungulates is more advanced than that of other mammalian groups. The next logical step will be the construction of models that explicitly incorporate measurements from demography and population genetics. Such an approach should prove most productive in choosing

Table 24-3 The behavioral and ecological classification of African antelopes and buffaloes. (Based on Jarman, 1974.)

Social organization	Feeding style	Size (average weight in kg)	Antipredator behavior	Examples
Class A Singly or in pairs, sometimes with offspring. Group size 1–3. Small permanent home range.	Select particular parts of a wide range of plant species. Most diversified diets of all species.	1–20	Freeze, lie down, or run to cover and freeze. Too small to outrun most predators and cannot utilize mass counterattack.	Dik-dik (*Madoqua*), duiker (*Cephalophus*)
Class B Commonly several female-offspring units associate. Group size 1–12, usually 3–6. Permanent home range within ranges of single males.	Feed entirely on range of grass species or on browse, selecting particular plant parts.	15–100	Same as in Class A species.	Reedbucks (*Redunca*), vaal rhebuck (*Pelea*), oribi (*Ourebia*), lesser kudu (*Tragelaphus imberbis*)
Class C Larger herds, with 6 to several hundred members, varying according to region and season. In breeding season a few males hold territories to exclude other males, many of which travel singly or in bachelor herds. Kob males display on leks.	Feed selectively on parts of a diversity of grass *and* browse species.	20–200	Diverse. In heavy cover, freeze or run (when detected). In open habitats, run; sometimes "explode" in all directions, reunite later. Utilize alarm behavior of fellows to provide better predator detection.	Kob, waterbuck, puku, and lechwe (*Kobus*); springbuck (*Antidorcas*); gazelles (*Gazella*); impala (*Aepyceros*); greater kudu (*Tragelaphus strepsiceros*)
Class D During sedentary periods, when forage is ample, societies are organized as in Class C species. During migrations conducted in order to follow changing food supply, herds often coalesce into "superherds" of thousands.	Feed on diversity of grasses, being selective concerning parts eaten. Since this resource is patchily distributed in space and time, the antelopes migrate in aggregations at the appropriate season.	100–250	Flee from large predators, but may face enemies as a group and even attack them en masse.	Wildebeest or gnu (*Connochaetes*), hartebeest (*Alcelaphus*), topi and blesbok (*Damaliscus*)
Class E Large, relatively stable herds of females and young are accompanied by multiple males organized into dominance hierarchies. These groups commonly contain hundreds and even 1 or 2 thousand members. Bachelor herds also occur. Superherds do not form during migrations.	Feed unselectively on many parts of a wide array of grasses and browse; much of the material has low nutritional value.	200–700	Defensive formations and communal attacks a common response, even to large predators. Group answers distress call of young.	Buffalo (*Syncerus caffer*), probably eland (*Taurotragus*), beisa oryx (*Oryx beisa*), and gemsbok (*Oryx gazella*)

among the intricate hypotheses advanced by Geist, Jarman, and other mammalogists to explain sexual dimorphism in the ungulates. The differences among species in this one trait is enormous. Some of the cases offer no great problem. In Jarman's solitary Class A species, for example, the males closely resemble the females. The monomorphism is obviously favored both by the stability of the home ranges, offering little opportunity for males to contend for harems, and by the need to avoid predators through stealth and concealment. Among the more social ungulates, belonging to Jarman's Classes C-E, some species are strongly dimorphic and others monomorphic, in some cases with the females mimicking the males or vice versa (see Estes, 1974). Geist has accounted for the variation with the following hypothesis. When the food supply varies from year to year, with sustained periods of abundance, the females are able to breed with a minimum of interference. They will not be territorial or otherwise display much aggression within their tightly knit herds. The males will then be freed to compete for females, who have now become the limiting resource. In the language of population biology, such species are r selectionists, and there is a strong tendency for males to differ from females in ways that relate purely to intrasexual selection. When the food supply is still more patchily distributed, however, to the degree that it becomes a fine-grained resource, selection for sexual selection will collapse. The species are more nearly K selectionists. Since the females do not reproduce in spurts, it is not profitable for males to expend large amounts of energy controlling harems by the exclusion of rivals. And since energy is not so readily available during the breeding seasons, the females will find it profitable to avoid the superfluous attentions of small males. They are likely to become more aggressive and even malelike in appearance, to the extent of having penislike tufts of hair as in the wildebeest. This process could account for the approach to monomorphism in such forms as the bison, African buffalo, reindeer, springbuck, gazelles, and eland. It also helps to explain the permanent incorporation of multiple males into the female herds in Jarman's Class E antelope species.

A competing and equally plausible hypothesis has been advanced by Estes (1974). This author views monomorphism as a means of maintaining cohesion within mixed-species herds during migrations. The trait is often associated with striking markings and body conformations that are peculiar to each species and make recognition still easier. In the case of nonterritorial species, such as the eland and buffalo, there has even been opposing selection pressure for continued growth of the males, since rank order is based upon size. The result is a pronounced variation in the size of adult males belonging to the herds.

The remainder of this chapter is devoted to a series of natural history sketches of species that in aggregate range across the entire spectrum of social stages. These examples have also been selected to provide the greatest possible phylogenetic array, from the morphologically primitive tragulids to the advanced and specialized wildebeest and African elephant.

Chevrotains (Tragulidae)

The behavior of the chevrotains, or mouse deer, is of extraordinary interest because of their primitive position within the Ruminantia, the ungulate suborder containing the largest number of species and greatest diversity of social systems. The five living species are secretive, forest-dwelling animals seldom observed in the wild, and information on their behavior is unfortunately fragmentary.

Tragulids resemble nothing so much as large mice, being in many respects convergent to the acouchis and other large forest-dwelling caviimorph rodents of South America. Their movements are swift and agile. R. A. Sterndale (1884) said that "they trip about most daintily on the tips of their toes, and look as if a puff of wind would blow them away." Males are outwardly very similar to females, except for the possession of a pair of small tusks. The social organization appears to be simple in nature. Dorst (1970) reports that water chevrotains (*Hyemoschus aquaticus*), the only African tragulids, occur alone or in pairs. Males of the Asiatic *Tragulus* evidently maintain territories or at least aggressively protect females within their domain. Katherine Ralls (personal communication) has observed captive *T. napu* males mark their enclosures with scent from the intermandibular glands. They also smear the secretion over the backs of females. Strange males have been observed to fight when placed in the same enclosures, slashing at each other with their tusks (see Figure 24-1). However, males forced to live together in groups are seldom antagonistic, a condition that Ralls believes may be due to inbreeding over several generations. This view is in accord with that of Davis (1965), who observed no hostility between a father and son of *T. javanicus*, even when the latter copulated with a female previously associated with the older male.

The Vicuña (*Vicugna vicugna*)

High in the central Andes of western South America, above the limit of cultivated crops, lies a treeless pastoral zone, the puna. While scanning the bleak rolling grasslands of the puna a traveler may be startled by a prolonged screech. The cry attracts his gaze to a racing troop of fifty gazelle-like mammals, bright cinnamon in color—vicuñas! As they gallop up a barren slope he sees that a single large vicuña pursues them closely. The pursuer charges at one straggler, then another, as if to nip its heels. But suddenly the aggressor halts, stands tall with slender neck and stout tail erect, stares at a line of llamas in the distance, and whistles a high trill. Then it gallops away to join a band of several vicuñas, some obviously young, which graze close by.

Figure 24-1 Fighting between males of the mouse deer *Tragulus napu.* (Photograph by Karen Minkowski; by courtesy of Katherine Ralls.)

Thus begins Carl B. Koford's classic account (1957) of the vicuña, one of the first of the studies of a vertebrate species to integrate social behavior and ecology in the modern way. The individual described in the passage is a male that is driving a herd of bachelors out of his territory and away from his harem of females and young. *Vicugna vicugna*, a member of the camel family, is notable for the fact that its males are the most strictly territorial mammals known. It is also one of the few ungulate species in which year-round harems are the norm. The details of its social life were worked out by Koford during a year's visit to several localities in the Peruvian Andes, and they have been confirmed and extended by William L. Franklin in a second excellent study conducted from 1967 to 1971 in Peru's Pampa Galeras National Vicuña Reserve.

The basic social unit is the territorial family group, consisting of the male and his harem (see Figure 24-2). At Huaylarco Koford found such bands to contain an average of 1 male, 4 females, and 2 juveniles, with the maximum numbers of females and juveniles ranging upward to 18 and 9, respectively. At the Pampa Galeras Reserve each group occupied both a feeding territory in which it fed and reproduced and a smaller sleeping territory in which it spent the nights. The holdings of six groups studied by Franklin varied from 7 to 30 hectares and averaged 17 hectares. Sometimes roads and streambeds serve as convenient physical barriers to separate the territories, but more often there is an invisible line recognized by the vicuñas alone. The males approach to within 2 or 3 meters of one another at this line and exchange threat displays. If one steps across he is promptly chased back.

The territories are dotted with large piles of dung that are treated

Figure 24-2 Societies of the vicuña, a small member of the camel family found on the barren plains of the high Andes. A territorial family group is arrayed in the foreground. The single dominant male faces the observer in a hostile pose, making himself appear as large as possible by standing erect on a rock with his head and tail held high. Behind him his harem, composed of ten females and three young, rests and feeds. In repose the vicuña pulls the legs under the body to conserve heat; the effect is enhanced by the bib of white fur that cloaks the chest and upper parts of the forelegs. The female on the far right is "spitting," an expulsion of air that expresses irritation or hostility toward another animal. In the distance to the left can be seen a nonterritorial herd of bachelor males. Such groups form and break up casually while wandering

from place to place in search of the best forage, and their members are always ready to take over the territory of a resident male if he weakens or disappears. Some of the plants of the harsh Andean environment are shown in the foreground. They include the grasses *Calamagrostis vicunarum* (far left) and *Festuca rigescens* (center), the lettucelike malvaceous *Nototriche transandica* in the lower lefthand corner, the composites *Baccharis microphylla* and *Lepidophyllum quadrangulare* just behind and to the right of the *Nototriche,* and the legume *Astragulus peruvianus* in the lower righthand corner. All but the *Lepidophyllum* are eaten by the vicuñas. (Drawing by Sarah Landry; based on Koford, 1957, and Franklin, 1973.)

in a ritual manner. All members of the family visit the piles regularly to sniff them, to knead them with the forefeet, and to add feces and urine. Franklin believes that these scent posts do not serve as warning signals. When the family group is temporarily absent, wandering males and family groups enter territories without hesitation. It is more likely that the dung heaps are used primarily to keep the residents *in*, by serving as guideposts to the territorial boundaries. When females and young accidentally step over these lines, they are promptly chased back into their own territory by the resident male. But however it is identified, the ultimate function of the territory is probably defense of the food supply, which in the barren puna environment appears to be the limiting resource through most or all of the year. The food-limitation hypothesis is strengthened by the fact that the size of the territory is greatest in areas with the least density of edible plants. Indeed, the very strictness of this limiting factor may have been responsible for the evolution of the unusual territorial system of the vicuña.

The male vicuña watches his little band at all times and leads it from one point in the territory to another. In times of danger he emits the screechlike alarm trill, consisting of several descending whistles delivered over about 4 seconds, and interposes himself between the source of the threat and the group. A nonterritorial male acquires a territory by taking over an unoccupied site or land abandoned by another male. At first he grazes and rests quietly, maintaining, as it were, a low profile. Then after a few days he begins to test neighboring males with aggressive encounters. By this means he appears to learn the precise limits of the land that can be safely occupied. Having thus consolidated his position, he sets about acquiring females to build a family group. A few females are available throughout the year in the form of solitary yearlings, as well as unattached groups and older individuals somehow deprived of mates.

At the season of birth, in March, the sex ratio of the newborn vicuña "crias" is close to parity. Within six months, however, the proportion of juvenile males starts to plummet. By the following March male yearlings are relatively scarce; at Pampa Galeras Reserve Franklin counted only 7 for every 100 females. The reason for the decline is the increasing aggressiveness of the adult male. At first some of the mothers try to protect their sons. Occasionally they even try to leave the group with their sons but are driven back by the adult male. Eventually they acquiesce and the young males are forced to leave. As the next birth season approaches, each yearling female becomes the target of aggression by both the adult male and her own mother. In effect, she occupies an untenable position at the bottom of the dominance hierarchy, and in time she is also forced to leave. The number of adult members in the territorial family group represents an equilibrium between recruitment of such expelled individuals, together with females who have lost their harem master, and

loss by death and emigration. It is clear that in the severe vicuña patriarchy, the male exercises a large part of the control leading to this number.

A second principal social unit is the nonterritorial male herd. This bachelor group usually contains from 15 to 25 members, but the total range is from 2 to 100, and solitary wanderers are common. The males aggregate loosely, with individuals coming and going in an evidently casual manner. The all-male groups wander widely along the fringes of the family territories, pausing to rest and feed. Individuals frequently test the defenses of the territorial males by deliberate intrusions and challenges—always ready to take over on a minute's notice if the resident weakens or leaves.

The Blue Wildebeest (*Connochaetes taurinus*)

The blue wildebeest, or brindled gnu, symbolizes the almost vanished glory of African wildlife. Regarded by zoologists as an aberrant form of antelope, it has been the most abundant ungulate of the African short grasslands. Its great migratory herds, containing thousands of individuals, once stretched to the horizon. Even today as many as a million wildebeest populate the Serengeti Plains. The wildebeest dominates the ecology of its range. It thrives best on pastures of colonial grasses such as Bermuda grass (*Cynodon dactylon*), which can withstand constant trampling and grazing and in fact benefit from the manuring of the animals feeding on them. Thus the wildebeest to a large extent creates its own optimum environment. It is an excellent example of Jarman's Class D species, in which unattached groups of females and their offspring pass in and out of the breeding territories of the males. But, as shown by the careful studies of R. D. Estes (1969, 1975a), an even greater distinction of the species is the great flexibility of its social system, which is finely adjusted to the highly variable environment of the African plains.

Under consistently favorable grazing conditions, wildebeest populations are organized into resident herds of either females and calves (nursery herds) or bachelor males. In Tanzania's Ngorongoro Crater, the nursery herds contain an average of ten members and apparently occupy a consistent home range of at most several hundred hectares. They also appear to be relatively stable in composition and closed to outsiders, since strange cows attempting to join them are often harassed. During the dry season this arrangement is altered. The nursery herds begin to aggregate in the moister, low-lying areas to which the suitable forage becomes increasingly restricted. At first the herds return at night to their home ranges, but eventually they come to spend all of their time in the new feeding areas. Simultaneously their numbers are swelled by an influx of bachelor herds and some of the territorial males. In regions with permanently drier conditions, wildebeest exist year-round in large aggregations that migrate from one site of suitable forage to another. In fact, permanent sedentary and migratory populations are the two poles of wildebeest social organization adapted respectively to very stable and very fluctuating environments. All intermediate stages are conceivable and do in fact occur. It is also possible for sedentary populations to bud off from migratory ones when local conditions become favorable, as reported for example in Rhodesia's Wankie National Park and in southern Botswana.

Blue wildebeest are well adapted to conduct mass migrations. They travel in single file along traditional game paths, leaving behind a scent from the interdigital glands of the hooves so strong that even a human being can follow them by smell alone. Their tolerated individual distance is less than in most other ungulates, permitting them to crowd closely together when occasion demands.

Superimposed upon the mostly female-centered herd system is the territorial organization of the solitary males. Where the vicuña male defends a territory to protect his harem and its food supply, the wildebeest bull defends it solely for purposes of courtship. The defense and the sexual advertisement associated with it are conducted throughout the year and are greatly intensified during the brief rutting season. In sedentary populations territories are moderate in size, averaging about 100–150 meters in diameter. But in the midst of migratory herds, where the males must shift their location at frequent intervals, the territories are often compressed to a diameter of 20 meters or less. During the most stringent period of the dry season, when the population is constantly on the move, territorial behavior is sometimes attenuated or lost altogether for brief periods. Only about half of the adult bulls are able to maintain territories in any season; the remainder are relegated to the bachelor herds.

The territorial advertisement displays of male wildebeest are among the most elaborate and spectacular to be found within the vertebrates. In the first place they employ all of the basic repertory of the alcelaphine antelopes: head-up posture, pawing and ritual defecating, kneeling, and horning. Much of the action takes place on the stamping ground, a patch of bare ground at or near the center of the territory. Often the males roll and wallow on the ground. The action probably serves not only as a visual display but as a means of impregnating the body with odors of feces and urine. Wildebeest bulls also engage daily in the unique "challenge ritual" (Estes, 1969). Every male makes the round of all his territorial neighbors, performing the ceremony with each in turn for an average of 7 minutes. At least 45 minutes of the day is required to communicate with all of them. The apparent function of the challenge ritual is to reaffirm the male's property rights while testing those of his neighbors. The territorial owner seems to recognize his neighbors personally. The exchanges are marked by what can be reasonably called mutual respect and restraint, and fighting is extremely rare. Real combat and injury usually occur at another time—when a male is first establishing its domain, in other words when it is still a stranger. About 30 distinct behavior patterns are employed in the ritual. They are used in almost every conceivable permutation, by either partner and at any moment in the ceremony. The displays include lateral posturing; ritualized grazing and grooming; cavorting, which includes head shaking, bucking, leaping, running about, and spinning; "pretended" alarm signals, in which one or both of the animals raise their heads, look away from each other, and stamp; urine testing; and the various general alcelaphine displays mentioned earlier (see Figure 24-3). Another peculiar feature of the challenge ritual is that the encounters take place anywhere in the territory and not just on the stamping ground or along the boundaries.

Although the histories of individual herd members have not been worked out in detail, the general life cycle is known. Before the calving season starts, young males are excluded from the nursery herds and begin to band together. By the time of the rutting season, four months later, all but a very few of the yearling males have joined the bachelor herds. While rejection by the mother and other females is a factor, the main force causing separation is the aggression of the territorial males, who treat the yearlings as rivals. Young females are treated more tolerantly, and it is possible that membership in the nursery herds is based at least to some extent on kinship through the female lines.

The African Elephant (*Loxodonta africana*)

The largest of land mammals is also distinguished by one of the most advanced social organizations. The African elephant is remarkable in the closeness and intimacy of the ties formed between the females, the power of the matriarch who rules over the family group, and the length of time these individual associations endure. This conception of elephant sociobiology is of recent vintage. The essential facts were inferred by Laws and Parker (1968) from demographic data and confirmed in direct behavioral observations by Hubert and Ursula Hendrichs (1971), who devoted two years to studying a population on the Serengeti Plains. More recently Iain Douglas-Hamilton (1972, 1973) has conducted a four-and-a-half-year study at Lake Manyara National Park, Tanzania, during which he came to recognize 414 of the approximately 500 elephants present and recorded an impressive amount of detail on their individual relationships and the histories of family groups. The following account is based to a large extent on the Douglas-Hamilton study.

The African elephant occurs today through most of sub-Saharan Africa exclusive of the Cape, but as recently as Roman times it ranged

Figure 24-3 The social organization of the blue wildebeest, or gnu, is depicted in this scene from the Serengeti Plains of Tanzania. In the foreground two males perform the challenge ritual, a daily exchange by which each reasserts his territorial rights and challenges those of his neighbors. The male on the left cavorts in front of his rival, who has just finished digging at the ground with his horns in another display of the ritual. The bulls appear to know each other, with the result that the challenge ritual lasts an average of only 7 minutes and almost never results in real fighting and injury. The exchange may take place on any site within the territory of either bull including the stamping ground, an example of which is seen in the center foreground. To the right, a nursery herd feeds and rests while in transit through the territory of

one of the males. Any female in estrus is likely to be mated by the resident bull at this time. Two male calves play in a manner anticipating the elaborate aggressive rituals and combat that will consume so much of their adult lives. Other solitary bulls are seen standing on their territories, one beyond each of the two displaying individuals and another pair along the territorial boundary at the acacia tree in the right background. Other nursery herds graze to the left; in the center background is a loose herd of nonterritorial bachelor males. The dominant ground vegetation is bermuda grass, a tough colonial species that thrives under heavy grazing and manuring by the wildebeest. (Drawing by Sarah Landry; based on Estes, 1969 and personal communication.)

north to the shores of the Mediterranean and Syria. Possibly several hundred populations now exist, each comprised of 1000 to 8000 individuals inhabiting an area of 1300-2600 square kilometers. Elephants are exclusively vegetarian, browsing on a great variety of plants. Within a 12-hour period one animal was seen to sample no less than 64 species of plants belonging to 28 families. As suitable vegetation grows scarce in a particular locality, the animals turn increasingly to the consumption of grass, but they cannot thrive indefinitely on this secondary food. Elephants can have a devastating effect on their environment. They strip trees of bark and branches, killing many. At higher population densities they eventually turn dry forests into parkland. A few bulls have the ability to push over larger trees, providing meals for themselves and their companions. The seeds of acacia and other trees and bushes pass through the digestive tract unharmed and sprout from the dung, so that in time an equilibrium is attained between the size of the elephant populations and the thickness of the vegetation on which they live.

Each population is organized into a two- or three-tiered hierarchy of social groupings. The most important grouping directly above the individual is the *family unit*, a tightly knit herd of 10-20 females and their offspring led by a powerful matriarch. At Manyara each unit contained an average of 3.4 female-offspring groups. Members appear never to wander from their unit for distances greater than a kilometer during intervals longer than a day. The matriarch is generally the oldest individual—and hence the largest and strongest, since elephants continue growing past maturity. Because of her age, the adult females around her are likely to include not only her daughters but also her granddaughters, and the female-female bonds can be assumed to last as long as 50 years. The matriarch rallies the others and leads them from one place to another. She takes the forward position when confronting danger and the rear position during retreats. When she grows old and feeble a younger cow gradually takes her place. But in cases where the matriarch dies suddenly the effect is traumatic. The survivors mill around her body in panic, disorganized and seemingly unable to retreat or to mount a proper defense. Hunters have long known that when the leader is shot, the rest of the herd can easily be brought down in rapid succession. For this reason, Laws and Parker recommended that when culling is made necessary by population pressure, entire family units should be removed and not just individuals picked at random.

The second level in social organization is the *kinship group*, an ensemble of family units that remain near one another and whose members show some degree of personal familiarity. It is probable that such groups originate when family units divide by fission. That the units do split is indicated by the fact that few contain more than 20 individuals, even though most are constantly growing. Douglas-Hamilton witnessed the process of division in the largest unit at

Manyara, which contained 22 members. Over a period of a year 2 young cows, an adolescent female, and 2 calves moved increasing distances from the remainder of the unit. After the adolescent female calved for the first time, the two subgroups remained apart for varying periods. Then one day the matriarch led the original family unit southward for a distance of 15 kilometers, producing the first major spatial separation of the two groups. When the parental unit returned to the original site, the derivative group rejoined it and continued to stay nearby. If this case history proves to be typical, the description of such complexes as kinship groups will be justified.

It is possible that population growth, expanding the assemblages of ultra-stable female groups, produces even larger social complexes which are coextensive with the local populations themselves. Such "clans" contain perhaps 100-250 individuals. During migrations as many as a thousand elephants form mobile aggregations that are evidently unorganized above the level of the kinship group. At Manyara, family units occupied home ranges 14 to 52 square kilometers in extent, through which they wandered in irregular patterns. The ranges overlapped greatly and there was no overt territorial behavior, possibly a result of the kinship ties of adjacent groups.

The degree of cooperation and altruism displayed within the family group is extraordinary. Young calves of both sexes are treated equally, and each is permitted to suckle from any nursing mother in the group. Adolescent cows serve as "aunts," restraining the calves from running ahead and nudging them awake from naps. When Douglas-Hamilton felled a young bull with an anesthetic dart, the adult cows rushed to his aid and tried to raise him to his feet. Similar behavior has been observed frequently by elephant hunters. In its adaptive value the response is basically similar to the raising of injured dolphins by their fellow school members. Because of the great bulk of the animal, a fallen elephant will soon suffocate from its own weight or overheat from lying still in the sun. Finally, the matriarch is exceptionally altruistic. She is ready to expose herself to danger while protecting her herd, and she is the most courageous individual when the group assembles in the characteristic circular defense formation (see Figure 24-4).

While still in the company of their mothers, young bulls anticipate their future roles by rushing at one another in mock charges and play-fighting. In adolescence they begin to be pushed away by the cows and at the age of 13 years, when almost grown, they are repeatedly chased away until they leave altogether. Adult males live alone or in loose bands and disperse more widely than the females. When in groups they compete for position in a dominance hierarchy, with the outcome usually being settled on the basis of size. The struggles become most strenuous in the presence of estrous females, but even then they seldom result in serious injury. Coalitions of the kind seen in higher primates appear to exist among the male elephant groups.

Hendrichs and Hendrichs observed a "protected threat" maneuver very similar to that reported independently in the hamadryas baboon by Kummer (see Chapter 26). That is, smaller bulls were able to dominate middle-sized ones by the mere proximity of senior bulls. The largest animals intimidated the small bulls less than they did the middle-sized animals, which were evidently more likely to be treated as rivals.

African elephants communicate mostly by visual signals produced with the forward part of the body. Hostility is expressed by a graded series of composite postures and movements. At lowest intensity the animal "stands tall," increasing its apparent size by lifting its head up to peer over its tusks, with its ears cocked forward. According to the Hendrichs, elephants convey a higher-intensity threat by moving toward the enemy, lifting the ears with a loud crack, and extending the trunk jerkily forward. When displaying toward a smaller rival the elephant may employ the "forward trunk swish," in which the trunk is rolled up and then suddenly unfurled toward the opponent. At the same time it emits a blast of air or trumpet call. A few individuals hurl bunches of grass, branches, and other objects in the direction of the rival. The use of the trunk illustrates the importance of context in elephant communication. When accompanied by an erect stance and a forward posture of the ears, a trunk extension is almost certainly a signal of hostility. But the trunk can also be held out simply to test the air or as a friendly gesture. When two elephants meet after a temporary separation they perform a greeting ceremony closely similar to that of the wolf and African wild dog. Each places the tip of its trunk into the mouth of the other, with the smaller animal ordinarily taking the initiative. The behavior could be a ritualized feeding movement. Calves often probe the mouths of their mothers to sample the food being eaten.

The ultimate aggressive act by an African elephant is the full charge, one of the awesome spectacles of nature. It is probably directed only at dangerous predators, including man. In a serious attack hostile displays are minimal and little warning of any kind is given:

One unknown young female with new-born calf disappeared to the right. After a 60 second interval, a large female (size category 5), with ears fully extended, charged silently out of the bush into which the young female and calf had vanished. She forced one tusk into the side of my landrover behind the cab without checking her stride. The vehicle was turned through 90°. Now other elephants appeared, which prevented any observation of the first cow, but from the damage it appears that she had withdrawn her tusk and dealt one more blow. The new elephants, with a calf of about 3 years among the foremost, came running from the right-hand side and went straight into the attack without any hesitation, but this time the action was mingled with loud, continuous trumpeting. A second fully adult female used her head to butt and afterwards press down upon the roof of the cab. She leaned heavily sideways against the vehicle and her tusks scraped the bodywork behind the door. A third large female charged from the front and drove her left tusk through one of the headlights. She withdrew it rapidly and thrust again penetrating past the radiator until about $3\frac{1}{2}$ feet of the tusk were buried in the car. She jerked up her head, let it return, and began to push. The car was moved backwards for about 35 yards until it hit a small tree. The third cow and the others now retired for about 30 yards where they stopped and formed a tight circle, still trumpeting, and facing outward with ears spread out and heads lifted. Within the next minute the group dissolved into the bush.   (Douglas-Hamilton, 1972)

The hearing of elephants is evidently about as acute as that of human beings, and in captivity they can easily be trained to respond to the human voice. Fully trained Indian elephants are able to obey as many as 24 separate verbal commands from their mahouts. In the free-living African elephant, vocal communication is as rich and frequent as visual communication. The sounds can be roughly classified as growls, trumpets, squeals, and shrieks, but these vary greatly in intensity and the context in which they are emitted. Growling, which sounds like a deep, rolling r, is one of the commonest and most versatile elephant sounds. A growl can carry as far as a kilometer, and its usual function seems to be the maintenance of contacts between individuals and families. But it also serves as a mildly aggressive signal between cows and calves when the young animals try to push their way to water holes dug by the adults. Calves growl while play-fighting. Another form of growling is combined with trumpeting during the more serious aggressive displays between adults. Some anecdotal evidence suggests that individual members of a group are able to recognize one another by minor variations in the quality of the sounds.

Chemical communication is also well developed, which is perhaps surprising in such a gigantic mammal. Douglas-Hamilton saw a separated individual track its family unit by following a two-hour-old trail with the tip of its trunk. Bulls frequently check the sexual condition of cows by putting the tips of their trunks to the females' genital openings. A major mystery is provided by the temporal gland, which is located between the ear and eye and periodically secretes a viscous, strongly smelling liquid. The secretion is released in greatest quantities when the animals are excited or under stress, which suggests that the gland may be under autonomic control. It is functional in both sexes, whereas in the Asiatic elephant it is functional only in the male. Like Asiatic elephants, *Loxodonta* rub the secretion against trees and on the earth, but the purpose is unclear. There is no evidence that males mark and defend territories, even though the flow of the liquid does seem to increase with population density. On the basis of numerous field observations, Douglas-Hamilton has hypothesized that the secretion serves multiple communicative functions—in trail marking, individual recognition, alarm, and perhaps social spacing.

Figure 24-4 The two basic social groups of the African elephant are illustrated in this drawing. In the left foreground a family unit faces the observer in a tightly grouped defensive formation. Alertness and mild hostility are indicated by the erect stance of the animals, the forward position of their ears, and the extension of their trunks. The family unit consists entirely of cows and young elephants in various stages of growth. The matriarch is the second individual from the left; her greater age is revealed by her more wrinkled skin and tattered ears. Several infants and young juveniles belonging to both sexes have shifted to protected positions in the rear of the group. To the far right can be seen a cow about three-quarters grown, and next to her one about half grown. The larger individuals are all adult cows. If the group is forced to retreat, the matriarch will cover the rear, continuing to face the enemy and perhaps making mock or real charges. When she moves away she will

move no faster than the smallest, slowest calf. The family units constitute the central social grouping of elephants. These associations of individual cows, which are strongly dependent on the matriarchs, often last
for decades. To the rear right is a loosely organized herd of bull elephants,
two of whom are contending for dominance. The ranking males become
the temporary consorts of estrous females in the cow herd. In the right
foreground is an acacia tree recently broken down by a feeding elephant.
This form of damage thins the vegetation. In regions supporting dense
elephant populations dry forests are often converted into parklands of
the kind shown in this illustration. (Drawing by Sarah Landry; based
on Douglas-Hamilton, 1972 and personal communication, together with
photographs by Peter Haas.)

The studies of Eisenberg, McKay, and their associates in Ceylon indicate that the social behavior of the Asiatic elephant (*Elephas maximus*) is basically similar to that of the African elephant. In particular, the stable groups are family units containing 8 to 21 cows and young; the units are led by a matriarch; calves nurse from any lactating female in the group; and males begin to depart when they are about 5 to 7 years old. Some differences have been noted, however. Males over 14 years of age exhibit the phenomenon of musth, a temporary state in which they become exceptionally aggressive and sexually active while secreting large quantities of temporal gland liquid. The males rub the secretion on tree trunks, evidently as a means of signaling their presence and mood. Bull elephants can breed when not in musth, but the condition clearly increases their chances of achieving dominance among rivals and permits more ready access to estrous females. It would be interesting to know whether the secretions vary enough to impart individual odor "signatures."

Chapter 25 **The Carnivores**

Among the mammalian orders the carnivores are surpassed only by the primates in the intricacy and variety of their social behavior. A majority of the 253 living species, which include dogs, cats, bears, raccoons, mongooses, and related forms, are wholly "solitary." This means that a society is comprised exclusively of the mother and her unweaned young, and adult males and females associate only during the breeding season. From this base several forms of more complex organization have evolved. One grade commonly encountered, for example in the jackals, raccoon dogs, foxes, and some mongooses, is characterized by pair bonding: the male remains with the female for extended periods of time and assists in some manner with the care and protection of the young. The coati *Nasua narica* represents another grade, distinguished by bands of females and offspring that are accompanied by males during the mating season. Many mongoose species possess a still higher form of organization, in which families headed by bonded male-female pairs cooperate during hunting. Sea otters display still another kind of organization. True to their marine environment, they gather like seals at safe places to live in loosely organized herds. There the males fight among themselves, and courtship and mating take place. Lions, the only cats with an advanced form of social organization, form prides of females to which one or two dominant males attach themselves in a nearly parasitic existence. Finally, at what might be called the summit of carnivore social evolution, packs of wolves and African hunting dogs display degrees of coordination and altruism attained elsewhere only by social insects and a few of the Old World monkeys and apes.

Social behavior is diversified not only within the Carnivora as a whole but also within single families and genera (Table 25-1). The high evolutionary lability of individual social traits is comparable to that seen elsewhere in the mammals, making it difficult to represent trends by conventional phylogenetic diagrams. The carnivores are more social as a whole than the great majority of other mammalian orders. Not only are a higher percentage of the species organized above the elementary female-offspring unit, but more are in or near the highest evolutionary grades. But an even greater interest lies in the fact that most of carnivore social behavior serves to increase the efficiency of predation. This peculiarity has two consequences. First, in accordance with the ecological efficiency rule, carnivores live in far less dense populations than herbivores and their home ranges are correspondingly larger. Consequently, territories are spatiotemporal and in some cases consist of little more than broadly overlapping networks of traplines marked by scent posts. Second, being at the top of the energy pyramid, the largest carnivores are not themselves subject to significant predation. Lions, tigers, and wolves are the premier "top carnivores" usually cited by ecologists to illustrate this category. They present the results of a significant evolutionary experiment. Their social adaptations are almost certain to be keyed pri-

Table 25-1 Families and genera of living carnivores (order Carnivora) and their principal sociobiological traits. Selected references are also given for each genus; for more general reviews see Eisenberg (1966), Kleiman (1967), Ewer (1973), and Kleiman and Eisenberg (1973).

Kind of carnivore	Sociobiology	References
SUPERFAMILY CANOIDEA		
FAMILY CANIDAE (dogs, foxes, and allied forms)		
SUBFAMILY CANINAE		
Canis ("true" dogs, including wolves, coyotes, jackals). 7 species. North America, Eurasia, Africa.	*Diverse.* Mated pairs of jackals defend territories. Wolves form packs of up to 20 which are normally extended family groups; see description elsewhere in this chapter.	Murie (1944), Banks et al. (1967), Scott(1967),Snow(1967),Woolpy and Ginsburg (1967), Woolpy (1968a,b), Fox (1969, 1971), Mech (1970), H. and Jane van Lawick-Goodall (1971), Ewer (1973), Wolfe and Allen (1973)
Alopex (arctic fox). 1 species. Circumpolar.	*Mated pairs,* occasionally solitary.	Kleiman (1967), MacPherson (1969)
Chrysocyon (maned wolf). 1 species. Southern South America.	*Solitary.*	Langguth (1969), Kleiman (1972b)
Dusicyon (Paraguayan fox, Chiloe fox, and related forms). 10 species. South America.	*Solitary.*	Housse (1949), Kleiman (1967)
Fennecus (fennec fox). 1 species. North Africa to Arabia.	*Mated pairs.*	Gauthier-Pilters (1967)
Nyctereutes (raccoon dog). 1 species. Eastern USSR, China, Japan.	*Mated pairs.*	Seitz (1955)
Urocyon (gray foxes). 2 species. North America.	*Mated pairs.*	Lord (1961)
Vulpes (foxes). 10 species. Europe, Asia, and Africa.	*Mated pairs;* males may keep more than one female, which are possibly mother and daughters or sisters. Territorial.	Vincent (1958), Ables (1969), Kilgore (1969), Ewer (1973)
Atelocynus (small-eared zorros), *Cerdocyon* (crab-eating foxes)	*Unknown.*	
SUBFAMILY SIMOCYONINAE		
Lycaon (African hunting dog). 1 species. Africa.	*Highly coordinated packs.* See elsewhere in this chapter.	Kühme (1965a,b), Estes and Goddard (1967), H. and Jane van Lawick-Goodall (1971), van Lawick (1974), Estes (1975b)
Cuon (dhole or red dog). 1 species. Southern USSR to Java.	*Packs.* Hunt in groups.	Keller (1973), Kleiman and Eisenberg (1973)
Speothos (bush dog). 1 species. Central and South America.	*Packs.* Hunt rodents and other prey in small groups.	Kleiman (1972b)
SUBFAMILY OTOCYONINAE		
Otocyon (bat-eared fox). Africa.	*Packs.* Hunt small game and insects in small groups.	Kleiman (1967)
FAMILY URSIDAE (bears and giant pandas). 6 genera, 8 species.	*Solitary,* probably generally territorial; prolonged association between mother and cubs. See description elsewhere in this chapter.	Krott and Krott (1963), Perry (1966, 1969), Ewer (1973), Poelker and Hartwell (1973)
FAMILY PROCYONIDAE (raccoons and allies)		
Procyon (raccoons). 6 species, New World.	*Solitary.* Common raccoon (*P. lotor*) is solitary although yearlings may den together. Home ranges overlap greatly. No evidence has been found of territorial defense, although dominance hierarchies form at feeding stations.	Stuewer (1943), Sharp and Sharp (1956), Bider et al. (1968), Ewer (1973), Barash (1974b)
Ailurus (lesser or red panda). 1 species. Sikkim to China.	*Unknown.* Mutually tolerant in captivity and thus may form bands in nature.	Ewer (1973)
Bassaricyon (olingos). 2 species. Mexico to South America.	*Unknown.*	

Table 25-1   (*continued*).

Kind of carnivore	Sociobiology	References
Bassariscus (ring-tailed "cats" or cacomistles). 2 species. Oregon to Central America.	*Solitary.*	Richardson (1942)
Nasua (coatis). 3 species. Central and South America.	*Social.* Small bands of females and offspring, joined during breeding season by males. See elsewhere in this chapter.	Kaufmann (1962)
Nasuella (little coati). 1 species. South America.	*Unknown.*	
Potos (kinkajous). 1 species. Mexico to South America.	*Solitary.*	Poglayen-Neuwall (1962, 1966)
FAMILY MUSTELIDAE (badgers, otters, skunks, weasels). 25 genera, 70 species. Worldwide except for Australia and Oceania.	*Diverse.* The great majority of species appear to be solitary except for universal mother-offspring associations and male visits during the breeding season. In the sea otter *Enhydra lutris*, large, weakly organized herds of mixed sexes form where kelp and rocks provide sleeping sites; courtship, pairing, and male fighting occur within the herds. Other otters appear to be solitary. In the European badger *Meles meles*, a mated pair and their offspring occupy the same burrow system for as much as 2 generations; the burrow sites are "traditional" and sometimes last for hundreds of years, across many badger generations. The American badger *Taxidea taxus* is solitary.	Eisenberg (1966), Lockie (1966), Verts (1967), Erlinge (1968), Kenyon (1969), Ewer (1973)
SUPERFAMILY FELOIDEA		
FAMILY VIVERRIDAE (civets, genets, mongooses). 36 genera, 75 species. Old World except for Australia and Oceania.	*Diverse.* Viverrinae are nocturnal and evidently solitary, but the Herpestinae are generally much more social than even the mustelids. Pair bonds are typically formed by herpestines, and at least in *Crossarchus* and *Suricata* the female is larger and dominant over the male. In some species, e.g., in *Crossarchus*, *Helogale*, *Mungos*, and *Suricata*, several families den and forage together; in the banded mongoose *M. mungo*, the extreme case, packs of 30 to 40 individuals are common. In other species, e.g., in *Cynictis* and *Herpestes*, families den together but forage separately. In *Helogale* several generations may coexist and forage together.	Ewer (1963, 1973), Wemmer (1972), Albignac (1973), Rasa (1973)
FAMILY HYAENIDAE (hyenas, aardwolves). 3 genera, 4 species. Africa to India.	*Social.* Aardwolves (*Proteles cristatus*) feed primarily on termites and have been seen alone and in small packs, which may be family groups. Spotted hyenas (*Crocuta crocuta*) form "clans" of *10–100* individuals which are territorial; females are larger than males and dominant to them.	Eisenberg (1966), Kruuk (1972), Kruuk and Sands (1972)
FAMILY FELIDAE (cats). 4 genera, 37 species. Worldwide except Australia and Oceania.	*Diverse.* Most species solitary, although the juveniles often assist the mother in hunting prey; in the cheetah (*Acinonyx jubatus*) the male apparently also joins the female until the birth of the cubs. Lions (*Panthera leo*) form matrifocal packs accompanied by one or two males; see description elsewhere in this chapter.	Eaton (1969, 1970), Schaller (1970, 1972), Eisenberg and Lockhart (1972), Bertram (1973), Eloff (1973), Ewer (1973), Kleiman and Eisenberg (1973), Muckenhirn and Eisenberg (1973)

marily or exclusively to hunting prey, and as such they can be contrasted profitably with the adaptations of rodents, antelopes, and other herbivores whose social systems represent to some extent devices for avoiding these same predators.

The species to be described in the following sections represent the best-studied paradigms of most of the carnivore social grades. Because several of the species concerned are also "big game" and popular zoo animals, interest in them has been more intense and field studies more careful than usual. Zoologists are consequently in a better position to consider the ecological basis of their social evolution.

The Black Bear (*Ursus americanus*)

Bears have long been considered to be exclusively solitary. In an admirable field study conducted in northern Minnesota, L. L. Rogers (1974) showed that although this is approximately the case in the American black bear, individual relationships are far more intimate and prolonged than had been suspected. In brief, females depend on the exclusive occupancy of feeding territories to breed, and in this sense they are solitary. But they also permit their female offspring to share subdivisions of the territories and bequeath their rights to these offspring when they move away or die. In order to learn these facts, Rogers trapped and tagged 94 individuals over a four-year period. With the aid of radio-telemetry he was able to trace the histories of 7 female cubs from birth to maturity.

During the mating season, from mid-May to late July, adult females defend exclusive territories, which in Minnesota average 15 square kilometers and range from 10 to 25 square kilometers in extent. There appears to be a clear cut-off point below which reproduction becomes difficult. Two females possessing territories of only 7 square kilometers did not produce litters, while a third left the area after having a single cub. As the end of the summer approached, aggressiveness toward intruders waned, even though most of the females remained within their territories.

Nine families monitored by Rogers broke up during the first three weeks of June, when the cubs were 16 to 17 months old. Each of the female yearlings then remained in the mother's territory, utilizing a subdivision of her own for a period of at least two years. In one case four young females lived close to older females that were probably siblings from previous litters. The ranges of both the mother and the young females tended to remain separate, despite the fact that the entire ensemble represented the mother's original mating territory. When a mother bear was killed, one of her daughters took sole possession of a 15-square-kilometer sector of the territory. She gave birth to a litter in the following winter and raised it in the inherited area. In another case a three-year-old female became the exclusive occupant of the eastern portion of her mother's territory when the latter shifted her site 2.4 kilometers to the west. Her sister, who acquired the smaller western portion, grew more slowly and failed to produce a litter. The mother made the move in the first place to occupy the former territory of a deceased neighbor. Her presence caused the neighbor's three-year-old daughter to move into the western half of the dead bear's former territory. The displaced daughter shared this portion with a five-year-old, who was probably a sibling from a previous litter. She was dominated by the older bear and did not reproduce the following winter.

Male black bears take no part in this inheritance system. They disperse from the maternal territories as subadults. During the mating season the fully mature males enter the female territories and displace one another by aggressive interactions, especially when they meet in the immediate vicinity of the females. Later, as their testosterone levels drop, they withdraw from the females and assemble in peaceful feeding aggregations wherever the richest food supplies are to be found. In the late fall they return to the female territories to den.

The Coati (*Nasua narica*)

Coatis resemble elongated raccoons with tapering snouts and mobile, expressive tails. They are the most social of the American Procyonidae. The term "coatimundi" is often used to refer to one of these animals, but technically it is supposed to refer to a solitary coati—which zoologists have now shown to be almost invariably a male. *Nasua narica* is the northernmost species of the genus, ranging from Arizona south to Panama. Its ecology and social behavior were investigated on Barro Colorado Island by Kaufmann (1962), and additional information on the same population was supplied by Smythe (1970a).

Although the coati and the black bear represent independently evolving lines, the sociobiology of the first can be conveniently thought of as one step beyond that of the second. In essence it differs only in that several female-offspring groups cohere as stable bands. The home ranges of the bands overlap widely, but the core areas are occupied exclusively. Kaufmann's six bands varied in total membership from 4 to 13 individuals, with 1 to 4 adult females forming the nucleus. One lone adult female and something in excess of 12 solitary adult males were also observed in the study area. Although the composition of the bands remained nearly constant over prolonged periods of time, they frequently broke up into casual and temporary subgroups during the daytime foraging trips. Associations formed on the basis of varying combinations of individuals, so that a diagram of the splittings and regroupings would resemble a loosely braided cord. The most stable combinations within the bands were individual females and their cubs. It is likely but not yet proved that the females are closely related, perhaps at the level of sisters and first cousins. Bands undoubtedly multiply by simple fission, with one or more females departing to colonize new core areas.

Relationships among the band members are relatively loose. Mutual grooming occurs, being most frequent between mothers and their young, next most frequent between other members of different ages, and least common between age-peers. There is no clear-cut dominance hierarchy, although juveniles tend to prevail over all other members except their own mothers. As the "spoiled brats" of the coati societies, they belligerently chitter, squeal, and play-wrestle with their siblings. Sometimes they attack other coatis for no apparent reason other than that these animals come too close while the juveniles are eating or being groomed. Their dominance is based on the vigorous support of their mothers, who rush to their aid in disputes. The influence is lasting, because the youngsters are able to intimidate other coatis even when the mothers are temporarily absent. On such occasions they are occasionally supported by other adult females.

There is little evidence of cooperation or altruism in coati social behavior. Food items are the object of scrambling competition. Although small prey such as mice and lizards are often flushed into the open by the combined activity of several individuals, they are eaten by the first coati that can seize them. The winner holds the others at bay with nose-up squealing and aggressive rushes. The wrangling continues until every scrap is consumed. Once Kaufmann saw a mother allow one of her cubs to share a land crab, but then only after she had eaten most of it herself. When a coati is busy digging out the burrow of a lizard or tarantula, it threatens any others who try to join it. Leadership is all but absent. Juveniles tend to follow their mothers, but the troop as a whole moves with whichever members appear the most strongly motivated. No coati is specialized to be the sentinel. All of the troop members scatter at the first sign of danger—every coati, as it were, for itself.

Through most of the year the adult males lead solitary lives. When two individuals meet in the forest they exchange nose-up squealing, growling, and other hostile displays that sometimes lead to chases and fighting. In the Barro Colorado population a dominance hierarchy seems to exist, in the sense that the disputes are usually brief and the winners predictable in advance. When males meet family bands, hostility also breaks out. In most cases the bands take the initiative, with the male making an unhurried retreat. Only during the mating period at the start of the dry season (January–March) are the males able to approach the families in peace.

As illustrated in Figure 25-1, the reproductive cycle of the Barro Colorado coatis is intimately tied to the food supply. Mating occurs when a large amount of fruit is ripening on the trees. By the time the young emerge from the maternal nests and start to forage with their bands, there is such a surplus of fruit that much of it is left to rot on the ground. All of the coatis, including the still solitary males, become principally frugivores. Toward the end of the wet season, as the supply of fruit dwindles, the bands of females and young turn increasingly to the capture of invertebrates and small vertebrates in the litter of the forest floor, while the males prey not only on these animals but also on agoutis, spiny rats, and probably other vertebrates. The male populations appear to be ultimately food limited. At the time of lowest fruit fall they extend their foraging time well into the hours of darkness. They fight more, and the condition of their pelage deteriorates. The significance of this sexual difference is unclear. The ecological partition could result from some altruistic tendency of the males to shift to whatever other foods are available, leaving the pick of the crop for their offspring. But it is more likely, or at least more plausible with reference to current genetic theory, that the pursuit of larger prey is due primarily or exclusively to natural selection based on individual survival of the males. Perhaps the concerted action of the bands crops the smaller prey items to a level below that which can sustain an adult using

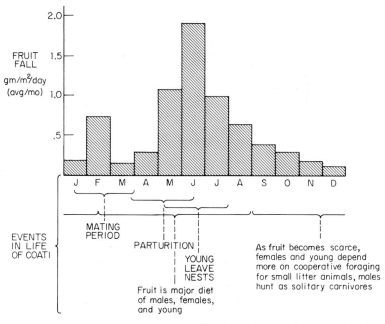

Figure 25-1 The relation between food supply and the annual cycle of social behavior in the coati. (Modified from Smythe, 1970a; coati data partially from Kaufmann, 1962.)

solitary foraging methods. As a result the males use their slightly larger size (they are 10 percent heavier than the females) to capture rodents and other larger prey.

The Lion (*Panthera leo*)

To the zoocentric human mind the lion has long enjoyed an exalted status: king of beasts, symbol of the sun, even animal god. The Egyptian pharaoh Rameses II took lions with him into battle, and kings from Amenhotep II to Saint Louis have traditionally hunted them for sport. But only within the last ten years has *Panthera leo* been made the subject of intensive zoological studies. For three years, from 1966 into 1969, George Schaller followed lion prides over the grasslands of Tanzania's Serengeti Park, "a boundless region with horizons so wide that one can see clouds between the legs of an ostrich," where heat waves at noon transform "distant granite boulders into visions of castles and zebra into lean Giacometti sculptures." Schaller logged 149,000 kilometers of travel while keeping the lions under observation for a total of 2900 hours. Subsequently Brian Bertram followed the same prides for an additional four years, confirming Schaller's results and acquiring valuable new insights into the ecological basis of their social behavior. Few animal populations have been studied for so long in the wild. As in Lynn Rogers' black bears, Iain Douglas-Hamilton's elephants, and Jane van Lawick-Goodall's chimpanzees, a new level of resolution has been attained, in which free-ranging individuals were tracked from birth through socialization, parturition, and death, and their idiosyncrasies and personal alliances recorded in clinical detail.

The core of a lion pride is a closed sisterhood of several adult females, related to one another at least as closely as cousins and associated for most or all of their lives within fixed territories passed from one generation to the next. In the prides most closely monitored by Schaller the average number of individuals per pride was 15 with a variation of 4 to 37. The degree of cooperation that the female members display is one of the most extreme recorded for mammal species other than man. The lionesses often stalk prey by fanning out and then rushing simultaneously from different directions. Their young, like calves of the African elephant, are maintained in something approaching a crèche: each lactating female prefers to nurse her own cubs but will permit those of other pride members to suckle. A single cub may wander to three, four, or five nursing females in succession in order to obtain a full meal. The adult males, in contrast, exist as partial parasites on the females. Young males almost invariably leave the prides in which they were born, wandering either singly or in groups. (A minority of the young females also become nomads.) When the opportunity arises these males attach themselves to a new pride, sometimes by aggressively displacing the resident males. Male

bands both inside and outside the prides typically consist of brothers, or at least of individuals who have been associated through much of their lives. The pride males permit the females to lead them from one place to another, and they depend on them to hunt and kill most of the prey. Once the animal is downed, the males move in and use their superior size to push the lionesses and cubs aside and to eat their fill. Only after they have finished do the others gain full access to the prey (see Figure 25-2). Males also respond more aggressively to strangers, especially to other males who attempt to intrude into the pride domain. The larger the size of the brotherhood, the longer its members are able to maintain possession of a pride before being driven out by rivals.

What is the significance of this peculiar social structure, in a group of mammals (the cat family Felidae) otherwise celebrated for its solitary habits? Schaller convincingly argues that the prides evolved primarily because group hunting is a superior means of catching large herbivorous mammals in open terrain. His data show that several lions stalking together are generally twice as successful at catching prey as are solitary lions. They are also capable of bringing down exceptionally large and dangerous prey, particularly giraffes and adult male buffalos, which are virtually inaccessible to single individuals. Schaller further found that cubs are better protected from leopards and nomadic male lions when their mother belongs to a group. For both these reasons, prides are far more successful at rearing litters than mothers living alone.

A loose dominance order exists among the lions and lionesses, based entirely on strength. Each lion seems to know the fighting potential of every other. The result is a tense peace broken only by noisy sporadic clashes that are intimidating but ordinarily do little damage. However, real fighting occurs, especially as an outgrowth of quarreling over the spoils, and the big cats show little restraint when they start to slash and bite. The best strategy for a pride member is to anticipate the attacks and to stay out of harm's way. Sometimes lionesses are able to force male lions to back off by launching concerted attacks. Occasionally lions even kill each other. Schaller recorded several fights between males that resulted in death. He also witnessed a case of the killing and cannibalism of cubs after one of the resident males died and the territory was invaded by other prides.

Wolves and Dogs (Canidae)

Three species of canids hunt in packs: the wolf (with its derivative the domestic dog), the African wild dog, and the dhole of Asia. Mass predation requires the highest degree of cooperation and coordinated movement, which redound in all other aspects of social life. Pack hunting permits relatively small animals to exploit large, difficult prey. Bourlière (1963) and other zoologists have noted that predatory mammals hunt mostly animals their own size and smaller. By weight of numbers alone, the pack-hunting canids have been able to break this restriction. Their counterparts among the marine mammals are the killer whales, which attack much larger whales in coordinated groups. Among the insects, the socioecological analogs are the army ants, which employ group foraging and mass assaults to subdue colonies of other social insects, including those of ants. And according to prevailing theory, primitive man was the analog among the primates (see Chapter 27).

Two behavioral traits basic to the Canidae seem to have made it easy for pack hunting to evolve on multiple occasions (Kleiman and Eisenberg, 1973). There is first the unique form of the pair bond, in which the male provisions both the female and her young, so that large litters can be reared whenever sufficient prey are available. Packs have formed in the most social species by an extension of this economic system to hold groups of related families together. Second, canids, unlike the majority of cats and other carnivorous mammals, pursue their prey in the open instead of relying on stealth and ambush. It is easier for cooperative pack hunting to evolve from such an initial hunting strategy.

The wolf, *Canis lupus*, is the northern representative of the pack hunters. Before being largely exterminated by man it ranged throughout North America south to the highlands of Mexico, and from Eurasia to Arabia, India, and southern China. It is larger in size than all but the most massive breeds of domestic dogs. Adults weigh 35-45 kilograms on the average and in extreme instances reach 80 kilograms, with males being slightly heavier than females. In other words, wolves are as large as small human adults. They also occupy the top of the food web. Over 50 percent of their food items consist of mammals the size of beavers or larger. Typical prey in North America include beavers, deer, moose, caribou, elk, mountain sheep, and, in the vicinity of settled areas, cattle, sheep, cats, and dogs (but seldom if ever human beings). Smaller prey, from mice to ptarmigans, add variety to the diet at all seasons but undoubtedly become more important in times of hardship. When packs sight an animal, they stalk and chase it as a coordinated unit. Smaller prey can be secured and disabled by the canine teeth of a single wolf. Larger prey must be literally torn down by concerted slashing and pulling on the part of the pack. Even group efforts frequently fail. The fleetest prey, such as deer and mountain sheep, often outrun the wolves, while an adult moose can fend off a large pack indefinitely if it stands its ground. Among 131 moose detected by wolves on Isle Royale as David Mech watched, only 6 were finally killed and eaten. Most of the remainder fled before the pack could close in, while the rest either stood at bay until the pack gave up or simply outran the wolves in straight pursuit (see Figure 25-3). The literature contains many accounts of successful hunts that were made possible only because of concerted action. Usually the prey was either cornered or else flushed from impregnable positions by an onslaught from several directions. At least three observers, Murie (1944), Crisler (1956), and Kelsall (1968), have witnessed wolves driving caribou toward other members of the pack lying in wait. Kelsall saw a pack of five wolves wait quietly as a minor band of caribou moved into a small clump of stunted spruce. When the caribou were out of sight an adult wolf walked just uphill from the spruce and concealed itself directly in the path being followed by the caribou. The other four wolves simultaneously circled the spruce, spread out along its downhill side, and began a stealthy "drive" through it. The goal was evidently to move the caribou toward the wolf waiting uphill.

The large size and specialized predatory habits of wolves dictate that they exist in low population densities and occupy relatively immense home ranges. On Isle Royale, Michigan, and in Algonquin Park, Ontario, as many as 40 wolves per thousand square kilometers have been counted, but the more common figures in Canada and Alaska are between 4 and 10 wolves. Because most packs contain between 5 and 15 members (the record is 36 from south-central Alaska), it is reasonable to suppose that the home range of a pack is on the order of 1000 square kilometers. Actual estimates from the field vary from approximately 100 to 10,000 square kilometers, with the majority falling between 300 and 1000 square kilometers (see Mech, 1970: Table 18, p. 165). The wolves move ceaselessly over their domains in search of prey. They commonly remain in the vicinity of kills for a period of several days to rest and to feed before heading off again. Although certain trails are repeatedly used during segments of their journeys, the overall pattern of movement has a random quality, and no grand circuit is followed. Running at the steady, tireless trot of the marathoner, the wolves can travel more than 100 kilometers in a 24-hour period. When hunted by man over hard snow in Finland, packs have covered as much as 200 kilometers in a day (Pulliainen, 1965). The work of Durwood L. Allen, David Mech, and their associates on Isle Royale has revealed that the packs are territorial, but the form of the territory is usually spatiotemporal and home ranges overlap considerably. It appears that one pack avoids using an area through which another pack has traveled a few hours or days previously. Undoubtedly the smell of scent in urine is an important sign employed by the wolves, although the sound of howling might also result in further separation. On occasion packs meet and fight. Wolfe and Allen (1973) recorded an encounter between the largest pack on Isle Royale and a pack of 4, during which 1 of the 4 was killed. At certain times the larger pack imposed territorial dominance on the smaller, but there were also quiescent periods during which the home ranges overlapped broadly.

The details of social behavior have been reviewed by Mech (1970), one of the principal observers of free-living packs, and Fox (1971),

Figure 25-2 In the Serengeti Park, a pride of lions devours a newly killed buffalo. The two males, who are brothers, have already eaten their fill and wandered away, permitting the remainder of the pride to approach and feed. The latter group consists of the lionesses, two three-year-old males, a juvenile about 18 months old, and two cubs 5 months in age. In the background two black-backed jackals and a group of vultures wait for a chance to share in the remains. A herd of wildebeest can also be seen. The adult male to the rear displays a relaxed open-mouthed face, while his companion stares at an unidentified object past the observer. Two of the lionesses snarl at each other during one of the

frequent low-keyed aggressive exchanges that occur between pride members at the kills. One of the young males, temporarily displaced during the jostling, crouches behind the kill. In the dominance hierarchy of the pride, cubs are at the bottom, and they suffer a high mortality rate from malnutrition due to an inability to eat fully before the prey is consumed. (Drawing by Sarah Landry; based on Schaller, 1972, in consultation with Brian Bertram.)

Figure 25-3 A wolf pack surrounds a moose on Isle Royale in Lake Superior. By standing its ground the moose successfully held the wolves at bay for five minutes, after which they gave up. (From Mech, 1970.)

who has studied the socialization process in captive animals. Mech's account is the more detailed and has the added advantage of being collated with current knowledge of the ecology of the species. A new pack is formed when a mated pair leaves its parental group to produce a litter on its own. As the family grows, separate linear dominance orders form among the males and females, respectively, with the founding pair occupying the alpha positions for at least a time. Dominance is expressed in priority of access to food, favored resting places, and mates. It is not absolute, however. An "ownership zone" exists within about half a meter of any wolf's mouth, and food in the zone is not disputed by higher-ranking animals. Rank begins to be established early in life, when puppies play-fight. It is reinforced in maturity by repeated exchanges of hostile and submissive displays. Fights usually end quickly by the submission of one of the contenders. But occasionally, especially during the breeding season, all-out battles erupt that result in serious injury. Cliques of wolves have been seen to gang up on individuals during these disputes. The alpha male is the center of constant attention, in every sense the lord and master of the pack. He is the leader in most chases and reacts first and most strongly to intruders. Other members normally defer to him during the greeting ceremony, during which one wolf tenderly nips, licks, and smells the mouth of another. The ceremony appears to be a ritualized version of food-begging movements by puppies. Although conducted most commonly following a separation, it is on many occasions directed spontaneously at the alpha male. Sometimes whole groups crowd around the leader in this act of friendly obeisance.

The alpha male also has greater access to estrous females, but this privilege is not absolute (Woolpy, 1968). The leader and other dominant males each show preference for particular females. The females in turn choose among the males, indicating their readiness to mate by standing still and moving their tails aside.

As Schenkel (1947, 1967) first demonstrated, wolves employ a rich repertory of facial expressions, tail positions, and body postures to express the nuances of rank and hostile intention. The presentation of the jugular region was interpreted by Lorenz in *King Solomon's Ring* (1952) to be a submissive signal, but this now appears to be an error. Lorenz said, "Every second you expect violence and await with bated breath the moment when the winner's teeth will rip the jugular vein of the loser. But your fears are groundless, for it will not happen. In this particular situation, the victor will definitely not close in on his less fortunate rival. You can see that he would like to, but he just cannot!" According to Schenkel (1967), the opposite is true. The dominant animal exposes his throat to the subordinate, who evidently does not dare to carry through this advantage. The early observations may simply have confused the roles of the two animals, although the matter is still far from completely resolved.

The visual repertory is supplemented by a comparable array of barks, howling, and other vocalizations. Although wolf pheromones have been poorly studied, they appear to be produced in five sites in the anal region alone: the genital glands, precaudal glands, anal glands, urine, and feces. Odors appear to be used in territorial marking, communication of which foods have recently been eaten (through the process of "snuffling" the lips of another animal with the nose), and the identification of the stage of the estrous cycle in the female. They are also used to augment communication in dominance interactions. A higher-ranking animal first sniffs the anal area of the lower-ranking animal, then presents its own posterior for inspection.

The evidence now seems overwhelming that the domestic dog originated entirely from the wolf, without receiving any detectable infusion of genes from jackals, coyotes, or other species of *Canis*. In fact *Canis familiaris*, the domestic dog, cannot really be separated as a valid biological species from the ancestral *Canis lupus*. Possibly the only universal diagnostic trait of the domestic dog is the sickle-shaped or curly tail, which is found in all breeds and is easily distinguishable from the drooping tail carriage of wolves and other wild dogs. The intensely social nature of wolves, their eagerness to express submission by groveling and ritual licking, their readiness to follow the leadership of a dominant animal, and their habit of hunting in packs, preadapted them to become symbiotic companions of man. Carbon-dated archaeological remains indicate that this event had occurred by 12,000 B.P., when populations of hunter-gatherers were spreading behind the retreating edge of the final continental ice sheet. How could wolf cubs young enough to be socialized and still requiring milk have been incorporated into human society? J. P. Scott (1968) has offered the following ingenious and entirely plausible hypothesis:

Scavenging wolves would have come around the hunting camps, looking for offal and attempting to steal stored supplies of meat. The hunters may, on occasion, have even hunted wolves and dug the young cubs out of their dens. Some of these may have been brought home alive and escaped the soup pot by attracting the attention of a woman who had lost her baby and was suffering discomfort from persistent lactation. Such a wolf cub could be very easily reared on the breast by a human mother for a few weeks, after which it could subsist on scraps and bits of cooked food. In a time of ample meat supplies there would have been plenty to go around. The adopted cub would have become rapidly attached to human beings, as wolf cubs do today, if taken at the right time, and it would have been friendly and playful with the children. By the time it was three months old it would have been largely self-sufficient, living on scraps of food and becoming a member of the human group. And unless human behavior has changed markedly, the foster mother would have become strongly attached to it.

The social behavior so well marked in the wolf is carried to further heights in the African wild dog *Lycaon pictus*, appropriately called by Hediger "the super beast of prey." The species is one of the

Figure 25-4 The "super beasts of prey" and most highly social canids: a pack of wild dogs on the Serengeti Plains of Tanzania. Most of the adults are just returning from a successful hunt. In the foreground an adult prepares to regurgitate some of the fresh meat to pups who tumble out of the den. On the left the mother dog performs the greeting cere- mony to the dominant male. In a moment she, too, will be fed by regurgitation. In the distance can be seen herds of zebras and wildebeest, which are among the largest animals attacked by the dogs. The excep- tionally populous litter is another trait of this species. Only one or two females produce a litter in a given year, and the remaining adults partici-

pate fully in the care and upbringing of the young animals. The exceptional altruism and cooperativeness of the species is associated with the habit of hunting in packs, a technique that increases the efficiency of capturing prey during daylight chases and makes possible the killing of animals much larger than the individual dogs. (Drawing by Sarah Landry; based on Estes and Goddard, 1967, and Hugo van Lawick-Goodall in van Lawick-Goodall and van Lawick-Goodall, 1971, in consultation with Richard D. Estes.)

scarcest yet most wide-ranging mammals of Africa. It occurs in most habitats other than extreme desert and dense forest. One pack of five has even been seen on the summit of Mt. Kilimanjaro (5895 meters), evidently the altitude record for mammals generally. One of the strictest of carnivores, the wild dog usually hunts prey approximately its own size, such as Grant's and Thomson's gazelles, impala, and the calves of wildebeest. But it also attacks and consumes much larger animals, including adult wildebeest and zebras. The hunts are almost always conducted in a tight group. Lasting an average of only 30 minutes and usually ending in success, they are scenes of unparalleled ferocity. The pack leader selects the target while still at a distance and leads the others toward it in a determined sprint. Gazelles flee when the dogs approach to within 200-300 meters. The predators rely on a combination of speed, endurance, and numbers to capture even the fleetest animals. Running at 55 kilometers per hour and in bursts at 65 kilometers per hour, the dogs overtake most quarries within the first 3 kilometers. Occasionally they hold a 50 kilometer per hour pace for 5 kilometers or more. They do not, as once thought, run in relays. One dog, usually a member of the leadership "cadre," holds the lead throughout, while the others string out behind it for as much as a kilometer or more. The advantages of group chasing are twofold. Some of the prey run in wide circles or in zigzag patterns in attempts to throw off pursuers at their heels. Other members of the pack running behind are able to cut across the curve and close the distance. Once the animal is seized, all the members of the pack rush in to immobilize it, quickly tearing it to pieces by yanking in all directions. Gazelles can be killed and eaten within 10 minutes following capture. A bull wildebeest or zebra might require more than an hour, but it is still remarkable that a creature the size of a German shepherd can take such oversized prey at all.

Our knowledge of the social behavior of the African wild dog is quite recent, stemming from field studies in the Serengeti National Park by Kühme (1965), Estes and Goddard (1967), and Hugo van Lawick (1974, and in H. and J. van Lawick-Goodall, 1971). During hundreds of hours of observations these zoologists found a degree of cooperation and altruism unmatched by any other animals except elephants and chimpanzees. As soon as the pack has eaten its fill it returns to the den to regurgitate to the pups, their mother, and any other adults who remained behind (see Figure 25-4). Even when the prey is not large enough to feed all of the dogs to repletion, the hunters still share their booty. Sick and crippled adults are thus cared for indefinitely. At the kill juveniles are given precedence by the adults, a complete reversal of the procedure in lions and wolves. Communal behavior is developed to such a degree that when a litter of nine pups watched by Estes and Goddard was orphaned at the age of five weeks, they were reared by the eight remaining members of the pack, all of which happened to be males.

Despite the savagery displayed in the hunts, wild dogs are relaxed and egalitarian in relations with one another. No individual distance is observed, and the pack members sometimes lie in heaps to keep warm. Females vie with one another for the privilege of nursing the pups, although the mother normally retains first rights. Separate dominance orders exist among the males and females, but they are so subtle in expression as to be easily overlooked by human observers. Threats are especially difficult to recognize. Instead of snarling and bristling like a wolf, the wild dog assumes a posture resembling that taken during stalking. The head is lowered to the level of the shoulder or below, the tail hangs quietly, and the dog either stands rigidly while facing its opponent or walks stiffly toward it. Submission, in contrast, is an elaborate and conspicuous performance. It grades insensibly into the greeting ceremony, by which the animals reestablish contact and on other occasions initiate pack chases. In potentially tense situations, especially following a kill, the dogs seem to compete with one another in making submissive displays. Their lips draw back in a rictuslike grin, the forepart of the body is lowered, and the tail is lifted over the back. The animals excitedly twitter back and forth as each tries to burrow beneath the other. There is an effort, to use Estes' expression, to be underdog instead of top dog. When ritualized begging behavior in the form of face licking and mouth snuffling is added, the performance turns into the full-fledged greeting ceremony.

With the gestures of obeisance prevailing, it is no wonder that the meaning of aggression and dominance within the wild dog society has remained obscure. The uncertainty is greatest in the relationships between females. In one case observed by Hugo van Lawick (in H. and J. van Lawick-Goodall, 1971), a linear order existed among four bitches. The current mother was at the bottom of the ranking and often harassed by the others, who appeared to be motivated by an intense interest in the pups. This relationship is puzzling, but it takes on a possibly sinister significance in light of van Lawick's subsequent discovery (1974) of open hostility and infanticide between two bitches who littered at the same time. When "Angel" became pregnant at the same time that "Havoc" was raising a litter, she was persistently driven away by this more dominant female. After Angel's pups were born they were systematically caught and killed by Havoc until only one remained. The survivor, "Solo," was finally adopted by Havoc and allowed to play with her own pups, albeit in a subordinate role where it was often the target of aggression. Thereafter Havoc prevented Angel from approaching Solo.

An intriguing picture of wild dog reproduction is now emerging. In any given year only one or two of the females produce a litter. Parturition, or at least success in bringing a litter to weaning, may depend on the position of the female in the dominance hierarchy. But whether or not this is really the case, it is indisputably true that the pack as a whole invests in only one or at most two litters at a time. These litters are relatively enormous in size, averaging about 10 pups in the wild and ranging to as many as 16. Most are born during the rainy season, when most herbivores are also born. The significance of the trait can be inferred, I believe, by comparing wild dogs with army ants. Both are extreme carnivores that use mass forays to conquer prey too large or otherwise too formidable for single predators. Probably as an ultimate consequence of this specialization, both the dogs and the ants are nomadic, shifting from site to site on an almost daily basis. Not to do so would be to reduce the food supply within striking range of the core area to below the maintenance level. Army ants are notable among social insects for the high degree of synchronization in their brood development, which is made possible by extraordinary bursts of oviposition over short periods of time and at regularly spaced intervals. These insects are nomadic only when the brood is in the larval stage. Thus synchronization of brood development means that the colony can remain safely in one well-entrenched home site for long stretches of time, when all of the young are in the egg and pupal stages. The wild dogs also benefit from synchronization but in a different way. When a litter is born the pack is tied down to one spot until the pups are large and strong enough to join the nomadic marches. If each bitch had a litter of the usual canid size and independently of the others, the pack would

be forced to spend much longer periods of time in one place. Therefore it can be reasonably suggested that large litters by single females has as its raison d'être the synchronization of development, which permits the pack to be nomadic on a maximum number of days in each year.

Although wild dogs are nomadic, they appear to stay within certain very large defined areas. The pack monitored by Kühme remained within an area of 50 square kilometers during February when game was densest, but by May, when game became scarce, the dogs were covering 150-200 square kilometers in long marches. Other observations suggest that over a period of years the total range covered by a single pack can extend to thousands of square kilometers. On the infrequent occasions when packs meet, their interactions vary greatly. Sometimes they react in an apparently friendly manner, but just as often they avoid each other or one group chases the other. The urine marking so characteristic of other canids is weakly developed in the wild dog. Dominant females mark the denning area a great deal, and on two occasions van Lawick saw small intruding packs chased from a den neighborhood. Thus it is possible that territorial behavior in the strict sense is limited to this one site during the two months of each year that pups are being raised. It may also be true that packs repel one another in ways too subtle for detection.

Chapter 26 **The Nonhuman Primates**

The living primate species can be profitably viewed as a kind of *scala naturae* that proceeds from near the phylogenetic base of placental mammals upward in small steps through increasing anatomical specialization, behavioral complexity, and social organization. It embraces the following taxonomic sequence: the tree shrews, the tarsiers, the lemuroids, the New World monkeys, the Old World monkeys, the anthropoid apes, and finally man. As T. H. Huxley said in 1876, "Perhaps no order of mammals presents us with so extraordinary a series of gradations as this—leading us insensibly from the crown and summit of the animal creation down to creatures from which there is but a step, as it seems, to the lowest, smallest and least intelligent of the placental mammals." In modern terms the *scala* must be interpreted as a series of evolutionary grades transecting a branching phylogenetic tree rather than literal steps leading from ancestors to descendants among the living forms (see Hill, 1972). But the precise definition of the grades remains one of the key problems of current primate social studies, and special attention will be devoted to it here.

The Distinctive Social Traits of Primates

First let us consider the underlying biological qualities that have contributed to the primates' remarkable social evolution. In 1932 Solly Zuckerman proposed in *The Social Life of Monkeys and Apes* that the binding force of primate sociality is sexual attraction. He had been led to this view by observations of a newly formed group of hamadryas baboons in the London Zoological Gardens. The males fought for possession of females while engaging in intense sexual activity. But the really unique feature that Zuckerman believed he saw was the uninterrupted sexual life of monkeys, apes, and man generally. Even if a given species possesses a breeding season, Zuckerman claimed, the variation in activity does not affect the sexual nature of the social bonds, "since there is no implication that the sexual stimulus holding individuals together is ever totally absent. Seasonal diminution of the reproductive activity of all the animals in a group would not disturb its intrinsic sexual basis, because society would hold together as long as its members were to any extent sexually potent." For the next 25 years this theory dominated thought in primate sociobiology. As late as 1959 Sahlins could still say that "it was the development of the physiological capacity to mate during much of, if not throughout, the menstrual cycle, and at all seasons, that impelled the formation of year round heterosexual groups among monkeys and apes. Within the primate order, a new level of social integration emerges, one that surpasses that of other mammals whose mating periods, and hence heterosexual grouping, are very limited in duration and by season."

Zuckerman's theory is wrong. It was disproved by the field studies of primate biology that began to flourish in the late 1950's and have accelerated to the present time. Primates have been found to possess distinct breeding seasons which are sharply marked even in a large percentage of species with very cohesive societies (Lancaster and Lee, 1965; Hill, 1972). Many of the fine details of social interaction have proved to be wholly dissociated from reproductive behavior. The important partial correlates of advanced sociality include the presence or absence of territory, the strategy of defense against predators, and other nonsexual phenomena. Ironically, one of the more persuasive pieces of evidence came from Hans Kummer's later studies of free-living hamadryas baboons in Ethiopia. Kummer found that subadult males start herding females even before they set up separate groups and long before sexual activity begins. They try to kidnap and mother infants less than six months of age. Eventually they adopt juvenile females and use threat to condition them to stay close. The one-male unit is thus created well in advance of sexual activity. Kummer believes that the bond has evolved as a transferred form of the mother-child relationship. The same conclusion was reached independently by N. A. Tikh, who studied hamadryas in compounds at the Sukhumi Station on the Black Sea (see Bowden, 1966).

The Zuckerman theory constituted the first—and perhaps the last—of the great unitary explanations of primate social evolution. The subsequent accretion of facts has revealed a large degree of idiosyncrasy in species characteristics, leading to the belief that the evolutionary grade attained by a particular species is at least partially determined by the peculiarities of the immediate environment to which it is adapted. So much can be explained, then, by viewing primate evolution in the same way that has proved successful in studies of social insects, birds, ungulates, and a few other vertebrate taxa. But the question remains as to why some of the primate species have attained higher evolutionary grades than other vertebrate groups. Surely a large brain is an essential concomitant, because the animals of greatest interest are the larger cercopithecoid monkeys and apes. But we do not know to what extent intelligence was a preadaptation that biased them toward complex societies, and to what extent it was a postadaptive device implementing the improvement of the social organization in response to some external selection pressure.

Preadaptation cannot yet be teased apart from postadaptation. The best that can be done is to tie them together in a logical but hypothetical sequence of cause and effect that accounts for the most distinctive features of primate social life recognized by specialists. The scheme represented in Figure 26-1 postulates certain basic primate qualities to be evolutionary prime movers. Following the method outlined in Chapter 3, I have classified them as stemming either from phylogenetic inertia or from the major adaptive shift of primates to arboreal life. Both of these influences, the inertial and postadaptive, triggered chains of other adaptations which together consititute the diagnostic social qualities of the primates.

The basic systems of mammalian reproduction and heredity are ultraconservative. An evolving mammalian population cannot easily alter the pituitary-gonadal endocrine system, substitute haplodiploidy for the XY sex-determination mechanism, or dispense with maternal care based on lactation. Consequently the reproductive and genetic systems are inertial in their effects. Because of them certain ancient mammalian traits continue to prevail throughout the primates. There is a tendency for males to be polygynous and aggressive toward one another, although pair bonding and pacific associations are permissible minority strategies (Washburn et al., 1968). Where long-term sexual alliances are not the rule, the strongest and most enduring bonds are between the mother and her offspring, to an extent that matrilines can be said to be the heart of the society. Mothers are the principal socializing force in early life. In at least some of the aggressively organized species they exert an influence on the identity of the peers and social rank of their sons and daughters. Their influence may even extend to later generations (Kawamura, 1967; Marsden, 1968; Missakian, 1972).

The second class of ultimate determinants of primate social behavior consists of the basic postadaptive traits, shown as the righthand side of Figure 26-1. The vast majority of arboreal animals, from insects to squirrels, are small and have no difficulty moving through the canopies of trees. The surfaces of trunks, limbs, and even leaves are broad enough in proportion to their bodies to be navigated as though they were uneven extensions of the ground. However, most primates, particularly the phylogenetically more advanced prosimians, monkeys, and apes, are unusual in being *large* arboreal animals. The ultimate reason why they filled the large-size categories is unknown, but the immediate physiological consequences of this adaptive shift are clear. For animals that must judge distances and the strength of supports with precision, vision is the paramount sense. Visual acuity in primates has been enhanced by moving the eyes to the front part of the head, making stereoscopic vision possible, and adding color vision, which increases the power of discrimination of objects within the variegated foliage. Cartmill (1974) has suggested that the tendency to prey on small insects made these changes even more advantageous. Sound has taken on added significance as the only means of detecting other animals through dense foliage. At the same time the sense of smell has declined in importance. A large animal can depend less on the tracking of odors in the irregular air currents of the canopy. It moves too rapidly and must follow pathways through the branches too irregular to permit exact orientation along the active odor spaces emitted by other animals. As a consequence the primates have come to depend heavily on visual and auditory signals in their communication systems. The trend has been carried much further in the generally larger Old World monkeys and apes than in the prosimians and New World monkeys.

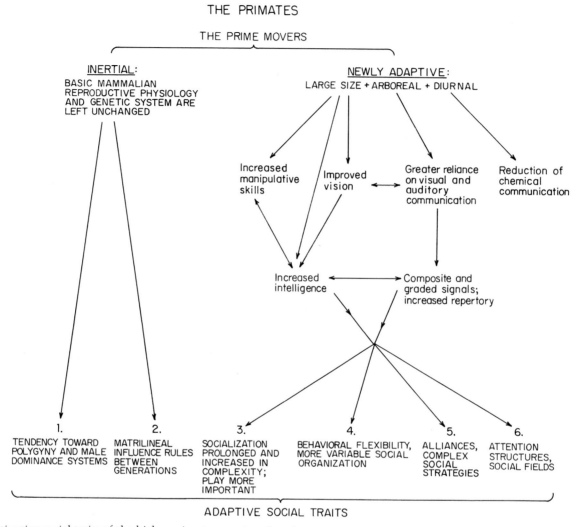

THE PRIMATES

THE PRIME MOVERS

INERTIAL:
BASIC MAMMALIAN REPRODUCTIVE PHYSIOLOGY AND GENETIC SYSTEM ARE LEFT UNCHANGED

NEWLY ADAPTIVE:
LARGE SIZE + ARBOREAL + DIURNAL

Increased manipulative skills

Improved vision

Greater reliance on visual and auditory communication

Reduction of chemical communication

Increased intelligence

Composite and graded signals; increased repertory

1. TENDENCY TOWARD POLYGYNY AND MALE DOMINANCE SYSTEMS

2. MATRILINEAL INFLUENCE RULES BETWEEN GENERATIONS

3. SOCIALIZATION PROLONGED AND INCREASED IN COMPLEXITY; PLAY MORE IMPORTANT

4. BEHAVIORAL FLEXIBILITY, MORE VARIABLE SOCIAL ORGANIZATION

5. ALLIANCES, COMPLEX SOCIAL STRATEGIES

6. ATTENTION STRUCTURES, SOCIAL FIELDS

ADAPTIVE SOCIAL TRAITS

Figure 26-1 The distinctive social traits of the higher primates are viewed as the outcome of conservative mammalian qualities ("inertial" forces) and adaptation to arboreal life. Even phyletic lines that are now terrestrial have retained the evolutionary advances made by their arboreal ancestors.

As Bernhard Rensch (1956, 1960) has argued on various occasions, large body size is crudely correlated in mammals with greater intelligence, seemingly as an inevitable result of the increase in absolute brain size. Thus the higher primates gained some component of their intelligence in the simple process of becoming large. Their mental capacity has been enhanced still more by the method of using the hands and feet to grasp branches during locomotion and rest. Both the New World monkeys and the Old World monkeys and apes have gone further by developing a "precision grip" as distinct from the more primitive "power grip" (Napier, 1960). Instead of merely closing the hand around the object, whether for support or feeding, they invest some amount of separate control in the index finger and thumb, permitting the fine manipulation of food particles and the grooming of fur. In general, the larger the primate the more dexterous the manipulation. Chimpanzees are more skilled than macaques and baboons, which in turn are superior to langurs and guenons. Man represents the culmination of this evolutionary trend.

Intelligence is the prerequisite for the most complex societies in the vertebrate style. Individual relationships are personalized, finely graduated, and rapidly changing. There is a premium on the precise expression of mood. Higher primates have extended the basic mammalian tendency away from the use of elementary sign stimuli and toward the perception of gestalt, that is, toward the simultaneous summation of complex sets of signals. In vision, for example, a bird or fish may respond to a single patch of color or the correct performance of one movement of the head—and to virtually nothing else. The monkey or ape more consistently tends to act on the appearance of the entire body, the posture, and the history of previous encounters with the individual confronted. There is also a tendency to utilize information from more than one sensory modality. At close range, visual and auditory signals are compatible and can be blended with tactile cues to form composite signals that convey messages redundantly and with greater exactitude (Marler, 1965). R. J. Andrew (1963a) has pointed out that the deep grunts commonly used by Old World monkeys and apes during close social encounters are particularly well suited to this purpose. The sounds are rich in overtones and therefore highly personalized, allowing the identification of individuals by voice alone. They are generated by the upper part of the respiratory tract, so that in addition to unique messages they carry redundant information concerning visual signals based on the shape of the mouth, position of the tongue, and other muscular postures that determine the expression of the face. Increasing sophistication in the employment of such composite signals among primitive hominids could have set the stage for the origin of human speech. Another probable consequence was the use of the face for personal recognition. Van Lawick-Goodall, Schaller, and others have documented the striking, humanlike variation in the facial features of chimpanzees and gorillas. It is easy for human observers to recognize individuals at a glance and even to guess their parentage with a high level of accuracy. These and other special qualities of primate communication have been extensively discussed in reviews by Andrew (1963b, 1972), Altmann et al. (1967), Anthoney (1968), Moynihan (1969), Wickler (1969b), and van Hooff (1972).

In addition to monitoring multiple signals, higher primates evaluate the behavior of many individuals within the society simultaneously. The animal lives in a *social field* in which it responds to multiple individuals simultaneously, in ways that take differing relationships into account and often entail compromise. Observers of free-living societies of Old World monkeys and apes have noted the use of behavioral strategies that manipulate the social field. Kummer (1967), for example, described the "protected threat" tactic of the hamadryas baboon. A female competing with a rival moves next to the overlord male, where she is in a better position to intimidate and to resist attack. If she is threatened, the male is much more likely to drive her rival away than to punish her. As a result she is more likely to advance in social rank. Alliances are also commonplace, especially between mothers and their grown offspring. Alloparental care leads to coalitions between adults as well as to the more rapid extension of social contacts by developing young. In troops of macaques and baboons adult males, not necessarily related, back one another up during aggressive encounters. The rank of an individual depends not only on his personal prowess but on the strength and dependability of his allies (Altmann, 1962a; Hall and DeVore, 1965). The dominant female of a troop of bonnet macaques studied by Simonds (1965) relied on the assistance of the dominant male to win encounters with every other member of the troop. But when her protector fell in status following the loss of a canine tooth and defeat in a major fight, the female was no longer able to dominate the other males.

Chance (1967) and Chance and Jolly (1970) have conceptualized the organization of individual social fields in terms of the *attention structures* of whole societies. Among the species of Old World monkeys and apes two categories of attention structure can be roughly distinguished. Centripetal societies, possessed by macaques, baboons, and most other cercopithecoid species, are organized around a dominant male. The members watch the male predominantly, shift their positions according to his approach or departure, and adjust their aggressive behavior toward others according to his responses. When the group is attacked from the outside, the dominant male and his allies lead the defense or retreat. The more pronounced the dominance structure, the stronger the centripetal orientation. When aggression breaks out inside the group, the members tend to move toward the cohort of dominant males, and this is sometimes true even when some of the males are the aggressors. Acentric societies, the second type, are exemplified by the patas, langurs, and gibbons. Although the attention structure varies in detail among species, all acentric societies are characterized by the tendency of the females and young to separate from the males during aggressive episodes. In other words, the society fragments in the face of tension. During peaceful moments the patas male lives mostly on the fringe of his little troop, serving principally as a watchdog. When a predator threatens, he runs to a low tree or other prominent position and threatens, while the females and young take refuge in another direction. Chance and Jolly view attention structure as basic and its analysis as the key to the understanding of primate societies. But in fact attention structure is just one more parameter, compounded of a multiplicity of behaviors and evolving as an adaptation to special features of the environment. As such it can be fed into certain models of social organization along with other parameters such as age structure, group size, and signal transmission rates. Loy (1971) has also criticized the classification of attention structure as being oversimplified, pointing out that not all species can be fitted comfortably into the dichotomy. Chimpanzees, identified by Chance and Jolly as centripetal, are actually much too loosely organized to make this specification useful. They occur in weakly structured, frequently changing social groups that include heterosexual bands without dependent young, bands made up solely of adult males, bands of mothers and young alone, and in fact bands of nearly all possible sex-age combinations. Males of rhesus monkeys, to cite another example, play a minor role in dictating group activities, far less than in the case of the baboons, which are the real paradigms of centripetal arrangement. Despite the shortcomings of their scheme, Chance and Jolly are correct in calling attention to the higher level of organization that self-assembles out of the higher primates' tendency to operate in complex social fields.

Social fields and attention structures enrich the roles played by individuals. DeVore (in Hall and DeVore, 1965) found that old anubis baboon males can maintain dominance and leadership when well past their physical prime, because they remain respected members of the "central hierarchy." Thelma Rowell (1969a), on the basis of separate studies of the same species, reasoned that other members of the troop benefit when they accord respect and prestige to declining but experienced leaders. Because primates are the chief predators of other primates, and man in particular has hunted African species through periods of evolutionary time, it is probably advantageous for individuals to utilize the special knowledge accumulated by the oldest, wiliest members of the group.

All of the distinctive primate traits just cited tighten the moment-to-moment adjustment by individuals to fluctuations in the environment—this in the most general terms is the key primate behavioral adaptation. To shift quickly and precisely from one response to another according to subtle changes in the social field requires that the structure of the society itself be malleable. The primate literature is filled with accounts of primate social malleability; Hans Kummer, Thelma Rowell, and others have stressed that it is one of the most distinctive phenomena seen in free-ranging societies. The anubis baboon is a particularly instructive species. On the savanna of Kenya's Nairobi Park, DeVore observed a definite marching order among the members of troops moving from one location to another. The dominant males accompanied females with small infants near the center, juveniles flanked these individuals near the center, and other adult males and females formed the van and rear. When a potential predator appeared the dominant males at once moved to the front to meet it. Rowell (1966a) discovered a different organization in anubis baboon troops in the forests of Uganda. Here troops progressed and communicated more like arboreal primate species than other baboons. The movements were less regular, with no consistent marching order. While moving through thicker vegetation, the baboons communicated more with grunts and showed a much greater concern for stragglers than did baboons on the savannas. Aggressive interactions among the males were also less frequent, and Rowell could find no evidence of the dominance hierarchies that are the hallmark of the savanna troops. Forest troops are casual about their sleeping places and generally avoid other troops. But on the open savanna of the

Amboseli Reserve, where groves of sleeping trees are scarce, the anubis baboons tolerate the presence of other groups, and large sleeping aggregations sometimes form similar to those of the hamadryas baboon. At the Awash Falls of Ethiopia a troop of anubis baboons has penetrated into terrain otherwise occupied by hamadryas. By studying this transferred group in detail, Nagel (1973) was able to differentiate to some extent the genetic and learned components of the baboons' social behavior, in the special sense of determining which differences between the species persist when the two forms are placed in a similar environment. The anubis troop gathered in a common sleeping place and separated for foraging, the hamadryas-like pattern Rowell had seen in Uganda. They also resembled the hamadryas in the length of the foraging routes they followed and in the amount of time spent foraging in wooded areas. But they retained the one-level social organization characteristic of anubis elsewhere instead of shifting to the hamadryas two-level harem system.

Kummer has described an experiment that illustrates how dramatically one feature of social behavior can change if sufficient stress is exerted. When a hamadryas female is placed in a group of anubis baboons, she quickly alters her social responses from the hamadryas forms to those of her new associates. Within half an hour she starts fleeing from attacking males like an anubis female rather than moving toward them. The reverse experiment is even more suggestive. An anubis female inserted into a hamadryas troop learns within one hour to approach the attacking male, thus conforming to the harem system that characterizes this species as opposed to her own. The adaptation is imperfect, however. After learning the behavior perfectly, the majority of the anubis females escape the herding male and stay away for good. This inability to make a total adjustment might be sufficient by itself to explain why the anubis troop at the Awash Falls failed to shift to a hamadryas organization even when surrounded by hamadryas societies in an altered environment.

The Ecology of Social Behavior in Primates

The principal organizing concept in the study of primate societies has been the theory that social parameters are fixed in each species as an adaptation to the particular environment in which the species lives. The parameters include size, demographic structure, home-range size and stability, and attention structure. Because this theory is rudimentary and still lacking in formal structure, it is most quickly grasped when traced historically. Its seeds were laid by C. Ray Carpenter (1934, 1942b, 1952, 1954), the first to recognize clearly that group size, demography, and various social behaviors are diagnostic traits of species. Carpenter proposed that the sex-age structure tends toward a steady state. Each primate species can be characterized by a "central grouping tendency," which is the array of the median

Table 26-1   A synopsis of the living primates. Species are cited that have been the objects of significant amounts of sociobiological research. (Higher classification based on Simpson, 1945; lower classification and geographical distributions based on Napier and Napier, 1967.)

ORDER PRIMATES

SUBORDER PROSIMII
(THE PROSIMIANS)

Family Tupaiidae. Tree shrews.
Anathana (1 species). Madras tree shrews. Wooded areas in southern India.
Dendrogale (2 species). Smooth-tailed tree shrews. Forests, Vietnam to Borneo.
Ptilocerus (1 species). Feather-tailed tree shrews. Forests, Malaya to Borneo.
Tupaia (12 species; see especially *T. glis*). Tree shrews. Forests of southeastern Asia and fringing islands. (Martin, 1968; Sorenson, 1970.)
Urogale (1 species). Philippine tree shrews. Mindanao, Philippine Islands.

Family Lemuridae. Lemurs.
Cheirogaleus (3 species). Dwarf lemurs. Forests, Madagascar.
Hapalemur (2 species). Gentle lemurs. Forests, Madagascar.
Lemur (5 species; see especially the ring-tailed lemur *L. catta*). True lemurs. Forests, Madagascar and Comoro Islands. (Petter, 1962, 1970; Petter-Rousseaux, 1962; Jolly, 1966, 1972b; Klopfer and Jolly, 1970; Klopfer, 1972.)
Lepilemur (1 species). Sportive lemurs. Forests, Madagascar. (Petter, 1962; Petter-Rousseaux, 1962; Charles-Dominique and Hladik, 1971.)
Microcebus (2 species). Mouse lemurs. Forests, Madagascar. (Petter, 1962; Petter et al., 1971; Martin, 1972, 1973.)
Phaner (1 species). Fork-marked dwarf lemurs. Forests, Madagascar. (Petter et al., 1971.)

Family Indriidae. Indrises.
Avahi (1 species). Avahis. Forests, Madagascar.
Indri (1 species). Indrises. Forests, Madagascar.

Propithecus (2 species; see especially Verreaux's sifaka *P. verreauxi*). Sifakas. Forests, Madagascar. (Petter, 1962; Jolly, 1966, 1972b.)

Family Daubentoniidae. Aye-ayes.
Daubentonia (1 species). Aye-ayes. Forests, Madagascar. (Petter and Petter, 1967; Petter and Peyrieras, 1970.)

Family Lorisidae. Lorises and galagos.
Arctocebus (1 species). Angwantibos. Forests of West Africa. (Charles-Dominique, 1971.)
Galago subgenus *Euoticus* (2 species). Needle-nailed galagos. Forests of Fernando Póo and tropical Africa. (Charles-Dominique, 1971.)
Galago subgenus *Galago* (3 species; see especially *G. senegalensis*, the bushbaby). Galagos and bushbabies. Forests and woodland savannas of Africa between 13°N and 27°S. (Sauer and Sauer, 1963; Doyle et al., 1967, 1969; Charles-Dominique, 1971; Rosenson, 1973.)
Galago subgenus *Galagoides* (1 species, *G. demidovii*). Dwarf galagos. Forests of Fernando Póo and tropical Africa east to the Rift Valley. (Vincent, 1968; Struhsaker, 1970b; Charles-Dominique, 1971, 1972; Charles-Dominique and Martin, 1972.)
Loris (1 species). Slender lorises. Forests, southern India and Ceylon. (Subramoniam, 1957; Petter and Hladik, 1970.)
Nycticebus (2 species). Slow lorises. Forests, India and Cambodia to Borneo.
Perodicticus (1 species). Potto. Forests in Africa. (Blackwell and Menzies, 1968.)

Family Tarsiidae. Tarsiers.
Tarsius (3 species). Tarsiers. Forests on Sumatra, Borneo, Celebes, Philippines, and surrounding islands.

SUBORDER ANTHROPOIDEA
(MONKEYS, APES, AND MAN)

SUPERFAMILY CEBOIDEA (= "PLATYRRHINI")
(NEW WORLD MONKEYS)

Family Callithricidae. Marmosets and tamarins.
Callimico (1 species, *C. goeldii*). Goeldi's marmosets. Forests of the Upper Amazon.
Callithrix (8 species). Marmosets. Forests of Brazil south of the Amazon, thence southward to Paraguay.
Cebuella (1 species). Pygmy marmoset, *C. pygmaea*. Upper Amazon Valley. (Christen, 1974.)
Leontideus (3 species). Golden lion marmosets (or tamarins). Forests, Brazil.

Saguinus subgenus *Marikina* (4 species). True bare-faced tamarins. Forests, upper Amazon.
Saguinus subgenus *Oedipomidas* (2 species; see especially the rufous-naped, or Geoffroy's, tamarin *S. geoffroyi*). Crested bare-faced tamarins, pinchés. Forests, Panama to Colombia. (Moynihan, 1970b.)
Saguinus subgenus *Saguinus* (16 species). Tamarins. Forests, Amazon Basin.

Table 26-1 (*continued*).

Family Cebidae. New World Monkeys.

Alouatta (5 species; see especially the mantled howler *A. villosa* [= *A. palliata*]). Howlers. Tropical forests of Central and South America. (Carpenter, 1934, 1965; Collias and Southwick, 1952; Altmann, 1959; Chivers, 1969; Richard, 1970.)

Aotus (1 species, the douroucouli, or night monkey *A. trivirgatus*). Tropical forests of Central and South America.

Ateles (4 species; see especially the black-handed spider monkey *A. geoffroyi*). Spider monkeys. Tropical forests. Mexico to Amazon Basin. (Carpenter, 1935; Eisenberg and Kuehn, 1966; Richard, 1970.)

Brachyteles (1 species). Woolly spider monkeys. Forests of south-eastern Brazil.

Cacajao (3 species). Uakaris. Forests, upper Amazon.

Callicebus (3 species; see especially the dusky titi *C. moloch*). Titis.

Forests, Amazon-Orinoco basins to southeastern Brazil. (Moynihan, 1966, 1969; Mason, 1968, 1971.)

Cebus (4 species). Capuchins. Tropical forests of Central and South America. (Bernstein, 1965; Mason, 1971; Oppenheimer, 1968, 1973.)

Chiropotes (2 species). Bearded sakis. Forests of the Amazon-Orinoco basins.

Lagothrix (2 species). Woolly monkeys. Forests of the Orinoco-Amazon basins.

Pithecia (2 species). Sakis. Forests of the Amazon-Orinoco basins.

Saimiri (2 species; see especially the common squirrel monkey *S. sciureus*). Squirrel monkeys. Tropical forests of Central and South America. (Ploog, 1967; Baldwin, 1969, 1971; Rosenblum and Cooper, eds., 1968; Mason, 1971.)

SUPERFAMILY CERCOPITHECOIDEA (= "CATARRHINI")
(OLD WORLD MONKEYS AND APES)

Family Cercopithecidae. Old World monkeys.

Cercocebus (5 species; see especially the gray-cheeked mangabey *C. albigena* and white-collared mangabey *C. torquatus*). Mangabeys. Tropical forests of Africa. (Struhsaker, 1969.)

Cercopithecus subgenus *Allenopithecus* (1 species). Allen's swamp monkey. Forests, Congo.

Cercopithecus subgenus *Cercopithecus* (21 species; see especially the vervet, or grivet, *C. aethiops*, Sykes' monkey *C. albogularis*, the blue monkey *C. mitis*, and spot-nosed guenon *C. nictitans*). Guenons. Widespread in forest and savanna woodland of sub-Saharan Africa. (Haddow, 1952; Booth, 1962; Struhsaker, 1967a–d, 1969, 1970a; Gartlan and Brain, 1968; Bourlière et al., 1970; Aldrich-Blake, 1970; Hunkeler et al., 1972; McGuire, 1974.)

Cercopithecus subgenus *Miopithecus* (1 species, *C. talapoin*). Talapoin, or mangrove monkey. Forests of West Central Africa. (Gautier–Hion, 1970, 1973.)

Colobus subgenus *Colobus* (2 species). Black and white colobus. Forests, Abyssinia and Senegal to Tanzania. (Haddow, 1952; Ullrich, 1961; Marler, 1969; Sabater Pi, 1973.)

Colobus subgenus *Piliocolobus* (2 species). Red colobus. Forests of West, Central, and East Africa.

Colobus subgenus *Procolobus* (1 species). Olive colobus. Forest, West Africa. (Booth, 1957.)

Cynopithecus (1 species, *C. niger*). Celebes black ape. Forests, Celebes and Batjan Island of Moluccas.

Erythrocebus (1 species, *E. patas*). Patas monkey. Ground-dwelling in savanna and grassland of sub-Saharan Africa. (Hall, 1965, 1967, 1968a; Hall and Mayer, 1967; Struhsaker and Gartlan, 1970.)

Macaca (12 species; see especially the Japanese monkey *M. fuscata*, the crab-eating monkey *M. fascicularis* [= *M. irus*], the rhesus monkey, or macaque, *M. mulatta*, the pig-tailed macaque *M.*

nemestrina, the bonnet monkey *M. radiata*, the stump-tailed macaque *M. speciosa*, and Barbary ape *M. sylvanus*). Forests and open habitats in North Africa and Asia from Afghanistan, Tibet, and Japan to the Philippines and Celebes. (Carpenter, 1942a; Sugiyama, 1960; Altmann, 1962a, 1965; Izawa and Nishida, 1963; Southwick, ed., 1963; Mizuhara, 1964; Furuya, 1965, 1969; Koford, 1965; Simonds, 1965; Southwick et al., 1965; Bernstein, 1969a,b; Bernstein and Sharpe, 1966; Nishida, 1966; Yamada, 1966; Kaufmann, 1967; Blurton Jones and Trollope, 1968; Bertrand, 1969; Crook, 1970b; Sackett, 1970; Deag and Crook, 1971; Lindburg, 1971; Sugiyama, 1971; Rowell, 1972; Carpenter, ed., 1973; Deag, 1973; Itoigawa, 1973.)

Mandrillus (2 species). Mandrill *M. sphinx* and drill *M. leucophaeus*. Forests, Fernando Póo and West Africa. (Gartlan, 1970; Sabater Pi, 1972.)

Nasalis (1 species). Proboscis monkey *N. larvatus*. Forests and mangrove swamps, Borneo. (Kern, 1964.)

Papio (5 species, possibly well-marked geographic races of a single species: the olive baboon *P. anubis* found just south of the Sahara across Africa, the hamadryas, or sacred, baboon *P. hamadryas* of eastern Ethiopia and Somalia, the Guinea baboon *P. papio* of the western tip of the African bulge (Guinea, Senegal, Sierra Leone), the yellow baboon *P. cynocephalus* found across Africa from Somalia to Angola just south of the range of *P. anubis*, and the chacma baboon *P. ursinus* of South Africa). The ranges of the 5 forms are contiguous. Ground-dwelling, savannas, grasslands, and open forests. (Bolwig, 1958; Washburn and DeVore, 1961; Hall and DeVore, 1965; Bowden, 1966; Rowell, 1966, 1972; Hall, 1968b; Kummer, 1968, 1971; Crook, 1970b; Ransom, 1971; Dunbar and Nathan, 1972.)

Table 26-1 (*continued*).

Presbytis (14 species; see especially the hanuman langur *P. entellus* and Nilgiri langur *P. johnii*). Langurs. Forests and mangrove swamps, India, Bhutan, and southwestern China to Borneo. (Jay, 1965; Ripley, 1967, 1970; Sugiyama, 1967; Bernstein, 1968; Yoshiba, 1968; Poirier, 1970.)

Pygathrix (1 species). Douc langurs. Forests, Laos, Vietnam, and Hainan.

Rhinopithecus (2 species). Snub-nosed langurs. Vietnam and western China.

Simias (1 species). Pagai Island langurs. Forests of Mentawai Islands, off coast of Sumatra.

Theropithecus (1 species, the gelada *T. gelada*). Ground-dwelling, grassy mountain slopes of Ethiopia. (Crook, 1966, 1970b; Crook and Aldrich-Blake, 1968.)

Family Hylobatidae. The lesser apes: gibbons and siamangs.

Hylobates (6 species; see especially the white-handed gibbon *H. lar*). Gibbons. Forests, Thailand, southern China, and Tenasserim to Borneo. (Carpenter, 1940; Bernstein and Schusterman, 1964; Ellefson, 1968; Chivers, 1973.)

Symphalangus (1 species, the siamang *S. syndactylus*). Forests of Sumatra and the Malay Peninsula. (Chivers, 1973.)

Family Pongidae. The great apes: chimpanzees, gorillas, and orang-utans.

Gorilla (1 species, *G. gorilla*; three subspecies are usually recognized: the western lowland gorilla *G. g. gorilla*, the eastern lowland gorilla *G. g. graueri*, and the eastern highland gorilla *G. g. beringei*). Ground-dwelling in forests from Nigeria and Cameroun to the mountains of East Africa. (Schaller, 1963, 1965a,b; Fossey, 1972.)

Pan (2 species, the chimpanzee *P. troglodytes* and pygmy chimpanzee *P. paniscus*). Ground-dwelling in tropical forests. *P. troglodytes* ranges from Guinea and Sierra Leone across Africa to as far east as Lake Victoria and Lake Tanganyika; *P. paniscus* is limited to the region between the Congo and Lualaba rivers. (Yerkes and Yerkes, 1929; Nissen, 1931; Kortlandt, 1962; Goodall, 1965; Reynolds and Reynolds, 1965; Izawa and Itani, 1966; Itani and Suzuki, 1967; van Lawick-Goodall, 1967, 1968a,b, 1971; Nishida, 1968, 1970; Nishida and Kawanaka, 1972; Sugiyama, 1968, 1969, 1973; Suzuki, 1969; Izawa, 1970; Okano et al., 1973.)

Pongo (1 species, *P. pygmaeus*). Orang-utans. Forests, Sumatra and Borneo. (Schaller, 1961; Davenport, 1967; MacKinnon, 1974; Rodman, 1973.)

SUPERFAMILY HOMINOIDEA
(MAN)

Family Hominidae. Man
Homo (1 species). Man.

---

numbers in each sex-age category calculated from a sample of societies. Thus, the median numbers for two of the first species he studied were as follows:

Howler (*Alouatta villosa*), 51 troops:
3 adult males + 8 adult females + 4 juveniles + 3 infants + unknown number of males living alone

White-handed gibbon (*Hylobates lar*), 21 troops (families):
1 adult male + 1 adult female + 3 juveniles + 1 infant + unknown number of males and females living temporarily alone

Carpenter called the average ratio of adult males to adult females the "socionomic sex ratio." He postulated that this and other social characteristics represent adaptations to the environment, although he was unclear about the specific processes involved.

Carpenter was further aware that social life offers some degree of protection against predators. Once on Barro Colorado Island he saw a juvenile howling monkey attacked by an ocelot. The youngster emitted a distress call, and three adult males immediately rushed to its aid while roaring loudly. M. R. A. Chance (1955, 1961) independently generalized this notion, hypothesizing that aggregation in monkeys and apes is in general a device to reduce predation. He noted that more than one strategy is available to the members of the society. They can stand and fight in the fashion of baboon males, or flee together where cover is adequate, in the manner of gibbon families. In 1963 Irven DeVore added an important new insight. Impressed by what he had seen of the anubis baboon in Kenya, he suggested that a shift to a terrestrial existence carries with it a tendency to evolve larger, more organized societies. Since food is scarcer, the group must occupy a larger home range. And being more exposed to predators during their long forays through open country, group members are likely to be more numerous and well organized. The adults, especially the adult males, are forced to fight when caught away from the protection of trees. As a result there is a tendency for them to evolve more aggressive behavior. In the case of baboons the males are notably larger and possess stout canine teeth, which are employed as fangs during combat. And perhaps inevitably, the aggressive modus vivendi extends into the internal structure of the society itself, intensifying the dominance systems by which the adults of both sexes are partially organized. DeVore's view received support from the observations of the Altmanns (1970), who found 11 circumstances under which anubis baboons at Amboseli draw close together, the majority of which are clearly related to defense. Baboons cluster (1) when predators are encountered; (2) when nearby baboon groups emit predator alarm calls; (3) during false alarms; (4) when Masai cattle or other baboon groups approach closely; (5) when the troop forages in heavy undergrowth; (6) when the troop is about to go through an opening in the foliage; (7) when traveling along an unfamiliar route; (8) when using the shade of a tree or a water hole; (9) at or just before moving from one locality to another; (10) just before climbing the sleeping trees; (11) during morning and evening "social hours."

With the idea in mind of social behavior as a direct ecological adaptation, the next logical step for primatologists was to undertake a careful comparison of species occupying different habitats. Phyllis Jay (1965) showed that some arboreal and leaf-eating colobine monkeys, particularly the langurs (*Presbytis*) of Asia and *Colobus* of Africa, differ consistently from the ground-dwelling macaques in ways that seem to fit them to their environment. They occupy small but well-defined territories, which may be defended stoutly against other groups of the same species. This trait is consistent with the more even, dependable distribution of food and parallels a similar correlation within the birds. But colobine males are less powerful in comparison with the females and less aggressive than macaques and baboons, characteristics that evidently reflect the tendency of the monkeys to flee into the trees rather than face predators head-on.

Two other authors, K. R. L. Hall (1965) and John F. Eisenberg (1966), considered a wider range of primate species but felt that the correlations were either too weak or the data too few to conclude more than the elementary generalization made by Jay. Hall was nevertheless optimistic about the ultimate outcome, prophesying that when sustained investigations are carried out, "it is not improbable that the perspective may revolutionize some of the conventional concepts of this branch of comparative study, while, in the process, demonstrating beyond doubt the unreality of making any social behaviour comparisons of these animals without a detailed knowledge of the ecological circumstances of their natural life." But at this point J. H. Crook and J. S. Gartlan (1966) became impatient and decided to try to force the issue. What they undertook was to classify all primates, including prosimians, into five evolutionary grades of social behavior. Then in effect they searched for partial correlates in the habitat and diet of those species for which even fragmentary data are available. This approach was later extended and refined somewhat by Crook

(1970b, 1971) and Denham (1971), but the original version of the scheme, presented in Table 26-2, still deserves attention for its directness and clarity. The classification does not include the essentially solitary tree shrews of the family Tupaiidae, which I have discussed previously (Chapter 16). The value of the Crook-Gartlan approach lies in its objectivity. When such a matrix is devised, its assumptions are disclosed and the necessary degree of arbitrariness in the division of categories is easily surmised. Data can be added and new kinds of analyses applied without reworking the original sources.

Let us examine the conclusions drawn by Crook and Gartlan and then the weaknesses of their particular analysis. Grade I is formed almost exclusively of prosimians, most of whose behavior can safely be regarded as primitive. The member species are nocturnal, forest-dwelling insectivores that live as solitary individuals or mated pairs in territories. It is noteworthy that the only phylogenetically higher species listed in this grade is the night monkey *Aotus trivirgatus*, a ceboid which shows signs of having become nocturnal only secondarily. Grade II represents a short step to small family groups with a single male, and it is correlated with the large ecological shift to diurnality and a largely vegetarian diet. Grades III and IV are distinguished by the tolerant attitude of multiple males toward one another and the trait, closely associated with it, of larger group size. The overall ecological correlation is poor at best. Terrestrial, open-country primate species tend to fall into Grades III or IV, but so do many arboreal, forest-dwelling species. Grade V is a curious variant of Grade II in which the basic social units are dominated by one male or (in the case of hamadryas baboons) two cooperating males. The most distinctive feature of Grade V is the markedly larger size and behavioral differentiation of the male. In the hamadryas and gelada the units typically aggregate into large foraging and sleeping groups. All three species placed in Grade V are inhabitants of the driest, most barren habitats of Africa.

Two difficulties plague the Crook-Gartlan analysis. First, the correlations are very weak and uncertain, a fact verifiable by simple inspection. This problem has been worsened by the addition of new data, especially in the case of the New World monkeys. The ceboid species span Grades I through III, and they vary enormously among themselves in group size, sex-age distribution, and dominance relations. Yet all are arboreal forest dwellers exhibiting less than major variation in diet. Moynihan (personal communication), who recently reviewed the group, could find almost no ecological correlates at all. It might be significant that the night monkey has remained in or reverted to grade I, a simpler state often associated with nocturnal habits, while a tendency of spider monkeys (*Ateles*) to form fission-fusion groups can be explained as an adaptation to exploit patchy food sources. Perhaps—believers would say surely—other correlations exist among the ceboids, but they are not at the level expressed in

Table 26-2 The first attempt by Crook and Gartlan (1966) to arrange all primate societies into evolutionary grades and to correlate the grades with the ecology of particular species.

	Grade I	Grade II	Grade III	Grade IV	Grade V
Taxonomy					
Species	*Aotus trivirgatus* *Microcebus* sp. *Cheirogaleus* sp. *Phaner* sp. *Daubentonia* sp. *Lepilemur* sp. *Galago* sp.	*Hapalemur griseus* *Indri* sp. *Propithecus* sp. *Avahi* sp. *Callicebus moloch* *Hylobates* sp.	*Lemur* sp. *Alouatta villosa* *Saimiri sciureus* *Colobus* sp. *Cercopithecus ascanius* *Gorilla gorilla*	*Macaca mulatta*, etc. *Presbytis entellus* *Cercopithecus aethiops* *Papio cynocephalus* *Pan troglodytes*	*Erythrocebus patas* *Papio hamadryas* *Theropithecus gelada*
Ecology					
Habitat	Forest	Forest	Forest–forest fringe	Forest fringe, tree savanna	Grassland or arid savanna
Diet	Mostly insects	Fruit or leaves	Fruit or fruit and leaves. Stems, etc.	Vegetarian-omnivore; occasionally carnivorous in *Papio* and *Pan*	Vegetarian-omnivore; *P. hamadryas* occasionally also carnivorous
Behavior and Sociobiology					
Diurnal activity	Nocturnal	Crepuscular or diurnal	Diurnal	Diurnal	Diurnal
Size of groups	Usually solitary	Very small groups	Small to occasionally large parties	Medium to large groups; *Pan* groups inconstant in size	Medium to large groups; variable size in *T. gelada* and probably *P. hamadryas*
Reproductive units	Pairs where known	Small family parties based on single male	Multimale groups	Multimale groups	One-male groups
Male motility between groups	—	Probably slight	Yes—where known	Yes in *M. fuscata* and *C. aethiops*, otherwise not observed	Not observed
Sex dimorphism and social role differentiation	Slight	Slight	Slight—size and behavioral dimorphism marked in *Gorilla*; color contrasts in *Lemur*	Marked dimorphism and role differentiation in *Papio* and *Macaca*	Marked dimorphism; social role differentiation
Population dispersion	Limited information suggests territories	Territories with display, marking, etc.	Territories known in *Alouatta, Lemur*; home ranges in *Gorilla* with some group avoidance probable	Territories with display in *C. aethiops*; home ranges with avoidance or group combat in others; extensive group mixing in *Pan*	Home ranges in *E. patas, P. hamadryas* and *T. gelada* show much congregation in feeding and sleeping; *T. gelada* in poor feeding conditions shows group dispersal

the Crook-Gartlan analysis. It has become fashionable among some primatologists to say that ecology and not phylogeny determines the social systems of particular species. But considerable phylogenetic inertia exists and much more is likely to come to light when comparative studies become more detailed. Eisenberg et al. (1972) have pointed out that Madagascan lemuroids such as *Lemur* and *Propithecus* are characterized by multimale groups with more males than females, dominance of females over males, and the frequent segregation of troops into all-male and all-female subgroups. These traits are not shared with any other known primate species, despite the fact that the lemuroids are ecologically similar to a great many of them. Struhsaker (1969) found a similar phylogenetic conservatism in some aspects of the social behavior of cercopithecoid monkeys in Africa. The savanna-dwelling *Erythrocebus patas*, for example, is anatomically closely related to the forest-dwelling guenons of the genus *Cercopithecus*. It is also quite close to them in social structure, so that placing it next to the hamadryas and gelada in Grade V is probably incorrect. On the other hand, the vervet *C. aethiops* is socially very different from other *Cercopithecus*, despite the fact that it is ecologically similar.

Of equal importance, the Crook-Gartlan format lacks a true dependent variable. It is constructed in the spirit of multiple regression analysis yet does not follow the correct procedure. What is needed is to define the grades of social evolution according to one intuitively satisfying dependent variable, then seek to document as fully as possible other variables that can be partially correlated with it. The dependent variable can be a single trait or an index based on several traits. In the Crook-Gartlan study no such variable is defined, and the analysis shifts implicitly from one trait to another as one proceeds upward through the grades. Crook and Gartlan seem to give secondary roles to certain social traits, such as the degrees of sexual dimorphism and group dispersion, that other authors might regard as paramount.

In a subsequent synthesis of the subject, Eisenberg and his coworkers (1972) went far to correct the methodological flaw. As shown in Table 26-3, the key trait selected by these authors is the degree of male involvement in social life. This variable is not only satisfying by itself but also reasonably well correlated with other social traits such as group size, the nature of the dominance system, and territoriality. Working with more data than had been available to previous authors, Eisenberg et al. recognized an intermediate social category, the age-graded-male troop. Some species that appear to be organized into multimale societies do not in practice adhere strictly to that pattern. Younger, weaker males may be tolerated but only in a subordinate status. After a time they take over the dominant position or else leave the troop altogether. Societies in this evolutionary grade do not contain ranking males of approximately the same age. Con-

sequently there are no alliances and cliques of the kind that form the central hierarchies in baboon and macaque troops.

Although the matrix of Table 26-3 provides a more efficient and heuristic system than that in the original Crook-Gartlan scheme, the correlations remain disturbingly weak. Insectivores remain in the bottom grade. Terrestrial and semiterrestrial species are still characterized by the most advanced social organization, and the same is true of omnivores. Little more can be extracted. Within single evolutionary grades it is possible to define subgrades based on additional social characteristics and to correlate them with certain aspects of the preferred niche. Thus folivores (leaf eaters) have smaller home ranges than frugivores (fruit eaters), and they are more likely to employ individual calling or troop chorusing to maintain spacing between adjacent groups.

The ecological analysis of social evolution in primates has not progressed as rapidly as its earliest proponents might have hoped. Yet the multiple regression approach initiated by Crook and Gartlan is on the right track and can be expected to yield new insights as the variables are increased and enriched by new data. At the same time it should be borne in mind that multiple regression analysis can never prove causal relations; it can only provide clues about their existence. A second, parallel effort that can result in a new leap forward is the construction of evolutionary hypotheses on the basis of models of population biology. This method, the necessary principles of which were given in Chapter 4, is already well advanced in the social insects. Deductive reasoning of the correct kind, based on population biology, can be expected to complement the multiple regression method. In fact it is destined to press well beyond it by suggesting the existence of parameters and mathematical relationships not easily identifiable by wholly inductive methods.

A case in point is the very simple but promising start in model building made by Denham (1971), who stresses the crucial parameter of food distribution. His approach accords with current ecological theory and can be extended as follows. Earlier in this book (Chapter 3) it was argued that greater predictability of food in space and time promotes the evolution of territoriality. When the resources are dense and easily defensible, and when food is the limiting resource, the optimum strategy is double defense—by means of the monogamous pair bond. If the quality of the environment is not only predictable but also uniform from locality to locality, so that its variation is kept under the polygyny threshold, the tendency toward monogamy will be reinforced. This latter factor could explain one ecological difference between the "solitary" species of the first grade shown in Table 26-3, which include a high proportion of insectivores, and the pair-bonding species of the second grade, most or all of which are primarily vegetarian. The explanation is the hypothesis, both reasonable and testable, that the plant items vary in quantity and quality less

Table 26-3   The arrangement of primate societies into evolutionary grades and their ecological correlates by Eisenberg et al. (1972 and personal communication). The grades are based on the degree of male involvement, given as the column headings. (Copyright © 1972 by the American Association for the Advancement of Science.)

Solitary species	Parental family	Minimal adult ♂ tolerance[a] (unimale troop)[b]	Intermediate ♂ tolerance[c] (age-graded-male troop)[b]	Highest ♂ tolerance[d] (multimale troop)[b]
A. Insectivore-frugivore	A. Frugivore-insectivore Callithricidae (Hapalidae)	A. Arboreal folivore Colobinae	A. Arboreal folivore Colobinae	A. Arboreal frugivore Indriidae
Tupaiidae	*Saguinus oedipus*	*Colobus guereza*	*Presbytis cristatus*	*Propithecus verreauxi*
Tupaia glis	*Cebuella pygmaea*	*Presbytis senex*	*Presbytis entellus*	Lemuridae
Lemuridae	*Callithrix jacchus*	*Presbytis johnii*	Cebidae	*Lemur catta*
Microcebus murinus	Cebidae	*Presbytis entellus*	*Alouatta villosa*	B. Semiterrestrial
Cheirogaleus major	*Callicebus moloch*	B. Arboreal frugivore	B. Arboreal frugivore	frugivore-omnivore
Daubentoniidae	*Aotus trivirgatus*	Cebidae	Cebidae	Cercopithecidae
Daubentonia	B. Folivore-frugivore	*Cebus capucinus*	*Ateles geoffroyi*	*Cercopithecus aethiops*
madagascariensis	Indriidae	Cercopithecidae	*Saimiri sciureus*	*Macaca fuscata*
Lorisidae	*Indri indri*	*Cercopithecus mitis*	Cercopithecidae	*Macaca mulatta*
Loris tardigradus	Hylobatidae	*Cercopithecus campbelli*	*Cercopithecus talapoin*	*Macaca radiata*
Perodicticus potto	*Hylobates lar*	*Cercocebus albigena*	C. Semiterrestrial	*Papio cynocephalus*
B. Folivore	*Symphalangus*	C. Semiterrestrial frugivore	frugivore-omnivore	*Papio ursinus*
Lemuridae	*syndactylus*	Cercopithecidae	Cercopithecidae	*Papio anubis*
Lepilemur mustelinus		*Erythrocebus patas*	*Cercopithecus aethiops*	*Macaca sinica*
		Theropithecus gelada	*Cercocebus torquatus*	Pongidae
		Mandrillus leucophaeus	*Macaca sinica*	*Pan troglodytes*
		Papio hamadryas	D. Terrestrial folivore-	
			frugivore	
			Pongidae	
			Gorilla gorilla	

[a]Troop with one adult male and strong intolerance to maturing males.
[b]"Troop" refers to the basic social grouping of adult females and their dependent or semidependent offspring.
[c]Troop typically showing age-graded-male series.
[d]Troop with several mature, adult males and age-graded series of males.

from territory to territory than do the insects. The same hypothesis is consistent with the fact that leaf-eating species defend smaller territories and use more conspicuous vocal displays than do otherwise similar fruit-eating species. Higher social grades can be expected to evolve in accordance with the Horn principle, which states that when food becomes sufficiently patchy in space and unpredictable in time, the optimum strategy is to abandon feeding territories and to join groupings larger than the family (see Chapter 3). As Crook, Denham, and others have pointed out, this may be the ultimate causation of greater group size in open-country species of Old World monkeys and apes. These primates live in an exceptionally patchy and unpredictable environment. The same principle can be extended to the forest-dwelling species in higher social grades, if further data reveal their food items to be similarly distributed. Tropical forests, contrary to the popular conception, are seasonal and usually strongly so. Many potential food sources within the forests, including the buds, flowers,

and fruit of *particular* plant species, are not only seasonal but also patchy and unpredictable through time. Finally, predation plays an unquestioned auxiliary role in evolution, forcing species into one defensive strategy or another and thereby helping to shape group size and organization.

In the remainder of this chapter, we will view the full range of primate sociality by considering individual species that represent each of the evolutionary grades. Since the sequence will proceed upward through the grades rather than through phylogenetic groupings, the reader will note some curious taxonomic juxtapositions. For example, the anthropoid apes are distributed from one end of the array to the other. The mostly solitary orang-utan must be sociobiologically classed with the primitive prosimians, while gibbons are grouped with marmosets, titis, and the New World monkeys. The gorilla possesses age-graded-male societies with a reasonably complex organization, but it is still well behind the chimpanzee, which by all reasonable standards occupies the pinnacle of nonhuman primate social evolution. The apes are extreme in this display of diversity, but each of the remaining major phylogenetic groups spans several evolutionary grades as well.

The Lesser Mouse Lemur (*Microcebus murinus*)

The galagos, pottos, mouse lemurs, and other nocturnal prosimians are among the most primitively social primates. Because of the difficulty of studying these animals in the field, data on most of the species are still fragmentary and inconclusive. Thanks largely to the work of Petter (1962) and R. D. Martin (1973), the population structure and behavior of one species, the lesser mouse lemur, has become sufficiently clear to serve as a paradigm for the lowest evolutionary grade. *Microcebus murinus* is the smallest and most widespread of all the prosimians of Madagascar, inhabiting nearly all forested areas along the coast. It is wholly nocturnal, spending the day in nests constructed from dry leaves in bushes and tree holes. Although primarily arboreal, it descends readily to the ground to cross gaps in the foliage and to forage through the leaf litter. The lesser mouse lemur is the most nearly omnivorous of all the known primates. Its diet includes fruits, flowers, and leaves of a variety of trees, bushes, and vines. It also eats insects, spiders, probably tree frogs and chameleons, and possibly mollusks. Individuals collect sap from holes that they score in live bark with rotating movements of their front teeth. In this last habit the species is convergent to the smallest of the New World primates, the pygmy marmoset.

Possibly as a result of its catholic diet, the lesser mouse lemur has a small home range, evidently 50 meters or less in diameter. The ranges tend to be exclusive, at least within sexes, and it is reasonable to hypothesize some form of territorial defense. Individuals placed by Petter into the same small enclosure all at the same time were compatible. But when one was allowed to occupy the space first, it subsequently attacked all newcomers. Even within the compatible groups males fought with each other whenever females came into estrus.

Martin's data show that the species is dispersed into localized population nuclei. Each nucleus contains a core of high density characterized by a proportion of 4 adult females to each adult male. At Mandena, Madagascar, the cores were located at sites containing a high frequency of the two preferred species of food trees. Since the sex ratio at birth is 1/1, it follows that males either emigrate or suffer higher mortality early in life. In fact, the surplus males are found concentrated in nests along the periphery of the core area. Females often nest in groups, evidently mating and rearing young compatibly. The number of females per nest in 1968 ranged from 1 to 15, with a median of 4. When females were in estrus, they were often accom-

panied by a single male. When the females passed out of estrus, the males evidently became more tolerant toward one another, and two and even three sometimes collected in the same female nest. Martin suggests that groups of females are often mothers and daughters. Sons, however, tend to be displaced to the periphery of the favored habitats, where they await an opportunity to join the dominant, breeding males. Communal nesting might be a consequence of the limited number of nest sites, perhaps abetted by kin selection. In any case the mouse lemur must still be regarded as an essentially solitary animal. No evidence exists of organized social life within the nests. Of equal importance, the lemurs forage wholly on their own. A similar sociobiological pattern has been demonstrated in the dwarf galago (*Galago demidovii*) of West Africa by Charles-Dominique and Martin (1970).

The communication system of *Microcebus* has not been well studied. Preliminary observations by Petter have revealed the existence of a rich vocal repertory, which includes defense cries by adults and distress calls by infants and juveniles. Chemical communication is also prominent. When entering a new area, the adults smear urine on branches with their feet, while at the time of female estrus the males appear to use a special genital secretion to mark their territories.

The Orang-utan (*Pongo pygmaeus*)

Until recently the orang-utan was the least known of the great apes, the mysterious Old Man of the rain forests of Sumatra and Borneo, seldom more than glimpsed in the wild. Careful studies have now been completed by David Horr, P. S. Rodman (1973), and J. R. MacKinnon (1974). Rodman and his assistants logged the quite respectable observation time of 1639 hours, in the course of which they came to know 11 orangs individually. They were able to follow their subjects continuously for hours or days through the nearly trackless forests of Borneo's Kutai Reserve.

As the orangs' unusual body form testifies, they are exclusively arboreal, relying extensively on brachiation to move through various levels in the rain forests from the canopy to near the ground. Although primarily frugivores, they also consume some leaves, bits of bark, and bird eggs. Their natural population densities had been previously judged to be 0.4 individuals per square kilometer or less. At the Kutai Reserve, which contains some of the least disturbed lowland habitats left in Asia, Rodman found the density to be closer to 3 individuals per square kilometer. Nuclear groups consist of females and their offspring, and are sometimes accompanied by an adult male. Solitary males are common, but juveniles or adult females are encountered alone only on rare occasions. Orang group size seldom if ever exceeds 4 individuals. At the Kutai Reserve, the following composition of seven such groups was recorded: adult female + infant male; adult female + infant male; adult female + infant, sex unknown; adult female + juvenile male; juvenile female alone; adult male alone; adult male alone. Occasionally these units met to form secondary groupings, the largest of which contained 6 individuals. Two of the temporary combinations were seen on multiple occasions and appeared to be based on kin ties. The others were passive aggregations brought together by the common attraction of a fruiting tree. The contacts were facilitated by a broad overlapping of the home ranges.

The orang society can be viewed as incorporating the loose fission-fusion structure so strikingly elaborated in the chimpanzee. But this is very elementary in form, and in most other respects the orang-utan is much closer to solitary prosimians such as the mouse lemurs. Specifically, females tend to aggregate, and males visit them only in order to copulate. As juvenile females mature they disperse slowly from the mother's home range. Males disperse for great distances and wander a good deal before settling into home ranges of their own.

Social interactions among the orang-utans are few in kind and far simpler than in the other anthropoid apes. They are virtually limited to relations between mothers and their offspring and the brief, simple encounters between adult males and females. Aggression within the society is quite rare, and nothing resembling a dominance system has been established in studies to date. During their lengthy period of observation at the Kutai Reserve, Rodman and his coworkers recorded only one clear instance of open hostility—when one adult female drove another from a fruit tree.

But the adult males, wandering mostly in solitude, probably do repel one another in the vicinity of the females. Although direct confrontations have not yet been observed, a few pieces of indirect evidence suggest that such intrasexual conflict does exist. Sexual dimorphism is strongly developed, with the males averaging twice the size of the females and possessing large extensible vocal pouches. The males use their pouches to deliver the "long call," a loud, throaty scream that can be heard by human beings for as much as a kilometer away. They sound the call most frequently when they have separated from their temporary female consorts for a short period of time. The function seems to be to reestablish contact. But the males also call on occasion when they are with the females. Since the display is evidently designed for long-distance communication, its second function may be to threaten away rivals. Finally, it is surely significant that no more than one adult male is ever seen in the company of a receptive female.

The Dusky Titi (*Callicebus moloch*)

Titis are small ceboid monkeys that are relatively common throughout the rain forests of the Amazon-Orinoco Basin. The dusky titi, the best studied of the three living species, possesses one of the simplest familial forms of society. It shares this evolutionary grade with many other ceboids, including the "typical" marmosets of the genus *Callithrix*, Goeldi's marmoset (*Callimico goeldii*), the pygmy marmoset (*Cebuella pygmaea*), the tamarin *Saguinus oedipus*, and among the Cebidae proper, the night monkey (*Aotus trivirgatus*) and the monk saki (*Pithecia monachus*). Among the Old World forms the same basic organization occurs in the gibbons and the siamang.

The dusky titi is small in comparison with the remaining primates, ranging from about 280 to 400 millimeters in length exclusive of the tail and weighing between 500 and 600 grams. It prefers low forest canopy, thickets, and undergrowth, through which it runs and leaps with quick, nervous movements. Occasionally it travels short distances over the ground. At Hacienda Barbascal, Colombia, Mason (1971) found groups to consist of a mated pair and one or two immature offspring. The populations are dense, each family occupying a roughly circular territory only about 50 meters in diameter. The territories are strictly defended. On frequent occasions, and especially during the early morning, the families confront one another at regular sites along the territorial boundary, exchange displays, and depart without making significant physical contact.

Cohesion within the titi family is close. The members forage together in tight groups and come into frequent physical contact. Tail twining, depicted in Figure 26-2, is an intimate signal commonly exchanged between individuals while they rest together. Mason has described the formation of the sex bond between individuals introduced to each other in large outdoor enclosures. Both sexes are wary of the approach of strangers, but the females are more strongly and persistently cautious. When the bond is finally achieved, it is evidently lifelong, with the female showing the more pronounced signs of attachment. Breaking the bond may have adverse psychogenic effects. Titis, unlike more gregarious and loosely organized ceboids such as the capuchins and squirrel monkeys, become silent and withdrawn in captivity. Only a minority survive more than a few weeks.

The communication system of the titi is surprisingly rich (Moynihan, 1966). In addition to odors and tactile signals, a wide range of visual displays is used. The acoustic repertory is one of the most diverse known in the animal kingdom. In order merely to verbalize the basic titi notes Moynihan came close to exhausting the English vocabulary: whistles, chuck notes, chirrups, moans, resonating notes, trills, pumping notes, squeaks, gobbling, and purring. These and intermediate sounds are given alone or in combinations of one to three to produce an almost endless concatenation of songs. The signals are further characterized by gradations in quality and intensity and by apparent shifts in meaning due at least in part to changes in context. More than one phrase or song may be used to convey what is evidently the same meaning, and elements are often combined with touch and visual displays. In Figure 26-2 is shown an example of a composite display used in territorial defense and other aggressive exchanges.

Figure 26-2 Communication in the dusky titi, a marmosetlike primate with elementary societies based on close pair bonding. At the left a mated pair engage in tail twining, a common form of tactile exchange. The adult in the right figure assumes the extreme "arch posture," an aggressive display. The hands are lifted off the perch and hang down, the lips are protruded as the mouth opens, the fur is erected, and the tail is curled and sometimes lashed back and forth. The arch posture is often accompanied by a variety of vocalizations. (From Moynihan, 1966.)

What is the explanation for the titi's expanded repertory? Moynihan hypothesizes that it has resulted from the unusually narrow and specific "acoustic niche" occupied by the dusky titi. The species is surrounded by birds and other monkeys, such as capuchins and howlers, that utter a great variety of trills, chuck notes, whistles, and other sounds more or less resembling the titi phrases. By elaborating their songs, and then employing them redundantly and in conjunction with visual and other forms of displays, the titis can greatly restrict their communication channel and insure privacy of communication even in the noisiest forest. Moynihan further believes that the dusky titi is subject to relatively little predation. If this is true, counterselection against loud, frequent communication has been reduced, and the auditory system has been freed to seek its highest potential level. In Moynihan's words, the system may well represent "the maximum elaboration and complexity which can be attained by a species-specific, and presumably largely 'innate,' language in particularly favourable circumstances." This hypothesis poses a new and interesting challenge, and the case of the dusky titi illustrates how much we have to learn about the meaning of communication in New World primates generally.

The White-Handed Gibbon (*Hylobates lar*)

The six species of gibbons and their close relative the siamang (*Symphalangus syndactylus*) are the smallest of the great apes. As exemplified by the white-handed gibbon, the commonest and best studied member of the group, they show a remarkable convergence in social behavior to the dusky titi and other monogamous New World primates. The white-handed gibbon, *Hylobates lar*, ranges from Indochina west to the Mekong River and south to Malaya and Sumatra. It is intensely arboreal in habit, depending on brachiation through the branches of trees for approximately 90 percent of its locomotion. It prefers the closed canopy of dense forest, where it can travel quickly from tree to tree. The gibbon occasionally descends to clumps of low bushes during feeding and all the way to the ground to drink from streams, although the great bulk of its liquid is obtained from eating fruit and licking bark and leaves after rain. In keeping with their monogamy the sexes are similar in appearance and size, both ranging between 4 and 8 kilograms in total body weight. Gibbon troops defend territories 100–120 hectares in extent (Ellefson, 1968).

Most of what we know about gibbon social behavior still originates from the classic field study of C. Ray Carpenter (1940) conducted near Chiengmai, Thailand. Carpenter mounted a full-scale expedition in order to remain with the gibbons over a period of months. He employed recording equipment to make the first precise study of primate vocal communication under natural conditions. It is notable for historical purposes that one of Carpenter's assistants was Sherwood L. Washburn, then a graduate student. Washburn later joined the faculty at the University of California at Berkeley, where, independently of S. A. Altmann at Harvard and the scientists of the Japanese Monkey Center, he and his associates played a key role in reviving field studies of primate social behavior during the 1950's.

Carpenter found that the *Hylobates lar* society is identical to the family. There are two to six members, the mated pair plus up to four offspring. Occasionally an aging male is also retained in the group. Solitary individuals are sometimes encountered in the forests. They are evidently either aged individuals or young adults still in search of mates and territories. The family stays close together, and dominance is weak or altogether absent. The female plays an equal role in territorial defense and in precoital sexual behavior (see also Bernstein and Schusterman, 1964). The mother takes care of the infant, allowing it to cling to her belly when very young, nursing and playing with it, and leading it about when the youngster begins to travel on its own. The male's relation to the infant is also close. He frequently inspects, manipulates, and grooms it. Play sessions are frequent, during which the youngster is permitted to be the mock aggressor. When a young gibbon calls in alarm, the male quickly swings to its aid. He sometimes breaks up play between infants and juveniles that has become too rough. In a captive group of the dark-handed gibbon *H. agilis* assembled by Carpenter, a lone male allowed a small juvenile to adopt him. Thereafter he carried the smaller animal in the maternal position during much of the day. This observation suggests that not only is paternal care close under normal circumstances, but the male is also prepared to assume the role of the mother when she falls ill or dies.

The origin of new gibbon groups has never been observed in nature, but its course can be safely inferred from circumstantial evidence. As Berkson et al. (1971) have noted, young gibbons become aggressive at puberty, and adults placed close together are very hostile. Young adults tend to be excluded, especially at feeding sessions. It is probable that as relations between the parents and young adults become more abrasive, the offspring scatter to form families of their own. Carpenter observed one such pair that might have been in the process of forming an incestuous union, although their sexes and origin could not be ascertained. They remained close together at all times and often stayed well apart from the rest of the family. Berkson and his co-workers observed the formation of one pair from among a group of adults assembled as strangers in an outdoor enclosure.

The gibbon family follows a precise daily cycle of activity which can be summarized as follows from Carpenter's Chiengmai data:

1. Dawn at 5:30 to 6:30: awaken.
2. 6:00–7:30 or 8:00: exchange calls with neighboring gibbon families and engage in general activity.
3. 7:30–8:30 or 9:00: progress through the territory.
4. 8:30–11:00: feed.
5. 11:00–11:30: progress to a place for the midday rest.
6. 11:30–2:30 or 3:00: "siesta" with some play and other general activity, especially by the young.
7. 2:30–4:30 or 5:00: feed and progress through the territory.
8. 5:00–5:30 or 6:00: direct movement to the place where the night will be spent.
9. 6:00–sundown: settle for the night.
10. Sundown–dawn: sleep or at least rest quietly. Gibbons do not build nests, but they do select trees with dense tops and a central location within the territory as sleeping "lodges."

Gibbon communication is frequent and complex. Grooming, which is performed with the hands, feet, and teeth, is a prominent part of social life. Individuals invite others to groom them by taking a supine posture while holding their arms level with or above the shoulders and head. They also emit characteristic invitatory grunts that change to squeaking, accompanied by withdrawal of the mouth corners, during the actual grooming bouts (Andrew, 1963). Using his sound recording apparatus, Carpenter distinguished nine categories of vocalizations in the free-ranging gibbons of the Chiengmai area. The most striking are the celebrated territorial calls, which carry over distances

of kilometers. Adults of either sex, but especially the females, emit a series of hoots with rising inflection, rising pitch, and increasing tempo. The call reaches a climax, then abruptly tapers off to two or three notes of lower pitch. The entire song takes 12 to 22 seconds. The males also use an abbreviated version, incorporating the first notes of the full song, which are repeated over and over. The same vocalizations are employed when the family is surprised by a hunter or some other potential enemy. The gibbons emit a special assembly or searching call when a member of the group is separated, and a chatter or series of clucks to lead others during group progression. Still other vocalizations and correlated postures and facial expressions are employed during greeting, play, and various levels of threat directed toward other members of the group.

The Mantled Howler (*Alouatta villosa*)

The howlers of the genus *Alouatta* are among the largest and most conspicuous of the New World monkeys. The mantled howler *A. villosa*, frequently referred to in the literature by the synonymous name *A. palliata*, is the most widespread of the five species, occurring from the coastal forests of Mexico through Central America to the Pacific coastal forests of South America as far south as the equator. The species is of special sociobiological interest because a high level of individual tolerance permits the formation of large multimale societies that may or may not be age-graded in nature. In this regard the species is convergent to the lemurs, as well as to the macaques and some other cercopithecids. The mantled howler and its congeners are also famous for the loud daily chorusing by which the males space their troops apart. The inaugural study of the species, its basic conclusions still unchanged, was conducted by Carpenter (1934, 1965) in Panama. Important data on ecology and behavior have been added by Collias and Southwick (1952), Altmann (1959), Bernstein (1964b), Chivers (1969), and Alison Richard (1970).

Mantled howlers are among the most impressive animals of the tropical American forests. Exceeding 5 kilograms in weight, the adults are robust, their heads set down and forward on the shoulders to give them a hunched appearance. Long black fur covers all of the body except for the deeply pigmented face and soles of the hands and feet. The voice box is swollen as part of the specialization for long-distance calling. Sexual dimorphism is marked. The males are 30 percent heavier than the females and their laryngeal swellings are larger and covered by beards.

Mantled howler groups are populous by primate standards. In 1932–33, when the Barro Colorado Island population studied by Carpenter was at or near saturation, the groups contained from 4 to 35 members with a median number of 18 individuals distributed according to sex and age as follows: 3 adult males, 8 adult females,

4 juveniles, and 3 infants. By the early 1950's, following a disastrous yellow fever epizooitic, the average group size had been halved and the average number of adult males had dropped to about 1. Ten years later, the population and original sex-age distribution had been mostly restored. Thus we have the unusual circumstance of a species that appears to alternate between multimale and unimale organization according to the density of the population as a whole. Solitary males are occasionally encountered in the tree tops, evidently in the process of transferring from one group to another. They follow troops for days, accepting threats and rebuffs until they are finally accepted. The process of group multiplication is not known, but it probably consists of elementary fission.

The howler troops are territorial, but the method by which they exclude one another is unconventional. Each day, at frequent intervals but especially during the early morning, the males of neighboring troops roar back and forth at one another. These thunderous sounds are the loudest produced by any animal in the tropical American forests and carry for a kilometer or more. They are apparently sufficient by themselves to maintain the spacing of the troops. Chivers observed an increase in calling between troops as they moved toward each other during daily rounds of foraging, followed by a mutual withdrawal and decrease in the vocalization. Because the troops do not meet at the territorial boundaries to threaten and fight in the manner of dusky titis, their home ranges overlap to some extent. However, as Collias, Southwick, and Chivers have documented, the overlap decreases with an increase in population density. Following the epizooitic the overlap was extensive. As density rebounded, the overlap diminished until it was no greater than in species that defend territorial boundaries through overt aggression. The method is quite effective, if a bit hard on the ears of human observers.

Conflict within troops is uncommon. It is signaled by a baring of teeth and cackling vocalization and almost never entails fighting. Aggression is especially rare between females; an observer may spend hundreds of hours without seeing a single episode of overt hostility. Dominance orders are correspondingly weakly defined. For this reason, and the difficulty of determining the true age of adults, it has not yet been established whether the troops are age-graded-male, with one dominant individual controlling younger animals, or whether the troops contain multiple high-ranking males (J. F. Eisenberg, personal communication). The latter alternative seems more probable on the basis of prima facie evidence. Males cooperate closely in the defense of young, and they share estrous females with no sign of hostility.

Allogrooming is also rare. This statistic supports the general hypothesis that the behavior functions in large measure as a conciliatory device, so that among primate species the less aggressively organized the society the less its members need to groom one another (see Chapter 9). Communication within the troops is primarily vocal. The

territorial roars of the males, together with the equivalent terrierlike barks of the females, are also employed as warning signals when a human being or larger predator is sighted. The remaining repertory is comparable in richness to those of most other New World primates. Special sounds are employed to lead progressions through trees, to direct the attention of the troop to strange situations, and to invite and orient play. Infants cry for help when lost, and a mother wails in a characteristic manner when her infant falls or is otherwise separated from her.

The Ring-Tailed Lemur (*Lemur catta*)

The true lemurs, comprised of five species in the Madagascan genus *Lemur*, represent the pinnacle of social evolution within the Prosimii. As such they provide a separate natural experiment that can be compared with the attainment of higher evolutionary grades in the ceboid and cercopithecoid monkeys and apes. Alison Jolly, the chronicler of the ring-tailed lemur, describes its general appearance as follows. "Its fur is a brilliant light gray, its face a black-and-white mask, its tail ringed with about fourteen circles of black and white. Its black skin shows on nose, palms, soles, and genitalia. Your first impression of an *L. catta* troop is a series of tails dangling straight down among the branches like enormous fuzzy striped caterpillars. Later, with difficulty, you put together the patches of light and shade into a set of curved gray backs, of black-and-white spotted faces, of amber eyes. By this time, if the troop does not know you, they are already clicking to each other, and first one and then a chorus begin to mob you with high, outraged barks. The troop is quite willing to click and bark for an hour at a time in the yapping soprano of twenty ill-bred little terriers." These animals display only the weakest sexual dimorphism, the adult males being somewhat heavier in the head and shoulders than the females. Observers also find it difficult to tell individual lemurs apart, in contrast to the anthropoid apes and the larger cercopithecoid monkeys, which show marked individual variation.

Lemur catta inhabits the dry gallery and mixed deciduous forests of southern and western Madagascar. It is the most terrestrial member of the genus, spending up to 20 percent of its time on the ground, four times more than the otherwise ecologically similar sifaka (*Propithecus verreauxi*) and almost as much as the "terrestrial" baboons. But it never strays far from the trees, to which it sprints at the slightest alarm. The lemur is exclusively vegetarian, feeding on the leaves, fruit, and seeds of a variety of tree species and a few ground plants. It obeys a strict cycle of diurnal activity. The troop begins to stir before dawn. No later than 8:30 A.M., the exact time depending on the temperature and weather conditions, it enters a period of sunning, feeding, and travel. Commonly two lengthy progressions occur during the morning, the first leading to feeding grounds in lower vegetation strata and the second to the place of the noon siesta. After further wandering and feeding in the afternoon, the troop returns to the feeding trees. There is a tendency to cover the same routes for three or four days in succession, then shift to a different part of the home range.

Like the mantled howlers of Barro Colorado Island, Jolly's population of lemurs at the Berenty Reserve changed markedly in group composition and territorial occupancy over a period of several years. In 1963–64 there were two troops, consisting of 21 and 24 individuals, respectively. Adult males and females were equally numerous, and their total population was approximately matched by the combined numbers of juveniles and infants. Two or more subordinate males formed the "Drones' Club." They trailed the main group during progressions and tended to feed and take siestas by themselves. The troops avoided one another consistently and occupied mostly exclusive ranges. Fighting was rare. In 1970 the same population had subdivided into four troops with an average total membership of 11 adults and young. Now the home ranges overlapped widely, and feeding and drinking sites were shared on a time plan. Contacts and fighting were much more frequent, while the subordinate drone males often lagged so far behind as to be out of sight. Jolly (1972b) believes that these changes were due to one or more bad years that restricted the number of usable sites and forced the troops together. However, the observed subdivision of the troops cannot be readily explained in this way.

The lemur society is aggressively organized. Exchanges range from simple visual threats and cuffs to full-scale "jump-fights" during which the animals sometimes rake one another with long downward slashes of their canines. Adult females are dominant over adult males, a reversal of an otherwise nearly universal primate pattern. The female hierarchy is loose and at least partly nontransitive, while that of the males is strictly linear. Aggression among the males reaches its maximum during the April breeding season. Yet oddly, male dominance seems to have no influence on access to estrous females. Jolly saw a female copulate with three males in succession, while one subordinate male accomplished three out of six observed matings. Perhaps dominance determines which of the males remain with the troop over long stretches of time, and which succeed in staying close to the troop during the short breeding season. Leadership is divorced from dominance. During a group progression first one and then another adult takes the van. Occasionally the troop splits into fractions that move in different directions until finally some begin to mew loudly, a signal that brings the lemurs back together.

Lemur catta resembles the Old World monkeys and apes in some aspects of its communication system while differing strikingly in others. Play is well developed and largely concerned with mock aggression among juveniles. Grooming is also a prominent form of

interaction but is peculiar in being mutual between pairs and unrelated in any obvious way to rank. Chemical communication is much more strongly developed than in the monkeys and apes and is employed principally during aggressive encounters. Both females and males mark small, vertical branches with genital secretions. They stand on their hands, hold on to the branch with their feet as high as possible, and rub their genitalia up and down in short strokes. The males also employ palmar marking, smearing an odorous secretion on branches by rubbing the surfaces with their forearms and hands. Brachial glands, which occur high on the male's chest, and conspicuous antebrachial organs on the forearms also produce odorous substances (see Figure 26-3). The male places the forearm glands against the chest glands, appearing to mix their secretions. During aggressive encounters the tail is pulled repeatedly between the forearms and waved in the air in a way that wafts the scent toward the opponents. Full-scale encounters between males entail a flurry of chemical, visual, and vocal signaling. They are usually initiated by transfer of secretions to the tail and sometimes lead to spectacular "stink fights":

> At the same time the *Lemur* stares toward another animal. He draws his upper lip forward and down, so that it covers the points of his canines and protrudes somewhat below the lower jaw. This gives the front part of the lemur's muzzle a square houndlike appearance with two lowered flaring lips, but the lips are tense and do not droop like a hound's. This expression probably flares the nostrils. He may either squeal or purr while tail-marking. He then stands on all fours with tail arched over his back, its tip just above his head. He quivers the tail violently in the vertical plane, shaking its odor forward. Tail-waving is always directed toward another animal which may be right in front of him or more than 3 m away . . . A stink-fight is a long series of palmar-marking, tail-marking, and tail-waving directed by two males toward each other. The animals stand between 3 and 10 m apart. First one marks, then the other, with pauses between. Occasionally both tail-wave simultaneously, the two arched backs and tails reflecting each other like a heraldic design. The more aggressive male gradually moves forward, the other retreats, although they are often not close enough to supplant each other in one leap. (Jolly, 1966)

The remainder of the lemur's repertory consists chiefly of vocal displays with a rich admixture of visual signals. The functional categories are comparable to those in the ceboid and cercopithecoid primates, but many of the specific displays give every appearance of having been independently evolved.

The Hamadryas Baboon (*Papio hamadryas*)

The hamadryas baboon, also called the sacred baboon, is a large, diurnal, almost exclusively terrestrial cercopithecoid monkey. It ranges across the arid acacia savanna and grassland in the region surrounding the mouth of the Red Sea—eastern Abyssinia, southern Somalia, and southwestern Arabia. Because hamadryas baboons hy-

Figure 26-3 The encounter of two ring-tailed lemur troops at the Berenty Reserve in Madagascar. The habitat is a riverside gallery forest, dominated in the foreground by a large tamarind tree (*Tamarindus indica*). The arboreal troop on the left is stirring into activity after a noontime siesta. One male faces the observer with a threat stare, his antebrachial gland visible on the inside of the left forearm. A second male behind him has begun to move down the tree trunk in the direction of the other troop. Directly to his rear two adults engage in mutual grooming, while other members stay clumped together in rest or in the early moments of arousal. The troop on the ground has begun its after-

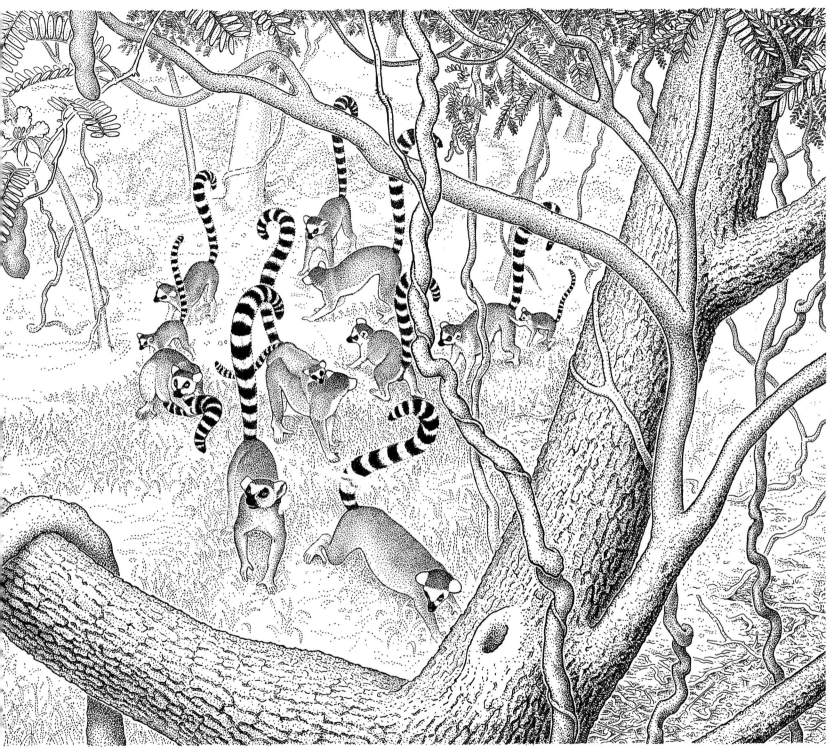

noon progression to a feeding site. Two adults to the left and front have spotted the group in the tree and are staring and barking in their direction. One of these, a male, draws its tail over the antebrachial glands in preparation for a hostile display. He is ready for a stink fight, during which the tail will be jerked back and forth to waft the scent toward the opponents. Well to the rear and in the center of this picture, two subordinate males of the "Drones' Club" trail the second troop. (Drawing by Sarah Landry; based on data from Alison Jolly, 1966 and personal communication.)

bridize extensively with anubis baboons, there is some question as to whether they really have the status of a full species (*Papio hamadryas*) or merely constitute a local subspecies (*Papio papio hamadryas*) of one unified baboon species. The former designation still seems the more prudent, especially in view of the strong morphological traits that distinguish this animal. The face is fleshy pink instead of black as in all other baboons. The males are twice the size of the females, their appearance made still more striking by a large mane of wavy gray hair. This dimorphism is related to the feature of hamadryas behavior that makes the species uniquely interesting: the extreme dominance of the adult male over females, who are kept forcibly in a permanent harem. This relationship influences virtually every other aspect of the social organization.

The sociobiology of the hamadryas baboon has been painstakingly documented by Hans Kummer over a 15-year period, first with captive animals and then in the field in Ethiopia (see especially Kummer, 1968, 1971). The species is totally social. Only one solitary individual, an adult male, was seen during months of observation over a substantial part of the species' range. The peculiarities of hamadryas organization are best understood by comparison with the more "conventional" system of other kinds of baboons. The basic unit of the *Papio anubis* society, as shown by DeVore, Hall, and others, is the *group*, an assembly of females, offspring, and multiple males. Aside from mothers and their dependents, no other distinct level of organization exists, at least not in the savanna population. The males are organized into dominance hierarchies. The group is ruled by a "central hierarchy" of dominant males who cooperate in defense and the control of subordinates. Access to females is determined to a large extent by rank, and possession is mostly limited to the time of estrus. In contrast, hamadryas males maintain permanent possession of females, and the societies are organized into three levels. The basic social element is the *one-male unit*, consisting of a mature male and the harem of females permanently associated with him. A limited number of one-male units combine in a *band*, the members of which stay together during part of the foraging expeditions and cooperate in the defense of food finds against other bands. The bands in turn collect at sleeping rocks to spend the night together more or less amicably. This sleeping unit, the *troop*, contains as many as 750 individuals in regions where suitable shelters are scarce, and as few as 12 where they are common. Finally, bachelor males, who constitute about 20 percent of the population, form little bands of their own.

The harems contain from one to as many as ten adult females. At their physical peak most males control from two to five of these adult consorts. The relationship is easily the most "sexist" known in all of the primates. The male herds the females, never letting them stray too far, associate with strangers, or quarrel too vigorously with one another. He employs forms of aggression that vary from a simple hostile stare or slap to a sharp bite on the neck (see Figure 26-4).

The chastised female responds by running to the male. The following three episodes from Kummer's 1968 protocols are typical:

A fight breaks out on the sleeping rock. As soon as it begins, Smoke looks up, advances quickly to the farthest of his females and hits her gently on the head with his hand.

A male, during the daily march, looks back for one of his females in oestrus. As she appears from behind a small ridge he lunges at her. Uttering a staccato cough, she runs toward him.

A male, having just arrived at the sleeping rock, turns suddenly and rushes 30 meters back along the on-coming column. An adult female from the farthest party runs toward him and receives a bite on the back of the neck. Squealing, she follows the male up to the sleeping rock where his other females are waiting.

Such events occur at frequent intervals. The females are also aggressive toward one another. An individual never confronts a rival unless the male is nearby. As a result the male almost invariably assists one or the other. The goal of this "protected threat" tactic is access to the male himself. Fighting is especially intense when two rivals attempt to groom him at the same time.

Since the males sequester their harems with such jealousy, they are also responsible for most of the interactions with other hamadryas units. Young male leaders tend to initiate band movement by moving out with their families closely in tow. Older male leaders then either follow or remain seated, and their actions decide the issue for the band as a whole. When preparing to change position the males notify one another with special gestures. Fighting between the bands is also conducted by the males. It consists almost entirely of spectacular bluffing, during which the opponents fence at each other with open jaws and slap swiftly back and forth with their hands. Film analysis shows that in spite of appearances, physical contact seldom occurs. Only when one male turns and flees is he apt to receive a scratch on the anal region. Fights also end when one animal turns his head to expose the side of his neck. This surrender ritual stops the aggression of the winner instantly.

It is remarkable that in the face of all this jealousy and rage some overlord males tolerate the presence of a follower male. The attachment begins when a subadult male associates with estrous females in the harem. The overlord not only tolerates this intrusion, he allows the youngster to copulate with the females. Soon the subadult male accepts the older male as a leader, running to him like a female when threatened and following the unit out to the feeding sites. At this stage he shows the first, rudimentary tendency to form a harem of his own, by kidnapping infant males and females and holding on to them for periods of up to 30 minutes. As he matures, he ceases paying close attention to the overlord's females and begins to adopt and to mother juvenile females of his own. Thus the sexual bonds are formed, and reinforced by disciplinary aggression, long before copulation can be attempted. Now the two males, each with his own harem, constitute a team. As the older individual weakens with age, his mates stray and the harem grows smaller, but he can still count on the cooperation and support of his younger partner. It is not yet known whether the follower male is ordinarily a relative, perhaps even a son, or a stranger. Harems are also sometimes formed by young adult males who work in solitude to adopt juvenile females.

The extreme polygynous system evolved in the hamadryas, representing an extension of a trend already evident in other baboons, demands an ecological explanation. Kummer (1971) has interpreted the social structure as an adaptation to the patchy, unpredictable food resources of the Ethiopian semidesert. The fusion-fission principle, by which the baboons form feeding groups of all sizes from clusters of bands down to the one-male unit, permits a fuller exploitation of food patches that vary greatly in extent from place to place and day to day. Kummer's concept of this mechanism appears to be true as far as it goes. But if we ask why permanent harems are part of the system, no answer follows from the ecological explanation. It is necessary to return to more basic theory of the kind developed in Chapter 15. The Orians-Verner model does not apply, since feeding territories are not maintained by the baboons. The inapplicability of the hypothesis is further indicated by the fact that the extended adoption procedures by which harems are built reduce the opportunities for female choice. Yet the energy expenditure devoted to acquiring harems is so extraordinarily great that females must be a limiting resource in male multiplication. How can this be the case in an environment with unusually poor food resources? The answer may lie in the pattern of fluctuation in the food supply rather than in its average quantity. Kummer has pointed out that in anubis baboons, the vervets, and Indian langurs, the highest proportion of females to males within groups occurs in populations that occupy environments with the greatest fluctuation in food availability. Although data are still insufficient, it seems likely that such populations would be found to vary more in size if monitored over periods of years. In other words, they pass through more frequent episodes of precipitous decline followed by temporary rapid population growth. If this is indeed the case, females can be expected to function as a limiting resource during good times, so that a moderate amount of reproductive effort will result in a large increase in personal fitness.

The Eastern Mountain Gorilla
(*Gorilla gorilla beringei*)

The gorilla commands attention because it is the largest of the primates, its great males attaining a height of nearly 2 meters and a weight of 180 kilograms or more. But this "amiable vegetarian," as George Schaller called it, has sociobiological peculiarities that would make it worthy of attention even if it were a midget. The gorilla is the one anthropoid ape species organized into age-graded-male troops. Its social life is also one of the most muted in all the higher primates. Although the groups are cohesive and follow one another's movements closely, dominance behavior is very low-keyed and overt agression nearly nonexistent. Territorial spacing is either absent or extremely subtle and erratic, while sexual behavior is so rare that it has been observed in the wild only on a handful of occasions.

The species is comprised of isolated populations scattered across equatorial Africa. The easternmost fringe of the range is occupied by gorillas which are distinguished by longer hair and stronger development of the silverback trait in the males. They are collectively denoted as the subspecies *Gorilla g. beringei*, or, in the vernacular, the eastern mountain gorillas. Their range covers the Virunga Volcanoes and the Mt. Kahuzi district, which includes the mountains north and east of Lake Kivu and the surrounding lowlands. The gorillas are surprisingly adaptable, thriving in a diversity of habitats from lowland rain forest to the thick bamboo stands, *Hagenia* parkland, and *Lobelia-Senecio* groves of the high mountains. Troops have been observed to climb through the mountain forests to as high as 4115 meters, where the temperature drops below freezing each night. The common denominator is a preference for humid environments and low, verdant vegetation. At lower elevations the apes prefer secondary growth over primary forest, a preference that brings them into frequent contact with human beings.

Gorillas are exclusively diurnal. Like their closest phylogenetic allies, the chimpanzees, they build arboreal nests where they spend the night. They are also total vegetarians, eating the leaves, blossoms, shoots, fruit, and bark of many kinds of plants. Among the *Hagenia* trees of the eastern highlands they are surrounded by a virtually unlimited food supply. During the bamboo growing season gorillas supplement their diet with large quantities of shoots. Although they have repeated opportunities in the wild to feed on such animal materials as termite colonies and the remains of birds and small antelopes, there is no evidence of their ever doing so. Yet, curiously, gorillas accept meat in captivity.

The key published work on the wild mountain gorilla is by Schaller (1963, 1965a,b). A newer, even more prolonged study by Dian Fossey has begun to add valuable supplementary information, but at the time of the present review had only been the subject of a preliminary report (Fossey, 1972). The social organization of at least this population of the species is now quite clear. The mountain gorillas live in groups of 2 to 30. In Schaller's overall census data silver-backed males, that is, those about ten years in age or older, constituted 13.1 percent of the group populations, black-backed (young adult) males 9.4 percent, adult females 34.1 percent, and infants and juveniles the remainder. A "typical" troop might consist of one silver-backed male, 0–2 black-backed males, about a half dozen females, and a comparable number of immature individuals. Lone males are relatively common,

Figure 26-4 Social behavior in the hamadryas baboon. The scene is the arid grassland of the Danakil Plain near the low foothills of the Ahmar Mountains, seen along the horizon. In the early morning, a large group of baboons departs from the communal sleeping rock (left background) on the way to the feeding and watering sites. The procession is beginning to break up into the basic social units, which consist of single males and their harems of females and offspring. Aggressive interactions are frequent and animated. The two males in the foreground threaten each other, the one on the right using a hostile stare while his opponent responds with a more intense gaping display. This exchange might escalate into a ritual fight, with rapid boxing and mouth fencing. The females directly behind the two males crouch, make fear faces, and

scream; otherwise they stay out of the conflict. About 2 meters behind the right male, a younger follower male watches the exchange. Although he is teamed with the overlord and has been trying to acquire a harem of his own under the protection of this older partner, he is not likely to join the fight. Directly to his rear another overlord bites the neck of one of his females as punishment for straying too far. Her response will be to run closer to him. At the far left, to the rear of the mother carrying a young infant, two young bachelor males move along the procession in a social formation of their own. (Drawing by Sarah Landry; based on Kummer, 1968 and personal communication.)

and Fossey observed one small troop composed entirely of bachelors. When these individuals are taken into account, the overall ratio in the population is approximately 1 male to 1.5 females. Some of the solitary males actively follow troops, giving the impression of being in leisurely transit from one group to another.

The gorilla troops are demographically stable, and each occupies a home range that changes only slightly over a period of weeks. The home ranges of Fossey's four groups on the slopes of Mount Visoke drifted substantially during two years but still retained constant alignments relative to one another. The home ranges overlap extensively, and neither Schaller nor Fossey detected signs of territorial defense. It is nonetheless clear that some kind of spacing occurs, because the centers of the home ranges are spread out at regular intervals rather than randomly distributed. Groups respond in variable and unpredictable ways when they meet. Usually the encounters are peaceful; the groups continue to feed or progress in full view of each other without visible excitement and sometimes even mingle for a few minutes. But mutual aggression and aversion also occur on occasion. Schaller saw the dominant male of one group charge silently at the dominant male of a second group. The two stared at each other, at times with their brow ridges almost touching. The two groups separated later in the day, sooner than in the case of most encounters. Overt aggressiveness was also noted in a second case when a female, a juvenile, and an infant made incipient charges toward an approaching group. Schaller hypothesized that the members of adjacent groups know one another as individuals, and that much of the variability of the intergroup responses stems from the histories of previous encounters as remembered by the gorillas themselves. Fossey has stressed the importance of the personal idiosyncrasies of the dominant males, who control the movements of the group. One of her groups was under the control of Whinny, a silver-backed male given his name because of his inability to vocalize properly. When Whinny died, the leadership passed to the group's second silverback, Uncle Bert, who clamped down on the group activities "like a gouty headmaster." Where the group had previously accepted Fossey's presence calmly, under Uncle Bert's command they changed to chest beating, whacking at foliage, hiding, and other signs of alarm. Soon the group retreated into a more remote area higher up on Mount Visoke. Also, they kept away from an all-male group trying to make contact—an avoidance behavior that can by itself account for the observed spacing of the home ranges observed in this case. Further evidence of active spacing, perhaps even territorial advertisement, is provided by the loud hoot calls of the mountain gorilla. Consisting of a prolonged chain of *hoo hoo hoos*, they are emitted only by silver-backed males and only during exchanges with other groups or solitary males in the vicinity. The distance between two silverbacks calling varies from as little as 6 meters to a kilometer or more (Fossey 1972).

Mountain gorillas are organized into age-graded-male troops. The core of each troop is the silver-backed male, the adult females, and the young. Extra males, including both subordinate silverbacks and black-backed individuals, remain at the periphery. In spite of this form of dispersion, and the general slow tempo of gorilla social life, the groups are strongly cohesive. The cluster of individuals seldom exceeds 70 meters in diameter, and the dominant male is always within easy vocal range of the other troop members.

Dominance is well marked but subtle in expression. Rank is loosely correlated with size, so that the big silverback is usually at the top, with the somewhat smaller black-backed males dominating the females and young. If more than one silverback is present, the hierarchy is linear and influenced by age, with young and conspicuously aged males taking subordinate positions. Most dominance interactions consist of a mere acknowledgment of precedence. When two animals meet on a narrow trail the subordinate gives the right-of-way; subordinates also yield their sitting place if approached by superiors. Sometimes the dominant animal intimidates the subordinate by starting at it. At most it snaps its mouth or taps the body of the other animal with the back of its hand. Higher levels of aggression within the troop are quite rare. Schaller saw females grapple, scream at each other, and engage in mock biting, but the actions never resulted in visible injury. Even aggression directed at intruders is minimal, consisting mostly of bluffing on the part of the dominant male, who advances to the front of the troop. During 3000 hours of observation with the Mount Visoke gorillas, Fossey was treated to less than 5 minutes of hostility—entirely defensive in nature and never more than a bluff.

If anything, sexual behavior is even lower keyed. Schaller witnessed only two copulations, both involving a dominant silver-backed male. Allogrooming is much less common than in the chimpanzee and most other primate species. It is principally directed from adults to immature individuals or practiced between the young themselves. Allogrooming between adults is quite rare; it was witnessed by Fossey but never by Schaller.

Gorillas communicate principally through the audiovisual channel. There are 16 or 17 distinct vocal displays, including the long-distance hooting of the silverbacks, and perhaps a somewhat smaller repertory of distinguishable facial expressions and postures. It is a matter of considerable interest that this great ape, with its presumed higher level of intelligence, employs a communication system no richer than that of the majority of other social primates, or for that matter other social mammals and birds generally. It is only when we come to its nearest relative, the chimpanzee, that a new evolutionary grade in social behavior is approached. It has been stated that if the sociobiology of gorillas is not more advanced than that of other Old World monkeys and apes, it is at least qualitatively different in some important respects. The cumulative data of Schaller and Fossey now seem to contradict this notion. Gorilla life is certainly much quieter, slower in pace, and in some ways more subtle, but it does not seem to have diverged in any basic way from the mode of the great majority of other Old World species.

The Chimpanzee (*Pan troglodytes*)

By reference to most intuitive criteria chimpanzees are the socially most advanced of the nonhuman primates. They are organized into moderately large societies within which casual groups form, break up, and re-form with extraordinary fluidity. Although the societies are cohesive and occupy stable home ranges, their meetings are amicable and they readily exchange adult females—not males as in other primate species. These two special qualities, great flexibility and openness, are enhanced by the intelligence and individuality in behavior of the troop members. The chimpanzee life cycle is characterized by a long period of socialization and loose but enduring ties between the mother and her adult offspring. Finally, the males are unique among nonhuman primates in the amount of cooperation they display while hunting animals and in the subsequent begging and sharing of the meat.

The common chimpanzee *Pan troglodytes* ranges through equatorial Africa from Sierra Leone and Guinea on the Atlantic coast eastward to Lake Tanganyika and Lake Victoria. A second form, the pygmy chimpanzee, is restricted to a limited area between the Congo and Lualaba rivers and has been variously classified as either a full species (*Pan paniscus*) or subspecies (*Pan troglodytes paniscus*). The common chimpanzee occurs widely in many forested habitats, from rain forest to savanna-forest mosaics, at every elevation from sea level to 3000 meters. It is semiterrestrial, spending 20–50 percent of its time on the ground on ordinary days. It forages during the day and builds sleeping nests in trees to spend the night. The chimpanzee is truly omnivorous, feeding to a large extent on fruit but also on the leaves, bark, and seeds of a wide variety of plant species. It collects termites and ants and at frequent intervals kills and eats small baboons and monkeys.

Pioneering field studies of chimpanzee social behavior were conducted by Nissen (1931) and Kortlandt (1962). In recent years three major studies have advanced our knowledge much further: at the Budongo Forest, near Lake Albert in Uganda, by Vernon and Frances Reynolds and by Izawa, Itani, Nishida, Sugiyama, Suzuki, and other researchers of the Kyoto University project staff; in the region of the Kabogo and Mahali Mountains, east of Lake Tanganyika, Tanzania, by the Kyoto University team; and at the Gombe Stream National Park in Tanzania by Jane van Lawick-Goodall and her associates. These efforts have benefited by the technique of habituation introduced by van Lawick-Goodall. The observer presents himself

openly to the wild chimpanzees, allowing the animals to grow accustomed to his presence over a period of days or weeks. Given enough time the method can be totally successful. The chimpanzees not only come to behave in what seems to be a normal fashion among themselves, they virtually accept the human being as a strange but benevolent member of the group.

The results of the Japanese studies (see especially Izawa, 1970; Nishida and Kawanaka, 1972; and Sugiyama, 1968, 1973) indicate that the basic social unit of chimpanzees is a loose consociation of about 30–80 individuals that occupy a persistent and reasonably well-defined home range over a period of years. The home ranges are moderately large in size, 5–20 square kilometers at Budongo and about 10 square kilometers near the Mahali Mountains, and they partially overlap one another (see Figure 26-6). At the Gombe Stream National Park van Lawick-Goodall estimated the total population to be about 150 individuals. However, only 38 visited her research station with any consistency during 1964–65, so that regional differentiation cannot be ruled out at this locality. The duration of groups and their fidelity to the home ranges evidently persist over chimpanzee generations. Thus, contrary to earlier surmises, the societies are of demographic and not the casual form as defined in Chapter 6. When groups meet, for example at common feeding sites, they often travel together for short periods of time without evident antagonism. However, Sugiyama twice witnessed behavior at Budongo that resembles the territorial display of other kinds of primates. When two groups met, they mingled excitedly. They used exaggerated movements to eat leaves and fruits seldom touched under ordinary circumstances, ran over the ground, and clambered through the branches while shouting and barking. After about an hour of these noisy displays, each group withdrew toward the exclusive portions of its home range. Groups periodically exchange members during their brief encounters. Nishida and Kawanaka (1972) noted that the migrants at Budongo were mostly adult females, especially those that had become sexually receptive. Some females with children also transferred, but in each case they eventually returned to their home group.

In sum, chimpanzee populations appear to be organized along conventional lines. The temporary mingling of neighboring groups is unusual. But the apparent familiarity with neighbors as individuals is not at all out of the ordinary, having been documented in other species of mammals and birds. The transfer of females rather than males is peculiar, but its genetic consequences are the same as the all-male exchanges of other species.

The fluidity of chimpanzee internal organization is truly exceptional, however. The entire group is seldom seen together in the same place. The members gather during migrations from one part of the home range to another in search of special foods. For example, one group moved north in September at Budongo to find the juicy fruits

Figure 26-5 Gorillas have the most relaxed, amiable societies of any of the great apes and larger Old World monkeys. In the scene depicted here, a troop of mountain gorillas rests and feeds in a *Hypericum* forest 3000 meters high in the Virunga Volcanoes of Uganda. The dominant silverback male stands in the left foreground. To his right are two adult females and a pair of two-year-old twins, who try to push each other off a branch in play resembling the human game "king of the castle." In the left rear three juveniles play "follow the leader." To the dominant male's left are another group of females; one cradles a nursing year-old infant, another grooms a three-year-old, while (on the far right) a third

carries a two-year-old on her back while she feeds. Behind this group a black-backed male rests in a hunched sitting position while beyond him two black-backed males and a female forage and another silverback male rests in a prone position. In the upper righthand corner can be seen a solitary male who watches the troop from afar. Notice the large amount of variation in facial features, which is believed to be used by the gorillas themselves in recognizing individual group members. The *Hypericum* forest is rich in wild celery and *Galium* vines, both of which are major foods of the gorillas. (Drawing by Sarah Landry; based on Schaller, 1965a, b, and personal communication, and Fossey, 1972.)

of *Garcinia* plants. But most of the time much smaller parties form and reform within such groups in almost kaleidoscopic fashion. Except for the continued association of offspring with mothers, sometimes well past the time of weaning, the associations have no consistent demographic structure. They are in fact true casual groups of the sort common in human societies (see Chapter 6). An example is provided in Figure 26-7. Parties that discover fruit trees call in other chimpanzees by the carnival display, first described by the missionary Thomas S. Savage in 1844 as "hooting, screaming, and drumming with sticks on old logs." In fact, the chimpanzees spank tree trunks and buttresses with their hands, while running excitedly about, brachiating from branch to branch, and barking, whooping, and crying. The

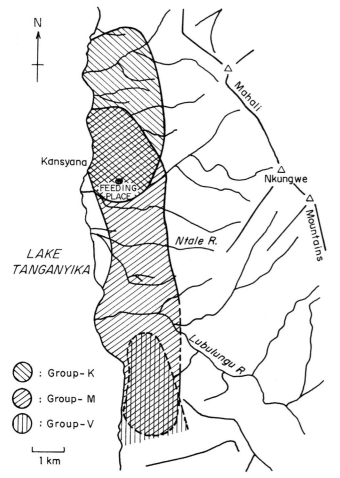

Figure 26-6 The home ranges of three groups of chimpanzees near the Mahali Mountains. A common feeding place occurs in the overlap zone of two of the home ranges. (Redrawn from Sugiyama, 1973.)

sounds can be heard more than a kilometer away, and parties within hearing distance often respond by running in the direction of the sound. The display is used in other contexts: when a party divides and one unit moves away, when a party starts to travel after resting or feeding, and sometimes with no apparent external stimulus. It serves to establish and strengthen ties within the group and perhaps, like the booming of howler monkeys and gibbons, to space the troops. When parties of the same group meet, it is usual for greeting ceremonies to be performed, especially between the adult males. A male newly arrived at a fruiting tree already occupied by a party may beat the buttresses of a tree while calling out. The male of the first party then approaches the newcomer, and the two mutually embrace and groom each other before settling down to eat. Sometimes the newcomer approaches the previous occupant directly and stretches out his hand. The other touches his hand, then the two embrace and groom.

Cooperation within chimpanzee parties is extraordinary in both kind and degree. Most of the time party members feed on fruit and other vegetable items in separate actions. But if the supply is limited—for example if a human observer offers fruit and only the males are venturesome enough to pick it up—the chimps beg from one another and share the food. Cooperation of a different, more significant kind is displayed by chimpanzees while hunting animals. The cumulative observations of Suzuki (1971), Teleki (1973), and others have shown that predation on larger animals such as baboons is an infrequent but quite normal form of specialized behavior. The readiness to pursue, shown always by adult males, is conveyed by changes in posture, behavior, and facial expression. Other chimpanzees respond to these signals with alert, excited movements that often culminate in simultaneous pursuit. According to Geza Teleki, predatory interest and intent are shown by a set or blank facial expression. The chimpanzee becomes unusually quiet and stares fixedly at the target prey. Its posture is tensed, and hair is partially erected all over its body. Ordinarily only mature males engage in the hunt, although on one occasion two females were seen to capture and kill a pair of young pigs. A notable aspect of the pursuit is the complete silence on the part of the chimpanzees until actual seizure is attempted. Such restraint on the part of one of the noisiest of all animals is most unusual.

Teleki distinguishes three modes of pursuit. In the first, the chimpanzees mingle with the prey and seize a victim with sudden, explosive movements. The second technique is a running pursuit. When the prey is a young baboon, capture can sometimes be achieved only after a battle with the adult males who rush out to defend it. The third mode, the most interesting of all, consists of stalking maneuvers in which the prey is helplessly treed or otherwise trapped. Each of the three modes entails some amount of teamwork on the part of the hunters.

At the Gombe Stream National Park the best opportunities for predation come when chimpanzees and anubis baboons mingle at feeding sites. For long stretches of time the interactions are neutral or at most mildly aggressive. Immature chimpanzees and baboons occasionally even play together. Then the mood changes quickly to presage a scene of grisly violence:

Two other chimpanzees, Charlie and Goliath, finish their bananas at this point and lie down a few yards away. Mike and Hugo also finish, and both begin to groom Hugh. Two more adult baboons and several juveniles join Mandrill and Sif [adult baboons], so the mixed group (all are within a circle of 10 yards' diameter) now includes 5 mature male chimpanzees and 7 baboons of various ages. Thor is the only infant [baboon] present. All seem quite relaxed: several chimpanzees groom mutually, and Salty baboon begins to groom Sif. Mike suddenly *waas* and arm-threatens Mandrill again at 11:02; Mandrill wheels and returns the threat by eyelidding at Mike, who in turn slaps the baboon's nose; Mandrill leaps backward, and calms down rapidly . . . At 11:07 Mandrill takes infant Thor from his mother, who does not pause in grooming the adult male. The baboons are only a yard away from the chimpanzees. Then at 11:09, several male chimpanzees—Mike, Hugh, Hugo, and Charlie—suddenly pounce on Mandrill, grab Thor from his hands, and

begin immediately to tear the infant apart as they huddle in a tight cluster. The baboons—including mother Sif—scatter quickly. The only baboon who stays near the chimpanzees is Mandrill, who *barks* repeatedly as he pushes with both hands against one chimpanzee's hunched back, but to no apparent avail. Thor is soon halved by Mike and Hugh, and other chimpanzees follow as these two walk away from the scene, climb nearby trees, and begin to consume the infant.   (Teleki, 1973)

This is a case of explosive seizure. The coordination among the male chimpanzees becomes still more apparent when the pursuit is conducted over the ground at a run, and obvious when the third mode of stalking and encirclement is employed. The following account by Teleki (1973) describes the initiation of the last kind of maneuver.

Figan, who has been sitting idly in a tree, suddenly drops to the ground at 12:32 P.M. and hurries silently across the open slope toward a small baboon cluster—an adult male, a female, and a young juvenile. Almost simultaneously, Rix and Worzle descend from other trees to join Figan, and all stop about 5 yards away and watch the same baboons. Figan, standing slightly ahead of the others, slowly begins to approach the juvenile, who *gecks* (yaks) loudly; the male baboon immediately joins the juvenile, and both stand their ground and face the watching chimpanzees. Figan stops again about 3 yards distant. At this moment Hugh, Charlie and Mike quickly cross the slope toward the baboons with hair

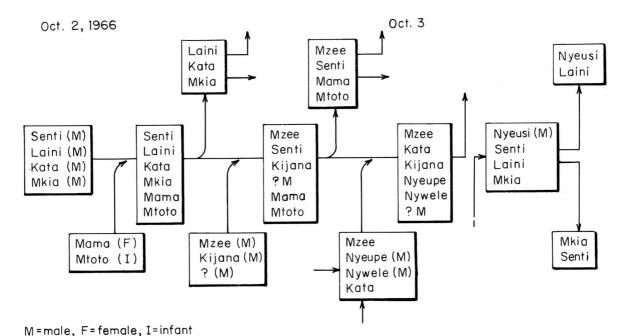

M=male, F=female, I=infant

Figure 26-7 The composition of chimpanzee parties changes in kaleidoscopic fashion, as illustrated by this history of the gathering of elements of one troop in the Budongo Forest. Although such groupings are very short-lived, the regional troops of which they are a part may persist for generations. (From Sugiyama, 1973.)

Figure 26-8 Temporary resting parties of chimpanzees in the Gombe Stream National Park. Left: three adult males on the left (Worzel, Charlie, and Hugo) are accompanied by two adult females (Sophie, with a female infant, and Melissa). Right: in a second group, two infants play in the middle, one with a typical "play-face," while a juvenile grooms an adult. (Photographs by Peter Marler and Richard Zigmond.)

on end; the juvenile baboon *screeches* and *gecks* loudly, the male lunges and eyelids; Charlie stands, arm-waves, and swaggers toward the cluster.

This episode ended when the baboons were able to retreat, after which the chimpanzee males quickly dispersed. Van Lawick-Goodall (1968a) witnessed another episode in which the roles played by the males were even more distinct. The same Figan began the action by stalking a juvenile baboon up the trunk of a palm tree. Within moments other males that had been resting and grooming nearby stood and approached the tree. Some moved to its base while others dispersed to adjacent trees that threatened to serve as alternate routes of escape. The baboon did indeed jump from one tree to another,

whereupon one chimpanzee stationed close by began to climb quickly toward it. The baboon then was able to escape by jumping 20 feet to the ground and running to the protection of its troop nearby.

The distribution of the meat is an equally complicated procedure. As van Lawick-Goodall and Teleki showed, various sequences of begging signals are used. The requesting animal may peer intently while placing its face close to the face of the meat eater or to the meat it is holding, or it may reach out and touch the meat itself or the chin and lips of the other animal. Alternatively, it extends an open hand with palm upward beneath the chin of the meat eater. Often a soft whimper or *hoo* accompanies these gestures. The individuals observed to beg belong to both sexes and all ages above two years. The meat eater sometimes rejects the request by pulling its booty away, moving to another position, or signaling refusal. Occasionally it acquiesces by allowing the other animal to chew directly on the meat or to remove small pieces with its hands. On four occasions during one year Teleki observed chimpanzees actually tear off pieces of meat and hand them over to supplicants.

Dominance behavior is well developed in the chimpanzee. A low-ranking individual gives way to a high-ranking one when they meet on a branch or when both approach the same piece of food. Its subordinate status is further signified when it detours around another animal or conciliates it by reaching out to touch it on the lips, thigh, or genital area. But these interactions are subtle. Overt threats and retreats are uncommon. Sugiyama witnessed only 31 such exchanges during 360 hours of observation time, the Reynolds saw 17 "quarrels" in 300 hours, and Jane van Lawick-Goodall recorded 72 aggressive incidents in the first two years of her stay at the Gombe Stream National Park. The great majority of hostile acts involve adult males. Yet curiously in view of this fact, the dominance system appears to have no influence on access to females. Chimpanzee females are essentially promiscuous. They often copulate with more than one male in rapid succession, yet without provoking interference from nearby males. Once van Lawick-Goodall saw seven males mount the same female, one after the other, with less than two minutes separating each of the first five copulations. On occasions the females themselves seek contact. An estrous female in Sugiyama's Budongo troop stopped grooming a dominant male, approached a young adult male on a nearby branch, copulated with him, and then resumed her ministrations to the first male. A second notable feature of chimpanzee domi-nance is that rank has little to do with allogrooming patterns. Chimpanzees groom one another regularly, seeming to use the behavior for mutual reassurance. Allogrooming occurs during a high percentage of those occasions, for example, when mothers and their offspring rejoin after a prolonged absence or when two parties of the same regional group meet during foraging excursions. Sometimes a dominant animal briefly grooms a subordinate that has approached it for reassurance, but in most cases it gives a mere token touch or pat.

Leadership, defined narrowly as the initiation of group movement, is well developed among chimpanzees. Ordinarily the dominant male of a party leads all the others. When the party is progressing rapidly from one food tree to another, the leader takes the front position. On other occasions it remains near the center or rear. Regardless of position it seldom loses control, because when it moves the rest move and when it halts, they halt also.

The rich communication system of chimpanzees has been described in detail by van Lawick-Goodall (1968b, 1971). It consists to a large extent of composite signals comprised of vocalizations, facial expressions, and body postures and movements. Touch, including allogrooming, is also frequently employed but is far poorer in signal diversity than the audiovisual system. Like human beings, the chimpanzee appears to make very little use of chemical signals. Yet it must be admitted that this subject has not been explicitly investigated with appropriate behavioral and chemical tests.

Chapter 27 **Man:
From Sociobiology
to Sociology**

Let us now consider man in the free spirit of natural history, as though we were zoologists from another planet completing a catalog of social species on Earth. In this macroscopic view the humanities and social sciences shrink to specialized branches of biology; history, biography, and fiction are the research protocols of human ethology; and anthropology and sociology together constitute the sociobiology of a single primate species.

Homo sapiens is ecologically a very peculiar species. It occupies the widest geographical range and maintains the highest local densities of any of the primates. An astute ecologist from another planet would not be surprised to find that only one species of *Homo* exists. Modern man has preempted all the conceivable hominid niches. Two or more species of hominids did coexist in the past, when the *Australopithecus* man-apes and possibly an early *Homo* lived in Africa. But only one evolving line survived into late Pleistocene times to participate in the emergence of the most advanced human social traits.

Modern man is anatomically unique. His erect posture and wholly bipedal locomotion are not even approached in other primates that occasionally walk on their hind legs, including the gorilla and chimpanzee. The skeleton has been profoundly modified to accommodate the change: the spine is curved to distribute the weight of the trunk more evenly down its length; the chest is flattened to move the center of gravity back toward the spine; the pelvis is broadened to serve as an attachment for the powerful striding muscles of the upper legs and reshaped into a basin to hold the viscera; the tail is eliminated, its vertebrae (now called the coccyx) curved inward to form part of the floor of the pelvic basin; the occipital condyles have rotated far beneath the skull so that the weight of the head is balanced on them; the face is shortened to assist this shift in gravity; the thumb is enlarged to give power to the hand; the leg is lengthened; and the foot is drastically narrowed and lengthened to facilitate striding. Other changes have taken place. Hair has been lost from most of the body. It is still not known why modern man is a "naked ape." One plausible explanation is that nakedness served as a device to cool the body during the strenuous pursuit of prey in the heat of the African plains. It is associated with man's exceptional reliance on sweating to reduce body heat; the human body contains from two to five million sweat glands, far more than in any other primate species.

The reproductive physiology and behavior of *Homo sapiens* have also undergone extraordinary evolution. In particular, the estrous cycle of the female has changed in two ways that affect sexual and social behavior. Menstruation has been intensified. The females of some other primate species experience slight bleeding, but only in women is there a heavy sloughing of the wall of the "disappointed womb" with consequent heavy bleeding. The estrus, or period of female "heat," has been replaced by virtually continuous sexual activity. Copulation is initiated not by response to the conventional primate signals of estrus, such as changes in color of the skin around the female sexual organs and the release of pheromones, but by extended foreplay entailing mutual stimulation by the partners. The traits of physical attraction are, moreover, fixed in nature. They include the pubic hair of both sexes and the protuberant breasts and buttocks of women. The flattened sexual cycle and continuous female attractiveness cement the close marriage bonds that are basic to human social life.

At a distance a perceptive Martian zoologist would regard the globular head as a most significant clue to human biology. The cerebrum of *Homo* was expanded enormously during a relatively short span of evolutionary time (see Figure 27-1). Three million years ago *Australopithecus* had an adult cranial capacity of 400-500 cubic centimeters, comparable to that of the chimpanzee and gorilla. Two million years later its presumptive descendant *Homo erectus* had a capacity of about 1000 cubic centimeters. The next million years saw an increase to 1400-1700 cubic centimeters in Neanderthal man and 900-2000 cubic centimeters in modern *Homo sapiens.* The growth in intelligence that accompanied this enlargement was so great that it cannot yet be measured in any meaningful way. Human beings can be compared among themselves in terms of a few of the basic components of intelligence and creativity. But no scale has been invented that can objectively compare man with chimpanzees and other living primates.

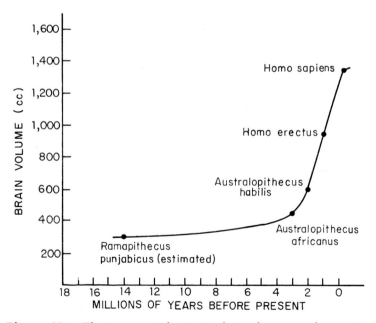

Figure 27-1 The increase in brain size during human evolution. (Redrawn from Pilbeam, 1972.)

We have leaped forward in mental evolution in a way that continues to defy self-analysis. The mental hypertrophy has distorted even the most basic primate social qualities into nearly unrecognizable forms. Individual species of Old World monkeys and apes have notably plastic social organizations; man has extended the trend into a protean ethnicity. Monkeys and apes utilize behavioral scaling to adjust aggressive and sexual interactions; in man the scales have become multidimensional, culturally adjustable, and almost endlessly subtle. Bonding and the practices of reciprocal altruism are rudimentary in other primates; man has expanded them into great networks where individuals consciously alter roles from hour to hour as if changing masks.

It is the task of comparative sociobiology to trace these and other human qualities as closely as possible back through time. Besides adding perspective and perhaps offering some sense of philosophical ease, the exercise will help to identify the behaviors and rules by which individual human beings increase their Darwinian fitness through the manipulation of society. In a phrase, we are searching for the human biogram (Count, 1958; Tiger and Fox, 1971). One of the key questions, never far from the thinking of anthropologists and biologists who pursue real theory, is to what extent the biogram represents an adaptation to modern cultural life and to what extent it is a phylogenetic vestige. Our civilizations were jerrybuilt around the biogram. How have they been influenced by it? Conversely, how much flexibility is there in the biogram, and in which parameters particularly? Experience with other animals indicates that when organs are hypertrophied, phylogeny is hard to reconstruct. This is the crux of the problem of the evolutionary analysis of human behavior. In the remainder of the chapter, human qualities will be discussed insofar as they appear to be general traits of the species. Then current knowledge of the evolution of the biogram will be reviewed, and finally some implications for the planning of future societies will be considered.

Plasticity of Social Organization

The first and most easily verifiable diagnostic trait is statistical in nature. The parameters of social organization, including group size, properties of hierarchies, and rates of gene exchange, vary far more among human populations than among those of any other primate species. The variation exceeds even that occurring between the remaining primate species. Some increase in plasticity is to be expected. It represents the extrapolation of a trend toward variability already apparent in the baboons, chimpanzees, and other cercopithecoids. What is truly surprising, however, is the extreme to which it has been carried.

Why are human societies this flexible? Part of the reason is that the members themselves vary so much in behavior and achievement. Even in the simplest societies individuals differ greatly. Within a small tribe of !Kung Bushmen can be found individuals who are acknowledged as the "best people"—the leaders and outstanding specialists among the hunters and healers. Even with an emphasis on sharing goods, some are exceptionally able entrepreneurs and unostentatiously acquire a certain amount of wealth. !Kung men, no less than men in advanced industrial societies, generally establish themselves by their mid-thirties or else accept a lesser status for life. There are some who never try to make it, live in run-down huts, and show little pride in themselves or their work (Pfeiffer, 1969). The ability to slip into such roles, shaping one's personality to fit, may itself be adaptive. Human societies are organized by high intelligence, and each member is faced by a mixture of social challenges that taxes all of his ingenuity. This baseline variation is amplified at the group level by other qualities exceptionally pronounced in human societies: the long, close period of socialization; the loose connectedness of the communication networks; the multiplicity of bonds; the capacity, especially within literate cultures, to communicate over long distances and periods of history; and from all these traits, the capacity to dissemble, to manipulate, and to exploit. Each parameter can be altered easily, and each has a marked effect on the final social structure. The result could be the observed variation among societies.

The hypothesis to consider, then, is that genes promoting flexibility in social behavior are strongly selected at the individual level. But note that variation in social organization is only a possible, not a necessary consequence of this process. In order to generate the amount of variation actually observed to occur, it is necessary for there to be multiple adaptive peaks. In other words, different forms of society within the same species must be nearly enough alike in survival ability for many to enjoy long tenure. The result would be a statistical ensemble of kinds of societies which, if not equilibrial, is at least not shifting rapidly toward one particular mode or another. The alternative, found in some social insects, is flexibility in individual behavior and caste development, which nevertheless results in an approach toward uniformity in the statistical distribution of the kinds of individuals when all individuals within a colony are taken together. In honeybees and in ants of the genera *Formica* and *Pogonomyrmex*, "personality" differences are strongly marked even within single castes. Some individuals, referred to by entomologists as the elites, are unusually active, perform more than their share of lifetime work, and incite others to work through facilitation. Other colony members are consistently sluggish. Although they are seemingly healthy and live long lives, their per-individual output is only a small fraction of that of the elites. Specialization also occurs. Certain individuals remain with the brood as nurses far longer than the average, while others concentrate on nest building or foraging. Yet somehow

the total pattern of behavior in the colony converges on the species average. When one colony with its hundreds or thousands of members is compared with another of the same species, the statistical patterns of activity are about the same. We know that some of this consistency is due to negative feedback. As one requirement such as brood care or nest repair intensifies, workers shift their activities to compensate until the need is met, then change back again. Experiments have shown that disruption of the feedback loops, and thence deviation by the colony from the statistical norms, can be disastrous. It is therefore not surprising to find that the loops are both precise and powerful (Wilson, 1971a).

The controls governing human societies are not nearly so strong, and the effects of deviation are not so dangerous. The anthropological literature abounds with examples of societies that contain obvious inefficiencies and even pathological flaws—yet endure. The slave society of Jamaica, compellingly described by Orlando Patterson (1967), was unquestionably pathological by the moral canons of civilized life. "What marks it out is the astonishing neglect and distortion of almost every one of the basic prerequisites of normal human living. This was a society in which clergymen were the 'most finished debauchees' in the land; in which the institution of marriage was officially condemned among both masters and slaves; in which the family was unthinkable to the vast majority of the population and promiscuity the norm; in which education was seen as an absolute waste of time and teachers shunned like the plague; in which the legal system was quite deliberately a travesty of anything that could be called justice; and in which all forms of refinements, of art, of folkways, were either absent or in a state of total disintegration. Only a small proportion of whites, who monopolized almost all of the fertile land in the island, benefited from the system. And these, no sooner had they secured their fortunes, abandoned the land which the production of their own wealth had made unbearable to live in, for the comforts of the mother country." Yet this Hobbesian world lasted for nearly two centuries. The people multiplied while the economy flourished.

The Ik of Uganda are an equally instructive case (Turnbull, 1972). They are former hunters who have made a disastrous shift to cultivation. Always on the brink of starvation, they have seen their culture reduced to a vestige. Their only stated value is *ngag*, or food; their basic notion of goodness (*marangik*) is the individual possession of food in the stomach; and their definition of a good man is *yakw ana marang*, "a man who has a full belly." Villages are still built, but the nuclear family has ceased to function as an institution. Children are kept with reluctance and from about three years of age are made to find their own way of life. Marriage ordinarily occurs only when there is a specific need for cooperation. Because of the lack of energy, sexual activity is minimal and its pleasures are considered to be about on the same level as those of defecation. Death is treated with relief or amusement, since it means more *ngag* for survivors. Because the unfortunate Ik are at the lowest sustainable level, there is a temptation to conclude that they are doomed. Yet somehow their society has remained intact and more or less stable for at least 30 years, and it could endure indefinitely.

How can such variation in social structure persist? The explanation may be lack of competition from other species, resulting in what biologists call ecological release. During the past ten thousand years or longer, man as a whole has been so successful in dominating his environment that almost any kind of culture can succeed for a while, so long as it has a modest degree of internal consistency and does not shut off reproduction altogether. No species of ant or termite enjoys this freedom. The slightest inefficiency in constructing nests, in establishing odor trails, or in conducting nuptial flights could result in the quick extinction of the species by predation and competition from other social insects. To a scarcely lesser extent the same is true for social carnivores and primates. In short, animal species tend to be tightly packed in the ecosystem with little room for experimentation or play. Man has temporarily escaped the constraint of interspecific competition. Although cultures replace one another, the process is much less effective than interspecific competition in reducing variance.

It is part of the conventional wisdom that virtually all cultural variation is phenotypic rather than genetic in origin. This view has gained support from the ease with which certain aspects of culture can be altered in the space of a single generation, too quickly to be evolutionary in nature. The drastic alteration in Irish society in the first two years of the potato blight (1846-1848) is a case in point. Another is the shift in the Japanese authority structure during the American occupation following World War II. Such examples can be multiplied endlessly—they are the substance of history. It is also true that human populations are not very different from one another genetically. When Lewontin (1972b) analyzed existing data on nine blood-type systems, he found that 85 percent of the variance was composed of diversity within populations and only 15 percent was due to diversity between populations. There is no a priori reason for supposing that this sample of genes possesses a distribution much different from those of other, less accessible systems affecting behavior.

The extreme orthodox view of environmentalism goes further, holding that in effect there is no genetic variance in the transmission of culture. In other words, the capacity for culture is transmitted by a single human genotype. Dobzhansky (1963) stated this hypothesis as follows: "Culture is not inherited through genes, it is acquired by learning from other human beings . . . In a sense, human genes have surrendered their primacy in human evolution to an entirely new,

nonbiological or superorganic agent, culture. However, it should not be forgotten that this agent is entirely dependent on the human genotype." Although the genes have given away most of their sovereignty, they maintain a certain amount of influence in at least the behavioral qualities that underlie variations between cultures. Moderately high heritability has been documented in introversion-extroversion measures, personal tempo, psychomotor and sports activities, neuroticism, dominance, depression, and the tendency toward certain forms of mental illness such as schizophrenia (Parsons, 1967; Lerner, 1968). Even a small portion of this variance invested in population differences might predispose societies toward cultural differences. At the very least, we should try to measure this amount. It is not valid to point to the absence of a behavioral trait in one or a few societies as conclusive evidence that the trait is environmentally induced and has no genetic disposition in man. The very opposite could be true.

In short, there is a need for a discipline of anthropological genetics. In the interval before we acquire it, it should be possible to characterize the human biogram by two indirect methods. First, models can be constructed from the most elementary rules of human behavior. Insofar as they can be tested, the rules will characterize the biogram in much the same way that ethograms drawn by zoologists identify the "typical" behavioral repertories of animal species. The rules can be legitimately compared with the ethograms of other primate species. Variation in the rules among human cultures, however slight, might provide clues to underlying genetic differences, particularly when it is correlated with variation in behavioral traits known to be heritable. Social scientists have in fact begun to take this first approach, although in a different context from the one suggested here. Abraham Maslow (1954, 1972) postulated that human beings respond to a hierarchy of needs, such that the lower levels must be satisfied before much attention is devoted to the higher ones. The most basic needs are hunger and sleep. When these are met, safety becomes the primary consideration, then the need to belong to a group and receive love, next self-esteem, and finally self-actualization and creativity. The ideal society in Maslow's dream is one which "fosters the fullest development of human potentials, of the fullest degree of humanness." When the biogram is freely expressed, its center of gravity should come to rest in the higher levels. A second social scientist, George C. Homans (1961), has adopted a Skinnerian approach in an attempt to reduce human behavior to the basic processes of associative learning. The rules he postulates are the following:

1. If in the past the occurrence of a particular stimulus-situation has been the occasion on which a man's activity has been rewarded, then the more similar the present stimulus-situation is to the past one, the more likely the man is at the present time to emit this activity or one similar to it.

2. The more often within a given period of time a man's activity rewards the behavior of another, the more often the other will perform the behavior.

3. The more valuable to a man a unit of the activity another gives him, the more often he behaves in the manner rewarded by the activity of the other.

4. The more often a man has in the recent past received a rewarding activity from another, the less valuable any further unit of that activity becomes to him.

Maslow the ethologist and visionary seems a world apart from Homans the behaviorist and reductionist. Yet their approaches are reconcilable. Homans' rules can be viewed as comprising some of the enabling devices by which the human biogram is expressed. His operational word is *reward*, which is in fact the set of all interactions defined by the emotive centers of the brain as desirable. According to evolutionary theory, desirability is measured in units of genetic fitness, and the emotive centers have been programmed accordingly. Maslow's hierarchy is simply the order of priority in the goals toward which the rules are directed.

The other indirect approach to anthropological genetics is through phylogenetic analysis. By comparing man with other primate species, it might be possible to identify basic primate traits that lie beneath the surface and help to determine the configuration of man's higher social behavior. This approach has been taken with great style and vigor in a series of popular books by Konrad Lorenz (*On Aggression*), Robert Ardrey (*The Social Contract*), Desmond Morris (*The Naked Ape*), and Lionel Tiger and Robin Fox (*The Imperial Animal*). Their efforts were salutary in calling attention to man's status as a biological species adapted to particular environments. The wide attention they received broke the stifling grip of the extreme behaviorists, whose view of the mind of man as a virtually equipotent response machine was neither correct nor heuristic. But their particular handling of the problem tended to be inefficient and misleading. They selected one plausible hypothesis or another based on a review of a small sample of animal species, then advocated the explanation to the limit. The weakness of this method was discussed earlier in a more general context (Chapter 2) and does not need repetition here.

The correct approach using comparative ethology is to base a rigorous phylogeny of closely related species on many biological traits. Then social behavior is treated as the dependent variable and its evolution deduced from it. When this cannot be done with confidence (and it cannot in man) the next best procedure is the one outlined in Chapter 7: establish the lowest taxonomic level at which each character shows significant intertaxon variation. Characters that shift from species to species or genus to genus are the most labile. We cannot safely extrapolate them from the cercopithecoid monkeys and apes to man. In the primates these labile qualities include group size, group cohesiveness, openness of the group to others, involvement of the male in parental care, attention structure, and the intensity and form of territorial defense. Characters are considered conservative if they remain constant at the level of the taxonomic family or throughout the order Primates, and they are the ones most likely to have persisted in relatively unaltered form into the evolution of *Homo*. These conservative traits include aggressive dominance systems, with males generally dominant over females; scaling in the intensity of responses, especially during aggressive interactions; intensive and prolonged maternal care, with a pronounced degree of socialization in the young; and matrilineal social organization. This classification of behavioral traits offers an appropriate basis for hypothesis formation. It allows a qualitative assessment of the probabilities that various behavioral traits have persisted into modern *Homo sapiens*. The possibility of course remains that some labile traits are homologous between man and, say, the chimpanzee. And conversely, some traits conservative throughout the rest of the primates might nevertheless have changed during the origin of man. Furthermore, the assessment is not meant to imply that conservative traits are more genetic—that is, have higher heritability—than labile ones. Lability can be based wholly on genetic differences between species or populations within species. Returning finally to the matter of cultural evolution, we can heuristically conjecture that the traits proven to be labile are also the ones most likely to differ from one human society to another on the basis of genetic differences. The evidence, reviewed in Table 27-1, is not inconsistent with this basic conception. Finally, it is worth special note that the comparative ethological approach does not in any way predict man's unique traits. It is a general rule of evolutionary studies that the direction of quantum jumps is not easily read by phylogenetic extrapolation.

Barter and Reciprocal Altruism

Sharing is rare among the nonhuman primates. It occurs in rudimentary form only in the chimpanzee and perhaps a few other Old World monkeys and apes. But in man it is one of the strongest social traits, reaching levels that match the intense trophallactic exchanges of termites and ants. As a result only man has an economy. His high intelligence and symbolizing ability make true barter possible. Intelligence also permits the exchanges to be stretched out in time, converting them into acts of reciprocal altruism (Trivers, 1971). The conventions of this mode of behavior are expressed in the familiar utterances of everyday life:

"Give me some now; I'll repay you later."

Table 27-1 General social traits in human beings, classified according to whether they are unique, belong to a class of behaviors that are variable at the level of the species or genus in the remainder of the primates (labile), or belong to a class of behaviors that are uniform through the remainder of the primates (conservative).

Evolutionarily labile primate traits	Evolutionarily conservative primate traits	Human traits
		SHARED WITH SOME OTHER PRIMATES
Group size		Highly variable
Group cohesiveness		Highly variable
Openness of group to others		Highly variable
Involvement of male in parental care		Strong
Attention structure		Centripetal on leading males
Intensity and form of territorial defense		Highly variable, but territoriality is general
		SHARED WITH ALL OR ALMOST ALL OTHER PRIMATES
	Aggressive dominance systems, with males dominant over females	Consistent with other primates, although variable
	Scaling of responses, especially in aggressive interactions	Consistent with other primates
	Prolonged maternal care; pronounced socialization of young	Consistent with other primates
	Matrilineal organization	Mostly consistent with other primates
		UNIQUE
		True language, elaborate culture
		Sexual activity continuous through menstrual cycle
		Formalized incest taboos and marriage exchange rules with recognition of kinship networks
		Cooperative division of labor between adult males and females

"Come to my aid this time, and I'll be your friend when you need me."

"I really didn't think of the rescue as heroism; it was only what I would expect others to do for me or my family in the same situation."

Money, as Talcott Parsons has been fond of pointing out, has no value in itself. It consists only of bits of metal and scraps of paper by which men pledge to surrender varying amounts of property and services upon demand; in other words it is a quantification of reciprocal altruism.

Perhaps the earliest form of barter in early human societies was the exchange of meat captured by the males for plant food gathered by the females. If living hunter-gatherer societies reflect the primitive state, this exchange formed an important element in a distinctive kind of sexual bond.

Fox (1972), following Lévi-Strauss (1949), has argued from ethnographic evidence that a key early step in human social evolution was the use of women in barter. As males acquired status through the control of females, they used them as objects of exchange to cement alliances and bolster kinship networks. Preliterate societies are characterized by complex rules of marriage that can often be interpreted directly as power brokerage. This is particularly the case where the elementary negative marriage rules, proscribing certain types of unions, are supplemented by positive rules that direct which exchanges must be made. Within individual Australian aboriginal societies two moieties exist between which marriages are permitted. The men of each moiety trade nieces, or more specifically their sisters' daughters. Power accumulates with age, because a man can control the descendants of nieces as remote as the daughter of his sister's daughter. Combined with polygyny, the system insures both political and genetic advantage to the old men of the tribe.

For all its intricacy, the formalization of marital exchanges between tribes has the same approximate genetic effect as the haphazard wandering of male monkeys from one troop to another or the exchange of young mature females between chimpanzee populations. Approximately 7.5 percent of marriages contracted among Australian aborigines prior to European influence were intertribal, and similar rates have been reported in Brazilian Indians and other preliterate societies (Morton, 1969). It will be recalled (Chapter 4) that gene flow of the order of 10 percent per generation is more than enough to counteract fairly intensive natural pressures that tend to differentiate populations. Thus intertribal marital exchanges are a major factor in creating the observed high degree of genetic similarity among populations. The ultimate adaptive basis of exogamy is not gene flow per se but rather the avoidance of inbreeding. Again, a 10 percent gene flow is adequate for the purpose.

The microstructure of human social organization is based on sophisticated mutual assessments that lead to the making of contracts.

As Erving Goffman correctly perceived, a stranger is rapidly but politely explored to determine his socioeconomic status, intelligence and education, self-perception, social attitudes, competence, trustworthiness, and emotional stability. The information, much of it subconsciously given and absorbed, has an eminently practical value. The probe must be deep, for the individual tries to create the impression that will gain him the maximum advantage. At the very least he maneuvers to avoid revealing information that will imperil his status. The presentation of self can be expected to contain deceptive elements:

Many crucial facts lie beyond the time and place of interaction or lie concealed within it. For example, the "true" or "real" attitudes, beliefs, and emotions of the individual can be ascertained only indirectly, through his avowals or through what appears to be involuntary expressive behavior. Similarly, if the individual offers the others a product or service, they will often find that during the interaction there will be no time or place immediately available for eating the pudding that the proof can be found in. They will be forced to accept some events as conventional or natural signs of something not directly available to the senses.   (Goffman, 1959)

Deception and hypocrisy are neither absolute evils that virtuous men suppress to a minimum level nor residual animal traits waiting to be erased by further social evolution. They are very human devices for conducting the complex daily business of social life. The level in each particular society may represent a compromise that reflects the size and complexity of the society. If the level is too low, others will seize the advantage and win. If it is too high, ostracism is the result. Complete honesty on all sides is not the answer. The old primate frankness would destroy the delicate fabric of social life that has built up in human populations beyond the limits of the immediate clan. As Louis J. Halle correctly observed, good manners have become a substitute for love.

Bonding, Sex, and Division of Labor

The building block of nearly all human societies is the nuclear family (Reynolds, 1968; Leibowitz, 1968). The populace of an American industrial city, no less than a band of hunter-gatherers in the Australian desert, is organized around this unit. In both cases the family moves between regional communities, maintaining complex ties with primary kin by means of visits (or telephone calls and letters) and the exchange of gifts. During the day the women and children remain in the residential area while the men forage for game or its symbolic equivalent in the form of barter and money. The males cooperate in bands to hunt or deal with neighboring groups. If not actually blood relations, they tend at least to act as "bands of brothers." Sexual bonds are carefully contracted in observance with tribal customs and are intended to be permanent. Polygamy, either covert or explicitly sanctioned by custom, is practiced predominantly by the males. Sexual behavior is nearly continuous through the menstrual cycle and marked by extended foreplay. Morris (1967a), drawing on the data of Masters and Johnson (1966) and others, has enumerated the unique features of human sexuality that he considers to be associated with the loss of body hair: the rounded and protuberant breasts of the young woman, the flushing of areas of skin during coition, the vaso-dilation and increased erogenous sensitivity of the lips, soft portions of the nose, ear, nipples, areolae, and genitals, and the large size of the male penis, especially during erection. As Darwin himself noted in 1871, even the naked skin of the woman is used as a sexual releaser. All of these alterations serve to cement the permanent bonds, which are unrelated in time to the moment of ovulation. Estrus has been reduced to a vestige, to the consternation of those who attempt to practice birth control by the rhythm method. Sexual behavior has been largely dissociated from the act of fertilization. It is ironic that religionists who forbid sexual activity except for purposes of procreation should do so on the basis of "natural law." Theirs is a misguided effort in comparative ethology, based on the incorrect assumption that in reproduction man is essentially like other animals.

The extent and formalization of kinship prevailing in almost all human societies are also unique features of the biology of our species. Kinship systems provide at least three distinct advantages. First, they bind alliances between tribes and subtribal units and provide a conduit for the conflict-free emigration of young members. Second, they are an important part of the bartering system by which certain males achieve dominance and leadership. Finally, they serve as a homeostatic device for seeing groups through hard times. When food grows scarce, tribal units can call on their allies for altruistic assistance in a way unknown in other social primates. The Athapaskan Dogrib Indians, a hunter-gatherer people of the northwestern Canadian arctic, provide one example. The Athapaskans are organized loosely by the bilateral primary linkage principle (June Helm, 1968). Local bands wander through a common territory, making intermittent contacts and exchanging members by intermarriage. When famine strikes, the endangered bands can coalesce with those temporarily better off. A second example is the Yanomamö of South America, who rely on kin when their crops are destroyed by enemies (Chagnon, 1968).

As societies evolved from bands through tribes into chiefdoms and states, some of the modes of bonding were extended beyond kinship networks to include other kinds of alliances and economic agreements. Because the networks were then larger, the lines of communication longer, and the interactions more diverse, the total systems became vastly more complex. But the moralistic rules underlying these arrangements appear not to have been altered a great deal. The average individual still operates under a formalized code no more elaborate than that governing the members of hunter-gatherer societies.

Role Playing and Polyethism

The superman, like the super-ant or super-wolf, can never be an individual; it is the society, whose members diversify and cooperate to create a composite well beyond the capacity of any conceivable organism. Human societies have effloresced to levels of extreme complexity because their members have the intelligence and flexibility to play roles of virtually any degree of specification, and to switch them as the occasion demands. Modern man is an actor of many parts who may well be stretched to his limit by the constantly shifting demands of his environment. As Goffman (1961) observed, "Perhaps there are times when an individual does march up and down like a wooden soldier, tightly rolled up in a particular role. It is true that here and there we can pounce on a moment when an individual sits fully astride a single role, head erect, eyes front, but the next moment the picture is shattered into many pieces and the individual divides into different persons holding the ties of different spheres of life by his hands, by his teeth, and by his grimaces. When seen up close, the individual, bringing together in various ways all the connections he has in life, becomes a blur." Little wonder that the most acute inner problem of modern man is identity.

Roles in human societies are fundamentally different from the castes of social insects. The members of human societies sometimes cooperate closely in insectan fashion, but more frequently they compete for the limited resources allocated to their role-sector. The best and most entrepreneurial of the role-actors usually gain a disproportionate share of the rewards, while the least successful are displaced to other, less desirable positions. In addition, individuals attempt to move to higher socioeconomic positions by changing roles. Competition between classes also occurs, and in great moments of history it has proved to be a determinant of societal change.

A key question of human biology is whether there exists a genetic predisposition to enter certain classes and to play certain roles. Circumstances can be easily conceived in which such genetic differentiation might occur. The heritability of at least some parameters of intelligence and emotive traits is sufficient to respond to a moderate amount of disruptive selection. Dahlberg (1947) showed that if a single gene appears that is responsible for success and an upward shift in status, it can be rapidly concentrated in the uppermost socioeconomic classes. Suppose, for example, there are two classes, each beginning with only a 1 percent frequency of the homozygotes of the upward-mobile gene. Suppose further that 50 percent of the homozygotes in the lower class are transferred upward in each generation. Then in only ten generations, depending on the relative sizes of the groups, the upper class will be comprised of as many as 20 percent homozygotes or more and the lower class of as few as 0.5 percent or less. Using a similar argument, Herrnstein (1971b) proposed that as environmental opportunities become more nearly equal within societies, socioeconomic groups will be defined increasingly by genetically based differences in intelligence.

A strong initial bias toward such stratification is created when one human population conquers and subjugates another, a common enough event in human history. Genetic differences in mental traits, however slight, tend to be preserved by the raising of class barriers, racial and cultural discrimination, and physical ghettos. The geneticist C. D. Darlington (1969), among others, postulated this process to be a prime source of genetic diversity within human societies.

Yet despite the plausibility of the general argument, there is little evidence of any hereditary solidification of status. The castes of India have been in existence for 2000 years, more than enough time for evolutionary divergence, but they differ only slightly in blood type and other measurable anatomical and physiological traits. Powerful forces can be identified that work against the genetic fixation of caste differences. First, cultural evolution is too fluid. Over a period of decades or at most centuries ghettos are replaced, races and subject people are liberated, the conquerors are conquered. Even within relatively stable societies the pathways of upward mobility are numerous. The daughters of lower classes tend to marry upward. Success in commerce or political life can launch a family from virtually any socioeconomic group into the ruling class in a single generation. Furthermore, there are many Dahlberg genes, not just the one postulated for argument in the simplest model. The hereditary factors of human success are strongly polygenic and form a long list, only a few of which have been measured. IQ constitutes only one subset of the components of intelligence. Less tangible but equally important qualities are creativity, entrepreneurship, drive, and mental stamina. Let us assume that the genes contributing to these qualities are scattered over many chromosomes. Assume further that some of the traits are uncorrelated or even negatively correlated. Under these circumstances only the most intense forms of disruptive selection could result in the formation of stable ensembles of genes. A much more likely circumstance is the one that apparently prevails: the maintenance of a large amount of genetic diversity within societies and the loose correlation of some of the genetically determined traits with success. This scrambling process is accelerated by the continuous shift in the fortunes of individual families from one generation to the next.

Even so, the influence of genetic factors toward the assumption of certain *broad* roles cannot be discounted. Consider male homosexuality. The surveys of Kinsey and his coworkers showed that in the 1940's approximately 10 percent of the sexually mature males in the United States were mainly or exclusively homosexual for at least three years prior to being interviewed. Homosexuality is also exhibited by comparably high fractions of the male populations in many if not most other cultures. Kallmann's twin data indicate the probable existence of a genetic predisposition toward the condition. Accordingly, Hutchinson (1959) suggested that the homosexual genes may

possess superior fitness in heterozygous conditions. His reasoning followed lines now standard in the thinking of population genetics. The homosexual state itself results in inferior genetic fitness, because of course homosexual men marry much less frequently and have far fewer children than their unambiguously heterosexual counterparts. The simplest way genes producing such a condition can be maintained in evolution is if they are superior in the heterozygous state, that is, if heterozygotes survive into maturity better, produce more offspring, or both. An interesting alternative hypothesis has been suggested to me by Herman T. Spieth (personal communication) and independently developed by Robert L. Trivers (1974). The homosexual members of primitive societies may have functioned as helpers, either while hunting in company with other men or in more domestic occupations at the dwelling sites. Freed from the special obligations of parental duties, they could have operated with special efficiency in assisting close relatives. Genes favoring homosexuality could then be sustained at a high equilibrium level by kin selection alone. It remains to be said that if such genes really exist they are almost certainly incomplete in penetrance and variable in expressivity, meaning that which bearers of the genes develop the behavioral trait and to what degree depend on the presence or absence of modifier genes and the influence of the environment.

Other basic types might exist, and perhaps the clues lie in full sight. In his study of British nursery children Blurton Jones (1969) distinguished two apparently basic behavioral types. "Verbalists," a small minority, often remained alone, seldom moved about, and almost never joined in rough-and-tumble play. They talked a great deal and spent much of their time looking at books. The other children were "doers." They joined groups, moved around a great deal, and spent much of their time painting and making objects instead of talking. Blurton Jones speculated that the dichotomy results from an early divergence in behavioral development persisting into maturity. Should it prove general it might contribute fundamentally to diversity within cultures. There is no way of knowing whether the divergence is ultimately genetic in origin or triggered entirely by experiential events at an early age.

Communication

All of man's unique social behavior pivots on his use of language, which is itself unique. In any language words are given arbitrary definitions within each culture and ordered according to a grammar that imparts new meaning above and beyond the definitions. The fully symbolic quality of the words and the sophistication of the grammar permit the creation of messages that are potentially infinite in number. Even communication about the system itself is made possible. This is the essential nature of human language. The basic attributes can be broken down, and other features of the transmission process itself can be added, to make a total of 16 design features (C. F. Hockett, reviewed by Thorpe, 1972a). Most of the features are found in at least rudimentary form in some other animal species. But the productivity and richness of human languages cannot be remotely approached even by chimpanzees taught to employ signs in simple sentences. The development of human speech represents a quantum jump in evolution comparable to the assembly of the eucaryotic cell.

Even without words human communication would be the richest known. The study of nonverbal communication has become a flourishing branch of the social sciences. Its codification is made difficult by the auxiliary role so many of the signals play to verbal communication. Categories of these signals are often defined inconsistently, and classifications are rarely congruent (see, for example, Renský, 1966; Crystal, 1969; Lyons, 1972). In Table 27-2 a composite arrangement is presented that I hope is both free of internal contradiction and consistent with current usage. The number of nonvocal signals, including all facial expressions, body postures and movement, and touch, probably number somewhat in excess of 100. Brannigan and Humphries (1972) have made a list of 136, which they believe is close to exhaustive. The number is consistent with the wholly independent estimate of Birdwhistle (1970), who believes that although the human face is capable of as many as 250,000 expressions, less than 100 sets of the expressions comprise distinct, meaningful symbols. Vocal paralanguage, insofar as it can be separated from the prosodic modifications of true speech, has not been cataloged so painstakingly. Grant (1969) recognized 6 distinct sounds, but several times this number would probably be distinguished by a zoologist accustomed to preparing ethograms of other primate species. In summary, all paralinguistic signals taken together almost certainly exceed 150 and may be close to 200. This repertory is larger than that of the majority of other mammals and birds by a factor of three or more, and it exceeds

Table 27-2   The modes of human communication.

I. Verbal Communication (Language): the utterance of words and sentences
II. Non-verbal Communication
  A. *Prosody:* tone, tempo, rhythm, loudness, pacing, and other qualities of voice that modify the meaning of verbal utterances
  B. *Paralanguage:* signals separate from words used to supplement or to modify language
    1. Vocal paralanguage: grunts, giggles, laughs, sobs, cries, and other nonverbal sounds
    2. Nonverbal paralanguage: body posture, motion, and touch (kinesic communication); possibly also chemical communication

slightly the total repertories of both the rhesus monkey and chimpanzee.

Another useful distinction in the analysis of human paralanguage can be made between signals that are prelinguistic, defined as having been in service before the evolutionary origin of true language, and those that are postlinguistic. The postlinguistic signals are most likely to have originated as pure auxiliaries to speech. One approach to the problem is through the phylogenetic analysis of the relevant properties of primate communication. Hooff (1972), for example, has established the homologues of smiling and laughing in facial expressions of the cercopithecoid monkeys and apes, thus classifying these human behaviors among our most primitive and universal signals.

Human language, as Marler (1965) argued, probably stemmed from richly graded vocal signals not unlike those employed by the rhesus monkey and chimpanzee, as opposed to the more discrete sounds characterizing the repertories of some of the lower primates. Human infants can utter a wide variety of vocalizations resembling those of macaques, baboons, and chimpanzees. But very early in their development they convert to the peculiar sounds of human speech. Multiple plosives, fricatives, nasals, vowels, and other sounds are combined to create the 40 or so basic phonemes. The human mouth and upper respiratory tract have been strongly modified to permit this vocal competence (see Figure 27-2). The crucial changes are associated with man's upright posture, which may have provided the initial but still incomplete impetus toward the present modification. With the face directed fully forward, the mouth gave way to the upper pharyngeal space at a 90-degree angle. This configuration helped to push the rear of the tongue back until it formed part of the forward wall of the upper pharyngeal tract. Simultaneously the pharyngeal space and the epiglottis were both considerably lengthened.

These two principal changes, the shift in tongue position and lengthening of the pharyngeal tract, were responsible for the versatility in sound production. When air is forced upward through the vocal cords it generates a buzzing noise that can be varied in intensity and duration but not in the all-important qualities of tone that produce phoneme differentiation. The latter effect is achieved as the air passes up through the pharyngeal tract and mouth cavity and out through the mouth. These structures together form an air tube which, like any cylinder, serves as a resonator. When its position and shape are altered, the tube emphasizes different combinations of frequencies emanating from the vocal cords. The result, illustrated in Figure 27-2, is the sounds we distinguish as phonemes (see also Lenneberg, 1967, and Denes and Pinson, 1973).

However, the great advance in language acquisition did not come from the ability to form many sounds. After all, it is theoretically possible for a highly intelligent being to speak only a *single* word and still communicate rapidly. It need only be programmed like a

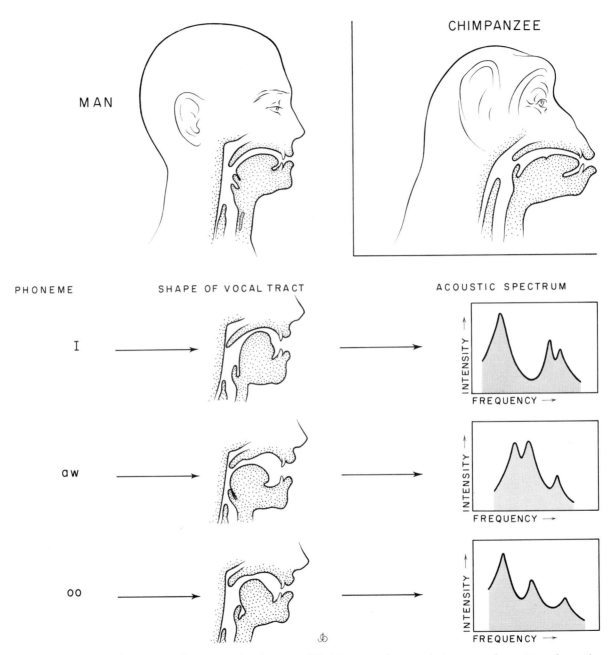

Figure 27-2 The human vocal apparatus has been modified in a way that greatly increases the variety of sounds that can be produced. The versatility was an essential accompaniment of the evolution of human speech. The upper diagrams show the ways in which man differs from the chimpanzee and other nonhuman primates: the angulation between the mouth and the upper respiratory tract is increased, the pharyngeal space is lengthened, and the back half of the tongue has come to form the front wall of the long tract above the vocal cords. The lower diagrams illustrate how movement of the tongue changes the shape of the air space to generate different sounds. (Modified from Howells, 1973, and Denes and Pinson, 1973.)

digital computer. Variation in loudness, duration, and pacing could be added to increase the transmission rate still more. It will be recalled that a single chemical substance, if modulated perfectly under ideal conditions, can generate up to 10,000 bits per second, far in excess of the capacity of human speech. Human languages gain their power instead from syntax, the dependence of meaning on the linear ordering of words. Each language possesses a grammar, the set of rules governing syntax. To truly understand the nature and origin of grammar would be to understand a great deal about the construction of the human mind. It is possible to distinguish three competing models that attempt to describe the known rules:

First Hypothesis: *Probabilistic left-to-right model.* The explanation favored by extreme behavioristic psychologists is that the occurrence of a word is Markovian, meaning that its probability is determined by the immediately preceding word or string of words. The developing child learns which words to link together in each appropriate circumstance.

Second Hypothesis: *Learned deep-structure model.* There exist a limited number of formal principles by which phrases of words are combined and juxtaposed to create various meanings. The child more or less unconsciously learns the deep structure of his own culture. Although the principles are finite in number, the sentences that can be generated from them are infinite in number. Animals cannot speak simply because they lack the necessary level of cognitive or intellectual ability, not because of the absence of any special "language faculty."

Third Hypothesis: *Innate deep-structure model.* The formal principles exist as suggested in hypothesis number two, but they are partially or wholly genetic. In other words, at least some of the principles emerge by maturation in an invariant manner. A corollary of this proposition is that much of the deep structure of grammar is widespread if not universal in mankind, notwithstanding the profound differences in surface structure and word meaning that exist between languages. A second corollary is that animals cannot speak because they lack this inborn language faculty, which is a qualitatively unique human property and not simply an outcome of man's quantitatively superior intelligence. The innate deep-structure model is the one that has come to be associated most prominently with the name of Noam Chomsky, and appears to be currently favored by most psycholinguists.

The probabilistic left-to-right model has already been eliminated, at least in its extreme version. The number of transitional probabilities a child would have to learn in order to compute in a language such as English is enormous, and there is simply not enough time in childhood to master them all (Miller, Galanter, and Pribram, 1960). Grammatical rules are actually learned very rapidly and in a predict-

able sequence, with the child passing through forms of construction that anticipate the adult form while differing significantly from it (Brown, 1973). This kind of ontogeny is typical of the maturation of innate components of animal behavior. Nevertheless, the similarity cannot be taken as conclusive evidence of a genetic program general to humanity.

The ultimate resolution of the problem, as Roger Brown and other developmental psycholinguists have stressed, cannot be achieved until deep grammar itself has been securely characterized. This is a relatively new area of investigation, scarcely dating beyond Chomsky's *Syntactic Structures* (1957). From the beginning it has been marked by a complicated, rapidly shifting argumentation. The basic ideas have been presented in recent reviews by Slobin (1971) and Chomsky (1972). Here it will suffice to define the main processes recognized by the new linguistic analysis. *Phrase structure grammar*, which is exemplified in Figure 27-3, consists of the rules by which sentences are built up in a hierarchical manner. Phrases can be thought of as modules that are substituted for other, equivalent modules or added *de novo* into sentences to change meanings. These elements cannot be split and the parts interchanged without creating serious difficulties. In the example "The boy hit the ball," "the ball" is intuitively such a unit. It can be easily taken out and replaced with some other phrase such as "the shuttlecock" or simply the word "it." The combination "hit the" is not such a unit. Despite the fact that the two words are juxtaposed, they cannot be easily replaced without creating difficulties for the construction of the entire remainder of the sentence. By observing the rules we all know subconsciously, the sentence can be expanded by the insertion of appropriately selected phrases: *After taking his position,* the *little* boy *swung twice and finally* hit the ball *and ran to first base.*

In short, phrase structure grammar decrees the ways in which phrases can be formed. It generates what has been called the deep structure of the word strings as opposed to the surface structure, or the mere order in which the individual words appear. But of course the sequences in which phrases and terminal words appear are crucial to the meaning of the sentence. "The boy hit the ball" is very different from "What did the boy hit?" even though the deep (phrase) structure is similar. The rules by which the deep structures are converted into surface structures by the assembling of phrases are called *transformational grammar.* A transformation is an operation that converts one phrase structure into another. Among the most basic operations are substitutions ("what" for "the ball"), displacement (placing "what" before the verb), and permutation (switching the positions of related words).

The psycholinguists have described, for English, both phrase structure and transformational grammar. The evidence does not appear

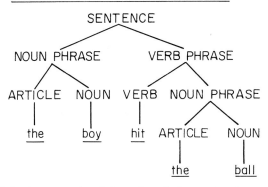

RULES OF PHRASE STRUCTURE GRAMMAR

1. SENTENCE ————————→ NOUN PHRASE + VERB PHRASE

2. NOUN PHRASE ————→ ARTICLE + NOUN

3. VERB PHRASE ————→ VERB + NOUN PHRASE

4. ARTICLE ————————→ the, a

5. NOUN ———————————→ boy, girl, ball

6. VERB ———————————→ hit

TREE OF PHRASE STRUCTURES

Figure 27-3 An example of the rules of phrase structure grammar in the English language. The simple sentence "The boy hit the ball" is seen to consist of a hierarchy of phrases. At each level one phrase can be substituted for another of equivalent composition, but the phrases cannot be split and their elements interchanged. (Based on Slobin, 1971.)

to be adequate, however, to choose between hypotheses two and three, in other words to decide whether the grammars are innately programmed or whether they are learned. The basic operations of transformation occur in all known human languages. However, this observation by itself does not establish that the precise rules of transformation are the same.

Is there a universal grammar? This question is difficult to answer because most attempts to generalize the rules of deep grammar have been based on the semantic content of one particular language. Students of the subject seldom confront the problem as if it were genuinely scientific, in a way that would reveal how concrete and soluble it might be. In fact, natural scientists are easily frustrated by the diffuse, oblique quality of much of the psycholinguistic literature, which often seems unconcerned with the usual canons of proposition and evidence. The reason is that many of the writers, including Chomsky, are structuralists in the tradition of Lévi-Strauss and Piaget.

They approach the subject with the implicit world view that the processes of the human mind are indeed structured, and also discrete, enumerable, and evolutionarily unique with no great need to be referred to the formulations of other scientific disciplines. The analysis is nontheoretical in the sense that it fails to argue from postulates that can be tested and extended empirically. Some psychologists, including Roger Brown and his associates and Fodor and Garrett (1966), have adduced testable propositions and pursued them with mixed results, but the trail of speculation on deep grammar has not been easy to follow even for these skillful experimentalists.

Like poet naturalists, the structuralists celebrate idiosyncratic personal visions. They argue from hidden premises, relying largely on metaphor and exemplification, and with little regard for the method of multiple competing hypotheses. Clearly, this discipline, one of the most important in all of science, is ripe for the application of rigorous theory and properly meshed experimental investigation.

A key question that the new linguistics may never answer is when human language originated. Did speech appear with the first use of stone tools and the construction of shelters by the *Australopithecus* man-apes, over two million years ago? Or did it await the emergence of fully modern *Homo sapiens*, perhaps even the development of religious rites in the past 100,000 years? Lieberman (1968) believes that the date was relatively recent. He interprets the Makapan *Australopithecus* restored by Dart to fall close to the chimpanzee in the form of its palate and pharyngeal tract. If he is right, this early hominid might not have been able to articulate the sounds of human speech. The same conclusion has been drawn with respect to the anatomy and vocal capacity of the Neanderthal man (Lieberman et al., 1972), which if true places the origin of language in the latest stages of speciation in the genus *Homo*. Other theoretical aspects of the evolutionary origin of human speech have been discussed by Jane Hill (1972) and I. G. Mattingly (1972). Lenneberg (1971) has hypothesized that the capacity for mathematical reasoning originated as a slight modification of linguistic ability.

Culture, Ritual, and Religion

The rudiments of culture are possessed by higher primates other than man, including the Japanese monkey and chimpanzee (Chapter 7), but only in man has culture thoroughly infiltrated virtually every aspect of life. Ethnographic detail is genetically underprescribed, resulting in great amounts of diversity among societies. Underprescription does not mean that culture has been freed from the genes. What has evolved is the capacity for culture, indeed the overwhelming tendency to develop one culture or another. Robin Fox (1971) put the argument in the following form. If the proverbial experiments of the pharaoh Psammetichos and James IV of Scotland had worked, and children reared in isolation somehow survived in good health,

I do not doubt that they *could* speak and that, theoretically, given time, they or their offspring would invent and develop a language despite their never having been taught one. Furthermore, this language, although totally different from any known to us, would be analyzable by linguists on the same basis as other languages and translatable into all known languages. But I would push this further. If our new Adam and Eve could survive and breed—still in total isolation from any cultural influences— then eventually they would produce a society which would have laws about property, rules about incest and marriage, customs of taboo and avoidance, methods of settling disputes with a minimum of bloodshed, beliefs about the supernatural and practices relating to it, a system of social status and methods of indicating it, initiation ceremonies for young men, courtship practices including the adornment of females, systems of symbolic body adornment generally, certain activities and associations set aside for men from which women were excluded, gambling of some kind, a tool- and weapon-making industry, myths and legends, dancing, adultery, and various doses of homicide, suicide, homosexuality, schizophrenia, psychosis and neuroses, and various practitioners to take advantage of or cure these, depending on how they are viewed.

Culture, including the more resplendent manifestations of ritual and religion, can be interpreted as a hierarchical system of environmental tracking devices. In Chapter 7 the totality of biological responses, from millisecond-quick biochemical reactions to gene substitutions requiring generations, was described as such a system. At that time culture was placed within the scheme at the slow end of the time scale. Now this conception can be extended. To the extent that the specific details of culture are nongenetic, they can be decoupled from the biological system and arrayed beside it as an auxiliary system. The span of the purely cultural tracking system parallels much of the slower segment of the biological tracking system, ranging from days to generations. Among the fastest cultural responses in industrial civilizations are fashions in dress and speech. Somewhat slower are political ideology and social attitudes toward other nations, while the slowest of all include incest taboos and the belief or disbelief in particular high gods. It is useful to hypothesize that cultural details are for the most part adaptive in a Darwinian sense, even though some may operate indirectly through enhanced group survival (Washburn and Howell, 1960; Masters, 1970). A second proposition worth considering, to make the biological analogy complete, is that the rate of change in a particular set of cultural behaviors reflects the rate of change in the environmental features to which the behaviors are keyed.

Slowly changing forms of culture tend to be encapsulated in ritual. Some social scientists have drawn an analogy between human ceremonies and the displays of animal communication. This is not correct. Most animal displays are discrete signals conveying limited meaning. They are commensurate with the postures, facial expressions, and elementary sounds of human paralanguage. A few animal displays, such as the most complex forms of sexual advertisement and nest changing in birds, are so impressively elaborate that they have occasionally been termed ceremonies by zoologists. But even here the comparison is misleading. Most human rituals have more than just an immediate signal value. As Durkheim stressed, they not only label but reaffirm and rejuvenate the moral values of the community.

The sacred rituals are the most distinctively human. Their most elementary forms are concerned with magic, the active attempt to manipulate nature and the gods. Upper Paleolithic art from the caves of Western Europe shows a preoccupation with game animals. There are many scenes showing spears and arrows embedded in the bodies of the prey. Other drawings depict men dancing in animal disguises or standing with heads bowed in front of animals. Probably the function of the drawings was sympathetic magic, based on the quite logical notion that what is done with an image will come to pass with the real thing. This anticipatory action is comparable to the intention movements of animals, which in the course of evolution have often been ritualized into communicative signals. The waggle dance of the honeybee, it will be recalled, is a miniaturized rehearsal of the flight from the nest to the food. Primitive man might have understood the meaning of such complex animal behavior easily. Magic was, and still is in some societies, practiced by special people variously called shamans, sorcerers, or medicine men. They alone were believed to have the secret knowledge and power to deal effectively with the supernatural, and as such their influence sometimes exceeded that of the tribal headmen.

Formal religion *sensu stricto* has many elements of magic but is focused on deeper, more tribally oriented beliefs. Its rites celebrate the creation myths, propitiate the gods, and resanctify the tribal moral codes. Instead of a shaman controlling physical power, there is a priest who communes with the gods and curries their favor through obeisance, sacrifice, and the proffered evidences of tribal good behavior. In more complex societies, polity and religion have always blended naturally. Power belonged to kings by divine right, but high priests often ruled over kings by virtue of the higher rank of the gods.

It is a reasonable hypothesis that magic and totemism constituted direct adaptations to the environment and preceded formal religion in social evolution. Sacred traditions occur almost universally in human societies. So do myths that explain the origin of man or at the very least the relation of the tribe to the rest of the world. But belief in high gods is not universal. Among 81 hunter-gatherer societies surveyed by Whiting (1968), only 28, or 35 percent, included high gods in their sacred traditions. The concept of an active, moral God who created the world is even less widespread. Furthermore, this concept most commonly arises with a pastoral way of life. The greater

Table 27-3 The religious beliefs of 66 agrarian societies, partitioned according to the percentage of subsistence derived from herding. (From *Human Societies* by G. and Jean Lenski. Copyright © 1970 by McGraw-Hill Book Company. Used with permission.)

Percentage of subsistence from herding	Percentage of societies believing in an active, moral creator God	Number of societies
36–45	92	13
26–35	82	28
16–25	40	20
6–15	20	5

the dependence on herding, the more likely the belief in a shepherd god of the Judaeo-Christian model (see Table 27-3). In other kinds of societies the belief occurs in 10 percent or less of the cases. Also, the God of monotheistic religions is always male. This strong patriarchal tendency has several cultural sources (Lenski, 1970). Pastoral societies are highly mobile, tightly organized, and often militant, all features that tip the balance toward male authority. It is also significant that herding, the main economic base, is primarily the responsibility of men. Because the Hebrews were originally a herding people, the Bible describes God as a shepherd and the chosen people as his sheep. Islam, one of the strictest of all monotheistic faiths, grew to early power among the herding people of the Arabian peninsula. The intimate relation of the shepherd to his flock apparently provides a microcosm which stimulates deeper questioning about the relation of man to the powers that control him.

An increasingly sophisticated anthropology has not given reason to doubt Max Weber's conclusion that more elementary religions seek the supernatural for the purely mundane rewards of long life, abundant land and food, the avoidance of physical catastrophes, and the defeat of enemies. A form of group selection also operates in the competition between sects. Those that gain adherents survive; those that cannot, fail. Consequently, religions, like other human institutions, evolve so as to further the welfare of their practitioners. Because this demographic benefit applies to the group as a whole, it can be gained in part by altruism and exploitation, with certain segments profiting at the expense of others. Alternatively, it can arise as the sum of generally increased individual fitnesses. The resulting distinction in social terms is between the more oppressive and the more beneficent religions. All religions are probably oppressive to some degree, especially when they are promoted by chiefdoms and states. The tendency is intensified when societies compete, since religion can be effectively harnessed to the purposes of warfare and economic exploitation.

The enduring paradox of religion is that so much of its substance is demonstrably false, yet it remains a driving force in all societies. Men would rather believe than know, have the void as purpose, as Nietzsche said, than be void of purpose. At the turn of the century Durkheim rejected the notion that such force could really be extracted from "a tissue of illusions." And since that time social scientists have sought the psychological Rosetta stone that might clarify the deeper truths of religious reasoning. In a penetrating analysis of this subject, Rappaport (1971) proposed that virtually all forms of sacred rites serve the purposes of communication. In addition to institutionalizing the moral values of the community, the ceremonies can offer information on the strength and wealth of tribes and families. Among the Maring of New Guinea there are no chiefs or other leaders who command allegiance in war. A group gives a ritual dance, and individual men indicate their willingness to give military support by whether they attend the dance or not. The strength of the consortium can then be precisely determined by a head count. In more advanced societies military parades, embellished by the paraphernalia and rituals of the state religion, serve the same purpose. The famous potlatch ceremonies of the Northwest Coast Indians enable individuals to advertise their wealth by the amount of goods they give away. Rituals also regularize relationships in which there would otherwise be ambiguity and wasteful imprecision. The best examples of this mode of communication are the *rites de passage*. As a boy matures his transition from child to man is very gradual in a biological and psychological sense. There will be times when he behaves like a child when an adult response would have been more appropriate, and vice versa. The society has difficulty in classifying him one way or the other. The *rite de passage* eliminates this ambiguity by arbitrarily changing the classification from a continuous gradient into a dichotomy. It also serves to cement the ties of the young person to the adult group that accepts him.

To sanctify a procedure or a statement is to certify it as beyond question and imply punishment for anyone who dares to contradict it. So removed is the sacred from the profane in everyday life that simply to repeat it in the wrong circumstance is a transgression. This extreme form of certification, the heart of all religions, is granted to the practices and dogmas that serve the most vital interests of the group. The individual is prepared by the sacred rituals for supreme effort and self-sacrifice. Overwhelmed by shibboleths, special costumes, and the sacred dancing and music so accurately keyed to his emotive centers he has a "religious experience." He is ready to reassert allegiance to his tribe and family, perform charities, consecrate his life, leave for the hunt, join the battle, die for God and country. *Deus vult* was the rallying cry of the First Crusade. God wills it, but the summed Darwinian fitness of the tribe was the ultimate if unrecognized beneficiary.

It was Henri Bergson who first identified a second force leading to the formalization of morality and religion. The extreme plasticity of human social behavior is both a great strength and a real danger. If each family worked out rules of behavior on its own, the result would be an intolerable amount of tradition drift and growing chaos. To counteract selfish behavior and the "dissolving power" of high intelligence, each society must codify itself. Within broad limits virtually any set of conventions works better than none at all. Because arbitrary codes work, organizations tend to be inefficient and marred by unnecessary inequities. As Rappaport succinctly expressed it, "Sanctification transforms the arbitrary into the necessary, and regulatory mechanisms which are arbitrary are likely to be sanctified." The process engenders criticism, and in the more literate and self-conscious societies visionaries and revolutionaries set out to change the system. Reform meets repression, because to the extent that the rules have been sanctified and mythologized, the majority of the people regard them as beyond question, and disagreement is defined as blasphemy.

This leads us to the essentially biological question of the evolution of indoctrinability (Campbell, 1972). Human beings are absurdly easy to indoctrinate—they *seek* it. If we assume for argument that indoctrinability evolves, at what level does natural selection take place? One extreme possibility is that the group is the unit of selection. When conformity becomes too weak, groups become extinct. In this version selfish, individualistic members gain the upper hand and multiply at the expense of others. But their rising prevalence accelerates the vulnerability of the society and hastens its extinction. Societies containing higher frequencies of conformer genes replace those that disappear, thus raising the overall frequency of the genes in the metapopulation of societies. The spread of the genes will occur more rapidly if the metapopulation (for example, a tribal complex) is simultaneously enlarging its range. Formal models of the process, presented in Chapter 5, show that if the rate of societal extinction is high enough relative to the intensity of the counteracting individual selection, the altruistic genes can rise to moderately high levels. The genes might be of the kind that favors indoctrinability even at the expense of the individuals who submit. For example, the willingness to risk death in battle can favor group survival at the expense of the genes that permitted the fatal military discipline. The group-selection hypothesis is sufficient to account for the evolution of indoctrinability.

The competing, individual-level hypothesis is equally sufficient. It states that the ability of individuals to conform permits them to enjoy the benefits of membership with a minimum of energy expenditure and risk. Although their selfish rivals may gain a momentary advantage, it is lost in the long run through ostracism and repression. The conformists perform altruistic acts, perhaps even to the extent of risking their lives, not because of self-denying genes selected at

the group level but because the group is occasionally able to take advantage of the indoctrinability which on other occasions is favorable to the individual.

The two hypotheses are not mutually exclusive. Group and individual selection can be reinforcing. If war requires spartan virtues and eliminates some of the warriors, victory can more than adequately compensate the survivors in land, power, and the opportunity to reproduce. The average individual will win the inclusive fitness game, making the gamble profitable, because the summed efforts of the participants give the average member a more than compensatory edge.

Ethics

Scientists and humanists should consider together the possibility that the time has come for ethics to be removed temporarily from the hands of the philosophers and biologicized. The subject at present consists of several oddly disjunct conceptualizations. The first is *ethical intuitionism*, the belief that the mind has a direct awareness of true right and wrong that it can formalize by logic and translate into rules of social action. The purest guiding precept of secular Western thought has been the theory of the social contract as formulated by Locke, Rousseau, and Kant. In our time the precept has been rewoven into a solid philosophical system by John Rawls (1971). His imperative is that justice should be not merely integral to a system of government but rather the object of the original contract. The principles called by Rawls "justice as fairness" are those which free and rational persons would choose if they were beginning an association from a position of equal advantage and wished to define the fundamental rules of the association. In judging the appropriateness of subsequent laws and behavior, it would be necessary to test their conformity to the unchallengeable starting position.

The Achilles heel of the intuitionist position is that it relies on the emotive judgment of the brain as though that organ must be treated as a black box. While few will disagree that justice as fairness is an ideal state for disembodied spirits, the conception is in no way explanatory or predictive with reference to human beings. Consequently, it does not consider the ultimate ecological or genetic consequences of the rigorous prosecution of its conclusions. Perhaps explanation and prediction will not be needed for the millennium. But this is unlikely—the human genotype and the ecosystem in which it evolved were fashioned out of extreme unfairness. In either case the full exploration of the neural machinery of ethical judgment is desirable and already in progress. One such effort, constituting the second mode of conceptualization, can be called *ethical behaviorism.* Its basic proposition, which has been expanded most fully by J. F. Scott (1971), holds that moral commitment is entirely learned, with operant conditioning being the dominant mechanism. In other words, children simply internalize the behavioral norms of the society. Opposing this theory is the *developmental-genetic conception* of ethical behavior. The best-documented version has been provided by Lawrence Kohlberg (1969). Kohlberg's viewpoint is structuralist and specifically Piagetian, and therefore not yet related to the remainder of biology. Piaget has used the expression "genetic epistemology" and Kohlberg "cognitive-developmental" to label the general concept. However, the results will eventually become incorporated into a broadened developmental biology and genetics. Kohlberg's method is to record and classify the verbal responses of children to moral problems. He has delineated six sequential stages of ethical reasoning through which an individual may progress as part of his mental maturation. The child moves from a primary dependence on external controls and sanctions to an increasingly sophisticated set of internalized standards (see Table 27-4). The analysis has not yet been directed to the question of plasticity in the basic rules. Intracultural variance has not been measured, and heritability therefore not assessed. The

difference between ethical behaviorism and the current version of developmental-genetic analysis is that the former postulates a mechanism (operant conditioning) without evidence and the latter presents evidence without postulating a mechanism. No great conceptual difficulty underlies this disparity. The study of moral development is only a more complicated and less tractable version of the genetic variance problem (see Chapters 2 and 7). With the accretion of data the two approaches can be expected to merge to form a recognizable exercise in behavioral genetics.

Even if the problem were solved tomorrow, however, an important piece would still be missing. This is the *genetic evolution of ethics*. In the first chapter of this book I argued that ethical philosophers intuit the deontological canons of morality by consulting the emotive centers of their own hypothalamic-limbic system. This is also true of the developmentalists, even when they are being their most severely objective. Only by interpreting the activity of the emotive centers as a biological adaptation can the meaning of the canons be deciphered. Some of the activity is likely to be outdated, a relic of adjustment to the most primitive form of tribal organization. Some of it may prove to be *in statu nascendi*, constituting new and quickly changing adaptations to agrarian and urban life. The resulting confusion will be reinforced by other factors. To the extent that unilaterally altruistic genes have been established in the population by group selection, they will be opposed by allelomorphs favored by individual selection. The conflict of impulses under their various controls is likely to be widespread in the population, since current theory predicts that the genes will be at best maintained in a state of balanced polymorphism (Chapter 5). Moral ambivalency will be further intensified by the circumstance that a schedule of sex- and age-dependent ethics can impart higher genetic fitness than a single moral code which is applied uniformly to all sex-age groups. The argument for this statement is the special case of the Gadgil-Bossert distribution in which the contributions of social interactions to survivorship and fertility schedules are specified (see Chapter 4). Some of the differences in the Kohlberg stages could be explained in this manner. For example, it should be of selective advantage for young children to be self-centered and relatively disinclined to perform altruistic acts based on personal principle. Similarly, adolescents should be more tightly bound by age-peer bonds within their own sex and hence unusually sensitive to peer approval. The reason is that at this time greater advantage accrues to the formation of alliances and rise in status than later, when sexual and parental morality become the paramount determinants of fitness. Genetically programmed sexual and parent-offspring conflict of the kind predicted by the Trivers models (Chapters 15 and 16) are also likely to promote age differences in the kinds and degrees of moral commitment. Finally, the moral standards of individuals during early phases of colony growth should

Table 27-4 The classification of moral judgment into levels and stages of development. (Based on Kohlberg, 1969.)

Level	Basis of moral judgment	Stage of development
I	Moral value is defined by punishment and reward	1. Obedience to rules and authority to avoid punishment 2. Conformity to obtain rewards and to exchange favors
II	Moral value resides in filling the correct roles, in maintaining order and meeting the expectations of others	3. Good-boy orientation: conformity to avoid dislike and rejection by others 4. Duty orientation: conformity to avoid censure by authority, disruption of order, and resulting guilt
III	Moral value resides in conformity to shared standards, rights, and duties	5. Legalistic orientation: recognition of the value of contracts, some arbitrariness in rule formation to maintain the common good 6. Conscience or principle orientation: primary allegiance to principles of choice, which can overrule law in cases where the law is judged to do more harm than good

differ in many details from those of individuals at demographic equilibrium or during episodes of overpopulation. Metapopulations subject to high levels of r extinction will tend to diverge genetically from other kinds of populations in ethical behavior (Chapter 5).

If there is any truth to this theory of innate moral pluralism, the requirement for an evolutionary approach to ethics is self-evident. It should also be clear that no single set of moral standards can be applied to all human populations, let alone all sex-age classes within each population. To impose a uniform code is therefore to create complex, intractable moral dilemmas—these, of course, are the current condition of mankind.

Esthetics

Artistic impulses are by no means limited to man. In 1962, when Desmond Morris reviewed the subject in *The Biology of Art*, 32 individual nonhuman primates had produced drawings and paintings in captivity. Twenty-three were chimpanzees, 2 were gorillas, 3 were orang-utans, and 4 were capuchin monkeys. None received special training or anything more than access to the necessary equipment. In fact, attempts to guide the efforts of the animals by inducing imitation were always unsuccessful. The drive to use the painting and drawing equipment was powerful, requiring no reinforcement from the human observers. Both young and old animals became so engrossed with the activity that they preferred it to being fed and sometimes threw temper tantrums when stopped. Two of the chimpanzees studied extensively were highly productive. "Alpha" produced over 200 pictures, while the famous "Congo," who deserves to be called the Picasso of the great apes, was responsible for nearly 400. Although most of the efforts consisted of scribbling, the patterns were far from random. Lines and smudges were spread over a blank page outward from a centrally located figure. When a drawing was started on one side of a blank page the chimpanzee usually shifted to the opposite side to offset it. With time the calligraphy became bolder, starting with simple lines and progressing to more complicated multiple scribbles. Congo's patterns progressed along approximately the same developmental path as those of very young human children, yielding fan-shaped diagrams and even complete circles. Other chimpanzees drew crosses.

The artistic activity of chimpanzees may well be a special manifestation of their tool-using behavior. Members of the species display a total of about ten techniques, all of which require manual skill. Probably all are improved through practice, while at least a few are passed as traditions from one generation to the next. The chimpanzees have a considerable facility for inventing new techniques, such as the use of sticks to pull objects through cage bars and to pry open boxes. Thus the tendency to manipulate objects and to explore their uses appears to have an adaptive advantage for chimpanzees.

The same reasoning applies a fortiori to the origin of art in man. As Washburn (1970) pointed out, human beings have been hunter-gatherers for over 99 percent of their history, during which time each man made his own tools. The appraisal of form and skill in execution were necessary for survival, and they probably brought social approval as well. Both forms of success paid off in greater genetic fitness. If the chimpanzee Congo could reach the stage of elementary diagrams, it is not too hard to imagine primitive man progressing to representational figures. Once that stage was reached, the transition to the use of art in sympathetic magic and ritual must have followed quickly. Art might then have played a reciprocally reinforcing role in the development of culture and mental capacity. In the end, writing emerged as the idiographic representation of language.

Music of a kind is also produced by some animals. Human beings consider the elaborate courtship and territorial songs of birds to be beautiful, and probably ultimately for the same reasons they are of use to the birds. With clarity and precision they identify the species, the physiological condition, and the mental set of the singer. Richness of information and precise transmission of mood are no less the standards of excellence in human music. Singing and dancing serve to draw groups together, direct the emotions of the people, and prepare them for joint action. The carnival displays of chimpanzees described in earlier chapters are remarkably like human celebrations in this respect. The apes run, leap, pound the trunks of trees in drumming motions, and call loudly back and forth. These actions serve at least in part to assemble groups at common feeding grounds. They may resemble the ceremonies of earliest man. Nevertheless, fundamental differences appeared in subsequent human evolution. Human music has been liberated from iconic representation in the same way that true language has departed from the elementary ritualization characterizing the communication of animals. Music has the capacity for unlimited and arbitrary symbolization, and it employs rules of phrasing and order that serve the same function as syntax.

Territoriality and Tribalism

Anthropologists often discount territorial behavior as a general human attribute. This happens when the narrowest concept of the phenomenon is borrowed from zoology—the "stickleback model," in which residents meet along fixed boundaries to threaten and drive one another back. But earlier, in Chapter 12, I showed why it is necessary to define territory more broadly, as any area occupied more or less exclusively by an animal or group of animals through overt defense or advertisement. The techniques of repulsion can be as explicit as a precipitous all-out attack or as subtle as the deposit of a chemical secretion at a scent post. Of equal importance, animals respond to their neighbors in a highly variable manner. Each species is characterized by its own particular behavioral scale. In extreme cases the scale may run from open hostility, say, during the breeding season or when the population density is high, to oblique forms of advertisement or no territorial behavior at all. One seeks to characterize the behavioral scale of the species and to identify the parameters that move individual animals up and down it.

If these qualifications are accepted, it is reasonable to conclude that territoriality is a general trait of hunter-gatherer societies. In a perceptive review of the evidence, Edwin Wilmsen (1973) found that these relatively primitive societies do not differ basically in their strategy of land tenure from many mammalian species. Systematic overt aggression has been reported in a minority of hunter-gatherer peoples, for example the Chippewa, Sioux, and Washo of North America and the Murngin and Tiwi of Australia. Spacing and demographic balance were implemented by raiding parties, murder, and threats of witchcraft. The Washo of Nevada actively defended nuclear portions of their home ranges, within which they maintained their winter residences. Subtler and less direct forms of interaction can have the same result. The !Kung Bushmen of the Nyae Nyae area refer to themselves as "perfect" or "clean" and other !Kung people as "strange" murderers who use deadly poisons.

Human territorial behavior is sometimes particularized in ways that are obviously functional. As recently as 1930 Bushmen of the Dobe area in southwestern Africa recognized the principle of exclusive family land-holdings during the wet season. The rights extended only to the gathering of vegetable foods; other bands were allowed to hunt animals through the area (R. B. Lee in Wilmsen, 1973). Other hunter-gatherer peoples appear to have followed the same dual principle: more or less exclusive use by tribes or families of the richest sources of vegetable foods, opposed to broadly overlapping hunting ranges. Thus the original suggestion of Bartholomew and Birdsell (1953) that *Australopithecus* and the primitive *Homo* were territorial remains a viable hypothesis. Moreover, in obedience to the rule of ecological efficiency, the home ranges and territories were probably large and population density correspondingly low. This rule, it will be recalled, states that when a diet consists of animal food, roughly ten times as much area is needed to gain the same amount of energy yield as when the diet consists of plant food. Modern hunter-gatherer bands containing about 25 individuals commonly occupy between 1000 and 3000 square kilometers. This area is comparable to the home range of a wolf pack but as much as a hundred times greater than that of a troop of gorillas, which are exclusively vegetarian.

Hans Kummer (1971), reasoning from an assumption of territoriality, provided an important additional insight about human behavior. Spacing between groups is elementary in nature and can be achieved by a relatively small number of simple aggressive techniques. Spacing and dominance within groups is vastly more complex, being tied to all the remainder of the social repertory. Part of man's problem is that his intergroup responses are still crude and primitive, and

inadequate for the extended extraterritorial relationships that civilization has thrust upon him. The unhappy result is what Garrett Hardin (1972) has defined as tribalism in the modern sense:

Any group of people that perceives itself as a distinct group, and which is so perceived by the outside world, may be called a tribe. The group might be a race, as ordinarily defined, but it need not be; it can just as well be a religious sect, a political group, or an occupational group. The essential characteristic of a tribe is that it should follow a double standard of morality—one kind of behavior for in-group relations, another for out-group.

It is one of the unfortunate and inescapable characteristics of tribalism that it eventually evokes counter-tribalism (or, to use a different figure of speech, it "polarizes" society).

Fearful of the hostile groups around them, the "tribe" refuses to concede to the common good. It is less likely to voluntarily curb its own population growth. Like the Sinhalese and Tamils of Ceylon, competitors may even race to outbreed each other. Resources are sequestered. Justice and liberty decline. Increases in real and imagined threats congeal the sense of group identity and mobilize the tribal members. Xenophobia becomes a political virtue. The treatment of nonconformists within the group grows harsher. History is replete with the escalation of this process to the point that the society breaks down or goes to war. No nation has been completely immune.

Early Social Evolution

Modern man can be said to have been launched by a two-stage acceleration in mental evolution. The first occurred during the transition from a larger arboreal primate to the first man-apes (*Australopithecus*). If the primitive hominid *Ramapithecus* is in the direct line of ancestry, as current opinion holds, the change may have required as much as ten million years. *Australopithecus* was present five million years ago, and by three million years B.P. it had speciated into several forms, including possibly the first primitive *Homo* (Tobias, 1973). As shown in Figure 27-1, the evolution of these intermediate hominids was marked by an accelerating increase in brain capacity. Simultaneously, erect posture and a striding, bipedal locomotion were perfected, and the hands were molded to acquire the precision grip. These early men undoubtedly used tools to a much greater extent than do modern chimpanzees. Crude stone implements were made by chipping, and rocks were pulled together to form what appear to be the foundations of shelters.

The second, much more rapid phase of acceleration began about 100,000 years ago. It consisted primarily of cultural evolution and must have been mostly phenotypic in nature, building upon the genetic potential in the brain that had accumulated over the previous millions of years. The brain had reached a threshold, and a wholly new, enormously more rapid form of mental evolution took over.

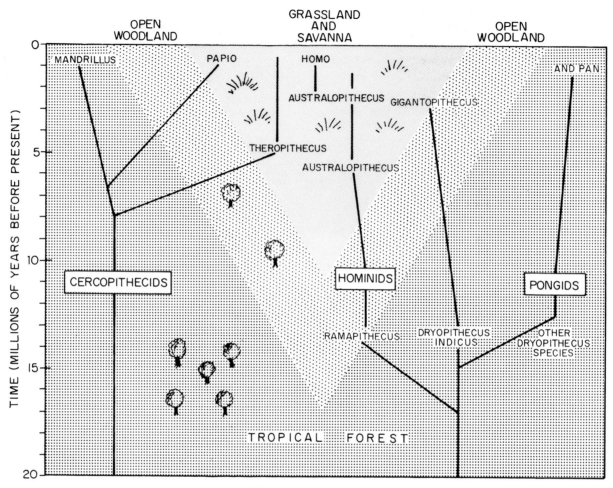

Figure 27-4 This simplified phylogeny of the Old World higher primates shows that only three existing groups have shifted from the forest to the savanna. They are the baboons (*Papio*), the gelada monkey (*Theropithecus gelada*), and man. (Based on Napier and Napier, 1967, and Simons and Ettel, 1970.)

This second phase was in no sense planned, and its potential is only now being revealed.

The study of man's origins can be referred to two questions that correspond to the dual stages of mental evolution:

———What features of the environment caused the hominids to adapt differently from other primates and started them along their unique evolutionary path?

———Once started, why did the hominids go so far?

The search for the prime movers of early human evolution has extended over more than 25 years. Participants in the search have included Dart (1949, 1956), Bartholomew and Birdsell (1953), Etkin (1954), Washburn and Avis (1958), Washburn et al. (1961), Rabb et al. (1967), Reynolds (1968), Schaller and Lowther (1969), C. J. Jolly (1970), and Kortlandt (1972). These writers have concentrated on two indisputably important facts concerning the biology of *Australopithecus* and early *Homo*. First, the evidence is strong that *Australopithecus africanus*, the species most likely to have been the direct ancestor of *Homo*, lived on the open savanna. The wear pattern of sand grains taken from the Sterkfontein fossils suggests a dry climate, while the pigs, antelopes, and other mammals found in association with the hominids are of the kind usually specialized for existence in grasslands. The australopithecine way of life came as the result of a major habitat shift. The ancestral *Ramapithecus* or an even more antecedent form lived in forests and was adapted for progression through trees by arm swinging. Only a very few other large-bodied primates have been able to join man in leaving the forest to spend most of their lives on the ground in open habitats (Figure 27-4). This is not to say that bands of *Australopithecus africanus* spent all of their lives running about in the open. Some of them might have carried their game into caves and even lived there in permanent residence, although the evidence pointing to this often quoted trait is still far from conclusive (Kurtén, 1972). Other bands could have retreated at night to the protection of groves of trees, in the manner of modern baboons. The important point is that much or all of the foraging was conducted on the savanna.

The second peculiar feature of the ecology of early men was the degree of their dependence on animal food, evidently far greater than in any of the living monkeys and apes. The *Australopithecus* were catholic in their choice of small animals. Their sites contain the remains of tortoises, lizards, snakes, mice, rabbits, porcupines, and other small, vulnerable prey that must have abounded on the savanna. The man-apes also hunted baboons with clubs. From analysis of 58 baboon skulls, Dart estimated that all had been brought down by blows to the head, 50 from the front and the remainder from behind. The *Australopithecus* also appear to have butchered larger animals, including the giant sivatheres, or horned giraffes, and dinotheres, elephantlike forms with tusks that curved downward from the lower jaws. In early Acheulean times, when *Homo erectus* began employing stone axes, some of the species of large African mammals became extinct. It is reasonable to suppose that this impoverishment was due to excessive predation by the increasingly competent bands of men (Martin, 1966).

What can we deduce from these facts about the life of early man? Before an answer is attempted, it should be noted that very little can be inferred directly from comparisons with other living primates. Geladas and baboons, the only open-country forms, are primarily vegetarian. They represent a sample of at most six species, which differ too much from one another in social organization to provide a baseline for comparison. The chimpanzees, the most intelligent and socially sophisticated of the nonhuman primates, are forest-dwelling and mostly vegetarian. Only during their occasional ventures into predation do they display behavior that can be directly correlated with ecology in a way that has meaning for human evolution. Other notable features of chimpanzee social organization, including the rapidly shifting composition of subgroups, the exchange of females between groups, and the intricate and lengthy process of socialization (see Chapter 26), may or may not have been shared by primitive man. We cannot argue either way on the basis of ecological correlation. It is often stated in the popular literature that the life of chimpanzees reveals a great deal about the origin of man. This is not necessarily true. The manlike traits of chimpanzees could be due to evolutionary

convergence, in which case their use in evolutionary reconstructions would be misleading.

The best procedure to follow, and one which I believe is relied on implicitly by most students of the subject, is to extrapolate backward from living hunter-gatherer societies. In Table 27-5 this technique is made explicit. Utilizing the synthesis edited by Lee and DeVore (1968; see especially J. W. M. Whiting, pp. 336-339), I have listed the most general traits of hunter-gatherer peoples. Then I have evaluated the lability of each behavioral category by noting the amount of variation in the category that occurs among the nonhuman primate species. The less labile the category, the more likely that the trait displayed by the living hunter-gatherers was also displayed by early man.

What we can conclude with some degree of confidence is that primitive men lived in small territorial groups, within which males were dominant over females. The intensity of aggressive behavior and the nature of its scaling remain unknown. Maternal care was prolonged, and the relationships were at least to some extent matrilineal. Speculation on remaining aspects of social life is not supported either way by the lability data and is therefore more tenuous. It is likely that the early hominids foraged in groups. To judge from the behavior of baboons and geladas, such behavior would have conferred some protection from large predators. By the time *Australopithecus* and early *Homo* had begun to feed on large mammals, group hunting almost certainly had become advantageous and even necessary, as in the African wild dog. But there is no compelling reason to conclude that men did the hunting while women stayed at home. This occurs today in hunter-gatherer societies, but comparisons with other primates offer no clue as to *when* the trait appeared. It is certainly not essential to conclude a priori that males must be a specialized hunter class. In chimpanzees males do the hunting, which may be suggestive. But in lions, it will be recalled, the females are the providers, often working in groups and with cubs in tow, while the males usually hold back. In the African wild dog both sexes participate. This is not to suggest that male group hunting was not an early trait of hominids, only that there is no strong independent evidence to support the hypothesis.

This brings us to the prevailing theory of the origin of human sociality. It consists of a series of interlocking models that have been fashioned from bits of fossil evidence, extrapolations back from extant hunter-gatherer societies, and comparisons with other living primate species. The core of the theory can be appropriately termed the *autocatalysis model.* It holds that when the earliest hominids became bipedal as part of their terrestrial adaptation, their hands were freed, the manufacture and handling of artifacts was made easier, and intelligence grew as part of the improvement of the tool-using habit. With mental capacity and the tendency to use artifacts increas-

Table 27-5 Social traits of living hunter-gatherer groups and the likelihood that they were also possessed by early man.

Traits that occur generally in living hunter-gatherer societies	Variability of trait category among nonhuman primates	Reliability of concluding early man had the same trait through homology
Local group size:		
Mostly 100 or less	Highly variable but within range of 3–100	Very probably 100 or less but otherwise not reliable
Family as the nuclear unit	Highly variable	Not reliable
Sexual division of labor:		
Women gather, men hunt	Limited to man among living primates	Not reliable
Males dominant over females	Widespread although not universal	Reliable
Long-term sexual bonding (marriage) nearly universal; polygyny general	Highly variable	Not reliable
Exogamy universal, abetted by marriage rules	Limited to man among living primates	Not reliable
Subgroup composition changes often (fission-fusion principle)	Highly variable	Not reliable
Territoriality general, especially marked in rich gathering areas	Occurs widely, but variable in pattern	Probably occurred; pattern unknown
Game playing, especially games that entail physical skill but not strategy	Occurs generally, at least in elementary form	Very reliable
Prolonged maternal care; pronounced socialization of young; extended relationships between mother and children, especially mothers and daughters	Occurs generally in higher cercopithecoids	Very reliable

ing through mutual reinforcement, the entire materials-based culture expanded. Cooperation during hunting was perfected, providing a new impetus for the evolution of intelligence, which in turn permitted still more sophistication in tool using, and so on through cycles of causation. At some point, probably during the late *Australopithecus* period or the transition from *Australopithecus* to *Homo*, this autocatalysis carried the evolving populations to a certain threshold of competence, at which the hominids were able to exploit the antelopes, elephants, and other large herbivorous mammals teeming around them on the African plains. Quite possibly the process began when the hominids learned to drive big cats, hyenas, and other carnivores from their kills (see Figure 27-5). In time they became the primary hunters themselves and were forced to protect their prey from other predators and scavengers. The autocatalysis model usually includes the proposition that the shift to big game accelerated the process of mental evolution. The shift could even have been the impetus that led to the origin of early *Homo* from their australo-

pithecine ancestors approximately two million years ago. Another proposition is that males became specialized for hunting. Child care was facilitated by close social bonding between the males, who left the domiciles to hunt, and the females, who kept the children and conducted most of the foraging for vegetable food. Many of the peculiar details of human sexual behavior and domestic life flow easily from this basic division of labor. But these details are not essential to the autocatalysis model. They are added because they are displayed by modern hunter-gatherer societies.

Although internally consistent, the autocatalysis model contains a curious omission—the triggering device. Once the process started, it is easy to see how it could be self-sustaining. But what started it? Why did the earliest hominids become bipedal instead of running on all fours like baboons and geladas? Clifford Jolly (1970) has proposed that the prime impetus was a specialization on grass seeds. Because the early pre-men, perhaps as far back as *Ramapithecus*, were the largest primates depending on grain, a premium was set on the ability to manipulate objects of very small size relative to the hands. Man, in short, became bipedal in order to pick seeds. This hypothesis is by no means unsupported fantasy. Jolly points to a number of convergent features in skull and dental structure between man and the gelada, which feeds on seeds, insects, and other small objects. Moreover, the gelada is peculiar among the Old World monkeys and apes in sharing the following epigamic anatomical traits with man: growth of hair around the face and neck of the male and conspicuous fleshy adornments on the chest of the female. According to Jolly's model, the freeing of the hands of the early hominids was a preadaptation that permitted the increase in tool use and the autocatalytic concomitants of mental evolution and predatory behavior.

Later Social Evolution

Autocatalytic reactions in living systems never expand to infinity. Biological parameters normally change in a rate-dependent manner to slow growth and eventually bring it to a halt. But almost miraculously, this has not yet happened in human evolution. The increase in brain size and the refinement of stone artifacts indicate a gradual improvement in mental capacity throughout the Pleistocene. With the appearance of the Mousterian tool culture of *Homo sapiens neanderthalensis* some 75,000 years ago, the trend gathered momentum, giving way in Europe to the Upper Paleolithic culture of *Homo s. sapiens* about 40,000 years B.P. Starting about 10,000 years ago agriculture was invented and spread, populations increased enormously in density, and the primitive hunter-gatherer bands gave way locally to the relentless growth of tribes, chiefdoms, and states. Finally, after A.D. 1400 European-based civilization shifted gears again, and knowledge and technology grew not just exponentially but superexponentially (see Figures 27-6, 27-7).

There is no reason to believe that during this final sprint there has been a cessation in the evolution of either mental capacity or the predilection toward special social behaviors. The theory of population genetics and experiments on other organisms show that substantial changes can occur in the span of less than 100 generations, which for man reaches back only to the time of the Roman Empire. Two thousand generations, roughly the period since typical *Homo sapiens* invaded Europe, is enough time to create new species and to mold them in major ways. Although we do not know how much mental evolution has actually occurred, it would be false to assume that modern civilizations have been built entirely on capital accumulated during the long haul of the Pleistocene.

Since genetic and cultural tracking systems operate on parallel tracks, we can bypass their distinction for the moment and return to the question of the prime movers in later human social evolution in its broadest sense. Seed eating is a plausible explanation to account

for the movement of hominids onto the savanna, and the shift to big-game hunting might account for their advance to the *Homo erectus* grade. But was the adaptation to group predation enough to carry evolution all the way to the *Homo sapiens* grade and farther, to agriculture and civilization? Anthropologists and biologists do not consider the impetus to have been sufficient. They have advocated the following series of additional factors, which can act singly or in combination.

Sexual Selection

Fox (1972), following a suggestion by Chance (1962), has argued that sexual selection was the auxiliary motor that drove human evolution all the way to the *Homo* grade. His reasoning proceeds as follows. Polygyny is a general trait in hunter-gatherer bands and may also have been the rule in the early hominid societies. If so, a premium would have been placed on sexual selection involving both epigamic display toward the females and intrasexual competition among the males. The selection would be enhanced by the constant mating provocation that arises from the female's nearly continuous sexual receptivity. Because of the existence of a high level of cooperation within the band, a legacy of the original *Australopithecus* adaptation, sexual selection would tend to be linked with hunting prowess, leadership, skill at tool making, and other visible attributes that contribute to the success of the family and the male band. Aggressiveness was constrained and the old forms of overt primate dominance replaced by complex social skills. Young males found it profitable to fit into the group, controlling their sexuality and aggression and awaiting their turn at leadership. As a result the dominant male in hominid societies was most likely to possess a mosaic of qualities that reflect the necessities of compromise: "controlled, cunning, cooperative, attractive to the ladies, good with the children, relaxed, tough, eloquent, skillful, knowledgeable and proficient in self-defense and hunting." Since positive feedback occurs between these more sophisticated social traits and breeding success, social evolution can proceed indefinitely without additional selective pressures from the environment.

Multiplier Effects in Cultural Innovation and in Network Expansion

Whatever its prime mover, evolution in cultural capacity was implemented by a growing power and readiness to learn. The network of contacts among individuals and bands must also have grown. We can postulate a critical mass of cultural capacity and network size in which it became advantageous for bands actively to enlarge both. In other words, the feedback became positive. This mechanism, like sexual selection, requires no additional input beyond the limits of

Figure 27-5 At the threshold of autocatalytic social evolution two million years ago, a band of early men (*Homo habilis*) forages for food on the African savanna. In this speculative reconstruction the group is in the act of driving rival predators from a newly fallen dinothere. The great elephantlike creature had succumbed from exhaustion or disease, its end perhaps hastened by attacks from the animals closing in on it. The men have just entered the scene. Some drive away the predators by variously shouting, waving their arms, brandishing sticks, and throwing rocks, while a few stragglers, entering from the left, prepare to join the fray. To the right a female sabertooth cat (*Homotherium*) and her two grown cubs have been at least temporarily intimidated and are backing away. Their threat faces reveal the extraordinary gape of their jaws. In the left foreground, a pack of spotted hyenas (*Crocuta*) has also retreated but is ready to rush back the moment an opening is provided.

The men are quite small, less than 1.5 meters in height, and individually no match for the large carnivores. According to prevailing theory, a high degree of cooperation was therefore required to exploit such prey; and it evolved in conjunction with higher intelligence and the superior ability to use tools. In the background can be seen the environment of the Olduvai region of Tanzania as it may have looked at this time. The area was covered by rolling parkland and rimmed to the east by volcanic highlands. The herbivore populations were dense and varied, as they are today. In the left background are seen three-toed horses (*Hipparion*), while to the right are herds of wildebeest and giant horned giraffelike creatures called sivatheres. (Drawing by Sarah Landry; prepared in consultation with F. Clark Howell. The reconstruction of *Homotherium* was based in part on an Aurignacian sculpture; see Rousseau, 1971.)

social behavior itself. But unlike sexual selection, it probably reached the autocatalytic threshold level very late in human prehistory.

Increased Population Density and Agriculture

The conventional view of the development of civilization used to be that innovations in farming led to population growth, the securing of leisure time, the rise of a leisure class, and the contrivance of civilized, less immediately functional pursuits. The hypothesis has been considerably weakened by the discovery that !Kung and other hunter-gatherer peoples work less and enjoy more leisure time than most farmers. Primitive agricultural people generally do not produce surpluses unless compelled to do so by political or religious authorities (Carneiro, 1970). Ester Boserup (1965) has gone so far as to suggest the reverse causation: population growth induces societies to deepen their involvement and expertise in agriculture. However, this explanation does not account for the population growth in the first place. Hunter-gatherer societies remained in approximate demographic equilibrium for hundreds of thousands of years. Something else tipped a few of them into becoming the first farmers. Quite possibly the crucial events were nothing more than the attainment of a certain level of intelligence and lucky encounters with wild-growing food plants. Once launched, agricultural economies permitted higher population densities which in turn encouraged wider networks of social contact, technological advance, and further dependence on farming. A few innovations, such as irrigation and the wheel, intensified the process to the point of no return.

Warfare

Throughout recorded history the conduct of war has been common among tribes and nearly universal among chiefdoms and states. When Sorokin analyzed the histories of 11 European countries over periods of 275 to 1,025 years, he found that on the average they were engaged in some kind of military action 47 percent of the time, or about one year out of every two. The range was from 28 percent of the years in the case of Germany to 67 percent in the case of Spain. The early chiefdoms and states of Europe and the Middle East turned over with great rapidity, and much of the conquest was genocidal in nature. The spread of genes has always been of paramount importance. For

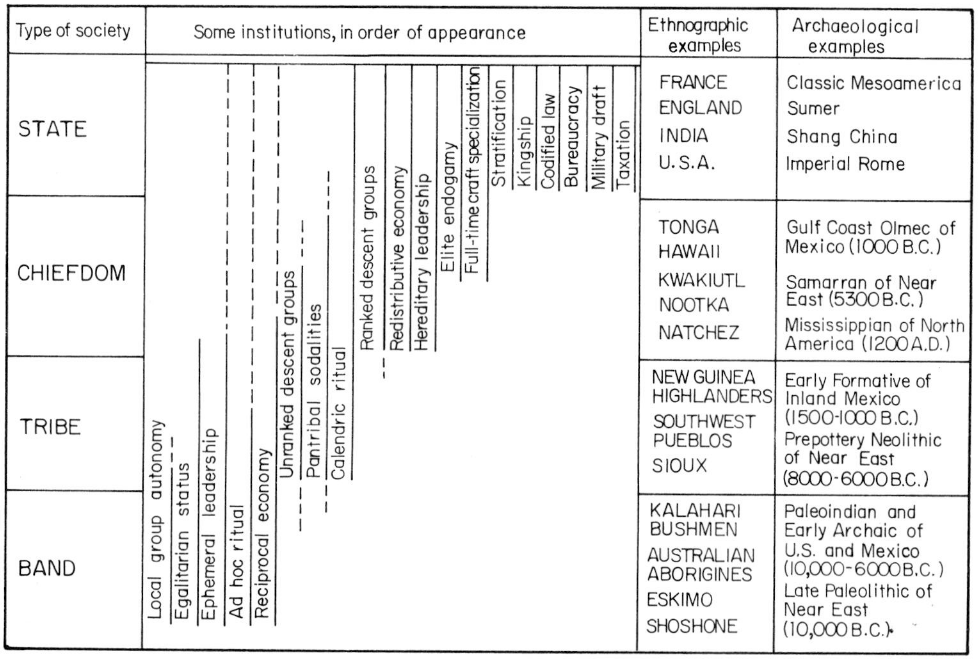

Type of society	Some institutions, in order of appearance		Ethnographic examples	Archaeological examples
STATE		Ranked descent groups / Redistributive economy / Hereditary leadership / Elite endogamy / Full-time craft specialization / Stratification / Kingship / Codified law / Bureaucracy / Military draft / Taxation	FRANCE ENGLAND INDIA U.S.A.	Classic Mesoamerica / Sumer / Shang China / Imperial Rome
CHIEFDOM			TONGA HAWAII KWAKIUTL NOOTKA NATCHEZ	Gulf Coast Olmec of Mexico (1000 B.C.) / Samarran of Near East (5300 B.C.) / Mississippian of North America (1200 A.D.)
TRIBE	Local group autonomy / Egalitarian status / Ephemeral leadership / Ad hoc ritual / Reciprocal economy	Unranked descent groups / Pantribal sodalities / Calendric ritual	NEW GUINEA HIGHLANDERS SOUTHWEST PUEBLOS SIOUX	Early Formative of Inland Mexico (1500-1000 B.C.) / Prepottery Neolithic of Near East (8000-6000 B.C.)
BAND			KALAHARI BUSHMEN AUSTRALIAN ABORIGINES ESKIMO SHOSHONE	Paleoindian and Early Archaic of U.S. and Mexico (10,000-6000 B.C.) / Late Paleolithic of Near East (10,000 B.C.)

Figure 27-6 The four principal types of societies in ascending order of sociopolitical complexity, with living and extinct examples of each. A few of the sociopolitical institutions are shown, in the approximate order in which they are interpreted to have arisen. (From Flannery, 1972. Reproduced, with permission, from "The Cultural Evolution of Civilizations," *Annual Review of Ecology and Systematics,* Vol. 3, p. 401. Copyright © 1972 by Annual Reviews, Inc. All rights reserved.)

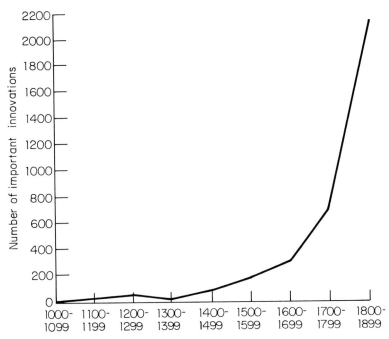

Figure 27-7 The number of important inventions and discoveries, by century, from A.D. 1000 to A.D. 1900. (From Lenski, 1970; after Ogburn and Nimkoff, 1958. Compiled from L. Darmstaedter and R. DuBois Reymond, *4000 Jahre-Pionier-Arbeit in den Exacten Wissenschaften*, Berlin, J. A. Stargart, 1904.)

example, after the conquest of the Midianites Moses gave instructions identical in result to the aggression and genetic usurpation by male langur monkeys:

Now kill every male dependent, and kill every woman who has had intercourse with a man, but spare for yourselves every woman among them who has not had intercourse.   (Numbers 31)

And centuries later, von Clausewitz conveyed to his pupil the Prussian crown prince a sense of the true, biological joy of warfare:

Be audacious and cunning in your plans, firm and persevering in their execution, determined to find a glorious end, and fate will crown your youthful brow with a shining glory, which is the ornament of princes, and engrave your image in the hearts of your last descendants.

The possibility that endemic warfare and genetic usurpation could be an effective force in group selection was clearly recognized by Charles Darwin. In *The Descent of Man* he proposed a remarkable model that foreshadowed many of the elements of modern group-selection theory:

Now, if some one man in a tribe, more sagacious than the others, invented a new snare or weapon, or other means of attack or defence,

the plainest self-interest, without the assistance of much reasoning power, would prompt the other members to imitate him; and all would thus profit. The habitual practice of each new art must likewise in some slight degree strengthen the intellect. If the invention were an important one, the tribe would increase in number, spread, and supplant other tribes. In a tribe thus rendered more numerous there would always be a rather greater chance of the birth of other superior and inventive members. If such men left children to inherit their mental superiority, the chance of the birth of still more ingenious members would be somewhat better, and in a very small tribe decidedly better. Even if they left no children, the tribe would still include their blood-relations, and it has been ascertained by agriculturists that by preserving and breeding from the family of an animal, which when slaughtered was found to be valuable, the desired character has been obtained.

Darwin saw that not only can group selection reinforce individual selection, but it can oppose it—and sometimes prevail, especially if the size of the breeding unit is small and average kinship correspondingly close. Essentially the same theme was later developed in increasing depth by Keith (1949), Bigelow (1969), and Alexander (1971). These authors envision some of the "noblest" traits of mankind, including team play, altruism, patriotism, bravery on the field of battle, and so forth, as the genetic product of warfare.

By adding the additional postulate of a threshold effect, it is possible to explain why the process has operated exclusively in human evolution (Wilson, 1972a). If any social predatory mammal attains a certain level of intelligence, as the early hominids, being large primates, were especially predisposed to do, one band would have the capacity to consciously ponder the significance of adjacent social groups and to deal with them in an intelligent, organized fashion. A band might then dispose of a neighboring band, appropriate its territory, and increase its own genetic representation in the metapopulation, retaining the tribal memory of this successful episode, repeating it, increasing the geographic range of its occurrence, and quickly spreading its influence still further in the metapopulation. Such primitive cultural capacity would be permitted by the possession of certain genes. Reciprocally, the cultural capacity might propel the spread of the genes through the genetic constitution of the metapopulation. Once begun, such a mutual reinforcement could be irreversible. The only combinations of genes able to confer superior fitness in contention with genocidal aggressors would be those that produce either a more effective technique of aggression or else the capacity to preempt genocide by some form of pacific maneuvering. Either probably entails mental and cultural advance. In addition to being autocatalytic, such evolution has the interesting property of requiring a selection episode only very occasionally in order to proceed as swiftly as individual-level selection. By current theory, genocide or genosorption strongly favoring the aggressor need take place only once every few generations to direct evolution. This alone could push truly altruistic genes to a high frequency within the bands (see Chapter 5). The turnover of tribes and chiefdoms estimated from atlases of early European and Mideastern history (for example, the atlas by McEvedy, 1967) suggests a sufficient magnitude of differential group fitness to have achieved this effect. Furthermore, it is to be expected that some isolated cultures will escape the process for generations at a time, in effect reverting temporarily to what ethnographers classify as a pacific state.

Multifactorial Systems

Each of the foregoing mechanisms could conceivably stand alone as a sufficient prime mover of social evolution. But it is much more likely that they contributed jointly, in different strengths and with complex interaction effects. Hence the most realistic model may be fully cybernetic, with cause and effect reciprocating through sub-cycles that possess high degrees of connectivity with one another. One such scheme, proposed by Adams (1966) for the rise of states and urban societies, is presented in Figure 27-8. Needless to say, the equations needed to translate this and similar models have not been written, and the magnitudes of the coefficients cannot even be guessed at the present time.

In both the unifactorial and multifactorial models of social evolution, an increasing internalization of the controls is postulated. This shift is considered to be the basis of the two-stage acceleration cited earlier. At the beginning of hominid evolution, the prime movers were external environmental pressures no different from those that have guided the social evolution of other animal species. For the moment, it seems reasonable to suppose that the hominids underwent two adaptive shifts in succession: first, to open-country living and seed eating, and second, after being preadapted by the anatomical and mental changes associated with seed eating, to the capture of

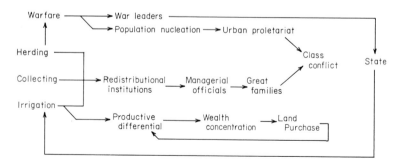

Figure 27-8 A multifactorial model of the origin of the state and urban society. (From Flannery, 1972; based on Adams, 1966. Reproduced, with permission, from "The Cultural Evolution of Civilizations," *Annual Review of Ecology and Systematics*, Vol. 3, p. 408. Copyright © 1972 by Annual Reviews, Inc. All rights reserved.)

large mammals. Big-game hunting induced further growth in mentality and social organization that brought the hominids across the threshold into the autocatalytic, more nearly internalized phase of evolution. This second stage is the one in which the most distinctive human qualities emerged. In stressing this distinction, however, I do not wish to imply that social evolution became independent of the environment. The iron laws of demography still clamped down on the spreading hominid populations, and the most spectacular cultural advances were impelled by the invention of new ways to control the environment. What happened was that mental and social change came to depend more on internal reorganization and less on direct responses to features in the surrounding environment. Social evolution, in short, had acquired its own motor.

The Future

When mankind has achieved an ecological steady state, probably by the end of the twenty-first century, the internalization of social evolution will be nearly complete. About this time biology should be at its peak, with the social sciences maturing rapidly. Some historians of science will take issue with this projection, arguing that the accelerating pace of discoveries in these fields implies a more rapid development. But historical precedents have misled us before: the subjects we are talking about are more difficult than physics or chemistry by at least two orders of magnitude.

Consider the prospects for sociology. This science is now in the natural history stage of its development. There have been attempts at system building but, just as in psychology, they were premature and came to little. Much of what passes for theory in sociology today is really labeling of phenomena and concepts, in the expected manner of natural history. Process is difficult to analyze because the fundamental units are elusive, perhaps nonexistent. Syntheses commonly consist of the tedious cross-referencing of differing sets of definitions and metaphors erected by the more imaginative thinkers (see for example Inkeles, 1964, and Friedrichs, 1970). That, too, is typical of the natural history phase.

With an increase in the richness of descriptions and experiments, sociology is drawing closer each day to cultural anthropology, social psychology, and economics, and will soon merge with them. These disciplines are fundamental to sociology *sensu lato* and are most likely to yield its first phenomenological laws. In fact, some viable qualitative laws probably already exist. They include tested statements about the following relationships: the effects of hostility and stress upon ethnocentrism and xenophobia (LeVine and Campbell, 1972); the positive correlation between and within cultures of war and combative sports, resulting in the elimination of the hydraulic model of aggressive drive (Sipes, 1973); precise but still specialized models of promotion and opportunity within professional guilds (White, 1970); and, far from least, the most general models of economics.

The transition from purely phenomenological to fundamental theory in sociology must await a full, neuronal explanation of the human brain. Only when the machinery can be torn down on paper at the level of the cell and put together again will the properties of emotion and ethical judgment come clear. Simulations can then be employed to estimate the full range of behavioral responses and the precision of their homeostatic controls. Stress will be evaluated in terms of the neurophysiological perturbations and their relaxation times. Cognition will be translated into circuitry. Learning and creativeness will be defined as the alteration of specific portions of the cognitive machinery regulated by input from the emotive centers. Having cannibalized psychology, the new neurobiology will yield an enduring set of first principles for sociology.

The role of evolutionary sociobiology in this enterprise will be twofold. It will attempt to reconstruct the history of the machinery and to identify the adaptive significance of each of its functions. Some of the functions are almost certainly obsolete, being directed toward such Pleistocene exigencies as hunting and gathering and intertribal warfare. Others may prove currently adaptive at the level of the individual and family but maladaptive at the level of the group—or the reverse. If the decision is taken to mold cultures to fit the requirements of the ecological steady state, some behaviors can be altered experientially without emotional damage or loss in creativity. Others cannot. Uncertainty in this matter means that Skinner's dream of a culture predesigned for happiness will surely have to wait for the new neurobiology. A genetically accurate and hence completely fair code of ethics must also wait.

The second contribution of evolutionary sociobiology will be to monitor the genetic basis of social behavior. Optimum socioeconomic systems can never be perfect, because of Arrow's impossibility theorem and probably also because ethical standards are innately pluralistic. Moreover, the genetic foundation on which any such normative system is built can be expected to shift continuously. Mankind has never stopped evolving, but in a sense his populations are drifting. The effects over a period of a few generations could change the identity of the socioeconomic optima. In particular, the rate of gene flow around the world has risen to dramatic levels and is accelerating, and the mean coefficients of relationship within local communities are correspondingly diminishing. The result could be an eventual lessening of altruistic behavior through the maladaption and loss of group-selected genes (Haldane, 1932; Eshel, 1972). It was shown earlier that behavioral traits tend to be selected out by the principle of metabolic conservation when they are suppressed or when their original function becomes neutral in adaptive value. Such traits can largely disappear from populations in as few as ten generations, only two or three centuries in the case of human beings. With our present inadequate understanding of the human brain, we do not know how many of the most valued qualities are linked genetically to more obsolete, destructive ones. Cooperativeness toward groupmates might be coupled with aggressivity toward strangers, creativeness with a desire to own and dominate, athletic zeal with a tendency to violent response, and so on. In extreme cases such pairings could stem from pleiotropism, the control of more than one phenotypic character by the same set of genes. If the planned society—the creation of which seems inevitable in the coming century—were to deliberately steer its members past those stresses and conflicts that once gave the destructive phenotypes their Darwinian edge, the other phenotypes might dwindle with them. In this, the ultimate genetic sense, social control would rob man of his humanity.

It seems that our autocatalytic social evolution has locked us onto a particular course which the early hominids still within us may not welcome. To maintain the species indefinitely we are compelled to drive toward total knowledge, right down to the levels of the neuron and gene. When we have progressed enough to explain ourselves in these mechanistic terms, and the social sciences come to full flower, the result might be hard to accept. It seems appropriate therefore to close this book as it began, with the foreboding insight of Albert Camus:

A world that can be explained even with bad reasons is a familiar world. But, on the other hand, in a universe divested of illusions and lights, man feels an alien, a stranger. His exile is without remedy since he is deprived of the memory of a lost home or the hope of a promised land.

This, unfortunately, is true. But we still have another hundred years.

Glossary, Bibliography, and Index

Glossary

For a rapid comprehension of sociobiology the equivalent of a college-level course in biology is desirable. Also, some training in elementary mathematics, particularly calculus and probability theory, is needed to make the most technical chapters readily understandable. However, *Sociobiology: The New Synthesis* has been written with the broadest possible audience in mind, and most of it can be read with full understanding by any intelligent person whether or not he or she has had formal training in science. To this end the following glossary has been stocked with the elementary terms of biology and mathematics that are most frequently used in the book. The reader will also find that it contains more technical expressions limited to sociobiology, including a few that appear only sparingly in the present book but are encountered consistently in the literature cited.

Absenteeism The practice of certain animals, such as tree shrews, of nesting away from their offspring and visiting them from time to time only to provide them with food and a minimum of additional care.

Active space The space within which the concentration of a pheromone (or any other behaviorally active chemical substance) is at or above threshold concentration. The active space of a pheromone is, in fact, the chemical signal itself.

Aculeate Pertaining to the Aculeata, or stinging Hymenoptera, a group including the bees, ants, and many of the wasps.

Adaptation In evolutionary biology, any structure, physiological process, or behavioral pattern that makes an organism more fit to survive and to reproduce in comparison with other members of the same species. Also, the evolutionary process leading to the formation of such a trait.

Adaptive Pertaining to any trait, anatomical, physiological, or behavioral, that has arisen by the evolutionary process of adaptation (*q.v.*).

Adaptive radiation The process of evolution in which species multiply, diverge into different ecological niches (for example, species that are predators on different kinds of prey, occupants of different habitats, and so forth), and come to occupy the same or at least overlapping ranges.

Age polyethism The regular changing of labor roles by members of a society as they age.

Aggregation A group of individuals of the same species, comprised of more than just a mated pair or a family, gathered in the same place but not internally organized or engaged in cooperative behavior. To be distinguished from a true society, *q.v.*

Aggression A physical act or threat of action by one individual that reduces the freedom or genetic fitness of another.

Agonistic   Referring to any activity related to fighting, whether aggression or conciliation and retreat.

Agonistic buffering   The use of infants by adults to inhibit aggression by other adults; reported among male macaques and a few other monkeys.

Alarm-defense system   Defensive behavior that also functions as an alarm signaling device within the colony. Examples include the use by certain ant species of chemical defensive secretions that double as alarm pheromones.

Alarm pheromone   A chemical substance exchanged among members of the same species that induces a state of alertness or alarm in the face of a common threat.

Alarm-recruitment system   A communication system that rallies others to some particular place to aid in the defense of the society. An example is the odor trail system of lower termites, which is used to recruit colony members to the vicinity of intruders and breaks in the nest wall.

Alate   Winged. Sometimes also used as a noun to refer to a reproductive social insect still bearing wings.

Allele   A particular form of a gene, distinguishable from other forms or alleles of the same gene.

Allodapine   A ceratinine bee belonging to *Allodape* or one of a series of closely related genera, all of which are either primitively eusocial or socially parasitic. Excluded from this informal taxonomic category is *Ceratina*, the only other major living genus of the tribe Ceratinini.

Allogrooming   Grooming directed at another individual, as opposed to self-grooming, which is directed at one's own body.

Allometry   Any size relation between two body parts that can be expressed by $y = bx^a$, where a and b are fitted constants. In the special case of isometry, $a = 1$, and the relative proportions of the body parts therefore remain constant with change in total body size. In all other cases ($a \neq 1$) the relative proportions change as the total body size is varied. Allometry is important in the differentiation of castes of the social insects, especially ants.

Allomone   A chemical substance released by one species that serves as a communicative signal to another species. (Contrast with pheromone.)

Alloparent   An individual that assists the parents in care of the young.

Alloparental care   The assistance by individuals other than the parents in the care of offspring. The behavior may be shown either by females (allomaternal care) or by males (allopaternal care).

Allopatric   Referring to populations, particularly species, that occupy different geographical ranges. (Contrast with sympatric.)

Allozygous   Referring to two genes on the same chromosome locus that are different or at least whose identity is not due to common descent. (Contrast with autozygous.)

Alpha   Referring to the highest-ranking individual within a dominance hierarchy.

Altricial   Pertaining to young animals that are helpless for a substantial period following birth; used especially with reference to birds. (Contrast with precocial.)

Altruism   Self-destructive behavior performed for the benefit of others. (See discussion in Chapter 5.)

Ameba   Any one of a large number of single-celled organisms, especially in the phylum Sarcodina, characterized by its ability to change shape frequently through the protrusion and retraction of soft extensions of cytoplasm called pseudopodia.

Amphibian   Any member of the vertebrate class Amphibia, such as a salamander, frog, or toad.

Analog signal   Same as graded signal (*q.v.*).

Analogue   Referring to structures, physiological processes, or behaviors that are similar owing to convergent evolution as opposed to common ancestry; hence, displaying analogy. (Opposed to homologue.)

Analogy   A resemblance in function, and often in appearance also, between two structures, physiological processes, or behaviors that is due to evolution rather than to common ancestry. (Contrast with homology.)

Anisogamy   The condition in which the female sex cell (ovum) is larger than the male sex cell (sperm). (Contrast with isogamy.)

Annual   Referring to a life cycle, or the species possessing the life cycle, that is completed in one growing season.

Antennation   Touching with the antennae. The movement can serve as a sensory probe or as a tactile signal to another insect.

Antisocial factor   Any selection pressure that tends to inhibit or to reverse social evolution.

Aposematism The advertisement by dangerous animals of their identity. Thus the most venomous wasps, coral fishes, and snakes are also often the most brightly colored.

Arachnid A member of the Class Arachnida, such as a spider, mite, or scorpion.

Arena An area used consistently for communal courtship displays. Same as lek.

Army ant A member of an ant species that shows both nomadic and group-predatory behavior. In other words, the nest site is changed at relatively frequent intervals, in some cases daily, and the workers forage in groups. (Same as legionary ant.)

Arthropod Any member of the phylum Arthropoda, such as a crustacean, spider, millipede, centipede, or insect.

Artiodactyl Any mammal belonging to the order Artiodactyla, hence an ungulate with an even number of toes in each hoof. The commonest artiodactyls include the pigs, deer, camels, and antelopes. (Contrast with perissodactyl.)

Asexual reproduction Any form of reproduction that does not involve the actual fusion of sex cells (syngamy), such as budding and parthenogenesis.

Assembly The calling together of the members of a society for any communal activity.

Assortative mating The nonrandom pairing of individuals who resemble each other in one or more traits. (Contrast with disassortative mating.)

Aunt Any female who assists a parent in caring for the young.

Australopithecine Pertaining to the "man-apes," primates belonging to the genus *Australopithecus*, which were primitive forms that lived during the Pleistocene Epoch and were ancestral to modern men (genus *Homo*). Australopithecines possessed postures and dentition similar to those of modern men but brains not much larger than those of modern apes.

Australopithecus See australopithecine.

Autocatalysis Any process the rate of which is increased by its own products. Thus autocatalytic reactions, fed by positive feedback, tend to accelerate until the ingredients are exhausted or some external constraint is imposed.

Automimicry The imitation by one sex or life stage of communication in another sex or life stage of the same species. An example is the imitation by the males of some monkey species of female sexual signals, which they appear to employ in appeasement rituals.

Autozygous Referring to two or more alleles on the same locus that are identical by common descent.

Auxiliaries Female social insects, especially bees, wasps, and ants, that associate with other females of the same generation and become workers.

Band The term applied to groups of certain social mammals, including coatis and human beings.

Behavioral biology The scientific study of all aspects of behavior, including neurophysiology, ethology, comparative psychology, sociobiology, and behavioral ecology.

Behavioral scale See behavioral scaling.

Behavioral scaling The range of forms and intensities of a behavior that can be expressed in an adaptive fashion by the same society or individual organism. For example, a society may be organized into individual territories at low densities but shift to a dominance system at high densities. (See discussion in Chapter 2.)

Biomass The weight of a set of plants, animals, or both. The set is chosen for convenience; it can be, for example, a colony of insects, a population of wolves, or an entire forest.

Bit The basic quantitative unit of information; specifically, the amount of information required to control, without error, which of two equiprobable alternates is to be chosen by the receiver.

Bivouac The mass of army ant workers within which the queen and brood find refuge. Also, the site where the mass is located.

Bonding Any close relationship formed between two or more individuals.

Brood Any young animals that are being cared for by adults. In social insects in particular, the immature members of a colony collectively, including eggs, nymphs, larvae, and pupae. In the strict sense eggs and pupae are not members of the society, but they are nevertheless referred to as part of the brood.

Brood cell A special chamber or pocket built to house immature stages of insects.

Brood parasitism In birds, the insertion of eggs by one species (such as a European cuckoo) into the nest of another species, with the result that the host rears the brood of the parasite as if it were its own.

Budding The reproduction of organisms by the direct growth of a new individual from the body of an old one. Also, the multiplication of insect colonies by fission. (See colony fission.)

Callow workers In colonies of social insects, the newly emerged adult workers, whose exoskeletons are still relatively soft and lightly pigmented.

Canid A member of the mammal family Canidae, such as a wolf, domestic dog, or jackal.

Carnivore An animal that eats fresh meat.

Carrying capacity Usually symbolized by K, the largest number of organisms of a particular species that can be maintained indefinitely in a given part of the environment.

Carton In entomology, the chewed vegetable fibers used by many kinds of ants, wasps, and other insects to construct nests.

Caste Broadly defined, as in ergonomic theory (Chapter 14), any set of individuals of a particular morphological type, or age group, or both, that performs specialized labor in the colony. More narrowly defined, any set of individuals in a given colony that are both morphologically distinct from other individuals and specialized in behavior.

Casual society (or group) A temporary group formed by individuals within a society. The casual society is unstable, being open to new members and losing old ones at a high rate. Examples include feeding groups of monkeys within a troop and groups of playing children. (Contrast with demographic society.)

Central nervous system Often abbreviated as the CNS, that part of the nervous system which is condensed and centrally located; for example, the brain and spinal cord of vertebrates, and the brain and ladderlike chain of ganglia in insects.

Cercopithecoid Pertaining to the Old World monkeys and apes (classified as the superfamily Cercopithecoidea by many authors.)

Ceremony A highly evolved and complex display used to conciliate others and to establish and maintain social bonds.

Character In taxonomy and a few other fields of biology, the word *character* is commonly used a a synonym for trait. A particular trait possessed by one individual and not another, or by one species and not another, is often called a character state.

Character convergence The process whereby two newly evolved species interact in such a way that one or both converges in one or more traits toward the other. (Contrast with character displacement.)

Character displacement The process whereby two newly evolved species interact in such a way as to cause one or both of them to diverge still further in evolution. (Contrast with character convergence.)

Chorus A group of calling anurans (frogs or toads) or insects.

Chromosome A complex, often rodlike structure found in the nucleus of a cell, bearing part of the basic genetic units (genes) of the cell.

Circadian rhythm A rhythm in behavior, metabolism, or some other activity that recurs about every 24 hours. (The prefix *circa-* refers to the lack of precision in the timing.)

Clade A species or set of species representing a distinct branch in a phylogenetic tree. (Contrast with evolutionary grade.)

Cladogram A phylogenetic tree that depicts only the splitting of species and groups of species through evolutionary time.

Class In systems of classification of organisms, the category below the phylum and above the order; a group of related, similar orders. Examples of classes are the Insecta (all true, six-legged insects) and Aves (all birds).

Claustral colony founding The procedure during which queens of ants and other social hymenopterans (or royal pairs in the case of termites) seal themselves off in cells and rear the first generation of workers on nutrients obtained mostly or entirely from their own storage tissues, including fat bodies and histolyzed wing muscles.

Cleptobiosis The relation in which one species robs the food stores or scavenges in the refuse piles of another species, but does not nest in close association with it.

Cleptoparasitism The parasitic relation in which a female seeks out the prey or stored food of another female, usually belonging to a different species, and appropriates it for the rearing of her own offspring.

Cline A pattern of gradual genetic change in a population from one part of its geographic range to another. Many mammal species, for example, show clines of increasing size toward the colder portions of their ranges.

Clone A population of individuals all derived asexually from the same single parent.

Clutch The number of eggs laid by a female at one time.

Coefficient of consanguinity Same as coefficient of kinship (q.v.).

Coefficient of kinship Symbolized by F_{IJ} or f_{IJ}, the probability that a pair of alleles drawn at random from the same locus on two individuals are identical by virtue of common descent. Also called the coefficient of consanguinity.

Coefficient of relationship Also known as the degree of relatedness, and symbolized by r, the coefficient of relationship is the fraction of genes identical by descent between two individuals.

Colony A society which is highly integrated, either by physical union of the members, division of the members into specialized zooids or castes, or both. In vernacular usage the term colony is on occasion applied to almost any group of organisms, especially a group of nesting birds or a cluster of rodents living in dens.

Colony fission The multiplication of colonies by the departure of one or more reproductive forms, accompanied by groups of workers, from the parental nest, leaving behind comparable units to perpetuate the "parental" colony. This mode of colony multiplication is referred to occasionally as hesmosis in ant literature and sociotomy in termite literature. Swarming in honeybees can be regarded as a special form of colony fission.

Colony odor The odor found on the bodies of social insects that is peculiar to a given colony. By smelling the colony odor of another member of the same species, an insect is able to determine whether it is a nestmate. (See nest odor and species odor.)

Comb (of cells or cocoons) A layer of brood cells or cocoons crowded together in a regular arrangement. Combs are a characteristic feature of the nests of many species of social wasps and bees.

Commensalism Symbiosis in which members of one species are benefited while those of the other species are neither benefited nor harmed.

Communal Applied to the condition or to the group showing it in which members of the same generation cooperate in nest building but not in brood care.

Communication Action on the part of one organism (or cell) that alters the probability pattern of behavior in another organism (or cell) in an adaptive fashion. (See discussion in Chapter 8.)

Compartmentalization The manner and extent to which subgroups of societies act as discrete units.

Competition The active demand by two or more organisms (or two or more species) for a common resource.

Composite signal A signal composed of two or more simpler signals.

Compound nest A nest containing colonies of two or more species of social insects, up to the point where the galleries of the nest run together and the adults sometimes intermingle but in which the broods of the species are still kept separate. (See mixed nest.)

Connectedness The number and direction of communication links within and between societies.

Conspecific Belonging to the same species.

Control According to strict sociobiological usage, particularly in primate studies, control is intervention by one or more individuals to reduce or halt aggression between other members of the group.

Conventional behavior According to the hypothesis proposed by V. C. Wynne-Edwards, any behavior by which members of a population reveal their presence and allow others to assess the density of the population. A more elaborate form of such behavior is referred to as an epideictic display.

Coordination Interaction among individuals or subgroups such that the overall effort of the group is divided among these units without leadership being assumed by any one of them.

Core area The area of heaviest regular use within the home range.

Cormidium A group of zooids (individual members) of a siphonophore colony that can separate from the remainder of the colony and live an independent existence. The cormidium is the unit of organization between the zooid and the complete colony.

Coterie The basic society of prairie dogs (a kind of rodent). The coterie consists of a small group of individuals that occupy communal burrows.

Counteracting selection The operation of selection pressures on two or more levels of organization, for example on the individual, family, and population, in such a way that certain genes are favored at one level but disfavored at another. (Contrast with reinforcing selection.)

Court The area defended by individual males within a lek, or communal display area; especially in birds. Also, the group of workers in an insect colony that surrounds the queen, especially in honeybees; the composition of such a court, also called a retinue, changes constantly.

Darling effect   See Fraser Darling effect.

Darwinism   The theory of evolution by natural selection, as originally propounded by Charles Darwin. The modern version of this theory still recognizes natural selection as the central process and for this reason is often called neo-Darwinism.

Dealate   Referring to an individual that has shed its wings, usually after mating; used both as an adjective and a noun.

Dealation   The removal of the wings by the queens (and also males in the termites) during or immediately following the nuptial flight and prior to colony foundation.

Dear enemy phenomenon   The recognition of territorial neighbors as individuals, with the result that aggressive interactions are kept at a minimum. The more intense forms of aggression are reserved for strangers.

Deme   A local population within which breeding is completely random. Hence the largest population unit that can be analyzed by the simpler models of population genetics.

Demographic society   A society that is stable enough through time, usually owing to its being relatively closed to newcomers, for the demographic processes of birth and death to play a significant role in its composition. (Contrast with casual group.)

Demography   The rate of growth and the age structure of populations, and the processes that determine these properties.

Dendrogram   A diagram showing evolutionary change in a biological trait, including the branching of the trait into different forms due to the multiplication of the species possessing it.

Density dependence   The increase or decrease of the influence of a physiological or environmental factor on population growth as the density of the population increases.

Deterministic   In mathematics, referring to a fixed relationship between two or more variables, without taking into account the effect of chance on the outcome of particular cases. (Contrast with stochastic.)

Developmental cycle   The period from the birth of the egg to the eclosion of the adult insect. (Applied to social wasp studies.)

Dialect   In the study of animal behavior, local geographic variants of bird songs, honeybee waggle dances, and other displays used in communication.

Dimorphism   In caste systems, the existence in the same colony of two different forms, including two size classes, not connected by intermediates.

Diploid   With reference to a cell or to an organism, having a chromosome complement consisting of two copies (called homologues) of each chromosome. A diploid cell or organism usually arises as the result of the union of two sex cells, each bearing just one copy of each chromosome. Thus, the two homologues in each chromosome pair in a diploid cell are of separate origin, one derived from the mother and the other from the father. (Contrast with haploid.)

Direct role   A behavior or set of behaviors displayed by a subgroup of the society that benefits other subgroups and therefore the society as a whole. (Contrast with indirect role; see discussion in Chapter 14.)

Directional selection   Selection that operates against one end of the range of variation and hence tends to shift the entire population toward the opposite end. (Contrast with disruptive and stabilizing selection.)

Disassortative mating   The nonrandom pairing of individuals that differ from each other in one or more traits.

Discrete signal   A signal used in communication that is turned either on or off, without significant intermediate gradations. (Contrast with graded signal.)

Displacement activity   The performance of a behavioral act, usually in conditions of frustration or indecision, that is not directly relevant to the situation at hand.

Display   A behavior pattern that has been modified in the course of evolution to convey information. A display is a special kind of signal, which in turn is broadly defined as *any* behavior that conveys information regardless of whether it serves other functions.

Disruptive selection   Selection that operates against the middle of the range of variation and hence tends to split populations. (Contrast with directional and stabilizing selection.)

Distraction display   A performance by a parent that draws the attention of predators away from its offspring.

DNA (deoxyribonucleic acid)   The basic hereditary material of all kinds of organisms. In higher organisms, including animals, the great bulk of DNA is located within the chromosomes.

Dominance hierarchy The physical domination of some members of a group by other members, in relatively orderly and long-lasting patterns. Except for the highest- and lowest-ranking individuals, a given member dominates one or more of its companions and is dominated in turn by one or more of the others. The hierarchy is initiated and sustained by hostile behavior, albeit sometimes of a subtle and indirect nature. (See discussion in Chapter 13.)

Dominance order Same as dominance hierarchy (*q.v.*).

Dominance system Same as dominance hierarchy (*q.v.*).

Driver ants African legionary ants belonging to the genus *Dorylus* and, less frequently, other members of the tribe Dorylini.

Drone A male social bee, especially a male honeybee or bumblebee.

Duetting The rapid and precise exchange of notes between two individuals, especially mated birds.

Dulosis The relation in which workers of a parasitic (dulotic) ant species raid the nests of another species, capture brood (usually in the form of pupae), and rear them as enslaved nestmates.

Dynamic selection Same as directional selection.

Eclosion Emergence of the adult (imago) insect from the pupa; less commonly, the hatching of an egg.

Ecological pressure See under prime movers.

Ecology The scientific study of the interaction of organisms with their environment, including both the physical environment and the other organisms that live in it.

Ecosystem All of the organisms of a particular habitat, such as a lake or a forest, together with the physical environment in which they live.

Effective population number The number of individuals in an ideal, randomly breeding population with a 1/1 sex ratio that would have the same rate of heterozygosity decrease as the real population under consideration.

Elite Referring to an insect colony member displaying greater than average initiative and activity.

Emery's rule The rule that species of social parasites are very similar to their host species and therefore presumably closely related to them phylogenetically. (First suggested by Carlo Emery.)

Emigration The movement of an individual or society from one nest site to another.

Empathic learning See observational learning.

Enculturation The transmission of a particular culture, especially to the young members of the society.

Endemic Referring to a species that is native to a particular place and found nowhere else.

Endocrine gland Any gland, such as the adrenal or pituitary gland of vertebrates, that secretes hormones into the body through the blood or lymph. (Opposed to exocrine gland.)

Endocrinology The scientific study of endocrine glands and hormones.

Entomology The scientific study of insects.

Environmentalism In biology, the form of analysis that stresses the role of environmental influences in the development of behavioral or other biological traits. Also, the point of view that such influences tend to be paramount in behavioral development.

Epideictic display In theory at least, a display by which members of a population reveal their presence and allow others to assess the density of the population. An extreme form of "conventional behavior" as postulated by V. C. Wynne-Edwards.

Epidermis The outer layer of living cells in the skin.

Epigamic Any trait related to courtship and sex other than the essential organs and behavior of copulation.

Epigamic selection See under sexual selection.

Epizootic The spread of a disease through a population of animals; the equivalent of an epidemic in human beings.

Ergatogyne Any form morphologically intermediate between the worker and the queen in an insect society.

Ergonomics The quantitative study of work, performance, and efficiency. (See discussion in Chapter 14.)

Estrous cycle The repeated series of changes in reproductive physiology and behavior that culminates in the estrus, or time of heat.

Estrus The period of heat, or maximum sexual receptivity, in the female. Ordinarily the estrus is also the time of the release of eggs in the female.

Ethocline A series of different behaviors observed among related species and interpreted to represent stages in an evolutionary trend.

Ethology The study of whole patterns of animal behavior in natural environments, stressing the analysis of adaptation and the evolution of the patterns.

Eusocial Applied to the condition or to the group possessing it in which individuals display all of the following three traits: cooperation in caring for the young; reproductive division of labor, with more or less sterile individuals working on behalf of individuals engaged in reproduction; and overlap of at least two generations of life stages capable of contributing to colony labor. "Eusocial" is the formal equivalent of the expressions "truly social" or "higher social," which are commonly used with less exact meaning in the study of social insects.

Eutherian Pertaining to the placental mammals (*q.v.*).

Evolution Any gradual change. Organic evolution, often referred to as evolution for short, is any genetic change in organisms from generation to generation; or more strictly, a change in gene frequencies within populations from generation to generation. (See discussion in Chapter 4.)

Evolutionary biology The collective disciplines of biology that treat the evolutionary process and the characteristics of populations of organisms, as well as ecology, behavior, and systematics.

Evolutionary convergence The evolutionary acquisition of a particular trait or set of traits by two or more species independently.

Evolutionary grade The evolutionary level of development in a particular structure, physiological process, or behavior occupied by a species or group of species. The evolutionary grade is distinguished from the phylogeny of a group, which is the relationship of species by descent.

Exocrine gland Any gland, such as the salivary gland, that secretes to the outside of the body or into the alimentary tract. Exocrine glands are the most common sources of pheromones, the chemical substances used in communication by most kinds of animals. (Opposed to endocrine gland.)

Exoskeleton The hardened outer body layer of insects and other arthropods that functions both as a protective covering and as a skeletal attachment for muscles.

Exponential growth Growth, especially in the number of organisms of a population, that is a simple function of the size of the growing entity; the larger the entity, the faster it grows.

F, f Symbol of the inbreeding coefficient (*q.v.*).

F_{IJ}, f_{IJ} Symbol of the coefficient of kinship (*q.v.*).

Facilitation See social facilitation.

Family In sociobiology, the word is given the conventional meaning of parents and offspring, together with other kin who are closely associated with them. In taxonomy, the family is the category below the order and above the genus; a group of related, similar genera. Examples of taxonomic families include the Formicidae, including all of the ants; and the Felidae, including all of the cats.

Fitness See genetic fitness.

Fixation In population genetics, the complete prevalence of one gene form (allele), resulting in the complete exclusion of another.

Flagellate A member of the phylum Mastigophora; a unicellular organism that propels itself by flagella, which are whiplike motile organs.

Floaters Individuals unable to claim a territory and hence forced to wander through less suitable surrounding areas.

Folivore An animal that eats leaves.

Food chain A portion of a food web, most frequently a simple sequence of prey species and the predators that consume them.

Food web The complete set of food links between species in a community; a diagram indicating which ones are the eaters and which are consumed.

Founder effect The genetic differentiation of an isolated population due to the fact that by chance alone its founders contained a set of genes statistically different from those of other populations.

Fraser Darling effect The stimulation of reproductive activity by the presence and activity of other members of the species in addition to the mating pair.

Frequency curve A curve plotted on a graph to display a particular frequency distribution (*q.v.*).

Frequency distribution The array of numbers of individuals showing differing values of some variable quantity; for example, the numbers of animals of different ages, or the numbers of nests containing different numbers of young.

Frugivore An animal that eats fruit.

Gamete The mature sexual reproductive cell: the egg or the sperm.

Gametogenesis The specialized series of cellular divisions that leads to the production of sex cells (gametes).

Gaster A special term occasionally applied to the metasoma, or terminal major body part behind the "waist," of ants and other hymenopterans.

Gene The basic unit of heredity.

Gene flow The exchange of genes between different species (an extreme case referred to as hybridization) or between different populations of the same species.

Gene pool All the genes—hence, hereditary material—in a population.

Genetic drift Evolution (change in gene frequencies) by chance processes alone.

Genetic fitness The contribution to the next generation of one genotype in a population relative to the contributions of other genotypes. By definition, this process of natural selection leads eventually to the prevalence of the genotypes with the highest fitnesses.

Genetic load The average loss of genetic fitness (*q.v.*) in an entire population due to the presence of individuals less fit than others.

Genome The complete genetic constitution of an organism.

Genotype The genetic constitution of an individual organism, designated with reference either to a single trait or to a set of traits. (Contrast with phenotype.)

Gens (plural: gentes) In the European cuckoo *Cuculus canorus*, a group of females within a population that lay their eggs primarily in the nest of a single host species. Their eggs mimic those of the host.

Genus (plural: genera) A group of related, similar species. Examples include *Apis* (the four species of honeybees) and *Canis* (wolves, domestic dogs, and their close relatives).

Geographic race See subspecies.

Gonad An organ that produces sex cells; either an ovary (female gonad) or testis (male gonad).

Grade See evolutionary grade.

Graded signal A signal that varies in intensity or frequency or both, thereby transmitting quantitative information about such variables as mood of the sender, distance of the target, and so forth.

Grooming The cleaning of the body surface by licking, nibbling, picking with the fingers, or other kinds of manipulation. When the action is directed toward one's own body, it is called self-grooming; when directed at another individual, it is referred to as allogrooming.

Group Any set of organisms, belonging to the same species, that remain together for a period of time while interacting with one another to a distinctly greater degree than with other conspecific organisms. The term is also frequently used in a loose taxonomic sense to refer to a set of related species; thus a genus, or a division of a genus, would be an example of a taxonomic "group."

Group effect An alteration in behavior or physiology within a species brought about by signals that are directed in neither space nor time. A simple example is social facilitation, in which there is an increase of an activity merely from the sight or sound (or other form of stimulation) coming from other individuals engaged in the same activity.

Group predation The hunting and retrieving of living prey by groups of cooperating animals. This behavior pattern is developed, for example, in army ants and wolves.

Group selection Selection that operates on two or more members of a lineage group as a unit. Defined broadly, group selection includes both kin selection and interdemic selection (*q.v.*).

Gynandromorphism The existence of both male and female sexual organs in the same individual.

Habitat The organisms and physical environment in a particular place.

Haplodiploidy The mode of sex determination in which males are derived from haploid (hence, unfertilized) eggs and females from diploid, usually fertilized eggs.

Haploid Having a chromosome complement consisting of just one copy of each chromosome. Sex cells are typically haploid. (Contrast with diploid.)

Harem A group of females guarded by a male, who prevents other males from mating with them.

Harvesting ants Ant species that store seeds in their nests. Many taxonomic groups have developed this habit independently in evolution.

Hemimetabolous Undergoing development that is gradual and lacks a sharp separation into larval, pupal, and adult stages. Termites, for example, are hemimetabolous. (Opposed to holometabolous.)

Heritability The fraction of variation of a trait within a population—more precisely, the fraction of its variance, which is the statistical measure—due to heredity as opposed to environmental influences. A heritability score of one means that all the variation is genetic in basis; a heritability score of zero means that all the variation is due to the environment. (See Chapter 4.)

Hermaphroditism The coexistence of both female and male sex organs in the same individual.

Heterozygous Referring to a diploid organism having different alleles of a given gene on the pair of homologous chromosomes carrying that gene. (See chromosome.)

Hierarchy In general, a system of two or more levels of units, the higher levels controlling at least to some extent the activities of the lower levels in order to integrate the group as a whole. In dominance systems within societies, a hierarchy is the sequence of dominant and dominated individuals.

Holometabolous Undergoing a complete metamorphosis during development, with distinct larval, pupal, and adult stages. The Hymenoptera, for example, are holometabolous. (Opposed to hemimetabolous.)

Home range The area that an animal learns thoroughly and patrols regularly. The home range may or may not be defended; those portions that are defended constitute the territory.

Homeostasis The maintenance of a steady state, especially a physiological or social steady state, by means of self-regulation through internal feedback responses.

Hominid Pertaining to man, including early man. A term derived from the family Hominidae, the taxonomic group that includes modern man and his immediate predecessors.

Homo The genus of true men, including several extinct forms (*H. habilis, H. erectus, H. neanderthalensis*) as well as modern man (*H. sapiens*), who are or were primates characterized by completely erect stature, bipedal locomotion, reduced dentition, and above all by an enlarged brain size.

Homogamy Same as assortative mating (*q.v.*).

Homologue Referring to a structure, physiological process, or behavior that is similar to another owing to common ancestry; hence, displaying homology. In genetics a homologue is one chromosome belonging to a group of chromosomes having the same overall genetic composition. (See diploid.)

Homology A similarity between two structures that is due to inheritance from a common ancestor. The structures are said to be homologous. (Contrast with analogy.)

Homopteran A member of, or pertaining to, the insect order Homoptera, which includes the aphids, jumping plant lice, treehoppers, spittlebugs, whiteflies, and related groups.

Homozygous Referring to a diploid organism possessing identical alleles of a given gene on both homologous chromosomes. An organism can be a homozygote with respect to one gene and at the same time a heterozygote with respect to another gene.

Honeybee A member of the genus *Apis*. Unless qualified otherwise, a honeybee is more particularly a member of the domestic species *A. mellifera*, and the term is usually applied to the worker caste.

Honeydew A sugar-rich fluid derived from the phloem sap of plants and passed as excrement through the guts of sap-feeding aphids and other insects. Honeydew is a principal food of many kinds of ants.

Hormone Any substance, secreted by an endocrine gland into the blood or lymph, that affects the physiological activity of other organs in the body; hormones can also influence the nervous system, and through it, the behavior of the organism.

Hymenopteran Pertaining to the insect order Hymenoptera; also, a member of the order, such as a wasp, bee, or ant.

Imago The adult insect. In termites, the term is usually applied only to adult primary reproductives.

Imitation The copying of a novel or otherwise improbable act.

Inbreeding The mating of kin. The degree of inbreeding is measured by the fraction of genes that will be identical owing to common descent. (See inbreeding coefficient; and contrast with outcrossing.)

Inbreeding coefficient Symbolized by f or F, the probability that both alleles (gene forms) on one locus on a pair of chromosomes are identical by virtue of common descent.

Inclusive fitness The sum of an individual's own fitness plus all its influence on fitness in its relatives other than direct descendants; hence the total effect of kin selection with reference to an individual.

Indirect role A behavior or set of behaviors that benefits only the subgroups that display it and is neutral or even destructive to other subgroups of the society. (Opposed to direct role; see discussion in Chapter 14.)

Individual distance   The fixed minimal distance an animal attempts to keep between itself and other members of the species.

Inquilinism   The relation in which a socially parasitic species of insect spends its entire life cycle in the nests of its host species. Workers either are lacking or, if present, are usually scarce and degenerate in behavior. This condition is sometimes referred to loosely as "permanent parasitism."

Insect society   In the strict sense, a colony of eusocial insects (ants, termites, eusocial wasps, or eusocial bees). In the broad sense adopted in this book, any group of presocial or eusocial insects.

Instar   Any period between molts during the course of development of an insect or other arthropod.

Instinct   Behavior that is highly stereotyped, more complex than the simplest reflexes, and usually directed at particular objects in the environment. Learning may or may not be involved in the development of instinctive behavior; the important point is that the behavior develops toward a narrow, predictable end product. (See discussion in Chapter 2.)

Intention movement   The preparatory motions that an animal goes through prior to a complete behavioral response; for example, the crouch before the leap, the snarl before the bite, and so forth.

Intercompensation   The effect caused by the domination of some density-dependent factors over others in population control. If the leading factor, say food shortage, is eliminated, a second factor, for example disease, takes over. This compensation follows a sequence that is peculiar to each species.

Interdemic selection   The selection of entire breeding populations (demes) as the basic unit. One of the extreme forms of group selection, to be contrasted with kin selection (q.v.).

Intrasexual selection   See under sexual selection.

Intrinsic rate of increase   Symbolized by r, the fraction by which a population is growing in each instant of time.

Invertebrate zoology   The scientific study of invertebrate animals.

Invertebrates   All kinds of animals lacking a vertebral column, from protozoans to insects and starfish. (See vertebrates.)

Isogamy   The condition in which the male and female sex cells are of the same size. (Contrast with anisogamy.)

Iteroparity   The production of offspring by an organism in successive groups. (Contrast with semelparity.)

K   Symbol for the carrying capacity of the environment (q.v.).

K extinction   The regular extinction of populations when they are at or near the carrying capacity of the environment (when there are K individuals in the population). (Contrast with r extinction.)

K selection   Selection favoring superiority in stable, predictable environments in which rapid population growth is unimportant. (Contrast with r selection.)

Kin selection   The selection of genes due to one or more individuals favoring or disfavoring the survival and reproduction of relatives (other than offspring) who possess the same genes by common descent. One of the extreme forms of group selection. (Contrast with interdemic selection.)

King   In sociobiology, the male who accompanies the queen (egg-laying female) in termite colonies and inseminates her from time to time.

Kinopsis   Attraction to other members of a society by the sight of their movement alone.

Kinship   Possession of a common ancestor in the not too distant past. Kinship is measured precisely by the coefficient of kinship and coefficient of relationship (q.v.).

Lability   In this book, the term is used with reference to evolutionary lability: the ease and speed with which particular categories of traits evolve. Thus, territorial behavior is usually highly labile, and maternal behavior much less so.

Labium   The lower "lip," or lowermost mouthpart-bearing segment of insects, located just below the mandibles and the maxillae. In zoology generally, any lip or liplike structure.

Langur   An Asiatic monkey belonging to the genus *Presbytis*.

Larva   An immature stage that is radically different in form from the adult; characteristic of many aquatic and marine invertebrate animals and the holometabolous insects, including the Hymenoptera. In the termites, the term is used in a special sense to designate an immature individual without any external trace of wing buds or soldier characteristics.

Leadership As narrowly used in sociobiology, leadership means only the role of leading other members of the society when the group progresses from one place to another.

Legionary ant See army ant.

Lek An area used consistently for communal courtship displays.

Lestobiosis The relation in which colonies of a small species of social insect nest in the walls of the nests of a larger species and enter the chambers of the larger species to prey on brood or to rob the food stores.

Life cycle The entire span of the life of an organism (or of a society) from the moment it originates to the time that it reproduces.

Lineage group A group of species allied by common descent.

Locus The location of a gene on the chromosome.

Logistic growth Growth, especially in the number of organisms constituting a population, that slows steadily as the entity approaches its maximum size. (Compare with exponential growth.)

Macaque Any monkey belonging to the genus *Macaca*, such as the rhesus monkey (*Macaca mulatta*).

Major worker A member of the largest worker subcaste, especially in ants. In ants the subcaste is usually specialized for defense, so that an adult belonging to it is often also referred to as a soldier. (See media worker and minor worker.)

Mammal Any animal of the class Mammalia, characterized by the production of milk by the female mammary glands and the possession of hair for body covering.

Mammalogy The scientific study of mammals.

Marsupial A mammal belonging to the subclass Metatheria; most marsupials, such as opossums and kangaroos, have a pouch (the marsupium) that contains the milk glands and shelters the young.

Mass communication The transfer among groups of individuals of information of a kind that cannot be transmitted from a single individual to another. Examples include the spatial organization of army ant raids, the regulation of numbers of worker ants on odor trails, and certain aspects of the thermoregulation of nests.

Mass provisioning The act of storing all of the food required for the development of a larva at the time the egg is laid. (Opposed to progressive provisioning.)

Matrifocal Pertaining to a society in which most of the activities and personal relationships are centered on the mothers.

Matrilineal Passed from the mother to her offspring, as for example access to a territory or status within a dominance system.

Maturation The automatic development of a behavioral pattern, which becomes increasingly complex and precise as the animal matures. Unlike learning, maturation does not require experience to occur.

Mean The numerical average.

Media worker In polymorphic ant series involving three or more worker subcastes, an individual belonging to the medium-sized subcaste(s). (See minor worker and major worker.)

Meiosis The cellular processes that lead to the formation of sex cells (gametes). In particular, a diploid cell divides twice to form four daughter cells; but the chromosomes are replicated only once, so that the four products are haploid (with only one complement of chromosomes each).

Melittology The scientific study of bees.

Melittophile An organism that must spend at least part of its life cycle with bee colonies.

Mesosoma The middle of the three major divisions of the insect body. In most insects it is the strict equivalent of the thorax, but in higher Hymenoptera it includes the propodeum, the first segment of the abdomen fused to the thorax.

Metacommunication Communication about communication. A metacommunicative signal imparts information about how other signals should be interpreted. Thus a play invitation signal indicates that subsequent threat displays should be taken as play and not as serious hostility.

Metapopulation A set of populations of organisms belonging to the same species and existing at the same time; by definition each population occupies a different area.

Metasoma The hindmost of the three principal divisions of the insect body. In most insect groups it is the strict equivalent of the abdomen. In the higher Hymenoptera it is composed only of some of the abdominal segments, since the first segment (the "propodeum") is fused with the thorax and has therefore become part of the mesosoma.

Metazoan Referring to any or all of the multicellular animals with the exception of the sponges.

Microevolution A small amount of evolutionary change, consisting of minor alterations in gene proportions, chromosome structure, or chromosome numbers. (A larger amount of change would be referred to as macroevolution or simply as evolution.)

Migrant selection Selection based on the different abilities of individuals of different genetic constitution to migrate. For example, if new populations are founded more consistently by individuals with gene *A* as opposed to those bearing gene *a*, gene *A* is said to be favored by migrant selection.

Minima In ants, a minor worker.

Minor worker A member of the smallest worker subcaste, especially in ants. Same as minima. (See media worker and major worker.)

Mixed nest A nest containing colonies of two or more species of social insects, in which mixing of both the adults and brood occurs. (See compound nest.)

Mixed-species flocks Groups of birds belonging to two or more species that travel and forage together.

Mobbing The joint assault on a predator too formidable to be handled by a single individual in an attempt to disable it or at least to drive it from the vicinity.

Molt (moult) The casting off of the outgrown skin or exoskeleton in the process of growth of an insect or other arthropod. Also the cast-off skin itself. The word is further used as an intransitive verb to designate the performance of the behavior.

Monogamy The condition in which one male and one female join to rear at least a single brood.

Monogyny In animals generally, the tendency of each male to mate with only a single female. In social insects, the term also means the existence of only one functional queen in the colony. (Opposed to polygyny.)

Monomorphism In entomology, the existence within an insect species or colony of only a single worker subcaste. (Opposed to polymorphism.)

Monophasic allometry Polymorphism in which the allometric regression line has a single slope; in ants the use of the term also implies that the relation of some of the body parts measured is nonisometric.

Morphogenetic Pertaining to the development of anatomical structures during the growth of an organism.

Multiplier effect In sociobiology, the amplification of the effects of evolutionary change in behavior when the behavior is incorporated into the mechanisms of social organization.

Mutation In the broad sense, any discontinuous change in the genetic constitution of an organism. In the narrow sense, the word refers usually to a "point mutation," a change along a very narrow portion of the nucleic acid sequence.

Mutation pressure Evolution (change in gene frequencies) by different mutation rates alone.

Mutualism Symbiosis in which both species benefit from the association. (Contrast with commensalism and parasitism.)

Myrmecioid complex One of the two major taxonomic groups of ants; the name is based on the subfamily Myrmeciinae, one of the constituent taxa. It should not be confused with the subfamily Myrmicinae, which belongs to the poneroid complex. (See Chapter 20.)

Myrmecology The scientific study of ants.

Myrmecophile An organism that must spend at least part of its life cycle with ant colonies.

Myrmecophytes Higher plants that live in an obligatory mutualistic relationship with ants.

Nasus The snoutlike organ possessed by soldiers of some species in the termite subfamily Nasutitermitinae. The nasus is used to eject poisonous or sticky fluid at intruders.

Nasute soldier A soldier termite possessing a nasus (*q.v.*).

Natural selection The differential contribution of offspring to the next generation by individuals of different genetic types but belonging to the same population. This is the basic mechanism proposed by Charles Darwin and is generally regarded today as the main guiding force in evolution. (See discussion in Chapter 4.)

Necrophoresis Transport of dead members of the colony away from the nest. A highly developed, stereotyped behavior in ants.

Neoteinic A supplementary reproductive termite. Used either as a noun or as an adjective (e.g., neoteinic reproductive).

Nest odor The distinctive odor of a nest, by which its inhabitants are able to distinguish the nest from those belonging to other societies or a least from the surrounding environment. In some cases the animals, e.g., honeybees and some ants, can orient toward the nest by means of the odor. It is possible that the nest odor is the same as the colony odor in some cases. The nest odor of honeybees is often referred to as the hive aura or hive odor.

Nest parasitism   The relation, found in some termites, in which colonies of one species live in the walls of the nests of a second, host species and feed directly on the carton material of which they are constructed.

Net reproductive rate   Symbolized by R_0, the average number of female offspring produced by each female during her entire lifetime.

Neurophysiology   The scientific study of the nervous system, especially the physiological processes by which it functions.

Niche   The range of each environmental variable, such as temperature, humidity, and food items, within which a species can exist and reproduce. The preferred niche is the one in which the species performs best, and the realized niche is the one in which it actually comes to live in a particular environment.

Nomadic phase   The period in the activity cycle of an army ant colony during which the colony forages more actively for food and moves frequently from one bivouac site to another. At this time the queen does not lay eggs, and the bulk of the brood is in the larval stage. (Opposed to statary phase.)

Nomadism   The relatively frequent movement by an entire society from one nest site or home range to another.

Nomogram   A graph that lines up two scales (for example the Celsius and Fahrenheit temperature scales) and matches them point for point.

Nuptial flight   The mating flight of the winged queens and males of an insect society.

Nymph   In general entomology, the young stage of any insect species with hemimetabolous development. In termites, the term is used in a slightly more restricted sense to designate immature individuals that possess external wing buds and enlarged gonads and that are capable of developing into functional reproductives by further molting.

Observational learning   Unrewarded learning that occurs when one animal watches the activities of another. Same as empathic learning.

Odor trail   A chemical trace laid down by one animal and followed by another. The odorous material is referred to either as the trail pheromone or as the trail substance.

Oligogyny   The occurrence in a single colony of social insects of from two to several functional queens. A special case of polygyny.

Omnivore   An animal that eats both animal and vegetable materials.

Ontogeny   The development of a single organism through the course of its life history. (Contrast with phylogeny.)

Opportunistic species   Species specialized to exploit newly opened habitats. Such species usually are able to disperse for long distances and to reproduce rapidly; in other words they are r selected (q.v.).

Optimal yield   The highest rate of increase that a population can sustain in a given environment. Theoretically, there exists a particular size, less than the carrying capacity, at which this yield is realized.

Order   In taxonomy, the category below the class and above the family; a group of related, similar families. Examples of orders include the Hymenoptera, which includes the wasps, ants, and bees; and the Primates, which includes the monkeys, apes, man, and other primates.

Organism   Any living creature.

Ornithology   The scientific study of birds.

Outcrossing   The pairing of unrelated individuals. (Contrast with inbreeding.)

Ovariole   One of the egg tubes that, together, form the ovary in female insects.

Pair bonding   A close and long-lasting association formed between a male and female; in animals at least, pair bonding serves primarily for the cooperative rearing of young.

Palpation   Touching with the labial or maxillary palps. The movement can serve as a sensory probe or as a tactile signal to another insect.

Panmictic   Referring to a population in which mating is completely random; a panmictic population is often referred to as a deme.

Parabiosis   The utilization of the same nest and sometimes even the same odor trails by colonies of different species of ants, which nevertheless keep their brood separate.

Parameter   In strict, mathematical usage a parameter is a quantity that can be held constant in a model while other quantities are being varied to study their relationships, but changed in value in other particular versions of the same model. Thus r, the rate of population increase, is a parameter that can be held constant at a particular value when varying N (number of organisms) and t (time), but changed to some new value in other versions of the same population-growth model. The term parameter is also used loosely to designate any variable property that exerts an effect upon a system.

Parasitism   Symbiosis in which members of one species exist at the expense of members of another species, usually without going so far as to cause their deaths.

Parasocial   See presocial.

Parental investment   Any behavior toward offspring that increases the chances of the offspring's survival at the cost of the parent's ability to invest in other offspring.

Parthenogenesis   The production of an organism from an unfertilized egg.

Partially claustral colony founding   The procedure during which an ant queen founds the colony by isolating herself in a chamber but still occasionally leaves to forage for part of her food supply.

Path analysis   A graphical mode of analysis used to determine the inbreeding coefficient.

Patroling   The act of investigating the nest interior. Worker honeybees are especially active in patroling and are thereby quick to respond as a group to contingencies when they arise in the nest.

Peck order   A term sometimes applied to a dominance order, especially in birds.

Pedicel   The "waist" of the ant. It is made up of either one segment (the petiole) or two segments (the petiole plus the postpetiole).

Perissodactyl   Any mammal belonging to the order Perissodactyla, hence an ungulate with an odd number of toes in its hooves, such as a tapir or rhinoceros. (Contrast with artiodactyl.)

Permeability   The degree of openness of a society to new members.

Phenodeviant   An individual of a scarce, aberrant kind due to the segregation of unusual combinations of individually common genes.

Phenotype   The observable properties of an organism as they have developed under the combined influences of the genetic constitution of the individual and the effects of environmental factors. (Contrast with genotype.)

Pheromone   A chemical substance, usually a glandular secretion, that is used in communication within a species. One individual releases the material as a signal and another responds after tasting or smelling it.

Philopatry   The tendency of animals to remain at certain places or at least to return to them for feeding and resting.

Phyletic group   A group of species related to one another through common descent.

Phylogenetic group   Same as phyletic group.

Phylogenetic inertia   See under prime movers.

Phylogeny   The evolutionary history of a particular group of organisms; also, the diagram of the "family tree" that shows which species (or groups of species) gave rise to others. (Contrast with ontogeny.)

Phylum   In taxonomy, a high-level category just beneath the kingdom and above the class; a group of related, similar classes. Examples of phyla include the Arthropoda, or all the crustaceans, spiders, insects, and related forms; and the Chordata, which includes the vertebrates, tunicates, and related forms.

Physiology   The scientific study of the functions of organisms and of the individual organs, tissues, and cells of which they are composed. In its broadest sense physiology also encompasses most of molecular biology and biochemistry.

Placenta   The organ, found in most mammals, that provides for the nourishment of the fetus and elimination of the fetal waste products. It is formed by the union of membranes from the fetus and the mother.

Placental   Pertaining to mammals belonging to the subclass Eutheria, a group that is characterized by the presence of a placenta in the female and that contains the great majority of living mammal species. (Contrast with marsupial.)

Plasmodium   A stage in the life cycle of true slime molds (Myxomycetales) in which a mass of tissue containing multiple nuclei but no distinct cell boundaries grows and spreads by nuclear division and accretion of cytoplasm. (Opposed to pseudoplasmodium, q.v.)

Pleiotropism   The control of more than one phenotypic characteristic, for example eye color, courtship behavior, or size, by the same gene or set of genes.

Plesiobiosis   The close proximity of two or more nests, accompanied by little or no direct communication between the colonies inhabiting them.

Pod   A school of fish in which the bodies of the individuals actually touch. Also, a group of whales.

Point mutation   A mutation resulting from a small, localized alteration in the chemical structure of a gene.

Pollen storers   Bumblebee species that store pollen in abandoned cocoons. From time to time the adult females remove the pollen from the cocoons and feed it into a larval cell in the form of a liquid mixture of pollen and honey. (Opposed to pouch makers.)

Polyandry   The acquisition by a female of more than one male as a mate. In the narrower sense of zoology, polyandry usually means that the males also cooperate with the female in raising a brood.

Polydomous   Pertaining to single colonies that occupy more than one nest.

Polyethism   Division of labor among members of a society. In social insects a distinction can be made between caste polyethism, in which morphological castes are specialized to serve different functions, and age polyethism, in which the same individual passes through different forms of specialization as it grows older. These two kinds of castes are referred to as physical castes and temporal castes, respectively.

Polygamy   The acquisition, as part of the normal life cycle, of more than one mate. Polygyny: more than one female to a male. Polyandry: more than one male to a female. In the narrower sense of zoology, polygamy usually also implies a relationship in which the partners cooperate to raise a brood.

Polygenes   Genes affecting the same trait but located on two or more loci on the chromosomes.

Polygyny   In animals generally, the tendency of each male to mate with two or more females. In strict usage, the male also cooperates to some extent in rearing the young. In social insects, the term also means the coexistence in the same colony of two or more egg-laying queens. When multiple queens found a colony together, the condition is referred to as primary polygyny. When supplementary queens are added after colony foundation, the condition is referred to as secondary polygyny. The coexistence of only two or several queens is sometimes called oligogyny. (Opposed to monogyny.)

Polymorphism   In social insects, the coexistence of two or more functionally different castes within the same sex. In ants it is possible to define polymorphism somewhat more precisely as the occurrence of nonisometric relative growth occurring over a sufficient range of size variation within a normal mature colony to produce individuals of distinctly different proportions at the extremes of the size range. In genetics, polymorphism is the maintenance of two or more forms of gene on the same locus at higher frequencies than would be expected by mutation and immigration alone.

Poneroid complex   One of the two major taxonomic groups of ants; the name is based on the subfamily Ponerinae, one of the constituent taxa. (See Chapter 20.)

Pongid   Any anthropoid ape other than a gibbon or siamang; the larger living apes (chimpanzee, gorilla, and orang-utan) together with certain fossil forms constitute the family Pongidae.

Population   A set of organisms belonging to the same species and occupying a clearly delimited space at the same time. A group of populations of the same species, each of which by definition occupies a different area, is sometimes called a metapopulation.

Postadaptation   An adaptation in the strict sense, meaning an evolutionary change in some trait that occurred in response to a particular selection pressure from the environment and did not accidentally precede it. (Contrast with preadaptation.)

Pouch makers   Bumblebee species that build special wax pouches adjacent to groups of larvae and fill them with pollen. (Opposed to pollen storers.)

Preadaptation   Any previously existing anatomical structure, physiological process, or behavior pattern that makes new forms of evolutionary adaptation more likely. (Contrast with adaptation and postadaptation.)

Precocial   Referring to young animals who are able to move about and forage at a very early age; especially in birds. (Contrast with altricial.)

Predator   Any organism that kills and eats other organisms.

Preferred niche   See under niche.

Presocial   Especially in insects, applied to the condition or to the group possessing it in which individuals display some degree of social behavior short of eusociality. Presocial species are either subsocial, i.e., the parents care for their own nymphs and larvae; or else parasocial, i.e., one or two of the following three traits are shown: cooperation in care of young, reproductive division of labor, and overlap of generations of life stages that contribute to colony labor.

Primary reproductive   In termites, the colony-founding type of queen or male derived from the winged adult.

Primate   Any member of the order Primates, such as a lemur, monkey, ape, or man.

Prime movers   The ultimate factors that determine the direction and velocity of evolutionary change. There are two kinds of prime movers: phylogenetic inertia, which includes basic genetic mechanisms and prior adaptations that make certain changes more likely or less likely; and ecological pressure, the set of all environmental influences that constitute the agents of natural selection. (See discussion in Chapter 3.)

Primer pheromone A pheromone (chemical signal) that acts to alter the physiology of the receiving organism in some way and eventually causes the organism to respond differently. (Contrast with releaser pheromone.)

Primitive Referring to a trait that appeared first in evolution and gave rise to other, more "advanced" traits later. Primitive traits are often but not always less complex than the advanced ones.

Progressive provisioning The feeding of a larva in repeated meals. (Opposed to mass provisioning.)

Prosimian Any primate, such as a lemur or tarsier, belonging to the primitive suborder Prosimii.

Protease An enzyme that catalyzes the digestion of proteins.

Protistan Referring to the kingdom Protista, which embraces most of what used to be included in the old phylum Protozoa, including the flagellates, amebas, ciliates, and a few other unicellular organisms.

Protozoa A group of single-celled organisms classified by some zoologists as a single phylum; it includes the flagellates, amebas, and ciliates.

Proximate causation The conditions of the environment or internal physiology that trigger the responses of an organism. They are to be distinguished from the environmental forces, referred to as the ultimate causation, that led to the evolution of the response in the first place.

Pseudergate A special caste found in the lower termites, comprised of individuals who either have regressed from nymphal stages by molts that reduced or eliminated the wing buds, or else were derived from larvae by undergoing nondifferentiating molts. Pseudergates serve as the principal elements of the worker caste, but remain capable of developing into other castes by further molting.

Pseudoplasmodium The motile, sluglike organism formed by the aggregation of the amebas of cellular slime molds.

Pupa The inactive developmental stage of the holometabolous insects (including the Hymenoptera) during which development into the final adult form is completed.

Pupate In insects, to change from a larva into a pupa.

Quasisocial Applied to the condition or to the group showing it in which members of the same generation use the same composite nest and cooperate in brood care.

Queen A member of the reproductive caste in semisocial or eusocial insect species. The existence of a queen caste presupposes the existence also of a worker caste at some stage of the colony life cycle. Queens may or may not be morphologically different from workers.

Queen substance Originally, the set of pheromones by which the queen honeybee continuously attracts the workers and controls their reproductive activities. The term is commonly used in a narrower sense to designate *trans*-9-keto-2-decenoic acid, the most potent component of the pheromone mixture. But it may also be defined more broadly in line with the original usage as any pheromone or set of pheromones used by a queen to control the reproductive behavior of the workers or of other queens.

Queenright Referring to a colony, especially a honeybee colony, that contains a functional queen.

r The symbol used to designate either the intrinsic rate of increase of a population, or the degree of relationship of two individuals.

r extinction The extinction of entire populations shortly after colonization and while they are in an early stage of growth and expansion. (Contrast with *K* extinction.)

r selection Selection favoring rapid rates of population increase, especially prominent in species that specialize in colonizing short-lived environments or undergo large fluctuations in population size. (Contrast with *K* selection.)

Race See subspecies.

Ramapithecus A small primate that lived in the Old World approximately 15 million years ago; its dental characteristics make it a likely candidate as one of the direct ancestors of man-apes (*Australopithecus*), which in turn gave rise to true men (*Homo*).

Realized niche See under niche.

Recessive In genetics, referring to an allele the phenotype of which is suppressed when it occurs in combination with a dominant allele.

Reciprocal altruism The trading of altruistic acts by individuals at different times. For example, one person saves a drowning person in exchange for the promise (or at least the expectation) that his altruistic act will be repaid if the circumstances are reversed at some time in the future.

Recombination The repeated formation of new combinations of genes through the processes of meiosis and fertilization that occur in the typical sexual cycle of most kinds of organisms.

Recruitment A special form of assembly by which members of a society are directed to some point in space where work is required.

Redirected activity The direction of some behavior, such as an act of aggression, away from the primary target and toward another, less appropriate object.

Reinforcing selection The operation of selection pressures on two or more levels of organization, for example on the individual, family, and population, in such a way that certain genes are favored at all levels and their spread through the population is accelerated.

Releaser A sign stimulus used in communication. Often the term is used broadly to mean any sign stimulus.

Releaser pheromone A pheromone (chemical signal) that is quickly perceived and causes a more or less immediate response. (Contrast with primer pheromone.)

Replete An individual ant whose crop is greatly distended with liquid food, to the extent that the abdominal segments are pulled apart and the intersegmental membranes are stretched tight. Repletes usually serve as living reservoirs, regurgitating food on demand to their nestmates.

Reproductive effort The effort required to reproduce, measured in terms of the decrease in the ability of the organism to reproduce at later times.

Reproductive success The number of surviving offspring of an individual.

Reproductive value Symbolized by v_x, the relative number of female offspring remaining to be born to each female of age x.

Reproductivity effect In social insects, the relation in which the rate of production of new individuals per colony member drops as the colony size increases.

Retinue The group of workers in an insect colony that surrounds the queen, especially in honeybees; the composition of the retinue changes constantly. Also known as the court.

Ritualization The evolutionary modification of a behavior pattern that turns it into a signal used in communication or at least improves its efficiency as a signal.

Role A pattern of behavior displayed by certain members of a society that has an effect on other members. (See discussion in Chapter 14.)

Royal cell In honeybees, the large, pitted, waxen cell constructed by the workers to rear queen larvae. In some species of termites, the special cell in which the queen is housed.

Royal jelly A material, supplied by workers to female larvae in royal cells, that is necessary for the transformation of larvae into queens. Royal jelly is secreted primarily by the hypopharyngeal glands and consists of a rich mixture of nutrient substances, many of them possessing a complex chemical structure.

Scaling See behavioral scaling.

School A group of fish or fishlike animals such as squid that swim together in an organized fashion; all or most of the school members are typically in the same stage of the life cycle.

Sclerite A portion of the insect body wall bounded by sutures.

Selection pressure Any feature of the environment that results in natural selection; for example, food shortage, the activity of a predator, or competition from other members of the same sex for a mate can cause individuals of different genetic types to survive to different average ages, to reproduce at different rates, or both.

Self-grooming Grooming directed at one's own body. Opposed to allogrooming, or the grooming of another individual.

Selfishness In the strict usage of sociobiology, behavior that benefits the individual in terms of genetic fitness at the expense of the genetic fitness of other members of the same species. (Compare with altruism and spite.)

Sematectonic Referring to communication by means of constructed objects. Examples include the sand pyramids of male ghost crabs and various portions of the nest structures of social insects.

Semelparity The production of offspring by an organism in one group all at the same time. (Contrast with iteroparity.)

Semiotics The scientific study of communication.

Semisocial In social insects, applied to the condition or to the group showing it in which members of the same generation cooperate in brood care and there is also a reproductive division of labor, i.e., some individuals are primarily egg layers and some are primarily workers.

Sensory physiology The study of sensory organs and the ways in which they receive stimuli from the environment and transmit them to the nervous system.

Sex determination The process by which the sex of an individual is determined. For example, the presence of a Y chromosome in a human embryo causes the fetus to develop into a male; while the fertilization of wasp and ant eggs causes them to develop into females.

Sex ratio The ratio of males to females (for example, 3 males/1 female) in a population, a society, a family, or any other group chosen for convenience.

Sexual dimorphism Any consistent difference between males and females beyond the basic functional portions of the sex organs.

Sexual selection The differential ability of individuals of different genetic types to acquire mates. Sexual selection consists of epigamic selection, based on choices made between males and females, and intra-sexual selection, based on competition between members of the same sex.

Sib A close kinsman, especially a brother or sister.

Sign stimulus The single stimulus, or one out of a very few such crucial stimuli, by which an animal distinguishes key objects such as enemies, potential mates, and suitable nesting places.

Signal In sociobiology, any behavior that conveys information from one individual to another, regardless of whether it serves other functions as well. A signal specially modified in the course of evolution to convey information is called a display.

Social drift Random divergence in the behavior and mode of organization of societies.

Social facilitation An ordinary pattern of behavior that is initiated or increased in pace or frequency by the presence or actions of another animal.

Social homeostasis The maintenance of steady states at the level of the society either by control of the nest microclimate or by the regulation of the population density, behavior, and physiology of the group members as a whole.

Social insect In the strict sense, a "true social insect" is one that belongs to a eusocial species: in other words, it is an ant, a termite, or one of the eusocial wasps or bees. In the broad sense, a "social insect" is one that belongs to either a presocial or eusocial species.

Social parasitism The coexistence of two species of animals, of which one is parasitically dependent on the societies of the other.

Social releaser See releaser.

Sociality The combined properties and processes of social existence.

Socialization The total modification of behavior in an individual due to its interaction with other members of the society, including its parents.

Society A group of individuals belonging to the same species and organized in a cooperative manner. The diagnostic criterion is reciprocal communication of a cooperative nature, extending beyond mere sexual activity. (See further discussion in Chapter 2.)

Sociobiology The systematic study of the biological basis of all social behavior. (See discussion in Chapter 1.)

Sociocline A series of different social organizations observed among related species and interpreted to represent stages in an evolutionary trend.

Sociogram The full description, taking the form of a catalog, of all the social behaviors of a species.

Sociology The study of human societies.

Sociotomy Same as colony fission.

Soldier A member of a worker subcaste specialized for colony defense.

Song In the study of animal behavior, any elaborate vocal signal.

Speciation The processes of the genetic diversification of populations and the multiplication of species.

Species The basic lower unit of classification in biological taxonomy, consisting of a population or series of populations of closely related and similar organisms. The somewhat more narrowly defined "biological species" consists of individuals that are capable of interbreeding freely with one another but not with members of other species under natural conditions.

Species odor The odor found on the bodies of social insects that is peculiar to a given species. It is possible that the species odor is merely the less distinctive component of a larger mixture comprising the colony odor (q.v.).

Spermatheca The receptacle in a female insect in which the sperm are stored.

Sphecology The scientific study of wasps.

Spite In the strict terminology of evolutionary biology, behavior that lowers the genetic fitnesses of both the perpetrator and the individual toward which the behavior is directed.

Stable age distribution The condition in which the proportions of individuals belonging to different age groups remain constant for generation after generation.

Stabilizing selection   Selection that operates against the extremes of variation in a population and hence tends to stabilize the population around the mean. (Contrast with directional selection and disruptive selection.)

Statary phase   The period in the activity cycle of an army ant colony during which the colony is relatively quiescent and does not move from site to site. At this time the queen lays the eggs and the bulk of the brood is in the egg and pupal stages. (Opposed to nomadic phase.)

Steady state   An apparently unchanging condition due to a balance between the synthesis (or arrival) and degradation (or departure) of all relevant components of a system.

Stochastic   Referring to the properties of mathematical probability. A stochastic model takes into account variations in outcome that are due to chance alone. (Contrast with deterministic.)

Straight run   The middle run made by a honeybee worker during the waggle dance and the element that contains most of the symbolical information concerning the location of the target outside the hive. The dancing bee makes a straight run, then loops back to the left (or right), then makes another straight run, then loops back in the opposite direction, and so on—the three basic movements together form the characteristic figure-eight pattern of the waggle dance.

Stridulation   The production of sound by rubbing one part of the body surface against another. A common form of communication in insects.

Subsocial   In the study of social insects, applied to the condition or to the group showing it in which the adults care for their nymphs or larvae for some period of time. (See also presocial.)

Subspecies   A subdivision of a species. Usually defined narrowly as a geographical race: a population or series of populations occupying a discrete range and differing genetically from other geographical races of the same species.

Superfamily   In taxonomy, the category between the family and the order; thus an order consists of a set of one or more superfamilies. Examples of superfamilies include the Apoidea, or all of the bees; and the Formicoidea, or all of the ants.

Superorganism   Any society, such as the colony of a eusocial insect species, possessing features of organization analogous to the physiological properties of a single organism. The insect colony, for example, is divided into reproductive castes (analogous to gonads) and worker castes (analogous to somatic tissue); it may exchange nutrients by trophallaxis (analogous to the circulatory system), and so forth.

Supplementary reproductive   A queen or male termite that takes over as the functional reproductive after the removal of the primary reproductive of the same sex. Supplementary reproductives are adultoid, mymphoid, or workerlike in form.

Surface pheromone   A pheromone with an active space restricted so close to the body of the sending organism that direct contact, or something approaching it, must be made with the body in order to perceive the pheromone. Examples include the colony odors of many species of social insects.

Swarming   In honeybees, the normal method of colony reproduction, in which the queen and a large number of workers depart suddenly from the parental nest and fly to some exposed site. There they cluster while scout workers fly in search of a suitable new nest cavity. In ants and termites, the term "swarming" is often applied to the mass exodus of reproductive forms from the nests at the beginning of the nuptial flight.

Symbiont   An organism that lives in symbiosis with another species.

Symbiosis   The intimate, relatively protracted, and dependent relationship of members of one species with those of another. The three principal kinds of symbiosis are commensalism, mutualism, and parasitism (q.v.).

Sympatric   Referring to populations, particularly species, the geographical ranges of which at least partially overlap. (Contrast with allopatric.)

Symphile   A symbiont, in particular a solitary insect or other kind of arthropod, that is accepted to some extent by an insect colony and communicates with it amicably. Most symphiles are licked, fed, or transported to the host brood chambers, or treated to a combination of these actions.

Syngamy   The final step in fertilization, in which the nuclei of the sex cells meet and fuse.

Tandem running   A form of communication, used by the workers of certain ant species during exploration or recruitment, in which one individual follows closely behind another, frequently contacting the abdomen of the leader with its antennae.

Taxis   The movement of an organism in a fixed direction with reference to a single stimulus. Thus, a phototaxis is a movement toward or away from a light, a geotaxis is a movement up or down in response to gravity, and so forth.

Taxon (plural: taxa)   Any group of organisms representing a particular unit of classification, such as all the members of a given subspecies or of a species, genus, and so on. Thus *Homo sapiens*, the species of man, is one taxon; so is the order Primates, embracing all the species of monkeys, apes, men, and other kinds of primates.

Taxon cycle A cycle in which a species spreads while adapted to one habitat and restricts its range and splits into two or more species while adapting to another habitat. Taxon cycles have been especially noted in large systems of islands; dispersal commonly occurs in open habitats and restriction and speciation in forests.

Taxonomy The science of classification, especially of organisms.

Temporal polyethism Same as age polyethism (*q.v.*).

Temporary social parasitism In social insects, parasitism in which a queen of one species enters an alien nest, usually belonging to another species, kills or renders infertile the resident queen, and takes her place. The population of the colony then becomes increasingly dominated by the offspring of the parasite queen as the host workers die off from natural causes.

Termitarium A termite nest. Also, an artificial nest used in the laboratory to house termites.

Termitology The scientific study of termites.

Termitophile An organism that must spend at least part of its life cycle with termite colonies.

Territory An area occupied more or less exclusively by an animal or group of animals by means of repulsion through overt defense or advertisement.

Time-energy budget The amounts of time and energy allotted by animals to various activities.

Total range The entire area covered by an individual in its lifetime.

Tradition A specific form of behavior, or a particular site used for breeding or some other function, passed from one generation to the next by learning.

Tradition drift Social drift (random divergence in social behavior) that is based purely on differences in experience and hence passed on as part of tradition.

Trail pheromone A substance laid down in the form of a trail by one animal and followed by another member of the same species.

Trail substance Same as trail pheromone.

Troop In sociobiology, a group of lemurs, monkeys, apes, or some other kind of primate.

Trophallaxis In social insects, the exchange of alimentary liquid among colony members and guest organisms, either mutually or unilaterally. In stomodeal (oral) trophallaxis the material originates from the mouth; in proctodeal (anal) trophallaxis it originates from the anus.

Trophic Pertaining to food.

Trophic egg An egg, usually degenerate in form and inviable, that is fed to other members of the colony.

Trophic level The position of a species in a food chain, determined by which species it consumes and which consume it.

Trophic parasitism The intrusion of one species into the social system of another, for example by utilization of the trail system, in order to steal food.

Trophobiosis The relationship in which ants receive honeydew from aphids and other homopterans, or the caterpillars of certain lycaenid and riodinid butterflies, and provide these insects with protection in return. The insects providing the honeydew are referred to as trophobionts.

Ultimate causation The conditions of the environment that render certain traits adaptive and others nonadaptive; hence the adaptive traits tend to be retained in the population and are "caused" in this ultimate sense. (Contrast with proximate causation.)

Umwelt A German expression (loosely translated, "the world around me") used to indicate the total sensory input of an animal. Each species, including man, has its own distinctive *Umwelt*.

Unicolonial Pertaining to a population of social insects in which there are no behavioral colony boundaries. Thus the entire local population consists of one colony. (Opposed to multicolonial.)

Variance The most commonly used statistical measure of variation (dispersion) of a trait within a population. It is the mean squared deviation of all individuals from the sample mean.

Vertebrate zoology The scientific study of vertebrate animals.

Vertebrates Animals having a vertebral column ("backbone"), including all of the fishes, amphibians, reptiles, birds, and mammals.

Viscosity In sociobiology and population genetics, the slowness of individual dispersal and hence the low rate of gene flow.

Waggle dance The dance whereby workers of various species of honeybees (genus *Apis*) communicate the location of food finds and new nest sites. The dance is basically a run through a figure-eight pattern, with the middle, transverse line of the eight containing the information about the direction and distance of the target. (See Chapter 8.)

Ward A group of prairie dog societies (coteries) separated from others by some kind of physical barrier, such as a stream or ridge.

Worker A member of the nonreproductive, laboring caste in semisocial and eusocial insect species. The existence of a worker caste presupposes the existence also of royal (reproductive) castes. In termites, the term is used in a more restricted sense to designate individuals in the family Termitidae that completely lack wings and have reduced pterothoraces, eyes, and genital apparatus.

Xenobiosis The relation in which colonies of one species live in the nests of another species and move freely among the hosts, obtaining food from them by regurgitation or other means but still keeping their brood separate.

Zoology The scientific study of animals.

Zoosemiotics The scientific study of animal communication.

Zygote The cell created by the union of two gametes (sex cells), in which the gamete nuclei are also fused. The earliest stage of the diploid generation.

Bibliography

Ables, E. D. 1969. Home-range studies of red foxes (*Vulpes vulpes*). *Journal of Mammalogy*, 50(1): 108–120.

Ackerman, R., and P. D. Weigl. 1970. Dominance relations of red and grey squirrels. *Ecology*, 51(2): 332–334.

Adams, R. McC. 1966. *The evolution of urban society: early Mesopotamia and prehispanic Mexico.* Aldine Publishing Co., Chicago. xii + 191 pp.

Ader, R., and P. M. Conklin. 1963. Handling of pregnant rats: effects on emotionality of their offspring. *Science*, 142: 411–412.

Adler, N. T. 1969. Effects of the male's copulatory behavior on successful pregnancy of the female rat. *Journal of Comparative and Physiological Psychology*, 69(4): 613–622.

Adler, N. T., J. A. Resko, and R. W. Goy. 1970. The effect of copulatory behavior on hormonal change in the female rat prior to implantation. *Physiology and Behavior*, 5(9): 1003–1007.

Adler, N. T., and S. R. Zoloth. 1970. Copulatory behavior can inhibit pregnancy in female rats. *Science*, 168: 1480–1482.

Albignac, R. 1973. *Mammiferes Carnivores.* Faune de Madagascar, no. 36. Centre National de Recherche Scientifique, Paris. 206 pp.

Alcock, J. 1969. Observational learning in three species of birds. *Ibis*, 111(3): 308–321.

——— 1972. The evolution of the use of tools by feeding animals. *Evolution*, 26(3): 464–473.

Aldrich-Blake, F. P. G. 1970. Problems of social structure in forest monkeys. In J. H. Crook, ed. (*q.v.*), *Social behaviour in birds and mammals: essays on the social ethology of animals and man*, pp. 79–101.

Alexander, B. K., and Jennifer Hughes. 1971. Canine teeth and rank in Japanese monkeys (*Macaca fuscata*). *Primates*, 12(1): 91–93.

Alexander, R. D. 1961. Aggressiveness, territoriality, and sexual behavior in field crickets (Orthoptera: Gryllidae). *Behaviour*, 17(2,3): 130–223.

——— 1962. Evolutionary change in cricket acoustical communication. *Evolution*, 16(4): 443–467.

——— 1968. Arthropods. In T. A. Sebeok, ed. (*q.v.*), *Animal communication: techniques of study and results of research*, pp. 167–216.

——— 1971. The search for an evolutionary philosophy of man. *Proceedings of the Royal Society of Victoria*, 84(1): 99–120.

——— 1974. The evolution of social behavior. *Annual Review of Ecology and Systematics*, 5: 325–383.

Alexander, R. D., and T. E. Moore. 1962. The evolutionary relationships of 17-year and 13-year cicadas, and three new species (Homoptera, Cicadidae, *Magicicada*). *Miscellaneous Publications, Museum of Zoology, University of Michigan, Ann Arbor*, 21: 1–57.

Alexander, T. R. 1964. Observations on the feeding behavior of *Bufo marinus* (Linné). *Herpetologica*, 20(4): 255–259.

Alexopoulos, C. J. 1963. The Myxomycetes II. *Botanical Review*, 29(1): 1–78.

Alibert, J. 1968. Influence de la société et de l'individu sur la trophallaxie chez *Calotermes flavicollis* Fabr. et *Cubitermes fungifaber* (Isoptera). In R. Chauvin and C. Noirot, eds. (*q.v.*), *L'effet de groupe chez les animaux*, pp. 237–288.

Allee, W. C. 1926. Studies in animal aggregations: causes and effects of bunching in land isopods. *Journal of Experimental Zoology*, 45: 255–277.

———— 1931. *Animal aggregations: a study in general sociology.* University of Chicago Press, Chicago. ix + 431 pp.

———— 1938. *The social life of animals.* W. W. Norton, New York. 293 pp.

———— 1942. Group organization among vertebrates. *Science*, 95: 289–293.

Allee, W. C., N. E. Collias, and Catherine Z. Lutherman. 1939. Modification of the social order in flocks of hens by the injection of testosterone propionate. *Physiological Zoology*, 12(4): 412–440.

Allee, W. C., and J. C. Dickinson, Jr. 1954. Dominance and subordination in the smooth dogfish *Mustelus canis* (Mitchill). *Physiological Zoology*, 27(4): 356–364.

Allee, W. C., A. E. Emerson, O. Park, T. Park, and K. P. Schmidt. 1949. *Principles of Animal Ecology.* W. B. Saunders Co., Philadelphia. xii + 837 pp.

Allee, W. C., and A. M. Guhl. 1942. Concerning the group survival value of the social peck order. *Anatomical Record*, 84(4): 497–498.

Alpert, G. D., and R. D. Akre. 1973. Distribution, abundance, and behavior of the inquiline ant *Leptothorax diversipilosus*. *Annals of the Entomological Society of America*, 66(4): 753–760.

Altmann, Margaret. 1956. Patterns of herd behavior in free-ranging elk of Wyoming, *Cervus canadensis nelsoni*. *Zoologica, New York*, 41(2): 65–71.'

———— 1958. Social integration of the moose calf. *Animal Behaviour*, 6(3,4): 155–159.

———— 1960. The role of juvenile elk and moose in the social dynamics of their species. *Zoologica, New York*, 45(1): 35–39.

———— 1963. Naturalistic studies of maternal care in moose and elk. In Harriet L. Rheingold, ed. (*q.v.*), *Maternal behavior in mammals*, pp. 233–253.

Altmann, S. A. 1956. Avian mobbing behavior and predator recognition. *Condor*, 58(4): 241–253.

———— 1959. Field observations on a howling monkey society. *Journal of Mammalogy*, 40(3): 317–330.

———— 1962a. A field study of the sociobiology of rhesus monkeys, *Macaca mulatta*. *Annals of the New York Academy of Sciences*, 102(2): 338–435.

———— 1962b. Social behavior of anthropoid primates: analysis of recent concepts. In E. L. Bliss, ed., *Roots of behavior*, pp. 277–285. Harper and Brothers, New York. xi + 339 pp.

———— 1965a. Sociobiology of rhesus monkeys: II, stochastics of social communication. *Journal of Theoretical Biology*, 8(3): 490–522.

———— ed. 1965b. *Japanese monkeys, a collection of translations*, selected by K. Imanishi. The Editor, Edmonton. v + 151 pp.

———— ed. 1967a. *Social communication among primates.* University of Chicago Press, Chicago. xiv + 392 pp.

———— 1967b. Preface. In S. A. Altmann, ed. (*q.v.*), *Social communication among primates*, pp. ix–xii.

———— 1967c. *The structure of primate social communication.* In S. A.

Altmann, ed. (*q.v.*), *Social communication among primates*, pp. 325–362.

———— 1969. Review of *Social organization of hamadryas baboons: a field study*, by H. Kummer. *American Anthropologist*, 71(4): 781–783.

Altmann, S. A., and Jeanne Altmann. 1970. *Baboon ecology: African field research.* University of Chicago Press, Chicago. viii + 220 pp.

Alverdes, F. 1927. *Social life in the animal world.* Harcourt, Brace, London. ix + 216 pp.

Amadon, D. 1964. The evolution of low reproductive rates in birds. *Evolution*, 18(1): 105–110.

Anderson, P. K. 1961. Density, social structure, and nonsocial environment in house-mouse populations and the implications for regulation of numbers. *Transactions of the New York Academy of Sciences*, 2d ser., 23(5): 447–451.

———— 1970. Ecological structure and gene flow in small mammals. *Symposia of the Zoological Society of London*, 26: 299–325.

Anderson, S., and J. K. Jones, Jr., eds. 1967. *Recent mammals of the world: a synopsis of families.* Ronald Press Co., New York. viii + 453 pp.

Anderson, W. W., and C. E. King. 1970. Age-specific selection. *Proceedings of the National Academy of Sciences, U.S.A.*, 66(3): 780–786.

Andrew, R. J. 1956. Intention movements of flight in certain passerines, and their uses in systematics. *Behaviour*, 10(1,2): 179–204.

———— 1961a. The motivational organisation controlling the mobbing calls of the blackbird (*Turdus merula*): I, effects of flight on mobbing calls. *Behaviour*, 17(2,3): 224–246.

———— 1961b. The motivational organisation controlling the mobbing calls of the blackbird (*Turdus merula*): II, the quantitative analysis of changes in the motivation of calling. *Behaviour*, 17(4): 288–321.

———— 1961c. The motivational organisation controlling the mobbing calls of the blackbird (*Turdus merula*): III, changes in the intensity of mobbing due to changes in the effect of the owl or to the progressive waning of mobbing. *Behaviour*, 18(1,2): 25–43.

———— 1961d. The motivational organisation controlling the mobbing calls of the blackbird (*Turdus merula*): IV, a general discussion of the calls of the blackbird and certain other passerines. *Behaviour*, 18(3): 161–176.

———— 1962. Evolution of intelligence and vocal mimicking. *Science*, 137: 585–589.

———— 1963a. Trends apparent in the evolution of vocalization in the Old World monkeys and apes. *Symposia of the Zoological Society of London*, 10: 89–107.

———— 1963b. The origin and evolution of the calls and facial expressions of the primates. *Behaviour*, 20(1,2): 1–109.

———— 1969. The effects of testosterone on avian vocalizations. In R. A. Hinde, ed. (*q.v.*), *Bird vocalizations: their relations to current problems in biology and psychology. Essays presented to W. H. Thorpe*, pp. 97–130.

———— 1972. The information potentially available in mammal displays. In R. A. Hinde, ed. (*q.v.*), *Non-verbal communication*, pp. 179–206.

Anthoney, T. R. 1968. The ontogeny of greeting, grooming, and sexual motor patterns in captive baboons (superspecies *Papio cynocephalus*). *Behaviour*, 31(4): 358–372.

Anthony, H. E. 1916. Habits of *Aplodontia*. *Bulletin of the American Museum of Natural History*, 35: 53–63.

Arata, A. A. 1967. Muroid, glirioid, and dipodoid rodents. In S. Anderson and J. K. Jones, Jr., eds. (*q.v.*), *Recent mammals of the world: a synopsis of families*, pp. 226–253.

Araujo, R. L. 1970. Termites of the Neotropical region. In K. Krishna and Frances M. Weesner, eds. (*q.v.*), *Biology of termites*, vol. 2, pp. 527–576.

Archer, J. 1970. Effects of population density on behaviour in rodents. In J. H. Crook, ed. (*q.v.*), *Social behaviour in birds and mammals: essays on the social ethology of animals and man*, pp. 169–210.

Armstrong, E. A. 1947. *Bird display and behaviour: an introduction to the study of bird psychology*, 2d ed. Lindsay Drummond, London. 431 pp. (Reprinted by Dover, New York, 1965, 431 pp.)

———— 1955. *The wren.* Collins, London. viii + 312 pp.

———— 1971. Social signalling and white plumage. *Ibis*, 113(4): 534.

Arnoldi, K. V. 1932. Biologische Beobachtungen an der neuen paläarktischen Sklavenhalterameise *Rossomyrmex proformicarum* K. Arn., nebst einigen Bemerkungen über die Beförderungsweise der Ameisen. *Zeitschrift für Morphologie und Ökologie der Tiere*, 24(2): 319–326.

Aronson, L. R., Ethel Tobach, D. S. Lehrman, and J. S. Rosenblatt, eds. 1970. *Development and evolution of behavior: essays in memory of T. C. Schneirla.* W. H. Freeman, San Francisco. xviii + 656 pp.

Ashmole, N. P. 1963. The regulation of numbers of tropical oceanic birds. *Ibis*, 103b(3): 458–473.

Ashmole, N. P., and H. Tovar S. 1968. Prolonged parental care in royal terns and other birds. *Auk*, 85(1): 90–100.

Assem, J. van den. 1967. Territory in the three-spined stickleback (*Gasterosteus aculeatus*). *Behaviour*, supplement 16. 164 pp.

———— 1971. Some experiments on sex ratio and sex regulation in the pteromalid *Lariophagus distinguendus*. *Netherlands Journal of Zoology*, 21(4): 373–402.

Auclair, J. L. 1963. Aphid feeding and nutrition. *Annual Review of Entomology*, 8: 439–490.

Ayala, F. J. 1968. Evolution of fitness: II, correlated effects of natural selection on the productivity and size of experimental populations of *Drosophila serrata*. *Evolution*, 22(1): 55–65.

Baerends, G. P., and J. M. Baerends-van Roon. 1950. An introduction to the study of the ethology of cichlid fishes. *Behaviour*, supplement 1. viii + 242 pp.

Baikov, N. 1925. *The Manchurian tiger.* [Cited by G. B. Schaller, 1967 (*q.v.*).]

Baker, A. N. 1971. *Pyrosoma spinosum* Herdman, a giant pelagic tunicate new to New Zealand waters. *Records of the Dominion Museum, Wellington, New Zealand*, 7(12): 107–117.

Baker, E. C. S. 1929. *The fauna of British India*, vol. 6, *Birds*, 2d ed. Taylor and Francis, London. xxxv + 499 pp.

Baker, H. G., and G. L. Stebbins, eds. 1965. *The genetics of colonizing species.* Academic Press, New York. xv + 588 pp.

Baker, R. H. 1971. Nutritional strategies of myomorph rodents in North American grasslands. *Journal of Mammalogy*, 52: 800–805.

Bakker, R. T. 1968. The superiority of dinosaurs. *Discovery*, 3(2): 11–22.

———— 1971. Ecology of the brontosaurs. *Nature, London*, 229: 172–174.

Bakko, E. B., and L. N. Brown. 1967. Breeding biology of the white-tailed prairie dog, *Cynomys leucurus*, in Wyoming. *Journal of Mammalogy*, 48(1): 100–112.

Baldwin, J. D. 1969. The ontogeny of social behaviour of squirrel monkeys (*Saimiri sciureus*) in a seminatural environment. *Folia Primatologica*, 11(1,2): 35–79.

———— 1971. The social organization of a semifree-ranging troop of squirrel monkeys (*Saimiri sciureus*). *Folia Primatologica*, 14(1,2): 23–50.

Banks, E. M., D. H. Pimlott, and B. E. Ginsburg, eds. 1967. *Ecology and behavior of the wolf.* Symposium of the Animal Behavior Society. *American Zoologist*, 7(2): 220–381.

Banta, W. C. 1973. Evolution of avicularia in cheilostome Bryozoa. In R. S. Boardman, A. H. Cheetham, and W. A. Oliver, Jr., eds. (*q.v.*), *Animal colonies: development and function through time*, pp. 295–303.

Barash, D. P. 1973. The social biology of the Olympic marmot. *Animal Behaviour Monographs*, 6(3): 172–245.

———— 1974a. The evolution of marmot societies: a general theory. *Science*, 185: 415–420.

———— 1974b. Neighbor recognition in two "solitary" carnivores: the raccoon (*Procyon lotor*) and the red fox (*Vulpes fulva*). *Science*, 185: 794–796.

Bardach, J. E., and J. H. Todd. 1970. Chemical communication in fish. In J. W. Johnston, Jr., D. G. Moulton, and A. Turk, eds. (*q.v.*), *Advances in chemoreception*, vol. 1, *Communication by chemical signals*, pp. 205–240.

Barksdale, A. W. 1969. Sexual hormones of *Achlya* and other fungi. *Science*, 166: 831–837.

Barlow, G. W. 1967. Social behavior of a South American leaf fish, *Polycentrus schomburgkii*, with an account of recurring pseudo-female behavior. *American Midland Naturalist*, 78(1): 215–234.

———— 1968. Ethological units of behavior. In D. Ingle, ed., *The central nervous system and fish behavior*, pp. 217–232. University of Chicago Press, Chicago. viii + 272 pp.

———— 1974a. Contrasts in social behavior between Central American cichlid fishes and coral-reef surgeon fishes. *American Zoologist*, 14(1): 9–34.

———— 1974b. Hexagonal territories. *Ecology.* (In press.)

Barlow, G. W., and R. F. Green. 1969. Effect of relative size of mate on color patterns in a mouthbreeding cichlid fish, *Tilapia melanotheron*. *Communications in Behavioral Biology*, ser. A, 4(1–3): 71–78.

Barlow, J. C. 1967. Edentates and pholidotes. In S. Anderson and J. K. Jones, Jr., eds. (*q.v.*), *Recent mammals of the world: a synopsis of families*, pp. 178–191.

Barnes, H. 1962. So-called anecdysis in *Balanus balanoides* and the effect of breeding upon the growth of the calcareous shell of some common barnacles. *Limnology and Oceanography,* 7(4): 462–473.

Barnes, R. D. 1969. *Invertebrate zoology,* 2d ed. W. B. Saunders Co., Philadelphia. x + 743 pp.

Barnett, S. A. 1958. An analysis of social behaviour in wild rats. *Proceedings of the Zoological Society of London,* 130(1): 107–152.

———— 1963. *A study in behaviour: principles of ethology and behavioural physiology, displayed mainly in the rat.* Methuen, London. xiii + 288 pp.

Barrai, I., L. L. Cavalli-Sforza, and M. Mainardi. 1964. Testing a model of dominant inheritance for metric traits in man. *Heredity,* 19(4): 651–668.

Barrington, E. 1965. *The biology of Hemichordata and Protochordata.* Oliver and Boyd, Edinburgh. 176 pp.

Barth, R. H., Jr. 1970. Pheromone-endocrine interactions in insects. In G. K. Benson and J. G. Phillips, eds., *Hormones and the environment,* pp. 373–404. Memoirs of the Society for Endocrinology no. 18. Cambridge University Press, Cambridge. xvi + 629 pp.

Bartholomew, G. A. 1952. Reproductive and social behavior of the northern elephant seal. *University of California Publications in Zoology,* 47(15): 369–472.

———— 1959. Mother-young relations and the maturation of pup behaviour in the Alaskan fur seal. *Animal Behaviour,* 7(3,4): 163–171.

———— 1970. A model for the evolution of pinniped polygyny. *Evolution,* 24(3): 546–559.

Bartholomew, G. A., and J. B. Birdsell. 1953. Ecology and the protohominids. *American Anthropologist,* 55: 481–498.

Bartholomew, G. A., and N. E. Collias. 1962. The role of vocalization in the social behaviour of the northern elephant seal. *Animal Behaviour,* 10(1,2): 7–14.

Bartholomew, G. A., and V. A. Tucker. 1964. Size, body temperature, thermal conductance, oxygen consumption, and heart rate in Australian varanid lizards. *Physiological Zoology,* 37(4): 341–354.

Bartlett, D., and J. Bartlett. 1974. Beavers—master mechanics of pond and stream. *National Geographic,* 145(5) (May): 716–732.

Bartlett, D. P., and G. W. Meier. 1971. Dominance status and certain operants in a communal colony of rhesus monkeys. *Primates,* 12(3,4): 209–219.

Bartlett, P. N., and D. M. Gates. 1967. The energy budget of a lizard on a tree trunk. *Ecology,* 48(2): 315–322.

Bastock, Margaret. 1956. A gene mutation which changes a behavior pattern. *Evolution,* 10(4): 421–439.

Bastock, Margaret, and A. Manning. 1955. The courtship of *Drosophila melanogaster. Behaviour,* 8(2,3): 85–111.

Bateman, A. J. 1948. Intra-sexual selection in *Drosophila. Heredity,* 2(3): 349–368.

Bates, B. C. 1970. Territorial behavior in primates: a review of recent field studies. *Primates,* 11(3): 271–284.

Bateson, G. 1955. A theory of play and fantasy. *Psychiatric Research Reports* (American Psychiatric Association), 2: 39–51.

———— 1963. The role of somatic change in evolution. *Evolution,* 17(4): 529–539.

Bateson, P. P. G. 1966. The characteristics and context of imprinting. *Biological Reviews, Cambridge Philosophical Society,* 41: 177–220.

Batra, Suzanne W. T. 1964. Behavior of the social bee, *Lasioglossum zephyrum,* within the nest (Hymenoptera: Halictidae). *Insectes Sociaux,* 11(2): 159–185.

———— 1966. The life cycle and behavior of the primitively social bee, *Lasioglossum zephyrum* (Halictidae). *Kansas University Science Bulletin* (Lawrence), 46(10): 359–422.

———— 1968. Behavior of some social and solitary halictine bees within their nests: a comparative study (Hymenoptera: Halictidae). *Journal of the Kansas Entomological Society,* 41(1): 120–133.

Batzli, G. O., and F. A. Pitelka. 1971. Condition and diet of cycling populations of the California vole, *Microtus californicus. Journal of Mammalogy,* 52(1): 141–163.

Bayer, F. M. 1973. Colonial organization in octocorals. In R. S. Boardman, A. H. Cheetham, and W. A. Oliver, Jr., eds. (q.v.), *Animal colonies: development and function through time,* pp. 69–93.

Beach, F. A. 1940. Effects of cortical lesions upon the copulatory behavior of male rats. *Journal of Comparative Psychology,* 29(2): 193–244.

———— 1945. Current concepts of play in animals. *American Naturalist,* 79: 523–541.

———— 1964. Biological bases for reproductive behavior. In W. Etkin, ed. (q.v.), *Social behavior and organization among vertebrates,* pp. 117–142.

Beatty, H. 1951. A note on the behavior of the chimpanzee. *Journal of Mammalogy,* 32(1): 118.

Beaumont, J. de. 1958. Le parasitisme social chez les guêpes et les bourdons. *Mitteilungen der Schweizerischen Entomologischen Gesellschaft,* 31(2): 168–176.

Beebe, W. 1922. *A monograph of the pheasants,* vol. 4. Witherby, London. xv + 242 pp.

———— 1926. The three-toed sloth *Bradypus cuculliger cuculliger* Wagler. *Zoologica, New York,* 7(1): 1–67.

———— 1947. Notes on the hercules beetle, *Dynastes hercules* (Linn.), at Rancho Grande, Venezuela, with special reference to combat behavior. *Zoologica, New York,* 32(2): 109–116.

Beilharz, R. G., and P. J. Mylrea. 1963. Social position and movement orders of dairy heifers. *Animal Behaviour,* 11(4): 529–533.

Beklemishev, W. N. 1969. *Principles of comparative anatomy of invertebrates,* vol. 1, Promorphology, trans. by J. M. MacLennan, ed. by Z. Kabata. University of Chicago Press, Chicago. xxx + 490 pp.

Bekoff, M. 1972. The development of social interaction, play, and metacommunication in mammals: an ethological perspective. *Quarterly Review of Biology,* 47(4): 412–434.

Bell, P. R., ed. 1959. *Darwin's biological work: some aspects reconsidered.* Cambridge University Press, Cambridge. xiii + 342 pp.

Bell, R. H. V. 1971. A grazing ecosystem in the Serengeti. *Scientific American,* 225(1) (July): 86–93.

Belt, T. 1874. *The naturalist in Nicaragua.* John Murray, London. xvi + 403 pp.

Bendell, J. F., and P. W. Elliot. 1967, *Behaviour and the regulation of numbers of the blue grouse.* Canadian Wildlife Report Series no. 4. Dept. of Indian Affairs and Northern Development, Ottawa. 76 pp.

Benois, A. 1972. Étude écologique de *Camponotus vagus* Scop. (= *pubescens* Fab.) (Hymenoptera, Formicidae) dans la région d'Antibes: nidification et architecture des nids. *Insectes Sociaux,* 19(2): 111–129.

———— 1973. Incidence des facteurs écologiques sur le cycle annuel et l'activité saissonnière de la fourmi d'Argentine *Iridomyrmex humilis* Mayr (Hymenoptera, Formicidae) dans la région d'Antibes. *Insectes Sociaux,* 20(3): 267–296.

Benson, W. W. 1971. Evidence for the evolution of unpalatability through kin selection in the Heliconiinae (Lepidoptera). *American Naturalist,* 105(943): 213–226.

Benson, W. W., and T. C. Emmel. 1973. Demography of gregariously roosting populations of the nymphaline butterfly *Marpesia berania* in Costa Rica. *Ecology,* 54(2): 326–335.

Bequaert, J. C. 1935. Presocial behavior among the Hemiptera. *Bulletin of the Brooklyn Entomological Society,* 30(5): 177–191.

Bergson, H. 1935. *The two sources of morality and religion,* trans. by R. A. Audra, C. Brereton, and W. H. Carter. Henry Holt, New York. viii + 308 pp.

Berkson, G., B. A. Ross, and S. Jatinandana. 1971. The social behavior of gibbons in relation to a conservation program. In L. A. Rosenblum, ed. (*q.v.*), *Primate behavior: developments in field and laboratory research,* vol. 2, pp. 225–255.

Berkson, G., and R. J. Schusterman. 1964. Reciprocal food sharing of gibbons. *Primates,* 5(1,2): 1–10.

Berndt, R., and H. Sternberg. 1969. Alters- und Geschlechtsunterschiede in der Dispersion des Trauerschnäppers (*Ficedula hypoleuca*). *Journal für Ornithologie,* 110(1): 22–26.

Bernstein, I. S. 1964a. A comparison of New and Old World monkey social organizations and behavior. *American Journal of Physical Anthropology,* 22(2): 233–238.

———— 1964b. A field study of the activities of howler monkeys. *Animal Behaviour,* 12(1): 92–97.

———— 1965. Activity patterns in a cebus monkey group. *Folia Primatologica,* 3(2,3): 211–224.

———— 1966. Analysis of a key role in a capuchin (*Cebus albifrons*) group. *Tulane Studies in Zoology,* 13(2): 49–54.

———— 1967. Intertaxa interactions in a Malayan primate community. *Folia Primatologica,* 7(3,4): 198–207.

———— 1968. The lutong of Kuala Selangor. *Behaviour,* 32(1–3): 1–16.

———— 1969a. Introductory techniques in the formation of pigtail monkey troops. *Folia Primatologica,* 10(1,2): 1–19.

———— 1969b. Spontaneous reorganization of a pigtail monkey group. *Proceedings of the Second International Congress of Primatology, Atlanta, Georgia,* 1: 48–51.

Bernstein, I. S., and R. J. Schusterman. 1964. The activities of gibbons in a social group. *Folia Primatologica,* 2(3): 161–170.

Bernstein, I. S., and L. G. Sharpe. 1966. Social roles in a rhesus monkey group. *Behaviour,* 26(1,2): 91–104.

Beroza, M., ed. 1970. *Chemicals controlling insect behavior.* Academic Press, New York. xii + 170 pp.

Berry, Kristin H. 1971. Social behavior of the chuckwalla, *Sauromalus obesus.* Herpetological Abstracts of the American Society of Ichthyologists and Herpetologists, 51st Annual Meeting, Los Angeles, pp. 2–3.

Bertram, B. C. R. 1970. The vocal behaviour of the Indian hill mynah, *Gracula religiosa. Animal Behaviour Monographs,* 3(2): 79–192.

———— 1973. Lion population regulation. *East African Wildlife Journal,* 11(3,4): 215–225.

Bertram, G. C. L., and C. K. R. Bertram. 1964. Manatees in the Guianas. *Zoologica, New York,* 49(2): 115–120.

Bertrand, Mireille. 1969. *The behavioral repertoire of the stumptail macaque: a descriptive and comparative study.* Bibliotheca Primatologica, no. 11. S. Karger, Basel. xii + 273 pp.

Bess, H. A. 1970. Termites of Hawaii and the Oceanic islands. In K. Krishna and Frances M. Weesner, eds. (*q.v.*), *Biology of termites,* vol. 2, pp. 449–476.

Best, J. B., A. B. Goodman, and A. Pigon. 1969. Fissioning in planarians: control by the brain. *Science,* 164: 565–566.

Betz, Barbara J. 1932. The population of a nest of the hornet *Vespa maculata. Quarterly Review of Biology,* 7(2): 197–209.

Bick, G. H., and Juanda C. Bick. 1965. Demography and behavior of the damselfly, *Argia apicalis* (Say), (Odonata: Coenagriidae). *Ecology,* 46(4): 461–472.

Bider, J. R., P. Thibault, and R. Sarrazin. 1968. Schemes dynamiques spatio-temporels de l'activité de *Procyon lotor* en relation avec le comportement. *Mammalia,* 32(2): 137–163.

Bieg, D. 1972. The production of males in queenright colonies of *Trigona* (*Scaptotrigona*) *postica. Journal of Apicultural Research,* 11(1): 33–39.

Bierens de Haan, J. A. 1940. *Die tierischen Instinkte und ihr Umbau durch Erfahrung: eine Einführung in die allgemeine Tierpsychologie.* E. J. Brill, Leyden. xi + 478 pp.

Bigelow, R. 1969. *The dawn warriors: man's evolution toward peace.* Atlantic Monthly Press, Little, Brown, Boston. xi + 277 pp.

Birch, H. G., and G. Clark. 1946. Hormonal modification of social behavior: II, the effects of sex-hormone administration on the social dominance status of the female-castrate chimpanzee. *Psychosomatic Medicine,* 8(5): 320–321.

Birdwhistle, R. L. 1970. *Kinesics and context: essays on body motion and communication.* University of Pennsylvania Press, Philadelphia. xiv + 338 pp.

Bishop, J. W., and L. M. Bahr. 1973. Effects of colony size on feeding by *Lophopodella carteri* (Hyatt). In R. S. Boardman, A. H. Cheetham, and W. A. Oliver, Jr., eds. (*q.v.*), *Animal colonies: development and function through time,* pp. 433–437.

Black-Cleworth, Patricia. 1970. The role of electrical discharges in the non-reproductive social behaviour of *Gymnotus carapo* (Gymnotidae, Pisces). *Animal Behaviour Monographs,* 3(1): 1–77.

Blackwell, K. F., and J. I. Menzies. 1968. Observations on the biology of the potto (*Perodicticus potto,* Miller). *Mammalia,* 32(3): 447–451.

Blair, W. F. 1968. Amphibians and reptiles. In T. A. Sebeok, ed. (q.v.), *Animal communication: techniques of study and results of research*, pp. 289–310.

Blair, W. F., and W. E. Howard. 1944. Experimental evidence of sexual isolation between three forms of mice of the cenospecies *Peromyscus maniculatus*. *Contributions from the Laboratory of Vertebrate Biology, University of Michigan, Ann Arbor*, 26: 1–19.

Blest, A. D. 1963. Longevity, palatability and natural selection in five species of New World saturniid moth. *Nature, London*, 197(4873): 1183–1186.

Blum, M. S. 1966. Chemical releasers of social behavior: VIII, citral in the mandibular gland secretion of *Lestrimelitta limao* (Hymenoptera: Apoidea: Melittidae). *Annals of the Entomological Society of America*, 59(5): 962–964.

Blum M. S., and E. O. Wilson. 1964. The anatomical source of trail substances in formicine ants. *Psyche, Cambridge*, 71(1): 28–31.

Blurton Jones, N. G. 1969. An ethological study of some aspects of social behaviour of children in nursery school. In D. Morris, ed. (q.v.), *Primate ethology: essays on the socio-sexual behavior of apes and monkeys*, pp. 437–463.

———— ed. 1972. *Ethological studies of child behaviour*. Cambridge University Press, Cambridge. x + 400 pp.

Blurton Jones, N. G., and J. Trollope. 1968. Social behavior of stump-tailed macaques in captivity. *Primates*, 9(4): 365–394.

Boardman, R. S., and A. H. Cheetham. 1973. Degrees of colony dominance in stenolaemate and gymnolaemate Bryozoa. In R. S. Boardman, A. H. Cheetham, and W. A. Oliver, Jr., eds. (q.v.), *Animal colonies: development and function through time*, pp. 121–220.

Boardman, R. S., A. H. Cheetham, and W. A. Oliver, Jr., eds. 1973. *Animal colonies: development and function through time*. Dowden, Hutchinson, and Ross, Stroudsburg, Pa. xiii + 603 pp.

Bodot, Paulette. 1969. Composition des colonies de termites: ses fluctuations au cours du temps. *Insectes Sociaux*, 16(1): 39–53.

Boice, R., and D. W. Witter. 1969. Hierarchical feeding behaviour in the leopard frog (*Rana pipiens*). *Animal Behaviour*, 17(3): 474–479.

Bolton, B. 1974. A revision of the palaeotropical arboreal ant genus *Cataulacus* F. Smith (Hymenoptera: Formicidae). *Bulletin of the British Museum of Natural History, Entomology*, 30(1): 1–105.

Bolwig, N. 1958. A study of the behaviour of the chacma baboon, *Papio ursinus*. *Behaviour*, 14(1,2): 136–163.

Bonner, J. T. 1955. *Cells and societies*. Princeton University Press, Princeton, N. J. iv + 234 pp.

———— 1958. *The evolution of development*. Cambridge University Press, Cambridge. 102 pp.

———— 1965. *Size and cycle: an essay on the structure of biology*. Princeton University Press, Princeton, N. J. viii + 219 pp.

———— 1967. *The cellular slime molds*, 2d ed. Princeton University Press, Princeton, N. J. xii + 205 pp.

———— 1970. The chemical ecology of cells in the soil. In E. Sondheimer and J. B. Simeone, eds. (q.v.), *Chemical ecology*, pp. 1–19.

———— 1974. *On development: the biology of form*. Harvard University Press, Cambridge. viii + 282 pp.

Boorman, S. A., and P. R. Levitt. 1972. Group selection on the boundary of a stable population. *Proceedings of the National Academy of Sciences, U.S.A.*, 69(9): 2711–2713.

———— 1973a. Group selection on the boundary of a stable population. *Theoretical Population Biology*, 4(1): 85–128.

———— 1973b. A frequency-dependent natural selection model for the evolution of social cooperation networks. *Proceedings of the National Academy of Sciences, U.S.A.*, 70(1): 187–189.

Booth, A. H. 1957. Observations on the natural history of the olive colobus monkey, *Procolobus verus* (van Beneden). *Proceedings of the Zoological Society of London*, 129(3): 421–430.

———— 1960. *Small mammals of West Africa*. Longmans, Green, London. 68 pp. [Cited by Bradbury, 1975 (q.v.).]

Booth, Cynthia. 1962. Some observations on behavior of Cercopithecus monkeys. *Annals of the New York Academy of Sciences*, 102(2): 477–487.

Borgmeier, T. 1955. *Die Wanderameisen der Neotropischen Region (Hym. Formicidae)*. Studia Entomologica, no. 3. Editora Vozes, Petrópolis, Rio de Janeiro. 716 pp.

Boserup, Ester. 1965. *The conditions of agricultural growth*. Aldine Publishing Co., Chicago. 124 pp.

Bossert, W. H. 1967. Mathematical optimization: are there abstract limits on natural selection? In P. S. Moorhead and M. M. Kaplan, eds. (q.v.), *Mathematical challenges to the Neo-Darwinian interpretation of evolution*, pp. 35–46.

———— 1968. Temporal patterning in olfactory communication. *Journal of Theoretical Biology*, 18(2): 157–170.

Bossert, W. H., and E. O. Wilson. 1963. The analysis of olfactory communication among animals. *Journal of Theoretical Biology*, 5(3): 443–469.

Bouillon, A. 1970. Termites of the Ethiopian region. In K. Krishna and Frances M. Weesner, eds. (q.v.), *Biology of termites*, vol. 2, pp. 153–280.

Bourlière, F. 1955. *The natural history of mammals*. G. G. Harrap, London. xxii + 363 pp. + xi.

———— 1963. Specific feeding habits of African carnivores. *African Wildlife*, 17(1): 21–27.

Bourlière, F., C. Hunkeler, and M. Bertrand. 1970. Ecology and behavior of Lowe's guenon (*Cercopithecus campbelli lowei*) in the Ivory Coast. In J. R. Napier and P. H. Napier, eds. (q.v.), *Old World monkeys: evolution, systematics, and behavior*, pp. 297–350.

Bovbjerg, R. V. 1956. Some factors affecting aggressive behavior in crayfish. *Physiological Zoology*, 29(2): 127–136.

———— 1960. Behavioral ecology of the crab, *Pachygrapsus crassipes*. *Ecology*, 41(4): 668–672.

———— 1970. Ecological isolation and competitive exclusion in two crayfish (*Orconectes virilis* and *Orconectes immunis*). *Ecology*, 51(2): 225–236.

Bovbjerg, R. V., and Sandra L. Stephen. 1971. Behavioral changes in crayfish with increased population density. *Bulletin of the Ecological Society of America*, 52(4): 37–38.

Bowden, D. 1966. Primate behavioral research in the USSR: the Sukhumi Medico-Biological Station. *Folia Primatologica*, 4(4): 346–360.

Boyd, H. 1953. On encounters between wild white-fronted geese in winter flocks. *Behaviour*, 5(1): 85–129.

Bradbury, J. 1975. Social organization and communication. In W. Wimsatt, ed., *Biology of bats*, vol. 3. Academic Press, New York. (In press.)

Bragg, A. N. 1955–56. In quest of the spadefoots. *New Mexico Quarterly*, 25(4): 345–358.

Brandon, R. A., and J. E. Huheey. 1971. Movements and interactions of two species of *Desmognathus* (Amphibia: Plethodontidae). *American Midland Naturalist*, 86(1): 86–92.

Brannigan, C. R., and D. A. Humphries. 1972. Human non-verbal behaviour, a means of communication. In N. Blurton Jones, ed. (*q.v.*), *Ethological studies of child behaviour*, pp. 37–64.

Brattstrom, B. H. 1962. Call order and social behavior in the foam-building frog, *Engystomops pustulosus*. *American Zoologist*, 2(3): 394.

——— 1973. Social and maintenance behavior of the echidna, *Tachyglossus aculeatus*. *Journal of Mammalogy*, 54(1): 50–70.

——— 1974. The evolution of reptilian social behavior. *American Zoologist*, 14(1): 35–49.

Braun, R. 1958. Das Sexualverhalten der Krabbenspinne *Diaea dorsata* (F.) und der Zartspinne *Anyphaena accentuata* (Walck.) als Hinweis auf ihre systematische Eingliederung. *Zoologischer Anzeiger*, 160(7,8): 119–134.

Brauns, H. 1926. A contribution to the knowledge of the genus *Allodape*, St. Farg. & Serv. Order Hymenoptera; section Apidae (Anthophila). *Annals of the South African Museum*, 23(3): 417–434.

Breder, C. M., Jr. 1959. Studies on social groupings in fishes. *Bulletin of the American Museum of Natural History*, 117(6): 393–482.

——— 1965. Vortices and fish schools. *Zoologica, New York*, 50(2): 97–114.

Breder, C. M., Jr., and C. W. Coates. 1932. A preliminary study of population stability and sex ratio of *Lebistes*. *Copeia*, 1932(3): 147–155.

Brémond, J.-C. 1968. Recherches sur la sémantique et les éléments vecteurs d'information dans les signaux acoustiques du Rouge-gorge (*Erithacus rubecula* L.). *La Terre et la Vie*, 115(2): 109–220.

Brereton, J. L. G. 1962. Evolved regulatory mechanisms of population control. In G. W. Leeper, ed. (*q.v.*), *The evolution of living organisms*, pp. 81–93.

——— 1971. Inter-animal control of space. In A. H. Esser, ed. (*q.v.*), *Behavior and environment: the use of space by animals and men*, pp. 69–91.

Brian, M. V. 1952a. Interaction between ant colonies at an artificial nest-site. *Entomologist's Monthly Magazine*, 88: 84–88.

——— 1952b. The structure of a dense natural ant population. *Journal of Animal Ecology*, 21(1): 12–24.

——— 1955. Food collection by a Scottish ant community. *Journal of Animal Ecology*, 24(2): 336–351.

——— 1956a. The natural density of *Myrmica rubra* and associated ants in West Scotland. *Insectes Sociaux*, 3(4): 473–487.

——— 1956b. Segregation of species of the ant genus *Myrmica*. *Journal of Animal Ecology*, 25(2): 319–337.

——— 1965. Caste differentiation in social insects. *Symposia of the Zoological Society of London*, 14: 13–38.

——— 1968. Regulation of sexual production in an ant society. In R. Chauvin and C. Noirot, eds. (*q.v.*), *L'effet de groupe chez les animaux*, pp. 61–76.

Brian, M. V., G. Elmes, and A. F. Kelly. 1967. Populations of the ant *Tetramorium caespitum* Latreille. *Journal of Animal Ecology*, 36(2): 337–342.

Brien, P. 1953. Étude sur les Phylactolemates. *Annales de la Société Royale Zoologique de Belgique*, 84(2): 301–444.

Broadbooks, H. E. 1965. Ecology and distribution of the pikas of Washington and Alaska. *American Midland Naturalist*, 73(2): 299–335.

——— 1970. Home ranges and territorial behavior of the yellow-pine chipmunk, *Eutamias amoenus*. *Journal of Mammalogy*, 51(2): 310–326.

Brock, V. E., and R. H. Riffenburgh. 1960. Fish schooling: a possible factor in reducing predation. *Journal du Conseil, Conseil Permanent International pour l'Exploration de la Mer*, 25: 307–317.

Bro Larsen, Ellinor. 1952. On subsocial beetles from the salt-marsh, their care of progeny and adaptation to salt and tide. *Transactions of the Ninth International Congress of Entomology, Amsterdam, 1951*, 1: 502–506.

Bromley, P. T. 1969. Territoriality in pronghorn bucks on the National Bison Range, Moiese, Montana. *Journal of Mammalogy*, 50(1): 81–89.

Bronson, F. H. 1963. Some correlates of interaction rate in natural populations of woodchucks. *Ecology*, 44(4): 637–643.

——— 1967. Effects of social stimulation on adrenal and reproductive physiology of rodents. In M. L. Conalty, ed., *Husbandry of laboratory animals*, pp. 513–542. Academic Press, New York.

——— 1969. Pheromonal influences on mammalian reproduction. In M. Diamond, ed., *Perspectives in reproduction and sexual behavior*, pp. 341–361. Indiana University Press, Bloomington. x + 532 pp.

——— 1971. Rodent pheromones. *Biology of Reproduction*, 4(3): 344–357.

Bronson, F. H., and B. E. Eleftheriou. 1963. Adrenal responses to crowding in *Peromyscus* and C57BL/10J mice. *Physiological Zoology*, 36(2): 161–166.

Brooks, R. J., and E. M. Banks. 1973. Behavioural biology of the collared lemming (*Dicrostonyx groenlandicus* [Traill]): an analysis of acoustic communication. *Animal Behaviour*, 6(1): 1–83.

Brothers, D. J., and C. D. Michener. 1974. Interactions in colonies of primitively social bees: III, ethometry of division of labor in *Lasioglossum zephyrum* (Hymenoptera: Halictidae). *Journal of Comparative Physiology*, 90(2): 129–168.

Brower, L. P. 1969. Ecological chemistry. *Scientific American*, 220(2) (February): 22–29.

Brown, B. A., Jr. 1974. Social organization in male groups of white-tailed deer. In V. Geist and F. Walther, eds. (q.v.), *The behaviour of ungulates and its relation to management*, vol. 1, pp. 436–446.

Brown, D. H., D. K. Caldwell, and Melba C. Caldwell. 1966. Observations on the behavior of wild and captive false killer whales, with notes on associated behavior of other genera of captive delphinids. *Contributions in Science, Los Angeles County Museum*, 95: 1–32.

Brown, D. H., and K. S. Norris. 1956. Observations of captive and wild cetaceans. *Journal of Mammalogy*, 37(3): 311–326.

Brown, E. S. 1959. Immature nutfall of coconuts in the Solomon Islands: II, changes in ant populations, and their relation to vegetation. *Bulletin of Entomological Research*, 50(3): 523–558.

Brown, J. C. 1964. Observations on the elephant shrews (Macroscelididae) of equatorial Africa. *Proceedings of the Zoological Society of London*, 143(1): 103–119.

Brown, J. H. 1971. Mechanisms of competitive exclusion between two species of chipmunks. *Ecology*, 52(2): 305–311.

Brown, J. L. 1963. Aggressiveness, dominance and social organization in the Steller jay. *Condor*, 65(6): 460–484.

———— 1964. The evolution of diversity in avian territorial systems. *Wilson Bulletin*, 76(2): 160–169.

———— 1966. Types of group selection. *Nature, London*, 211(5051): 870.

———— 1969. Territorial behavior and population regulation in birds: a review and re-evaluation. *Wilson Bulletin*, 81(3): 293–329.

———— 1970a. Cooperative breeding and altruistic behaviour in the Mexican jay, *Aphelocoma ultramarina. Animal Behaviour*, 18(2): 366–378.

———— 1970b. The neural control of aggression. In C. H. Southwick, ed. (q.v.) *Animal aggression: selected readings*, pp. 164–186.

———— 1972. Communal feeding of nestlings in the Mexican jay (*Aphelocoma ultramarina*): interflock comparisons. *Animal Behaviour*, 20(2): 395–403.

———— 1974. Alternate-routes to sociality in jays—with a theory for the evolution of altruism and communal breeding. *American Zoologist*, 14(1): 63–80.

Brown, J. L., R. W. Hunsperger, and H. E. Rosvold. 1969. Interaction of defence and flight reactions produced by simultaneous stimulation at two points in the hypothalamus of the cat. *Experimental Brain Research*, 8: 130–149.

Brown, L. H. 1966. Observations on some Kenya eagles. *Ibis*, 108(4): 531–572.

Brown, R. 1973. *A first language: the early stages.* Harvard University Press, Cambridge. xxii + 437 pp.

Brown, R. G. B. 1962. The aggressive and distraction behaviour of the western sandpiper *Ereunetes mauri. Ibis*, 104(1): 1–12.

Brown, W. L. 1952a. Contributions toward a reclassification of the Formicidae: I, tribe Platythyreini. *Breviora, Cambridge, Mass.*, 6: 1–6.

———— 1952b. Revision of the ant genus *Serrastruma. Bulletin of the Museum of Comparative Zoology, Harvard*, 107(2): 67–86.

———— 1955. A revision of the Australian ant genus *Notoncus* Emery, with notes on the other genera of Melophorini. *Bulletin of the Museum of Comparative Zoology, Harvard*, 113(6): 471–494.

———— 1957. Predation of arthropod eggs by the ant genera *Proceratium* and *Discothyrea. Psyche, Cambridge*, 64(3): 115.

———— 1958. General adaptation and evolution. *Systematic Zoology*, 7(4): 157–168.

———— 1960. Contributions toward a reclassification of the Formicidae: III, tribe Amblyoponini (Hymenoptera). *Bulletin of the Museum of Comparative Zoology, Harvard*, 122(4): 145–230.

———— 1964. Revision of *Rhoptromyrmex. Pilot Register of Zoology*, cards nos. 11–19.

———— 1965. Contributions to a reclassification of the Formicidae: IV, tribe Typhlomyrmecini (Hymenoptera). *Psyche, Cambridge*, 72(1): 65–78.

———— 1968. An hypothesis concerning the function of the metapleural glands in ants. *American Naturalist*, 102(924): 188–191.

———— 1973. A comparison of the Hylean and Congo-West African rain forest ant faunas. In Betty J. Meggers, E. S. Ayensu, and W. D. Duckworth, eds., *Tropical forest ecosystems in Africa and South America: a comparative review*, pp. 161–185. Smithsonian Institution Press, Washington, D.C. viii + 350 pp.

———— 1975. Contributions toward a reclassification of the Formicidae: V, Ponerinae, tribes Platythyreini, Cerapachyini, Cylindromyrmecini, Acanthostichini, and Aenictogitini. *Search, Ithaca, Entomology.* (In press.)

Brown, W. L., T. Eisner, and R. H. Whittaker. 1970. Allomones and kairomones: transspecific chemical messengers. *BioScience*, 20(1): 21–22.

Brown, W. L., W. H. Gotwald, and J. Lévieux. 1970. A new genus of ponerine ants from West Africa (Hymenoptera: Formicidae) with ecological notes. *Psyche, Cambridge*, 77(3): 259–275.

Brown, W. L., and W. W. Kempf. 1960. A world revision of the ant tribe Basicerotini. *Studia Entomologica*, n.s. 3(1–4): 161–250.

———— 1969. A revision of the Neotropical dacetine ant genus *Acanthognathus* (Hymenoptera: Formicidae). *Psyche, Cambridge*, 76(2): 87–109.

Brown, W. L., and E. O. Wilson. 1959. The evolution of the dacetine ants. *Quarterly Review of Biology*, 34: 278–294.

Bruce, H. M. 1966. Smell as an exteroceptive factor. *Journal of Animal Science*, supplement 25: 83–89.

Brun, R. 1952. Das Zentralnervensystem von *Teleutomyrmex schneideri* Kutt. ♀ (Hym. Formicid.). *Mitteilungen der Schweizerischen Entomologischen Gesellschaft*, 25(2): 73–86.

Bruner, J. S. 1968. *Processes of cognitive growth: infancy.* Clark University Press, with Barre Publishers, Barre, Mass. vii + 75 pp.

Buck, J. B. 1938. Synchronous rhythmic flashing of fireflies. *Quarterly Review of Biology*, 13(3): 301–314.

Buckley, Francine G. 1967. Some notes on flock behaviour in the blue-crowned hanging parrot *Loriculus galgulus* in captivity. *Pavo* (Indian Journal of Ornithology), 5(1,2): 97–99.

Buckley, Francine G., and P. A. Buckley. 1972. The breeding ecology of royal terns *Sterna (Thalasseus) maxima maxima. Ibis*, 114: 344–359.

Buckley, P. A., and Francine G. Buckley. 1972. Individual egg and chick recognition by adult royal terns (*Sterna maxima maxima*). *Animal Behaviour*, 20(3): 457–462.

Buechner, H. K. 1950. Life history, ecology, and range use of the pronghorn antelope in Trans-Pecos, Texas. *American Midland Naturalist*, 43(2): 257–354.

—— 1961. Territorial behavior in Uganda kob. *Science*, 133: 698–699.

—— 1963. Territoriality as a behavioral adaptation to environment in Uganda kob. *Proceedings of the Sixteenth International Congress of Zoology, Washington, D.C.*, 3: 59–63.

Buechner, H. K., and H. D. Roth. 1974. The lek system in Uganda kob antelope. *American Zoologist*, 14(1): 145–162.

Buettner-Janusch, J., and R. J. Andrew. 1962. The use of the incisors by primates in grooming. *American Journal of Physical Anthropology*, 20(1): 127–129.

Bullis, H. R., Jr. 1961. Observations on the feeding behavior of white-tip sharks on schooling fishes. *Ecology*, 42(1): 194–195.

Bullock, T. H. 1973. Seeing the world through a new sense: electroreception in fish. *American Scientist*, 61(3): 316–325.

Bunnell, P. 1973. Vocalizations in the territorial behavior of the frog *Dendrobates pumilio*. *Copeia*, 1973, no. 2: pp. 277–284.

Bünzli, G. H. 1935. Untersuchungen über coccidophile Ameisen aus den Kaffeefeldern von Surinam. *Mitteilungen der Schweizerischen Entomologischen Gesellschaft*, 16(6,7): 453–593.

Burchard, J. E., Jr. 1965. Family structure in the dwarf cichlid *Apistogramma trifasciatum* Eigenmann and Kennedy. *Zeitschrift für Tierpsychologie*, 22(2): 150–162.

Buren, W. F. 1968. A review of the species of *Crematogaster*, sensu stricto, in North America (Hymenoptera, Formicidae): II, descriptions of new species. *Journal of the Georgia Entomological Society*, 3(3): 91–121.

Burghardt, G. M. 1970. Chemical perception in reptiles. In J. W. Johnston, Jr., D. G. Moulton, and A. Turk, eds. (*q.v.*) *Advances in chemoreception*, vol. 1, *Communication by chemical signals*, pp. 241–308.

Burnet, F. M. 1971. "Self-recognition" in colonial marine forms and flowering plants in relation to the evolution of immunity. *Nature, London*, 232(5308): 230–235.

Burt, W. H. 1943. Territoriality and home range concepts as applied to mammals. *Journal of Mammalogy*, 24(3): 346–352.

Burton, Frances D. 1972. The integration of biology and behavior in the socialization of *Macaca sylvana* of Gibraltar. In F. E. Poirier, ed. (q.v.), *Primate socialization*, pp. 29–62.

Busnel, R.-G., and A. Dziedzic. 1966. Acoustic signals of the pilot whale *Globicephala melaena* and of the porpoises *Delphinus delphis* and *Phocoena phocoena*. In K. S. Norris, ed. (*q.v.*), *Whales, dolphins, and porpoises*, pp. 607–646.

Bustard, H. R. 1970. The role of behavior in the natural regulation of numbers in the gekkonid lizard *Gehyra variegata*. *Ecology*, 51(4): 724–728.

Butler, Charles. 1609. *The feminine monarchie: on a treatise concerning bees, and the due ordering of them*. Joseph Barnes, Oxford.

Butler, C. G. 1954a. *The world of the honeybee*. Collins, London. xiv + 226 pp.

—— 1954b. The method and importance of the recognition by a colony of honeybees (*A. mellifera*) of the presence of its queen. *Transactions of the Royal Entomological Society of London*, 105(2): 11–29.

—— 1967. Insect pheromones. *Biological Reviews, Cambridge Philosophical Society*, 42(1): 42–87.

—— 1969. Some pheromones controlling honeybee behaviour. *Proceedings of the Seventh Congress of the International Union for the Study of Social Insects, Bern*, pp. 19–32.

Butler, C. G., and D. H. Calam. 1969. Pheromones of the honey bee—the secretion of the Nassanoff gland of the worker. *Journal of Insect Physiology*, 15(2): 237–244.

Butler, C. G., R. K. Callow, and J. R. Chapman. 1964. 9-Hydroxydec-trans-2-enoic acid, a pheromone stabilizing honeybee swarms. *Nature, London*, 201 (4920): 733.

Butler, C. G., D. J. C. Fletcher, and Doreen Watler. 1969. Nest-entrance marking with pheromones by the honeybee, *Apis mellifera* L., and by a wasp, *Vespula vulgaris* L. *Animal Behaviour*, 17(1): 142–147.

Butler, C. G., and J. B. Free. 1952. The behaviour of worker honeybees at the hive entrance. *Behaviour*, 4(4): 262–292.

Butler, C. G., and J. Simpson. 1967. Pheromones of the queen honeybee (*Apis mellifera* L.) which enable her workers to follow her when swarming. *Proceedings of the Royal Entomological Society of London*, ser. A, 42(10–12): 149–154.

Butler, R. A. 1954. Incentive conditions which influence visual exploration. *Journal of Experimental Psychology*, 48: 17–23.

—— 1965. Investigative behavior. In A. M. Schrier, H. F. Harlow, and F. Stollnitz, eds., *Behavior of non-human primates: modern research trends*, vol. 2, pp. 463–493. Academic Press, New York. xv + pp. 287–595.

Cairns, J., Jr., M. L. Dahlberg, K. L. Dickson, Nancy Smith, and W. T. Waller. 1969. The relationship of fresh-water protozoan communities to the MacArthur-Wilson equilibrium model. *American Naturalist*, 103(933): 439–454.

Calaby, J. H. 1956. The distribution and biology of the genus *Ahamitermes* (Isoptera). *Australian Journal of Zoology*, 4(2): 111–124.

—— 1960. Observations on the banded ant-eater *Myrmecobius f. fasciatus* Waterhouse (Marsupialia), with particular reference to its food habits. *Proceedings of the Zoological Society of London*, 135(2): 183–207.

Caldwell, D. K., Melba C. Caldwell, and D. W. Rice. 1966. Behavior of the sperm whale, *Physeter catodon* L. In K. S. Norris, ed. (*q.v.*) *Whales, dolphins, and porpoises*, pp. 679–716.

Caldwell, L. D., and J. B. Gentry. 1965. Interactions of *Peromyscus* and *Mus* in a one-acre field enclosure. *Ecology*, 46(1,2): 189–192.

Caldwell, Melba C., and D. K. Caldwell. 1966. Epimeletic (care-giving) behavior in Cetacea. In K. S. Norris, ed. (*q.v.*) *Whales, dolphins, and porpoises*, pp. 755–788.

———— 1972. Behavior of marine mammals: sense and communication. In S. H. Ridgway, ed. (*q.v.*), *Mammals of the sea: biology and medicine*, pp. 419–502.

Calhoun, J. B. 1962a. *The ecology and sociology of the Norway rat.* U.S. Department of Health, Education, and Welfare, Public Health Service Document no. 1008. Superintendent of Documents, U.S. Government Printing Office, Washington, D.C. viii + 288 pp.

———— 1962b. Population density and social pathology. *Scientific American*, 206(2) (February): 139–148.

———— 1971. Space and the strategy of life. In A. H. Esser, ed. (*q.v.*), *Behavior and environment: the use of space by animals and men*, pp. 329–387.

Callow, R. K., J. R. Chapman, and Patricia N. Paton. 1964. Pheromones of the honeybee: chemical studies of the mandibular gland secretion of the queen. *Journal of Apicultural Research*, 3(2): 77–89.

Campbell, B. G., ed. 1972. *Sexual selection and the descent of man 1871–1971.* Aldine Publishing Co., Chicago. x + 378 pp.

Campbell, D. T. 1972. On the genetics of altruism and the counter-hedonic components in human culture. *Journal of Social Issues*, 28(3): 21–37.

Camus, A. 1955. *The myth of Sisyphus.* Vintage Books, Alfred A. Knopf, New York. viii + 151 pp.

Candland, D. K., and A. I. Leshner. 1971. Formation of squirrel monkey dominance order is correlated with endocrine output. *Bulletin of the Ecological Society of America*, 52(4): 54.

Capranica, R. R. 1968. The vocal repertoire of the bullfrog (*Rana catesbeiana*). *Behaviour*, 31(3): 302–325.

Carl, E. A. 1971. Population control in arctic ground squirrels. *Ecology*, 52(3): 395–413.

Carne, P. B. 1966. Primitive forms of social behaviour, and their significance in the ecology of gregarious insects. *Proceedings of the Ecological Society of Australia*, 1: 75–78.

Carneiro, R. L. 1970. A theory of the origin of the state. *Science*, 169: 733–738.

Carpenter, C. C. 1971. Discussion of Session I: Territoriality and dominance. In A. H. Esser, ed. (*q.v.*), *Behavior and environment: the use of space by animals and men*, pp. 46–47.

Carpenter, C. R. 1934. A field study of the behavior and social relations of howling monkeys. *Comparative Psychology Monographs*, 10(2): 1–168.

———— 1935. Behavior of red spider monkeys in Panama. *Journal of Mammalogy*, 16(3): 171–180.

———— 1940. A field study in Siam of the behavior and social relations of the gibbon (*Hylobates lar*). *Comparative Psychology Monographs*, 16(5): 1–212.

———— 1942a. Sexual behavior of free ranging rhesus monkeys (*Macaca mulatta*): II, periodicity of estrus, homosexual, autoerotic and nonconformist behavior. *Journal of Comparative Psychology*, 33(1): 143–162.

———— 1942b. Characteristics of social behavior in non-human primates.

Transactions of the New York Academy of Sciences, 2d ser., 4(8): 248–258.

———— 1952. Social behavior of non-human primates. In P.-P. Grassé, ed. (*q.v.*), *Structure et physiologie des sociétés animales*, pp. 227–246.

———— 1954. Tentative generalizations on the grouping behavior of non-human primates. *Human Biology*, 26(3): 269–276.

———— 1965. The howlers of Barro Colorado Island. In I. DeVore, ed. (*q.v.*), *Primate behavior: field studies of monkeys and apes*, pp. 250–291.

———— ed. 1973. *Behavioral regulators of behavior in primates.* Bucknell University Press, Lewisburg, Pa. 303 pp.

Carr, A., and H. Hirth. 1961. Social facilitation in green turtle siblings. *Animal Behaviour*, 9(1,2): 68–70.

Carr, A., and L. Ogren. 1960. The ecology and migrations of sea turtles: 4, the green turtle in the Caribbean Sea. *Bulletin of the American Museum of Natural History*, 121(1): 1–48.

Carr, W. J., R. D. Martorano, and L. Krames. 1970. Responses of mice to odors associated with stress. *Journal of Comparative and Physiological Psychology*, 71(2): 223–228.

Carrick, R. 1963. Ecological significance of territory in the Australian magpie, *Gymnorhina tibicen. Proceedings of the Thirteenth International Ornithological Congress*, 2: 740–753.

Carrick, R., S. E. Csordas, Susan E. Ingham, and K. Keith. 1962. Studies on the southern elephant seal, *Mirounga leonina* (L.), III, IV. *C.S.I.R.O. Wildlife Research, Canberra, Australia*, 7(2): 119–197.

Cartmill, M. 1974. Rethinking primate origins. *Science*, 184: 436–443.

Castle, G. B. 1934. The damp-wood termites of western United States, genus *Zootermopsis* (formerly, *Termopsis*). In C. A. Kofoid et al., eds. (*q.v.*), *Termites and termite control*, pp. 273–310.

Castoro, P. L., and A. M. Guhl. 1958. Pairing behavior related to aggressiveness and territory. *Wilson Bulletin*, 70(1): 57–69.

Caughley, G. 1964. Social organization and daily activity of the red kangaroo and the grey kangaroo. *Journal of Mammalogy*, 45(3): 429–436.

Cavalli-Sforza, L. L. 1971. Similarities and dissimilarities of sociocultural and biological evolution. In F. R. Hodson, D. G. Kendall, and P. Tautu, eds., *Mathematics in the archaeological and historical sciences*, pp. 535–541. Edinburgh University Press, Edinburgh. vii + 565 pp.

Cavalli-Sforza, L. L., and W. F. Bodmer. 1971. *The genetics of human populations.* W. H. Freeman, San Francisco. xvi + 965 pp.

Cavalli-Sforza, L. L., and M. W. Feldman. 1973. Models for cultural inheritance: I, group mean and within group variation. *Theoretical Population Biology*, 4(1): 42–55.

Chagnon, N. A. 1968. *Yanomamö: the fierce people.* Holt, Rinehart and Winston, New York. xviii + 142 pp.

Chalmers, N. R. 1968. The social behaviour of free living mangabeys in Uganda. *Folia Primatologica*, 8(3,4): 263–281.

———— 1972. Comparative aspects of early infant development in some captive cercopithecines. In F. E. Poirier, ed. (*q.v.*), *Primate socialization*, pp. 63–82.

Chalmers, N. R., and Thelma E. Rowell. 1971. Behaviour and female reproductive cycles in a captive group of mangabeys. *Folia Primatologica*, 14(1): 1–14.

Chance, M. R. A. 1955. The sociability of monkeys. *Man*, 55(176): 162–165.

—— 1961. The nature and special features of the instinctive social bond of primates. In S. L. Washburn, ed. (*q.v.*), *Social life of early man*, pp. 17–33.

—— 1962. Social behaviour and primate evolution. In M. F. Ashley Montagu, ed., *Culture and the evolution of man*, pp. 84–130. Oxford University Press, New York. xiii + 376 pp.

—— 1967. Attention structure as the basis of primate rank orders. *Man*, 2(4): 503–518.

Chance, M. R. A., and C. J. Jolly. 1970. *Social groups of monkeys, apes and men*. E. P. Dutton, New York. 224 pp.

Charles-Dominique, P. 1971. Eco-ethologie des prosimiens du Gabon. *Biologia Gabonica*, 7(2): 121–228.

—— 1972. Ecologie et vie sociale de *Galago demidovii* (Fischer 1808; Prosimii). *Zeitschrift für Tierpsychologie*, supplement 9: 7–42.

Charles-Dominique, P., and C. M. Hladik. 1971. Le *Lepilemur* du Sud de Madagascar: *écologie*, alimentation et vie sociale. *La Terre et la Vie*, 118(1): 3–66.

Charles-Dominique, P., and R. D. Martin. 1970. Evolution of lorises and lemurs. *Nature, London*, 227 (5255): 257–260.

—— 1972. Behaviour and ecology of nocturnal prosimians. *Zeitschrift für Tierpsychologie*, supplement 9. 91 pp.

Chase, I. D. 1973. A working paper on explanations of hierarchy in animal societies. (Unpublished manuscript, cited by permission of the author.)

—— 1974. Models of hierarchy formation in animal societies. *Behavioral Science*. (In press.)

Chauvin, R. 1960. Les substances actives sur le comportement à l'intérieur de la ruche. *Annales de l'Abeille*, 3(2): 185–197.

—— ed. 1968. *Traité de biologie de l'abeille*, 5 vols. Vol. 1, *Biologie et physiologie générales*. xvi + 547 pp. Vol. 2, *Système nerveux, comportement et régulations sociales*. viii + 566 pp. Vol. 3, *Les produits de la ruche*. viii + 400 pp. Vol. 4, *Biologie appliquée*. viii + 434 pp. Vol. 5, *Histoire, ethnographie et folklore*. viii + 152 pp. Masson et Cie, Paris.

Chauvin, R., and C. Noirot, eds. 1968. *L'effet de groupe chez les animaux*. Colloques Internationaux no. 173. Centre National de la Recherche Scientifique, Paris. 390 pp.

Cheetham, A. H. 1973. Study of cheilostome polymorphism using principal components analysis. In G. P. Larwood, ed. (*q.v.*), *Living and fossil Bryozoa: recent advances in research*, pp. 385–409.

Chepko, Bonita Diane. 1971. A preliminary study of the effects of play deprivation on young goats. *Zeitschrift für Tierpsychologie*, 28(5): 517–526.

Cherrett, J. M. 1972. Some factors involved in the selection of vegetable substrate by *Atta cephalotes* (L.) (Hymenoptera: Formicidae) in tropical rain forest. *Journal of Animal Ecology*, 41: 647–660.

Cherry, C. 1957. *On human communication*. John Wiley & Sons, New York. xvi + 333 pp.

Chiang, H. C., and O. Stenroos. 1963. Ecology of insect swarms: II, occurrence of swarms of *Anarete* sp. under different field conditions (Cecidomyiidae, Diptera). *Ecology*, 44(3): 598–600.

Chitty, D. 1967a. The natural selection of self-regulatory behaviour in animal populations. *Proceedings of the Ecological Society of Australia*, 2: 51–78.

—— 1967b. What regulates bird populations? *Ecology*, 48(4): 698–701.

Chivers, D. J. 1969. On the daily behaviour and spacing of free-ranging howler monkey groups. *Folia Primatologica*, 10(1): 48–102.

—— 1973. An introduction to the socio-ecology of Malayan forest primates. In R. P. Michael and J. H. Crook, eds. (*q.v.*), *Comparative ecology and behaviour of primates*, pp. 101–146.

Chomsky, N. 1957. *Syntactic structures*. Mouton, The Hague. 118 pp.

—— 1972. *Language and mind*, enlarged ed. Harcourt, Brace, Jovanovich, New York. xii + 194 pp.

Christen, Anita. 1974. Fortpflanzungsbiologie und Verhalten bei *Cebuella pygmaea* und *Tamarin tamarin* (Primates, Platyrrhina, Callithricidae). *Zeitschrift für Tierpsychologie*, supplement 14. 79 pp.

Christian, J. J. 1955. Effect of population size on the adrenal glands and reproductive organs of male mice. *American Journal of Physiology*, 182(2): 292–300.

—— 1961. Phenomena associated with population density. *Proceedings of the National Academy of Sciences, U.S.A.*, 47(4): 428–449.

—— 1968. Endocrine-behavioral negative feed-back responses to increased population density. In R. Chauvin and C. Noirot, eds. (*q.v.*), *L'effet de groupe chez les animaux*, pp. 289–322.

—— 1970. Social subordination, population density, and mammalian evolution. *Science*, 168: 84–90.

Christian, J. J., and D. E. Davis. 1964. Endocrines, behavior, and population. *Science*, 146: 1550–1560.

Clark, Eugenie. 1972. The Red Sea's garden of eels. *National Geographic*, 142(5) (November): 724–735.

Clark, L. R., P. W. Geier, R. D. Hughes, and R. F. Morris. 1967. *The ecology of insect populations in theory and practice*. Methuen, London. xiv + 232 pp.

Clarke, T. A. 1970. Territorial behavior and population dynamics of a pomacentrid fish, the garibaldi, *Hypsypops rubicunda*. *Ecological Monographs*, 40(2): 189–212.

Clausen, C. P. 1940. *Entomophagous insects*. McGraw-Hill Book Co., New York. x + 688 pp.

Clausen, J. A., ed. 1968. *Socialization and society*. Little, Brown, Boston. xvi + 400 pp.

Clausewitz, C. von. 1960. *Principles of war*, trans. by H. W. Gatzke from the Appendix of *Vom Kriege*, 1832. Stackpole Co., Harrisburg, Pa. iv + 82 pp.

Clemente, Carmine D., and D. B. Lindsley, eds. 1967. *Brain function*, vol. 5, *Aggression and defense: neural mechanisms and social patterns*. University of California Press, Berkeley. xv + 361 pp.

Cleveland, L. R., S. R. Hall, Elizabeth P. Sanders, and Jane Collier. 1934.

The wood-feeding roach *Cryptocercus*, its Protozoa, and the symbiosis between Protozoa and roach. *Memoirs of the American Academy of Arts and Sciences*, 17(2): 185–342.

Clough, G. C. 1971. Behavioral responses of Norwegian lemmings to crowding. *Bulletin of the Ecological Society of America*, 52(4): 38.

——— 1972. Biology of the Bahaman hutia, *Geocapromys ingrahami*. *Journal of Mammalogy*, 53(4): 807–823.

Coates, A. G., and W. A. Oliver, Jr. 1973. Coloniality in zoantharian corals. In R. S. Boardman, A. H. Cheetham, and W. A. Oliver, Jr., eds. (q.v.), *Animal colonies: development and function through time*, pp. 3–27.

Cody, M. L. 1966. A general theory of clutch size. *Evolution*, 20(2): 174–184.

——— 1969. Convergent characteristics in sympatric species: a possible relation to interspecific competition and aggression. *Condor*, 71(3): 223–239.

——— 1971. Finch flocks in the Mohave Desert. *Theoretical Population Biology*, 2(2): 142–158.

——— 1974. *Competition and the structure of bird communities.* Princeton University Press, Princeton, N.J. x + 318 pp.

Cody, M. L., and J. H. Brown. 1970. Character convergence in Mexican finches. *Evolution*, 24(2): 304–310.

Coe, M. J. 1962. Notes on the habits of the Mount Kenya hyrax (*Procavia johnstoni mackinderi* Thomas). *Proceedings of the Zoological Society of London*, 138(4): 639–644.

——— 1967. Co-operation of three males in nest construction by *Chiromantis rufescens* Gunther (Amphibia: Rhacophoridae). *Nature, London*, 214(5083): 112–113.

Cohen, D. 1967. Optimization of seasonal migratory behaviour. *American Naturalist*, 101(917): 1–17.

Cohen, J. E. 1969a. Grouping in a vervet monkey troop. *Proceedings of the Second International Congress of Primatology, Atlanta, Georgia (U.S.A.), 1968*, 1: 274–278.

——— 1969b. Natural primate troops and a stochastic population model. *American Naturalist*, 103(933): 455–477.

——— 1971. *Casual groups of monkeys and men: stochastic models of elemental social systems.* Harvard University Press, Cambridge. xiii + 175 pp.

Cole, A. C. 1968. *Pogonomyrmex harvester ants: a study of the genus in North America.* University of Tennessee Press, Knoxville. x + 222 pp.

Cole, L. C. 1954. The population consequences of life history phenomena. *Quarterly Review of Biology*, 29(2): 103–137.

Collias, N. E. 1943. Statistical analysis of factors which make for success in initial encounters between hens. *American Naturalist*, 77(773): 519–538.

——— 1950. Social life and the individual among vertebrate animals. *Annals of the New York Academy of Sciences*, 51(6): 1076–1092.

Collias, N. E., and Elsie C. Collias. 1967. A field study of the red jungle fowl in north-central India. *Condor*, 69(4): 360–386.

——— 1969. Size of breeding colony related to attraction of mates in a tropical passerine bird. *Ecology*, 50(3): 481–488.

Collias, N. E., Elsie C. Collias, D. Hunsaker, and Lory Minning. 1966. Locality fixation, mobility and social organization within an unconfined population of red jungle fowl. *Animal Behaviour*, 14(4): 550–559.

Collias, N. E., and L. R. Jahn. 1959. Social behavior and breeding success in Canada geese (*Branta canadensis*) confined under semi-natural conditions. *Auk*, 76(4): 478–509.

Collias, N. E., and C. H. Southwick. 1952. A field study of population density and social organization in howling monkeys. *Proceedings of the American Philosophical Society, Philadelphia*, 96(2): 143–156.

Collias, N. E., J. K. Victoria, and R. J. Shallenberger. 1971. Social facilitation in weaverbirds: importance of colony size. *Ecology*, 52(2): 823–828.

Colombel, P. 1970a. Recherches sur la biologie et l'éthologie d'*Odontomachus haematodes* L. (Hym. Formicoïdea Poneridae): étude des populations dans leur milieu naturel. *Insectes Sociaux*, 17(3): 183–198.

——— 1970b. Recherches sur la biologie et l'éthologie d'*Odontomachus haematodes* L. (Hym. Formicoïdea Poneridae): biologie des reines. *Insectes Sociaux*, 17(3): 199–204.

Comfort, A. 1971. Likelihood of human pheromones. *Nature, London*, 230(5294): 432–433, 479.

Conder, P. J. 1949. Individual distance. *Ibis*, 91(4): 649–655.

Connell, J. H. 1961. The influence of interspecific competition and other factors on the distribution of the barnacle *Chthamalus stellatus*. *Ecology*, 42(4): 710–723.

——— 1963. Territorial behavior and dispersion in some marine invertebrates. *Researches on Population Ecology*, 5(2): 87–101.

Cook, S. F., and K. G. Scott. 1933. The nutritional requirements of *Zootermopsis* (*Termopsis*) *angusticollis*. *Journal of Cellular and Comparative Physiology*, 4(1): 95–110.

Cooper, K. W. 1957. Biology of eumenine wasps: V, digital communication in wasps. *Journal of Experimental Zoology*, 134(3): 469–509.

Corliss, J. O. 1961. *The ciliated Protozoa: characterization, classification, and guide to the literature.* Pergamon, New York. 310 pp.

Corning, W. C., J. A. Dyal, and A. O. D. Willows, eds. 1973. *Invertebrate learning*, 2 vols. Vol. 1, *Protozoans through annelids.* xvii + 296 pp. Vol. 2, *Arthropods and gastropod mollusks.* xiii + 284 pp. Plenum Press, New York.

Cott, H. B. 1957. *Adaptive coloration in animals.* Methuen, London. xxxii + 508 pp.

Coulson, J. C. 1966. The influence of the pair-bond and age on the breeding biology of the kittiwake gull *Rissa tridactyla*. *Journal of Animal Ecology*, 35(2): 269–279.

Coulson, J. C., and E. White. 1956. A study of colonies of the kittiwake *Rissa tridactyla* (L.). *Ibis*, 98(1): 63–79.

——— 1960. The effect of age and density of breeding birds on the time of breeding of the kittiwake *Rissa tridactyla*. *Ibis*, 102(1): 71–86.

Count, E. W. 1958. The biological basis of human sociality. *American Anthropologist*, 60(6): 1049–1085.

Cousteau, J.-Y., and P. Diolé. 1972. Killer whales have fearsome teeth and a strange gentleness to man. *Smithsonian*, 3(3) (June): 66–73. (Reprinted in modified form from J.-Y. Cousteau, *The whale: mighty monarch of the sea*, Doubleday, Garden City, N.Y., 1972.)

Cowdry, E. V., ed. 1930. *Human biology and racial welfare*. Hoeber, New York. 612 pp.

Craig, G. B. 1967. Mosquitoes: female monogamy induced by male accessory gland substance. *Science*, 156: 1499–1501.

Craig, J. V., and A. M. Guhl. 1969. Territorial behavior and social interactions of pullets kept in large flocks. *Poultry Science*, 48(5): 1622–1628.

Craig, J. V., L. L. Ortman, and A. M. Guhl. 1965. Genetic selection for social dominance ability in chickens. *Animal Behaviour*, 13(1): 114–131.

Crane, Jocelyn. 1949. Comparative biology of salticid spiders at Rancho Grande, Venezuela: IV, an analysis of display. *Zoologica, New York*, 34(4): 159–214.

———— 1957. Imaginal behavior in butterflies of the family Heliconiidae: changing social patterns and irrelevant actions. *Zoologica, New York*, 42(4): 135–145.

Creighton, W. S. 1953. New data on the habits of *Camponotus* (*Myrmaphaenus*) *ulcerosus* Wheeler. *Psyche, Cambridge*, 60(2): 82–84.

Creighton, W. S., and R. H. Crandall. 1954. New data on the habits of *Myrmecocystus melliger* Forel. *Biological Review, City College of New York*, 16(1): 2–6.

Creighton, W. S., and R. E. Gregg. 1954. Studies on the habits and distribution of *Cryptocerus texanus* Santschi (Hymenoptera: Formicidae). *Psyche, Cambridge*, 61(2): 41–57.

Crisler, Lois. 1956. Observations of wolves hunting caribou. *Journal of Mammalogy*, 37(3): 337–346.

Crisp, D. J., and P. S. Meadows. 1962. The chemical basis of gregariousness in cirripedes. *Proceedings of the Royal Society*, ser. B, 156: 500–520.

Crook, J. H. 1961. The basis of flock organisation in birds. In W. H. Thorpe and O. L. Zangwill, eds. (*q.v.*), *Current problems in animal behaviour*, pp. 125–149.

———— 1964. The evolution of social organization and visual communication in the weaver birds (Ploceinae). *Behaviour*, supplement 10. 178 pp.

———— 1965. The adaptive significance of avian social organizations. *Symposia of the Zoological Society of London*, 14: 181–218.

———— 1966. Gelada baboon herd structure and movement: a comparative report. *Symposia of the Zoological Society of London*, 18:237–258.

———— 1970a. Introduction—social behaviour and ethology. In J. H. Crook, ed. (*q.v.*), *Social behaviour in birds and mammals: essays on the social ethology of animals and man*, pp. xxi–xl.

———— 1970b. The socio-ecology of primates. In J. H. Crook, ed. (*q.v.*), *Social behaviour in birds and mammals: essays on the social ethology of animals and man*, pp. 103–166.

———— ed. 1970c. *Social behaviour in birds and mammals: essays on the social ethology of animals and man*. Academic Press, New York. xl + 492 pp.

———— 1971. Sources of cooperation in animals and man. In J. F. Eisenberg

and W. S. Dillon, eds. (*q.v.*), *Man and beast: comparative social behaviour*, pp. 237–272.

———— 1972. Sexual selection, dimorphism, and social organization in the primates. In B. G. Campbell, ed. (*q.v.*), *Sexual selection and the descent of man, 1871-1971*, pp. 231–281.

Crook, J. H., and P. Aldrich-Blake. 1968. Ecological and behavioural contrasts between sympatric ground dwelling primates in Ethiopia. *Folia Primatologica*, 8(3,4): 192–227.

Crook, J. H., and P. A. Butterfield. 1970. Gender role in the social system of quelea. In J. H. Crook, ed. (*q.v.*), *Social behaviour in birds and mammals: essays on the social ethology of animals and man*, pp. 211–248.

Crook, J. H., and J. S. Gartlan. 1966. Evolution of primate societies. *Nature, London*, 210(5042): 1200–1203.

Crovello, T. J., and C. S. Hacker. 1972. Evolutionary strategies in life table characteristics among feral and urban strains of *Aedes aegypti* (L.). *Evolution*, 26(2): 185–196.

Crow, J. F., and M. Kimura. 1965. Evolution in sexual and asexual populations. *American Naturalist*, 99(909): 439–450.

———— 1970. *An introduction to population genetics theory*. Harper & Row, New York. xiv + 591 pp.

Crow, J. F., and A. P. Mange. 1965. Measurement of inbreeding from the frequency of marriages between persons of the same surname. *Eugenics Quarterly*, 12: 199–203.

Crowcroft, P. 1957. *The life of the shrew*. Max Reinhardt, London. viii + 166 pp.

Crystal, D. 1969. *Prosodic systems and intonation in English*. Cambridge University Press, London. viii + 381 pp.

Cullen, Esther. 1957. Adaptations in the kittiwake to cliff-nesting. *Ibis*, 99(2): 275–302.

Cullen, J. M. 1960. Some adaptations in the nesting behaviour of terns. *Proceedings of the Twelfth International Ornithological Congress, Helsinki*, 1: 153–157.

Curio, E. 1963. Probleme des Feinderkennens bei Vögeln. *Proceedings of the Thirteenth International Ornithological Congress, Ithaca, New York*, 1: 206–239.

Curtis, Helena. 1968a. *Biology*. Worth Publishers, New York. 854 pp.

———— 1968b. *The marvelous animals: an introduction to the Protozoa*. Natural History Press, Garden City, N.Y. xvi + 189 pp.

Curtis, H. J. 1971. Genetic factors in aging. *Advances in Genetics*, 16: 305–325.

Curtis, R. F., J. A. Ballantine, E. B. Keverne, R. W. Bonsall, and R. P. Michael. 1971. Identification of primate sexual pheromones and the properties of synthetic attractants. *Nature, London*, 232(5310): 396–398.

Daanje, A. 1950. On locomotory movements in birds and the intention movements derived from them. *Behaviour*, 3(1): 48–98.

Dagg, Anne I., and D. E. Windsor. 1971. Olfactory discrimination limits in gerbils. *Canadian Journal of Zoology*, 49(3): 283–285.

Dahl, E., H. Emanuelsson, and C. von Mecklenburg. 1970. Pheromone transport and reception in an amphipod. *Science*, 170: 739–740.

Dahlberg, G. 1947. *Mathematical methods for population genetics.* S. Karger, New York. 182 pp.

Dalke, P. D., D. B. Pyrah, D. C. Stanton, J. E. Crawford, and E. Schlatterer. 1963. Ecology, productivity, and management of sage grouse in Idaho. *Journal of Wildlife Management*, 27(4): 810–841.

Dambach, M. 1963. Vergleichende Untersuchungen über das Schwarmverhalten von *Tilapia*-Jungfischen (Cichlidae, Teleostei). *Zeitschrift für Tierpsychologie*, 20(3): 267–296.

Dane, B., and W. G. Van der Kloot. 1964. An analysis of the display of the goldeneye duck (*Bucephala clangula* [L.]). *Behaviour*, 22(3,4): 282–328.

Dane, B., C. Walcott, and W. H. Drury. 1959. The form and duration of the display actions of the goldeneye (*Bucephala clangula*). *Behaviour*, 14(4): 265–281.

Darling, F. F. 1937. *A herd of red deer.* Oxford University Press, London. x + 215 pp. (Reprinted as a paperback, Doubleday, Garden City, N.Y., 1964. xiv + 226 pp.)

—— 1938. *Bird flocks and the breeding cycle: a contribution to the study of avian sociality.* Cambridge University Press, Cambridge. x + 124 pp.

Darlington, C. D. 1969. *The evolution of man and society.* Simon and Schuster, New York. 753 pp.

Darlington, P. J. 1971. Interconnected patterns of biogeography and evolution. *Proceedings of the National Academy of Sciences, U.S.A.*, 68(6): 1254–1258.

Dart, R. A. 1949. The predatory implemental technique of *Australopithecus. American Journal of Physical Anthropology*, n.s. 7: 1–38.

—— 1953. The predatory transition from ape to man. *International Anthropological and Linguistic Review*, 1(4): 201–213.

—— 1956. Cultural status of the South African man-apes. *Report of the Smithsonian Institution, Washington, D.C., 1955*, pp. 317–338.

Darwin, C. 1871. *The descent of man, and selection in relation to sex*, 2 vols. Appleton, New York. Vol. 1: vi + 409 pp.; vol. 2: viii + 436 pp.

Dasmann, R. F., and R. D. Taber. 1956. Behavior of Columbian black-tailed deer with reference to population ecology. *Journal of Mammalogy*, 37(2): 143–164.

Davenport, R. K. 1967. The orang-utan in Sabah. *Folia Primatologica*, 5(4): 247–263.

Davis, D. E. 1942. The phylogeny of social nesting habits in the Crotophaginae. *Quarterly Review of Biology*, 17(2): 115–134.

—— 1946. A seasonal analysis of mixed flocks of birds in Brazil. *Ecology*, 27(2): 168–181.

—— 1957. Aggressive behavior in castrated starlings. *Science*, 126: 253.

—— 1958. The role of density in aggressive behaviour of house mice. *Animal Behaviour*, 6(3,4): 207–210.

—— 1964. The physiological analysis of aggressive behavior. In W. Etkin, ed. (*q.v.*), *Social behavior and organization among vertebrates*, pp. 53–74.

Davis, J. A., Jr. 1965. A preliminary report of the reproductive behavior of the small Malayan chevrotain, *Tragulus javanicus*, at New York Zoo. *International Zoo Yearbook*, 5: 42–48.

Davis, R. B., C. F. Herreid, and H. L. Short. 1962. Mexican free-tailed bat in Texas. *Ecological Monographs*, 32(4): 311–346.

Davis, R. M. 1972. Behavior of the Vlei rat, *Otomys irroratus. Zoologica Africana*, 7: 119–140.

Davis, R. T., R. W. Leary, Mary D. C. Smith, and R. F. Thompson. 1968. Species differences in the gross behaviour of nonhuman primates. *Behaviour*, 31(3,4): 326–338.

Davis, W. H., R. W. Barbour, and M. D. Hassell. 1968. Colonial behavior of *Eptesicus fuscus. Journal of Mammalogy*, 49(1): 44–50.

Deag, J. M. 1973. Intergroup encounters in the wild Barbary macaque *Macaca sylvanus* L. In R. P. Michael and J. H. Crook, eds. (*q.v.*), *Comparative ecology and behaviour of primates*, pp. 315–373.

Deag, J. M., and J. H. Crook. 1971. Social behaviour and "agonistic buffering" in the wild Barbary macaque *Macaca sylvana* L. *Folia Primatologica*, 15(3,4): 183–200.

Deegener, P. 1918. *Die Formen der Vergesellschaftung im Tierreiche: ein systematisch-soziologischer Versuch.* Veit, Leipzig. 420 pp.

DeFries, J. C., and G. E. McClearn. 1970. Social dominance and Darwinian fitness in the laboratory mouse. *American Naturalist*, 104(938): 408–411.

Delage-Darchen, Bernadette. 1972. Une fourmi de Côte-d'Ivoire: *Melissotarsus titubans* Del., n. sp. *Insectes Sociaux*, 19(3): 213–226.

Deleurance, É.-P. 1948. Le comportement reproducteur est indépendant de la présence des ovaires chez *Polistes* (Hyménoptères Vespides). *Compte Rendu de l'Académie des Sciences, Paris*, 227(17): 866–867.

—— 1952. Le polymorphisme social et son déterminisme chez les Guêpes. In P.-P. Grassé, ed. (*q.v.*), *Structure et physiologie des sociétés animales*, pp. 141–155.

—— 1957. Contribution à l'étude biologique des *Polistes* (Hyménoptères Vespoides): I, l'activité de construction. *Annales des Sciences Naturelles, Zoologie*, 11th ser., 19(1,2): 91–222.

Deligne, J. 1965. Morphologie et fonctionnement des mandibules chez les soldats des termites. *Biologia Gabonica*, 1(2): 179–186.

Denenberg, V. H., ed. 1972. *The development of behavior.* Sinauer Associates, Sunderland, Mass. ix + 483 pp.

Denenberg, V. H., and K. M. Rosenberg. 1967. Nongenetic transmission of information. *Nature, London*, 216(5115): 549–550.

Denes, P. B., and E. N. Pinson. 1973. *The speech chain: the physics and biology of spoken language*, rev. ed. Anchor Press, Doubleday, Garden City, N.Y. xviii + 217 pp.

Denham, W. W. 1971. Energy relations and some basic properties of primate social organization. *American Anthropologist*, 73: 77–95.

Deutsch, J. A. 1957. Nest building behaviour of domestic rabbits under semi-natural conditions. *British Journal of Animal Behaviour*, 5(2): 53–54.

DeVore, B. I. 1963a. Mother-infant relations in free-ranging baboons. In Harriet L. Rheingold, ed. (*q.v.*), *Maternal behavior in mammals*, pp. 305–335.

———— 1963b. A comparison of the ecology and behavior of monkeys and apes. In S. L. Washburn, ed. (q.v.), *Classification and human evolution*, pp. 301–319.

———— ed. 1965. *Primate behavior: field studies of monkeys and apes.* Holt, Rinehart and Winston, New York. xiv + 654 pp.

———— 1971. The evolution of human society. In J. F. Eisenberg and W. S. Dillon, eds. (q.v.), *Man and beast: comparative social behavior,* pp. 297–311.

———— 1972. Quest for the roots of society. In P. R. Marler, ed. (q.v.), *The marvels of animal behavior,* pp. 393–408.

DeVore, B. I., and S. L. Washburn. 1960. Baboon behavior. 16-mm sound color film. University Extension, University of California, Berkeley. 30 min.

Diamond, J. M., and J. W. Terborgh. 1968. Dual singing in New Guinea birds. *Auk,* 85(1): 62–82.

Dingle, H. 1972a. Migration strategies of insects. *Science,* 175: 1327–1335.

———— 1972b. Aggressive behavior in stomatopods and the use of information theory in the analysis of animal communication. In H. E. Winn and B. L. Olla, eds., *Behavior of marine animals: current perspectives in research,* vol. 1, *Invertebrates,* pp. 126–156. Plenum Press, New York. xxix + 244 pp.

Dingle, H., and R. L. Caldwell. 1969. The aggressive and territorial behaviour of the mantis shrimp *Gonodactylus bredini* Manning (Crustacea: Stomatopoda). *Behaviour,* 33(1,2): 115–136.

Dixon, K. L. 1956. Territoriality and survival in the plain titmouse. *Condor,* 58(3): 169–182.

Dobrzański, J. 1961. Sur l'éthologie guerrière de *Formica sanguinea* Latr. (Hyménoptère, Formicidae). *Acta Biologiae Experimentalis, Warsaw,* 21: 53–73.

———— 1965. Genesis of social parasitism among ants. *Acta Biologiae Experimentalis, Warsaw,* 25(1): 59–71.

———— 1966. Contribution to the ethology of *Leptothorax acervorum* (Hymenoptera: Formicidae). *Acta Biologiae Experimentalis, Warsaw,* 26(1): 71–78.

Dobzhansky, T. 1963. Anthropology and the natural sciences—the problem of human evolution. *Current Anthropology,* 4: 138, 146–148.

Dobzhansky, T., H. Levene, and B. Spassky. 1972. Effects of selection and migration on geotactic and phototactic behaviour of *Drosophila,* III. *Proceedings of the Royal Society,* ser. B, 180: 21–41.

Dobzhansky, T., R. C. Lewontin, and Olga Pavlovsky. 1964. The capacity for increase in chromosomally polymorphic and monomorphic populations of *Drosophila pseudoobscura. Heredity,* 19(4): 597–614.

Dobzhansky, T., and Olga Pavlovsky. 1971. Experimentally created incipient species of *Drosophila. Nature, London,* 230(5292): 289–292.

Dobzhansky, T., and B. Spassky. 1962. Selection for geotaxis in monomorphic and polymorphic populations of *Drosophila pseudoobscura. Proceedings of the National Academy of Sciences, U.S.A.,* 48(10): 1704–1712.

Dodson, C. H. 1966. Ethology of some bees of the tribe Euglossini (Hymenoptera: Apidae). *Journal of the Kansas Entomological Society,* 39(4): 607–629.

Doetsch, R. N., and T. M. Cook. 1973. *Introduction to bacteria and their ecobiology.* University Park Press, Baltimore, Md. xii + 371 pp.

Donisthorpe, H. St. J. K. 1915. *British ants, their life-history and classification.* William Brendon and Son, Plymouth, England. xv + 379 pp.

Dorst, J. 1970. *A field guide to the larger mammals of Africa.* Houghton Mifflin Co., Boston. 287 pp.

Douglas-Hamilton, I. 1972. On the ecology and behaviour of the African elephant: the elephants of Lake Manyara. Ph.D. Thesis, Oriel College, Oxford University, Oxford. xiv + 268 pp.

———— 1973. On the ecology and behaviour of the Lake Manyara elephants. *East African Wildlife Journal,* 11(3,4): 401–403.

Downes, J. A. 1958. Assembly and mating in the biting Nematocera. *Proceedings of the Tenth International Congress of Entomology, Montreal, 1956,* 2: 425–434.

Downhower, J. F., and K. B. Armitage. 1971. The yellow-bellied marmot and the evolution of polygamy. *American Naturalist,* 105(944): 355–370.

Doyle, G. A., Annette Anderson, and S. K. Bearder. 1969. Maternal behaviour in the lesser bushbaby (*Galago senegalensis moholi*) under semi-natural conditions. *Folia Primatologica,* 11(3): 215–238.

Doyle, G. A., Annette Pelletier, and T. Bekker. 1967. Courtship, mating and parturition in the lesser bushbaby (*Galago senegalensis moholi*) under semi-natural conditions. *Folia Primatologica,* 7(2): 169–197.

Drabek, C. M. 1973. Home range and daily activity of the round-tailed ground squirrel, *Spermophilus tereticaudus neglectus. American Midland Naturalist,* 89(2): 287–293.

Dreher, J. J., and W. E. Evans. 1964. Cetacean communication. In W. N. Tavolga, ed. (q.v.), *Marine bio-acoustics,* pp. 373–393.

Drury, W. H., Jr. 1962. Breeding activities, especially nest building, of the yellowtail (*Ostinops decumanus*) in Trinidad, West Indies. *Zoologica, New York,* 47(1): 39–58.

Dubost, G. 1965a. Quelques renseignements biologiques sur *Potamogale velox. Biologia Gabonica,* 1(3): 257–272.

———— 1965b. Quelques traits remarquables du comportement de *Hyaemoschus aquaticus* (Tragulidae, Ruminantia, Artiodactyla). *Biologia Gabonica,* 1(3): 282–287.

———— 1970. L'organisation spatiale et sociale de *Muntiacus reevesi* Ogilby 1839 en semi-liberté. *Mammalia,* 34(3): 331–335.

Ducke, A. 1910. Révision des guêpes sociales polygames d'Amérique. *Annales Historico-Naturales Musei Nationales Hungarici,* 8(2): 449–544.

———— 1914. Über Phylogenie und Klassifikation der sozialen Vespiden. *Zoologische Jahrbücher, Abteilungen Systematik, Ökologie und Geographie der Tiere,* 36(2,3): 303–330.

Duellman, W. E. 1966. Aggressive behavior in dendrobatid frogs. *Herpetologica,* 22(3): 217–221.

———— 1967. Social organization in the mating calls of some Neotropical anurans. *American Midland Naturalist,* 77(1): 156–163.

Dumas, P. C. 1956. The ecological relations of sympatry in *Plethodon dunni* and *Plethodon vehiculum. Ecology,* 37(3): 484–495.

DuMond, F. V. 1968. The squirrel monkey in a seminatural environment. In L. A. Rosenblum and R. W. Cooper, eds. (*q.v.*), *The squirrel monkey*, pp. 87–146.

Dunaway, P. B. 1968. Life history and populational aspects of the eastern harvest mouse. *American Midland Naturalist*, 79(1): 48–67.

Dunbar, M. J. 1960. The evolution of stability in marine environments: natural selection at the level of the ecosystem. *American Naturalist*, 94(875): 129–136.

———— 1972. The ecosystem as a unit of natural selection. In E. S. Deevey, ed., *Growth by intussusception: ecological essays in honor of G. Evelyn Hutchinson*, pp. 114–130. Transactions of the Academy, vol. 44. Connecticut Academy of Arts and Sciences, New Haven. 442 pp.

Dunbar, R. I. M., and M. F. Nathan. 1972. Social organization of the Guinea baboon, *Papio papio*. *Folia Primatologica*, 17(5,6): 321–334.

Dunford, C. 1970. Behavioral aspects of spatial organization in the chipmunk, *Tamias striatus*. *Behaviour*, 36(3): 215–231.

Dunn, E. R. 1941. Notes on *Dendrobates auratus*. *Copeia*, 1941, no. 2, pp. 88–93.

Eaton, R. L. 1969. Cooperative hunting by cheetahs and jackals and a theory of domestication of the dog. *Mammalia*, 33(1): 87–92.

———— 1970. Group interactions, spacing and territoriality in cheetahs. *Zeitschrift für Tierpsychologie*, 27(4): 481–491.

———— ed. 1973. *The world's cats*, vol. 1. World Wildlife Safari, Winston, Oreg.

Eberhard, A. 1972. Inhibition and activation of bacterial luciferase synthesis. *Journal of Bacteriology*, 109(3): 1101–1105.

Eberhard, Mary Jane West. 1969. The social biology of polistine wasps. *Miscellaneous Publications, Museum of Zoology, University of Michigan, Ann Arbor*, 140: 1–101.

Eberhard, W. G. 1972. Altruistic behavior in a sphecid wasp: support for kin-selection theory. *Science*, 172: 1390–1391.

Edmondson, W. T. 1945. Ecological studies of sessile Rotatoria: II, dynamics of populations and social structures. *Ecological Monographs*, 15(2): 141–172.

Ehrlich, P. R., and Anne H. Ehrlich. 1973. Coevolution: heterotypic schooling in Caribbean reef fishes. *American Naturalist*, 107(953): 157–160.

Ehrlich, S. 1966. Ecological aspects of reproduction in nutria *Myocastor coypus* Mol. *Mammalia*, 30(1): 142–152.

Ehrman, Lee. 1964. Genetic divergence in M. Vetukhiv's experimental populations of *Drosophila pseudoobscura*: 1, rudiments of sexual isolation. *Genetical Research, Cambridge*, 5(1): 150–157.

———— 1966. Mating success and genotype frequency in *Drosophila*. *Animal Behaviour*, 14(2,3): 332–339.

Eibl-Eibesfeldt, I. 1950. Über die Jugendentwicklung des Verhaltens eines männlichen Dachses (*Meles meles* L.) unter besonderer Berücksichtigung des Spieles. *Zeitschrift für Tierpsychologie*, 7(3): 327–355.

———— 1953. Zur Ethologie des Hamsters (*Cricetus cricetus* L.). *Zeitschrift für Tierpsychologie*, 10(2): 204–254.

———— 1955. Über Symbiosen, Parasitismus und andere besondere zwischenartliche Beziehungen tropischer Meeresfische. *Zeitschrift für Tierpsychologie*, 12(2): 203–219.

———— 1962. Freiwasserbeobachtungen zur Deutung des Schwarmverhaltens verschiedener Fische. *Zeitschrift für Tierpsychologie*, 19(2): 163–182.

———— 1966. Das Verteidigen der Eiablageplätze bei der Hood-Meerechse (*Amblyrhynchus cristatus venustissimus*). *Zeitschrift für Tierpsychologie*, 23(5): 627–631.

———— 1970. *Ethology: the biology of behavior*. Holt, Rinehart and Winston, New York. xiv + 530 pp.

Eickwort, G. C., and Kathleen R. Eickwort. 1971. Aspects of the biology of Costa Rican halictic bees: II, *Dialictus umbripennis* and adaptations of its caste structure to different climates. *Journal of the Kansas Entomological Society*, 44(3): 343–373.

———— 1972. Aspects of the biology of Costa Rican halictine bees: IV, *Augochlora* (*Oxystoglossella*). *Journal of the Kansas Entomological Society*, 45(1): 18–45.

———— 1973a. Aspects of the biology of Costa Rican halictine bees: V, *Augochlorella edentata* (Hymenoptera: Halictidae). *Journal of the Kansas Entomological Society*, 46(1): 3–16.

———— 1973b. Notes on the nests of three wood-dwelling species of *Augochlora* from Costa Rica (Hymenoptera: Halictidae). *Journal of the Kansas Entomological Society*, 46(1): 17–22.

Eimerl, S., and I. DeVore. 1965. *The primates*. Time-Life Books, Chicago. 200 pp.

Eisenberg, J. F. 1962. Studies on the behavior of *Peromyscus maniculatus gambelii* and *Peromyscus californicus parasiticus*. *Behaviour*, 19(3): 177–207.

———— 1963. The behavior of heteromyid rodents. *University of California Publications in Zoology*, 69. iv + 100 pp.

———— 1966. The social organization of mammals. *Handbuch der Zoologie*, 10(7): 1–92.

———— 1967. A comparative study in rodent ethology with emphasis on evolution of social behavior, I. *Proceedings of the United States National Museum, Washington, D.C.*, 122(3597): 1–51.

———— 1968. Behavior patterns. In J. A. King, ed. (*q.v.*), *Biology of Peromyscus (Rodentia)*, pp. 451–495.

———— 1972. The elephant: life at the top. In P. R. Marler, ed. (*q.v.*), *The marvels of animal behavior*, pp. 191–207.

Eisenberg, J. F., and W. Dillon, eds. 1971. *Man and beast: comparative social behavior*. Smithsonian Institution Press, Washington, D.C. 401 pp.

Eisenberg, J. F., and E. Gould. 1966. The behavior of *Solenodon paradoxus* in captivity with comments on the behavior of other Insectivora. *Zoologica, New York*, 51(1): 49–58.

———— 1970. *The tenrecs: a study in mammalian behavior and evolution*. Smithsonian Institution Press, Washington, D.C. vi + 138 pp.

Eisenberg, J. F., and Devra G. Kleiman. 1972. Olfactory communication in mammals. *Annual Review of Ecology and Systematics*, 3: 1–32.

Eisenberg, J. F., and R. E. Kuehn. 1966. The behavior of *Ateles geoffroyi* and related species. *Smithsonian Miscellaneous Collections*, 151(8). iv + 63 pp.

Eisenberg, J. F., and M. Lockhart. 1972. An ecological reconnaissance of Wilpattu National Park, Ceylon. *Smithsonian Contributions to Zoology*, 101. vi + 118 pp.

Eisenberg, J. F., N. A. Muckenhirn, and R. Rudran. 1972. The relation between ecology and social structure in primates. *Science*, 176: 863–874.

Eisenberg, R. M. 1966. The regulation of density in a natural population of the pond snail *Lymnaea elodes*. *Ecology*, 47(6): 889–906.

Eisner, T. 1957. A comparative morphological study of the proventriculus of ants (Hymenoptera: Formicidae). *Bulletin of the Museum of Comparative Zoology, Harvard*, 116(8): 439–490.

—— 1970. Chemical defense against predation in arthropods. In E. Sondheimer and J. B. Simeone, eds. (q.v.), *Chemical ecology*, pp. 157–217.

Eisner, T., and J. Meinwald. 1966. Defensive secretions of arthropods. *Science*, 153: 1341–1350.

Elder, W. H., and Nina L. Elder. 1970. Social groupings and primate associations of the bushbuck (*Tragelaphus scriptus*). *Mammalia*, 34(3): 356–362.

Ellefson, J. O. 1968. Territorial behavior in the common white-handed gibbon, *Hylobates lar* Linn. In Phyllis C. Jay, ed. (q.v.), *Primates: studies in adaptation and variability*, pp. 180–199.

Ellis, Peggy E. 1959. Learning and social aggregation in locust hoppers. *Animal Behaviour*, 7(1,2): 91–106.

Ellison, L. N. 1971. Territoriality in Alaskan spruce grouse. *Auk*, 88(3): 652–664.

Eloff, F. 1973. Ecology and behavior of the Kalahari lion. In R. L. Eaton, ed., (q.v.) *The world's cats*, vol. 1, pp. 90–126.

Emerson, A. E. 1938. Termite nests—a study of the phylogeny of behavior. *Ecological Monographs*, 8(2): 247–284.

—— 1956a. Regenerative behavior and social homeostasis in termites. *Ecology*, 37(2): 248–258.

—— 1956b. Ethospecies, ethotypes, taxonomy, and evolution of *Apicotermes* and *Allognathotermes* (Isoptera, Termitidae). *American Museum Novitates*, no. 1771. 31 pp.

—— 1967. Cretaceous insects from Labrador: 3, a new genus and species of termite (Isoptera: Hodotermitidae). *Psyche, Cambridge*, 74(4): 276–289.

—— 1969. A revision of the Tertiary fossil species of the Kalotermitidae (Isoptera). *American Museum Novitates*, no. 2359. 57 pp.

—— 1971. Tertiary fossil species of the Rhinotermitidae (Isoptera), phylogeny of genera, and reciprocal phylogeny of associated Flagellata (Protozoa) and the Staphylinidae (Coleoptera). *Bulletin of the American Museum of Natural History*, 146(3): 243–303.

Emery, C. 1909. Über den Ursprung der dulotischen, parasitischen und myrmekophilen Ameisen. *Biologisches Centralblatt*, 29(11): 352–362.

Emlen, J. M. 1970. Age specificity and ecological theory. *Ecology*, 51(4): 588–601.

Emlen, J. T. 1938. Midwinter distribution of the American crow in New York State. *Ecology*, 19(2): 264–275.

—— 1940. The midwinter distribution of the crow in California. *Condor*, 42(6): 287–294.

Emlen, J. T., and G. B. Schaller. 1960. Distribution and status of the mountain gorilla (*Gorilla gorilla beringei*). *Zoologica, New York*, 45(5): 41–52.

Emlen, S. T. 1968. Territoriality in the bullfrog, *Rana catesbeiana. Copeia*, 1968, no. 2, pp. 240–243.

—— 1971. The role of song in individual recognition in the indigo bunting. *Zeitschrift für Tierpsychologie*, 28(3): 241–246.

—— 1972. An experimental analysis of the parameters of bird song eliciting species recognition. *Behaviour*, 41(1,2): 130–171.

Enders, R. K. 1935. Mammalian life histories from Barro Colorado Island, Panama. *Bulletin of the Museum of Comparative Zoology, Harvard*, 78(4): 385–502.

Erickson, J. G. 1967. Social hierarchy, territoriality, and stress reactions in sunfish. *Physiological Zoology*, 40(1): 40–48.

Erlinge, S. 1968. Territoriality of the otter *Lutra lutra* L. *Oikos*, 19(1): 81–98.

Ernst, E. 1959. Beobachtungen beim Spritzakt der *Nasutitermes*-Soldaten. *Revue Suisse de Zoologie*, 66(2): 289–295.

—— 1960. Fremde Termitenkolonien in *Cubitermes*-Nestern. *Revue Suisse de Zoologie*, 67(2): 201–206.

Errington, P. L. 1963. *Muskrat populations*. Iowa State University Press, Ames, Iowa. x + 665 pp.

Esch, H. 1967a. The evolution of bee language. *Scientific American*, 216(4) (April): 96–104.

—— 1967b. Die Bedeutung der Lauterzeugung für die Verständigung der stachellosen Bienen. *Zeitschrift für Vergleichende Physiologie*, 56(2): 199–220.

—— 1967c. The sounds produced by swarming honey bees. *Zeitschrift für Vergleichende Physiologie*, 56(4): 408–411.

Esch, H., Ilse Esch, and W. E. Kerr. 1965. Sound: an element common to communication of stingless bees and to dances of the honey bee. *Science*, 149: 320–321.

Eshel, I. 1972. On the neighbor effect and the evolution of altruistic traits. *Theoretical Population Biology*, 3(3): 258–277.

Espinas, A. 1878. *Des sociétés animales: étude de psychologie comparée*. Librairie Germer Ballière, Paris. (Reprinted by Stechert, Hafner, New York, 1924). 389 pp.

Espmark, Y. 1971. Mother-young relationship and ontogeny of behaviour in reindeer (*Rangifer tarandus* L.). *Zeitschrift für Tierpsychologie*, 29(1): 42–81.

Esser, A. H., ed. 1971. *Behavior and environment: the use of space by animals and men*. Proceedings of an international symposium held at the 1968 meeting of the American Association for the Advancement of Science in Dallas, Texas. Plenum Press, New York. xvii + 411 pp.

Estes, R. D. 1966. Behaviour and life history of the wildebeest (*Connochaetes taurinus* Burchell). *Nature, London,* 212(5066): 999–1000.

——— 1967. The comparative behavior of Grant's and Thomson's gazelles. *Journal of Mammalogy,* 48(2): 189–209.

——— 1969. Territorial behavior of the wildebeest (*Connochaetes taurinus* Burchell, 1823). *Zeitschrift für Tierpsychologie,* 26(3): 284–370.

——— 1974. Social organization of the African Bovidae. In V. Geist and F. Walther, eds. (*q.v.*), *The behaviour of ungulates and its relation to management,* pp. 166–205.

——— 1975a. *The behavior of African mammals,* vol. 1, *Ungulates.* (In preparation.)

——— 1975b. *The behavior of African mammals,* vol. 2, *Carnivores.* (In preparation.)

Estes, R. D., and J. Goddard. 1967. Prey selection and hunting behavior of the African wild dog. *Journal of Wildlife Management,* 31(1): 52–70.

Etkin, W. 1954. Social behavior and the evolution of man's mental faculties. *American Naturalist,* 88(840): 129–142.

——— ed. 1964. *Social behavior and organization among vertebrates.* University of Chicago Press, Chicago. xii + 307 pp.

Ettershank, G. 1966. A generic revision of the world Myrmicinae related to *Solenopsis* and *Pheidologeton* (Hymenoptera: Formicidae). *Australian Journal of Zoology,* 14: 73–171.

Evans, H. E. 1958. The evolution of social life in wasps. *Proceedings of the Tenth International Congress of Entomology, Montreal, 1956,* 2: 449–457.

——— 1964. Observations on the nesting behavior of *Moniaecera asperata* (Fox) (Hymenoptera, Sphecidae, Crabroninae) with comments on communal nesting in solitary wasps. *Insectes Sociaux,* 11(1): 71–78.

——— 1966. *The comparative ethology and evolution of the sand wasps.* Harvard University Press, Cambridge. xvi + 526 pp.

Evans, H. E., and Mary Jane West Eberhard. 1970. *The wasps.* University of Michigan Press, Ann Arbor. vi + 265 pp.

Evans, L. T. 1951. Field study of the social behavior of the black lizard, *Ctenosaura pectinata. American Museum Novitates,* 1493. 26 pp.

——— 1953. Tail display in an iguanid lizard, *Liocephalus carinatus coryi. Copeia,* 1953, no. 1, pp. 50–54.

Evans, Mary Alice, and H. E. Evans. 1970. *William Morton Wheeler, biologist.* Harvard University Press, Cambridge. xii + 363 pp.

Evans, S. M. 1973. A study of fighting reactions in some nereid polychaetes. *Animal Behaviour,* 21(1): 138–146.

Evans, W. E., and J. Bastian. 1969. Marine mammal communication: social and ecological factors. In H. T. Andersen, ed., *The biology of marine mammals,* pp. 425–475. Academic Press, New York. 511 pp.

Ewer, Rosalie F. 1959. Suckling behaviour in kittens. *Behaviour,* 15(1,2): 146–162.

——— 1963a. The behaviour of the meerkat, *Suricata suricatta* (Schreber). *Zeitschrift für Tierpsychologie,* 20(5): 570–607.

——— 1963b. A note on the suckling behaviour of the viverrid, *Suricata suricatta* (Schreber). *Animal Behaviour,* 11(4): 599–601.

——— 1967. The behaviour of the African giant rat (*Cricetomys gambianus* Waterhouse). *Zeitschrift für Tierpsychologie,* 24(1): 6–79.

——— 1968. *Ethology of mammals.* Plenum Press, New York. xiv + 418 pp.

——— 1971. The biology and behaviour of a free-living population of black rats (*Rattus rattus*). *Animal Behaviour Monographs,* 4(3): 125–174.

——— 1973. *The carnivores.* Cornell University Press, Ithaca, N.Y. xvi + 494 pp.

Ewing, L. S. 1967. Fighting and death from stress in a cockroach. *Science,* 155: 1035–1036.

Faber, W. 1967. Beiträge zur Kenntnis sozialparasitischer Ameisen: 1, *Lasius* (*Austrolasius* n.sg.) *reginae* n.sp., eine neue temporär sozialparasitische Erdameise aus Österreich (Hym. Formicidae). *Pflanzenschutz-Berichte,* 36(5–7): 73–107.

Fabricius, E., and K. Gustafson. 1953. Further aquarium observations on the spawning behaviour of the char, *Salmo alpinus* L. *Reports of the Institute of Freshwater Research, Drottningholm,* 35: 58–104.

Fady, J.-C. 1969. Les jeux sociaux: le compagnon de jeux chez les jeunes. Observations chez *Macaca irus. Folia Primatologica,* 11(1,2): 134–143.

Fagen, R. M. 1972. An optimal life-history strategy in which reproductive effort decreases with age. *American Naturalist,* 106(948): 258–261.

——— 1973. The paradox of play. (Unpublished manuscript.)

——— 1974. Selective and evolutionary aspects of animal play. *American Naturalist,* 108(964): 850–858.

Falconer, D. S. 1960. *Introduction to quantitative genetics.* Ronald Press, New York. x + 365 pp.

Falls, J. B. 1969. Functions of territorial song in the white-throated sparrow. In R. A. Hinde, ed. (*q.v.*), *Bird vocalizations: their relations to current problems in biology and psychology: essays presented to W. H. Thorpe,* pp. 207–232.

Fara, J. W., and R. H. Catlett. 1971. Cardiac response and social behaviour in the guinea-pig (*Cavia porcellus*). *Animal Behaviour,* 19(3): 514–523.

Farentinos, R. C. 1971. Some observations on the play behavior of the Steller sea lion (*Eumetopias jubata*). *Zeitschrift für Tierpsychologie,* 28(4): 428–438.

Fedigan, Linda M. 1972. Roles and activities of male geladas (*Theropithecus gelada*). *Behaviour,* 41(1,2): 82–90.

Fenner, F. 1965. Myxoma virus and *Oryctolagus cuniculus*: two colonizing species. In H. G. Baker and G. L. Stebbins, eds. (*q.v.*), *The genetics of colonizing species,* pp. 485–499.

Fiedler, K. 1954. Vergleichende Verhaltensstudien an Seenadeln,

Schlangennadeln und Seepferdchen (Syngnathidae). *Zeitschrift für Tierpsychologie*, 11(3): 358–416.

Fielder, D. R. 1965. A dominance order for shelter in the spiny lobster *Jasus lalandei* (H. Milne-Edwards). *Behaviour*, 24(3,4): 236–245.

Findley, J. S. 1967. Insectivores and dermopterans. In S. Anderson and J. K. Jones, Jr., eds. (q.v.), *Recent mammals of the world: a synopsis of families*, pp. 87–108.

Fiscus, C. H., and K. Niggol. 1965. Observations of cetaceans off California, Oregon, and Washington. *Special Scientific Report, U.S. Department of the Interior, Fish and Wildlife Service*, 498. 27 pp.

Fishelson, L. 1964. Observations on the biology and behaviour of Red Sea coral fishes. *Bulletin of the Sea Fisheries Research Station, Haifa, Israel*, 37: 11–26.

Fishelson, L., D. Popper, and N. Gunderman. 1971. Diurnal cyclic behaviour of *Pempheris oualensis* Cuv. & Val. (Pempheridae, Teleostei). *Journal of Natural History*, 5: 503–506.

Fisher, A. E. 1964. Chemical stimulation of the brain. *Scientific American*, 210(6) (June): 60–68.

Fisher, J. 1954. Evolution and bird sociality. In J. Huxley, A. C. Hardy, and E. B. Ford, eds., *Evolution as a process*, pp. 71–83. George Allen & Unwin, London. 376 pp. (Reprinted as a paperback, Collier Books, New York, 1963. 416 pp.)

Fisher, R. A. 1930. *The genetical theory of natural selection*. Clarendon Press, Oxford. xiv + 272 pp.

Flanders, S. E. 1956. The mechanisms of sex-ratio regulation in the (parasitic) Hymenoptera. *Insectes Sociaux*, 3(2): 325–334.

Flannery, K. V. 1972. The cultural evolution of civilizations. *Annual Review of Ecology and Systematics*, 3: 399–426.

Fleay, D. H. 1935. Breeding of *Dasyurus viverrinus* and general observations on the species. *Journal of Mammalogy*, 16(1): 10–16.

Floody, O. R., and D. W. Pfaff. 1974. Steroid hormones and aggressive behavior: approaches to the study of hormone-sensitive brain mechanisms for behavior. In S. H. Frazier, ed., *Aggression*. Research Publications, Association for Research in Nervous and Mental Disease, vol. 52 (1972 Symposium on Aggression). Waverly Press, Boston. (In press.)

Fodor, J., and M. Garrett. 1966. Some reflections on competence and performance. In J. Lyons and R. J. Wales, eds., *Psycholinguistic papers*, pp. 133–163. Edinburgh University Press, Edinburgh. 243 pp.

Forbes, S. A. 1906. The corn-root aphis and its attendant ant. *Bulletin, U.S. Department of Agriculture, Division of Entomology*, 60: 29–39.

Ford, E. B. 1971. *Ecological genetics*, 3d ed. Chapman & Hall, London. xx + 410 pp.

Forel A. 1874. *Les fourmis de la Suisse*. Société Helvétique des Sciences Naturelles, Zurich. iv + 452 pp.

——— 1898. La parabiose chez les fourmis. *Bulletin de la Société Vaudoise des Sciences Naturelles* (Lausanne), 34(130): 380–384.

Fossey, Dian. 1972. Living with mountain gorillas. In P. R. Marler, ed. (q.v.), *The marvels of animal behavior*, pp. 209–229.

Foster, J. B., and A. I. Dagg. 1972. Notes on the biology of the giraffe. *East African Wildlife Journal*, 10(1): 1–16.

Fox, M. W. 1969. The anatomy of aggression and its ritualization in Canidae: a developmental and comparative study. *Behaviour*, 35(3,4): 242–258.

——— 1971. *Behaviour of wolves, dogs and related canids*. Jonathan Cape, London. 214 pp.

——— 1972. Socio-ecological implications of individual differences in wolf litters: a developmental and evolutionary perspective. *Behaviour*, 46(3,4): 298–313.

Fox, R. 1971. The cultural animal. In J. F. Eisenberg and W. S. Dillon, eds. (q.v.), *Man and beast: comparative social behavior*, pp. 263–296.

——— 1972. Alliance and constraint: sexual selection in the evolution of human kinship systems. In B. G. Campbell, ed. (q.v.), *Sexual selection and the descent of man 1871–1971*, pp. 282–331.

Frädrich, H. 1965. Zur Biologie und Ethologie des Warzenschweines (*Phacochoerus aethiopicus* Pallas), unter Berücksichtigung des Verhaltens anderer Suiden. *Zeitschrift für Tierpsychologie*, 22(3): 328–374, 22(4): 375–393.

——— 1974. A comparison of behaviour in the Suidae. In V. Geist and F. Walther, eds. (q.v.), *The behaviour of ungulates and its relation to management*, vol. 1, pp. 133–143.

Francoeur, A. 1973. Révision taxonomique des espèces nearctiques du groupe *fusca*, genre *Formica* (Formicidae, Hymenoptera). *Memoires de la Société Entomologique du Québec*, 3: 1–316.

Frank, F. 1957. The causality of microtine cycles in Germany. *Journal of Wildlife Management*, 21(2): 113–121.

Franklin, I., and R. C. Lewontin. 1970. Is the gene the unit of selection? *Genetics*, 65(4): 707–734.

Franklin, W. L. 1973. High, wild world of the vicuña. *National Geographic*, 143(1) (January): 76–91.

——— 1974. The social behaviour of the vicuña. In V. Geist and F. Walther, eds. (q.v.), *The behaviour of ungulates and its relation to management*, vol. 1, pp. 477–487.

Franzisket, L. 1960. Experimentelle Untersuchung über die optische Wirkung der Streifung beim Preussenfisch (*Dascyllus aruanus*). *Behaviour*, 15(1,2): 77–81.

Fraser, A. F. 1968. *Reproductive behaviour in ungulates*. Academic Press, New York. x + 202 pp.

Free, J. B. 1955a. The behaviour of egg-laying workers of bumblebee colonies. *British Journal of Animal Behaviour*, 3(4): 147–153.

——— 1955b. The division of labour within bumblebee colonies. *Insectes Sociaux*, 2(3): 195–212.

——— 1956. A study of the stimuli which release the food begging and offering responses of worker honeybees. *British Journal of Animal Behaviour*, 4(3): 94–101.

——— 1959. The transfer of food between the adult members of a honeybee community. *Bee World*, 40(8): 193–201.

——— 1961a. The social organization of the bumble-bee colony. A lecture given to The Central Association of Bee-keepers on 18th January 1961. North Hants Printing and Publishing Co., Fleet, Hants, England. 11 pp.

——— 1961b. Hypopharyngeal gland development and division of labour in honey-bee (*Apis mellifera* L.) colonies. *Proceedings of the Royal Entomological Society of London*, ser. A, 36(1–3): 5–8.

——— 1969. Influence of the odour of a honeybee colony's food stores on the behaviour of its foragers. *Nature, London*, 222(5195): 778.

Free, J. B., and C. G. Butler. 1959. *Bumblebees.* New Naturalist, Collins, London. xiv + 208 pp.

Freeland, J. 1958. Biological and social patterns in the Australian bulldog ants of the genus *Myrmecia. Australian Journal of Zoology*, 6(1): 1–18.

Fretwell, S. D. 1972. *Populations in a seasonal environment.* Princeton University Press, Princeton, N.J. xxiii + 217 pp.

Friedlaender, J. S. 1971. Isolation by distance in Bougainville. *Proceedings of the National Academy of Sciences, U.S.A.*, 68(4): 704–707.

Friedlander, C. P. 1965. Aggregation in *Oniscus asellus* Linn. *Animal Behaviour*, 13(2,3): 342–346.

Friedrichs, R. W. 1970. *A sociology of sociology.* Free Press, Collier-Macmillan, New York. xxxiv + 429 pp.

Frisch, K. von. 1954. *The dancing bees: an account of the life and senses of the honey bee*, trans. by Dora Ilse. Methuen, London. xiv + 183 pp.

——— 1965. *Tanzsprache und Orientierung der Bienen.* Springer-Verlag, Berlin. vii + 578 pp.

——— 1967. *The dance language and orientation of bees*, trans. by L. E. Chadwick. Belknap Press of Harvard University Press, Cambridge. xiv + 566 pp.

Frisch, K. von, and R. Jander, 1957. Über den Schwänzeltanz der Bienen. *Zeitschrift für Vergleichende Physiologie*, 40(3): 239–263.

Frisch, K. von, and G. A. Rösch. 1926. Neue Versuche über die Bedeutung von Duftorgan und Pollenduft für die Verständigung im Bienenvolk. *Zeitschrift für Vergleichende Physiologie*, 4(1): 1–21.

Frisch, O. von. 1966a. Versuche über die Herzfrequenzänderung von Jungvögeln bei Fütterungs- und Schreckreizen. *Zeitschrift für Tierpsychologie*, 23(1): 52–55.

——— 1966b. Herzfrequenzänderung bei Drückreaktionen junger Nestflüchter. *Zeitschrift für Tierpsychologie*, 23(4): 497–500.

Fry, C. H. 1972a. The biology of African bee-eaters. *Living Bird*, 11: 75–112.

——— 1972b. The social organization of bee-eaters (Meropidae) and co-operative breeding in hot-climate birds. *Ibis*, 114(1): 1–14.

Fry, W. G., ed. 1970. *The biology of the Porifera.* Symposia of the Zoological Society of London no. 25. Academic Press, New York. xxviii + 512 pp.

Furuya, Y. 1963. On the Gagyusan troop of Japanese monkeys after the first separation. *Primates*, 4(1): 116–118.

——— 1965. Social organization of the crabeating monkey. *Folia Primatologica*, 6(3,4): 285–336.

——— 1969. On the fission of troops of Japanese monkeys: II, general view of the troop fission of Japanese monkeys. *Primates*, 10(1): 47–70.

Gadgil, M. 1971. Dispersal: population consequences and evolution. *Ecology*, 52(2): 253–261.

Gadgil, M., and W. H. Bossert. 1970. Life history consequences of natural selection. *American Naturalist*, 104(935): 1–24.

Galton, F. 1871. Gregariousness in cattle and men. *Macmillan's Magazine, London*, 23: 353.

Gander, F. F. 1929. Experiences with wood rats, *Neotoma fuscipes macrotis. Journal of Mammalogy*, 10(1): 52–58.

Garattini, S., and E. B. Sigg, eds. 1969. *Aggressive behaviour.* Proceedings of the Symposium on the Biology of Aggressive Behaviour, Milan, May 1968. Excerpta Medica, Amsterdam. 369 pp.

Garcia, J., B. K. McGowan, F. R. Ervin, and R. A. Koelling. 1968. Cues: their relative effectiveness as a function of the reinforcer. *Science*, 160: 794–795.

Garstang, W. 1946. The morphology and relations of the Siphonophora. *Quarterly Journal of Microscopical Science*, n.s. 87(2): 103–193.

Gartlan, J. S. 1968. Structure and function in primate society. *Folia Primatologica*, 8(2): 89–120.

——— 1969. Sexual and maternal behavior of the vervet monkey, *Cercopithecus aethiops. Journal of Reproduction and Fertility*, supplement 6: 137–150.

——— 1970. Preliminary notes on the ecology and behavior of the drill, *Mandrillus leucophaeus* Ritgen, 1824. In J. R. Napier and P. H. Napier, eds. (q.v.), *Old World monkeys: evolution, systematics, and behavior*, pp. 445–480.

Gartlan, J. S., and C. K. Brain. 1968. Ecology and social variability in *Cercopithecus aethiops* and *C. mitis*. In Phyllis C. Jay, ed. (q.v.), *Primates: studies in adaptation and variability*, pp. 253–292.

Gaston, A. J. 1973. The ecology and behaviour of the long-tailed tit. *Ibis*, 115(3): 330–351.

Gates, D. M. 1970. Animal climates (where animals must live). *Environmental Research*, 3(2): 132–144.

Gauss, C. H. 1961. Ein Beitrag zur Kenntnis des Balzverhaltens einheimischer Molche. *Zeitschrift für Tierpsychologie*, 18(1): 60–66.

Gauthier-Pilters, Hilde. 1959. Einige Beobachtungen zum Droh-, Angriffs- und Kampverhalten des Dromedarhengstes, sowie über Geburt und Verhaltensentwicklung des Jungtiers, in der nordwestlichen Sahara. *Zeitschrift für Tierpsychologie*, 16(5): 593–604.

——— 1967. The fennec. *African Wildlife*, 21(2): 117–125.

——— 1974. The behaviour and ecology of camels in the Sahara, with special reference to nomadism and water management. In V. Geist and F. Walther, eds. (q.v.), *The behaviour of ungulates and its relation to management*, vol. 2, pp. 542–551.

Gautier-Hion, A. 1970. L'organisation sociale d'une bande de talapoins (*Miopithecus talapoin*) dans le nord-est du Gabon. *Folia Primatologica*, 12(2): 116–141.

——— 1973. Social and ecological features of talapoin monkey—comparisons with sympatric cercopithecines. In R. P. Michael and J. H. Crook, eds. (q.v.), *Comparative ecology and behaviour of primates*, pp. 147–170.

Gay, F. J. 1966. A new genus of termites (Isoptera) from Australia. *Journal of the Entomological Society of Queensland*, 5: 40–43.

Gay, F. J., and J. H. Calaby. 1970. Termites of the Australian region. In K. Krishna and Frances M. Weesner, eds. (*q.v.*), *Biology of termites*, vol. 2, pp. 393–448.

Gehlbach, F. 1971. Discussion. In A. H. Esser, ed. (*q.v.*), *Behavior and environment: the use of space by animals and men*, p. 211.

Geist, V. 1963. On the behaviour of the North American moose (*Alces alces andersoni* Peterson 1950) in British Columbia. *Behaviour*, 20(3,4): 377–416.

——— 1971a. *Mountain sheep: a study in behavior and evolution*. University of Chicago Press, Chicago. xvi + 383 pp.

——— 1971b. The relation of social evolution and dispersal in ungulates during the Pleistocene, with emphasis on the Old World deer and the genus *Bison*. *Quarternary Research*, 1(3): 283–315.

——— 1974. On the relationship of social evolution and ecology in ungulates. *American Zoologist*, 14(1): 205–220.

Geist, V., and F. Walther, eds. 1974. *The behaviour of ungulates and its relation to management*, 2 vols. IUCN Publications, n.s., no. 24. International Union for the Conservation of Nature and Natural Resources, Morges, Switzerland. Vol. 1, pp. 1–511; vol. 2, pp. 512–940.

Gerking, S. D. 1953. Evidence for the concepts of home range and territory in stream fishes. *Ecology*, 34(2): 347–365.

Gersdorf, E. 1966. Beobachtungen über das Verhalten von Vogelschwärmen. *Zeitschrift für Tierpsychologie*, 23(1): 37–43.

Gervet, J. 1956. L'action des températures différentielles sur la monogynie fonctionnelle chez les *Polistes* (Hyménoptères Vespides). *Insectes Sociaux*, 3(1): 159–176.

——— 1962. Étude de l'effet de groupe sur la ponte dans la société polygyne de *Polistes gallicus*. *Insectes Sociaux*, 9(3): 231–263.

Getz, L. L. 1972. Social structure and aggressive behavior in a population of *Microtus pennsylvanicus*. *Journal of Mammalogy*, 53(2): 310–317.

Ghent, A. W. 1960. A study of the group-feeding behaviour of larvae of the jack pine sawfly, *Neodiprion pratti banksianae* Roh. *Behaviour*, 16(1,2): 110–148.

Ghent, R. L., and N. E. Gary. 1962. A chemical alarm releaser in honey bee stings (*Apis mellifera* L.). *Psyche, Cambridge*, 69(1): 1–6.

Ghiselin, M. T. 1969. The evolution of hermaphroditism among animals. *Quarterly Review of Biology*, 44(2): 189–208.

Gibb, J. A. 1966. Tit predation and the abundance of *Ernarmonia conicolana* (Heyl.) on Weeting Heath, Norfolk, 1962–63. *Journal of Animal Ecology*, 35(1): 43–53.

Gibson, J. B., and J. M. Thoday. 1962. Effects of disruptive selection: VI, a second chromosome polymorphism. *Heredity*, 17(1): 1–26.

Giesel, J. T. 1971. The relations between population structure and rate of inbreeding. *Evolution*, 25(3): 491–496.

Gilbert, J. J. 1963. Contact chemoreception, mating behaviour, and sexual isolation in the rotifer genus *Brachionus*. *Journal of Experimental Biology*, 40(4): 625–641.

——— 1966. Rotifer ecology and embryological induction. *Science*, 151: 1234–1237.

——— 1973. Induction and ecological significance of gigantism in the rotifer *Asplancha sieboldi*. *Science*, 181: 63–66.

Gilbert, L. E., and M. C. Singer. 1973. Dispersal and gene flow in a butterfly species. *American Naturalist*, 107(953): 58–72.

Gill, J. C., and W. Thomson. 1956. Observations on the behaviour of suckling pigs. *British Journal of Animal Behaviour*, 4(2): 46–51.

Gilliard, E. T. 1962. On the breeding behavior of the cock-of-the-rock (Aves, *Rupicola rupicola*). *Bulletin of the American Museum of Natural History*, 124(2): 31–68.

Ginsburg, B., and W. C. Allee. 1942. Some effects of conditioning on social dominance and subordination in inbred strains of mice. *Physiological Zoology*, 15(4): 485–506.

Glancey, B. M., C. E. Stringer, C. H. Craig, P. M. Bishop, and B. B. Martin. 1970. Pheromone may induce brood tending in the fire ant, *Solenopsis saevissima. Nature, London*, 226(5248): 863–864.

Glass, Lynn W., and R. V. Bovbjerg. 1969. Density and dispersion in laboratory populations of caddisfly larvae (*Cheumatopsyche*, Hydropsychidae). *Ecology*, 50(6): 1082–1084.

Goddard, J. 1967. Home range, behaviour, and recruitment rates of two black rhinoceros populations. *East African Wildlife Journal*, 5: 133–150.

——— 1973. The black rhinoceros. *Natural History*, 82(4): 58–67.

Godfrey, J. 1958. Social behaviour in four bank vole races. *Animal Behaviour*, 6(1,2): 117.

Goffman, E. 1959. *The presentation of self in everyday life*. Doubleday Anchor Books, Doubleday, Garden City, N.Y. xvi + 259 pp.

——— 1961. *Encounters: two studies in the sociology of interaction*. Bobbs-Merrill, Indianapolis. 152 pp.

——— 1969. *Strategic interaction*. University of Pennsylvania Press, Philadelphia. x + 145 pp.

Goin, C. J. 1949. The peep order in peepers: a swamp water serenade. *Quarterly Journal of the Florida Academy of Sciences* (Gainesville), 11(2,3): 59–61.

Goin, C. J., and Olive B. Goin. 1962. *Introduction to herpetology*. W. H. Freeman, San Francisco. 341 pp.

Goin, Olive B., and C. J. Goin. 1962. Amphibian eggs and the montane environment. *Evolution*, 16(3): 364–371.

Gompertz, T. 1961. The vocabulary of the great tit. *British Birds*, 54(10): 369–394; 54(11,12): 409–418.

Goodall, Jane, 1965. Chimpanzees of the Gombe Stream Reserve. In I. DeVore, ed. (*q.v.*), *Primate behavior: field studies of monkeys and apes*, pp. 425–481.

Gosling, L. M. 1974. The social behaviour of Coke's hartebeest (*Alcelaphus buselaphus cokei*). In V. Geist and F. Walther, eds. (*q.v.*), *The behaviour of ungulates and its relation to management*, vol. 1, pp. 488–511.

Goss-Custard, J. D. 1970. Feeding dispersion in some overwintering wading birds. In J. H. Crook, ed. (*q.v.*), *Social behaviour in birds and mammals: essays on the social ethology of animals and man,* pp. 3–35.

Gösswald, K. 1933. Weitere Untersuchungen über die Biologie von *Epimyrma gösswaldi* Men. und Bemerkungen über andere parasitische Ameisen. *Zeitschrift für Wissenschaftliche Zoologie,* 144(2): 262–288.

———— 1953. Histologische Untersuchungen an den arbeiterlosen Ameise *Teleutomyrmex schneideri* Kutter (Hym. Formicidae). *Mitteilungen der Schweizerischen Entomologischen Gesellschaft,* 26(2): 81–128.

Gösswald, K., and W. Kloft. 1960. Neuere Untersuchungen über die sozialen Wechselbeziehungen im Ameisenvolk, durchgeführt mit Radio-Isotopen. *Zoologische Beiträge,* 5(2,3): 519–556.

Gottesman, I. I. 1968. A sampler of human behavioral genetics. *Evolutionary Biology,* 2: 276–320.

Gottschalk, L. A., S. M. Kaplan, Goldine C. Gleser, and Carolyn Winget. 1961. Variations in the magnitude of anxiety and hostility with phases of the menstrual cycle. *Psychosomatic Medicine,* 23(5): 448.

Gotwald, W. H. 1971. Phylogenetic affinity of the ant genus *Cheliomyrmex* (Hymenoptera: Formicidae). *Journal of the New York Entomological Society,* 79(3): 161–173.

Gotwald, W. H., and W. L. Brown. 1966. The ant genus *Simopelta* (Hymenoptera: Formicidae). *Psyche, Cambridge,* 73(4): 261–277.

Gotwald, W. H., and J. Lévieux. 1972. Taxonomy and biology of a new West African ant belonging to the genus *Amblyopone* (Hymenoptera: Formicidae). *Annals of the Entomological Society of America,* 65(2): 383–396.

Gramza, A. F. 1967. Responses of brooding nighthawks to a disturbance stimulus. *Auk,* 84(1): 72–86.

Grandi, G. 1961. Studi di un entomologo sugli imenotteri superiori. *Bollettino dell'Istituto di Entomologia della Università degli studi di Bologna,* 25. 659 pp.

Grant, E. C. 1969. Human facial expression. *Man,* 4(4): 525–536.

Grant, P. R. 1966. The coexistence of two wren species of the genus *Thryothorus. Wilson Bulletin,* 78(3): 266–278.

———— 1968. Polyhedral territories of animals. *American Naturalist,* 102(923): 75–80.

———— 1970. Experimental studies of competitive interaction in a two-species system: II, the behaviour of *Microtus, Peromyscus* and *Clethrionomys* species. *Animal Behaviour,* 18(3): 411–426.

———— 1972. Convergent and divergent character displacement. *Biological Journal of the Linnaean Society,* 4(1): 39–68.

Grant, T. R. 1973. Dominance and association among members of a captive and a free-ranging group of grey kangaroos (*Macropus giganteus*). *Animal Behaviour,* 21(3): 449–456.

Grant, W. C., Jr. 1955. Territorialism in two species of salamanders. *Science,* 121: 137–138.

Grassé, P.-P. 1952a. *Traité de zoologie,* vol. 1, pt. 1, *Phylogenie; protozoaires: généralités, flagellés.* Masson et Cie, Paris.

———— ed. 1952b. *Structure et physiologie des sociétés animales.* Colloques Internationaux no. 34. Centre National de la Recherche Scientifique, Paris. 359 pp.

———— 1959. La reconstruction du nid et les coordinations inter-individuelles chez *Bellicositermes natalensis* et *Cubitermes* sp. La théorie de la stigmergie: essai d'interprétation du comportement des termites constructeurs. *Insectes Sociaux,* 6(1): 41–83.

———— 1967. Nouvelles expériences sur le termite de Müller (*Macrotermes mülleri*) et considérations sur la théorie de la stigmergie. *Insectes Sociaux,* 14(1): 73–102.

Grassé, P.-P., and C. Noirot. 1958. Construction et architecture chez les termites champignonnistes (Macrotermitinae). *Proceedings of the Tenth International Congress of Entomology, Montreal, 1956,* 2: 515–520.

Gray, B. 1971a. Notes on the biology of the ant species *Myrmecia dispar* (Clark) (Hymenoptera: Formicidae). *Insectes Sociaux,* 18(2): 71–80.

———— 1971b. Notes on the field behaviour of two ant species *Myrmecia desertorum* Wheeler and *Myrmecia dispar* (Clark) (Hymenoptera: Formicidae). *Insectes Sociaux,* 18(2): 81–94.

Greaves, T. 1962. Studies of foraging galleries and the invasion of living trees by *Coptotermes acinaciformis* and *C. brunneus* (Isoptera). *Australian Journal of Zoology,* 10(4): 630–651.

Green, R. G., C. L. Larson, and J. F. Bell. 1939. Shock disease as the cause of the periodic decimation of the snowshoe hare. *American Journal of Hygiene,* ser. B, 30: 83–102.

Greenberg, B. 1946. The relation between territory and social hierarchy in the green sunfish. *Anatomical Record,* 94(3): 395.

———— 1947. Some relations between territory, social hierarchy, and leadership in the green sunfish (*Lepomis cyanellus*). *Physiological Zoology,* 20(3): 267–299.

Greer, A. E., Jr. 1971. Crocodilian nesting habits and evolution. *Fauna,* 2: 20–28.

Griffin, D. J. G., and J. C. Yaldwyn. 1970. Giant colonies of pelagic tunicates (*Pyrosoma spinosum*) from SE Australia and New Zealand. *Nature, London,* 226(5244): 464–465.

Groos, K. 1896. *Die Spiele der Thiere.* Gustav Fischer, Jena. xvi + 359 pp. (Translated as *The play of animals,* Appleton, New York, 1898.)

Groot, A. P. de. 1953. Protein and amino acid requirements of the honeybee (*Apis mellifica* L.). *Physiologia Comparata et Oecologia,* 3(2,3): 197–285.

Grubb, P., and P. A. Jewell. 1966. Social grouping and home range in feral Soay sheep. *Symposia of the Zoological Society of London,* 18: 179–210.

Guhl, A. M. 1950. Social dominance and receptivity in the domestic fowl. *Physiological Zoology,* 23(4): 361–366.

———— 1958. The development of social organization in the domestic chick. *Animal Behaviour,* 6(1,2): 92–111.

———— 1964. Psychophysiological interrelations in the social behavior of chickens. *Psychological Bulletin,* 61(4): 277–285.

———— 1968. Social inertia and social stability in chickens. *Animal Behaviour,* 16(2,3): 219–232.

Guhl, A. M., N. E. Collias, and W. C. Allee. 1945. Mating behavior and the social hierarchy in small flocks of white leghorns. *Physiological Zoology,* 18(4): 365–390.

Guhl, A. M., and Gloria J. Fischer. 1969. The behaviour of chickens. In E. S. E. Hafez, ed. (*q.v.*), *The behaviour of domestic animals*, pp. 515–553.

Guiglia, Delfa. 1972. *Les guêpes sociales (Hymenoptera Vespidae) d'Europe occidentale et septentrionale.* Masson et Cie, Paris. viii + 181 pp.

Guiler, E. R. 1970. Observations on the Tasmanian devil, *Sarcophilus harrisii* (Marsupalia: Dasyuridae), I, II. *Australian Journal of Zoology*, 18(1): 49–70.

Guiton, P. 1959. Socialisation and imprinting in brown leghorn chicks. *Animal Behaviour*, 7(1,2): 26–41.

Gundlach, H. 1968. Brutfürsorge, Brutpflege, Verhaltensontogenese und Tagesperiodik beim europäischen Wildschwein (*Sus scrofa* L.). *Zeitschrift für Tierpsychologie*, 25(8): 955–995.

Gurney, J. H. 1913. *The Gannet: a bird with a history.* Witherby, London. li + 567 pp.

Guthrie, R. D. 1971. A new theory of mammalian rump patch evolution. *Behaviour*, 38(1,2): 132–145.

Guthrie-Smith, H. 1925. *Bird life on island and shore.* Blackwood, Edinburgh. xix + 195 pp.

Gwinner, E. 1966. Über einige Bewegungsspiele des Kolkraben (*Corvus corax* L.). *Zeitschrift für Tierpsychologie*, 23(1): 28–36.

Haartman, L. von. 1954. Die Trauerfliegenschnäpper: III, die Nahrungsbiologie. *Acta Zoologica Fennica*, 83: 1–96.

———— 1956. Territory in the pied flycatcher *Muscicapa hypoleuca*. *Ibis*, 98(3): 460–475.

———— 1969. Nest-site and evolution of polygamy in European passerine birds. *Ornis Fennica*, 46(1): 1–12.

Haas, A. 1960. Vergleichende Verhaltsstudien zum Paarungsschwarm solitärer Apiden. *Zeitschrift für Tierpsychologie*, 17(4): 402–416.

Haddow, A. J. 1952. Field and laboratory studies on an African monkey, *Cercopithecus ascanius schmidti* Matschie. *Proceedings of the Zoological Society of London*, 122(2): 297–394.

Haeckel, E. 1888. *Report on the Siphonophorae collected by H.M.S. Challenger during the years 1873–76.* Scientific Results of the Voyage of H.M.S. Challenger, Zoology, vol. 28. Eyre and Spottiswoode, London. xii + 380 pp.

Hafez, E. S. E., ed. 1969. *The behaviour of domestic animals*, 2d ed. Williams & Wilkins Co., Baltimore. xii + 647 pp.

Haga, R. 1960. Observations on the ecology of the Japanese pika. *Journal of Mammalogy*, 41(2): 200–212.

Hahn, M. E., and P. Tumolo. 1971. Individual recognition in mice: how is it mediated? *Bulletin of the Ecological Society of America*, 52(4): 53–54.

Hailman, J. P. 1960. Hostile dancing and fall territory of a color-banded mockingbird. *Condor*, 62(6): 464–468.

Haldane, J. B. S. 1932. *The causes of evolution.* Longmans, Green, London. vii + 234 pp. (Reprinted as a paperback, Cornell University Press, Ithaca, N.Y., 1966. vi + 235 pp.)

———— 1955. Animal communication and the origin of human language. *Science Progress, London*, 43(171): 385–401.

Haldane, J. B. S., and H. Spurway. 1954. A statistical analysis of communication in "Apis mellifera" and a comparison with communication in other animals. *Insectes Sociaux*, 1(3): 247–283.

Hall, E. T. 1966. *The hidden dimension.* Doubleday, Garden City, N.Y. (Reprinted as a paperback, Anchor Books, Doubleday, Garden City, N.Y., 1969. xii + 217 pp.)

Hall, J. R. 1970. Synchrony and social stimulation in colonies of the black-headed weaver *Ploceus cucullatus* and Vieillot's black weaver *Melanopteryx nigerrimus.* *Ibis*, 112(1): 93–104.

Hall, K. R. L. 1960. Social vigilance behaviour of the chacma baboon, *Papio ursinus*. *Behaviour*, 16(3,4): 261–294.

———— 1963a. Variations in the ecology of the chacma baboon (*P. ursinus*). *Symposia of the Zoological Society of London*, 10: 1–28.

———— 1963b. Tool-using performances as indicators of behavioral adaptability. *Current Anthropology*, 4(5): 479–487.

———— 1965. Social organization of the old-world monkeys and apes. *Symposia of the Zoological Society of London*, 14: 265–289.

———— 1967. Social interactions of the adult male and adult females of a patas monkey group. In S. A. Altmann, ed. (*q.v.*), *Social communication among primates*, pp. 261–280.

———— 1968a. Behaviour and ecology of the wild patas monkey, *Erythrocebus patas*, in Uganda. In Phyllis C. Jay, ed. (*q.v.*), *Primates: studies in adaptation and variability*, pp. 32–119.

———— 1968b. Experiment and quantification in the study of baboon behavior in its natural habitat. In Phyllis C. Jay, ed. (*q.v.*), *Primates: studies in adaptation and variability*, pp. 120–130.

Hall, K. R. L., and I. DeVore. 1965. Baboon social behavior. In I. DeVore, ed. (*q.v.*), *Primate behavior: field studies of monkeys and apes*, pp. 53–110.

Hall, K. R. L., and Barbara Mayer. 1967. Social interactions in a group of captive patas monkeys (*Erythrocebus patas*). *Folia Primatologica*, 5(3): 213–236.

Halle, L. J. 1971. International behavior and the prospects for human survival. In J. F. Eisenberg and W. S. Dillon, eds. (*q.v.*), *Man and beast: comparative social behavior*, pp. 353–368.

Hamilton, T. H. 1962. Species relationships and adaptations for sympatry in the avian genus *Vireo*. *Condor*, 64(1): 40–68.

Hamilton, T. H., and R. H. Barth, Jr. 1962. The biological significance of season change in male plumage appearance in some New World migratory bird species. *American Naturalist*, 96(888): 129–144.

Hamilton, W. D. 1964. The genetical theory of social behaviour, I, II. *Journal of Theoretical Biology*, 7(1): 1–52.

———— 1966. The moulding of senescence by natural selection. *Journal of Theoretical Biology*, 12(1): 12–45.

———— 1967. Extraordinary sex ratios. *Science*, 156: 477–488.

———— 1970. Selfish and spiteful behaviour in an evolutionary model. *Nature, London*, 228(5277): 1218–1220.

———— 1971a. Geometry for the selfish herd. *Journal of Theoretical Biology*, 31(2): 295–311.

———— 1971b. Selection of selfish and altruistic behavior in some extreme models. In J. F. Eisenberg and W. S. Dillon, eds. (q.v.), *Man and beast: comparative social behavior*, pp. 57–91.

———— 1972. Altruism and related phenomena, mainly in social insects. *Annual Review of Ecology and Systematics*, 3: 193–232.

Hamilton, W. J., III, and W. M. Gilbert. 1969. Starling dispersal from a winter roost. *Ecology*, 50(5): 886–898.

Hangartner, W. 1969a. Structure and variability of the individual odor trail in *Solenopsis geminata* Fabr. (Hymenoptera, Formicidae). *Zeitschrift für Vergleichende Physiologie*, 62(1): 111–120.

———— 1969b. Carbon dioxide, a releaser for digging behavior in *Solenopsis geminata* (Hymenoptera: Formicidae). *Psyche, Cambridge*, 76(1): 58–67.

Hangartner, W., J. M. Reichson, and E. O. Wilson. 1970. Orientation to nest material by the ant, *Pogonomyrmex badius* (Latreille). *Animal Behaviour*, 18(2): 331–334.

Hanks, J., M. S. Price, and R. W. Wrangham. 1969. Some aspects of the ecology and behaviour of the defassa waterbuck (*Kobus defassa*) in Zambia. *Mammalia*, 33(3): 471–494.

Hansen, E. W. 1966. The development of maternal and infant behavior in the rhesus monkey. *Behaviour*, 27(1,2): 107–149.

Hardin, G. 1956. Meaninglessness of the word protoplasm. *Scientific Monthly*, 82(3): 112–120.

———— 1972. Population skeletons in the environmental closet. *Bulletin of the Atomic Scientists*, 28(6) (June): 37–41.

Hardy, A. C. 1960. Was man more aquatic in the past? *New Scientist*, 7: 642–645.

Harlow, H. F. 1959. The development of learning in the rhesus monkey. *American Scientist*, 47(4): 459–479.

Harlow, H. F., M. K. Harlow, R. O. Dodsworth, and G. L. Arling. 1966. Maternal behavior of rhesus monkeys deprived of mothering and peer associations in infancy. *Proceedings of the American Philosophical Society*, 110(1): 58–66.

Harlow, H. F., and R. R. Zimmerman. 1959. Affectional responses in the infant monkey. *Science*, 130: 421–432.

Harrington, J. R. 1971. Olfactory communication in *Lemur fuscus*. Ph.D. thesis, Duke University, Durham, N.C. [Cited by Thelma Rowell, 1972 (q.v.).]

Harris, G. W., and R. P. Michael. 1964. The activation of sexual behaviour by hypothalamic implants of oestrogen. *Journal of Physiology*, 171(2): 275–301.

Harris, M. P. 1970. Territory limiting the size of the breeding population of the oystercatcher (*Haematopus ostralegus*)—a removal experiment. *Journal of Applied Ecology*, 39(3): 707–713.

Harris, V. T. 1952. An experimental study of habitat selection by prairie and forest races of the deermouse, *Peromyscus maniculatus*. *Contributions from the Laboratory of Vertebrate Biology, University of Michigan, Ann Arbor*, no. 56. 53 pp.

Harris, W. V. 1970. Termites of the Palearctic region. In K. Krishna and Frances M. Weesner, eds. (q.v.), *Biology of termites*, vol. 2, pp. 295–313.

Harrison, C. J. O. 1965. Allopreening as agonistic behaviour. *Behaviour*, 34(3,4): 161–209.

Harrison, G. A., and A. J. Boyce, eds. 1972. *The structure of human populations*. Clarendon Press, Oxford University Press, Oxford. xvi + 447 pp.

Hartley, P. H. T. 1949. Biology of the mourning chat in winter quarters. *Ibis*, 91(3): 393–413.

———— 1950. An experimental analysis of interspecific recognition. *Symposia of the Society for Experimental Biology*, 4: 313–336.

Hartman, W. D., and H. M. Reiswig. 1971. The individuality of sponges. *Abstracts with Programs, Geological Society of America*, 3(7): 593.

———— 1973. The individuality of sponges. In R. S. Boardman, A. H. Cheetham, and W. A. Oliver, Jr., eds. (q.v.), *Animal colonies: development and function through time*, pp. 567–584.

Hartshorne, C. 1958. Some biological principles applicable to song-behavior. *Wilson Bulletin*, 70(1): 41–56.

Harvey, P. A. 1934. Life history of *Kalotermes minor*. In C. A. Kofoid et al., eds. (q.v.), *Termites and termite control*, pp. 217–233.

Haskell, P. T. 1970. The hungry locust. *Science Journal* (January), pp. 61–67.

Haskins, C. P. 1939. *Of ants and men*. Prentice-Hall, New York. vii + 244 pp.

———— 1970. Researches in the biology and social behavior of primitive ants. In L. R. Aronson et al., eds. (q.v.), *Development and evolution of behavior: essays in memory of T. C. Schneirla*, pp. 355–388.

Haskins, C. P., and Edna F. Haskins. 1950. Notes on the biology and social behavior of the archaic ponerine ants of the genera *Myrmecia* and *Promyrmecia*. *Annals of the Entomological Society of America*, 43(4): 461–491.

———— 1951. Note on the method of colony foundation of the ponerine ant *Amblyopone australis* Erichson. *American Midland Naturalist*, 45(2): 432–445.

———— 1965. *Pheidole megacephala* and *Iridomyrmex humilis* in Bermuda—equilibrium or slow replacement? *Ecology*, 46(5): 736–740.

Haskins, C. P., and R. M. Whelden. 1954. Note on the exchange of ingluvial food in the genus *Myrmecia*. *Insectes Sociaux*, 1(1): 33–37.

Haskins, C. P., and P. A. Zahl. 1971. The reproductive pattern of *Dinoponera grandis* Roger (Hymenoptera, Ponerinae) with notes on the ethology of the species. *Psyche, Cambridge*, 78(1,2): 1–11.

Hasler, A. D. 1966. *Underwater guideposts: homing of salmon*. University of Wisconsin Press, Madison. xii + 155 pp.

———— 1971. Orientation and fish migration. *Fish Physiology*, 6: 429–510.

Hassell, M. P. 1966. Evaluation of parasite or predator responses. *Journal of Animal Ecology*, 35(1): 65–75.

Haubrich, R. 1961. Hierarchical behaviour in the South African clawed frog, *Xenopus laevis* Daudin. *Animal Behaviour*, 9(1,2): 71–76.

Hay, D. A. 1972. Recognition by *Drosophila melanogaster* of individuals from other strains or cultures: support for the role of olfactory cues in selective mating. *Evolution*, 26(2): 171–176.

Haydak, M. H. 1935. Brood rearing by honeybees confined to a pure carbohydrate diet. *Journal of Economic Entomology*, 28(4): 657–660.

———— 1945. The language of the honeybee. *American Bee Journal*, 85: 316–317.

Hazlett, B. A. 1966. Social behavior of the Paguridae and Diogenidae of Curaçao. *Studies on the Fauna of Curaçao and Other Caribbean Islands* (The Hague), 23: 1–143.

———— 1970. The effect of shell size and weight on the agonistic behavior of a hermit crab. *Zeitschrift für Tierpsychologie*, 27(3): 369–374.

Hazlett, B. A., and W. H. Bossert. 1965. A statistical analysis of the aggressive communications systems of some hermit crabs. *Animal Behaviour*, 13(2,3): 357–373.

Healey, M. C. 1967. Aggression and self-regulation of population size in deermice. *Ecology*, 48(3): 377–392.

Heatwole, H. 1965. Some aspects of the association of cattle egrets with cattle. *Animal Behaviour*, 13(1): 79–83.

Hediger, H. 1941. Biologische Gesetzmässigkeiten im Verhalten von Wirbeltieren. *Mitteilungen der Naturforschenden Gesellschaft Bern, 1940*, pp. 37–55.

———— 1950. *Wildtiere in Gefangenschaft—ein Grundriss der Tiergartenbiologie.* Benno Schwabe, Basle. (Reprinted as *Wild animals in captivity: an outline of the biology of zoological gardens*, trans. by G. Sircom, Butterworth Scientific Publications, London. 207 pp.)

———— 1955. *Studies of the psychology and behaviour of captive animals in zoos and circuses*, trans. by G. Sircom. Criterion Books, New York. vii + 166 pp. (Reprinted as *The psychology and behaviour of animals in zoos and circuses*, Dover, New York, 1968. vii + 166 pp.)

Heimburger, N. 1959. Das Markierungsverhalten einiger Caniden. *Zeitschrift für Tierpsychologie*, 16(1): 104–113.

Heinroth, O., and Magdalena Heinroth. 1928. *Die Vögel Mitteleuropas*, vol. 3. Hugo Bermühler Verlag, Berlin-Lichterfelde. x + 286 pp.

Heldmann, G. 1936a. Ueber die Entwicklung der polygynen Wabe von *Polistes gallica* L. *Arbeiten über Physiologische und Angewandte Entomologie aus Berlin-Dahlem*, 3: 257–259.

———— 1936b. Über das Leben auf Waben mit mehreren überwinterten Weibchen von *Polistes gallica* L. *Biologisches Zentralblatt*, 56(7,8): 389–401.

Heller, H. C. 1971. Altitudinal zonation of chipmunks (*Eutamias*): interspecific aggression. *Ecology*, 52(2): 312–329.

Helm, June. 1968. The nature of Dogrib socioterritorial groups. In R. B. Lee and I. DeVore, eds. (q.v.), *Man the hunter*, pp. 118–125.

Helversen, D. von, and W. Wickler. 1971. Über den Duettgesang des afrikanischen Drongo *Dicrurus adsimilis* Bechstein. *Zeitschrift für Tierpsychologie*, 29(3): 301–321.

Hendrichs, H., and Ursula Hendrichs. 1971. *Dikdik und Elefanten.* R. Piper, Munich. 173 pp.

Hendrickson, J. R. 1958. The green sea turtle, *Chelonia mydas* (Linn.) in Malaya and Sarawak. *Proceedings of the Zoological Society of London*, 130(4): 455–534.

Henry, C. S. 1972. Eggs and repagula of *Ululodes* and *Ascaloptynx* (Neuroptera: Ascalaphidae): a comparative study. *Psyche, Cambridge*, 79(1,2): 1–22.

Hensley, M. M., and J. B. Cope. 1951. Further data on removal and repopulation of the breeding birds in a spruce-fir forest community. *Auk*, 68(4): 483–493.

Hergenrader, G. L., and A. D. Hasler. 1967. Seasonal changes in swimming rates of yellow perch in Lake Mendota as measured by sonar. *Transactions of the American Fisheries Association*, 96(4): 373–382.

Herrnstein, R. J. 1971a. Quantitative hedonism. *Journal of Psychiatric Research*, 8: 399–412.

———— 1971b. I.Q. *Atlantic Monthly*, 228(3) (September): 43–64.

Hess, E. H. 1958. "Imprinting" in animals. *Scientific American*, 198(3) (March): 81–90.

Highton, R., and T. Savage. 1961. Functions of the brooding behavior in the female red-backed salamander, *Plethodon cinereus. Copeia*, 1961, no. 1, pp. 95–98.

Hildén, O., and S. Vuolanto. 1972. Breeding biology of the red-necked phalarope *Phalaropus lobatus* in Finland. *Ornis Fennica*, 49(3,4): 57–85.

Hill, C. 1946. Playtime at the zoo. *Zoo Life* (Zoological Society of London), 1(1): 24–26.

Hill, Jane H. 1972. On the evolutionary foundations of language. *American Anthropologist*, 74(3): 308–317.

Hill, W. C. O. 1972. *Evolutionary biology of the primates.* Academic Press, New York. x + 233 pp.

Hinde, R. A. 1952. The behaviour of the great tit (*Parus major*) and some other related species. *Behaviour*, supplement 2. x + 201 pp.

———— 1954. Factors governing the changes in strength of a partially inborn response, as shown by the mobbing behaviour of the chaffinch (*Fringilla coelebs*): I, the nature of the response, and an examination of its course. *Proceedings of the Royal Society*, ser. B, 142: 306–331.

———— 1956. The biological significance of the territories of birds. *Ibis*, 98(3): 340–369.

———— 1958. Alternative motor patterns in chaffinch song. *Animal Behaviour*, 6(3,4): 211–218.

———— ed. 1969. *Bird vocalizations: their relations to current problems in biology and psychology: essays presented to W. H. Thorpe.* Cambridge University Press, Cambridge. xvi + 394 pp.

———— 1970. *Animal behaviour: a synthesis of ethology and comparative psychology*, 2d ed. McGraw-Hill Book Co., New York. xvi + 876 pp.

———— ed. 1972. *Non-verbal communication.* Cambridge University Press, Cambridge. xiii + 423 pp.

———— 1974. *Biological bases of human social behaviour.* McGraw-Hill Book Co., New York. xvi + 462 pp.

Hinde, R. A., and Lynda M. Davies. 1972a. Changes in mother-infant relationship after separation in *Rhesus* monkeys. *Nature, London*, 239(5366): 41–42.

———— 1972b. Removing infant rhesus from mother for 13 days compared with removing mother from infant. *Journal of Child Psychology and Psychiatry*, 13: 227–237.

Hinde, R. A., and Yvette Spencer-Booth. 1967. The behaviour of socially living rhesus monkeys in their first two and a half years. *Animal Behaviour*, 15(1): 169–196.

———— 1969. The effect of social companions on mother-infant relations in rhesus monkeys. In D. Morris, ed. (*q.v.*), *Primate ethology: essays on the socio-sexual behavior of apes and monkeys*, pp. 343–364.

———— 1971. Effects of brief separation from mother on rhesus monkeys. *Science*, 173: 111–118.

Hingston, R. W. G. 1929. *Instinct and intelligence*. Macmillan Co., New York. xv + 296 pp.

Hirsch, J. 1963. Behavior genetics and individuality understood. *Science*, 142: 1436–1442.

Hjorth, I. 1970. Reproductive behaviour in Tetraonidae with special references to males. *Viltrevy*, 7(4): 183–596.

Hochbaum, H. A. 1955. *Travels and traditions of waterfowl*. University of Minnesota Press, Minneapolis. xii + 301 pp.

Hockett, C. F. 1960. Logical considerations in the study of animal communication. In W. E. Lanyon and W. N. Tavolga, eds. (*q.v.*), *Animal sounds and communication*, pp. 392–430.

Hockett, C. F., and S. A. Altmann. 1968. A note on design features. In T. A. Sebeok, ed. (*q.v.*), *Animal communication: techniques of study and results of research*, pp. 61–72.

Hocking, B. 1970. Insect associations with the swollen thorn acacias. *Transactions of the Royal Entomological Society of London*, 122(7): 211–255.

Hodjat, S. H. 1970. Effects of crowding on colour, size and larval activity of *Spodoptera littoralis* (Lepidoptera: Noctuidae). *Entomologia Experimentalis et Applicata*, 13: 97–106.

Hoesch, W. 1960. Zum Brutverhalten des Laufhühnchens *Turnix sylvatica lepurana*. *Journal für Ornithologie*, 101(3): 265–295.

Hoese, H. D. 1971. Dolphin feeding out of water in a salt marsh. *Journal of Mammalogy*, 52(1): 222–223.

Hoffer, E. 1882–83. *Die Hummeln Steiermarks: Lebensgeschichte und Beschreibung Derselben*, two parts. Leuschner and Lubensky, Graz. Part I: 92 pp; part 2: 98 pp. (Behavioral descriptions are in the first part, published in 1882.)

Hoffmeister, D. F. 1967. Tubulidentates, proboscideans, and hyracoideans. In S. Anderson and J. K. Jones, Jr., eds. (*q.v.*), *Recent mammals of the world: a synopsis of families*, pp. 355–365.

Hogan-Warburg, A. J. 1966. Social behavior of the ruff, *Philomachus pugnax* (L.). *Ardea*, 54(3,4): 109–229.

Höhn, E. O. 1969. The phalarope. *Scientific American*, 220(6) (June): 104–111.

Holgate, P. 1967. Population survival and life history phenomena. *Journal of Theoretical Biology*, 14(1): 1–10.

Hölldobler, B. 1962. Zur Frage der Oligogynie bei *Camponotus ligniperda* Latr. und *Camponotus herculeanus* L. (Hym. Formicidae). *Zeitschrift für Angewandte Entomologie*, 49(4): 337–352.

———— 1967a. Verhaltensphysiologische Untersuchungen zur Myrmecophilie einiger Staphylinidenlarven. *Verhandlungen der Deutschen Zoologischen Gesellschaft, Heidelberg, 1967*, pp. 428–434.

———— 1967b. Zur Physiologie der Gast-Wirt-Beziehungen (Myrmecophilie) bei Ameisen: I, das Gastverhältnis der *Atemeles*- und *Lomechusa*-Larven (Col. Staphylinidae) zu *Formica* (Hym. Formicidae). *Zeitschrift für Vergleichende Physiologie*, 56(1): 1–21.

———— 1969a. Host finding by odor in the myrmecophilic beetle *Atemeles pubicollis* Bris. (Staphylinidae). *Science*, 166: 757–758.

———— 1969b. Orientierungsmechanismen des Ameisengastes *Atemeles* (Coleoptera, Staphylinidae) bei der Wirtssuche. *Verhandlungen der Deutschen Zoologischen Gesellschaft, Würzburg, 1969*, pp. 580–585.

———— 1970. Zur Physiologie der Gast-Wirt-Beziehungen (Myrmecophilie) bei Ameisen: II, das Gastverhältnis des imaginalen *Atemeles pubicollis* Bris. (Col. Staphylinidae) zu *Myrmica* und *Formica* (Hym. Formicidae). *Zeitschrift für Vergleichende Physiologie*, 66(2): 215–250.

———— 1971a. Recruitment behavior in *Camponotus socius* (Hym. Formicidae). *Zeitschrift für Vergleichende Physiologie*, 75(2): 123–142.

———— 1971b. Sex pheromone in the ant *Xenomyrmex floridanus*. *Journal of Insect Physiology*, 17(8): 1497–1499.

———— 1971c. Communication between ants and their guests. *Scientific American*, 224(3) (March): 86–93.

———— 1973. Chemische Strategie beim Nahrungserwerb der Diebsameise (*Solenopsis fugax* Latr.) und der Pharaoameise (*Monomorium pharaonis* L.). *Oecologia, Berlin*, 11: 371–380.

Hölldobler, B., M. Möglich, and U. Maschwitz. 1974. Communication by tandem running in the ant *Camponotus sericeus*. *Journal of Comparative Physiology*, 90(2): 105–127.

Hölldobler, K. 1953. Beobachtungen über die Koloniengründung von *Lasius umbratus umbratus* Nyl. *Zeitschrift für Angewandte Entomologie*, 34(4): 598–606.

Holling, C. S. 1959. Some characteristics of simple types of predation and parasitism. *Canadian Entomologist*, 91(7): 385–398.

Holmes, R. T. 1966. Breeding ecology and annual cycle adaptations of the red-backed sandpiper (*Calidris alpina*) in northern Alaska. *Condor*, 68(1): 3–46.

———— 1970. Differences in population density, territoriality, and food supply of dunlin on arctic and subarctic tundra. In A. Watson, ed. (*q.v.*), *Animal populations in relation to their food resources*, pp. 303–319.

Holst, D. von. 1969. Sozialer Stress bei Tupajas (*Tupaia belangeri*) *Zeitschrift für Vergleichende Physiologie*, 63(1): 1–58.

———— 1972a. Renal failure as the cause of death in *Tupaia belangeri* exposed to persistent social stress. *Journal of Comparative Physiology*, 78(3): 236–273.

———— 1972b. Die Funktion der Nebennieren männlicher *Tupaia belangeri*. *Journal of Comparative Physiology*, 78(3): 289–306.

Homans, G. C. 1961. *Social behavior: its elementary forms*. Harcourt, Brace & World, New York. xii + 404 pp.

Hooff, J. A. R. A. M. van. 1972. A comparative approach to the phylogeny of laughter and smiling. In R. A. Hinde, ed. (*q.v.*), *Non-verbal communication*, pp. 209–241.

Hooker, Barbara I. 1968. Birds. In T. A. Sebeok, ed. (*q.v.*), *Animal communication: techniques of study and results of research*, pp. 311–337.

Hooker, T., and Barbara I. Hooker. 1969. Duetting. In R. A. Hinde, ed. (*q.v.*), *Bird vocalizations: their relations to current problems in biology and psychology*, pp. 185–205.

Horn, E. G. 1971. Food competition among the cellular slime molds (Acrasieae). *Ecology*, 52(3): 475–484.

Horn, H. S. 1968. The adaptive significance of colonial nesting in the Brewer's blackbird (*Euphagus cyanocephalus*). *Ecology*, 49(4): 682–694.

Horwich, R. H. 1972. The ontogeny of social behavior in the gray squirrel (*Sciurus carolinensis*). *Zeitschrift für Tierpsychologie*, supplement 8. 103 pp.

Houlihan, R. T. 1963. The relationship of population density to endocrine and metabolic changes in the California vole (*Microtus californicus*). *University of California Publications in Zoology*, 65: 327–362.

Housse, R. P. R. 1949. Las zorros de Chile o chacales americanos. *Anales de la Academia Chilena de Ciencias Naturales, Santiago*, 34(1): 33–56.

Houston, D. B. 1974. Aspects of the social organization of moose. In V. Geist and F. Walther, eds. (*q.v.*), *The behaviour of ungulates and its relation to management*, vol. 2, pp. 690–696.

Howard, H. E. 1920. *Territory in bird life*. John Murray, London. xiii + 308 pp. (Reprinted with an introduction by J. Huxley and J. Fisher, Collins, London. 1948. 224 pp.)

——— 1940. *A waterhen's worlds*. Cambridge University Press, Cambridge. ix + 84 pp.

Howard, W. E. 1960. Innate and environmental dispersal of individual vertebrates. *American Midland Naturalist*, 63(1): 152–161.

Howells, W. W. 1973. *Evolution of the genus Homo*. Addison-Wesley, Reading, Mass. 188 pp.

Howse, P. E. 1964. The significance of the sound produced by the termite *Zootermopsis angusticollis* (Hagen). *Animal Behaviour*, 12(2,3): 284–300.

——— 1970. *Termites: a study in social behaviour*. Hutchinson University Library, London. 150 pp.

Hoyt, C. P., G. O. Osborne, and A. P. Mulcock. 1971. Production of an insect sex attractant by symbiotic bacteria. *Nature, London*, 230(5294): 472–473.

Hrdy, Sarah Blaffer. 1974. The care and exploitation of non-human primate infants by conspecifics other than the mother. *Advances in the Study of Behavior*. (In press.)

Hubbard, H. G. 1897. The ambrosia beetles of the United States. *Bulletin of the United States Department of Agriculture*, n.s. 7: 9–30.

Hubbard, J. A. E. B. 1973. Sediment-shifting experiments: a guide to functional behavior in colonial corals. In R. S. Boardman, A. H. Cheetham, and W. A. Oliver, Jr., eds. (*q.v.*), *Animal colonies: development and function through time*, pp. 31–42.

Huber, P. 1802. Observations on several species of the genus Apis, known by the name of humble-bees, and called Bombinatrices by Linnaeus. *Transactions of the Linnean Society of London*, 6: 214–298.

——— 1810. *Recherches sur les moeurs des fourmis indigènes*. J. J. Paschoud, Paris. xvi + 328 pp.

Hughes, R. L. 1962. Reproduction of the macropod marsupial *Potorous tridactylus* (Kerr). *Australian Journal of Zoology*, 10(2): 193–224.

Hunkeler, C., F. Bourlière, and M. Bertrand. 1972. Le comportement social de la mone de Lowe (*Cercopithecus campbelli lowei*). *Folia Primatologica*, 17(3): 218–236.

Hunsaker, D. 1962. Ethological isolating mechanisms in the *Sceloporus torquatus* group of lizards. *Evolution*, 16(1): 62–74.

Hunsaker, D., and T. C. Hahn. 1965. Vocalization of the South American tapir, *Tapirus terrestris*. *Animal Behaviour*, 13(1): 69–78.

Hunter, J. R. 1969. Communication of velocity changes in jack mackerel (*Trachurus symmetricus*) schools. *Animal Behaviour*, 17(3): 507–514.

Hutchinson, G. E. 1948. Circular causal systems in ecology. *Annals of the New York Academy of Sciences*, 50(4): 221–246.

——— 1951. Copepodology for the ornithologist. *Ecology*, 32(3): 571–577.

——— 1959. A speculative consideration of certain possible forms of sexual selection in man. *American Naturalist*, 93(869): 81–91.

——— 1961. The paradox of the plankton. *American Naturalist*, 95(882): 137–145.

Hutchinson, G. E., and S. D. Ripley. 1954. Gene dispersal and the ethology of the Rhinocerotidae. *Evolution*, 8(2): 178–179.

Hutt, Corinne. 1966. Exploration and play in children. *Symposia of the Zoological Society of London*, 18: 61–81.

Huxley, J. S. 1914. The courtship-habits of the great crested grebe (*Podiceps cristatus*); with an addition to the theory of sexual selection. *Proceedings of the Zoological Society of London*, 35: 491–562. (Reprinted as *The courtship of the great crested grebe*, with a foreword by Desmond Morris, Cape Editions, Grossman, London, 1968.)

——— 1923. Courtship activities in the red-throated diver (*Colymbus stellatus* Pontopp.); together with a discussion of the evolution of courtship in birds. *Journal of the Linnean Society of London, Zoology*, 35(234): 253–292.

——— 1934. A natural experiment on the territorial instinct. *British Birds*, 27(10): 270–277.

——— 1938. The present standing of the theory of sexual selection. In G. R. de Beer, ed., *Evolution: essays on aspects of evolutionary biology presented to Professor E. S. Goodrich on his seventieth birthday*, pp. 11–42. Clarendon Press, Oxford. viii + 350 pp.

——— 1966. Introduction. In J. S. Huxley, ed., A discussion on ritualization of behaviour in animals and man, pp. 249–271. *Philosophical Transactions of the Royal Society of London*, ser. B, 251(772): 247–526.

Hyman, Libbie H. 1940. *The invertebrates: Protozoa through Ctenophora*. McGraw-Hill Book Co., New York. xii + 726 pp.

——— 1951a. *The invertebrates*, vol. 2, *Platyhelminthes and Rhynchocoela: the acoelomate Bilateria*. McGraw-Hill Book Co., New York. viii + 550 pp.

——— 1951b. *The invertebrates*, vol. 3, *Acanthocephala, Aschelminthes, and Entoprocta: the pseudocoelomate Bilateria*. McGraw-Hill Book Co., New York. vii + 572 pp.

——— 1959. *The invertebrates*, vol. 5, *Smaller coelomate groups: Chaetognatha, Hemichordata, Pogonophora, Phoronida, Ectoprocta, Brachiopoda, Sipunculida, the coelomate Bilateria*. McGraw-Hill Book Co., New York. viii + 783 pp.

Ihering, H. von. 1896. Zur Biologie der socialen Wespen Brasiliens. *Zoologischer Anzeiger*, 19(516): 449–453.

Imaizumi, Y., and N. E. Morton. 1969. Isolation by distance in Japan and Sweden compared with other countries. *Human Heredity*, 19: 433–443.

Imaizumi, Y., N. E. Morton, and D. E. Harris. 1970. Isolation by distance in artificial populations. *Genetics*, 66(3): 569–582.

Imanishi, K. 1958. Identification: a process of enculturation in the subhuman society of *Macaca fuscata. Primates*, 1(1): 1–29. (In Japanese with English introduction.)

——— 1960. Social organization of subhuman primates in their natural habitat. *Current Anthropology*, 1(5,6): 393–407.

——— 1963. Social behavior in Japanese monkeys, *Macaca fuscata.* In C. H. Southwick, ed. (*q.v.*), *Primate social behavior: an enduring problem*, pp. 68–81. (Originally published in Japanese in *Psychologia*, 1[1]: 47–54, 1957.)

Immelmann, K. 1966. Beobachtungen an Schwalbenstaren. *Journal für Ornithologie*, 107(1): 37–69.

——— 1972. Sexual and other long-term aspects of imprinting in birds and other species. *Advances in the Study of Behavior*, 4: 147–174.

Inhelder, E. 1955. Zur Psychologie einiger Verhaltensweisen—besonders des Spiels—von Zootieren. *Zeitschrift für Tierpsychologie*, 12(1): 88–144.

Inkeles, A. 1964. *What is sociology? An introduction to the discipline and profession.* Prentice-Hall, Englewood Cliffs, N.J. viii + 120 pp.

Innis, Anne C. 1958. The behaviour of the giraffe, *Giraffa camelopardalis*, in the eastern Transvaal. *Proceedings of the Zoological Society of London*, 131(2): 245–278.

Ishay, J., H. Bytinski-Salz, and A. Shulov. 1967. Contributions to the bionomics of the Oriental hornet (*Vespa orientalis* Fab.). *Israel Journal of Entomology*, 2: 45–106.

Ishay, J., and R. Ikan. 1969. Gluconeogenesis in the Oriental hornet *Vespa orientalis* F. *Ecology*, 49(1): 169–171.

Ishay, J., and E. M. Landau. 1972. *Vespa* larvae send out rhythmic hunger signals. *Nature, London*, 237(5353): 286–287.

Istock, C. A. 1967. The evolution of complex life cycle phenomena: an ecological perspective. *Evolution*, 21(3): 592–605.

Itani, J. 1958. On the acquisition and propagation of a new habit in the natural group of the Japanese monkey at Takasaki-Yama. *Primates*, 1(2): 84–98. (In Japanese with English summary.)

——— 1959. Paternal care in the wild Japanese monkey, *Macaca fuscata fuscata. Primates*, 2(1): 61–93.

——— 1966. Social organization of chimpanzees. *Shizen*, 21(8): 17–30. [Cited by K. Izawa, 1970 (*q.v.*).]

——— 1972. A preliminary essay on the relationship between social organization and incest avoidance in nonhuman primates. In F. E. Poirier, ed. (*q.v.*), *Primate socialization*, pp. 165–171.

Itani, J., and A. Suzuki. 1967. The social unit of chimpanzees. *Primates*, 8(4): 355–381.

Itoigawa, N. 1973. Group organization of a natural troop of Japanese monkeys and mother-infant interactions. In C. R. Carpenter, ed.

(*q.v.*), *Behavioral regulators of behavior in primates*, pp. 229–250.

Ivey, M. E., and Judith M. Bardwick. 1968. Patterns of affective fluctuation in the menstrual cycle. *Psychosomatic Medicine*, 30(3): 336–345.

Iwata, K. 1967. Report of the fundamental research on the biological control of insect pests in Thailand: II, the report on the bionomics of subsocial wasps of Stenogastrinae (Hymenoptera, Vespidae). *Nature and Life in Southeast Asia*, 5: 259–293.

——— 1969. On the nidification of *Ropalidia* (*Anthreneida*) *taiwana koshunensis* Sonan in Formosa (Hymenoptera, Vespidae). *Kontyû*, 37: 367–372.

Izawa, K. 1970. Unit groups of chimpanzees and their nomadism in the savanna woodland. *Primates*, 11(1): 1–46.

Izawa, K., and J. Itani. 1966. Chimpanzees in Kasakati Basin, Tanganyika: I, ecological study in the rainy season 1963–1964. *Kyoto University African Studies*, 1: 73–156.

Izawa, K., and T. Nishida. 1963. Monkeys living in the northern limit of their distribution. *Primates*, 4(2): 67–88.

Jackson, J. A. 1970. A quantitative study of the foraging ecology of downy woodpeckers. *Ecology*, 51(2): 318–323.

Jackson, L. A., and J. N. Farmer. 1970. Effects of host fighting behavior on the course of infection of *Trypanosoma duttoni* in mice. *Ecology*, 51(4): 672–679.

Jacobson, M. 1972. *Insect sex pheromones.* Academic Press, New York. xii + 382 pp.

Jameson, D. L. 1954. Social patterns in the leptodactylid frogs *Syrrhophus* and *Eleutherodactylus. Copeia*, 1954, no. 1, pp. 36–38.

——— 1957. Life history and phylogeny in the salientians. *Systematic Zoology*, 6(2): 75–78.

Janzen, D. H. 1967. Interaction of the bull's-horn acacia (*Acacia cornigera* L.) with an ant inhabitant (*Pseudomyrmex ferruginea* F. Smith) in eastern Mexico. *Kansas University Science Bulletin*, 47(6): 315–558.

——— 1969. Allelopathy by myrmecophytes: the ant *Azteca* as an allelopathic agent of *Cecropia. Ecology*, 50(1): 147–153.

——— 1970. Altruism by coatis in the face of predation by *Boa constrictor. Journal of Mammalogy*, 51(2): 387–389.

——— 1972. Protection of *Barteria* (Passifloraceae) by *Pachysima* ants (Pseudomyrmecinae) in a Nigerian rain forest. *Ecology*, 53(5): 884–892.

Jardine, N., and R. Sibson. 1971. *Mathematical taxonomy.* John Wiley & Sons, New York. xviii + 286 pp.

Jarman, P. J. 1974. The social organisation of antelope in relation to their ecology. *Behaviour*, 58(3,4): 215–267.

Jarman, P. J., and M. V. Jarman. 1973. Social behaviour, population structure and reproductive potential in impala. *East African Wildlife Journal*, 11(3,4): 329–338.

Jay, Phyllis C. 1963. Mother-infant relations in langurs. In Harriet L. Rheingold, ed. (*q.v.*), *Maternal behavior in mammals*, pp. 282–304.

——— 1965. The common langur of North India. In I. DeVore, ed. (*q.v.*), *Primate behavior: field studies of monkeys and apes*, pp. 197–249.

——— ed. 1968. *Primates: studies in adaptation and variability.* Holt, Rinehart and Winston, New York. xiv + 529 pp.

Jeanne, R. L. 1972. Social biology of the Neotropical wasp *Mischocyttarus drewseni*. *Bulletin of the Museum of Comparative Zoology, Harvard*, 144(3): 63–150.

——— 1975. The adaptiveness of social wasp nest architecture. *Quarterly Review of Biology*. (In press.)

Jehl, J. R. 1970. Sexual selection for size differences in two species of sandpipers. *Evolution*, 24(2): 311–319.

Jenkins, D. 1961. Social behaviour in the partridge *Perdix perdix*. *Ibis*, 103a(2): 155–188.

Jenkins, D., A. Watson, and G. R. Miller. 1963. Population studies on red grouse, *Lagopus lagopus scoticus* (Lath.) in north-east Scotland. *Journal of Animal Ecology*, 32: 317–376.

Jenkins, T. M., Jr. 1969. Social structure, position choice and micro-distribution of two trout species (*Salmo trutta* and *Salmo gairdneri*) resident in mountain streams. *Animal Behaviour Monographs*, 2(2): 55–123.

Jennings, H. S. 1906. *Behavior of the lower organisms*. Columbia University Press, New York. xvi + 366 pp.

Jennrich, R. I., and F. B. Turner. 1969. Measurement of non-circular home range. *Journal of Theoretical Biology*, 22(2): 227–237.

Jewell, P. A. 1966. The concept of home range in mammals. In P. A. Jewell and Caroline Loizos, eds. (*q.v.*), *Play, exploration and territory in mammals*, pp. 85–109.

Jewell, P. A., and Caroline Loizos, eds. 1966. *Play, exploration and territory in mammals*. Symposia of the Zoological Society of London no. 18. Academic Press, New York. xiii + 280 pp.

Johnsgard, P. A. 1967. Dawn rendezvous on the lek. *Natural History*, 76(3) (March): 16–21.

Johnson, C. 1964. The evolution of territoriality in the Odonata. *Evolution*, 18(1): 89–92.

Johnson, C. G. 1969. *Migration and dispersal of insects by flight*. Methuen, London. xxii + 766 pp.

Johnson, N. K. 1963. Biosystematics of sibling species of flycatchers in the *Empidonax hammondii-oberholseri-wrightii* complex. *University of California Publications in Zoology*, 66(2): 79–238.

Johnston, J. W., Jr., D. G. Moulton, and A. Turk, eds. 1970. *Advances in chemoreception*, vol. 1, *Communication by chemical signals*. Appleton-Century-Crofts, New York. x + 412 pp.

Johnston, Norah C., J. H. Law, and N. Weaver. 1965. Metabolism of 9-ketodec-2-enoic acid by worker honeybees (*Apis mellifera* L.). *Biochemistry*, 4: 1615–1621.

Jolicoeur, P. 1959. Multivariate geographical variation in the wolf *Canis lupus* L. *Evolution*, 13(3): 283–299.

Jolly, Alison. 1966. *Lemur behavior: a Madagascar field study*. University of Chicago Press, Chicago. xiv + 187 pp.

——— 1972a. *The evolution of primate behavior*. Macmillan Co., New York. xiii + 397 pp.

——— 1972b. Troop continuity and troop spacing in *Propithecus verreauxi* and *Lemur catta* at Berenty (Madagascar). *Folia Primatologica*, 17(5,6): 335–362.

Jolly, C. J. 1970. The seed-eaters: a new model of hominid differentiation based on a baboon analogy. *Man*, 5(1): 5–26.

Jones, J. K., Jr., and R. R. Johnson. 1967. Sirenians. In S. Anderson and J. K. Jones, Jr., eds. (*q.v.*), *Recent mammals of the world: a synopsis of families*, pp. 366–373.

Jones, T. B., and A. C. Kamil. 1973. Tool-making and tool-using in the northern blue jay. *Science*, 180: 1076–1078.

Jonkel, C. J., and I. McT. Cowan. 1971. The black bear in the spruce-fir forest. *Wildlife Monographs*, 27: 1–57.

Joubert, S. C. J. 1974. The social organization of the roan antelope *Hippotragus equinus* and its influence on the spatial distribution of herds in the Kruger National Park. In V. Geist and F. Walther, eds. (*q.v.*), *The behaviour of ungulates and its relation to management*, vol. 2, pp. 661–675.

Jullien, J. 1885. Monographie des bryozoaires d'eau douce. *Bulletin de la Société Zoologique de France*, 10: 91–207.

Kahl, M. P. 1971. Social behavior and taxonomic relationships of the storks. *Living Bird*, 10: 151–170.

Kaiser, P. 1954. Über die Funktion der Mandibeln bei den Soldaten von *Neocapritermes opacus* (Hagen). *Zoologischer Anzeiger*, 152(9,10): 228–234.

Kalela, O. 1954. Über den Revierbesitz bei Vögeln und Säugetieren als populationsökologischer Faktor. *Annales Zoologici Societatis Zoologicae Botanicae Fennicae "Vanamo"* (Helsinki), 16(2): 1–48.

——— 1957. Regulation of reproductive rate in subarctic populations of the vole *Clethrionomys rufocanus* (Sund.). *Annales Academiae Scientiarum Fennicae* (Suomalaisen Tiedeakatemian Toimituksia), ser. A (IV, Biologica), 34: 1–60.

Kalleberg, H. 1958. Observations in a stream tank of territoriality and competition in juvenile salmon and trout (*Salmo salar* L. and *S. trutta* L.). *Reports of the Institute of Freshwater Research, Drottningholm*, 39: 55–98.

Kallmann, F. J. 1952. Twin and sibship study of overt male homosexuality. *American Journal of Human Genetics*, 4(2): 136–146.

Kalmijn, A. J. 1971. The electric sense of sharks and rays. *Journal of Experimental Biology*, 55(2): 371–383.

Kalmus, H. 1941. Defence of source of food by bees. *Nature, London*, 148(3747): 228.

Karlin, S. 1969. *Equilibrium behavior of population genetic models with non-random mating*. Gordon and Breach, New York. 163 pp.

Karlin, S., and J. McGregor. 1972. Polymorphisms for genetic and ecological systems with weak coupling. *Theoretical Population Biology*, 3(2): 210–238.

Karlson, P., and A. Butenandt. 1959. Pheromones (ectohormones) in insects. *Annual Review of Entomology*, 4: 39–58.

Kästle, W. 1963. Zur Ethologie des Grasanolis (*Norops auratus*)

(Daudin). *Zeitschrift für Tierpsychologie*, 20(1): 16–33.

———— 1967. Soziale Verhaltensweisen von Chamäleonen aus der *pumilis*- und *bitaeniatus*-Gruppe. *Zeitschrift für Tierpsychologie*, 24(3): 313–341.

Kaston, B. J. 1936. The senses involved in the courtship of some vagabond spiders. *Entomologica Americana*, n.s. 16(2): 97–167.

———— 1965. Some little known aspects of spider behavior. *American Midland Naturalist*, 73(2): 336–356.

Kaufman, I. C., and L. A. Rosenblum. 1967. Depression in infant monkeys separated from their mothers. *Science*, 155: 1030–1031.

Kaufmann, J. H. 1962. Ecology and social behavior of the coati, *Nasua narica*, on Barro Colorado Island, Panama. *University of California Publications in Zoology*, 60(3): 95–222.

———— 1966. Behavior of infant rhesus monkeys and their mothers in a free-ranging band. *Zoologica, New York*, 51(1): 17–27.

———— 1967. Social relations of adult males in a free-ranging band of rhesus monkeys. In S. A. Altmann, ed. (*q.v.*), *Social communication among primates*, pp. 73–98.

———— 1974a. Social ethology of the whiptail wallaby, *Macropus parryi*, in northeastern New South Wales. *Animal Behaviour*, 22(2): 281–369.

———— 1974b. The ecology and evolution of social organization in the kangaroo family (Macropodidae). *American Zoologist*, 14(1): 51–62.

———— 1974c. Habitat use and social organization of nine sympatric species of macropodid marsupials. *Journal of Mammalogy*, 55(1): 66–80.

Kaufmann, J. H., and Arleen B. Kaufmann. 1971. Social organization of whiptail wallabies, *Macropus parryi*. *Bulletin of the Ecological Society of America*, 52(4): 54–55.

Kaufmann, K. W. 1970. A model for predicting the influence of colony morphology on reproductive potential in the phylum Ectoprocta. *Biological Bulletin, Marine Biological Laboratory, Woods Hole*, 139(2): 426.

———— 1971. The form and function of the avicularia of *Bugula* (phylum Ectoprocta). *Postilla*, 151: 1–26.

———— 1973. The effect of colony morphology on the life-history parameters of colonial animals. In R. S. Boardman, A. H. Cheetham, and W. A. Oliver, Jr., eds. (*q.v.*), *Animal colonies: development and function through time*, pp. 221–222.

Kaufmann, T. 1965. Ecological and biological studies on the West African firefly *Luciola discicollis* (Coleoptera: Lampyridae). *Annals of the Entomological Society of America*, 58(4): 414–426.

Kawai, M. 1958. On the system of social ranks in a natural troop of Japanese monkeys: I, basic rank and dependent rank. *Primates*, 1–2: 111–130. (In Japanese; translated in S. A. Altmann, ed., 1965b [*q.v.*].)

———— 1965a. Newly acquired pre-cultural behavior of the natural troop of Japanese monkeys on Koshima Islet. *Primates*, 6(1): 1–30.

———— 1965b. On the system of social ranks in a natural troop of Japanese monkeys: I, basic rank and dependent rank. In S. A. Altmann, ed. (*q.v.*), *Japanese monkeys*, pp. 66–85.

Kawamura, S. 1954. A new type of action expressed in the feeding behavior of the Japanese monkey in its wild habitat. *Organic Evolution*, 2(1): 10–13. (In Japanese; cited by K. Imanishi, 1963 [*q.v.*].)

———— 1958. Matriarchal social ranks in the Minoo-B troop: a study of the rank system of Japanese monkeys. *Primates*, 1–2: 149–156. (In Japanese; translated in S. A. Altmann, ed., 1965b [*q.v.*].)

———— 1963. The process of sub-culture propagation among Japanese macaques. In C. H. Southwick, ed. (*q.v.*), *Primate social behavior: an enduring problem*, pp. 82–90. (Originally published in Japanese in *Journal of Primatology*, 1959, 2[1]: 43–60.)

———— 1967. Aggression as studied in troops of Japanese monkeys. In Carmine D. Clemente and D. B. Lindsley, eds., (*q.v.*), *Brain function*, vol. 5, *Aggression and defense, neural mechanisms and social patterns*, pp. 195–223.

Kawanabe, H. 1958. On the significance of the social structure for the mode of density effect in a salmon-like fish, "Ayu," *Plecoglossus altivelis* Temminck et Schlegel. *Memoirs of the College of Science, University of Kyoto*, ser. B, 25(3): 171–180.

Keenleyside, M. H. A. 1955. Some aspects of the schooling behaviour of fish. *Behaviour*, 8(2,3): 183–248.

———— 1972. The behaviour of *Abudefduf zonatus* (Pisces, Pomacentridae). *Animal Behaviour*, 20(4): 763–774.

Keith, A. 1949. *A new theory of human evolution*. Philosophical Library, New York. x + 451 pp.

Keller, R. 1973. Einige Beobachtungen zum Verhalten des Dekkan-Rothundes (*Cuon alpinus dukhunensis* Sykes) im Kanha-Nationalpark. *Vierteljahrsschrift der Naturforschenden Gesellschaft in Zurich*, 118(1): 129–135.

Kelsall, J. P. 1968. *The migratory barren-ground caribou of Canada*. Department of Indian Affairs and Northern Development, Ottawa. 340 pp.

Kemp, G. A., and L. B. Keith. 1970. Dynamics and regulation of red squirrel (*Tamiasciurus hudsonicus*) populations. *Ecology*, 51(5): 763–779.

Kemper, H., and Edith Döhring. 1967. *Die sozialen Faltenwespen Mitteleuropas*. Paul Parey, Berlin. 180 pp.

Kempf, W. W. 1951. A taxonomic study on the ant tribe Cephalotini (Hymenoptera: Formicidae). *Revista de Entomologia, Rio de Janeiro*, 22(1–3): 1–244.

———— 1958. New studies of the ant tribe Cephalotini (Hym. Formicidae). *Studia Entomologica*, 1(1,2): 1–168.

———— 1959. A revision of the Neotropical ant genus *Monacis* Roger (Hym., Formicidae). *Studia Entomologica*, 2(1–4): 225–270.

Kendeigh, S. C. 1952. *Parental care and its evolution in birds*. Illinois Biological Monographs, 22(1–3). x + 356 pp.

Kennedy, J. M., and K. Brown. 1970. Effects of male odor during infancy on the maturation, behavior, and reproduction of female mice. *Developmental Psychobiology*, 3(3): 179–189.

Kenyon, K. W. 1969. *The sea otter in the eastern Pacific Ocean*. North American Fauna no. 68. U.S. Bureau of Sport Fisheries and Wildlife, Washington, D.C. xiii + 352 pp.

Kern, J. A. 1964. Observations on the habits of the proboscis monkey, *Nasalis larvatus* (Wurmb), made in the Brunei Bay area, Borneo. *Zoologica, New York*, 49(3): 183–192.

Kerr, W. E., A. Ferreira, and N. Simões de Mattos. 1963. Communication among stingless bees—additional data (Hymenoptera: Apidae). *Journal of the New York Entomological Society*, 71: 80–90.

Kerr, W. E., S. F. Sakagami, R. Zucchi, V. de Portugal-Araújo, and J. M. F. de Camargo. 1967. Observações sôbre a arquitetura dos ninhos e comportamento de algumas espécies de abelhas sem ferrão das vizinhanças de Manaus, Amazonas (Hymenoptera, Apoidea). *Atas do Simpósio sôbre a Biota Amazônica, Conselho Nacional de Pesquisas, Rio de Janeiro*, 5 (Zoology): 255–309.

Kessel, E. L. 1955. The mating activities of balloon flies. *Systematic Zoology*, 4(3): 97–104.

Kessler, S. 1966. Selection for and against ethological isolation between *Drosophila pseudoobscura* and *Drosophila persimilis*. *Evolution*, 20(4): 634–645.

Keyfitz, N. 1968. *Introduction to the mathematics of population.* Addison-Wesley Publishing Co., Reading, Mass. xiv + 450 pp.

Kiley, Marthe. 1972. The vocalizations of ungulates, their causation and function. *Zeitschrift für Tierpsychologie*, 31(2): 171–222.

Kiley-Worthington, Marthe. 1965. The waterbuck (*Kobus defassa* Ruppell 1835 and *K. ellipsiprimnus* Ogilby 1833) in East Africa: spatial distribution: a study of the sexual behaviour. *Mammalia*, 29(2): 176–204.

Kilgore, D. L. 1969. An ecological study of the swift fox (*Vulpes velox*) in the Oklahoma Panhandle. *American Midland Naturalist*, 81(2): 512–534.

Kilham, L. 1970. Feeding behavior of downy woodpeckers: I, preference for paper birches and sexual differences. *Auk*, 87(3): 544–556.

King, C. E. 1964. Relative abundance of species and MacArthur's model. *Ecology*, 45(4): 716–727.

King, C. E., and W. W. Anderson. 1971. Age-specific selection: II, the interaction between *r* and *K* during population growth. *American Naturalist*, 105(942): 137–156.

King, J. A. 1955. Social behavior, social organization, and population dynamics in a black-tailed prairiedog town in the Black Hills of South Dakota. *Contributions from the Laboratory of Vertebrate Biology, University of Michigan, Ann Arbor*, no. 67. 123 pp.

——— 1956. Social relations of the domestic guinea pig living under semi-natural conditions. *Ecology*, 37(2): 221–228.

——— 1957. Relationships between early social experience and adult aggressive behavior in inbred mice. *Journal of Genetic Psychology*, 90: 151–166.

——— 1968. Psychology. In J. A. King, ed. (q.v.), *Biology of Peromyscus* (*Rodentia*), pp. 496–542.

——— ed. 1968. *Biology of Peromyscus* (*Rodentia*). Special Publication no. 2. American Society of Mammalogists, Stillwater, Oklahoma. xiii + 593 pp.

King, J. L. 1967. Continuously distributed factors affecting fitness. *Genetics*, 55(3): 483–492.

Kinsey, K. P. 1971. Social organization in a laboratory colony of wood rats, *Neotoma fuscipes*. In A. H. Esser, ed. (q.v.), *Behavior and environment: the use of space by animals and men*, pp. 40–45.

Kislak, J. W., and F. A. Beach. 1955. Inhibition of aggressiveness by ovarian hormones. *Endocrinology*, 56(6): 684–692.

Kitchener, D. J. 1972. The importance of shelter to the quokka, *Setonix brachyurus* (Marsupialia), on Rottnest Island. *Australian Journal of Zoology*, 20(3): 281–299.

Kittredge, J. S., Michelle Terry, and F. T. Takahashi. 1971. Sex pheromone activity of the molting hormone, crustecdysone, on male crabs. *Fishery Bulletin*, 69(2): 337–343.

Kleiber, M. 1961. *The fire of life: an introduction to animal energetics.* John Wiley & Sons, New York. 454 pp.

Kleiman, Devra G. 1967. Some aspects of social behavior in the Canidae. *American Zoologist*, 7(2): 365–372.

——— 1971. The courtship and copulatory behaviour of the green acouchi, *Myoprocta pratti. Zeitschrift für Tierpsychologie*, 29(3): 259–278.

——— 1972a. Maternal behaviour of the green acouchi (*Myoprocta pratti* Pocock), a South American caviomorph rodent. *Behaviour*, 43(3,4): 48–84.

——— 1972b. Social behavior of the maned wolf (*Chrysocyon brachyurus*) and bush dog (*Speothos venaticus*): a study in contrast. *Journal of Mammalogy*, 53(4): 791–806.

Kleiman, Devra G., and J. F. Eisenberg. 1973. Comparisons of canid and felid social systems from an evolutionary perspective. *Animal Behaviour*, 21(4): 637–659.

Klingel, H. 1965. Notes on the biology of the plains zebra *Equus quagga boehmi* Matschie. *East African Wildlife Journal*, 3: 86–88.

——— 1967. Soziale Organisation und Verhalten freilebender Steppenzebras. *Zeitschrift für Tierpsychologie*, 24(5): 580–624.

——— 1968. Soziale Organisation und Verhaltensweisen vom Hartmann- und Bergzebras (*Equus zebra hartmannae* und *E. z. zebra*). *Zeitschrift für Tierpsychologie*, 25(1): 76–88.

——— 1972. Social behaviour of African Equidae. *Zoologica Africana*, 7: 175–186.

Klopfer, P. H. 1957. An experiment on empathic learning in ducks. *American Naturalist*, 91(856): 61–63.

——— 1961. Observational learning in birds: the establishment of behavioral modes. *Behaviour*, 17(1): 71–80.

——— 1970. Sensory physiology and esthetics. *American Scientist*, 58(4): 399–403.

——— 1972. Patterns of maternal care in lemurs: II, effects of group size and early separation. *Zeitschrift für Tierpsychologie*, 30(3): 277–296.

Klopfer, P. H., and Alison Jolly. 1970. The stability of territorial boundaries in a lemur troop. *Folia Primatologica*, 12(3): 199–208.

Klopman, R. B. 1968. The agonistic behavior of the Canada goose (*Branta canadensis canadensis*): I, attack behavior. *Behaviour*, 30(4): 287–319.

Kluijver, H. N., and L. Tinbergen. 1953. Territory and the regulation of density in titmice. *Archives Néerlandaises de Zoologie, Leydig,* 10(3): 265–289.

Kneitz, G. 1964. Saisonales Trageverhalten bei *Formica polyctena* Foerst. (Formicidae, Gen. *Formica*). *Insectes Sociaux,* 11(2): 105–129.

Knerer, G., and C. E. Atwood. 1966. Nest architecture as an aid in halictine taxonomy (Hymenoptera: Halictidae). *Canadian Entomologist,* 98(12): 1337–1339.

Knerer, G., and Cécile Plateaux-Quénu. 1967a. Sur la production continue ou périodique de couvain chez les Halictinae (Insectes Hyménoptères). *Compte Rendu de l'Académie des Sciences, Paris,* 264(4): 651–653.

—— 1967b. Sur la production de mâles chez les Halictinae (Insectes Hyménoptères) sociaux. *Compte Rendu de l'Académie des Sciences, Paris,* 264(8): 1096–1099.

—— 1967c. Usurpation de nids étrangers et parasitisme facultatif chez *Halictus scabiosae* (Rossi) (Insecte Hyménoptère). *Insectes Sociaux,* 14(1): 47–50.

Koenig, Lilli. 1960. Das Aktionssystem des Siebenschläfers (*Glis glis* L.). *Zeitschrift für Tierpsychologie,* 17(4): 427–505.

Koenig, O. 1962. *Kif-Kif.* Wollzeilen-Verlag, Vienna. [Cited by W. Wickler, 1972a (*q.v.*).]

Kofoid, C. A., S. F. Light, A. C. Horner, M. Randall, W. B. Herms, and E. E. Bowe, eds. 1934. *Termites and termite control,* 2d ed., rev. University of California Press, Berkeley. xxvii + 795 pp.

Koford, C. B. 1957. The vicuña and the puna. *Ecological Monographs,* 27(2): 153–219.

—— 1963. Rank of mothers and sons in bands of rhesus monkeys. *Science,* 141: 356–357.

—— 1965. Population dynamics of rhesus monkeys on Cayo Santiago. In I. DeVore, ed. (*q.v.*), *Primate behavior: field studies of monkeys and apes,* pp. 160–174.

—— 1967. Population changes in rhesus monkeys: Cayo Santiago 1960–1964. *Tulane Studies in Zoology, New Orleans,* 13(1): 1–7.

Kohlberg, L. 1969. Stage and sequence: the cognitive-developmental approach to socialization. In D. A. Goslin, ed., *Handbook of socialization theory and research,* pp. 347–480. Rand McNally Co., Chicago. xiii + 1182 pp.

Köhler, W. 1927. *The mentality of apes,* trans. by Ella Winter, 2d ed. Kegan Paul, Trench, and Trubner, London. viii + 336 pp.

Konijn, T. M., J. G. C. van de Meene, J. T. Bonner, and D. S. Barkley. 1967. The acrasin activity of adenosine-3′,5′-cyclic phosphate. *Proceedings of the National Academy of Sciences, U.S.A.,* 58(3): 1152–1154.

Konishi, M. 1965. The role of auditory feedback in the control of vocalization in the white-crowned sparrow. *Zeitschrift für Tierpsychologie,* 22(7): 770–783.

Koopman, K. F., and E. L. Cockrum. 1967. Bats. In S. Anderson and J. K. Jones, Jr., eds. (*q.v.*), *Recent mammals of the world: a synopsis of families,* pp. 109–150.

Kortlandt, A. 1940. Eine Übersicht der angeboren Verhaltungsweisen des Mittel-Europäischen Kormorans (*Phalacrocorax carbo sinensis*

[Shaw & Nodd.]), ihre Funktion, ontogenetische Entwicklung und phylogenetische Herkunft. *Archives Néerlandaises de Zoologie, Leydig,* 4(4): 401–442.

—— 1962. Chimpanzees in the wild. *Scientific American,* 206(5) (May): 128–138.

—— 1972. *New perspectives on ape and human evolution.* Stichting voor Psychobiologie, Universiteit van Amsterdam, The Netherlands. 100 pp.

Kortlandt, A., and M. Kooij. 1963. Protohominid behaviour in primates (preliminary communication). *Symposia of the Zoological Society of London,* 10: 61–88.

Krames, L., W. J. Carr, and B. Bergman. 1969. A pheromone associated with social dominance among male rats. *Psychonomic Science,* 16(1): 11–12.

Krebs, C. J. 1964. The lemming cycle at Baker Lake, Northwest Territories, during 1959–62. Arctic Institute of North America, Technical Paper no. 15. [Cited by D. Chitty, 1967a (*q.v.*).]

—— 1972. *Ecology: the experimental analysis of distribution and abundance.* Harper & Row, New York. x + 694 pp.

Krebs, C. J., M. S. Gaines, B. L. Keller, Judith H. Myers, and R. H. Tamarin. 1973. Population cycles in small rodents. *Science,* 179: 35–44.

Krebs, C. J., B. L. Keller, and R. H. Tamarin. 1969. *Microtus* population biology: demographic changes in fluctuating populations of *M. ochrogaster* and *M. pennsylvanicus* in southern Indiana. *Ecology,* 50(4): 587–607.

Krebs, J. R. 1971. Territory and breeding density in the great tit, *Parus major* L. *Ecology,* 52(1): 2–22.

Krieg, H. 1939. Begegnungen mit Ameisenbären und Faultieren in freier Wildbahn. *Zeitschrift für Tierpsychologie,* 2(3): 282–292.

Krishna, K. 1970. Taxonomy, phylogeny, and distribution of termites. In K. Krishna and Frances M. Weesner, eds. (*q.v.*) *Biology of termites,* vol. 2, pp. 127–152.

Krishna, K., and Frances M. Weesner, eds. 1969. *Biology of termites,* vol. 1. Academic Press, New York. xiii + 598 pp.

—— 1970. *Biology of termites,* vol. 2. Academic Press, New York. xiv + 643 pp.

Krott, P., and Gertraud Krott. 1963. Zum Verhalten des Braunbären (*Ursus arctos* L. 1758) in den Alpen. *Zeitschrift für Tierpsychologie,* 20(2): 160–206.

Kruuk, H. 1964. Predators and anti-predator behaviour of the black-headed gull (*Larus ridibundus*). *Behaviour,* supplement 11. 129 pp.

—— 1972. *The spotted hyena: a study of predation and social behavior.* University of Chicago Press, Chicago. xvi + 335 pp.

Kruuk, H., and W. A. Sands. 1972. The aardwolf (*Proteles cristatus* Sparrman) 1783 as predator of termites. *East African Wildlife Journal,* 10(3): 211–227.

Kühlmann, D. H. H., and H. Karst. 1967. Freiwasserbeobachtungen zum Verhalten von Tobiasfischschwärmen (*Ammodytidae*) in der westlichen Ostsee. *Zeitschrift für Tierpsychologie,* 24(3): 282–297.

Kühme, W. 1963. Ergänzende Beobachtungen an afrikanischen Elefanten

(*Loxodonta africana* Blumenbach 1797) in Freigehege. *Zeitschrift für Tierpsychologie*, 20(1): 66–79.

——— 1965a. Freilandstudien zur Soziologie des Hyänenhundes (*Lycaon pictus lupinus* Thomas 1902). *Zeitschrift für Tierpsychologie*, 22(5): 495–541.

——— 1965b. Communal food distribution and division of labour in African hunting dogs. *Nature, London*, 205(4970): 443–444.

Kullenberg, B. 1956. Field experiments with chemical sexual attractants on aculeate Hymenoptera males. *Zoologiska Bidrag från Uppsala*, 31: 253–354.

Kullmann, E. 1968. Soziale Phaenomene bei Spinnen. *Insectes Sociaux*, 15(3): 289–297.

Kummer, H. 1967. Tripartite relations in hamadryas baboons. In S. A. Altmann, ed. (*q.v.*), *Social communication among primates*, pp. 63–71.

——— 1968. *Social organization of hamadryas baboons: a field study.* University of Chicago Press, Chicago. viii + 189 pp.

——— 1971. *Primate societies: group techniques of ecological adaptation.* Aldine-Atherton, Chicago. 160 pp.

Kunkel, P., and Irene Kunkel. 1964. Beiträge zur Ethologie des Hausmeerschweinchens *Cavia aperea f. porcellus* (L.). *Zeitschrift für Tierpsychologie*, 21(5): 602–641.

Kurtén, B. 1972. *Not from the apes.* Vintage Books, Random House, New York. viii + 183 pp.

Kutter, H. 1923. Die Sklavenräuber *Strongylognathus huberi* For. ssp. *alpinus* Wheeler. *Revue Suisse de Zoologie*, 30(15): 387–424.

——— 1950. Über eine neue, extrem parasitische Ameise, 1. *Mitteilungen der Schweizerischen Entomologischen Gesellschaft*, 23(2): 81–94.

——— 1956. Beiträge zur Biologie palaearktischer *Coptoformica* (Hym. Form.). *Mitteilungen der Schweizerischen Entomologischen Gesellschaft*, 29(1): 1–18.

——— 1957. Zur Kenntnis schweizerischer Coptoformicaarten (Hym. Form.), 2. *Mitteilungen der Schweizerischen Entomologischen Gesellschaft*, 30(1): 1–24.

——— 1969. *Die sozialparasitischen Ameisen der Schweiz.* Naturforschenden Gesellschaft in Zürich, Neujahrsblatt, 1969. 62 pp.

Lack, D. 1954. *The natural regulation of animal numbers.* Oxford University Press, Oxford. viii + 343 pp.

——— 1966. *Population studies of birds.* Oxford University Press, Oxford. v + 341 pp.

——— 1968. *Ecological adaptations for breeding in birds.* Methuen, London. xii + 409 pp.

La Follette, R. M. 1971. Agonistic behaviour and dominance in confined wallabies, *Wallabia rufogrisea frutica. Animal Behaviour*, 19(1): 93–101.

Lamprecht, J. 1970. Duettgesang beim Siamang, *Symphalangus syndactylus* (Hominoidea, Hylobatinae). *Zeitschrift für Tierpsychologie*, 27(2): 186–204.

Lancaster, D. A. 1964. Life history of the Boucard tinamou in British Honduras: II, breeding biology. *Condor*, 66(4): 253–276.

Lancaster, Jane B. 1971. Play-mothering: the relations between juvenile females and young infants among free-ranging vervet monkeys (*Cercopithecus aethiops*). *Folia Primatologica*, 15(3,4): 161–182.

Lancaster, Jane B., and R. B. Lee. 1965. The annual reproductive cycle in monkeys and apes. In I. DeVore, ed. (*q.v.*), *Primate behavior: field studies of monkeys and apes*, pp. 486–513.

Landau, H. G. 1951. On dominance relations and the structure of animal societies: I, effect of inherent characteristics; II, some effects of possible social factors. *Bulletin of Mathematical Biophysics*, 13(1): 1–19; 13(4): 245–262.

——— 1953. On dominance relations and the structure of animal societies: III, the condition for a score structure. *Bulletin of Mathematical Biophysics*, 15(2): 143–148.

——— 1965. Development of structure in a society with a dominance relation when new members are added successively. *Bulletin of Mathematical Biophysics*, special issue 27: 151–160.

Lang, E. M. 1961. Beobachtungen am indischen Panzernashorn (*Rhinoceros unicornis*). *Zoologischer Garten, Leipzig*, n.s. 25: 369–409.

Lange, R. 1960. Über die Futterweitergabe zwischen Angehörigen verschiedener Waldameisenstaaten. *Zeitschrift für Tierpsychologie*, 17(4): 389–401.

——— 1967. Die Nahrungsverteilung unter den Arbeiterinnen des Waldameisenstaates. *Zeitschrift für Tierpsychologie*, 24(5): 513–545.

Langguth, A. 1969. Die südamerikanischen Canidae unter besonderer Berücksichtigung des Mähnenwolfes, *Chrysocyon brachyurus* (Illiger). *Zeitschrift für Wissenschaftlichen Zoologie*, 179(1): 1–187.

Langlois, T. H. 1936. A study of the small-mouth bass, *Micropterus dolomieu* (Lacepede) in rearing ponds in Ohio. *Bulletin of the Ohio Biological Survey*, 6: 189–225.

Lanyon, W. E. 1956. Territory in the meadowlarks, genus *Sturnella. Ibis*, 98(3): 485–489.

Lanyon, W. E., and W. N. Tavolga, eds. 1960. *Animal sounds and communication.* Publication no. 7. American Institute of Biological Sciences, Washington, D.C. ix + 443 pp.

Larwood, G. P., ed. 1973. *Living and fossil Bryozoa: recent advances in research.* Academic Press, New York. xviii + 634 pp.

Lasiewski, R. C., and W. R. Dawson. 1967. A re-examination of the relation between standard metabolic rate and body weight in birds. *Condor*, 69(1): 13–23.

La Val, R. K. 1973. Observations on the biology of *Tadarida brasiliensis cynocephala* in southeastern Louisiana. *American Midland Naturalist*, 89(1): 112–120.

Law, J. H., E. O. Wilson, and J. A. McCloskey. 1965. Biochemical polymorphism in ants. *Science*, 149: 544–546.

Lawick, H. van. 1974. *Solo: the story of an African wild dog.* Houghton Mifflin Co., Boston. 159 pp.

Lawick, H. van, and Jane van Lawick-Goodall. 1971. *Innocent killers.* Houghton Mifflin Co., Boston. 222 pp.

Lawick-Goodall, Jane van. 1967. *My friends the wild chimpanzees.* National Geographic Society, Washington, D.C. 204 pp.

——— 1968a. The behaviour of free-living chimpanzees in the Gombe Stream Reserve. *Animal Behaviour Monographs,* 1(3): 161–311.

——— 1968b. A preliminary report on expressive movements and communication in the Gombe Stream chimpanzees. In Phyllis C. Jay, ed. (*q.v.*), *Primates: studies in adaptation and variability,* pp. 313–374.

——— 1969. Mother-offspring relationships in free-ranging chimpanzees. In D. Morris, ed. (*q.v.*), *Primate ethology: essays on the socio-sexual behavior of apes and monkeys,* pp. 364–436.

——— 1970. Tool-using in primates and other vertebrates. *Advances in the Study of Behavior,* 3: 195–249.

——— 1971. *In the shadow of man.* Houghton Mifflin Co., Boston. xx + 297 pp.

Laws, R. M. 1974. Behaviour, dynamics and management of elephant populations. In V. Geist and F. Walther, eds. (*q.v.*), *The behaviour of ungulates and its relation to management,* vol. 2, pp. 513–529.

Laws, R. M., and I. S. C. Parker. 1968. Recent studies on elephant populations in East Africa. *Symposia of the Zoological Society of London,* 21: 319–359.

Layne, J. N. 1954. The biology of the red squirrel, *Tamiasciurus hudsonicus loquax* (Bangs), in central New York. *Ecological Monographs,* 24(3): 227–267.

——— 1958. Observations on freshwater dolphins in the upper Amazon. *Journal of Mammalogy,* 39(1): 1–22.

——— 1967. Lagomorphs. In S. Anderson and J. K. Jones, Jr., eds. (*q.v.*), *Recent mammals of the world: a synopsis of families,* pp. 192–205.

Layne, J. N., and D. K. Caldwell. 1964. Behavior of the Amazon dolphin, *Inia geoffrensis* (Blainville), in captivity. *Zoologica, New York,* 49(2): 81–108.

Le Boeuf, B. J. 1972. Sexual behavior in the northern elephant seal *Mirounga angustirostris. Behaviour,* 41(1,2): 1–26.

——— 1974. Male-male competition and reproductive success in elephant seals. *American Zoologist,* 14(1): 163–176.

Le Boeuf, B. J., and R. S. Peterson. 1969a. Social status and mating activity in elephant seals. *Science,* 163: 91–93.

——— 1969b. Dialects in elephant seals. *Science,* 166: 1654–1656.

Le Boeuf, B. J., R. J. Whiting, and R. F. Gantt. 1972. Perinatal behavior of northern elephant seal females and their young. *Behaviour,* 43(1–4): 121–156.

Lechleitner, R. R. 1958. Certain aspects of behavior of the black-tailed jack rabbit. *American Midland Naturalist,* 60(1): 145–155.

Lecomte, J. 1956. Über die Bildung von "Strassen" durch Sammelbienen, deren Stock um 180° gedreht wurde. *Zeitschrift für Bienenforschung,* 3: 128–133.

Le Cren, E. D., and M. W. Holdgate, eds. 1962. *The exploitation of natural animal populations.* John Wiley & Sons, New York. xiv + 399 pp.

Lederer, E. 1950. Odeurs et parfums des animaux. *Fortschritte der Chemie Organischer Naturstoffe,* 6: 87–153.

Ledoux, A. 1950. Recherche sur la biologie de la fourmi fileuse (*Oecophylla longinoda* Latr.). *Annales des Sciences Naturelles,* 11th ser., 12(3,4): 313–461.

Lee, K. E., and T. G. Wood. 1971. *Termites and soils.* Academic Press, New York. x + 251 pp.

Lee, R. B. 1968. What hunters do for a living, or how to make out on scarce resources. In R. B. Lee and I. DeVore, eds. (*q.v.*), *Man the hunter,* pp. 30–48.

Lee, R. B., and I. DeVore, eds. 1968. *Man the hunter.* Aldine Publishing Co., Chicago. xvi + 415 pp.

Leeper, G. W., ed. 1962. *The evolution of living organisms.* Melbourne University Press, Parksville, Victoria. x + 459 pp.

Lees, A. D. 1966. The control of polymorphism in aphids. *Advances in Insect Physiology,* 2: 207–277.

Lehrman, D. S. 1964. The reproductive behavior of ring doves. *Scientific American,* 211(5) (November): 48–54.

——— 1965. Interaction between internal and external environments in the regulation of the reproductive cycle of the ring dove. In F. A. Beach, ed., *Sex and behavior,* pp. 355–380. John Wiley & Sons, New York. xvi + 592 pp.

Lehrman, D. S., and J. S. Rosenblatt. 1971. The study of behavioral development. In H. Moltz, ed. (*q.v.*), *The ontogeny of vertebrate behavior,* pp. 1–27.

Leibowitz, Lila. 1968. Founding families. *Journal of Theoretical Biology,* 21(2): 153–169.

Leigh, E. G. 1970. Sex ratio and differential mortality between the sexes. *American Naturalist,* 104(954): 205–210.

——— 1971. *Adaptation and diversity: natural history and the mathematics of evolution.* Freeman, Cooper, San Francisco. 288 pp.

Lein, M. R. 1972. Territorial and courtship songs of birds. *Nature, London,* 237(5349): 48–49.

——— 1973. The biological significance of some communication patterns of wood warblers (Parulidae). Ph.D. thesis, Harvard University, Cambridge. 252 pp.

Le Masne, G. 1953. Observations sur les relations entre le couvain et les adultes chez les fourmis. *Annales des Sciences Naturelles,* 11th ser., 15(1): 1–56.

——— 1956a. Recherches sur les fourmis parasites *Plagiolepis grassei* et l'évolution des *Plagiolepis* parasites. *Compte Rendu de l'Académie Sciences, Paris,* 243(7): 673–675.

——— 1956b. La signification des reproducteurs aptères chez la fourmi *Ponera eduardi* Forel. *Insectes Sociaux,* 3(2): 239–259.

——— 1965. Les transports mutuels autour des nids de *Neomyrma rubida* Latr.: un nouveau type de relations inter-spécifiques chez les fourmis? *Comptes Rendus du Cinquième Congrès de l'Union Internationale pour l'Étude des Insectes Sociaux, Toulouse, 1965,* pp. 303–322.

Lemon, R. E. 1967. The response of cardinals to songs of different dialects. *Animal Behaviour,* 15(4): 538–545.

——— 1968. The relation between organization and function of song in cardinals. *Behaviour,* 32(1–3): 158–178.

——— 1971a. Differentiation of song dialects in cardinals. *Ibis*, 113(3): 373–377.

——— 1971b. Vocal communication by the frog *Eleutherodactylus martinicensis*. *Canadian Journal of Zoology*, 49(2): 211–217.

Lemon, R. E., and A. Herzog. 1969. The vocal behavior of cardinals and pyrrhuloxias in Texas. *Condor*, 71(1): 1–15.

Lengerken, H. von. 1954. *Die Brutfürsorge- und Brutpflegeinstinkte der Käfer*, 2d ed. Akademische Verlagsgessellschaft M.B.H., Leipzig. 383 pp.

Lenneberg, E. H. 1967. *Biological foundations of language*. John Wiley & Sons, New York. xviii + 489 pp.

——— 1971. Of language knowledge, apes, and brains. *Journal of Psycholinguistic Research*, 1(1): 1–29.

Lenski, G. 1970. *Human societies: a macrolevel introduction to sociology*. McGraw-Hill Book Co., New York. xvi + 525 pp.

Lent, P. C. 1966. Calving and related social behavior in the barren-ground caribou. *Zeitschrift für Tierpsychologie*, 23(6): 701–756.

Leopold, A. S. 1944. The nature of heritable wildness in turkeys. *Condor*, 46(4): 133–197.

Lerner, I. M. 1954. *Genetic homeostasis*. Oliver and Boyd, London. vii + 134 pp.

——— 1958. *The genetic basis of selection*. John Wiley & Sons, New York. xvi + 298 pp.

——— 1968. *Heredity, evolution, and society*. W. H. Freeman, San Francisco. xviii + 307 pp.

Leshner, A. I., and D. K. Candland. 1971. Adrenal determinants of squirrel monkey dominance orders. *Bulletin of the Ecological Society of America*, 52(4): 54.

Leuthold, R. H. 1968a. A tibial gland scent-trail and trail-laying behavior in the ant *Crematogaster ashmeadi* Mayr. *Psyche, Cambridge*, 75(3): 233–248.

——— 1968b. Recruitment to food in the ant *Crematogaster ashmeadi*. *Psyche, Cambridge*, 75(4): 334–350.

Leuthold, W. 1966. Variations in territorial behavior of Uganda kob *Adenota kob thomasi* (Neumann 1896). *Behaviour*, 27(3,4): 215–258.

——— 1970. Observations on the social organization of impala (*Aepyceros melampus*). *Zeitschrift für Tierpsychologie*, 27(6): 693–721.

——— 1974. Observations on home range and social organization of lesser kudu, *Tragelaphus imberbis* (Blyth, 1869). In V. Geist and F. Walther, eds. (*q.v.*), *The behaviour of ungulates and its relation to management*, vol. 1, pp. 206–234.

Lévieux, J. 1966. Note préliminaire sur les colonnes de chasse de *Megaponera foetens* F. (Hyménoptère Formicidae). *Insectes Sociaux*, 13(2): 117–126.

——— 1971. Mise en évidence de la structure des nids et de l'implantation des zones de chasse de deux espèces de *Camponotus* (Hym. Form.) à l'aide de radio-isotopes. *Insectes Sociaux*, 18(1): 29–48.

——— 1972. Le role des fourmis dans les réseaux trophiques d'une savane préforestière de Côte-d'Ivoire. *Annales de l'Université d'Abidjan*, 5(1): 143–240.

Levin, B. R., and W. L. Kilmer. 1974. Interdemic selection and the evolution of altruism: a computer simulation study. *Evolution*. (In press.)

Levin, B. R., M. L. Petras, and D. I. Rasmussen. 1969. The effect of migration on the maintenance of a lethal polymorphism in the house mouse. *American Naturalist*, 103(934): 647–661.

Levin, M. D., and S. Glowska-Konopacka. 1963. Responses of foraging honeybees in alfalfa to increasing competition from other colonies. *Journal of Apicultural Research*, 2(1): 33–42.

LeVine, R. A., and D. T. Campbell. 1972. *Ethnocentrism: theories of conflict, ethnic attitudes, and group behavior*. John Wiley & Sons, New York. x + 310 pp.

Levins, R. 1965. The theory of fitness in a heterogeneous environment: IV, the adaptive significance of gene flow. *Evolution*, 18(4): 635–638.

——— 1968. *Evolution in changing environments: some theoretical explorations*. Princeton University Press, Princeton, N.J. ix + 120 pp.

——— 1970. Extinction. In M. Gerstenhaber, ed., *Some mathematical questions in biology*, pp. 77–107. Lectures on Mathematics in the Life Sciences, vol. 2. American Mathematical Society, Providence, R.I. vii + 156 pp.

Lévi-Strauss, C. 1949. *Les structures élémentaires de la parenté*. Presses Universitaires de France, Paris. xiv + 639 pp. (*The elementary structures of kinship*, rev. ed., trans. by J. H. Bell and J. R. von Sturmer and ed. by R. Needham, Beacon Press, Boston, 1969. xlii + 541 pp.)

Lewontin, R. C. 1965. Selection for colonizing ability. In H. G. Baker and G. L. Stebbins, eds. (*q.v.*), *The genetics of colonizing species*, pp. 77–94.

——— 1972a. Testing the theory of natural selection. (Review of R. Creed, ed., *Ecological genetics and evolution*, Blackwell Scientific Publications, Oxford, 1971.) *Nature, London*, 236(5343): 181–182.

——— 1972b. The apportionment of human diversity. *Evolutionary Biology*, 6: 381–398.

Lewontin, R. C., and L. C. Dunn. 1960. The evolutionary dynamics of a polymorphism in the house mouse. *Genetics*, 45(6): 705–722.

Lewontin, R. C., and J. L. Hubby. 1966. A molecular approach to the study of genic heterozygosity in natural populations: II, amount of variation and degree of heterozygosity in natural populations of *Drosophila pseudoobscura*. *Genetics*, 54(2): 595–609.

Leyhausen, P. 1956. Verhaltensstudien an Katzen. *Zeitschrift für Tierpsychologie*, supplement 2. vi + 120 pp.

——— 1965. The communal organization of solitary mammals. *Symposia of the Zoological Society of London*, 14: 249–263.

——— 1971. Dominance and territoriality as complemented in mammalian social structure. In A. H. Esser, ed. (*q.v.*), *Behavior and environment: the use of space by animals and men*, pp. 22–33.

Lidicker, W. Z., Jr. 1962. Emigration as a possible mechanism permitting the regulation of population density below carrying capacity. *American Naturalist*, 96(886): 29–33.

——— 1965. Comparative study of density regulation in confined populations of four species of rodents. *Researches on Population Ecology*, 7(2): 57–72.

Lidicker, W. Z., Jr., and B. J. Marlow. 1970. A review of the dàsyurid marsupial genus *Antechinomys* Krefft. *Mammalia*, 34(2): 212–227.

Lieberman, P. 1968. Primate vocalizations and human linguistic ability. *Journal of the Acoustic Society of America*, 44:1574–1584.

Lieberman, P., E. S. Crelin, and D. H. Klatt. 1972. Phonetic ability and related anatomy of the newborn and adult human, Neanderthal man, and the chimpanzee. *American Anthropologist*, 74(3): 287–307.

Ligon, J. D. 1968. Sexual differences in foraging behavior in two species of *Dendrocopos* woodpeckers. *Auk*, 85(2): 203–215.

Lill, A. 1968. An analysis of sexual isolation in the domestic fowl: I, the basis of homogamy in males; II, the basis of homogamy in females. *Behaviour*, 30(2,3): 107–145.

Lilly, J. C. 1961. *Man and dolphin.* Doubleday, New York. (Reprinted as a paperback, Pyramid Books, New York, 1969. 191 pp.)

——— 1967. *The mind of the dolphin: a nonhuman intelligence.* Doubleday, New York. (Reprinted as a paperback, Avon Books, Hearst Corporation, New York, 1969. 286 pp.)

Lin, N. 1963: Territorial behavior in the cicada killer wasp, *Sphecius speciosus* (Drury) (Hymenoptera: Sphecidae), I. *Behaviour*, 20(1,2): 115–133.

——— 1964. Increased parasitic pressure as a major factor in the evolution of social behavior in halictine bees. *Insectes Sociaux*, 11(2): 187–192.

Lin, N., and C. D. Michener. 1972. Evolution of sociality in insects. *Quarterly Review of Biology*, 47(2): 131–159.

Lindauer, M. 1952. Ein Beitrag zur Frage der Arbeitsteilung im Bienenstaat. *Zeitschrift für Vergleichende Physiologie*, 34(4): 299–345.

——— 1954. Temperaturregulierung und Wasserhaushalt im Bienenstaat. *Zeitschrift für Vergleichende Physiologie*, 36(4): 391–432.

——— 1955. Schwarmbienen auf Wohnungssuche. *Zeitschrift für Vergleichende Physiologie*, 37(4): 263–324.

——— 1961. *Communication among social bees.* Harvard University Press, Cambridge. ix + 143 pp.

——— 1970. Lernen und Gedächtnis—Versuche an der Honigbiene. *Naturwissenschaften*, 57: 463–467.

Lindauer, M., and W. E. Kerr. 1958. Die gegenseitige Verständigung bei den stachellosen Bienen. *Zeitschrift für Vergleichende Physiologie*, 41(4): 405–434.

——— 1960. Communication between the workers of stingless bees. *Bee World*, 41: 29–41, 65–71.

Lindburg, D. G. 1971. The rhesus monkey in North India: an ecological and behavioral study. In L. A. Rosenblum, ed. (*q.v.*), *Primate behavior: developments in field and laboratory research*, vol. 2, pp. 1–106.

Lindhard, E. 1912. Humlebien som Husdyr. Spredte Traek af nogle danske Humlebiarters Biologi. *Tidsskrift for Landbrukets Planteavl* (Copenhagen), 19: 335–352.

Linsdale, J. M., and L. P. Tevis, Jr. 1951. *The dusky-footed wood rat.* University of California Press, Berkeley. vii + 664 pp.

Linsdale, J. M., and P. Q. Tomich. 1953. *A herd of mule deer.* University of California Press, Berkeley. xiii + 567 pp.

Linsenmair, K. E. 1967. Konstruktion und Signalfunktion der Sandpyramide der Reiterkrabbe *Ocypode saratan* Forsk. (Decapoda Brachyura Ocypodidae). *Zeitschrift für Tierpsychologie*, 24(4): 403–456.

——— 1972. Die Bedeutung familienspezifischer "Abzeichen" für den Familienzusammenhalt bei der sozialen Wüstenassel *Hemilepistus reaumuri* Audouin u. Savigny (Crustacea, Isopoda, Oniscoidea). *Zeitschrift für Tierpsychologie*, 31(2): 131–162.

Linsenmair, K. E., and Christa Linsenmair. 1971. Paarbildung and Paarzusammenhalt bei der monogamen Wüstenassel *Hemilepistus reaumuri* (Crustacea, Isopoda, Oniscoidea). *Zeitschrift für Tierpsychologie*, 29(2): 134–155.

Linzey, D. W. 1968. An ecological study of the golden mouse, *Ochrotomys nuttalli*, in the Great Smoky Mountains National Park. *American Midland Naturalist*, 79(2): 320–345.

Lipton, J. 1968. *An exaltation of larks or, the venereal game.* Grossman, New York. 118 pp.

Lissmann, H. W. 1958. On the function and evolution of electric organs in fish. *Journal of Experimental Biology*, 35(1): 156–191.

Littlejohn, M. J., and J. J. Loftus-Hills. 1968. An experimental evaluation of premating isolation in the *Hyla ewingi* complex (Anura: Hylidae). *Evolution*, 22(4): 659–663.

Llewellyn, L. M., and F. H. Dale. 1964. Notes on the ecology of the opossum in Maryland. *Journal of Mammalogy*, 45(1): 113–122.

Lloyd, J. A., J. J. Christian, D. E. Davis, and F. H. Bronson. 1964. Effects of altered social structure on adrenal weights and morphology in populations of woodchucks (*Marmota monax*). *General and Comparative Endocrinology*, 4(3): 271–276.

Lloyd, J. E. 1966. *Studies on the flash communication system in* Photinus *fireflies.* Miscellaneous Publications, Museum of Zoology, University of Michigan, Ann Arbor, 130. 95 pp.

——— 1973. Fireflies of Melanesia: bioluminescence, mating behavior, and synchronous flashing (Coleoptera: Lampyridae). *Annals of the Entomological Society of America*, 2(6): 991–1008.

Lloyd, M., and H. S. Dybas. 1966a. The periodical cicada problem: I, population ecology. *Evolution*, 20(2): 133–149.

——— 1966b. The periodical cicada problem: II, evolution. *Evolution*, 20(4): 466–505.

Lockie, J. D. 1966. Territory in small carnivores. In P. A. Jewell and Caroline Loizos, eds. (*q.v.*), *Play, exploration and territory in mammals*, pp. 143–165.

Loconti, J. D., and L. M. Roth. 1953. Composition of the odorous secretion of *Tribolium castaneum*. *Annals of the Entomological Society of America*, 46(2): 281–289.

Loizos, Caroline. 1966. Play in mammals. In P. A. Jewell and Caroline Loizos, eds. (*q.v.*), *Play, exploration and territory in mammals*, pp. 1–9.

——— 1967. Play behaviour in higher primates: a review. In D. Morris, ed. (*q.v.*), *Primate ethology: essays on the socio-sexual behavior of apes and monkeys*, pp. 226–282.

Łomnicki, A., and L. B. Slobodkin. 1966. Floating in *Hydra littoralis*. *Ecology*, 47(6): 881–889.

Lord, R. D., Jr. 1961. A population study of the gray fox. *American Midland Naturalist*, 66(1): 87–109.

Lorenz, K. Z. 1935. Der Kumpan in der Umwelt des Vogels. *Journal für Ornithologie*, 83(2): 137–213.

——— 1950. The comparative method in studying innate behaviour patterns. *Symposia of the Society for Experimental Biology*, 4: 221–268.

——— 1952. *King Solomon's ring, new light on animal ways*. Methuen, London. xxii + 202 pp.

——— 1956. Plays and vacuum activities. In M. Autuori et al., *L'instinct dans le comportement des animaux et de l'homme*, pp. 633–645. Masson et Cie, Paris. 796 pp.

——— 1970. *Studies in animal and human behaviour*, vol. 1, trans. by R. Martin. Harvard University Press, Cambridge. xx + 403 pp.

——— 1971. *Studies in animal and human behaviour*, vol. 2, trans. by R. Martin. Harvard University Press, Cambridge. xxiv + 366 pp.

Low, R. M. 1971. Interspecific territoriality in a pomacentrid reef fish, *Pomacentrus flavicauda* Whitley. *Ecology*, 52(4): 648–654.

Lowe, Mildred E. 1956. Dominance-subordinance relationships in the crawfish *Cambarellus shufeldti*. *Tulane Studies in Zoology, New Orleans*, 4(5): 139–170.

Lowe, V. P. W. 1966. Observations on the dispersal of red deer on Rhum. In P. A. Jewell and Caroline Loizos, eds. (*q.v.*), *Play, exploration and territory in mammals*, pp. 211–228.

Lowe, V. T. 1963. Observations on the painted snipe. *Emu*, 62(4): 221–237.

Loy, J. 1970. Behavioral responses of free-ranging rhesus monkeys to food shortage. *American Journal of Physical Anthropology*, 33(2): 263–271.

——— 1971. On the primate biogram. (Review of M. R. A. Chance and C. J. Jolly, *Social groups of monkeys, apes and men*, Dutton, New York, 1970.) *Science*, 172: 680–681.

Lüscher, M. 1952. Die Produktion und Elimination von Ersatzgeschlechtstieren bei der Termite *Kalotermes flavicollis* Fabr. *Zeitschrift für Vergleichende Physiologie*, 34(2): 123–141.

——— 1961a. Air-conditioned termite nests. *Scientific American*, 205(1) (July): 138–145.

——— 1961b. Social control of polymorphism in termites. In J. S. Kennedy, ed., *Insect polymorphism*, pp. 57–67. Symposium of the Royal Entomological Society of London, no. 1. Royal Entomological Society, London. 115 pp.

Lüscher, M., and B. Müller. 1960. Ein spurbildendes Sekret bei Termiten. *Naturwissenschaften*, 47(21): 503.

Lush, J. L. 1947. Family merit and individual merit as bases for selection, I, II. *American Naturalist*, 81(799): 241–261; 81(800): 362–379.

Lyons, J. 1972. Human language. In R. A. Hinde, ed. (*q.v.*), *Non-verbal communication*, pp. 49–85.

MacArthur, R. H. 1962. Some generalized theorems of natural selection. *Proceedings of the National Academy of Sciences, U.S.A.*, 48(11): 1893–1897.

——— 1965. Ecological consequences of natural selection. In T. H. Waterman and H. J. Morowitz, eds., *Theoretical and mathematical biology*, pp. 388–397. Blaisdell Publishing Co., New York. xvii + 426 pp.

——— 1971. Patterns of terrestrial bird communities. In D. S. Farner, J. R. King, and K. C. Parkes, eds., *Avian biology*, vol. 1, pp. 189–221. Academic Press, New York. xix + 586 pp.

——— 1972. *Geographical ecology: patterns in the distribution of species*. Harper & Row, New York. xviii + 269 pp.

MacArthur, R. H., and E. O. Wilson. 1967. *The theory of island biogeography*. Princeton University Press, Princeton, N.J. xi + 203 pp.

MacCluer, Jean W., J. Van Neel, and N. A. Chagnon. 1971. Demographic structure of a primitive population: a simulation. *American Journal of Physical Anthropology*, 35(2): 193–207.

MacFarland, C. 1972. Goliaths of the Galapagos. *National Geographic*, 142(5) (November): 633–649.

Machlis, L., W. H. Nutting, and H. Rapoport. 1968. The structure of sirenin. *Journal of the American Chemical Society*, 90: 1674–1676.

MacKay, D. M. 1972. Formal analysis of communicative processes. In R. A. Hinde, ed. (*q.v.*), *Non-verbal communication*, pp. 3–25.

Mackerras, M. Josephine, and Ruth H. Smith. 1960. Breeding the short-nosed marsupial bandicoot, *Isoodon macrourus* (Gould) in captivity. *Australian Journal of Zoology*, 8(3): 371–382.

Mackie, G. O. 1963. Siphonophores, bud colonies, and superorganisms. In E. C. Dougherty, ed., *The lower Metazoa: comparative biology and phylogeny*, pp. 329–337. University of California Press, Berkeley. xi + 478 pp.

——— 1964. Analysis of locomotion in a siphonophore colony. *Proceedings of the Royal Society*, ser. B, 159: 366–391.

——— 1973. Coordinated behavior in hydrozoan colonies. In R. S. Boardman, A. H. Cheetham, and W. A. Oliver, Jr., eds. (*q.v.*), *Animal colonies: development and function through time*, pp. 95–106.

MacKinnon, J. 1970. Indications of territoriality in mantids. *Zeitschrift für Tierpsychologie*, 27(2): 150–155.

——— 1974. The behaviour and ecology of wild orang-utans (*Pongo pygmaeus*). *Animal Behaviour*, 22(1): 3–74.

MacMillan, R. E. 1964. Population ecology, water relations, and social behavior of a southern California semidesert rodent fauna. *University of California Publications in Zoology*, 71. 59 pp.

MacPherson, A. H. 1969. *The dynamics of Canadian arctic fox populations*. Canadian Wildlife Report Series no. 8. Dept. of Indian Affairs and Northern Development, Ottawa. 52 pp.

Mainardi, D. 1964. Interazione tra preferenze sessuali delle femmine e predominanza sociale dei maschi nel determinismo della selezione sessuale nel topo (*Mus musculus*). *Rendiconti Accademia Nazionale Lincei, Roma*, 37: 484–490.

Mainardi, D., M. Marsan, and A. Pasquali. 1965. Causation of sexual preferences of the house mouse. The behaviour of mice reared by parents whose odour was artificially altered. *Atti Societa Italiana Scienze Naturale Museo Civico Storia Naturale, Milano*, 54: 325–338.

Malécot, G. 1948. *Les mathématiques de l'hérédité*. Masson et Cie, Paris. vi + 63 pp.

Mann, T. 1964. *The biochemistry of semen and of the male reproductive tract*. Methuen, London. xxiii + 493 pp.

Manning, A. 1967. *An introduction to animal behavior*. Addison-Wesley Publishing Co., Reading, Mass. viii + 208 pp.

Marchal, P. 1896. La reproduction et l'évolution des guêpes sociales. *Archives de Zoologie Expérimentale et Générale*, 3d ser., 4: 1–100.

———— 1897. La castration nutriciale chez les Hyménoptères sociaux. *Compte Rendu de la Société de Biologie, Paris*, pp. 556–557.

Markin, G. P. 1970. Food distribution within laboratory colonies of the Argentine ant, *Iridomyrmex humilis* (Mayr). *Insectes Sociaux*, 17(2): 127–157.

Markl, H. 1968. Die Verständigung durch Stridulationssignale bei Blattschneiderameisen: II, Erzeugung und Eigenschaften der Signale. *Zeitschrift für Vergleichende Physiologie*, 60(2): 103–150.

Marler, P. R. 1956. Behaviour of the chaffinch, *Fringilla coelebs*. *Behaviour*, supplement 5. vii + 184 pp.

———— 1957. Specific distinctiveness in the communication signals of birds. *Behaviour*, 11(1): 13–39.

———— 1959. Developments in the study of animal communication. In P. R. Bell, ed. (q.v.), *Darwin's biological work: some aspects reconsidered*, pp. 150–206.

———— 1960. Bird songs and mate selection. In W. E. Lanyon and W. N. Tavolga, eds. (q.v.), *Animal sounds and communication*, pp. 348–367.

———— 1961. The logical analysis of animal communication. *Journal of Theoretical Biology*, 1(3): 295–317.

———— 1965. Communication in monkeys and apes. In I. DeVore, ed. (q.v.), *Primate behavior: field studies of monkeys and apes*, pp. 544–584.

———— 1967. Animal communication signals. *Science*, 157: 769–774.

———— 1969. *Colobus guereza*: territoriality and group composition. *Science*, 163: 93–95.

———— 1970. Vocalizations of East African monkeys: I, red *Colobus*. *Folia Primatologica*, 13(2,3): 81–91.

———— ed. 1972. *The marvels of animal behavior*. National Geographic Society, Washington, D.C. 422 pp.

———— 1973. A comparison of vocalizations of red-tailed monkeys and blue monkeys, *Cercopithecus ascanius* and *C. mitis*, in Uganda. *Zeitschrift für Tierpsychologie*, 33(3): 223–247.

Marler, P. R., and W. J. Hamilton III. 1966. *Mechanisms of animal behavior*. John Wiley & Sons, New York. xi + 771 pp.

Marler, P. R., and P. Mundinger. 1971. Vocal learning in birds. In H. Moltz, ed. (q.v.), *The ontogeny of vertebrate behavior*, pp. 389–450.

Marler, P. R., and M. Tamura. 1964. Culturally transmitted patterns of vocal behavior in sparrows. *Science*, 146: 1483–1486.

Marr, J. N., and L. E. Gardner, Jr. 1965. Early olfactory experience and later social behavior in the rat: preference, sexual responsiveness, and care of the young. *Journal of Genetic Psychology*, 107: 167–174.

Marsden, H. M. 1968. Agonistic behaviour of young rhesus monkeys after changes induced in social rank of their mothers. *Animal Behaviour*, 16(1): 38–44.

———— 1971. Intergroup relations in rhesus monkeys (*Macaca mulatta*). In A. H. Esser, ed. (q.v.), *Behavior and environment: the use of space by animals and men*, pp. 112–113.

Marshall, A. J. 1954. *Bower-birds, their displays and breeding cycles*. Clarendon Press of Oxford University Press, Oxford. x + 208 pp.

Martin, M. M., Mary J. Gieselmann, and Joan Stadler Martin. 1973. Rectal enzymes of attine ants, α-amylase and chitinase. *Journal of Insect Physiology*, 19(7): 1409–1416.

Martin, M. M., and Joan Stadler Martin. 1971. The presence of protease activity in the rectal fluid of primitive attine ants. *Journal of Insect Physiology*, 17(10): 1897–1906.

Martin, P. S. 1966. Africa and Pleistocene overkill. *Nature, London*, 212(5060): 339–342.

Martin, R. D. 1968. Reproduction and ontogeny in tree shrews (*Tupaia belangeri*) with reference to their general behavior and taxonomic relationships. *Zeitschrift für Tierpsychologie*, 25(4): 409–495; 25(5): 505–532.

———— 1972. Adaptive radiation and behaviour of the Malagasy lemurs. *Philosophical Transactions of the Royal Society of London*, ser. B, 264: 295–352.

———— 1973. A review of the behaviour and ecology of the lesser mouse lemur (*Microcebus murinus* J. F. Miller 1777). In R. P. Michael and J. H. Crook, eds. (q.v.), *Comparative ecology and behaviour of primates*, pp. 1–68.

Martinez, D. R., and E. Klinghammer. 1970. The behavior of the whale *Orcinus orca*: a review of the literature. *Zeitschrift für Tierpsychologie*, 27(7): 828–839.

Martof, B. S. 1953. Territoriality in the green frog, *Rana clamitans*. *Ecology*, 34(1): 165–174.

Maschwitz, U. 1964. Gefahrenalarmstoffe und Gefahrenalarmierung bei sozialen Hymenopteren. *Zeitschrift für Vergleichende Physiologie*, 47(6): 596–655.

———— 1966a. Alarm substances and alarm behavior in social insects. *Vitamins and hormones*, 24: 267–290.

———— 1966b. Das Speichelsekret der Wespenlarven und seine biologische Bedeutung. *Zeitschrift für Vergleichende Physiologie*, 53(3): 228–252.

Maschwitz, U., R. Jander, and D. Burkhardt. 1972. Wehrsubstanzen und Wehrverhalten der Termite *Macrotermes carbonarius*. *Journal of Insect Physiology*, 18(9): 1715–1720.

Maschwitz, U., K. Koob, and H. Schildknecht. 1970. Ein Beitrag zur Funktion der Metathoracaldrüse der Ameisen. *Journal of Insect Physiology*, 16(2): 387–404.

Maslow, A. H. 1936. The role of dominance in the social and sexual behavior of infra-human primates: IV, the determination of hierarchy in pairs and in a group. *Journal of Genetic Psychology*, 49(1): 161–198.

———— 1940. Dominance-quality and social behavior in infra-human primates. *Journal of Social Psychology*, 11: 313–324.

—— 1954. *Motivation and personality.* Harper, New York. 411 pp.

—— 1972. *The farther reaches of human nature.* Viking Press, New York. xxii + 423 pp.

Mason, J. W. 1968. Organization of the multiple endocrine responses to avoidance in the monkey. *Psychosomatic Medicine*, 30(5): 774–790.

Mason, W. A. 1960. The effects of social restriction on the behavior of rhesus monkeys: I, free social behavior. *Journal of Comparative and Physiological Psychology*, 53(6): 582–589.

—— 1965. The social development of monkeys and apes. In I. DeVore, ed. (q.v.), *Primate behavior: field studies of monkeys and apes*, pp. 514–543.

—— 1968. Use of space in *Callicebus* groups. In Phyllis C. Jay, ed. (q.v.), *Primates: studies in adaptation and variability*, pp. 200–216.

—— 1971. Field and laboratory studies of social organization in *Saimiri* and *Callicebus*. In L. A. Rosenblum, ed. (q.v.), *Primate behavior: developments in field and laboratory research*, vol. 2, pp. 107–137.

Mason, W. A., and G. Berkson. 1962. Conditions influencing vocal responsiveness of infant chimpanzees. *Science*, 137: 127–128.

Masters, R. D. 1970. Genes, language, and evolution. *Semiotica*, 2(4): 295–320.

Masters, W. H., and Virginia E. Johnson. 1966. *Human sexual response.* Little, Brown, Boston. xiii + 366 pp.

Mather, K., and B. J. Harrison. 1949. The manifold effect of selection. *Heredity*, 3(1): 1–52; 3(2): 131–162.

Mathew, D. N. 1964. Observations on the breeding habits of the bronze-winged jaçana, *Metopidius indicus* (Latham). *Journal of the Bombay Natural History Society*, 61(2): 295–302.

Mathewson, Sue F. 1961. Gonadotrophic control of aggressive behavior in starlings. *Science*, 134: 1522–1523.

Matthews, L. H. 1971. *The life of mammals*, vol. 2. Universe Books, New York. 440 pp.

Matthews, R. W. 1968a. *Microstigmus comes*: sociality in a sphecid wasp. *Science*, 160: 787–788.

—— 1968b. Nesting biology of the social wasp *Microstigmus comes*. *Psyche, Cambridge*, 75(1): 23–45.

Mattingly, I. G. 1972. Speech cues and sign stimuli. *American Scientist*, 60(3): 327–337.

Mautz, D., R. Boch, and R. A. Morse. 1972. Queen finding by swarming honey bees. *Annals of the Entomological Society of America*, 65(2): 440–443.

May, R. M. 1973. *Stability and complexity in model ecosystems.* Princeton University Press, Princeton, N.J. x + 235 pp.

Maynard Smith, J. 1956. Fertility, mating behaviour, and sexual selection in *Drosophila subobscura*. *Journal of Genetics*, 54(2): 261–279.

—— 1964. Group selection and kin selection. *Nature, London*, 201(4924): 1145–1147.

—— 1965. The evolution of alarm calls. *American Naturalist*, 99(904): 59–63.

—— 1971. What use is sex? *Journal of Theoretical Biology*, 30(2): 319–335.

Maynard Smith, J., and G. R. Price. 1973. The logic of animal conflict. *Nature, London*, 246(5427): 15–18.

Maynard Smith, J., and M. G. Ridpath. 1972. Wife sharing in the Tasmanian native hen, *Tribonyx mortierii*: a case of kin selection? *American Naturalist*, 106(950): 447–452.

Mayr, E. 1935. Bernard Altum and the territory theory. *Proceedings of the Linnaean Society of New York* (1933–34), nos. 45, 46, pp. 24–38.

—— 1960. The emergence of evolutionary novelties. In S. Tax, ed., *Evolution after Darwin*, vol. 1, *The evolution of life, its origin, history, and future*, pp. 349–380. University of Chicago Press, Chicago. viii + 629 pp.

—— 1963. *Animal species and evolution.* Belknap Press of Harvard University Press, Cambridge. xiv + 797 pp.

—— 1969. *Principles of systematic zoology.* McGraw-Hill Book Co., New York. xi + 428 pp.

—— 1970. *Populations, species, and evolution.* Belknap Press of Harvard University Press, Cambridge. xv + 453 pp.

Mazokhin-Porshnyakov, G. A. 1969. Die Fähigkeit der Bienen, visuelle Reize zu generalisieren. *Zeitschrift für Vergleichende Physiologie*, 65(1): 15–28.

McBride, A. F., and D. O. Hebb. 1948. Behavior of the captive bottle-nose dolphin, *Tursiops truncatus*. *Journal of Comparative and Physiological Psychology*, 41: 111–123.

McBride, G. 1958. Relationship between aggressiveness and egg production in the domestic hen. *Nature, London*, 181(4612): 858.

—— 1963. The "teat order" and communication in young pigs. *Animal Behaviour*, 11(1): 53–56.

McBride, G., I. P. Parer, and F. Foenander. 1969. The social organization and behaviour of the feral domestic fowl. *Animal Behaviour Monographs*, 2(3): 125–181.

McCann, C. 1934. Observations on some of the Indian langurs. *Journal of the Bombay Natural History Society*, 36(3): 618–628.

McClearn, G. E. 1970. Behavioral genetics. *Annual Review of Genetics*, 4: 437–468.

McClearn, G. E., and J. C. DeFries. 1973. *Introduction to behavioral genetics.* W. H. Freeman, San Francisco. x + 349 pp.

McClintock, Martha. 1971. Menstrual synchrony and suppression. *Nature, London*, 229(5282): 244–245.

McCook, H. C. 1879. Combats and nidification of the pavement ant, *Tetramorium caespitum*. *Proceedings of the Academy of Natural Sciences of Philadelphia*, 31: 156–161.

McDonald, A. L., N. W. Heimstra, and D. K. Damkot. 1968. Social modification of agonistic behaviour in fish. *Animal Behaviour*, 16(4): 437–441.

McEvedy, C. 1967. *The Penguin atlas of ancient history.* Penguin Books, Baltimore, Md. 96 pp.

McFarland, W. N., and S. A. Moss. 1967. Internal behavior in fish schools. *Science*, 156: 260–262.

McGrew, W. C., and Caroline E. G. Tutin. 1973. Chimpanzee tool use in dental grooming. *Nature, London*, 241(5390): 477–478.

McGuire, M. T. 1974. The St. Kitts vervet. *Contributions to Primatology*, 1. xii + 199 pp.

McHugh, T. 1958. Social behavior of the American buffalo (*Bison bison bison*). *Zoologica, New York*, 43(1): 1–40.

McKay, F. E. 1971. Behavioral aspects of population dynamics in unisexual-bisexual *Poeciliopsis* (Pisces: Poeciliidae). *Ecology*, 52(5): 778–790.

McKay, G. M. 1973. Behavior and ecology of the Asiatic elephant in southeastern Ceylon. *Smithsonian Contributions to Zoology*, 125. iv + 113 pp.

McKnight, T. L. 1958. The feral burro in the United States: distribution and problems. *Journal of Wildlife Management*, 22(2): 163–179.

McLaren, I. A. 1967. Seals and group selection. *Ecology*, 48(1): 104–110.

McLaughlin, C. A. 1967. Aplodontoid, sciuroid, geomyoid, castoroid, and anomaluroid rodents. In S. Anderson and J. K. Jones, Jr., eds. (*q.v.*), *Recent mammals of the world: a synopsis of families*, pp. 210–225.

McManus, J. J. 1970. Behavior of captive opossums, *Didelphis marsupialis virginiana*. *American Midland Naturalist*, 84(1): 144–169.

McNab, B. K. 1963. Bioenergetics and the determination of home range size. *American Naturalist*, 97(894): 133–140.

Mead, Margaret. 1963. Socialization and enculturation. *Current Anthropology*, 4(1): 184–188.

Mech, L. D. 1970. *The wolf: the ecology and behavior of an endangered species*. Natural History Press, Garden City, N.Y. xx + 384 pp.

Medawar, P. B. 1952. *An unsolved problem of biology*. H. K. Lewis, London. 24 pp. (Reprinted in P. B. Medawar, *The uniqueness of the individual*, pp. 44–70, Methuen, London, 1957. 191 pp.)

Medler, J. T. 1957. Bumblebee ecology in relation to the pollination of alfalfa and red clover. *Insectes Sociaux*, 4(3): 245–252.

Meier, G. W. 1965. Other data on the effects of social isolation during rearing upon adult reproductive behaviour in the rhesus monkey (*Macaca mulatta*). *Animal Behaviour*, 13(2,3): 228–231.

Menzel, E. W., Jr. 1966. Responsiveness to objects in free-ranging Japanese monkeys. *Behaviour*, 26(1,2): 130–149.

———— 1971. Communication about the environment in a group of young chimpanzees. *Folia Primatologica*, 15(3,4): 220–232.

Menzel, R. 1968. Das Gedächtnis der Honigbiene für Spektralfarben: I, kurzzeitiges und langzeitiges Behalten. *Zeitschrift für Vergleichende Physiologie*, 60(1): 82–102.

Merfield, F. G., and H. Miller. 1956. *Gorillas were my neighbours*. Longmans, London.

Merrell, D. J. 1953. Selective mating as a cause of gene frequency changes in laboratory populations of *Drosophila melanogaster*. *Evolution*, 7(4): 287–296.

———— 1968. A comparison of the estimated size and the "effective size" of breeding populations of the leopard frog, *Rana pipiens*. *Evolution*, 22(2): 274–283.

Mertz, D. B. 1971a. Life history phenomena in increasing and decreasing populations. In G. P. Patil, E. C. Pielou, and W. E. Waters, eds., *Statistical ecology*, vol. 2, *Sampling and modeling biological popula-*

tions and population dynamics, pp. 361–399. Pennsylvania State University Press, University Park, Pa.

———— 1971b. The mathematical demography of the California condor population. *American Naturalist*, 105(945): 437–453.

Mesarović, M. D., D. Macko, and Y. Takahara. 1970. *Theory of hierarchical, multilevel systems*. Academic Press, New York. xiii + 294 pp.

Mewaldt, L. R. 1964. Effects of bird removal on winter population of sparrows. *Bird Banding*, 35(3): 184–195.

Meyerriecks, A. J. 1960. *Comparative breeding behavior of four species of North American herons*. Publication no. 2. The Nuttall Ornithological Club, Cambridge, Mass. viii + 158 pp.

———— 1972. *Man and birds: evolution and behavior*. Pegasus, Bobbs-Merrill Co., Indianapolis. xii + 209 pp.

Michael, R. P. 1966. Action of hormones on the cat brain. In R. A. Gorski and R. E. Whalen, eds., *Brain and behavior*, vol. 3, *The brain and gonadal function*, pp. 81–98. University of California Press, Berkeley. xv + 289 pp.

Michael, R. P., and J. H. Crook, eds. 1973. *Comparative ecology and behaviour of primates*. Academic Press, New York. xvi + 847 pp.

Michael, R. P., and Patricia P. Scott. 1964. The activation of sexual behaviour by the subcutaneous administration of oestrogen. *Journal of Physiology*, 171(2): 254–274.

Michener, C. D. 1958. The evolution of social behavior in bees. *Proceedings of the Tenth International Congress of Entomology, Montreal, 1956*, 2: 441–447.

———— 1961a. Probable parasitism among Australian bees of the genus *Allodapula* (Hymenoptera, Apoidea, Ceratinini). *Annals of the Entomological Society of America*, 54(4): 532–534.

———— 1961b. Observations on the nests and behavior of *Trigona* in Australia and New Guinea (Hymenoptera, Apidae). *American Museum Novitates*, 2026. 46 pp.

———— 1962. Biological observations on the primitively social bees of the genus *Allodapula* in the Australian region (Hymenoptera, Xylocopinae). *Insectes Sociaux*, 9(4): 355–373.

———— 1964a. Reproductive efficiency in relation to colony size in hymenopterous societies. *Insectes Sociaux*, 11(4): 317–341.

———— 1964b. The bionomics of *Exoneurella*, a solitary relative of *Exoneura* (Hymenoptera: Apoidea: Ceratinini). *Pacific Insects*, 6(3): 411–426.

———— 1965. The life cycle and social organization of bees of the genus *Exoneura* and their parasite, *Inquilina* (Hymenoptera: Xylocopinae). *Kansas University Science Bulletin*, 46(9): 317–358.

———— 1966a. Interaction among workers from different colonies of sweat bees (Hymenoptera, Halictidae). *Animal Behaviour*, 14(1): 126–129.

———— 1966b. The bionomics of a primitively social bee, *Lasioglossum versatum* (Hymenoptera: Halictidae). *Journal of the Kansas Entomological Society*, 39(2): 193–217.

———— 1966c. Evidence of cooperative provisioning of cells in *Exomalopsis* (Hymenoptera: Anthophoridae). *Journal of the Kansas Entomological Society*, 39(2): 315–317.

—— 1966d. Parasitism among Indoaustralian bees of the genus *Allodapula* (Hymenoptera: Ceratinini). *Journal of the Kansas Entomological Society*, 39(4): 705–708.

—— 1969. Comparative social behavior of bees. *Annual Review of Entomology*, 14: 299–342.

—— 1970. Social parasites among African allodapine bees (Hymenoptera, Anthophoridae, Ceratinini). *Journal of the Linnean Society, London, Zoology*, 49(3): 199–215.

—— 1971. Biologies of African allodapine bees. *Bulletin of the American Museum of Natural History*, 145(3): 221–301.

—— 1973. The Brazilian honeybee. *BioScience*, 23(9): 523–533.

—— 1974. *The social behavior of the bees: a comparative study.* Belknap Press of Harvard University Press, Cambridge. xii + 404 pp.

Michener, C. D., and D. J. Brothers. 1974. Were workers of eusocial Hymenoptera initially altruistic or oppressed? *Proceedings of the National Academy of Sciences, U.S.A.*, 71(3): 671–674.

Michener, C. D., D. J. Brothers, and D. R. Kamm. 1971. Interactions in colonies of primitively social bees: artificial colonies of *Lasioglossum zephyrum*. *Proceedings of the National Academy of Sciences, U.S.A.*, 68(6): 1241–1245.

Michener, C. D., and W. B. Kerfoot. 1967. Nests and social behavior of three species of *Pseudaugochloropsis* (Hymenoptera: Halictidae). *Journal of the Kansas Entomological Society*, 40(2): 214–232.

Milkman, R. D. 1967. Heterosis as a major cause of heterozygosity in nature. *Genetics*, 55(3): 493–495.

—— 1970. The genetic basis of natural variation in *Drosophila melanogaster*. *Advances in Genetics*, 15: 55–114.

Miller, E. M. 1969. Caste differentiation in the lower termites. In K. Krishna and Frances M. Weesner, eds. (*q.v.*), *Biology of termites*, vol. 1, pp. 283–310.

Miller, G. A., E. Galanter, and K. H. Pribram. 1960. *Plans and the structure of behavior*. Henry Holt, New York. xii + 226 pp.

Miller, N. E. 1948. Theory and experiment relating psychoanalytic displacement to stimulus-response generalization. *Journal of Abnormal and Social Psychology*, 43(2): 155–178.

Miller, R. S. 1964. Ecology and distribution of pocket gophers (Geomyiidae) in Colorado. *Ecology*, 45(2): 256–272.

—— 1967. Pattern and process in competition. *Advances in Ecological Research*, 4: 1–74.

Miller, R. S., and W. J. D. Stephen. 1966. Spatial relationships in flocks of sandhill cranes (*Grus canadensis*). *Ecology*, 47(2): 323–327.

Millikan, G. C., and R. I. Bowman. 1967. Observations on Galápagos tool-using finches in captivity. *Living Bird*, 6: 23–41.

Milstead, W. W., ed. 1967. *Lizard ecology: a symposium*. University of Missouri Press, Columbia. xi + 300 pp.

Milum, V. G. 1955. Honey bee communication. *American Bee Journal*, 95(3): 97–104.

Minchin, A. K. 1937. Notes on the weaning of a young koala (*Phascolarctos cinereus*). *Records of the South Australian Museum, Adelaide*, 6(1): 1–3.

Minks, A. K., W. L. Roelofs, F. J. Ritter, and C. J. Persoons. 1973. Reproductive isolation of two tortricid moth species by different ratios of a two-component sex attractant. *Science*, 180: 1073.

Missakian, Elizabeth A. 1972. Genealogical and cross-genealogical dominance relations in a group of free-ranging rhesus monkeys (*Macaca mulatta*) on Cayo Santiago. *Primates*, 13(2): 169–180.

Mitchell, G. D. 1969. Paternalistic behavior in primates. *Psychological Bulletin*, 71: 399–417.

Mitchell, R. 1970. An analysis of dispersal in mites. *American Naturalist*, 104(939): 425–431.

Mizuhara, H. 1964. Social changes of Japanese monkey troops in the Takasakiyama. *Primates*, 5(1,2): 27–52.

Moffat, C. B. 1903. The spring rivalry of birds. Some views on the limit to multiplication. *Irish Naturalist*, 12(6): 152–166.

Mohnot, S. M. 1971. Some aspects of social changes and infant-killing in the Hanuman langur, *Presbytis entellus* (Primates: Cercopithecidae), in western India. *Mammalia*, 35: 175–198.

Mohr, H. 1960. Zum Erkennen von Raubvögeln, insbesondere von Sperber und Baumfalk, durch Kleinvögeln. *Zeitschrift für Tierpsychologie*, 17(6): 686–699.

Möhres, F. P. 1957. Elektrische Entladungen im Dienste der Revierabgrenzung bei Fischen. *Naturwissenschaften*, 44(15): 431–432.

Moltz, H. 1971a. The ontogeny of maternal behavior in some selected mammalian species. In H. Moltz, ed. (*q.v.*), *The ontogeny of vertebrate behavior*, pp. 263–313.

—— ed. 1971b. *The ontogeny of vertebrate behavior*. Academic Press, New York. xi + 500 pp.

Moment, G. 1962. Reflexive selection: a possible answer to an old puzzle. *Science*, 136: 262–263.

Montagner, H. 1963. Etude préliminaire des relations entre les adultes et le couvain chez les guêpes sociales du genre *Vespa*, au moyen d'un radio-isotope. *Insectes Sociaux*, 10(2): 153–165.

—— 1966. Le mécanisme et les conséquences des comportements trophallactiques chez les guêpes du genre *Vespa*. Thesis, Faculté des Sciences de l'Université de Nancy, France. 143 pp.

—— 1967. Comportements trophallactiques chez les guêpes sociales. Sound, color film produced by Service du Film Recherche Scientifique; 96, Boulevard Raspail, Paris. No. B2053, 19 min.

Montagu, M. F. Ashley. 1968a. The new litany of "innate depravity," or original sin revisited. In M. F. Ashley Montagu, ed. (*q.v.*), *Man and aggression*, pp. 3–17.

—— ed. 1968b. *Man and aggression*. Oxford University Press, Oxford. xiv + 178 pp. (2d ed., 1973.)

Montgomery, G. G., and M. E. Sunquist. 1974. Impact of sloths on neotropical forest energy flow and nutrient cycling. In F. B. Golley and E. Medina, eds., *Tropical ecological systems: trends in terrestrial and aquatic research*, vol. 2, *Ecology studies, analysis and synthesis*. Springer-Verlag, New York. (In press.)

Moore, B. P. 1964. Volatile terpenes from *Nasutitermes* soldiers (Isoptera, Termitidae). *Journal of Insect Physiology*, 10(2): 371–375.

—— 1968. Studies on the chemical composition and function of the cephalic gland secretion in Australian termites. *Journal of Insect Physiology*, 14(1): 33–39.

——— 1969. Biochemical studies in termites. In K. Krishna and Frances M. Weesner, eds. (*q.v.*), *Biology of termites*, vol. 1, pp. 407–432.

Moore, J. C. 1956. Observations of manatees in aggregations. *American Museum Novitates*, 1811. 24 pp.

Moore, N. W. 1964. Intra- and interspecific competition among dragonflies (Odonata): an account of observations and field experiments on population control in Dorset, 1954–60. *Journal of Animal Ecology*, 33(1): 49–71.

Moore, W. S., and F. E. McKay. 1971. Coexistence in unisexual-bisexual species complexes of *Poeciliopsis* (Pisces: Poeciliidae). *Ecology*, 52(5): 791–799.

Moorhead, P. S., and M. M. Kaplan, eds. 1967. *Mathematical challenges to the neo-Darwinian interpretation of evolution*. Wistar Institute Symposium Monograph no. 5. Wistar Institute Press, Philadelphia. xi + 140 pp.

Moreau, R. E. 1960. Conspectus and classification of the ploceine weaverbirds. *Ibis*, 102(2): 298–321; 102(3): 443–471.

Morgan, C. L. 1896. *An introduction to comparative psychology*. Walter Scott, London. xvi + 382 pp.

——— 1922. *Emergent evolution*. Holt, New York. (3d ed., 1931. xii + 313 pp.)

Morgan, Elaine. 1972. *The descent of woman*. Stein and Day, New York. 258 pp.

Morimoto, R. 1961a. On the dominance order in *Polistes* wasps: I, studies on the social Hymenoptera in Japan XII. *Science Bulletin of the Faculty of Agriculture, Kyushu University*, 18(4): 339–351.

——— 1961b. On the dominance order in *Polistes* wasps: II, studies on the social Hymenoptera in Japan XIII. *Science Bulletin of the Faculty of Agriculture, Kyushu University*, 19(1): 1–17.

Morris, C. 1946. *Signs, language, and behavior*. Prentice-Hall, Englewood Cliffs, N.J. xiv + 365 pp.

Morris, D. 1957. "Typical intensity" and its relation to the problem of ritualization. *Behaviour*, 11(1): 1–12.

——— 1962. *The biology of art*. Alfred Knopf, New York. 176 pp.

——— 1967a. *The naked ape: a zoologist's study of the human animal*. McGraw-Hill Book Co., New York. 252 pp.

——— ed. 1967b. *Primate ethology: essays on the socio-sexual behavior of apes and monkeys*. Aldine Publishing Co., Chicago. x + 374 pp. (Reprinted as a paperback, Anchor Books, Doubleday, Garden City, N.Y., 1969. vii + 471 pp.)

Morrison, B. J., and W. F. Hill. 1967. Socially facilitated reduction of the fear response in rats raised in groups or in isolation. *Journal of Comparative and Physiological Psychology*, 63: 71–76.

Morse, D. H. 1967. Foraging relationships of brown-headed nuthatches and pine warblers. *Ecology*, 48(1): 94–103.

——— 1970. Ecological aspects of some mixed-species foraging flocks of birds. *Ecological Monographs*, 40(1): 119–168.

Morse, R. A., and N. E. Gary. 1961. Colony response to worker bees confined with queens (*Apis mellifera* L.). *Bee World*, 42(8): 197–199.

Morse, R. A., and F. M. Laigo. 1969. Apis dorsata *in the Philippines (including an annotated bibliography)*. Monograph of the Philippine Association of Entomologists, Inc. (University of the Philippines, Laguna, P.I.), no. 1. 96 pp.

Morton, N. E. 1969. Human population structure. *Annual Review of Genetics*, 3: 53–74.

Morton, N. E., Shirley Yee, D. E. Harris, and Ruth Lew. 1971. Bioassay of kinship. *Theoretical Population Biology*, 2(4): 507–524.

Mörzer Bruyns, W. F. J. 1971. *Field guide of whales and dolphins*. C. A. Mees, Amsterdam. 258 pp.

Mosebach-Pukowski, Erna. 1937. Über die Raupengesellschaften von *Vanessa io* und *Vanessa urticae*. *Zeitschrift für Morphologie und Ökologie der Tiere*, 33(3): 358–380.

Moyer, K. E. 1969. Internal impulses to aggression. *Transactions of the New York Academy of Sciences*, 31(2): 104–114.

——— 1971. *The physiology of hostility*. Markham, Chicago. x + 194 pp.

Moynihan, M. H. 1958. Notes on the behavior of some North American gulls: II, non-aerial hostile behavior of adults. *Behaviour*, 12(1,2): 95–182.

——— 1960. Some adaptations which help to promote gregariousness. *Proceedings of the Twelfth International Ornithological Congress, Helsinki*, pp. 523–541.

——— 1962. The organization and probable evolution of some mixed species flocks of Neotropical birds. *Smithsonian Miscellaneous Collections*, 143(7). 140 pp.

——— 1964. Some behavior patterns of platyrrhine monkeys: I, the night monkey (*Aotus trivirgatus*). *Smithsonian Miscellaneous Collections*, 146(5). iv + 84 pp.

——— 1966. Communication in the titi monkey, *Callicebus*. *Journal of Zoology, London*, 150(1): 77–127.

——— 1968. Social mimicry: character convergence versus character displacement. *Evolution*, 22(2): 315–331.

——— 1969. Comparative aspects of communication in New World primates. In D. Morris, ed. (*q.v.*), *Primate ethology: essays on the socio-sexual behavior of apes and monkeys*, pp. 306–342.

——— 1970a. Control, suppression, decay, disappearance and replacement of displays. *Journal of Theoretical Biology*, 29(1): 85–112.

——— 1970b. Some behavior patterns of platyrrhine monkeys: II, *Saguinus geoffroyi* and some other tamarins. *Smithsonian Contributions to Zoology*, 28. iv + 77 pp.

——— 1973. The evolution of behavior and the role of behavior in evolution. *Breviora*, 415. 29 pp.

——— 1974. Conservatism of displays and comparable stereotyped patterns among cephalopods. (Unpublished manuscript.)

Muckenhirn, N. A., and J. F. Eisenberg. 1973. Home ranges and predation in the Ceylon leopard. In R. L. Eaton, ed. (*q.v.*), *The world's cats*, vol. 1, pp. 142–175.

Mueller, H. C. 1971. Oddity and specific searching image more important than conspicuousness in prey selection. *Nature, London*, 233(5318): 345–346.

Mukinya, J. G. 1973. Density, distribution, population structure and social organization of the black rhinoceros in Masai Mara Game Reserve. *East African Wildlife Journal*, 11(3,4): 385–400.

Müller, D. G., L. Jaenicke, M. Donike, and T. Akintobi. 1971. Sex attractant in a brown alga: chemical structure. *Science*, 171: 815–817.

Müller-Schwarze, D. 1968. Play deprivation in deer. *Behaviour*, 31(3): 144–162.

—— 1969. Complexity and relative specificity in a mammalian pheromone. *Nature, London*, 223(5205): 525–526.

—— 1971. Pheromones in black-tailed deer (*Odocoileus hemionus columbianus*). *Animal Behaviour*, 19(1): 141–152.

Müller-Velten, H. 1966. Über den Angstgeruch bei der Hausmaus. *Zeitschrift für Vergleichende Physiologie*, 52(4): 401–429.

Murchison, C. 1935. The experimental measurement of a social hierarchy in *Gallus domesticus*: IV, loss of body weight under conditions of mild starvation as a function of social dominance. *Journal of General Psychology*, 12: 296–312.

Murdoch, W. W. 1966. Population stability and life history phenomena. *American Naturalist*, 100(910): 5–11.

Murie, A. 1944. *The wolves of Mount McKinley*. Fauna of the National Parks of the United States, Fauna Series no. 5. U.S. Department of the Interior, Washington, D.C. xix + 238 pp.

Murphy, G. I. 1968. Patterns in life history. *American Naturalist*, 102(927): 391–403.

Murray, B. G. 1967. Dispersal in vertebrates. *Ecology*, 48(6): 975–978.

—— 1971. The ecological consequences of interspecific territorial behavior in birds. *Ecology*, 52(3): 414–423.

Murton, R. K. 1968. Some predator-prey relationships in bird damage and population control. In R. K. Murton and E. N. Wright, eds., *The problems of birds as pests*, pp. 157–169. Academic Press, New York.

Murton, R. K., A. J. Isaacson, and N. J. Westwood. 1966. The relationships between wood-pigeons and their clover food supply and the mechanism of population control. *Journal of Applied Ecology*, 3(1): 55–96.

Myers, Judith H., and C. J. Krebs. 1971. Genetic, behavioral, and reproductive attributes of dispersing field voles *Microtus pennsylvanicus* and *Microtus ochrogaster*. *Ecological Monographs*, 41(1): 53–78.

Myers, K., C. S. Hale, R. Mykytowycz, and R. L. Hughes. 1971. The effects of varying density and space on sociality and health in animals. In A. H. Esser, ed. (*q.v.*), *Behavior and environment: the use of space by animals and men*, pp. 148–187.

Mykytowycz, R. 1958–60. Social behaviour of an experimental colony of wild rabbits, *Oryctolagus cuniculus* (L.), I, II, III. *Commonwealth Scientific and Industrial Research Organization, Wildlife Research*, Canberra, 3: 7–25; 4: 1–13; 5: 1–20.

—— 1962. Territorial function of chin gland secretion in the rabbit, *Oryctolagus cuniculus* (L.). *Nature, London*, 193(4817): 799.

—— 1964. Territoriality in rabbit populations. *Australian Natural History*, 14(10): 326–329.

—— 1965. Further observations on the territorial function and histology of the submandibular cutaneous (chin) glands in the rabbit, *Oryctolagus cuniculus* (L.). *Animal Behaviour*, 13(4): 400–412.

—— 1968. Territorial marking by rabbits. *Scientific American*, 218(5) (May): 116–126.

Mykytowycz, R., and M. L. Dudziński. 1972. Aggressive and protective behaviour of adult rabbits *Oryctolagus cuniculus* (L.) towards juveniles. *Behaviour*, 43(1–4): 97–120.

Myton, Becky. 1974. Utilization of space by *Peromyscus leucopus* and other small mammals. *Ecology*, 55(2): 277–290.

Nagel, U. 1973. A comparison of anubis baboons, hamadryas baboons and their hybrids at a species border in Ethiopia. *Folia Primatologica*, 19(2,3): 104–165.

Nakamura, E. L. 1972. Development and use of facilities for studying tuna behavior. In H. E. Winn and B. L. Olla, eds., *Behavior of marine animals: current perspectives in research*, vol. 2, *Vertebrates*, pp. 245–277. Plenum Press, New York.

Napier, J. R. 1960. Studies of the hands of living primates. *Proceedings of the Zoological Society of London*, 134(4): 647–657.

Napier, J. R., and P. H. Napier. 1967. *A handbook of living primates*. Academic Press, New York. xiv + 456 pp.

—— eds. 1970. *Old World monkeys: evolution, systematics, and behavior*. Academic Press, New York. xvi + 660 pp.

Narise, T. 1968. Migration and competition in *Drosophila*: I, competition between wild and vestigial strains of *Drosophila melanogaster* in a cage and migration-tube population. *Evolution*, 22(2): 301–306.

Naylor, A. F. 1959. An experimental analysis of dispersal in the flour beetle, *Tribolium confusum*. *Ecology*, 40(3): 453–465.

Neal, E. 1948. *The badger*. Collins, London. xvi + 158 pp.

Nedel, J. O. 1960. Morphologie und Physiologie der Mandibeldrüse einiger Bienen-arten (Apidae). *Zeitschrift für Morphologie und Ökologie der Tiere*, 49(2): 139–183.

Neel, J. V. 1970. Lessons from a "primitive" people. *Science*, 170: 815–822.

Neill, W. T. 1971. *The last of the ruling reptiles: alligators, crocodiles, and their kin*. Columbia University Press, New York. xvii + 486 pp.

Nel, J. J. C. 1968. Aggressive behaviour of the harvester termites *Hodotermes mossambicus* (Hagen) and *Trinervitermes trinervoides* (Sjöstedt). *Insectes Sociaux*, 15(2): 145–156.

Nelson, J. B. 1965. The behaviour of the gannet. *British Birds*, 58(7): 233–288; 58(8): 313–336.

Nero, R. W. 1956. A behavior study of the red-winged blackbird: I, mating and nesting activities. *Wilson Bulletin*, 68(1): 5–37.

Neuweiler, G. 1969. Verhaltensbeobachtungen an einer indischen Flughundkolonie (*Pteropus g. giganteus* Brünn). *Zeitschrift für Tierpsychologie*, 26(2): 166–199.

Neville, M. K. 1968. Ecology and activity of Himalayan foothill rhesus monkeys (*Macaca mulatta*). *Ecology*, 49(1): 110–123.

Nice, Margaret M. 1937. Studies in the life history of the song sparrow: I, a population study of the song sparrow. *Transactions of the Linnaean Society of New York*, 4. vi + 247 pp.

—— 1941. The role of territory in bird life. *American Midland Naturalist*, 26(3): 441–487.

———— 1943. Studies in the life history of the song sparrow: II, the behavior of the song sparrow and other passerines. *Transactions of the Linnaean Society of New York*, 6. viii + 328 pp.

Nicholls, D. G. 1970. Dispersal and dispersion in relation to the birthsite of the southern elephant seal, *Mirounga leonina* (L.), of Macquarie Island. *Mammalia*, 34(4): 598–616.

Nicholson, A. J. 1954. An outline of the dynamics of animal populations. *Australian Journal of Zoology*, 2(1): 9–65.

Nicholson, E. M. 1929. Report on the "British Birds" census of heronries, 1928. *British Birds*, 22(12): 334–372.

Nicolai, J. 1964. Der Brutparasitismus der Viduinae als ethologisches Problem: Prägungsphänomene als Faktoren der Rassen- und Artbildung. *Zeitschrift für Tierpsychologie*, 21(2): 129–204.

———— 1969. Beobachtungen an Paradieswitwen (*Steganura paradisaea* L., *Steganura obtusa* Chapin) und der Strohwitwe (*Tetraenura fischeri* Reichenow) in Ostafrika. *Journal für Ornithologie*, 110(4): 421–447.

Nielsen, H. T. 1964. Swarming and some other habits of *Mansonia perturbans* and *Psorophora ferox* (Diptera: Culicidae). *Behaviour*, 24(1,2): 67–89.

Niemitz, C., and A. Krampe. 1972. Untersuchungen zum Orientierungsverhalten der Larven von *Necrophorus vespillo* F. (Silphidae Coleoptera). *Zeitschrift für Tierpsychologie*, 30(5): 456–463.

Nietzsche, F. 1956. *The birth of tragedy* and *The genealogy of morals: an attack*, trans. by Francis Golffing. Anchor Books, Doubleday, Garden City, N.Y. xii + 299 pp.

Nisbet, I. C. T. 1973. Courtship-feeding, egg-size and breeding success in common terns. *Nature, London*, 241(5385): 141–142.

Nishida, T. 1966. A sociological study of solitary male monkeys. *Primates*, 7(2): 141–204.

———— 1968. The social group of wild chimpanzees in the Mahali Mountains. *Primates*, 9(2): 167–227.

———— 1970. Social behavior and relationship among wild chimpanzees of the Mahali Mountains. *Primates*, 11(1): 47–87.

Nishida, T., and K. Kawanaka. 1972. Inter-unit-group relationships among wild chimpanzees of the Mahali Mountains. *Kyoto University African Studies*, 7: 131–169.

Nishiwaki, M. 1972. General biology. In S. H. Ridgway, ed. (*q.v.*), *Mammals of the sea: biology and medicine*, pp. 3–204.

Nissen, H. W. 1931. A field study of the chimpanzee: observations of chimpanzee behavior and environment in western French Guinea. *Comparative Psychology Monographs*, 8(1). vi + 122 pp.

Nixon, H. L., and C. R. Ribbands. 1952. Food transmission within the honeybee community. *Proceedings of the Royal Society*, ser. B, 140: 43–50.

Noble, G. A. 1962. Stress and parasitism: II, effect of crowding and fighting among ground squirrels on their coccidia and trichomonads. *Experimental Parasitology*, 12(5): 368–371.

Noble, G. K. 1931. *The biology of the Amphibia*. McGraw-Hill Book Co., New York. xiii + 577 pp.

———— 1939. The role of dominance in the social life of birds. *Auk*, 56(3): 263–273.

Nogueira-Neto, P. 1950. Notas bionomicas sobre Meliponíneos (Hymenoptera, Apoidea): IV, colonias mistas e questões relacionadas. *Revista de Entomologia, Rio de Janeiro*, 21(1,2): 305–367.

———— 1970a. *A criação de abelhas indígenas sem ferrão* (*Meliponinae*). Editora Chácaras e Quintais, São Paulo. 365 pp.

———— 1970b. Behavior problems related to the pillages made by some parasitic stingless bees (Meliponinae, Apidae). In L. R. Aronson et al., eds. (*q.v.*), *Development and evolution of behavior: essays in memory of T. C. Schneirla*, pp. 416–434.

Noirot, C. 1958–59. Remarques sur l'écologie des termites. *Annales de la Société Royale Zoologique de Belgique*, 89(1): 151–169.

———— 1969a. Glands and secretions. In K. Krishna and Frances M. Weesner, eds. (*q.v.*), *Biology of termites*, vol. 1, pp. 89–123.

———— 1969b. Formation of castes in the higher termites. In K. Krishna and Frances M. Weesner, eds. (*q.v.*), *Biology of termites*, vol. 1, pp. 311–350.

Noirot, Elaine. 1972. The onset of maternal behavior in rats, hamsters, and mice: a selective review. *Advances in the Study of Behavior*, 4: 107–145.

Nolte, D. J., I. Dési, and Beryl Meyers. 1969. Genetic and environmental factors affecting chiasma formation in locusts. *Chromosoma, Berlin*, 27(2): 145–155.

Nolte, D. J., S. H. Eggers, and I. R. May. 1973. A locust pheromone: locustol. *Journal of Insect Physiology*, 19(8): 1547–1554.

Nolte, D. J., I. R. May, and B. M. Thomas. 1970. The gregarisation pheromone of locusts. *Chromosoma, Berlin*, 29(4): 462–473.

Nordeng, H. 1971. Is the local orientation of anadromous fishes determined by pheromones? *Nature, London*, 233(5319): 411–413.

Nørgaard, E. 1956. Environment and behaviour of *Theridion saxatile*. *Oikos* (Acta Oecologica Scandinavica), 7(2): 159–192.

Norris, K. S., ed. 1966. *Whales, dolphins, and porpoises*. University of California Press, Berkeley. xvi + 789 pp.

———— 1967. Aggressive behavior in Cetacea. In Carmine D. Clemente and D. B. Lindsley, eds. (*q.v.*), *Brain function*, vol. 5, *Aggression and defense, neural mechanisms and social patterns*, pp. 225–241.

Norris, K. S., and J. H. Prescott. 1961. Observations on Pacific cetaceans of Californian and Mexican waters. *University of California Publications in Zoology*, 63(4): 291–402.

Norris, Maud J. 1968. Some group effects on reproduction in locusts. In R. Chauvin and C. Noirot, eds. (*q.v.*), *L'effet de groupe chez les animaux*, pp. 147–161.

Northrop, F. S. C. 1959. *The logic of the sciences and the humanities*. Meridian Books, New York. xiv + 402 pp.

Norton-Griffiths, M. N. 1969. The organisation, control and development of parental feeding in the oystercatcher (*Haematopus ostralegus*). *Behaviour*, 34(2): 55–114.

Nottebohm, F. 1967. The role of sensory feedback in the development of avian vocalizations. *Proceedings of the Fourteenth International Ornithological Congress, Oxford, 1966*, pp. 265–280.

———— 1970. Ontogeny of bird song. *Science*, 167: 950–956.

Novick, A. 1969. *The world of bats*. Holt, Rinehart and Winston, New York. 171 pp.

Nutting, W. L. 1969. Flight and colony foundation. In K. Krishna and Frances M. Weesner, eds. (*q.v.*), *Biology of termites*, vol. 1, pp. 233–282.

O'Connell, C. P. 1960. Use of fish schools for conditioned response experiments. *Animal Behaviour*, 8(3,4): 225–227.

O'Donald, P. 1972. Sexual selections by variations in fitness at breeding time. *Nature, London*, 237(5354): 349–351.

O'Farrell, T. P. 1965. Home range and ecology of snowshoe hares in interior Alaska. *Journal of Mammalogy*, 46(3): 406–418.

Ogburn, W. F., and M. Nimkoff. 1958. *Sociology*, 3d ed. Houghton Mifflin Co., Boston. x + 756 pp.

Ohba, S. 1967. Chromosomal polymorphism and capacity for increase under near optimal conditions. *Heredity*, 22(2): 169–185.

Okano, T., C. Asami, Y. Haruki, M. Sasaki, N. Itoigawa, S. Shinohara, and T. Tsuzuki. 1973. Social relations in a chimpanzee colony. In C. R. Carpenter, ed. (*q.v.*), *Behavioral regulators of behavior in primates*, pp. 85–105.

Økland, F. 1934. Utvandring og overvintring hos den røde skogmaur (*Formica rufa* L.). *Norsk Entomologisk Tidsskrift*, 3(5):316–327.

Oliver, J. A. 1956. Reproduction in the king cobra, *Ophiophagus hannah* Cantor. *Zoologica, New York*, 41(4): 145–152.

Oppenheimer, J. R. 1968. Behavior and ecology of the white-faced monkey, *Cebus capucinus*, on Barro Colorado Island, C.Z. Ph.D. thesis, University of Illinois, Urbana. viii + 181 pp.

———— 1973. Social and communicative behavior in the *Cebus* monkey. In C. R. Carpenter, ed. (*q.v.*), *Behavioral regulators of behavior in primates*, pp. 251–271.

Ordway, Ellen. 1965. Caste differentiation in *Augochlorella* (Hymenoptera, Halictidae). *Insectes Sociaux*, 12(4): 291–308.

———— 1966. The bionomics of *Augochlorella striata* and *A. persimilis* in eastern Kansas (Hymenoptera: Halictidae). *Journal of the Kansas Entomological Society*, 39(2): 270–313.

Orians, G. H. 1961a. Social stimulation within blackbird colonies. *Condor*, 63(4): 330–337.

———— 1961b. The ecology of blackbird (*Agelaius*) social systems. *Ecological Monographs*, 31(3): 285–312.

———— 1969. On the evolution of mating systems in birds and mammals. *American Naturalist*, 103(934): 589–603.

Orians, G. H., and G. M. Christman. 1968. A comparative study of the behavior of red-winged, tricolored, and yellow-headed blackbirds. *University of California Publications in Zoology*, 84. 81 pp.

Orians, G. H., and G. Collier. 1963. Competition and blackbird social systems. *Evolution*, 17(4): 449–459.

Orians, G. H., and Mary F. Willson. 1964. Interspecific territories of birds. *Ecology*, 45(4): 736–745.

Orr, R. T. 1967. The Galapagos sea lion. *Journal of Mammalogy*, 48(1): 62–69.

Ostrom, J. H. 1972. Were some dinosaurs gregarious? *Palaeogeography, Palaeoclimatology, Palaeoecology*, 11: 287–301.

Otte, D. 1970. *A comparative study of communicative behavior in grasshoppers*. Miscellaneous Publications, Museum of Zoology, University of Michigan, Ann Arbor, 141. 168 pp.

———— 1972. Simple versus elaborate behavior in grasshoppers: an analysis of communication in the genus *Syrbula*. *Behaviour*, 42(3,4): 291–322.

Otto, D. 1958. Über die Arbeitsteilung im Staate von *Formica rufa rufopratensis minor* Gössw. und ihre verhaltensphysiologischen Grundlagen, ein Beitrag zur Biologie der Roten Waldameise. *Wissenschaftliche Abhandlungen der Deutschen Akademie der Landwirtschaftswissenschaften zu Berlin*, 30: 1–169.

Owen, D. F. 1963. Similar polymorphisms in an insect and a land snail. *Nature, London*, 198(4876): 201–203.

Owen-Smith, R. N. 1971. Territoriality in the white rhinoceros (*Ceratotherium simum*) Burchell. *Nature, London*, 231(5301): 294–296.

———— 1974. The social system of the white rhinoceros. In V. Geist and F. Walther, eds. (*q.v.*), *The behaviour of ungulates and its relation to management*, vol. 1, pp. 341–351.

Packard, R. L. 1967. Octodontoid, bathyergoid, and ctenodactyloid rodents. In S. Anderson and J. K. Jones, Jr., eds. (*q.v.*), *Recent mammals of the world: a synopsis of families*, pp. 273–290.

———— 1968. An ecological study of the fulvous harvest mouse in eastern Texas. *American Midland Naturalist*, 79(1): 68–88.

Packer, W. C. 1969. Observations on the behavior of the marsupial *Setonix brachyurus* (Quoy and Gaimard) in an enclosure. *Journal of Mammalogy*, 50(1): 8–20.

Pagès, Elisabeth. 1965. Notes sur les pangolins du Gabon. *Biologia Gabonica*, 1(3): 209–238.

———— 1970. Sur l'écologie et le adaptation de l'orcyterope et des pangolins sympatriques du Gabon. *Biologia Gabonica*, 6(1): 27–92.

———— 1972a. Comportamente agressif et sexuel chez les pangolins arboricoles (*Manis tricuspis* et *M. longicaudata*). *Biologia Gabonica*, 8(1): 3–62.

———— 1972b. Comportamente maternale et developement du jeune chez un pangolin arboricole (*M. tricuspis*). *Biologia Gabonica*, 8(1): 63–120.

Paine, R. T. 1966. Food web complexity and species diversity. *American Naturalist*, 100(910): 65–75.

Pardi, L. 1940. Ricerche sui Polistini: I, poliginia vera ed apparente in *Polistes gallicus* (L.). *Processi Verbali della Società Toscana di Scienze Naturali in Pisa*, 49: 3–9.

———— 1948. Dominance order in *Polistes* wasps. *Physiological Zoology*, 21(1):1–13.

Pardi, L., and M. T. M. Piccioli. 1970. Studi sulla biologia di *Belonogaster* (Hymenoptera, Vespidae): 2, differenziamento castale incipiente in *B. griseus* (Fab.). *Monitore Zoologico Italiana*, n.s., supplement 3, pp. 235–265.

Parker, G. A. 1970a. Sperm competition and its evolutionary consequences in the insects. *Biological Reviews, Cambridge Philosophical Society*, 45: 525–568.

———— 1970b. The reproductive behaviour and the nature of sexual selection in *Scatophaga stercoraria* L. (Diptera: Scatophagidae): IV, epigamic competition and competition between males for the possession of females. *Behaviour*, 37(1,2): 113–139.

Parr, A. E. 1927. A contribution to the theoretical analysis of the schooling behaviour of fishes. *Occasional Papers of the Bingham Oceanographic Collection*, 1: 1–32.

Parsons, P. A. 1967. *The genetic analysis of behaviour*. Methuen, London. x + 174 pp.

Passera, L. 1968. Observations biologiques sur la fourmi *Plagiolepis grassei* Le Masne Passera parasite social de *Plagiolepis pygmaea* Latr. (Hym. Formicidae). *Insectes Sociaux*, 15(4): 327–336.

Pastan, I. 1972. Cyclic AMP. *Scientific American*, 227(2) (August): 97–105.

Patterson, I. J. 1965. Timing and spacing of broods in the black-headed gull *Larus ridibundus*. *Ibis*, 107(4): 433–459.

Patterson, O. 1967. *The sociology of slavery: an analysis of the origins, development and structure of Negro slave society in Jamaica*. Fairleigh Dickinson University Press, Cranbury, N.J. 310 pp.

Patterson, R. G. 1971. Vocalization in the desert tortoise, *Gopherus agassizi*. M.A. thesis, California State University, Fullerton. [Cited by B. H. Brattstrom, 1974 (q.v.).]

Pavlov, I. P. 1928. *Lectures on conditioned reflexes*. International Publishers, New York. 414 pp.

Payne, R. S., and S. McVay. 1971. Songs of humpback whales. *Science*, 173: 585–597.

Peacock, A. D., and A. T. Baxter. 1950. Studies in Pharaoh's ant, *Monomorium pharaonis* (L.): 3, life history. *Entomologist's Monthly Magazine*, 86: 171–178.

Peacock, A. D., I. C. Smith, D. W. Hall, and A. T. Baxter. 1954. Studies in Pharaoh's ant, *Monomorium pharaonis* (L): 8, male production by parthenogenesis. *Entomologist's Monthly Magazine*, 90: 154–158.

Pearson, O. P. 1948. Life history of mountain viscachas in Peru. *Journal of Mammalogy*, 29(4): 345–374.

—— 1966. The prey of carnivores during one cycle of mouse abundance. *Journal of Animal Ecology*, 35(1): 217–233.

—— 1971. Additional measurements of the impact of carnivores on California voles (*Microtus californicus*). *Journal of Mammalogy*, 52(1): 41–49.

Peek, F. W. 1971. Seasonal change in the breeding behavior of the male red-winged blackbird. *Wilson Bulletin*, 83(4): 383–395.

Peek, J. M., R. E. LeResche, and D. R. Stevens. 1974. Dynamics of moose aggregations in Alaska, Minnesota, and Montana. *Journal of Mammalogy*, 55(1): 126–137.

Pérez, J. 1899. *Les abeilles*. Librairie Hachette et Cie, Paris. viii + 348 pp.

Perry, R. 1966. *The world of the polar bear*. University of Washington Press, Seattle. xi + 195 pp.

—— 1967. *The world of the wolves*. Cassell, London. xi + 162 pp.

—— 1969. *The world of the giant panda*. Taplinger, New York. ix + 136 pp.

Peters, D. S. 1973. *Crossocerus dimidiatus* (Fabricius, 1781), eine weitere soziale Crabroninen-art. *Insectes Sociaux*, 20(2): 103–108.

Peterson, R. L. 1955. *North American moose*. University of Toronto Press, Toronto. xi + 280 pp.

Peterson, R. S. 1968. Social behavior in pinnipeds. In R. J. Harrison, ed., *The behavior and physiology of pinnipeds*, pp. 3–53. Appleton-Century-Crofts, New York. 411 pp.

Peterson, R. S., and G. A. Bartholomew. 1967. *The natural history and behavior of the California sea lion*. Special Publication no. 1. American Society of Mammalogists, Stillwater, Okla. xii + 79 pp.

Petit, Claudine. 1958. Le déterminisme génétique et psycho-physiologique de la compétition sexuelle chez *Drosophila melanogaster*. *Bulletin Biologique de la France et de la Belgique*, 92(3): 248–329.

Petit, Claudine, and Lee Ehrman. 1969. Sexual selection in *Drosophila*. *Evolutionary Biology*, 3:177–223.

Petter, F. 1961. Répartition géographique et écologie des rongeurs désertiques (du Sahara occidental à l'Iran oriental). *Mammalia*, 25 (special number): 1–219.

Petter, J.-J. 1962a. Recherches sur l'écologie et l'éthologie des lémuriens malgaches. *Mémoires du Muséum National d'Histoire Naturelle, Paris*, ser. A (Zoology), 27(1): 1–146.

—— 1962b. Ecological and behavioral studies of Madagascar lemurs in the field. *Annals of the New York Academy of Sciences*, 102(2): 267–281.

—— 1970. "Domaine vital" et "territoire" chez les lémuriens malgaches. In G. Richard, ed. (q.v.), *Territoire et domaine vital*, pp. 107–114.

Petter, J.-J., and C. M. Hladik. 1970. Observations sur le domaine vital et la densité de population de *Loris tardigradus* dans les forêts de Ceylan. *Mammalia*, 34(3): 394–409.

Petter, J.-J., and Arlette Petter. 1967. The aye-aye of Madagascar. In S. A. Altmann, ed. (q.v.), *Social communication among primates*, pp. 195–205.

Petter, J.-J., and A. Peyrieras. 1970. Nouvelle contribution à l'étude d'un lémurien malgache, le aye-aye (*Daubentonia madagascariensis* E. Geoffroy). *Mammalia*, 34(2): 167–193.

Petter, J.-J., A. Schilling, and G. Pariente. 1971. Observations éco-éthologiques sur deux lémuriens malgaches nocturnes: *Phaner furcifer* et *Microcebus coquereli*. *La Terre et la Vie*, 118(3): 287–327.

Petter-Rousseaux, Arlette. 1962. Recherches sur la biologie de la reproduction des primates inférieurs. *Mammalia*, 26, supplement 1. 88 pp.

Pfeffer, P. 1967. Le mouflon de Corse (*Ovis ammon musimom* Schreber 1782); position systématique, écologie et éthologie comparées. *Mammalia*, 31, supplement. 262 pp.

Pfeffer, P., and H. Genest. 1969. Biologie comparée d'une population de mouflons de Corse (*Ovis ammon musimon*) du Parc Naturel du Caroux. *Mammalia*, 33(2): 165–192.

Pfeiffer, J. E. 1969. *The emergence of man*. Harper & Row, New York. xxiv + 477 pp.

Pfeiffer, W. 1962. The fright reaction of fish. *Biological Reviews, Cambridge Philosophical Society*, 37(4): 495–511.

Phillips, P. J. 1973. Evolution of holopelagic Cnidaria: colonial and non-colonial strategies. In R. S. Boardman, A. H. Cheetham, and W. A. Oliver, Jr., eds. (*q.v.*), *Animal colonies: development and function through time*, pp. 107–118.

Pianka, E. R. 1970. On *r*- and *K*-selection. *American Naturalist*, 104(940): 592–597.

Piccioli, M. T. M., and L. Pardi. 1970. Studi della biologia di *Belonogaster* (Hymenoptera, Vespidae): 1, sull'etogramma di *Belonogaster griseus* (Fab.). *Monitore Zoologico Italiana*, n.s., supplement 3, pp. 197–225.

Pickles, W. 1940. Fluctuations in the populations, weights and biomasses of ants at Thornhill, Yorkshire, from 1935 to 1939. *Transactions of the Royal Entomological Society of London*, 90(17): 467–485.

Pielou, E. C. 1969. *An introduction to mathematical ecology*. Wiley-Interscience, New York. viii + 286 pp.

Pilbeam, D. 1972. *The ascent of man: an introduction to human evolution*. Macmillan Co., New York. x + 207 pp.

Pilleri, G., and J. Knuckey. 1969. Behaviour patterns of some Delphinidae observed in the western Mediterranean. *Zeitschrift für Tierpsychologie*, 26(1): 48–72.

Pilters, Hilde. 1954. Untersuchungen über angeborene Verhaltensweisen bei Tylopoden, unter besonderer Berücksichtigung der neuweltlichen Formen. *Zeitschrift für Tierpsychologie*, 11(2): 213–303.

Pisarski, B. 1966. Etudes sur les fourmis du genre *Strongylognathus* Mayr (Hymenoptera, Formicidae). *Annales Zoologici, Warsaw*, 23(22): 509–523.

Pitcher, T. J. 1973. The three-dimensional structure of schools in the minnow, *Phoxinus phoxinus* (L.). *Animal Behaviour*, 21(4): 673–686.

Pitelka, F. A. 1942. Territoriality and related problems in North American hummingbirds. *Condor*, 44(5): 189–204.

———— 1957. Some aspects of population structure in the short-term cycle of the brown lemming in northern Alaska. *Cold Spring Harbor Symposia on Quantitative Biology*, 22: 237–251.

———— 1959. Numbers, breeding schedule, and territoriality in pectoral sandpipers of northern Alaska. *Condor*, 61(4): 233–264.

Plateaux-Quénu, Cécile. 1961. Les sexués de remplacement chez les insectes sociaux. *Année Biologique*, 37(5,6): 177–216.

———— 1972. *La biologie des abeilles primitives*. Les grand problèmes de la biologie, no. 11. Masson, Paris. 200 pp.

———— 1973. Construction et évolution annuelle du nid d'*Evylaeus calceatus* Scopoli (Hym., Halictinae) avec quelques considérations sur la division du travail dans les sociétés monogynes et digynes. *Insectes Sociaux*, 20(3): 297–320.

Plath, O. E. 1922. Notes on *Psithyrus*, with records of two new American hosts. *Biological Bulletin, Marine Biological Laboratory, Woods Hole*, 43(1): 23–44.

———— 1934. *Bumblebees and their ways*. Macmillan Co., New York. xvi + 201 pp.

Platt, J. R. 1964. Strong inference. *Science*, 146: 347–353.

Plempel, M. 1963. Die chemischen Grundlagen der Sexualreaktion bei Zygomyceten. *Planta*, 59: 492–508.

Ploog, D. W. 1967. The behavior of squirrel monkeys (*Saimiri sciureus*) as revealed by sociometry, bioacoustics, and brain stimulation. In S. A. Altmann, ed. (*q.v.*), *Social communication among primates*, pp. 149–184.

Poelker, R. J., and H. D. Hartwell. 1973. Black bear of Washington. *Biological Bulletin*, Washington State Game Department, 14: 1–180.

Poglayen-Neuwall, I. 1962. Beiträge zu einem Ethogramm des Wickelbären (*Potos flavus* Schreber). *Zeitschrift für Säugetierkunde, Berlin*, 27(1): 1–44.

———— 1966. On the marking behavior of the kinkajou (*Potos flavus* Schreber). *Zoologica, New York*, 51(4): 137–142.

Poirier, F. E. 1968. The Nilgiri langur (*Presbytis johnii*) mother-infant dyad. *Primates*, 9(1,2): 45–68.

———— 1969a. Behavioral flexibility and intergroup variation among Nilgiri langurs (*Presbytis johnii*) of South India. *Folia Primatologica*, 11(1,2): 119–133.

———— 1969b. The Nilgiri langur (*Presbytis johnii*) troop: its composition, structure, function, and change. *Folia Primatologica*, 10(1,2): 20–47.

———— 1970a. The Nilgiri langur (*Presbytis johnii*) of South India. In L. A. Rosenblum, ed. (*q.v.*), *Primate behavior: developments in field and laboratory reserach*, vol. 1, pp. 251–383.

———— 1970b. Dominance structure of the Nilgiri langur (*Presbytis johnii*) of South India. *Folia Primatologica*, 12(3): 161–186.

———— ed. 1972a. *Primate socialization*. Random House, New York. x + 260 pp.

———— 1972b. Introduction. In F. E. Poirier, ed. (*q.v.*), *Primate socialization*, pp. 3–28.

Pontin, A. J. 1961. Population stabilization and competition between the ants *Lasius flavus* (F.) and *L. niger* (L.). *Journal of Animal Ecology*, 30(1): 47–54.

———— 1963. Further considerations of competition and the ecology of the ants *Lasius flavus* (F.) and *L. niger* (L.). *Journal of Animal Ecology*, 32(3): 565–574.

Poole, T. B. 1966. Aggressive play in polecats. *Symposia of the Zoological Society of London*, 18: 23–44.

Porter, W. P., and D. M. Gates. 1969. Thermodynamic equilibria of animals with environment. *Ecological Monographs*, 39(3): 227–244.

Porter, W. P., J. W. Mitchell, W. A. Beckman, and C. B. DeWitt. 1973. Behavioral implications of mechanistic ecology: thermal and behavioral modeling of desert ectotherms and their microenvironment. *Oecologia, Berlin*, 13(1): 1–54.

Powell, G. C., and R. B. Nickerson. 1965. Aggregations among juvenile king crabs (*Paralithodes camtschatica*, Tilesius), Kodiak, Alaska. *Animal Behaviour*, 13(2,3): 374–380.

Priesner, E. 1968. Die interspezifischen Wirkungen der Sexuallockstoffe der Saturniidae (Lepidoptera). *Zeitschrift für Vergleichende Physiologie*, 61(3): 263–297.

Pringle, J. W. S. 1951. On the parallel between learning and evolution. *Behaviour*, 3(3): 174–215.

Prior, R. 1968. *The roe deer of Cranborne Chase: an ecological survey.* Oxford University Press, Oxford. xvi + 222 pp.

Prokopy, R. J. 1972. Evidence for a marking pheromone deterring repeated oviposition in apple maggot flies. *Environmental Entomology*, 1(3): 326–332.

Pukowski, Erna. 1933. Ökologische Untersuchungen an *Necrophorus* F. *Zeitschrift für Morphologie und Ökologie der Tiere*, 27(3):518–586.

Pulliainen, E. 1965. Studies on the wolf (*Canis lupus* L.) in Finland. *Annales Zoologici Fennici, Helsinki*, 2(4):215–259.

Pulliam, R., B. Gilbert, P. Klopfer, D. McDonald, Linda McDonald, and G. Millikan. 1972. On the evolution of sociality, with particular reference to *Tiaris olivacea*. *Wilson Bulletin*, 84(1): 77–89.

Quastler, H. 1958. A primer on information theory. In H. P. Yockey, R. L. Platzman, and H. Quastler, eds. (q.v.), *Symposium on information theory in biology*, pp. 3–49.

Quilliam, T. A., ed. 1966. The mole: its adaptation to an underground environment. *Journal of Zoology, London*, 149(1): 31–114.

Quimby, D. C. 1951. The life history and ecology of the jumping mouse, *Zapus hudsonius*. *Ecological Monographs*, 21(1): 61–95.

Rabb, G. B., and Mary S. Rabb. 1963. On the behavior and breeding biology of the African pipid frog *Hymenochirus boettigeri*. *Zeitschrift für Tierpsychologie*, 20(2): 215–241.

Rabb, G. B., J. H. Woolpy, and B. E. Ginsburg. 1967. Social relationships in a group of captive wolves. *American Zoologist*, 7(2): 305–311.

Radakov, D. V. 1973. *Schooling in the ecology of fish*, trans. by H. Mills. Halsted Press, Wiley, New York. viii + 173 pp.

Rahm, U. 1961. Verhalten der Schuppentiere (Pholidota). *Handbuch der Zoologie*, 8(10): 32–48.

———— 1969. Notes sur le cri du *Dendrohyrax dorsalis* (Hyracoidea). *Mammalia*, 33(1): 68–79.

Raignier, A. 1972. Sur l'origine des nouvelle sociétés des fourmis voyageuses africaines (Hyménoptères Formicidae, Dorylinae). *Insectes Sociaux*, 19(3): 153–170.

Raignier, A., and J. Van Boven. 1955. Etude taxonomique, biologique et biométrique des *Dorylus* du sous-genre *Anomma* (Hymenoptera Formicidae). *Annales du Musée Royal du Congo Belge, Tervuren* (Belgium), n.s. 4 (Sciences Zoologiques) 2: 1–359.

Ralls, Katherine. 1971. Mammalian scent marking. *Science*, 171: 443–449.

Rand, A. L. 1941. Development and enemy recognition of the curve-billed thrasher *Toxostoma curvirostre*. *Bulletin of the American Museum of Natural History*, 78: 213–242.

———— 1953. Factors affecting feeding rates of anis. *Auk*, 70(1): 26–30.

———— 1954. Social feeding behavior of birds. *Fieldiana, Zoology* (Chicago): 36(1): 1–71.

Rand, A. S. 1967a. The adaptive significance of territoriality in iguanid lizards. In W. W. Milstead, ed. (q.v.), *Lizard ecology: a symposium*, pp. 106–115.

———— 1967b. Ecology and social organization in the iguanid lizard *Anolis lineatopus*. *Proceedings of the United States National Museum, Smithsonian Institution*, 122: 1–79.

Rand, A. S., and E. E. Williams. 1970. An estimation of redundancy and information content of anole dewlaps. *American Naturalist*, 104(935): 99–103.

Ransom, T. W. 1971. Ecology and social behavior of baboons (*Papio anubis*) at the Gombe National Park. Ph.D. thesis, University of California, Berkeley.

Ransom, T. W., and B. S. Ransom. 1971. Adult male-infant relations among baboons (*Papio anubis*). *Folia Primatologica*, 16(3,4): 179–195.

Ransom, T. W., and Thelma E. Rowell. 1972. Early social development of feral baboons. In F. E. Poirier, ed. (q.v.), *Primate socialization*, pp. 105–144.

Rappaport, R. A. 1971. The sacred in human evolution. *Annual Review of Ecology and Systematics*, 2:23–44.

Rasa, O. Anne E. 1973. Marking behaviour and its social significance in the African dwarf mongoose, *Helogale undulata rufula*. *Zeitschrift für Tierpsychologie*, 32(3): 293–318.

Rasmussen, D. I. 1964. Blood group polymorphism and inbreeding in natural populations of the deer mouse *Peromyscus maniculatus*. *Evolution*, 18(2): 219–229.

Ratcliffe, F. N., F. J. Gay, and T. Greaves. 1952. *Australian termites, the biology, recognition, and economic importance of the common species.* Commonwealth Scientific and Industrial Research Organization, Melbourne. 124 pp.

Rau, P. 1933. *The jungle bees and wasps of Barro Colorado Island (with notes on other insects).* Published by the author, Kirkwood, St. Louis County, Mo. 324 pp.

Rawls, J. 1971. *A theory of justice.* Belknap Press of Harvard University Press, Cambridge. xvi + 607 pp.

Ray, C., W. A. Watkins, and J. J. Burns. 1969. The underwater song of *Erignathus* (bearded seal). *Zoologica, New York*, 54(2): 79–83.

Regnier, F. E., and E. O. Wilson. 1968. The alarm-defence system of the ant *Acanthomyops claviger*. *Journal of Insect Physiology*, 14(7): 955–970.

———— 1969. The alarm-defence system of the ant *Lasius alienus*. *Journal of Insect Physiology*, 15(5): 893–898.

———— 1971. Chemical communication and "propaganda" in slave-maker ants. *Science*, 172: 267–269.

Reid, M. J., and J. W. Atz. 1958. Oral incubation in the cichlid fish *Geophagus jurupari* Heckel. *Zoologica, New York*, 43(5): 77–88.

Renner, M. 1960. Das Duftorgan der Honigbiene und die physiologische Bedeutung ihres Lockstoffes. *Zeitschrift für Vergleichende Physiologie*, 43(4): 411–468.

Renner, M., and Margot Baumann. 1964. Über Komplexe von subepidermalen Drüsenzellen (Duftdrüsen?) der Bienenkönigin. *Naturwissenschaften*, 51(3): 68–69.

Rensch, B. 1956. Increase of learning ability with increase of brain size. *American Naturalist*, 90(851): 81–95.

——— 1960. *Evolution above the species level.* Columbia University Press, New York. xvii + 419 pp.

Renský, M. 1966. The systematics of paralanguage. *Travaux linguistiques de Prague*, 2:97–102.

Ressler, R. H., R. B. Cialdini, M. L. Ghoca, and Suzanne M. Kleist. 1968. Alarm pheromone in the earthworm *Lumbricus terrestris*. *Science*, 161: 597–599.

Rettenmeyer, C. W. 1962. The behavior of millipeds found with neotropical army ants. *Journal of the Kansas Entomological Society*, 35(4): 377–384.

——— 1963a. The behavior of Thysanura found with army ants. *Annals of the Entomological Society of America*, 56(2): 170–174.

——— 1963b. Behavioral studies of army ants. *Kansas University Science Bulletin*, 44(9): 281–465.

Reynolds, H. C. 1952. Studies on reproduction in the opossum (*Didelphis virginiana virginiana*). *University of California Publications in Zoology*, 52(3): 223–284.

Reynolds, V. 1965. Some behavioural comparisons between the chimpanzee and the mountain gorilla in the wild. *American Anthropologist*, 67(3): 691–706.

——— 1966. Open groups in hominid evolution. *Man*, 1(4): 441–452.

——— 1968. Kinship and the family in monkeys, apes and man. *Man*, 3(2): 209–233.

Reynolds, V., and Frances Reynolds. 1965. Chimpanzees of the Budongo Forest. In I. DeVore, ed. (*q.v.*), *Primate behavior: field studies of monkeys and apes*, pp. 368–424.

Rheingold, Harriet L. 1963a. Maternal behavior in the dog. In Harriet Rheingold, ed. (*q.v.*), *Maternal behavior in mammals*, pp. 169–202.

——— ed. 1963b. *Maternal behavior in mammals.* John Wiley & Sons, New York. viii + 349 pp.

Rhijn, J. G. van. 1973. Behavioural dimorphism in male ruffs, *Philomachus pugnax* (L.). *Behaviour*, 47(3,4): 153–229.

Ribbands, C. R. 1953. *The behaviour and social life of honeybees.* Bee Research Association, London. 352 pp.

Rice, D. W. 1967. Cetaceans. In S. Anderson and J. K. Jones, Jr., eds. (*q.v.*), *Recent mammals of the world: a synopsis of families*, pp. 291–324.

Rice, D. W., and K. W. Kenyon. 1962. Breeding cycles and behavior of Laysan and black-footed albatrosses. *Auk*, 79(4): 517–567.

Richard, Alison. 1970. A comparative study of the activity patterns and behavior of *Alouatta villosa* and *Ateles geoffroyi*. *Folia Primatologica*, 12(4): 241–263.

Richard, G., ed. 1970. *Territoire et domaine vital.* Masson et Cie, Paris. viii + 125 pp.

Richards, Christina M. 1958. The inhibition of growth in crowded *Rana pipiens* tadpoles. *Physiological Zoology*, 31(2): 138–151.

Richards, K. W. 1973. Biology of *Bombus polaris* Curtis and *B. hyperboreus* Schönherr at Lake Hazen, Northwest Territories (Hymenoptera: Bombini). *Quaestiones Entomologicae*, 9: 115–157.

Richards, O. W. 1927a. The specific characters of the British humblebees (Hymenoptera). *Transactions of the Entomological Society of London*, 75(2): 233–268.

——— 1927b. Sexual selection and allied problems in the insects. *Biological Reviews, Cambridge Philosophical Society*, 2(4): 298–364.

——— 1965. Concluding remarks on the social organization of insect communities. *Symposia of the Zoological Society of London*, 14: 169–172.

——— 1969. The biology of some W. African social wasps (Hymenoptera: Vespidae, Polistinae). *Memorie Società Entomologica Italiana*, 48(1B): 79–93.

——— 1971. The biology of the social wasps (Hymenoptera, Vespidae). *Biological Reviews, Cambridge Philosophical Society*, 46(4): 483–528.

Richards, O. W., and Maud J. Richards. 1951. Observations on the social wasps of South America (Hymenoptera Vespidae). *Transactions of the Royal Entomological Society of London*, 102(1): 1–170.

Richardson, W. B. 1942. Ring-tailed cats (*Bassariscus astutus*): their growth and development. *Journal of Mammalogy*, 23(1): 17–26.

Richter-Dyn, Nira, and N. S. Goel. 1972. On the extinction of colonizing species. *Theoretical Population Biology*, 3(4): 406–433.

Ride, W. D. L. 1970. *A guide to the native mammals of Australia.* Oxford University Press, Oxford. xiv + 249 pp.

Ridgway, S. H., ed. 1972. *Mammals of the sea: biology and medicine.* C. C. Thomas, Springfield, Ill. xiv + 812 pp.

Ridpath, M. G. 1972. The Tasmanian native hen, *Tribonyx mortierii*, I–III. *Commonwealth Scientific and Industrial Research Organization, Wildlife Research, East Melbourne*, 17(1): 1–118.

Riemann, J. G., Donna J. Moen, and Barbara J. Thorson. 1967. Female monogamy and its control in houseflies. *Journal of Insect Physiology*, 13(3): 407–418.

Ripley, Suzanne. 1967. Intertroop encounters among Ceylon gray langurs (*Presbytis entellus*). In S. A. Altmann, ed. (*q.v.*), *Social communication among primates*, pp. 237–253.

——— 1970. Leaves and leaf-monkeys. In J. R. Napier and P. H. Napier, eds. (*q.v.*), *Old World monkeys: evolution, systematics, and behavior*, pp. 481–509.

Ripley, S. D. 1952. Territory and sexual behavior in the great Indian rhinoceros, a speculation. *Ecology*, 33(4): 570–573.

——— 1958. Comments on the black and square-lipped rhinoceros species in Africa. *Ecology*, 39(1): 172–174.

——— 1959. Competition between sunbird and honeyeater species in the Moluccan Islands. *American Naturalist*, 93(869): 127–132.

——— 1961. Aggressive neglect as a factor in interspecific competition in birds. *Auk*, 78(3): 366–371.

Roberts, Pamela. 1971. Social interactions of *Galago crassicaudatus*. *Folia Primatologica*, 14(3,4): 171–181.

Roberts, R. B., and C. H. Dodson. 1967. Nesting biology of two communal bees, *Euglossa imperialis* and *Euglossa ignita* (Hymenoptera: Apidae), including description of larvae. *Annals of the Entomological Society of America*, 60(5): 1007–1014.

Robertson, A., D. J. Drage, and M. H. Cohen. 1972. Control of aggregation in *Dictyostelium discoideum* by an external periodic pulse of cyclic adenosine monophosphate. *Science*, 175: 333–335.

Robertson, D. R. 1972. Social control of sex reversal in a coral-reef fish. *Science*, 177: 1007–1009.

Robins, C. R., C. Phillips, and Fanny Phillips. 1959. Some aspects of the behavior of the blennioid fish *Chaenopsis ocellata* Poey. *Zoologica, New York*, 44(2): 77–84.

Robinson, D. J., and I. McT. Cowan. 1954. An introduced population of the gray squirrel (*Sciurus carolinensis* Gmelin) in British Columbia. *Canadian Journal of Zoology*, 32(3): 261–282.

Rodman, P. S. 1973. Population composition and adaptive organisation among orang-utans of the Kutai Reserve. In R. P. Michael and J. H. Crook, eds. (*q.v.*), *Comparative ecology and behaviour of primates*, pp. 171–209.

Roe, Anne, and G. G. Simpson, eds. 1958. *Behavior and evolution.* Yale University Press, New Haven, Conn. vii + 557 pp.

Roe, F. G. 1970. *The North American buffalo: a critical study of the species in the wild state*, 2d ed. University of Toronto Press, Toronto. xi + 991 pp.

Roelofs, W. L., and A. Comeau. 1969. Sex pheromone specificity: taxonomic and evolutionary aspects in Lepidoptera. *Science*, 165: 398–400.

——— 1971. Sex attractants in Lepidoptera. *Proceedings of the Second International Congress of Pesticide Chemistry, IUPAC, Tel Aviv, Israel*, pp. 91–114.

Rogers, L. L. 1974. Movement patterns and social organization of black bears in Minnesota. Ph.D. thesis, University of Minnesota, Minneapolis.

Rood, J. P. 1970. Ecology and social behavior of the desert cavy (*Microcavia australis*). *American Midland Naturalist*, 83(2): 415–454.

Rood, J. P., and F. H. Test. 1968. Ecology of the spiny rat, *Heteromys anomalus*, at Rancho Grande, Venezuela. *American Midland Naturalist*, 79(1): 89–102.

Roonwal, M. L. 1970. Termites of the Oriental region. In K. Krishna and Frances M. Weesner, eds. (*q.v.*), *Biology of termites*, vol. 2, pp. 315–391.

Ropartz, P. 1966. Contribution à l'étude du déterminisme d'un effet de groupe chez les souris. *Comptes Rendus de l'Académie des Sciences, Paris*, 263: 2070–2072.

——— 1968. Olfaction et comportement social chez les rongeurs. *Mammalia*, 32(4): 550–569.

Rose, R. M., J. W. Holaday, and I. S. Bernstein. 1971. Plasma testosterone, dominance rank and aggressive behaviour in male rhesus monkeys. *Nature, London*, 231(5302): 366–368.

Rosen, M. W. 1959. *Water flow about a swimming fish.* Station Technical Publications, NOTS TP 2298. U.S. Naval Ordnance Test Station, China Lake, Calif. iv + 94 pp. [Cited by C. M. Breder, 1965 (*q.v.*).]

Rosen, M. W., and N. E. Cornford. 1971. Fluid friction of fish slimes. *Nature, London*, 234(5323): 49–51.

Rosenblatt, J. S. 1965. The basis of synchrony in the behavioral inter-

action between the mother and her offspring in the laboratory rat. In B. M. Foss, ed., *Determinants of infant behaviour*, vol. 3, pp. 3–45. Methuen, London. xiii + 264 pp.

——— 1972. Learning in newborn kittens. *Scientific American*, 227(6) (December): 18–25.

Rosenblatt, J. S., and D. S. Lehrman. 1963. Maternal behavior of the laboratory rat. In Harriet L. Rheingold, ed. (*q.v.*), *Maternal behavior in mammals*, pp. 8–57.

Rosenblum, L. A., ed. 1970. *Primate behavior: developments in field and laboratory research*, vol. 1. Academic Press, New York. xii + 400 pp.

——— 1971a. The ontogeny of mother-infant relations in macaques. In H. Moltz, ed. (*q.v.*), *The ontogeny of vertebrate behavior*, pp. 315–367.

——— ed. 1971b. *Primate behavior: developments in field and laboratory research*, vol. 2. Academic Press, New York. xi + 267 pp.

Rosenblum, L. A., and R. W. Cooper, eds. 1968. *The squirrel monkey.* Academic Press, New York. xii + 451 pp.

Rosenson, L. M. 1973. Group formation in the captive greater bushbaby (*Galago crassicaudatus crassicaudatus*). *Animal Behaviour*, 21(1): 67–77.

Rothballer, A. B. 1967. Aggression, defense and neurohumors. In Carmine D. Clemente and D. B. Lindsley, eds. (*q.v.*), *Brain function*, vol. 5, *Aggression and defense, neural mechanisms and social patterns*, pp. 135–170.

Roubaud, E. 1916. Recherches biologiques sur les guêpes solitaires et sociales d'Afrique: la genèse de la vie sociale et l'évolution de l'instinct maternel chez les vespides. *Annales des Sciences Naturelles*, 10th ser. (Zoologie), 1: 1–160.

Roughgarden, J. 1971. Density-dependent natural selection. *Ecology*, 52(3): 453–468.

——— 1974. Species packing and competition function with illustrations from coral reef fish. *Theoretical Population Biology*, 5(2): 163–186.

Rousseau, M. 1971. Un machairodonte dans l'art Aurignacien? *Mammalia*, 35(4): 648–657.

Rovner, J. S. 1968. Territoriality in the sheet-web spider *Linyphia triangularis* (Clerck) (Araneae, Linyphiidae). *Zeitschrift für Tierpsychologie*, 25(2): 232–242.

Rowell, Thelma E. 1963. Behaviour and female reproductive cycles of rhesus macaques. *Journal of Reproduction and Fertility*, 6: 193–203.

——— 1966a. Forest living baboons in Uganda. *Journal of Zoology, London*, 149(3): 344–364.

——— 1966b. Hierarchy in the organization of a captive baboon group. *Animal Behaviour*, 14(4): 430–443.

——— 1967. A quantitative comparison of the behaviour of a wild and a caged baboon troop. *Animal Behaviour*, 15(4): 499–509.

——— 1969a. Long-term changes in a population of Ugandan baboons. *Folia Primatologica*, 11(4): 241–254.

——— 1969b. Variability in the social organization of primates. In D. Morris, ed. (*q.v.*), *Primate ethology, essays on the socio-sexual behavior of apes and monkeys*, pp. 283–305.

——— 1970. Baboon menstrual cycles affected by social environment. *Journal of Reproduction and Fertility*, 21: 133–141.

——— 1971. Organization of caged groups of *Cercopithecus* monkeys. *Animal Behaviour*, 19(4): 625–645.

——— 1972. *Social behaviour of monkeys.* Penguin Books, Harmondsworth, Middlesex. 203 pp.

Rowell, Thelma E., N. A. Din, and A. Omar. 1968. The social development of baboons in their first three months. *Journal of Zoology, London*, 155(4): 461–483.

Rowell, Thelma E., R. A. Hinde, and Yvette Spencer-Booth. 1964. "Aunt"-infant interaction in captive rhesus monkeys. *Animal Behaviour*, 12(2,3): 219–226.

Rowley, I. 1965. The life history of the superb blue wren, *Malurus cyaneus. Emu*, 64(4): 251–297.

Ruelle, J. E. 1970. A revision of the termites of the genus *Macrotermes* from the Ethiopian Region (Isoptera: Termitidae). *Bulletin of the British Museum of Natural History, Entomology*, 24: 365–444.

Rumbaugh, D. M. 1970. Learning skills of anthropoids. In L. A. Rosenblum, ed. (q.v.), *Primate behavior: developments in field and laboratory research*, vol. 1, pp. 1–70.

Russell, Eleanor. 1970. Observations on the behaviour of the red kangaroo (*Megaleia rufa*) in captivity. *Zeitschrift für Tierpsychologie*, 27(4): 385–404.

Ryan, E. P. 1966. Pheromone: evidence in a decapod crustacean. *Science*, 151: 340–341.

Ryland, J. S. 1970. *Bryozoans.* Hutchinson University Library, London. 175 pp.

Saayman, G. S. 1971a. Behaviour of the adult males in a troop of free-ranging chacma baboons (*Papio ursinus*). *Folia Primatologica*, 15(1,2): 36–57.

——— 1971b. Grooming behaviour in a troop of free-ranging chacma baboons (*Papio ursinus*). *Folia Primatologica*, 16(3,4): 161–178.

Saayman, G. S., C. K. Tayler, and D. Bower. 1973. Diurnal activity cycles in captive and free-ranging Indian Ocean bottlenose dolphins (*Tursiops aduncus* Ehrenburg). *Behaviour*, 44(3,4): 212–233.

Sabater Pi, J. 1972. Contribution to the ecology of *Mandrillus sphinx* Linnaeus 1758 of Rio Muni (Republic of Equatorial Guinea). *Folia Primatologica*, 17(4): 304–319.

——— 1973. Contribution to the ecology of *Colobus polykomos satanas* (Waterhouse, 1838) of Rio Muni, Republic of Equatorial Guinea. *Folia Primatologica*, 19(2,3): 193–207.

Sackett, G. P. 1970. Unlearned responses, differential rearing experiences, and the development of social attachments by rhesus monkeys. In L. A. Rosenblum, ed. (q.v.), *Primate behavior: developments in field and laboratory research*, vol. 1, pp. 111–140.

Sade, D. S. 1965. Some aspects of parent-offspring and sibling relations in a group of rhesus monkeys, with a discussion of grooming. *American Journal of Physical Anthropology*, 23(1): 1–17.

——— 1967. Determinants of dominance in a group of free-ranging rhesus monkeys. In S. A. Altmann, ed. (q.v.), *Social communication among primates*, pp. 99–114.

Sadleir, R. M. F. S. 1965. The relationship between agonistic behaviour and population changes in the deermouse, *Peromyscus maniculatus* (Wagner). *Journal of Animal Ecology*, 34(2): 331–352.

Sahlins, M. D. 1959. The social life of monkeys, apes and primitive man. In J. N. Spuhler, ed., *The evolution of man's capacity for culture*, pp. 54–73. Wayne State University Press, Detroit, Mich. 79 pp.

Saint Girons, M.-C. 1967. Etude du genre *Apodemus* Kaup, 1829 en France (suite et fin). *Mammalia*, 31(1): 55–100.

Sakagami, S. F. 1954. Occurrence of an aggressive behaviour in queenless hives, with considerations on the social organization of honeybee. *Insectes Sociaux*, 1(4): 331–343.

——— 1960. Ethological peculiarities of the primitive social bees, *Allodape* Lepeletier and allied genera. *Insectes Sociaux*, 7(3): 231–249.

——— 1971. Ethosoziologischer Vergleich zwischen Honigbienen und stachellosen Bienen. *Zeitschrift für Tierpsychologie*, 28(4): 337–350.

Sakagami, S. F., and Y. Akahira. 1960. Studies on the Japanese honeybee, *Apis cerana cerana* Fabricius: 8, two opposing adaptations in the post-stinging behavior of honeybees. *Evolution*, 14(1): 29–40.

Sakagami, S. F., and K. Fukushima. 1957. *Vespa dybowskii* André as a facultative temporary social parasite. *Insectes Sociaux*, 4(1): 1–12.

Sakagami, S. F., and K. Hayashida. 1968. Bionomics and sociology of the summer matrifilial phase in the social halictine bee, *Lasioglossum duplex. Journal of the Faculty of Science, Hokkaido University*, 6th ser. (Zoology), 16(3): 413–513.

Sakagami, S. F., and S. Laroca. 1963. Additional observations on the habits of the cleptobiotic stingless bees, the genus *Lestrimelitta* Friese (Hymenoptera, Apoidea). *Journal of the Faculty of Science, Hokkaido University*, 6th ser. (Zoology), 15(2): 319–339.

Sakagami, S. F., and C. D. Michener. 1962. *The nest architecture of the sweat bees (Halictinae): a comparative study of behavior.* University of Kansas Press, Lawrence. 135 pp.

Sakagami, S. F., Maria J. Montenegro, and W. E. Kerr. 1965. Behavior studies of the stingless bees, with special reference to the oviposition process: 5, *Melipona quadrifasciata anthidioides* Lepeletier. *Journal of the Faculty of Science, Hokkaido University*, 6th ser. (Zoology), 15(4): 578–607.

Sakagami, S. F., and Y. Oniki. 1963. Behavior studies of the stingless bees, with special reference to the oviposition process: 1, *Melipona compressipes manaosensis* Schwarz. *Journal of the Faculty of Science, Hokkaido University*, 6th ser. (Zoology), 15(2): 300–318.

Sakagami, S. F., and K. Yoshikawa. 1968. A new ethospecies of *Stenogaster* wasps from Sarawak, with a comment on the value of ethological characters in animal taxonomy. *Annotationes Zoologicae Japonensis*, 41(2): 77–84.

Sakagami, S. F., and R. Zucchi. 1965. Winterverhalten einer neotropischen Hummel, *Bombus atratus*, innerhalb des Beobachtungskastens: ein Beitrag zur Biologie der Hummeln. *Journal of the Faculty of Science, Hokkaido University*, 6th ser. (Zoology), 15(4): 712–762.

Sale, P. F. 1972. Effect of cover on agonistic behavior of a reef fish: a possible spacing mechanism. *Ecology*, 53(4): 753–758.

Salt, G. 1936. Experimental studies in insect parasitism: 4, the effect of superparasitism on populations of *Trichogramma evanescens*. *Journal of Experimental Biology*, 13: 363–375.

Sanders, C. J. 1970. The distribution of carpenter ant colonies in the spruce-fir forests of northeastern Ontario. *Ecology*, 51(5): 865–873.

——— 1971. Sex pheromone specificity and taxonomy of budworm moths (Choristoneura). *Science*, 171: 911–913.

Sanders, C. J., and F. B. Knight. 1968. Natural regulation of the aphid *Pterocomma populifoliae* on bigtooth aspen in northern lower Michigan. *Ecology*, 49(2): 234–244.

Sands, W. A. 1957. The soldier mandibles of the Nasutitermitinae (Isoptera, Termitidae). *Insectes Sociaux*, 4(1): 13–24.

——— 1972. The soldierless termites of Africa (Isoptera: Termitidae). *Bulletin of the British Museum of Natural History, Entomology*, supplement 18. 244 pp.

Santschi, F. 1920. Fourmis du genre *Bothriomyrmex* Emery (systématique et moeurs). *Revue Zoologique Africaine*, 7(3): 201–224.

Sauer, E. G. F., and Eleonore M. Sauer. 1963. The South-West African bush-baby of the *Galago senegalensis* group. *Journal of the South West Africa Scientific Society*, 16: 5–35. [Synopsis in J. R. Napier and P. H. Napier, 1967 (*q.v.*).]

——— 1972. Zur Biologie der kurzohrigen Elefantenspitzmaus. *Zeitschrift des Kölner Zoo*, 15(4): 119–139.

Savage, T. S., and J. Wyman. 1843–1844. Observations on the external characters and habits of the Troglodytes Niger, Geoff. and on its organization. *Boston Journal of Natural History*, 4(3): 362–376; 4(4): 377–386.

Schaller, G. B. 1961. The orang-utan in Sarawak. *Zoologica, New York*, 46(2): 73–82.

——— 1963. *The mountain gorilla: ecology and behavior*. University of Chicago Press, Chicago. xviii + 431 pp.

——— 1965a. The behavior of the mountain gorilla. In I. DeVore, ed. (*q.v.*), *Primate behavior: field studies of monkeys and apes*, pp. 324–367.

——— 1965b. *The year of the gorilla*. Ballantine Books, New York. 285 pp.

——— 1967. *The deer and the tiger: a study of wildlife in India*. University of Chicago Press, Chicago. ix + 370 pp.

——— 1970. This gentle and elegant cat. *Natural History*, 79(6): 30–39.

——— 1972. *The Serengeti lion: a study of predator-prey relations*. University of Chicago Press, Chicago. xiii + 480 pp.

Schaller, G. B., and G. R. Lowther. 1969. The relevance of carnivore behavior to the study of early hominids. *Southwestern Journal of Anthropology*, 25(4): 307–341.

Scheffer, V. B. 1958. *Seals, sea lions, and walruses: a review of the Pinnipedia*. Stanford University Press, Stanford, Calif. x + 179 pp.

Schein, M. W., and M. H. Fohrman. 1955. Social dominance relationships in a herd of dairy cattle. *British Journal of Animal Behaviour*, 3(2): 45–55.

Schenkel, R. 1947. Ausdrucks-Studien an Wölfen. Gefangenschafts-Beobachtungen. *Behaviour*, 1(2): 81–129.

——— 1966a. Zum Problem der Territorialität und des Markierens bei Säugern—am Beispiel des Schwarzen Nashorns und des Löwens. *Zeitschrift für Tierpsychologie*, 23(5): 593–626.

——— 1966b. Play, exploration and territoriality in the wild lion. *Symposia of the Zoological Society of London*, 18: 11–22.

——— 1967. Submission: its features and function in the wolf and dog. *American Zoologist*, 7(2): 319–329.

Scherba, G. 1964. Species replacement as a factor affecting distribution of *Formica opaciventris* Emery (Hymenoptera: Formicidae). Journal of the New York Entomological Society, 72: 231–237.

Scheven, J. 1958. Beitrag zur Biologie der Schmarotzerfeldwespen *Sulcopolistes atrimandibularis* Zimm., *S. semenowi* F. Morawitz und *S. sulcifer* Zimm. *Insectes Sociaux*, 5(4): 409–437.

Schevill, W. E. 1964. Underwater sounds of cetaceans. In W. N. Tavolga, ed. (*q.v.*), *Marine bio-acoustics*, pp. 307–316.

Schevill, W. E., and W. A. Watkins. 1962. *Whale and porpoise voices: a phonograph record*. Contribution no. 1320. Woods Hole Oceanographic Institution, Woods Hole, Mass. 24 pp.

Schiller, P. H. 1952. Innate constituents of complex responses in primates. *Psychological Review*, 59(3): 177–191.

——— 1957. Innate motor action as a basis of learning. In Claire H. Schiller, trans. and ed., *Instinctive behavior: the development of a modern concept*, pp. 264–287. International Universities Press, New York. xix + 328 pp.

Schjelderup-Ebbe, T. 1922. Beiträge zur Sozialpsychologie des Haushuhns. *Zeitschrift für Psychologie*, 88(3–5): 225–252.

——— 1923. Weitere Beiträge zur Sozial- und Individualpsychologie des Haushuhns. *Zeitschrift für Psychologie*, 92(1,2): 60–87.

——— 1935. Social behavior of birds. In C. A. Murchison, ed., *A handbook of social psychology*, pp. 947–972. Clark University Press, Worcester, Mass. xii + 1195 pp.

Schloeth, R. 1961. Das Sozialleben des Camargue-Rindes. *Zeitschrift für Tierpsychologie*, 18(5): 575–627.

Schmid, B. 1939. Psychologische Beobachtungen und Versuche an einem jungen, männlichen Ameisenbären (*Myrmecophaga tridactylus* L.). *Zeitschrift für Tierpsychologie*, 2(2): 117–126.

Schneider, D. 1969. Insect olfaction: deciphering system for chemical messages. *Science*, 163: 1031–1037.

Schneirla, T. C. 1933. Studies on army ants in Panama. *Journal of Comparative Psychology*, 15(2): 267–299.

——— 1938. A theory of army-ant behavior based upon the analysis of activities in a representative species. *Journal of Comparative Psychology*, 25(1): 51–90.

——— 1940. Further studies on the army-ant behavior pattern. Mass-organization in the swarm-raiders. *Journal of Comparative Psychology*, 29(3): 401–460.

——— 1946. Problems in the biopsychology of social organization. *Journal of Abnormal and Social Psychology*, 41(4): 385–402.

——— 1956. A preliminary survey of colony division and related processes in two species of terrestrial army ants. *Insectes Sociaux*, 3(1): 49–69.

——— 1971. *Army ants: a study in social organization*, ed. by H. R. Topoff. W. H. Freeman, San Francisco. xxii + 349 pp.

Schneirla, T. C., and R. Z. Brown. 1952. Sexual broods and the production of young queens in two species of army ants. *Zoologica, New York*, 37(1): 5–32.

Schneirla, T. C., and G. Piel. 1948. The army ant. *Scientific American*, 178(6) (June): 16–23.

Schneirla, T. C., J. S. Rosenblatt, and Ethel Tobach. 1963. Maternal behavior in the cat. In Harriet L. Rheingold, ed. (*q.v.*), *Maternal behavior in mammals*, pp. 122–168.

Schoener, T. W. 1965. The evolution of bill size differences among sympatric congeneric species of birds. *Evolution*, 19(2): 189–213.

——— 1967. The ecological significance of sexual dimorphism in size in the lizard *Anolis conspersus*. *Science*, 155: 474–477.

——— 1968a. Sizes of feeding territories among birds. *Ecology*, 49(1): 123–141.

——— 1968b. The *Anolis* lizards of Bimini: resource partitioning in a complex fauna. *Ecology*, 49(4): 704–726.

——— 1971. Theory of feeding strategies. *Annual Review of Ecology and Systematics*, 2: 369–404.

——— 1973. Population growth regulated by intraspecific competition for energy or time: some simple representations. *Theoretical Population Biology*, 4(1): 56–84.

Schoener, T. W., and Amy Schoener. 1971a. Structural habitats of West Indian *Anolis* lizards: 1, lowland Jamaica. *Breviora*, 368. 53 pp.

——— 1971b. Structural habitats of West Indian *Anolis* lizards: 2, Puerto Rican uplands. *Breviora*, 375. 39 pp.

Schopf, T. J. M. 1973. Ergonomics of polymorphism: its relation to the colony as the unit of natural selection in species of the phylum Ectoprocta. In R. S. Boardman, A. H. Cheetham, and W. A. Oliver, Jr., eds. (*q.v.*), *Animal colonies: development and function through time*, pp. 247–294.

Schremmer, F. 1972. Beobachtungen zur Biologie von *Apoica pallida* (Olivier, 1791), einer neotropischen sozialen Faltenwespe (Hymmenoptera, Vespidae). *Insectes Sociaux*, 19(4): 343–357.

Schull, W. J., and J. V. Neel. 1965. *The effects of inbreeding on Japanese children*. Harper & Row, New York. xii + 419 pp.

Schultz, A. H. 1958. The occurrence and frequency of pathological and teratological conditions and of twinning among non-human primates. *Primatologia, Handbuch der Primatenkunde*, 1: 965–1014.

Schultze-Westrum, T. 1965. Innerartliche Verständigung durch Düfte beim Gleitbeutler *Petaurus breviceps papuanus* Thomas (Marsupialia, Phalangeridae). *Zeitschrift für Vergleichende Physiologie*, 50(2): 151–220.

Schusterman, R. J., and R. G. Dawson. 1968. Barking, dominance, and territoriality in male sea lions. *Science*, 160: 434–436.

Schwarz, H. F. 1948. Stingless bees (Meliponidae) of the western hemisphere. *Bulletin of the American Museum of Natural History*, 90. xvii + 546 pp.

Scott, J. F. 1971. *Internalization of norms: a sociological theory of moral commitment*. Prentice-Hall, Englewood Cliffs, N.J. xviii + 237 pp.

Scott, J. P. 1967. The evolution of social behavior in dogs and wolves. *American Zoologist*, 7(2): 373–381.

——— 1968. Evolution and domestication of the dog. *Evolutionary Biology*, 2: 243–275.

Scott, J. P., and E. Fredericson. 1951. The causes of fighting in mice and rats. *Physiological Zoology*, 24(4): 273–309.

Scott, J. P., and J. L. Fuller. 1965. *Genetics and the social behavior of the dog*. University of Chicago Press, Chicago. xviii + 468 pp.

Scott, J. W. 1942. Mating behavior of the sage grouse. *Auk*, 59(4): 477–498.

——— 1950. A study of the phylogenetic or comparative behavior of three species of grouse. *Annals of the New York Academy of Sciences*, 51(6): 1062–1073.

Scudo, F. M. 1967. The adaptive value of sexual dimorphism: 1, anisogamy. *Evolution*, 21(2): 285–291.

Seay, B. 1966. Maternal behavior in primiparous and multiparous rhesus monkeys. *Folia Primatologica*, 4(2): 146–168.

Sebeok, T. A. 1962. Coding in the evolution of signalling behavior. *Behavioral Science*, 7(4): 430–442.

——— 1963. Communication among social bees; porpoises and sonar; man and dolphin. *Language*, 39(3): 448–466.

——— 1965. Animal communication. *Science*, 147: 1006–1014.

——— ed. 1968. *Animal communication: techniques of study and results of research*. Indiana University Press, Bloomington. xviii + 686 pp.

Seemanova, Eva. 1972. (Quoted by *Time*, October 9, 1972, p. 58.)

Seitz, A. 1955. Untersuchungen über angeborene Verhaltensweisen bei Caniden: III, Beobachtungen an Marderhunden (*Nyctereutes procyonoides* Gray). *Zeitschrift für Tierpsychologie*, 12(3): 463–489.

Sekiguchi, K., and S. F. Sakagami. 1966. Structure of foraging population and related problems in the honeybee, with considerations on the division of labour in bee colonies. *Report of the Hokkaido National Agricultural Experiment Station* (Hitsujigaoka, Sapporo, Japan), no. 69. 65 pp.

Selander, R. K. 1965. On mating systems and sexual selection. *American Naturalist*, 99(906): 129–141.

——— 1966. Sexual dimorphism and differential niche utilization in birds. *Condor*, 68(2): 113–151.

——— 1972. Sexual selection and dimorphism in birds. In B. Campbell, ed. (*q.v.*), *Sexual selection and the descent of man, 1871–1971*, pp. 180–230.

Selous, E. 1927. *Realities of bird life*. Constable, London.

Selye, H. 1956. *The stress of life*. McGraw-Hill Book Co., New York. xviii + 324 pp.

Seton, E. T. 1909. *Life-histories of northern animals: an account of the mammals of Manitoba*, 2 vols. Charles Scribner's Sons, New York. Vol. 1: xxx + 673 pp.; vol. 2: xii + 590 pp.

Sexton, O. J. 1960. Some aspects of the behavior and of the territory of a dendrobatid frog, *Prostherapis trinitatis. Ecology,* 41(1): 107–115.

——— 1962. Apparent territorialism in *Leptodactylus insularum* Barbour. *Herpetologica,* 18(3): 212–214.

Shank, C. C. 1972. Some aspects of behaviour in a population of feral goats (*Capra hircus* L.). *Zeitschrift für Tierpsychologie,* 30(5): 488–528.

Shannon, C. E., and W. Weaver. 1949. *The mathematical theory of communication.* University of Illinois Press, Urbana. 117 pp.

Sharp, W. M., and Louise H. Sharp. 1956. Nocturnal movement and behavior of wild raccoons at a winter feeding station. *Journal of Mammalogy,* 37(2): 170–177.

Shaw, Evelyn. 1962. The schooling of fishes. *Scientific American,* 206(6) (June): 128–138.

——— 1970. Schooling in fishes: critique and review. In L. R. Aronson, Ethel Tobach, D. S. Lehrman, and J. S. Rosenblatt, eds. (*q.v.*), *Development and evolution of behavior: essays in memory of T. C. Schneirla,* pp. 452–480.

Shearer, D., and R. Boch. 1965. 2-Heptanone in the mandibular gland secretion of the honey-bee. *Nature, London,* 206(4983): 530.

Shepher, J. 1972. [A news report of his studies of marriage in Israeli kibbutzes in "Science and the citizen," *Scientific American,* 227(6) (December): 43.]

Shettleworth, Sara J. 1972. Constraints on learning. *Advances in the Study of Behavior,* 4: 1–68.

Shillito, Joy F. 1963. Field observations on the growth, reproduction and activity of a woodland population of the common shrew *Sorex araneus* L. *Proceedings of the Zoological Society of London,* 140(1): 99–114.

Shoemaker, H. H. 1939. Social hierarchy in flocks of the canary. *Auk,* 56(4): 381–406.

Shorey, H. H. 1970. Sex pheromones of Lepidoptera. In D. L. Wood, R. M. Silverstein, and M. Nakajima, eds. (*q.v.*), *Control of insect behavior by natural products,* pp. 249–284.

Short, L. 1961. Interspecies flocking of birds of montane forest in Oaxaca, Mexico. *Wilson Bulletin,* 73(4): 341–347.

Shuleikin, V. V. 1968. *Marine physics.* Nauka Publishing House, Moscow. [Cited by D. V. Radakov, 1973 (*q.v.*).]

Siegel, R. W., and L. W. Cohen. 1962. The intracellular differentiation of cilia. *American Zoologist,* 2(4): 558.

Sikes, Sylvia K. 1971. *The natural history of the African elephant.* Elsevier, New York. xxvi + 397 pp.

Silberglied, R. E., and O. R. Taylor. 1973. Ultraviolet differences between the sulfur butterflies, *Colias eurytheme* and *C. philodice,* and a possible isolating mechanism. *Nature, London,* 241(5389): 406–408.

Silén, L. 1942. Origin and development of the cheilo-ctenostomatous stem of Bryozoa. *Zoologiska Bidrag,* Uppsala, 22: 1–59.

——— 1975. Polymorphism. In R. M. Woollacott, ed., *The biology of bryozoans.* Academic Press, New York. (In press.)

Silverstein, R. M. 1970. Attractant pheromones of Coleoptera. In M. Beroza, ed. (*q.v.*), *Chemicals controlling insect behavior,* pp. 21–40.

Simberloff, D. S., and E. O. Wilson. 1969. Experimental zoogeography of islands: the colonization of empty islands. *Ecology,* 50(2): 278–296.

Simmons, J. A., E. G. Wever, and J. M. Pylka. 1971. Periodical cicada: sound production and hearing. *Science,* 171: 212–213.

Simmons, K. E. L. 1951. Interspecific territorialism. *Ibis,* 93(3): 407–413.

——— 1955. Studies on great crested grebes. *Avicultural Magazine,* 61(1): 3–13; 61(2): 93–102; 61(3): 131–146; 61(4): 181–201; 61(5): 235–253; 61(6): 294–316.

——— 1970. Ecological determinants of breeding adaptations and social behaviour in two fish-eating birds. In J. H. Crook, ed. (*q.v.*), *Social behaviour in birds and mammals,* pp. 37–77.

Simon, H. A. 1962. The architecture of complexity. *Proceedings of the American Philosophical Society,* 106(6): 467–482.

Simonds, P. E. 1965. The bonnet macaque in South India. In I. DeVore, ed. (*q.v.*), *Primate behavior: field studies of monkeys and apes,* pp. 175–196.

Simons, E. L., and P. C. Ettel. 1970. Gigantopithecus. *Scientific American,* 222(1) (January): 76–85.

Simpson, G. G. 1944. *Tempo and mode in evolution.* Columbia University Press, New York. xviii + 237 pp.

——— 1945. The principles of classification and a classification of mammals. *Bulletin of the American Museum of Natural History,* 85. xvi + 350 pp.

——— 1953. *The major features of evolution.* Columbia University Press, New York. xx + 434 pp.

——— 1961. *Principles of animal taxonomy.* Columbia University Press, New York. xii + 247 pp.

Simpson, T. L. 1973. Coloniality among the Porifera. In R. S. Boardman, A. H. Cheetham, and W. A. Oliver, Jr., eds. (*q.v.*), *Animal colonies: development and function through time,* pp. 549–565.

Sinclair, A. R. E. 1970. Studies of the ecology of the East African buffalo. Ph.D. thesis, Oxford University, Oxford. [Cited by H. Kruuk, 1972 (*q.v.*).]

Sipes, R. G. 1973. War, sports and aggression: an empirical test of two rival theories. *American Anthropologist,* 75(1): 64–86.

Skaife, S. H. 1953. Subsocial bees of the genus *Allodape* Lep. & Serv. *Journal of the Entomological Society of South Africa,* 16(1): 3–16.

——— 1954a. The black-mound termite of the Cape, *Amitermes atlanticus* Fuller. *Transactions of the Royal Society of South Africa,* 34(1): 251–271.

——— 1954b. Caste differentiation among termites. *Transactions of the Royal Society of South Africa,* 34(2): 345–353.

——— 1955. *Dwellers in darkness.* Longmans, Green, London. x + 134 pp.

Skinner, B. F. 1966. The phylogeny and ontogeny of behavior. *Science,* 153: 1205–1213.

Skutch, A. F. 1935. Helpers at the nest. *Auk,* 52(3): 257–273.

——— 1959. Life history of the groove-billed ani. *Auk,* 76(3): 281–317.

——— 1961. Helpers among birds. *Condor,* 63(3): 198–226.

Sladen, F. W. L. 1912. *The humble-bee, its life-history and how to domes-*

ticate it, with descriptions of all the British species of Bombus and Psithyrus. Macmillan Co., London. xiii + 283 pp.

Slijper, E. J. 1962. *Whales.* Hutchinson, London. 475 pp.

Slobin, D. 1971. *Psycholinguistics.* Scott, Foresman, Glenview, Ill. xii + 148 pp.

Slobodkin, L. B., and A. Rapoport. 1974. An optimal strategy of evolution. *Quarterly Review of Biology,* 49(3): 181–200.

Smith, C. C. 1968. The adaptive nature of social organization in the genus of tree squirrels *Tamiasciurus. Ecological Monographs,* 38(1): 31–63.

Smith, E. A. 1968. Adoptive suckling in the grey seal. *Nature, London,* 217(5130): 762–763.

Smith, H. M. 1943. Size of breeding populations in relation to egg-laying and reproductive success in the eastern red-wing (*Agelaius p. phoeniceus*). *Ecology,* 24(2): 183–207.

Smith, M. R. 1936. *Distribution of the Argentine ant in the United States and suggestions for its control or eradication.* U.S. Department of Agriculture, Circular no. 387. 39 pp.

Smith, N. G. 1968. The advantages of being parasitized. *Nature, London,* 219(5155): 690–694.

Smith, W. J. 1963. Vocal communication in birds. *American Naturalist,* 97(893): 117–125.

———— 1969a. Messages of vertebrate communication. *Science,* 165: 145–150.

———— 1969b. Displays of *Sayornis phoebe* (Aves, Tyrannidae). *Behaviour,* 33(3,4): 283–322.

Smith, W. J., Sharon L. Smith, Elizabeth C. Oppenheimer, Jill G. de Villa, and F. A. Ulmer. 1973. Behavior of a captive population of black-tailed prairie dogs: annual cycle of social behavior. *Behaviour,* 46(3,4): 189–220.

Smyth, M. 1968. The effects of removal of individuals from a population of bank voles (*Clethrionomys glareolus*). *Journal of Animal Ecology,* 37(1): 167–183.

Smythe, N. 1970a. The adaptive value of the social organization of the coati (*Nasua narica*). *Journal of Mammalogy,* 51(4): 818–820.

———— 1970b. On the existence of "pursuit invitation" signals in mammals. *American Naturalist,* 104(938): 491–494.

Snow, Carol J. 1967. Some observations on the behavioral and morphological development of coyote pups. *American Zoologist,* 7(2): 353–355.

Snow, D. W. 1958. *A study of blackbirds.* Allen and Unwin, London. 192 pp.

———— 1961. The natural history of the oilbird, *Steatornis caripensis,* in Trinidad, W.I.: 1, general behavior and breeding habits. *Zoologica, New York,* 46(1): 27–48.

———— 1963. The evolution of manakin displays. *Proceedings of the Thirteenth International Ornithological Congress, Ithaca, 1962,* pp. 553–561.

Snyder, N. 1967. An alarm reaction of aquatic gastropods to intraspecific extract. *Memoirs of the Cornell University Agricultural Experiment Station,* 403: 1–222.

Snyder, R. L. 1961. Evolution and integration of mechanisms that regulate population growth. *Proceedings of the National Academy of Sciences, U.S.A.,* 47(4): 449–455.

Sody, H. J. V. 1959. Das javanische Nashorn, *Rhinoceros sondaicus. Zeitschrift für Säugetierkunde,* 24(3,4): 109–240.

Solomon, M. E. 1969. *Population dynamics.* St. Martin's Press, New York. 60 pp.

Sondheimer, E., and J. B. Simeone, eds. 1970. *Chemical ecology.* Academic Press, New York. xvi + 336 pp.

Sorenson, M. W. 1970. Behavior of tree shrews. In L. A. Rosenblum, ed. (*q.v.*), *Primate behavior: developments in field and laboratory research,* vol. 1, pp. 141–193.

Sorokin, P. 1957. *Social and cultural dynamics.* Porter Sargent, Boston. 719 pp.

Soulié, J. 1960a. Des considérations écologiques peuvent-elles apporter une contribution à la connaissance du cycle biologique des colonies de *Cremastogaster* (Hymenoptera-Formicoidea). *Insectes Sociaux,* 7(3): 283–295.

———— 1960b. La "sociabilité" des *Cremastogaster* (Hymenoptera-Formicoidea). *Insectes Sociaux,* 7(4): 369–376.

———— 1964. Le contrôle par les ouvrières de la monogynie des colonies chez *Sphaerocrema striatula* (Myrmicidae, Cremastogastrini). *Insectes Sociaux,* 11(4): 383–388.

Southern, H. N. 1948. Sexual and aggressive behaviour in the wild rabbit. *Behaviour,* 1(3,4): 173–194.

Southwick, C. H., ed. 1963. *Primate social behavior: an enduring problem.* Van Nostrand Co., Princeton, N.J. viii + 191 pp.

———— 1967. An experimental study of intragroup agonistic behavior in rhesus monkeys (*Macaca mulatta*). *Behaviour,* 28(1,2): 182–209.

———— 1969. Aggressive behaviour of rhesus monkeys in natural and captive groups. In S. Garattini and E. B. Sigg, eds. (*q.v.*), *Aggressive behaviour,* pp. 32–43.

———— ed. 1970. *Animal aggression: selected readings.* Van Nostrand Reinhold, New York. xii + 229 pp.

Southwick, C. H., Mirza Azhar Beg, and M. R. Siddiqi. 1965. Rhesus monkeys in North India. In I. DeVore, ed. (*q.v.*), *Primate behavior: field studies of monkeys and apes,* pp. 111–159.

Southwick, C. H., and M. R. Siddiqi. 1967. The role of social tradition in the maintenance of dominance in a wild rhesus group. *Primates,* 8(4): 341–353.

Sowls, L. K. 1974. Social behaviour of the collared peccary, *Dicotyles tajacu* (L.). In V. Geist and F. Walther, eds. (*q.v.*), *The behaviour of ungulates and its relation to management,* vol. 1, pp. 144–165.

Sparks, J. H. 1965. On the role of allopreening invitation behaviour in reducing aggression among red avadavats, with comments on its evolution in the Spermestidae. *Proceedings of the Zoological Society of London,* 145(3): 387–403.

———— 1969. Allogrooming in primates: a review. In D. Morris, ed. (*q.v.*), *Primate ethology: essays on the socio-sexual behavior of apes and monkeys,* pp. 190–225.

Spencer-Booth, Yvette. 1968. The behaviour of group companions towards rhesus monkey infants. *Animal Behaviour,* 16(4): 541–557.

——— 1970. The relationships between mammalian young and conspecifics other than mothers and peers: a review. *Advances in the Study of Behavior*, 3: 119–194.

Spencer-Booth, Yvette, and R. A. Hinde. 1967. The effects of separating rhesus monkey infants from their mothers for six days. *Journal of Child Psychology and Psychiatry*, 7: 179–197.

——— 1971. The effects of thirteen days maternal separation on infant rhesus monkeys compared with those of shorter and repeated separations. *Animal Behaviour*, 19(3): 595–605.

Spieth, H. T. 1968. Evolutionary implications of sexual behavior in *Drosophila*. *Evolutionary Biology*, 2: 157–193.

Spradbery, J. P. 1965. The social organization of wasp communities. *Symposia of the Zoological Society of London*, 14: 61–96.

——— 1973. *Wasps: an account of the biology and natural history of solitary and social wasps*. Sidgwick and Jackson, London, xvi + 408 pp.

Stains, H. J. 1967. Carnivores and pinnipeds. In S. Anderson and J. K. Jones, Jr., eds. (*q.v.*), *Recent mammals of the world: a synopsis of families*, pp. 325–354.

Stamps, Judy A. 1973. Displays and social organization in female *Anolis aeneus*. *Copeia*, 1973, no. 2, pp. 264–272.

Starr, R. C. 1968. Cellular differentiation in *Volvox*. *Proceedings of the National Academy of Sciences, U.S.A.*, 59(4): 1082–1088.

Starrett, A. 1967. Hystricoid, erethizontoid, cavioid, and chinchilloid rodents. In S. Anderson and J. K. Jones, Jr., eds. (*q.v.*), *Recent mammals of the world: a synopsis of families*, pp. 254–272.

Stefanski, R. A. 1967. Utilization of the breeding territory in the black-capped chickadee. *Condor*, 69(3): 259–267.

Steiner, A. L. 1971. Play activity of Columbian ground squirrels. *Zeitschrift für Tierpsychologie*, 28(3): 247–261.

Stenger, Judith. 1958. Food habits and available food of ovenbirds in relation to territory size. *Auk*, 75(3): 335–346.

Stenger, Judith, and J. B. Falls. 1959. The utilized territory of the ovenbird. *Wilson Bulletin*, 71(2): 125–140.

Stephens, J. S., Jr., R. K. Johnson, G. S. Key, and J. E. McCosker. 1970. The comparative ecology of three sympatric species of California blennies of the genus *Hypsoblennius* Gill (Teleostomi, Blenniidae). *Ecological Monographs*, 40(2): 213–233.

Sterba, G. 1962. *Freshwater fishes of the world*. Pet Library, Cooper Square, New York. 877 pp.

Sterndale, R. A. 1884. *Natural history of the Mammalia of India and Ceylon*. Calcutta. [Cited by L. H. Matthews, 1971 (*q.v.*).]

Števčić, Z. 1971. Laboratory observations on the aggregations of the spiny spider crab (*Maja squinado* Herbst). *Animal Behaviour*, 19(1): 18–25.

Stevenson, Joan G. 1969. Song as a reinforcer. In R. A. Hinde, ed. (*q.v.*), *Bird vocalizations: their relation to current problems in biology and psychology*, pp. 49–60.

Stewart, R. E., and J. W. Aldrich. 1951. Removal and repopulation of breeding birds in a spruce-fir forest community. *Auk*, 68(4): 471–482.

Steyn, J. J. 1954. The pugnacious ant (*Anoplolepis custodiens* Smith) and its relation to the control of citrus scales at Letaba. *Memoirs of the Entomological Society of South Africa*, no. 3. iii + 96 pp.

Stiles, F. G. 1971. Time, energy, and territoriality of the Anna hummingbird (*Calypte anna*). *Science*, 173: 818–821.

Stiles, F. G., and L. L. Wolf. 1970. Hummingbird territoriality at a tropical flowering tree. *Auk*, 87(3): 467–491.

Stimson, J. 1970. Territorial behavior of the owl limpet, *Lottia gigantea*. *Ecology*, 51(1): 113–118.

Stirling, I. 1971. Studies on the behaviour of the South Australian fur seal, *Arctocephalus forsteri* (Lesson), 1, 2. *Australian Journal of Zoology*, 19(3): 243–273.

——— 1972. Observations on the Australian sea lion, *Neophoca cinerea* (Peron). *Australian Journal of Zoology*, 20(3): 271–279.

Stones, R. C., and C. L. Hayward. 1968. Natural history of the desert woodrat, *Neotoma lepida*. *American Midland Naturalist*, 80(2): 458–476.

Struhsaker, T. T. 1967a. Behavior of vervet monkeys (*Cercopithecus aethiops*). *University of California Publications in Zoology*, 82. 64 pp.

——— 1967b. Social structure among vervet monkeys (*Cercopithecus aethiops*). *Behaviour*, 29(2–4): 83–121.

——— 1967c. Auditory communication among vervet monkeys (*Cercopithecus aethiops*). In S. A. Altmann, ed. (*q.v.*), *Social communication among primates*, pp. 281–324.

——— 1967d. Ecology of vervet monkeys (*Cercopithecus aethiops*) in the Masai-Amboseli Game Reserve, Kenya. *Ecology*, 48(6): 891–904.

——— 1969. Correlates of ecology and social organization among African cercopithecines. *Folia Primatologica*, 11(1,2): 80–118.

——— 1970a. Phylogenetic implications of some vocalizations of *Cercopithecus* monkeys. In J. R. Napier and P. H. Napier eds. (*q.v.*), *Old World monkeys: evolution, systematics, and behavior*, pp. 365–444.

——— 1970b. Notes on *Galagoides demidovii* in Cameroon. *Mammalia*, 34(2): 207–211.

Struhsaker, T. T., and J. S. Gartlan. 1970. Observations on the behaviour and ecology of the patas monkey (*Erythrocebus patas*) in the Wazas Reserve, Cameroon. *Journal of Zoology, London*, 161(1): 49–63.

Struhsaker, T. T., and P. Hunkeler. 1971. Evidence of tool-using by chimpanzees in the Ivory Coast. *Folia Primatologica*, 15(3,4): 212–219.

Stuart, A. M. 1960. Experimental studies on communication in termites. Ph.D. thesis, Harvard University, Cambridge, Mass. 95 pp.

——— 1963. Studies on the communication of alarm in the termite *Zootermopsis nevadensis* (Hagen), Isoptera. *Physiological Zoology*, 36(1): 85–96.

——— 1969. Social behavior and communication. In K. Krishna and Frances M. Weesner, eds. (*q.v.*), *Biology of termites*, vol. 1, pp. 193–232.

——— 1970. The role of chemicals in termite communication. In J. W. Johnston, D. G. Moulton, and A. Turk, eds. (*q.v.*), *Advances in chemoreception*, vol. 1, *Communication by chemical signals*, pp. 79–106.

Stuewer, F. W. 1943. Raccoons: their habits and management in Michigan. *Ecological Monographs*, 13(2): 203–257.

Stumper, R. 1950. Les associations complexes des fourmis. Commensalisme, symbiose et parasitisme. *Bulletin Biologique de la France et de la Belgique*, 84(4): 376–399.

Subramoniam, Swarna. 1957. Some observations on the habits of the slender loris, *Loris tardigradus* (Linnaeus). *Journal of the Bombay Natural History Society*, 54(2): 387–398.

Sudd, J. H. 1963. How insects work in groups. *Discovery*, June, pp. 15–19.

———— 1967. *An introduction to the behaviour of ants.* Arnold, London. viii + 200 pp.

Sugiyama, Y. 1960. On the division of a natural troop of Japanese monkeys at Takasakiyama. *Primates*, 2(2): 109–148.

———— 1967. Social organization of hanuman langurs. In S. A. Altmann, ed. (*q.v.*), *Social communication among primates*, pp. 221–236.

———— 1968. Social organization of chimpanzees in the Budongo Forest, Uganda. *Primates*, 9(3): 225–258.

———— 1969. Social behavior of chimpanzees in the Budongo Forest, Uganda. *Primates*, 10(3,4): 197–225.

———— 1971. Characteristics of the social life of bonnet macaques (*Macaca radiata*). *Primates*, 12(3,4): 247–266.

———— 1972. Social characteristics and socialization of wild chimpanzees. In F. E. Poirier, ed. (*q.v.*), *Primate socialization*, pp. 145–163.

———— 1973. Social organization of wild chimpanzees. In C. R. Carpenter, ed. (*q.v.*), *Behavioral regulators of behavior in primates*, pp. 68–80.

Summers, F. M. 1938. Some aspects of normal development in the colonial ciliate *Zoothamnium alternans*. *Biological Bulletin, Marine Biological Laboratory, Woods Hole*, 74(1): 117–129.

Suzuki, A. 1969. An ecological study of chimpanzees in a savanna woodland. *Primates*, 10(2): 103–148.

———— 1971. Carnivory and cannibalism observed among forest-living chimpanzees. *Journal of the Anthropological Society of Nippon*, 79(1): 30–48.

Sved, J. A., T. E. Reed, and W. F. Bodmer. 1967. The number of balanced polymorphisms that can be maintained in a natural population. *Genetics*, 55(3): 469–481.

Szlep, Raja, and T. Jacobi. 1967. The mechanism of recruitment to mass foraging in colonies of *Monomorium venustum* Smith, *M. subopacum* ssp. *phoenicium* Em., *Tapinoma israelis* For. and *T. simothi* v. *phoenicium* Em. *Insectes Sociaux*, 14(1): 25–40.

Taber, F. W. 1945. Contribution on the life history and ecology of the nine-banded armadillo. *Journal of Mammalogy*, 26(3): 211–226.

Talbot, Mary. 1943. Population studies of the ant, *Prenolepis imparis* Say. *Ecology*, 24(1): 31–44.

———— 1957. Population studies of the slave-making ant *Leptothorax duloticus* and its slave *Leptothorax curvispinosus*. *Ecology*, 38(3): 449–456.

———— 1967. Slave-raids of the ant *Polyergus lucidus* Mayr. *Psyche, Cambridge*, 74(4): 299–313.

Talbot, Mary, and C. H. Kennedy. 1940. The slave-making ant, *Formica sanguinea subintegra* Emery, its raids, nuptial flights and nest structure. *Annals of the Entomological Society of America*, 33(3): 560–577.

Talmadge, R. V., and G. D. Buchanan. 1954. The armadillo: a review of its natural history, ecology, anatomy, and reproductive physiology. *Rice Institute Pamphlet, Houston*, 41(2): 1–135. [Cited by J. F. Eisenberg, 1966 (*q.v.*).]

Tavistock, H. W. 1931. The food-shortage theory. *Ibis*, 13th ser., 1: 351–354.

Tavolga, Margaret C. 1966. Behavior of the bottlenose dolphin (*Tursiops truncatus*); social interactions in a captive colony. In K. S. Norris, ed. (*q.v.*), *Whales, dolphins and porpoises*, pp. 718–730.

Tavolga, Margaret C., and F. S. Essapian. 1957. The behavior of the bottle-nosed dolphin (*Tursiops truncatus*): mating, pregnancy, parturition, and mother-infant behavior. *Zoologica, New York*, 42(1): 11–31.

Tavolga, W. N., ed. 1964. *Marine bio-acoustics.* Pergamon, New York. xiv + 413 pp.

Tayler, C. K., and G. S. Saayman. 1973. Imitative behaviour by Indian Ocean bottlenose dolphins (*Tursiops aduncus*) in captivity. *Behaviour*, 44(3,4): 286–298.

Taylor, L. H. 1939. Observations on social parasitism in the genus *Vespula* Thomson. *Annals of the Entomological Society of America*, 32(2): 304–315.

Teleki, G. 1973. *The predatory behavior of wild chimpanzees.* Bucknell University Press, Lewisburg, Pa. 232 pp.

Tembrock, G. 1968. Land mammals. In T. A. Sebeok, ed. (*q.v.*), *Animal communication: techniques of study and results of research*, pp. 338–404.

Tener, J. S. 1954. A preliminary study of the musk-oxen of Fosheim Peninsula, Ellesmere Island, N.W.T. *Canada Wildlife Service, Wildlife Management Bulletin*, 1st ser., no. 9. 34 pp. [Cited by L. D. Mech, 1970 (*q.v.*).]

———— 1965. *Muskoxen in Canada: a biological and taxonomic review.* Department of Northern Affairs and National Resources, Ottawa. 166 pp.

Test, F. H. 1954. Social aggressiveness in an amphibian. *Science*, 120: 140–141.

Tevis, L. 1950. Summer behavior of a family of beavers in New York State. *Journal of Mammalogy*, 31(1): 40–65.

Thaxter, R. 1892. On the Myxobacteriaceae, a new order of Schizomycetes. *Botanical Gazette*, 17: 389–406.

Theodor, J. L. 1970. Distinction between "self" and "not-self" in lower invertebrates. *Nature, London*, 227(5259): 690–692.

Thielcke, G. 1965. Gesangsgeographische Variation des Gartenbaumläufers (*Certhia brachydactyla*) in Hinblick auf das Artbildungsproblem. *Zeitschrift für Tierpsychologie*, 22(5): 542–566.

———— 1969. Geographic variation in bird vocalizations. In R. A. Hinde, ed. (*q.v.*), *Bird vocalizations: their relation to current problems in biology and psychology: essays presented to W. H. Thorpe*, pp. 311–339.

Thielcke, G., and Helga Thielcke. 1970. Die sozialen Funktionen verschiedener Gesangsformen des Sonnenvogels (*Leiothrix lutea*). *Zeitschrift für Tierpsychologie*, 27(2): 177–185.

Thiessen, D. D. 1964. Population density, mouse genotype, and endocrine function in behavior. *Journal of Comparative and Physiological Psychology*, 57(3): 412–416.

——— 1973. Footholds for survival. *American Scientist*, 61(3): 346–351.

Thiessen, D. D., H. C. Friend, and G. Lindzey. 1968. Androgen control of territorial marking in the Mongolian gerbil. *Science*, 160: 432–433.

Thiessen, D. D., K. Owen, and G. Lindzey. 1971. Mechanisms of territorial marking in the male and female Mongolian gerbils (*Meriones unguiculatus*). *Journal of Comparative and Physiological Psychology*, 77(1): 38–47.

Thiessen, D. D., and P. Yahr. 1970. Central control of territorial marking in the Mongolian gerbil. *Physiology and Behavior*, 5: 275–278.

Thines, G., and B. Heuts. 1968. The effect of submissive experiences on dominance and aggressive behaviour of *Xiphophorus* (Pisces, Poeciliidae). *Zeitschrift für Tierpsychologie*, 25(2): 139–154.

Thoday, J. M. 1953. Components of fitness. *Symposia of the Society for Experimental Biology*, 7: 96–113.

——— 1964. Genetics and integration of reproductive systems. *Symposia of the Royal Entomological Society of London*, 2: 108–119.

Thompson, W. L. 1960. Agonistic behavior in the house finch: 2, factors in aggressiveness and sociality. *Condor*, 62(5): 378–402.

Thompson, W. R. 1957. Influence of prenatal maternal anxiety on emotionality in young rats. *Science*, 125: 698–699.

——— 1958. Social behavior. In Anne Roe and G. G. Simpson, eds. (*q.v.*), *Behavior and evolution*, pp. 291–310.

Thorpe, W. H. 1954. The process of song-learning in the chaffinch as studied by means of the sound spectrograph. *Nature, London*, 173(4402): 465–469.

——— 1961. *Bird-song: the biology of vocal communication and expression in birds*. Cambridge University Press, Cambridge. xii + 143 pp.

——— 1963a. *Learning and instinct in animals*, 2d ed. Methuen, London. xii + 558 pp.

——— 1963b. Antiphonal singing in birds as evidence for avian auditory reaction time. *Nature, London*, 197(4869): 774–776.

——— 1972a. The comparison of vocal communication in animals and man. In R. A. Hinde, ed. (*q.v.*), *Non-verbal communication*, pp. 27–47.

——— 1972b. Vocal communication in birds. In R. A. Hinde, ed. (*q.v.*), *Non-verbal communication*, pp. 153–176.

Thorpe, W. H., and M. E. W. North. 1965. Origin and significance of the power of vocal imitation: with special reference to the antiphonal singing of birds. *Nature, London*, 208(5007): 219–222.

——— 1966. Vocal imitation in the tropical bou-bou shrike *Laniarius aethiopicus major* as a means of establishing and maintaining social bonds. *Ibis*, 108(3): 432–435.

Thorpe, W. H., and O. L. Zangwill, eds. 1961. *Current problems in animal behaviour*. Cambridge University Press, Cambridge. xiv + 424 pp.

Tiger, L. 1969. *Men in groups*. Random House, New York. xx + 254 pp.

Tiger, L., and R. Fox. 1971. *The imperial animal*. Holt, Rinehart and Winston, New York. xi + 308 pp.

Tinbergen, L. 1960. The natural control of insects in pinewoods: I, factors influencing the intensity of predation by songbirds. *Archives Néerlandaises de Zoologie, Leydig*, 13(3): 265–336.

Tinbergen, N. 1939. Field observations of East Greenland birds: II, the behavior of the snow bunting (*Plectrophenax nivalis subnivalis* [Brehm]) in spring. *Transactions of the Linnaean Society of New York*, 5: 1–94.

——— 1951. *The study of instinct*. Clarendon Press of Oxford University Press, Oxford. xii + 228 pp.

——— 1952. "Derived" activities; their causation, biological significance, origin, and emancipation during evolution. *Quarterly Review of Biology*, 27(1): 1–32.

——— 1953. *The herring gull's world: a study of the social behaviour of birds*. Collins, London. xvi + 255 pp.

——— 1959. Comparative studies of the behaviour of gulls (Laridae): a progress report. *Behaviour*, 15(1,2): 1–70.

——— 1960. The evolution of behavior in gulls. *Scientific American*, 203(6) (December): 118–130.

——— 1967. Adaptive features of the black-headed gull *Larus ridibundus* L. *Proceedings of the Fourteenth International Ornithological Congress, Oxford, 1966*, pp. 43–59.

Tinbergen, N., M. Impekoven, and D. Franck. 1967. An experiment on spacing-out as a defence against predation. *Behaviour*, 28(3,4): 307–321.

Tinkle, D. W. 1965. Population structure and effective size of a lizard population. *Evolution*, 19(4): 569–573.

——— 1967. The life and demography of the side-blotched lizard, *Uta stansburiana*. *Miscellaneous Publications, Museum of Zoology, University of Michigan, Ann Arbor*, 132. 182 pp.

——— 1969. The concept of reproductive effort and its relation to the evolution of life histories of lizards. *American Naturalist*, 103(933): 501–516.

Tobias, P. V. 1973. Implications of the new age estimates of the early South African hominids. *Nature, London*, 246(5428): 79–83.

Todd, J. H. 1971. The chemical language of fishes. *Scientific American*, 224(5) (May): 99–108.

Todt, D. 1970. Die antiphonen Paargesänge des ostafrikanischen Grassängers *Cisticola hunteri prinioides* Neumann. *Journal für Ornithologie*, 111(3,4): 332–356.

Tokuda, K., and G. D. Jensen. 1968. The leader's role in controlling aggressive behavior in a monkey group. *Primates*, 9(4): 319–322.

Tordoff, H. B. 1954. Social organization and behavior in a flock of captive, nonbreeding red crossbills. *Condor*, 56(6): 346–358.

Tretzel, E. 1966. Artkennzeichnende und reaktionsauslösende Komponenten im Gesang der Heidelerche (*Lullula arborea*). *Verhandlungen der Deutschen Zoologischen Gesellschaft, Jena, 1965*, pp. 367–380.

Trivers, R. L. 1971. The evolution of reciprocal altruism. *Quarterly Review of Biology*, 46(4): 35–57.

———— 1972. Parental investment and sexual selection. In B. Campbell, ed. (q.v.), *Sexual selection and the descent of man, 1871–1971*, pp. 136–179.

———— 1974. Parent-offspring conflict. *American Zoologist*, 14(1): 249–264.

———— 1975. Haplodiploidy and the evolution of the social insects. *Science*. (In press.)

Trivers, R. L., and D. E. Willard. 1973. Natural selection of parental ability to vary the sex ratio of offspring. *Science*, 179: 90–92.

Troughton, E. L. 1966. *Furred animals of Australia*, 8th ed., rev. Livingston Publishing Co., Wynnewood, Pa. xxxii + 376 pp.

Truman, J. W., and Lynn M. Riddiford. 1974. Hormonal mechanisms underlying insect behaviour. *Advances in Insect Physiology*, 10: 297–352.

Trumler, E. 1959. Das "Rossigkeitsgesicht" und ähnliches Ausdrucksverhalten bei Einhufern. *Zeitschrift für Tierpsychologie*, 16(4): 478–488.

Tschanz, B. 1968. Trottellummen. *Zeitschrift für Tierpsychologie*, supplement 4. 103 pp.

Tsumori, A. 1967. Newly acquired behavior and social interactions of Japanese monkeys. In S. A. Altmann, ed. (q.v.), *Social communication among primates*, pp. 207–219.

Tsumori, A., M. Kawai, and R. Motoyoshi. 1965. Delayed response of wild Japanese monkeys by the sand-digging method: 1, case of the Koshima troop. *Primates*, 6(2): 195–212.

Tucker, D., and N. Suzuki. 1972. Olfactory responses to Schreckstoff of catfish. *Proceedings of the Fourth International Symposium on Olfaction and Taste, Starnberg, Germany*, pp. 121–127.

Turnbull, C. M. 1968. The importance of flux in two hunting societies. In R. B. Lee and I. DeVore, eds. (q.v.), *Man the hunter*, pp. 132–137.

———— 1972. *The mountain people*. Touchstone Books, Simon and Schuster, New York. 309 pp.

Turner, C. D., and J. T. Bagnara. 1971. *General endocrinology*, 5th ed. W. B. Saunders Co., Philadelphia. x + 659 pp.

Turner, E. R. A. 1964. Social feeding in birds. *Behaviour*, 24(1,2): 1–46.

Turner, F. B., R. I. Jennrich, and J. D. Weintraub. 1969. Home ranges and body size of lizards. *Ecology*, 50(6): 1076–1081.

Tyler, Stephanie. 1972. The behaviour and social organization of the New Forest ponies. *Animal Behaviour Monographs*, 5(2): 85–196.

Ullrich, W. 1961. Zur Biologie und Soziologie der Colobusaffen (*Colobus guereza caudatus* Thomas 1885). *Zoologische Garten, Leipzig*, n.s. 25(6): 305–368.

Urquhart, F. A. 1960. *The monarch butterfly*. University of Toronto Press, Toronto, xxiv + 361 pp.

Uzzell, T. 1970. Meiotic mechanisms of naturally occurring unisexual vertebrates. *American Naturalist*, 104(939): 433–445.

Valone, J. A., Jr. 1970. Electrical emissions in *Gymnotus carapo* and their relation to social behavior. *Behaviour*, 37(1,2): 1–14.

Vandenbergh, J. G. 1967. The development of social structure in free-ranging rhesus monkeys. *Behaviour*, 29(2–4): 179–194.

———— 1971. The effects of gonadal hormones on the aggressive behaviour of adult golden hamsters (*Mesocricetus auratus*). *Animal Behaviour*, 19(3): 589–594.

Van Denburgh, J. 1914. The gigantic land tortoises of the Galapagos Archipelago. *Proceedings of the California Academy of Sciences, San Francisco*, 4th ser. 2(1): 203–374.

Van Deusen, H. M., and J. K. Jones, Jr. 1967. Marsupials. In S. Anderson and J. K. Jones, Jr., eds. (q.v.), *Recent mammals of the world: a synopsis of families*, pp. 61–86.

Van Valen, L. 1971. Group selection and the evolution of dispersal. *Evolution*, 25(4): 591–598.

Varley, Margaret, and D. Symmes. 1966. The hierarchy of dominance in a group of macaques. *Behaviour*, 27(1,2): 54–75.

Vaughan, T. A. 1972. *Mammalogy*. W. B. Saunders Co., Philadelphia. viii + 463 pp.

Velthuis, H. H. V., and J. van Es. 1964. Some functional aspects of the mandibular glands of the queen honeybee. *Journal of Apicultural Research*, 3(1): 11–16.

Verheyen, R. 1954. *Monographie éthologique de l'hippopotame* (Hippopotamus amphibius *Linné*). Institut des Parcs Nationaux du Congo Belge. Exploration du Parc National Albert, Brussels. 91 pp.

Verner, J. 1965. Breeding biology of the long-billed marsh wren. *Condor*, 67(1): 6–30.

Verner, J., and Gay H. Engelsen. 1970. Territories, multiple nest building, and polygyny in the long-billed marsh wren. *Auk*, 87(3): 557–567.

Verner, J., and Mary F. Willson. 1966. The influence of habitats on mating systems of North American passerine birds. *Ecology*, 47(1): 143–147.

Vernon, W., and R. Ulrich. 1966. Classical conditioning of pain-elicited aggression. *Science*, 152: 668–669.

Verron, H. 1963. Rôle des stimuli chimiques dans l'attraction sociale chez *Calotermes flavicollis* (Fabr.). *Insectes Sociaux*, 10(2): 167–184; 10(3): 185–296; 10(4): 297–335.

Verts, B. J. 1967. *The biology of the striped skunk*. University of Illinois Press, Urbana. xiv + 218 pp.

Verwey, J. 1930. Die Paarungsbiologie des Fischreihers. *Zoologische Jahrbücher, Abteilungen Physiologie*, 48: 1–120.

Vince, Margaret A. 1969. Embryonic communication, respiration and the synchronization of hatching. In R. A. Hinde, ed. (q.v.), *Bird vocalizations: their relations to current problems in biology and psychology*, pp. 233–260.

Vincent, F. 1968. La sociabilité du galago de Demidoff. *La Terre et la Vie*, 115(1): 51–56.

Vincent, R. E. 1958. Observations of red fox behavior. *Ecology*, 39(4): 755–757.

Voeller, B. 1971. Developmental physiology of fern gametophytes: relevance for biology. *BioScience*, 21(6): 266–270.

Vos, A. de, P. Brokx, and V. Geist. 1967. A review of social behavior of the North American cervids during the reproductive period. *American Midland Naturalist*, 77(2): 390–417.

Vuilleumier, F. 1967. Mixed species flocks in Patagonian forests, with remarks on interspecies flock formation. *Condor*, 69(4): 400–404.

Waddington, C. H. 1957. *The strategy of the genes: a discussion of some aspects of theoretical biology*. George Allen and Unwin, London. x + 262 pp.

Wahlund, S. 1928. Zusammensetzung von Populationen und Korrelationserscheinungen vom Standpunkt der Vererbungslehre aus betrachtet. *Hereditas*, 11: 65–106.

Walker, E. P., ed. 1964. *Mammals of the world*, vol. 3, *A classified bibliography*. Johns Hopkins Press, Baltimore. ix + 769 pp.

Wallace, B. 1958. The average effect of radiation-induced mutations on viability in *Drosophila melanogaster*. *Evolution*, 12(4): 532–556.

——— 1968. *Topics in population genetics*. W. W. Norton, New York. x + 481 pp.

——— 1973. Misinformation, fitness, and selection. *American Naturalist*, 107(953): 1–7.

Wallis, D. I. 1961. Food-sharing behaviour of the ants *Formica sanguinea* and *Formica fusca*. *Behaviour*, 17(1): 17–47.

Waloff, Z. 1966. *The upsurges and recessions of the desert locust plague: an historical survey*. Anti-Locust Memoir no. 8. Anti-Locust Research Centre, London. 111 pp.

Walther, F. R. 1964. Verhaltensstudien an der Gattung *Tragelaphus* De Blainville, 1816, in Gefangenschaft, unter besonderer Berücksichtigung des Sozialverhaltens. *Zeitschrift für Tierpsychologie*, 21(4): 393–467.

——— 1969. Flight behaviour and avoidance of predators in Thomson's gazelle (*Gazella thomsoni* Guenther 1884). *Behaviour*, 34(3): 184–221.

Ward, P. 1965. Feeding ecology of the black-faced dioch *Quelea quelea* in Nigeria. *Ibis*, 107(2): 173–214.

Waring, G. H. 1970. Sound communications of black-tailed, white-tailed, and Gunnison's prairie dogs. *American Midland Naturalist*, 83(1): 167–185.

Warren, J. M., and R. J. Maroney. 1958. Competitive social interaction between monkeys. *Journal of Social Psychology*, 48: 223–233.

Washburn, S. L., ed. 1961. *Social life of early man*. Viking Fund Publications in Anthropology no. 31. Aldine Publishing Co., Chicago. ix + 299 pp.

——— ed. 1963. *Classification and human evolution*. Viking Fund Publications in Anthropology no. 37. Aldine Publishing Co., Chicago. viii + 371 pp.

——— 1970. Comment on: "A possible evolutionary basis for aesthetic appreciation in men and apes." *Evolution*, 24(4): 824–825.

——— 1971. On understanding man. *Rehovot, Weizmann Institute of Science*, 6(2): 22–29.

Washburn, S. L., and Virginia Avis. 1958. Evolution of human behavior. In Anne Roe and G. G. Simpson, eds. (*q.v.*), *Behavior and evolution*, pp. 421–436.

Washburn, S. L., and I. DeVore. 1961. The social life of baboons. *Scientific American*, 204(6) (June): 62–71.

Washburn, S. L., and D. A. Hamburg. 1965. The implications of primate research. In I. DeVore, ed. (*q.v.*), *Primate behavior: field studies of monkeys and apes*, pp. 607–622.

Washburn, S. L., and R. S. Harding. 1970. Evolution of primate behavior. In F. O. Schmitt, ed., *Neural and behavioral evolution. Neurosciences: second study program*, pp. 39–47. Rockefeller University Press, New York. 1068 pp.

Washburn, S. L., and F. C. Howell. 1960. Human evolution and culture. In S. Tax, ed., *Evolution after Darwin*, vol. 2, *Evolution of man*, pp. 33–56. University of Chicago Press, Chicago. viii + 473 pp.

Washburn, S. L., Phyllis C. Jay, and Jane B. Lancaster. 1968. Field studies of Old World monkeys and apes. *Science*, 150: 1541–1547.

Wasmann, E. 1915. Neue Beiträge zur Biologie von *Lomechusa* und *Atemeles*, mit kritischen Bemerkungen über das echte Gastverhältnis. *Zeitschrift für Wissenschaftliche Zoologie*, 114(2): 233–402.

Watson, A. 1967. Population control by territorial behaviour in red grouse. *Nature, London*, 215(5107): 1274–1275.

——— ed. 1970. *Animal populations in relation to their food resources*. Blackwell Scientific Publications, Oxford. xx + 477 pp.

Watson, A., and D. Jenkins. 1968. Experiments on population control by territorial behaviour in red grouse. *Journal of Animal Ecology*, 37(3): 595–614.

Watson, A., and R. Moss. 1971. Spacing as affected by territorial behavior, habitat and nutrition in red grouse (*Lagopus l. scoticus*). In A. H. Esser, ed. (*q.v.*), *Behavior and environment: the use of space by animals and men*, pp. 92–111.

Watson, J. A. L., J. J. C. Nel, and P. H. Hewitt. 1972. Behavioural changes in founding pairs of the termite, *Hodotermes mossambicus*. *Journal of Insect Physiology*, 18(2): 373–387.

Watts, C. R., and A. W. Stokes. 1971. The social order of turkeys. *Scientific American*, 224(6) (June): 112–118.

Wautier, V. 1971. Un phénomène social chez les coléoptères: le grégarisme des *Brachinus* (Caraboïdea Brachinidae). *Insectes Sociaux*, 18(3): 1–84.

Way, M. J. 1953. The relationship between certain ant species with particular reference to biological control of the coreid, *Theraptus* sp. *Bulletin of Entomological Research*, 44(4): 669–691.

——— 1954a. Studies of the life history and ecology of the ant *Oecophylla longinoda* Latreille. *Bulletin of Entomological Research*, 45(1): 93–112.

——— 1954b. Studies on the association of the ant *Oecophylla longinoda* (Latr.) (Formicidae) with the scale insect *Saissetia zanzibarensis* Williams (Coccidae). *Bulletin of Entomological Research*, 45(1): 113–134.

——— 1963. Mutualism between ants and honeydew-producing Homoptera. *Annual Review of Entomology*, 8: 307–344.

Weber, M. 1964. *The sociology of religion*, trans. by E. Fischoff, with an introduction by T. Parsons. Beacon Press, Boston. lxx + 304 pp.

Weber, N. A. 1943. Parabiosis in Neotropical "ant gardens." *Ecology*, 24(3): 400–404.

———— 1944. The Neotropical coccid-tending ants of the genus *Acropyga* Roger. *Annals of the Entomological Society of America*, 37(1): 89–122.

———— 1966. Fungus-growing ants. *Science*, 153: 587–604.

———— 1972. *Gardening ants: the attines.* Memoirs of the American Philosophical Society no. 92. American Philosophical Society, Philadelphia. xx + 146 pp.

Wecker, S. C. 1963. The role of early experience in habitat selection by the prairie deer mouse, *Peromyscus maniculatus bairdi. Ecological Monographs*, 33(4): 307–325.

Weeden, Judith Stenger. 1965. Territorial behavior of the tree sparrow. *Condor*, 67(3): 193–209.

Weeden, Judith Stenger, and J. B. Falls. 1959. Differential responses of male ovenbirds to recorded songs of neighboring and more distant individuals. *Auk*, 76(3): 343–351.

Weesner, Frances M. 1970. Termites of the Nearctic region. In K. Krishna and Frances M. Weesner, eds. (*q.v.*), *Biology of termites*, vol. 2, pp. 477–525.

Weir, J. S. 1959. Egg masses and early larval growth in *Myrmica. Insectes Sociaux*, 6(2): 187–201.

Weismann, A. 1891. *Essays upon heredity and kindred biological problems*, 2d ed. Clarendon Press, Oxford. xv + 471 pp.

Weiss, P. A. 1970. *Life, order, and understanding: a theme in three variations.* Graduate Journal, University of Texas, supplement 8. 157 pp.

Weiss, R. F., W. Buchanan, Lynne Altstatt, and J. P. Lombardo. 1971. Altruism is rewarding. *Science*, 171: 1262–1263.

Welch, B. L., and Annemarie S. Welch. 1969. Aggression and the biogenic amine neurohumors. In S. Garattini and E. B. Sigg, eds. (*q.v.*), *Aggressive behaviour*, pp. 188–202.

Weller, M. W. 1968. The breeding biology of the parasitic black-headed duck. *Living Bird*, 7: 169–207.

Wemmer, C. 1972. Comparative ethology of the large-spotted genet, *Genetta tigrina*, and related viverrid genera. Ph.D. thesis, University of Maryland, College Park.

Wesson, L. G. 1939. Contributions to the natural history of *Harpagoxenus americanus* (Hymenoptera: Formicidae). *Transactions of the American Entomological Society*, 65: 97–122.

———— 1940. Observations on *Leptothorax duloticus. Bulletin of the Brooklyn Entomological Society*, 35(3): 73–83.

West, Mary Jane. 1967. Foundress associations in polistine wasps: dominance hierarchies and the evolution of social behavior. *Science*, 157: 1584–1585.

West, Mary Jane, and R. D. Alexander. 1963. Sub-social behavior in a burrowing cricket *Anurogryllus muticus* (De Geer): Orthoptera: Gryllidae. *Ohio Journal of Science*, 63(1): 19–24.

Weygoldt, P. 1972. Geisselskorpione und Geisselspinnen (*Uropygi* und *Amblypygi*). *Zeitschrift des Kölner Zoo*, 15(3): 95–107.

Wharton, C. H. 1950. Notes on the life history of the flying lemur. *Journal of Mammalogy*, 31(3): 269–273.

Wheeler, W. M. 1904. A new type of social parasitism among ants. *Bulletin of the American Museum of Natural History*, 20(30): 347–375.

———— 1910. *Ants: their structure, development and behavior.* Columbia University Press, New York. xxv + 663 pp.

———— 1916. The Australian ants of the genus *Onychomyrmex. Bulletin of the Museum of Comparative Zoology, Harvard*, 60(2): 45–54.

———— 1918. A study of some ant larvae with a consideration of the origin and meaning of social habits among insects. *Proceedings of the American Philosophical Society*, 57: 293–343.

———— 1921. A new case of parabiosis and the "ant gardens" of British Guiana. *Ecology*, 2(2): 89–103.

———— 1922. Ants of the American Museum Congo Expedition, a contribution to the myrmecology of Africa: VII, keys to the genera and subgenera of ants; VIII, a synonymic list of the ants of the Ethiopian region; IX, a synonymic list of the ants of the Malagasy Region. *Bulletin of the American Museum of Natural History*, 45(1): 631–1055.

———— 1923. *Social life among the insects.* Harcourt, Brace, New York. vii + 375 pp.

———— 1925. A new guest-ant and other new Formicidae from Barro Colorado Island, Panama. *Biological Bulletin, Marine Biological Laboratory, Woods Hole*, 49(3): 150–181.

———— 1927a. *Emergent evolution and the social.* Kegan Paul, Trench, Trubner, London. 57 pp.

———— 1927b. The physiognomy of insects. *Quarterly Review of Biology*, 2(1): 1–36.

———— 1928. *The social insects: their origin and evolution.* Harcourt, Brace, New York. xviii + 378 pp.

———— 1930. Social evolution. In E. V. Cowdry, ed. (*q.v.*), *Human biology and racial welfare*, pp. 139–155.

———— 1933. *Colony-founding among ants, with an account of some primitive Australian species.* Harvard University Press, Cambridge. x + 179 pp.

———— 1934. A second revision of the ants of the genus *Leptomyrmex* Mayr. *Bulletin of the Museum of Comparative Zoology, Harvard*, 77(3): 69–118.

———— 1936. Ecological relations of ponerine and other ants to termites. *Proceedings of the American Academy of Arts and Sciences*, 71(3): 159–243.

Whitaker, J. O., Jr. 1963. A study of the meadow jumping mouse, *Zapus hudsonius* (Zimmerman) in central New York. *Ecological Monographs*, 33(3): 215–254.

White, H. C. 1970. *Chains of opportunity: system models of mobility in organizations.* Harvard University Press, Cambridge. xvi + 418 pp.

White, J. E. 1964. An index of the range of activity. *American Midland Naturalist*, 71(2): 369–373.

White, Sheila J., and R. E. C. White. 1970. Individual voice production in gannets. *Behaviour*, 37(1,2): 40–54.

Whitehead, G. K. 1972. *The wild goats of Great Britain and Ireland.* David and Charles, Newton Abbot, U.K. 184 pp.

Whiting, J. W. M. 1968. Discussion, "Are the hunter-gatherers a cultural type?" In R. B. Lee and I. DeVore, eds. (q.v.), *Man the hunter,* pp. 336–339.

Whittaker, R. H., and P. P. Feeny. 1971. Allelochemics: chemical interactions between species. *Science,* 171: 757–770.

Whitten, W. K., and F. H. Bronson. 1970. The role of pheromones in mammalian reproduction. In J. W. Johnston, Jr., D. G. Moulton, and A. Turk, eds. (q.v.), *Advances in chemoreception,* vol. 1, *Communication by chemical signals,* pp. 309–325.

Wickler, W. 1962. Ei-Attrapen und Maulbrüten bei afrikanischen Cichliden. *Zeitschrift für Tierpsychologie,* 19(2): 129–164.

——— 1963. Zur Klassifikation der Cichlidae, am Beispiel der Gattungen *Tropheus, Petrochromis, Haplochromis* und *Hemihaplochromis* n. gen. (Pisces, Perciformes). *Senckenbergiana Biologica,* 44(2): 83–96.

——— 1967a. Vergleichende Verhaltensforschung und Phylogenetik. In G. Heberer, ed., *Die Evolution der Organismen,* vol. 1, pp. 420–508. G. Fischer, Stuttgart. xvi + 754 pp.

——— 1967b. Specialization of organs having a signal function in some marine fish. *Studies in Tropical Oceanography, Miami,* 5: 539–548.

——— 1969a. Zur Soziologie des Brabantbuntbarsches, *Tropheus moorei* (Pisces, Cichlidae). *Zeitschrift für Tierpsychologie,* 26(8): 967–987.

——— 1969b. Socio-sexual signals and their intra-specific imitation among primates. In D. Morris, ed. (q.v.), *Primate ethology: essays on the socio-sexual behavior of apes and monkeys,* pp. 89–189.

——— 1972a. *The sexual code: the social behavior of animals and men.* Doubleday, Garden City, N.Y. xxxi + 301 pp. (Translated from *Sind Wir Sünder?,* Droemer Knaur, Munich, 1969.)

——— 1972b. Aufbau und Paarspezifität des Gesangsduettes von *Laniarius funebris* (Aves, Passeriformes, Laniidae). *Zeitschrift für Tierpsychologie,* 30(5): 464–476.

——— 1972c. Deuttieren zwischen artverschiedenen Vögeln im Freiland. *Zeitschrift für Tierpsychologie,* 31(1): 98–103.

Wickler, W., and Uta Seibt. 1970. Das Verhalten von *Hymenocera picta* Dana, einer Seesterne fressenden Garnele (Decapoda, Natantia, Gnathophyllidae). *Zeitschrift für Tierpsychologie,* 27(3): 352–368.

Wickler, W., and Dagmar Uhrig. 1969a. Verhalten und ökologische Nische der Gelbflügelfledermaus, *Lavia frons* (Geoffroy) (Chiroptera, Megadermatidae). *Zeitschrift für Tierpsychologie,* 26(6): 726–736.

——— 1969b. Bettelrufe, Antwortszeit und Rassenunterschiede im Begrüssungsduett des Schmuckbartvogels *Trachyphonus d'arnaudii. Zeitschrift für Tierpsychologie,* 26(6): 651–661.

Wiegert, R. G. 1974. Competition: a theory based on realistic, general equations of population growth. *Science,* 185: 539–542.

Wiener, N. 1948. Time, communication, and the nervous system. *Annals of the New York Academy of Sciences,* 50(4): 197–220.

Wilcox, R. S. 1972. Communication by surface waves: mating behavior of a water strider (Gerridae). *Journal of Comparative Physiology,* 80(3): 255–266.

Wiley, R. H. 1973. Territoriality and non-random mating in sage grouse, *Centrocercus urophasianus. Animal Behaviour Monographs,* 6(2): 85–169.

——— 1974. Evolution of social organization and life history patterns among grouse (Aves: Tetraonidae). *Quarterly Review of Biology,* 49(3): 201–227.

Wille, A., and C. D. Michener. 1973. The nest architecture of stingless bees with special reference to those of Costa Rica (Hymenoptera: Apidae). *Revista de Biología Tropical* (Universidad de Costa Rica, San José), 21 (supplement 1): 1–278.

Wille, A., and E. Orozco. 1970. The life cycle and behavior of the social bee *Lasioglossum* (*Dialictus*) *umbripenne* (Hymenoptera: Halictidae). *Revista de Biología Tropical* (Universidad de Costa Rica, San José), 17(2): 199–245.

Williams, C. B. 1964. *Patterns in the balance of nature and related problems in quantitative biology.* Academic Press, New York. vii + 324 pp.

Williams, Elizabeth, and J. P. Scott. 1953. The development of social behavior patterns in the mouse in relation to natural periods. *Behaviour,* 6(1): 35–65.

Williams, E. C. 1941. An ecological study of the floor fauna of the Panama rain forest. *Bulletin of the Chicago Academy of Science,* 6(4): 63–124.

Williams, E. E. 1972. The origin of faunas, evolution of lizard congeners in a complex island fauna: a trial analysis. *Evolutionary Biology,* 6: 47–89.

Williams, F. X. 1919. Philippine wasp studies: II, descriptions of new species and life history studies. *Bulletin of the Experiment Station, Hawaiian Sugar Planters' Association, Entomology Series,* 14: 19–184.

Williams, G. C. 1957. Pleiotropy, natural selection, and evolution of senescence. *Evolution,* 11(4): 398–411.

——— 1964. Measurement of consociation among fishes and comments on the evolution of schooling. *Publications of the Museum, Michigan State University, East Lansing, Biological Series,* 2(7): 351–383.

——— 1966a. *Adaptation and natural selection: a critique of some current evolutionary thought.* Princeton University Press, Princeton, N.J. x + 307 pp.

——— 1966b. Natural selection, the costs of reproduction, and a refinement of Lack's principle. *American Naturalist,* 100(916): 687–690.

Williams, G. C., and J. B. Mitton. 1973. Why reproduce sexually? *Journal of Theoretical Biology,* 39(3): 545–554.

Williams, G. C., and Doris C. Williams. 1957. Natural selection of individually harmful social adaptations among sibs with special reference to social insects. *Evolution,* 11(1): 32–39.

Williams, H. W., M. W. Sorenson, and P. Thompson. 1969. Antiphonal calling of the tree shrew *Tupaia palawanensis. Folia Primatologica,* 11(3): 200–205.

Williams, T. R. 1972. The socialization process: a theoretical perspective. In F. E. Poirier, ed. (q.v.), *Primate socialization,* pp. 206–260.

Willis, E. O. 1966. The role of migrant birds at swarms of army ants. *Living Bird*, 5: 187–231.

——— 1967. The behavior of bicolored antbirds. *University of California Publications in Zoology*, 79. 127 pp.

Wilmsen, E. N. 1973. Interaction, spacing behavior, and the organization of hunting bands. *Journal of Anthropological Research*, 29(1): 1–31.

Wilson, A. P. 1968. Social behavior of free-ranging rhesus monkeys with an emphasis on aggression. Ph.D. thesis, University of California, Berkeley. [Cited by J. H. Crook, 1970b (*q.v.*).]

Wilson, A. P., and C. Boelkins. 1970. Evidence for seasonal variation in aggressive behaviour by *Macaca mulatta*. *Animal Behaviour*, 18(4): 719–724.

Wilson, E. O. 1953. The origin and evolution of polymorphism in ants. *Quarterly Review of Biology*, 28(2): 136–156.

——— 1955a. A monographic revision of the ant genus *Lasius*. *Bulletin of the Museum of Comparative Zoology, Harvard*, 113(1): 1–205.

——— 1955b. Ecology and behavior of the ant *Belonopelta deletrix* Mann. *Psyche, Cambridge*, 62(2): 82–87.

——— 1957. The organization of a nuptial flight of the ant *Pheidole sitarches* Wheeler. *Psyche, Cambridge*, 64(2): 46–50.

——— 1958a. The beginnings of nomadic and group-predatory behavior in the ponerine ants. *Evolution*, 12(1): 24–31.

——— 1958b. Observations on the behavior of the cerapachyine ants. *Insectes Sociaux*, 5(1): 129–140.

——— 1958c. Studies on the ant fauna of Melanesia: I, the tribe Leptogenyini; II, the tribes Amblyoponini and Platythyreini. *Bulletin of the Museum of Comparative Zoology, Harvard*, 118(3): 101–153.

——— 1958d. A chemical releaser of alarm and digging behavior in the ant *Pogonomyrmex badius* (Latreille). *Psyche, Cambridge*, 65(2,3): 41–51.

——— 1959a. Communication by tandem running in the ant genus *Cardiocondyla*. *Psyche, Cambridge*, 66(3): 29–34.

——— 1959b. Adaptive shift and dispersal in a tropical ant fauna. *Evolution*, 13(1): 122–144.

——— 1959c. Source and possible nature of the odor trail of fire ants. *Science*, 129: 643–644.

——— 1961. The nature of the taxon cycle in the Melanesian ant fauna. *American Naturalist*, 95(882): 169–193.

——— 1962a. Chemical communication among workers of the fire ant *Solenopsis saevissima* (Fr. Smith): 1, the organization of mass-foraging; 2, an information analysis of the odour trail; 3, the experimental induction of social responses. *Animal Behaviour*, 10(1,2): 134–164.

——— 1962b. Behavior of *Daceton armigerum* (Latreille), with a classification of self-grooming movements in ants. *Bulletin of the Museum of Comparative Zoology, Harvard*, 127(7): 403–422.

——— 1963. Social modifications related to rareness in ant species. *Evolution*, 17(2): 249–253.

——— 1964. The true army ants of the Indo-Australian area (Hymenoptera: Formicidae: Dorylinae). *Pacific Insects*, 6(3): 427–483.

——— 1966. Behaviour of social insects. In P. T. Haskell, ed., *Insect behaviour*, pp. 81–96. Symposium of the Royal Entomological Society of London, no. 3. Royal Entomological Society, London. 113 pp.

——— 1968a. The ergonomics of caste in the social insects. *American Naturalist*, 102(923): 41–66.

——— 1968b. Chemical systems. In T. A. Sebeok, ed. (*q.v.*), *Animal communication: techniques of study and results of research*, pp. 75–102.

——— 1969. The species equilibrium. In G. M. Woodwell, ed., *Diversity and stability in ecological systems*, pp. 38–47. Brookhaven Symposia in Biology no. 22. Biology Department, Brookhaven National Laboratory, Upton, N.Y. vii + 264 pp.

——— 1970. Chemical communication within animal species. In E. Sondheimer and J. B. Simeone, eds. (*q.v.*), *Chemical ecology*, pp. 133–155.

——— 1971a. *The insect societies*. Belknap Press of Harvard University Press, Cambridge. x + 548 pp.

——— 1971b. Competitive and aggressive behavior. In J. F. Eisenberg and W. Dillon, eds. (*q.v.*), *Man and beast: comparative social behavior*, pp. 183–217.

——— 1972a. On the queerness of social evolution. *Bulletin of the Entomological Society of America*, 19(1): 20–22.

——— 1972b. Animal communication. *Scientific American*, 227(3) (September): 52–60.

——— 1973. Group selection and its significance for ecology. *BioScience*, 23(11): 631–638.

——— 1974a. The soldier of the ant *Camponotus (Colobopsis) fraxinicola* as a trophic caste. *Psyche, Cambridge*, 81(1): 182–188.

——— 1974b. *Leptothorax duloticus* and the beginnings of slavery in ants. *Evolution*. (In press.)

——— 1974c. Aversive behavior and competition within colonies of the ant *Leptothorax curvispinosus* Mayr (Hymenoptera: Formicidae). *Annals of the Entomological Society of America*, 67(5): 777–780.

——— 1974d. The population consequences of polygyny in the ant *Leptothorax curvispinosus* Mayr (Hymenoptera: Formicidae). *Annals of the Entomological Society of America*, 67(5): 781–786.

Wilson, E. O., and W. H. Bossert. 1963. Chemical communication among animals. *Recent Progress in Hormone Research*, 19: 673–716.

——— 1971. *A primer of population biology*. Sinauer Associates, Sunderland, Mass. 192 pp.

Wilson, E. O., and W. L. Brown. 1956. New parasitic ants of the genus *Kyidris*, with notes on ecology and behavior. *Insectes Sociaux*, 3(3): 439–454.

——— 1958. Recent changes in the introduced population of the fire ant *Solenopsis saevissima* (Fr. Smith). *Evolution*, 12(2): 211–218.

Wilson, E. O., F. M. Carpenter, and W. L. Brown. 1967. The first Mesozoic ants. *Science*, 157: 1038–1040.

Wilson, E. O., T. Eisner, W. R. Briggs, R. E. Dickerson, R. L. Metzenberg, R. D. O'Brien, M. Susman, and W. E. Boggs. 1973. *Life on earth*. Sinauer Associates, Sunderland, Mass. xiv + 1053 pp.

Wilson, E. O., T. Eisner, G. C. Wheeler, and Jeanette Wheeler. 1956.

Aneuretus simoni Emery, a major link in ant evolution. *Bulletin of the Museum of Comparative Zoology, Harvard,* 115(3): 81–99.

Wilson, E. O., and F. E. Regnier. 1971. The evolution of the alarm-defense system in the formicine ants. *American Naturalist,* 105(943): 279–289.

Wilson, E. O., and R. W. Taylor. 1964. A fossil ant colony: new evidence of social antiquity. *Psyche, Cambridge,* 71(2): 93–103.

—— 1967. The ants of Polynesia (Hymenoptera: Formicidae). *Pacific Insects Monograph,* 14. 109 pp.

Wilsson, L. 1971. Observations and experiments on the ethology of the European beaver (*Castor fiber* L.). *Viltrevy,* 8(3): 115–266.

Wing, M. W. 1968. Taxonomic revision of the Nearctic genus *Acanthomyops* (Hymenoptera: Formicidae). *Memoirs, Cornell University Agricultural Experiment Station,* 405: 1–173.

Winn, H. E. 1964. The biological significance of fish sounds. In W. N. Tavolga, ed. (*q.v.*), *Marine bio-acoustics,* pp. 213–231.

Winterbottom, J. M. 1943. On woodland bird parties in northern Rhodesia. *Ibis,* 85(4): 437–442.

—— 1949. Mixed bird parties in the Tropics, with special reference to northern Rhodesia. *Auk,* 66(3): 258–263.

Wolf, L. L., and F. R. Hainsworth. 1971. Time and energy budgets of territorial hummingbirds. *Ecology,* 52(6): 980–988.

Wolf, L. L., and F. G. Stiles. 1970. Evolution of pair cooperation in a tropical hummingbird. *Evolution,* 24(4): 759–773.

Wolfe, M. L., and D. L. Allen. 1973. Continued studies of the status, socialization, and relationships of Isle Royale wolves, 1967 to 1970. *Journal of Mammalogy,* 54(3): 611–633.

Wood, D. H. 1970. An ecological study of *Antechinus stuartii* (Marsupialia) in a south-east Queensland rain forest. *Australian Journal of Zoology,* 18(2): 185–207.

—— 1971. The ecology of *Rattus fuscipes* and *Melomys cervinipes* (Rodentia: Muridae) in a south-east Queensland rain forest. *Australian Journal of Zoology,* 19(4): 371–392.

Wood, D. L., R. M. Silverstein, and M. Nakajima, eds. 1970. *Control of insect behavior by natural products.* Academic Press, New York. x + 345 pp.

Wood-Gush, D. G. M. 1955. The behaviour of the domestic chicken: a review of the literature. *British Journal of Animal Behaviour,* 3(3): 81–110.

Woolfenden, G. E. 1973. Nesting and survival in a population of Florida scrub jays. *Living Bird,* 12: 25–49.

—— 1974a. Florida scrub jay helpers at the nest. *Auk.* (In press.)

—— 1974b. The effect and source of Florida scrub jay helpers. (Unpublished manuscript.)

Woollacott, R. M., and R. L. Zimmer. 1972. Origin and structure of the brood chamber in *Bugula neritina* (Bryozoa). *Marine Biology,* 16: 165–170.

Woolpy, J. H. 1968a. The social organization of wolves. *Natural History,* 77(5): 46–55.

—— 1968b. Socialization of wolves. *Science and Psychoanalysis,* 12: 82–94.

Woolpy, J. H., and B. E. Ginsburg. 1967. Wolf socialization: a study of temperament in a wild social species. *American Zoologist,* 7(2): 357–363.

Wortis, R. P. 1969. The transition from dependent to independent feeding in the young ring dove. *Animal Behaviour Monographs,* 2(1): 1–54.

Wright, S. 1931. Evolution in Mendelian populations. *Genetics,* 16(2): 97–158.

—— 1943. Isolation by distance. *Genetics,* 28(2): 114–138.

—— 1945. Tempo and mode in evolution: a critical review. *Ecology,* 26(4): 415–419.

—— 1969. *Evolution and the genetics of populations,* vol. 2, *The theory of gene frequencies.* University of Chicago Press, Chicago. vii + 511 pp.

Wünschmann, A. 1966. Einige Gefangenschaftsbeobachtungen an Breitstirn-Wombats (*Lasiorhinus latifrons* Owen 1845). *Zeitschrift für Tierpsychologie,* 23(1): 56–71.

Wüst, Margarete. 1973. Stomodeale und proctodeale Sekrete von Ameisenlarven und ihre biologische Bedeutung. *Proceedings of the Seventh Congress of the International Union for the Study of Social Insects, London,* pp. 412–417.

Wynne-Edwards, V. C. 1962. *Animal dispersion in relation to social behaviour.* Oliver and Boyd, Edinburgh. xi + 653 pp.

—— 1971. Space use and the social community in animals and men. In A. H. Esser, ed. (*q.v.*), *Behavior and environment: the use of space by animals and men,* pp. 267–280.

Yamada, M. 1958. A case of acculturation in a society of Japanese monkeys. *Primates,* 1(2): 30–46. (In Japanese.)

—— 1966. Five natural troops of Japanese monkeys in Shodoshima Island: 1, distribution and social organization. *Primates,* 7(3): 315–362.

Yamanaka, M. 1928. On the male of a paper wasp, *Polistes fadwigae* Dalla Torre. *Science Reports of the Tôhoku Imperial University, Sendai, Japan,* 6th ser. (Biology), 3(3): 265–269.

Yamane, S. 1971. Daily activities of the founding queens of two *Polistes* species, *P. snelleni* and *P. biglumis* in the solitary stage (Hymenoptera, Vespidae). *Kontyû,* 39: 203–217.

Yasuno, M. 1965. Territory of ants in the Kayano grassland at Mt. Hakkôda. *Science Reports of the Tôhoku University, Sendai, Japan,* 6th ser. (Biology), 31(3): 195–206.

Yeaton, R. I. 1972. Social behavior and social organization in Richardson's ground squirrel (*Spermophilus richardsonii*) in Saskatchewan. *Journal of Mammalogy,* 53(1): 139–147.

Yeaton, R. I., and M. L. Cody. 1974. Competitive release in island song sparrow populations. *Theoretical Population Biology,* 5(1): 42–58.

Yerkes, R. M. 1943. *Chimpanzees: a laboratory colony.* Yale University Press, New Haven. xv + 321 pp.

Yerkes, R. M., and Ada M. Yerkes. 1929. *The great apes: a study of anthropoid life.* Yale University Press, New Haven. xix + 652 pp.

Yockey, H. P., R. L. Platzman, and H. Quastler, eds. 1958. *Symposium* on information theory in biology. Pergamon Press, New York. xii + 418 pp.

Yoshiba, K. 1968. Local and intertroop variability in ecology and social behavior of common Indian langurs. In Phyllis C. Jay, ed. (q.v.), *Primates: studies in adaptation and variability*, pp. 217–242.

Yoshikawa, K. 1963. Introductory studies on the life economy of polistine wasps: 2, superindividual stage; 3, dominance order and territory. *Journal of Biology, Osaka City University*, 14: 55–61.

———— 1964. Predatory hunting wasps as the natural enemies of insect pests in Thailand. *Nature and Life in Southeast Asia* (Tokyo), 3: 391–398.

Yoshikawa, K., R. Ohgushi, and S. F. Sakagami. 1969. Preliminary report on entomology of the Osaka City University 5th Scientific Expedition to Southeast Asia, 1966, with descriptions of two new genera of stenogastrine wasps by J. van der Vecht. *Nature and Life in Southeast Asia* (Tokyo), 6: 153–182.

Young, C. M. 1964. An ecological study of the common shelduck (*Tadorna tadorna* L.) with special reference to the regulation of the Ythan population. Ph.D. thesis, Aberdeen University, Aberdeen. [Cited by J. R. Krebs, 1971 (q.v.).]

Zajonc, R. B. 1971. Attraction, affiliation, and attachment. In J. F. Eisenberg and W. S. Dillon, eds. (q.v.), *Man and beast: comparative social behavior*, pp. 141–179.

Zarrow, M. X., J. E. Philpott, V. H. Denenberg, and W. B. O'Connor. 1968. Localization of ^{14}C-4-corticosterone in the two day old rat and a consideration of the mechanism involved in early handling. *Nature, London*, 218(5148): 1264–1265.

Zimmerman, J. L. 1971. The territory and its density dependent effect in *Spiza americana*. *Auk*, 88(3): 591–612.

Zucchi, R., S. F. Sakagami, and J. M. F. de Camargo. 1969. Biological observations on a Neotropical parasocial bee, *Eulaema nigrita*, with a review of the biology of Euglossinae: a comparative study. *Journal of the Faculty of Science, Hokkaido University*, 6th ser. (Zoology), 17: 271–380.

Zuckerman, S. 1932. *The social life of monkeys and apes*. Harcourt, Brace, New York. xii + 356 pp.

Zumpe, Doris. 1965. Laboratory observations on the aggressive behaviour of some butterfly fishes (*Chaetodontidae*). *Zeitschrift für Tierpsychologie*, 22(2): 226–236.

Zwölfer, H. 1958. Zur Systematik, Biologie und Ökologie unterirdisch lebender Aphiden (Homoptera, Aphidoidea) (Anoeciinae, Tetraneurini, Pemphigini und Fordinae): IV, ökologische und systematische Erörterungen. *Zeitschrift für Angewandte Entomologie*, 43(1): 1–52.

Index

A